AF780227

10th Anniversary of *Plants*—Recent Advances and Perspectives—Volume I

10th Anniversary of *Plants*—Recent Advances and Perspectives—Volume I

Editors

Milan Stankovic
Paula Baptista
Petronia Carillo

MDPI

Basel • Beijing • Wuhan • Barcelona • Belgrade • Novi Sad • Cluj • Manchester

Editors
Milan Stankovic
University of Kragujevac
Kragujevac
Serbia

Paula Baptista
Polytechnic Institute
of Bragança
Bragança
Portugal

Petronia Carillo
University of Campania
"Luigi Vanvitelli"
Caserta
Italy

Editorial Office
MDPI
St. Alban-Anlage 66
4052 Basel, Switzerland

This is a reprint of articles from the Special Issue published online in the open access journal *Plants* (ISSN 2223-7747) (available at: https://www.mdpi.com/journal/plants/special_issues/10th_anniversary_plants).

For citation purposes, cite each article independently as indicated on the article page online and as indicated below:

Lastname, A.A.; Lastname, B.B. Article Title. *Journal Name* **Year**, *Volume Number*, Page Range.

Volume I
ISBN 978-3-0365-8418-8 (Hbk)
ISBN 978-3-0365-8419-5 (PDF)
doi.org/10.3390/books978-3-0365-8419-5

Set
ISBN 978-3-0365-8372-3 (Hbk)
ISBN 978-3-0365-8373-0 (PDF)

© 2023 by the authors. Articles in this book are Open Access and distributed under the Creative Commons Attribution (CC BY) license. The book as a whole is distributed by MDPI under the terms and conditions of the Creative Commons Attribution-NonCommercial-NoDerivs (CC BY-NC-ND) license.

Contents

About the Editors

Milan Stankovic

Milan Stanković is an associate professor at the Department of Biology and Ecology, Faculty of Science, University of Kragujevac, Republic of Serbia and Head of Department of Biology and Ecology (2016-). His scientific and teaching careeer started at the Department of Biology and Ecology, Faculty of Science, University of Kragujevac (2008). He acquired his PhD degree in Plant Science (2012) at the same university and completed his postdoctoral research at the Université François-Rabelais de Tours, France. At the Faculty of Science, he was appointed as an assistant professor (2013), as well as an associate professor (2019), and he teaches several BSc, MSc, and PhD courses on plant science. His current research is focused on plant biology, ecology, and phytochemistry. Dr. Stanković is the (co-)author of over 300 references including articles in peer-reviewed journals, edited books, book chapters, conference papers, meeteng abstracts, etc. He is currently working as an associate editor of *Plants* (2012-).

Paula Baptista

Paula Baptista is currently an associate professor at the Polytechnic Institute of Bragança (Portugal), the coordinator of the topic "Sustainable Agriculture and Innovative Agro-food Chains" of the Mountain Research Center (CIMO, http://cimo.ipb.pt/web/), and the general secretary of the International Organisation for Biological Control (IOBC-WPRS). More recently, she has become a member of the Plant Health (PLH) Panel of EFSA. She holds a bachelor's degree in agricultural engineering (1996, University of Trás-os-Montes), a master's degree in quality control (2000, University of Porto), and a PhD in science (2007, University of Minho). Her main interests are focused on agricultural microbiology and biotechnology, and in particular on the exploitation of beneficial microorganisms to improve plant health and yield. Her specific lines of research include the identification and characterization of microbial biological control agents, and the study of the molecular bases of plant(s)–pathogen(s)–biocontrol agent(s) interactions using '-omics' approaches. Her research has led to the publication of more than 150 ISI papers.

Petronia Carillo

Petronia Carillo has been a full professor of agronomy at the University of Campania "Luigi Vanvitelli" in Caserta, Italy, since 2020, where she teaches agronomy, plant physiology, and post-harvest physiology. She was an associate professor of plant physiology (2010–2020) and permanent researcher of plant physiology (1999–2010) at the Second University of Naples, Italy. She studied at University Federico II of Naples, Italy, where she received her Ph.D. in plant physiology (1996) and MS (1992). She was a guest scientist at the Botanical Institute, Ruprecht-Karls-University of Heidelberg, Germany (4 months, 2000), and Max Planck Institute of Molecular Plant Physiology of Golm-Potsdam Germany (2–3 months a year from 2001 to 2010 and shorter visits from 2011 to 2020). She currently studies the metabolic and physiological responses of species of agronomic interest to nutrient deficiency, salt stress, and the type of cultivation (conventional, biological, hydroponic, eustress, microgravity, and the use of biostimulants).

plants

Editorial

10th Anniversary of Plants—Recent Advances and Further Perspectives

Milan Stanković

Department of Biology and Ecology, Faculty of Science, University of Kragujevac, Radoja Domanovića 12, 34000 Kragujevac, Serbia; mstankovic@kg.ac.rs

Published for the first time in 2012, *Plants* will celebrate its 10th anniversary. To mark this significant milestone and celebrate the achievements made throughout the years, we intend to publish a Special Issue entitled "10th Anniversary of *Plants*—Recent Advances and Perspectives". In the past decade, the continuous support of the authors, editors, and reviewers, as well as the readers, has resulted in noteworthy success and the achievement of a common goal, as well as the sustained reputation of *Plants* in the world of science. In parallel with the development of our journal, great success has also been achieved across the field of plant science itself, from the molecular to the ecosystem level, and many new findings are based on new methodological approaches. Apart from the fact that this Special Issue will serve as a celebration of the anniversary, it should also serve as a guide for discoveries in plant science and thus for the development of the journal.

This anniversary Special Issue contains 101 papers, the majority comprising articles (77 papers), followed by reviews (20 papers), communications (3 papers), and one protocol. The number of citations for this 10th anniversary Special Issue papers should be emphasized on this occasion; in the first year, this value has already reached almost 400, thereby confirming the importance and scientific impact of this collection as well as predicting an enviable future. Over 600 authors from all over the world contributed to the published papers of this Special Issue. Considering the satisfactory quality and diversity of the submissions, seven papers received the "featured" status, including four articles [1–4] and two reviews [5,6]. This status is awarded under very strict criteria defined by the publication policy of *Plants*. "Feature papers represent the most advanced research with significant potential for high impact in the field. A Feature Paper should be a substantial original Article that involves several techniques or approaches, provides an outlook for future research directions and describes possible research applications. Feature papers are submitted upon individual invitation or recommendation by the scientific editors and must receive positive feedback from the reviewers." The number of papers with this status in relation to the total number of papers indicates the enviable scientific quality and impact of this Special Issue. The *Plants* editorial team practices the "Editor's Choice" option. "Editor's Choice articles are based on recommendations by the scientific editors of MDPI journals from around the world. Editors select a small number of articles recently published in the journal that they believe will be particularly interesting to readers, or important in the respective research area. The aim is to provide a snapshot of some of the most exciting work published in the various research areas of the journal." This status was deservedly awarded to four papers, two articles [7,8], one review paper [9], as well as one protocol-type paper in this Special Issue [10]. An indispensable opportunity was to recognize the best paper from the 10th anniversary Special Issue. This complex task was assigned to the editors in charge of the issue. Scientific rigor, significance, citation, and originality were assessed in detail for all papers in the Special Issue. After that, the paper with the highest marks was announced [11].

During the establishment of the 10th anniversary Special Issue of *Plants*, no specific topic was defined, but all potential papers which were within the aims and scope of

Citation: Stanković, M. 10th Anniversary of Plants—Recent Advances and Further Perspectives. *Plants* **2023**, *12*, 1696. https://doi.org/10.3390/plants12081696

Received: 14 March 2023
Accepted: 8 April 2023
Published: 18 April 2023

Copyright: © 2023 by the author. Licensee MDPI, Basel, Switzerland. This article is an open access article distributed under the terms and conditions of the Creative Commons Attribution (CC BY) license (https://creativecommons.org/licenses/by/4.0/).

Plants were considered, and, based on that, the Special Issue was entitled "10th Anniversary of *Plants*—Recent Advances and Further Perspectives". The published papers are characterized by great diversity concerning the topics in plant sciences, reflecting recent developments and the main trends, particularly in plant molecular biology and physiology, genetics, and phytochemistry. Though a complex task, based on the main topic, objects of research, as well as the contribution of the results, it was possible to categorize the papers into several groups. One group of papers is dedicated to molecular biology and the physiology of plants, applying modern molecular methodological approaches, cell biology, microbiology, etc., where molecular and physiological processes, interactions, stress resistance, etc. are elucidated. This group includes articles [1,2,7,12–26], communications [27,28], and review papers [5,6,11,29–34] with significant results applicable both in science and practice. The biology and ecology of plant secondary metabolites, their identification, and biological and therapeutic activity from different aspects of the phytochemistry of edible, aromatic, medicinal, or potentially medicinal plants are represented in a significant number of articles [3,8,35–49], as well as review papers [9,50,51]. The morphology and systematics of plants, as well as taxonomic methods, are also the subject of significant articles [52–56] and one review paper [57]. A similar number of papers are devoted to the scientific and practical aspects of ecology and the environment, comprising several articles [4,58–61] and one review [62]. The diversity of the topics of the Special Issue indicates the significant representation of agricultural plants in the trends of plant science. Physiology, molecular biology, and genetics, with a special aspect on the biotechnological approach, are covered in an impressive number of papers, such as articles [63–94], communications [95], and reviews [96–101], as well as the protocol described [10].

The abovementioned information illustrates the twofold significance of this Special Issue. It is a culmination of ten years of efforts to improve the quality of *Plants* and, at the same time, serves as a general overview of the achievements of plant sciences in the current period. I would like to take this opportunity to thank all the authors on behalf of the editorial office, especially for their interest in participating and willingness to share their experiences in science, as well as for the high-quality contributions submitted. We would like to thank the numerous reviewers for their valuable comments, which contributed to the quality of the published articles and thus to the overall quality of the Special Issue.

In terms of gratitude, it is imperative to note that the realization of the Special Issue was only possible thanks to the cooperation and dedication of the Special Issue editorial team—Prof. Dr. Milan Stanković (University of Kragujevac, Kragujevac, Serbia), Prof. Dr. Paula Baptista (Mountain Research Centre—CIMO, Bragança, Portugal), and Prof. Dr. Petronia Carillo (University of Campania Luigi Vanvitelli, Caserta, Italy). I would also like to thank the *Plants* editorial office, especially Ms. Sumi Sun, for their collaboration and guidance during the initiation, review, and editing process of the issue.

It has been my great pleasure to invest time and effort in editing this special issue, as well as to contribute over the past decade in the positions of assistant and associate editor of *Plants*.

Conflicts of Interest: The authors declare no conflict of interest.

References

1. Han, M.; Xu, X.; Xiong, Y.; Wei, H.; Yao, K.; Huang, T.; Long, Y.; Su, T. Genome-Wide Survey and Expression Analyses of Hexokinase Family in Poplar (*Populus trichocarpa*). *Plants* **2022**, *11*, 2025. [CrossRef] [PubMed]
2. Marchica, A.; Cotrozzi, L.; Lorenzini, G.; Nali, C.; Pellegrini, E. Antioxidants and Phytohormones Act in Coordination to Regulate Sage Response to Long Term Ozone Exposure. *Plants* **2022**, *11*, 904. [CrossRef] [PubMed]
3. Vitalini, S.; Iriti, M.; Vaglia, V.; Garzoli, S. Chemical Investigation and Dose-Response Phytotoxic Effect of Essential Oils from Two Gymnosperm Species (*Juniperus communis* var. *saxatilis* Pall. and *Larix decidua* Mill.). *Plants* **2022**, *11*, 1510. [CrossRef]
4. Chauhan, S.; Yadav, G.; Babu, S. Ecological Networks in Urban Forest Fragments Reveal Species Associations between Native and Invasive Plant Communities. *Plants* **2022**, *11*, 541. [CrossRef] [PubMed]
5. Bertaccini, A. Plants and Phytoplasmas: When Bacteria Modify Plants. *Plants* **2022**, *11*, 1425. [CrossRef]

6. Abideen, Z.; Hanif, M.; Munir, N.; Nielsen, B.L. Impact of Nanomaterials on the Regulation of Gene Expression and Metabolomics of Plants under Salt Stress. *Plants* **2022**, *11*, 691. [CrossRef]
7. Chiusano, M.L.; Incerti, G.; Colantuono, C.; Termolino, P.; Palomba, E.; Monticolo, F.; Benvenuto, G.; Foscari, A.; Esposito, A.; Marti, L.; et al. *Arabidopsis thaliana* Response to Extracellular DNA: Self Versus Nonself Exposure. *Plants* **2021**, *10*, 1744. [CrossRef]
8. D'Agostino, G.; Giambra, B.; Palla, F.; Bruno, M.; Badalamenti, N. The Application of the Essential Oils of *Thymus vulgaris* L. and *Crithmum maritimum* L. as Biocidal on Two *Tholu Bommalu* Indian Leather Puppets. *Plants* **2021**, *10*, 1508. [CrossRef]
9. Motolinía-Alcántara, E.A.; Castillo-Araiza, C.O.; Rodríguez-Monroy, M.; Román-Guerrero, A.; Cruz-Sosa, F. Engineering Considerations to Produce Bioactive Compounds from Plant Cell Suspension Culture in Bioreactors. *Plants* **2021**, *10*, 2762. [CrossRef]
10. Pepe, M.; Hesami, M.; Jones, A.M.P. Machine Learning-Mediated Development and Optimization of Disinfection Protocol and Scarification Method for Improved In Vitro Germination of Cannabis Seeds. *Plants* **2021**, *10*, 2397. [CrossRef]
11. Shelp, B.J.; Aghdam, M.S.; Flaherty, E.J. γ-Aminobutyrate (GABA) Regulated Plant Defense: Mechanisms and Opportunities. *Plants* **2021**, *10*, 1939. [CrossRef] [PubMed]
12. Vitti, A.; Pagán, I.; Bochicchio, B.; De Stradis, A.; Piazzolla, P.; Scopa, A.; Nuzzaci, M. *Cucumber mosaic virus* Is Unable to Self-Assemble in Tobacco Plants When Transmitted by Seed. *Plants* **2022**, *11*, 3217. [CrossRef] [PubMed]
13. Zagorchev, L.; Du, Z.; Shi, Y.; Teofanova, D.; Li, J. *Cuscuta australis* Parasitism-Induced Changes in the Proteome and Photosynthetic Parameters of *Arabidopsis thaliana*. *Plants* **2022**, *11*, 2904. [CrossRef]
14. Piotrowska, J.; Jodoi, Y.; Trang, N.H.; Wawrzynska, A.; Takahashi, H.; Sirko, A.; Maruyama-Nakashita, A. The C-Terminal Region of SLIM1 Transcription Factor Is Required for Sulfur Deficiency Response. *Plants* **2022**, *11*, 2595. [CrossRef] [PubMed]
15. Roach, T.; Baur, T.; Kranner, I. β-Cyclocitral Does Not Contribute to Singlet Oxygen-Signalling in Algae, but May Down-Regulate Chlorophyll Synthesis. *Plants* **2022**, *11*, 2155. [CrossRef] [PubMed]
16. Panara, F.; Fasano, C.; Lopez, L.; Porceddu, A.; Facella, P.; Fantini, E.; Daddiego, L.; Perrella, G. Genome-Wide Identification and Spatial Expression Analysis of Histone Modification Gene Families in the Rubber Dandelion *Taraxacum kok-saghyz*. *Plants* **2022**, *11*, 2077. [CrossRef]
17. Pramsohler, M.; Lichtenberger, E.; Neuner, G. Seasonal Xylem Sap Acidification Is Governed by Tree Phenology, Temperature and Elevation of Growing Site. *Plants* **2022**, *11*, 2058. [CrossRef]
18. AL-Juhani, W.; Al Thagafi, N.T.; Al-Qthanin, R.N. Gene Losses and Plastome Degradation in the Hemiparasitic Species *Plicosepalus acaciae* and *Plicosepalus curviflorus*: Comparative Analyses and Phylogenetic Relationships among Santalales Members. *Plants* **2022**, *11*, 1869. [CrossRef]
19. Sakata, N.; Haraguchi, T.; Masuo, S.; Ishiga, T.; Ishiga, Y. *Pseudomonas cannabina* pv. *alisalensis* Virulence Factors Are Involved in Resistance to Plant-Derived Antimicrobials during Infection. *Plants* **2022**, *11*, 1742. [CrossRef]
20. Kadoglidou, K.; Irakli, M.; Boutsika, A.; Mellidou, I.; Maninis, N.; Sarrou, E.; Georgiadou, V.; Tourvas, N.; Krigas, N.; Moysiadis, T.; et al. Metabolomic Fingerprinting and Molecular Characterization of the Rock Samphire Germplasm Collection from the Balkan Botanic Garden of Kroussia, Northern Greece. *Plants* **2022**, *11*, 573. [CrossRef]
21. Tomasella, M.; Natale, S.; Petruzzellis, F.; Di Bert, S.; D'Amico, L.; Tromba, G.; Nardini, A. No Evidence for Light-Induced Embolism Repair in Cut Stems of Drought-Resistant Mediterranean Species under Soaking. *Plants* **2022**, *11*, 307. [CrossRef] [PubMed]
22. Acharya, B.R.; Sandhu, D.; Dueñas, C.; Ferreira, J.F.S.; Grover, K.K. Deciphering Molecular Mechanisms Involved in Salinity Tolerance in Guar (*Cyamopsis tetragonoloba* (L.) Taub.) Using Transcriptome Analyses. *Plants* **2022**, *11*, 291. [CrossRef]
23. Scheible, N.; Yoon, G.M.; McCubbin, A.G. Calmodulin Domain Protein Kinase PiCDPK1 Regulates Pollen Tube Growth Polarity through Interaction with RhoGDI. *Plants* **2022**, *11*, 254. [CrossRef]
24. Sala, F.; Herbei, M.V.; Rujescu, C. RWLMod—Potential Model to Study Plant Tolerance in Drought Stress Conditions. *Plants* **2021**, *10*, 2576. [CrossRef] [PubMed]
25. van Voorthuizen, M.J.; Song, J.; Novák, O.; Jameson, P.E. Plant Growth Regulators INCYDE and TD-K Underperform in Cereal Field Trials. *Plants* **2021**, *10*, 2309. [CrossRef]
26. Im, G.; Choi, D. *AIP1*, Encoding the Small Subunit of Acetolactate Synthase, Is Partially Responsible for Resistance to Hypoxic Stress in *Arabidopsis thaliana*. *Plants* **2021**, *10*, 2251. [CrossRef] [PubMed]
27. Kong, Q.; Low, P.M.; Lim, A.R.Q.; Yang, Y.; Yuan, L.; Ma, W. Functional Antagonism of WRI1 and TCP20 Modulates *GH3.3* Expression to Maintain Auxin Homeostasis in Roots. *Plants* **2022**, *11*, 454. [CrossRef]
28. Bunce, J. Carboxylation Capacity Can Limit C_3 Photosynthesis at Elevated CO_2 throughout Diurnal Cycles. *Plants* **2021**, *10*, 2603. [CrossRef]
29. Soltabayeva, A.; Dauletova, N.; Serik, S.; Sandybek, M.; Omondi, J.O.; Kurmanbayeva, A.; Srivastava, S. Receptor-like Kinases (LRR-RLKs) in Response of Plants to Biotic and Abiotic Stresses. *Plants* **2022**, *11*, 2660. [CrossRef]
30. Bandurska, H. Drought Stress Responses: Coping Strategy and Resistance. *Plants* **2022**, *11*, 922. [CrossRef]
31. Matilla, A.J. Exploring Breakthroughs in Three Traits Belonging to Seed Life. *Plants* **2022**, *11*, 490. [CrossRef] [PubMed]
32. Shen, G.; Sun, W.; Chen, Z.; Shi, L.; Hong, J.; Shi, J. Plant GDSL Esterases/Lipases: Evolutionary, Physiological and Molecular Functions in Plant Development. *Plants* **2022**, *11*, 468. [CrossRef]
33. Zhang, N.; Zabotina, O.A. Critical Determinants in ER-Golgi Trafficking of Enzymes Involved in Glycosylation. *Plants* **2022**, *11*, 428. [CrossRef] [PubMed]

34. Blasio, F.; Prieto, P.; Pradillo, M.; Naranjo, T. Genomic and Meiotic Changes Accompanying Polyploidization. *Plants* **2022**, *11*, 125. [CrossRef]
35. Ognyanov, M.; Denev, P.; Petkova, N.; Petkova, Z.; Stoyanova, M.; Zhelev, P.; Matev, G.; Teneva, D.; Georgiev, Y. Nutrient Constituents, Bioactive Phytochemicals, and Antioxidant Properties of Service Tree (*Sorbus domestica* L.) Fruits. *Plants* **2022**, *11*, 1832. [CrossRef]
36. Păltinean, R.; Ielciu, I.; Hanganu, D.; Niculae, M.; Pall, E.; Angenot, L.; Tits, M.; Mocan, A.; Babotă, M.; Frumuzachi, O.; et al. Biological Activities of Some Isoquinoline Alkaloids from *Fumaria schleicheri* Soy. Will. *Plants* **2022**, *11*, 1202. [CrossRef] [PubMed]
37. Jamil, M.; Wang, J.Y.; Yonli, D.; Ota, T.; Berqdar, L.; Traore, H.; Margueritte, O.; Zwanenburg, B.; Asami, T.; Al-Babili, S. *Striga hermonthica* Suicidal Germination Activity of Potent Strigolactone Analogs: Evaluation from Laboratory Bioassays to Field Trials. *Plants* **2022**, *11*, 1045. [CrossRef]
38. Kim, M.-K.; Park, G.; Ji, Y.; Lee, Y.-G.; Choi, M.; Go, S.-H.; Son, M.; Jang, Y.-P. Design of Experiments-Based Optimization of Flavonoids Extraction from *Daphne genkwa* Flower Buds and Flavonoids Contents at Different Blooming Stages. *Plants* **2022**, *11*, 925. [CrossRef]
39. Mazzara, E.; Torresi, J.; Fico, G.; Papini, A.; Kulbaka, N.; Dall'Acqua, S.; Sut, S.; Garzoli, S.; Mustafa, A.M.; Cappellacci, L.; et al. A Comprehensive Phytochemical Analysis of Terpenes, Polyphenols and Cannabinoids, and Micromorphological Characterization of 9 Commercial Varieties of *Cannabis sativa* L. *Plants* **2022**, *11*, 891. [CrossRef]
40. Piccini, C.; Cantini, C.; Cai, G.; Pinto, D.C.G.A.; Silva, A.M.S.; Romi, M.; Dias, M.C. Chemical Profiling of Two Italian *Olea europaea* (L.) Varieties Subjected to UV-B Stress. *Plants* **2022**, *11*, 680. [CrossRef]
41. Zuzarte, M.; Sousa, C.; Cavaleiro, C.; Cruz, M.T.; Salgueiro, L. The Anti-Inflammatory Response of *Lavandula luisieri* and *Lavandula pedunculata* Essential Oils. *Plants* **2022**, *11*, 370. [CrossRef]
42. Cornara, L.; Ambu, G.; Alberto, A.; Trombetta, D.; Smeriglio, A. Characterization of Ingredients Incorporated in the Traditional Mixed-Salad of the Capuchin Monks. *Plants* **2022**, *11*, 301. [CrossRef] [PubMed]
43. Teneva, D.; Pencheva, D.; Petrova, A.; Ognyanov, M.; Georgiev, Y.; Denev, P. Addition of Medicinal Plants Increases Antioxidant Activity, Color, and Anthocyanin Stability of Black Chokeberry (*Aronia melanocarpa*) Functional Beverages. *Plants* **2022**, *11*, 243. [CrossRef] [PubMed]
44. Andonova, T.; Muhovski, Y.; Fidan, H.; Slavov, I.; Stoyanova, A.; Dimitrova-Dyulgerova, I. Chemical Compounds, Antitumor and Antimicrobial Activities of Dry Ethanol Extracts from *Koelreuteria paniculata* Laxm. *Plants* **2021**, *10*, 2715. [CrossRef] [PubMed]
45. Letaief, T.; Garzoli, S.; Laghezza Masci, V.; Mejri, J.; Abderrabba, M.; Tiezzi, A.; Ovidi, E. Chemical Composition and Biological Activities of Tunisian *Ziziphus lotus* Extracts: Evaluation of Drying Effect, Solvent Extraction, and Extracted Plant Parts. *Plants* **2021**, *10*, 2651. [CrossRef]
46. Šic Žlabur, J.; Radman, S.; Fabek Uher, S.; Opačić, N.; Benko, B.; Galić, A.; Samirić, P.; Voća, S. Plant Response to Mechanically-Induced Stress: A Case Study on Specialized Metabolites of Leafy Vegetables. *Plants* **2021**, *10*, 2650. [CrossRef]
47. Smeriglio, A.; Denaro, M.; Trombetta, D.; Ragusa, S.; Circosta, C. New Insights on *Euphorbia dendroides* L. (*Euphorbiaceae*): Polyphenol Profile and Biological Properties of Hydroalcoholic Extracts from Aerial Parts. *Plants* **2021**, *10*, 1621. [CrossRef]
48. Ariyawansa, H.A.; Tsai, I.; Wang, J.-Y.; Withee, P.; Tanjira, M.; Lin, S.-R.; Suwannarach, N.; Kumla, J.; Elgorban, A.M.; Cheewangkoon, R. Molecular Phylogenetic Diversity and Biological Characterization of *Diaporthe* Species Associated with Leaf Spots of *Camellia sinensis* in Taiwan. *Plants* **2021**, *10*, 1434. [CrossRef]
49. Setzer, W.N.; Duong, L.; Pham, T.; Poudel, A.; Nguyen, C.; Mentreddy, S.R. Essential Oils of Four Virginia Mountain Mint (*Pycnanthemum virginianum*) Varieties Grown in North Alabama. *Plants* **2021**, *10*, 1397. [CrossRef]
50. Ahmad, A.; Ahmad, N.; Anis, M.; Faisal, M.; Alatar, A.A.; Abdel-Salam, E.M.; Meena, R.P.; Sivanesan, I. Biotechnological Advances in Pharmacognosy and In Vitro Manipulation of *Pterocarpus marsupium* Roxb. *Plants* **2022**, *11*, 247. [CrossRef]
51. Arif, Y.; Singh, P.; Bajguz, A.; Hayat, S. Phytocannabinoids Biosynthesis in Angiosperms, Fungi, and Liverworts and Their Versatile Role. *Plants* **2021**, *10*, 1307. [CrossRef]
52. Arnelas, I.; Pérez-Collazos, E.; López-Martínez, J.; Devesa, J.A.; Catalán, P. Molecular Systematics of *Valerianella* Mill. (Caprifoliaceae): Challenging the Taxonomic Value of Genetically Controlled Carpological Traits. *Plants* **2022**, *11*, 1276. [CrossRef] [PubMed]
53. Arévalo-Rodrigues, G.; Hurbath, F.; Prado, E.; Galvão, I.; Cordeiro, I.; Demarco, D. Adnate Leaf-Base and the Origin of Ribs in Succulent Stems of *Euphorbia* L. *Plants* **2022**, *11*, 1076. [CrossRef] [PubMed]
54. Martín-Gómez, J.J.; Porceddu, M.; Bacchetta, G.; Cervantes, E. Seed Morphology in Species from the *Silene mollissima* Aggregate (Caryophyllaceae) by Comparison with Geometric Models. *Plants* **2022**, *11*, 901. [CrossRef] [PubMed]
55. Sala, F.; Dobrei, A.; Herbei, M.V. Leaf Area Calculation Models for Vines Based on Foliar Descriptors. *Plants* **2021**, *10*, 2453. [CrossRef] [PubMed]
56. Paiva, E.A.S.; Couy-Melo, G.A.; Ballego-Campos, I. Colleters, Extrafloral Nectaries, and Resin Glands Protect Buds and Young Leaves of *Ouratea castaneifolia* (DC.) Engl. (Ochnaceae). *Plants* **2021**, *10*, 1680. [CrossRef]
57. Cervantes, E.; Martín-Gómez, J.J.; Gutiérrez del Pozo, D.; Tocino, Á. Seed Geometry in the Vitaceae. *Plants* **2021**, *10*, 1695. [CrossRef] [PubMed]
58. Sciandrello, S.; Cambria, S.; Giusso del Galdo, G.; Minissale, P.; Puglisi, M.; Tavilla, G.; Tomaselli, V. Ecological Features and Conservation of *Urtica rupestris* Guss. (Urticaceae): A Narrow Endemic Species of Sicily. *Plants* **2023**, *12*, 164. [CrossRef]

59. Rahman, I.U.; Afzal, A.; Iqbal, Z.; Alzain, M.N.; Al-Arjani, A.-B.F.; Alqarawi, A.A.; Abd_Allah, E.F.; Ali, N.; Sakhi, S.; Khan, M.A.; et al. Classification and Characterization of the Manoor Valley's (Lesser Himalaya) Vegetation from the Subtropical-Temperate Ecotonal Forests to the Alpine Pastures along Ecological Variables. *Plants* **2022**, *11*, 87. [CrossRef]
60. Hatzilazarou, S.; El Haissoufi, M.; Pipinis, E.; Kostas, S.; Libiad, M.; Khabbach, A.; Lamchouri, F.; Bourgou, S.; Megdiche-Ksouri, W.; Ghrabi-Gammar, Z.; et al. GIS-Facilitated Seed Germination and Multifaceted Evaluation of the Endangered *Abies marocana* Trab. (Pinaceae) Enabling Conservation and Sustainable Exploitation. *Plants* **2021**, *10*, 2606. [CrossRef]
61. Libiad, M.; Khabbach, A.; El Haissoufi, M.; Anestis, I.; Lamchouri, F.; Bourgou, S.; Megdiche-Ksouri, W.; Ghrabi-Gammar, Z.; Greveniotis, V.; Tsiripidis, I.; et al. Agro-Alimentary Potential of the Neglected and Underutilized Local Endemic Plants of Crete (Greece), Rif-Mediterranean Coast of Morocco and Tunisia: Perspectives and Challenges. *Plants* **2021**, *10*, 1770. [CrossRef] [PubMed]
62. Fasani, E.; Li, M.; Varotto, C.; Furini, A.; DalCorso, G. Metal Detoxification in Land Plants: From Bryophytes to Vascular Plants. STATE of the Art and Opportunities. *Plants* **2022**, *11*, 237. [CrossRef] [PubMed]
63. Herrera, S.; Lora, J.; Hormaza, J.I.; Rodrigo, J. Self-Incompatibility in Apricot: Identifying Pollination Requirements to Optimize Fruit Production. *Plants* **2022**, *11*, 2019. [CrossRef] [PubMed]
64. Cano, A.; Giraldo-Acosta, M.; García-Sánchez, S.; Hernández-Ruiz, J.; Arnao, M.B. Effect of Melatonin in Broccoli Postharvest and Possible Melatonin Ingestion Level. *Plants* **2022**, *11*, 2000. [CrossRef]
65. Hajjar, G.; Quellec, S.; Challois, S.; Bousset-Vaslin, L.; Joly, G.; Langrume, C.; Deleu, C.; Leport, L.; Musse, M. Characterization of the Water Shortage Effects on Potato Tuber Tissues during Growth Using MRI Relaxometry and Biochemical Parameters. *Plants* **2022**, *11*, 1918. [CrossRef]
66. Catalano, C.; Abbate, L.; Fatta Del Bosco, S.; Motisi, A.; Carimi, F.; De Michele, R.; Mercati, F.; D'Onghia, A.M.; Carra, A. Different Cell Types Affect the Transition from Juvenile to Mature Phase in Citrus Plants Regenerated through Somatic Embryogenesis. *Plants* **2022**, *11*, 1811. [CrossRef]
67. Candan, F.; Markushin, Y.; Ozbay, G. Uptake and Presence Evaluation of Nanoparticles in *Cicer arietinum* L. by Infrared Spectroscopy and Machine Learning Techniques. *Plants* **2022**, *11*, 1569. [CrossRef]
68. Alam, M.A.; Rahman, M.; Ahmed, S.; Jahan, N.; Khan, M.A.-A.; Islam, M.R.; Alsuhaibani, A.M.; Gaber, A.; Hossain, A. Genetic Variation and Genotype by Environment Interaction for Agronomic Traits in Maize (*Zea mays* L.) Hybrids. *Plants* **2022**, *11*, 1522. [CrossRef]
69. Fernández-López, D.J.; Fernández-Fernández, J.I.; Martínez-Mora, C.; Bleda-Sánchez, J.A.; Ruiz-García, L. Productiveness and Berry Quality of New Wine Grape Genotypes Grown under Drought Conditions in a Semi-Arid Wine-Producing Mediterranean Region. *Plants* **2022**, *11*, 1363. [CrossRef]
70. Caeiro, A.; Caeiro, S.; Correia, S.; Canhoto, J. Induction of Somatic Embryogenesis in Tamarillo (*Solanum betaceum* Cav.) Involves Increases in the Endogenous Auxin Indole-3-Acetic Acid. *Plants* **2022**, *11*, 1347. [CrossRef]
71. Márquez Gutiérrez, R.; Cherubino Ribeiro, T.H.; de Oliveira, R.R.; Benedito, V.A.; Chalfun-Junior, A. Genome-Wide Analyses of MADS-Box Genes in *Humulus lupulus* L. Reveal Potential Participation in Plant Development, Floral Architecture, and Lupulin Gland Metabolism. *Plants* **2022**, *11*, 1237. [CrossRef] [PubMed]
72. Di Martino, C.; Torino, V.; Minotti, P.; Pietrantonio, L.; Del Grosso, C.; Palmieri, D.; Palumbo, G.; Crawford, T.W., Jr.; Carfagna, S. Mycorrhized Wheat Plants and Nitrogen Assimilation in Coexistence and Antagonism with Spontaneous Colonization of Pathogenic and Saprophytic Fungi in a Soil of Low Fertility. *Plants* **2022**, *11*, 924. [CrossRef] [PubMed]
73. Fan, Q.; Bibi, S.; Vallad, G.E.; Goss, E.M.; Hurlbert, J.C.; Paret, M.L.; Jones, J.B.; Timilsina, S. Identification of Genes in *Xanthomonas euvesicatoria* pv. *rosa* That Are Host Limiting in Tomato. *Plants* **2022**, *11*, 796. [CrossRef] [PubMed]
74. Tominaga, T.; Yao, L.; Saito, H.; Kaminaka, H. Conserved and Diverse Transcriptional Reprogramming Triggered by the Establishment of Symbioses in Tomato Roots Forming *Arum*-Type and *Paris*-Type Arbuscular Mycorrhizae. *Plants* **2022**, *11*, 747. [CrossRef]
75. Burducea, M.; Lobiuc, A.; Dirvariu, L.; Oprea, E.; Olaru, S.M.; Teliban, G.-C.; Stoleru, V.; Poghirc, V.A.; Cara, I.G.; Filip, M.; et al. Assessment of the Fertilization Capacity of the Aquaculture Sediment for Wheat Grass as Sustainable Alternative Use. *Plants* **2022**, *11*, 634. [CrossRef] [PubMed]
76. Ghosh, A.; Saha, I.; Fujita, M.; Debnath, S.C.; Hazra, A.K.; Adak, M.K.; Hasanuzzaman, M. Photoactivated TiO_2 Nanocomposite Delays the Postharvest Ripening Phenomenon through Ethylene Metabolism and Related Physiological Changes in *Capsicum* Fruit. *Plants* **2022**, *11*, 513. [CrossRef]
77. Salinier, J.; Lefebvre, V.; Besombes, D.; Burck, H.; Causse, M.; Daunay, M.-C.; Dogimont, C.; Goussopoulos, J.; Gros, C.; Maisonneuve, B.; et al. The INRAE Centre for Vegetable Germplasm: Geographically and Phenotypically Diverse Collections and Their Use in Genetics and Plant Breeding. *Plants* **2022**, *11*, 347. [CrossRef]
78. Mijailovic, N.; Richet, N.; Villaume, S.; Nesler, A.; Perazzolli, M.; Aït Barka, E.; Aziz, A. D-Tagatose-Based Product Triggers Sweet Immunity and Resistance of Grapevine to Downy Mildew, but Not to Gray Mold Disease. *Plants* **2022**, *11*, 296. [CrossRef]
79. Bochicchio, R.; Labella, R.; Vitti, A.; Nuzzaci, M.; Logozzo, G.; Amato, M. Root Morphology, Allometric Relations and Rhizosheath of Ancient and Modern Tetraploid Wheats (*Triticum durum* Desf.) in Response to Inoculation with *Trichoderma harzianum* T-22. *Plants* **2022**, *11*, 159. [CrossRef]

80. Tadmor, Y.; Raz, A.; Reikin-Barak, S.; Ambastha, V.; Shemesh, E.; Leshem, Y.; Crane, O.; Stern, R.A.; Goldway, M.; Tchernov, D. et al. Metamitron, a Photosynthetic Electron Transport Chain Inhibitor, Modulates the Photoprotective Mechanism of Apple Trees. *Plants* **2021**, *10*, 2803. [CrossRef]
81. Maloupa, E.; Karapatzak, E.; Ganopoulos, I.; Karydas, A.; Papanastasi, K.; Kyrkas, D.; Yfanti, P.; Nikisianis, N.; Zahariadis, A. Kosma, I.S.; et al. Molecular Authentication, Phytochemical Evaluation and Asexual Propagation of Wild-Growing *Rosa canina* L. (Rosaceae) Genotypes of Northern Greece for Sustainable Exploitation. *Plants* **2021**, *10*, 2634. [CrossRef] [PubMed]
82. Mohamed, A.H.; Omar, A.A.; Attya, A.M.; Elashtokhy, M.M.A.; Zayed, E.M.; Rizk, R.M. Morphological and Molecular Characterization of Some Egyptian Six-Rowed Barley (*Hordeum vulgare* L.). *Plants* **2021**, *10*, 2527. [CrossRef] [PubMed]
83. Friel, J.; Bombarely, A.; Fornell, C.D.; Luque, F.; Fernández-Ocaña, A.M. Comparative Analysis of Genotyping by Sequencing and Whole-Genome Sequencing Methods in Diversity Studies of *Olea europaea* L. *Plants* **2021**, *10*, 2514. [CrossRef] [PubMed]
84. Benakashani, F.; Gonzalez-Andujar, J.L.; Soltani, E. Differences in Germination of ACCase-Resistant Biotypes Containing Isoleucine-1781-Leucine Mutation and Susceptible Biotypes of Wild Oat (*Avena sterilis* ssp. *ludoviciana*). *Plants* **2021**, *10*, 2350 [CrossRef]
85. Kishor, D.S.; Alavilli, H.; Lee, S.-C.; Kim, J.-G.; Song, K. Development of SNP Markers for White Immature Fruit Skin Color in Cucumber (*Cucumis sativus* L.) Using QTL-seq and Marker Analyses. *Plants* **2021**, *10*, 2341. [CrossRef]
86. Deng, P.; Xie, X.; Long, F.; Zhang, L.; Li, Y.; Zhao, Z.; Yang, S.; Wang, Y.; Fan, R.; Li, Z. Trait Variations and Probability Grading Index System on Leaf-Related Traits of *Eucommia ulmoides* Oliver Germplasm. *Plants* **2021**, *10*, 2280. [CrossRef]
87. Egamberdieva, D.; Ma, H.; Alaylar, B.; Zoghi, Z.; Kistaubayeva, A.; Wirth, S.; Bellingrath-Kimura, S.D. Biochar Amendments Improve Licorice (*Glycyrrhiza uralensis* Fisch.) Growth and Nutrient Uptake under Salt Stress. *Plants* **2021**, *10*, 2135. [CrossRef]
88. Alharby, H.F.; Nahar, K.; Al-Zahrani, H.S.; Hakeem, K.R.; Hasanuzzaman, M. Enhancing Salt Tolerance in Soybean by Exogenous Boron: Intrinsic Study of the Ascorbate-Glutathione and Glyoxalase Pathways. *Plants* **2021**, *10*, 2085. [CrossRef]
89. Smulders, L.; Benítez, E.; Moreno, B.; López-García, Á.; Pozo, M.J.; Ferrero, V.; de la Peña, E.; Alcalá Herrera, R. Tomato Domestication Affects Potential Functional Molecular Pathways of Root-Associated Soil Bacteria. *Plants* **2021**, *10*, 1942. [CrossRef]
90. Dhungana, S.K.; Park, J.-H.; Oh, J.-H.; Kang, B.-K.; Seo, J.-H.; Sung, J.-S.; Kim, H.-S.; Shin, S.-O.; Baek, I.-Y.; Jung, C.-S. Quantitative Trait Locus Mapping for Drought Tolerance in Soybean Recombinant Inbred Line Population. *Plants* **2021**, *10*, 1816. [CrossRef]
91. Hong, W.-J.; Jiang, X.; Choi, S.-H.; Kim, Y.-J.; Kim, S.-T.; Jeon, J.-S.; Jung, K.-H. A Systemic View of Carbohydrate Metabolism in Rice to Facilitate Productivity. *Plants* **2021**, *10*, 1690. [CrossRef] [PubMed]
92. Van Nguyen, T.; Kim, S.-W.; Min, C.-W.; Gupta, R.; Lee, G.-H.; Jang, J.-W.; Rathi, D.; Shin, H.-W.; Jung, J.-Y.; Jo, I.-H.; et al. Optimization of Protein Isolation and Label-Free Quantitative Proteomic Analysis in Four Different Tissues of Korean Ginseng. *Plants* **2021**, *10*, 1409. [CrossRef] [PubMed]
93. Carella, A.; Gianguzzi, G.; Scalisi, A.; Farina, V.; Inglese, P.; Bianco, R.L. Fruit Growth Stage Transitions in Two Mango Cultivars Grown in a Mediterranean Environment. *Plants* **2021**, *10*, 1332. [CrossRef] [PubMed]
94. Gruner, P.; Miedaner, T. Perennial Rye: Genetics of Perenniality and Limited Fertility. *Plants* **2021**, *10*, 1210. [CrossRef] [PubMed]
95. Pérez-Jiménez, M.; Pérez-Tornero, O. Comparison of Four Systems to Test the Tolerance of 'Fortune' Mandarin Tissue Cultured Plants to *Alternaria alternata*. *Plants* **2021**, *10*, 1321. [CrossRef]
96. Pathirana, R.; Carimi, F. Management and Utilization of Plant Genetic Resources for a Sustainable Agriculture. *Plants* **2022**, *11*, 2038. [CrossRef]
97. Chaisiri, C.; Liu, X.; Lin, Y.; Luo, C. *Diaporthe citri*: A Fungal Pathogen Causing Melanose Disease. *Plants* **2022**, *11*, 1600. [CrossRef]
98. Riviello-Flores, M.d.l.L.; Cadena-Iñiguez, J.; Ruiz-Posadas, L.d.M.; Arévalo-Galarza, M.d.L.; Castillo-Juárez, I.; Soto Hernández, M.; Castillo-Martínez, C.R. Use of Gamma Radiation for the Genetic Improvement of Underutilized Plant Varieties. *Plants* **2022**, *11*, 1161. [CrossRef]
99. Neupane, D.; Lohaus, R.H.; Solomon, J.K.Q.; Cushman, J.C. Realizing the Potential of *Camelina sativa* as a Bioenergy Crop for a Changing Global Climate. *Plants* **2022**, *11*, 772. [CrossRef]
100. Panzeri, D.; Guidi Nissim, W.; Labra, M.; Grassi, F. Revisiting the Domestication Process of African *Vigna* Species (Fabaceae): Background, Perspectives and Challenges. *Plants* **2022**, *11*, 532. [CrossRef]
101. Opačić, N.; Radman, S.; Fabek Uher, S.; Benko, B.; Voća, S.; Šic Žlabur, J. Nettle Cultivation Practices—From Open Field to Modern Hydroponics: A Case Study of Specialized Metabolites. *Plants* **2022**, *11*, 483. [CrossRef] [PubMed]

Disclaimer/Publisher's Note: The statements, opinions and data contained in all publications are solely those of the individual author(s) and contributor(s) and not of MDPI and/or the editor(s). MDPI and/or the editor(s) disclaim responsibility for any injury to people or property resulting from any ideas, methods, instructions or products referred to in the content.

plants

Article

Perennial Rye: Genetics of Perenniality and Limited Fertility

Paul Gruner and Thomas Miedaner *

State Plant Breeding Institute, University of Hohenheim, 70593 Stuttgart, Germany; Paul.Gruner@uni-hohenheim.de
* Correspondence: miedaner@uni-hohenheim.de; Tel.: +49-711-4592-2690

Abstract: Perenniality, the ability of plants to regrow after seed set, could be introgressed into cultivated rye by crossing with the wild relative and perennial *Secale strictum*. However, studies in the past showed that *Secale cereale* × *Secale strictum*-derived cultivars were also characterized by reduced fertility what was related to so called chromosomal multivalents, bulks of chromosomes that paired together in metaphase I of pollen mother cells instead of only two chromosomes (bivalents). Those multivalents could be caused by ancient translocations that occurred between both species. Genetic studies on perennial rye are quite old and especially the advent of molecular markers and genome sequencing paved the way for new insights and more comprehensive studies. After a brief review of the past research, we used a basic QTL mapping approach to analyze the genetic status of perennial rye. We could show that for the trait perennation 0.74 of the genetic variance in our population was explained by additively inherited QTLs on chromosome 2R, 3R, 4R, 5R and 7R. Fertility on the other hand was with 0.64 of explained genetic variance mainly attributed to a locus on chromosome 5R, what was most probably the self-incompatibility locus *S5*. Additionally, we could trace the *Z* locus on chromosome 2R by high segregation distortion of markers. Indications for chromosomal co-segregation, like multivalents, could not be found. This study opens new possibilities to use perennial rye as genetic resource and for alternative breeding methods, as well as a valuable resource for comparative studies of perennation across different species.

Keywords: *Secale cereale*; *Secale montanum*; *Secale strictum*; QTL mapping; molecular marker; self-incompatibility; fertility; seed set

Citation: Gruner, P.; Miedaner, T. Perennial Rye: Genetics of Perenniality and Limited Fertility. *Plants* **2021**, *10*, 1210. https://doi.org/10.3390/plants10061210

Academic Editor: Ioannis Ganopoulos

Received: 6 May 2021
Accepted: 11 June 2021
Published: 14 June 2021

Publisher's Note: MDPI stays neutral with regard to jurisdictional claims in published maps and institutional affiliations.

Copyright: © 2021 by the authors. Licensee MDPI, Basel, Switzerland. This article is an open access article distributed under the terms and conditions of the Creative Commons Attribution (CC BY) license (https://creativecommons.org/licenses/by/4.0/).

1. Introduction

In 2019, rye was grown on about 4.2 million hectares worldwide resulting in an overall production of about 12.8 million tons [1]. This data, however, must be mainly based on annual rye (*Secale cereale* L.) because only a limited number of perennial rye (*Secale cereale* × *Secale strictum*) cultivars were available and to our knowledge the last breeding efforts were done more than a decade ago. According to the original habitat of *Secale strictum* Presl. (syn. *Secale montanum* Guss.) in dry, stony or sandy mountain areas or as weed in or along cultivated fields [2], the breeding goal for perennial rye was to develop cultivars for poor and sandy soils that could be used as forage [3–5] or as soil cover and (especially winter) pasture for solely grazed areas [6] or combined as fodder and grain crop in low-input farming [7,8]. Beside high drought (and cold) tolerance more traits of *S. strictum* were interesting for breeding: large root systems, high tillering capacity, weed suppression, pest resistances, tolerance to (heavy) metals like nickel, zinc, aluminum and manganese and high protein content of kernels [2,9–11].

Secale cereale × *Secale strictum* crosses aiming for establishing perennial rye cultivars were made by several groups worldwide. Reimann-Phillipp in Germany released two tetraploid varieties 'Permontra' (winter type) and 'Sopertra' (spring type) and produced diploid breeding material that did not reach official variety registration [8], but was used in this study. Kruppa and Kotvics in Hungary released the varieties 'Kriszta' and 'Perenne' [3,5] and Myers in Australia produced the (non-registered) cultivar 'Black

Mountain' [6]. The Letbridge Research and Developmental Centre in Canada developed the cultivar 'ACE-1' based on selections derived from Reimann-Philipp [4].

Compared to the annual *S. cereale*, all breeding efforts were confronted with fragile rachis (brittle ear), low grain yields, loose stands especially in the years after first harvest, low fertility, long periods of flowering and ripening and high ergot infections arising as consequence of the latter two in combination with wet weather conditions [4–8,10]. Crossing barriers between *S. cereale* and *S. strictum* were the main reason for the low fertility and cytological studies of pollen mother cells (PMCs) revealed abnormalities. In metaphase I (or anaphase) of the PMCs, where the sister chromatids usually pair together (= seven bivalents), often a multivalent of six chromatids (= three chromosomes in ring or line formation) plus four bivalents (= four chromosomes) was found and this indicated (two) translocations on the three multivalent-forming chromosomes (Figure 1a) [12–16].

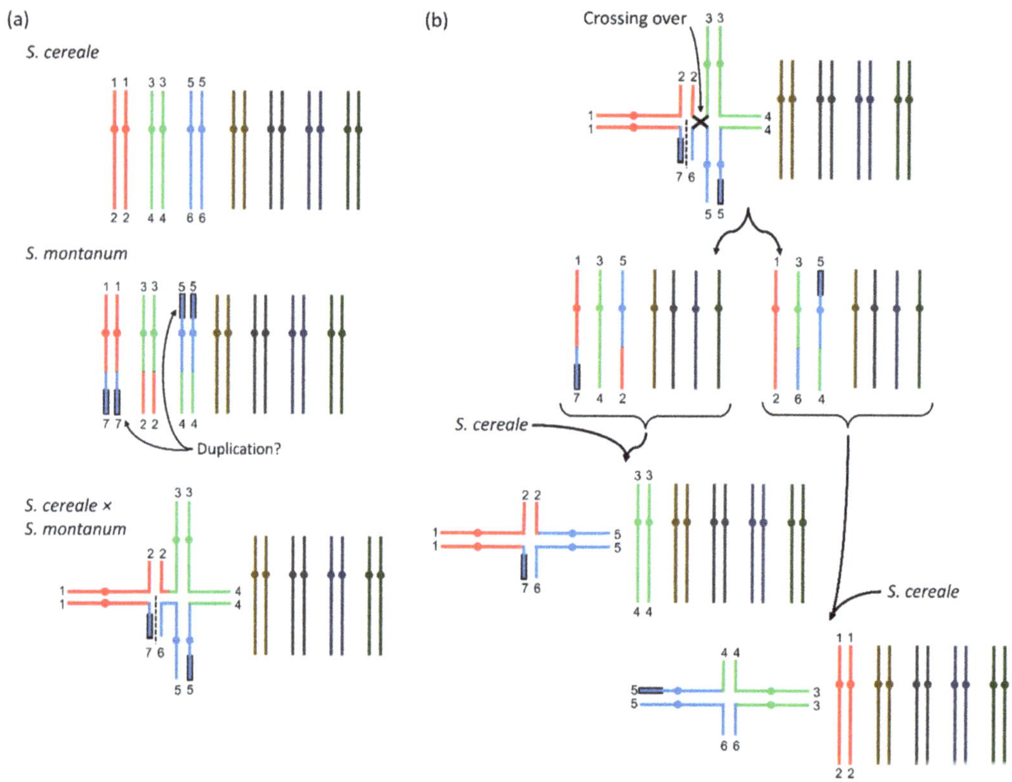

Figure 1. Visualization of multivalents in *S. cereale* × *S. strictum* progenies according to results from Stutz [17]. (**a**) In *S. cereale* × *S. strictum* a multivalent of six chromatids could be detected in metaphase I of pollen mother cells. This was caused by translocations between three chromosomes of the two parental species (in color). Following [17], the numbers indicate the chromosomal assignment, the black framing of the chromosomes indicates a potential duplication and the dashed line the separation point for chromosomal line configurations that were also observed. (**b**) In an F_2-generation, in rare cases also multivalents with four chromatids could be observed. This could be explained by a crossing over within the ring conformation (top). When the resulting gametes (two possible configurations) were then combined with *S. cereale* gametes, two differently composed four-chromosome multivalents could be formed and another two different compositions would be possible when they would be combined with the S. strictum gametes. Here, only the combination with *S. cereale* gametes is shown and more details can be found in [17].

With low frequency, even more chromosomal constitutions could be identified in the progenies [12–14,16], however, some of the results had to be questioned [17] because *S. strictum* accessions that were used showed multivalents also for plants of wild *S. strictum*, which was in disagreement with the other authors listed before. The phenomenon of multivalent formations and explanations for the different constitutions were best described in Stutz [18] and Dierks and Reimann-Philipp [19]. The important link between low fertility and chromosomal translocations was the assumption that DNA of all seven chromosomes was required for a viable gamete. The arbitrary segregation of the six multivalent chromosomes would only result in fertile gametes when all three chromosomes of a single parental species would segregate together. If the arbitrary segregation of the chromosomes of the multivalent was considered, Dierks and Reimann-Philipp [19] calculated that functional chromosomal constitutions could be expected in 25 percent of the cases only. Additional fertility problems of *S. strictum* and potential fertile abnormal chromosomal constitutions may had influenced this theoretical ratio in a way that fertility assessment of microgametes (pollen) and macrogametes (kernels/inflorescences) often exceeded the theoretical expectation calculated from chromosomal segregation ratios [12,19].

Aside of cytogenetically caused non-fertility in the *S. cereale* × *S. strictum* progenies, we added another fertility-related factor in our experiment by crossing a self-incompatible *S. cereale* × *S. strictum* genotype with a self-fertile inbred line. The cultivated rye *S. cereale* is generally a self-incompatible cross-pollinating species, but self-fertile genotypes (resulting in inbred lines) were developed by recurrent selection of partially self-fertile plants detected in extremely large populations (N > 50,000) [20]. The self-incompatibility in rye has been referred to gametophytic mechanisms and interaction of two loci named *S* and *Z* [21] and two loci on chromosome 1R and 2R have been referred to those genes [22–24]. The genes for self-fertility (or pseudocompatibility) have been referred to the same loci [25,26] and self-fertility can be interpreted as special allele *Sf* of the self-incompatibility locus [27,28]. Aside of the *S* and *Z* loci (= *S1* and *S2* loci), a further self-fertility locus *S5* was located on chromosome 5R [28,29]. Even more dominant self-fertility genes were found on chromosomes 1R, 4R, 5R and 6R [29].

Based on microscopic studies, the chromosomes involved in multivalent formation were referred to 2R, 6R and 7R [8] and due to the reasons listed before, the identification of any gene thereon would be challenging. For the perennial habit (perenniality), Dierks and Reimann-Philipp [19] concluded that a major (dominant) gene *P* was located on one of the respective chromosomes. The fertility problems of all the breeding attempts with *S. cereale* × *S. strictum* progenies did support this hypothesis and the only disagreement with this theory coming from Stutz [18], could be regarded to the misinterpretation of the perennial phenotype. Other traits like the fragile rachis ("brittle ears") typical for *S. strictum*, or spring/winter type (vernalization requirement for flower induction) when crossed with spring-type *S. cereale* were inherited independently from the multivalent according to Dierks and Reimann-Philipp [19], however, also close correlation of perenniality and fragile rachis had been observed in experiments where only 1% of the F_2-plants showed perenniality without fragile rachis [20].

Nevertheless, recombination was also observed between the homologous chromosomal segments of the multivalent. Thus, if the perenniality gene *P* would be located at a (distal) chromosomal position where still recombination occurred, it must be possible to identify perennial *S. cereale* recombinants. If there would be no crossing over between a *P*-carrying *S. strictum* chromosome and a *S. cereale* chromosome resulting in viable gametes, the mapping of the perenniality gene would be impossible with classical crossing experiments and consequently, the breeding of fertile perennial rye varieties would require homozygosity for all three translocated *S. strictum* chromosomes [18,19]. Even further, any introgression of *S. cereale* would disturb this configuration and lead again to a high percentage of non-viable gametes and hence reduced fertility. However, Stutz [18] found another multivalent constitution by analyzing a F_2 generation. There, single genotypes showed pollen mother cells with only four chromatids (=2 chromosomes) and five remaining bivalents in metaphase I. This could be explained by a crossing over across the six-chromatid ring multivalent in the F_1 plants (Figure 1b). These configurations would be especially valuable for genetic studies and breeding, because two possible configurations of four-chromosome multivalents with different chromosomes involved could theoretically occur (Figure 1b) and thus it would allow to identify the respective chromosomes and with further crossing and recombination steps also the gene loci.

To our knowledge, the last genetic studies of perennial rye were at the methodological level of chromosome microscopy, but perenniality (and fertility) was also studied in other cereals (rice, sorghum, maize, wheat) and their perennial wild relatives by means of molecular markers [30–36] and sequence based genomics and transcriptomics [37–40]. All those studies showed that the genetics of perenniality was highly complex. Several QTLs and even more gene candidates, of which often several could be located in one single QTL were found. Interestingly, almost all loci that were identified independently in the separate species could be connected through syntenic gene motifs (DNA and protein sequences) across the species, showing that this trait was highly conserved between the species.

For this study, a F_2 population originating from crossing an annual and self-fertile inbred line with a perennial and self-incompatible plant from an improved perennial population that was originally derived from breeding material of Reimann-Philipp [8] was phenotypically assessed for perenniality and fertility and genomically with an Infinium iSelect 10K SNP-chip. At first, we present the phenotypes for both traits separately and the relation between them, then results from studying the pure marker data in regard to multivalents and other abnormalities and finally the QTL-mapping results for both traits.

2. Results

2.1. Phenotype

For both traits, perenniality and fertility, a high amount of genetic variance was observed (Table 1). Perenniality (1–9) was assessed at two sites, but the maximum value of 9 was only recorded once for a single plant and the maximum calculated BLUE was 8 (Figure 2). Most of the phenotypic variance was explained by the genotype, but also the genotype–location interaction was high (Table 1).

Table 1. Variance components (Var.comp.) with standard errors (St. Error) and entry-mean heritability (H^2) for the traits perenniality (scored in 1-9 scale) and fertility (scored from 0 to 100%). Fertility was assessed in one location and hence some factors could not be assessed (n.a.).

	Perenniality (1–9)		Fertility (0–100%)	
	Var. Comp.	St. Error	Var. Comp.	St. Error
Location (L)	0.22	0.39	n.a.	n.a.
Replicate	0.08	0.09	2.6	4.4
Genotype (G)	3.72	0.51	313.7	38.9
G×L	1.18	0.21	n.a.	n.a.
Residual	1.39	0.10	94.9	9.5
H^2	0.81		0.87	

The lowest observation for fertility was 10%, but only a few plants showed this low fertility and all estimated BLUES were higher than 20% (Figure 2). Most of the phenotypic variance was explained by the genotype but no genotype–location interaction could be calculated. For both traits, the entry-mean heritability was high (0.81 and 0.87) but the phenotypic distributions were non-normal (Figure 2). Correlation between both traits was only moderate (0.37). This and the observation of highly perennating genotypes combined with high fertility (Figure 2) indicated an incomplete linkage of both traits. The opposite off-correlation extreme, no perenniality and low fertility, could not be found indicating that still the annual genotypes were the most fertile.

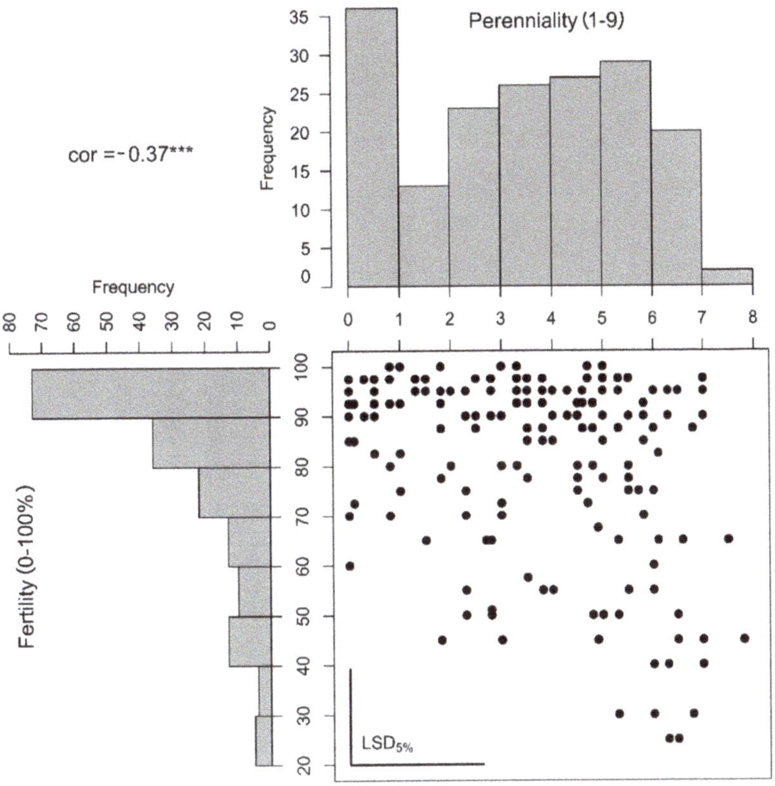

Figure 2. Histograms and correlation plot based on the estimated genotypic means (best linear unbiased estimators, BLUES) of perenniality (scored from 1 to 9) and non-fertility (scored from 0 to 100%). The calculated correlation (cor) was estimated to be −0.37 and highly significant with $\alpha \leq 1\%$ (***). The least-significant differences (LSD) on $\alpha \leq 5\%$ significance level were plotted as bars in the bottom-left corner of the correlation plot.

2.2. Marker Studies

The study of pure marker data had two purposes. Firstly, this was the basis for QTL mapping and secondly, we could use it to find indications for chromosomal abnormalities like multivalents that would had directly influenced fertility. We could successfully apply the Infinium iSelect SNP chip that was initially developed for *S. cereale*, for a *S. cereale* × *S. strictum* population. After filtering, 2641 markers remained from the 10K SNP assay and entered into the analysis. Caused by the few possible recombination events generally observable in F_2-generations, 2314 markers were intercorrelated with one (=redundant), many in several combinations, resulting in 789 unique markers. For those, the "A" marker allele had a frequency of 25.5% on average, the "B" allele of 22.8%, the heterozygous of 51.5% and the missing values 0.1%. A linkage map could be constructed and except for some chromosomal regions, it covered the full genome when compared with overlapping markers (n = 812) from a previously published linkage map [41] (Figure 3). Some chromosomal regions (on 2R, 3R, 5R and 7R) were less covered with molecular markers compared to the previously published map (Figure 3). However, the map from Bauer et al. [41] had many more markers (N = 87,820) and no (large) differences in the marker order of both maps were observed so that our map was still considered sufficient for the linkage mapping.

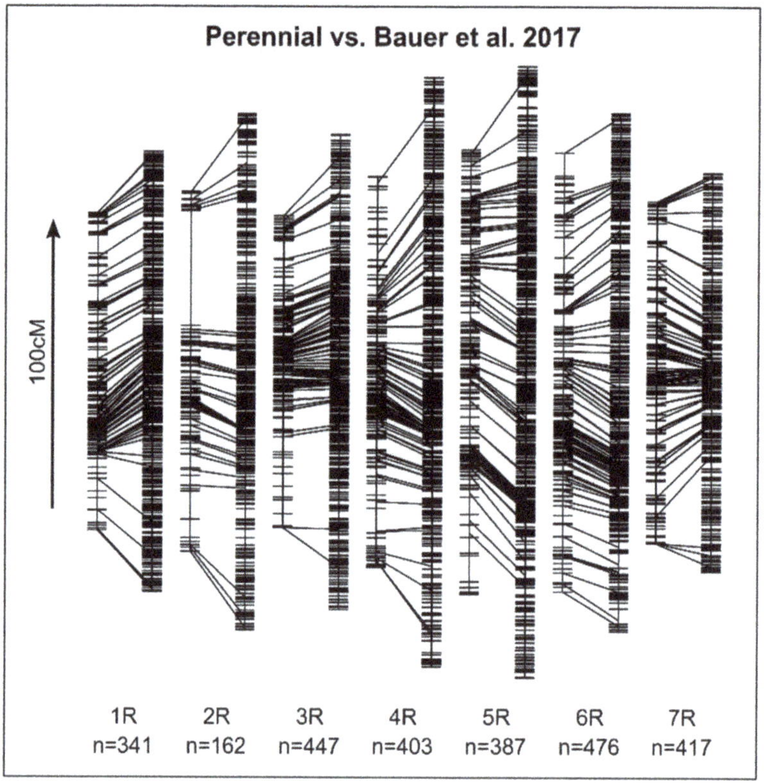

Figure 3. Chromosome-wise comparison of a linkage map constructed from the marker data analyzed here (Perennial, left) with a previously published linkage map [41] (right). Overlapping markers of the respective linkage groups (1R–7R) are connected by lines. For comparisons of the absolute sizes in cM, an arrow indicating the chromosomal order with a length of 100 cM is displayed in the left of the plot. The number of markers in each linkage group of the constructed linkage map (Perennial) is reported (n).

The largest gap on the top arm of chromosome 2R was also characterized by high segregation distortion, where the "B" marker alleles were reduced to zero resulting in a 1:1 ratio for the "A" and "H" allele. The full reduction of the "B" marker allele at zero could not be visualized in Figure 4, because we filtered the marker data set in a first step (before linkage map construction). If we used a linkage map based on other material like the one from Bauer et al. [41], the gap would be flanked by the markers isotig16940 at 114.7cM and C3277_855 at 124.3 cM. We referred this locus to the self-incompatibility locus *Z* [23,24] what is probably equal to the self-fertility locus *S2* [25,26]. The missing of the "B" allele and segregation ratio of 1:1 for the "A" and "H" allele agreed well with a self-fertility model with pollen compatibility [28] and previously reported segregation ratios [25]. Segregation distortion to a lower extend could also be observed at the top arms of chromosome 4R and 6R, but it was considered irrelevant for following results. The successful construction of a linkage map was also an indicator that the linkage of markers was not influenced by chromosomal segregation abnormalities like multivalents. We further correlated all markers with each other and did not find any high correlation between markers of certain chromosomes (Figure S1) like it would be expected if only a certain parental chromosome (chromatid) combination would result in fertile gametes as proposed for example by Stutz [18] (Figure 1a).

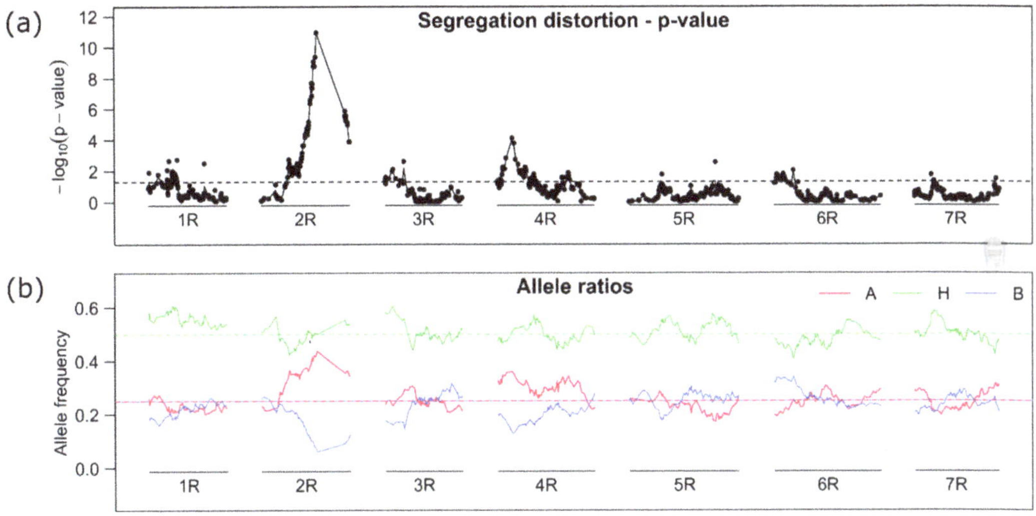

Figure 4. (**a**) Test of markers allele frequencies for distortion from the expected 1:2:1 (A, H, B) ratio of markers on the seven linkage groups (1R–7R). The $-\log_{10}$ of a Chi-square test is displayed. The horizontal dashed line gives a (unadjusted) global 5% threshold level. (**b**) The allele frequency for the respective marker alleles (A, H, B) are displayed (red, green, blue) along the chromosomes. The horizontal lines give the expected frequencies 0.25 for the homozygous state (A, B) and 0.5 for the heterozygous state (H). The lines were drawn by connecting single marker-based estimates.

2.3. Mapping

To detect markers associated with perenniality or fertility, we tested different procedures (scans) without marker cofactors and with differently selected cofactors and results were presented in Table 2 and Figures S2–S7, marker sequences can be found in Table S1. The cofactors and cofactor selection methods only marginally influenced the mapping results. Difference in detected loci between the methods could only be observed for the markers that explained the least variance in single marker fits (QTL-F1a, QTL-F1b, QTL-P2, QTL-P2).

For perenniality, three QTL were consistently found by all methods (Table 2). They were located on chromosome 4R (QTL-P4), chromosome 5R (QTL-P5) and chromosome 7R (QTL-P7). Those QTLs also explained most of the genetic variance, each 0.16, 0.23 and 0.24, respectively. The effects for the first two were mainly codominant with effect sizes of 1.15 and 1.34, whereas the latter was dominant (d effect = cd effect) or even over-dominant (d effect > cd effect), i.e., having a codominant effect of 0.64 and a dominant effect 1.61. All QTL (QTL-P2 to QTL-P7) were highly additive as the explained genetic variance of a model with markers from all QTL together was 0.74 and exactly the same as when explained variances of the single marker fits would be summed up. When combinations of the three QTLs explaining most of the variance (QTL-P4, QTL-P5, QTL-P7) were compared (Figure 5a), the genotypes having all three wild type alleles (H or B) had the highest perenniality followed by genotypes with two wild type alleles of which the combinations of QTL-P4 with QTL-P5 and QTL-P5 with QTL-P7 were on average higher than the combination of QTL-P4 with QTL-P7 (Figure 5a). However, for all combinations also genotypes with only little or even no perennation could be found (Figure S8).

Table 2. QTL mapping results for the traits "perenniality" (P) and "non-fertility" (F). For each locus (QTL) the markers with the smallest *p*-values are reported. Two markers for a locus are reported when different methods (single marker regression without cofactors and with two different sets of cofactors) resulted in a different marker with smallest *p*-value, but an overlapping locus.

Name	Marker	Chr.	Pos.	CI L	CI R	Pos. [41]	Pval c	Pval d	Pval cd+d	Eff. cd	SE cd	Eff. d	SE d	n_A	n_B	n_H	p_G	p_{CovG}
								Perenniality										
QTL-P2	Contig1964 [a]	2	121.0	79.2 [a]	123.3 [a]	153.6 *	6.4×10^{-1}	1.7×10^{-4}	3.6×10^{-5}	−0.69	0.42	1.32	0.30	62	18	96	0.07	0.00
QTL-P3	isotig21327 [b,c]	3	71.1	66.9 [c]	73.3 [c]	100.5	1.0×10^{-2}	2.9×10^{-1}	3.0×10^{-1}	0.58	0.19	0.34	0.27	45	44	87	0.04	−0.08
QTL-P4	C9941_1700 [a]	4	72.0	70.3 [a]	75.4 [a]	89.0 *	5.0×10^{-8}	8.9×10^{-1}	6.4×10^{-7}	1.15	0.18	0.07	0.45	51	40	85	0.16	0.32
	Contig1605 [b,c]	4	74.6	72.0 [c]	75.4 [c]	94.7 *	1.5×10^{-7}	7.6×10^{-1}	5.2×10^{-6}	1.11	0.18	0.14	0.39	51	41	84	0.16	0.33
QTL-P5	C14811_2522 [a,b,c]	5	77.0	71.9 [a]	80.8 [a]	88.8	8.6×10^{-11}	9.2×10^{-1}	6.4×10^{-8}	1.34	0.17	−0.04	0.29	40	49	87	0.23	0.39
QTL-P7	isotig17332 [a,b,c]	7	0.6	0.0 [a]	2.3 [a]	1.2 *	3.4×10^{-4}	5.5×10^{-7}	4.8×10^{-8}	0.64	0.18	1.61	0.27	50	36	90	0.24	0.09
				Contig1964 + isotig21327 + C9941_1700 + C14811_2522 + isotig17332													0.74	0.80
								Fertility										
QTL-F1a	isotig11337 [b]	1	29.7	27.2 [b]	30.6 [b]	62.7 *	9.1×10^{-4}	2.0×10^{-1}	3.4×10^{-4}	−7.2	0.23	3.6	0.30	37	36	103	0.06	0.14
	isotig22259 [c]	1	33.9	30.6 [c]	34.8 [c]	68.3 *	1.3×10^{-2}	1.4×10^{-1}	2.5×10^{-3}	−5.8	0.24	4.2	0.30	37	32	107	0.04	0.11
QTL-F1b	Contig1017 [a]	1	63.2	59.3 [a]	66.6 [a]	91.6 *	1.7×10^{-4}	3.0×10^{-3}	3.2×10^{-6}	−7.7	0.22	8.1	0.29	37	38	101	0.12	0.17
QTL-F4	isotig03456 [a]	4	78.2	69.2 [a]	87.7 [a]	103.1 *	2.7×10^{-5}	7.7×10^{-3}	1.7×10^{-6}	−7.3	0.20	−7.2	0.29	53	38	85	0.13	0.29
	Contig1437 [b,c]	4	86.2	83.3 [c]	87.1 [c]	126.5 *	1.6×10^{-5}	2.4×10^{-1}	1.3×10^{-5}	−7.9	0.20	−3.2	0.29	37	41	82	0.11	0.25
QTL-F5 (*S5*)	C28789_183 [a,b,c]	5	41.7	40.6 [a]	42.0 [a]	44.9 *	2.8×10^{-37}	5.5×10^{-15}	9.5×10^{-33}	−18.4	0.14	14.9	0.20	48	38	90	0.64	0.35
					isotig11337 + Contig1017 + isotig03456 + C28789_183												0.77	0.71

The letter superscripts indicate in which model the respective marker passed the threshold and was the most significant: [a] single marker regression without cofactors, [b] single marker regression with marker cofactors selected from a first run and [c] single marker regression with marker cofactors selected by information criterion based cofactor selection. The superscript also indicates the method used for confidence interval estimation. For each marker the following parameters are reported: the chromosome (Chr.) and position (Pos.) based on a population specific linkage map, based on the linkage map published by [41] where * indicates that this marker could not be found on the public map and the position from the most closely linked marker was taken instead, confidence limits (CI left and CI right) based on $\log_{10}$(*p*-value) drop in 1, *p*-values (Pval) for codominant (c) effects, dominant (d) effects and the combination thereof (cd + d) as well as the effect sizes (Effect) with respective standard errors (SE), the counts (n) for the ABH-marker alleles, the explained genetic variance (p_G), and the explained genetic covariance (p_{CovG}). For estimation of the respective parameters, each marker was used as fixed effect (cd + d) in a separate model and only for calculation of total explained variance a model was fitted combining cd + d effects of all markers simultaneously.

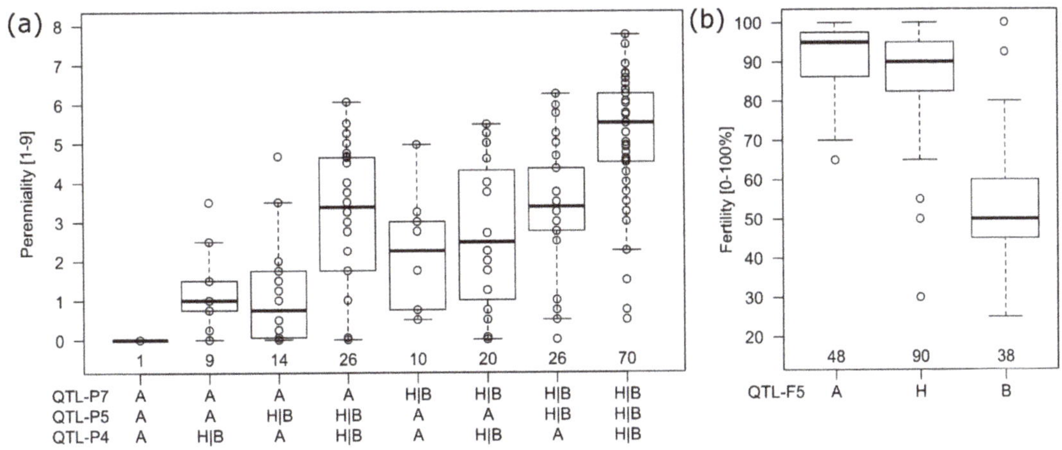

Figure 5. Genotypic means for perenniality (**a**) and fertility (**b**) clustered by marker alleles A, H and B (x-axis) of the most significant markers in the respective QTLs. For perenniality (**a**) the H and B allele was combined in one group (H | B) and all genotype means were additionally plotted as dots in the boxplot display (box at first and third quartile with median in center). The number of genotypes with the respective marker allele combinations is written above the x-axis within the plot.

The trait fertility was mainly explained by a major QTL (QTL-F5) on chromosome 5R explaining 0.64 of the genetic variance (Table 2, Figure 5b). The codominant effect was −18.4 and the dominant effect with 14.9 almost as high as the codominant. The negative effect size indicated that the A allele (inbred line) was leading to higher fertility (Figure 5b). More QTLs (QTL-F1a, QTL-F1b, QTL-F4) on chromosome 1R and 4R could be found. The additional QTLs explained only 0.06 to 0.13 of the genetic variance, had smaller (negative) effect sizes and the (positive) dominant effects of QTL-F1a and QTL-F1b again indicated the fertility to be dominantly inherited by the parental A allele for those two loci.

Additionally to this basic QTL mapping approach, we studied epistatic effects in terms of marker–marker interactions (Figures S9–S14) and to detect those, all possible combinations with cd-cd, d-d and cd-d marker interactions were tested in a model with the same markers as single (cd and d) main effects simultaneously. As this resulted in many more tests, we adjusted the global significance threshold. With this stricter threshold no significant marker–marker interactions could be found, except of some single neighboring markers on chromosome 4R with cd-d interaction being significant. Nevertheless, for both traits several chromosomal regions showed high p-values for marker–marker interactions, but only a few were associated with the main QTLs we found. We concluded that with the given study set-up (population size and phenotypic error) we could not infer any useful conclusion regarding epistasis. Anyway, a detailed visualization can be found in Figures S9–S14. We also estimated explained covariance of both traits for the respective QTLs, but the explained covariance estimated for each QTL (Table 2) was generally low and the largest influence on the explained genetic covariance for a single marker were QTL-F4, QTL-F5, QTL-P4 and QTL-P5 with values between 0.25 and 0.39 each.

3. Discussion

Our study revealed new insights into perenniality and fertility of *S. cereale* × *S. strictum* progenies and we could show that perenniality was a complex trait with several QTLs involved compared to fertility (or non-fertility) which was mainly caused by a self-fertility allele at a self-incompatibility locus coming from the perennial parent. We could not find any indications for abnormalities in chromosomal segregation (multivalents) and their relation to low fertility, but as this was intensively discussed in previous studies we will discuss it in more detail here. It will be followed by a discussion of the results from mapping fertility and perenniality. Due to the scarce genomic-related literature on perenniality in rye, we addressed the usefulness of results from other species (species sytheny) and finally ended in future breeding perspectives.

3.1. Multivalents

We could not find an ultimate explanation why we could not prove the presences of multivalents with molecular markers. It was most reasonable, that the perennial genotype that we had used as parent did not result in multivalents (anymore), because it belonged to an improved perennial population from Reimann-Phillipp, who continued (also privately) with perennial rye breeding after releasing his research works on this topic and his major selection criterium was high fertility. Stutz [18] showed a possible way out of multivalents (Figure 1b) where crossing overs between the chromosomes could reduce the initial six-chromatid multivalent into a four-chromatid multivalent and it may be possible that following recombination events would have even led to chromosomal constitutions without any multivalent. Additionally, recombination was observed between the chromosomes of the multivalent [18,19] what additionally could had caused translocated chromosomal segments to become smaller and reduced the lethality for gametes with certain combinations of translocated chromosomes. Even further, not only the genetic resource itself but also the perennial F_1 plant was chosen based on high fertility from crosses that had been made with several perennial plants to develop the population under study. Even more could the (indirect) selection against the "B" allele at the self-incompatibility loci *Z* have influenced the constitution of the respective population as it would be located on one of the multivalent forming chromosomes [8]. To further clarify this issue, again the chromosomes in metaphase I of the pollen mother cells must be (microscopically) studied. Unfortunately, the material used here was not maintained as inbred lines, but remaining kernels of the same cross are currently developed into inbred lines so that more seeds for future studies combining both, molecular markers and microscopy of metaphase chromosomes in pollen mother cells would be available. If we could show, that the crossing barrier in terms of multivalent formation was overcome (or not an issue at all), it would allow (the three multivalent forming chromosomes of) *S. strictum* to be used as new (secondary) rye breeding pool, what may be especially interesting in terms of disease or drought resistance. Surprisingly, in a diversity study based on molecular markers [42] a single variety Gonello (KWS Lochow), that was characterized by high frost tolerance, was related to *S. strictum* showing that this wild species may had already been used in annual rye breeding.

3.2. Fertility

The locus QTL-F5 could be identical with the *S5* locus [25,26] and the high amount of explained genetic variance of this locus proved it to be the main reason for non-fertility. It was surprising, that even though we did not place isolation bags on the heads of the plants, we most probably detected a self-incompatibility locus (with a segregating self-fertility allele) to explain most of the genetic variance for fertility. The reason could be that the experiment was flowering 2–3 weeks later than the rye stand growing on the experimental station so that the main pollen cloud could not pollinate the experiment. Additionally, the genotypes were grown as single plants in a distance of each other (about 0.25 m) so that also within the experiment cross pollination was reduced. Still, for confirmation of limited self-fertility it would be necessary to use isolation bags preventing any cross-pollination in future experiments. In this study, no genotype was completely sterile. In rye, there were several loci known for self-incompatibility, three of them were located on chromosome 1R, 2R and 5R and reported most often [23–27] but also loci on 4R and 6R have been proposed [29]. When self-fertility is interpreted as a special allele *Sf* at those loci, we hypothesised that the expression of *Sf* lead to an universal (or complementary) structure resulting in successful pollination. However, the pollination was only successful when *Sf* was expressed in both, pollen and stigma. Important for the segregation of the respective alleles at both loci was that a pollen structure would be defined by one haplotype (n = 1x) and consequently no pollen with the "B" allele at locus *Z* could fertilize the macrogamete.

This resulted in a 1:1 ratio of the "A" and "H" allele at the Z locus, what was proven by marker segregation. The pistil surface on the other side was in diploid stage and here a single (dominant) "A" allele must had expressed the self-fertility as we could show by the respective effect sizes (Table 2, Figure 5b). We visualized our hypothesis in Figure 6. However, we never (microscopically) studied the growth of the pollen tube on the stigma and eventually also other mechanisms could prevent the fertilization of the macrogamete.

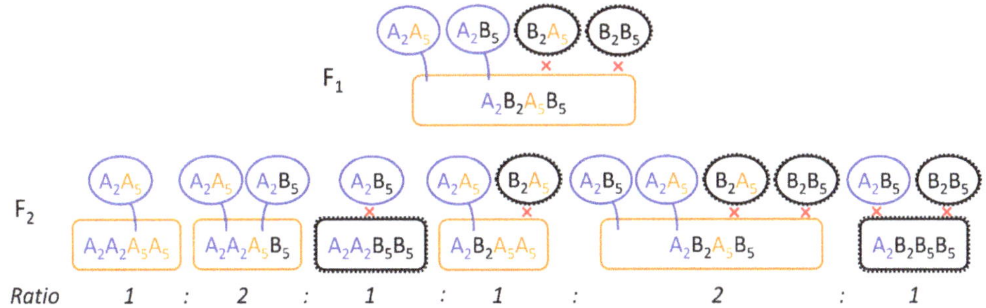

Figure 6. Visualization of self-fertility hypothesis for locus *S2* (Z) and *S5* (subscript 2 and 5) with the respective parental alleles for the inbred line (A) and self-incompatible genotype (B). The hypothesis displayed here was that both, pollen and stigma, carry proteins or any structure (black) on the surface that was responsible for the self-incompatibility mechanism. Those structures were expressed by a respective allele at each (multiallelic) locus. The *Z* locus was causal for the pollen (circles) and the *S* locus for the stigma (rounded rectangles). Self-fertility (*Sf*) could have been expressed by a mutation at each locus (pollen = blue, stigma = orange) that resulted in no or a special (protein) structure causing, if matched with a similar or complementary structure on the opposite surface (= two colored surfaces), successful pollination. Because the pollen was haploid and the stigma diploid, the *Z* alleles segregated 1:1 (A_2A_2:A_2B_2) and the *S5* alleles 1:2:1 (A_5A_5:A_5B_5:B_5B_5) and the ratio of the combination of both loci is displayed in the figure. The *Sf* allele for the stigma (*S5*) was dominant.

Previous studies were mainly based on marker segregation [25,27] and in further inbreeding generations, the alleles for the *S5* locus would theoretically deviate more and more from a 1:1 ratio (of randomly segregating parental alleles), because certain genotypes could not be self-pollinated (Figure 6). In the F_3 generation, therefore, we would expect an average ratio of the *S5* alleles of A:H:B = 3:2:1 (= 2:1 ratio of parental alleles A and B). Due to excess of pollen, the macrogametes could be successfully self-pollinated even when the *Z* locus was heterozygous so that the interaction between both loci could (principally) not be validated on marker basis. If we assume that there were only two target sides (pollen or stigma surface) affected by expressed *Sf* alleles, it may even be possible that there were more (stigma-based) self-incompatibility loci segregating in this population, that could not be detected because they were masked by the *Z* and the *S5* locus. Further, the high amount of explained genetic variation of the *S5* locus indicated that the lack of fertility was mainly dependent on self-incompatibility and not on chromosomal abnormalities and the additional QTLs we detected could also be additional self-incompatibility loci. For the identification of reasons for fertility other than self-incompatibility, additional studies of pollen vitality or germination, as done in early studies [12], could also help to better assess non-fertility. In this study, (extremely) sterile plants could also be caused by pollen sterility instead of self-incompatibility that we did not investigate. Unintentionally, the self-incompatibility loci were also selection factors for perenniality. As discussed before, the unintentional selection at the *Z* locus may have even helped to select crosses with "advanced" perennial genotypes. The second self-incompatibility locus *S5* could additionally help to shorten the distance to QTL-P5, because it explained 0.35 of the genetic covariance between non-fertility and perenniality (Table 2). Though, in following generations the most-fertile and most perennating genotypes must be selected. A random continuous self-pollination could otherwise result in a reduction of perennial genotypes.

3.3. Perenniality

To the best of our knowledge, this is the first mapping study for perenniality in rye. We could show, that perenniality was not caused by a single major gene like it was proposed by Dierks and Reimann-Phillip [19] and five QTLs could be located on chromosomes 2R, 3R, 4R, 5R and 7R that in combination explained 0.74 of the genetic variance. All identified QTLs were highly additive so that three of them, QTL-P4, QTL-P5 and QTL-P7 with the highest explained genetic variance, each ranging from 0.16 to 0.24 may be most interesting for breeding. However, a fixation of three QTLs simultaneously would require large population sizes and for the marker-defined clusters still a large variation between the genotypes was observed (Figure 5a). Future (backcross) populations with defined marker combinations must proof potential redundancy of the identified QTLs (genes) and whether the number of QTLs for perenniality could be reduced to still reach a high trait level. We additionally tried to estimate marker–marker interaction but concluded that it was not significant. Here, the adjustment of the genome-wide error might had been too strict. With the given method for adjustment [43] it was computationally not possible to enter all marker–marker interactions into a PCA and calculate the effective marker number from it directly. Thus, we used the number of all possible combinations of effective markers from the single marker fit as effective marker number (=$q_{eff}(q_{eff} - 1)$) for the global-threshold adjustment of the marker–marker interactions. Additionally, the incremental fit of the fixed model effects reduced the power for the (lastly fitted) marker–marker effects and hence most probably, the procedure was generally over-adjusted. An example for epistasis could be found in rice where a dominant complementary gene action for *Rhz2* and *Rhz3* was concluded [31].

Aside from the genetic complexity, the trait was also dependent on the environment. This was shown by large genotype–location interaction (Table 1) and could be explained by the expression of the perennial phenotype by several genes, of which each could be differently affected by environment. If we would split up the trait perenniality (and the environment) into more factors, like number of axillary buds (shoots), general number of tillers or sensitivity to vernalization, we may probably gain a better understanding of the different genetic and environmental factors. In this study, multi-environment trials were limited by the use of the F_2 generation that we had chosen to reduce the influence of (at this point) unknown fertility reasons. With further inbreeding generations, more seeds would be available to replicate a single genotype (line) in several environments. To still not neglect this issue, we vegetatively cloned the single genotypes, resulting in plants strong enough to be tested in two (ecologically highly) different environments.

3.4. Species Synteny

To our knowledge, this was the first mapping study of perenniality in rye, but in other cereals perenniality has already been mapped and even further, certain loci could be connected across species (e.g., in rice, sorghum, teosinte, *Leymus*-wildrye) by comparing gene, marker or protein sequences [30–32,35,37,38,40,44,45]. Because rye was generally shown to be highly syntenic to some of those species [46,47], it could be interesting to compare our results with the other species. However, on a phenotypic level, the perenniality in our study differed from the other crops, because there, perennial genotypes were simultaneously characterized by rhizomatous growth. On a genomic level, it was difficult to make proper comparisons, because so far for rye only reference genome sequences for the annual crop (*S. cereale*) had been published [47,48] and functional perennation genes could principally only be found in genomes of perennial species (genotypes). If we expect similar genes for perenniality across cereals, the comparison of genomic sequences from perennation-associated QTLs in perennial species could be a valuable resource to dissect perennation from related traits and to filter the vast amount of potential gene candidates. The sequence of perennial rice, *Oryza longistaminata,* was already published [49] and perennial sorghum, *Sorghum propinquum,* was mentioned to be sequenced [45]. To our knowledge, research on perennation in the closest rye relatives, wheat (*Triticum aestivum*) and barley (*Hordeum vulgare*) ended with crosses between crop and wild relative [50,51]. In terms of basic knowledge of perenniality, studies of the model plant *Arabidopsis thaliana* or better its perennial relative *Arabis alpina* were also useful. A perenniality-related flowering gene was identified [52] and the study showed that flowers of *Arabis alpina* developed from the main shot and the axillary buds (meristems) were most important for the perennial life cycle. Another study [53] showed that the axillary buds that were developed at a distance from the shoot apical meristem (SAM) and before the onset of vernalization, remained dormant during flowering and this was the prerequisite for perenniality. Caused by vernalization and the following determination of the SAM into flowers, also axillary buds in proximity were developed and their fate of vegetative growth (in the first flowering period) was determined. This showed, that one key mechanism for perenniality was the conservation of axillary buds and consequently the isolation or even protection from plant hormones triggering flower induction and later senescence. The fact, that axillary buds were developed before flowering showed that plant resources were invested into the vegetative growth (instead of seed set) before flowering and seed yield reduction due to perenniality would mainly be caused by less (dense) tillers compared to annual plants. An important question would be, to what extent the roots were affected by the senescence of flowering shoots and how the plant can reach new nutrients, especially when plants cannot spread out by rhizomatous growth like in perennial rye. It was also shown that environmental factors like the duration of cold treatment influenced the percentage of plants that showed senescence after seed set [53].

3.5. Breeding Perspectives

Compared to the trait fertility, which could be reduced to a single self-incompatibility locus, perenniality remained a complex topic in which we, as a first step, identified QTLs for perenniality in rye. The trait also showed high genotype-environment interaction. In comparison to our experimental set up with intensively cherished single-plant growing practice, more farming-like practices with field plots previously showed worse results in the degree of perenniality. When perennial genotypes were grown in large-drilled yield plots, the regrowth in the second year was much lower compared to a single plant growing practice and the yield of the perennial progenies was reduced compared to annual rye sown a second time in the same plot [54] illustrating that intense intra-plot competition negatively affects perenniality. Perennial rye varieties were better suited for dual purpose use (grain + biomass), grazing or mixed cropping [3,5,54] and it is probable that such practical challenges would also appear in other perennial cereals like wheat or barley as soon as they reach higher breeding progress. In wheat, the progress was challenged by the hexaploid karyotype of wheat combined with the diploid or tetraploid karyotype of the perennial (*Thinopyrum*, *Leymus*) species [35,51]. Recombination between the genomes was lacking and as a solution amphiploid wheat was discussed to serve as potential new breeding pool [55], but has previously also been found as impractical [56]. Generally, all species comparisons highlighted that perenniality must be well defined [57] and also for rye the differences in the phenotype could be refined for more detailed studies. Especially with a focus on future breeding efforts, the identification of perenniality-related plant structures like dormant buds or shoots in the young rye plants could be a powerful selection criterion and allow the trait perenniality to be assessed even before flowering and hence to shorten the breeding cycle.

Aside from breeding perennial varieties, perenniality may be interesting as an additional tool for rye breeding. The conservation of single plants over several years without development of inbred lines would allow to implement new breeding strategies, especially for population breeding. By means of polycross or (incomplete) diallel methods [58,59] the combining ability of a single genotype (plant) could be estimated from its offspring after crossing with a tester population and based on it, superior (perennial) plants could be intercrossed after the two-year testing procedure to build up a new improved population. Such a methodology was so far only possible by special short-day cultivation practices keeping clones of the plants in a vegetative state alive over several years that was in practice not very successful [60,61]. Later on, in vitro propagation of single rye plants was tried, but after the long storage period under cold conditions the plants quickly switched into the generative stage resulting in one or a few tillers only and thus restricting seed availability. Beside this special breeding purpose, the largest potential of perennial rye was to use it as genetic resource for trait introgression into cultivated (hybrid) rye, because we observed a good resistance to leaf rust (*Puccinia recondita*) and stem rust (*P. graminis* f.sp. *secalis*).

4. Materials and Methods

4.1. Breeding Material

Breeding material of a self-incompatible perennial rye population originally derived from a cross *S. cereale* × *S. strictum* was received from Reimann-Philipp who selected for high levels of fertility and perennation over several cycles. Several plants thereof were crossed plant-wise with a self-fertile breeding line of the hybrid rye program of the University of Hohenheim (L301-N) in the greenhouse in 2014. From the resulting F_1 plants (sown in autumn 2014), the most fertile single plant with high perenniality was self-pollinated under an isolation bag (in 2015) resulting in the F_2 population 'L301-N × 84/1′ that derived from a single gamete of the perennial rye population. We had to analyze this F_2 population and could not self the population subsequently (e.g., by single-seed descent), because otherwise the non-fertile plants would have been eliminated equaling a strong (natural) selection and producing a bias to our mapping population given the correlation of fertility and perenniality.

4.2. Field Trials

In autumn 2015, the F_2 population was sown in trays and transplanted in pots after having 2–3 leaves. When the plants (= genotypes) produced enough axillary shoots (tillers), 200 plants were vegetative cloned (ripped apart) resulting in four clones per genotype. Two clones of each genotype were planted as replicates in the field located in Stuttgart-Hohenheim (48°42′54″ N 9°11′22″ E, 389 m above sea level, mean annual temperature 10.1 °C, mean annual precipitation 691 mm) and two clones at 'Oberer Lindenhof' (48°28′26″ N 9°18′18″ E, 720 m above sea level, mean annual temperature 6.8 °C, mean annual precipitation 942 mm) close to Würtingen (St. Johann, Germany). We used spaced planting with a distance of 0.25 m from each neighboring plant. The genotypes plus some checks were grown in two randomized complete blocks at each location. The plants flowered in 2016, seeds were harvested subsequently and the plants cut by hand to 20–30 cm above the ground, so that perenniality (plant regrowth) could be assessed some weeks later according to the following.

Perenniality was assessed on a 0 to 9 scale, indicating the amount of new arising tillers in relation to the remaining stubble from the initial shoots. Zero indicated no regrowth of the plant and nine a full regrowth resulting in about the same number of tillers as remaining stubbles from harvest. To prevent false scorings due to germinated seeds that fell out by threshing, the plots were cleaned with a commercial leaf blower after harvest. In Hohenheim, the seed set or better non-seed set ("Schartigkeit") of the heads was visually scored as percentage (0–100%) of florets that did not develop seeds (in comparison to the seed set of non-perennial self-fertile rye grown as check). Consequently, fully sterile plants were scored with 100% non-seed set and fully fertile (like standard self-fertile rye) were rated with 0% non-seed set. To simplify the wording throughout the paper, we calculated the fertility by 100% minus non-seed set and used this term instead.

4.3. Marker Analysis and Linkage Map Construction

From the F_2 population, 182 genotypes and the respective parents were used for marker analysis. The DNA was extracted from about 4–5 cm long, dried leaf samples using the Macherey and Nagel (Düren, Germany) NucleoSpin 96 Plant II DNA extraction kit. Marker analysis was done with a proprietary rye 10K Infinium iSelect SNP chip at

KWS SAAT SE and Co. KGaA, Grimsehlstr. 31, 37555 Einbeck, Germany. The SNPs of this assay were partially overlapping with the 5k-SNP assay of Martis et al. [46] and the 600k-SNP assay of Bauer et al. [41]. Only segregating markers were kept and coded as ABH (parent1, parent2, heterozygous). When one of the parents had a missing or heterozygous marker allele, the other parent was used as reference for ABH coding. If marker alleles were heterozygous and/or missing for both parents, the markers were dropped from the data frame. Additionally, markers with more than 10 percent missing values were dropped, resulting in 2641 markers of which 2314 were correlated with one (= redundant) in several combinations so that only a single marker thereof (with least missing values) was kept for further analysis, resulting in 789 unique markers. Those markers had 0.1% missing values on average and were imputed with the "imputeByFlanks" function from the R package ABHgenotypeR [62]. After this, still seven missing values remained in the data. Those were imputed manually. For linkage map construction we used the non-imputed data and included the redundant markers as they were located at the same position anyway, but allowed a better comparison of markers with the already published linkage map from Bauer et al. [41]. The linkage map was constructed using the R package ASmap and the "mst.map" function with "Kosambi" method [63,64]. Based on marker data, two genotypes were identified as being identical, each to a further one. The duplicates (one genotype per duplicate) were removed for linkage map construction and marker studies. For mapping, both duplicates (two genotypes per duplicate) were removed.

4.4. Phenotypic Analysis and Mapping Procedure

Phenotypic means were calculated by using mixed models implemented in the software ASReml for R [64,65]. Given by the experimental structure the model (1) was $y_{ijkl} = \mu + g_i + l_j + r_{jk} + (gl)_{ij} + e_{ijkl}$, with the observation y_{ijkl} explained by the overall intercept μ and effects for the ith genotype g, which was modelled as fixed for best linear unbiased estimators (BLUEs) and as random to estimate variance components, and the further random effects for the jth location l, the kth replicate (= cloned plant) r nested within the location, the genotype–location interaction $(gl)_{ij}$ and the error term e_{ijkl}. As the fertility was only assessed in a single location, the location effect and the respective interactions were not fitted for this trait. The heritability was calculated by $H^2 = \sigma_g{}^2/(\sigma_g{}^2 + av.VD/2)$, where $\sigma_g{}^2$ is the genetic variance and $av.VD$ the mean variance of a difference of two BLUEs [66].

The QTL-mapping procedure was based on single marker regression fitting each marker as codominant (first) and as dominant (second) fixed effect. The codominant (cd) markers were coded as 0, 1, 2 (A, H, B) and the dominant (d) as 0, 1, 0. For the trait perenniality, also random effects for the (cd and d) marker-location interactions were fitted. For each (fixed) effect the p-value was extracted from a Wald-test statistic of the fixed effects. Additionally, a p-value for the combination of both effects was calculated by adding up the (incremental) Wald-statistics of both effects and calculating p-values based on Chi-square statistics with and adjusted number of degrees of freedom (df = 2). The global significance threshold was calculated by the simpleM method [43]. Comparable with Bonferroni correction, the defined genome-wide significance threshold $\alpha = 0.05$ was divided by the number of tests (markers) q in order to obtain the SNP-wise significance level. Due to correlation of the markers, the tests were not independent and α was divided by an effective number of markers q_{eff} instead. The effective number of markers q_{eff} was calculated by running a principal component analysis (PCA) based on the marker data and the number of eigenvalues that explained 99.5% of the variation of the SNP data. PCA was based on a correlation matrix of the markers coded in 0, 1, 2 (A, H, B) and calculated by using the "eigen" function in R [64]. The procedure was initially developed for genome-ide association studies (GWAS) [43] and we considered the linkage mapping as special case of GWAS resulting in an even smaller number q_{eff} due to the high linkage of markers, compared to GWAS studies.

To adjust for masking effects of genes located on other chromosomes than the single marker in the fit, we additionally run two scans with models including markers (cd and

d) as fixed cofactors fitted in the model sequence before the actual marker under testing (cofactor-based mapping, CM). The two additional scans differed in terms of the cofactor selection. For the first additional scan CM1, from each chromosome the most significant markers were chosen, that passed the global significance threshold in a cofactor-free scan (compare [67]). For the second additional scan (CM2), a simultaneous forward- and backward-selection procedure was used to identify the respective cofactors. For simplicity reasons and computational speed, this procedure was based on a linear model including only the BLUEs and markers in cd coding. The procedure was implemented as "steps" function in R [64]. We started the procedure with a Null model (only intercept) and used the (Schwarz) Bayesian information criterion (BIC) [68] by setting k = log($N_{genotypes}$). Similar to the single marker regression without cofactors, *p*-values for cd and d marker effects were extracted from the Wald-test statistics.

For all markers identified as significant we calculated effect sizes plus standard errors from the estimated coefficients and coefficient error variance. The explained genetic variance p_G was calculated as reduction in the genetic variance estimated in the phenotypic model (1) by additional marker effects, divided by the full genetic variance estimated in model (1). We additionally ran a bivariate model including both, perenniality and non-fertility in one model. We fitted random effects with unstructured variance-covariance matrix for the genotype and the residual and with diagonal variance-covariance matrix for the replicate. Only data with records from both traits was used and thus no location effect was fitted. Similar to p_G, the explained genetic covariance p_{CovG} was calculated as reduction in the genetic covariance estimated in the bivariate phenotypic model by additional marker effects divided by the full genetic covariance estimated in in the bivariate phenotypic model.

To detect potential epistasis of marker1 (m1) with marker2 (m2), a scan with all marker–marker interactions and possible d and cd combinations was run. The (incremental) sequence of fixed effects was the following: $m1_{cd} + m2_{cd} + m1_d + m2_d + m1_{cd}:m2_{cd} + m1_d:m2_d + m1_d:m2_{cd}$. The $m1_d:m2_{cd}$ interaction was tested also vice versa ($m2_d:m1_{cd}$). No additional cofactors were used in this run, but for the trait perenniality all fixed marker effects were additionally fitted as random marker-location interactions. We defined the q_{eff} for global significance threshold of epistasis effects as product of q_{eff} and (q_{eff}-1) as this would be the number of all possible combinations. Due to the high number of marker–marker combinations tested (n = 621,732) we did not run a PCA and not deduced q_{eff} specifically for the interactions.

5. Conclusions

We could show that perenniality in rye was quantitatively inherited and we identified five QTLs with medium effect sizes of which some acted dominantly. This quantitative nature could be challenging for breeding of perennial rye and if breeders aim for perennial varieties, future studies should investigate the impact of the environment and to what extent selection on perenniality-related plant structures or with molecular markers could shorten breeding cycles. Reduced fertility was related to cytogenetic causes in previous studies. We could show, that in our material it was mainly related to self-incompatibility by identifying a locus that explained 0.64 of the genetic variance and is most probably the *S5* locus, known from other studies. If the self-incompatibility is considered consequently, reduced fertility should not be a concern in future breeding programs.

Supplementary Materials: The following are available online at https://www.mdpi.com/article/10.3390/plants10061210/s1, Table S1 DNA sequences of the reported markers; Figure S1 Pearson correlation of markers; Figure S2 LOD scores for mapping the phenotype perenniality by single marker regression; Figure S3 LOD scores for mapping the phenotype perenniality by single marker regression and using additional markers from cofactor selection as cofactors; Figure S4 LOD scores for mapping the phenotype perenniality by single marker regression and using additional markers from a first run as cofactors; Figure S5 LOD scores for mapping the phenotype fertility by single marker regression; Figure S6 LOD scores for mapping the phenotype fertility by single marker regression and using additional markers from cofactor selection as cofactors; Figure S7 LOD scores for mapping the phenotype fertility by single marker regression and using additional markers from a first fit as cofactors; Figure S8 Genotype means for the trait perenniality clustered by marker alleles; Figure S9 Display of codominant-codominant marker–marker interaction of mapping perenniality; Figure S10 Display of dominant-dominant marker–marker interaction of mapping perenniality; Figure S11 Display of codominant-dominant marker–marker interaction of mapping perenniality; Figure S12 Display of codominant-codominant marker–marker interaction of mapping fertility; Figure S13 Display of dominant-dominant marker–marker interaction of mapping fertility; Figure S14 Display of codominant-dominant marker–marker interaction of mapping fertility.

Author Contributions: P.G. analyzed the data and wrote the manuscript. T.M. conceived the study, supervised the project and edited the manuscript. All authors have read and agreed to the published version of the manuscript.

Funding: Marker analyses were funded by a special budget of the University of Hohenheim (TG77). The first author was funded by the German Federal Ministry for Economic Affairs and Energy (BMWi, grant # IGF-Nr. 246 EBG) via AiF (Arbeitsgemeinschaft industrieller Forschungsvereinigungen "Otto von Guericke" e.V. Cologne) and Gemeinschaft zur Förderung von Pflanzeninnovation e. V., Bonn, Germany (GFPi), Bonn (G163/19 AiF), in the framework of the CORNET (Collective Research Networking) program ProtectRye.

Institutional Review Board Statement: Not applicable.

Informed Consent Statement: Not applicable.

Data Availability Statement: The SNP marker chip we used in this study was proprietary to KWS SAAT SE and Co. KGaA and thus the marker data is not publicly available. Marker sequences for the most important markers can be found in the Supplementary Table S1 and were additionally referenced on a previously published linkage map [41]. Seeds for perennial rye genotypes can be requested from the corresponding author. The material used in this study was not maintained.

Acknowledgments: We cordially thank Mark Raith for conducting the field trials and doing the assessments and the whole technical staff of the rye group for assistance.

Conflicts of Interest: The authors declare no conflict of interest.

References

1. FAO. FAOSTAT. Available online: http://www.fao.org/faostat/en/#data/QC (accessed on 13 January 2021).
2. Frederiksen, S.; Petersen, G. A Taxonomic Revision of *Secale* (*Triticeae*, *Poaceae*). *Nord. J. Bot.* **1998**, *18*, 399–420. [CrossRef]
3. Füle, L.; Galli, Z.; Kotvics, G.; Heszky, L. Forage Quality of 'Perenne', a New Perennial Rye Variety (Secale cereale x Secale montanum). In Genetic Variation for Plant Breeding, Proceedings of the 17th EUCARPIA Congress, Tulln, Austria, 8–11 September 2004; BOKU—University of Natural Resources and Applied Life Sciences: Vienna, Austria; Tulln, Austria, 2004; pp. 435–438.
4. Acharya, S.N.; Mir, Z.; Moyer, J.R. ACE-1 Perennial Cereal Rye. *Can. J. Plant Sci.* **2004**, *84*, 819–821. [CrossRef]
5. Halász, E.; Sipos, T. Perennial Rye (*Secale cereanum*) Breeding in Hungary, at University of Debrecen. *An. Univ. din Oradea Fasc. Protecţia Mediu.* **2007**, *12*, 57–62.
6. Oram, R.N. *Secale montanum*—A Wider Role in Australasia? *N. Z. J. Agric. Res.* **1996**, *39*, 629–633. [CrossRef]
7. Reimann-Philipp, R. Perennial Spring Rye as a Crop Alternative. *J. Agron. Crop Sci.* **1986**, *157*, 281–285. [CrossRef]
8. Reimann-Philipp, R. Breeding Perennial Rye. *Plant Breed. Rev.* **1995**, *13*, 265–292.
9. Kotvics, G. Investigation on increasing the protein content of *Secale cereale* L. In *Protein Growth by Plant Breeding*; Akadémiai Kiadó: Budapest, Hungary, 1970; pp. 89–98.
10. Andersen, M.R.; Depuit, E.J.; Abernethy, R.H.; Kleinman, L.H. Value Mountain Rye for Suppression of Annual Bromegrasses on Semiarid Mined Lands. *J. Range Manag.* **1992**, *45*, 345–351. [CrossRef]
11. Tang, Z.X.; Ross, K.; Ren, Z.L.; Yang, Z.J.; Zhang, H.Y.; Chikmawati, T.; Gustafson, M.; Secale, J.P. *Wild Crop Relatives: Genomic and Breeding Resources: Cereals*; Kole, C., Ed.; Springer: Berlin/Heidelberg, Germany, 2011; pp. 367–396. ISBN 978-3-642-14228-4.
12. Schiemann, E.; Nürnberg-Krüger, U. Neue Untersuchungen an *Secale africanum* Stapf. *Naturwissenschaften* **1952**, *39*, 136–137. [CrossRef]
13. Riley, R. The Cytogenetics of the Differences between Some *Secale* Species. *J. Agric. Sci.* **1955**, *46*, 377–383. [CrossRef]
14. Khush, G.S.; Stebbins, G.L. Cytogenetic and Evolutionary Studies in *Secale*. I. Some New Data on the Ancestry of *S. Cereale*. *Am. J. Bot.* **1961**, *48*, 723–730. [CrossRef]
15. Stutz, H.C. On the Origin of Cultivated Rye. *Am. J. Bot.* **1972**, *59*, 59–70. [CrossRef]
16. Singh, R.J.; Röbbelen, G. Identification by Giemsa Technique of the Translocations Separating Cultivated Rye from Three Wild Species of *Secale*. *Chromosoma* **1977**, *59*, 217–225. [CrossRef]

17. Singh, R.J. Cross Compatibility, Meiotic Pairing and Fertility in 5 *Secale* Species and Their Interspecific Hybrids. *Cereal Res. Commun.* **1977**, *5*, 67–75.
18. Stutz, H.C. A Cytogenetic Analysis of the Hybrid *Secale cereale* L. × *Secale montanum* Guss. and Its Progeny. *Genetics* **1957**, *42*, 199–221. [CrossRef] [PubMed]
19. Dierks, W.; Reimann-Philipp, R. Die Züchtung eines perennierenden Roggens als Möglichkeit zur Verbesserung der Roggenzuchtmethodik und zur Schaffung eines mehrfach nutzbaren Grünfutter- und Körnerroggens. *Z. Pflanzenzucht* **1966**, 343–368.
20. Ossent, H.P. Perennierender Kulturroggen. *Der Züchter* **1930**, *2*, 221–227. [CrossRef]
21. Lundqvist, A. Self-Incompatibility in Rye I. Genetic Control in the Diploid. *Hereditas* **1956**, *42*, 293–348. [CrossRef]
22. Wricke, G.; Wehling, P. Linkage between an Incompatibility Locus and a Peroxidase Isozyme Locus (*Prx 7*) in Rye. *Theor. Appl. Genet.* **1985**, *71*, 289–291. [CrossRef]
23. Gertz, A.; Wricke, G. Linkage between the Incompatibility Locus Z and a β-Glucosidase Locus in Rye. *Plant Breed.* **1989**, *102*, 255–259. [CrossRef]
24. Hackauf, B.; Wehling, P. Approaching the Self-Incompatibility Locus Z in Rye (*Secale cereale* L.) via Comparative Genetics. *Theor. Appl. Genet.* **2005**, *110*, 832–845. [CrossRef]
25. Fuong, F.T.; Voylokov, A.V.; Smirnov, V.G. Genetic Studies of Self-Fertility in Rye (*Secale cereale* L.). 2. The Search for Isozyme Marker Genes Linked to Self-Incompatibility Loci. *Theor. Appl. Genet.* **1993**, *87*, 619–623. [CrossRef]
26. Voylokov, A.V.; Korzun, V.; Börner, A. Mapping of Three Self-Fertility Mutations in Rye (*Secale cereale* L.) Using RFLP, Isozyme and Morphological Markers. *Theor. Appl. Genet.* **1998**, *97*, 147–153. [CrossRef]
27. Voylokov, A.V.; Fuong, F.T.; Smirnov, V.G. Genetic Studies of Self-Fertility in Rye (*Secale cereale* L.). 1. The Identification of Genotypes of Self-Fertile Lines for the *Sf* Alleles of Self-Incompatibility Genes. *Theor. Appl. Genet.* **1993**, *87*, 616–618. [CrossRef] [PubMed]
28. Do Canto, J.; Studer, B.; Lubberstedt, T. Overcoming Self-Incompatibility in Grasses: A Pathway to Hybrid Breeding. *Theor. Appl. Genet.* **2016**, *129*, 1815–1829. [CrossRef]
29. Melz, G.; Kaczmarek, J.; Szigat, G. Genetical Analysis of Rye (*Secale cereale* L.). Location of Self-Fertility Genes in Different Inbred Lines. *Genet. Pol.* **1990**, *31*, 1–7.
30. Paterson, A.H.; Schertz, K.F.; Lin, Y.R.; Liu, S.C.; Chang, Y.L. The Weediness of Wild Plants: Molecular Analysis of Genes Influencing Dispersal and Persistence of Johnsongrass, *Sorghum halepense* (L.) Pers. *Proc. Natl. Acad. Sci. USA* **1995**, *92*, 6127–6131. [CrossRef]
31. Hu, F.Y.; Tao, D.Y.; Sacks, E.; Fu, B.Y.; Xu, P.; Li, J.; Yang, Y.; McNally, K.; Khush, G.S.; Paterson, A.H.; et al. Convergent Evolution of Perenniality in Rice and Sorghum. *Proc. Natl. Acad. Sci. USA* **2003**, *100*, 4050–4054. [CrossRef] [PubMed]
32. Westerbergh, A.; Doebley, J. Quantitative Trait Loci Controlling Phenotypes Related to the Perennial versus Annual Habit in Wild Relatives of Maize. *Theor. Appl. Genet.* **2004**, *109*, 1544–1553. [CrossRef]
33. Chen, Z.; Hu, F.; Xu, P.; Li, J.; Deng, X.; Zhou, J.; Li, F.; Chen, S.; Tao, D. QTL Analysis for Hybrid Sterility and Plant Height in Interspecific Populations Derived from a Wild Rice Relative, *Oryza longistaminata*. *Breed. Sci.* **2009**, *59*, 441–445. [CrossRef]
34. Zhao, J.; Li, J.; Xu, P.; Zhou, J.; Hu, F.; Deng, X.; Deng, W.; Tao, D. A New Gene Controlling Hybrid Sterility between *Oryza sativa* and *Oryza longistaminata*. *Euphytica* **2012**, *187*, 339–344. [CrossRef]
35. Yun, L.; Larson, S.R.; Mott, I.W.; Jensen, K.B.; Staub, J.E. Genetic Control of Rhizomes and Genomic Localization of a Major-Effect Growth Habit QTL in Perennial Wildrye. *Mol. Genet. Genom.* **2014**, *289*, 383–397. [CrossRef]
36. Fan, Z.; Wang, K.; Rao, J.; Cai, Z.; Tao, L.-Z.; Fan, Y.; Yang, J. Interactions among Multiple Quantitative Trait Loci Underlie Rhizome Development of Perennial Rice. *Front. Plant Sci.* **2020**, *11*, 1757. [CrossRef]
37. Jang, C.S.; Kamps, T.L.; Skinner, D.N.; Schulze, S.R.; Vencill, W.K.; Paterson, A.H. Functional Classification, Genomic Organization, Putatively Cis-Acting Regulatory Elements, and Relationship to Quantitative Trait Loci of *Sorghum* Genes with Rhizome-Enriched Expression. *Plant Physiol.* **2006**, *142*, 1148–1159. [CrossRef]
38. Hu, F.; Wang, D.; Zhao, X.; Zhang, T.; Sun, H.; Zhu, L.; Zhang, F.; Li, L.; Li, Q.; Tao, D.; et al. Identification of Rhizome-Specific Genes by Genome-Wide Differential Expression Analysis in *Oryza longistaminata*. *BMC Plant Biol.* **2011**, *11*, 1–14. [CrossRef]
39. He, R.; Salvato, F.; Park, J.-J.; Kim, M.-J.; Nelson, W.; Balbuena, T.S.; Willer, M.; Crow, J.A.; May, G.D.; Soderlund, C.A.; et al. A Systems-Wide Comparison of Red Rice (*Oryza longistaminata*) Tissues Identifies Rhizome Specific Genes and Proteins That Are Targets for Cultivated Rice Improvement. *BMC Plant Biol.* **2014**, *14*, 1–21. [CrossRef]
40. Zhang, Y.; Zhang, S.; Liu, H.; Fu, B.; Li, L.; Xie, M.; Song, Y.; Li, X.; Cai, J.; Wan, W.; et al. Genome and Comparative Transcriptomics of African Wild Rice *Oryza longistaminata* Provide Insights into Molecular Mechanism of Rhizomatousness and Self-Incompatibility. *Mol. Plant* **2015**, *8*, 1683–1686. [CrossRef] [PubMed]
41. Bauer, E.; Schmutzer, T.; Barilar, I.; Mascher, M.; Gundlach, H.; Martis, M.M.; Twardziok, S.O.; Hackauf, B.; Gordillo, A.; Wilde, P. Towards a Whole-genome Sequence for Rye (*Secale cereale* L.). *Plant J.* **2017**, *89*, 853–869. [CrossRef]
42. Bolibok-Brągoszewska, H.; Targońska, M.; Bolibok, L.; Kilian, A.; Rakoczy-Trojanowska, M. Genome-Wide Characterization of Genetic Diversity and Population Structure in *Secale*. *BMC Plant Biol.* **2014**, *14*, 1–15. [CrossRef] [PubMed]
43. Gao, X.; Starmer, J.; Martin, E.R. A Multiple Testing Correction Method for Genetic Association Studies Using Correlated Single Nucleotide Polymorphisms. *Genet. Epidemiol.* **2008**, *32*, 361–369. [CrossRef]
44. Zhang, T.; Zhao, X.; Huang, L.; Liu, X.; Zong, Y.; Zhu, L.; Yang, D.; Fu, B. Tissue-Specific Transcriptomic Profiling of *Sorghum propinquum* Using a Rice Genome Array. *PLoS ONE* **2013**, *8*, e60202. [CrossRef] [PubMed]

45. Kong, W.; Kim, C.; Goff, V.H.; Zhang, D.; Paterson, A.H. Genetic Analysis of Rhizomatousness and Its Relationship with Vegetative Branching of Recombinant Inbred Lines of *Sorghum bicolor* × *S. propinquum*. *Am. J. Bot.* **2015**, *102*, 718–724. [CrossRef]
46. Martis, M.M.; Zhou, R.; Haseneyer, G.; Schmutzer, T.; Vrána, J.; Kubaláková, M.; König, S.; Kugler, K.G.; Scholz, U.; Hackauf, B. Reticulate Evolution of the Rye Genome. *Plant Cell* **2013**, *25*, 3685–3698. [CrossRef]
47. Li, G.; Wang, L.; Yang, J.; He, H.; Jin, H.; Li, X.; Ren, T.; Ren, Z.; Li, F.; Han, X.; et al. A High-Quality Genome Assembly Highlights Rye Genomic Characteristics and Agronomically Important Genes. *Nat. Genet.* **2021**. [CrossRef]
48. Rabanus-Wallace, M.T.; Hackauf, B.; Mascher, M.; Lux, T.; Wicker, T.; Gundlach, H.; Baez, M.; Houben, A.; Mayer, K.F.X.; Guo, L.; et al. Chromosome-Scale Genome Assembly Provides Insights into Rye Biology, Evolution and Agronomic Potential. *Nat. Genet.* **2021**. [CrossRef]
49. Reuscher, S.; Furuta, T.; Bessho-Uehara, K.; Cosi, M.; Jena, K.K.; Toyoda, A.; Fujiyama, A.; Kurata, N.; Ashikari, M. Assembling the Genome of the African Wild Rice *Oryza longistaminata* by Exploiting Synteny in Closely Related *Oryza* Species. *Commun. Biol.* **2018**, *1*, 1–10. [CrossRef]
50. von Bothmer, R.; Flink, J.; Jakobsen, N.; Kotimäki, M.; Landström, T. Interspecific Hybridization with Cultivated Barley (*Hordeum vulgare* L.). *Hereditas* **1983**, *99*, 219–244. [CrossRef]
51. Jones, T.A.; Zhang, X.-Y.; Wang, R.R.-C. Genome Characterization of MT-2 Perennial and OK-906 Annual Wheat × Intermediate Wheatgrass Hybrids. *Crop Sci.* **1999**, *39*, 1041–1043. [CrossRef]
52. Wang, R.; Farrona, S.; Vincent, C.; Joecker, A.; Schoof, H.; Turck, F.; Alonso-Blanco, C.; Coupland, G.; Albani, M.C. PEP1 Regulates Perennial Flowering in *Arabis alpina*. *Nature* **2009**, *459*, 423–427. [CrossRef] [PubMed]
53. Ponraj, U.; Theres, K. Keep a Distance to Be Different: Axillary Buds Initiating at a Distance from the Shoot Apical Meristem Are Crucial for the Perennial Lifestyle of *Arabis alpina*. *New Phytol.* **2020**, *227*, 116–131. [CrossRef] [PubMed]
54. Miedaner, T.; Longin, F. St. John's rye—Efinitely too good for goats and sheep. In *Neglected Cereals—From Ancient Grains to Superfood*; ERLLING Verlag GmbH Co. KG: Clenze, Germany, 2017; pp. 111–122.
55. Curwen-McAdams, C.; Jones, S.S. Breeding Perennial Grain Crops Based on Wheat. *Crop Sci.* **2017**, *57*, 1172–1188. [CrossRef]
56. Banks, P.M.; Xu, S.J.; Wang, R.R.-C.; Larkin, P.J. Varying Chromosome Composition of 56-Chromosome Wheat × *Thinopyrum intermedium* Partial Amphiploids. *Genome* **1993**, *36*, 207–215. [CrossRef] [PubMed]
57. Munné-Bosch, S. Do Perennials Really Senesce? *Trends Plant Sci.* **2008**, *13*, 216–220. [CrossRef]
58. Falconer, D.S.; Mackay, T.F. *Introduction to Quantitative Genetics*, 4th ed.; Pearson-Prentice Hall: Harlow, UK, 2008.
59. Wellensiek, S.J. Rational Methods for Breeding Cross-Fertilizers. *Meded. Landbouwhogesch.* **1947**, *7*, 1–36.
60. Adam, J. Der Weg Zur Züchtung einer neuen Roggensorte. In *Berichte Über Die Arbeitstagungen der "Arbeitsgemeinschaft der Saatzuchtleiter"*; Bundesanstalt für Alpine Landwirtschaft: Admont, Austria, 1954; pp. 137–157.
61. Jungfer, E. Kurztagbehandelte Klone in der Roggenzüchtung. *Der Züchter* **1955**, *25*, 255–262. [CrossRef]
62. Reuscher, S.; Furuta, T. ABHgenotypeR: Easy Visualization of ABH Genotypes R package version 1.0.1. Available online: https://CRAN.R-project.org/package=ABHgenotypeR (accessed on 6 May 2021).
63. Taylor, J.; Butler, D. R Package ASMap: Efficient Genetic Linkage Map Construction and Diagnosis. *J. Stat. Softw.* **2017**, *79*, 1–29. [CrossRef]
64. *R Core Team R: A Language and Environment for Statistical Computing*; R Foundation for Statistical Computing: Vienna, Austria, 2019.
65. Butler, D. asreml: asreml() fits the linear mixed model. R package version 3.0. Available online: https://www.vsni.co.uk/ (accessed on 6 May 2021).
66. Piepho, H.-P.; Möhring, J. Computing Heritability and Selection Response from Unbalanced Plant Breeding Trials. *Genetics* **2007**, *177*, 1881–1888. [CrossRef] [PubMed]
67. Malosetti, M.; van Eeuwijk, F.A.; Boer, M.P.; Casas, A.M.; Elía, M.; Moralejo, M.; Bhat, P.R.; Ramsay, L.; Molina-Cano, J.-L. Gene and QTL Detection in a Three-Way Barley Cross under Selection by a Mixed Model with Kinship Information Using SNPs. *Theor. Appl. Genet.* **2011**, *122*, 1605–1616. [CrossRef]
68. Schwarz, G. Estimating the Dimension of a Model. *Ann. Stat.* **1978**, *6*, 461–464. [CrossRef]

plants

MDPI

Review

Phytocannabinoids Biosynthesis in Angiosperms, Fungi, and Liver

Yamshi Arif [1], Priyanka Singh [1], Andrzej Bajguz [2,*] and Shamsul Hayat [1]

[1] Department of Botany, Plant Physiology Section, Faculty of Life Sciences, Aligarh Muslim University, Aligarh 202002, India; yamshiarifalig@gmail.com (Y.A.); singh.priyanka8156@gmail.com (P.S.); hayat_68@yahoo.co.in (S.H.)

[2] Department of Biology and Plant Ecology, Faculty of Biology, University of Bialystok, Ciolkowskiego 1J, 15-245 Bialystok, Poland

* Correspondence: abajguz@uwb.edu.pl; Tel.: +48-85-7388361

Abstract: Phytocannabinoids are a structurally diverse class of bioactive naturally occurring compounds found in angiosperms, fungi, and liverworts and produced in several plant organs such as the flower and glandular trichrome of *Cannabis sativa*, the scales in *Rhododendron*, and oil bodies of liverworts such as *Radula* species; they show a diverse role in humans and plants. Moreover, phytocannabinoids are prenylated polyketides, i.e., terpenophenolics, which are derived from isoprenoid and fatty acid precursors. Additionally, targeted productions of active phytocannabinoids have beneficial properties via the genes involved and their expression in a heterologous host. Bioactive compounds show a remarkable non-hallucinogenic biological property that is determined by the variable nature of the side chain and prenyl group defined by the enzymes involved in their biosynthesis. Phytocannabinoids possess therapeutic, antibacterial, and antimicrobial properties; thus, they are used in treating several human diseases. This review gives the latest knowledge on their role in the amelioration of abiotic (heat, cold, and radiation) stress in plants. It also aims to provide synthetic and biotechnological approaches based on combinatorial biochemical and protein engineering to synthesize phytocannabinoids with enhanced properties.

Keywords: abiotic stress; cell homeostasis; heterologous host synthetic approach; terpenophenolics

Citation: Arif, Y.; Singh, P.; Bajguz, A.; Hayat, S. Phytocannabinoids Biosynthesis in Angiosperms, Fungi, and Liverworts and Their Versatile Role. *Plants* **2021**, *10*, 1307. https://doi.org/10.3390/plants10071307

Academic Editors: Milan S. Stankovic, Paula Baptista and Petronia Carillo

Received: 6 June 2021
Accepted: 25 June 2021
Published: 28 June 2021

Publisher's Note: MDPI stays neutral with regard to jurisdictional claims in published maps and institutional affiliations.

Copyright: © 2021 by the authors. Licensee MDPI, Basel, Switzerland. This article is an open access article distributed under the terms and conditions of the Creative Commons Attribution (CC BY) license (https://creativecommons.org/licenses/by/4.0/).

1. Introduction

Phytocannabinoids are meroterpenoids bearing a resorcinol core with an isoprenyl, alkyl, or aralkyl para-positioned side chain, or alkyl group usually containing an odd number of carbon atoms—cannabinoids that have an even number of carbon atoms in the side chain are rare. Phytocannabinoids can be obtained from angiosperms (flowering plants), fungi, and liverworts (Figure 1). The first phytocannabinoid was isolated from the *Cannabis sativa* family Cannabaceae, but it has a long controversial history of its use and abuse [1,2]. From *C. sativa* more than 113 phytocannabinoids were isolated and classified into several groups such as cannabidiols (CBDs), cannabigerols (CBGs), cannabicyclols (CBLs), cannabidiols (CBNDs), cannabinols (CBNs), cannabitriols (CBTs), cannabichromenes (CBCs), (−)-Δ^9-*trans*-tetrahydrocannabinol (Δ^9-THC) and miscellaneous cannabinoids [1,3–5]. Compounds obtained from *C. sativa* predominately generate alkyl-type phytocannabinoids with a monoterpene isoprenyl and the pentyl side chain [4,6]. In *C. sativa*, CBD, CBG, CBC, cannabichromevarine (CBCV), and Δ^9-THC are the most abundant cannabinoids in their respective acidic form. The acidic form of the cannabinoid (C22, "pre-cannabinoids") is the final step of the cannabinoid biosynthetic pathway. Oxidation, decarboxylation, and cyclization lead to the development of modified phytocannabinoid via spontaneous breakdown or conversion product. The conversion mainly occurs due to the poor oxidative stability of phytocannabinoids, especially with the alkyl group. *C. sativa* produces the most common phytocannabinoids. In addition to this, the brains of mammals have receptors that respond to the *C. sativa* cannabinoid, so they were termed as cannabinoid receptor types 1 and 2 (CB_1R and CB_2R) and thus participated in the endocannabinoid system [1,3,4,7,8].

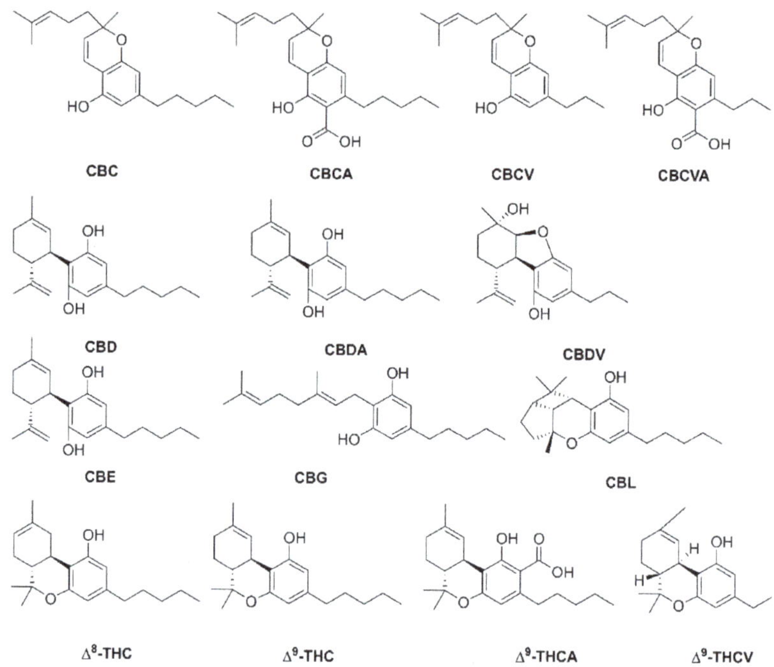

Figure 1. Structure of phytocannabinoids in *Cannabis sativa*. Abbreviations: CBC, cannabichromene; CBCA, cannabichromenic acid; CBCV, cannabichromevarine; CBCVA, cannabichromevarinic acid; CBD, cannabidiol; CBDA, cannabidiolic acid; CBDV, cannabidivarine, CBE, cannabielsoin; CBG, cannabigerol; CBL, cannabicyclol; Δ^8-THC, Δ^8-tetrahydrocannabinol; Δ^9-THC, Δ^9-tetrahydrocannabinol; Δ^9-THCA, Δ^9-tetrahydrocannabinolic acid; Δ^9-THCV, Δ^9-tetrahydrocannabivarinic acid.

The Endocannabinoid system in humans and animals revealed that it participates in the regulation of biological functions such as memory, brain system, mood and addiction along with cellular and metabolic processes such as glycolysis, lipolysis, and the energy balance system [7,9]. Other angiosperms such as *Helichrysum umbraculigerum* (Asteraceae) native to South Africa, *Amorpha fruticosa* (Fabaceae), and *Glycyrrhiza foetida* (Fabaceae) contains a bioactive compound bearing a cannabinoid backbone (Figure 2); they are characterized as prenylated bibenzyl derivatives because the aralkyl side chain occurs [1,10].

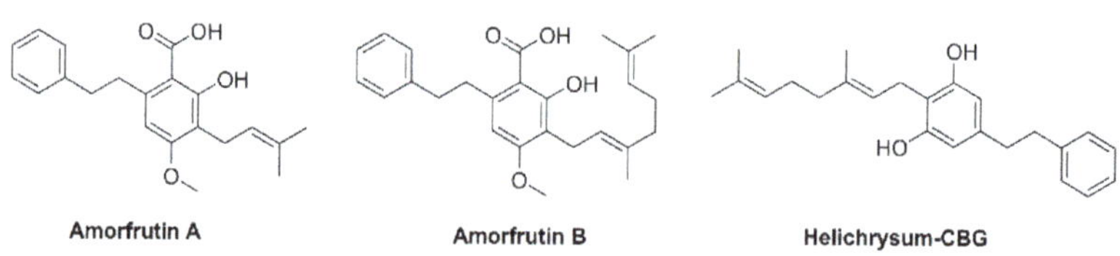

Figure 2. Structure of phytocannabinoids in *Helichrysum* and *Glycyrrhiza* plants.

Many *Rhododendron* species (family Ericaceae) such as *Rh. dauricum* native to Northeastern Asia, *Rh. adamsii* found in Eastern Siberia and Mongolia, *Rh. rubiginosum* var. *rubiginosum* native to Southwest China, and *Rh. anthopogonoides* grown in Southern China, all generate active monoterpenoids that have a cannabinoid backbone. Phytocannabinoids are CBC types with an orcinol or methyl group side chain (Figure 3). *Rh. dauricum* particularly produces cannabinoids bearing sesquiterpene moiety such as daurichromenic acid (DCA), grifolic acid (GFA), confluentin, and rhododaurichromenic acid [11–13]. *Rh. adamsii* produces cannabigerorcinic acid, DCA, cannabigerorcinic acid methylase, chromane, and chromene monoterpenoids; *Rh. rubiginosum* produces cannabinoid rubiginosins A–G [14,15]. *Rh. anthopogonoides* contains chromane/chromene derivatives such as cannabiorcicyclolic acid, cannabiorcichromenic acid, anthopogochromenic acid, and anthopogocyclolic acid (Figure 3) [16].

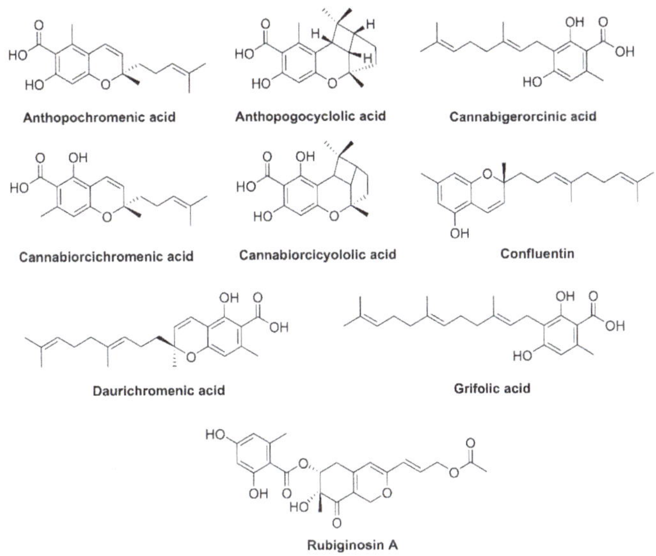

Figure 3. Structure of phytocannabinoids in *Rhododendron* plants.

Liverworts, such as *Radula marginata*, *R. perrottetii*, and *R. laxirameae* native to New Zealand, produce active cannabinoids with bibenzyl backbones such as lunularic acid and its dimeric form—vittatin (Figure 4) [17–20]. Some fungi, e.g., *Albatrellus* (Albatrellaceae, mycorrhizal fungi) species, also produce GFA along with its derivative confluentin, grifolin, and neogrifolin (Figure 4). Additionally, *Cylindrocarpon olidum* generates cannabiorcichromenic acid and halogenated cannabinoid, i.e., 8-chlorocannabiorcichromenic acid (Figure 4) [1,21].

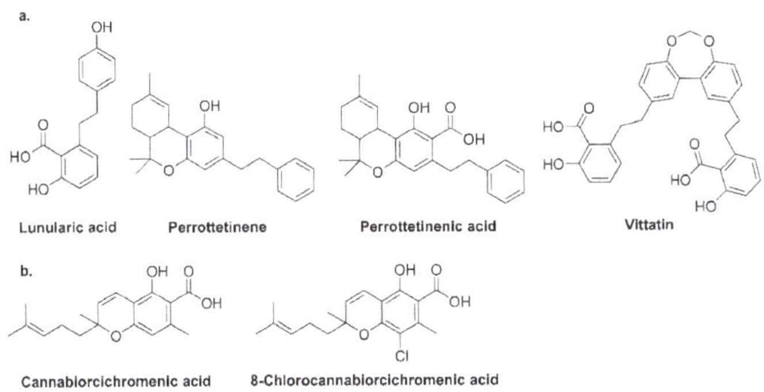

Figure 4. Structure of phytocannabinoids in (**a**) liverworts and (**b**) fungi.

This review focuses on the biosynthesis of different active phytocannabinoids in several cellular compartments of *C. sativa*, *Rhododendron*, and *Radula* species. In this topic framework, the most crucial criterion is the synthetic and biotechnological techniques for the production of phytocannabinoids. The current review highlights the multi-faceted role of different active phytocannabinoids in humans and plants. Interestingly, this review briefly highlights the antimicrobial, antibacterial, and antibiotic properties of phytocannabinoid based on recent papers. Additionally, the role of phytocannabinoids in ameliorating pathogenic attack, and environmental stresses, e.g., cold, heat, and UV radiation, is also briefly assessed.

2. Phytocannabinoid Biosynthesis Sites

In *C. sativa*, phytocannabinoids are stored in glandular trichomes, located all over the aerial part of the plant, so root surface and root tissues do not keep phytocannabinoid Female flowers possess a high density of phytocannabinoid [22,23]. Glandular trichomes have balloon shaped secretory vessicle which store cannabinoid. High temperature or herbivory leads to trichome rupture, which releases the sticky contents on the plant parts with viscous and non-crystallizing properties [24,25]. Higher temperatures increase cannabinoid production. Furthermore, cannabinoid production is raised in the cannabis flower after UV-B exposure. Nevertheless, phytocannabinoids act as a sun shield that absorbs lethal UV radiation [26]. *Rhododendron* genus lepidote consists of small leaves surrounded by glandular scales on both the abaxial and adaxial surfaces. These scales have lipophilic globules that contain major bioactive compounds such as cannabinoids, terpenes; the apoplastic space of the glandular scale also contains cannabinoids such as DCA in the *Rhododendron* [27]. Liverworts have oil bodies that are membrane-bound cellular structures that contain cannabinoids, aromatic oil, and terpenoid (*cis* configuration), mostly sesquiterpenoids and diterpenoids. Oil bodies are odiferous bitter, pungent compounds, which make them biologically active. Furthermore, these possess several ecological advantages such as tolerance from temperature, light, or radiation [28,29].

3. Biosynthesis of Phytocannabinoids

This section focuses on detailed events undergone in the production of several phytocannabinoids inside *C. sativa*, *Rhododendron*, and liverworts. Moreover, biosynthesis of phytocannabinoids via biotechnological approaches in a heterologous host and synthetic methods are discussed.

3.1. *Cannabis sativa*

Phytocannabinoids are prenylated polyketides, i.e., terpenophenolic compounds, which are derived from isoprenoid and fatty acid precursors. Phytocannabinoid biosynthesis occurs in different cellular compartments: gland cells cytosol, the plastids, and the extracellular storage cavity. In the cytosol, oxidative cleavage of fatty acid such as palmitic acid yields hexanoic acid; it further synthesizes olivetolic acid (OA). The next step is prenylation of phenolic moiety (the polyketide derivatives, 5-pentenyl resorcinolic acid, and OA) with the terpenoid geranyl pyrophosphate (GPP). This step originates from the methylerythritol-4-phosphate (MEP) pathway in plastids. Cyclization (oxidative) and storage of the final products take place outside the gland cells. Transport proteins and vesicle trafficking participate in mobilizing intermediates across the morphologically highly specialized interface between the gland cells and storage cavity [30–32].

In *C. sativa*, phytocannabinoids biosynthesis is divided into three important places: cytosol (for polyketide pathway), plastids (MEP pathway for prenylation), and apoplastic spaces (oxidocyclization and storage) (Figure 5). Inside cytosol, biosynthesis of phytocannabinoids participates in the integration of major steps in polyketide and isoprenoid metabolism. Fatty acids (C18) are sequentially desaturated, peroxygenated, and cleaved into the hexanoic acid (C6) and C12 product via enzyme desaturase, lipoxygenase (LOX), and hydroperoxide lyases, respectively. Hexanoic acid is converted into thioester hexanoyl-

CoA; this reaction is catalyzed by acyl-activated enzyme 1 (AAE1). Later, hexanoyl-CoA and malonyl-CoA (C2 donor) together via the action of olivetol synthase (OLS) and olivetolic acid cyclase (OAC) synthesizes OA. Moreover, it was reported that OAC is the dimeric (α+β) barrel protein, it is the first plant enzyme that catalyzes (C2-C7) intramolecular aldol condensation along with carboxylate retention; OAC contains distinctive active-site bearing the pentyl-binding hydrophobic pocket and polyketide binding site, whereas it is devoid of aromatase and thioesterase activities [31,33,34].

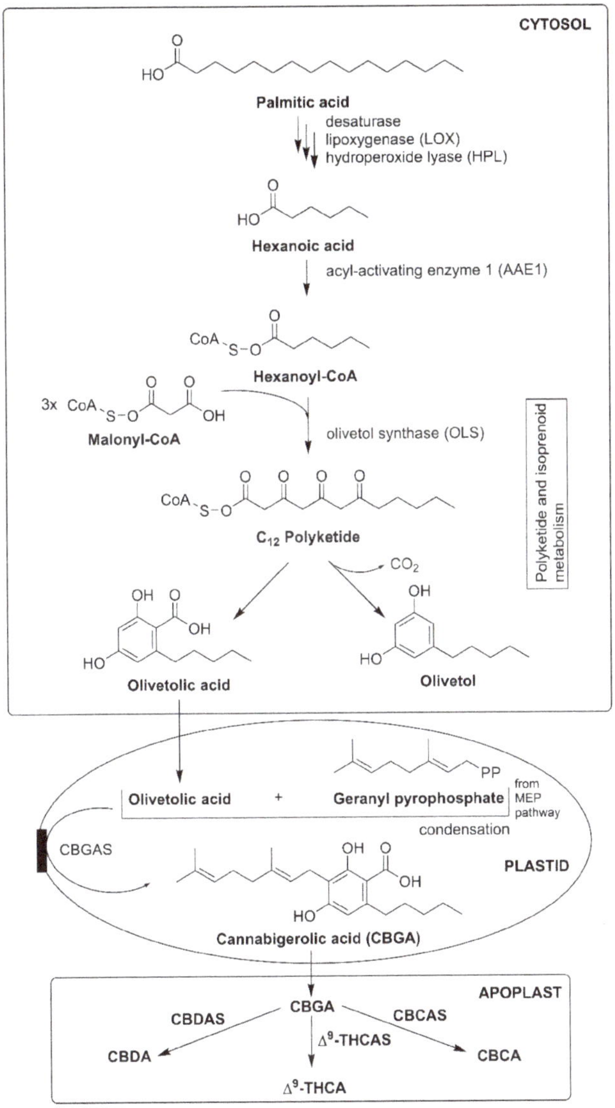

Figure 5. Phytocannabinoids biosynthesis in *Cannabis sativa*. Abbreviations: CBCA, cannabichromenic acid; CBCAS, cannabichromenic acid synthase; CBDA, cannabidiolic acid; CBDAS, cannabidiolic acid synthase; Δ^9-THCA, Δ^9-tetrahydrocannabinolic acid; Δ^9-THCAS, Δ^9-tetrahydrocannabinolic acid synthase.

Inside plastids, the MEP pathway synthesizes GPP. It prenylates OA, which forms the intermediate branch-point and first cannabinoid which is cannabigerolic acid (CBGA), and this reaction is catalyzed by cannabigerolic acid synthase (CBGAS). CBGA is an essential cannabinoid because it acts as the precursor of several cannabinoids with an alkylic pentyl side chain. In contrast, CBGAS is a transmembrane aromatic prenyltransferase (PT) that transfers plastid signals. Then CBGA is converted into Δ^9-THCA and cannabidiolic acid (CBDA) with the help of two enzymes which are CBDAS and Δ^9-tetrahydrocannabinolic acid synthase (Δ^9-THCAS). This conversion continues by reducing oxygen (O_2) into hydrogen peroxide (H_2O_2) via oxidative cyclization reactions. Additionally, CBDAS and Δ^9-THCAS are necessary flavoprotein enzymes that are dependent on O_2 (electron acceptor) [1,30,31].

Another important enzyme, cannabichromenic acid synthase (CBCAS), dependent from FAD and O_2, takes part in the synthesis of cannabichromenic (CBCA). Additionally, enzymes Δ^9-THCAS and CBCAS have high sequence similarity at about a 96% nucleotide level. Both remain active inside resin space, which shows that CBCAS participates as an O_2 dependent flavoprotein that converts CBGA to cannabichromenic acid (CBCA) with H_2O_2 as the side product via an oxidocyclization reaction. Active cannabinoid Δ^9-THCA, CBDA, and CBCA with a pentyl side chain are synthesized in the apoplastic cannabis space. Furthermore, these active phytocannabinoids undergo decarboxylation and spontaneous rearrangement reactions on exposure to heat, radiation, or during storage. Some phytocannabinoid having unknown C1-C4 alkyl side chain are synthesized from acetyl-CoA, propanoyl-CoA, or pentanoyl-CoA [1,5,31,35].

3.2. *Rhododendron*

DCA and its derivative are produced and stored inside specialized glandular scales in *Rh. dauricum*. DCA utilizes carbon atoms from acetyl-CoA and farnesyl-CoA, with two significant intermediates, i.e., orsellinic acid (OSA) and GFA. DCA biosynthesis in *Rhododendron* is split between the cytosol, plastid, and apoplastic spaces (Figure 6) [11,36].

The biosynthesis of DCA starts in the cytosol with polyketide formation, type III polyketide synthase (PKS) helps in acetyl-CoA chain extension. Then another enzyme, orcinol synthase (ORS), catalyzes orcinol, OSA, triacetic acid, tetracetic acid, lactone and phloroacetophenone, where malonyl-CoA (three units) act as a carbon donor. Furthermore, tetraketide cyclase catalyzes OSA from ORS [36,37]. OA is transported to the plastid via a transporter, which is still unknown. Inside the plastids, the MEP pathway derives farnesyl-CoA. The inhibition of the MEP pathway via clomazone decreases OSA and DCA synthesis. In contrast, inhibition of the mevalonate-dependent pathway via mevastatin led to an increment in OSA and DCA biosynthesis. Moreover, aromatic farnesyltransferase *Rh. dauricum* prenyltransferase (PT) helps in regiospecific farnesylation; this enzyme moderates sequence identity with UbiA aromatic PTs that lie within chloroplasts. Geranyl-CoA and geranylgeranyl-CoA serve as the alternative prenyl donors used by PT, but their activity rate is 13%, and 2.5% of the activity acquired by farnesyl-CoA. GFA is synthesized as the intermediate within the plastids. Then, within apoplastic spaces, an oxidocyclization reaction takes place via DCA synthase (DCAS) forming CBC scaffold; reaction moves forward by H_2O_2 release. Like Δ^9-THCAS and CBDAS, DCAS is active enzymatically outside apoplastic spaces and dependent on O_2. In *Rh. dauricum*, DCA decarboxylated forms produce confluentin; spontaneous decarboxylation occurs via heat, irradiation, and during storage, similar to the decarboxylation acidic to neutral phytocannabinoids in *C. sativa* trichomes [5,11,37].

Apoplastic spaces serve as storage for many metabolites, essential oils, DCA, and confluentin. Moreover, GFA and DCA act as phytotoxic compounds in *Rh. dauricum* cell culture, as they induce cell death. Similarly, H_2O_2 formed as a side product in DCA biosynthesis also increases cell death by enhancing apoptosis-related reactions. However, to overcome cell death, autotoxicity, and cell damage, DCA storage occurs in the apoplast, and H_2O_2 is released to participate in the plant-defense system and provide plant immunity. In *Rh. dauricum*, transport proteins and vesicle trafficking mechanisms are still not well understood and remain a valuable and exciting approach for further future investigations [1,2,37].

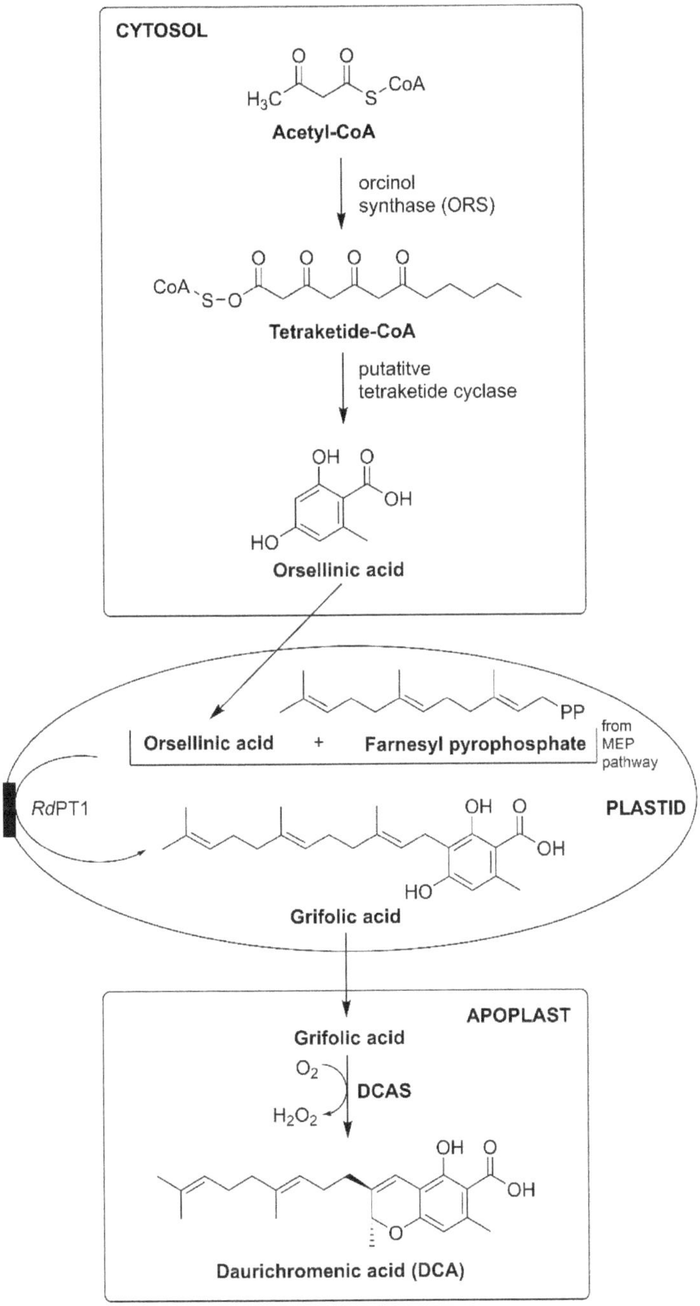

Figure 6. Phytocannabinoids biosynthesis in Rhododendron. Abbreviations: DCAS, daurichromenic acid synthase; MEP, methylerythritol-4-phosphate; PT, prenyltransferase.

3.3. *Liverworts*

It was reported that *Radula marignata* possesses enzymes for GPP biosynthesis and helps in the biosynthesis of bibenzyl cannabinoid. Moreover, the production of bibenzyl CBGA analog (i.e., lunularic acid, perrottetinenic acid, and perrottetinene) needs precursor stilbene acid or dihydrostilbene acid, which is very rare, and compounds of this type were found in *Hydrangea macrophylla* var. *thunbergii* and liverworts such as *Marchantia polymorpha* and *Convolvulus hystrix* (Figure 7) [1,17,38].

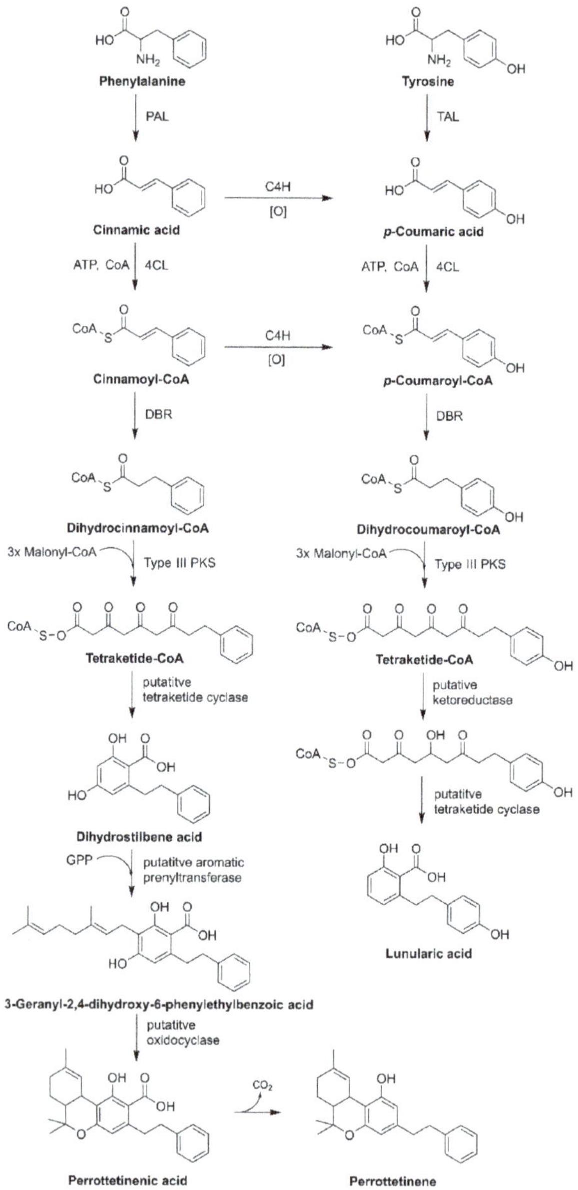

Figure 7. Phytocannabinoids biosynthesis in liverworts. Abbreviations: C4H, cinnamate 4-hydroxylase; 4CL, 4-coumarate:CoA ligase; PAL, phenylalanine ammonia-lyase; TAL, tyrosine ammonia-lyase.

Stilbene acid is synthesized from type III PKS via coumaroyl-CoA or dihydrocoumaroyl-CoA, CoA-activated precursors. Starter molecules are extended by malonyl-CoA with decarboxylation, followed by a condensation reaction, which produces polyketide intermediate, which synthesizes different core structures. Hydrangic acid is the starter molecule that acts as coumaroyl-CoA. It is extended utilizing malonyl-CoA (three units) as a C2 donor to synthesize tetraketide intermediate. These reactions are catalyzed via stilbene synthase (STS)-type PKS enzymes. Ketoreductase (KR) leads to polyketide reduction followed by STS-like C2 to C7 intramolecular aldol condensation, here retention of the carboxylic group produces hydrangic acid. KR is involved in the loss of the C5-hydroxyl group on the aromatic ring structure of hydrangic acid, contrary to the stilbene acid structure. In *R. marginata*, the precursor of stilbene acid or dihydrostilbene is derived from the coumaroyl CoA or cinnamoyl-CoA; type III PKS enzyme helps in chain elongation, later putative tetraketide cyclase or (dihydrostilbene acid cyclase, DHAC) helps in cyclization [1,18,38,39].

The lunularic acid precursor, prelunularic acid is produced by the type III PKS, named bibenzyl synthase (BBS). Moreover, BBS catalyzes the reaction where dihydrocoumaroyl-CoA serves as the starter molecule for the extension utilizing malonyl-CoA (3-units), which serves as the carbon donor. Later, cyclization occurs on reduced polyketide, with a lack of C5-hydroxyl group on the aromatic ring structure. Furthermore, type III PKS plays a crucial role in the bibenzyl cannabinoid and its analog synthesis by also catalyzing the carboxylate retaining reaction mechanism. KR is involved in the cyclization and assists proper ring formation. Thus, after cyclization, lunularic acid is synthesized [40–42].

Perrottetinenic acid (PA) is synthesized in *R. marginate* [43]. The transcriptomic approach of liverworts bears the mRNA encoding for type III PKS (responsible for chain elongation), which were later recognized as STS. Furthermore, it also exhibits a 60% homology of the amino acid sequence to stilbene-carboxylate synthase in *Marchantia polymorpha*. Other enzymes, such as double-bond reductase (DBR)—aromatic PT and oxidocyclase (perrottetinenic acid synthase, PAS), are also responsible for the production of PA. DBR catalyzes compounds that are precursors of phenylpropanoid, and generates dihydrocinnamoyl-CoA. The production of bibenzyl phytocannabinoid is dispersed across the liverworts cells in the same way as in cannabis and *Rhododendron*. Therefore, DHAC and DBR reside in the cytosol, PT is localized in the plastids and PAS inside the oil body. Signal peptides act as the crucial indicators for the encoding genes selection [18,39,41,44]. Another bibenzyl *cis*-THC, (−)-*cis*-perrottetinene (*cis*-PET) was isolated from the liverwort *R. perrottetii* [45]. The *cis* configuration in the cyclohexene ring in *cis*-PET is comparable with Δ^9-*trans*-THC. PET resembles Δ^9-THC in its 3D shape, and can bind to many of the same cannabinoid receptors (CBRs) as Δ^9-THC. Interestingly, PET also reduces the level of prostaglandins in the brain—a compound with inflammatory properties that increase in response to Δ^9-THC and may be responsible for adverse effects [17].

3.4. Application of Biotechnological Approaches to Phytocannabinoids Production in Heterologous Hosts

Phytocannabinoids can be generated in different heterologous hosts such as fungi, bacteria, and plants with the help of biotechnological techniques. Several investigations reported that in vitro culture of phytocannabinoids biosynthesis in *C. sativa*, with the help of explants and micropropagation, is a widely used biotechnological approach for phytocannabinoid production. Moreover, apart from *C. sativa*, *Nicotiana benthamiana* also emerged as the favorable heterologous host for the production of phytocannabinoids. It exhibits the production of several proteins and bioactive compounds and has glandular trichomes that help overcome cell death, autotoxicity, and cell damage caused due to intermediates produced during phytocannabinoids biosynthesis. In recent research, it was found that a major biotechnological tool for phytocannabinoids production and inducing genetic modification is established via the micropropagation technique. Additionally, cell suspension culture, hairy root, and adventitious root culture also produce a small quantity of cannabinoids. In *Saccharomyces cerevisiae*, galactose produces phytocannabinoids such as CBG, CBDA, Δ^9-THCA, and minor phytocannabinoids such as cannabidivarinic acid (CBDVA) and Δ^9-tetrahydrocannabidivarinic acid (Δ^9-THCVA) [46–51].

3.5. Production of Phytocannabinoids through Synthetic Approaches

Phytocannabinoids are terpenophenolic compounds that are produced by polyketide and the MEP pathway. In the heterologous biosynthesis approach, two important pathways that are optimizing and engineering provide abundant precursors for cannabinoid production. Additionally, these pathways are linked via aromatic PT, ubiquitous in plants, animals, bacteria, and fungi [52,53]. Thus, this helps in the production of aromatic metabolites such as coumarin, flavonoid, and phenylpropanoid at different reaction spectra [54]. As discussed, biosynthesis of phytocannabinoid is different in plants, fungi, and liverworts; thus, the recent techniques which emerge out to be beneficial is by using the aromatic PT-based approach to generate novel phytocannabinoids in the heterologous hosts formed on combinational utilization of module over several species [52–54].

Similar to *Humulus lupulus*, inside trichomes of *C. sativa*, chalcone isomerases such as proteins (CHILs) are expressed. Additionally, CHILs are polyketide binding proteins, and their co-expression in heterologous cannabinoid production mechanism plays a pivotal role in augmenting biosynthesis. Apart from combinatorial biochemistry techniques, enzyme-based approaches also play a remarkable role in phytocannabinoid production. Furthermore, some non-natural precursors such as pentanoic acid, hexanoic acid, heptanoic acid are incorporated inside the generated cannabinoid. Derivatization of phytocannabinoid diversifies their functionality. Glycosylation of phytocannabinoid reduces cell damage and autotoxicity and increases cell stability and life; in a therapeutic context, glycosylated phytocannabinoid enhances absorption, distribution, metabolism and, excretion (ADME) features. Halogenation of cannabinoids (like DCA) by co-expression of halogenase AscD derived from *Fusarium* exhibits several antifungal, antibacterial, antitumor, antiparasitic properties. Thus, the synthetic biology technique serves as a new insight into the production and designing of phytocannabinoids [55–58].

4. Phytocannabinoid Storage and Maintenance of Cell Homeostasis

The last step of phytocannabinoid biosynthesis and their storage both occur in different cellular compartments like in C. sativa storage organs are resin present inside glandular trichomes, in Rh. dauricum they storage takes place inside apoplast of glandular scale while in R. marginata oil bodies are the storage organs [1]. In *C. sativa*, nuclear magnetic resonance (NMR) based metabolomics of trichomes revealed storage of phytocannabinoids along with terpenes, sugars, organic acid, amino acid, and choline [6]. Sugars, organic acids, and choline, which present in appropriate stoichiometric ratios, give rise to natural deep eutectic solvents (NADES) [59]. NMR-based techniques revealed that constituents for NADES display proton-associated intermolecular interactions, forming aggregates to form larger structures in the liquid phase [60]. NADES is a distinguished solvent bearing this property when the water content is less than 40%; this unique property of NADES incorporates them in the third membrane-less solvent phase inside biological systems [61,62]. Δ^9-THCA and CBDA support NADES mediated solubilization in trichomes and oil bodies as they are virtually insoluble in water. In *C. sativa* and *Rh. dauricum*, oxidocyclase releases H_2O_2. Furthermore, it participates in a plant-defense system and provides plant immunity. NADES stabilizes the activity of oxidocyclases, whereas it also serves as a biological solvent where the biosynthetic catalytic enzyme remains active and functional. NADES plays a crucial role in a plant cell to maintain cell homeostasis. Thus, it plays a chief role in stabilizing phytocannabinoid biosynthesis enzymes, their storage and further induces cellular homeostasis. Recently, cytochrome P450 (CYP) enzymes, CYP79A1, CYP71E1, and P450 oxidoreductase involved in the biosynthesis of the cyanogenic glucoside dhurrin, which was stabilized NADES. Biosynthesis and appropriate storage of cannabinoids in heterologous hosts need to be engineered to reduce autotoxicity and apoptosis [63,64].

5. Biotechnology and In Vitro Propagation of Cannabis

Cannabis is an essential plant as it yields phytocannabinoid, which has multiple applications. Cannabis cultivation is regulated in several countries, so alternative biotechnological and in vitro tissue culture approaches require attention. Meanwhile, these approaches are beneficial to preserve cultivars or clones of plants with specific metabolites such as phytocannabinoid. Micropropagation generates genetically stable plants; thus, this method is beneficial for the clonal multiplication of cannabis. Nevertheless, micropropagation of the cannabis plant is performed via adventitious buds and axillary buds located on the nodal segment. Another method that is synthetic seed technology (encapsulation of axillary bud or nodal segment in calcium alginate seeds) has also gained importance in plant clonal propagation. It displays homogenous growth and development of plants and genetic stability even after long storage [65,66].

Cannabis is not a well-known recalcitrant plant for transformation. The plant's regeneration capacity is low and is completely dependent on plant tissue, age, explants, and a combination of plant growth regulators (PGRs), i.e., indole-3-acetic acid (IAA), indole-3-butyric acids, 1-naphthaleneacetic acid (NAA), 2,4-dichlorophenoxyacetic acid, and kinetin. Moreover, successful transformation is performed via *Agrobacterium tumefaciens*; cells that do not regenerate the shoots are undifferentiated and can be transformed via phosphomannose isomerize and calorimetric assays, which showed efficacious expression of the transgene [67]. Cannabis transformation using shoot tip explants regenerates a high amount of cannabinoid after infection with *Agrobacterium tumefaciens* [68,69]. For the successful regeneration of cannabis, thidiazuron (behavior as cytokinin-like compound) with kinetin (represent cytokinins) has not only the effect on the shoots but also shows a high yielding capacity of phytocannabinoids. Dicamba (3,6-dichloro-2-methoxybenzoic acid) herbicide is also known for inducing the regeneration of shoots from calli. Moreover, a most potent and efficient transformation of cannabis was carried out using 1–2 cm of hypocotyl explants and by supplementation of zeatin and 6-benzylaminopurine [65].

Another important tissue culture technique used for the production of bioactive cannabinoids showing pharmaceutical effects is hairy root culture. In this system, both *Agrobacterium tumefaciens* and *Agrobacterium rhizogenes* are used for transformation to yield cannabinoids by establishing a hairy root system. Meanwhile, the most responsive tissue for infection in this system is hypocotyl, which produces large quantities of hypocotyls or engineered plants for the production of bioactive compounds that are industrially valuable; for example, Δ^9-THCA is produced by expressing Δ^9-THCAS. The hairy root technique is obtained by the increased growth rate, independent of a hormone, and has the same metabolic potential as the original plant organ. In the hairy root system, cannabis callus was cultured on a full-strength B5 medium feed with an NAA or IAA displayed increase in the accumulation of phytocannabinoids. Rhizogenesis induction in cannabis undifferentiated cells is crucial, as it can be performed on calli overexpressing some important transcription factors (TFs) and genes which are involved in cannabinoid synthesis. Additionally, cannabis hairy root culture is an efficient method implemented with adsorbents to reduce toxicity problems [65,67].

Cannabis cell suspension culture transformation with the genes involved in the cannabinoid synthesis pathway offers a great opportunity to enhance bioactive phytocannabinoid production possessing pharmacological potential. Additionally, increased production via cannabis cell suspension culture can be obtained through TF. TF have cascade mechanism (that is when the master regulator of cannabinoid synthesis is identified; there it can be expressed inducibly or constitutively inside cell suspension culture). Thus, TF constitutes potent and efficient tool involved in plant metabolic engineering [65].

TF, belonging to the MYB family, was involved in cannabinoid production and tolerant oxidative stress. TF expression in an inducible manner confers toxicity tolerance caused due to the accumulation of phytocannabinoids during the growth of transformed cells. Genetic engineering and plant cell suspension culture were elicited to induce the production of specific cannabinoids. Additionally, in cannabis suspension cells, induction with biotic and abiotic elicitors did not cause an increment in phytocannabinoid synthesis. Non-essential elements, such as silicon, play a key role in mitigating biotic and abiotic stress. Additionally, silicon has a stimulatory effect on the production of cannabis secondary metabolites like cannabinoids. Cyclodextrins are cyclic oligosaccharides bearing five or more than five α-D-glucopyranose residues; these form an inclusion complex structure with lipophilic compounds like cannabinoids. In plant cell suspension culture, cyclodextrins are used for the production of non-polar compounds such as stilbenes and other cannabinoids. Moreover, cyclodextrins improve cannabinoid and other metabolite solubilities in an aqueous environment. Cyclodextrins possess a similar structure to alkyl-derived oligosaccharide released from the plant cell wall during a fungal infection; thus they act as elicitors for the production of secondary metabolites. More investigations are needed on the effect of cyclodextrins on the synthesis of non-polar cannabinoids in cannabis suspension cultures [65,70].

6. Phytocannabinoids and Their Derivative Role and Bioactivity in Humans and Animals

Phytocannabinoids are used as medicines since antiquity. They possess antimicrobial, antibacterial, and antifungal activity and serve as remarkable antibiotics (Table 1). Additionally, cannabinoids show little antifungal power because fungi can metabolize cannabinoids except a few, such as *Phomopsis ganjae*. An increasing number of developing countries are relaxing their legislation around phytocannabinoids and cannabis-derived products; legalized cannabis-derived product industries are increasing worldwide. They are likely to constitute a projected US$ 57 billion market by the year 2027 [71].

Table 1. Bioactivity and effect of different phytocannabinoid in animals and humans.

Phytocannabinoids	Bioactivity in Animals and Human	References
Δ^9-THC	pleiotropic effects such as analgesic, muscle relaxation, and pain tolerance	[1]
	increases weight and appetite; improves sleep and depression; alters mood, behavior, feeling, and thoughts	[72,73]
	cures neuropathic pain, spasticity, dysphoria, and anxiety disorders	[74,75]
	antiemetic activity, i.e., prevents vomiting in cancer patient during chemotherapy	[76]
	displays antiglaucoma activity and reduces intraocular pressure	[76]
Δ^9-THCA	neuroprotective and antitumor activity	[77]
Δ^9-THCV	non-psychoactive effect; cures obesity	[75]
	effective against metabolic disorders, pancreatic disease, and hepatosteatosis syndrome	[1]
CBC	non-psychotropic and anti-inflammatory activity	[77,78]
CBD	cures memory loss, obesity, convulsive disorder, and rheumatoid arthritis	[79]
	cures epilepsy; exhibits antipsychotic, antinausea, and antianxiety activity	[80]
CBDA	anti-inflammatory and antihyperalgesia effect	[81,82]
CBG	non-psychotic activity	[75]
CBC, CBL, and DCA	cures HIV and cancer; possess immune boosting activity	[83]
CBG and amofrutin	anti-inflammatory activity	[84]
CBL	anti-inflammatory, anti-microbial, antipsychotic, and antiallergic activity	[5]
GFA and DCA	anti-microbial activity	[85]
PET	increases analgesia, catalepsy, hypolocomotion and hypothermia	[28]

6.1. Cannabinoid Receptors in Humans and Their Role

CB_1R (cannabinoid receptor type 1, first cloned in 1990) mRNA is highly expressed in the brain. It is also found in the heart, lung, ovary, adrenal gland, thymus, testes, prostate, tonsils, bone marrow, and uterus. CB_1R is also widely distributed inside non-neuronal tissues and inside various cells, tissues and also co-exist with other CBRs [86,87]. CB_1R is found inside the brain and displays high protein density and expression inside brain parts such as the cerebellar molecular layer, substantia nigra pars reticulate, globus pallidus external and internal, olfactory bulb, olfactory nucleus (anterior), hippocampus and layers II–III, Va and VI of the cerebral cortex. Additionally, in humans, the highest CB_1R level is found inside the cingulated gyrus, motor cortices and frontal secondary somatosensory. CB_1R is found in moderate levels inside the hypothalamus and ventral striatum and low levels in the brainstem; other respiratory control centers lack CB_1R. Moreover, this is the main reason CB_1R has a low effect on respiratory and cardiovascular activities [88]. Another important receptor, CB_2R, has high density inside immune cells and tissues, whereas a low amount is expressed inside the brain. Inside the brain, most of the presynaptic CBRs are observed as CB_1R, whereas it was displayed that presynaptic inhibitory CB_2R in GABAergic terminals of the hippocampus [86,87]. Phytocannabinoid applies its strong physio-psychotropic and psychotogenic actions along with the modulation of fast synaptic transmission in the brain showing its action on receptor CB_1R, present in the synaptic region. Additionally, fast synaptic activity encloses signaling via GABA and glutamate (activate ionotropic receptors). However, CB_1R contributes to psychoactivity and neurophysiology along with phytocannabinoid [89]. CB_2R is found inside pyramidal cells (II/III) of the medial prefrontal cortex; when it activates, it causes IP3-dependent opening of calcium-activated chloride channels and ultimately resulted in the inactivation of neuronal firing [90,91].

G-protein-coupled receptors (GPR), such as GPR18 and GPR55, serve as potent phytocannabinoid receptors. GPR18 is a "deorphanized" receptor present in cells and tissues of the thymus, spleen, small intestine, lymph nodes, leukocytes, and gametes [92–95]. Phytocannabinoids (e.g., Δ^9-THC) are agonists to GPR18. It plays a potent role in signalizing toxin-sensitive G-protein, Akt, PI3K, and p42/44 mitogen-activated protein kinase [96,97]. Additionally, cannabinoid (e.g., Δ^9-THC) induces β-arrestin recruitment in the cells which are transfected with GPR18 [86,95]. GPR55 is the important "deorphanized" metabotropic receptor that interacts with several active phytocannabinoids; it is present in the central nervous system (CNS), adipose tissues, adrenal gland, immune cells, small intestine, osteoblasts and osteoclast [98,99]. GPR55 displays a low sequence with both CB_1R and CB_2R (about 15%); GPRs interact with $G\alpha12$ or $G\alpha13$, which results in activation of several pathways such as Rho-associated protein kinase, Ras homolog gene family member A (RhoA), p38, MAPK/ERK pathways [91,100]. Some phytocannabinoids potently activate GPR5. Moreover, GPR55 is the most complex CBR. The same ligand acts as an agonist, antagonist, or neutral; it also possesses the allosteric modulator site [86,87]. Phytocannabinoids serve as weak and partial agonists on the GPR55, but via the help of some cannabinoids (CBD, cannabidivarine, CBDV), they inhibit lysophosphatidylinositol linked GPR55 activation. Thus, GPR55, along with cannabinoids, can participate in curing cancer cells, obesity, inflammation, neuropathic pain, osteoporosis, and neuromodulation. Transient receptor potential (TRP) superfamily channels have 27 polymodal sensor cation channels classified into six broad types in humans [101]. Among these six families, four groups interact with cannabinoids and other derivatives. Thus, these participate in boosting the immune system, with analgesia (pain reliever), and nociception (processing of sensory nervous system); they also modulate inflammations and pain sensations [102–104].

Phytocannabinoids exerts strong therapeutic potential in humans assessed by the meta-analysis of several clinical trials; it possesses a strong effect on health conditions such as nausea, vomiting, insomnia (sleeping disorder), depression, anxiety, paraplegia (paralyzes in a lower limb due to spinal cord injury), psychosis (emotional metal disorder), appetite stimulation in several syndromes [86]. Additionally, cannabinoids play a significant role in ailments of human diseases and syndromes such as Tourette syndrome (nervous system disorder such as continuous movements), AIDS, and treating intraocular pressure of the eye in glaucoma [105]. The endocannabinoid system (internal lipid retrograde neurotransmitters which bind on CBR) along with phytocannabinoid is involved in modulating pharmacological, physiological, biological, and cognitive processes [106]. Due to the limited space, in this review, we only name diseases and ailments cured by phytocannabinoids that are liver encephalopathy, hepatic disease, asthma, respiratory tract changes, bronchospasm, bone remodeling and metabolism, osteoarthritis, and osteoporosis. Phytocannabinoids may also be used to treat diseases related to CNS such as brain trauma, stroke, brain aging, neuroinflammation, and neurodegradation. Additionally, they have served as an anti-solid depressant to cure patients with suicidal tendencies; apart from all this, they are a strong pain reliever and immunomodulator. Phytocannabinoids are also given to pregnant women to avoid miscarriage and preimplantation embryo development [86,105,107].

Phytocannabinoids have several medicinal properties in curing diseases and disorders and assist as an important template for chemists to form novel and synthetic medicine [108]. Several investigations display the potential of synthetic cannabinoids in human ailments. Δ^9-THC and CBD together or alone form synthetic cannabinoids such as nabiximols, Sativex®, nabilone, dronabinol, levonantradol, and other synthetic Δ^9-THC-analogs. These synthetic cannabinoids are used to cure cancer pain, neuropathic pain, and spasticity caused by sclerosis [75,105].

Phytocannabinoids have diverse therapeutic potential, but they have several adverse risks, mainly due to fewer trials. Certainly, there is a necessity to support research and investigation on phytocannabinoid legally and ethically; globally, the legalization of cannabinoids encounters several controversies by society, researchers, and health practitioners. Many countries such as Canada and the US also support phytocannabinoid production for medical use. Colorado's state is employing private support from the medical cannabinoid (marijuana) industry to endorse future research on phytocannabinoids and cannabis [108,109].

6.2. *Bioactivity of Phytocannabinoids*

Phytocannabinoids are of great clinical use in humans and mammals, but the plant sources do not accumulate at high levels and are also regarded as endangered species. This enhances the demand for the biosynthesis of phytocannabinoids via biotechnological approaches to understand their bioactivity and role in medication. The prerequisite and establishment for such biotechnological approaches should be proper and deep knowledge of the pathways and mechanisms involved in phytocannabinoid production in *C. sativa*, liverworts and *Rhododendron* species [1,2,78,105].

6.2.1. Neutral Cannabinoids

In humans, phytocannabinoids play a potent role in signal transduction. They show crucial interaction with the G-protein coupled cannabinoid receptors (GPCRs such as CB_1R and CB_2R), peroxisome proliferator-activated receptor γ (PPARγ), and TRP ion channel. CB_1R is the most abundant GPCR located in CNS, whereas CB_2R is found in immune cells and tissues. Thus, cannabinoids play a crucial role in signaling, immune, and CNS proper functioning [72,74]. In humans, Δ^9-THC displays pleiotropic effects such as analgesic (relieve pain), relaxation, pain tolerance, and dysphoria (anxiety disorder); thus this reveals Δ^9-THC displays agonistic effects by the activation of CB_1R of β-arrestin 2 recruitment and signaling; also Δ^9-THC is the essential psychoactive (alters mood, behavior, feeling and thoughts by affecting CNS) bioactive constituent [73,104]. Δ^9-THC is found in the drug dronabinol (Marinol®), and the constituent of sesame oil plays potent antiemetic (prevents vomiting) for cancer patients receiving chemotherapy. Additionally, it also stimulates appetite in a patient with AIDS (acquired immunodeficiency syndrome); chronic administration of doses increases weight and appetite. Δ^9-THC is also given to patients with insomnia and depression as it improves sleep. In humans, Δ^8-THC decreases intraocular pressure (fluid pressure inside the eye) and exhibits antiglaucoma activity [76]. Phytocannabinoid CBG possesses non-psychotic activity that has a lower affinity for both CB_1R and CB_2R, several times lower than Δ^9-THC. CBG has remarkable activity towards ligand-gated cation channels of superfamily TRP. CBG serves as an agonist of TRP type ankyrin 1 (TRPA1) and TRP type vanilloid 1 (TRPV1), whereas it acts as an antagonistic of TRP type melastatin 8 (TRPM8). Δ^9-tetrahydrocannabivarinic acid (Δ^9-THCV) serves as a non-psychoactive that is antagonistic to CB_1R. In particular, Δ^9-THCV potentially acts against obesity-linked glucose intolerance. In humans, it acts as the potential phytocannabinoid for treating metabolic disorders, hepatosteatosis syndrome, and obesity. It also helps recover fasting plasma glucose and pancreatic β cell functioning. Additionally, Δ^9-THCV reduces adiponectin and apolipoprotein [75].

CBD plays a potent pharmacological role in humans by affecting CNS and some peripheral portions, as it is highly antagonistic to CB_1R and CB_2R. Additionally, CBD also acts as an essential allosteric modulator of receptor μ-opioid (pain reliever), thus used to relieve pain [75]. In humans, doses in the range of 10 to 700 mg CBD are not toxic and are delivered to patients in the form of Epidiolex® with epilepsy treatment-resistant, interlinked with CDKL5 deficiency disorder and many other disorders and syndromes [80]. CBD does not act as a psychoactive compound and can be delivered in patients receiving pharmacotherapy. Furthermore, CBD reduces Δ^9-THC-elicited psychotic disorders and decreases the deleterious effect of Δ^9-THC on memory which is hippocampus-dependent. CBD acts as the potential cannabinoid to cure obesity, convulsive disorder, and rheumatoid arthritis. Furthermore, CBD also possesses antipsychotic, antinausea, and antianxiety properties [79].

CBC acts as the non-psychotropic which does not interact with CB_1R and CB_2R [5,78]. Additionally, CBC inhibits endocannabinoid inactivation and further activates TRPA1, which helps recover intestinal inflammation and has several protective roles [110,111]. Additionally, CBC also possesses anti-inflammatory activity for curing lipopolysaccharide-enhanced edema. In particular, till 2021, no CBC study involves humans [112].

6.2.2. Cannabinoid Acid

In *C. sativa*, CBDA, CBCA, CBDA, CBGA, and Δ^9-THCA belong to the cannabinoid acid group, which does not possess cannabimimetic or psychotropic bioactivity [5]. Continuous decarboxylation of Δ^9-THCA leads to a low level of Δ^9-THC [113]. Δ^9-THCA has more affinity than Δ^9-THC and binds to PPARγ, whereas it has a low affinity towards CB_1R and CB_2R. Additionally, Δ^9-THCA possesses the neuroprotective property [77]. Phytocannabinoids of cannabis and their derivatives hamper the level of tumor necrosis factor α after stimulating lipopolysaccharide in culture supernatants of some macrophages. Additionally, phytocannabinoids, i.e., cannabinoid acid, also display their effect on phospholipase activity specific to phosphatidylcholine; thus, this effects signaling. Cannabinoid acids, e.g., CBDA, possess antihyperalgesic (i.e., reduce abnormally enhance pain) and anti-inflammatory properties [114]. In mice, cannabinoid acid (i.e., CBDA, and Δ^9-THCA) treatment increased antihyperalgesic and anti-inflammatory properties with inflammations induced by carrageenan. Additionally, in mice, oral administration of cannabinoids increases the antihyperalgesic property before carrageenan. In intestinal segments of house musk shrews, CBDA enhances tissue reduction in the resting state with the help of non-neuronal-associated pathways or pathways that are independent or are not associated with CB_1R or CB_2R [82,115].

6.2.3. Bibenzyl Cannabinoids

Analogs of bioactive bibenzyl (aralkyl) compound such as CBG from few liverworts have a phenethyl side chain; so bibenzyl CBG displays reduced affinity to cannabinoids receptors [116]. Additionally, it has an affinity to TRPA1 and TRP 1-4, which are ionotropic receptors, and shows a strong association to TRPM8 [10]. Bibenzyl phytocannabinoids belonging to the class amofrutin were extracted from *Glycyrrhiza foetida* and *Amorpha fruticosa*. They possess anti-inflammatory properties and are a strong activator of PPARγ [84]. Bibenzyl cannabinoid perrottetinene produced by *Radula* species shows a structural resemblance to cannabis Δ^9-THC, distinguishing it by having an aromatic side chain instead of a pentyl.

In mammals, perrottetinene acts as the psychoactive compound that is agonistic to CB_1R Furthermore, it enhances behavioral tetrads of analgesia, catalepsy, hypolocomotion, and hypothermia. Interestingly, enantiomers produced in liverworts are the major hallmark of their bioactivity and metabolic responses [28].

6.2.4. Rhododendron Cannabinoids

Rhododendron produces phytocannabinoids in the form of prenylated orcinoids. They are generally classified in CBC or CBL type as they have chromane or chromene scaffold Prenylated orcinoids and their derivatives play a potent role in building and boosting up the immune system in mammals and humans. *Rhododendron* cannabinoids possess anti-cancer property; hence can be recommended to a cancer patient. It also has antithrombotic activity. Furthermore, it also displays anti-inflammatory, antimicrobial, and antipsychotic properties and has very low toxicity in humans when administered [5,37,83,117]. Other essential phytocannabinoids of the *Rhododendron* species, DCA and rhododaurichromenic acids, serve as the potent bioactive compounds in curing HIV as they possess anti-HIV bioactivities. It was found that when DCA was administered in infected H9 cells, it showed a half-maximal effective concentration (EC_{50}) of 15 nM, much lower than the azidothymidine (a drug used to cure HIV), which has EC_{50} of 44 nM. Moreover, DCA with potent anti-HIV bioactivity has a therapeutic index (TI) of 3701 and EC_{50} of 5.67 ng mL^{-1} [81] Moreover, rhododaurichromanic acid possesses anti-HIV properties with TI of 92 and EC_{50} of 370 ng mL^{-1}. Active phytocannabinoids from *Rh. anthopogonoides* such as cannabiorcichromenic acid (CBC-type), anthopogocyclolic acid, and anthopogochromenic acid, which are CBL-type possess antiallergic activity as they inhibit the release of histamine [5,16,117]

7. Antibacterial and Antimicrobial Property of Phytocannabinoids

Cannabis fibers are used in textile industries as they are rich in active phytocannabinoids, which provide high biomass in less time. Hemp fibers, also called bast fibers, are rich in Δ^9-THC, cellulose, lignin, and other metabolites, so they are used in industries, animal bedding, and used as a substitute for glass fibers. Hemp fibers also possess high antibacterial properties because of the presence of phytocannabinoids, so they are used as antibacterial finishing agents and in surgical industries. Recently it was found that cannabis powder contains a high amount of cannabinoid content which possesses antibacterial properties against *Escherichia coli*. Interestingly, phytocannabinoids have high pharmacological, antibacterial, and antimicrobial activities [65,78,105,118,119].

Cannabis extract contains antibacterial property against *Bacillus subtilis* and *Staphylococcus aureus*, which are Gram-positive bacteria; also against Gram-negative bacteria *E. coli* *Pseudomonas aeruginosa*. In contrast, it displays no activity against fungal infections caused by *Aspergillus niger* and *Candida albicans*. Phytocannabinoids, e.g., Δ^9-THC, CBG, CBN, CBD and CBC, possess antibiotic properties against methicillin-resistant *Staphylococcus aureus* CBD and Δ^9-THC display bactericidal activity against streptococci and staphylococci. GFA and DCA also show antimicrobial properties against Gram-positive bacteria. Moreover, the prenyl group shows structural similarities to bioactive monoterpenes [58,65,85,108,120].

8. Phytocannabinoids in Stress Tolerance

Phytocannabinoids have a diverse role in humans and also exhibit antimicrobial, antibacterial, and antibiotic activity. Apart from this, they also have some other biological beneficial properties in mitigating biotic and abiotic stress in the plant. Cannabis trichomes possess phytocannabinoid in large quantities. High temperatures or herbivory cause trichome rupture and the release of the phytocannabinoid content, which protects the plant from desiccation and high-temperature stress. Therefore, they enhance plant tolerance to heat stress [2,79]. It was also reported that phytocannabinoid production was enhanced in cannabis flowers after UV-B-induced stress. Thus, phytocannabinoids serve as a sun shield against destructive UV-B radiation [26]. Liverworts oil bodies possess phytocannabinoid, so they are not damaged by fungi and bacteria, insect larvae and adults, slugs, snails, and small mammals. Significantly, phytocannabinoids inside oil bodies provide tolerance against several abiotic stresses such as cold temperature, heat, excessive light, and UV radiation [26,29,58].

Additionally, phytocannabinoids provide resilience to desiccation. Liverworts are unable to produce abscisic acid (ABA); besides this, they have phytocannabinoid, i.e., lunularic acid that shows the same activity as ABA, which is a plant hormone. There is a lack of knowledge about the role of phytocannabinoids in biotic and abiotic stress mitigation. Thus, more in-depth studies are still required to understand the mechanism involved in stress tolerance via phytocannabinoids. However, studies must understand their role in other plants' stress amelioration [1].

9. Phytocannabinoid Foe

Why is cannabis cultivation illegal in many countries and what are the disadvantages associated with phytocannabinoid consumption? Cannabis is the third most popular substance consumed after alcohol and tobacco. The USA and New Zealand are at the top rates (42%) of consuming it; however, it is the most illicit drug used globally. Its consumption is escalating among teenagers, and they are consuming it in the form of hashish and marijuana [86,121,122]. It was reported that almost everyone who tried cannabis (marijuana or hashish) becomes a habitual abuser and dependent on cannabis; thus, this enhanced the cannabis use disorder [123,124]. Cannabis elicits a plethora of biological, metabolic, and physical responses. Nevertheless, everything depends on its utilization and what dose it is consumed in (what, when, and how). Most commonly used cannabis marijuana is prepared by drying leaves and flowering buds of the cannabis plant. Moreover, hashish which is eventually stronger than marijuana, is the concentrated resin from the female cannabis plant and is directly consumed via chewing and smoking. Kief is a trichomes mass obtained from flowers and leaves, which is used to make hashish. Additionally, hashish contains a mixture of several phytocannabinoids and terpenes [125]. Smoking or dabbing marijuana is the most common way in the consumption of cannabinoids. Phytocannabinoid delays the onset effect by half to two hour circulating level; circulating level is greater when they are smoked or intravenously consumed. In particular, the delay time in the onset of psychobiological effect is due to the biosynthesis of bioactive phytocannabinoid metabolite or induction in endocannabinoid availability or to reach receptors to show its effect. Cannabis consumption produces two kinds of effects: short-term (acute) and long-term (chronic) [86]. In humans, acute consumption of cannabis effect is classified into two groups: (i) physiological effect, which displays therapeutic potential, and (ii) recreation effects which show the effect on brain identical to psychotic state (psychotomimetic) [126,127]. Acute physiological effects include relaxation, hyperlocomotion, resting of heart, and increase in thirst, appetite, and food palatability. Acute psychotomimetic effects encompass enhanced sociability, euphoria (sense of great happiness), relaxation, hallucinations, delusions, conceptual disorganization, paranoia, social withdrawal, blurred vision, and decreased attention [128,129]. Furthermore, it also leads to anxiety, panic attacks, and dysphoria; all of these effects are similar to schizophrenia. Long-term (chronic) consumption of cannabis is related to addiction risk which depends on age, gender, lifestyle, frequency of intake, dosage, and genetic makeup [122]. Long-term cannabis consumption increases various alterations inside the human body (mainly heart, brain, and nervous system) and develops the serious risk of several disorders such as schizophrenia, neuropsychiatric syndrome, cerebellar infarction, vasoconstriction, and several fatal cardiovascular problems. Additionally, cannabis consumption has a detrimental effect on metabolic and biological processes; but these disorders are low if consumption of alcohol is low; apart from this, cannabis consumption confers useful effects on the risk of cardiometabolic factors such as insulin sensitivity, glycemia, lipid amount. Nevertheless, it remains to be demonstrated whether long-term cannabis consumption in humans functions as an activator or deactivator of CBRs [79,105,108,109,122,126].

10. Conclusions and Future Prospects

Phytocannabinoids are bioactive naturally occurring terpenoids that were earlier thought to be exclusive to *C. sativa* but have now also been produced in *Rhododendron* species, some legumes, the liverwort genus *Radula*, and some fungi. Bioactive phytocannabinoids show remarkable non-hallucinogenic biological properties determined by the variable nature of the side chain and prenyl group defined by the enzymes involved in their biosynthesis. The present review focused on genes and enzymes involved in biosynthesis across several plant species such as cannabis, *Rhododendron* and liverworts were discussed here. Moreover, these species are used in the combinatorial fashion to construct a required new bioactive phytocannabinoid structure established on integrating peculiar prenyl moieties, side chains, and unique cyclized core structures. Meanwhile, phytocannabinoids biosynthesis involves a large collection of enzymes that can potently develop bioactive phytocannabinoids via genetic engineering. This review gives a better understanding of the diverse role of phytocannabinoids in humans, plants, microbiology, and biotechnology (Figure 8). In particular, active phytocannabinoids play a crucial role in treating several diseases in humans. They possess antibacterial and antimicrobial properties in several industries. Recent studies also appraise phytocannabinoids role in cold, heat, and radiation stress tolerance. Phytocannabinoids also protect the plant from pathogens and herbivory. Moreover, the review also focuses on the foe of improper use of phytocannabinoids and why it is banned in many countries.

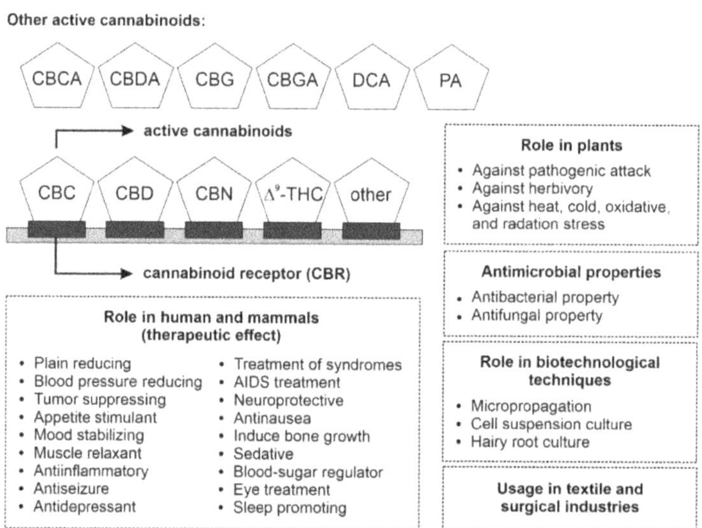

Figure 8. Selective role of active phytocannabinoids in humans, mammals, plants, biotechnology, and industries. Abbreviations: CBC, cannabichromene; CBCA, cannabichromenic acid; CBD, cannabidiol; CBDA, cannabidiolic acid; CBG, cannabigerol; CBGA, cannabigerolic acid; CBN, cannabinol; CBR, cannabinoid receptor; DCA, daurichromenic acid; PA, perrottetinenic acid; Δ^9-THC, Δ^9-tetrahydrocannabinol.

The hypothesis is that phytocannabinoids have versatile use and are beneficial for humans and plants if appropriately used. Further investigations are needed on genes and enzymes involved in the biosynthetic pathway in different plant species. Comparative study between type III PKS from cannabis and liverwort might explain their specificity to particular molecular structure and insight of genetic and protein engineering. Semi-synthetic techniques and chemical modification integration can be used for the biosynthesis of different types of cannabinoids—it also modulates their bioactivity and bioavailability. The availability of naturally occurring cannabinoids provides an insight into scrutinizing their role in plant stress tolerance, including individual and combinatorial exogenous effects. Lastly, quantitative chemical profiling can also give a deep knowledge of the occurrence and importance of NADES and their other possible role in plants and their role in reducing autotoxicity.

Author Contributions: Conceptualization, S.H.; methodology, S.H.; writing—original draft preparation, Y.A. and P.S.; writing—review and editing, S.H. and A.B.; visualization, A.B.; supervision, S.H. and A.B. All authors have read and agreed to the published version of the manuscript.

Funding: This research received no external funding.

Institutional Review Board Statement: Not applicable.

Informed Consent Statement: Not applicable.

Data Availability Statement: Not applicable.

Acknowledgments: Yamshi Arif acknowledges the receipt of Inspire fellowship (IF 180778) by the Department of Science & Technology, Government of India, New Delhi.

Conflicts of Interest: The authors declare no conflict of interest.

Abbreviations

AAE1	acyl-activated enzyme 1
ABA	abscisic acid
ADME	absorption, distribution, metabolism, excretion
BBS	bibenzyl synthase
C4H	cinnamate 4-hydroxylase
4CL	4-coumarate:CoA ligase
CBR	cannabinoid receptor
CB_1R	cannabinoid receptor type 1
CB_2R	cannabinoid receptor type 2
CBC	cannabichromene
CBCA	cannabichromenic acid
CBCAS	cannabichromenic acid synthase
CBCV	cannabichromevarine
CBCVA	cannabichromevarinic acid
CBD	cannabidiol
CBDA	cannabidiolic acid
CBDAS	cannabidiolic acid synthase
CBDV	cannabidivarine
CBDVA	cannabidivarinic acid
CBE	cannabielsoin
CBG	cannabigerol
CBGA	cannabigerolic acid
CBGAS	cannabigerolic acid synthase
CBL	cannabicyclol
CBN	cannabinol
CBND	cannabidiol
CBT	cannabitriol
CHIL	chalcone isomerases-like protein
CNS	central nervous system
CYP	cytochrome P450
DBR	double-bond reductase
DCA	daurichromenic acid
DCAS	daurichromenic acid synthase
DHAC	dihydrostilbene acid cyclase
Dicamba	3,6-dichloro-2-methoxybenzoic acid

EC_{50}	effective concentration
GFA	grifolic acid
GPP	geranyl pyrophosphate
GPR	G-protein-coupled receptor
GPCR	G-protein-coupled cannabinoid receptor
IAA	indole-3-acetic acid
KR	ketoreductase
LOX	lipoxygenase
MEP	methylerythritol-4-phosphate
NAA	1-naphthaleneacetic acid
NADES	natural deep eutectic solvent
OA	olivetolic acid
OAC	olivetolic acid cyclase
OLS	olivetol synthase
ORS	orcinol synthase
OSA	orsellinic acid
PA	perrottetinenic acid
PAL	phenylalanine ammonia-lyase
PAS	perrottetinenic acid synthase
PET	perrottetinene
PGR	plant growth regulator
PKS	polyketide synthase
PPARγ	peroxisome proliferator-activated receptor γ
PT	prenyltransferase
STS	stilbene synthase
TAL	tyrosine ammonia-lyase
TF	transcription factor
TI	therapeutic index
TRP	transient receptor potential
TRPA1	TRP type ankyrin 1
TRPM8	TRP type melastanin 8
TRPV1	TRP type vanilloid 1
Δ^8-THC	Δ^8-tetrahydrocannabinol
Δ^9-THC	Δ^9-tetrahydrocannabinol
Δ^9-THCA	Δ^9-tetrahydrocannabinolic acid
Δ^9-THCAS	Δ^9-tetrahydrocannabinolic acid synthase
Δ^9-THCV	Δ^9-tetrahydrocannabivarinic acid
Δ^9-THCVA	Δ^9-tetrahydrocannabidivarinic acid

References

1. Gülck, T.; Møller, B.L. Phytocannabinoids: Origins and biosynthesis. *Trends Plant Sci.* **2020**, *25*, 985–1004. [CrossRef] [PubMed]
2. Hanuš, L.O.; Meyer, S.M.; Muñoz, E.; Taglialatela-Scafati, O.; Appendino, G. Phytocannabinoids: A unified critical inventory. *Nat. Prod. Rep.* **2016**, *33*, 1357–1392. [CrossRef]
3. ElSohly, M.A.; Slade, D. Chemical constituents of marijuana: The complex mixture of natural cannabinoids. *Life Sci.* **2005**, *78*, 539–548. [CrossRef] [PubMed]
4. Berman, P.; Futoran, K.; Lewitus, G.M.; Mukha, D.; Benami, M.; Shlomi, T.; Meiri, D. A new ESI-LC/MS approach for comprehensive metabolic profiling of phytocannabinoids in *Cannabis*. *Sci. Rep.* **2018**, *8*, 14280. [CrossRef] [PubMed]
5. Pollastro, F.; Caprioglio, D.; Del Prete, D.; Rogati, F.; Minassi, A.; Taglialatela-Scafati, O.; Munoz, E.; Appendino, G. Cannabichromene. *Nat. Prod. Commun.* **2018**, *13*, 1189–1194. [CrossRef]
6. Happyana, N.; Kayser, O. Monitoring metabolite profiles of *Cannabis sativa* L. trichomes during flowering period using ^{1}H NMR-based metabolomics and real-time PCR. *Planta Med.* **2016**, *82*, 1217–1223. [CrossRef] [PubMed]
7. Chanda, D.; Neumann, D.; Glatz, J.F.C. The endocannabinoid system: Overview of an emerging multi-faceted therapeutic target. *Prostaglandins Leukot. Essent. Fatty Acids* **2019**, *140*, 51–56. [CrossRef]
8. Flores-Sanchez, I.J.; Verpoorte, R. Secondary metabolism in *Cannabis*. *Phytochem. Rev.* **2008**, *7*, 615–639. [CrossRef]
9. Dei Cas, M.; Casagni, E.; Casiraghi, A.; Minghetti, P.; Fornasari, D.M.M.; Ferri, F.; Arnoldi, S.; Gambaro, V.; Roda, G. Phytocannabinoids profile in medicinal cannabis oils: The impact of plant varieties and preparation methods. *Front. Pharmacol.* **2020**, *11*, 570616. [CrossRef]

10. Pollastro, F.; De Petrocellis, L.; Schiano-Moriello, A.; Chianese, G.; Heyman, H.; Appendino, G.; Taglialatela-Scafati, O. Amorfrutin-type phytocannabinoids from *Helichrysum umbraculigerum*. *Fitoterapia* **2017**, *123*, 13–17. [CrossRef]
11. Taura, F.; Iijima, M.; Kurosaki, F. Daurichromenic acid and grifolic acid: Phytotoxic meroterpenoids that induce cell death in cell culture of their producer *Rhododendron dauricum*. *Plant Signal. Behav.* **2018**, *13*, e1422463. [CrossRef] [PubMed]
12. Taura, F.; Iijima, M.; Lee, J.-B.; Hashimoto, T.; Asakawa, Y.; Kurosaki, F. Daurichromenic acid-producing oxidocyclase in the young leaves of *Rhododendron dauricum*. *Nat. Prod. Commun.* **2014**, *9*, 1329–1332. [CrossRef]
13. Kashiwada, Y.; Yamazaki, K.; Ikeshiro, Y.; Yamagishi, T.; Fujioka, T.; Mihashi, K.; Mizuki, K.; Cosentino, L.M.; Fowke, K.; Morris-Natschke, S.L.; et al. Isolation of rhododaurichromanic acid B and the anti-HIV principles rhododaurichromanic acid A and rhododaurichromenic acid from *Rhododendron dauricum*. *Tetrahedron* **2001**, *57*, 1559–1563. [CrossRef]
14. Rogachev, A.; Komarova, N.; Korchagina, D.; Dolgikh, M.; Sorokina, I.; Baev, D.; Tolstikova, T.; Fomenko, V.; Salakhutdinov, N. Some prenylated phenols of *Rhododendron adamsii*: Isolation, modification and pharmacological tests. *Chem. Sustain. Dev.* **2009**, *17*, 185–193.
15. Yang, Y.-X.; Wang, J.-X.; Wang, Q.; Li, H.-L.; Tao, M.; Luo, Q.; Liu, H. New chromane and chromene meroterpenoids from flowers of *Rhododendron rubiginosum* Franch. var. *rubiginosum*. *Fitoterapia* **2018**, *127*, 396–401. [CrossRef] [PubMed]
16. Iwata, N.; Kitanaka, S. New cannabinoid-like chromane and chromene derivatives from *Rhododendron anthopogonoides*. *Chem. Pharm. Bull.* **2011**, *59*, 1409–1412. [CrossRef] [PubMed]
17. Chicca, A.; Schafroth, M.A.; Reynoso-Moreno, I.; Erni, R.; Petrucci, V.; Carreira, E.M.; Gertsch, J. Uncovering the psychoactivity of a cannabinoid from liverworts associated with a legal high. *Sci. Adv.* **2018**, *4*, eaat2166. [CrossRef]
18. Hussain, T.; Espley, R.V.; Gertsch, J.; Whare, T.; Stehle, F.; Kayser, O. Demystifying the liverwort *Radula marginata*, a critical review on its taxonomy, genetics, cannabinoid phytochemistry and pharmacology. *Phytochem. Rev.* **2019**, *18*, 953–965. [CrossRef]
19. Métoyer, B.; Lebouvier, N.; Hnawia, E.; Herbette, G.; Thouvenot, L.; Asakawa, Y.; Nour, M.; Raharivelomanana, P. Chemotypes and biomarkers of seven species of new Caledonian liverworts from the Bazzanioideae subfamily. *Molecules* **2018**, *23*, 1353. [CrossRef] [PubMed]
20. Pryce, R.J. Lunularic acid, a common endogenous growth inhibitor of liverworts. *Planta* **1971**, *97*, 354–357. [CrossRef]
21. Quaghebeur, K.; Coosemans, J.; Toppet, S.; Compernolle, F. Cannabiorci- and 8-chlorocannabiorcichromenic acid as fungal antagonists from *Cylindrocarpon olidum*. *Phytochemistry* **1994**, *37*, 159–161. [CrossRef]
22. Elhendawy, M.A.; Wanas, A.S.; Radwan, M.M.; Azzaz, N.A.; Toson, E.S.; ElSohly, M.A. Chemical and biological studies of *Cannabis sativa* roots. *Med. Cannabis Cannabinoids* **2019**, *1*, 104–111. [CrossRef] [PubMed]
23. Happyana, N.; Agnolet, S.; Muntendam, R.; Van Dam, A.; Schneider, B.; Kayser, O. Analysis of cannabinoids in laser-microdissected trichomes of medicinal *Cannabis sativa* using LCMS and cryogenic NMR. *Phytochemistry* **2013**, *87*, 51–59. [CrossRef]
24. Livingston, S.J.; Quilichini, T.D.; Booth, J.K.; Wong, D.C.J.; Rensing, K.H.; Laflamme-Yonkman, J.; Castellarin, S.D.; Bohlmann, J.; Page, J.E.; Samuels, A.L. Cannabis glandular trichomes alter morphology and metabolite content during flower maturation. *Plant J.* **2019**, *101*, 37–56. [CrossRef]
25. Mahlberg, P.G.; Kim, E.-S. Cuticle development on glandular trichomes of *Cannabis sativa* (Cannabaceae). *Am. J. Bot.* **1991**, *78*, 1113–1122. [CrossRef]
26. Eichhorn Bilodeau, S.; Wu, B.-S.; Rufyikiri, A.-S.; MacPherson, S.; Lefsrud, M. An update on plant photobiology and implications for cannabis production. *Front. Plant Sci.* **2019**, *10*, 296. [CrossRef]
27. Iijima, M.; Munakata, R.; Takahashi, H.; Kenmoku, H.; Nakagawa, R.; Kodama, T.; Asakawa, Y.; Abe, I.; Yazaki, K.; Kurosaki, F.; et al. Identification and characterization of daurichromenic acid synthase active in anti-HIV biosynthesis. *Plant Physiol.* **2017**, *174*, 2213–2230. [CrossRef] [PubMed]
28. Asakawa, Y.; Ludwiczuk, A. Chemical constituents of bryophytes: Structures and biological activity. *J. Nat. Prod.* **2017**, *81*, 641–660. [CrossRef]
29. He, X.; Sun, Y.; Zhu, R.-L. The oil bodies of liverworts: Unique and important organelles in land plants. *Crit. Rev. Plant Sci.* **2013**, *32*, 293–302. [CrossRef]
30. Geissler, M.; Volk, J.; Stehle, F.; Kayser, O.; Warzecha, H. Subcellular localization defines modification and production of Δ^9-tetrahydrocannabinolic acid synthase in transiently transformed *Nicotiana benthamiana*. *Biotechnol. Lett.* **2018**, *40*, 981–987. [CrossRef] [PubMed]
31. Luo, X.; Reiter, M.A.; d'Espaux, L.; Wong, J.; Denby, C.M.; Lechner, A.; Zhang, Y.; Grzybowski, A.T.; Harth, S.; Lin, W.; et al. Complete biosynthesis of cannabinoids and their unnatural analogues in yeast. *Nature* **2019**, *567*, 123–126. [CrossRef]
32. Rodziewicz, P.; Loroch, S.; Marczak, Ł.; Sickmann, A.; Kayser, O. Cannabinoid synthases and osmoprotective metabolites accumulate in the exudates of *Cannabis sativa* L. glandular trichomes. *Plant Sci.* **2019**, *284*, 108–116. [CrossRef] [PubMed]
33. Gagne, S.J.; Stout, J.M.; Liu, E.; Boubakir, Z.; Clark, S.M.; Page, J.E. Identification of olivetolic acid cyclase from *Cannabis sativa* reveals a unique catalytic route to plant polyketides. *Proc. Natl. Acad. Sci. USA* **2012**, *109*, 12811–12816. [CrossRef] [PubMed]
34. Raharjo, T.J.; Chang, W.-T.; Choi, Y.H.; Peltenburg-Looman, A.M.G.; Verpoorte, R. Olivetol as product of a polyketide synthase in *Cannabis sativa* L. *Plant Sci.* **2004**, *166*, 381–385. [CrossRef]
35. Laverty, K.U.; Stout, J.M.; Sullivan, M.J.; Shah, H.; Gill, N.; Holbrook, L.; Deikus, G.; Sebra, R.; Hughes, T.R.; Page, J.E.; et al. A physical and genetic map of *Cannabis sativa* identifies extensive rearrangements at the THC/CBD acid synthase loci. *Genome Res.* **2019**, *29*, 146–156. [CrossRef] [PubMed]

36. Saeki, H.; Hara, R.; Takahashi, H.; Iijima, M.; Munakata, R.; Kenmoku, H.; Fuku, K.; Sekihara, A.; Yasuno, Y.; Shinada, T.; et al. An aromatic farnesyltransferase functions in biosynthesis of the anti-HIV meroterpenoid daurichromenic acid. *Plant Physiol.* **2018**, *178*, 535–551. [CrossRef]
37. Taura, F.; Iijima, M.; Yamanaka, E.; Takahashi, H.; Kenmoku, H.; Saeki, H.; Morimoto, S.; Asakawa, Y.; Kurosaki, F.; Morita, H. A novel class of plant type III polyketide synthase involved in orsellinic acid biosynthesis from *Rhododendron dauricum*. *Front. Plant Sci.* **2016**, *7*, 1452. [CrossRef] [PubMed]
38. Katsuyama, Y.; Kita, T.; Funa, N.; Horinouchi, S. Curcuminoid biosynthesis by two type III polyketide synthases in the herb *Curcuma longa*. *J. Biol. Chem.* **2009**, *284*, 11160–11170. [CrossRef]
39. Hussain, T.; Plunkett, B.; Ejaz, M.; Espley, R.V.; Kayser, O. Identification of putative precursor genes for the biosynthesis of cannabinoid-like compound in *Radula marginata*. *Front. Plant Sci.* **2018**, *9*, 537. [CrossRef]
40. Akiyama, T.; Shibuya, M.; Liu, H.-M.; Ebizuka, Y. *p*-Coumaroyltriacetic acid synthase, a new homologue of chalcone synthase, from *Hydrangea macrophylla* var. *thunbergii*. *Eur. J. Biochem.* **1999**, *263*, 834–839. [CrossRef]
41. Austin, M.B.; Noel, J.P. The chalcone synthase superfamily of type III polyketide synthases. *Nat. Prod. Rep.* **2002**, *20*, 79–110. [CrossRef]
42. Imoto, S.A.; Ohta, Y. Intracellular localization of lunularic acid and prelunularic acid in suspension cultured cells of *Marchantia polymorpha*. *Plant Physiol.* **1985**, *79*, 751–755. [CrossRef] [PubMed]
43. Toyota, M.; Shimamura, T.; Ishii, H.; Renner, M.; Braggins, J.; Asakawa, Y. New bibenzyl cannabinoid from the New Zealand liverwort *Radula marginata*. *Chem. Pharm. Bull.* **2002**, *50*, 1390–1392. [CrossRef]
44. Yang, X.; Matsui, T.; Kodama, T.; Mori, T.; Zhou, X.; Taura, F.; Noguchi, H.; Abe, I.; Morita, H. Structural basis for olivetolic acid formation by a polyketide cyclase from *Cannabis sativa*. *FEBS J.* **2016**, *283*, 1088–1106. [CrossRef]
45. Toyota, M.; Kinugawa, T.; Asakawa, Y. Bibenzyl cannabinoid and bisbibenzyl derivative from the liverwort *Radula perrottetii*. *Phytochemistry* **1994**, *37*, 859–862. [CrossRef]
46. Fuentes, P.; Armarego-Marriott, T.; Bock, R. Plastid transformation and its application in metabolic engineering. *Curr. Opin. Biotechnol.* **2018**, *49*, 10–15. [CrossRef]
47. Kodym, A.; Leeb, C.J. Back to the roots: Protocol for the photoautotrophic micropropagation of medicinal *Cannabis*. *Plant Cell Tissue Organ Cult.* **2019**, *138*, 399–402. [CrossRef]
48. Lange, K.; Schmid, A.; Julsing, M.K. Δ^9-Tetrahydrocannabinolic acid synthase: The application of a plant secondary metabolite enzyme in biocatalytic chemical synthesis. *J. Biotechnol.* **2016**, *233*, 42–48. [CrossRef] [PubMed]
49. Lata, H.; Chandra, S.; Khan, I.A.; ElSohly, M.A. Micropropagation of *Cannabis sativa* L. -An update. In *Cannabis sativa L. Botany and Biotechnology*; Chandra, S., Lata, H., ElSohly, M.A., Eds.; Springer International Publishing: Cham, Switzerland, 2017; pp. 285–297. [CrossRef]
50. Wróbel, T.; Dreger, M.; Wielgus, K.; Słomski, R. The application of plant in vitro cultures in cannabinoid production. *Biotechnol. Lett.* **2017**, *40*, 445–454. [CrossRef] [PubMed]
51. Andersen-Ranberg, J.; Kongstad, K.T.; Nielsen, M.T.; Jensen, N.B.; Pateraki, I.; Bach, S.S.; Hamberger, B.; Zerbe, P.; Staerk, D.; Bohlmann, J.; et al. Expanding the landscape of diterpene structural diversity through stereochemically controlled combinatorial biosynthesis. *Angew. Chem. Int. Ed.* **2016**, *55*, 2142–2146. [CrossRef]
52. Chen, R.; Gao, B.; Liu, X.; Ruan, F.; Zhang, Y.; Lou, J.; Feng, K.; Wunsch, C.; Li, S.-M.; Dai, J.; et al. Molecular insights into the enzyme promiscuity of an aromatic prenyltransferase. *Nat. Chem. Biol.* **2017**, *13*, 226–234. [CrossRef]
53. Kannangara, R.; Siukstaite, L.; Borch-Jensen, J.; Madsen, B.; Kongstad, K.T.; Staerk, D.; Bennedsen, M.; Okkels, F.T.; Rasmussen, S.A.; Larsen, T.O.; et al. Characterization of a membrane-bound C-glucosyltransferase responsible for carminic acid biosynthesis in *Dactylopius coccus* Costa. *Nat. Commun.* **2017**, *8*, 1987. [CrossRef]
54. Yazaki, K.; Sasaki, K.; Tsurumaru, Y. Prenylation of aromatic compounds, a key diversification of plant secondary metabolites. *Phytochemistry* **2009**, *70*, 1739–1745. [CrossRef] [PubMed]
55. Ban, Z.; Qin, H.; Mitchell, A.J.; Liu, B.; Zhang, F.; Weng, J.-K.; Dixon, R.A.; Wang, G. Noncatalytic chalcone isomerase-fold proteins in *Humulus lupulus* are auxiliary components in prenylated flavonoid biosynthesis. *Proc. Natl. Acad. Sci. USA* **2018**, *115*, E5223–E5232. [CrossRef] [PubMed]
56. Gribble, G.W. Biological activity of recently discovered halogenated marine natural products. *Mar. Drugs* **2015**, *13*, 4044–4136. [CrossRef] [PubMed]
57. Hansen, E.H.; Osmani, S.A.; Kristensen, C.; Møller, B.L.; Hansen, J. Substrate specificities of family 1 UGTs gained by domain swapping. *Phytochemistry* **2009**, *70*, 473–482. [CrossRef]
58. Khan, B.A.; Warner, P.; Wang, H. Antibacterial properties of hemp and other natural fibre plants: A review. *BioResources* **2014**, *9*, 3642–3659. [CrossRef]
59. Choi, Y.H.; van Spronsen, J.; Dai, Y.; Verberne, M.; Hollmann, F.; Arends, I.W.C.E.; Witkamp, G.-J.; Verpoorte, R. Are natural deep eutectic solvents the missing link in understanding cellular metabolism and physiology? *Plant Physiol.* **2011**, *156*, 1701–1705. [CrossRef]
60. Dai, Y.; Witkamp, G.-J.; Verpoorte, R.; Choi, Y.H. Tailoring properties of natural deep eutectic solvents with water to facilitate their applications. *Food Chem.* **2015**, *187*, 14–19. [CrossRef]
61. Courchaine, E.M.; Lu, A.; Neugebauer, K.M. Droplet organelles? *EMBO J.* **2016**, *35*, 1603–1612. [CrossRef]

62. Hammond, O.S.; Bowron, D.T.; Edler, K.J. The effect of water upon deep eutectic solvent nanostructure: An unusual transition from ionic mixture to aqueous solution. *Angew. Chem. Int. Ed.* **2017**, *56*, 9782–9785. [CrossRef] [PubMed]
63. Knudsen, C.; Bavishi, K.; Viborg, K.M.; Drew, D.P.; Simonsen, H.T.; Motawia, M.S.; Møller, B.L.; Laursen, T. Stabilization of dhurrin biosynthetic enzymes from *Sorghum bicolor* using a natural deep eutectic solvent. *Phytochemistry* **2020**, *170*, 112214. [CrossRef] [PubMed]
64. Knudsen, C.; Gallage, N.J.; Hansen, C.C.; Møller, B.L.; Laursen, T. Dynamic metabolic solutions to the sessile life style of plants. *Nat. Prod. Rep.* **2018**, *35*, 1140–1155. [CrossRef] [PubMed]
65. Andre, C.M.; Hausman, J.-F.; Guerriero, G. *Cannabis sativa*: The plant of the thousand and one molecules. *Front. Plant Sci.* **2016**, *7*, 19. [CrossRef] [PubMed]
66. Lata, H.; Chandra, S.; Techen, N.; Khan, I.A.; ElSohly, M.A. Molecular analysis of genetic fidelity in *Cannabis sativa* L. plants grown from synthetic (encapsulated) seeds following in vitro storage. *Biotechnol. Lett.* **2011**, *33*, 2503–2508. [CrossRef] [PubMed]
67. Gai, Q.-Y.; Jiao, J.; Luo, M.; Wei, Z.-F.; Zu, Y.-G.; Ma, W.; Fu, Y.-J. Establishment of hairy root cultures by *Agrobacterium rhizogenes* mediated transformation of *Isatis tinctoria* L. For the efficient production of flavonoids and evaluation of antioxidant activities. *PLoS ONE* **2015**, *10*, e0119022. [CrossRef]
68. Feeney, M.; Punja, Z.K. Tissue culture and *Agrobacterium*-mediated transformation of hemp (*Cannabis sativa* L.). *In Vitro Cell. Dev. Biol. Plant* **2003**, *39*, 578–585. [CrossRef]
69. Lata, H.; Chandra, S.; Khan, I.; ElSohly, M. High frequency plant regeneration from leaf derived callus of high Δ^9-tetrahydrocannabinol yielding *Cannabis sativa* L. *Planta Med.* **2010**, *76*, 1629–1633. [CrossRef]
70. Hazekamp, A.; Verpoorte, R. Structure elucidation of the tetrahydrocannabinol complex with randomly methylated β-cyclodextrin. *Eur. J. Pharm. Sci.* **2006**, *29*, 340–347. [CrossRef]
71. Pellechia, T. Legal Cannabis Industry Poised for Big Growth, in North America and around the World. *Forbes* **2018**. Available online: https://www.forbes.com/sites/thomaspellechia/2018/03/01/double-digit-billions-puts-north-america-in-the-worldwide-cannabis-market-lead/?sh=356dc9565109 (accessed on 15 April 2021).
72. Booth, J.K.; Bohlmann, J. Terpenes in *Cannabis sativa*—From plant genome to humans. *Plant Sci.* **2019**, *284*, 67–72. [CrossRef]
73. Laprairie, R.B.; Bagher, A.M.; Kelly, M.E.M.; Dupré, D.J.; Denovan-Wright, E.M. Type 1 cannabinoid receptor ligands display functional selectivity in a cell culture model of striatal medium spiny projection neurons. *J. Biol. Chem.* **2014**, *289*, 24845–24862. [CrossRef]
74. Kendall, D.A.; Yudowski, G.A. Cannabinoid receptors in the central nervous system: Their signaling and roles in disease. *Front. Cell. Neurosci.* **2017**, *10*, 294. [CrossRef]
75. Pryce, G.; Riddall, D.R.; Selwood, D.L.; Giovannoni, G.; Baker, D. Neuroprotection in experimental autoimmune encephalomyelitis and progressive multiple sclerosis by cannabis-based cannabinoids. *J. Neuroimmune Pharmacol.* **2014**, *10*, 281–292. [CrossRef]
76. Hippalgaonkar, K.; Gul, W.; ElSohly, M.A.; Repka, M.A.; Majumdar, S. Enhanced solubility, stability, and transcorneal permeability of δ-8-tetrahydrocannabinol in the presence of cyclodextrins. *AAPS PharmSciTech* **2011**, *12*, 723–731. [CrossRef]
77. Nadal, X.; Del Río, C.; Casano, S.; Palomares, B.; Ferreiro-Vera, C.; Navarrete, C.; Sánchez-Carnerero, C.; Cantarero, I.; Bellido, M.L.; Meyer, S.; et al. Tetrahydrocannabinolic acid is a potent PPARγ agonist with neuroprotective activity. *Br. J. Pharmacol.* **2017**, *174*, 4263–4276. [CrossRef]
78. Turner, S.E.; Williams, C.M.; Iversen, L.; Whalley, B.J. Molecular pharmacology of phytocannabinoids. In *Phytocannabinoids: Unraveling the Complex Chemistry and Pharmacology of Cannabis sativa*; Kinghorn, A.D., Falk, H., Gibbons, S., Kobayashi, J.I., Eds.; Springer International Publishing: Cham, Switzerland, 2017; pp. 61–101. [CrossRef]
79. Huestis, M.A.; Solimini, R.; Pichini, S.; Pacifici, R.; Carlier, J.; Busardò, F.P. Cannabidiol adverse effects and yoxicity. *Curr. Neuropharmacol.* **2019**, *17*, 974–989. [CrossRef]
80. Devinsky, O.; Verducci, C.; Thiele, E.A.; Laux, L.C.; Patel, A.D.; Filloux, F.; Szaflarski, J.P.; Wilfong, A.; Clark, G.D.; Park, Y.D.; et al. Open-label use of highly purified CBD (Epidiolex®) in patients with CDKL5 deficiency disorder and Aicardi, Dup15q, and Doose syndromes. *Epilepsy Behav.* **2018**, *86*, 131–137. [CrossRef] [PubMed]
81. Rok Lee, Y.; Wang, X. A short synthetic route to biologically active (±)-daurichromenic acid as highly potent anti-HIV agent. *Org. Biomol. Chem.* **2005**, *3*, 3955. [CrossRef] [PubMed]
82. Verhoeckx, K.C.M.; Korthout, H.A.A.J.; van Meeteren-Kreikamp, A.P.; Ehlert, K.A.; Wang, M.; van der Greef, J.; Rodenburg, R.J.T.; Witkamp, R.F. Unheated *Cannabis sativa* extracts and its major compound THC-acid have potential immuno-modulating properties not mediated by CB1 and CB2 receptor coupled pathways. *Int. Immunopharmacol.* **2006**, *6*, 656–665. [CrossRef] [PubMed]
83. Costa, M.; Dias, T.A.; Brito, A.; Proença, F. Biological importance of structurally diversified chromenes. *Eur. J. Med. Chem.* **2016**, *123*, 487–507. [CrossRef]
84. Weidner, C.; Wowro, S.J.; Freiwald, A.; Kawamoto, K.; Witzke, A.; Kliem, M.; Siems, K.; Müller-Kuhrt, L.; Schroeder, F.C.; Sauer, S. Amorfrutin B is an efficient natural peroxisome proliferator-activated receptor gamma (PPARγ) agonist with potent glucose-lowering properties. *Diabetologia* **2013**, *56*, 1802–1812. [CrossRef]
85. Ali, E.M.M.; Almagboul, A.Z.I.; Khogali, S.M.E.; Gergeir, U.M.A. Antimicrobial activity of *Cannabis sativa* L. *Chin. Med.* **2012**, *3*, 61–64. [CrossRef]
86. Solymosi, K.; Kofalvi, A. Cannabis: A treasure trove or Pandora's Box? *Mini Rev. Med. Chem.* **2017**, *17*, 1223–1291. [CrossRef] [PubMed]

87. Morales, P.; Hurst, D.P.; Reggio, P.H. Molecular targets of the phytocannabinoids: A complex picture. In *Phytocannabinoids Unraveling the Complex Chemistry and Pharmacology of Cannabis sativa*; Kinghorn, A.D., Falk, H., Gibbons, S., Kobayashi, J.i., Eds. Springer International Publishing: Cham, Switzerland, 2017; pp. 103–131. [CrossRef]
88. Lenkey, N.; Kirizs, T.; Holderith, N.; Máté, Z.; Szabó, G.; Vizi, E.S.; Hájos, N.; Nusser, Z. Tonic endocannabinoid-mediated modulation of GABA release is independent of the CB1 content of axon terminals. *Nat. Commun.* **2015**, *6*, 6557. [CrossRef] [PubMed]
89. Herkenham, M.; Lynn, A.B.; Johnson, M.R.; Melvin, L.S.; de Costa, B.R.; Rice, K.C. Characterization and localization of cannabinoid receptors in rat brain: A quantitative in vitro autoradiographic study. *J. Neurosci.* **1991**, *11*, 563–583. [CrossRef] [PubMed]
90. Fezza, F.; Bari, M.; Florio, R.; Talamonti, E.; Feole, M.; Maccarrone, M. Endocannabinoids, related compounds and their metabolic routes. *Molecules* **2014**, *19*, 17078–17106. [CrossRef] [PubMed]
91. Mukhopadhyay, P.; Baggelaar, M.; Erdelyi, K.; Cao, Z.; Cinar, R.; Fezza, F.; Ignatowska-Janlowska, B.; Wilkerson, J.; van Gils, N.; Hansen, T.; et al. The novel, orally available and peripherally restricted selective cannabinoid CB2 receptor agonist LEI-101 prevents cisplatin-induced nephrotoxicity. *Br. J. Pharmacol.* **2016**, *173*, 446–458. [CrossRef] [PubMed]
92. Alexander, S.P.H. So what do we call GPR18 now? *Br. J. Pharmacol.* **2012**, *165*, 2411–2413. [CrossRef]
93. Caldwell, M.D.; Hu, S.S.-J.; Viswanathan, S.; Bradshaw, H.; Kelly, M.E.M.; Straiker, A. A GPR18-based signalling system regulates IOP in murine eye. *Br. J. Pharmacol.* **2013**, *169*, 834–843. [CrossRef]
94. McHugh, D.; Page, J.; Dunn, E.; Bradshaw, H.B. Δ^9-Tetrahydrocannabinol and *N*-arachidonyl glycine are full agonists at GPR18 receptors and induce migration in human endometrial HEC-1B cells. *Br. J. Pharmacol.* **2012**, *165*, 2414–2424. [CrossRef] [PubMed]
95. Console-Bram, L.; Brailoiu, E.; Brailoiu, G.C.; Sharir, H.; Abood, M.E. Activation of GPR18 by cannabinoid compounds: A tale of biased agonism. *Br. J. Pharmacol.* **2014**, *171*, 3908–3917. [CrossRef] [PubMed]
96. Lu, V.B.; Puhl, H.L., 3rd; Ikeda, S.R. N-Arachidonyl glycine does not activate G protein-coupled receptor 18 signaling via canonical pathways. *Mol. Pharmacol.* **2013**, *83*, 267–282. [CrossRef]
97. Offertáler, L.; Mo, F.-M.; Bátkai, S.; Liu, J.; Begg, M.; Razdan, R.K.; Martin, B.R.; Bukoski, R.D.; Kunos, G. Selective ligands and cellular effectors of a G protein-coupled endothelial cannabinoid receptor. *Mol. Pharmacol.* **2003**, *63*, 699–705. [CrossRef]
98. Baker, D.; Pryce, G.; Davies, W.L.; Hiley, C.R. In silico patent searching reveals a new cannabinoid receptor. *Trends Pharmacol. Sci* **2006**, *27*, 1–4. [CrossRef] [PubMed]
99. Sawzdargo, M.; Nguyen, T.; Lee, D.K.; Lynch, K.R.; Cheng, R.; Heng, H.H.Q.; George, S.R.; O'Dowd, B.F. Identification and cloning of three novel human G protein-coupled receptor genes GPR52, ΨGPR53 and GPR55: GPR55 is extensively expressed in human brain. *Mol. Brain Res.* **1999**, *64*, 193–198. [CrossRef]
100. Shore, D.M.; Reggio, P.H. The therapeutic potential of orphan GPCRs, GPR35 and GPR55. *Front. Pharmacol.* **2015**, *6*, 69. [CrossRef] [PubMed]
101. Nilius, B.; Szallasi, A. Transient receptor potential channels as drug targets: From the science of basic research to the art of medicine. *Pharmacol. Rev.* **2014**, *66*, 676–814. [CrossRef]
102. Köfalvi, A. Alternative interacting sites and novel receptors for cannabinoid ligands. In *Cannabinoids and the Brain*; Köfalvi, A. Ed.; Springer: Boston, MA, USA, 2008; pp. 131–160. [CrossRef]
103. Pertwee, R.G.; Howlett, A.C.; Abood, M.E.; Alexander, S.P.H.; Di Marzo, V.; Elphick, M.R.; Greasley, P.J.; Hansen, H.S.; Kunos, G.; Mackie, K.; et al. International union of basic and clinical pharmacology. LXXIX. Cannabinoid receptors and their ligands Beyond CB_1 and CB_2. *Pharmacol. Rev.* **2010**, *62*, 588–631. [CrossRef] [PubMed]
104. Storozhuk, M.V.; Zholos, A.V. TRP channels as novel targets for endogenous ligands: Focus on endocannabinoids and nociceptive signalling. *Curr. Neuropharmacol.* **2018**, *16*, 137–150. [CrossRef]
105. Whiting, P.F.; Wolff, R.F.; Deshpande, S.; Di Nisio, M.; Duffy, S.; Hernandez, A.V.; Keurentjes, J.C.; Lang, S.; Misso, K.; Ryder, S. et al. Cannabinoids for medical use. *JAMA* **2015**, *313*, 2456. [CrossRef]
106. Owens, B. Drug development: The treasure chest. *Nature* **2015**, *525*, S6–S8. [CrossRef]
107. Tzadok, M.; Uliel-Siboni, S.; Linder, I.; Kramer, U.; Epstein, O.; Menascu, S.; Nissenkorn, A.; Yosef, O.B.; Hyman, E.; Granot, D. et al. CBD-enriched medical cannabis for intractable pediatric epilepsy. *Seizure* **2016**, *35*, 41–44. [CrossRef] [PubMed]
108. Alexander, S.P.H. Therapeutic potential of cannabis-related drugs. *Prog. Neuropsychopharmacol. Biol. Psychiatry* **2016**, *64*, 157–166. [CrossRef] [PubMed]
109. Eisenstein, M. Medical marijuana: Showdown at the cannabis corral. *Nature* **2015**, *525*, S15–S17. [CrossRef] [PubMed]
110. Izzo, A.A.; Capasso, R.; Aviello, G.; Borrelli, F.; Romano, B.; Piscitelli, F.; Gallo, L.; Capasso, F.; Orlando, P.; Di Marzo, V. Inhibitory effect of cannabichromene, a major non-psychotropic cannabinoid extracted from *Cannabis sativa*, on inflammation-induced hypermotility in mice. *Br. J. Pharmacol.* **2012**, *166*, 1444–1460. [CrossRef]
111. Jadoon, K.A.; Ratcliffe, S.H.; Barrett, D.A.; Thomas, E.L.; Stott, C.; Bell, J.D.; O'Sullivan, S.E.; Tan, G.D. Efficacy and safety of cannabidiol and tetrahydrocannabivarin on glycemic and lipid parameters in patients with type 2 diabetes: A randomized double-blind, placebo-controlled, parallel group pilot study. *Diabetes Care* **2016**, *39*, 1777–1786. [CrossRef]
112. Valliere, M.A.; Korman, T.P.; Woodall, N.B.; Khitrov, G.A.; Taylor, R.E.; Baker, D.; Bowie, J.U. A cell-free platform for the prenylation of natural products and application to cannabinoid production. *Nat. Commun.* **2019**, *10*, 565. [CrossRef] [PubMed]
113. Moreno-Sanz, G. Can you pass the acid test? Critical review and novel therapeutic perspectives of Δ^9-tetrahydrocannabinolic acid A. *Cannabis Cannabinoid Res.* **2016**, *1*, 124–130. [CrossRef]

114. Rock, E.M.; Limebeer, C.L.; Parker, L.A. Effect of cannabidiolic acid and Δ^9-tetrahydrocannabinol on carrageenan-induced hyperalgesia and edema in a rodent model of inflammatory pain. *Psychopharmacology* **2018**, *235*, 3259–3271. [CrossRef]
115. McPartland, J.M.; MacDonald, C.; Young, M.; Grant, P.S.; Furkert, D.P.; Glass, M. Affinity and efficacy studies of tetrahydrocannabinolic acid A at cannabinoid receptor types one and two. *Cannabis Cannabinoid Res.* **2017**, *2*, 87–95. [CrossRef]
116. Fuhr, L.; Rousseau, M.; Plauth, A.; Schroeder, F.C.; Sauer, S. Amorfrutins are natural PPARγ agonists with potent anti-inflammatory properties. *J. Nat. Prod.* **2015**, *78*, 1160–1164. [CrossRef]
117. Iwata, N.; Kitanaka, S. Tetracyclic chromane derivatives from *Rhododendron anthopogonoides*. *J. Nat. Prod.* **2010**, *73*, 1203–1206. [CrossRef]
118. Khan, B.A.; Wang, J.; Warner, P.; Wang, H. Antibacterial properties of hemp hurd powder against *E. coli*. *J. Appl. Polym. Sci.* **2014**, *132*, 41588. [CrossRef]
119. Wahby, I.; Caba, J.M.; Ligero, F. Agrobacteriuminfection of hemp (*Cannabis sativa* L.): Establishment of hairy root cultures. *J. Plant Interact.* **2013**, *8*, 312–320. [CrossRef]
120. Appendino, G.; Gibbons, S.; Giana, A.; Pagani, A.; Grassi, G.; Stavri, M.; Smith, E.; Rahman, M.M. Antibacterial cannabinoids from *Cannabis sativa*: A structure-activity study. *J. Nat. Prod.* **2008**, *71*, 1427–1430. [CrossRef]
121. Degenhardt, L.; Chiu, W.-T.; Sampson, N.; Kessler, R.C.; Anthony, J.C.; Angermeyer, M.; Bruffaerts, R.; de Girolamo, G.; Gureje, O.; Huang, Y.; et al. Toward a global view of alcohol, tobacco, cannabis, and cocaine use: Findings from the WHO World Mental Health Surveys. *PLoS Med.* **2008**, *5*, e141. [CrossRef]
122. Volkow, N.D.; Baler, R.D.; Compton, W.M.; Weiss, S.R.B. Adverse health effects of marijuana use. *N. Engl. J. Med.* **2014**, *370*, 2219–2227. [CrossRef] [PubMed]
123. Agrawal, A.; Lynskey, M.T.; Bucholz, K.K.; Kapoor, M.; Almasy, L.; Dick, D.M.; Edenberg, H.J.; Foroud, T.; Goate, A.; Hancock, D.B.; et al. DSM-5 cannabis use disorder: A phenotypic and genomic perspective. *Drug Alcohol Depend.* **2014**, *134*, 362–369. [CrossRef] [PubMed]
124. Hasin, D.S.; O'Brien, C.P.; Auriacombe, M.; Borges, G.; Bucholz, K.; Budney, A.; Compton, W.M.; Crowley, T.; Ling, W.; Petry, N.M.; et al. DSM-5 criteria for substance use disorders: Recommendations and rationale. *Am. J. Psychiatry* **2013**, *170*, 834–851. [CrossRef] [PubMed]
125. Pierre, J.M.; Gandal, M.; Son, M. Cannabis-induced psychosis associated with high potency "wax dabs". *Schizophr. Res.* **2016**, *172*, 211–212. [CrossRef] [PubMed]
126. Ashton, C.H. Pharmacology and effects of cannabis: A brief review. *Br. J. Psychiatry* **2001**, *178*, 101–106. [CrossRef] [PubMed]
127. Rodondi, N.; Pletcher, M.J.; Liu, K.; Hulley, S.B.; Sidney, S. Marijuana use, diet, body mass index, and cardiovascular risk factors (from the CARDIA Study). *Am. J. Cardiol.* **2006**, *98*, 478–484. [CrossRef]
128. Moosmann, B.; Roth, N.; Auwärter, V. Finding cannabinoids in hair does not prove cannabis consumption. *Sci. Rep.* **2015**, *5*, 14906. [CrossRef] [PubMed]
129. Sewell, R.A.; Schnakenberg, A.; Elander, J.; Radhakrishnan, R.; Williams, A.; Skosnik, P.D.; Pittman, B.; Ranganathan, M.; D'Souza, D.C. Acute effects of THC on time perception in frequent and infrequent cannabis users. *Psychopharmacology* **2013**, *226*, 401–413. [CrossRef] [PubMed]

plants

MDPI

Communication

Comparison of Four Systems to Test the Tolerance of 'Fortune' Mandarin Tissue Cultured Plants to *Alternaria alternata*

Margarita Pérez-Jiménez and Olaya Pérez-Tornero *

Equipo de Mejora Genética de Cítricos, Instituto Murciano de Investigación y Desarrollo Agrario y Alimentario (IMIDA), 30150 Murcia, Spain; margarita.perez3@carm.es
* Correspondence: olalla.perez@carm.es

Abstract: Alternaria brown spot is a severe disease that affects leaves and fruits on susceptible mandarin and mandarin-like cultivars, and is produced by *Alternaria alternata*. Consequently, there is an urge to obtain new cultivars resistant to *A. alternata*, and mutation breeding together with tissue culture can help shorten the process. However, a protocol for the in vitro selection of resistant citrus genotypes is lacking. In this study, four methods to evaluate the sensitivity to *Alternaria* of mandarin 'Fortune' explants in in vitro culture were tested. The four tested systems consisted of: (1) the addition of the mycotoxin, produced by *A. alternata* in 'Fortune', to the propagation culture media, (2) the addition of the *A. alternata* culture filtrate to the propagation culture media, (3) the application of the mycotoxin to the intact shoot leaves, and (4) the application of the mycotoxin to the previously excised and wounded leaves. After analyzing the results, only the addition of the *A. alternata* culture filtrate to the culture media and the application of the mycotoxin to the wounded leaves produced symptoms of infection. However, the addition of the fungus culture filtrate to the culture media produced results, which might indicate that, in addition to the mycotoxin, many other unknown elements that can affect the plant growth and behavior could be found in the fungus culture filtrate. Therefore, the application of the toxin to the excised and wounded leaves seems to be the most reliable method to analyze sensitivity to *Alternaria* of 'Fortune' explants cultured in vitro.

Keywords: brown spot; ACT; fungus culture filtrate; mycotoxin

Citation: Pérez-Jiménez, M.; Pérez-Tornero, O. Comparison of Four Systems to Test the Tolerance of 'Fortune' Mandarin Tissue Cultured Plants to *Alternaria alternata*. *Plants* **2021**, *10*, 1321. https://doi.org/10.3390/plants10071321

Academic Editors: Milan S. Stankovic, Paula Baptista and Petronia Carillo

Received: 1 June 2021
Accepted: 25 June 2021
Published: 28 June 2021

Publisher's Note: MDPI stays neutral with regard to jurisdictional claims in published maps and institutional affiliations.

Copyright: © 2021 by the authors. Licensee MDPI, Basel, Switzerland. This article is an open access article distributed under the terms and conditions of the Creative Commons Attribution (CC BY) license (https://creativecommons.org/licenses/by/4.0/).

1. Introduction

Alternaria alternata is a fungus that can produce severe damage in economically important plants, including cereal crops, vegetables, and fruits. *A. alternata* is able to infect plants producing host-selective toxins during the germination of spores on plant surfaces. Taking these toxins into consideration, seven pathotypes can be distinguished [1]. *A. alternata* pv. citri is the specific pathotype for *Citrus reticulata* and its hybrids, and produces the host-specific Alternaria citri toxin (ACT), which provokes, in sensible mandarin trees, the emergence of Alternaria brown spot (ABS), a severe disease whose main features are lesions on leaves and immature fruits of the infected plants that can result in early fruit abscission [2]. This reduces the quality and the commercial value of fruits in the market, leading to huge economic losses globally every year [3].

'Fortune' mandarin ('Fina' clementine × 'Dancy' mandarin) is a *Citrus reticulata* hybrid. Although this cultivar is widely harvested in Spain due to the excellent characteristics of its fruits, it is sensible to ABS. Thus, obtaining new ABS-resistant cultivars with the 'Fortune' organoleptic and farming qualities would be desirable for the mandarin market. In this sense, mandarin breeding programs are trying to include these features in their new selections. However, citrus breeding programs using conventional methods involve several difficulties associated with long juvenile periods, high heterozygosity, polygenic traits, and complicated genetic systems [4]. To overcome these difficulties, breeders have developed mutation breeding techniques [5], which could lead to a new cultivar with the desirable qualities of 'Fortune', but excluding the sensitivity to ABS.

If mutation breeding is increasingly considered to be a powerful alternative for the generation of genetic variations for plant breeding in citrus, tissue culture offers the possibility of managing large populations in a limited space while allowing a more rigorous control of the environmental conditions [6] and reducing the time spent until the selection of a genotype can be performed. To make a selection in ex vitro plants, waiting until they are grown to evaluate them is imperative. However, tissue culture offers the possibility of making an early selection of plant material, which saves space, time, and human resources [7]. Hitherto, a tissue culture protocol for mutant selection has been developed in potato and apple for two different species of the genus *Alternaria* [8,9] and in sugarcane for *Fusarium sacchari* [10]. However, there is no in vitro selection protocol for genotypes that are resistant to *Alternaria alternata* in citrus, and this is key to evaluate mutants developed in vitro.

In this study, four systems aimed to test the tolerance of 'Fortune' mandarin tissue cultured plants, which are very sensitive to ABS, to *A. alternata* were compared. Sensitivity to an *A. alternata* culture filtrate, as well as the ACT, was tested.

2. Results

2.1. ACT or Culture Filtrate Addition to the Culture Media

After four weeks of culture, no effect of the ACT in the culture media was observed in the 'Fortune' shoots. Actually, no significant differences were found either in the proliferation rate, length, productivity, or leaf damage with any treatment (data not shown).

When the *A. alternata* culture filtrate was added to the culture media, 'Fortune' explants were significantly ($p < 0.0001$) affected by the filtrate volume in the culture media in all of the studied parameters. Proliferation rate decreased from 2.06 in the control to 1.25 when the fungus culture filtrate was 50% of the volume in the culture media, and significant differences were observed between the three concentrations (Figure 1). Likewise productivity decreased with the presence of the *A. alternata* culture filtrate in the culture media (Figure 1), although significant differences were not observed between the control and 25% of the culture filtrate. On the contrary, the addition of the culture filtrate to the culture media boosted the shoots average length from 11.29 mm in the control to 13.55 mm with 50% of the culture filtrate, and significant differences were not observed for 25% and 50% (Figure 1).

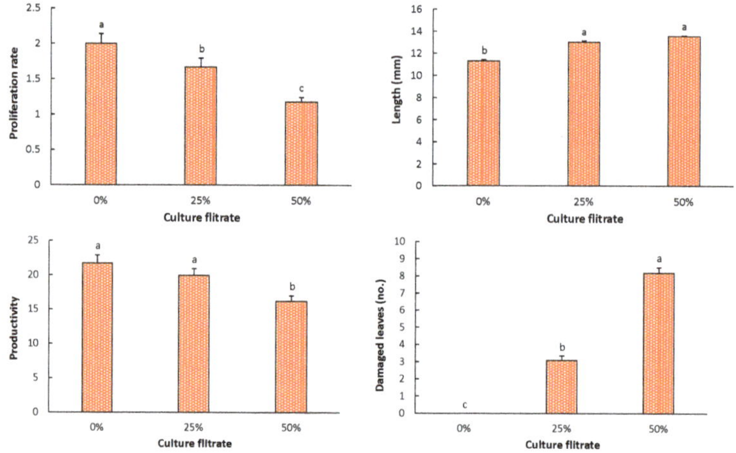

Figure 1. Proliferation rate, elongation, productivity, and number of damaged leaves in shoots of the mandarin cv. 'Fortune' exposed to different percentages of culture filtrate of *Alternaria alternata*. Data represent average ± SD values. Bars with different lower-case letters indicate a significant difference according to the LSD test ($p \leq 0.05$).

Finally, the culture of 'Fortune' shoots in vitro by adding the *A. alternata* culture filtrate was significantly affected ($p < 0.0001$) by the filtrate volume in the culture medium, producing damage in around three leaves by explant in shoots cultured with a 25% of the filtrate and around eight leaves by explant in the shoots in a 50% of the fungus filtrate (Figure 1).

2.2. ACT Application to Shoot Leaves

Regarding the ACT application to in vivo leaves, results were similar to those obtained with the ACT addition to the culture media, so there were no effects. Thus, after four weeks of culture, no significant differences were found between leaves in shoots grown in the control vessels (0 mL L^{-1} of ACT) and the rest of the treatments (Figure 2; data not shown).

Direct application of ACT in the 'Fortune' wounded excised leaves has a significant effect ($p < 0.0001$) over the percentage of damaged leaves. The mycotoxin caused a significant damage when the concentration of the ACT was 75 mg L^{-1} (Figures 3 and 4). At this concentration, almost 80% of the leaves per shoot were damaged. Although some leaves were damaged at 25 and 50 mg L^{-1} of ACT, no significant differences were found between these two treatments and the control.

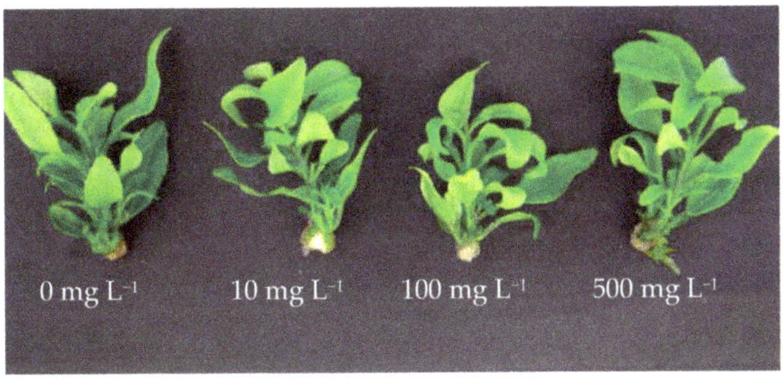

Figure 2. Shoots of the mandarin 'Fortune' grown at different concentrations of *Alternaria citri* toxin.

Figure 3. Excised and wounded leaves of the mandarin 'Fortune' exposed to 0 (**a**), 12.5 (**b**), 25 (**c**), 50 (**d**), and 75 (**e**) mg L^{-1} of *Alternaria citri* toxin and acetonitrile (**f**).

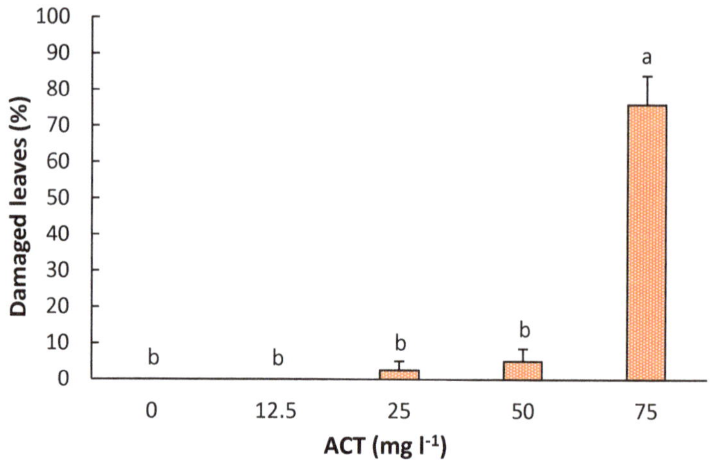

Figure 4. Percentage of damaged leaves exposed to a medium containing different concentrations of *Alternaria citri* toxin (ACT). Data represent average ± SD values. Bars with different lower-case letters indicate a significant difference by LSD test ($p \leq 0.05$).

3. Discussion

ABS is a severe disease on susceptible mandarin and mandarin-like cultivars [11]. Thus, the obtention of new genotypes that preserve good organoleptic qualities, along with a medium-to-high tolerance to the infection by *A. alternata*, will entail a huge advance in the mandarin culture. On the other hand, citrus breeding involves a costly and long technical effort that could be shortened (and made more cost-effective) with mutation breeding and tissue culture, a technique that has proved to be very efficient in the early selection of citrus genotypes [12]. In this study, we have tested four different methods to detect *A. alternata* susceptibility in vitro in order to set a fast and efficient protocol for plant selection.

The methods studied in this work can be divided into two categories: (1) modification of the culture media by the addition of *A. alternata* components (fungus culture filtrate or extracted ACT), and (2) the application of ACT to leaves excised from shoot culture. When ACT was added to the culture media, no differences between the control and the treatments were detected, so no toxicity emerged from the application of ACT at the used concentrations to the shoots in this experiment. Thus far, there is no information about the addition of ACT to media in citrus shoots in previous reports, and it is also scarce in other species. Only Chakraborty et al. (2020) [13] reported induced-tolerance in shoots of *Withania somnifera* to *A. alternata* by exposing calli to a media containing mycotoxin isolated from an *A. alternata* culture. The differences between this study and the current trial can be found in the concentration of the toxin and the sensitivity of certain tissues and species to the toxin. Thus, a higher sensitivity to the *A. alternata* toxin has been found in calli than in shoots, so a higher concentration is needed in shoots to replicate the effects produced in calli [14].

On the contrary, when the culture filtrate of *Alternaria* was added to the culture media, a toxic response was observed in the shoots (Figure 1). As it was revealed in this experiment, the capacity of in vitro shoot multiplication in citrus, as well as the number of damaged leaves, are clearly affected by any biotic [15] or abiotic stress [7]. However, the shoots length was boosted by the *A. alternata* culture filtrate in the culture media. This last result would suggest that the filtrate could contain many other unknown compounds, beyond ACT, that might be also playing positive or negative roles in the in vitro culture of 'Fortune', separately and/or synergistically with ACT. The fungal culture filtrate has been used by few authors for inducing resistance to some species of the *Alternaria* genus through tissue culture techniques [9,14,16,17]. The components of this filtrate were not analyzed in any of these studies; hence, the exact content of the culture media, and even the presence of mycotoxin and its concentration, were actually unknown. Nevertheless, pathogen culture filtrates are part of the selection protocol most commonly used for in vitro production and selection of disease-tolerant plants in many crops [18]. Moreover, although the exposition to the products of the fungus activity in the filtrate applicated via culture media has proved to be efficient for the induction of resistance to many pathogens, due to the big amount of components of the culture. Fungus culture would not probably be the best method to evaluate *A. alternata* resistance in vitro, since the infection by *A. alternata* comes into play during the germination of spores on plant surfaces and not through the absorption of the pathogen [19]. The simulation of an infection should be as similar as possible to the real process not to introduce new factors that can alter our results.

In contrast, the application of the toxin to the leaves would be more similar to the traditional and efficient ex vitro test and also to naturally-occurring infections. This application was tested in intact leaves in the shoot and in wounded leaves excised from the shoot. In the case of intact leaves, no damage was observed, and the opposite was noticed in the wounded leaves, since above 75% of the leaves presented damage with an application of a preparation containing 75 mg L^{-1} of ACT. Similar results were obtained by [8] with chemically-synthesized AM-toxin I of *A. alternata* in in vitro leaves obtained from apple mutant shoots, and the same method was used by [20] in *Vitis vinifera* to evaluate the capacity of vitis cultivars to resist the *Erysiphe necator* infection. This protocol had not been proven in vitro in citrus plants to test them against ACT before; however, it has been extendedly used ex vitro thanks to the studies made by [2] in *A. alternata* pv. *citri*.

4. Materials and Methods

4.1. In Vitro Material

Plant material was obtained from shoot cultures of 'Fortune' mandarin established in vitro and subcultured monthly on proliferation media composed of MS salts and vitamins [21] supplemented with 2 mg L^{-1} of 6-benzylaminopurine, 0.1 mg L^{-1} of indolebutyric acid, 0.6 mg L^{-1} of gibberellic acid, 30 g L^{-1} of sucrose and 7 g L^{-1} of agar (Hispanlab), as described by [22]. After adding plant growth regulators and adjusting the medium pH to 5.7 with 1 N NaOH, 100 mL of medium were dispensed in each of the 500-mL jars and sterilized in an autoclave at 121 °C for 21 min. Cultures were grown at 25 ± 1 °C with white light (5000 lux) and a 16-h photoperiod.

4.2. *A. alternata* Isolates and ACT Purification

Isolates of *A. alternata* pv. *citri* were obtained from fruits affected by ABS in a commercial plantation of 'Fortune' mandarin as described by Nemsa et al. (2012) [23]. Likewise, the fungus culture, pathogenicity studies, and ACT purification were conducted as described by del Río et al. (2018) [24]. Briefly, cultures in Petri dishes of the selected *A. alternata* isolates (15 day-old) were used to extract ACT. Cultures were dried in an oven at 60 °C and grounded before ACT extraction with acetonitrile (1 g 10 mL^{-1}) as a solvent, while stirring for 2 h.

4.3. *A. alternata* Culture Filtrate Preparation

The *A. alternata* culture filtrate preparation protocol was adapted from Kumar et al. (2008) (14). A virulent *A. alternata* pv. *citri* strain was inoculated initially in Petri plates containing PDA (potato dextrose agar) medium at 27 °C. After 10 days of culture, a mycelium portion of 5 × 5 mm was inoculated in an Erlenmeyer flask with liquid MS medium and 3% of sucrose. Cultures were maintained in agitation (125 rpm) and darkness at 25 ± 2 °C for 50 days. After that period, the mycelium was homogenized with a glass rod and filtered through a Whatman Grade 1 paper filter with a vacuum flask to obtain an *A. alternata* filtrate deprived from spores and mycelium. The fungus culture filtrate was centrifugated at 5000 rpm for 20 min and filtrated once again using a Whatman Grade 1 paper filter. The pH of the resulting filtrate was adjusted to 5.7, and the solution was stored at −20 °C in brown bottles.

4.4. ACT Addition to the Culture Media

Shoots of 'Fortune' from proliferation were cultured in glass tubes (150 × 20 mm) containing 15 mL of proliferation medium with different concentrations of ACT (0, 1, and 2 mg L^{-1}), previously filter-sterilized using a nylon filter with a membrane pore size of 0.45 μm, plus an additional control with water to know the possible effects of acetonitrile. The experiment consisted of 20 glass tubes per treatment. Medium sterilization was performed and culture conditions were as described previously.

After 4 weeks of treatment, the number of shoots (longer than 5 mm) per explant, their average length, and the number of damaged leaves were recorded. From these data, proliferation rate (no. shoots/explant) and productivity (shoot average length × proliferation rate) were calculated. The number of damaged leaves included not only the leaves that appeared chlorotic or with necrotic spots, but also the fallen leaves that remained on the culture media.

4.5. Culture-Filtrate Addition to Leaves Excised from Shoot Culture

Filtrate (1 mL) was previously cultured in PDA Petri dishes for 7 days at 27 °C to ensure sterility. Media was prepared as a proliferation media, but adding the corresponding percentage of filtrate to reach a final volume of 100 mL per vessel, after sterilization in the autoclave. Thus, 3 treatments were studied: 0%, 25%, and 50% of filtrate volume in the culture media. Medium sterilization and culture conditions were performed as described previously. The experiment consisted of 6 glass vessels per treatment with 6 shoots each. Evaluation criteria were similar to those of the previous experiment.

4.6. ACT Application to Shoot Leaves

ACT was applied to in vivo leaves or wounded leaves in different experiments. Firstly filter-sterilized ACT was diluted in acetonitrile to the concentrations of 0, 10, 100, and 500 mg L^{-1}, and an additional control with water was prepared in order to check the possible effects of acetonitrile. ACT was applied with a brush, previously sterilized in the autoclave, to the first and third fully-expanded in vivo leaves of 20 in vitro shoots of 'Fortune' per ACT treatment. The explants were cultured in vitro for 4 weeks after the application of the treatment, and after that time the number of damaged leaves was recorded.

Additionally, using the previously-described protocol, filter-sterilized ACT at the concentrations of 0, 12.5, 25, 50, and 75 mg L^{-1} was applied with a brush to the first 4 fully-expanded leaves of 'Fortune' explants. Leaves were first excised from the shoots and placed in sterile Petri plates with a humid filter paper. Subsequently, leaves were wounded in the abaxial side to improve mycotoxin penetration, and ACT was applied with a sterilized brush. Four repeats (plates) and 10 leaves per plate were used per treatment. The results were evaluated a week after the application of the treatment, and the percentage of damaged leaves was recorded.

4.7. Statistical Analysis and Data Presentation

Data were first tested for homogeneity of variance and normality of distribution. Significance was determined by analysis of variance (ANOVA), and the significance ($p < 0.05$) of any differences between mean values was tested by Duncan's new multiple range test using Statgraphics Centurion® XVI (StatPoint Technologies Inc., The Plains, VA, USA).

5. Conclusions

In the present study, four methods to evaluate the susceptibility of mandarin cultivar 'Fortune' explants to *A. alternata* pv. *citri* have been tested. After analyzing the results and the previous studies, the adaptation to the in vitro environment of the traditional method of excising and wounding leaves seems to be the most reliable approach. According to our results, a higher concentration of ACT could be necessary to obtain symptoms in the in vitro cultured shoots, and the fungus culture filtrate, although widely used, may contain many other components rather than ACT that might introduce noise in the results. In conclusion, the wounding in the leaves seems to be basic in order to infect the leaf surface. Thus, the selection of mandarin genotypes resistant to ABS can be better performed in vitro trough the excision of leaves of the studied shoots and after producing small wounds in the leaf surface. This first report about an in vitro protocol for the selection of mandarin genotypes resistant to *A. alternata* will provide a basic tool to produce and select mutants resistant to *A. alternata* in citrus.

Author Contributions: O.P.-T. conceived the study. O.P.-T. and M.P.-J. analyzed the experimental data, interpreted the results and prepared the original draft. O.P.-T. was responsible for funding acquisition and project administration. All authors have read and agreed to the published version of the manuscript.

Funding: This work was supported by the Seneca Foundation Project 08693/PI/08.

Institutional Review Board Statement: Not applicable.

Informed Consent Statement: Not applicable.

Data Availability Statement: Not applicable.

Acknowledgments: Authors want to thank F. Córdoba for his technical assistance in the laboratory, A. Lacasa and J. A. del Río for providing the virulent *A. alternata* pv. *citri* strain and the mycotoxin, respectively, and to Akimitsu Kazuya for his wise advises.

Conflicts of Interest: The authors declare no conflict of interest.

References

1. Wang, M.; Sun, X.; Yu, D.; Xu, J.; Chung, K.; Li, H. Genomic and transcriptomic analyses of the tangerine pathotype of *Alternaria alternata* in response to oxidative stress. *Sci. Rep.* **2016**, *1*, 32437. [CrossRef]
2. Kohmoto, K.; Scheffer, R.P.; Whiteside, J.O. Host-Selective Toxins from *Alternaria citri*. *Phytopathology* **1979**, *69*, 667–671. [CrossRef]
3. Bassimba, D.D.M. Epidemiología de la mancha marrón de las mandarinas causada por *Alternaria alternata*. Desarrollo de estrategias de control eficientes. Ph.D. Thesis, Universitat Politécnica de Valéncia, Valencia, Spain, 2015.
4. Kayim, M.; Koe, N.K. The effects of some carbohydrates on growth and somatic embryogenesis in citrus callus culture. *Sci. Hort.* **2006**, *109*, 29–34. [CrossRef]
5. Pérez-Jiménez, M.; Tallón, C.I.; Pérez-Tornero, O. Inducing Mutations in Citrus spp.: Sensitivity of different sources of plant material to gamma radiation. *Appl. Radiat. Isot.* **2020**, *157*, 109030. [CrossRef]
6. Pérez-Tornero, O.; Tallón, C.I.; Porras, I.; Navarro, J.M. Physiological and growth changes in micropropagated *Citrus macrophylla* explants due to salinity. *J. Plant Physiol.* **2009**, *166*, 1923–1933. [CrossRef] [PubMed]
7. Pérez-Jiménez, M.; Pérez-Tornero, O. Improved salt-tolerance in Citrus macrophylla mutant rootstocks. *Sci. Hortic.* **2020**, *259*, 108815. [CrossRef]
8. Saito, A.; Nakazawa, N.; Suzuki, M. Selection of mutants resistant to Alternaria blotch from in vitro cultured apple shoots irradiated with X- and γ-rays. *J. Plant Physiol.* **2001**, *158*, 391–400. [CrossRef]
9. Rodríguez, N.V.; Kowalski, B.; Rodríguez, L.G.; Caraballoso, I.B.; Suárez, M.A.; Pérez, M.O.; Quintana, C.R.; González, N.; Ramos, R.Q. In vitro and ex vitro Selection of Potato Plantlets for Resistance to Early Blight. *J. Phytopathol.* **2007**, *155*, 582–586. [CrossRef]
10. Mahlanza, T.; Rutherford, R.S.; Snyman, S.J.; Watt, M.P. In vitro generation of somaclonal variant plants of sugarcane for tolerance to *Fusarium sacchari*. *Plant Cell Rep.* **2013**, *32*, 249–262. [CrossRef]
11. Garganese, F.; Sanzani, S.M.; Di Rella, D.; Schena, L.; Ippolito, A. Pre- and postharvest application of altenative means to control Alternaria Brown spot of citrus. *Crop Prot.* **2019**, *121*, 7379. [CrossRef]
12. Pérez-Jiménez, M.; Pérez-Tornero, O. In vitro plant evaluation trial: Reliability test of salinity assays in citrus plants. *Plants* **2020**, *9*, 1352. [CrossRef]
13. Chakraborty, N.; Banerjee, M.; Acharya, K. In vitro selection of elite clone of *Withania somnifera* against leaf blight disease caused by *Alternaria alternata*. *Physiol. Mol. Plant Pathol.* **2020**, *112*, 101560. [CrossRef]
14. Kumar, J.V.; Kumari, B.D.R.; Sujatha, G.; Castaño, E. Production of plants resistant to *Alternaria carthami* via organogenesis and somatic embryogenesis of safflower cv. NARI-6 treated with fungal culture filtrates. *Plant Cell Tiss. Organ Cult.* **2008**, *93*, 85–96. [CrossRef]
15. Virk, G.S.; Nagpal, A. In Vitro selection of calli of *Citrus jambhiri* Lush for tolerance to culture filtrate of *Phytophthora parasitica* and their regeneration. *Physiol. Mol. Biol. Plants* **2011**, *17*, 41–47. [CrossRef]

16. Saxena, G.; Verma, P.C.; Rahman, L.; Banerjee, S.; Shukla, R.S.; Kumar, S. Selection of leaf blight-resistant *Pelargonium graveolens* plants regenerated from callus resistant to a culture filtrate of *Alternaria alternata*. *Crop Protect.* **2008**, *27*, 558–565. [CrossRef]
17. Grzebelus, E.; Kruk, M.; Macko-Podgórni, A.; Grzebelus, D. Response of carrot protoplasts and protoplast-derived aggregates to selection using a fungal culture filtrate of *Alternaria radicina*. *Plant Cell Tiss. Organ Cult.* **2013**, *115*, 209–222. [CrossRef]
18. Rai, M.K.; Kaliaa, R.K.; Singha, R.; Gangola, M.P.; Dhawana, A.K. Developing stress tolerant plants through *in vitro* selection—An overview of the recent progress. *Environ. Exp. Bot.* **2011**, *71*, 89–98. [CrossRef]
19. Vicent, A.; Asensi, M.J.; Badal, J.; Sanz, N.; Armengol, J.; Alfaro-Lassala, F.; Cuenca, F.; García-Jiménez, J. La Necrosis de Hojas y Frutos de Mandarino Fortune Causada por Alternaria. MAPAMA, docs. Available online: https://www.mapa.gob.es/app/MaterialVegetal/fichaMaterialVegetal.aspx?idFicha=159 (accessed on 25 March 2020).
20. Miazzi, M.M.; Hajjeh, H.R.; Faretra, F. An in vitro method to evaluate grapevine cultivars for *Erysiphe necator* susceptibility. *In Vitro Cell. Dev. Biol. Plant* **2010**, *46*, 363–367. [CrossRef]
21. Murashige, T.; Skoog, F. A revised medium for rapid growth and bioassays with tobacco tissue cultures. *Plant Physiol.* **1962**, *15*, 473–497. [CrossRef]
22. Tallón, C.I.; Córdoba, F.; Pérez-Tornero, O. Puesta a Punto de un Protocolo de Micropropagación de Mandarino 'Fortune'. SEFV **2011**. Abstract Book. Publicaciones de la Universidad Jaume I. ISBN: 978-84-8021-805-4. Colección 'Ciencias Experimentales' Nº 17. Available online: https://www.letrasavila.com/es/libros/sefv-2011-programa-de-sesiones-y-resumenes-de-conferencias-p_AGP0026172 (accessed on 24 June 2011).
23. Nemsa, I.; Hernández, M.A.; Lacasa, A.; Porras, I.; García-Li dón, A.; Cifuentes, D.; Bouzid, S.; Ortuño, A.; Del Río, J.A. Pathogenicity of Alternaria alternata on fruits and leaves of 'Fortune' mandarin (*Citrus clementina* × *Citrus tangerina*). *Can. J. Plant Pathol.* **2012**, *34*, 195–202. [CrossRef]
24. Diaz, L.; Del Rio, J.A.; Ortuño, A. Mechanism of the *Alternaria alternata* Pathogenicity in 'Fortune' Mandarin. *Horticulturae* **2018**, *4*, 54. [CrossRef]

plants

MDPI

Article

Fruit Growth Stage Transitions in Two Mango Cultivars Grown in a Mediterranean Environment

Alessandro Carella [1], Giuseppe Gianguzzi [1], Alessio Scalisi [1,2], Vittorio Farina [1], Paolo Inglese [1] and Riccardo Lo Bianco [1,*]

[1] Department of Agricultural, Food and Forest Sciences (SAAF), University of Palermo, 90133 Palermo, Italy; alessandro.carella@unipa.it (A.C.); giuseppe.gianguzzi@unipa.it (G.G.); alessio.scalisi@agriculture.vic.gov.au (A.S.); vittorio.farina@unipa.it (V.F.); paolo.inglese@unipa.it (P.I.)

[2] Tatura SmartFarm, Agriculture Victoria, 255 Ferguson Rd., Tatura, VIC 3616, Australia

* Correspondence: riccardo.lobianco@unipa.it

Abstract: Studying mango (*Mangifera indica* L.) fruit development represents one of the most important aspects for the precise orchard management under non-native environmental conditions. In this work, precision fruit gauges were used to investigate important eco-physiological aspects of fruit growth in two mango cultivars, Keitt (late ripening) and Tommy Atkins (early-mid ripening). Fruit absolute growth rate (AGR, mm day^{-1}), daily diameter fluctuation (ΔD, mm), and a development index given by their ratio (AGR/ΔD) were monitored to identify the prevalent mechanism (cell division, cell expansion, ripening) involved in fruit development in three ('Tommy Atkins') or four ('Keitt') different periods during growth. In 'Keitt', cell division prevailed over cell expansion from 58 to 64 days after full bloom (DAFB), while the opposite occurred from 74 to 85 DAFB. Starting at 100 DAFB, internal changes prevailed over fruit growth, indicating the beginning of the ripening stage. In Tommy Atkins (an early ripening cultivar), no significant differences in AGR/ΔD was found among monitoring periods, indicating that both cell division and expansion coexisted at gradually decreasing rates until fruit harvest. To evaluate the effect of microclimate on fruit growth the relationship between vapor pressure deficit (VPD) and ΔD was also studied. In 'Keitt', VPD was the main driving force determining fruit diameter fluctuations. In 'Tommy Atkins', the lack of relationship between VPD and ΔD suggest a hydric isolation of the fruit due to the disruption of xylem and stomatal flows starting at 65 DAFB. Further studies are needed to confirm this hypothesis.

Keywords: fruit development; fruit gauge; VPD; *Mangifera indica*; cell division; cell expansion; ripening

Citation: Carella, A.; Gianguzzi, G.; Scalisi, A.; Farina, V.; Inglese, P.; Bianco, R.L. Fruit Growth Stage Transitions in Two Mango Cultivars Grown in a Mediterranean Environment. *Plants* **2021**, *10*, 1332. https://doi.org/10.3390/plants10071332

Academic Editors: Milan S. Stankovic, Paula Baptista and Petronia Carillo

Received: 27 May 2021
Accepted: 25 June 2021
Published: 29 June 2021

Publisher's Note: MDPI stays neutral with regard to jurisdictional claims in published maps and institutional affiliations.

Copyright: © 2021 by the authors. Licensee MDPI, Basel, Switzerland. This article is an open access article distributed under the terms and conditions of the Creative Commons Attribution (CC BY) license (https://creativecommons.org/licenses/by/4.0/).

1. Introduction

Recently, tropical and sub-tropical crops like mango (*Mangifera indica* L.) have been introduced in Sicily. The region is characterized by Mediterranean climate with mild temperatures, long periods of summer drought and relatively wet winters. This environment can be considered a transition zone between the arid climate of North Africa and the temperate-humid climate of Central Europe [1]. The climate is influenced by the interactions between mid-latitudes and tropical phenomena that make it potentially vulnerable to climate change associated to an increase in temperature and a decrease in precipitation, so as to be identified as one of the most important hot-spots of the last decades [2,3]. Climate change has a major impact also on agriculture [4–6], especially if we consider that in these areas some growers are moving their business towards the production of newly introduced exotic fruits of tropical origin [7,8].

The possibility to cultivate mango in non-tropical or sub-tropical areas is subjected to the effect of temperature [9]. Mango cannot be cultivated in areas where the average temperature of the coldest month is less than 15 °C [10], while the optimum growing temperature ranges between 24 and 26 °C, reaching 30 to 33 °C for the flowering and fruit development stages [11,12]. These conditions are satisfied in the coastal areas of Sicily [13] where new orchards have been established [9].

Fruit growth is one of the most important parameters to evaluate the adaptation of a species to particular micro-climatic conditions, because the fruit represents the main sink organ of the plant and it can be considered an optimal indicator of its water and nutrient status [14]. The growth curve of mango fruits follows a sigmoidal pattern [15] and can be obtained non-destructively by measuring fruit length, width and thickness at short intervals along its developmental period [16]. The growth pattern is split in three developmental stages: cell division, cell expansion, and ripening [17,18]. Cell division is a very energy demanding process [19], due to the very fast cell division rate in fruit tissues. Carbohydrate intake is therefore crucial in this phase [20]. Carbohydrates translocated into the fruit are mainly imported from actively photosynthesizing leaves through the phloem [21].

Once this first stage is over, fruits start a linear growth, mainly characterized by the expansion of pulp cells due to water uptake led mainly by osmotic gradients. This stage is strongly influenced by the daily fluctuations of temperature and relative humidity, as well as vapor pressure deficit (VPD), which play an important role on fruit transpiration [22]. Specifically, daily VPD fluctuations drive fruit enlargement during the night and shrinkage during the day [18]. This occurs because during the day, transpiration reduces the xylem water potential, and consequently decreases the xylem flow to the fruit causing it to shrink in diameter. During the evening and the night, the water potential is restored and the fruit returns to its volume or increases it [23,24]. It is common, during this period that in situations of severe water stress, the plant takes water directly from the fruit via the xylem. This phenomenon is called backflow [25], as also documented in apple and kiwi [26–28]. In this regard, the variation in fruit diameter over a time interval represents the net contribution of the import of phloem flows, which are always positive, and xylem flows, which, depending on time of day, can be positive or negative [29]. When VPD is high, leaf and fruit transpiration flows will increase, determining some degree of plant dehydration and causing large fruit diameter fluctuations.

The last stage of development is fruit ripening, in which the fruit becomes physiologically and sexually mature enough to be separated from the mother plant [30]. At this stage internal and external changes in the fruit texture, flavor, and color are observed [31,32].

Fruit gauges are small dendrometers able to monitor continuously and very precisely the variations of fruit diameter, even at very short intervals. They are based on low-cost linear potentiometers connected to a data-logger device [29]. These instruments have already demonstrated good adaptability to different fruit species.

In kiwi, Morandi et al. [33] used these devices to monitor the development of the fruit in its final stage, determining the contributions of xylem and phloem flows to fruit growth. In another study, Morandi et al. [34] used fruit gauges to evaluate the influence of peach fruit transpiration on daily. The same devices were used to evaluate how the level of irrigation affects the daily growth pattern of pear [35], nectarine [36], orange [37], and olive [38,39] fruit.

Considering the increasing interest in mango cultivation in temperate areas, to date there is insufficient information about the growth and ripening stages of this fruit, as well as about its daily growth dynamics in Mediterranean environments.

The main aim of this trial was to acquire precise indications on the stages of fruit development of two mango cultivars ('Keitt' and 'Tommy Atkins') with reference to environmental parameters. The use of fruit gauges aimed to improve basic knowledge of the mango fruit physiology, but at the same time, provided useful information for the development of precise crop management resulting in quality and sustainable productions.

2. Materials and Methods

2.1. Orchard Characteristics and Plant Material

The experiment was carried out in a commercial orchard of the Cupitur Farm located in Caronia (38°03′ N, and 14°33′ E, 5 m a.s.l.) in northeastern Sicily (Italy) from July to October 2019. In that orchard, there are windbreaks made of cypress plants (*Cupressus sempervirens* L.), and nonwoven fabric windbreaks supported by wooden poles of 5 m high.

The study was performed using 15-year-old mango trees (*Mangifera indica* L.), three of the cultivar 'Tommy Atkins' (early- to mid-season ripening) and three of the cultivar 'Keitt' (late-season ripening) of similar size and crop load of 0.7 and 1.3 fruits cm^{-2} of trunk cross-sectional area, respectively. Both cultivars were grafted onto Gomera-3 mango rootstock. The planting density was 500 trees ha^{-1}, with a spacing of 5 × 4 m. Trees were trained to globe-shaped canopies, reaching 2.5–3 m in height. All the trees received the same conventional cultural cares. Trees were drip irrigated with a seasonal irrigation volume of 3300 m^3 ha^{-1}. Fertilization with N was carried out twice, at the beginning of vegetative growth in early spring, and at fruit set. P, K, and microelements were delivered with the irrigation water throughout the season. Two light pruning operations were carried out, one at the end of winter, before the vegetative resting period, and one after fruit harvest.

2.2. Environmental Conditions

The climate is Mediterranean [40], with average annual temperatures of 17–18 °C and average rainfalls of about 691 mm distributed across 77 days [13,41,42]. The experiment location falls into the upper thermos-Mediterranean lower sub-humid bioclimatic belt [13,43]. Data of average temperature and average humidity were acquired by a PCE-HT71 data-logger placed in the field. Data of daily temperature and relative humidity were used to calculate vapor pressure deficit (VPD).

2.3. Fruit Measurements and Experimental Design

Starting at 10 days after full bloom, fruit thickness, width, and length were measured in both cultivars using a digital caliper at two-week intervals.

Fruit diameter was monitored continuously, at 15-min intervals, with the fruit gauges described by Morandi et al. [29] connected to a CR-1000 data logger (Campbell Scientific, Inc., Logan, UT, USA). In each cultivar, measurements were carried out using 12 fruit gauges placed in four fruits from different portions of the canopy in each of three trees per cultivar. Fruit gauges were placed in the two cultivars at different times, as their timing of fruit development differs. Measurements started at 51 days after full bloom (DAFB) in the early-mid ripening 'Tommy Atkins', and at 58 DAFB in the late ripening 'Keitt'.

In 'Tommy Atkins', diameter changes were monitored in three different periods: I (20–27 July) from 51 to 58 DAFB; II (3–12 August) from 65 to 74 DAFB; III (24 August–7 September) from 86 to 100 DAFB. Measurements of 'Tommy Atkins' covered 30 days, corresponding to about 30% of its entire fruiting period.

In 'Keitt', diameter changes were monitored in four different periods: I (27 July–2 August) from 58 to 64 DAFB; II (12–23 August) from 74 to 85 DAFB; III (7–21 September) from 100 to 113 DAFB; IV (21 September–3 October) from 114 to 126 DAFB. Measurements of 'Keitt' covered 52 days, corresponding to about 35% of its entire fruiting period.

2.4. Fruit Development Parameters

Data recorded by the data-logger were processed by graphical analysis. Fruit absolute growth rate (AGR, mm day^{-1}) was estimated by calculating the slope of the diameter changes measured by fruit gauges in each monitored period.

The average daily fluctuation of fruit diameter (ΔD, mm) was also calculated for each development stage. It is mainly related to fruit water exchanges (via xylem and transpiration) and more evident during the cell expansion period. Finally, it was calculated a fruit development index, obtained from the ratio between AGR and ΔD. This index was useful to detect the shifts from cell division stage to cell expansion stage to ripening stage. The rationale behind this was that a relatively high index would be associated to cell division, where a low ΔD and a high AGR are expected (growth due to small and actively dividing cells); an intermediate index would be associated to cell expansion, where both high ΔD and AGR are expected (growth due mainly to cell water influx); a relatively low index would be associated to fruit ripening, where a medium to high ΔD and a low AGR are expected (very low growth and some cell water exchanges). To evaluate the influence of the environment on fruit development, VPD and ΔD were also related.

2.5. Statistical Analysis

Linear regression analysis was performed on fruit gauge data to estimate AGR in each fruit development stage, using Sigmaplot 12.0 (Systat Software Inc., Chicago, IL, USA) procedures. Sigmaplot regression analysis procedures were also used to test the relationships between ΔD and VPD and evaluate the influence of the environment on fruit growth. Differences of AGR, ΔD, and development index (AGR ΔD^{-1}) among measurement periods were tested using one-way analysis of variance (ANOVA). The means were compared by Tukey's multiple comparison test at the 0.05 significant level using Systat statistical software version 13 (Systat Software Inc.). As a temporal reference, it was used the central day of the interval of each period. For 'Tommy Atkins', the reference days were: 56 DAFB (I); 69 DAFB (II); 93 DAFB (III). For 'Keitt', the reference days were: 61 DAFB (I); 80 DAFB (II); 107 DAFB (III); 120 DAFB (IV).

3. Results and Discussion

3.1. Climate Data

During the experiment, the average temperature was 25.9 °C, with a maximum temperature of 35.9 °C reached on 22 July, and a minimum temperature of 18.0 °C reached on 3 October. The average relative humidity (RH) of the area was 68.1%, with a minimum value of 32.0% recorded on 8 August. The maximum RH was 98.9% recorded on 4 September during the rainfall events (>120 mm) that occurred in the period between the end of August and the beginning of September.

At the beginning of August, the weather was hot and dry, with relatively high VPD values reaching 1.5–2 kPa (on 8 August). VPD shown a constant pattern from 20 July to 28 August and from 7 September to the end of the trial (5 October), with daily minimum values of 0.64 kPa and maximum values of 2 kPa. Between 29 August and 6 September, a heavy rain event caused a rapid drop of daily VPD to 0.21 kPa (Figure 1).

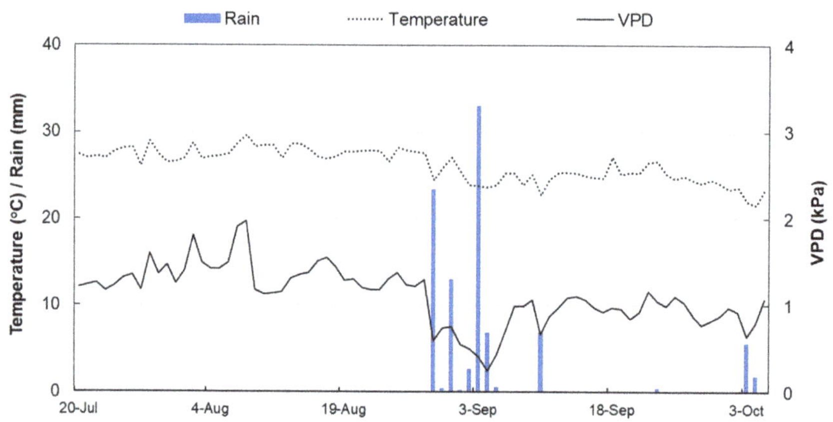

Figure 1. Trends of daily temperature, VPD, and rainfall at Caronia, northern Sicily (38°03′ N, and 14°33′ E, 5 m a.s.l.) during the trial period.

3.2. Fruit Growth

The typical sigmoidal fruit growth pattern was observed in fruits of both cultivars (Figure 2). In 'Keitt', the growth in length was the most rapid followed by width and finally by thickness, resulting in the characteristic flattened shape of the fruit (Figure 2A) [44]. In 'Tommy Atkins', although length was also the most rapidly growing fruit dimension, fruit width and thickness showed very similar trends, resulting in a rounder shape typical of this cultivar (Figure 2B).

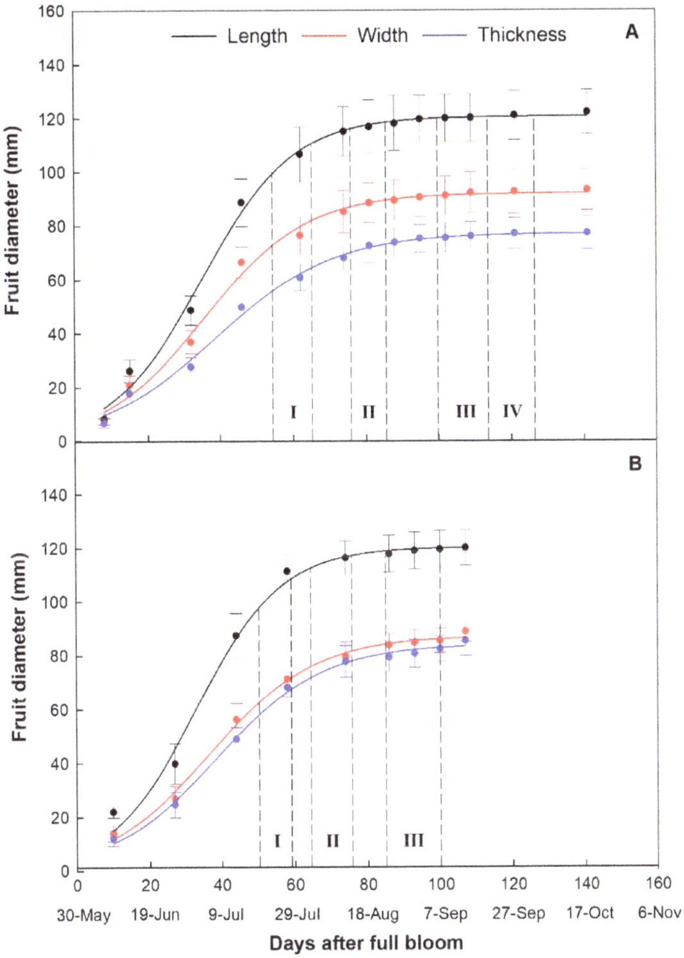

Figure 2. Fruit growth curve of 'Keitt' (**A**) in the four observation periods (I, II, III, IV), and 'Tommy Atkins' (**B**) in the three observation periods (I, II, III) monitored with a digital caliper. In 'Keitt', Length = $120.62/(1 + e^{(-(DAFB-35.11)/12.56)})$; $R^2 = 0.99$. Width = $92.04/(1 + e^{(-(DAFB-35.84)/14.01)})$; $R^2 = 0.99$. Thickness = $77.23/(1 + e^{(-(DAFB-39.01)/16.11)})$; $R^2 = 0.99$. In 'Tommy Atkins', Length = $120.31/(1 + e^{(-(DAFB-32.95)/11.81)})$; $R^2 = 0.99$. Width = $87.25/(1 + e^{(-(DAFB-36.66)/14.56)})$; $R^2 = 0.99$. Thickness = $84.13/(1 + e^{(-(DAFB-38.66)/14.56)})$; $R^2 = 0.99$.

The growth curves highlighted the expected difference in the length of the fruiting period in the two cultivars, lasting about 105 days in 'Tommy Atkins' and about 140 days in 'Keitt'. This difference consisted mostly in a longer ripening stage in 'Keitt' than in 'Tommy Atkins', because, even though both cultivars have similar polygalacturonase activity, 'Keitt' retains more total pectin at the beginning of the ripening stage than 'Tommy Atkins' [45].

The use of fruit gauges allowed for the detection of diameter variation and monitoring of growth rate during the 24 h (Figure 3). The fruit gauges recorded diameter fluctuations throughout the day, which continued with a similar fashion in all the monitoring periods. In fact, from about 18:00 to about 8:00 of the next day, there was a rapid increase of fruit diameter followed by a sharp decrease during the hottest hours, with a different net diameter increase (growth rate) depending on the period of observation. This is in line with what is explained by the model of Léchaudel et al. [46] and Faust [47], which emphasizes a negative correlation between transpiration and water potential in mango fruit. In detail the transpiration rate is maximum during the hottest hours, when xylem water potential reaches its minimum, while during the evening and the night water potential is restored and fruits gradually expand to reach sizes larger than the initial.

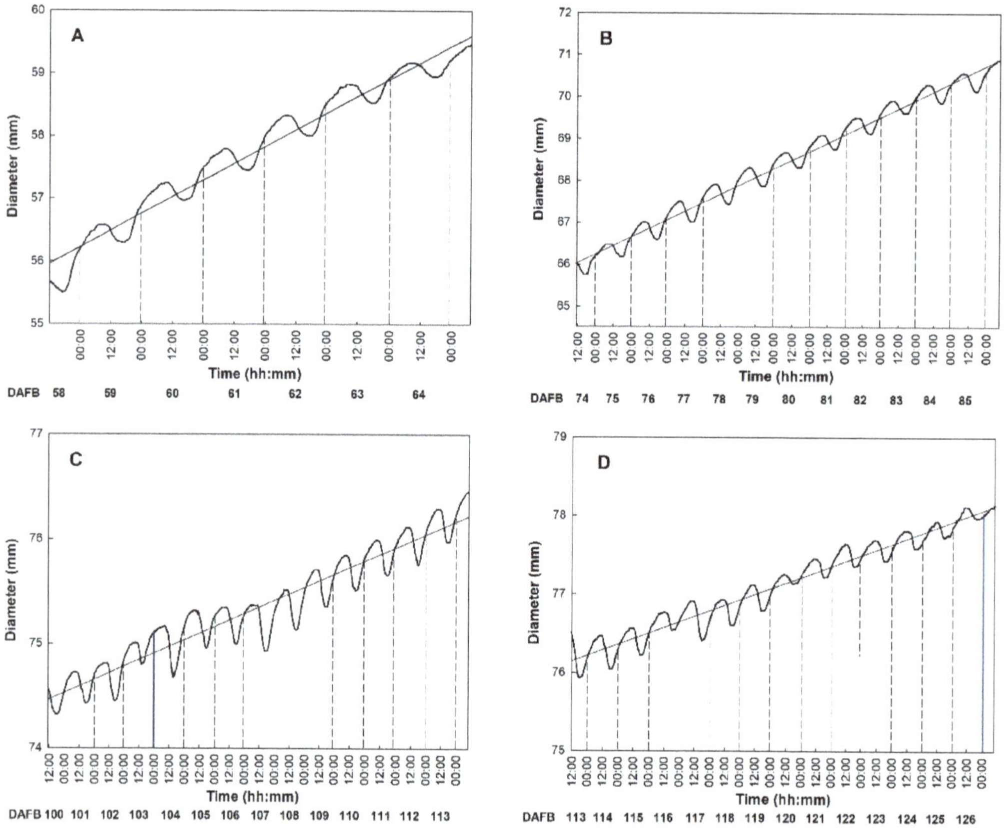

Figure 3. Trends of diametric growth of 'Keitt' mangos in the four observation periods monitored by fruit gauges every 15 min from 58 to 64 DAFB (**A**), from 74 to 85 DAFB (**B**), from 100 to 113 days DAFB (**C**), and from 114 to 126 DAFB (**D**). The blue lines indicate days of rainfall. Period I: Diameter = 0.535DAFB − 23,297; $R^2 = 0.957$; $p < 0.001$. Period (**II**): Diameter = 0.410DAFB − 17,864; $R^2 = 0.978$; $p < 0.0001$. Period (**III**): Diameter = 0.126DAFB − 5434; $R^2 = 0.910$; $p < 0.001$. Period (**III**): Diameter = 0.142DAFB − 6137; $R^2 = 0.948$; $p < 0.001$.

In 'Keitt', fruit diameter started shrinking at 8:30–9:30 and continued until 17:30–18:00. During this time of the day, fruits may lose water, and consequently volume in two ways: directly, because of the fruit transpiration, and indirectly, via xylem (backflow). This occurs when the leaf transpiration is rather high, especially during warm days and high VPD, and the water supply from the roots is insufficient to counteract the water shortage. In these cases, water might flow outward from the fruit according to the water potential gradient [25]. Thus, fruit shrinkage is caused by a negative water balance because water losses cannot be compensated by phloem inputs. From 18:00 onward fruit diameter increased until the morning, with the fruit generally reaching a larger size than the previous day (Figure 3). This size difference between consecutive days represents the net daily growth and can be mainly associated to an accumulation of dry matter, including a pool of organic and inorganic molecules that become part of the cellular structures of the fruit (Figure 4). Only in period III, between 91 and 96 DAFB (Figure 4 III), reduced and irregular diameter fluctuations were detected. This occurred in correspondence with intense rainfall events between 29 August and 3 September, i.e., when RH was >90% and VPD reached 0.21 kPa (Figure 1). According to Torres Ruiz et al. [48] supplying water when kiwi vines are experiencing stressing environmental conditions, e.g., very high transpiration rates, can influence fruit volume growth in the following days, causing a marked reduction of daily fluctuations. About two days later, RH, temperature and VPD values returned to the typical values of the period and fruits gradually resumed regular daily fluctuations and growth (Figure 4 III).

Similarly, in 'Tommy Atkins', a diametric decrease was observed during the warmest part of the day, starting between 8:30 and 9:15 and ending between 17:00 and 17:45. Subsequently, a constant increase in diameter was recorded during the observation periods (Figure 4). Only in period III, between 91 and 96 DAFB (Figure 4 III), reduced and irregular diameter fluctuations were detected. This occurred in correspondence with intense rainfall events between 29 August and 3 September, i.e., when RH was >90% and VPD reached 0.21 kPa (Figure 1). According to Torres Ruiz et al. [48], supplying water when kiwi vines are experiencing stressing environmental conditions, e.g., very high transpiration rates, can influence fruit volume growth in the following days, causing a marked reduction of daily fluctuations. About two days later, RH, temperature and VPD values returned to the typical values of the period and fruits gradually resumed regular daily fluctuations and growth (Figure 4 III).

3.3. Fruit Development

In both cultivars, fruit AGR decreased significantly with progressing development (Figure 5). Specifically, in 'Keitt', AGR was highest in period I, with values reaching 0.53 ± 0.05 mm day^{-1}. A significant and similar reduction was observed in periods III and IV, while AGR in period II was intermediate (Figure 5A). This suggests a substantial decrease in dry matter accumulation rate by the fruit during its development. It also indicates that during period I, fruits were still in an active growth phase, probably due to cell expansion mechanisms, strictly dependent on carbohydrate accumulation and water recall by osmosis. Subsequently, the reduction in growth confirmed that period II was a transition period between the end of cell expansion and the beginning of ripening. A similar decrease of AGR during mango fruit development was shown by Lechaudel in 'Lirfa' [49] and by Dambreville in 'Cogshall' and 'José' [50]. Following periods occurred at the time of full drupe maturation (Figure 5A).

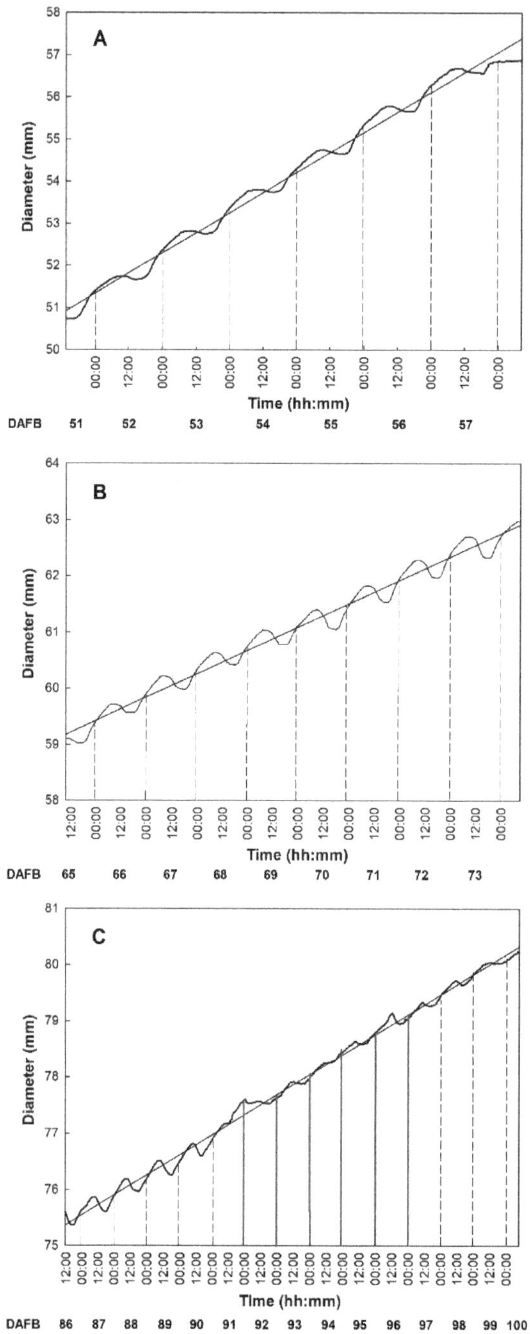

Figure 4. Trends of diametric growth of 'Tommy Atkins' mangos in the three observation periods monitored by fruit gauges every 15 min from 51 to 58 DAFB (**A**), from 65 to 74 DAFB (**B**), and from 86 to 100 DAFB (**C**). The blue lines indicate days of rainfall. Period I: Diameter = 1.010DAFB − 50,888; $R^2 = 0.991$; $p < 0.001$. Period II: Diameter = 0.417DAFB − 18,141; $R^2 = 0.989$; $p < 0.001$. Period (**III**): Diameter = 0.357DAFB – 15,513; $R^2 = 0.995$; $p < 0.001$.

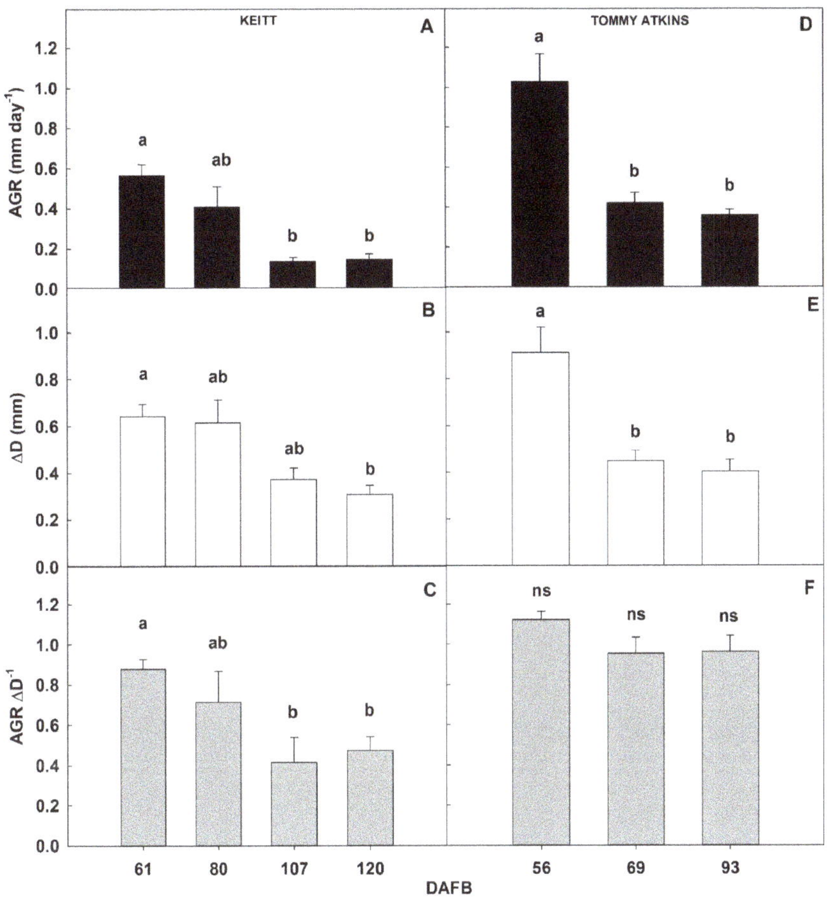

Figure 5. Fruit absolute growth rate (AGR), daily diameter fluctuations (ΔD), and development index (AGR ΔD^{-1}) of 'Keitt' (**A–C**) and 'Tommy Atkins' (**D–F**) mango during the monitoring periods (Keitt: 61, 80, 107, 120 DAFB. Tommy Atkins: 56, 69, 93 DAFB). DAFB = days after full bloom.

In 'Tommy Atkins', AGR was highest also in period I with an average of 1.01 ± 0.14 mm day^{-1}, showing a high net daily growth, also visible in the first period of the sigmoidal curve between 20 and 60 DAFB (Figure 2B). In subsequent periods, there was a significant and similar reduction in fruit growth of about 60%, with values of 0.42 ± 0.05 mm day^{-1} and 0.36 ± 0.02 mm day^{-1} in period II and III, respectively (Figure 5D). In this case, fruits monitored in period I were actively growing, most likely by cell expansion, while fruits at periods II and III were already at the ripening stage.

In 'Keitt', also ΔD showed a significant reduction over the four periods of fruit development, going from 0.64 ± 0.05 mm in period I, to 0.30 ± 0.04 mm in period IV (Figure 5B). This indicates that fruits monitored in periods III and IV were not at cell expansion stage as cell expansion depends mainly on water inflow that would have caused marked daily diameter fluctuations (Figure 5B). Decreases in ΔD could be associated with lower water exchanges from the fruit to either the atmosphere or the rest of the plant at more advanced developmental stages. These could be determined by the beginning of the ripening stage, in which the water exchanges between the fruit and the environment (transpiration) are gradually reduced, probably due to the thickening of the peel cuticle [51–53].

Also in 'Tommy Atkins', marked diameter fluctuations were detected in period I (0.91 ± 0.11 mm). However, these were more than halved in periods II and III dropping to 0.45 ± 0.04 and 0.40 ± 0.05 mm, respectively (Figure 5E). This is a second piece of evidence indicating that the fruit in period I was still in an active growth phase due to cell expansion mechanisms associated to water exchanges. The reduction of ΔD in period II and III could be due to a reduction in fruit transpiration, possibly driven by cuticular thickening phenomena during ripening [54], or to a lower xylem communication with the plant. Indeed, cuticular conductance values in mango can vary depending on the microclimatic conditions of fruit growth [24], thus limiting transpiration phenomena [55]. In this regard, Léchaudel et al. [56] showed that the rate of water accumulation in mangoes decreases when the dry matter of the fruit increases. All these phenomena (reduction of growth, reduction of the transpiration rate by cuticular thickening, xylem isolation) may be associated with the beginning of the ripening phase, which is also characterized by the activation of metabolic activities determining physiological, biochemical, and organoleptic changes [32,57].

In order to evaluate the effect of microclimate conditions on fruit growth dynamics, ΔD was related to VPD. In 'Keitt', a direct exponential relationship was found between the two parameters over the four monitoring periods, i.e., as VPD increased, the amplitude of daily fluctuations in fruit diameter rapidly increased (Figure 6A). In this case, VPD seems to be the main driving force determining fruit diameter fluctuations. Morandi et al. [23] found a similar behavior in peach fruit showing a direct relationship between transpiration rate (directly linked with diameter fluctuations) and VPD at cell division and cell expansion stages, demonstrating a tight coupling of fruit transpiration to environmental conditions during these developmental stages.

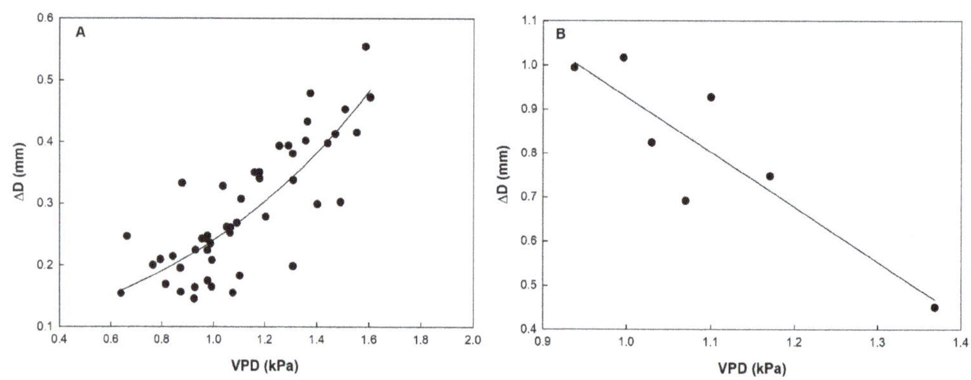

Figure 6. Relationship between vapor pressure deficit (VPD) and daily fluctuations in fruit diameter (ΔD) in 'Keitt' over the four monitoring periods (**A**), and 'Tommy Atkins' (**B**) mango at monitoring period I. Keitt: $\Delta D = 0.076 \times e^{(1.1579VPD)}$; $R^2 = 0.679$; $p < 0.001$. Tommy Atkins: $\Delta D = 2.175 - 1.248VPD$; $R^2 = 0.788$; $p = 0.008$.

In 'Tommy Atkins', on the other hand, a linear inverse relationship between ΔD and VPD was found only in period I (Figure 6B). Fruits of the Tommy Atkins cultivar have a relatively low rate of transpiration compared with other mango cultivars, due to a cuticle structure characterized by a relatively high wax content, a limited number of lenticels [51] and a high density of resin ducts [58]. This along with a possible leaf stomatal closure in response to high VPD may explain the decrease of daily diametrical oscillations as VPD increased. Already at period II, however, no relationship ($p = 0.083$) between the two parameters was found. This could support the hypothesis that in this period, the fruit was entering the ripening phase and was becoming isolated at the hydric level, i.e., interrupting xylem and stomatal flows, and non-dependent on VPD variations. This phenomenon was most evident in period III, where the absence of the relationship between ΔD and VPD was confirmed ($p = 0.278$), supporting the hypothesis that during ripening, fruits limit water exchanges with the atmosphere (no transpiration) and with the rest of the plant by xylem backflow. In fact, data recorded in period III showed no significant changes in diameter. In kiwi, this has been attributed to a drop in the daily transpiration rate, as fruits stop xylem inflow in the last phase limiting the formation of pressure gradients ideal for water movement [33]. According to Nordey [52], in 'Cogshal' mango, changes in xylem flow are related to a decrease in water conductivity of xylem vessels, driven by vessel occlusion. Also in fruits of other species, including kiwi [27,54], apple [26], cherry [59] and grapes [60], an occlusion of xylem vessels was observed as the conduction of these tissues during particular climatic conditions could imply large water losses by backflow [59]. As mentioned above, 'Tommy Atkins' fruits have very low cuticular conductance compared to other varieties, especially at the ripening stage, and this may explain the different behavior of 'Tommy Atkins' and 'Keitt' in response to VPD [61]. In particular, 'Keitt' presents greater proportion of amorphous zone than 'Tommy Atkins', due to a faster cuticular degradation, increasing fruit skin water permeability.

The ratio between AGR and ΔD, or development index, was calculated to understand what the prevailing mechanism between cell division and expansion in the growth model was, assuming that a large ΔD would be mainly due to intense water exchanges typical of the cell expansion mechanism. Specifically, in 'Keitt', the development index was significantly higher in period I than in periods III and IV, with intermediate values in period II (Figure 5C). Considering that both parameters followed similar decreasing trends and that AGR values were very low in periods III and IV, it can be stated that in period I, cell division may have prevailed over cell expansion, while in period III and IV internal changes prevailed over fruit growth, typical of the fruit ripening stage. Period II fell into an intermediate stage where growth gradually slowed down and cell expansion may have prevailed over cell division. The transition stage between periods I and III (65–99 DAFB) can be considered relatively long in this mid- to late-ripening cultivar (Figure 5C). In 'Tommy Atkins', AGR and ΔD showed the same trends resulting in no significant differences in development index among monitoring periods (Figure 5F). Since some growth was still present until fruit harvest (although at a slower rate than in period I), we can assume that both cell division and expansion coexisted at constant rates during all three monitoring periods. This indicates on one side, that this early-ripening cultivar keeps growing until the end of fruit development, and on the other side, that a period of greater cell division than cell expansion mechanism must occur before 56 DAFB, if ever.

4. Conclusions

This paper describes unpublished information about the growth of mango fruit in a Mediterranean environment and may represent a solid ground for further investigations. The use of fruit gauges on drupes allowed the monitoring of diametric variations over time in response to environmental and physiological conditions of two important cultivars such as Keitt and Tommy Atkins. Thanks to the identification of the different growth stages, and more importantly of the moments when either cell division or cell expansion prevails during fruit development, the optimal time for the application of specific management practices (fertilization, irrigation, fruit thinning) could be established.

The precise identification of the beginning of the ripening phase is also very useful to maximize fruit quality, for example by applying deficit irrigation with nearly no risk to compromise final fruit size.

The association between environmental conditions and fruit growth also indicated that 'Keitt' fruit growth takes advantage of increasing VPD, suggesting a good adaptation of this cultivar to Mediterranean environments. The same does not seem to be true for 'Tommy Atkins', acting as a genotype sensitive to quick changes in VPD and most likely to the dry conditions of Mediterranean summers. Further scientific work on this track will have to confirm or disprove the fruit hydric isolation hypothesis formulated in this study during the later fruit development stages.

Author Contributions: Conceptualization, R.L.B., V.F., P.I., A.C. and G.G.; methodology, A.S., R.L.B.; software, A.S.; validation, A.C., R.L.B. and G.G.; formal analysis, R.L.B.; investigation, A.C. and G.G.; resources, G.G. and V.F.; data curation, A.C., R.L.B. and G.G.; writing—original draft preparation, A.C.

and G.G.; writing—review and editing, R.L.B. and A.C.; visualization, R.L.B. and V.F.; supervision, P.I., R.L.B. and V.F.; project administration, V.F.; funding acquisition, V.F. All authors have read and agreed to the published version of the manuscript.

Funding: This research received no external funding.

Institutional Review Board Statement: Not applicable.

Informed Consent Statement: Not applicable.

Data Availability Statement: Data available on request.

Acknowledgments: We would like to express our gratitude to Pietro Cuccio for providing the experimental field of his farm "Cupitur s.r.l.".

Conflicts of Interest: The authors declare no conflict of interest.

References

1. Lionello, P.; Malanotte-Rizzoli, P.; Boscolo, R.; Alpert, P.; Artale, V.; Li, L.; Luterbacher, J.; May, W.; Trigo, R.; Tsimplis, M. The Mediterranean Climate: An Overview of the Main Characteristics and Issues. *Dev. Earth Environ. Sci.* **2006**, *4*, 1–26.
2. Giorgi, F. Climate Change Hot-Spots. *Geophys. Res. Lett.* **2006**, *33*. [CrossRef]
3. Giorgi, F.; Lionello, P. Climate Change Projections for the Mediterranean Region. *Glob. Planet. Chang.* **2008**, *63*, 90–104. [CrossRef]
4. Mariani, L.; Parisi, S.; Cola, G.; Failla, O. Climate Change in Europe and Effects on Thermal Resources for Crops. *Int. J. Biometeorol.* **2012**, *56*, 1123–1134. [CrossRef] [PubMed]
5. Mendelsohn, R. The Impact of Climate Change on Agriculture in Developing Countries. *J. Nat. Resour. Policy Res.* **2008**, *1*, 5–19. [CrossRef]
6. Nath, V.; Kumar, G.; Pandey, S.; Pandey, S. Impact of climate change on tropical fruit production systems and its mitigation strategies. In *Climate Change and Agriculture in India: Impact and Adaptation*; Springer: Berlin/Heidelberg, Germany, 2019; pp. 129–146.
7. Farina, V.; Barone, F.; Mazzaglia, A.; Lanza, C.M. Evaluation of Fruit Quality in Loquat Using Both Chemical and Sensory Analyses. In Proceedings of the Acta Horticulturae, International Society for Horticultural Science (ISHS), Leuven, Belgium, 31 January 2011; pp. 345–349.
8. Mossad, A.; El Helew, W.K.M.; Elsheshetawy, H.E.; Farina, V. Mass Modelling by Dimension Attributes for Mango (*Mangifera Indica* Cv. Zebdia) Relevant to Post-Harvest and Food Plants Engineering. *Agric. Eng. Int. CIGR J.* **2016**, *18*, 219–229.
9. Farina, V.; Tripodo, L.; Gianguzzi, G.; Sortino, G.; Giuffre, D.; Cicero, U.L.; Candia, R.; Collura, A. Innovative Techniques to Reduce Chilling Injuries in Mango (*Mangifera Indica* L.) Trees under Mediterranean Climate. *Chem. Eng. Trans.* **2017**, *58*, 823–828.
10. Ochse, J.; Dijkman, M.; Wehlburg, M. *Cultivo y Mejoramiento de Plantas Tropicales y Subtropicales*; Limusa-Wiley: Ciudad de México, Mexico, 1972.
11. Purseglove, J.W. *Tropical Crops. Dicotyledons 2*; Longmans: London, UK, 1968; pp. 333–719.
12. Chacko, E.K. *Physiology of Vegetative and Reproductive Growth in Mango (Mangifera Indica L.) Trees*; CSIRO: Melbourne, Australia, 1986; Volume 1, pp. 54–70.
13. Gianguzzi, L.; Papini, F.; Cusimano, D. Phytosociological Survey Vegetation Map of Sicily (Mediterranean Region). *J. Maps* **2016**, *12*, 845–851. [CrossRef]
14. Boini, A.; Bresilla, K.; Perulli, G.; Manfrini, L.; Corelli Grappadelli, L.; Morandi, B. Photoselective Nets Impact Apple Sap Flow and Fruit Growth. *Agric. Water Manag.* **2019**, *226*, 105738. [CrossRef]
15. Lizada, C. Mango. In *Biochemistry of Fruit Ripening*; Springer: Berlin/Heidelberg, Germany, 1993; pp. 255–271.
16. da Silva, A.C.; de Souza, A.P.; Leonel, S.; de Souza, M.E.; Ramos, D.P.; Tanaka, A.A. Growth and Flowering of Five Mango Cultivar under Subtropics Conditions of Brazil. *Am. J. Plant Sci.* **2014**, *2014*, 42759. [CrossRef]
17. Sergent, E. *El Cultivo Del Mango (*Mangifera Indica *L.): Botánica, Manejo y Comercialización*; CDCH UCV: Caracas, Venezuela, 1999; Volume 72, ISBN 980-00-1389-X.
18. Galán Saúco, V. *El Cultivo Del Mango*; Mundi-Prensa Libros: Madrid, Spain, 2009; ISBN 84-8476-367-6.
19. Li, M.; Feng, F.; Cheng, L. Expression Patterns of Genes Involved in Sugar Metabolism and Accumulation during Apple Fruit Development. *PLoS ONE* **2012**, *7*, e33055. [CrossRef] [PubMed]
20. Agustí, M. *Fruticultura*; Mundi-Prensa: Madrid, Spain, 2010; pp. 140–144; ISBN 978-84-8476-398-7.
21. Génard, M.; Dauzat, J.; Franck, N.; Lescourret, F.; Moitrier, N.; Vaast, P.; Vercambre, G. Carbon Allocation in Fruit Trees: From Theory to Modelling. *Trees* **2008**, *22*, 269–282. [CrossRef]
22. Lescourret, F.; Génard, M.; Habib, R.; Fishman, S. Variation in Surface Conductance to Water Vapor Diffusion in Peach Fruit and Its Effects on Fruit Growth Assessed by a Simulation Model. *Tree Physiol.* **2001**, *21*, 735–741. [CrossRef] [PubMed]
23. Morandi, B.; Rieger, M.; Grappadelli, L.C. Vascular Flows and Transpiration Affect Peach (Prunus Persica Batsch.) Fruit Daily Growth. *J. Exp. Bot.* **2007**, *58*, 3941–3947. [CrossRef] [PubMed]

24. Léchaudel, M.; Lopez-Lauri, F.; Vidal, V.; Sallanon, H.; Joas, J. Response of the Physiological Parameters of Mango Fruit (Transpiration, Water Relations and Antioxidant System) to Its Light and Temperature Environment. *J. Plant Physiol.* **2013**, *170*, 567–576. [CrossRef] [PubMed]
25. Higuchi, H.; Sakuratani, T. Water Dynamics in Mango (*Mangifera Indica* L.) Fruit during the Young and Mature Fruit Seasons as Measured by the Stem Heat Balance Method. *J. Jpn. Soc. Hortic. Sci.* **2006**, *75*, 11–19. [CrossRef]
26. Dražeta, L.; Lang, A.; Hall, A.J.; Volz, R.K.; Jameson, P.E. Causes and Effects of Changes in Xylem Functionality in Apple Fruit. *Ann. Bot.* **2004**, *93*, 275–282. [CrossRef]
27. Morandi, B.; Manfrini, L.; Zibordi, M.; Losciale, P.; Corelli Grappadelli, L. Effects of Drought Stress on the Growth, Water Relations and Vascular Flows of Young 'Summerkiwi' Fruit. In Proceedings of the XXVIII International Horticultural Congress on Science and Horticulture for People (IHC2010): International Symposium on CLIMWATER 2010: Horticultural Use of Water in a Changing Climate, Lisbon, Portugal, 22–27 August 2010; pp. 355–359.
28. Mazzeo, M.; Dichio, B.; Clearwater, M.J.; Montanaro, G.; Xiloyannis, C. Hydraulic Resistance of Developing Actinidia Fruit. *Ann. Bot.* **2013**, *112*, 197–205. [CrossRef] [PubMed]
29. Morandi, B.; Manfrini, L.; Zibordi, M.; Noferini, M.; Fiori, G.; Grappadelli, L.C. A Low-Cost Device for Accurate and Continuous Measurements of Fruit Diameter. *HortScience* **2007**, *42*, 1380–1382. [CrossRef]
30. Grierson, W. Fruit Development, Maturation and Ripening. In *Handbook of Plant and Crop Physiology*; CRC Press: Boca Raton, FL, USA, 2002; pp. 143–160.
31. Lakshminarayana, S. Respiration and Ripening Patterns in the Life Cycle of the Mango Fruit. *J. Hortic. Sci.* **1973**, *48*, 227–233. [CrossRef]
32. Giovannoni, J. Molecular Biology of Fruit Maturation and Ripening. *Annu. Rev. Plant Biol.* **2001**, *52*, 725–749. [CrossRef]
33. Morandi, B.; Losciale, B.P.; Manfrini, L.; Zibordi, M.; Studhalter, M.; Grappadelli, L.C. The Growth of the Kiwifruit in Its Final Stages. In *VI International Symposium on Kiwifruit*; ISHS: Leuven, Belgium, 2006; pp. 369–374.
34. Morandi, B.; Manfrini, L.; Losciale, P.; Zibordi, M.; Corelli Grappadelli, L. Changes in Vascular and Transpiration Flows Affect the Seasonal and Daily Growth of Kiwifruit (Actinidia Deliciosa) Berry. *Ann. Bot.* **2010**, *105*, 913–923. [CrossRef] [PubMed]
35. Morandi, B.; Anconelli, S.; Losciale, P.; Manfrini, L.; Pierpaoli, E.; Zibordi, M.; Grappadelli, L. How Irrigation Level Affects the Daily Growth Pattern of Young Pear Fruit of the Cv. 'Abbé Fetel'. *Acta Hortic.* **2014**, 163–169. [CrossRef]
36. Scalisi, A.; O'Connell, M.G.; Stefanelli, D.; Lo Bianco, R. Fruit and Leaf Sensing for Continuous Detection of Nectarine Water Status. *Front. Plant Sci.* **2019**, *10*, 805. [CrossRef]
37. Grilo, F.; Scalisi, A.; Pernice, F.; Morandi, B.; Lo Bianco, R. Recurrent Deficit Irrigation and Fruit Harvest Affect Tree Water Relations and Fruitlet Growth in 'Valencia'Orange. *Eur. J. Hortic. Sci* **2019**, *84*, 177–187. [CrossRef]
38. Scalisi, A.; Marino, G.; Marra, F.P.; Caruso, T.; Bianco, R.L. A Cultivar-Sensitive Approach for the Continuous Monitoring of Olive (*Olea Europaea* L.) Tree Water Status by Fruit and Leaf Sensing. *Front. Plant Sci.* **2020**, *11*, 11. [CrossRef] [PubMed]
39. Marino, G.; Scalisi, A.; Guzmán-Delgado, P.; Caruso, T.; Marra, F.P.; Lo Bianco, R. Detecting Mild Water Stress in Olive with Multiple Plant-Based Continuous Sensors. *Plants* **2021**, *10*, 131. [CrossRef]
40. KOPPEN, W. das Das Geographische System Der Klimat. In *Handbuch der klimatologie*; Gebrüder Borntraeger: Berlin, Germany, 1936; pp. 1–44.
41. Duro, A.; Piccione, V.; Scalia, C.; Zampino, D. Precipitazioni e Temperature Medie Mensili in Sicilia Relative al Sessantennio 1926–1985. In *5 Workshop Progetto Strategico Clima, Ambiente e Territorio nel Mezzogiorno CNR*; ISHS: Leuven, Belgium, 1996; pp. 17–103.
42. Drago, A.; Cartabellotta, D.; Lo Bianco, B.; Monterosso, I. Atlante Climatologico Della Sicilia. Seconda Edizione. *Riv. Ital. Di Agrometeorol.* **2005**, *2*, 67–83.
43. Farina, V.; Gianguzzi, G.; D'Asaro, A.; Mazzaglia, A.; Palazzolo, E. Fruit Production and Quality Evaluation of Four Litchi Cultivars (Litchi Chinensis Sonn.) Grown in Mediterranean Climate. *Fruits* **2017**, *72*, 203–211. [CrossRef]
44. Torres, A.C.; Galván, D.F.; Saúco, V.G. *Guía Descriptiva de Cultivares de Mango*; Consejería de Agricultura, Pesca y Alimentación: Seville, Spain, 1997; ISBN 84-86840-11-2.
45. Yahia, E.M. Mango (*Mangifera Indica* L.). In *Postharvest Biology and Technology of Tropical and Subtropical Fruits*; Elsevier: Amsterdam, The Netherlands, 2011; pp. 492–567e.
46. Léchaudel, M.; Joas, J. An Overview of Preharvest Factors Influencing Mango Fruit Growth, Quality and Postharvest Behaviour. *Braz. J. Plant Physiol.* **2007**, *19*, 287–298. [CrossRef]
47. Faust, M. *Physiology of Temperate Zone Fruit Trees*; John Wiley & Sons, Inc.: Hoboken, NJ, USA, 1989; ISBN 0-471-81781-3.
48. Torres-Ruiz, J.M.; Perulli, G.D.; Manfrini, L.; Zibordi, M.; Velasco, G.L.; Anconelli, S.; Pierpaoli, E.; Corelli-Grappadelli, L.; Morandi, B. Time of Irrigation Affects Vine Water Relations and the Daily Patterns of Leaf Gas Exchanges and Vascular Flows to Kiwifruit (Actinidia Deliciosa Chev.). *Agric. Water Manag.* **2016**, *166*, 101–110. [CrossRef]
49. Léchaudel, M.; Génard, M.; Lescourret, F.; Urban, L.; Jannoyer, M. Modeling Effects of Weather and Source–Sink Relationships on Mango Fruit Growth. *Tree Physiol.* **2005**, *25*, 583–597. [CrossRef]
50. Dambreville, A.; Normand, F.; Lauri, P.-E.; Guedon, Y. Modeling Developmental Growth Stages in Mango on the Basis of Absolute vs. Relative Growth Rates. In Proceedings of the International Symposium on Modelling in Fruit Research and Orchard Management, Montpellier, France, 2–5 June 2015; pp. 69–74.

51. Camacho-Vázquez, C.; Ruiz-May, E.; Guerrero-Analco, J.A.; Elizalde-Contreras, J.M.; Enciso-Ortiz, E.J.; Rosas-Saito, G.; López Sánchez, L.; Kiel-Martínez, A.L.; Bonilla-Landa, I.; Monribot-Villanueva, J.L. Filling Gaps in Our Knowledge on the Cuticle of Mangoes (Mangifera Indica) by Analyzing Six Fruit Cultivars: Architecture/Structure, Postharvest Physiology and Possible Resistance to Fruit Fly (Tephritidae) Attack. *Postharvest Biol. Technol.* **2019**, *148*, 83–96. [CrossRef]
52. Nordey, T.; Léchaudel, M.; Génard, M. The Decline in Xylem Flow to Mango Fruit at the End of Its Development Is Related to the Appearance of Embolism in the Fruit Pedicel. *Funct. Plant Biol.* **2015**, *42*, 668–675. [CrossRef] [PubMed]
53. Montanaro, G.; Dichio, B.; Xiloyannis, C. Significance of Fruit Transpiration on Calcium Nutrition in Developing Apricot Fruit. *J. Plant Nutr. Soil Sci.* **2010**, *173*, 618–622. [CrossRef]
54. Montanaro, G.; Dichio, B.; Xiloyannis, C.; Lang, A. Fruit Transpiration in Kiwifruit: Environmental Drivers and Predictive Model. *AoB Plants* **2012**, *2012*, pls036. [CrossRef] [PubMed]
55. Maguire, K.M.; Banks, N.H.; Opara, L.U. Factors Affecting Weight Loss of Apples. *Hortic. Rev.* **2001**, *25*, 197–234.
56. Lechaudel, M.; Génard, M.; Lescourret, F.; Urban, L.; Jannoyer, M. Leaf-to-Fruit Ratio Affects Water and Dry-Matter Content of Mango Fruit. *J. Hortic. Sci. Biotechnol.* **2002**, *77*, 773–777. [CrossRef]
57. Mukherjee, S.; Litz, R.E. Introduction: Botany and Importance. In *The Mango: Botany, Production and Uses*, 2nd ed.; CABI Publishing: Wallingford, UK, 2009; pp. 1–18.
58. Guillén, L.; Adaime, R.; Birke, A.; Velázquez, O.; Angeles, G.; Ortega, F.; Ruíz, E.; Aluja, M. Effect of Resin Ducts and Sap Content on Infestation and Development of Immature Stages of Anastrepha Obliqua and Anastrepha Ludens (Diptera: Tephritidae) in Four Mango (Sapindales: Anacardiaceae) Cultivars. *J. Econ. Entomol.* **2017**, *110*, 719–730.
59. Brüggenwirth, M.; Winkler, A.; Knoche, M. Xylem, Phloem, and Transpiration Flows in Developing Sweet Cherry Fruit. *Trees* **2016**, *30*, 1821–1830. [CrossRef]
60. Choat, B.; Gambetta, G.A.; Shackel, K.A.; Matthews, M.A. Vascular Function in Grape Berries across Development and Its Relevance to Apparent Hydraulic Isolation. *Plant Physiol.* **2009**, *151*, 1677–1687. [CrossRef] [PubMed]
61. González-Aguilar, G.; Sotelo-Mundo, R.; Báez-Sañudo, R.; Petit-Jiménez, D.; González-León, A. Permeability of Cuticular Membrane during the Ontogeny of *Mangifera Indica* L. In Proceedings of the VIII International Mango Symposium Conference, Sun City, South Africa, 5–10 February 2006; pp. 213–220.

plants

MDPI

Article

Essential Oils of Four Virginia Mountain Mint (*Pycnanthemum virginianum*) Varieties Grown in North Alabama

William N. Setzer [1,2,*], Lam Duong [3], Trang Pham [3], Ambika Poudel [2], Cuong Nguyen [1] and Srinivasa Rao Mentreddy [3,*]

1. Department of Chemistry, University of Alabama in Huntsville, Huntsville, AL 35899, USA; cvn0001@uah.edu
2. Aromatic Plant Research Center, 230 N 1200 E, Suite 100, Lehi, UT 84043, USA; apoudel@aromaticplant.org
3. Department of Biological and Environmental Sciences, Alabama A & M University, Normal, AL 35762, USA; lduong@bulldogs.aamu.edu (L.D.); trang.pham@bulldogs.aamu.edu (T.P.)

* Correspondence: wsetzer@chemistry.uah.edu (W.N.S.); srinivasa.mentreddy@aamu.edu (S.R.M.)

Abstract: Virginia mountain mint (*Pycnanthemum virginianum*) is a peppermint-flavored aromatic herb of the Lamiaceae and is mainly used for culinary, medicinal, aromatic, and ornamental purposes. North Alabama's climate is conducive to growing mint for essential oils used in culinary, confectionery, and medicinal purposes. There is, however, a need for varieties of *P. virginianum* that can be adapted and easily grown for production in North Alabama. Towards this end, four field-grown varieties with three harvesting times (M1H1, M1H2, M1H3; M2H1, M2H2, M2H3; M3H1, M3H2, M3H3, M4H1, M4H2, M4H3) were evaluated for relative differences in essential oil yield and composition. Thirty-day-old greenhouse-grown plants of the four varieties were transplanted on raised beds in the field at the Alabama A & M University Research Station in North Alabama. The plots were arranged in a randomized complete block with three replications. The study's objective was to compare the four varieties for essential oil yield and their composition at three harvest times, 135, 155, and 170 days after planting (DAP). Essential oils were obtained by hydrodistillation with continuous extraction with dichloromethane using a Likens–Nickerson apparatus and analyzed by gas chromatographic techniques. At the first harvest, the essential oil yield of the four varieties showed that M1H1 had a yield of 1.15%, higher than M2H1, M3H1, and M4H1 with 0.91, 0.76, and 1.03%, respectively. The isomenthone concentrations increased dramatically through the season in M1 (M1H1, M1H2, M1H3) by 19.93, 54.7, and 69.31%, and M3 (M3H1, M3H2, M3H3) by 1.81, 48.02, and 65.83%, respectively. However, it increased only slightly in M2 and M4. The thymol concentration decreased slightly but not significantly in all four varieties; the thymol in M2 and M4 was very high compared with M1 and M3. The study showed that mountain mint offers potential for production in North Alabama. Two varieties, M1 and M3, merit further studies to determine yield stability, essential oil yield, composition, and cultivation development practices.

Keywords: pulegone; isomenthone; menthone; thymol; *p*-cymene; chemotypes; seasonal variation; enantiomeric distribution

Citation: Setzer, W.N.; Duong, L.; Pham, T.; Poudel, A.; Nguyen, C.; Mentreddy, S.R. Essential Oils of Four Virginia Mountain Mint (*Pycnanthemum virginianum*) Varieties Grown in North Alabama. *Plants* **2021**, *10*, 1397. https://doi.org/10.3390/plants10071397

Academic Editors: Milan S. Stankovic, Paula Baptista and Petronia Carillo

Received: 4 June 2021
Accepted: 6 July 2021
Published: 8 July 2021

Publisher's Note: MDPI stays neutral with regard to jurisdictional claims in published maps and institutional affiliations.

Copyright: © 2021 by the authors. Licensee MDPI, Basel, Switzerland. This article is an open access article distributed under the terms and conditions of the Creative Commons Attribution (CC BY) license (https://creativecommons.org/licenses/by/4.0/).

1. Introduction

Discovered and named 'mountain mint' by the French Botanist Andre Micaux [1], *Pycnanthemum* Michx. is an herbaceous perennial belonging to the family Lamiaceae. The plants can grow up to 1 meter in height with delicate, angular stems and 4–6 cm long narrowly lanceolate leaves. The leaves are known for their mild mint flavor. The plants grow well in semi-shaded woodlands and along waterways on well-drained, light, sandy, loam clay soils with a pH ranging from slightly acidic to slightly alkaline [2].

Pycnanthemum Michx. has been reported to consist of an estimated 20 species according to World Flora Online [3], all of which occur in North America: ***Pycnanthemum albescens* Torr. & A. Gray**, *Pycnanthemum beadlei* (Small) Fernald, *Pycnanthemum californicum* Torr. ex

Durand, *Pycnanthemum clinopodioides* Torr. & A. Gray, ***Pycnanthemum curvipes* (Greene) E.Grant & Epling**, ***Pycnanthemum flexuosum* (Walter) Britton, Sterns & Poggenb.**, *Pycnanthemum floridanum* E.Grant & Epling, ***Pycnanthemum incanum* (L.) Michx.**, ***Pycnanthemum loomisii* Nutt.**, *Pycnanthemum monotrichum* Fernald, *Pycnanthemum montanum* Michx., ***Pycnanthemum muticum* (Michx.) Pers.**, ***Pycnanthemum nudum* Nutt.**, ***Pycnanthemum pycnanthemoides* (Leavenw.) Fernald**, *Pycnanthemum setosum* Nutt., ***Pycnanthemum tenuifolium* Schrad.**, *Pycnanthemum torreyi* Benth., ***Pycnanthemum verticillatum* (Michx.) Pers.**, ***Pycnanthemum virginianum* (L.) T. Durand & B.D. Jacks. ex B.L. Rob. & Fernald**, and *Pycnanthemum viridifolium* E.Grant & Epling. The species highlighted in **bold** have been recorded in the state of Alabama. Nearly 50% of the known species are found in Alabama, which underscores the adaptation of mountain mint to Alabama environments.

Several *Pycnanthemum* species have been used in Native American traditional medicine [1]. The Choctaw took a hot decoction of *P. albescens* leaves as a diaphoretic for colds. The Miwok used *P. californicum* as a treatment for colds. A poultice of leaves of *P. flexuosum* or *P. incanum* was used by the Cherokee to relieve headache, while a leaf infusion of these plants was taken for fevers. The Lakota took an infusion of the leaves of *P. virginianum* for coughs. In addition, *P. flexuosum* and *P. incanum* were used by the Cherokee for food, and the Chippewa used *P. virginianum* to season meat or broth. Today, mountain mint is popularly used as a mild-flavored tea, and the leaves and buds are often eaten in salads. Mountain mint tea is known to be curative, diaphoretic, and carminative [1,2]. The tea made of mountain mint leaves is used for treating menstrual disorders, mild headaches, fevers, colds, coughs, and indigestion [1–3]. Mountain mint has been shown to cause abortions if consumed by pregnant women [1].

The essential oil composition of mountain mints varies considerably among species, and the major components are carvacrol, menthone, isomenthone, β-elemene, limonene, piperitone (minty and camphor-like odor [4]), and germacrene D, characteristic of species in the Lamiaceae [5,6]. The uses of *Pycnanthemum* species are based on essential oil composition, e.g., *P. virginianum* rich in menthone and isomenthone have culinary and medicinal uses; *Pycnanthemum* species rich in β-elemene and low in pulegone are known to attract beneficial insects, mainly bees and butterflies, whereas species such as *P. muticum*, rich in pulegone, are insect repellants. If consumed, pulegone can be toxic to the liver, but it is apparently safe to rub *P. muticum* herb on clothes to deter chiggers, gnats, and ticks [7]. Species with little or no pulegone are used for making teas and infusions. The global market demand for mint essential oil was an estimated 177.88 million USD in 2018 with a predicted annual growth rate of 9.2% between 2014 and 2025 [8]. Thus, mountain mint rich in menthone and isomenthone may be suitable for pharmaceutical and nutraceutical, food & beverage, cosmetic, aromatherapy, and cleaning products markets [8]. About 50% of mountain mint species grow wild in Alabama and are adapted to shady environments and sandy marginal soils. Therefore, this crop merits consideration for evaluation as an alternative, niche-market cash crop in Alabama. Towards this end, four varieties of *P. virginianum* were evaluated for their leaf biomass and essential oil content and composition.

2. Results and Discussion

2.1. Fresh Leaf Biomass Yield

There were significant differences among varieties for fresh leaf biomass at 135 and 155 days after transplanting (DAP), but no such differences existed at 179 DAP (Figure 1). At 135 DAP, M1 and M2 had significantly more fresh leaf biomass than other varieties. At 155 DAP, Varieties M1 and M3 had similar leaf fresh weights and were significantly greater than that of M2 and M4. At 170 DAP, however, all varieties produced similar fresh leaf biomass. The fresh leaf biomass of all varieties decreased with age, and at 170 DAP, it was less than half of that produced at 135 DAP. M3 did not fare well at the first time of planting because the plants in two replications died, but this may be related to inadequate propagules and the harvesting plan. July, the month following planting, received the lowest amount of precipitation combined with high temperatures (see Supplementary Figure S1), which may have affected the poorly established plants of M3. However, this variety has shown consistently good growth in the greenhouse and in the field trials in progress.

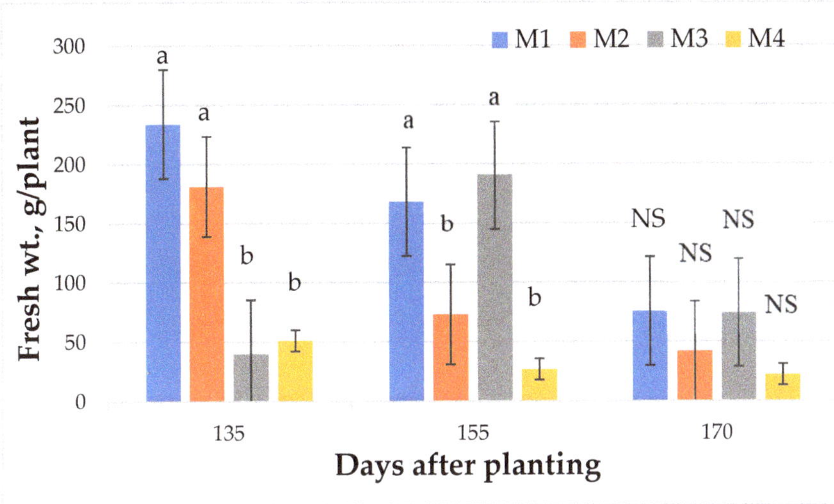

Figure 1. Changes with time in leaf fresh biomass of four mountain mint varieties grown in North Alabama, USA. Bars with same letter are not significantly different at $p \leq 0.05$. NS = not significant.

The fresh leaf biomass as a percentage of whole plant biomass varied with variety (Figure 2). At the first time of harvest (135 DAP), varieties M1 and M3 with leaf biomass of nearly 70% of the whole plant biomass were superior in leaf production than M1 and M2. At 155 DAP, all varieties had lower levels of leaf biomass as a percentage of whole-plant biomass with narrow differences between varieties. However, M3 had a higher percentage of leaf biomass than M4. At 170 DAP, the leaf biomass as a percentage of the whole plant was lower among all varieties relative to those at 155 or 135 DAP; the magnitude of differences between varieties increased. Thus, M3 partitioned a greater percentage of leaves compared to other varieties. M1 with 58% was superior to M2 and M4, which had a similar percentage of leaf biomass. In general, the M1 and M3 were consistently superior to M2 and M4. Seasonal effects were observed as all varieties had lower leaf biomass as the season advanced. Although the literature on the growth and yield of mountain mint is scarce, there is published research on variation in biomass accumulation and oil content in species belonging to the Lamiaceae [9–12]. Brar et al. [10] reported a larger leaf-to-stem ratio in *Mentha arvensis* L. Higher accumulations of fresh leaf biomass during the mid-season and decline as the crop matured has been observed in *Mentha piperita* L. [12] as well as *M. arvensis* [11], a trend observed in this study.

There were limited time-of-harvest × variety interactions for fresh leaf biomass. The interaction of M1 was significant with the first and second time of harvest, whereas the interaction of M2 and M3 was significant with the first and second time of harvest, respectively. All other interactions among time of harvest and varieties were not significantly different at $p \leq 0.05$.

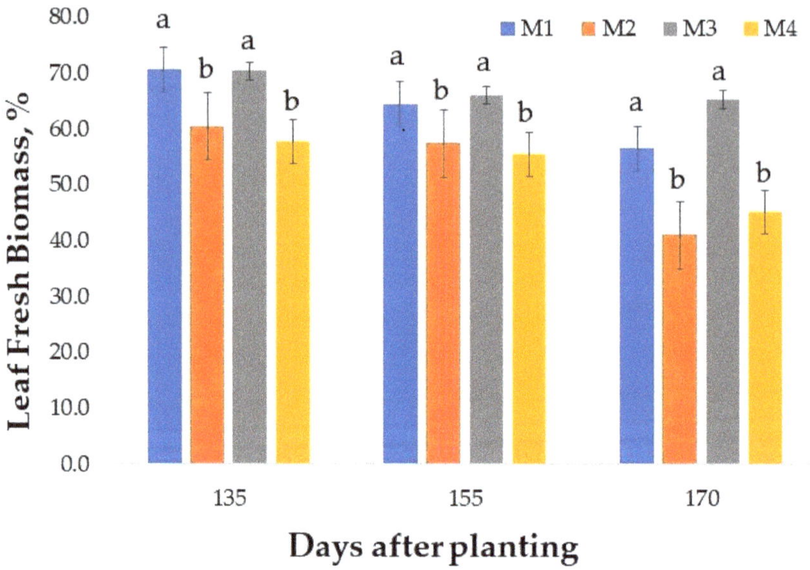

Figure 2. Variation among four mountain mint varieties for fresh leaf biomass as a percentage of whole plant biomass at three times of harvest. Alabama, USA. Bars with same letter are not significantly different at $p \leq 0.05$.

2.2. *Chemical Composition of Essential Oils*

The chemical compositions of the *P. virginianum* essential oils have been determined by gas chromatography–mass spectrometry (GC-MS) and quantified by gas chromatography-flame ionization detection (GC-FID). The essential oil compositions are compiled in Supplementary Tables S1–S4. The major components of the four varieties of *P. virginianum* over three harvest dates are summarized in Table 1.

Table 1. Major components of *Pycnanthemum virginianum* essential oils cultivated in North Alabama [a].

Compound	Percent Composition											
	M1H1 [b]	M1H2 [b]	M1H3 [b]	M2H1 [b]	M2H2 [b]	M2H3 [b]	M3H1 [c]	M3H2 [d]	M3H3 [d]	M4H1 [b]	M4H2 [b]	M4H3 [b]
1-Octen-3-ol	2.0 ± 0.2	2.0 ± 0.2	1.5 ± 0.2	2.6 ± 0.1	2.0 ± 0.4	2.2 ± 0.7	3.5 ± 0.4	1.8	1.9	3.2 ± 0.5	5.5 ± 0.7	5.8 ± 1.5
Myrcene	0.4 ± 0.1	0.4 ± 0.0	0.4 ± 0.0	3.1 ± 1.1	2.6 ± 0.9	1.9 ± 0.7	0.1 ± 0.1	0.4	0.4	1.3 ± 1.0	0.8 ± 0.5	0.5 ± 0.1
p-Cymene	0.1 ± 0.0	tr	tr	7.1 ± 0.8	8.7 ± 2.1	7.4 ± 0.7	0.2 ± 0.1	tr	0.3	5.0 ± 2.7	22.5 ± 2.4	18.9 ± 8.8
Limonene	1.9 ± 0.2	1.3 ± 0.1	1.1 ± 0.2	5.8 ± 1.2	5.5 ± 1.4	4.1 ± 1.0	1.0 ± 0.7	1.5	1.2	0.7 ± 0.3	0.6 ± 0.2	0.5 ± 0.1
γ-Terpinene	tr	tr	tr	2.4 ± 0.3	1.7 ± 0.6	0.9 ± 0.2	nd	nd	tr	3.9 ± 2.4	1.7 ± 0.5	1.2 ± 0.3
cis-Sabinene hydrate	tr	tr	tr	0.6 ± 0.1	0.6 ± 0.1	0.7 ± 0.1	tr	tr	tr	1.0 ± 0.2	2.5 ± 0.8	2.7 ± 1.0
Menthone	1.5 ± 0.6	4.7 ± 0.6	5.6 ± 0.9	5.4 ± 0.9	9.9 ± 0.1	12.9 ± 1.2	15.3 ± 15.1	3.8	5.6	0.3 ± 0.3	0.1 ± 0.0	0.9 ± 0.7
Isomenthone	19.9 ± 6.9	54.7 ± 3.8	69.3 ± 1.4	0.1 ± 0.0	0.4 ± 0.4	1.8 ± 1.3	1.8 ± 0.1	48.0	65.8	2.7 ± 3.2	0.4 ± 0.6	6.3 ± 5.9
trans-Isopulegone	1.6 ± 0.1	1.1 ± 0.1	0.9 ± 0.0	3.6 ± 0.2	4.1 ± 0.4	4.1 ± 0.8	1.6 ± 0.2	1.2	1.1	0.2 ± 0.4	nd	0.2 ± 0.3
cis-Piperitenol	nd	nd	nd	2.6 ± 0.2	2.3 ± 0.3	2.2 ± 0.4	nd	nd	0.1	tr	nd	0.1 ± 0.1
Pulegone	65.3 ± 8.4	29.8 ± 3.9	13.0 ± 1.8	15.3 ± 1.8	10.8 ± 1.4	10.1 ± 1.6	63.8 ± 11.2	37.3	16.1	11.8 ± 15.4	0.6 ± 0.9	2.3 ± 1.8
Thymol	0.4 ± 0.1	0.2 ± 0.2	tr	26.9 ± 4.1	22.1 ± 2.3	23.0 ± 1.3	0.3 ± 0.4	tr	0.6	53.8 ± 15.2	36.5 ± 5.4	30.9 ± 5.3
Carvacrol	tr	tr	tr	1.5 ± 0.2	1.4 ± 0.1	1.4 ± 0.1	tr	nd	tr	1.3 ± 0.4	1.1 ± 0.3	0.7 ± 0.1
Unidentified (RI 1345)	0.1 ± 0.0	0.1 ± 0.0	tr	7.5 ± 0.7	10.7 ± 0.8	10.7 ± 2.4	0.1 ± 0.1	nd	0.3	0.2 ± 0.2	nd	0.6 ± 0.6
(*E*)-β-Caryophyllene	1.2 ± 0.2	0.6 ± 0.5	1.3 ± 0.2	1.1 ± 0.1	1.2 ± 0.2	1.2 ± 0.2	0.8 ± 1.1	1.0	1.1	1.8 ± 0.2	1.7 ± 0.4	1.6 ± 0.7
Germacrene D	1.2 ± 0.4	0.7 ± 0.5	1.1 ± 0.0	1.8 ± 0.1	1.5 ± 0.2	1.4 ± 0.1	0.2 ± 0.3	0.7	1.0	1.4 ± 0.5	0.8 ± 0.0	1.0 ± 0.6

[a] M1 = Variety #1, M2 = Variety #2, M3 = Variety #3, M4 = Variety #4. H1 = Harvest #1 (30 September 2020, 135 days after transplanting), H2 = Harvest #2 (19 October 2020, 155 days after transplanting), H3 = Harvest #3 (4 November 2020, 170 days after transplanting). tr = trace (< 0.05%), nd = not detected. [b] Average of three different plants ± standard deviations. [c] Average of two plants (third plant died) ± standard deviations. [d] Only one plant survived.

The four varieties of *P. virginianum* showed different volatile chemical profiles and apparently define different chemical variations of this species. In order to discern the phytochemical differences between these varieties, a hierarchical cluster analysis was carried out using 16 of the most abundant essential oil components (1-octen-3-ol, myrcene, *p*-cymene, limonene, γ-terpinene, *cis*-sabinene hydrate, menthone, isomenthone, *trans*-isopulegone, *cis*-piperitenol, pulegone, thymol, carvacrol, unidentified (RI 1345), (*E*)-β-caryophyllene, and germacrene D) (Figure 3).

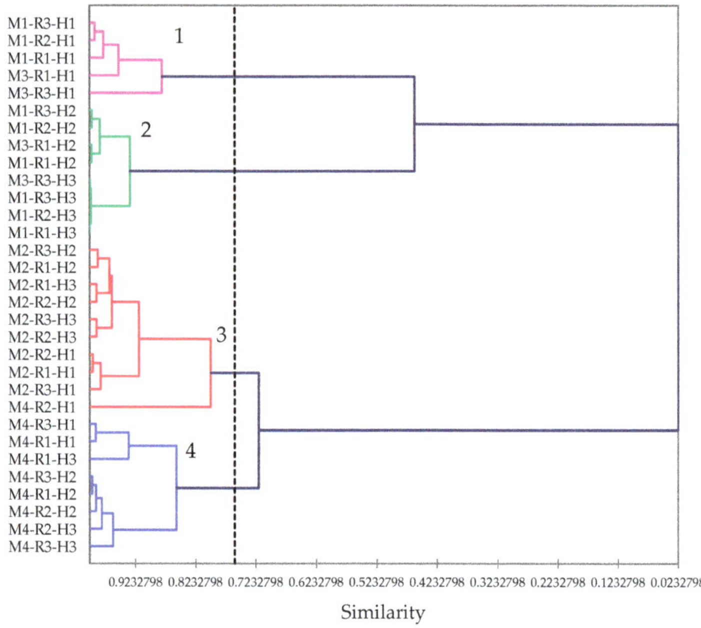

Figure 3. Dendrogram obtained by hierarchical cluster analysis of the 16 most abundant components of *Pycnanthemum virginianum* essential oils. 'M' represents the varieties, 'R' is the replicate of each variety, and 'H' is the harvesting times.

There are four clearly defined clusters based on the cluster analysis: Cluster 1 can be defined as a pulegone/menthone cluster and comprises five samples, three M1 samples, and two M3 samples, all collected from the first harvest. Cluster 2 is an isomenthone/pulegone cluster and is made up of a variety M1 and M3 samples from harvests two and three. Cluster 3 is a thymol/pulegone cluster made up chiefly of M2 samples. Cluster 4 is a thymol/*p*-cymene chemotype and is made up of variety M4 samples. Chemotype 1, rich in pulegone but also with high concentrations of menthone and isomenthone, is similar in composition to several samples of *Mentha pulegium* L. (pennyroyal) [13] and may, therefore, serve as a substitute herb for pennyroyal. Commercial *M. pulegium* essential oil contains around 84% pulegone (Aromatic Plant Research Center, Lehi, UT, USA). Note that these are early harvest samples of M1 and M3; later-harvested samples of M1 and M3 showed a preponderance of isomenthol with reduced concentrations of pulegone. The concentration of menthone also increased through the season for M1 (Figure 4). Varieties M2 and M3 also showed a seasonal decrease in pulegone and increasing isomenthone (Figures 5 and 6). The decrease in pulegone concentration with a concomitant increase in menthone and isomenthone concentrations is not surprising. Pulegone is the precursor in the biosynthesis of menthone and isomenthone [14,15]. There are several chemotypes of *Thymus vulgaris* L., but one chemotype is rich in thymol and *p*-cymene [16,17], and commercial thyme essential oil (doTERRA International, Pleasant Grove, UT, USA) is composed of 44% thymol and 10% *p*-cymene. Thus, the thymol/*p*-cymene chemotype (cluster 4) of *P. virginianum* could serve as a substitute for thyme. Variety M4 (i.e., cluster 4) showed a significant increase in *p*-cymene concentration coupled with a decrease in thymol concentration (Figure 7).

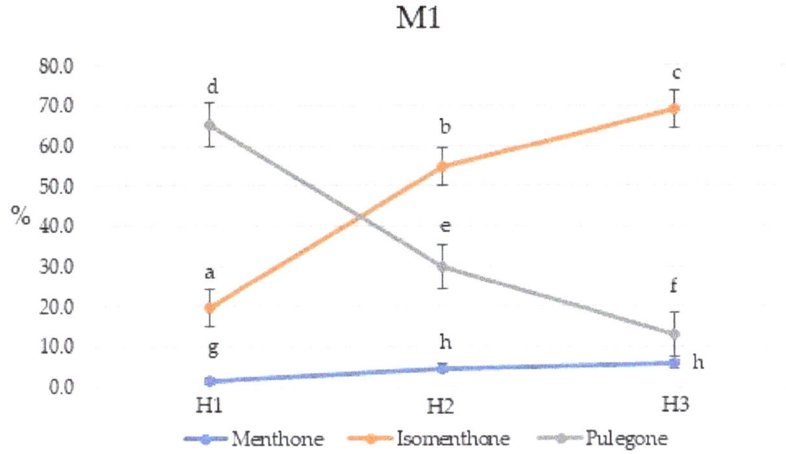

Figure 4. Seasonal variation in the menthone, isomenthone, and pulegone percent concentrations for *Pycnanthemum virginianum* variety M1. Percent concentrations with same letter are not significantly different at $p \leq 0.05$. H1, H2, and H3 are harvest times, 135, 155, and 170 days after planting, respectively.

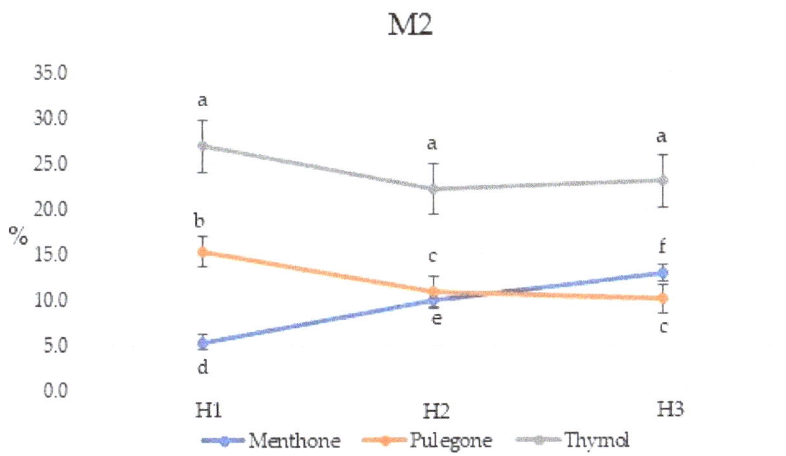

Figure 5. Seasonal variation in the menthone, pulegone, and thymol percent concentrations for *Pycnanthemum virginianum* variety M2. Percent concentrations with same letter are not significantly different at $p \leq 0.05$. H1, H2, and H3 are harvest times, 135, 155, and 170 days after planting, respectively.

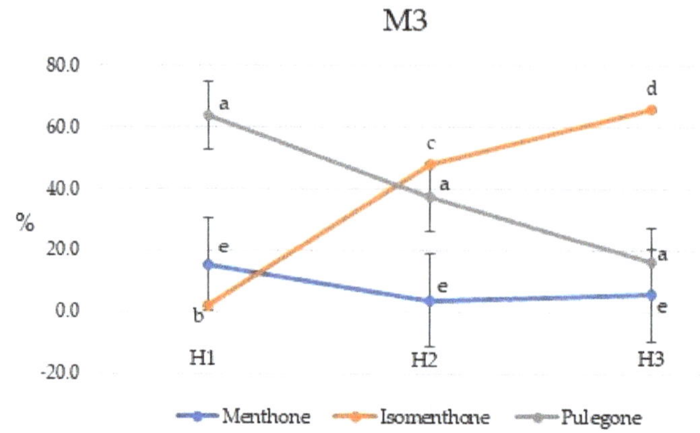

Figure 6. Seasonal variation in the menthone, isomenthone, and pulegone percent concentrations for *Pycnanthemum virginianum* variety M3. Percent concentrations with same letter are not significantly different at $p \leq 0.05$. H1, H2, and H3 are harvest times, 135, 155, and 170 days after planting, respectively.

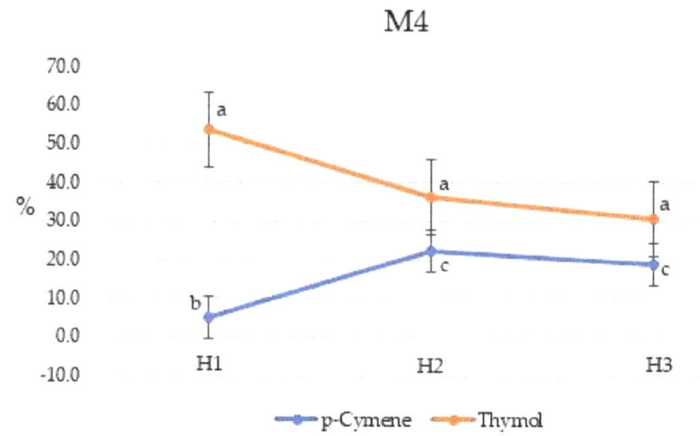

Figure 7. Seasonal variation in the *p*-cymene and thymol percent concentrations for *Pycnanthemum virginianum* variety M4. Percent concentrations with same letter are not significantly different at $p \leq 0.05$. H1, H2, and H3 are harvest times, 135, 155, and 170 days after planting, respectively.

There is currently very little information on the volatile phytochemistry of other *Pycnanthemum* species. The essential oil composition of *P. incanum*, cultivated in south Alabama, has been determined [18]. The major components in *P. incanum* oil were 1,8-cineole (30.7%), α-terpineol (16.9%), (*E*)-β-caryophyllene (11.0%), borneol (8.2%), and germacrene D (5.0%). *p*-Cymene (0.8%, menthone (0.2%), isomenthone (1.0%), pulegone (1.8%), and thymol (0.3%) were found in relatively low concentrations.

2.3. Chemical Composition of Essential Oils

The *P. virginianum* essential oils were analyzed by chiral gas chromatography–mass spectrometry in order to examine the enantiomeric distribution of the terpenoid components. The enantiomeric distributions are compiled in Supplementary Tables S5–S8; a summary of the enantiomeric distributions is shown in Table 2.

Table 2. Enantiomeric distribution of terpenoid constituents of *Pycnanthemum virginianum* [a].

Compound	M1	M2	M3	M4
α-Thujene	54:46	72:28	—	76:24
α-Pinene	24:76	58:42	23:77	72:28
Camphene	100:0	100:0	—	100:0
Sabinene	30:70	0:100	28:72	variable [b]
β-Pinene	46:54	54:46	46:54	31:69
α-Phellandrene	—	95:5	—	96:4
δ-3-Carene	—	100:0	—	100:0
α-Terpinene	—	100:0	—	100:0
Limonene	6:94	0:100	7:93	variable [c]
β-Phellandrene	—	0:100	—	0:100
cis-Sabinene hydrate	—	95:5	—	99:1
Linalool	—	100:0	—	variable [d]
trans-Sabinene hydrate	—	78:22	—	82:18
Menthone	0:100	0:100	0:100	0:100
Isomenthone	100:0	100:0	100:0	100:0
Borneol	—	0:100	—	0:100
Terpinen-4-ol	—	65:35	—	66:34
α-Terpineol	9:91	32:68	10:90	variable [e]
Pulegone	100:0	100:0	100:0	100:0
Piperitone	94:6	—	89:11	—
δ-Elemene	—	—	—	variable [f]
α-Copaene	—	—	—	100:0
trans-β-Elemene	15:85	20:80	16:84	6:94
(*E*)-β-Caryophyllene	100:0	100:0	100:0	100:0
Germacrene D	91:9	94:6	91:9	80:20
δ-Cadinene	0:100	0:100	0:100	0:100

[a] Average enantiomeric distributions, % (+)-enantiomer : % (−)-enantiomer, for each variety. [b] (+)-enantiomer ranged 0–72%. [c] Mostly (−)-enantiomer, but one sample with 49% (+)-limonene. [d] (+)-enantiomer ranged 19–100%. [e] Mostly (+)-enantiomer, but one sample with only 33% (+)-linalool. [f] (+)-enantiomer ranged 0–50%.

There was variability in several of the terpenoid constituents of *P. virginianum* essential oils. α-Thujene showed some variation in enantiomeric distribution between the different varieties; M1 had nearly a racemic mixture, whereas M2 and M4 were predominantly (+)-α-thujene. Camphene was exclusively the (−)-enantiomer in M1, M2, and M4, but it could not be measured in M3. Sabinene showed variation in enantiomeric distribution. M1 and M2 mostly showed the (−)-enantiomer, M3 was exclusively (−)-sabinene, and M4 showed considerable variation through the season. (+)-Linalool was the only enantiomer observed in M2, but the enantiomeric distribution in M4 varied through the season. (−)-α-Terpineol dominated the distribution in M1, M2, and M3, but (+)-α-terpineol was predominant in M4. The (+)-enantiomers were exclusively found for camphene, δ-3-carene, α-terpinene, isomenthone, pulegone, and (*E*)-β-caryophyllene, and the dominant enantiomers for α-phellandrene, *cis*-sabinene hydrate, *trans*-sabinene hydrate, terpinen-4-ol, piperitone, and germacrene D. The (−)-enantiomers were exclusive for β-phellandrene, menthone, and δ-cadinene, and the dominant enantiomer for limonene, δ-elemene, and trans-β-elemene. Consistent with the biosynthesis of monoterpenes in *Mentha* [14,15,19], (+)-pulegone is the apparent precursor of (−)-menthone and (+)-isomenthone in *Pycnanthemum*.

3. Materials and Methods

3.1. Cultivation of *Pycnanthemum virginianum* Varieties

The seeds of four varieties of *Pycnanthemum virginianum*, obtained from the USDA germplasm resource, were planted in seed germination flats filled with soilless potting mix and placed in a temperature-controlled greenhouse. The greenhouse was maintained at 26–28 °C/15–18 °C day/night air temperatures with 13 h daylength. The relative humidity

(RH) ranged 76–80%. The germinated plants were transplanted into 10-cm pots containing soilless mix and grown in the greenhouse for 1 year.

The propagules with root and rhizome from the 12-month-old potted plants were then transplanted onto raised beds (beds 50 cm wide and 15 cm high) covered with a black plastic sheet with drip tape underneath at the Alabama A&M Winfred Thomas Agricultural Research Station located in Hazel Green, AL (latitude 34°89′ N and longitude 86°56′ W). The soil at the experimental site is a Decatur silty loam (fine, kaolinitic, thermic Rhodic Paleudult). Before making the raised beds, a strip of land was rototilled. A mixture of organic manures, composted chicken manure, and vermicompost to provide an equivalent of 50 kg/ha of nitrogen was incorporated in the soil. Besides organic manure mix, the plants received soluble organic fertilizer at 3-week intervals through the drip irrigation method. The plants were maintained under soil moisture-stress-free conditions. The experimental design was a randomized complete block design with three replications (R). One plant per each variety per replication was harvested at 135, 155, and 170 days after planting (DAP) to assess fresh leaf biomass production and essential oil content and its composition. At each harvest time, the fresh leaves from the plants were collected early morning and placed immediately in a cooler with ice packs for transportation to the laboratory at the University of Alabama in Huntsville for further processing.

3.2. *Hydrodistillation of Pycnanthemum virginianum*

At each harvest, the fresh leaves of *P. virginianum* were chopped and hydrodistilled using a Likens–Nickerson apparatus with continuous extraction with dichloromethane to give the essential oils (Table 3).

Table 3. Hydrodistillation details of *Pycnanthemum virginianum* cultivated in North Alabama.

Sample [a]	Plant Mass (g)	Essential Oil Yield (mg)	% Yield	Essential Oil Color
M1H1R1	104.08	1127.6	1.083	pale yellow
M1H1R2	105.70	1164.6	1.102	pale yellow
M1H1R3	122.45	1548.9	1.265	pale yellow
M2H1R1	82.71	704.9	0.852	yellow
M2H1R2	70.22	580.0	0.826	yellow
M2H1R3	75.80	791.6	1.044	yellow
M3H1R1	40.13	302.3	0.753	pale yellow
M3H1R2	0.00	plant died	—	—
M3H1R3	74.15	572.1	0.772	pale yellow
M4H1R1	30.67	301.2	0.982	yellow
M4H1R2	35.31	345.8	0.979	yellow
M4H1R3	73.80	825.5	1.119	orange
M1H2R1	139.76	1330.7	0.952	pale yellow
M1H2R2	133.69	1093.3	0.818	pale yellow
M1H2R3	167.15	1862.5	1.114	pale yellow
M2H2R1	74.47	596.6	0.801	yellow
M2H2R2	59.86	463.1	0.774	yellow
M2H2R3	132.61	1003.9	0.757	yellow
M3H2R1	182.68	2157.9	1.181	pale yellow
M3H2R2	0.00	plant died	—	—
M3H2R3	0.00	plant died	—	—
M4H2R1	8.10	92.6	1.143	orange
M4H2R2	60.66	252.3	0.416	orange
M4H2R3	32.81	208.2	0.635	orange

Table 3. *Cont.*

Sample [a]	Plant Mass (g)	Essential Oil Yield (mg)	% Yield	Essential Oil Color
M1H3R1	100.14	582.0	0.581	pale yellow
M1H3R2	110.22	897.3	0.814	pale yellow
M1H3R3	59.71	556.9	0.933	pale yellow
M2H3R1	39.93	298.4	0.747	yellow
M2H3R2	53.23	510.8	0.960	yellow
M2H3R3	70.02	495.8	0.708	yellow
M3H3R1	0.00	plant died	—	—
M3H3R2	0.00	plant died	—	—
M3H3R3	110.07	945.5	0.859	yellow
M4H3R1	53.98	320.2	0.593	orange
M4H3R2	29.91	108.9	0.364	orange
M4H3R3	19.84	143.9	0.725	orange

[a] M indicates the *P. virginianum* variety; H is the harvest (H1 = 135 days after planting (DAP), H2 = 155 DAP, and H3 = 170 DAP); R is the number of replicates of each plant variety (three replicates of each variety).

3.3. Gas Chromatographic Analysis

The *P. virginianum* essential oils were analyzed by gas chromatography with flame ionization detection (GC-FID), gas chromatography–mass spectrometry (GC-MS), and chiral GC-MS as previously described [20]. The percent compositions were determined from raw peak areas from the GC-FID data without standardization. Essential oil components were identified by comparison of the MS fragmentation and retention indices with those in the databases [21–24]. The enantiomeric distributions were determined from raw peak areas from the chiral GC-MS without standardization.

3.4. Statistical Analysis

Each *P. virginianum* variety was analyzed using three replicate plants (when possible) for each harvest time. The data are expressed a means ± standard deviations. Analysis of variance was conducted by one-way ANOVA followed by the Tukey test using Minitab® 18 (Minitab Inc., State College, PA, USA). The interactions between harvesting time and variety were determined using two-way ANOVA. Differences at $p < 0.05$ were considered to be statistically significant. For the agglomerative hierarchical cluster (AHC) analysis, the 31 essential oil compositions were treated as operational taxonomic units (OTUs), and the concentrations (percentages) of 16 of the most abundant essential oil components (1-octen-3-ol, myrcene, *p*-cymene, limonene, γ-terpinene, *cis*-sabinene hydrate, menthone, isomenthone, *trans*-isopulegone, *cis*-piperitenol, pulegone, thymol, carvacrol, unidentified (RI 1345), (*E*)-β-caryophyllene, and germacrene D) were used to determine the chemical associations between the *P. virginianum* essential oil samples using XLSTAT Premium, version 2018.1.1.62926 (Addinsoft, Paris, France). Similarity was determined using Pearson correlation, and clustering was defined using the unweighted pair-group method with arithmetic mean (UPGMA).

4. Conclusions

The *Pycnanthemum virginianum* varieties in the study showed significant variation in fresh leaf biomass accumulation and essential oil composition, both between varieties and dramatic seasonal variations. Nevertheless, based on leaf biomass production and chemical profiles, two varieties, M1 and M3, merit further studies to determine yield stability, essential oil yield, composition, and development of cultivation practices for commercial production. For variety M1 at 130 DAP, the biomass yield was good, and the chemical profile was similar to pennyroyal. Variety M2 had a rather unique chemical profile, but the yield was inferior. M4 had a chemical profile similar to thyme, but the biomass yields were low. M3 did not fare well, with half the plants dying, but this may be related to the harvesting plan.

Supplementary Materials: The following are available online at https://www.mdpi.com/article/10.3390/plants10071397/s1, Figure S1: Huntsville, Alabama, weather data, 2020. Table S1: Essential oil composition of *Pycnanthemum virginianum* variety M1, Table S2: Essential oil composition of *Pycnanthemum virginianum* variety M2, Table S3: Essential oil composition of *Pycnanthemum virginianum* variety M3, Table S4: Essential oil composition of *Pycnanthemum virginianum* variety M4, Table S5: Enantiomeric distribution of terpenoid constituents of *Pycnanthemum virginianum*, variety M1, Table S6: Enantiomeric distribution of terpenoid constituents of *Pycnanthemum virginianum*, variety M2, Table S7: Enantiomeric distribution of terpenoid constituents of *Pycnanthemum virginianum*, variety M3, Table S8: Enantiomeric distribution of terpenoid constituents of *Pycnanthemum virginianum*, variety M4.

Author Contributions: Conceptualization, S.R.M.; methodology, S.R.M. and W.N.S.; formal analysis, W.N.S.; investigation, W.N.S., L.D., T.P., C.N., and A.P.; resources, S.R.M.; data curation, W.N.S.; writing—original draft preparation, W.N.S. and S.R.M.; writing—review and editing, all authors; supervision, S.R.M.; project administration, S.R.M.; funding acquisition, S.R.M. All authors have read and agreed to the published version of the manuscript.

Funding: This research was funded by the USDA/National Institute of Food and Agriculture (NIFA), Agriculture and the 1890 Capacity Building Grant Project, grant number: 2015-38821-24337.

Institutional Review Board Statement: Not applicable.

Informed Consent Statement: Not applicable.

Data Availability Statement: All of the data are available within this manuscript and the Supplementary Tables.

Acknowledgments: S.R.M. and L.D. thank Khadejah Scott, DeAnthony Price, and Suresh Kumar for their assistance with field trials. W.N.S. and A.P. participated in this work as part of the activities of the Aromatic Plant Research Center (APRC, https://aromaticplant.org/).

Conflicts of Interest: The authors declare no conflict of interest.

References

1. Moerman, D.E. *Native American Ethnobotany*; Timber Press, Inc.: Portland, OR, USA, 1998; ISBN 978-0-88192-453-4.
2. Foster, S.; Duke, J.A. *A Field Guide to Medicinal Plants*; Houghton Mifflin: Boston, MA, USA, 1990; ISBN 0-395-46722-5.
3. Bown, D. *Encyclopedia of Herbs & Their Uses*; Dorling Kindersley: London, UK, 1995; ISBN 978-0789401847.
4. Ravid, U.; Putievsky, E.; Katzir, I. Enantiomeric distribution of piperitone in essential oils of some *Mentha* spp., *Calamintha incána* (Sm.) Heldr. and *Artemisia judaica* L. *Flavour Fragr. J.* **1994**, *9*, 85–87. [CrossRef]
5. Karpiński, T.M. Essential oils of Lamiaceae family plants as antifungals. *Biomolecules* **2020**, *10*, 103. [CrossRef] [PubMed]
6. Ebadollahi, A.; Ziaee, M.; Palla, F. Essential oils extracted from different species of the Lamiaceae plant family as prospective bioagents against several detrimental pests. *Molecules* **2020**, *25*, 1556. [CrossRef] [PubMed]
7. Betz, S.; Knapke, D.; Schlosser, K.; Stravinsky, D.; Wells, L. The Herb Society of America's Notable Native 2016—Mountain Mints Pycnanthemum Michx. Available online: www.ncherbsociety.org/uploads/8/4/4/3/84437124/pycnanthemum_nn_2016.pdf (accessed on 31 March 2021).
8. Grand View Research. Mint Essential Oils Market Size, Share & Trends Analysis Report by Product (Cornmint, Peppermint, Spearmint, Dementholized Peppermint), by Application, by Usage (Direct, Indirect), and Segment Forecasts, 2019–2025. Available online: https://www.grandviewresearch.com/industry-analysis/mint-essential-oils-market (accessed on 22 June 2021).
9. Sims, C.A.; Juliani, H.R.; Mentreddy, S.R.; Simon, J.E. Essential oils in Holy basil (*Ocimum tenuiflorum* L.) as influenced by planting dates and harvest times in north Alabama. *J. Med. Act. Plants* **2014**, *2*, 33–41. [CrossRef]
10. Brar, S.K.; Gill, B.S.; Brar, A.S.; Kaur, T. Planting date and straw mulch affect biomass yield, oil yield and oil quality of Japanese mint (*Mentha arvensis* L.) harvested at successive intervals. *J. Essent. Oil Bear. Plants* **2014**, *17*, 676–695. [CrossRef]
11. Solomon, A.M.; Beemnet, M.K. Row spacing and harvesting age affect agronomic characteristics and essential oil yield of Japanese mint (*Mentha arvensis* L.). *Med. Aromat. Plant Sci. Biotechnol.* **2011**, *5*, 74–76.
12. Kassahun, B.M.; Teixeira da Silva, J.A.; Mekonnen, S.A. Agronomic characters, leaf and essential oil yield of peppermint (*Mentha piperiata* L.) as influenced by harvesting age and row spacing. *Med. Aromat. Plant Sci. Biotechnol.* **2011**, *5*, 49–53.
13. Domingues, P.M.; Santos, L. Essential oil of pennyroyal (*Mentha pulegium*): Composition and applications as alternatives to pesticides—New tendencies. *Ind. Crops Prod.* **2019**, *139*, 111534. [CrossRef]
14. Croteau, R.; Venkatachalam, K.V. Metabolism of monoterpenes: Demonstration that (+)-cis-isopulegone, not piperitenone, is the key intermediate in the conversion of (−)-isopiperitenone to (+)-pulegone in peppermint (*Mentha piperita*). *Arch. Biochem. Biophys.* **1986**, *249*, 306–315. [CrossRef]

15. Fuchs, S.; Beck, T.; Sandvoss, M.; Mosandl, A. Biogenetic studies in *Mentha* × *piperita*. 2. Stereoselectivity in the bioconversion of pulegone into menthone and isomenthone. *J. Agric. Food Chem.* **1999**, *47*, 3058–3062. [CrossRef] [PubMed]
16. Asllani, U.; Toska, V. Chemical composition of Albanian thyme oil (*Thymus vulgaris* L.). *J. Essent. Oil Res.* **2003**, *15*, 165–167. [CrossRef]
17. Satyal, P.; Murray, B.L.; McFeeters, R.L.; Setzer, W.N. Essential oil characterization of *Thymus vulgaris* from various geographical locations. *Foods* **2016**, *5*, 70. [CrossRef] [PubMed]
18. Lawson, S.K.; Satyal, P.; Setzer, W.N. The volatile phytochemistry of seven Native American aromatic medicinal plants. *Plants* **2021**, *10*, 1061. [CrossRef] [PubMed]
19. Lorenzo, D.; Paz, D.; Dellacassa, E.; Davies, P.; Vila, R.; Cañigueral, S. Essential oils of *Mentha pulegium* and *Mentha rotundifolia* from Uruguay. *Braz. Arch. Biol. Technol.* **2002**, *45*, 519–524. [CrossRef]
20. DeCarlo, A.; Johnson, S.; Okeke-Agulu, K.I.; Dosoky, N.S.; Wax, S.J.; Owolabi, M.S.; Setzer, W.N. Compositional analysis of the essential oil of *Boswellia dalzielii* frankincense from West Africa reveals two major chemotypes. *Phytochemistry* **2019**, *164*, 24–32. [CrossRef] [PubMed]
21. Adams, R.P. *Identification of Essential Oil Components by Gas Chromatography/Mass Spectrometry*, 4th ed.; Allured Publishing: Carol Stream, IL, USA, 2007; ISBN 978-1-932633-21-4.
22. *NIST17*; National Institute of Standards and Technology: Gaithersburg, MD, USA, 2017.
23. Mondello, L. *FFNSC 3*; Shimadzu Scientific Instruments: Columbia, MD, USA, 2016.
24. Satyal, P. Development of GC-MS Database of Essential Oil Components by the Analysis of Natural Essential Oils and Synthetic Compounds and Discovery of Biologically Active Novel Chemotypes in Essential Oils. Ph.D. Dissertation, University of Alabama in Huntsville, Huntsville, AL, USA, 2015.

plants

MDPI

Article

Optimization of Protein Isolation and Label-Free Quantitative Proteomic Analysis in Four Different Tissues of Korean Ginseng

Truong Van Nguyen [1], So-Wun Kim [1], Cheol-Woo Min [1], Ravi Gupta [2], Gi-Hyun Lee [1], Jeong-Woo Jang [1], Divya Rathi [1], Hye-Won Shin [1], Ju-Young Jung [1], Ick-Hyun Jo [3], Woo-Jong Hong [4], Ki-Hong Jung [4], Seungill Kim [5], Yu-Jin Kim [6,*] and Sun-Tae Kim [1,*]

1 Department of Plant Bioscience, Life and Industry Convergence Research Institute, Pusan National University, Miryang 50463, Korea; ng.truong.win@gmail.com (T.V.N.); kimsso12@nate.com (S.-W.K.); min0685@gmail.com (C.-W.M.); 7012nn@naver.com (G.-H.L.); com5222@naver.com (J.-W.J.); rathi.divya71@gmail.com (D.R.); angel981123@naver.com (H.-W.S.); xb01xb@naver.com (J.-Y.J.)
2 Department of Botany, School of Chemical and Life Science, Jamia Hamdard, New Delhi 110062, India; ravigupta07@ymail.com
3 Department of Herbal Crop Research, Rural Development Administration, Eumseong 27709, Korea; intron@korea.ac.kr
4 Graduate School of Biotechnology and Crop Biotech Institute, Kyung Hee University, Yongin 17104, Korea; hwj0602@nate.com (W.-J.H.); khjung2010@khu.ac.kr (K.-H.J.)
5 Department of Environmental Horticulture, University of Seoul, Seoul 02504, Korea; ksi2204@uos.ac.kr
6 Department of Life Science and Environmental Biochemistry, Life and Industry Convergence Research Institute, Pusan National University, Miryang 50463, Korea
* Correspondence: yjkim2020@pusan.ac.kr (Y.-J.K.); stkim71@pusan.ac.kr (S.-T.K.); Tel.: +82-55-350-5546 (Y.-J.K.); +82-55-350-5505 (S.-T.K.)

Citation: Van Nguyen, T.; Kim, S.-W.; Min, C.-W.; Gupta, R.; Lee, G.-H.; Jang, J.-W.; Rathi, D.; Shin, H.-W.; Jung, J.-Y.; Jo, I.-H.; et al. Optimization of Protein Isolation and Label-Free Quantitative Proteomic Analysis in Four Different Tissues of Korean Ginseng. *Plants* **2021**, *10*, 1409. https://doi.org/10.3390/plants10071409

Academic Editors: Milan S. Stankovic, Paula Baptista and Petronia Carillo

Received: 22 May 2021
Accepted: 8 July 2021
Published: 9 July 2021

Publisher's Note: MDPI stays neutral with regard to jurisdictional claims in published maps and institutional affiliations.

Copyright: © 2021 by the authors. Licensee MDPI, Basel, Switzerland. This article is an open access article distributed under the terms and conditions of the Creative Commons Attribution (CC BY) license (https://creativecommons.org/licenses/by/4.0/).

Abstract: Korean ginseng is one of the most valuable medicinal plants worldwide. However, our understanding of ginseng proteomics is largely limited due to difficulties in the extraction and resolution of ginseng proteins because of the presence of natural contaminants such as polysaccharides, phenols, and glycosides. Here, we compared four different protein extraction methods, namely, TCA/acetone, TCA/acetone–MeOH/chloroform, phenol–TCA/acetone, and phenol–MeOH/chloroform methods. The TCA/acetone–MeOH/chloroform method displayed the highest extraction efficiency, and thus it was used for the comparative proteome profiling of leaf, root, shoot, and fruit by a label-free quantitative proteomics approach. This approach led to the identification of 2604 significantly modulated proteins among four tissues. We could pinpoint differential pathways and proteins associated with ginsenoside biosynthesis, including the methylerythritol 4–phosphate (MEP) pathway, the mevalonate (MVA) pathway, UDP-glycosyltransferases (UGTs), and oxidoreductases (CYP450s). The current study reports an efficient and reproducible method for the isolation of proteins from a wide range of ginseng tissues and provides a detailed organ-based proteome map and a more comprehensive view of enzymatic alterations in ginsenoside biosynthesis.

Keywords: label-free proteomics; *Panax ginseng*; ginsenosides; cytochrome p450; UDP-glycosyltransferase; MEP pathway; MVA pathway; TCA/acetone; methanol/chloroform

1. Introduction

Ginseng (*Panax ginseng*) is a precious medicinal plant exhibiting significant economic values and pharmacological effects [1,2]. Owing to the presence of various bioactive compounds such as saponins, alkaloids, polysaccharides, free amino acids, and (poly)phenolics, ginseng has been proved to combat stress, improve the immune system, and maintain optimal oxidative status against aging, as well as assisting medical treatments related to central nervous system disorders, liver diseases, cardiovascular diseases, and cancer [1,3].

The world market of ginseng root and related products is worth USD 2084 million, suggesting a huge production and demand for ginseng products [2], and therefore, multiple studies at the genome [4], transcriptome [5], and metabolite [5] level have been conducted to understand the biology of this plant.

In addition, efforts have also been made to improve our understanding of ginseng at the protein level by utilizing proteomics approaches. Studies have focused on identifying stress-responsive and ginsenoside biosynthesis-related proteins, while some studies have concentrated on comparing and analyzing proteins from different ginseng parts and species [5–8]. However, a number of these studies used one or two tissues and were based on two-dimensional gel electrophoresis (2-DE) analysis, limiting the comprehensiveness of their proteome data [5,6,9]. Therefore, a systematic proteomics study using a wide range of tissues is necessary to provide a deeper understanding of ginseng.

Protein purification is a crucial step of the sample preparation, guaranteeing sufficient and high-quality proteins for proteome analysis [10]. TCA/acetone, phenol methanol, and methanol/chloroform precipitation methods have been developed for the isolation of plant proteins due to their efficiency in precipitating proteins and simultaneously removing interfering compounds [11]. A recent review [12] suggested that TCA/acetone precipitation displays high efficiency in the isolation of total proteins from a diversity of plant tissues while the phenol/methanol method effectively produces high-quality protein samples; minimizes protein degradation; and removes polysaccharides, ions, and nucleic acids. Besides, a study by Wessel and Flügge [13] pointed out that the methanol/chloroform precipitation can work well with different kinds of proteins, especially hydrophobic proteins, in the presence of detergents and with dilute samples. However, no single extraction method can reap the entire proteomes of a tissue or a plant species. Therefore, the combination of two or more approaches to integrate the strengths of each one for the isolation of proteins has been suggested [8,14]. A recent study by Wu [14] presented a protocol that was the combination of TCA/acetone precipitation and phenol extraction for the successful isolation of proteins from various recalcitrant tissues.

Advancements in proteomics approaches have facilitated the proteome analysis of various plants; however, difficulties in extracting relatively pure ginseng proteins have remained a primary obstacle [15]. Up to now, TCA/acetone method has been extensively used for extracting total ginseng proteins [7] while TCA precipitation and phenol extraction have been moderately employed to isolate ginseng proteins for 2-DE analysis [16]. Nonetheless, the efficiency of these methods has been tested on one or two ginseng tissues only, hindering their wide acceptability in ginseng proteome analysis [7,17,18]. Therefore, the development of a universal ginseng protein isolation method is a prerequisite for high-throughput ginseng proteome analysis.

Here, an attempt was made to first evaluate the efficiency and reproducibility of different protein extraction methods, namely TCA/acetone, TCA/acetone–MeOH/chloroform, phenol–TCA/acetone, and phenol–MeOH/chloroform, followed by utilizing the most effective approach for the comparative proteome analysis (Figure S1). Moreover, an attempt was also made to generate a relatively comprehensive proteome map of ginseng fruit, leaf, root, and shoot using a label-free quantitative proteomics approach (Figure 1). Furthermore, through the significantly modulated proteins, we generated a more comprehensive view of the ginsenoside biosynthesis. This in-depth study provides new insights into the protein complement of different ginseng tissues.

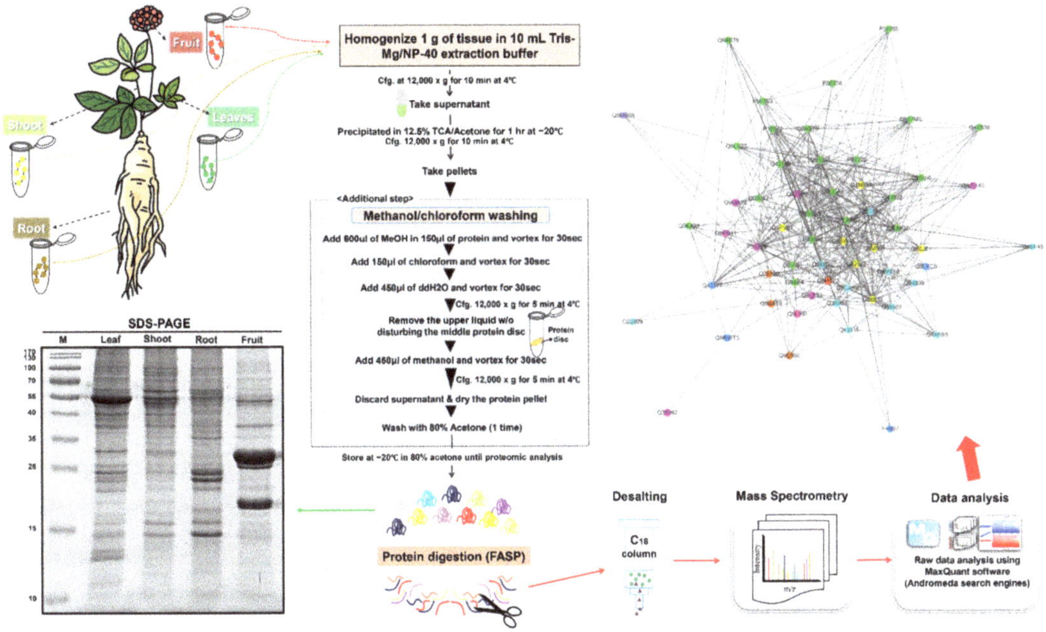

Figure 1. Workflow of the experiment. Ginseng samples were collected and homogenized in Tris–Mg/NP-40 buffer. After centrifugation at 16,000× *g* for 10 min at 4 °C, the supernatant was precipitated in 12.5% TCA/acetone at 4 °C for one hour. Protein pellets, obtained through centrifugation at 16,000× *g* for 10 min at 4 °C, were subsequently washed with methanol/chloroform, followed by trypsin digestion using the FASP method. The digested peptides were desalted and analyzed using a label-free quantitative proteomic approach. The obtained data were analyzed and annotated using MaxQuant, Perseus, and MapMan.

2. Results and Discussion

2.1. Optimization of Ginseng Protein Extraction Method

The medical value of *Panax ginseng* increases with its age, but for its use as medicine and commercial production, a growth period of 4–6 years is often required [18]. Therefore, in order to meet the practicality and enhance the reliability of the current study, fruit, leaf, root, and shoot samples were harvested from various 4-year-old *Panax ginseng* plants and pooled together before analysis. As ginseng leaves contain various natural contaminants such as lipids, saccharides, and various photosynthetic pigments, the extraction of proteins from ginseng leaves is more challenging than from other ginseng parts [19]. Therefore, we used ginseng leaves as a model sample for checking the protein extraction efficiency of four different extraction methods, namely TCA/acetone, TCA/acetone–MeOH/chloroform, phenol–TCA/acetone, and phenol–MeOH/chloroform (Figure S1). Eliminating interfering compounds is an initially crucial step in extracting proteins from plant samples. A review by Wu [12] revealed that finely powdered plant samples can be directly subjected to TCA/acetone but not to phenol. Therefore, to ensure the homogeneity of the samples, the finely ground ginseng samples were first homogenized with Tris–Mg/NP-40 extraction buffer, and the OS was subsequently extracted using four different abovementioned methods.

SDS-PAGE analysis of isolated proteins showed that using the TCA/acetone–MeOH/chloroform method produced more protein bands with a high resolution on the gel than the other tested methods (Figure 2A). Furthermore, the label-free quantitative proteomic analysis led to the identification of 36,145 peptides, corresponding to 4705 protein groups. The average numbers of peptides and unique peptides were 20,383 and 8256, 22,552 and 8919, 22,437 and 7919, and 22,437 and 8981 for TCA/acetone, TCA/acetone–

MeOH/chloroform, phenol–TCA/acetone, and phenol–MeOH/chloroform, respectively (Table S1; Figure S2A). The average sequence coverage was 13.24, 14.99, 16.40, and 15.26 (%) for TCA/acetone, TCA/acetone–MeOH/chloroform, phenol–TCA/acetone, and phenol–MeOH/chloroform, respectively (Table S1; Figure 2B). Filtering out by applying a cut-off value of 75% within three technical replicates of each sample led to the identification of 3049 proteins (Figure 2B), of which 2449, 2422, 2245, and 1883 proteins were identified when using phenol–MeOH/chloroform, TCA/acetone–MeOH/chloroform, TCA/acetone, and phenol–TCA/acetone extractions, respectively (Table S1; Figure 2B). Isoelectric point (Figure S2C), molecular weight (Figure S2D), and hydrophobicity (GRAVY) (Figure S2E) of most of these proteins were between 20 and 160 kDa, 4 and 12, and −2 and 1, respectively. Subcellular prediction analysis using CELLO2GO web-based software showed a relatively similar distribution of proteins isolated using the four different methods over 11 locations (Figure S2F). Since the numbers of proteins identified by each method were relatively similar, there was not a large difference in the molecular weight, isoelectric point, and hydrophobicity of proteins among the tested approaches.

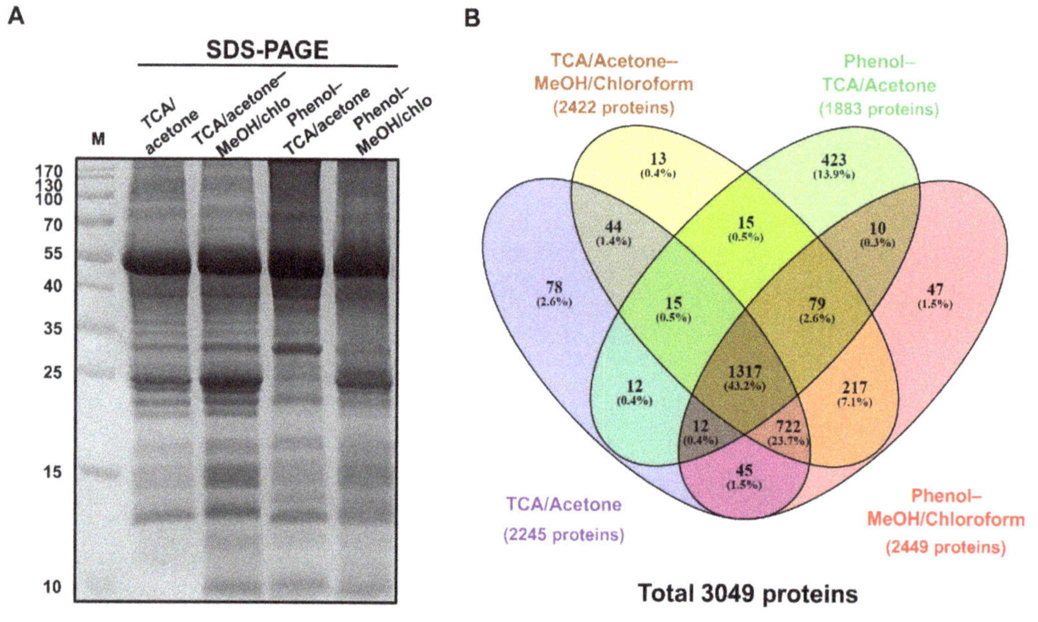

Figure 2. SDS-PAGE of proteins isolated from ginseng leaf using TCA/acetone, TCA/acetone–MeOH/chloroform, phenol–TCA/acetone, and phenol–MeOH/chloroform methods (**A**). Venn diagram showing the distribution of proteins isolated from ginseng leaves using four different protein extraction methods (**B**).

Common methods based on TCA/acetone precipitation and phenol extraction, which have successfully isolated ginseng proteins from one or two ginseng tissues for 2-DE analysis [17], might be no longer effective in extracting a wide range of ginseng tissues for label-free quantitative proteomic analysis. Alternatively, the idea of combining two extraction methods to incorporate the strengths of every single one for isolating proteins from different ginseng tissues has shown considerable potential. Particularly, a recent study by Li [8] showed that the combination of GdnHCl with methanol/chloroform precipitation led to improved extraction of proteins from ginseng cauline leaves, compared with GdnHCl lysate and Tris–HCl lysate methods. However, this combination still displayed certain limitations as the SDS-PAGE quality and the number of identified proteins were relatively modest [8]. In the current study, the TCA/acetone–MeOH/chloroform method maintained the advantages of both TCA/acetone precipitation, which allows extraction of total proteins [20], and MeOH/chloroform extraction, which efficiently removes remaining contaminants (especially lipids) without clear quantitative loss of proteins [13], resulting in a better extraction of ginseng proteins as observed on the SDS-PAGE (Figure 2A) and by the number of identified proteins (Figure 2B). An extraction method is considered to be effective when it reproducibly attains the most comprehensive proteome and simultaneously minimizes protein degradation and contaminants [20]. Therefore, although the phenol–MeOH/chloroform method produced a slightly higher number of identified proteins than the TCA/acetone–MeOH/chloroform, the poor gel profile and high toxicity to humans of phenol made it an unsuitable choice for our subsequent analysis. The TCA/acetone–MeOH/chloroform could produce a clear gel profile and a higher number of identified proteins, compared with the other tested methods; therefore, it was utilized to extract proteins from ginseng tissues for global identification.

2.2. Label-Free Quantification Using Four Different Ginseng Tissues

The LC-MS/MS analysis led to the identification of a total of 39,275 peptides, which corresponded to 4764 protein groups. A cut-off value of 75% was applied within four technical replicates of each tissue sample, leading to the identification of 3073 proteins (Figure 3A). Of these, 1434, 1958, 2137, and 2211 proteins were found to be in the fruit, root, leaf, and shoot samples, respectively. Subsequently, multiple ANOVA tests, controlled by Benjamini–Hochberg FDR threshold of 0.05, were applied on the identified proteins to demarcate 2604 differentially regulated proteins with fold change more than 1.5 (Table S2; Figure 3B). While 1179 proteins were common in all four tissues, 287, 18, 132, and 39 proteins were common in the leaf/shoot, leaf/root, shoot/root, and root/fruit samples, respectively (Figure 3B).

Sequential multi-scatter plot and principal component analysis (PCA) were thereafter performed to analyze the correlation and variations among the four ginseng tissues (Figure 3C,D). The PCA plot illustrates a clear separation among all of the four sample sets, demarcating the distinctness of the differential tissue proteomes (Figure 3C). While the root and leaf samples were separated in PC1 accounting for 42.8% of the total variation, the shoot and fruit samples were resolved in PC2 that accounted for 26.7% of the total variation. Furthermore, the multi-scatter plot with the Pearson correlation coefficients of the technical replicates in each sample set ranging from 0.931 to 0.965 indicated a strong correlation among the technical replicates of the same samples (Figure 3D).

Previously, ginseng proteomic studies were based primarily on 2-DE analysis, leading to the identification of a relatively low number of proteins (about 1000 proteins) in these studies [6,16]. The development of the shotgun techniques, coupled with advancements in MS, has significantly improved the number of proteins identified from various plant tissues [15]. A recent study combined GdnHCl with methanol/chloroform precipitation to extract proteins from ginseng cauline leaves, leading to the identification of 1366 proteins [8]. However, by applying basic fractionation, the number of proteins isolated using this method increased significantly to 3608 proteins [8]. In the current study, by using the TCA/acetone–MeOH/chloroform for protein extraction, followed by a label-free quantitative proteomic analysis, we successfully identified 4764 proteins from ginseng fruit, leaf, root, and shoot (Figure 3A). This is the first study on ginseng in which such a high number of identified proteins is reported from a wide range of tissues using only one extraction method without fractionation. However, further investigations comparing this method with different MS sample preparations such as single-pot solid-phase-enhanced sample preparation (SP3) [21], in-StageTip digestion (iST) [22], and the suspension trapping (S-Trap) filter [23] using various ginseng tissues might provide a deeper understanding of the sample preparation for ginseng proteomics.

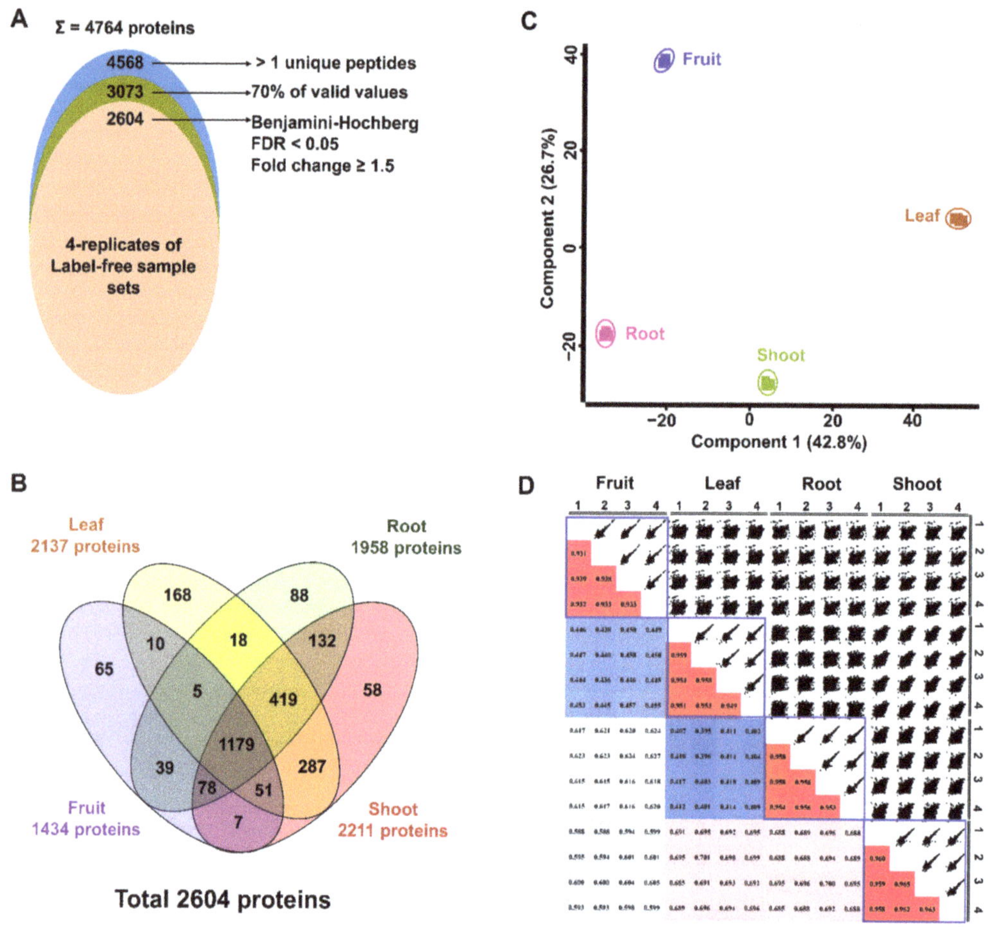

Figure 3. A total of 4764 protein groups were identified in this study. Out of these, 2604 proteins were significantly modulated among four tissues (**A**). Venn diagram showing the distribution of 2604 proteins (**B**). Principle component analysis of the differentially regulated proteins (**C**). Multi-scatter plots of label-free protein intensities between different technical replicates of the samples with Pearson correlation coefficient values (**D**).

2.3. *Functional Classification of Identified Proteins*

2.3.1. Functions of Commonly Identified Proteins among Four Tissues

For the further investigation of the significantly modulated proteins, we performed hierarchical clustering analysis (HCA) which separated all the identified proteins into four clusters based on $\log_2$ of the z-score normalized intensities among the technical replicates of each sample (Figure 4A). While Cluster 1 consisted of 265 proteins with high abundance in the shoot, Cluster 2 included 1104 proteins with increased abundance in the leaf. Clusters 3 and 4 contained 448 and 787 proteins, which were maximally accumulated in the fruit and root, respectively (Figure 4B)

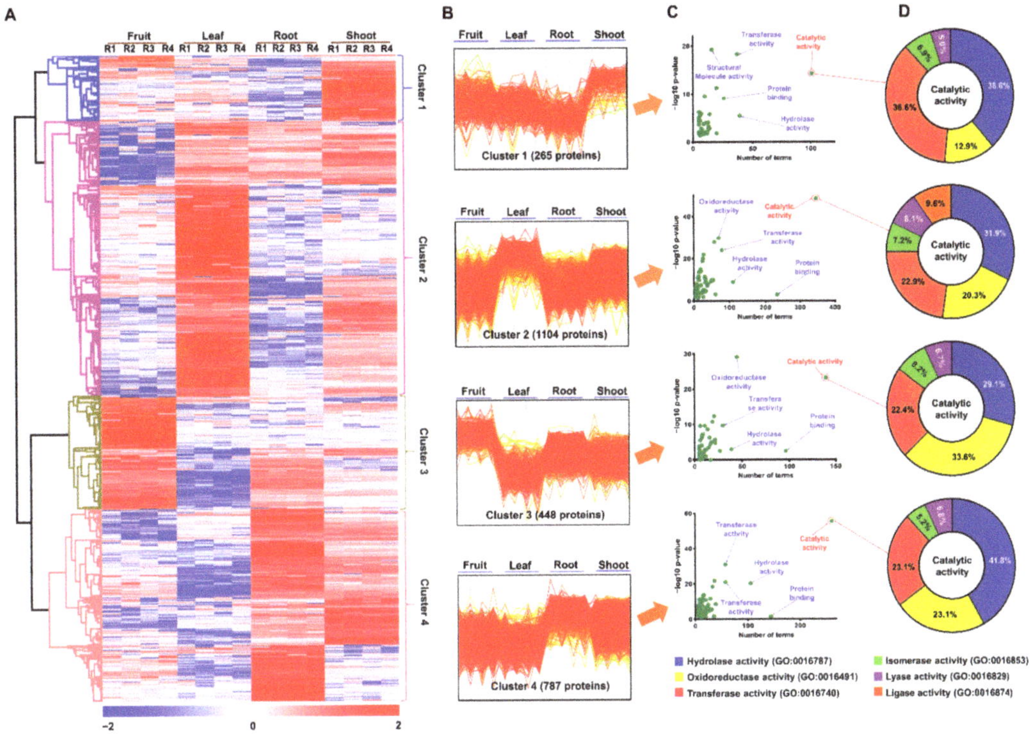

Figure 4. Expression profile of 2604 significantly modulated proteins identified by label-free quantitative proteome analysis. Hierarchical clustering (**A**) was carried out by Perseus software. Expression patterns of 4 protein clusters (**B**). Gene ontology analysis was performed for functional annotation of proteins in four clusters using AgriGO (ver. 2.0) (**C**,**D**).

For functional annotation of the identified proteins, we carried out gene ontology (GO) enrichment analysis via AgriGO through homolog identification of *P. ginseng* proteins in *A. thaliana* (TAIR10) [24] (Table S3). Notably, in the GO classification of molecular function, catalytic activity was the largest GO term in Clusters 1, 2, 3, and 4 with the involvement of 101 (38.1%), 347 (31.4%), 140 (31.3%), and 262 (33.3%) proteins, respectively (Figure 4C). Hydrolase activity, oxidoreductase activity, and transferase activity were the three main subgroups of catalytic activity found in all of the four clusters, while ligase activity was found in only Cluster 2 (Figure 4D). The metabolism overview of MapMan analysis indicated that most of the proteins related to the catalytic activity in Cluster 1 were involved in the biosynthesis of methionines, cellulose and precursors, phospholipids, flavonoids, and isoprenoids, which are more active in the shoot [25]. Meanwhile, the proteins associated with catalytic activity in Cluster 2 were majorly involved in the biosynthesis of various amino acids, photosynthesis, nucleotide metabolism (synthesis of purines and pyrimidines), CHO metabolism (synthesis of starch and sucrose), and the synthesis of secondary metabolites (flavonoids, isoprenoids, and phenylpropanoids), which take place predominantly in the leaf of plants [25]. By contrast, most of the proteins that belonged to the catalytic activity in Clusters 3 and 4 were mainly associated with the degradation of different molecules (such as amino acids, nucleotides, lipids, starch, sucrose, and flavonoids), glycolysis, and tricarboxylic acid cycle, which commonly occur in the fruit and root of plants [25] (Table S3). The result of MapMan analysis is consistent with the result from the HCA (Figure 4A,B).

2.3.2. Functions of Tissue-Specific Proteins

Among the 2604 identified proteins, 65, 168, 88, and 58 proteins were specifically identified in the fruit, leaf, root, and shoot, respectively (Table S2; Figure 5A). For understanding the functional significance of these proteins, the metabolic overview and cell function were analyzed using MapMan (Figure 5B), followed by an interactome analysis using STRING (v. 11.0) (Figure 5C).

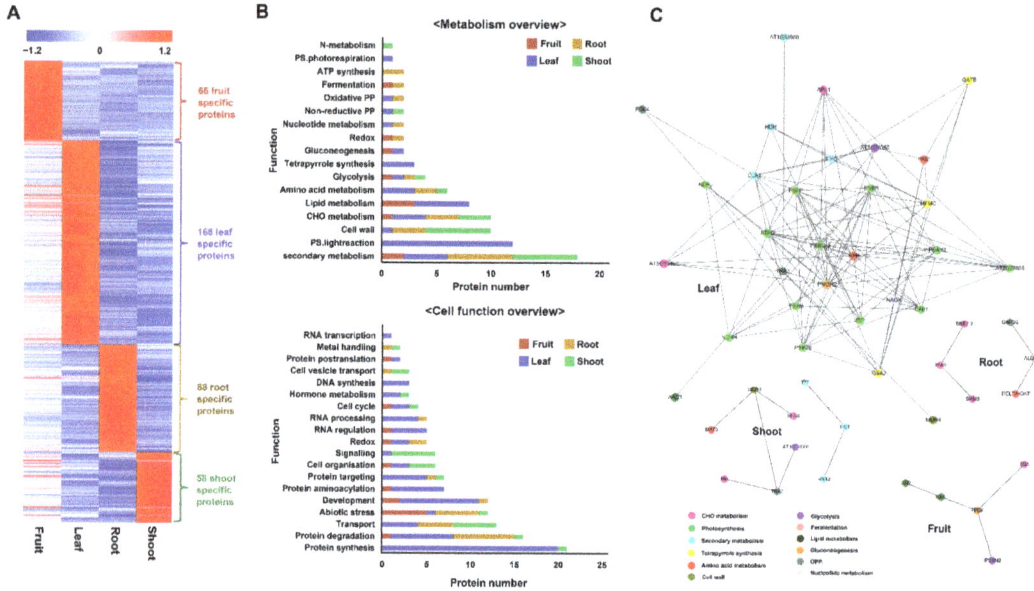

Figure 5. Overview of tissue-specific proteins (A). Functional annotation of tissue-specific proteins was carried out using MapMan (B). Protein–protein interaction networks of tissue-specific proteins related to metabolic processes were analyzed using STRING (ver. 11.0), coupled with Cytoscape (ver. 3.7.2) (C).

The metabolism overview of MapMan analysis revealed that among 65 proteins specific to the fruit, 10 proteins were classified into six metabolic groups, of these, lipid metabolism was the largest group, containing acyl-(acyl-carrier-protein) desaturase and 3-ketoacyl-acyl carrier protein synthase I involved in the fatty acid synthesis and elongation (Table S4). For the leaf-specific proteins, 13 groups accounting for 37 proteins were categorized; of these, the photosynthesis process with proteins associated with the light reaction of photosystems I and II and photorespiration was the major metabolism. Regarding the 88 root-specific proteins, 10 metabolic groups accounting for 21 proteins were sorted, of which secondary metabolism was the largest, containing six proteins. Differently, cell wall with six proteins associated with the formation and modification of the cell wall was the most dominant metabolic group of shoot-specific proteins (Table S4).

Furthermore, the cell function of MapMan analysis showed that six groups accounting for 14 fruit-specific proteins were categorized, of these, abiotic stress was the largest group, with five proteins. Protein synthesis, protein aminoacylation, and protein targeting were the most dominant groups associated with 38 proteins specific to the leaf. Meanwhile, the largest groups of proteins specific to the root were protein degradation and biotic stress. Transport and signaling were the most predominant groups related to 10 proteins included exclusively in the shoot (Figure 5B).

To have a global view of all possible interactions among specific proteins that were involved in the metabolisms of each sample set, protein–protein interaction networks were created. After STRING functional enrichment analysis, a total of 5, 29, 6, and 9 proteins uniquely stemming from the fruit, leaf, shoot, and root, respectively, showed interactions on the network (Figure 5C). Among these, photosynthesis was the primary metabolism influencing various activities in the leaf, while CHO metabolism and secondary metabolism were predominant metabolisms in the root and shoot, respectively. Gluconeogenesis was the metabolism linked to different metabolic activities in the fruit.

Tissue-specific proteins are important factors contributing to differences in anatomical characteristics and physiological functions among living tissues. Therefore, some studies have been conducted to identify and characterize tissue-specific proteins in various plants [26,27]. On *P. ginseng*, few studies have determined proteins specific to ginseng leaves and roots. A study by Seung [28] successfully identified and characterized root-specific RNase-like proteins (GMPs) in roots of wild ginseng, which might work as vegetative storage proteins promoting its survival in the natural habitat. Furthermore, Li [8] highlighted 878 and 1754 proteins specific to the roots and cauline leaves, respectively. The author also revealed that the cauline leaf-specific proteins were primarily associated with photosynthesis and related energy conversion while the proteins specific to the root were involved in the biosynthesis and modification of biomacromolecules [8]. The functional annotation and molecular processes highlighted in the leaf and root in the current study are were relatively consistent with the previous report [8]. However, as the present study was performed on all fruit, leaf, root, and shoot tissues, the number of overlapped proteins was significantly increased, while the number of tissue-specific proteins was also highlighted.

2.4. Decoding the Proteome Modulations in Association with Ginsenoside Biosynthesis

Ginsenosides, a well-known triterpenoid saponin type in the ginseng plant, are natural secondary metabolites of ginseng, exhibiting a diversity of medicinal effects [1]. Recently, more than 180 ginsenosides have been identified and categorized into three main types: protopanaxadiol (PPD) type, protopanaxatriol (PPT) type, and oleanane type, with the first two commonly existing in *P. ginseng* [29,30]. The biosynthesis of ginsenosides can be divided into three main stages: (1) the biosynthesis of the precursor isopentenyl pyrophosphate (IPP) and dimethylallyl pyrophosphate (DMAPP) through the MVA and MEP pathways, (2) the conversion of IPP and DMAPP into 2,3-oxidosqualene, and (3) the formation of ginsenosides and sterols from 2,3-oxidosqualene [31].

It is a fact that ginsenosides are unevenly distributed in different parts of ginseng. A few studies have confirmed that the total ginsenoside content of the ginseng leaf and fruit was higher than that of the root [32], yet there have been no studies elucidating the molecular mechanism for this difference. This study, for the first time, revealed a relatively comprehensive proteome profile of the ginseng fruit, leaf, root, and shoot, providing a new understanding of the molecular basis for the variation in the ginsenoside content among the four tissues. Our result identified a total of 67 proteins associated with the ginsenoside biosynthesis (Table S5). Of these, acetyl-CoA C-acetyltransferase (ACCT), hydroxymethylglutaryl-CoA synthase (HMGS), and diphosphomevalonate decarboxylase (MVD) related to the MVA pathway were more abundant in the shoot. Nine proteins associated with the MEP pathway, namely one protein of 1-deoxy-D-xylulose-5-phosphate synthase family (DXS), two proteins of 1-deoxy-D-xylulose 5-phosphate reductoisomerase family (DXR), one protein of 2-C-methyl-D-erythritol 4-phosphate cytidylyltransferase family (ispD), one protein of 4-diphosphocytidyl-2-C-methyl-D-erythritol kinase family (ispE), one protein of 2-C-methyl-D-erythritol 2,4-cyclodiphosphate synthase family (ispF), two proteins of (E)-4-hydroxy-3-methylbut-2-enyl-diphosphate synthase family (ispG), and one protein of 4-hydroxy-3-methylbut-2-enyl diphosphate reductase family (ispH), showed higher abundance in the leaf. In addition, 28 UGTs were identified, of which 5, 12, 6, and 5 proteins were highly accumulated in the fruit, leaf, root, and shoot, respectively. Furthermore, 22 CYP450s were also identified, of which 4, 5, 6, and 7 proteins were highly abundant in the fruit, leaf, root, and shoot, respectively. Proteins such as isopentenyl-diphosphate delta-isomerase (IDI) and beta-amyrin synthase (β–AS) were also identified with high abundance in the ginseng leaf sample, while cycloartenol synthase (CAS) was highly accumulated in shoot and leaf samples. Besides, 2, 3, 6, and 1 proteins related to the biosynthesis of ginsenosides were found to be specific to the fruit, leaf, root, and shoot respectively (Figure 6; Table S5).

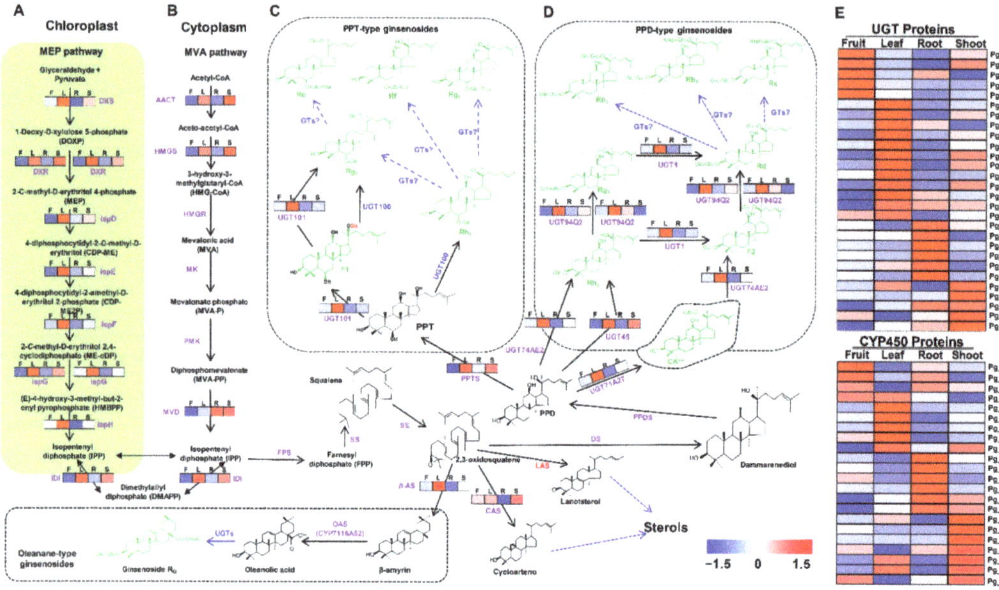

Figure 6. Expression profiles of identified proteins involved in the MEP (**A**) and the MVA (**B**) pathways. PPT-type (**C**) and PPD-type (**D**) ginsenosides. The abundance of UGTs and CYP450s related to ginsenoside biosynthesis (**E**). Color codes represent abundance patterns of identified proteins wherein red and blue indicate a high and low abundance of proteins in particular tissues, respectively. F—fruit, L—leaf, R—root, S—shoot.

Biosynthesis of IPP and DMAPP is essential to most living organisms. Depending on species, these precursors of isoprenoids can be synthesized through only the MVA pathway like some archaea and eukaryotes or only the MEP pathway like most bacteria or both of these pathways in most photosynthetic eukaryotes [33]. The MVA pathway is responsible for the conversion of acetyl-CoA into IPP and DMAPP, while the MEP pathway produces the IPP and DMAPP from glyceraldehyde and pyruvate (Figure 6A,B). In *P. ginseng*, studies based on phytochemical and inhibitor experiments, transcriptome, and genome sequencing revealed that the biosynthesis of IPP and DMAPP, the precursors of ginsenosides, has the involvement of both MVA and MEP pathways [24,31,34]. In addition, by conducting deep RNA sequencing on the 1–5-year-old ginseng root samples and five different tissues Xue [35] not only determined most genes related to the MVA and MEP pathways but also pointed out the relative expression of these genes among different aging samples and tissues. However, these genes are not directly involved in the reactions of the MVA and MEP pathways, but their products (enzymes) are. This means that the abundance pattern of these enzymes in the fruit, leaf, root, and shoot of ginseng might be a deciding factor for the differences in the biosynthesis of the IPP and DMAPP and subsequently of tissue specific CYP450s and UGTs, differentiating the types and concentration of ginsenosides in various parts of the ginseng plant [29,36]. In the present study, the higher abundance of proteins related to the MVA pathway (ACCT, HMGS, and MDV) was observed in the shoot, while all proteins associated with the MEP pathway (1 DXS, 2 DXR, ispD, ispE, ispF 2 ispG, and ispH) showed the highest abundance in the leaf. These findings suggest that the biosynthesis of ginsenosides in the shoots might have the major involvement of the MVA pathway, while the biosynthesis of ginsenosides in the leaves might rely majorly on the MEP pathway. The findings are logically suitable to the plastic location of the MEP pathway and in concordance with the results from the research of Xue [35] that all genes related to the MEP pathway had higher levels of gene expression in the leaf than the root of *P. ginseng*.

The present study also highlights the increased abundance of cycloartenol cyclase (CAS) in the shoot and β-amyrin synthase (β-AS) in the leaf. The precursors, IPP and DMAPP, are converted into several metabolic intermediates and then to 2,3-oxidosqualene, which in turn undergoes variable cyclization by oxidosqualene cyclases (OSCs), hydroxylation by CYP450s, and glycosylation by UGTs to form an array of ginsenosides (Figure 6) [33]. The formation of sterols from 2,3-oxidosqualene is catalyzed by two different OSCs, namely lanosterol synthase (LAS) and CAS, while β-AS is involved in the production of oleanane. The formation of sterols from 2,3-oxidosqualene is catalyzed by two different OSCs, namely lanosterol synthase (LAS) and CAS, while β-AS is involved in the production of oleanane [29,37]. Our findings led to speculation that the biosynthesis of sterols might be more active in the shoot samples, whereas the pathways involving the production of the triterpenoid oleanane are differentially active in leaf tissues.

The PPD- and PPT-type saponins make up a majority of ginsenosides in *P. ginseng*. The biosynthesis of PPD- and PPT-type saponins occurs when 2,3-oxidosqualene is converted into dammolarenediol by dammolarenediol synthase (DS), then into protopanaxadiol (PPD) by cytochrome P450 CYP716A47 (PPDS) before undergoing one more hydroxylation catalyzed by cytochrome P450 CYP716A53v2 (PPTS) to form protopanaxatriol (PPT). The PPD and PPT are subsequently glycosylated by different UGTs to form a diversity of PPD- and PPT-type ginsenosides [33,35]. In our study, six UGTs related to the biosynthesis of PPD-type saponins were found to be differentially accumulated in the leaf tissues, including UGT71A27 (Pg_S6256.3) catalyzing the biosynthesis of compound K from PPD, UGT45 (Pg_S5977.4) converting PPD into Rh2, UGT47AE2 (Pg_S4174.7) catalyzing the biosynthesis of Rh2 from PPD and the biosynthesis of F2 from compound K, UGT94Q2 (Pg_S6708.3 and Pg_S2289.21) catalyzing the conversion of ginsenoside Rh2 to ginsenoside Rg3 and triggering the biosynthesis of ginsenoside Rd from ginsenoside F2, and UGT1 (Pg_S4493.1) triggering C20–OH glycosylation of ginsenoside Rg3 to produce ginsenoside Rd and converting Rh2 to F2 [38,39]. Besides, two proteins participating in the formation of PPT-type saponins comprising PPTS (Pg_S1770.12) catalyzing the formation of PPT from PPD and UGT101 (Pg_S4157.4) catalyzing the biosynthesis of ginsenoside F1 from PPT and the conversion of ginsenoside F1 to ginsenoside Rg1 also displayed a high abundance in the leaf samples [40]. The increased abundance of these proteins in the leaf demonstrated that the biosynthesis of PPD- and PPT-type ginsenosides in the leaf tissues is promisingly higher than in the fruit, root, and shoot. Furthermore, the appearance of CYP450s and UGTs specific to the fruit, leaf, root, and shoot may explain the existence of ginsenosides that are specific to each tissue. These results are consistent with a previous study by Kang [32] showing that the total ginsenoside content in the leaf of *P. ginseng* is 12 times higher than that of the main root and that the leaves of *P. ginseng* contain a high amount of ginsenosides Rh1 and Rb3, whereas its main roots have a higher quantity of ginsenosides Rb1 and Rc.

3. Materials and Methods

3.1. Plant Materials

P. ginseng cv. Chunpoong was grown in a controlled growth chamber at the Department of Ginseng Research, National Institute of Horticultural and Herbal Science (NIHHS), Rural Development Administration (RDA), Eumseong, Korea (latitude 36°94, longitude 127°75). The average temperature and humidity of the greenhouse were maintained at 22.5 ± 2.5 °C and 50 ± 10%, respectively. Four-year-old leaves, shoots, roots, and fruits from five different ginseng plants were harvested and immediately stored at −80 °C for analysis.

3.2. Total Protein Extraction

The leaf samples (1 g) of 4-year-old plants from five different plants were pooled together, homogenized in 10 mL of Tris–Mg/NP-40 extraction buffer (0.5 M Tris–HCl, 2% (*v*/*v*) NP-40, 20 mM $MgCl_2$, 2% (*v*/*v*) β-mercaptoethanol, and 2% (*w*/*v*) polyvinylpolypyrrolidone, pH 8.3) and subjected to centrifugation at 16,000× *g* for 10 min at 4 °C. The OS was used for the subsequent extractions using trichloroacetic acid (TCA)/acetone, TCA/acetone–MeOH/chloroform, phenol–TCA/acetone, and phenol–MeOH/chloroform methods.

The TCA/acetone method was carried out as described previously [7,41]. Briefly, the OS was incubated with 4 volumes of 12.5% (*w*/*v*) TCA/acetone containing 0.07% (*v*/*v*) β-mercaptoethanol for 1 h at −20 °C and then centrifuged at 16,000× *g* for 10 min at 4 °C to obtain protein pellets. The TCA/acetone–MeOH/chloroform method was performed as described previously [13]. Briefly, the OS was first extracted using the TCA/acetone method, and the obtained proteins were then mixed with 4 volumes of methanol, then an equal volume of chloroform, and then 3 volumes of the sterile distilled water before being centrifuged at 16,000× *g* for 5 min to collect protein pellets. For the phenol–TCA/acetone, the OS was vigorously mixed with the same volume of saturated phenol and separated into two phases through centrifugation at 3500× *g* for 5 min at 4 °C. The lower phase containing proteins was incubated with 4 volumes of 12.5% (*w*/*v*) TCA/acetone containing 0.07% (*v*/*v*) β-mercaptoethanol for 1 h at −20 °C before being centrifuged at 16,000× *g* for 10 min at 4 °C to collect protein pellets. The phenol–MeOH/chloroform method was performed similarly to the phenol–TCA/acetone method with a slight difference: the lower phase yielded from the phenol extraction was incubated with 4 volumes of methanol and 1 volume of chloroform for 1 h at −20 °C prior to being centrifuged at 16,000× *g* for 10 min at 4 °C to collect protein pellets. The resulting pellets of these methods were finally washed with 80% acetone containing 0.07% (*v*/*v*) β-mercaptoethanol and then stored at −20 °C until further analysis.

The extraction of total proteins from ginseng fruits, leaves, roots, and shoots was conducted using the TCA/acetone–MeOH/chloroform as described above.

3.3. Label-Free Quantitative Proteome Analysis Using Q-Exactive Mass Spectrometer

Label-free quantitative proteomic analysis of ginseng fruit, leaf, root, and shoot samples was performed as described previously [7]. Briefly, the digested peptides, obtained from the in-solution trypsin digestion using the FASP method, coupled with a 30k spin filter (Merck Millipore, Darmstadt, Germany) [42], were desalted using C_{18} column (Oasis HLB 1 cc Vac Cartridge, 30 mg sorbent per cartridge, 30 μm, 100/pk, WAT094225, Waters, Ireland). Subsequently, the desalted peptides were dissolved in solvent A (water/ACN, 98:2 *v*/*v*; 0.1% formic acid), followed by the reversed-phase chromatography separation utilizing a UHPLC Dionex UltiMate 3000 (Thermo Fisher Scientific, Madison, WI, USA) instrument [43]. In the UHPLC, the sample was first trapped with an Acclaim PepMap 100 trap column (100 μm × 2 cm, nanoViper C18, 5 μm, 100 Å) and then washed with 98% solvent A for 6 min at a flow rate of 6 μL/min prior to being separated in an Acclaim PepMap 100 capillary column (75 μm × 15 cm, nanoViper C18, 3 μm, 100 Å) at a flow rate of 400 nL/min. As the UHPLC was running, the LC analytical gradient was increased gradually from 2% to 35% solvent B during the first 90 min and then from 35% to 95% in the next 10 min; finally, 90% and 5% solvent B were run for 5 min and 15 min, respectively. The integration of liquid chromatography–tandem mass spectrometry (LC-MS/MS) with an electrospray ionization source to the quadrupole-based mass spectrometer Q Exactive Orbitrap High-Resolution Mass Spectrometer (Thermo Fisher Scientific, Madison, WI, USA) allowed the resulting peptides to be electro-sprayed through a coated silica emitted tip (PicoTip emitter, New Objective, Massachusetts, USA) at an ion spray voltage of 2000 eV, generating the MS spectra with a resolution of 70,000 (200 *m*/*z*) in a mass range of 350–1800 *m*/*z*. For ion accumulation, 100 ms was set as the maximum injection time. The eluted samples, measured in a data-dependent mode for the 10 most abundant peaks

(Top15 method) with the high mass accuracy Orbitrap after ion activation/dissociation with Higher Energy C-trap Dissociation (HCD) at 27 collision energy in a 100–1650 m/z mass range, were used for MS/MS events (resolution of 17,500) [43]. The obtained proteomics data have been deposited to the ProteomeXchange Consortium via the PRIDE [44] partner repository with the dataset identifier PXD022914.

3.4. Data Processing Using MaxQuant Software

The MS spectra of ginseng fruit, leaf, root, and shoot samples were cross-referenced against the genome sequencing database (http://ginsengdb.snu.ac.kr/ (accessed on 1 April 2021)) maintained by Seoul National University [4]. Label-free quantification (LFQ) was performed using default precursor mass tolerances set by Andromeda, with 20 ppm for the first search and 4.5 ppm for the following ones. The search of the LFQ data was based on 0.5 Da of a product mass tolerance with a maximum of two missed tryptic digestions. Carbamidomethylation of cysteine residues was chosen for the fixed modifications, while acetylation of lysine residues and oxidation of methionine residues were selected for additional modifications. The false discovery rate (FDR), which was set at 1% for peptide identifications, was determined based on a reverse nonsense version of the original database.

The data processing for LFQ was performed using MaxLFQ, available as a part of the MaxQuant suite [45]. Subsequently, Perseus software (v. 1.6.14.0) [46] was employed for further statistical and graph analyses. The Perseus software enables performing missing value imputation of protein intensities from a normal distribution (width: 0.3, downshift: 1.8); HCA; and multiple-sample test (one-way ANOVA), controlled by Benjamini–Hochberg method based on an FDR threshold of 0.05, for identifying the significant differences in the protein abundance among the ginseng fruit, leaf, root, and shoot samples. Functional annotation of the identified proteins was undertaken, employing MapMan and AgriGO (v. 2.0) [47,48]. The interaction networks of differentially regulated proteins were predicted by STRING analysis (v. 11.0), coupled with Cytoscape (v. 3.7.2.0) [49]. Subcellular localization analysis was performed using CELLO2GO web-based software [50].

4. Conclusions

P. ginseng is a precious plant with immense medical and economic value; however, our knowledge about ginseng proteomics is still scanty. Here, a label-free quantitative proteomic analysis was employed to generate a comprehensive proteome map of the ginseng fruit, leaf, root, and shoot. To optimize the extraction of ginseng proteins, we first compared four different protein extraction methods, and we finally adopted the TCA/acetone–MeOH/chloroform method for further analysis. The increased abundance of most of the proteins related to the ginsenoside biosynthesis illustrated that the biosynthesis of ginsenosides in the leaves is probably higher than in the fruit, root, and shoot, while the tissue-specific CYP450s and UGTs might elucidate the existence of proteins specific to each tissue. In addition, the increased abundance of CAS in the shoot and β-AS in the leaf leads to speculation that the biosynthesis of sterols might be more active in the shoot samples, whereas the production of oleanane-type ginsenosides might be more active in the leaf tissues. Taken together, the results of the current study show that this efficient and reproducible method for the ginseng protein isolation, which plays a vital role in facilitating the development of ginseng proteomics, provides a relatively comprehensive picture of the ginsenoside biosynthesis and new insights into the protein complement of different ginseng tissues.

Supplementary Materials: The following are available online at https://www.mdpi.com/article/10.3390/plants10071409/s1, Figure S1: Diagram showing the protein extraction procedures using different protein extraction methods. Figure S2: In-depth proteome analysis of proteins isolated using four different protein extraction methods, namely, TCA/acetone, TCA/acetone–MeOH/chlo, Phenol–TCA/acetone, Phenol–MeOH/chlo. Table S1: Label-free proteomic analysis of proteins isolated from ginseng leaves using TCA/acetone, TCA/acetone–MeOH/chlo, Phenol–TCA/acetone, and Phenol–MeOH/chlo methods, Table S2: Label-free proteomic analysis of proteins isolated from ginseng fruit, leaf, root, and shoot tissues using TCA/acetone–MeOH/chlo, Table S3: Gene ontology (GO) enrichment analysis of proteins isolated from ginseng fruit, leaf, root, and shoot tissues, Table S4: Cell function and metabolism overview of proteins specific to fruit, leaf, root, and shoot using MapMan software, Table S5: List of proteins involved in ginsenoside biosynthesis.

Author Contributions: Conceptualization, S.-T.K. and Y.-J.K.; Methodology, S.-W.K., C.-W.M., H.W.S., J.-Y.J., R.G. and T.V.N.; Software, C.-W.M., R.G. and T.V.N.; Validation, S.-T.K., Y.-J.K. and T.V.N.; Formal Analysis, T.V.N.; Investigation, D.R., I.-H.J., W.-J.H., K.-H.J., S.K. and T.V.N.; Resources, G.-H.L., J.-W.J. and T.V.N.; Data Curation, C.-W.M., W.-J.H., R.G. and T.V.N.; Writing—Original Draft Preparation, T.V.N.; Writing—Review & Editing, S.-W.K., C.-W.M., D.R., I.-H.J., W.-J.H., K.-H.J., S.K. and R.G.; Visualization, G.-H.L., J.-W.J. and T.V.N.; Supervision, S.-T.K. and Y.-J.K.; Project Administration, S.-T.K. and Y.-J.K.; Funding Acquisition, S.-T.K. All authors have read and agreed to the published version of the manuscript.

Funding: This work was carried out with the support of the "Cooperative Research Program for Agriculture Science & Technology (Project No. PJ01492202)" Rural Development Administration, Republic of Korea.

Institutional Review Board Statement: Not applicable.

Informed Consent Statement: Not applicable.

Data Availability Statement: The data presented in this study are available in the article and Supplementary Materials.

Conflicts of Interest: The authors declare no conflict of interest.

References

1. Wee, J.J.; Park, K.M.; Chung, A.S. Biological activities of ginseng and its application to human health. In *Herbal Medicine: Biomolecular and Clinical Aspects*, 2nd ed.; Benzie, I.F.F., Wachtel, G., Eds.; CRC Press/Taylor & Francis: Oxfordshire, UK, 2011; pp. 157–174. [CrossRef]
2. Baeg, I.H.; So, S.H. The world ginseng market and the ginseng (Korea). *J. Ginseng Res.* **2013**, *37*, 1–7. [CrossRef] [PubMed]
3. Ru, W.; Wang, D.; Xu, Y.; He, X.; Sun, Y.E.; Qian, L.; Zhou, X.; Qin, Y. Chemical constituents and bioactivities of *Panax ginseng* (C. A. Mey.). *Drug Discov. Ther.* **2011**, *9*, 23–32. [CrossRef]
4. Jayakodi, M.; Choi, B.S.; Lee, S.C.; Kim, N.H.; Park, J.Y.; Jang, W.; Lakshmanan, M.; Mohan, S.V.G.; Lee, D.Y.; Yang, T.J. Ginseng Genome Database: An open-access platform for genomics of *Panax ginseng*. *BMC Plant Biol.* **2018**, *18*, 1–7. [CrossRef]
5. Lee, Y.S.; Park, H.S.; Lee, D.K.; Jayakodi, M.; Kim, N.H.; Koo, H.J.; Lee, S.C.; Kim, Y.J.; Kwon, S.W.; Yang, T.J. Integrated transcriptomic and metabolomic analysis of five *Panax ginseng* cultivars reveals the dynamics of ginsenoside biosynthesis. *Front. Plant Sci.* **2017**, *8*, 1048. [CrossRef]
6. Lei, X.; Wang, Z.; Song, J.; Liang, S.; Yao, L.; Hou, Z.; Wang, Y. Comparative Proteomic Analysis of *Panax ginseng* C. A. Meyer × *Panax quinquefolius* L. Leaves and Parental Lines. In *Active Ingredients from Aromatic and Medicinal Plants*, 1st ed.; El-Shemy, H., Ed.; InTechOpen: Rijeka, Croatia, 2017; pp. 127–142. [CrossRef]
7. Kim, S.W.; Gupta, R.; Min, C.W.; Lee, S.H.; Cheon, Y.E.; Meng, Q.F.; Jang, J.W.; Hong, C.E.; Lee, J.Y.; Jo, I.H.; et al. Label-free quantitative proteomic analysis of *Panax ginseng* leaves upon exposure to heat stress. *J. Ginseng Res.* **2019**, *43*, 143–153. [CrossRef]
8. Li, X.; Cheng, X.; Liao, B.; Xu, J.; Han, X.; Zhang, J.; Lin, Z.; Hu, L. Spatial protein expression of *Panax ginseng* by in-depth proteomic analysis for ginsenoside biosynthesis and transportation. *J. Ginseng Res.* **2020**, 1–8. [CrossRef]
9. Nam, M.H.; Kim, S.I.; Liu, J.R.; Yang, D.C.; Lim, Y.P.; Kwon, K.H.; Yoo, J.S.; Park, Y.M. Proteomic analysis of Korean ginseng (*Panax ginseng* C.A. Meyer). *J. Chromatogr. B Anal. Technol. Biomed. Life Sci.* **2005**, *815*, 147–155. [CrossRef]
10. Amer, H.E.A. Purification of Proteins: Between Meaning and Different Methods. In *Proteomics Technologies and Applications*, 1st ed.; Abdurakhmonov, I.Y., Ed.; InTechOpen: London, UK, 2019; pp. 1–13. [CrossRef]
11. Jiang, L.; He, L.; Fountoulakis, M. Comparison of protein precipitation methods for sample preparation prior to proteomic analysis. *J. Chromatogr. A* **2004**, *1023*, 317–320. [CrossRef] [PubMed]
12. Wu, X.; Gong, F.; Wang, W. Protein extraction from plant tissues for 2DE and its application in proteomic analysis. *Proteomics* **2014**, *14*, 645–658. [CrossRef] [PubMed]
13. Wessel, D.; Flügge, U.I. A method for the quantitative recovery of protein in dilute solution in the presence of detergents and lipids. *Anal. Biochem.* **1984**, *138*, 141–143. [CrossRef]
14. Wu, X.; Xiong, E.; Wang, W.; Scali, M.; Cresti, M. Universal sample preparation method integrating trichloroacetic acid/acetone precipitation with phenol extraction for crop proteomic analysis. *Nat. Protoc.* **2014**, *9*, 362–374. [CrossRef]
15. Ashwin, N.M.R.; Barnabas, L.; Ramesh Sundar, A.; Malathi, P.; Viswanathan, R.; Masi, A.; Agrawal, G.K.; Rakwal, R. Advances in proteomic technologies and their scope of application in understanding plant–pathogen interactions. *J. Plant Biochem. Biotechnol.* **2017**, *26*, 371–386. [CrossRef]

16. Ma, R.; Sun, L.; Chen, X.; Mei, B.; Chang, G.; Wang, M.; Zhao, D. Proteomic analyses provide novel insights into plant growth and ginsenoside biosynthesis in forest cultivated *Panax ginseng* (*F. Ginseng*). *Front. Plant Sci.* **2016**, *7*, 1–16. [CrossRef] [PubMed]
17. Sun, L.W.; Ma, P.T.; Li, D.N.; Lei, X.J.; Ma, R.; Qi, C. Protein extraction from the stem of *Panax ginseng* C. A. Meyer: A tissue of lower protein extraction efficiency for proteomic analysis. *Afr. J. Biotechnol.* **2011**, *10*, 4328–4333. [CrossRef]
18. Wang, K.; Jiang, S.; Sun, C.; Lin, Y.; Yin, R.; Wang, Y.; Zhang, M. The spatial and temporal transcriptomic landscapes of ginseng, *Panax ginseng* C. A. Meyer. *Sci. Rep.* **2015**, *5*, 1–12. [CrossRef] [PubMed]
19. Wang, H.; Peng, D.; Xie, J. Ginseng leaf-stem: Bioactive constituents and pharmacological functions. *Chin. Med.* **2009**, *4*, 20. [CrossRef]
20. Niu, L.; Zhang, H.; Wu, Z.; Wang, Y.; Liu, H.; Wu, X.; Wang, W. Modified TCA/acetone precipitation of plant proteins for proteomic analysis. *PLoS ONE* **2018**, *13*, 1–13. [CrossRef] [PubMed]
21. Mikulášek, K.; Konečná, H.; Potěšil, D.; Holánková, R.; Havliš, J.; Zdráhal, Z. SP3 Protocol for Proteomic Plant Sample Preparation Prior LC-MS/MS. *Front. Plant Sci.* **2021**, *12*, 635550. [CrossRef]
22. Sielaff, M.; Kuharev, J.; Bohn, T.; Hahlbrock, J.; Bopp, T.; Tenzer, S.; Distler, U. Evaluation of FASP, SP3, and iST Protocols for Proteomic Sample Preparation in the Low Microgram Range. *J. Proteome Res.* **2017**, *16*, 4060–4072. [CrossRef]
23. Ameni, G.; Pieper, R.; Yu, Y. S-Trap, an Ultrafast Sample-Preparation Approach for Shotgun Proteomics. *J. Proteome Res.* **2018**, *17*, 2917–2924. [CrossRef]
24. Jo, I.H.; Lee, J.; Hong, C.E.; Lee, D.J.; Bae, W.; Park, S.G.; Ahn, Y.J.; Kim, Y.C.; Kim, J.U.; Lee, J.W.; et al. Isoform sequencing provides a more comprehensive view of the *Panax ginseng* transcriptome. *Genes* **2017**, *8*, 228. [CrossRef]
25. Heldt, H.W. *Plant Biochemistry*, 3rd ed.; Elsevier Academic Press: San Diego, CA, USA, 2005; pp. 45–100.
26. Mooney, B.P.; Miernyk, J.A.; Michael Greenlief, C.; Thelen, J.J. Using quantitative proteomics of Arabidopsis roots and leaves to predict metabolic activity. *Physiol. Plant* **2006**, *128*, 237–250. [CrossRef]
27. Marcon, C.; Malik, W.A.; Walley, J.W.; Shen, Z.; Paschold, A.; Smith, L.G.; Piepho, H.P.; Briggs, S.P.; Hochholdinger, F. A high-resolution tissue-specific proteome and phosphoproteome atlas of maize primary roots reveals functional gradients along the root axes. *Plant Physiol.* **2015**, *168*, 233–246. [CrossRef] [PubMed]
28. Seung, I.K.; Kweon, S.M.; Kim, E.A.; Jin, Y.K.; Kim, S.; Jong, S.Y.; Young, M.P. Characterization of RNase-like major storage protein from the ginseng root by proteomic approach. *J. Plant Physiol.* **2004**, *161*, 837–845. [CrossRef]
29. Kim, Y.J.; Zhang, D.; Yang, D.C. Biosynthesis and biotechnological production of ginsenosides. *Biotechnol. Adv.* **2015**, *33*, 717–735. [CrossRef] [PubMed]
30. Yang, J.L.; Hu, Z.F.; Zhang, T.T.; Gu, A.D.; Gong, T.; Zhu, P. Progress on the studies of the key enzymes of ginsenoside biosynthesis. *Molecules* **2018**, *23*, 589. [CrossRef]
31. Zhao, S.; Wang, L.; Liu, L.; Liang, Y.; Sun, Y.; Wu, J. Both the mevalonate and the non-mevalonate pathways are involved in ginsenoside biosynthesis. *Plant Cell Rep.* **2014**, *33*, 393–400. [CrossRef]
32. Kang, O.J.; Kim, J.S. Comparison of ginsenoside contents in different parts of Korean ginseng (*Panax ginseng* C.A. Meyer). *Prev. Nutr. Food Sci.* **2016**, *21*, 389–392. [CrossRef]
33. Hoshino, Y.; Gaucher, E.A. On the origin of isoprenoid biosynthesis. *Mol. Biol. Evol.* **2018**, *35*, 2185–2197. [CrossRef]
34. Chen, S.; Luo, H.; Li, Y.; Sun, Y.; Wu, Q.; Niu, Y.; Song, J.; Lv, A.; Zhu, Y.; Sun, C.; et al. 454 EST analysis detects genes putatively involved in ginsenoside biosynthesis in *Panax ginseng*. *Plant Cell Rep.* **2011**, *30*, 1593–1601. [CrossRef]
35. Xue, L.; He, Z.; Bi, X.; Xu, W.; Wei, T.; Wu, S.; Hu, S. Transcriptomic profiling reveals MEP pathway contributing to ginsenoside biosynthesis in *Panax ginseng*. *BMC Genom.* **2019**, *20*, 1–14. [CrossRef] [PubMed]
36. Zhao, M.; Lin, Y.; Wang, Y.; Li, X.; Han, Y.; Wang, K.; Sun, C.; Wang, Y.; Zhang, M. Transcriptome analysis identifies strong candidate genes for ginsenoside biosynthesis and reveals its underlying molecular mechanism in *Panax ginseng* C.A. Meyer. *Sci. Rep.* **2019**, *9*, 1–10. [CrossRef] [PubMed]
37. Kang, K.B.; Jayakodi, M.; Lee, Y.S.; Nguyen, V.B.; Park, H.S.; Koo, H.J.; Choi, I.Y.; Kim, D.H.; Chung, Y.J.; Ryu, B.; et al. Identification of candidate UDP-glycosyltransferases involved in protopanaxadiol-type ginsenoside biosynthesis in *Panax ginseng*. *Sci. Rep.* **2018**, *8*, 11744. [CrossRef] [PubMed]
38. Jung, S.C.; Kim, W.; Park, S.C.; Jeong, J.; Park, M.K.; Lim, S.; Lee, Y.; Im, W.T.; Lee, J.H.; Choi, G.; et al. Two ginseng UDP-glycosyltransferases synthesize ginsenoside Rg3 and Rd. *Plant Cell Physiol.* **2014**, *55*, 2177–2188. [CrossRef] [PubMed]
39. Wang, P.; Wei, Y.; Fan, Y.; Liu, Q.; Wei, W.; Yang, C.; Zhang, L.; Zhao, G.; Yue, J.; Yan, X.; et al. Production of bioactive ginsenosides Rh2 and Rg3 by metabolically engineered yeasts. *Metab. Eng.* **2015**, *29*, 97–105. [CrossRef] [PubMed]
40. Wei, W.; Wang, P.; Wei, Y.; Liu, Q.; Yang, C.; Zhao, G.; Yue, J.; Yan, X.; Zhou, Z. Characterization of *Panax ginseng* UDP-glycosyltransferases catalyzing protopanaxatriol and biosyntheses of bioactive ginsenosides F1 and Rh1 in metabolically engineered yeasts. *Mol. Plant* **2015**, *8*, 1412–1424. [CrossRef]
41. Gupta, R.; Min, C.W.; Kim, S.W.; Wang, Y.; Agrawal, G.K.; Rakwal, R.; Kim, S.G.; Lee, B.W.; Ko, J.M.; Baek, I.Y.; et al. Comparative investigation of seed coats of brown- versus yellow-colored soybean seeds using an integrated proteomics and metabolomics approach. *Proteomics* **2015**, *15*, 1706–1716. [CrossRef]
42. Wiśniewski, J.R.; Zougman, A.; Nagaraj, N.; Mann, M. Universal sample preparation method for proteome analysis. *Nat. Methods* **2009**, *6*, 359–362. [CrossRef]

43. Min, C.W.; Jung, W.Y.; Park, H.J.; Moon, K.B.; Ko, H.; Sohn, J.H.; Jeon, J.H.; Kim, H.S.; Gupta, R.; Kim, S.T.; et al. Label-free quantitative proteomic analysis determines changes in amino acid and carbohydrate metabolism in three cultivars of Jerusalem artichoke tubers. *Plant Biotechnol. Rep.* **2019**, *13*, 111–122. [CrossRef]
44. Perez-Riverol, Y.; Csordas, A.; Bai, J.; Bernal-Llinares, M.; Hewapathirana, S.; Kundu, D.J.; Inuganti, A.; Griss, J.; Mayer, G.; Eisenacher, M.; et al. The PRIDE database and related tools and resources in 2019: Improving support for quantification data. *Nucleic Acids Res.* **2019**, *47*, D442–D450. [CrossRef]
45. Tyanova, S.; Temu, T.; Cox, J. The MaxQuant computational platform for mass spectrometry-based shotgun proteomics. *Nat. Protoc.* **2016**, *11*, 2301–2319. [CrossRef] [PubMed]
46. Tyanova, S.; Temu, T.; Sinitcyn, P.; Carlson, A.; Hein, M.Y.; Geiger, T.; Mann, M.; Cox, J. The Perseus computational platform for comprehensive analysis of (prote)omics data. *Nat. Methods* **2016**, 731–740. [CrossRef]
47. Usadel, B.; Poree, F.; Nagel, A.; Lohse, M.; Czedik-Eysenberg, A.; Stitt, M. A guide to using MapMan to visualize and compare Omics data in plants: A case study in the crop species, Maize. *Plant Cell Environ.* **2009**, *32*, 1211–1229. [CrossRef] [PubMed]
48. Tian, T.; Liu, Y.; Yan, H.; You, Q.; Yi, X.; Du, Z.; Xu, W.; Su, Z. AgriGO v2.0: A GO analysis toolkit for the agricultural community, 2017 update. *Nucleic Acids Res.* **2017**, *45*, W122–W129. [CrossRef] [PubMed]
49. Doncheva, N.T.; Morris, J.H.; Gorodkin, J.; Jensen, L.J. Cytoscape StringApp: Network Analysis and Visualization of Proteomics Data. *J. Proteome Res.* **2019**, *18*, 623–632. [CrossRef] [PubMed]
50. Yu, C.S.; Cheng, C.W.; Su, W.C.; Chang, K.C.; Huang, S.W.; Hwang, J.K.; Lu, C.H. CELLO2GO: A Web Server for Protein Subcellular Localization Prediction with Functional Gene Ontology Annotation. *PLoS ONE* **2014**, *9*, e99368. [CrossRef] [PubMed]

plants

MDPI

Article

Molecular Phylogenetic Diversity and Biological Characterization of *Diaporthe* Species Associated with Leaf Spots of *Camellia sinensis* in Taiwan

Hiran A. Ariyawansa [1,*,†], Ichen Tsai [1,2,3,†], Jian-Yuan Wang [1], Patchareeya Withee [4], Medsaii Tanjira [4], Shiou-Ruei Lin [5], Nakarin Suwannarach [6,7], Jaturong Kumla [6,7], Abdallah M. Elgorban [8] and Ratchadawan Cheewangkoon [4,7,*]

1 Department of Plant Pathology and Microbiology, College of Bioresources and Agriculture, National Taiwan University, Taipei 10617, Taiwan; ichenntsaii@gmail.com (I.T.); r07633020@gmail.com (J.-Y.W.)
2 Biodiversity and Climate Research Centre (BiK-F), 60325 Frankfurt am Main, Germany
3 Department of Biological Science, Goethe University Frankfurt, 60438 Frankfurt am Main, Germany
4 Department of Entomology and Plant Pathology, Faculty of Agriculture, Chiang Mai University, Chiang Mai 50200, Thailand; pwithee92@gmail.com (P.W.); t_medsaii@hotmail.com (M.T.)
5 Department of Tea Agronomy, Tea Research and Extension Station, Taoyuan 32654, Taiwan; tres226@ttes.gov.tw
6 Research Centre of Microbial Diversity and Sustainable Utilization, Faculty of Science, Chiang Mai University, Chiang Mai 50200, Thailand; suwan.462@gmail.com (N.S.); jaturong_yai@hotmail.com (J.K.)
7 Innovative Agriculture Research Center, Faculty of Agriculture, Chiang Mai University, Chiang Mai 50200, Thailand
8 Department of Botany and Microbiology, College of Science, King Saud University, Riyadh 11451, Saudi Arabia; aelgorban@ksu.edu.sa
* Correspondence: ariyawansa44@ntu.edu.tw (H.A.A.); ratchadawan.c@cmu.ac.th (R.C.)
† H.A.A. and I.T. contributed equally to this study.

Citation: Ariyawansa, H.A.; Tsai, I.; Wang, J.-Y.; Withee, P.; Tanjira, M.; Lin, S.-R.; Suwannarach, N.; Kumla, J.; Elgorban, A.M.; Cheewangkoon, R. Molecular Phylogenetic Diversity and Biological Characterization of *Diaporthe* Species Associated with Leaf Spots of *Camellia sinensis* in Taiwan. *Plants* **2021**, *10*, 1434. https://doi.org/10.3390/plants10071434

Academic Editors: Milan S. Stankovic, Paula Baptista and Petronia Carillo

Received: 28 June 2021
Accepted: 12 July 2021
Published: 14 July 2021

Publisher's Note: MDPI stays neutral with regard to jurisdictional claims in published maps and institutional affiliations.

Copyright: © 2021 by the authors. Licensee MDPI, Basel, Switzerland. This article is an open access article distributed under the terms and conditions of the Creative Commons Attribution (CC BY) license (https://creativecommons.org/licenses/by/4.0/).

Abstract: *Camellia sinensis* is one of the major crops grown in Taiwan and has been widely cultivated around the island. Tea leaves are prone to various fungal infections, and leaf spot is considered one of the major diseases in Taiwan tea fields. As part of a survey on fungal species causing leaf spots on tea leaves in Taiwan, 19 fungal strains morphologically similar to the genus *Diaporthe* were collected. ITS (internal transcribed spacer), *tef1-α* (translation elongation factor 1-α), *tub2* (beta-tubulin), and *cal* (calmodulin) gene regions were used to construct phylogenetic trees and determine the evolutionary relationships among the collected strains. In total, six *Diaporthe* species, including one new species, *Diaporthe hsinchuensis*, were identified as linked with leaf spot of *C. sinensis* in Taiwan based on both phenotypic characters and phylogeny. These species were further characterized in terms of their pathogenicity, temperature, and pH requirements under laboratory conditions. *Diaporthe tulliensis*, *D. passiflorae*, and *D. perseae* were isolated from *C. sinensis* for the first time. Furthermore, pathogenicity tests revealed that, with wound inoculation, only *D. hongkongensis* was pathogenic on tea leaves. This investigation delivers the first assessment of *Diaporthe* taxa related to leaf spots on tea in Taiwan.

Keywords: endophytes; foliar pathogens; pathogenicity; taxonomy

1. Introduction

Species of *Diaporthe* have been frequently reported as pathogens, endophytes, and saprobes of various types of hosts [1–17]. These taxa have been reported globally and cause various diseases on economically important plants and crops such as dieback of forest trees [2], leaf spots on tea and *Ixora* spp. [3,4], leaf and pod blights and seed decay in soybean [5], melanose and stem-end rot on *Citrus* spp. [6–10], leaf spot on common hop [11], twig blight and dieback of blueberry [12], trunk diseases of grapevines [13], branch canker on *Cinnamomum camphora* (Camphor Tree) [14], pear shoot canker [15], and stem canker on sunflower [16,17].

The genus *Diaporthe* (syn. *Phomopsis*) was introduced by Nitschke and is typified by *D. eres* [18–27]. It is categorized in the family Diaporthaceae and order Diaporthales [18–27]. The holomorphic name for *Diaporthe*/*Phomopsis* was complex. Therefore, following the nomenclature rules, Rossman et al. [28] recommended adopting the older sexual typified name *Diaporthe* over the younger asexual typified name *Phomopsis*. Currently, Index Fungorum (retrieved in April 2021) reported over 900 names under the genus *Phomopsis*, whereas *Diaporthe* contains over 1000 names. Species of *Diaporthe* were traditionally identified based on their phenotypic characters such as colony morphology, appearance of ascomata and conidiomata, variation in ascospores and conidiospores, and host affiliations [18–27]. However, due to improvements in DNA sequencing, various enigmatic taxa have been discovered, which has reformed our understanding of the natural classification of *Diaporthe*. Recent studies based on molecular phylogeny have continuously discovered that traditionally used morphological characters and host associations are not sufficient to determine the species of *Diaporthe* strains because they show variation under different environmental conditions [29]. Therefore, multilocus phylogenies based on DNA sequences of ITS (internal transcribed spacer), *tef1-α* (translation elongation factor 1-α), *tub2* (beta-tubulin), and *cal* (calmodulin) are more often used to determine the natural classification of *Diaporthe* species [1,26,29–31].

Foliar fungal pathogens that infect *Camellia sinensis* can lead to a notable reduction in their yield, resulting in a loss of income [3,32,33]. Tea leaves at all stages are susceptible to fungal diseases [3,32,33]. In total, 520 fungal species have been identified to occur on *Camellia* species, of which 303 were reported from *C. sinensis* according to the U.S. Department of Agriculture (USDA) database [3,4]. Various fungi are known to cause diseases of leaves, stems, and roots of *C. sinensis* and other *Camellia* species, for example brown blight caused by *Colletotrichum* species, gray blight by *Pestalotiopsis*-like taxa, leaf spots and dieback by *Diaporthe* species, and blister blight by *Exobasidium vexans* [3,32,33]. Our research group studies various foliar diseases allied with tea farms in Taiwan [32,33]. Our previous work has indicated that in addition to the major fungal diseases, such as brown blight due to *Colletotrichum* species complex and grey blight disease caused by *Pestalotiopsis*-like taxa, numerous other fungal species can potentially cause leaf spots on *C. sinensis* in Taiwan [32,33]. Therefore, the aim of this study was to fully characterize the diversity and phylogeny of *Diaporthe*-like strains originally isolated from the leaf spots of *C. sinensis* and evaluate their pathogenicity to the major tea cultivar Chin-Shin Oolong grown in Taiwan's tea fields. This study was conducted to obtain a better understanding of pathogen biology and determine the optimal temperature and pH for the mycelial growth of these fungi under laboratory conditions.

2. Results

2.1. Fungal Isolation

In total, 19 *Diaporthe*-like strains associated with leaf spots of *C. sinensis* were isolated from five different tea fields located in Taiwan (Figure 1 and Table S1).

2.2. Phylogenetic Evaluation of the Concatenated Data Matrix of Diaporthe Species

Prior to the multilocus gene analysis, alignments corresponding to the gene regions of ITS, *tef1-α*, *tub2*, and *cal* were analyzed using ML. Congruence between single loci made it conceivable to evaluate the phylogenies by concatenating the genes as it delivered a guarantee of gene orthology. The phylogenetic tree obtained from the concatenated gene datasets is in Figure 2.

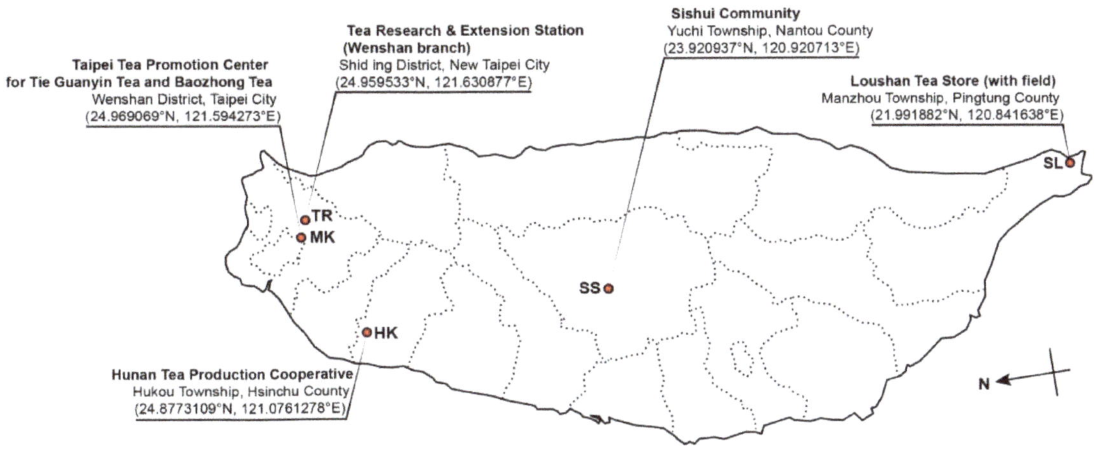

Figure 1. Tea fields surveyed in this study. Relevant geographical details are given for each indicated location.

The final phylogenetic tree was inferred using the combined gene data matrix of ITS, *tef1-α*, *tub2*, and *cal* gene regions from 244 strains of *Diaporthe* species and *Diaporthella corylina* CBS 121124 as the outgroup taxon. The final combined data matrix contained 1803 characters, including gaps (624 for ITS, 526 for *cal*, 381 for *tub2*, and 272 for *tef1-α*). The following priors were used in MrBayes for the five gene loci data partitions based on the results of MrModeltest: all partitions had Dirichlet base frequencies and GTR + I + G models with inverse gamma-distributed rates implemented for ITS and *tub2*, and HKY + I + G with inverse gamma-distributed rates for *cal* and *tef1-α*. The Bayesian analysis resulted in 90,000 trees after 90,000,000 generations. The first 20% of trees representing the burn-in phase of the analyses were discarded, while the remaining trees were used to calculate Bayesian posterior probabilities (PP) in the majority rule consensus tree. The best scoring RAxML tree had a likelihood value of: −4812.054120 and GTR + I + G was used as the evolutionary model.

The phylogenetic trees obtained from both ML and Bayesian analyses had a similar topology and branching pattern based on phenotypic characters and were supported by the molecular phylogenetic inference of strains generated in this study. The results were consistent with recent publications [3,12,13,30,34–36]. In the present phylogeny (Figure 2), out of 19 strains, 16 were assigned to previously described species, namely, *D. apiculatum* (4), *D. hongkongensis* (4), *D. tulliensis* (3), *D. passiflorae* (2), and *D. perseae* (3). Three strains in this study formed distinct clades with a highly supported sub-clade, which was identified as a novel species and named *D. hsinchuensis*.

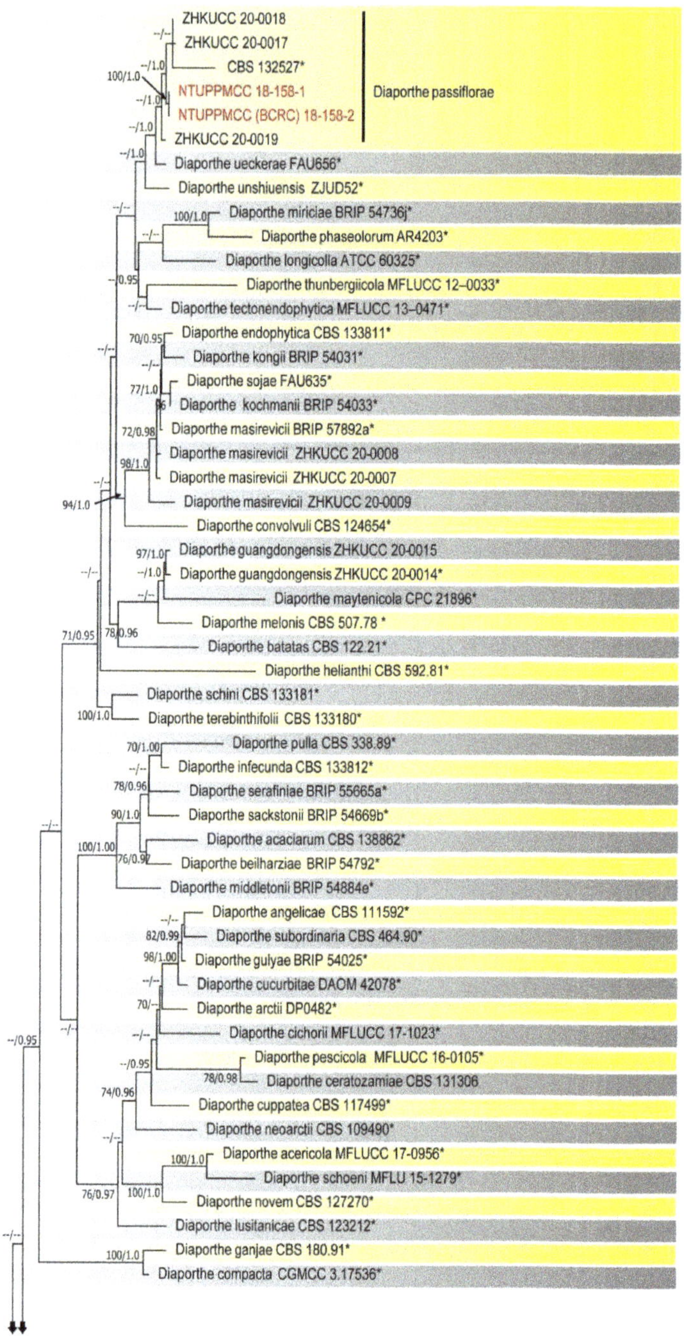

Figure 2. *Cont.*

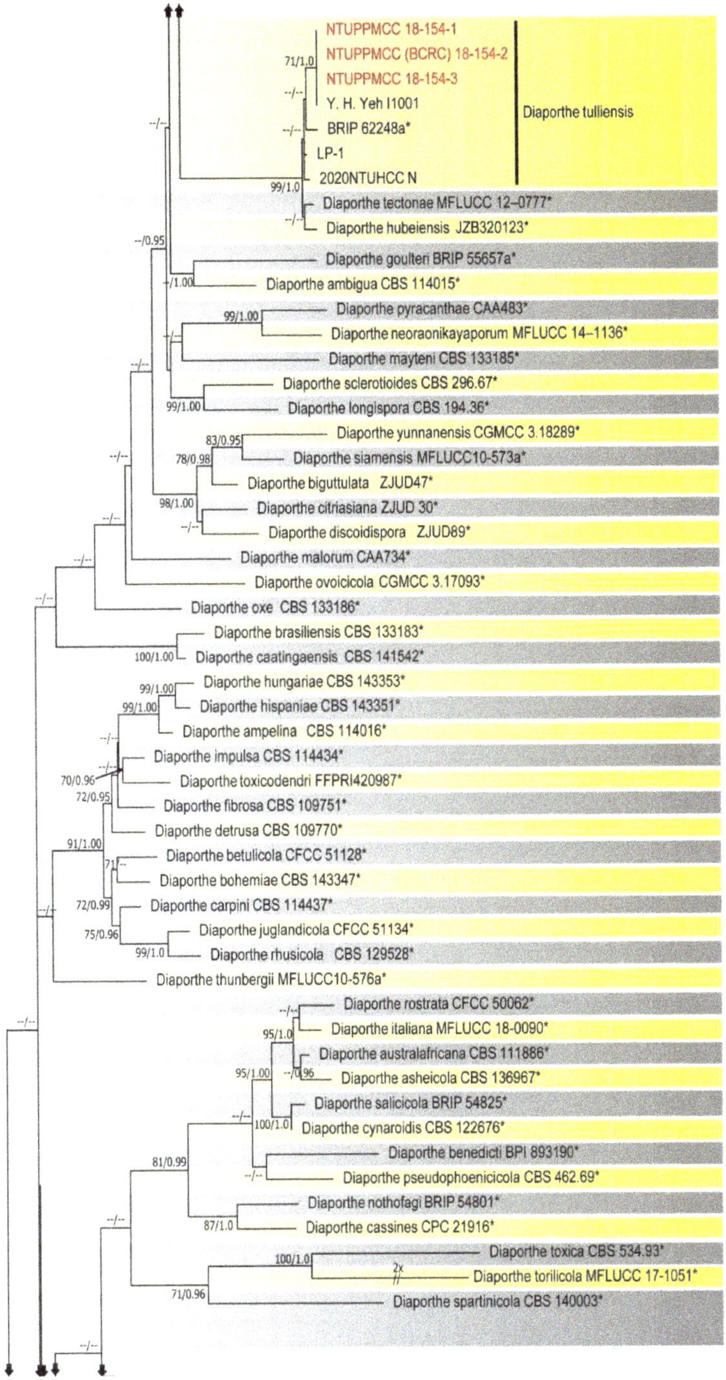

Figure 2. *Cont.*

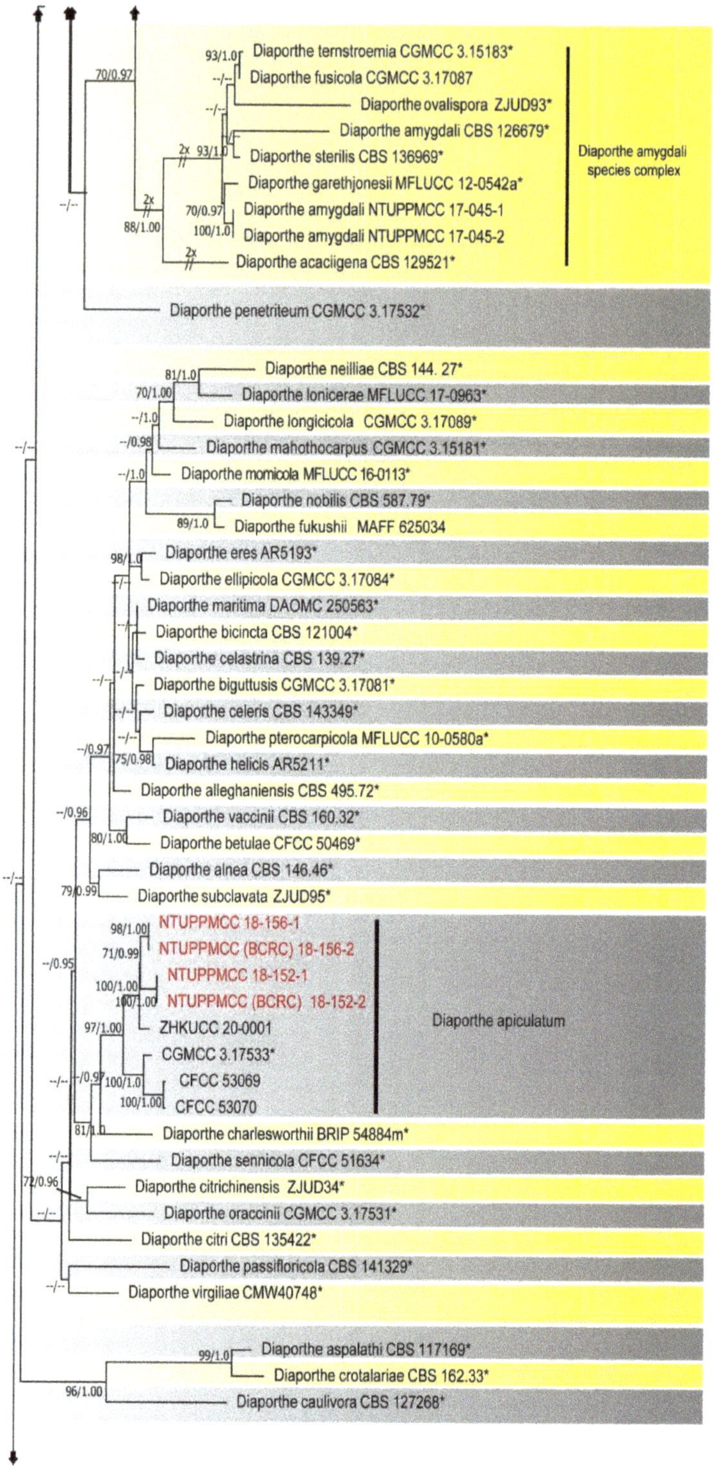

Figure 2. *Cont.*

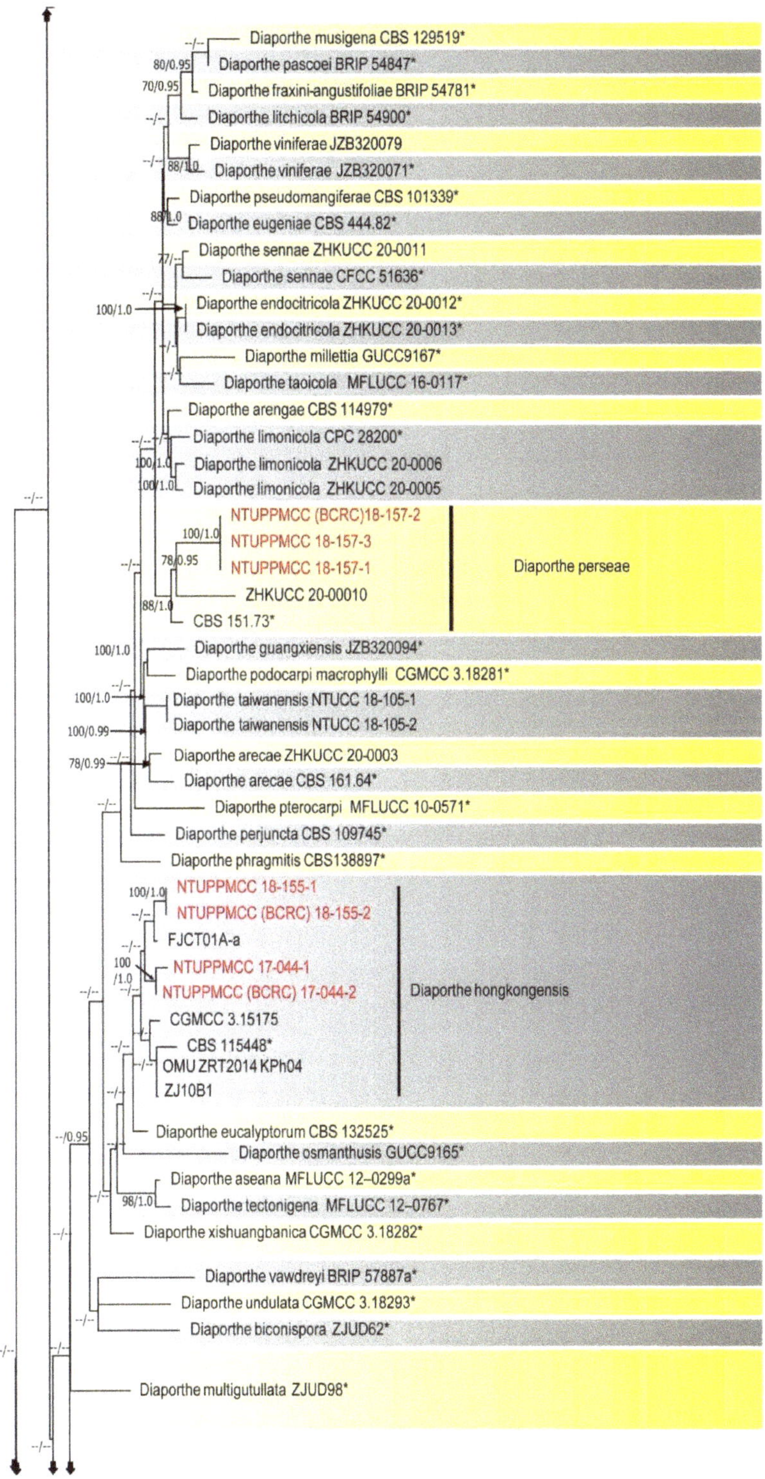

Figure 2. *Cont.*

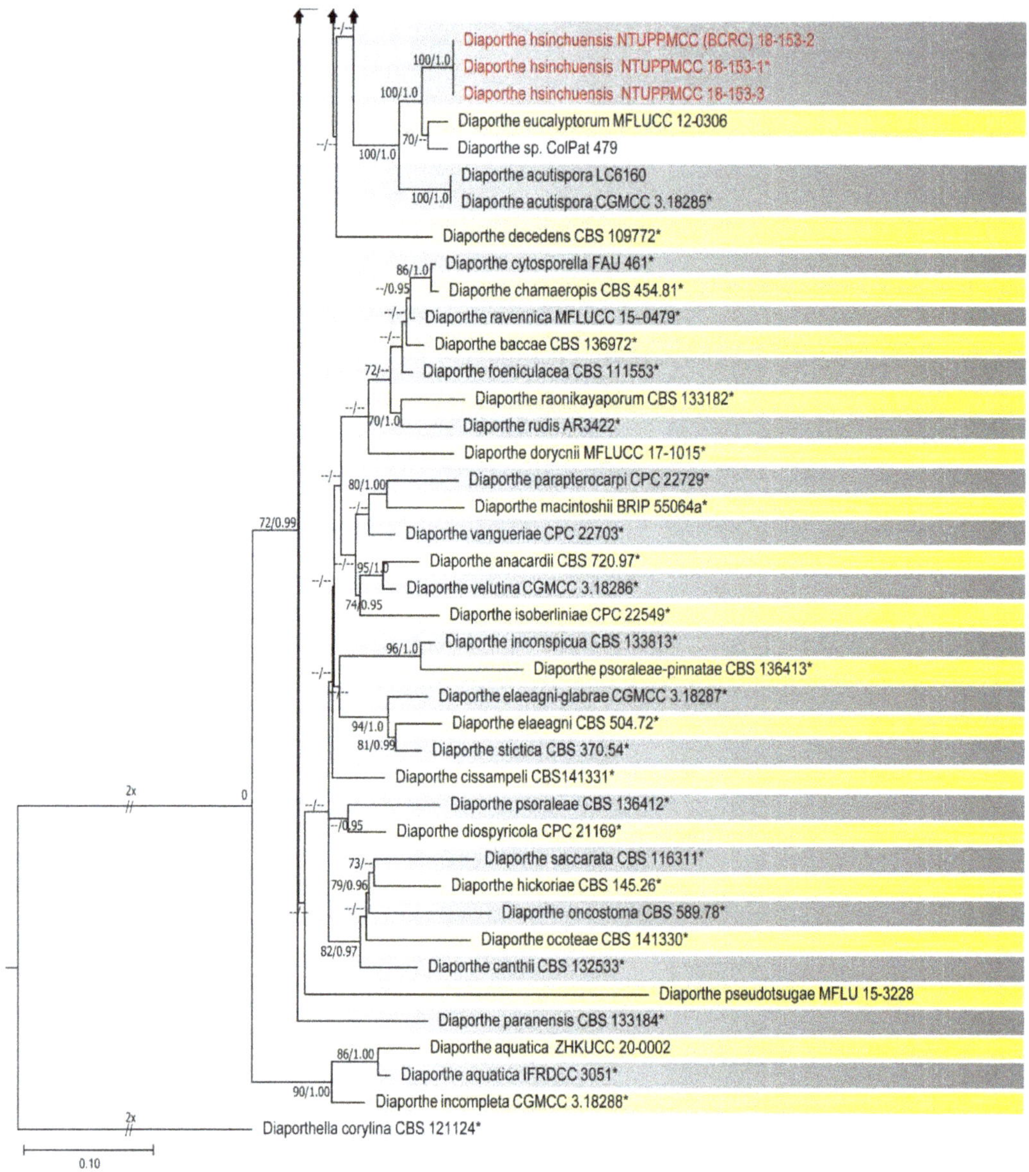

Figure 2. Phylogenetic tree of *Diaporthe* generated from a combined sequence dataset of ITS, *tef1-α*, *tub2*, and *cal* gene regions. Bootstrap values greater than 70% and Bayesian posterior probabilities greater than 0.95 were given below or above the nodes. Isolates obtained in this study are indicated in red, and the ex-type sequences are indicated by *. The strain codes were annotated after relevant species names. *Diaporthella corylina* CBS 121124 served as an outgroup taxon.

2.3. *Taxonomy Diaporthe hsinchuensis Ariyawansa and I. Tsai, sp. nov.*

MycoBank: MB840098, Figure 3

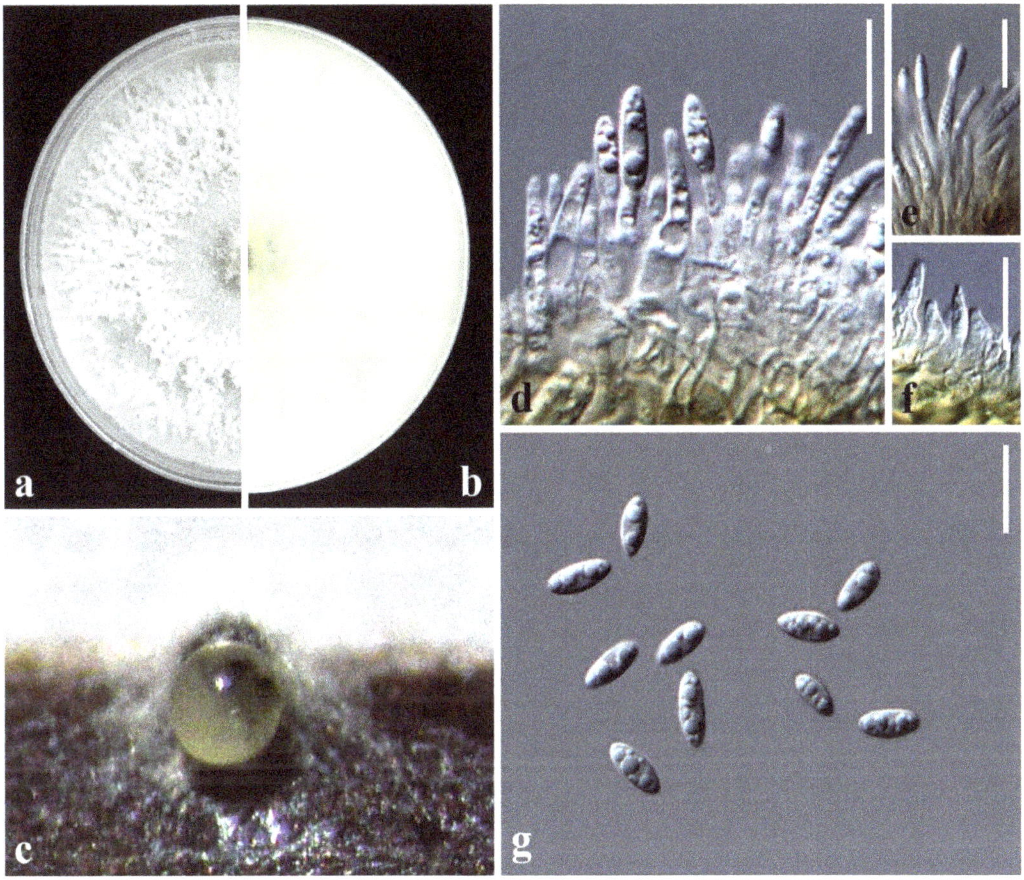

Figure 3. *Diaporthe hsinchuensis* NTUPPMCC 18-153-1: (**a**,**b**) surface and reverse side of colony on PDA; (**c**) conidiomata on PNA. (**d**–**f**); conidiogenous cells; (**g**) conidia. Scale bars = 10 μm.

Etymology: The epithet refers to Hsinchu, Taiwan, where this species was originally collected.

Description: Forms grey lesions on the tip of the tea (*Camellia sinensis*) leaf. *Sexual morph*: Undetermined. *Asexual morph*: *Conidiomata* pycnidial on PNA, globose, erumpent when mature, conidia exuding from the pycnidia in ivory white drops. *Conidiophores* cylindrical, phialidic, septate, and branched, 8–20 × 1–3 μm ($\overline{x} \pm$ SD = 14.1 ± 2.9 × 2.0 ± 0.4 μm, n = 30). *Alpha conidiogenous cells* hyaline, ovoid to ampulliform, cylindrical to subcylindrical tapering towards the apex, straight or curved, 1–8 × 1–4 μm ($\overline{x} \pm$ SD = 5.6 ± 1.9 × 2.4 ± 0.6 μm, n = 30). *Alpha conidia* hyaline, oval, or fusiform, unicellular, aseptate, 2–7-guttulate 6–9 × 2.5–4 μm ($\overline{x} \pm$ SD = 7.5 ± 0.9 × 3.2 ± 0.2 μm, n = 30). *Beta conidia* not observed. *Gamma conidia* not observed.

Colony characteristics: Colonies on PDA circular, edge entire, surface white, cottony, reverse yellowish white.

Material examination: Taiwan, Hsinchu County, Hukou Township, Hunan Tea Production Cooperative, on leaves of *Camellia sinensis* (Theaceae), 4 April 2018, Tsai Ichen, HK04-1 (NTUPPMH 18-153-1, holotype), ex-type culture NTUPPMCC 18-153-1; *ibid*., HK04-2 (NTUPPMH 18-153-2) = living culture NTUPPMCC 18-153-2.

Notes: The strains representing *Diaporthe hsinchuensis* clustered in a well-supported clade (ML/PP = 100/1.0) and formed a distinct linage sister to *D. eucalyptorum* (MFLUCC 12-0306), *D. acutispora* (LC6160 and LC6161), and *Diaporthe* sp. (ColPat479) (Figure 2) Unfortunately, morphological data were not available for *D. eucalyptorum* (MFLUCC 12-0306) or *Diaporthe* sp. (ColPat479). Furthermore, the ex-type strain of *D. eucalyptorum* (CBS 132525), which was used by Crous et al. [37] to introduce the species, appeared as a basal clade to strains containing *D. hongkongensis* in the present study in agreement with previous studies [2,30]. This may indicate that *D. eucalyptorum* (MFLUCC 12-0306), used by Senanayake et al. [38], is a different species, but further studies are required to confirm the taxonomic status of this strain. Based on the available data, we only compared the morphological features of *D. hsinchuensis* with *D. acutispora*. *D. hsinchuensis* differs from *D. acutispora* in its smaller conidiophores (8–20 × 1–3 μm versus 10–34.5 × 2–3 μm), relatively smaller alpha conidia (6–9 × 2.5–4 μm versus 7–10 × 2–3 μm), host (*Camellia sinensis* versus *Coffea* sp.), and geographical location (Taiwan versus China). In addition, *D. hsinchuensis* can be clearly differentiated based on nucleotide differences in ITS, *tef1-α*, *tub2*, and *cal* loci from its phylogenetically closely related species, *D. acutispora* (ITS 4%; *tef1-α* 14%; *tub2* 5%, *cal* 6%).

2.4. Growth Rate

All tested isolates were grown on PDA media for 7 days at 25 °C in the dark. The sizes of colonies were measured (mm), and their means were calculated and are presented in Figure 4. Isolate NTUPPMCC 18-153-1 (*D. hsinchuensis*) exhibited the widest diameter colony (70.33 mm on average), which, along with isolate NTUPPMCC 18-154-1 (*D. tulliensis*), presented significantly faster growth after 7 days of incubation. In contrast, strain NTUPPMCC 18-152-1 (*D. apiculatum*), 18-155-1 (*D. hongkongensis*), and 18-157-1 (*D. perseae*) had significantly slower growth compared to all the other isolates.

2.5. Temperature Effects

Fungal mycelial growth was detected for all the tested isolates between 10 to 45 °C and measured as colony diameter. Temperature regimes strongly influenced the growth of tested fungal strains ($p \leq 0.001$); the maximum growth was determined at 25–30 °C (mean 60.39 mm), while the minimum or no-growth was observed at 40–45 °C (mean 0.10 mm) for all the isolates (Figure 5, Table S2).

2.6. Optimal pH

The effects of pH on the mycelial growth of tested strains are presented in Figure 6 and Table S3. Results showed that, generally, all strains used in this study grow better in slightly acidic to alkaline medium (pH 5–10) compared with acidic medium (pH 3–5). Isolates NTUPPMCC 18-153-1 and 18-155-1 showed a relatively narrower optical pH range (pH 6–8 and 5–7, respectively) upon mycelial growth compared with the other isolates.

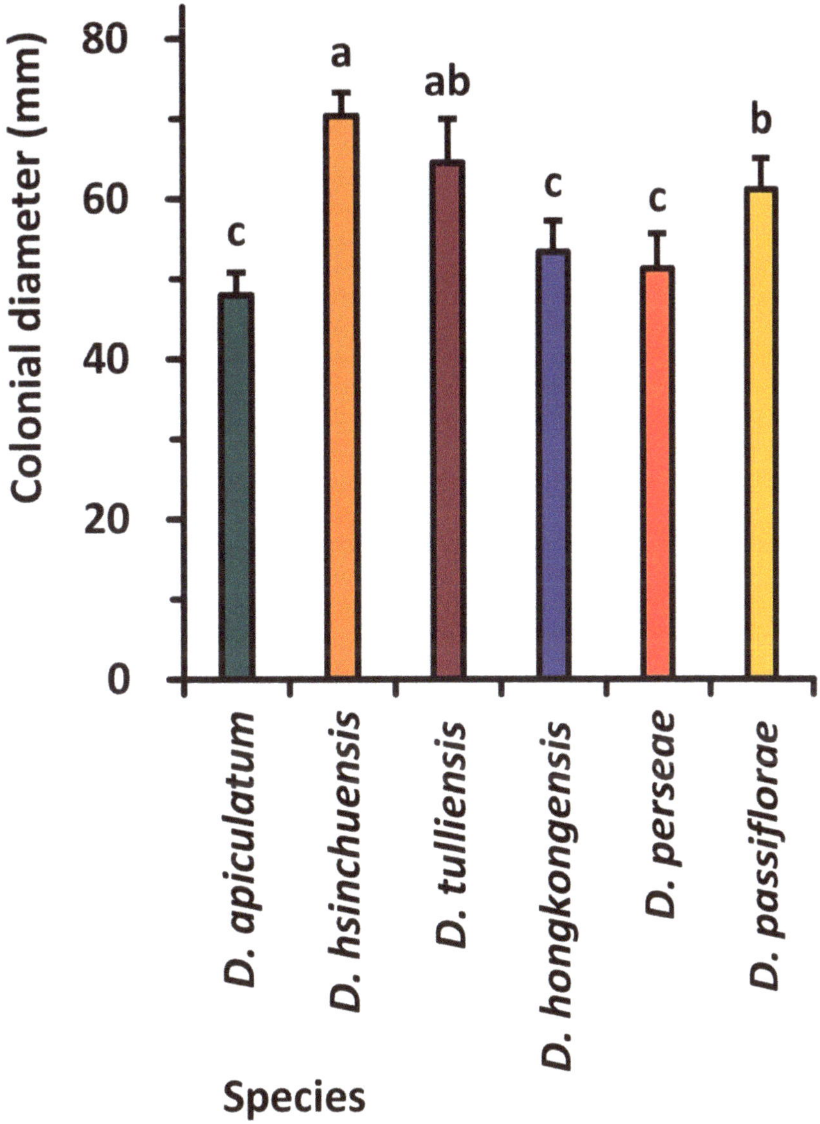

Figure 4. Comparison of mycelial growth rate of the six *Diaporthe* species based on perpendicular colonial diameter. According to Tukey's range test, data (mean ± standard deviation) with the same letters are not significantly different. Colors identify different taxa.

2.7. Pathogenicity

In total, all six selected isolates except NTUPPMCC 18-155-1 (*D. hongkongensis*) failed to cause symptoms on either wounded or unwounded inoculated sites. However, NTUPPMCC 18-155-1 caused symptoms on wounded inoculated sites on tea leaves (Figure 7), showing that the strain might enter tea leaves via wounds. The characteristic of the concentric circular fruiting-body formation and brown blight shared among observed symptoms (Figure 7d) were consistent with that observed in natural tea fields. The fungal strain capable of forming lesions on wounds in this experiment was re-isolated from the fruiting lesions with morphological characteristics identical to those of the original isolates.

Therefore, Koch's postulates were fulfilled and based on both the DNA sequence data and morphological evidence from these re-isolates, all were confirmed to be pathogenic to *C. sinensis*.

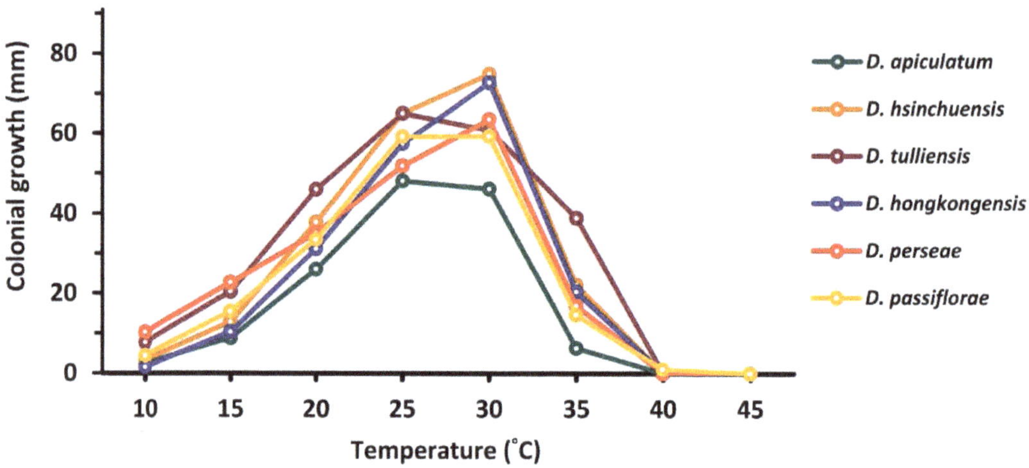

Figure 5. Temperature effect on mycelial growth according to the comparison of colonial growth (mm) of different species at each temperature, based on the mean values presented in Table S2.

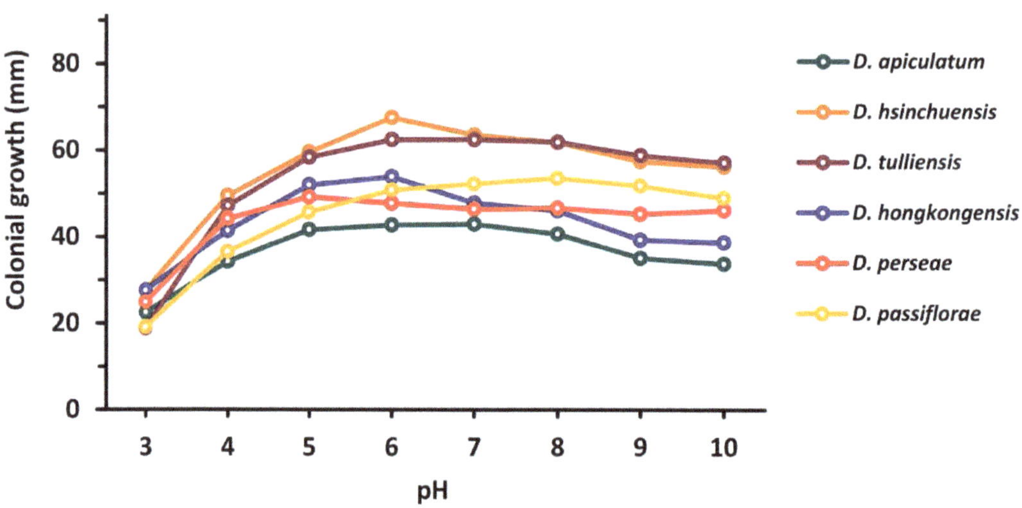

Figure 6. Optimal pH for mycelial growth of each species according to the comparison of colonial growth (mm diam), based on the mean values presented in Table S3.

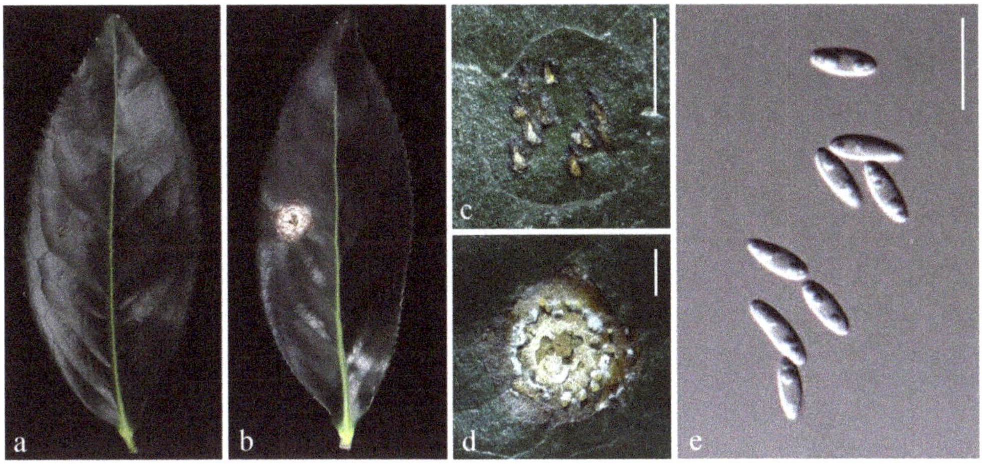

Figure 7. Symptoms caused by *D. hongkongensis*: (**a**,**c**) control leaf inoculated with PDA disk, symptoms absent; (**b**,**d**) symptom on tea leaf inoculated with *D. hongkongensis* 14 days after incubation (lesion size on average = 6.38 mm diam); (**e**) conidia obtained from artificially generated lesion. Scale bars: (**c**,**d**) = 2 mm; (**e**) = 10 µm.

3. Discussion

Tea is one of the major crops grown in Taiwan, and oolong tea produced by Taiwan is responsible for 25% of global oolong production (One Town One Product (OTOP) [32,33]. Foliar diseases of *C. sinensis* are of concern to tea farmers because leaves are the main part of the plant used to produce various teas. However, research regarding *Diaporthe* species causing foliar diseases of *C. sinensis* is rare, and twig blight caused by unknown *Phomopsis* species (asexual morph of *Diaporthe*) are the only diseases caused by *Diaporthe* mentioned in the list of plant diseases in Taiwan [39].

Accurately naming a fungus based on its molecular and phenotype characters that reflect phylogeny allows predictions about plant-associated organisms together with potential pathogenicity and appropriate control measures. In the present study, we identified six *Diaporthe* taxa linked with leaf spot in Taiwan tea fields, including a novel species. Various *Diaporthe* species have been identified as either pathogens or endophytes on *C. sinensis* such as *D. amygdali*, *D. apiculata*, *D. discoidispora*, *D. eres*, *D. foeniculacea*, *D. foeniculina*, *D. hongkongensis*, *D. incompleta*, *D. longicicola*, *D. masirevicii*, *D. nobilis*, *D. oraccinii*, *D. penetriteum*, *D. portugallica*, *D. tectonigena*, *D. ueckerae*, *D. velutina*, and *D. xishuangbanica* [40]. Among the tested isolates, *D. hongkongensis* fulfilled Koch's postulates and was identified as the only pathogen causing leaf spot on *C. sinensis* in the present study. *D. hongkongensis* has been recognized as a well-known phytopathogen affecting various hosts such as grapevine [41], peach [42], kiwifruit [43], dragon fruit [44], and pear [15] in recent studies. *D. hongkongensis* was also reported to be associated with both healthy and diseased tea leaves in China by Gao et al. [3]. However, in their study, Gao et al. [3] did not confirm the pathogenicity of *D. hongkongensis* to tea leaves by fulfilling Koch's postulates. In addition, there is no record of *D. hongkongensis* causing plant diseases in Taiwan. Therefore, to the best of our understanding, this is not only the first report of *D. hongkongensis* causing leaf spot on *C. sinensis*, but also a novel discovery of *D. hongkongensis* present in Taiwan.

Furthermore, in this study, we reported another species associated with tea leaf spot, namely, *D. apiculatum*, for the first time in Taiwan. Gao et al. [3] introduced *D. apiculatum* as a new species from the healthy and infected leaves of *C. sinensis* in China but did not confirm the pathogenicity of the fungal strains via Koch's postulates. However, in the present study, strains that we identified as *D. apiculatum* did not fulfill Koch's postulates after 14 days of incubation. Therefore, further studies are required to confirm whether this is a latent pathogen or a conditional pathogen, causing symptoms when the plant defense mechanisms are weak.

In the present study, *Diaporthe passiflorae*, *D. tulliensis*, and *D. perseae* were isolated from the leaf spots of *C. sinensis* for the first time. *D. passiflorae* was introduced by Crous et al. [37] as a potential endophyte on the fruit of *Passiflora edulis*. In contrast, Li et al. [45] identified the same species as a postharvest pathogen causing fruit rot of Kiwifruit in Sichuan Province, China. *D. perseae* was originally described as *Phomopsis perseae* Zerova from branches of dying *Persea gratissima* trees in Russia [29]. This species also has been identified as a pathogen of mango causing fruit rot in Malaysia [46] and as endophytes on *Citrus grandis* in China [30]. *D. tulliensis* has been recognized as an important pathogen of various hosts, such as kiwifruit [47], coffee [48], and grapes [13]. Interestingly, Huang et al. [49] and Chen and Kirschner [50] recently reported the same species as a pathogen causing leaf spot on *Parthenocissus tricuspidata* as well as an endophyte from *Nelumbo nucifera* in Taiwan, respectively. However, in the present study, none of the strains recognized as *D. passiflorae*, *D. perseae*, *D. tulliensis*, or the new species *D. hsinchuensis*, satisfied Koch's postulates after 14 days of incubation. Thus, they were not considered pathogenic to major tea cultivar Ching-Shin Oolong grown in Taiwan.

Only a few studies have shown the effect of environmental factors on the growth of *Diaporthe* species. The most recent study was carried out by Arciuolo et al. [51], who tested the optimal temperature and water activity required for mycelial growth, pycnidial conidiomata development, and asexual spore production of *D. eres*, the causal agent of hazelnut defects. Arciuolo et al. [51] concluded that the optimum temperature for mycelial growth of *D. eres* was observed at 20–25 °C and at 30 °C for pycnidia and cirrhi development. In another study, Hilário et al. [12] observed that four novel pathogens of *Diaporthe*, namely *D. crousii*, *D. phillipsii*, *D. rossmaniae*, and *D. vacuae*, causing twig blight and dieback of *Vaccinium corymbosum* in Portugal had an optimum temperature for mycelial growth at 20–25 °C. The present study provides insights into the effect of environmental factors such as temperature and pH on the mycelial growth of the *Diaporthe* strains isolated from *C. sinensis* in Taiwan under a controlled environment. Our results showed that all species identified in this study reached maximum colony diameter at 25–30 °C and preferred to grow under low acidic to alkaline rather than acidic medium. The observations in the present study are consistent with the results of Arciuolo et al. [51] and Hilário et al. [12]. In the future, these abiotic factors could be used as the basis for developing a predictive model for the infection of *Diaporthe* taxa of tea plants in Taiwan.

4. Conclusions

In conclusion, the research outcomes of the present study improve the understanding of *Diaporthe* species allied with leaf spots on tea leaves and deliver valuable data for effective disease management of *C. sinensis* in Taiwan. The study identified six *Diaporthe* species associated with leaf spots on *C. sinensis* in Taiwan tea fields. These species comprise one novel species and three new records in *Diaporthe* on *C. sinensis*. However, to gain better knowledge of the diversity, pathogenicity, distribution, and implications of *Diaporthe* taxa on tea plantations in Taiwan, extensive surveys on *C. sinensis* plantations and the collection of a larger number of samples should reflect the goals of future studies.

5. Materials and Methods

5.1. Sample Collection, Fungal Isolation, and Morphological Examination

Tea leaves with characteristic leaf spots were collected from five distinct tea fields located in major tea-growing areas in Taiwan (Figure 1). The samples were collected in resealable plastic bags and carried back to the laboratory. Pure cultures were obtained through a single conidium isolation method, as described in Udayanga et al. [26] and Ariyawansa et al. [4]. In brief, contents of the fruiting body were mounted in a drop of sterile distilled water on a flame-sterilized cavity slide and pipetted to thoroughly mix. The drop of spore suspension was spread evenly on a Petri dish of water agar (WA) and incubated at 25 °C in the dark for 12 h. Single germinating conidia were transferred to a Petri dish of potato dextrose agar (PDA; HiMedia Laboratories Pvt. Ltd., Mumbai, India) and incubated at 25 °C in the dark. Colonial characterization was carried out from isolates cultured on PDA (HiMedia Laboratories Pvt. Ltd., Mumbai, India). Conidiomatal growth was detected on WA with double-autoclaved pine needles placed on the surface of agar (PNA), corn meal agar (CMA; HiMedia Laboratories Pvt. Ltd., Mumbai, India), or PDA supplemented with 10% NaCl, which were incubated at 25 °C under continuous blue light for 7 to 14 days [33]. Microscopic characteristics were examined in distilled water, with 30 measurements taken from each structure using cellSense Standard software (XV Imaging, Version 3.17.0.16686) under an Olympus BX51 microscope (Olympus Corp., Tokyo, Japan) with differential interference contrast (DIC) illumination.

In this study, type specimens were deposited in the herbarium of the Department of Plant Pathology and Microbiology, National Taiwan University (NTUPPMH). Ex-type living cultures were deposited in the Department of Plant Pathology and Microbiology, National Taiwan University Culture Collection (NTUPPMCC), and the Bioresource Collection and Research Centre (BCRC).

5.2. DNA Extraction, PCR, and Sequencing

The growing mycelia of each isolate were gathered from cultures incubated at 25 °C in the dark for 7 to 14 days. Genomic DNA were extracted with EasyPure Genomic DNA Spin Kit (Bioman Scientific Co., Ltd., New Taipei, Taiwan) according to the manufacturer's guidelines (Bioman Scientific Co., Ltd., New Taipei, Taiwan). PCR amplifications of ITS, tef1-α, tub2, and cal gene regions were separately performed in 25 μL reaction mixtures, as described in Ariyawansa et al. [4]. The relevant primer pairs included in this study are in Table 1. The PCR products were visualized by electrophoresis on a 1.5% agarose gel stained with BioGreenTM Safe DNA Gel buffer (Bioman Scientific Co., Ltd., New Taipei, Taiwan). The sequence purified amplicons from each gene region were obtained using the Sanger sequencing method at the Genomics Co., Ltd., (New Taipei, Taiwan). Newly obtained sequences in the present study were deposited to NCBI GenBank.

Table 1. Gene regions and primer sequences used in this study.

Region	Primers	Sequence (5′→3′)	Reference
ITS	ITS5	GGA AGT AAA AGT CGT AAC AAG G	[52]
	ITS4	TCC GCT TAT TGA TAT GC	
tef1-α	EF1-728F	CAT CGA GAA GTT CGA GAA GG	[53]
	EF1-986R	TAC TTG AAG GAA CCC TTA CC	
tub2	Bt2a	GGT AAC CAA ATC GGT GCT TTC	[54]
	Bt2b	ACC CTC AGT GTA GTG ACC CTT GGC	
cal	CAL-228F	GAG TTC AAG GAG GCC TTC TCC C	[53]
	CAL-737R	CAT CTT TCT GGC CAT GG	

5.3. Strain Selection, Sequence Alignment, and Phylogenetic Analysis

An initial ITS-only tree containing all taxa currently recognized in *Diaporthe* was made to resolve the clades containing the isolates obtained in this study (data not shown). This analysis was further used to select the species to be included in the multilocus phylogenetic analyses. DNA sequence data of ITS, *tef1-α*, *tub*2, and *cal* loci were used to determine the phylogenetic placement of isolates. DNA sequences of each strain were searched against GenBank by nucleotide BLAST (BLASTn) to find the nearest matches. Strains in GenBank with 95–99% similarity to known *Diaporthe* species together with species previously described as pathogens on tea were included in the phylogenetic analysis following the recent publications of Ariyawansa et al. [4], Yang et al. [55], Dissanayake et al. [31,56], Gao et al. [57], Guarnaccia and Crous [8], and Manawasinghe et al. [13]. The strains used in the present study and their GenBank accession numbers are presented in Table S4.

Multiple sequence alignments were generated in MAFFT v. 6.864b with default parameters (http://mafft.cbrc.jp/alignment/server/index.html, accessed on 3 April 2021). The alignments for each gene were visually improved manually where necessary in MEGA v. 5 [58]. Single gene trees were first built for ITS, *tef1-α*, *tub2*, and *cal*, and finally subjected to a multilocus analysis. MrModeltest v. 2.3 [59] under the Akaike Information Criterion (AIC) implemented in PAUP v. 4.0b10 [60] was used to determine the individual selection of evolutionary models for phylogenetic analyses of each locus.

Two phylogenetic tree inference methods, maximum likelihood (ML) in RAxML [61], and Bayesian analyses in MrBayes v. 3.0b4 [62] were used to evaluate the phylogenetic relationships of the strains used in this study, as described in Ariyawansa et al. [4] and Tsai et al. [33]. Bootstrap values obtained via ML (MLB) and Bayesian posterior probabilities (BPP) that were equal to or greater than 70% or greater than 0.95, respectively, are indicated below or above each node (Figure 2). Topologies of the trees obtained from each gene were visually compared to confirm the similarity between the overall tree topology of the individual datasets and that of the tree obtained from the combined alignment. MEGA v. 5 [58] and FigTree v. 1.4 [63] software were used to assess and visualize phylogenetic trees and data files.

The principles of Genealogical Concordance Phylogenetic Species Recognition (GCPSR) were used to identify the species limits within *Diaporthe*-like taxa [64,65]. Dettman et al. [65] proposed that species should be differentiated when fulfilling one of the following two criteria: genealogical concordance or genealogical non-discordance. In other words, if clades exist in at least some of the trees, then they are recognized as genealogically concordant; clades are recognized as genealogically non-discordant if they are significantly supported with high statistical values (MLB $\geq$ 70, PP $\geq$ 0.95) in a single locus without conflict at or above this support level in any other single-locus trees. By following this criterion, poorly supported non-monophylies in one locus are eliminated without undermining well-supported monophylies in another locus.

Based on the outcomes of the phylogenetic analysis considering a single strain representing each taxon, a total of six isolates were randomly selected for the pathogenicity, mycelial growth, temperature, and pH tests.

5.4. Mycelial Growth Test

In total, six strains representing six taxa identified in this study, NTUPPMCC 18-152-1 (*Diaporthe apiculatum*), NTUPPMCC 18-055-1 (*D. hongkongensis*), NTUPPMCC 18-153-1 (*D. hsinchuensis*), NTUPPMCC 18-154-1 (*D. tulliensis*), NTUPPMCC 18-157-1 (*D. perseae*), and NTUPPMCC 18-158-1 (*D. passiflorae*), were used to determine the radial growth of mycelia. The growth rate was evaluated following a modified procedure of Huang et al. [66]. A 4 mm-diam mycelial disk was cut from the edge of a three-day-old PDA culture, placed centrally into a Petri dish (90 mm diam) of PDA (13 mL) and incubated at 25 °C in the dark for three days. Measurements were taken daily, based on the diameter of two perpendicular axes per fungal colony. The mycelial growth was determined on the final measurement. The test was conducted thrice with three replicates per trial.

5.5. Temperature and pH Effects on Mycelial Growth

The same isolates, volume of medium, inoculation method, and standard of measurement used in the mycelial growth test were used to test the effects of temperature and pH. Further details for each assessment are described below.

The effect of temperature on radial mycelial growth was checked on a daily basis and determined on the third day after inoculation at 10, 15, 20, 25, 30, 35, 40, and 45 °C in the dark. All single inoculations were conducted on Petri dishes of PDA. The test was performed three times, with three replicates per trial. The present protocol was modified from Keith et al. [67].

The optimal pH for radial mycelial growth was studied at pH 3, 4, 5, 6, 7, 8, 9, and 10. PDA plates were heated to mix prior to sterilization, and the pH values were adjusted with

1 M HCl and 1 M NaOH solutions [68,69]. The tested cultures were incubated at 25 °C in the dark for three days, and the colony sizes were measured on a daily basis. The test was conducted three times, with three replicates per trial.

5.6. Pathogenicity Assessment

The same isolates used in the above temperature and pH investigations were used in the pathogenicity assessment. The test was conducted on detached tea leaves (picked from fourth to sixth leaf below the apical bud) randomly collected from healthy branches of Ching-Shin Oolong gathered from a conventional tea field (ca. 15 years old) located in Pinglin District, New Taipei City (24°56′26.9″ N, 121°43′25.4″ E), as described in Tsai et al. [33]. In brief, fresh leaf-attached tea branches were thoroughly rinsed with tap water to remove dust, and excessive water was gently eliminated with tissue paper. A flat rack (164 × 114 × 26 mm^3) wrapped with sterile tissue paper was placed in a plastic box (320 × 240 × 70 mm^3), and the box was filled with 700 mL of sterilized distilled water. Four tea leaves were detached from the branch, surface sterilized with 75% ethanol and fixed to the rack at the foliar tip and base with rubber bands. The leaves were wounded by pinpricking with a sterile needle. A 4 mm-diam mycelial disk was cut from the margin of a seven-day-old colony on PDA and inoculated on a single wounded site [70]. In total, four tea leaves inoculated with PDA disks (4 mm diam) without mycelium served as controls. The box with the above contents was sealed with plastic wrap to maintain moisture and incubated at 26 ± 1 °C under a 12/12 h photocycle [33]. The pathogenicity was determined after 14 days of incubation. The test was conducted three times, with four replicates for each round per isolate.

5.7. Statistical Analysis

The statistical analysis was carried out by one-way analysis of variance (ANOVA) using SAS® University Edition v. 3.8, and the pairwise comparison was performed via Tukey's range test (α = 0.05).

Supplementary Materials: The following are available online at https://www.mdpi.com/article/10.3390/plants10071434/s1, Table S1: collection information of *Diaporthe* taxa used in this study, Table S2: mycelial growth (mm diam., with the origin diameter of inoculated mycelial plug deducted) of each isolate (NTUPPMCC) at every pH level. Data (mean ± standard deviation) with the same letters are not significantly different on the basis of Tukey's range test. Table S3: GenBank accession numbers of isolates included in the multilocus sequence analysis, Table S4: mycelial growth (mm diam., with the origin diameter of inoculated mycelial plug deducted) of individual isolate (NTUPPMCC) at each temperature. Data (mean ± standard deviation) with the same letters are not significantly different on the basis of Tukey's range test.

Author Contributions: Conceptualization, H.A.A.; methodology, I.T. and H.A.A.; investigation, I.T., P.W., M.T. and H.A.A.; data curation, H.A.A., I.T., J.-Y.W., P.W., M.T., S.-R.L., N.S., J.K., A.M.E., R.C.; writing—H.A.A. original draft preparation, I.T. and H.A.A.; writing—review and editing, all authors; funding acquisition, H.A.A., A.M.E. and R.C. All authors have read and agreed to the published version of the manuscript.

Funding: This study was funded by grants from the Higher Education Sprout Projects (grant number: NTUJP-108L7215 and NTUJP-110L7221) and The Researchers Supporting Project number (RSP-2021/56), King Saud University, Riyadh, Saudi Arabia.

Institutional Review Board Statement: Not applicable.

Informed Consent Statement: Not applicable.

Data Availability Statement: Publicly available datasets were analyzed in this study. This data can be found here: https://www.ncbi.nlm.nih.gov/ (accessed on 13 July 2021) and https://www.mycobank.org/page/Simple%20names%20search (accessed on 13 July 2021).

Acknowledgments: This research work was also partially supported by Chiang Mai University. The authors extend their appreciation to The Researchers supporting project number (RSP-2021/56), King Saud University, Riyadh, Saudi Arabia. We are grateful to C.P. Lin, Parichad Pakdeeniti, Panatda Kankavee, Chia-Yun Yen, Yu-Chen Lin, Vivienne Hsieh-Wu, Chia-Yi Wu, A.D. Ariyawansa, D.M.K. Ariyawansa, Ruwini Ariyawansa, Amila Gunasekara, and Oshen Kemika for their valuable ideas.

Conflicts of Interest: The authors declare no conflict of interest.

References

1. Udayanga, D.; Castlebury, L.A.; Rossman, A.Y.; Chukeatirote, E.; Hyde, K.D. Insights into the genus *Diaporthe*: Phylogenetic species delimitation in the *D.* eres species complex. *Fungal Divers.* **2014**, *67*, 203–229. [CrossRef]
2. Yang, Q.; Fan, X.L.; Guarnaccia, V.; Tian, C.M. High diversity of *Diaporthe* species associated with dieback diseases in China, with twelve new species described. *MycoKeys* **2018**, *39*, 97–149. [CrossRef] [PubMed]
3. Gao, Y.; Liu, F.; Cai, L. Unravelling *Diaporthe* species associated with *Camellia*. *Syst. Biodivers.* **2016**, *14*, 102–117. [CrossRef]
4. Ariyawansa, H.A.; Tsai, I.; Withee, P.; Tanjira, M.; Yen, C.Y.; Al-Rashed, S.; Elgorban, A.M.; Cheewangkoon, R. *Diaporthe taiwanensis*: A new taxon causing leaf spots and necrosis on *Ixora chinensis* in Taiwan. *Phytotaxa* **2020**, *461*, 155–165. [CrossRef]
5. Udayanga, D.; Castlebury, L.A.; Rossman, A.Y.; Chukeatirote, E.; Hyde, K.D. The *Diaporthe sojae* species complex: Phylogenetic re-assessment of pathogens associated with soybean, cucurbits and other field crops. *Fungal Biol.* **2015**, *119*, 383–407. [CrossRef]
6. Mondal, S.N.; Vicent, A.; Reis, R.F.; Timmer, L.W. Efficacy of pre-and postinoculation application of fungicides to expanding young citrus leaves for control of melanose, scab, and Alternaria brown spot. *Plant Dis.* **2007**, *91*, 1600–1606. [CrossRef]
7. Udayanga, D.; Castlebury, L.A.; Rossman, A.Y.; Hyde, K.D. Species limits in *Diaporthe*: Molecular re-assessment of *D.* citri, D. cytosporella, D. foeniculina and D. rudis. *Persoonia* **2014**, *32*, 83–101. [CrossRef]
8. Guarnaccia, V.; Crous, P.W. Emerging citrus diseases in Europe caused by species of *Diaporthe*. *IMA Fungus* **2017**, *8*, 317–334. [CrossRef]
9. Guarnaccia, V.; Crous, P.W. Species of *Diaporthe* on *Camellia* and *Citrus* in the Azores Islands. *Phytopathol. Mediterr.* **2018**, *57*, 307–319. [CrossRef]
10. Chaisiri, C.; Liu, X.Y.; Lin, Y.; Li, J.B.; Xiong, B.; Luo, C.X. Phylogenetic analysis and development of molecular tool for detection of *Diaporthe citri* causing melanose disease of citrus. *Plants* **2020**, *9*, 329. [CrossRef] [PubMed]
11. Allan-Perkins, E.; Li, D.W.; Schultes, N.; Yavuz, S.; LaMondia, J. The identification of a new species, *Diaporthe humulicola*, a pathogen causing Diaporthe leaf spot on common hop. *Plant Dis.* **2020**, *104*, 2377–2390. [CrossRef]
12. Hilário, S.; Amaral, I.A.; Gonçalves, M.F.; Lopes, A.; Santos, L.; Alves, A. *Diaporthe* species associated with twig blight and dieback of *Vaccinium corymbosum* in Portugal, with description of four new species. *Mycologia* **2020**, *112*, 293–308. [CrossRef]
13. Manawasinghe, I.S.; Dissanayake, A.J.; Li, X.; Liu, M.; Wanasinghe, D.N.; Xu, J.; Zhao, W.; Zhang, W.; Zhou, Y.; Hyde, K.D.; et al. High genetic diversity and species complexity of *Diaporthe* associated with grapevine dieback in China. *Front. Microbiol.* **2019**, *10*, 1936. [CrossRef]
14. Li, D.; Zhang, H.; Song, Q.; Liu, J.; Yang, Q.; Luan, F.; Li, D. First report of *Diaporthe eres* causing branch canker on *Cinnamomum camphora* (camphor tree) in Jiangxi Province, China. *Plant Dis.* **2021**. [CrossRef]
15. Guo, Y.S.; Crous, P.W.; Bai, Q.; Fu, M.; Yang, M.M.; Wang, X.H.; Du, Y.M.; Hong, N.; Xu, W.X.; Wang, G.P. High diversity of *Diaporthe* species associated with pear shoot canker in China. *Persoonia* **2020**, *45*, 132–162. [CrossRef]
16. Muntanola-Cvetkovic, M.; Mihaljcevic, M.; Petrov, M. On the identity of the causative agent of a serious *Phomopsis-Diaporthe* disease in sunflower plant. *Nova Hedwig.* **1981**, *34*, 417–435.
17. Thompson, S.M.; Tan, Y.P.; Young, A.J.; Neate, S.M.; Aitken, E.A.B.; Shivas, R.G. Stem cankers on sunflower (*Helianthus annuus*) in Australia reveal a complex of pathogenic *Diaporthe* (*Phomopsis*) species. *Persoonia* **2011**, *27*, 80–89. [CrossRef] [PubMed]
18. Díaz, G.A.; Latorre, B.A.; Lolas, M.; Ferrada, E.; Naranjo, P.; Zoffoli, J.P. Identification and characterization of *Diaporthe ambigua*, D. australafricana, D. novem, and D. rudis causing a postharvest fruit rot in kiwifruit. *Plant Dis.* **2017**, *101*, 1402–1410. [CrossRef]
19. Garcia-Reyne, A.; López-Medrano, F.; Morales, J.M.; Garcia-Esteban, C.; Martin, I.; Eraña, I.; Meije, Y.; Lalueza, A.; Alastruey-Izquierdo, A.; Rodríguez-Tudela, J.L.; et al. Cutaneous infection by *Phomopsis longicolla* in a renal transplant recipient from Guinea: First report of human infection by this fungus. *Transpl. Infect. Dis.* **2011**, *13*, 204–207. [CrossRef]
20. Mostert, L.; Crous, P.W.; Kang, J.C.; Phillips, A.J. Species of *Phomopsis* and a *Libertella* sp. occurring on grapevines with specific reference to South Africa: Morphological, cultural, molecular and pathological characterization. *Mycologia* **2001**, *93*, 146–167. [CrossRef]
21. Mostert, L.; Kang, J.C.; Crous, P.W.; Denman, S. *Phomopsis saccharata* sp. nov., causing a canker and die-back disease of *Protea repens* in South Africa. *Sydowia* **2001**, *53*, 227–235.
22. Muralli, T.S.; Suryanarayanan, T.S.; Geeta, R. Endophytic *Phomopsis* species: Host range and implications for diversity estimates. *Can. J. Microbiol.* **2006**, *52*, 673–680. [CrossRef]
23. Rehner, S.A.; Uecker, F.A. Nuclear ribosomal internal transcribed spacer phylogeny and host diversity in the coelomycete *Phomopsis*. *Can. J. Bot.* **1994**, *72*, 1666–1674. [CrossRef]
24. Santos, J.M.; Vrandečić, K.; Ćosić, J.; Duvnjak, T.; Phillips, A.J.L. Resolving the *Diaporthe* species occurring on soybean in Croatia. *Persoonia* **2011**, *27*, 9–19. [CrossRef] [PubMed]

25. Udayanga, D.; Liu, X.; McKenzie, E.H.; Chukeatirote, E.; Bahkali, A.H.; Hyde, K.D. The genus *Phomopsis*: Biology, applications, species concepts and names of common phytopathogens. *Fungal Divers.* **2011**, *50*, 189. [CrossRef]
26. Udayanga, D.; Liu, X.; Crous, P.W.; McKenzie, E.H.; Chukeatirote, E.; Hyde, K.D. A multi-locus phylogenetic evaluation of *Diaporthe* (*Phomopsis*). *Fungal Divers.* **2012**, *56*, 157–171. [CrossRef]
27. Van Rensburg, J.C.J.; Lamprecht, S.C.; Groenewald, J.Z.; Castlebury, L.A.; Crous, P.W. Characterisation of *Phomopsis* spp. associated with die-back of rooibos (*Aspalathus linearis*) in South Africa. *Stud. Mycol.* **2006**, *55*, 65–74. [CrossRef] [PubMed]
28. Rossman, A.Y.; Adams, G.C.; Cannon, P.F.; Castlebury, L.A.; Crous, P.W.; Gryzenhout, M.; Jaklitsch, W.M.; Mejia, L.C.; Stoykov, D.; Udayanga, D.; et al. Recommendations of generic names in Diaporthales competing for protection or use. *IMA Fungus* **2015**, *6*, 145–154. [CrossRef]
29. Gomes, R.R.; Glienke, C.; Videira, S.I.R.; Lombard, L.; Groenewald, J.Z.; Crous, P.W. *Diaporthe*: A genus of endophytic, saprobic and plant pathogenic fungi. *Persoonia* **2013**, *31*, 1–41. [CrossRef]
30. Dong, Z.; Manawasinghe, I.S.; Huang, Y.; Shu, Y.; Phillips, A.J.L.; Dissanayake, A.J.; Hyde, K.D.; Xiang, M.; Luo, M. Endophytic *Diaporthe* associated with *Citrus grandis* cv Tomentosa in China. *Front. Microbiol.* **2021**, *11*, 3621. [CrossRef]
31. Dissanayake, A.J.; Camporesi, E.; Hyde, K.D.; Wei, Z.; Yan, J.Y.; Li, X.H. Molecular phylogenetic analysis reveals seven new *Diaporthe* species from Italy. *Mycosphere* **2017**, *8*, 853–877. [CrossRef]
32. Ariyawansa, H.A.; Tsai, I.; Thambugala, K.M.; Chuang, W.Y.; Lin, S.R.; Hozzein, W.N.; Cheewangkoon, R. Species diversity of Pleosporalean taxa associated with *Camellia sinensis* (L.) Kuntze in Taiwan. *Sci. Rep.* **2020**, *10*, 12762. [CrossRef]
33. Tsai, I.; Chung, C.L.; Lin, S.R.; Hung, T.H.; Shen, T.L.; Hu, C.Y.; Hozzein, W.N.; Ariyawansa, H.A. Cryptic diversity, molecular systematics, and pathogenicity of genus *Pestalotiopsis* and allied genera causing gray blight disease of tea in Taiwan, with a description of a new *Pseudopestalotiopsis* species. *Plant Dis.* **2021**, *105*, 425–443. [CrossRef]
34. Dissanayake, A.J.; Chen, Y.Y.; Liu, J.K.J. Unravelling *Diaporthe* species associated with woody hosts from karst formations (Guizhou) in China. *J. Fungi* **2020**, *6*, 251. [CrossRef]
35. Guarnaccia, V.; Groenewald, J.Z.; Woodhall, J.; Armengol, J.; Cinelli, T.; Eichmeier, A.; Ezra, D.; Fontaine, F.; Gramaje, D.; Gutierrez-Aguirregabiria, A.; et al. *Diaporthe* diversity and pathogenicity revealed from a broad survey of grapevine diseases in Europe. *Persoonia* **2018**, *40*, 135–153. [CrossRef] [PubMed]
36. Long, H.; Zhang, Q.; Hao, Y.Y.; Shao, X.Q.; Wei, X.X.; Hyde, K.D.; Wang, Y.; Zhao, D.G. *Diaporthe* species in south-western China. *MycoKeys* **2019**, *57*, 113–127. [CrossRef] [PubMed]
37. Crous, P.W.; Summerell, B.A.; Shivas, R.G.; Burgess, T.I.; Decock, C.A.; Dreyer, L.L.; Granke, L.L.; Guest, D.I.; Hardy, G.S.J.; Hausbeck, M.K.; et al. Fungal planet description sheets: 107–127. *Persoonia* **2012**, *28*, 138–182. [CrossRef] [PubMed]
38. Senanayake, I.C.; Crous, P.W.; Groenewald, J.Z.; Maharachchikumbura, S.S.; Jeewon, R.; Phillips, A.J.; Bhat, J.D.; Perera, R.H.; Li, Q.R.; Li, W.J.; et al. Families of Diaporthales based on morphological and phylogenetic evidence. *Stud. Mycol.* **2017**, *86*, 217–296. [CrossRef]
39. Tzean, S.S.; Tzeng, K.C.; Chang, C.A.; Tsay, T.T.; Yen, S.F. *List of Plant Diseases in Taiwan*; Bureau of Animal and Plant Health Inspection and Quarantine, Council of Agriculture, Executive Yuan: Taipei, Taiwan, 2019.
40. Farr, D.F.; Rossman, A.Y. Fungal Databases, U.S. National Fungus Collections, ARS, USDA. Available online: https://nt.ars-grin.gov/fungaldatabases/ (accessed on 13 June 2021).
41. Dissanayake, A.J.; Liu, M.; Zhang, W.; Chen, Z.; Udayanga, D.; Chukeatirote, E.; Li, X.; Yan, J.; Hyde, K.D. Morphological and molecular characterisation of *Diaporthe* species associated with grapevine trunk disease in China. *Fungal Biol.* **2015**, *119*, 283–294. [CrossRef]
42. Zhang, Z.; Zhang, Z.B.; Huang, Y.T.; Wang, F.; Hu, W.H.; Dai, L.Y.; Zhong, J.; Liu, Y.; Zhu, J.Z. First report of *Diaporthe hongkongensis* causing fruit rot on peach (*Prunus persica*) in China. *Plant Dis.* **2021**. [CrossRef]
43. Erper, I.; Turkkan, M.; Ozean, M.; Luongo, L.; Belisario, A. Characterization of *Diaporthe hongkongensis* species causing stem-end rot on kiwifruit in Turkey. *J. Plant Pathol.* **2017**, *99*, 779–782.
44. Kee, Y.J.; Wong, K.L.; Zakaria, L.; Mohd, M.H. *Diaporthe* species causing stem gray blight of red-fleshed dragon fruit (*Hylocereus polyrhizus*) in Malaysia. *Sci. Rep.* **2021**, *11*, 3907. [CrossRef]
45. Li, H.; Yu, S.; Tang, W.; Miao, M.; Liu, Y. First report of *Diaporthe passiflorae* and *Diaporthe nobilis* causing a postharvest kiwifruit rot in Sichuan province, China. *Plant Dis.* **2019**, *103*, 771. [CrossRef]
46. Lim, L.; Mohd, M.H.; Zakaria, L. Identification and pathogenicity of *Diaporthe* species associated with stem-end rot of mango (*Mangifera indica* L.). *Eur. J. Plant Pathol.* **2019**, *155*, 687–696. [CrossRef]
47. Bai, Q.; Wang, G.P.; Hong, N.; Guo, Y.S.; Fu, M. First report of *Diaporthe tulliensis* and *Diaporthe actinidiae* causing kiwifruit stem canker in Hubei and Anhui provinces, China. *Plant Dis.* **2017**, *101*, 508–509. [CrossRef]
48. Gong, J.L.; Lu, Y.; Wu, W.H.; He, C.P.; Liang, Y.Q.; Huang, X.; Zheng, J.L.; Xi, J.G.; Tang, S.B.; Yi, K.X. First report of *Phomopsis heveicola* (anamorph of *Diaporthe tulliensis*) causing leaf blight of coffee (*Coffea arabica*) in China. *Plant Dis.* **2020**, *104*, 570. [CrossRef]
49. Huang, C.C.; Liu, H.H.; Wu, P.H.; Chang, H.X. First report of leaf spot caused by *Diaporthe tulliensis* on Boston ivy (*Parthenocissus tricuspidata*) in Taiwan. *Plant Dis.* **2021**. [CrossRef]
50. Chen, K.L.; Kirschner, R. Fungi from leaves of lotus (*Nelumbo nucifera*). *Mycol. Prog.* **2018**, *17*, 275–293. [CrossRef]
51. Arciuolo, R.; Camardo Leggieri, M.; Chiusa, G.; Castello, G.; Genova, G.; Spigolon, N.; Battilani, P. Ecology of *Diaporthe eres*, the causal agent of hazelnut defects. *PLoS ONE* **2021**, *16*, e0247563. [CrossRef]

52. White, T.J.; Bruns, T.; Lee, S.; Taylor, J. Amplification and Direct Sequencing of Fungal Ribosomal RNA Genes for Phylogenetics. In *PCR Protocols: A Guide to Methods and Applications*; Innis, M.A., Gelfand, D.H., Sninsky, J.J., White, T.J., Eds.; Academic Press: San Diego, CA, USA, 1990; pp. 315–322.
53. Carbone, I.; Kohn, L.M. A method for designing primer sets for speciation studies in filamentous ascomycetes. *Mycologia* **1999**, *91*, 553–556. [CrossRef]
54. Glass, N.L.; Donaldson, G.C. Development of primer sets designed for use with the PCR to amplify conserved genes from filamentous ascomycetes. *Appl. Environ. Microbiol.* **1995**, *61*, 1323–1330. [CrossRef]
55. Yang, Q.; Jiang, N.; Tian, C.M. Three new *Diaporthe* species from Shaanxi Province, China. *MycoKeys* **2020**, *67*, 1–18. [CrossRef]
56. Dissanayake, A.J.; Phillips, A.J.L.; Hyde, K.D.; Yan, J.Y.; Li, X.H. The current status of species in *Diaporthe*. *Mycosphere* **2017**, *8*, 1106–1156. [CrossRef]
57. Gao, Y.; Liu, F.; Duan, W.; Crous, P.W.; Cai, L. *Diaporthe* is paraphyletic. *IMA Fungus* **2017**, *8*, 153–187. [CrossRef] [PubMed]
58. Tamura, K.; Peterson, D.; Peterson, N.; Stecher, G.; Nei, M.; Kumar, S. MEGA5: Molecular evolutionary genetics analysis using maximum likelihood, evolutionary distance, and maximum parsimony methods. *Mol. Biol. Evol.* **2011**, *28*, 2731–2739. [CrossRef] [PubMed]
59. Nylander, J. *MrModeltest v2. Program*; Distributed by the Author; Evolutionary Biology Centre, Uppsala University: Uppsala, Sweden, 2004.
60. Swofford, D.L. *PAUP*: Phylogenetic Analysis Using Parsimony (and Other Methods), Version 4.0b10*; Sinauer Associates: Sunderland, MA, USA, 2002.
61. Silvestro, D.; Michalak, I. RaxmlGUI: A graphical front-end for RAxML. *Org. Divers. Evol.* **2012**, *12*, 335–337. [CrossRef]
62. Ronquist, F.; Huelsenbeck, J.P. MrBayes 3: Bayesian phylogenetic inference under mixed models. *Bioinformatics* **2003**, *19*, 1572–1574. [CrossRef]
63. Rambaut, A.; Drummond, A. *FigTree: Tree Figure Drawing Tool, Version 1.2.2*; Institute of Evolutionary Biology, University of Edinburgh: Edinburgh, Scotland, 2008.
64. Taylor, J.W.; Jacobson, D.J.; Kroken, S.; Kasuga, T.; Geiser, D.M.; Hibbett, D.S.; Fisher, M.C. Phylogenetic species recognition and species concepts in fungi. *Fungal Genet. Biol.* **2000**, *31*, 21–32. [CrossRef]
65. Dettman, J.R.; Jacobson, D.J.; Turner, E.; Pringle, A.; Taylor, J.W. Reproductive isolation and phylogenetic divergence in *Neurospora*: Comparing methods of species recognition in a model eukaryote. *Evolution* **2003**, *57*, 2721–2741. [CrossRef]
66. Huang, F.; Hou, X.; Dewdney, M.M.; Fu, Y.; Chen, G.; Hyde, K.D.; Li, H. *Diaporthe* species occurring on citrus in China. *Fungal Divers.* **2013**, *61*, 237–250. [CrossRef]
67. Keith, L.M.; Velasquez, M.E.; Zee, F.T. Identification and characterization of *Pestalotiopsis* spp. causing scab disease of guava, *Psidium guajava*, in Hawaii. *Plant Dis.* **2006**, *90*, 16–23. [CrossRef] [PubMed]
68. Abubakar, A.; Suberu, H.A.; Bello, I.M.; Abdulkadir, R.; Daudu, O.A.; Lateef, A.A. Effect of pH on mycelial growth and sporulation of *Aspergillus parasiticus*. *J. Plant Sci.* **2013**, *1*, 64–67. [CrossRef]
69. Saha, A.; Mandal, P.; Dasgupta, S.; Saha, D. Influence of culture media and environmental factors on mycelial growth and sporulation of *Lasiodiplodia theobromae* (Pat.) Griffon and Maubl. *J. Environ. Biol.* **2008**, *29*, 407–410.
70. Wang, Y.; Xiong, F.; Lu, Q.; Hao, X.; Zheng, M.; Wang, L.; Li, N.; Ding, C.; Wang, X.; Yang, Y. Diversity of *Pestalotiopsis*-like species causing gray blight disease of tea plants (*Camellia sinensis*) in China, including two novel *Pestalotiopsis* species, and analysis of their pathogenicity. *Plant Dis.* **2019**, *103*, 2548–2558. [CrossRef] [PubMed]

plants

MDPI

Article

The Application of the Essential Oils of *Thymus vulgaris* L. and *Crithmum maritimum* L. as Biocidal on Two *Tholu Bommalu* Indian Leather Puppets

Giulia D'Agostino, Belinda Giambra, Franco Palla, Maurizio Bruno * and Natale Badalamenti

Department of Biological, Chemical and Pharmaceutical Sciences and Technologies (STEBICEF), Università degli Studi di Palermo, Viale delle Scienze, Ed. 17, I-90128 Palermo, Italy; giuliadagostino@outlook.com (G.D.); info@belindagiambra.it (B.G.); franco.palla@unipa.it (F.P.); natale.badalamenti@unipa.it (N.B.)
* Correspondence: maurizio.bruno@unipa.it; Tel.: +39-091-23-897-531

Abstract: The chemical profile of the *Thymus vulgaris* (Lamiaceae) essential oil (EO) was investigated in order to evaluate its biological properties against microorganisms affecting two *Tholu Bommalu*, typical Indian leather puppets stored at the International Puppets Museum "Antonio Pasqualino" of Palermo, Italy. A GC–MS analysis, using both polar and apolar columns, was used to determine the chemical composition of the essential oil. The aim of this study was to evaluate the antimicrobial effectiveness of the *Thymus vulgaris* and *Crithmum maritimum* essential oils in vapor phase to disinfect heritage leather puppets. Pieces of leather artifacts that were affected by different bacterial colonies were exposed to EO under vacuum and static evaporation conditions. The results presented showed that the vaporization of essential oil was an efficient method in the disinfection of natural skins, eradicating microorganism in short times. *T. vulgaris* EO in the 50% solution showed excellent inhibitory activity against isolated bacteria with both methods, but the obtained results suggest that the vacuum method allowed for faster exposition of the artifacts to the biocide. Furthermore, the biocidal properties of the essential oil of a Sicilian accession of *Crithmum maritimum* (Apiaceae) aerial parts were compared and investigated. The results of the latter essential oil showed a poor activity against the isolated micro-organisms.

Keywords: *Thymus vulgaris*; *Crithmum maritimum*; leather artifacts; essential oils; anti-bacterial activity

Citation: D'Agostino, G.; Giambra, B.; Palla, F.; Bruno, M.; Badalamenti, N. The Application of the Essential Oils of *Thymus vulgaris* L. and *Crithmum maritimum* L. as Biocidal on Two *Tholu Bommalu* Indian Leather Puppets. *Plants* **2021**, *10*, 1508. https://doi.org/10.3390/plants10081508

Academic Editor: Milan S. Stankovic

Received: 6 July 2021
Accepted: 19 July 2021
Published: 22 July 2021

Publisher's Note: MDPI stays neutral with regard to jurisdictional claims in published maps and institutional affiliations.

Copyright: © 2021 by the authors. Licensee MDPI, Basel, Switzerland. This article is an open access article distributed under the terms and conditions of the Creative Commons Attribution (CC BY) license (https://creativecommons.org/licenses/by/4.0/).

1. Introduction

Biodeterioration of cultural heritage causes different alteration processes affecting the constitutive materials of artworks. In particular, fungi and bacteria are able to colonize different artworks comprised of natural materials through aerosol pollution, representing complex problems for conservation by causing a loss of mechanic resistance and deterioration of pictorial layers. Moreover, inadequate exposure to specific thermo-hygrometric parameters can increase the concentrations of microbial colonies, enhancing the deterioration process. For the purpose of inhibiting the biological colonization, different chemical biocides, such as permethrin or/and benzalkonium chloride (BAK), are frequently used. These products are usually toxic for humans and the environment. Hence, in the last decade, essential oils have been applied in order to combat cultural heritage biodeterioration as an eco-friendly solution [1–12]. To prevent biodeterioration caused by fungi and bacteria, objects must first be disinfected. The requirements for a disinfectant include the ability to inhibit the growth and metabolic activity of microorganisms without adversely affecting the material. Currently, fumigation with ethylene oxide is the most popular method for disinfecting fabrics, papers and leathers. However, this gas is an irritant and a dangerous human carcinogen, and its use should be avoided [13].

However, the disinfection process by using volatile compounds, such essential oils (EOs), can require a long time, so the purpose of this work was to find a way to accelerate this. During the evaporation process of a specific liquid, the number of molecules escaping from the liquid increases every second and those that condense back into the liquid also increase. Reaching thermodynamic equilibrium, the vapor phase exerts a pressure on its condensed phase, which is called "vapor pressure" [14]. The speed with which the evaporation occurs is proportional to the vapor pressure of the liquid. The purpose of a vacuum pump applied to a closed system is to maintain a lower pressure than that of the atmosphere. In order to make our disinfection process faster, a vacuum pump was connected to a closed system, ensuring a higher evaporation of the EO contained inside.

To investigate a suitable alternative to traditional synthetic biocides, we decided to analyze the biocidal properties of two different essential oils: *Thymus vulgaris* L. (Lamiaceae) and *Crithmum maritimum* L. (Apiaceae).

Thymus EO was selected due to its microbial properties, which have previously been reported in several studies [15–21]. In particular, the *Thymus vulgaris* essential oil has been shown to be quite effective against several microorganisms [22,23].

In addition, the essential oil of *Crithmum maritimum* has been largely investigated [24] Its antimicrobial properties have been studied [25,26], showing its antioxidant, anti-inflammatory, vermifuge and antifungal potentials. In particular, the essential oil's capacity to inhibit two important virulence factors in *Candida albicans*, *Cryptococcus neoformans* and several dermatophytes and *Aspergillus* spp has been demonstrated.

There are no reports on the use of *Thymus* essential oil for leather disinfection to date. Thus, the main purpose of this study was to undertake the first study of the application of the *Thymus* essential oil to heritage leather disinfection under a vacuum system Consequently, in the context of our ongoing research on endemic Sicilian plants [27–29] and the biological activity of essential oils [30–32], and in consideration of the important antibacterial properties of the essential oil *Thymus* demonstrated in the aforementioned articles, we decided to utilize the EO of *Thymus vulgaris* as natural biocide against bacteria affecting two *Tholu Bommalu*, typical Indian leather puppets, which were stored at the International Puppets Museum "Antonio Pasqualino". We exploited the high vacuum inside ad hoc chambers in order to speed up the disinfection process. The activity of the essential oil of *T. vulgaris* was then compared with that of Sicilian *Crithmum maritimum* EO. These results were compared with the results of using a conventional synthetic biocide, benzalkonium chloride, as a disinfectant.

2. Results and Discussion

2.1. Gas Chromatography and Mass Spectrometry (GC-MS) Analysis of the Essential Oil

The chemical composition of the essential oil of *Thymus vulgaris* was analyzed by GC-MS analysis and is reported in Table 1. Sixteen compounds, divided into three classes were identified, accounting for 97.97% of the total composition. In terms of compound classes, monoterpene hydrocarbons (49.96%) dominate the EO, with *p*-cymene as the most abundant compound (35.96%), followed by terpinen-4-ol (10.29%) and α-terpinene (8.85%). Oxygenated monoterpenes are also dominant (43.67%) with thymol (25.38%). In contrast sesquiterpene hydrocarbons accounted for only 4.34%; no oxygenated sesquiterpenes were identified in the chromatogram of the EO. Comparing different samples of *T. vulgaris* from Saudi Arabia [22] and France and Serbia [33] to our results, we found them to also be rich in not only thymol, *p*-cymene and α-terpineol but also in camphene and caryophyllene.

The chemical composition of the essential oil obtained from *Crithmum maritimum* was analyzed by GC-MS analysis and is reported in Table 2. Forty compounds, divided into four classes, were identified, accounting for 90.03% of the total oil. In terms of compound classes, monoterpene hydrocarbons (45.08%) dominate the EO, with β-myrcene as the most abundant compound (13.66%), followed by *p*-cymene (11.67%) and β-phellandrene (6.57%). Oxygenated monoterpenes are also dominant (40.03%) with thymol acetate (14.38%). In contrast, hydrocarbon and oxygenated sesquiterpenes accounted for only 1.94% and 0.50% respectively.

Table 1. Chemical composition of *T. vulgaris* essential oil.

No.	Compounds	LRI [a]	LRI [b]	Area (%)	Ident. [c]	Ref.
1	*β*-Pinene	974	1070	0.08	1, 2, 3	
2	*cis*-Carane	978	1033	0.90	1, 2	
3	*β*-Myrcene	990	1079	2.54	1, 2, 3	
4	3-Carene	1011	1129	0.48	1, 2	
5	*α*-Terpinene	1018	1140	8.85	1, 2, 3	
6	*p*-Cymene	1021	1224	35.96	1, 2, 3	
7	(3*E*)-3-Ethyl-2,5-dimethyl-1,3-hexadiene [d]	1031	976	0.04	1, 2	
8	Eucalyptol	1035	1211	1.56	1, 2, 3	
9	Terpinolene	1084	1264	1.11	1, 2	
10	*β*-Linalool	1098	1498	5.99	1, 2, 3	
11	*trans*-Pinocarveol	1139	1666	0.45	1, 2	
12	Terpinen-4-ol	1177	1612	10.29	1, 2	
13	Thymol	1267	2156	25.38	1, 2, 3	
14	*α*-Copaene	1376	1487	0.14	1, 2	[34]
15	Caryophyllene	1419	1598	4.13	1, 2, 3	
16	*δ*-Cadinene	1530	1722	0.07	1, 2	
	Monoterpene Hydrocarbons			49.96		
	Oxygenated Monoterpenes			43.67		
	Sesquiterpene Hydrocarbons			4.34		
	Total			97.97		

[a]: retention index on a HP-5MS apolar column; [b]: retention index on a DB-Wax polar column. [c]: 1 = retention index identical to bibliography; 2 = identification based on comparison of MS; 3 = retention time identical to authentic compounds; [d]: tentative identification. Compounds are classified in order of linear retention time of apolar column.

Table 2. Chemical composition of *C. maritimum* essential oil.

No.	Compounds	LRI [a]	LRI [b]	Area (%)	Ident. [c]	Ref.
1	*α*-Pinene	932	1019	5.51	1, 2, 3	
2	Camphene	944	1060	5.16	1, 2, 3	
3	Sabinene	966	1114	0.21	1, 2	
4	*β*-Pinene	974	1070	0.55	1, 2, 3	
5	*β*-Myrcene	990	1079	13.66	1, 2, 3	
6	*α*-Phellandrene	998	1166	0.88	1, 2	
7	*p*-Cymene	1021	1224	11.67	1, 2, 3	
8	*β*-Phellandrene	1022	1201	6.57	1, 2	
9	*γ*-Terpinene	1049	1233	0.87	1, 2,	
10	*β*-Linalool	1098	1498	0.43	1, 2, 3	
11	*cis*-2-Menthenol	1118	1527	0.39	1, 2	[35]
12	Camphor	1122	1501	0.27	1, 2, 3	[36]
13	3-Methylbutyl pentanoate	1134	1345	0.16	1, 2	
14	*trans*-2-Menthenol	1136	1561	0.56	1, 2	[37]
15	Camphene hydrate	1148	1583	0.09	1, 2	
16	Terpinen-4-ol	1177	1612	3.52	1, 2	
17	Pentyl pentanoate [d]	1185	1402	0.51	1, 2	
18	*trans*-3(10)-Caren-2-ol	1194	1525	0.18	1, 2	[38]
19	1-Methylhexyl butanoate	1197	1371	0.63	1, 2	[39]
20	Carvone	1218	1709	0.24	1, 2, 3	
21	*cis*-Carveol	1222	1820	0.19	1, 2	
22	Thymol methyl ether	1235	1587	1.23	1, 2	

Table 2. *Cont.*

No.	Compounds	LRI [a]	LRI [b]	Area (%)	Ident. [c]	Ref.
23	2-Methylcyclohexyl butyrate [d]	1241	1407	0.37	1, 2	
24	4,7-Dimethylbenzofuran	1260	1619	0.54	1, 2	[40]
25	Thymol	1267	2156	2.62	1, 2, 3	
26	Bornyl acetate	1270	1555	2.69	1, 2	
27	Piperitone	1277	1730	0.40	1, 2	
28	2-Undecanone	1295	1614	0.27	1, 2	
29	Thymol acetate	1352	1813	14.38	1, 2, 3	
30	4-Cyclohexyl-2-butanone [d]	1371	1823	0.10	1, 2	
31	Damascenone	1385	1810	0.48	1, 2	
32	α-Ionol	1395	1881	0.29	1, 2	
33	1,4-Dimethoxy-2-tert-butylbenzene	1408	1871	1.23	1, 2	[41]
34	α-Methylbenzyl propionate [d]	1412	1853	0.99	1, 2	
35	2-Methyl-6-(2-propenyl)phenol [d]	1423	1889	3.30	1, 2	
36	Geranyl acetone	1456	1859	0.30	1, 2	
37	Bornyl butyrate	1463	1760	1.30	1, 2	
38	γ-Cadinene	1521	1776	1.94	1, 2	
39	2,3,4-Trimethylacetophenone	1548	1882	5.35	1, 2	
	Monoterpene Hydrocarbons			45.08		
	Oxygenated Monoterpenes			40.03		
	Sesquiterpene Hydrocarbons			1.94		
	Oxygenated Sesquiterpenes			0.50		
	Others			2.48		
	Total			90.03		

[a]: retention index on a HP-5MS apolar column; [b]: retention index on a DB-Wax polar column. [c]: 1 = retention index identical to bibliography; 2 = identification based on comparison of MS; 3 = retention time identical to authentic compounds; [d]: tentative identification. Compounds are classified in order of linear retention time of apolar column.

2.2. Antimicrobial Activity

In order to test the antimicrobial activity of *T. vulgaris* and *C. maritimum* EOs against bacteria species isolated from Tholu Bommalu, an agar disk diffusion method (ADD) was performed (results are listed in Table 3). As far as the *T. vulgaris* EO was concerned, *Georgenia* sp. isolated colonies were the most susceptible to the oil; in this case, the antimicrobial activity was so high that the inhibition halos were confluent. Every bacterial colony showed a relevant sensitivity to both EO solutions (50% and 100%), which were able to produce inhibition halos of up to 33 mm. In contrast, the *C. maritimum* essential oil did not show inhibition halos, except in the case of *Bacillus* sp., although these inhibition halos were smaller than those of *T. vulgaris*.

Table 3. Antimicrobial activity of *T. vulgaris* EO using ADD method.

EO/Solvents	Conc. (%)	Inhibition Halos (mm) [a] × Bacterial Specie			
		Bacillus	*Georgenia*	*Streptomyces*	*Ornithinibacillus*
T. vulgaris EO	50	33	confl. [b]	42	39
	100	38	confl. [b]	46	54
C. maritimum EO	50	9	0	0	0
	100	10	0	0	0
Benzalkonium chloride	3	7	9	8	6
Pentane	100	0	0	0	0

[a] Inhibition halo diameter, including paper disk diameter (6 mm). Sensitive strains > 9 mm; not sensitive < 9 mm.
[b] Confluent: microbial growth on the entire surface of agar medium.

The antimicrobial activity of the *T. vulgaris* EO was displayed against all isolated colonies and clearly had a greater effect than that of the controls (benzalkonium chloride

(BAK), the reference biocide, and pentane, as shown in Figure 1). Completely in contrast was the antimicrobial activity of the *C. maritimum* EO, which produced inhibition halos smaller than those produced by the controls.

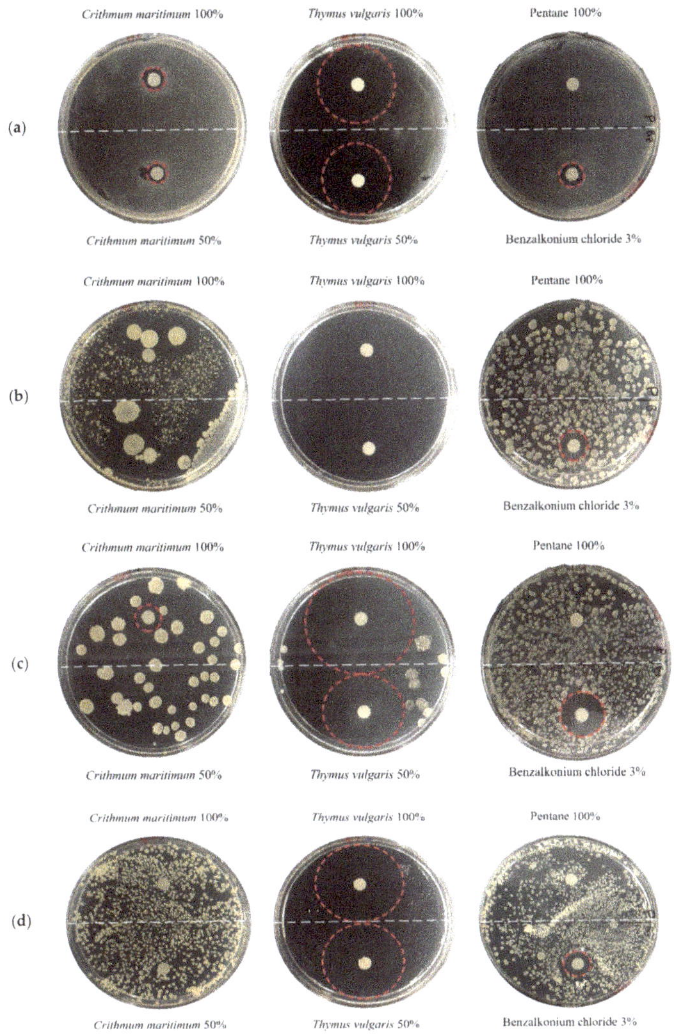

Figure 1. Inhibition halos highlighted using ADD method with *T. vulgaris* and *C. maritimum* EOs and control BAK against (**a**) *Bacillus* sp., (**b**) *Georgenia* sp., (**c**) *Ornithinibacillus* sp. and (**d**) *Streptomyces* sp.

The biocide properties of the *T. vulgaris* EO are in agreement with its chemical composition. Several studies highlighted how EOs rich in phenolic compounds, such as thymol or carvacrol, have a strong antimicrobial activity. In particular, thymol seems to be bioactive against more than twelve different bacterial colonies [21,42].

2.3. Use of a Vaccum Chamber for the Disinfection Process

Considering the huge presence of microbial colonies on *Tholu Bommalu* and the fragility of the parchment support, the aim was to find a disinfection process that was minimally invasive but also fast, allowing replicability in the future for the entire collection of Indian shadow puppets stored in the International Puppets Museum, "Antonio Pasqualino".

The study focused on observing the growth of isolated microbial colonies in response to their exposure to volatile compounds of the *Thymus vulgaris* EO, under both normal evaporation conditions and under vacuum inside a Plexiglas chamber (Figure 2).

Figure 2. *Tholu Bommalu* exposed to volatile compounds of EO inside the Plexiglas chamber under vacuum.

In order to track the microbial growth on *Tholu Bommalu* fragments affected by bacterial colonies, surface sampling was performed weekly both on the samples in the clean chamber and in the vacuum chamber. The exposure of the fragments to EO volatile compounds lasted a total of five weeks. *T. vulgaris* essential oil manifested a marked reduction in microbial growth over time. This was observed in both the group of samples. However, a clear difference was observed between the two methods, demonstrating that the exposure under vacuum allowed for faster inhibition of microbial growth.

Microbial in vitro cultures were performed on both nutrient agar (Oxoid, Thermo Fisher: 1.0 g Lab-Lemco powder; 5.0 g peptone; 5.0 g NaCl; 15.0 g agar/liter of dH_20) and Sabouraud dextrose agar (Oxoid, Thermo Fisher: 40.0 g dextrose; 10.0 g peptone; 15.0 g agar/liter of dH_2O). Nutrient agar supports a wide range of microorganisms, whereas Sabouraud is mainly used to cultivate fungi or filamentous bacteria.

The results from these different culture media were combined and evaluated, revealing the microbial growth trend over time for each fragment (as shown in Figure 3).

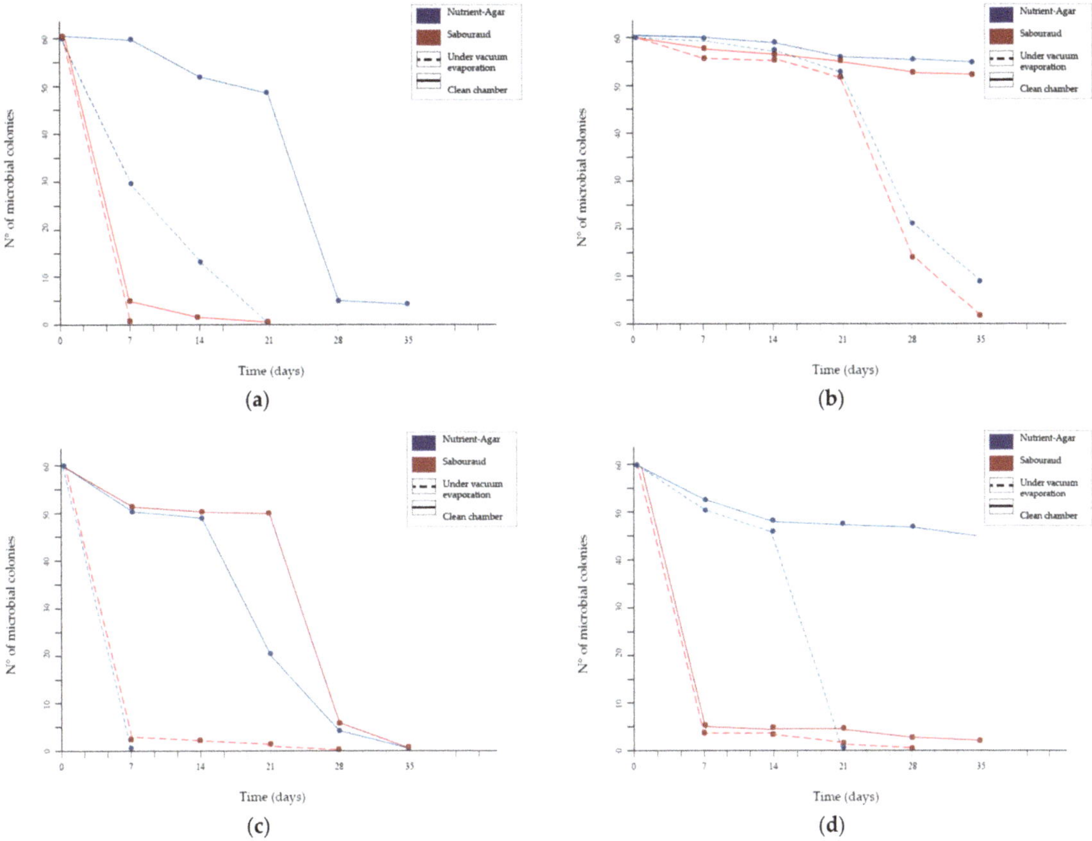

Figure 3. Microbial growth trends over time for fragments "A" (**a**); "B" (**b**); "C" (**c**) and "D" (**d**).

In particular, one of the clearest results was related to the colonies sampled from fragments "A" and "C" and grown on nutrient agar, which were significantly reduced in fragment "A" and completely absent in fragment "C" from the first week of exposure to EO volatile compounds. On the contrary, the same colonies exposed in the clean chamber without the vacuum showed confluent growths.

Colonies collected on surfaces of the "A" and "C" fragments showed differences in terms of the microbial load. Specifically, after only one week of exposure to EO volatile compounds under vacuum no microbial growth was observed, whereas microbial growth was still present under ambient conditions, although considerably reduced.

By the third week, the exposure to *T. vulgaris* EO volatile compounds under vacuum conditions resulted in the microbial colonies on "A" and "C" fragments being completely eradicated. In contrast, under environmental conditions, the microbial colonies were still present, revealing a considerable reduction in the microbial load only after the fourth week of the disinfection process.

In regard to fragments "B" and "D", the microbial colonies present on their surfaces immediately proved to be more resistant to the action of the EO, showing confluent growth in all cases. The first result of microbial growth inhibition was observed from the fourth week of exposure. Under vacuum conditions, the colonies present in "B" were significantly reduced on both nutrient and Sabouraud agar when compared to the same fragments exposed under environmental conditions. Regarding fragment "D", the microbial growth tested on Sabouraud agar was almost completely eradicated under vacuum conditions and was significantly reduced under environmental conditions. Microbial growth remained confluent on nutrient agar.

Reaching the fifth week of exposure to EO volatile compounds, microbial growth had almost completely stopped under vacuum conditions. The only exception was fragment "B", although in this case, the microbial load was still reduced. This result was completely in contrast with the exposure under environmental conditions, in which confluent growth was still present after 35 days of disinfection, although only for some taxa.

An important aspect, which must be emphasized, is that a non-topical and direct application on the leather fragments and on the entire artifact using only the vapor pressure of the EO did not cause decolorization processes. This aspect was checked with the aid of a colorimeter (NH300 Colorimeter, 3NH Shanghai Co., Ltd.), evaluating parameters such as total color differences (ΔE), a * (red-green), L * (lightness) and b * (yellow-blue) before and after the EO treatment.

Furthermore, to avoid discomfort and residual odor of the same on the artifact, the latter was subjected to three high vacuum cycles, maintaining stable thermo-hygrometric parameters, in order to eliminate the possible presence of some volatile components of the essential oil.

3. Materials and Methods

3.1. Essential Oils

The essential oil of *Thymus vulgaris* (100% pure, 100 mL) was purchased from Authentic Oil Co, Unit 1, Little Castle Farm, Raglan, Monmouthshire, NP15 2BX, UK. In this case, the aerial parts of *T. vulgaris* come from Daran, Karaman, Turkey.

The aerial parts of *Crithmum maritimum* were collected in Addaura (Mondello), Palermo, Sicily, Italy (38°11′31″N, 13°20′41″E, 2 m m.s.l.), in June 2020 and a voucher specimen was deposited in STEBICEF Department, University of Palermo (PAL 348/20).

A quantity of 300 g of the aerial parts of *C. maritimum* were subjected to hydrodistillation for 3 h using Clevenger's apparatus [43]. The oil (yield 2.08% (*v*/*w*)) was dried with anhydrous sodium sulfate, filtered and stored in a freezer at −20 °C until the time of analyses.

3.2. GC-MS Analysis of Essential Oil

Analyses of essential oils were performed according to the procedure reported by Rigano et al. [44]. EO analysis was performed using an Agilent 7000 C GC (Agilent Technologies, Inc., Santa Clara, CA, USA) system, fitted with a fused silica Agilent DB-Wax capillary column (30 m × 0.25 mm i.d.; 0.25 μm film thickness) and coupled to an Agilent triple quadrupole Mass Selective Detector MSD 5973 (Agilent Technologies, Inc., Santa Clara, CA, USA). The settings were as follows: ionization voltage, 70 eV; electron multiplier energy, 2000 V; transfer line temperature, 295 °C; solvent delay, 4 min. The other GC analysis was performed with a Shimadzu QP 2010 plus equipped with an AOC-20i autoinjector (Shimadzu, Kyoto, Japan), gas chromatograph equipped with a flame ionization detector (FID), a capillary column (HP-5MS) (30 m × 0.25 mm i.d.; film thickness, 0.25 μm) and a data processor. The oven program was as follows: temperature increase at 40 °C for 5 min at a rate of 2 °C/min up to 260 °C and then isothermal amplification for 20 min. Helium was used as the carrier gas (1 mL min^{-1}). The injector and detector temperatures were set at 250 °C and 290 °C, respectively. An amount of 1 μL of each oil solution (3% EO/hexane *v*/*v*) was injected with a split mode. Linear retention indices (LRI) were determined by using retention times of *n*-alkanes (C_8-C_{40}), and the peaks were identified by comparison with mass spectra and by comparison to their relative retention indices with WILEY275 (Wiley), NIST 17 (NIST, The National Institute of Standards and Technology, Gaithersburg, MD, USA), ADAMS (Allured Business Media, Carol Stream, IL, USA) and FFNSC2 (Shimadzu, Kyoto, Japan) libraries.

3.3. ADD Control Solutions

Benzalkonium chloride (3% v/v) (Sigma Aldrich, St. Louis, MO 68178 United States), the reference biocide, and pentane (Sigma Aldrich, 100%, St. Louis, MO 68178, USA), the solvent accounting for fifty percentage of the EO solution, were controls used in agar disc diffusion (ADD) assays [45,46]. ADD assays, performed on nutrient or Sabouraud agar media, were performed twice.

3.4. Microbial Taxa

Microbial patina on the leather substrata of *Tholu Bommalu* was sampled by sterile stubs. Distinctive colonies, in morphology and/or pigmentation, were isolated on the nutrient agar in vitro culture. Taxonomical identification was performed by molecular analysis of the 16S rDNA gene or ITS1-2 containing DNA region [47]. Specifically, *Bacillus* sp., *Georgenia* sp., *Ornithinibacillus* sp. and *Streptomyces* sp. were identified as the prevalent genera (Gram + bacteria).

3.5. Antibacterial Activity Assays

The ADD in vitro method was performed to evaluate the antibacterial activity of EO and BAK solutions. An amount of 10 µL of each bacterial broth culture (normalized to the concentration of 1 × 106 CFU/mL) was uniformly spread using a sterile Drigalsky spatula on the nutrient agar surface, and the surface was allowed to dry (1 h at 30 °C). A sterile paper disc (6 mm in diameter, Dutscher papier, FR) imbibed with 10 µL of the *T. vulgaris* essential oil (100% or 50%) or control solutions (BAK 3% v/v and pentane 100%,) was leaned on the agar medium (9 cm Petri dishes) surface. Due to the non-miscibility of essential oils in water, fifty percent of the EO solution was composed of pentane. After incubation at 30 °C for 24 h, inhibition halos (i.hs.) differing in diameter (mm) were revealed, reflecting the antimicrobial activity and categorized as a sensitive strain (i.h. > 9 mm) or a resistant strain (i.h. < 9 mm). *Georgenia* sp. colonies were the most susceptible isolated bacteria, showing a relevant sensitivity to the essential oil with an inhibition zone of up to 18 mm. At the other end of the spectrum, *Bacillus* sp. was the most resistant with an i.h. of 14–17 mm diameters. Table 2 shows the results of the antibacterial activity using the ADD method (Figure 1). Pentane, accounting for fifty percent of the EO solution, was applied as a control to evaluate the real biocide power of the diluted EO, and it did not show inhibition halos.

3.6. Vacuum Chamber

In order to maintain stable thermo-hygrometric parameters (13–18 °C; 50–60% RH), a saturated solution of $MgCl_2$ was placed inside the vacuum system. In addition, the chamber was filled with the *Thymus vulgaris* EO, calculating the right amount to saturate atmosphere using the equation of a perfect gas; a thermo-hygrometer; and the leather samples, placed on a reticulated support made of cardboard and nylon to allow the volatile compounds to easily reach all points of the samples. To create vacuum, a diaphragm vacuum pump (Type MZ 2C, Ser.No. 20635805) was connected by a tube to the top of the chamber and was kept running for about one minute and then disconnected.

4. Conclusions

The present work focused on determining the yield, the chemical composition and antimicrobial properties of *Thymus vulgaris* and *Crithmum maritimum* essential oils. The essential oil of *T. vulgaris* was characterized by a large presence of monoterpenes, *p*-cymene (35.96%), terpinen-4-ol (10.29%), α-terpinene (8.85%) and thymol (25.38%), while *C. maritimum* essential oil was dominated by β-myrcene (13.66%), followed by *p*-cymene (11.67%), β-phellandrene (6.57%) and thymol acetate (14.38%). In addition, we suggested an integrated approach to the most common disinfection processes, by using a vacuum chamber in order to allow the mechanism to act faster than usual. *Thymus* oil, in vapor phase, is a strong inhibitor for bacterial growth; every bacterial colony isolated showed significant sensitivity to both EO solutions (50%, 100%) and was able to produce inhibition halos up to

33 mm. All colonies under vacuum conditions were significantly reduced compared to the same ones exposed to environmental conditions. In contrast, the *C. maritimum* essential oil did not show inhibition halos. The potential use of commercial plant essential oils, together with the achievements reached during the in vitro and in situ applications to control the growth of bacterial taxa, led us to hypothesize their use as natural biocides, replacing the most common toxic biocide usually used in the conservation of cultural heritage.

Author Contributions: Conceptualization, M.B. Methodology, G.D., N.B,. and B.G. Validation, M.B. Formal analysis, G.D. and N.B. Investigation, F.P., N.B., and M.B. Resources, M.B. Writing—original draft preparation, G.D., N.B., F.P. and M.B. Writing—review and editing, M.B. All authors have read and agreed to the published version of the manuscript.

Funding: This research received no external funding.

Acknowledgments: The authors thank Giuseppe Gallo and Teresa Faddetta for the collaboration in microbial analysis and the Museo Internazionale delle Marionette, "A. Pasqualino" Palermo, Italy, for providing the artifact and active assistance. The results regarding the application of essential oils in contrast to microbial colonization of Indian leather puppets are part of G. D'Agostino's final examination in his master's degree in Conservation and Restoration of Cultural Heritage. He graduated cum laude and qualified as an Italian Restorer of Cultural Heritage (Italian Ministry of Culture) at the University of Palermo, Italy.

Conflicts of Interest: The authors declare no conflict of interest.

References

1. Casiglia, S.; Bruno, M.; Senatore, F. Volatile constituents of *Dianthus rupicola* Biv. from Sicily: Activity against microorganisms affecting cellulosic objects. *Nat. Prod. Res.* **2014**, *28*, 1739–1746. [CrossRef]
2. Casiglia, S.; Bruno, M.; Senatore, F. Activity against microorganisms affecting cellulosic objects of the volatile constituents of *Leonotis nepetaefolia* from Nicaragua. *Nat. Prod. Commun.* **2014**, *9*, 1637–1639. [CrossRef]
3. Casiglia, S.; Ben Jemia, M.; Riccobono, L.; Bruno, M.; Scandolera, E.; Senatore, F. Chemical composition of the essential oil of *Moluccella spinosa* L. (Lamiaceae) collected wild in Sicily and its activity on microorganisms affecting historical textiles. *Nat. Prod Res.* **2015**, *29*, 1201–1206. [CrossRef] [PubMed]
4. Casiglia, S.; Bruno, M.; Senatore, F.; Senatore, F. Composition of the essential oil of *Allium neapolitanum* Cirillo growing wild in Sicily and its activity on microorganisms affecting historical art crafts. *J. Oleo Sci.* **2015**, *64*, 1315–1320. [CrossRef] [PubMed]
5. Casiglia, S.; Riccobono, L.; Bruno, M.; Senatore, F.; Senatore, F. Chemical composition of the essential oil from *Pulicaria vulgaris* var. graeca (Sch.-Bip.) Fiori (Asteraceae) grown wild in Sicily and its antimicrobial activity. *Nat. Prod. Res.* **2016**, *30*, 259–267 [CrossRef] [PubMed]
6. Casiglia, S.; Bruno, M.; Senatore, F.; Senatore, F. Chemical composition of the essential oil of *Bupleurum fontanesii* Guss. ex Caruel (Apiaceae) growing wild in Sicily and its activity on microorganisms affecting historical art crafts. *Nat. Prod. Commun.* **2016**, *11* 105–108.
7. Casiglia, S.; Riccobono, L.; Bruno, M.; Rosselli, S.; Senatore, F.; Senatore, F. Chemical composition of the essential oil from Thapsia garganica L. (Apiaceae) grown wild in Sicily and its antimicrobial activity. *Nat. Prod. Res.* **2016**, *30*, 1042–1052. [CrossRef] [PubMed]
8. Casiglia, S.; Bruno, M.; Fontana, G.; Senatore, F. Chemical composition of the essential oil of *Mentha pulegium* L. (Linné) growing wild in Sicily and its activity on microorganisms affecting historical art crafts. *Nat. Prod. Commun.* **2017**, *12*, 311–315.
9. Casiglia, S.; Bruno, M.; Bramucci, M.; Quassinti, L.; Lupidi, G.; Fiorini, D.; Maggi, F. *Kundmannia sicula* (L.) DC growing in Sicily (Italy) as a rich source of germacrene D. *J. Essent. Oil Res.* **2017**, *29*, 437–442. [CrossRef]
10. Casiglia, S.; Bruno, M.; Senatore, F.; Rosselli, S. Chemical composition of essential oils of *Anthemis secundiramea* Biv. subsp. *secundiramea* (Asteraceae) collected wild in Sicily and their activity on microorganisms affecting historical art craft. *Nat. Prod. Res* **2019**, *33*, 970–979. [CrossRef]
11. Bruno, M.; Modica, A.; Catinella, G.; Canlı, C.; Arasoglu, T.; Çelik, S. Chemical composition of the essential oil of *Centaurea tomentella* Hand.-Mazz. and C. haussknechtii Boiss. (Asteraceae) collected wild in Turkey and its activity on microorganisms affecting historical art craft. *Nat. Prod. Res.* **2019**, *33*, 1092–1100. [CrossRef] [PubMed]
12. Cappitelli, F.; Cattò, C.; Villa, F. The control of cultural heritage microbial deterioration. *Microorganisms* **2020**, *8*, 154. [CrossRef]
13. Sequeira, S.; Cabrita, E.J.; Macedo, M.F. Antifungals on paper conservation: An overview. *Int. Biodeterior. Biodegrad.* **2012**, *74* 67–86. [CrossRef]
14. Matusiak, K.; Machnowski, W.; Wrzosek, H.; Polak, J.; Rajkowska, K.; Smigielski, K.; Kunicka-Styczynska, A. Application of *Cinnamomum zeylanicum* essential oil in vapour phase for heritage textiles disinfection. *Int. Biodeterior. Biodegrad.* **2018**, *131*, 88–96 [CrossRef]

15. Cimanga, K.; Kambu, K.; Tona, L.; Apers, S.; De Bruyne, T.; Hermans, N.; Tottè, J.; Pieters, L.; Vlietinck, A.J. Correlation between chemical composition and antibacterial activity of essential oils of some aromatic medicinal plants growing in the Democratic Republic of Congo. *J. Ethnopharmacol.* **2002**, *79*, 213–220.
16. Scaffaro, R.; Maio, A.; D'Arrigo, M.; Lo Presti, F.; Marino, A.; Bruno, M.; Nostro, A. Morpho-mechanical and antimicrobial properties of *Coridothymus capitatus* L. essential oil loaded into ultrafine poly (lactic acid) fibers prepared by electrospinning. *Future Microbiol.* **2020**, *15*, 1379–1392. [CrossRef] [PubMed]
17. De Feo, V.; Bruno, M.; Tahiri, B.; Napolitano, F.; Senatore, F. Chemical composition and antibacterial activity of the essential oils of *Thymus spinulosus* Ten. (Lamiaceae). *J. Agric. Food Chem.* **2003**, *51*, 3849–3853. [CrossRef]
18. De Martino, L.; Bruno, M.; Formisano, C.; De Feo, V.; Napolitano, F.; Rosselli, S.; Senatore, F. Chemical composition and antimicrobial activity of the essential oils from two species of *Thymus* growing wild in Southern Italy. *Molecules* **2009**, *14*, 4614–4624. [CrossRef]
19. Casiglia, S.; Bruno, M.; Scandolera, E.; Senatore, F. Influence of harvesting time on composition of the essential oil of *Thymus capitatus* (L.) Hoffmanns. & Link. growing wild in northern Sicily and its activity on microorganisms affecting historical art crafts. *Arab. J. Chem.* **2019**, *12*, 2704–2712.
20. Palla, F.; Bruno, M.; Mercurio, F.; Tantillo, A.; Rotolo, V. Essential oil as natural biocides in conservation of cultural heritage. *Molecules* **2020**, *25*, 730. [CrossRef]
21. Gagliano Candela, R.; Maggi, F.; Lazzara, G.; Rosselli, S.; Bruno, M. The essential oil of *Thymbra capitata* and its application as a biocide on stone and derived surfaces. *Plants* **2019**, *8*, 300. [CrossRef]
22. Al-Asmari, A.K.; Al-Faraidy, A.A.; Almuhaiza, M.S. Chemical composition of essential oil of *Thymus vulgaris* collected from Saudi Arabian market. *Asian Pac. J. Trop. Biomed.* **2017**, *7*, 147–150. [CrossRef]
23. De Carvalho, R.J.; Targino de Souza, G.; Gonçalves Honório, V.; Pereira de Sousa, J.; Conceição, M.L.; Maganani, M.; Leite de Souza, E. Comparative inhibitory effects of *Thymus vulgaris* L. essential oil against *Staphylococcus aureus*, *Listeria monoctogenes* and mesophilic starter co-culture in cheese-mimicking models. *Food Microbiol.* **2015**, *52*, 59–65. [CrossRef]
24. Pavela, R.; Maggi, F.; Cianfaglione, K.; Bruno, M.; Benelli, G. Larvicidal activity of essential oils of five Apiaceae taxa and some of their main constituents against *Culex quinquefasciatus*. *Chem. Biodivers.* **2017**, *14*. [CrossRef]
25. Alves-Silva, J.; Guerra, I.; Gonçalves, M.J.; Cavaleiro, C.; Cruz, M.T.; Figueirinha, A. Chemical composition of *Crithmum maritimum* L. essential oil and hydrodistillation residual water by GC-MS and HPLC-DAD-MS/MS, and their biological activities. *Ind. Crops Prod.* **2020**, *149*, 1–9. [CrossRef]
26. Meot-Duros, L.; Magné, C. Antioxidant activity and phenol content of *Crithmum maritimum* L. leaves. *Plant Physiol. Biochem.* **2009**, *47*, 37–41. [CrossRef] [PubMed]
27. Gagliano Candela, R.; Ilardi, V.; Badalamenti, N.; Bruno, M.; Rosselli, S.; Maggi, F. Essential oil compositions of *Teucrium fruticans*, *T. scordium* subsp. *scordioides* and *T. siculum* growing in Sicily and Malta. *Nat. Prod. Res.* **2020**, in press. [CrossRef] [PubMed]
28. Ilardi, V.; Badalamenti, N.; Bruno, M. Chemical composition of the essential oil from different vegetative parts of *Foeniculum vulgare* subsp. *piperitum* (Ucria) Coutinho (Umbelliferae) growing wild in Sicily. *Nat. Prod. Res.* **2020**, in press.
29. Badalamenti, N.; Ilardi, V.; Rosselli, S.; Bruno, M.; Maggi, F.; Leporini, M.; Falco, T.; Loizzo, M.R.; Tundis, R. *Ferulago nodosa* subsp. geniculata (Guss.) Troia & Raimondo from Sicily (Italy): Isolation of essential oil and evaluation of its bioactivity. *Molecules* **2020**, *25*, 3249.
30. Badalamenti, N.; Bruno, M.; Gagliano Candela, R.; Maggi, F. Chemical composition of the essential oil of *Elaeoselinum asclepium* (L.) Bertol subsp. *meoides* (Desf.) Fiori (Umbelliferae) collected wild in Central Sicily and its antimicrobial activity. *Nat. Prod. Res.* **2020**, in press. [CrossRef] [PubMed]
31. Di Napoli, M.; Maresca, V.; Varcamonti, M.; Bruno, M.; Badalamenti, N.; Basile, A.; Zanfardino, A. (+)-(*E*)-Chrysanthenyl acetate: A molecule with interesting biological properties contained in the *Anthemis secundiramea* (Asteraceae) flowers. *Appl. Sci.* **2020**, *10*, 6808. [CrossRef]
32. Badalamenti, N.; Ilardi, V.; Bruno, M.; Pavela, R.; Boukouvala, M.C.; Kavallieratos, N.G.; Maggi, F.; Canale, A.; Benelli, G. Chemical composition and broad-spectrum insecticidal activity of the flower essential oil from an ancient Sicilian food plant, *Ridolfia segetum*. *Agriculture* **2021**, *11*, 304. [CrossRef]
33. Satyal, P.; Murray, B.; McFeeters, R.; Setzer, W. Essential oil characterization of *Thymus vulgaris* from various geographical locations. *Foods* **2016**, *5*, 70. [CrossRef]
34. Salido, S.; Altarejos, J.; Nogueras, M.; Sánchez, A.; Pannecouque, C.; Witvrouw, M.; De Clercq, E. Chemical studies of essential oils of *Juniperus oxycedrus* ssp. *badia*. *J. Ethnopharmacol.* **2002**, *81*, 129–134. [PubMed]
35. Basta, A.; Tzakou, O.; Couladis, M. Composition of the leaves essential oil of *Melissa officinalis* s. l. from Greece. *Flavour Fragr. J.* **2005**, *20*, 642–644. [CrossRef]
36. Shang, C.; Hu, Y.; Deng, C.; Hu, K. Rapid determination of volatile constituents of *Michelia alba* flowers by gas chromatography—Mass spectrometry with solid-phase microextraction. *J. Chromatogr. A* **2001**, *942*, 283–288. [CrossRef]
37. Schmidt, J.M.; Noletto, J.A.; Vogler, B.; Setzer, W.N. Abaco Bush Medicine: Chemical composition of the essential oils of four aromatic medicinal plants from Abaco Island, Bahamas. *J. Herbs Spices Med. Plants* **2006**, *12*, 43–65. [CrossRef]
38. Iranshahi, M.; Amin, G.; Sourmaghi, M.S.; Shafiee, A.; Hadjiakhoondi, A. Sulphur-containing compounds in the essential oil of the root of *Ferula persica* Willd. var. *persica*. *Flavour Fragr. J.* **2006**, *21*, 260–261. [CrossRef]

39. Shiota, H. New esteric components in the volatiles of banana fruit (*Musa sapientum* L.). *J. Agric. Food Chem.* **1993**, *41*, 2056–2062. [CrossRef]
40. Varlet, V.; Knockaert, C.; Prost, C.; Serot, T. Comparison of odor-active volatile compounds of fresh and smoked salmon. *J. Agric. Food Chem.* **2006**, *54*, 3391–3401. [CrossRef]
41. Huang, B.; Qin, L.; Chu, Q.; Zhang, Q.; Gao, L.; Zheng, H. Comparison of headspace SPME with hydrodistillation and SFE for analysis of the volatile components of the roots of *Valeriana officinalis* var. latifolia. *Chromatographia* **2009**, *69*, 489–496. [CrossRef]
42. Naigre, R.; Kalck, P.; Roques, C.; Roux, I.; Michel, G. Comparison of antimicrobial properties of monoterpenes and their carbonylated products. *Planta Med.* **1996**, *62*, 275–277. [CrossRef]
43. EDQM (Council of Europe). *European Pharmacopoeia*, 6th ed.; EDQM: Strasbourg, France, 2008.
44. Rigano, D.; Formisano, C.; Rosselli, S.; Badalamenti, N.; Bruno, M. GC and GC—MS Analysis of volatile compounds from *Ballota nigra* subsp. uncinata collected in Aeolian Islands, Sicily (Southern Italy). *Nat. Prod. Commun.* **2020**, *15*, 1–7. [CrossRef]
45. Balouiri, M.; Sadiki, M.; Ibnsouda, S.K. Methods for in vitro evaluating antimicrobial activity: A review. *J. Pharm. Anal.* **2016**, *6*, 71–79. [CrossRef]
46. Palla, F.; Caruana, E.; Di Carlo, E.; Rotolo, V. Plant essential oils in controlling fungal colonization on wooden substrate. *Borziana* **2021**, *2*, 5–14.
47. Johnson, J.S.; Spakowicz, D.J.; Hong, B.-Y.; Petersen, L.M.; Demkowicz, P.; Chen, L.; Leopold, S.R.; Hanson, B.M.; Agresta, H.O.; Gerstein, M.; et al. Evaluation of 16S rRNA gene sequencing for species and strain-level microbiome analysis. *Nat. Commun.* **2019**, *10*, 5029. [CrossRef]

plants

Article

New Insights on *Euphorbia dendroides* L. (*Euphorbiaceae*): Polyphenol Profile and Biological Properties of Hydroalcoholic Extracts from Aerial Parts

Antonella Smeriglio [1,*], Marcella Denaro [1], Domenico Trombetta [1], Salvatore Ragusa [2] and Clara Circosta [1]

1 Department of Chemical, Biological, Pharmaceutical and Environmental Sciences, University of Messina, Viale Ferdinando Stagno d'Alcontres 31, 98166 Messina, Italy; marcella.denaro@unime.it (M.D.); domenico.trombetta@unime.it (D.T.); clara.circosta@unime.it (C.C.)
2 Department of Health Sciences, University "Magna Graecia" of Catanzaro, Viale Europa, 88100 Catanzaro, Italy; srgusa@unicz.it
* Correspondence: antonella.smeriglio@unime.it; Tel.: +39-0906774039

Abstract: *Euphorbia dendroides* L. is a rounded shrub commonly found in the Mediterranean area well-known, since ancient times, for its traditional use. The aim of the present study was to investigate the phytochemical profile as well as the antioxidant and anti-inflammatory properties of flower (FE), leaf (LE), fruit (FrE), and branch (BE) hydroalcoholic extracts. For this purpose, a preliminary phytochemical screening followed by RP-LC-DAD-ESI-MS analysis, as well as several in vitro cell-free colorimetric assays, were carried out. Moreover, the toxicity of the extracts was investigated by the brine shrimp lethality assay. All extracts showed a high content of polyphenols, in particular phenolic acids (chlorogenic acid 0.74–13.80 g/100 g) and flavonoids (rutin 0.05–2.76 g/100 g and isovitexin 8.02 in BE). All the extracts showed strong and concentration-dependent antioxidant and anti-inflammatory activity with, on average, the following order of potency: FE, LE, FrE, and BE. Interestingly, all the extracts investigated did not show any toxicity on *Artemia salina*. Moreover, the only LD_{50} found (BE, 8.82 mg/mL) is well above the concentration range, which has been shown the biological properties. Considering this, this study offers the first evidence of the possible use of the polyphenol extracts from the aerial parts of *E. dendroides* as promising antioxidant and anti-inflammatory agents.

Keywords: *Euphorbia dendroides* L.; aerial parts; polyphenols; antioxidant activity; anti-inflammatory activity; toxicity

Citation: Smeriglio, A.; Denaro, M.; Trombetta, D.; Ragusa, S.; Circosta, C. New Insights on *Euphorbia dendroides* L. (*Euphorbiaceae*): Polyphenol Profile and Biological Properties of Hydroalcoholic Extracts from Aerial Parts. *Plants* **2021**, *10*, 1621. https://doi.org/10.3390/plants10081621

Academic Editors: Milan S. Stankovic, Paula Baptista and Petronia Carillo

Received: 22 July 2021
Accepted: 3 August 2021
Published: 6 August 2021

Publisher's Note: MDPI stays neutral with regard to jurisdictional claims in published maps and institutional affiliations.

Copyright: © 2021 by the authors. Licensee MDPI, Basel, Switzerland. This article is an open access article distributed under the terms and conditions of the Creative Commons Attribution (CC BY) license (https://creativecommons.org/licenses/by/4.0/).

1. Introduction

Euphorbia (Euphorbiaceae) is the third most common genus among flowering plants [1]. It is widespread around the world with more than 2000 species and with an exceptional diversity such as shrubs, vines, and grassy plants [1]. *Euphorbia* is well known, since ancient times, for its therapeutic activity on several gastrointestinal ailments, infections, skin irritations, body pain, microbial illness, sensory disorders, and as an antidote against snake venom [2–4]. Numerous species are commonly cultivated for ornamental purposes, such as *E. pulcherrima* Willd., *E. fulgens* Karw, *E. milii* Des Moul., *E. milii* var. *splendens* Ursch and Leandri, *E. tirucalli* L., and *E. lactea* Roxb [5].

Moreover, *E. pekinensis* Rupr., *E. kansui* Liou, *E. lathyris* L., *E. humifusa* Willd., and *E. maculata* L. were described by the Chinese Pharmacopoeia for their application in traditional medicine against gonorrhea, migraine, edema, and warts [6].

All plants belonging to the *Euphorbia* genus are characterized by the presence of an irritating latex, which plays a pivotal role in the first defense mechanism against insects, pathogens, and herbivores. This latex is a rich source of phytochemicals, which have been extensively investigated over time [7–9].

More than 650 types of diterpenes and triterpenes with several biological properties such as antiproliferative, antimicrobial, antiviral, anti-inflammatory, vasoactive, cytotoxic, neuroprotective, and anticancer, have been identified in the *Euphorbia* genus, thus, supporting the traditional use of the *Euphorbia* spp. [4,7]. Ingenol mebutate, a diterpene isolated from *E. peplus* L., showed significant activity against the early stage of actinic keratosis following topical application [10]. These promising biological activities led to an increased interest in *Euphorbia* spp. [4,8,11]. However, despite this, there are many under-investigated species, such as *E. dendroides* L., characterized by deciduous leaves, for which only limited information is available today. Moreover, polyphenols, are under-looked compounds in this plant genus.

E. dendroides is a rounded shrub or a small tree, commonly found in Mediterranean areas (Spain, France, Italy, Balkans, Greece, Turkey, Northern Africa, etc.) along coasts and especially on rocks, cliffs, and arid calcareous soils. The few studies available are mainly focused on its latex with particular reference to terpenoids [12–16], while to the authors knowledge, only one study is currently available on extracts obtained from the whole plant which investigated the antioxidant and anti-proliferative activity of this species [17].

In our previous study, the phytochemical profile as well as the antioxidant and anti-acetylcholinesterase activities of *E. dendroides* latex, were investigated [9]. The latex extract showed interesting biological properties without any significant toxicity on *Artemia salina*, paving the way for new potential uses of *E. dendroides* latex for example as a safe, effective and environmentally friendly insecticide [9].

Considering this, and given the lack of studies concerning the polyphenol extracts of this plant, we decided to focus our attention on the aerial parts (leaf, branch, flower, and fruit), with the aim to extend the knowledge about the polyphenol profile as well as on the antioxidant and anti-inflammatory properties of this plant species, highlighting also its potential ecotoxicity.

2. Results

In the present study, the phytochemical profile as well as the antioxidant, anti-inflammatory, and toxicity of leaf (LE), branch (BE), flower (FE), and fruit (FrE) hydroalcoholic extracts of *E. dendroides* L. (Figure 1) were evaluated for the first time.

(a) **(b)**

Figure 1. *Euphorbia dendroides* L. (**a**) and particular of the aerial parts (**b**). Original photo by SR.

The extraction method adopted has allowed the obtainment of high extraction yields ranging from 13.70 to 24.21%, with flowers, which showed the best extraction yield, followed by leaves, branches, and fruits.

2.1. Phytochemical Screening and Characterization of Polyphenol Profile

Table 1 shows the results of the preliminary phytochemical screening carried out by several in vitro colorimetric assays.

Table 1. Phytochemical screening of the *E. dendroides* leaf (LE), branch (BE), flower (FE), and fruit (FrE) hydroalcoholic extracts. Results were expressed as the mean ± standard deviation of three independent experiments in triplicate (n = 3).

Extract	Total Phenols	Total Flavonoids	Vanillin Index	Proanthocyanidins
	g GAE [a]/100 g DE [b]	g QE [c]/100 g DE	g CE [d]/100 g DE	g CyE [e]/100 g DE
LE	55.91 ± 1.89	84.85 ± 1.50	2.08 ± 0.14 &	2.06 ± 0.02 &
BE	26.54 ± 1.25 §	46.93 ± 4.00 §	0.44 ± 0.04 §	1.68 ± 0.08 §
FE	64.75 ± 4.63 *	73.75 ± 7.06 °	0.92 ± 0.05 *	2.23 ± 0.04 *
FrE	48.76 ± 4.25	83.11 ± 3.13	0.67 ± 0.02	3.54 ± 0.02

[a] GAE, Gallic acid equivalents; [b] DE, Dry extract; [c] QE, Quercetin equivalents; [d] CE, Catechin equivalents; [e] CyE, Cyanidin equivalents; * $p < 0.001$ vs. LE, BE and FrE; § $p < 0.001$ vs. LE and FrE; ° $p < 0.001$ vs. LE and BE; & $p < 0.001$ vs. FrE.

All extracts, albeit with substantial differences, showed a high content of polyphenols. FE showed the highest content of total phenols expressed as g of gallic acid equivalents (GAE)/100 g of dry extract (DE) followed by LE, FrE, and BE. Comparable amounts of flavonoids, expressed as g of quercetin equivalents (QE)/100 g DE, were detected in LE and FrE, followed by FE. On the contrary, a consistent decrease of flavonoids content was observed in BE. The vanillin index and proanthocyanin assays were carried out in order to evaluate the flavonols and proanthocyanidins content. Moreover, these two assays, in particular their ratio (vanillin index/proanthocyanidins), allow the calculation of the polymerization degree. All extracts, which showed the highest proanthocyanidins content with respect to the flavonols content, showed a low polymerization degree (≤1). This allows for postulating the high presence of monomeric molecules.

These preliminary results were confirmed by the determination of the polyphenol profile by reverse-phase liquid chromatography coupled with diode array detection and electrospray ion trap mass spectrometry (RP-LC-DAD-ESI-MS) analysis (Figure 2), which identified the FE as the richest source of polyphenols (17.23 g/100 g DE), followed by LE, FrE, and BE (Table 2).

More than 10 compounds were identified and quantified in each extract by RP-LC-DAD-ESI-MS analysis (Figure 1; Table 2). Among them, a prevalence of chlorogenic acid and rutin was recorded in all the extracts investigated, with the exception of BE, in which the most abundant compound was isovitexin (8.02 g/100 g DE). However, the amount of both compounds changed significantly depending on the aerial part investigated. Indeed, chlorogenic acid, although it is more abundant in FE (13.80 g/100 g DE), is still expressed at high levels and in comparable amounts to RE and LE. On the contrary, the rutin content, which is always higher in FE (2.76 g/100 g DE), decreases significantly (~4 fold) in the other two extracts (RE and LE). A completely different polyphenol profile was discovered in BE, where chlorogenic acid content is clearly lower than other extracts (0.74 g/100 g DE) and rutin amount (0.05 g/100 g DE) is comparable to other flavonoids identified. Here, the most abundant compound was the isovitexin, a C-glycosylated flavonoid, generally abundant in the plant cortex.

2.2. Antioxidant and Anti-Inflammatory Activity

The antioxidant and free radical-scavenging activity of the *E. dendroides* extracts was evaluated by several cell-free colorimetric assays, based on different environments and reaction mechanisms (hydrogen atom transfer and electron transfer-based methods). As depicted in Figure 3, all extracts showed a concentration-dependent antioxidant and free radical-scavenging activity in each test performed.

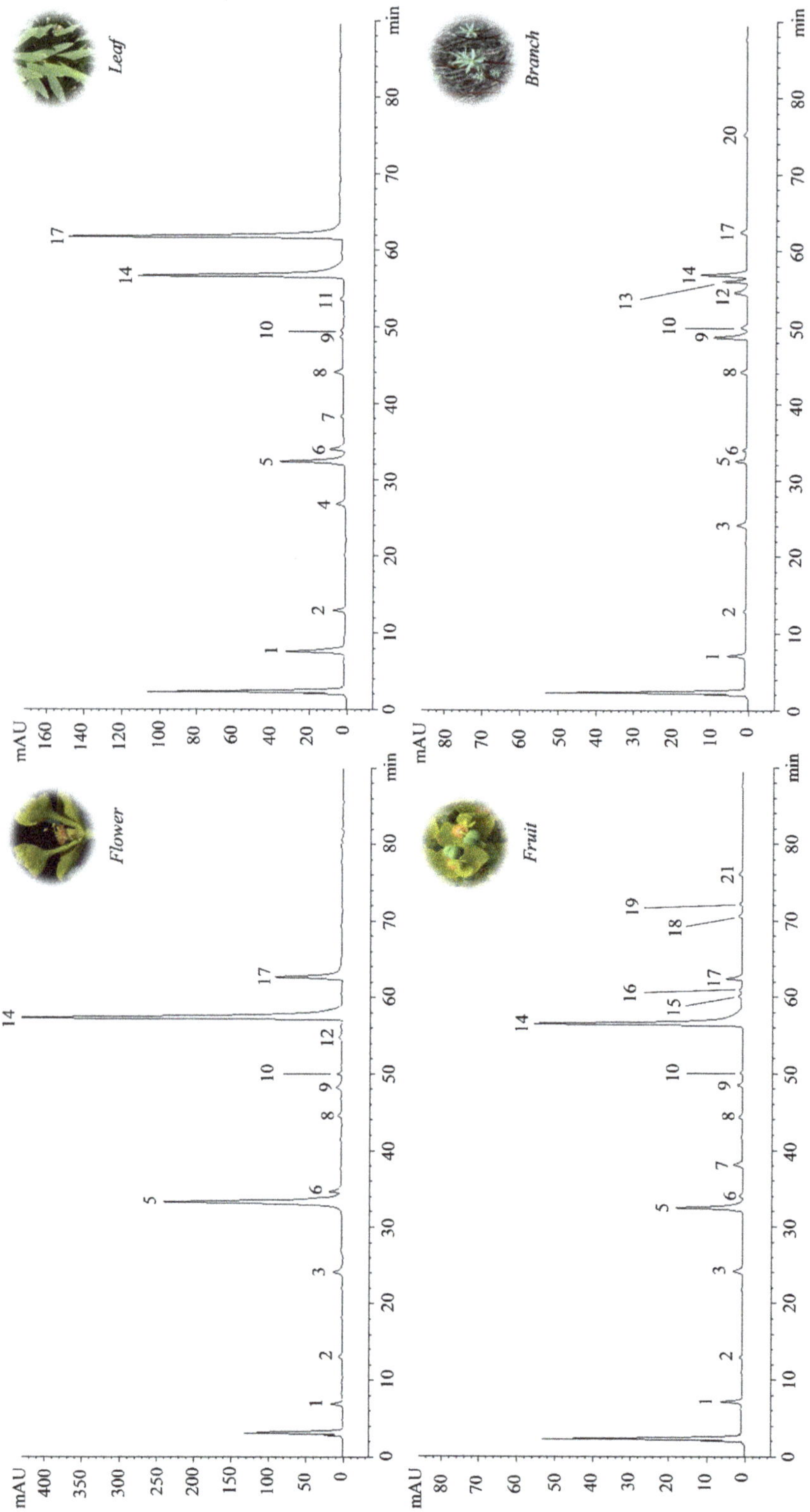

Figure 2. Representative LC-DAD chromatograms of *E. dendroides* flower (FE), leaf (LE), fruit (FrE), and branch (BE) extracts acquired at 260 nm. Peak numbers refer to compounds listed in Table 2.

Table 2. Polyphenol profile of *E. dendroides* flower (FE), leaf (LE), fruit (FrE), and branch (BE) extracts. Results were expressed as the mean ± standard deviation of three independent experiments in triplicate ($n = 3$).

n [a]	Compound	RT [b]	λ_{max} (nm)	[M-H]⁻	MW [c]	FE	LE	FrE	BE
							g/100 g DE [d]		
1	Gallic Acid	6.97	232;270	169	170	0.08 ± 0.01 *	0.01 ± 0.00 §	0.04 ± 0.00	0.04 ± 0.00
2	3,4 Dihydroxybenzoic acid	13.08	232	153	154	0.01 ± 0.00	0.01 ± 0.00	0.01 ± 0.00	0.01 ± 0.00
3	*p*-Hydroxybenzoic acid	23.97	232; 260	137	138	0.03 ± 0.00 *	-	0.01 ± 0.00	0.01 ± 0.00
4	Catechin	26.95	279	289	290	-	1.03 ± 0.01	-	-
5	Chlorogenic acid	32.40	290; 320	353	354	13.80 ± 0.58 *	11.66 ± 0.37	11.67 ± 0.42	0.74 ± 0.01 °
6	Caffeic acid	34.02	290; 325	179	180	0.02 ± 0.00 *	0.17 ± 0.01 §	0.01 ± 0.00 &	0.04 ± 0.00
7	Epicatechin	38.23	279	289	290	-	0.01 ± 0.00 $	0.12 ± 0.01	-
8	*p*-Cumaric acid	44.38	280; 300	163	164	0.01 ± 0.00	0.01 ± 0.00	0.02 ± 0.00 Ψ	0.01 ± 0.00
9	Eriodictyol-7-*O*-glucoside	48.31	232; 284	449	450	0.08 ± 0.00 *	0.04 ± 0.00 §	0.05 ± 0.00 &	0.06 ± 0.00
10	*m*-Cumaric acid	50.30	230; 274	163	164	0.01 ± 0.00	0.01 ± 0.00	0.01 ± 0.00	0.01 ± 0.00
11	Verbascoside	53.86	230; 330	623	624	-	0.01 ± 0.00	-	-
12	Isovitexin	54.91	284;328	431	432	0.02 ± 0.00 &	-	-	8.02 ± 0.22
13	Isoquercetin	56.37	262; 356	463	464	-	-	-	0.01 ± 0.00
14	Rutin	57.25	256;356	609	610	2.76 ± 0.11 *	0.65 ± 0.04 &	0.59 ± 0.03 &	0.05 ± 0.00
15	Roifolin	59.97	264; 336	577	578	-	-	0.01 ± 0.00	-
16	Kaempferol-3-*O*-rutinoside	60.95	266; 348	593	594	-	-	0.01 ± 0.00	-
17	Isorhamnetin-3-*O*-rutinoside	62.55	255; 356	623	624	0.41 ± 0.02 *	0.17 ± 0.01 §	0.04 ± 0.00 &	0.01 ± 0.00
18	Luteolin	70.77	268; 350	285	286	-	-	0.01 ± 0.00	-
19	Tiliroside	72.24	266; 314	593	594	-	-	0.01 ± 0.00	-
20	Isorhamnetin	75.71	230; 372	315	316	-	-	-	0.01 ± 0.00
21	Apigenin	75.96	267; 336	269	270	-	-	0.01 ± 0.00	-
	Total					**17.23**	**13.78**	**12.62**	**9.02**

[a] *n*. = peak number based on the elution order; [b] RT= retention time; [c] MW= Molecular weight; [d] DE= dry extract; * $p < 0.001$ vs. LE, FrE and BE; § $p < 0.001$ vs. FrE and BE; ° $p < 0.001$ vs. LE and FrE; & $p < 0.001$ vs. BE; $ $p < 0.001$ vs. FrE; Ψ $p < 0.001$ vs. FE, LE and BE.

However, the response of the extracts changes a lot depending on the aerial part considered. The oxygen radical absorbance capacity (ORAC) was identified as the main mechanism through which all the extracts exert their antioxidant activity, as can be observed from the lowest half-maximal inhibitory concentrations (IC_{50}) reported in Table 3. However, in this case, extracts did not show any statistically significant differences.

Table 3. The antioxidant and free radical-scavenging activity of *E. dendroides* flower (FE), leaf (LE), fruit (FrE), and branch (BE) hydroalcoholic extracts. Results were expressed as the half-inhibitory concentration (IC_{50} µg/mL) with confident limits (C.L.) at 95%.

Assay	FE	LE	FrE	BE	Standard [a]
			IC_{50} µg/mL (C.L.)		
ORAC	3.68 (1.30–5.46)	6.39 (5.32–7.58)	4.35 (2.15–8.79)	8.39 (6.87–10.25)	0.68 (0.38–1.27) *
TEAC	5.85 (4.90–6.98) °	22.31 (17.31–28.76) Ψ	13.88 (11.46–16.02)	14.17 (11.62–17.28)	2.89 (1.79–3.37) *
ICA	297.97 (147.01–403.96)	201.74 (167.03–243.68)	149.64 (76.74–291.77)	293.66 (247.85–347.93) &	6.62 (5.35–7.89) *
FRAP	10.36 (3.82–28.09) $	42.26 (33.19–53.81)	20.53 (10.35–40.70)	49.95 (13.12–68.38)	3.70 (1.65–6.29) *
DPPH	17.86 (15.75–20.26) °	31.75 (27.25–36.99)	25.72 (21.31–31.05)	30.07 (25.81–35.04)	3.95 (1.39–5.65) *
BCB	12.63 (9.52–16.75) &	24.01 (17.14–33.63)	14.62 (10.87–19.66) £	30.96 (21.16–45.30)	0.24 (0.17–0.32) *

[a] Trolox for ORAC, TEAC, FRAP and DPPH; Ethylenediaminetetraacetic acid (EDTA) for iron-chelating activity and butylhydroxytoluene (BHT) for β-carotene bleaching; ICA, Iron-chelating activity; BCB, β-carotene bleaching. * $p < 0.001$ vs. all *E. dendroides* extracts; ° $p < 0.001$ vs. LE, FrE, BE; Ψ $p < 0.001$ vs. FrE and BE; & $p < 0.001$ vs. LE and FrE; $ $p < 0.001$ vs. LE; £ $p < 0.001$ vs. BE.

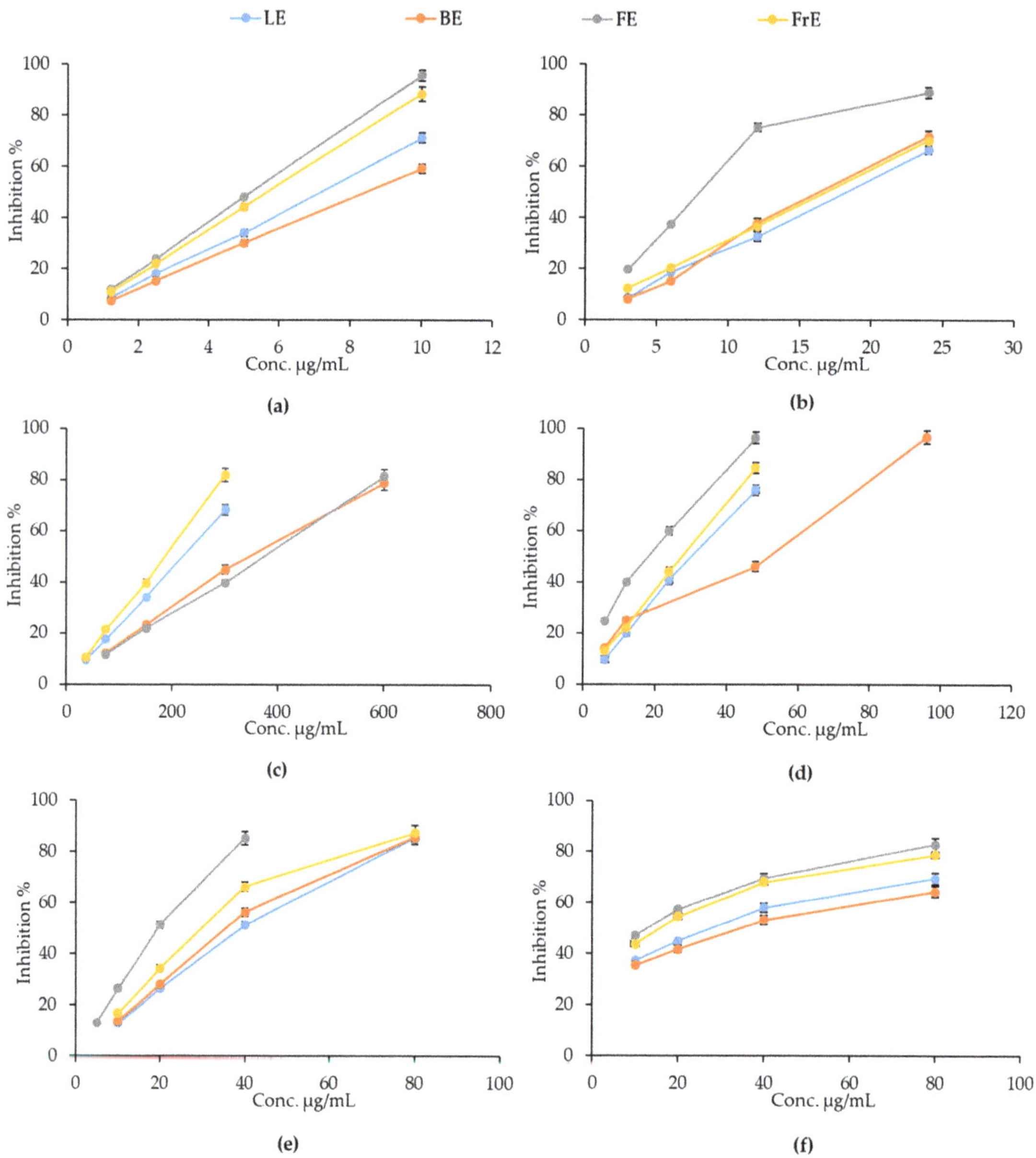

Figure 3. The antioxidant and free radical-scavenging activity of *E. dendroides* flower (FE), leaf (LE), fruit (FrE), and branch (BE) extracts towards ORAC (**a**); TEAC (**b**); Iron chelating-activity (**c**); FRAP (**d**); DPPH (**e**) and β-carotene bleaching (**f**) assay. Results were expressed as the mean inhibition percentage (%) ± standard deviation of three independent experiments ($n = 3$).

FE showed the most powerful antioxidant and free-radical scavenging activity in all assays carried out, with the exception of the iron-chelating activity, in which RE showed the best and statistically significant antioxidant activity with respect to all other *E. dendroides* extracts. In particular, statistically significant results, with respect to LE, FrE, and BE were observed for Trolox equivalent antioxidant capacity (TEAC) and DPPH assays. On the contrary, FE showed statistically significant results with respect to LE, and LE, and

FrE in ferric reducing antioxidant power (FRAP) and β-carotene bleaching (BCB) assays, respectively. All extracts showed statistically significant results with respect to Trolox, ethylenediaminetetraacetic acid (EDTA), and butylhydroxytoluene (BHT), used as reference compounds.

The anti-inflammatory properties of *E. dendroides* extracts were evaluated by two simple in vitro cell-free based assays, which evaluate the heat-induced protein denaturation (BSA denaturation assay) and the protease inhibitory activity. These assays allow simultaneous evaluation of enzymatic and non-enzymatic anti-inflammatory mechanisms. The results obtained showed that, first of all, the extracts show once again a concentration-dependent behavior, in accordance with what has already been seen in the antioxidant activity (Figure 4).

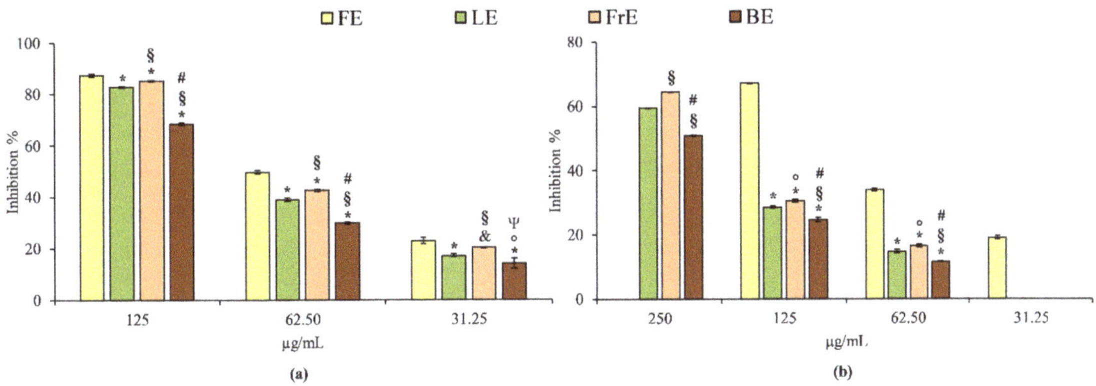

Figure 4. The anti-inflammatory activity of *E. dendroides* flower (FE), leaf (LE), fruit (FrE) and branch (BE) extracts towards BSA denaturation assay (**a**) and protease inhibition assay (**b**). Results were expressed as the mean inhibition percentage (%) ± standard deviation of three independent experiments ($n = 3$). * $p < 0.001$ vs. FE; § $p < 0.001$ vs. LE; # $p < 0.001$ vs. FrE; & $p < 0.05$; ° $p < 0.05$ vs. LE; Ψ $p < 0.05$ vs. FrE.

According to the antioxidant assays, FE is also the strongest anti-inflammatory agent among the extracts investigated both in BSA denaturation and protease inhibition assay. The order of potency, expressed as IC_{50} is the following: 62.88 μg/mL (C.L. 60.55–64.34 μg/mL) for FE followed by LE (IC_{50} 80.27 μg/mL; C.L. 78.48–82.66 μg/mL), FrE (IC_{50} 73.50 μg/mL; C.L.72.22–74.84 μg/mL) and BE (IC_{50} 104.80 μg/mL; C.L. 103.55–106.15 μg/mL) in the BSA denaturation assay, and 92.27 μg/mL (C.L. 90.45–94.05 μg/mL) for FE followed by FrE (IC_{50} 205.34 μg/mL; C.L. 203.89–207.23 μg/mL), LE (IC_{50} 219.02 μg/mL; C.L. 216.56–221.29 μg/mL) and BE (IC_{50} 253.84 μg/mL; C.L. 250.45–257.59 μg/mL) in the protease inhibition test. Diclofenac sodium, used as a reference compound in both assays, showed the following IC_{50} values: 42.37 μg/mL (C.L. 40.55–44.68) and 36.55 μg/mL (C.L. 34.88–38.79) in BSA denaturation and protease inhibition assay, respectively. Considering this, all *E. dendroides* extracts showed a different and statistically significant ($p < 0.001$) anti-inflammatory behavior in all tests carried out, both among them and with respect to the reference compound diclofenac sodium.

2.3. Brine Shirmp Letality Assay

The brine shrimp lethality assay is a useful and rapid screening tool to evaluate the toxicity of plant complexes or pure compounds [18]. *E. dendroides* extracts were tested at different concentrations ranging from 0.0001 to 10 mg/mL. No mortality was found at 24 h, while only weak toxicity (ranging from 0 to 10%, data not shown) was found after 48 h and only at the highest concentration (10 mg/mL) for FE, LE, and FrE. Despite all extracts not showing any significant toxicity over a broad range of concentrations, BE proved to be the most toxic extract among those investigated affecting the larvae viability with an LD_{50} value of 8.82 mg/mL (C.L. 7.05–9.16 mg/ml).

3. Discussion

The *Euphorbia* species are plants well known for their health effects and traditional use around the world [5]. Their biological activity is closely related to the phytochemical profile which, however, may vary according to different parameters such as species, grown conditions, the part of the plant investigated, as well as applied extraction methods.

Much research has been carried out in order to highlight the phytochemical profile of different *Euphorbia* spp. This has allowed the identification of numerous classes of secondary metabolites with marked health properties, including multi-drug resistance modulatory activity, antiproliferative, cytotoxic, antiviral, antimicrobial, antioxidant, and anti-inflammatory activity [5,9,13,14,17,19,20].

Among these compounds, the most representative are sesquiterpenes, diterpenes, triterpenes, steroids, cerebrosides, phenolic compounds, tannins, and coumarins [5,20].

In particular, it is well known that the *Euphorbia* genus is a rich source of jatrophane and modified jatrophane diterpenes, a wide range of structurally unique macrocyclic and polyoxygenated derivatives, which opened new frontiers for research studies on this genus [21]. Many of them have been isolated from the latex and aerial parts of E. *dendroides* and, for this reason, the phytochemical research on this species has mainly focused in recent years on this class of compounds [12,13,15,19].

To date, although polyphenols are among the most investigated compounds in the plant kingdom, studies concerning these secondary metabolites in the *Euphorbia* genus are rather scarce and even missing for some species.

In this study, the polyphenols profile of hydroalcoholic extracts of aerial parts (flower, leaf, fruit, and branch) of *E. dendroides* was investigated for the first time. Moreover, to date and to the authors' knowledge, only one study, which investigated the antiproliferative and antioxidant activities of two fractions (ethyl acetate and butanol) of whole plant methanol/water extract of *E. dendroides* is currently available [17].

The extraction method adopted in the present study allowed to obtain a very high extraction yield (13.70–24.21%) with respect to previous investigations (0.42–2.12%) [17]. Despite the polyphenols extraction from plant material is affected by several parameters such as chemical environment, the extraction process, and solvent polarity, as well as particle size and storage conditions [22], the results of the present study are in accordance with Ghout et al. [17] showing a total phenols content ranging from 26.54 to 64.75 g GAE/100 g DE with respect to 16.43 and 92.95 g GAE/100 g DE for butanol and ethyl acetate fractions of *E. dendroides* extract.

On the contrary, a greater content of flavonoids was found in the hydroalcoholic extract of flowers, leaves, fruits, and branches analysed in this study compared to the flavonoid content found in the two fractions mentioned above (46.93–84.85 g QE/100 g DE vs. 1.22 and 2.60 g QE/100 g DE) [17]. Moreover, extracts from aerial parts of *E. dendroides* showed a greater polyphenols content in comparison with the methanol latex extract, which showed a total phenols and flavonoids content equal to 4.75 g GAE/100 g and 1.47 g RE/100 g DE, respectively [9].

In the present study, the RP-LC-DAD-ESI-MS analysis allowed us to identify and quantify more than 10 compounds in each extract, according to previous results which identify ten [17] and fourteen compounds [9] in different *E. dendroides* extracts. According to Ghout et al. [17], chlorogenic acid is the most representative phenolic acid in all the extracts here investigated with the exception of the BE, in which the most abundant compound was the flavonoid isovitexin. However, expressing the results of Ghout et al. [17] as dry weight, the chlorogenic acid content found in FE, LE and FrE was about double. On the contrary, although it remains the second most abundant phenolic acid present in the extracts under consideration, the gallic acid content is about 100 times lower in the extracts analysed in the present study compared to those analysed previously [17]. Caffeic acid represent the second most abundant phenolic acid in LE, whereas comparable amount was found in the other extracts. This phenolic acid was identified also previously in the methanol latex extract [9] as well as in the methanol/water extract of whole plant [17]. Finally, according to previous results, other minor phenolic acids such as *p*-cumaric acid and *p*-hydroxybenzoic acid [9,17] were found. Beyond the phenolic acids, the extracts investigated in the present study showed, unlike those investigated previously, a high content of flavonoids as well, with rutin as the most abundant compound ranging from 200 to10,000 times higher with respect to previous results [17]. This flavonoid was identified also in other *Euphorbia* species such as *E. lathyris* and *E. geniculate*, in which also other quercetin derivatives such as quercetin-3-*O*-rhamnoside and quercetin-3-*O*-D-glucopyranoside as well as the aglycon (quercetin) were identified [20]. On the contrary, rutin is completely absent in the methanol latex extract previously investigated, characterized by the abundant presence of eryodictiol-7-*O*-glucoside, eriodyctiol, naringenin-7-*O*-glucoside, naringenin, and quercetin [9].

Two biological activities (antioxidant and anti-inflammatory) were investigated in the present study. According to previous results, all *E. dendroides* extracts showed concentration-dependent antioxidant [9,17] and anti-inflammatory activity.

The different in vitro colorimetric assays carried out, characterized by different environment and reaction mechanisms, allowed identification of the oxygen radical absorbance capacity as the main mechanism through which all the extracts exert their antioxidant activity. Despite all the analysed extracts showing significantly lower antioxidant activity compared to the reference standards, in accordance with what was previously observed, the IC_{50} values found in the present study are much lower than those previously highlighted in the literature for *E. dendroides* extracts [9,17], showing the most powerful antioxidant activity in the aerial part extracts.

On average, the extracts proved to be particularly active in tests based on hydrogen atoms-transfer reactions (ORAC and BCB), followed by electron and hydrogen atom-transfer-based assays (TEAC and DPPH), and finally on tests based on electrons transfer (FRAP). The iron-chelating activity was the lowest among those investigated, although the FrE showed interesting results. It is well-known that several flavonoids efficiently chelate trace metals such as iron and copper, which play an important role in oxygen metabolism avoiding the generation of highly aggressive secondary radical species [23]. The proposed binding sites for trace metals to flavonoids in order of potency are the following: catechol moiety in B ring, 3-hydroxyl, 4-oxo groups in the heterocyclic ring, and the 4-oxo, 5-hydroxyl groups between the heterocyclic and the A rings [23]. Considering this, certainly, rutin exerts the greatest contribution in terms of iron-chelating activity, although the presence of different flavonoids could enhance this activity. It is well-known that the antioxidant behaviour of plant extracts is strictly related to the quali-quantitative composition of the polyphenolic profile. In particular, the reducing ability depends on the number of free-hydroxyl groups on the base skeleton [9,24]. Considering this, among phenolic acids, chlorogenic acid plays a predominant role followed by gallic, caffeic, and dihydroxybenzoic acid. On the contrary, for flavonoids, the radical scavenging activity depends on the structure and the substituents of the heterocyclic and B rings and, in particular, by the catechol group in the B ring, which has the better electron-donating properties. Moreover, the 2,3-double bond conjugated with the 4-oxo group is responsible for electron delocalization. The presence of a 3-hydroxyl group in the heterocyclic ring also increases the radical-scavenging activity, while additional hydroxyl or methoxy groups at positions 3,5, and 7 of A and C rings seem to be less important [23]. Considering this, the order of potency of flavonoids is the following: flavonols, flavones, flavanols, and flavanones. Most of the isolated flavonoids from the *Euphorbia* genus are simple flavonols as well as O-, C-substituted and prenylated. The main glycosides are d-glucose, l-rhamnose, or glucorhamnose attached at either the C-3 or C-7 position. Structure–activity relationship studies showed that methylation of the hydroxyl groups on the C-3 or C-7 position decreases the activities while glycosylation loses the activity. In any case, the parent compound is essential in preserving biological activity [20]. Indeed, observing the average behaviour of the extracts in terms of antioxidant activity, it can be deduced that FE, the extract richer in rutin and isorhamnetin-3-*O*-rutinoside (both flavonols), shows the strongest antioxidant activity, followed by FrE, which is the most diversified extract in terms of flavonoid content, followed by LE and BE. BE in particular, shows the lowest antioxidant activity probably attributable to the very low concentration of rutin found compared to the other extracts analysed and to the high presence of isovitexin, which, as a C-glycosylated flavonoid, acts as a weak antioxidant [23].

The present study showed that *E. dendroides* extracts from aerial parts showed also strong and concentration-dependent anti-inflammatory properties both in enzymatic and non-enzymatic assays. It is well-known that polyphenols may exert anti-inflammatory effects through various mechanisms. Among the main ones that can be counted there is certainly the radical scavenging activity, due to the close connection between oxidative balance and inflammation, followed by modulation of the main enzymes involved in inflammation such as protease, COX-1 and COX-2, and phospholipase A2, as well as through complex signalling pathways that lead to a modulation of the release of important pro-inflammatory markers such as interleukins and nitric oxide [25].

One of the most important aspects of this study is that the extracts under examination showed such biological activities without showing any toxic effect on *Artemia salina*, as demonstrated previously for other leaf and flowering top extracts [26].

Moreover, this study demonstrates indirectly that the toxicity generally associated with *E. dendroides* is mainly due to the terpenes fraction in accordance with what was observed in our previous study on a methanol extract of *E. dendroides* latex, in which an LD_{50} ~350 times lower with respect to the most toxic extract investigated in the present study (0.025 mg/mL vs. 8.82 mg/mL) was found [9].

4. Materials and Methods

4.1. Chemicals

The 1,1-Diphenyl-2-picrylhydrazyl radical (DPPH), 2,20-azino-bis(3-ethylbenzothiazoline-6-sulfonic acid (ABTS), potassium persulfate ($K_2S_2O_8$), butylate-hydroxytoluene phenazine methosulphate, 6-Hydroxy-2,5,7,8-tetramethylchromane-2-carboxylic acid (Trolox) diclofenac sodium, butylated hydroxytoluene (BHT), 2,2′-Azobis(2-methylpropionamidine) dihydrochloride (AAPH), fluorescein sodium salt, sodium phosphate dibasic (Na_2HPO_4) potassium phosphate monobasic (KH_2PO_4), 2,4,6-Tris(2-pyridyl)-S-triazine (TPTZ), iron sulphate heptahydrate, sodium acetate, sodium carbonate, vanillin, Folin-Ciocalteu reagent β-carotene, linoleic acid, tween 80, ferrozine, iron (II) chloride, EDTA, fatty free BSA trypsin, tris-HCl, casein, potassium bichromate ($K_2Cr_2O_7$) were purchased from Sigma Aldrich (MSt. Louis, MO, USA). Methanol, acetonitrile, glacial acetic acid and phosphoric acid were HPLC grade and purchased from Merck (Darmstadt, Germany). Reference compounds were HPLC grade and purchased from Extrasynthese (Genay, France).

4.2. Plant Material Collection and Samples Preparation

Euphorbia dendroides L. (Euphorbiaceae) was collected and botanically identified by Prof. S. Ragusa in April 2019 in Messina (Italy), Masse locality (200 m a.s.l.) during flowering and fructification. A voucher specimen (19/04 ED) has been deposited within the herbarium of the Department ChiBioFarAm, University of Messina, Messina, Italy.

The aerial parts have been manually divided into leaves, flowers, fruits, and branches that have been air-dried in the dark at RT and then powdered by a blade mill (IKA® A11) with liquid nitrogen in order to block enzymatic activities and preserve the chemical features. Two grams of each frozen powder (leaves, flowers, fruits, and branches) were extracted with 40 mL of a hydroalcoholic mixture consisting of $EtOH/H_2O$ 80:20 *v/v* vortex-mixed for 3 min, and sonicated in an ice-cold bath for 5 min using a 3 mm titanium probe set at 200 W and 30% amplitude signal (Vibra Cell™ Sonics Materials, inc., Danbury CT, USA). Extracts were then centrifuged at 3000 × g for 10 min. Thereafter, the supernatant was filtered on filter paper and evaporated to dryness by a rotary evaporator at 37 °C. The procedure was repeated 3 times. Dry extracts were suspended and properly diluted in a hydroalcoholic mixture for phytochemical characterization and subsequent analyses.

4.3. Phytochemical Screening

4.3.1. Total Phenols Content

Total phenols content was determined using the Folin–Ciocalteu method according to Smeriglio et al. [27]. Briefly, 50 µL of sample solution (0.5–4.0 mg/mL) and 450 µL of deionized water were added to 500 µL of Folin–Ciocalteu reagent. After 3 min, sodium carbonate (500 µL, 10% *v*/*v*) was added and samples were left in the dark at RT for 1 hour, vortexing every 10 min. Absorbance was read at 785 nm with a UV-VIS spectrophotometer (Model UV-1601, Shimadzu, Kyoto, Japan) against a blank consisting of the same extraction hydroalcoholic mixture. Gallic acid was used as a reference compound (0.075–0.60 µg/mL). Results, which represent the average of three independent experiments in triplicate (n = 3) were expressed as g GAE/100 g DE.

4.3.2. Total Flavonoids Content

Total flavonoids content was determined as reported by Smeriglio et al. [9]. Briefly, 0.2 mL of sample solution (0.5–4.0 mg/mL) were mixed with 0.2 mL of $AlCl_3$ (2 mg/mL) and 1.2 mL of sodium acetate (50 mg/mL). After 2.5 h, the absorbance was recorded at 440 nm with a UV-VIS spectrophotometer (Model UV-1601, Shimadzu, Kyoto, Japan) against a blank consisting of the same extraction hydroalcoholic mixture. Quercetin was used as a reference compound (0.125–1.0 mg/mL). Results, which represent the average of three independent experiments in triplicate (n = 3), were expressed as g QE/100 g DE.

4.3.3. Vanillin Index Determination

Vanillin index was evaluated according to Smeriglio et al. [24]. Briefly, 2.0 mL of sample solution diluted in 0.5 M H_2SO_4 (absorbance ranging from 0.2 to 0.4) were loaded onto a conditioned Sep-Pak C18 cartridge (Waters, Milan, Italy). The column was washed with 2.0 ml of H_2SO_4 (5.0 mM), purged with air, and charged with methanol (5.0 mL) to elute the sample. Thereafter, 1 mL of the eluate was added to 6.0 mL of 4% vanillin methanol solution, and the mixture was conditioned in a water bath at 20 °C for 10 min. Chloridric acid (3.0 ml) was added and after incubation time (15 min), the absorbance was recorded at 500 nm with a UV-VIS spectrophotometer (Model UV-1601, Shimadzu, Kyoto, Japan) against a blank consisting of the same extraction hydroalcoholic mixture. Catechin was used as a reference compound (0.125–0.50 mg/mL). Results, which represent the average of three independent experiments in triplicate (n = 3), were expressed as g CE/100 g DE.

4.3.4. Proanthocyanidins Determination

The proanthocyanidins content was evaluated as described by Barreca et al. [28]. Briefly, 2.0 mL of sample solution, diluted 10 times with 0.05 M H_2SO_4, was loaded onto a conditioned Sep-Pak C18 cartridge (Waters, Milan, Italy), preconditioned with 5 mM H_2SO_4 (2.0 mL), and purged with air. The proanthocyanidins-reach fraction obtained was eluted with methanol (3.0 mL) and collected in a 100 ml flask shielded from light containing 9.5 mL of absolute ethanol. Thereafter, the mixture was added with 12.5 mL of $FeSO_4 \cdot 7H_2O$ solubilized in 37% HCl (300 mg/L) and placed to reflux for 50 min. After cooling by immersion in cold water (20 °C) for ten min, the absorbance was read at 550 nm with an UV-VIS spectrophotometer (Model UV-1601, Shimadzu, Kyoto, Japan) against a blank consisting of the same extraction hydroalcoholic mixture. The basal anthocyanins content of the sample was determined detracting the absorbance of a sample prepared under the same conditions but simply cooled in ice. Proanthocyanidins content was expressed as 5 times the amount of cyanidin formed by means of a cyanidin chloride (ε = 34,700) calibration curve. Results, which represent the average of three independent experiments in triplicate (n = 3), were expressed as g CyE/100 g DE.

4.4. Determination of Polyphenol Profile by RP-LC-DAD-ESI-MS Analysis

Polyphenols characterization of LE, FE, FrE, and BE was carried out according to Smeriglio et al. [29] by RP-LC-DAD-ESI-MS analysis. Separation was carried out by a

Luna Omega PS C18 column (150 mm × 2.1 mm, 5 µm; Phenomenex, Torrance, CA, United States) at 25 °C by using a mobile phase consisting of solvent A (0.1% formic acid) and solvent B (methanol) according to the following elution program: 0–3 min, 0% B; 3–9 min, 3% B; 9–24 min, 12% B; 24–30 min, 20% B; 30–33 min, 20% B; 33–43 min, 30% B; 43–63 min, 50% B; 63–66 min, 50% B; 66–76 min, 60% B; 76–81 min, 60% B; 81–86 min, 0% B and equilibrated 4 min for a total run time of 90 min. The injection volume was 5 µL. The UV-Vis spectra were recorded ranging from 190 to 600 nm and chromatograms were acquired at different wavelengths (260, 292, 330, and 370 nm) to identified all polyphenol classes. The experimental parameters of the mass spectrometer (ion trap, model 6320, Agilent Technologies, Santa Clara, CA, USA) operating in negative (ESI−) ionization mode were set as follows: capillary voltage 3.5 kV, nebulizer (N_2) pressure 40 psi, drying gas temperature 350 °C, drying gas flow 9 L/min and skimmer voltage 40 V. Acquisition was carried out in full-scan mode (90–1000 *m*/*z*). Data were acquired by Agilent ChemStation software version B.01.03 and Agilent trap control software version 6.2.

4.5. Antioxidant and Free-Radical Scavenging Activity

The antioxidant activity of *E. dendroides* extracts was evaluated by several in vitro colorimetric assays based on different mechanisms (electron, hydrogen and electron, and hydrogen transfer-based assays) and reaction environments. Absorbance was recorded by a UV-VIS spectrophotometer (Model UV-1601, Shimadzu, Kyoto, Japan). Results, which represent the average of three independent experiments in triplicate ($n = 3$), were expressed as inhibition percentage (%) of the oxidative/radical activity, calculating the IC_{50} with the respective C.L. at 95% by Litchfield and Wilcoxon's test using PHARM/PCS software version 4 (MCS Consulting, Wynnewood, PA, USA). All concentration ranges reported refer to final concentrations of *E. dendroides* extracts and reference compounds in the reaction mixture.

4.5.1. FRAP Assay

This assay was carried out according to Smeriglio et al. [30]. Fifty microliters of sample solution (2.5–100 µg/mL) or Trolox as reference compound (1.25–10 µg/mL) was added to fresh pre-warmed (37 °C) working FRAP reagent (1.5 mL) and incubated for 4 min at RT in the dark. Absorbance was recorded at 593 nm against a blank consisting of the extraction hydroalcoholic mixture

4.5.2. ORAC Assay

Oxygen radical absorbance capacity was evaluated according to Smeriglio et al. [31]. Briefly, 20 µL of sample solution (1.25–10 µg/mL) diluted in 75 mM phosphate buffer pH 7.4, were added to 120 µL of fresh fluorescein solution (117 nM) and incubated 15 min at 37 °C. After that, 60 µL of 40 mM AAPH solution was added to start the reaction, that was recorded every 30 s for 90 min (λ_{ex} 485; λ_{em} 520) by a fluorescence plate reader (FLUOstar Omega, BMG LABTECH, Ortenberg, Germany). A blank consisting of the extraction hydroalcoholic mixture diluted in phosphate buffer and Trolox as the reference compound (0.25–2.0 µg/mL) were included in each assay.

4.5.3. BCB Assay

The BCB assay was carried out using a β-carotene emulsion prepared according to Smeriglio et al. [32]. Briefly, 0.28 mL of sample solution (10–80 µg/mL) were added to 7 ml of β-carotene emulsion, whereas an emulsion without β-carotene was used as a negative control. The absorbance was recorded at starting time and during incubation at 50 °C every 20 min for 120 min at 470 nm. BHT was used as a reference compound (0.063–0.5 µg/mL).

4.5.4. TEAC Assay

The Trolox equivalent antioxidant capacity of samples was evaluated according to Monforte et al. [33]. The reaction mixture (4.3 mM $K_2S_2O_8$ and 1.7 mM ABTS solution,

1:5 v/v) was incubated for 12–16 h in the dark at RT and diluted just before the analyses until an absorbance of 0.7 ± 0.02 (734 nm). Fifty microliters of each sample (3.0–30 μg/mL) were added to 1 mL of the reagent and incubated at RT for 6 min. The absorbance was recorded at 734 nm against a blank consisting of the extraction hydroalcoholic mixture. Trolox was used as a reference compound (0.63–5.0 μg/mL).

4.5.5. ICA Assay

The iron-chelating activity was evaluated according to Bazzicalupo et al. [34]. Briefly, 50 μL of $FeCl_2 \cdot 4H_2O$ solution (2.0 mM) was added to 100 μL of the sample (0.075–0.60 mg/mL) and incubated at RT for 5 min. After that, 100 μL of 5 mM ferrozine and 3 mL of deionized water were added to the mixture and incubated for 10 min at RT. The absorbance was read at 562 nm against a blank consisting of the extraction hydroalcoholic mixture. EDTA was used as a reference compound (1.50–12.0 μg/mL).

4.5.6. DPPH Assay

The DPPH radical scavenging activity was evaluated according to Smeriglio et al. [32]. Briefly, 37.5 μL of sample solution (5–80 μg/ml) was added to fresh DPPH methanol solution (10^{-4} M), vortex-mixed for 10 s, and incubated in the dark at RT for 20 min. Absorbance was recorded at 517 nm against a blank consisting of the extraction hydroalcoholic mixture. Trolox was used as a reference compound (0.63–5.0 μg/mL).

4.6. *Anti-Inflammatory Activity*

The anti-inflammatory activity of *E. dendroides* extracts was evaluated by two simple in vitro colorimetric enzymatic and non-enzymatic assays. Absorbance was recorded by a multi-well plate reader (Multiskan GO; Thermo Scientific, MA, United States). Results, which represent the average of three independent experiments in triplicate (n = 3), were expressed as inhibition percentage (%) of the inflammatory/enzyme activity, calculating the IC_{50} with the respective C.L. at 95% by Litchfield and Wilcoxon's test using the PHARM/PCS software version 4 (MCS Consulting, Wynnewood, PA, USA). All concentration ranges following reported refer to final concentrations of *E. dendroides* extracts and reference compounds in the reaction mixture.

4.6.1. BSA Denaturation Assay

The ability of samples to inhibit the heat-induced bovine serum albumin denaturation was evaluated according to Saso et al. [35]. Briefly, 100 μL of 0.4 % fatty free BSA solution and 20 μL of PBS pH 5.3 were added into a 96-well plate. Therefore, 80 μL of sample solution (31.25–125 μg/mL) were added to the mixture. The absorbance was recorded at 595 nm at starting time and after incubation for 30 min at 70 °C. A blank consisting of the extraction hydroalcoholic mixture was used as a control. Diclofenac sodium was used as a reference compound (6.25–50 μg/mL).

4.6.2. Protease Inhibition Assay

The protease inhibitory activity of samples was evaluated according to Oyedapo and Famurewa [36]. Briefly, 200 μL of sample solution (31.25–250 μg/mL) were added to the reaction mixture consisting of 12 μL of trypsin (10 μg/mL) and 188 μL of 25 mM Tris-HCl buffer (pH 7.5). After that, 200 μL of 0.8% casein was added and the reaction mixture and incubated for 20 min at 37 °C in a water bath. At the end of the incubation time, 400 μL of perchloric acid was added to stop the reaction. The cloudy suspension was centrifuged at 3500 × g for 10 min and the absorbance of the supernatant was recorded at 280 nm against a blank consisting of the extraction hydroalcoholic mixture. Diclofenac sodium was used as a reference compound (6.25–50 μg/mL).

4.7. Brine Shrimp Lethality Assay

In order to investigate the toxicity of the extracts, a brine shrimp lethality assay was carried out according to Caputo et al. [37]. Eggs of *Artemia salina* were purchased from a local pet shop, placed in a hatcher chamber containing seawater, and incubated for 48 h at RT with continuous aeration and illumination. Two hundred microliters of each sample (0.0001 to 10 mg/mL) and $K_2Cr_2O_7$ as a reference compound (500 µg/mL) diluted in seawater were seeded in a 24-well plate. Ten nauplii per well were added and incubated for 48 h in the same conditions reported above.

Surviving larvae without abnormal swimming behavior were counted after 24 h and 48 h by a stereomicroscope (SMZ-171 Series, Motic, Seneco s.r.l.–Milano, Italy). One negative control (10 larvae treated with seawater only) were also evaluated. Three independent experiments (n = 10) were carried out for each treatment. Lethality was calculated using the following equation:

$$\% \, Lethality = 100 - [(slt \times 100)]/slcs \tag{1}$$

where *slt* were the survival larvae treated with extracts or $K_2Cr_2O_7$, whereas *slcs* were the survival larvae treated with seawater (negative control).

4.8. Statistical Analysis

Three independent experiments (n = 3 and n = 10) were carried out for the in vitro cell-free assays and brine shrimp lethality assay, respectively. Results were expressed as the mean ± standard deviation (S.D.). Data were analyzed by one-way analysis of variance (ANOVA) followed by Tukey's test and Student-Newman-Keuls Method by SigmaPlot12 (Systat Software, Inc., San Jose, CA, USA). Results were statistically significant for $p \leq 0.05$.

5. Conclusions

In conclusion, this is the first study in which extracts of the aerial parts (flowers, leaves, fruits, and branches) of *E. dendroides* were investigated. Phytochemical analyses showed a high content of polyphenols with chlorogenic acid and rutin as the most representative compounds, respectively for phenolic acids and flavonoids in regards to flowers, leaves, and fruits. On the contrary, chlorogenic acid and isovitexin were found as the most representative compounds for the branch extract. However, beyond the most representative compounds, small differences were found in the phytochemical profile of the extracts under examination, which certainly may contribute to the promising biological activity observed. On average, the flower extract showed the highest antioxidant and anti-inflammatory activity followed by fruits, leaves, and branches. Interestingly, all the extracts showed no toxicity, demonstrating indirectly that the toxicity, generally ascribed to this plant species, is due to the terpene component mainly present in the latex.

These results, which certainly require further cell-based studies to better investigate the biological properties investigated, show clear preliminary evidence of a possible use of the extracts as powerful antioxidant and anti-inflammatory agents.

Author Contributions: Conceptualization, C.C., A.S., S.R. and D.T.; methodology, C.C., A.S. and D.T.; software, D.T.; validation, C.C., A.S. and D.T.; formal analysis, A.S., M.D.; investigation, A.S., M.D.; resources, C.C. and D.T.; data curation, C.C., A.S. and D.T.; writing—original draft preparation, A.S., M.D.; writing—review and editing, C.C., A.S., S.R. and D.T.; visualization, C.C., S.R. and D.T.; supervision, C.C. and D.T.; project administration, C.C.; funding acquisition, C.C. and D.T. All authors have read and agreed to the published version of the manuscript.

Funding: This research received no external funding.

Institutional Review Board Statement: Not applicable.

Informed Consent Statement: Not applicable.

Data Availability Statement: The data presented in this study are available on request from the corresponding author.

Conflicts of Interest: The authors declare no conflict of interest.

References

1. Horn, J.W.; Vane, B.W.; Morawetz, J.J.; Riina, R.; Steinmann, V.W.; Berry, P.E.; Wurdack, K.J. Phyogenetic sand the evolution of major structural char-acters in the giant genus *Euphorbia* L. (Euphorbiaceae). *Mol. Phylogenet. E* **2012**, *63*, 305–326.
2. Nambudiri, N.S.; Nambudiri, V.E. *Euphorbia* peplus: 18th-centuryinsightsona 21st-century therapy. *JAMA Derm.* **2013**, *149*, 1081.
3. Gopi, K.; Renu, K.; Sannanaik Vishwanath, B.; Jayaraman, G. Protective effect of *Euphorbia hirta* and its components against snake venom induced lethality. *J. Ethnopharmacol.* **2015**, *165*, 180–190.
4. Kemboi, D.; Peter, X.; Langat, M.; Tembu, J. A Review of the Ethnomedicinal Uses, Biological Activities, and Triterpenoids of *Euphorbia* Species. *Molecules* **2020**, *25*, 4019.
5. Salehi, B.; Iriti, M.; Vitalini, S.; Antolak, H.; Pawlikowska, E.; Kręgiel, D.; Sharifi-Rad, J.; Oyeleye, S.I.; Ademiluyi, A.O.; Czopek, K.; et al. *Euphorbia*-Derived Natural Products with Potential for Use in Health Maintenance. *Biomolecules* **2019**, *9*, 337.
6. Liang, Q.L.; Dai, C.C.; Jiang, J.H.; Tang, Y.P.; Duan, J.A. A new cytotoxic casbane diterpene from *Euphorbia pekinensis*. *Fitoterapia* **2009**, *80*, 514.
7. Shi, Q.W.; Su, X.H.; Kiyota, H. Chemical and pharmacological research of the plants in genus *Euphorbia*. *Chem. Rev.* **2008**, *108*, 4295–4327.
8. Vasas, A.; Hohmann, J. *Euphorbia* Diterpenes: Isolation, Structure, Biological Activity, and Synthesis (2008−2012). *Chem. Rev.* **2014**, *114*, 8579–8612.
9. Smeriglio, A.; Ragusa, S.; Monforte, M.T.; D'Angelo, V.; Circosta, C. Phytochemical analysis and evaluation of antioxidant and anti-acetylcholinesterase activities of *Euphorbia dendroides* L. (Euphorbiaceae) latex. *Plant Biosyst.* **2019**, *153*, 498–505.
10. Berman, B. New developments in the treatment of actini ckeratos is: Focus on ingenol mebutategel. *Clin. Cosmet. Investig. Derm.* **2012**, *20*, 111–122.
11. Vasas, A.; Rédei, D.; Csupor, D.; Molnár, J.; Hohmann, J. Diterpenes from European *Euphorbia* species serving as prototypes for natural-product-based drug discovery. *Eur. J. Org. Chem.* **2012**, *2012*, 5115–5130.
12. Aljancić, I.S.; Pesić, M.; Milosavljević, S.M.; Todorović, N.M.; Jadranin, M.; Milosavljević, G.; Povrenović, D.; Banković, J.; Tanić, N.; Marković, I.D.; et al. Isolation and biological evaluation of jatrophane diterpenoids from *Euphorbia dendroides*. *J. Nat. Prod.* **2011**, *74*, 1613–1620.
13. Pešić, M.; Banković, J.; Aljančić, I.S.; Todorović, N.M.; Jadranin, M.; Vajs, V.E.; Tešević, V.V.; Vučković, I.; Momčilović, M.; Marković, I.D.; et al. New anti-cancer characteristics of jatrophane diterpenes from *Euphorbia dendroides*. *Food Chem. Toxicol.* **2011**, *49*, 3165–3173.
14. Jadranin, M.; Pešić, M.; Aljančić, I.S.; Milosavljević, S.M.; Todorović, N.M.; Podolski-Renić, A.; Banković, J.; Tanić, N.; kaMarković, I.; Vajsac, V.E.; et al. Jatrophane diterpenoids from the latex of *Euphorbia dendroides* and their anti-P-glycoprotein activity in human multi-drug resistant cancer cell lines. *Phytochemistry* **2013**, *86*, 208–217.
15. Esposito, M.; Nothias, L.F.; Nedev, H.; Gallard, J.F.; Leyssen, P.; Retailleau, P.; Costa, J.; Roussi, F.; Iorga, B.I.; Paolini, J.; et al. *Euphorbia dendroides* Latex as a Source of Jatrophane Esters: Isolation, Structural Analysis, Conformational Study, and Anti-CHIKV Activity. *J. Nat. Prod.* **2016**, *79*, 2873–2882.
16. Esposito, M.; Nim, S.; Nothias, L.F.; Gallard, J.F.; Rawal, M.K.; Costa, J.; Roussi, F.; Prasad, R.; Di Pietro, A.; Paolini, J.; et al. Evaluation of Jatrophane Esters from *Euphorbia* spp. as Modulators of Candida albicans Multidrug Transporters. *J. Nat. Prod.* **2017**, *80*, 479–487.
17. Ghout, A.; Zellagui, A.; Gherraf, N.; Demirtas, I.; Ayse Sahin, Y.; Boukhenaf, M.; Lahouel, M.; Nieto, G.; Akkal, S. Antiproliferative and Antioxidant Activities of Two Extracts of the Plant Species *Euphorbia dendroides* L. *Medicines* **2018**, *5*, 36.
18. Danabas, D.; Ates, M.; Ertit Tastan, B.; Cicek Cimen, I.C.; Unal, I.; Aksu, O.; Kutlu, B. Effects of Zn and ZnO Nanoparticles on *Artemia salina* and *Daphnia magna* Organisms: Toxicity, Accumulation and Elimination. *Sci. Total Environ.* **2020**, *711*, 134869.
19. Corea, G.; Fattorusso, E.; Lanzotti, V.; Taglialatela-Scafati, O.; Appendino, G.; Ballero, M.; Simon, P.N.; Dumontet, C.; Di Pietro, A. Modified jatrophane diterpenes as modulators of multidrug resistance from *Euphorbia dendroides* L. *Bioorg. Med. Chem.* **2003**, *11*, 5221–5227.
20. Magozwi, D.K.; Dinala, M.; Mokwana, N.; Siwe-Noundou, X.; Krause, R.W.M.; Sonopo, M.; McGaw, L.J.; Augustyn, W.A.; Tembu, V.J. Flavonoids from the Genus *Euphorbia*: Isolation, Structure, Pharmacological Activities and Structure-Activity Relationships. *Pharmaceuticals* **2021**, *14*, 428.
21. Fattahian, M.; Ghanadian, M.; Ali, Z.; Khan, I.A. Jatrophane and rearranged jatrophane-type diterpenes: Biogenesis, structure, isolation, biological activity and SARs (1984–2019). *Phytochem. Rev.* **2020**, *19*, 1–72.
22. Xu, B.J.; Chang, S.K. A comparative study on phenolic profiles and antioxidant activities of legumes as affected by extraction solvents. *J. Food Sci.* **2007**, *72*, S159–S166.
23. Pietta, P.G. Flavonoids as antioxidants. *J. Nat. Prod.* **2000**, *63*, 1035–1042.
24. Smeriglio, A.; Denaro, M.; Barreca, D.; D'Angelo, V.; Germanò, M.P.; Trombetta, D. Polyphenolic profile and biological activities of black carrot crude extract (*Daucus carota* L. ssp. sativus var. atrorubens Alef.). *Fitoterapia* **2018**, *124*, 49–57.

25. Hussain, T.; Tan, B.; Yin, Y.; Blachier, F.; Tossou, M.C.; Rahu, N. Oxidative Stress and Inflammation: What Polyphenols Can Do for Us? *Oxid. Med. Cell Longev.* **2016**, *2016*, 7432797.
26. Miceli, N.; Cavò, E.; Ragusa, M.; Cacciola, F.; Mondello, L.; Dugo, L.; Acquaviva, R.; Malfa, G.A.; Marino, A.; D'Arrigo, M.; et al. *Brassica incana* Ten. (Brassicaceae): Phenolic Constituents, Antioxidant and Cytotoxic Properties of the leaf an Flowering Top Extracts. *Molecules* **2020**, *25*, 1461.
27. Smeriglio, A.; Cornara, L.; Denaro, M.; Barreca, D.; Burlando, B.; Xiao, J.; Trombetta, D. Antioxidant and cytoprotective activities of an ancient Mediterranean citrus (*Citrus lumia* Risso) albedo extract: Microscopic observations and polyphenol characterization. *Food Chem.* **2019**, *279*, 347–355.
28. Barreca, D.; Laganà, G.; Leuzzi, U.; Smeriglio, A.; Trombetta, D.; Bellocco, E. Evaluation of the nutraceutical, antioxidant and cytoprotective properties of ripe pistachio (Pistachia vera L. variety Bronte) hulls. *Food Chem.* **2016**, *196*, 493–502.
29. Smeriglio, A.; Denaro, M.; D'Angelo, V.; Germanò, M.P.; Trombetta, D. Antioxidant, Anti-Inflammatory and Anti-Angiogenic Properties of *Citrus lumia* Juice. *Front. Pharm.* **2020**, *11*, 593506.
30. Smeriglio, A.; Denaro, M.; De Francesco, C.; Cornara, L.; Barreca, D.; Bellocco, E.; Ginestra, G.; Mandalari, M.; Trombetta, D. Feijoa Fruit Peel: Micro-morphological Features, Evaluation of Phytochemical Profile, and Biological Properties of Its Essential Oil. *Antioxidants* **2019**, *8*, 320.
31. Smeriglio, A.; Bonasera, S.; Germanò, M.P.; D'angelo, V.; Barreca, D.; Denaro, M.; Monforte, M.T.; Galati, E.M.; Trombetta, D. *Opuntia ficus-indica* (L.) Mill. fruit as source of betalains with antioxidant, cytoprotective, and anti-angiogenic properties. *Phytoter. Res.* **2019**, *33*, 1–12.
32. Smeriglio, A.; Denaro, M.; Barreca, D.; Calderaro, A.; Bisignano, C.; Ginestra, G.; Bellocco, E.; Trombetta, D. In Vitro evaluation of the antioxidant, cytoprotective, and antimicrobial properties of essential oil from *Pistacia vera* L. Variety Bronte Hull. *Int. J. Mol. Sci.* **2017**, *18*, E1212.
33. Monforte, M.T.; Smeriglio, A.; Germano, M.P.; Pergolizzi, S.; Circosta, C.; Galati, E.M. Evaluation of antioxidant, antiinflammatory, and gastroprotective properties of *Rubus fruticosus* L. fruit juice. *Phytoter. Res.* **2018**, *32*, 1–11.
34. Bazzicalupo, M.; Burlando, B.; Denaro, M.; Barreca, D.; Trombetta, D.; Smeriglio, A.; Cornara, L. Polyphenol Characterization and Skin-Preserving Properties of Hydroalcoholic Flower Extract from *Himantoglossum robertianum* (Orchidaceae). *Plants* **2019**, *8*, 502.
35. Saso, L.; Valentino, G.; Casini, M.L.; Grippa, E.; Gatto, M.T.; Leone, M.G.; Silvestrini, B. Inhibition of Heat-induced denaturation of albumin by nonsteroidal anti-inflammato drugs (NSAIDs): Pharmacological implications. *Arch. Pharm. Res.* **2001**, *24*, 150–158.
36. Oyedapo, O.O.; Famurewa, A.J. Antiprotease and membrane stabilizing activities of extracts of *Fagara zanthoxyloides*, *Olax subscorpioides* and *Tetrapeura tetraptera*. *Int. J. Pharm.* **1995**, *33*, 65–69.
37. Caputo, L.; Cornara, L.; Bazzicalupo, M.; De Francesco, C.; De Feo, V.; Trombetta, D.; Smeriglio, A. Chemical Composition and Biological Activities of Essential Oils from Peels of Three *Citrus* Species. *Molecules* **2020**, *25*, 1890.

plants

Article

Colleters, Extrafloral Nectaries, and Resin Glands Protect Buds and Young Leaves of *Ouratea castaneifolia* (DC.) Engl. (Ochnaceae)

Elder A. S. Paiva, Gabriel A. Couy-Melo and Igor Ballego-Campos *

Departamento de Botânica, Instituto de Ciências Biológicas, Universidade Federal de Minas Gerais, Belo Horizonte 31270-901, MG, Brazil; epaiva@icb.ufmg.br (E.A.S.P.); gabriel-couy@hotmail.com (G.A.C.-M.)
* Correspondence: igorballego@gmail.com

Abstract: Buds usually possess mechanical or chemical protection and may also have secretory structures. We discovered an intricate secretory system in *Ouratea castaneifolia* (Ochnaceae) related to the protection of buds and young leaves. We studied this system, focusing on the distribution, morphology, histochemistry, and ultrastructure of glands during sprouting. Samples of buds and leaves were processed following the usual procedures for light and electron microscopy. Overlapping bud scales protect dormant buds, and each young leaf is covered with a pair of stipules. Stipules and scales possess a resin gland, while the former also possess an extrafloral nectary. Despite their distinct secretions, these glands are similar and comprise secreting palisade epidermis. Young leaves also possess marginal colleters. All the studied glands shared some structural traits, including palisade secretory epidermis and the absence of stomata. Secretory activity is carried out by epidermal cells. Functionally, the activity of these glands is synchronous with the young and vulnerable stage of vegetative organs. This is the first report of colleters and resin glands for *O. castaneifolia*. We found evidence that these glands are correlated with protection against herbivores and/or abiotic agents during a developmental stage that precedes the establishment of mechanical defenses.

Keywords: calcium oxalate crystals; colleter; extrafloral nectaries; resin gland; bud protection; plant-environment interaction

Citation: Paiva, E.A.S.; Couy-Melo, G.A.; Ballego-Campos, I. Colleters, Extrafloral Nectaries, and Resin Glands Protect Buds and Young Leaves of *Ouratea castaneifolia* (DC.) Engl. (Ochnaceae). *Plants* **2021**, *10*, 1680. https://doi.org/10.3390/plants10081680

Academic Editors: Milan S. Stankovic, Paula Baptista and Petronia Carillo

Received: 14 July 2021
Accepted: 10 August 2021
Published: 16 August 2021

Publisher's Note: MDPI stays neutral with regard to jurisdictional claims in published maps and institutional affiliations.

Copyright: © 2021 by the authors. Licensee MDPI, Basel, Switzerland. This article is an open access article distributed under the terms and conditions of the Creative Commons Attribution (CC BY) license (https://creativecommons.org/licenses/by/4.0/).

1. Introduction

The botanical family Ochnaceae has a pantropical distribution, comprising 27 genera and approximately 500 species [1], with its center of diversity being the Neotropics. Most of the diversity of Neotropical taxa is in the Amazon Basin, with just a few extra-Amazonian distributions restricted to Andean forests or the Brazilian Cerrado and Atlantic Forest [2]. The *Ouratea* genus, with about 300 species, is the largest and most diverse of the family [2]. The genus is widely distributed across several phytogeographic domains of Central and South America [3,4]. Many species have been recently described as *Ouratea*, demonstrating the limited study of the group and the cause of many taxonomic controversies (see [4]).

Secretion and secretory products seem to be important features for species of *Ouratea*. Representatives of this genus possess a pair of conspicuous stipules, in which there is, at least for some species, an extrafloral nectary (EFN) on the abaxial face [5]. Furthermore, species of *Ouratea* are a rich source of flavonoids and biflavonoids, and show potential as constituents of medicines; triterpenes, diterpenes, steroids, monosaccharides, and triacylglycerides are also common in this plant group [6].

Plant secretions are related to several forms of plant–environment interactions. Floral and vegetative buds constitute a vulnerable portion of plants and, thus, physical and chemical protections have often been found in these meristematic regions. Shoot buds usually possess substances produced by glands as distinct as colleters, nectaries, resin-producing glands, elaiophores, or secretory cavities [7].

The protective roles of some bud secretions have long been studied, as attested by the reports made by Groom [8], who stated that "*many buds have a great protective auxiliary in the secretion which covers and fills them. This secretion consists of gummy mucilage or resin, or both together; it is secreted by the general epidermis, by colleters, or by "leaf-teeth"*." The substances present in bud secretions can protect vegetative and reproductive buds against several environmental stresses ([9,10] and references therein). These substances act complementary to the protection provided by enveloping shoot apical meristems by superimposing cataphylls and undeveloped leaves [11].

Damage to shoot apices presents a high cost because they are essential for plant growth [12]. Therefore, investments in protecting the apical meristem are advantageous since the resprouting ability is a key functional trait that enables plant populations to persist after the destruction of living tissues from disturbance [13].

Knowledge regarding chemical defenses in shoot buds of Ochnaceae is far from comprehensive, although nectar secretion in scales and stipules of some taxa is well known. Many authors have recently reported the presence of glands in the family [14–17] sometimes relating them to bud protection [16]. Nevertheless, many aspects of secretion in the family are still poorly understood.

Considering this knowledge gap, we investigated the anatomy, histochemistry, and ultrastructure of the colleters, nectaries, and resin glands present in the buds and developing leaves of *Ouratea castaneifolia* (DC.) Engl., an arboreal-shrub species of the Cerrado and Brazilian semi-deciduous forests [18]. This evergreen species possesses leaves with a long lifespan, which persist for more than two years [19]. We present novel anatomical and ultrastructural data regarding some glands of Ochnaceae, and further discuss functional aspects.

2. Results

2.1. Bud Dynamics and Structural Aspects

Adult individuals of *O. castaneifolia* exhibit rhythmic growth, with one event of shoot growth per year and a long bud dormancy period lasting about ten months. During this stationary phase, the vegetative buds are protected by overlapping bud scales covering the vegetative apices (Figure 1A). The buds re-establish meristematic activity and start a new phase of vegetative growth before the end of the dry season (April–September), resulting in new leaves. After bud burst, shoot elongation occurs for about two months, when leaves grow, differentiate, and become strongly coriaceous and hard.

Flowering occurs at the end of leaf sprouting, and the vegetative buds go into dormancy, remaining in a resting state until the next annual vegetative cycle. The leaf primordia are produced at the beginning of the vegetative growth period and can be found at the apical portion of buds, each protected by a pair of stipules. The scales and stipules are foliaceous, wide, long (0.5×0.8 cm and 0.6×1.2 cm, respectively), and overlap the apical meristem (Figure 1A). The deciduous scales fall during bud burst, and stipules show abscission at the end of leaf blade expansion. Mature leaves of *O. castaneifolia* are simple, sclerophyllous, and coriaceous, with long marginal teeth resembling spines.

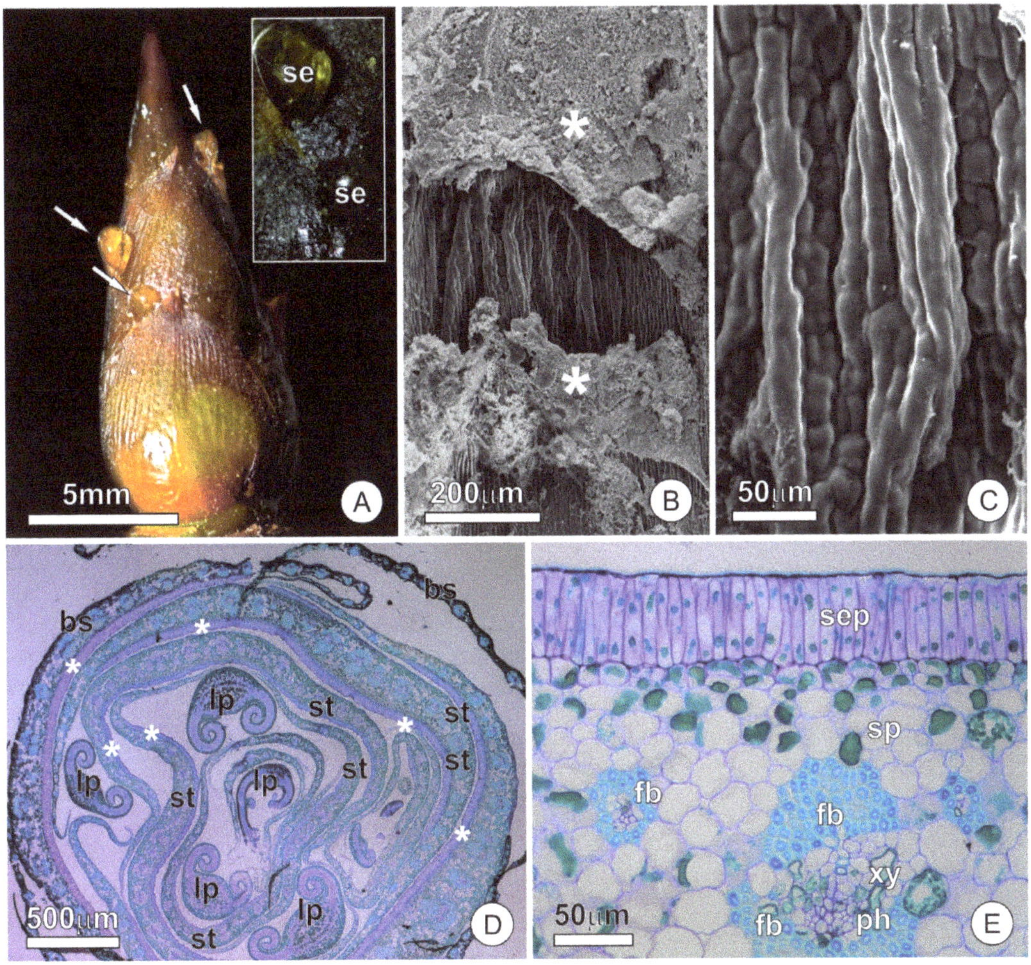

Figure 1. Resin glands of bud scales and stipules of *O. castaneifolia*: (**A**) Vegetative bud showing resin accumulation on the surface of bud scales (arrows). The insert shows released secretion towards the abaxial side of the bud scales; (**B**,**C**) The surface of resin-glands showing corrugated aspect and secretion residues (*). (**D**) Cross-section of a bud showing an overlapped arrangement of bud scales, stipules, and leaf primordia. Note the distinct epidermis at the abaxial face of bud scales and stipules (*); (**E**) Cross-section of a stipule showing adaxial secretory epidermis and the overall arrangement of the mesophyll and vascular tissue. Note the fiber cap that surrounds the vascular bundles; bs = bud scale, fb = fiber cap, lp = leaf primordium, ph = phloem, se = secretion, sep = secretory epidermis, st = stipule, xy = xylem.

2.2. Resin Glands

The median-basal portion of the adaxial surface of bud scales and stipules shows secretory epidermis related to resin secretion. The gland presents an irregular outline, with variable size among the different scales and stipules, sometimes occupying more than half of the adaxial surface. The surface of the gland is corrugated, with a smooth cuticle, and often covered by secretion residues. Stomata, pores, or cuticular ruptures were not observed on the gland surface (Figure 1B,C).

The resin gland possesses a secretory epidermis with columnar cells arranged in a palisade-like pattern (Figure 1D,E). The anticlinal surface of the secretory cells is about three times longer than the ordinary cells of the epidermis. The secretory cells have dense cytoplasm and large nuclei compared to other epidermal cells (Figure 1D). The mesophyll of scales and stipules have similar arrangements and cell types, being homogeneous and formed by round parenchyma cells (Figure 1D,E). Parenchyma cells of the portion of mesophyll underlying the gland do not have distinguishing characteristics (Figure 1D,E). The vascular bundles, arranged in the median portion of the mesophyll, are similar in scales and stipules; they are collateral and present an extensive cap of fibers on both abaxial and adaxial sides (Figure 1E). No change was observed in the vascular bundles towards the secretory region of the resin gland.

The secretory cells have a dense protoplast (Figure 2A), with a conspicuous nucleus and numerous organelles (Figure 2B). Mitochondria, dictyosomes, and the endoplasmic reticulum are the most abundant organelles, appearing scattered throughout the cytoplasm (Figure 2B–D). The endoplasmic reticulum is predominantly smooth, mainly appearing parallel to the plasma membrane and forming an extensive network permeating the entire cytoplasm (Figure 2B,C). Plastids are scarce and possess dense stroma and poorly developed inner membranes. The vacuome of secretory cells is inconspicuous and limited to small vacuoles. The presence of secretion, forming deposits of varying volumes throughout the cytosol, is striking in these cells (Figure 2C,D). The secretion observed in the cytosol is heterogeneous, with a peripheral portion strongly osmiophilic and a central region with a granular aspect (Figure 2D). The most striking feature in the cells of subglandular parenchyma is the presence of a large central vacuole, which has phenolic content (Figure 2E). Plastids, mitochondria, and endoplasmic reticulum are the predominant organelles in the extravacuolar cytoplasm (Figure 2E,F).

The ultrastructural analysis of the cuticle did not detect channels or ruptures that allow for the release of the secretion. Microdeposits of osmiophilic material are observed in the cuticle, similar to that observed in the protoplast (Figure 3A). The formation of subcuticular spaces is rarely observed and is mainly limited to a few cells (Figure 3A). Secretion residues, which spread throughout the adaxial surface of scales and stipules and disperse throughout all the structures they encompass, are present on the cuticular surface, sometimes forming lamellar structures (Figure 3B).

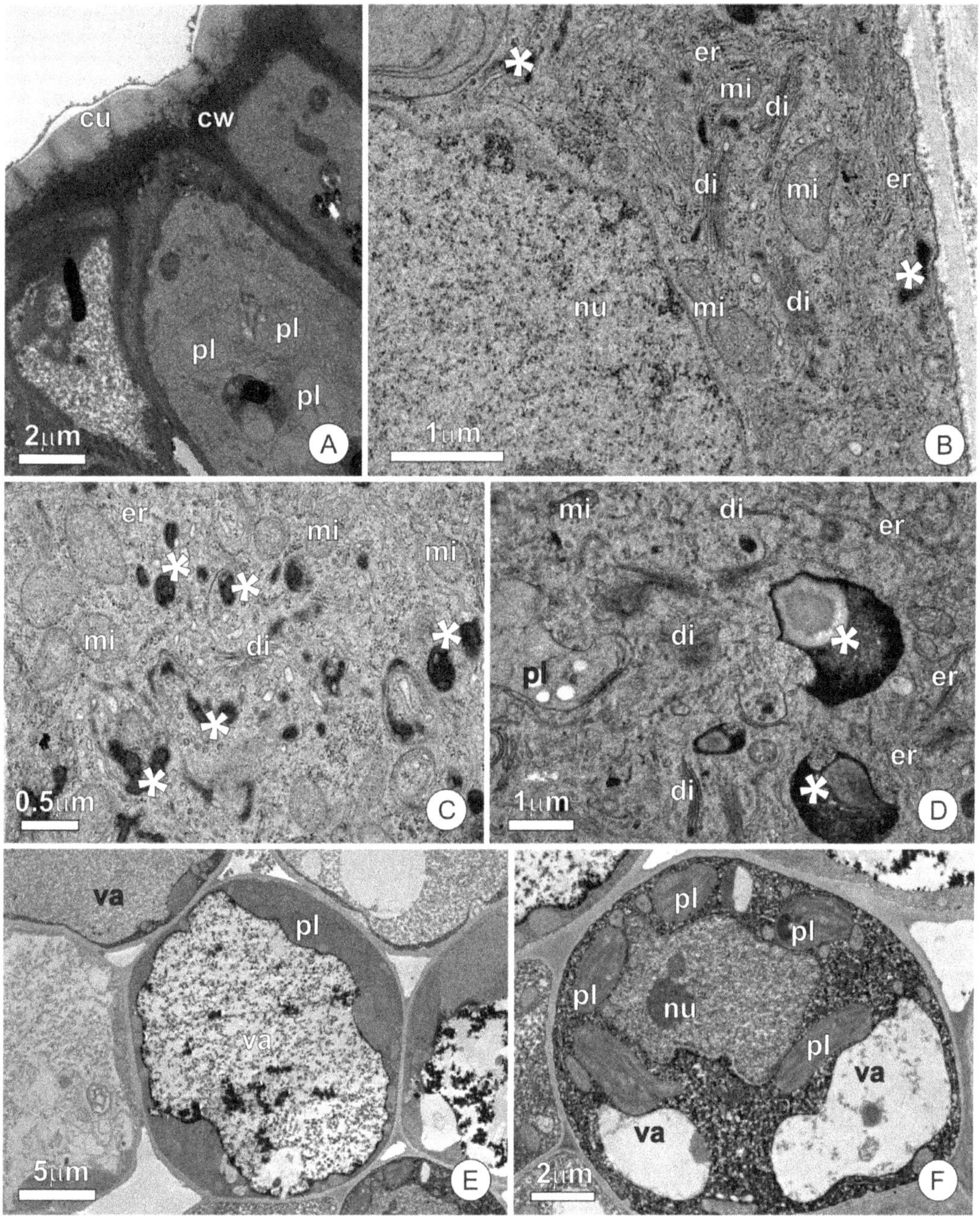

Figure 2. Ultrastructural aspects of the resin glands of *O. castaneifolia*: (**A**) Overall aspect of the secretory epidermis in cross-section. Note the dense protoplast of the cells. (**B–D**) Secretory cells showing organelle-rich protoplast with numerous mitochondria, dictyosomes, and segments of the smooth endoplasmic reticulum. Note the numerous deposits of osmiophilic secretion throughout the cytosol (*) (**E**,**F**). Cells of the subglandular parenchyma showing large vacuoles and phenolic contents. cu = cuticle, cw = cell wall, di = dictyosome, er = endoplasmic reticulum, mi = mitochondria, nu = nucleus, pl = plastid, va = vacuole.

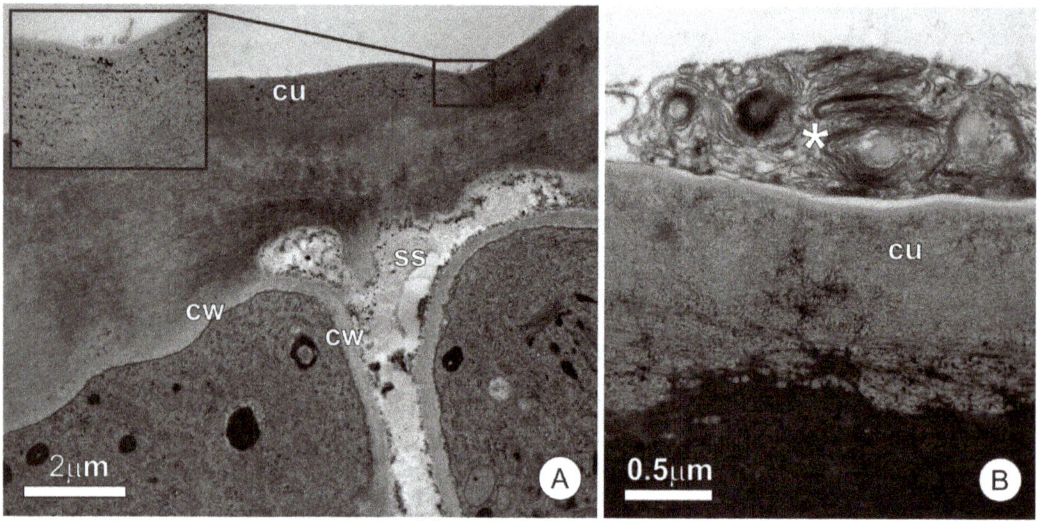

Figure 3. Cuticle structure in the secretory epidermis of resin glands: (**A**) Cross-section of the epidermic cells showing osmiophilic microdeposits inside the cuticle. A small subcuticular space can be observed. (**B**) Secretory residues are forming lamellar structures (*) on the surface of the cuticle. cu = cuticle, cw = cell wall, ss = subcuticular space.

2.3. Colleters

In the early stages of leaf differentiation, the marginal teeth present a colleter at the apex (Figure 4A,B). The colleters are long (up to 1 mm-long), pedunculate, and hyaline (Figure 4B). The peduncle region becomes sclerified and constitutes the marginal spines on mature leaves, leaving no evidence of the previous existence of the colleters (Figure 4C).

The colleters are formed in the early stages of leaf blade differentiation and are functional in young, unexpanded leaves (Figure 4D). Colleters persist in secretory activity throughout leaf expansion, but become senescent, turn brownish, and detach from the leaf after this stage. The secretory portion of colleters is conical, with a stomata-free epidermis covered by a smooth cuticle (Figure 4E–G). The fully-developed colleters present a secretory epidermis with columnar palisade cells that surround a parenchymatous central axis (Figure 4E,F).

In the secretory stage of colleters, the cells of the epidermal layer possess a dense and organelle-rich protoplast (Figure 5A–C). These cells display a large nuclei with uncondensed chromatin and conspicuous nucleoli (Figure 5B). The rough endoplasmic reticulum, mitochondria, and plastids complete the cytoplasmic organelles of the secretory cells (Figure 5B–D). Plastids have a poorly developed endomembrane system, and oil droplets that are similar to others observed free in the cytosol (Figure 5B). Dictyosomes are distributed throughout the cytoplasm, although they appear more numerous in the distal portion of cells (Figure 5C). The plasma membrane is sinuous, with the formation of irregular periplasmic spaces, within which the presence of amorphous and flocculated material can be observed (Figure 5C,E,F). This material is also accumulated in the intercellular spaces formed between palisade cells, in subcuticular spaces, and in large periplasmic spaces at the distal portion of the secretory cells (Figure 5E,F). The vacuome consists of small vacuoles, which are rare in most cells of the secretory epidermis. The parenchyma cells of the central axis present a set of organelles similar to those described for the secretory epidermis. However, these parenchyma cells possess few extravacuolar organelles due to the large central vacuoles filled with phenolics.

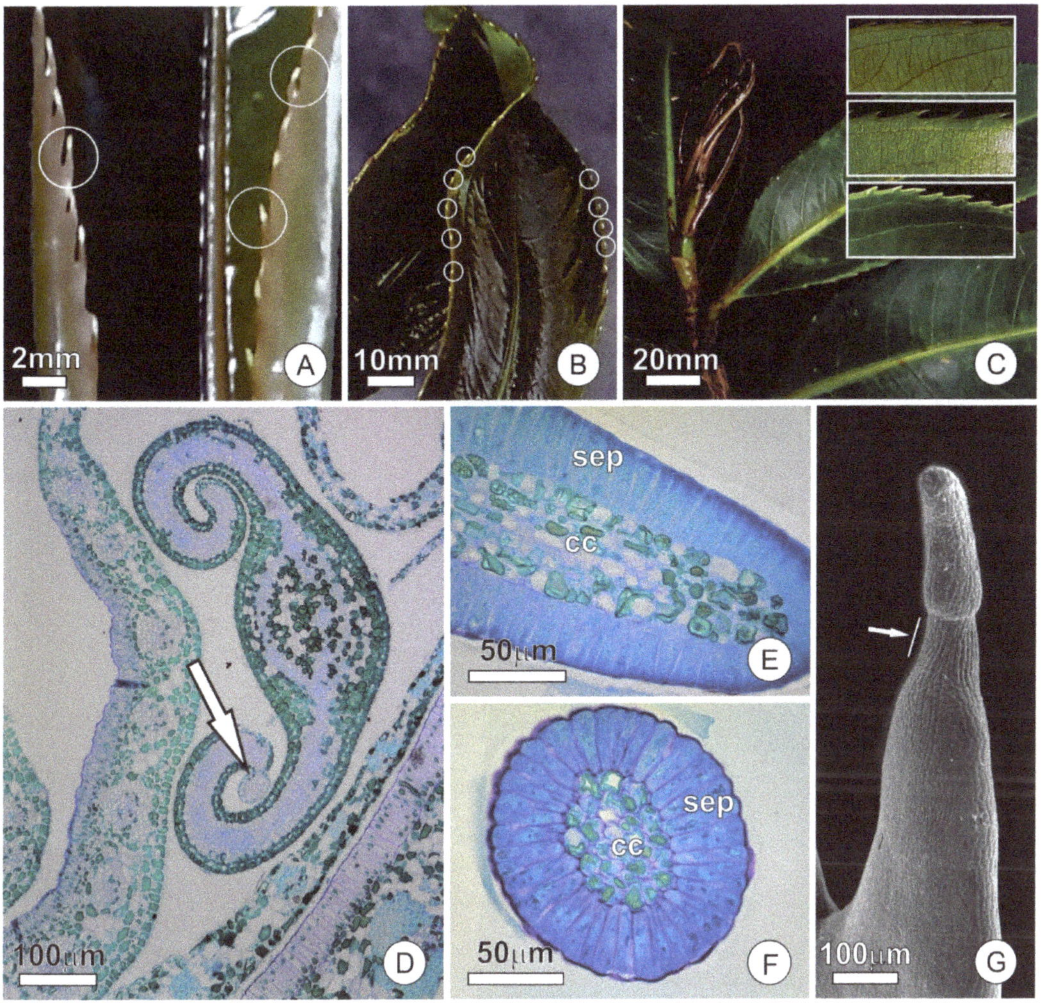

Figure 4. Distribution and structure of leaf colleters of *O. castaneifolia* leaves: (**A**,**B**) Young leaves, at the final stage of expansion, with hyaline colleters at the apex of the marginal teeth (circles). (**C**) Vegetative bud showing young leaves above mature leaves (from last sprouting event). Young leaves are shinny due to spread of colleter secretion; mature leaves present prominent marginal teeth. The inserts show the sequence of leaf maturation; notice the marginal teeth of the unexpanded young leaf (on top), followed by colleter abscission, intense sclerification and, finally, the fully-developed marginal spines (bottom). (**D**) Cross-section of a young leaf with involute ptyxis. Note the colleter at the leaf margin (arrow). (**E**) Longitudinal section of a colleter showing the secretory epidermis with columnar palisade cells and a central axis. (**F**) Cross-section of a colleter showing secretory epidermis surrounding the central axis. (**G**) Scanning electron micrograph of a marginal tooth in a young leaf. Note that the conical-shaped colleter at the apex is connected through a peduncle region (arrow). cc = central axis; sep = secretory epidermis.

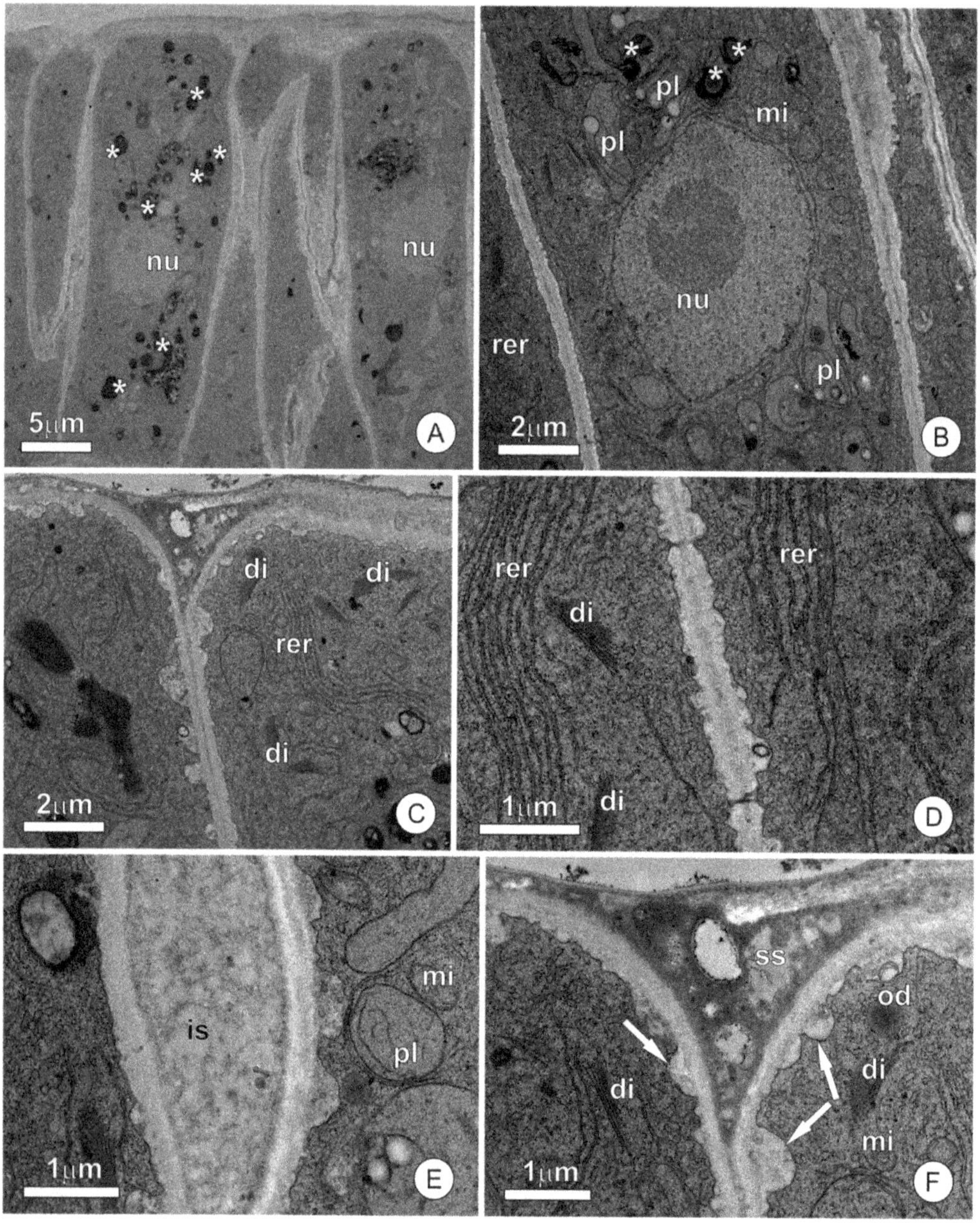

Figure 5. Ultrastructural aspects of the colleters of *O. castaneifolia*: (**A**) Overall aspect of the secretory epidermal cells showing dense protoplast and numerous osmiophilic inclusions (*); (**B–D**) Secretory cells showing large nuclei and organelle-rich cytoplasm with numerous mitochondria, plastids, and segments of the rough endoplasmic reticulum. Note oil droplets within the plastids and dispersed throughout the cytosol along with osmiophilic inclusions (*) in (**B**); (**E,F**) Details of the sinuous plasma membrane of secretory cells. Note the formation of irregular periplasmic spaces (arrows) and the accumulation of flocculated material within intercellular spaces and the subcuticular space. di = dictyosome, is = intercellular space, mi = mitochondria, nu = nucleus, od = oil droplet, pl = plastid, rer = rough endoplasmic reticulum, ss = subcuticular space.

2.4. Extrafloral Nectaries

A region of secretory cells stands out on the abaxial face of stipules (Figure 6A). This region is involved in the synthesis and release of nectar and constitutes an extrafloral nectary (EFN). The glandular surface is smaller than that of resin glands, slightly elongated in the axial direction, and also distinguished from the ordinary surface of the stipule by the absence of stomata, which are frequent in non-secretory portions (Figure 6A,B). Subcuticular spaces form conspicuous spaces on the secretory surface, which appear in different regions and reach large dimensions (Figure 6B,C), sometimes extending throughout the entire gland surface.

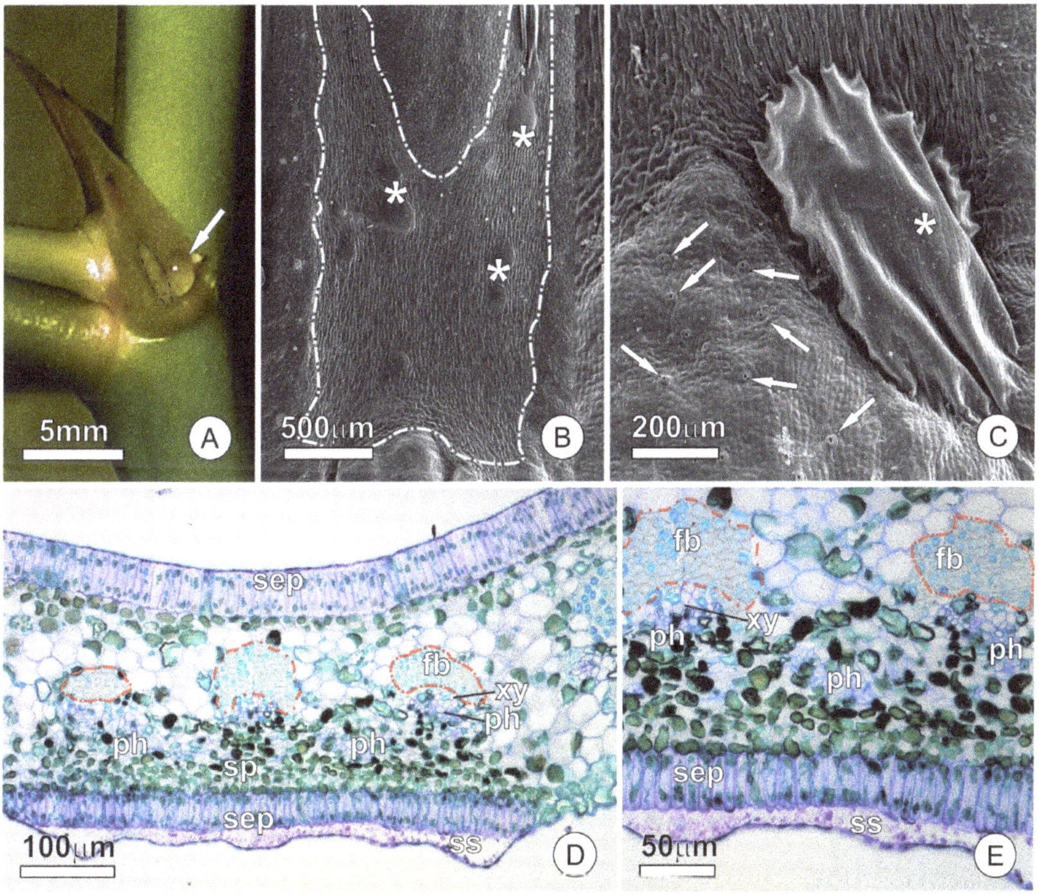

Figure 6. Structural aspects of extrafloral nectaries in stipules of *O. castaneifolia*: (**A**) Overview of a nectary (dashed line) in a stipule. Note the large nectar droplet (arrow). (**B**,**C**) Surface view of the nectary showing conspicuous subcuticular spaces (*) where nectar accumulates before being released. In (**C**), note the contrasting presence of stomata (arrows) on the ordinary surface of the stipule versus their absence over the nectary. (**D**) Cross-section of a stipule with a resin gland on the adaxial surface and a nectary on the abaxial surface. (**E**) Cross-section of a stipule showing the nectary portion. Note the dense arrangement of the subglandular parenchyma of the nectary and the presence of a large subcuticular space. A fiber cap (red dashed line), which is interrupted towards the nectary, surrounds the vascular bundles. fb = fiber cap, ph = phloem, sep = secretory epidermis, sp = subglandular parenchyma, ss = subcuticular space, xy = xylem.

The EFN consists of a uniseriate secretory epithelium, secretory parenchyma, and vascular tissues (Figure 6D). Epidermal cells are arranged in palisades similar to those described for the adaxial face (Figure 6E). The cells of the secretory parenchyma are smaller and the cytoplasm is denser than the other components of the mesophyll (Figure 6D,E). Vascular bundles in the vicinity of the nectary possess a gap on the abaxial fiber cap such that phloem cells make contact with the secretory parenchyma (Figure 6D,E). The secretory parenchyma of the EFN shows a remarkable presence of calcium oxalate (CaOx) crystals in the form of druses (Figure 7A–C). CaOx crystals are characteristically associated with the vascular bundles in both scales and stipules, especially in the parenchyma cells associated with the abaxial surface at the fiber cap limit. However, the crystals are more numerous where the cap of fibers is interrupted in the nectary region than in other areas of the stipule (Figure 7B–D).

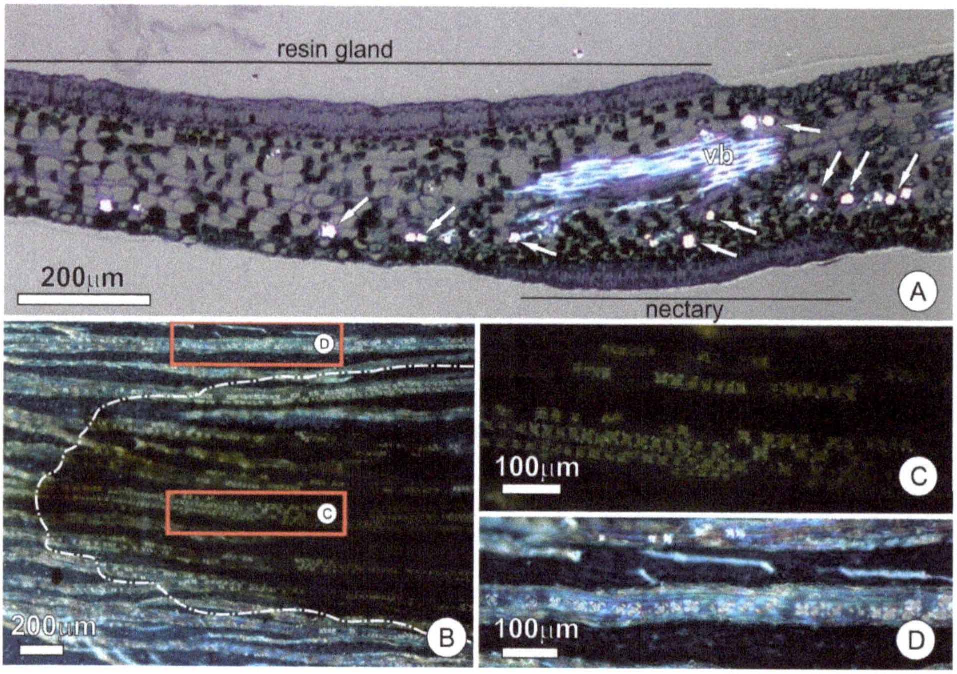

Figure 7. Distribution of calcium oxalate (CaOx) crystals in stipules of *O. castaneifolia*; all images were taken under polarized light: (**A**) Longitudinal section of a stipule showing greater accumulation of crystals (arrows) towards the abaxial surface and the nectary portion. Note that crystals are absent under the resin gland (adaxial face); (**B**–**D**) Surface view of the nectary portion (dashed line in (**B**)) showing the distribution of crystals. Note the numerous crystals in the nectary (**C**) in comparison to the area outside the nectary (**D**); The rectangles in (**B**) indicate the detailed areas in C and D. vb = vascular bundle.

The secretory cells present thin, pecto-cellulosic cell walls and cytoplasm rich in organelles, among which mitochondria, segments of the rough endoplasmic reticulum, dictyosomes, and plastids are the most representative (Figure 8A–D). Mitochondria have well-developed cristae and are distributed throughout the cytoplasm (Figure 8C,D). In the secretory stage, the dictyosomes appear inactive, with rare vesicles being produced (Figure 8D). Plastids have electron-lucent stroma, with a poorly developed inner membrane system with few grana thylakoids; plastoglobuli are dispersed in the stroma (Figure 8C,D), while starch is markedly absent. The few observed vacuoles are small and filled with a flocculated content (Figure 8C). Secretory cells of the epidermis were observed to connect via plasmodesmata (Figure 8C). Although secretory parenchyma cells have large vacuoles, the extravacuolar cytoplasm is organelle-rich and shows a composition similar to that of the cells of the secretory epidermis (Figure 8E,F).

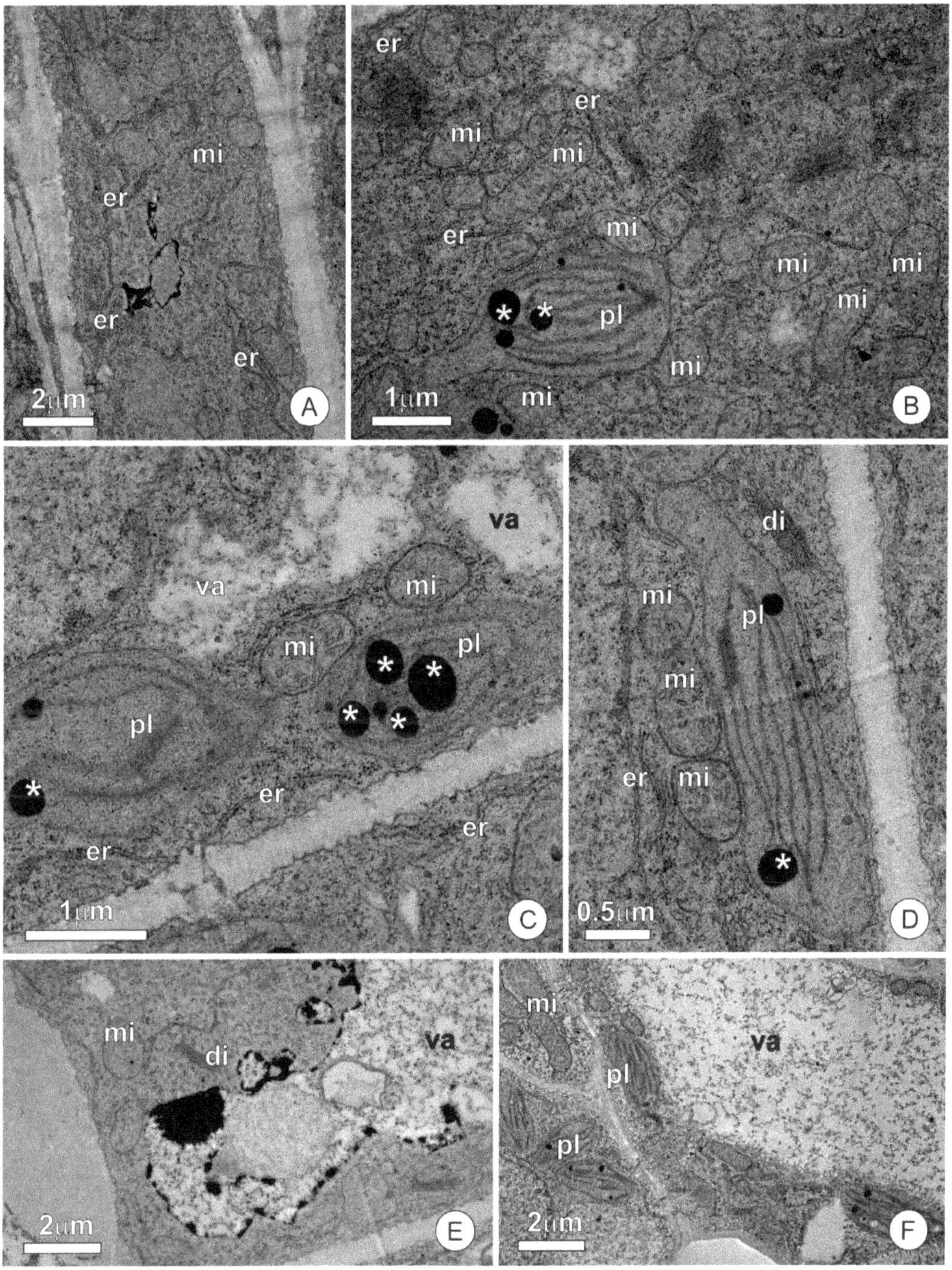

Figure 8. Ultrastructural aspects of the extrafloral nectaries of *O. castaneifolia*: (**A**) Overview of a secretory cell showing a dense protoplast; (**B–D**) Secretory cells showing organelle-rich cytoplasm with abundant mitochondria, segments of the rough endoplasmic reticulum, and plastids with a poorly developed inner membrane system. Note the numerous plastogobuli (*) dispersed in the stroma of the plastids and the small vacuoles filled with a flocculated content; (**E**,**F**) Cells of the subglandular parenchyma showing large vacuoles and extravacuolar cytoplasm rich in organelles. di = dictyosome, er = endoplasmic reticulum, mi = mitochondria, pl = plastid, va = vacuole.

2.5. Histochemistry and Sugar Analysis

Histochemical tests revealed a mixture of hydrophilic and lipophilic components, including terpenoids, mucilage, lipids, and proteins in both the resin-producing gland and colleters. Terpenoids were the most abundant and strongly marked by NADI reagent in the resin-producing glands, while mucilage was less conspicuous. Conversely, NADI reagent showed a weak reaction in the colleters, and both the protoplast and exudate marked strongly with Ruthenium Red. Differential coloration granted by the NADI reagent suggests that the terpene content is associated with essential oil production. Lipids and proteins were seen in both the protoplast and exudate of colleters but were absent in the resin-producing gland.

The secretion exuded by the EFNs tested positive for glucose by glucose strip tests, indicating a sugary secretion and confirming nectar release. Tests with Xylidine Ponceau indicated the presence of structural proteins in the protoplast of nectary cells, but other tests yielded negative results. The results for all histochemical tests performed are summarized in Table 1.

Table 1. Results for histochemical tests performed in the glands of *O. castaneifolia* buds and young leaves.

Test	Target Substance	Resin Gland		Colleter		Nectary	
		Protoplast	Secretion	Protoplast	Secretion	Protoplast	Secretion
NADI reagent	Terpenoids (essential oils)	+	+	−	−	−	−
Ruthenium Red	Mucilage	−	−	+	+	−	−
Sudan Red B	Lipids	−	−	+	+	−	−
Xylidine Ponceau	Proteins	−	−	+	+	+	−
Glucose strip-test	Sugars (glucose)	N/A	N/A	N/A	N/A	N/A	+

+ positive, − negative, or weak reaction, N/A = not applicable.

3. Discussion

3.1. Anatomy

The secretory portion of the studied glands share some similarities, mainly because epidermal cells are directly involved in the secretory process in all of them. The prevalence of epidermis in secretory processes is common to many other secretory structures of eudicotyledons, including colleters, nectaries, elaiophores, and other glands throughout distinct taxa [20–26].

Resin production by a patch of differentiated epithelium, as observed in *O. castaneifolia*, is uncommon. These secretions are often associated with trichomes, colleters, ducts, or cavities [7]. Buds of *Populus* spp. (Salicaceae) possess a palisade-like epidermis in the adaxial side of the stipules that secretes resin [7,20], as described here for *O. castaneifolia*. However, in *Populus*, the secretory epithelium is not restricted to a specific area, extending towards the entire adaxial surface, which is heavily ridged [20].

Nonetheless, the similarities between the secretory system in buds of *O. castaneifolia* and *Populus* species are worth mentioning. Apart from the stipular resin glands, *Populus* also possess specialized leaf teeth with resin-secreting glands and extrafloral nectaries (or hydathodes [7]). Thus, the glandular apparatus of these taxa might constitute an interesting case of convergence regarding bud protection within the Malpighiales.

The presence of a central axis in the colleters of *O. castaneifolia* that is very distinct from the epithelial cells indicates a mixed origin of this structure, encompassing both the protoderm and ground meristem. Therefore, such colleters can be considered as the "standard-type", following Thomas [21]. Standard colleters occur in several taxa of angiosperms, most notably the Rubiaceae and Apocynaceae [9,21,22,26]. Colleters or colleter-like glands (i.e., thick glandular hairs) have been reported in a few species of Ochnaceae, although usually associated with the inner base of stipules, sepals, or leaves [14–16]. Marginal glands, however, are commonly reported in *Sauvagesia* [15,27,28] and several additional genera of the subfamily Sauvagesioideae [14]. Recently, Rios et al. [17] also demonstrated marginal colleters in two species of *Luxemburgia*. Nonetheless, data on the anatomy, ultrastructure, and secretory activity of these structures is lacking, and the present description appears to be unprecedented, to the best of our knowledge.

The colleters of *O. castaneifolia* are very conspicuous due to their contrasting colors. However, these structures have not been described until now, and the reason seems to be the asynchrony between the phase in which they occur and that of interest for taxonomic studies. By the time most, if not all, *Ouratea* species bloom, the leaf is already wholly differentiated, and the colleters have already suffered abscission. Thus, in taxonomic analyses, which are mainly made of fertile material, colleters are not seen; this fact appears strikingly in the descriptions of new species, whose morphological descriptions are detailed yet do not register the presence of colleters. This gap in the reports of temporary secretory structures has also been reported for extrafloral nectaries [29]. Given that serrate leaves are a remarkable character for *Ouratea* [30], it seems reasonable to suppose that colleters, which occur at the apex of each marginal tooth, are a characteristic shared by several species of this genus. The report of marginal colleters in *Luxemburgia*, together with recent data showing a high correlation between leaf teeth and glands in eudicots [17], might corroborate this hypothesis.

Based on their structure, the nectaries of *O. castaneifolia* could be classified as embedded nectaries, i.e., totally embedded in tissues of other organs [31]. Nonetheless, they comprise slight specializations of the epidermis and subjacent tissue rather than conspicuous and distinct units enclosed in the mesophyll. The observed lack of bundle caps towards the nectary tissue is also noteworthy, as it exposes the phloem directly to the secretory parenchyma. While most vascularized nectaries rely on variable extensions of phloem, xylem, or both [31,32], the nectaries of *O. castaneifolia* are vascularized by direct contact with the vascular bundles. This, in turn, indicates the requirement of a steady and direct supply of pre-nectar solutions from phloem. Usually, extrafloral nectaries lack starch reserves [33], as we observed here. This remarkable absence of energetic reserves seems to reinforce the role of phloem as the source of pre-nectar.

3.2. Ultrastructure and Secretion Mechanism

The overall aspect of the protoplast, including dense cytoplasm, conspicuous nuclei, and numerous organelles, corroborates the secretory nature of the cells comprising the glands of *O. castaneifolia* [7,34,35]. Additionally, evidence of accumulated material (osmiophilic and granulated), either scattered throughout the cytoplasm or associated with vesicles and other organelles, corroborate an intense secretory process and a secretion of mixed nature, as also observed in histochemical tests.

The presence of abundant mitochondria observed in all studied glands likely reflects an intense metabolic activity with high-energy requirements [34,36], while other organelles are involved in specific types of secretory products [9,34,37]. In this sense, the presence of abundant active dictyosomes in the colleters and resin glands indicates polysaccharide synthesis related to mucilaginous secretory products, as commonly demonstrated in several glands secreting mucilage or mixed-secretions [7,9,37–39]. The presence of mucilaginous material, as revealed by histochemical tests, corroborates this view. However, in the resin glands, dictyosomes were also associated with osmiophilic material, indicating their involvement in resin synthesis. While this is less common, some authors have previously indicated the association of Golgi bodies with osmiophilic material in resin-secreting glands [40,41]. The osmiophilic nature and the positive reaction for terpenoids in histochemical tests suggest that this material comprises the terpene fraction of the secretion. Terpenoid synthesis in plants likely occurs at different cellular sites, so that a resinous substance might be composed of distinct portions produced after intercellular exchange between various compartments [7,42]. Plastids and the endoplasmic reticulum are usually the most common organelles associated with these type of secretions [7,9,37]; the abundant presence of these organelles in the colleters and resin glands of *O. castaneifolia* indicates that they are also involved in the resinous portion of the secretion. The presence of oil droplets and abundant, rough endoplasmic reticulum in colleters corroborates the occurrence of lipids and proteins in the secretion, as detected by histochemical tests.

In the case of the nectaries, the absence of osmiophilic inclusions, along with the inconspicuous activity of the Golgi apparatus and an abundance of endoplasmic reticulum is congruent with nectar secretion. According to Fahn [7], the endoplasmic reticulum is the dominant organelle in nectar-secreting cells, and the dictyosomes might be less developed during the secretory stage.

The secretory route in the colleters and resin glands is delineated by the presence of secretion products (lipophilic, granular, and amorphous inclusions) dispersed throughout the cytosol, periplasmic spaces, and subcuticular spaces, and is also included in the cuticle. In this sense, secretions produced in the various organelles involved are transported throughout the cytosol, potentially fusing and agglomerating before liberation in the periplasmic spaces. After this point, the secretion crosses the cell walls, usually accumulating in intercellular spaces and small subcuticular spaces before reaching the surface of the glands. Accumulation in the periplasmic space and other extracellular spaces indicates that a pressure-based model of secretion release is involved [43]. The presence of osmiophilic droplets in the colleters and resin-secreting glands indicates lipophilic material and is a typical feature of resin-secreting glands [7,44].

The ultrastructure of the secretory cells in the nectaries indicate a granulocrine secretion [7,34], in which the incoming pre-nectar is processed, transported in vesicles, and eliminated via fusion or invagination of the plasmalemma. The conspicuous subcuticular spaces observed in the nectaries of *O. castaneifolia* suggest cuticle rupture and nectar release in a cycling manner. This mechanism of nectar release is a common feature among stomata free nectaries, in which nectar can be released by repetitive cycles of cuticle detachments and rupture [45].

3.3. Functional Aspects

Secretions, such as nectar, resins, and mucilages, associated with EFNs, resin glands and colleters, respectively, are recognized for mediating plant–environment interactions. The resin-producing glands of *O. castaneifolia* are related to the protection of the bud itself, including the promeristem and all of the developing organs that it contains. In turn, the EFNs and colleters are related to protecting specific young organs, namely the developing leaves. The secretion observed in resin glands, in which we found essential oils in association with polysaccharides, is similar to those commonly observed in colleters as these structures also show mixed secretions with both hydrophilic and hydrophobic compounds. Therefore, resin glands act in the protection of buds in a similar way that typical colleters do, both providing a coverage of secretion that might protect against biotic and abiotic factors. In fact, from a functional point of view, these structures can be considered analogous. Although there are controversies about the definition of colleters, the functional aspect seems to be preponderant for recognizing these structures [46,47]. While the scales and stipules of *O. castaneifolia* have resin glands formed essentially by a secretory epithelium, Reinales and Parra-O [16] described the presence of standard colleters in scales and stipules for the clade comprising *Rhytidanthera*, *Godoya*, *Cespedesia*, and *Krukoviella*. It is important to note that these colleters and resin glands have similar secretory activity and most likely, perform the same function. The involvement of colleters in the protection of buds, especially those associated with stipules and scales, has been reported for several taxa [48]. Therefore, the evolution of the glandular system in vegetative buds of Ochnaceae proves to be an open and intriguing question.

The type of ptyxis showed by *O. castaneifolia*, and the arrangement of colleters at the leaf blade margin, seem to act in facilitating the spread of secretion throughout the leaf surface, on both sides, as suggested by Paiva [49]. Thus, these colleters seem to have an action directed at leaf blade protection. On the other hand, the meristem and young leaves in the phase that precedes the formation of colleters, are protected by the secretion of resin glands. In this way, there is no overlapping of functions but a complementarity between these two secretory structures.

There seems to be a correlation between the composition of the secretion of colleters and environmental factors. Tresmondi et al. [9] compared colleters of species from savanna environments with those from the forest and observed that resinous secretions prevail in the savanna environment, which is subject to greater luminous and water stresses. Considering that *O. castaneifolia* inhabits savanna (Brazilian Cerrado) and forest-edge environments, the presence of mixed secretion, both in the colleters and in the resin gland, seems to reflect a greater protection against desiccation.

Concerning mucilaginous secretions, such as that produced by colleters, Groom [8] argued that "hygroscopic substance like mucilage (and tannin) is an admirable means of controlling the water-supply of an organ for two reasons: first, the osmotic power of a solution increases with a rise of temperature; secondly, the osmotic power increases with the concentration of the solution. The result is that when a bud is in greatest danger of losing all its water—i.e., when the temperature is high and a considerable amount of water has been evaporated from the mucilage—the remaining water is held most firmly or a first supply of water is absorbed most fiercely". Similarly, resins are also likely to reduce water loss by cuticular transpiration or even reduce leaf temperature by increasing radiation reflectance in hot, arid conditions [7,50]. This protection against water loss is even more critical in young organs because their cuticle and vascular tissues are incipient, compromising adequate transport and water retention (see [49]). Additionally, due to its chemical composition, lipophilic substance such as essential oils and oleoresins are frequently associated with protection against pathogens and herbivores [7,23].

The occurrence of EFNs was reported for eight species of *Ouratea* [5], including *O. castaneifolia* [23,51,52]. In these reports, the location of the nectaries is the same, that is, on the abaxial face of the stipules or cataphylls. Thus, in all species of *Ouratea* with reports of EFNs, these structures are ephemeral and seem to be related exclusively to the protection of young organs, given the caducous nature of the stipules to which they are associated. According to Machado et al. [23], the EFNs of species of *Ouratea* effectively protect plants against herbivores; EFNs of *O. spectabilis* are visited by several ant species that significantly reduce damage by lepidopteran caterpillars.

In the studied EFNs of *O. castaneifolia*, the highest concentration of calcium oxalate crystals coincides with the vascularized portion of these structures. The presence of these crystals is associated with the control of cytosolic calcium levels [53], which seems to be an essential factor for nectar secretion [54]. It is not by chance that the presence of these crystals is frequently reported in the nectaries of different plant taxa [23,55–61]. Although the presence of these crystals is often linked to some protection against the action of herbivores [58], in *O. castaneifolia*, and in most of the taxa in which they occur, this seems unlikely. It is important to emphasize that the crystals occur in the deepest layers of the nectary, leaving the cells with dense protoplast, which are more nutritious and vulnerable to herbivory, exposed towards the gland surface.

4. Materials and Methods

4.1. Plant Material

Plant material was collected from three adult individuals of *O. castaneifolia* growing on the Campus of the Universidade Federal de Minas Gerais, Belo Horizonte (Brazil). The plants were observed and sampled during the years 2019 and 2020. Whole vegetative buds and the median portion of several isolated bud scales, stipules, and young developing leaves were obtained from each of these individuals and subjected to the procedures below. For each of the portions obtained, all individuals were sampled in each of the procedures, with at least three replicates per individual.

4.2. Light Microscopy

For microscopy analysis, whole buds and samples of bud scales, stipules, and young leaves were vacuum infiltrated with Karnovsky's fixative (paraformaldehyde 4% and glutaraldehyde 5% in phosphate buffer 0.1 M, pH 7.2; modified from [62]) for 5 min and left to set for 24 h in the same solution. Soon after, they were dehydrated in an increasing ethanol series (10–98%) and embedded in (2-hidroxiethyl)-methacrylate (Historesin embedding kit, Leica, Heidelberg, Germany). Transverse and longitudinal sections of the entire apex and fragments of stipules, bud scales, and young leaves were obtained using a rotary microtome (Hyrax M40, Carl Zeiss Mikroskopie, Jena, Germany). The 5–6 μm thick sections were mounted on glass slides and stained with toluidine blue (0.5% in acetate buffer 0.1 M, pH 4.7; modified from [63]). Analysis and image capture were performed using a light microscope (CX41RF, Olympus Scientific Solutions, Waltham, MA, USA) coupled to a digital camera (U-TV0.5XC-3, Olympus Scientific Solutions, Waltham, MA, USA) and a computer with an imaging software (LCmicro, Olympus Soft Imaging Solutions, Waltham, MA, USA).

4.3. Electron Microscopy

For scanning electron microscopy (SEM), whole buds and samples of bud scales, stipules, and young leaves were fixed in Karnovsky solution (paraformaldehyde 4% and glutaraldehyde 5% in phosphate buffer 0.1 M, pH 7.2; modified from [62]). Samples were left under vacuum for 5 min to improve infiltration, after which they were kept in the fixative for 24 h. The samples were then dehydrated in an increasing ethanol series (5–100%), critical-point dried (CPD030, Bal-Tec/Leica, Balzers, Liechtenstein), and coated with a palladium-gold alloy (MD20, Bal-Tec, Balzers, Liechtenstein). The samples were analyzed using a Quanta 200 scanning electron microscope (FEI Co., Eindhoven, The Netherlands).

For transmission electron microscopy (TEM), samples of bud scales, stipules, and young leaves were prepared to isolate fragments (1 × 1 mm) containing portions of the secretory glands. These samples were fixed in Karnovsky solution (paraformaldehyde 4% and glutaraldehyde 5% in 0.1 M phosphate buffer, pH 7.2; modified from [62]), infiltrated under vacuum for 5 min, and left in this fixative for 24 h. The fixed material was post-fixed in osmium tetroxide (1% in phosphate buffer 0.1 M, pH 7.2) for 2 h, dehydrated in an acetone series (30, 50, 70, 95, 100%), and embedded in low viscosity epoxy resin [64]. The material was then sectioned with an ultramicrotome (UC6, Leica, Deer-field, IL, USA) coupled with a diamond blade. The ultrathin sections (40–60 nm thick) were contrasted using a saturated solution of uranyl acetate and lead citrate [65]. The analysis was performed using a Tecnai G2-Spirit transmission electron microscope (Philips/FEI Co., Eindhoven, The Netherlands) at 80 Kv.

4.4. Histochemistry

Freshly collected samples of *O. castaneifolia* were used for histochemical tests. For each studied gland, samples were free-hand sectioned, subjected to histochemical tests, and mounted on glass slides. Sudan Red B (0.5%, in ethanol 95% and glycerin 1:1) was used for lipids (modified from [66]), Ruthenium Red (0.002%, aqueous solution) for mucilages [67], NADI reagent for oleoresins and essential oils [68], and Xylidine Ponceau (0.1% in acetic acid 3%) for proteins [69]. After treatment, the sections were briefly washed in the respective solvent of each test, and then finally washed in distilled water (for NADI test we used phosphate buffer 0.1 M, pH 7.2). An analysis was performed at the end of each test using a light microscope (CX41RF, Olympus Scientific Solutions, Waltham, MA, USA) coupled to a digital camera (U-TV0.5XC-3, Olympus Scientific Solutions, Waltham, MA, USA) and a computer with an imaging software (LCmicro, Olympus Soft Imaging Solutions, Waltham, MA, USA). Additionally, glucose strip tests (Alamar Tecno Científica, São Paulo, Brazil) were used to confirm the presence of sugars in the nectary secretion.

5. Conclusions

The vegetative buds of *O. castaneifolia* display a diverse secretory system comprised of resin-secreting glands, colleters, and extrafloral nectaries. There is marked synchrony of the secretory activity of these glands with the differentiation and expansion of young organs. Thus, it seems reasonable to assume that the secretory activity in these cases is correlated with the protection against herbivores and/or abiotic agents, since buds and young organs are vulnerable. Vegetative buds are vulnerable structures that have a high fitness value and are usually strongly defended. In *O. castaneifolia*, the defense system is expressed through mediators of plant–environment interactions, which prevail in young organs and act in a phase that precedes the development of mechanical defenses.

Author Contributions: Conceptualization, E.A.S.P.; methodology, I.B.-C., G.A.C.-M. and E.A.S.P.; validation, I.B.-C., G.A.C.-M. and E.A.S.P.; investigation, I.B.-C., G.A.C.-M. and E.A.S.P.; resources, E.A.S.P.; data curation, I.B.-C., G.A.C.-M. and E.A.S.P.; writing—original draft preparation, I.B.-C., G.A.C.-M. and E.A.S.P.; writing—review and editing, I.B.-C. and E.A.S.P.; visualization, I.B.-C., G.A.C.-M. and E.A.S.P.; supervision, E.A.S.P.; project administration, E.A.S.P.; funding acquisition, E.A.S.P. All authors have read and agreed to the published version of the manuscript.

Funding: This research was funded in part by the Coordenação de Aperfeiçoamento de Pessoal de Nível Superior—Brazil (CAPES), Finance Code 001. This work was also supported through a research grant from the Conselho Nacional de Desenvolvimento Científico e Tecnológico (CNPq, Brazil, process 305638/2018-1) to E.A.S.P.

Institutional Review Board Statement: Not applicable.

Informed Consent Statement: Not applicable.

Data Availability Statement: Not applicable.

Acknowledgments: The authors thank the funding agencies cited above and the Centro de Microscopia—Universidade Federal de Minas Gerais (CM-UFMG) for providing the equipment and technical support for electron microscopy analyses.

Conflicts of Interest: The authors declare no conflict of interest.

References

1. Schneider, J.V.; Bissiengou, P.; Amaral, M.C.E.; Tahir, A.; Fay, M.F.; Thines, M.; Sosef, M.S.M.; Zizka, G.; Chatrou, L.W. Phylogenetics, ancestral state reconstruction, and a new infrafamilial classification of the pantropical Ochnaceae (Medusagynaceae, Ochanceae s.str., Quiinaceae) based on five DNA regions. *Mol. Phylogenet. Evol.* **2014**, *78*, 199–214. [CrossRef] [PubMed]
2. Fiaschi, P.; Nicoletti de Fraga, C.; Yamamoto, K. Neotropical Ochnaceae s.l. (incl. Quiinaceae). In *Neotropikey—Interactive Key and Information Resources for Flowering Plants of the Neotropics*; Milliken, W., Klitgird, B., Baracat, A., Eds.; (2009 onwards); Available online: http://www.kew.org/science/tropamerica/neotropikey/families/Ochnaceae_s.l._(incl._Quiinaceae).htm (accessed on 12 March 2021).
3. Yamamoto, K. Estudos Taxonômicos Sobre *Ouratea parviflora* (DC.) Baill. (Ochnaceae) e Espécies Afins Ocorrentes em Floresta Atlântica Nas Regiões Sudeste e Sul do Brasil. Ph.D. Thesis, Universidade Estadual de Campinas, Campinas, Brasil, 1995.
4. Fraga, C.N.; Saavedra, M.M. A new cauliflorous white-flowered species of *Ouratea* (Ochnaceae) from the Brazilian Atlantic Forest. *Phytotaxa* **2014**, *167*, 119–126. [CrossRef]
5. Weber, M.G.; Porturas, L.D.; Keeler, K.H. World List of Plants with Extrafloral Nectaries. Available online: www.extrafloralnectaries.org (accessed on 16 March 2021).
6. Suzart, L.R.; Daniel, J.F.S.; Carvalho, M.G.; Kaplan, M.A.C. Biodiversidade flavonoídica e aspectos farmacológicos em espécies dos gêneros *Ouratea* e *Luxemburgia* (Ochnaceae). *Quim. Nova* **2007**, *30*, 984–987. [CrossRef]
7. Fahn, A. *Secretory Tissue in Plants*; Academic Press: London, UK, 1979.
8. Groom, P. On bud-protection in dicotyledons. *Trans. Linn. Soc. Lond. 2nd Ser. Bot.* **1892**, *3*, 255–266. [CrossRef]
9. Tresmondi, F.; Canaveze, Y.; Guimarães, E.; Machado, S.R. Colleters in Rubiaceae from forest and savanna: The link between secretion and environment. *Sci. Nat.* **2017**, *104*, 17. [CrossRef]
10. Voigt, D.; Kim, J.; Jantschke, A.; Varenberg, M. Robust, universal, and persistent bud secretion adhesion in horse-chestnut trees. *Sci. Rep.* **2020**, *10*, 16925. [CrossRef]
11. Campos, B.H.; Guimarães, E.; Canaveze, Y.; Machado, S.R. Epicormic bud protection traits vary along a latitudinal gradient in a neotropical savana. *Sci. Nat.* **2021**, *108*, 11. [CrossRef]
12. Coley, P.D.; Kursar, T.A. Anti-herbivore defences of young tropical leaves: Physiological constrains and ecological tradeoffs. In *Tropical Forest Plant Ecophysiology*; Mulkey, S.S., Chazdon, L.R., Eds.; Springer: Boston, MA, USA, 1996; pp. 305–336. [CrossRef]

13. Pausas, J.G.; Pratt, R.B.; Keeley, J.E.; Jacobsen, A.L.; Ramirez, A.R.; Vilagrosa, A.; Paula, S.; Kaneakua-Pia, I.N.; Davis, S.D. Towards understanding resprouting at the global scale. *New Phytol.* **2016**, *209*, 945–954. [CrossRef]
14. Amaral, M.C.E.; Bittrich, V. Ochnaceae. In *Flowering Plants. Eudicots. The Families and Genera of Vascular Plants*; Kubitzki, K., Ed.; Springer: Berlin/Heidelberg, Germany, 2014; Volume 11, pp. 253–268. [CrossRef]
15. Queiroz-Lima, A.; Amorim, A.M.; Cardoso, D.B. A new bristle-leaved species of *Sauvagesia* (Ochnaceae) endemic to the Espinhaço range, Brazil. *Syst. Bot.* **2017**, *42*, 346–350. [CrossRef]
16. Reinales, S.; Parra-O, C. Phylogenetic position and evolution of glandular structures of the unusual and narrowly distributed genus *Rhytidanthera* (Ochnaceae). *Bot. J. Linn. Soc.* **2020**, *193*, 84–99. [CrossRef]
17. Rios, A.B.M.; Menino, G.C.D.O.; Dalvi, V.C. Leaf teeth in eudicots: What can anatomy elucidate? *Bot. J. Linn. Soc.* **2020**, *193* 504–522. [CrossRef]
18. Lorenzi, H. *Árvores Brasileiras: Manual de Identificação e Cultivo de Plantas Arbóreas do Brasil*; Instituto Plantarum: Nova Odessa Brazil, 2000.
19. Barbosa-Campos, M.T.; Castro, S.A.B.; Kuster, V.C.; Santos, L.N.; Lemos-Filho, J.P.; Vale, F.H.A. How the long-life span leaves of *Ouratea castaneifolia* Engl. (Ochnaceae) differ in distinct light conditions. *Braz. J. Bot.* **2018**, *41*, 403–414. [CrossRef]
20. Curtis, J.D.; Lersten, N.R. Morphology, seasonal variation, and function of resin glands on buds and leaves of *Populus deltoides* (Salicaceae). *Am. J. Bot.* **1974**, *61*, 835–845. [CrossRef]
21. Thomas, V. Structural, functional and phylogenetic aspects of the colleter. *Ann. Bot.* **1991**, *68*, 287–305. [CrossRef]
22. Simões, A.O.; CASTRO, M.D.M.; Kinoshita, L.S. Calycine colleters of seven species of Apocynaceae (Apocynoideae) from Brazil. *Bot. J. Linn. Soc.* **2006**, *152*, 387–398. [CrossRef]
23. Machado, S.R.; Morellato, P.C.; Sajo, M.G.; Oliveira, P.S. Morphological patterns of extrafloral nectaries in woody plant species of the Brazilian cerrado. *Plant Biol.* **2008**, *10*, 660–673. [CrossRef]
24. Possobom, C.C.F.; Guimarães, E.; Machado, S.R. Structure and secretion mechanisms of floral glands in *Diplopterys pubipetala* (Malpighiaceae), a neotropical species. *Flora* **2015**, *211*, 26–39. [CrossRef]
25. Araújo, J.S.; Meira, R.M.S.A. Comparative anatomy of calyx and foliar glands of *Banisteriopsis* C. B. Rob. (Malpighiaceae). *Acta Bot. Brasil.* **2016**, *30*, 112–123. [CrossRef]
26. Ribeiro, J.C.; Ferreira, M.J.P.; Demarco, D. Colleters in Asclepiadoideae (Apocynaceae): Protection of meristems against desiccation and new functions assigned. *Int. J. Plant Sci.* **2017**, *178*, 465–477. [CrossRef]
27. Harley, R.M.; Giulietti, A.M.; Leite, K.R.B. Two new species and a new record of *Sauvagesia* (Ochnaceae) in the chapada diamantina of Bahia, Brazil. *Kew Bull.* **2005**, *60*, 571–580.
28. Cardoso, D.B.O.S. A new species of *Sauvagesia* (Ochnaceae) from the Espinhaço range of Minas Gerais, Brazil. *Brittonia* **2011**, *63* 150–155. [CrossRef]
29. Parkin, J. The extra-floral nectaries of *Hevea brasiliensis*, Müll.-Arg. (the Para Rubber Tree), an example of bud-scales serving as nectaries. *Ann. Bot.* **1904**, *18*, 217–226. [CrossRef]
30. Marinho, L.C.; Amorim, A.M.; Cardoso, D.B.O.S. Stirring up a wasp nest: Two new species of the taxonomically complex genus *Ouratea* (Ochnaceae). *Syst. Bot.* **2018**, *43*, 760–766. [CrossRef]
31. Elias, T.S. Extrafloral nectaries: Their structure and distribution. In *The Biology of Nectaries*; Bentley, B., Elias, T., Eds.; Columbia University Press: New York, NY, USA, 1983; pp. 174–203.
32. Frey-Wyssling, A. The phloem supply to the nectaries. *Acta. Bot. Neerl.* **1955**, *4*, 358–369. [CrossRef]
33. Paiva, E.A.S. Ultrastructure and post-floral secretion of the pericarpial nectaries of *Erythrina speciosa* (Fabaceae). *Ann. Bot.* **2009** *104*, 937–944. [CrossRef]
34. Fahn, A. Structure and function of secretory cells. *Adv. Bot. Res.* **2000**, *31*, 37–75. [CrossRef]
35. Oliveira, C.S.; Salino, A.; Paiva, E.A.S. Colleters in Thelypteridaceae: Unveiling mucilage secretion and its probable role in ferns *Flora* **2017**, *228*, 65–70. [CrossRef]
36. Mohan, J.S.S.; Inamdar, J.A. Ultrastructure and secretion of extrafloral nectaries of *Plumeria rubra* L. *Ann. Bot.* **1986**, *57*, 389–401 [CrossRef]
37. Fernandes, V.F.; Thadeo, M.; Dalvi, V.C.; Meira, R.M.S.A. Secretory structures in *Casearia sylvestris* Sw. (Salicaceae): Diversity mechanisms of secretion, and exudate complexity. *Int. J. Plant Sci.* **2018**, *178*, 288–301. [CrossRef]
38. Machado, S.R.; Paleari, L.M.; Paiva, E.A.S.; Rodrigues, T.M. Colleters on the inflorescence axis of *Croton glandulosus* (Euphorbiaceae): Structural and functional characterization. *Int. J. Plant Sci.* **2015**, *176*, 86–93. [CrossRef]
39. Ballego-Campos, I.; Paiva, E.A.S. Mucilage secretion in the inflorescences of *Aechmea blanchetiana*: Evidence of new functions of scales in bromeliaceae. *Flora* **2018**, *246–247*, 1–9. [CrossRef]
40. Fahn, A.; Benayoun, J. Ultrastructure of resin ducts in *Pinus halepensis* development, possible sites of resin synthesis, and mode of its elimination from the protoplast. *Ann. Bot.* **1976**, *40*, 857–863. [CrossRef]
41. Thomson, W.W.; Platt-Aloia, K.A.; Endress, A.G. Ultrastructure of oil gland development in the leaf of *Citrus sinensis* L. *Bot. Gaz.* **1976**, *137*, 330–340. [CrossRef]
42. McGarvey, D.J.; Croteau, R. Terpenoid metabolism. *Plant Cell* **1995**, *7*, 1015–1026. [CrossRef]
43. Paiva, E.A.S. How do secretory products cross the plant cell wall to be released? A new hypothesis involving cyclic mechanical actions of the protoplast. *Ann. Bot.* **2016**, *117*, 533–554. [CrossRef]

44. Sadala-Castilho, R.; Machado, S.R.; Sa-Haiad, B.; Lima, H.A. Oil-resin glands in Velloziaceae flowers: Structure, ontogenesis and secretion. *Plant Syst. Evol.* **2016**, *302*, 585–599. [CrossRef]
45. Paiva, E.A.S. How does the nectar of stomata-free nectaries cross the cuticle? *Acta. Bot. Bras.* **2017**, *31*, 525–530. [CrossRef]
46. Mayer, J.L.S.; Cardoso-Gustavson, P.; Appezzato-da-Gloria, B. Colleters in monocots: New record for orchidaceae. *Flora* **2011**, *206*, 185–190. [CrossRef]
47. Paiva, E.A.S.; Martins, L.C. Calycinal trichomes in *Ipomoea cairica* (Convolvulaceae): Ontogenesis, structure and functional aspects. *Aust. J. Bot.* **2011**, *59*, 91–98. [CrossRef]
48. Almeida, A.L.; Paiva, E.A.S. Colleters in *Mabea fistulifera* Mart. (Euphorbiaceae): Anatomy and biology of the secretory process. *Flora* **2019**, *258*, 151439. [CrossRef]
49. Paiva, E.A.S. Colleters in *Cariniana estrellensis* (Lecythidaceae): Structure, secretion and evidences for young leaf protection. *J. Torrey Bot. Soc.* **2012**, *139*, 1–8. [CrossRef]
50. Dell, B. Distribution and function of resins and glandular hairs in Western Australian plants. *J. R Soc. West. Aust.* **1977**, *59*, 119–123.
51. Oliveira, P.S.; Leitao-Filho, H.F. Extrafloral Nectaries: Their taxonomic distribution and abundance in the woody flora of cerrado vegetation in Southeast Brazil. *Biotropica* **1987**, *19*, 140–148. [CrossRef]
52. Oliveira, P.S.; Pie, M.R. Interaction between ants and plants bearing extrafloral nectaries in cerrado vegetation. *An. Soc. Entomol. Bras.* **1988**, *27*, 161–176. [CrossRef]
53. Paiva, E.A.S. Are calcium oxalate crystals a dynamic calcium store in plants? *New Phytol.* **2019**, *223*, 1707–1711. [CrossRef]
54. Paiva, E.A.S.; Ballego-Campos, I.; Gibernau, M. True nectar or stigmatic secretion? Structural evidence elucidates an old controversy regarding nectaries in *Anthurium*. *Am. J. Bot.* **2021**, *108*, 37–50. [CrossRef]
55. Schnell, R.; Cusset, G.; Quenum, M. Contribution a l'etude des glandes extra-florales chez quelques groupes de plantes tropicales. *Rev. Générale Bot.* **1963**, *70*, 269–342.
56. Metcalfe, C.R.; Chalk, L. *Anatomy of the Dicotyledons*, 2nd ed.; Claredon Press: Oxford, UK, 1979.
57. Konarska, A. Characteristics of flower nectaries of *Hedera helix* L. (Araliaceae). *Acta. Sci. Pol. Hortoru. Cultus* **2014**, *13*, 109–122.
58. Gish, M.; Mescher, M.C.; Moraes, C.M. Mechanical defenses of plant extrafloral nectaries against herbivory. *Commun. Integr. Biol. Biol.* **2016**, *9*, 3. [CrossRef]
59. Gonçalves-Souza, P.; Gonçalves, E.G.; Paiva, E.A.S. Extrafloral nectaries in *Philodendron* (Araceae): Distribution and structure. *Bot. J. Linn. Soc.* **2016**, *180*, 229–240. [CrossRef]
60. Pereira, P.S.; Gonçalves, L.A.; Silva, M.J.; Rezende, M.H. Extrafloral nectaries of four varieties of *Chamaecrista ramosa* (Vogel) H.S. Irwin & Barneby (Fabaceae): Anatomy, chemical nature, mechanisms of nectar secretion, and elimination. *Protoplasma* **2018**, *255*, 1635–1647. [CrossRef]
61. Mesquita-Neto, J.N.; Paiva, E.A.S.; Galetto, L.; Schlindwein, C. Nectar secretion of floral buds of *Tococa guianensis* mediates interactions with generalist ants that reduce florivory. *Front. Plant Sci.* **2020**, *11*, 627. [CrossRef] [PubMed]
62. Karnovsky, M.J. A formaldehyde glutaraldehyde fixative of high osmolality for use in electron microscopy. *J. Cell Biol.* **1965**, *27*, 137A–138A.
63. O'Brien, T.; Feder, N.; McCully, M.E. Polychromatic staining of plant cell walls by toluidine blue O. *Protoplasma* **1964**, *59*, 368–373. [CrossRef]
64. Spurr, A.R. A low-viscosity epoxy resin embedding medium for electron microscopy. *J. Ultrastruct. Res.* **1969**, *26*, 31–43. [CrossRef]
65. Roland, J.C. General preparation and staining of thin sections. In *Electron Microscopy of Plant Cells*; Hall, J., Ed.; Elsevier: New York, NY, USA, 1978; pp. 1–62.
66. Brundrett, M.C.; Kendrick, B.; Peterson, C.A. Efficient lipid staining in plant material with Sudan Red 7B or Fluoral Yellow 088 in polyethylene glycol-glycerol. *Biotech. Histochem.* **1991**, *66*, 111–116. [CrossRef]
67. Johansen, D.A. *Plant Microtechnique*; McGraw-Hill Book Co.: New York, NY, USA, 1940.
68. David, R.; Carde, J.P. Coloration différentielle des inclusions lipidiques et terpéniques des pseudophylles du Pin maritime au moyen du réactif Nadi. *C R Acad. Sci. Paris* **1964**, *258*, 1338–1340.
69. Vidal, B.C. Dichroism in collagen bundles stained with Xylidine-Ponceau 2R. *Ann. D Histochim.* **1970**, *15*, 289–296.

plants

MDPI

Article

A Systemic View of Carbohydrate Metabolism in Rice to Facilitate Productivity

Woo-Jong Hong [1,†], Xu Jiang [1,†], Seok-Hyun Choi [1], Yu-Jin Kim [2], Sun-Tae Kim [3], Jong-Seong Jeon [1] and Ki-Hong Jung [1,*]

[1] Graduate School of Biotechnology and Crop Biotech Institute, Kyung Hee University, Yongin 17104, Korea; hwj0602@khu.ac.kr (W.-J.H.); kangwuk97@khu.ac.kr (X.J.); lilchoi92@gmail.com (S.-H.C.); jjeon@khu.ac.kr (J.-S.J.)

[2] Department of Life Science and Environmental Biochemistry, Life and Industry Convergence Research Institute, Pusan National University, Miryang 50463, Korea; yjkim2020@pusan.ac.kr

[3] Department of Plant Bioscience, Life and Industry Convergence Research Institute, Pusan National University, Miryang 50463, Korea; stkim71@pusan.ac.kr

* Correspondence: khjung2010@khu.ac.kr

† These authors contributed equally to this work.

Citation: Hong, W.-J.; Jiang, X.; Choi, S.-H.; Kim, Y.-J.; Kim, S.-T.; Jeon, J.-S.; Jung, K.-H. A Systemic View of Carbohydrate Metabolism in Rice to Facilitate Productivity. *Plants* **2021**, *10*, 1690. https://doi.org/10.3390/plants10081690

Academic Editors: Milan S. Stankovic, Paula Baptista and Petronia Carillo

Received: 15 July 2021
Accepted: 13 August 2021
Published: 17 August 2021

Publisher's Note: MDPI stays neutral with regard to jurisdictional claims in published maps and institutional affiliations.

Copyright: © 2021 by the authors. Licensee MDPI, Basel, Switzerland. This article is an open access article distributed under the terms and conditions of the Creative Commons Attribution (CC BY) license (https://creativecommons.org/licenses/by/4.0/).

Abstract: Carbohydrate metabolism is an important biochemical process related to developmental growth and yield-related traits. Due to global climate change and rapid population growth, increasing rice yield has become vital. To understand whole carbohydrate metabolism pathways and find related clues for enhancing yield, genes in whole carbohydrate metabolism pathways were systemically dissected using meta-transcriptome data. This study identified 866 carbohydrate genes from the MapMan toolkit and the Kyoto Encyclopedia of Genes and Genomes database split into 11 clusters of different anatomical expression profiles. Analysis of functionally characterized carbohydrate genes revealed that source activity and eating quality are the most well-known functions, and they each have a strong correlation with tissue-preferred clusters. To verify the transcriptomic dissection, three pollen-preferred cluster genes were used and found downregulated in the *gori* mutant. Finally, we summarized carbohydrate metabolism as a conceptual model in gene clusters associated with morphological traits. This systemic analysis not only provided new insights to improve rice yield but also proposed novel tissue-preferred carbohydrate genes for future research.

Keywords: carbohydrate metabolism; microarray; crop; rice; productivity

1. Introduction

As the world's population increases and arable land decreases year by year, food security has become one of the most serious problems faced by all countries [1]. Rice (*Oryza sativa* L.) is not only a model crop plant but also the main staple cereal that supplies nearly half of the world's calorie consumption. Hence, improving its production is of great strategic significance for ensuring food security and sustainable agricultural development [2]. As a sessile and autophototrophic plant, rice generates carbohydrates by photosynthesis. These photoassimilates undergo a series of ordered metabolic processes and play a pivotal role in different developmental stages, including vegetative, reproductive, and ripening. Additionally, carbohydrate reserves in mature seeds provide the primary energy intake of mankind and contribute energy during its germination [3]. This source-sink coordination that runs through the entire plant life cycle reflects the importance of carbohydrate metabolism in rice productivity improvement.

Extensive research has provided evidence for the generation of more metabolic substrates by manipulating the potential of "source", resulting in increased rice yield. For instance, *OsDWARF4* mutation showed an erect leaf phenotype that may enhance light capture for photosynthesis and finally lead to enhanced grain yield [4,5]. High grain yield was also observed in SNU-SG1 rice with the stay-green phenotype [6]. In addition to these, several attempts have been made to evaluate sugar transporters and key enzymes due to their vital role in carbohydrate metabolic processes. There are two main steps in sucrose translocation: phloem loading and unloading [7]. In the apoplastic loading model, one of the phloem loading steps, sucrose moves to the apoplasmic region and is loaded into phloem via Sucrose Transporters (SUTs) and Sugar Will Eventually be Exported Transporters (SWEETs) [8–10]. In post-phloem unloading, many studies have focused on sugar signaling after sucrose conversion into hexose by *Hexokinase* (*HXK*) family genes [11] Moreover, the overexpression of *Grain Incomplete Filling 1* (*GIF1*), which encodes a cell wall invertase under the control of its native promoter, increases grain production [12] Similarly, in maize, the constitutive expression of *Cell Wall Invertase* (*CWINV*) elevates grain yield and starch content [13].

Great progress has been made in this field. However, there has not been any big success until now, such as the green revolution caused by the discovery and application of semi-dwarf rice cultivars [14]. One explanation could be the failure to establish giant "sink" cultivars with rich spikelets due to the grain-filling ability that could not match a large yield capacity [15]. Another explanation is that carbohydrate metabolism has been oversimplified [8]; recently, there have been several reports regarding its complexity. For example, there is considerable heterogeneity in phloem loading and transport even in one species [16], and invertase inhibitors capping invertase exist [17]. An understanding of the systematic perceptions of carbohydrate metabolism for further applications is still very limited

With the rise of bioinformatics and the establishment of high-throughput gene expression methods such as microarrays or next-generation sequencing technology, new technologies and methods have afforded systemic insights into various biological research fields. Recently, transcriptomic analyses of carbon partitioning during rice grain filling and the relationship between high temperature and grain filling have been carried out [18,19] Despite the importance of systematic insights on carbohydrate metabolism in tissues related to morphological traits, transcriptome analysis has only been focused on a type of tissue and developmental processes.

To provide systemic insights into carbohydrate metabolism in rice, a transcriptomic dissection of carbohydrate metabolism-related genes retrieved from the Kyoto Encyclopedia of Genes and Genomes (KEGG) database [20] and the MapMan toolkit [21], which cover genome-wide biological pathways, was performed. After clustering genes with meta expression profiles of anatomical samples, a functional enrichment analysis was performed and the results were validated by quantitative reverse transcription-polymerase chain reaction (qRT-PCR). Finally, a conceptual model of carbohydrate metabolism to enhance crop yield was constructed. This research can shed light on carbon metabolism and provide candidate genes to enhance crop yield of rice and other species.

2. Materials and Methods

2.1. Integration of Carbohydrate Metabolism Annotation Data

Carbohydrate metabolism-related genes were collected according to the annotation of the MapMan toolkit (version 3.6.0RC1) [21] and KEGG database (retrieved on 10 April 2021 [20]. First, 266 carbon metabolism genes were selected from the KEGG database. Carbohydrate metabolism-related genes with MapMan bincodes from the MapMan toolkit were selected next. In total, 787 genes had MapMan bincodes (1: photosynthesis; 2: major CHO; 3: minor CHO; 4: glycolysis; 6: gluconeogenesis/glyoxylate cycle 7: OPP; 8: TCA/org.trasnformation; 25: C1-metabolism; 34: transporters related to sugar or sucrose). Finally, 872 genes from the two data sources were selected, and 866 genes annotated by the Rice Genome Annotation Project (RGAP) [22] were chosen for further analysis.

2.2. Collection and Clustering of Microarray Data

Transcriptomic data were downloaded to analyze the anatomical expression patterns of carbohydrate metabolism-related genes. The data source mentioned was used in previ ous reports [23]. The detailed information is discussed below. For the analysis of anatomical expression profiles, anatomical data were retrieved from the Rice Oligonucleotide Array Database (ROAD) [24]. For heatmap analysis of cluster H genes, data were downloaded from RMEDB [25]. Multiple Experiment Viewer (MeV) is a widely used program for visualizing transcriptome data and performing statistical analysis [26]. MeV (version 4.9.0) was used to visualize the microarray data. For the dissection of transcriptome data, a k-means clustering (KMC) algorithm embedded in MeV was applied using the same method as with the identification of late pollen-preferred genes in rice [27]. Adobe Illustrator CS6 was used to edit the heatmap images.

2.3. Functional Classification via Literature Search

To find the previously characterized functional roles of the 866 carbohydrate metabolism-related genes in anatomical clusters, the Overview of Functionally characterized Genes in Rice Online (OGRO) database (http://qtaro.abr.affrc.go.jp/ogro/table (accessed on 14 April 2021)) was used [28]. Information for 1949 functionally characterized genes is available in this database. As in a previous study [29], information on the 866 genes was parsed, and data were summarized using Excel 365 (version 16.0.14228.20158). Count numbers for the characterized genes were visualized using R Studio (version 1.4.1106) and ggplot2 R package (version 3.3.3) [30]. The detailed, functionally characterized gene information of the 866 carbohydrate metabolism-related genes is listed in Table S1.

2.4. Gene Ontology (GO) Enrichment Analysis

GO enrichment is commonly used to interpret the functional roles of large-scale transcriptomic data [31]. This study used the ROAD to find the GO terminology for each cluster (http://ricephylogenomics-khu.org/road/go_analysis.php, temporary homepage for updating (accessed on 7 May 2021)). To perform the GO enrichment analysis, the following criteria were applied: query number > 2, hyper $p < 0.05$, and fold enrichment value (query number/query expected number) > 2 [32]. Significant GO terms and integrated cluster information were selected from the transcriptome data analysis with each selected GO term. Finally, these data were visualized via R Studio (version 1.4.1106) and ggplot2 R package (version 3.3.3).

2.5. KEGG Enrichment Analysis

KEGG enrichment analysis was performed using R Studio and the clusterProfiler package [33]. To use the enrichKEGG function in this package, input data consisting of cluster information and Rice Annotation Project Database ID (https://rapdb.dna.affrc.go.jp/ (accessed on 7 May 2021)) [34] were used. In addition, "dosa" was chosen as the organism code, and the results were filtered out by applying an adjusted p-value cutoff < 0.05. For the visualization of the results, the dot-plot function in the package was used, and the figure was modified with the ggplot2 package (version 3.3.3).

2.6. RNA Extraction and qRT-PCR

To isolate RNA, plants were grown in a paddy field condition, as reported previously [35]. Samples were immediately frozen in liquid nitrogen, and total RNA was isolated using a TRIzol reagent (Invitrogen, Waltham, MA, USA) combined with an RNase Plant Mini Kit (Qiagen, Hilton, Germany; http://www.qiagen.com (accessed on 7 May 2021)) and DNase treatment. First-strand cDNA was synthesized using the SuPrimeScript RT Premix (with oligo(dT), 2×; GeNet Bio, Daegu, Korea). A qRT-PCR was performed, as reported previously [36]. For the *gori* knockout mutant, anthers from a paddy field-grown plant were collected to extract RNA. All primers used in this study are listed in Table S2.

2.7. Construction of the Conceptual Carbohydrate Metabolism Model

To generate a conceptual model focusing on the source-sink communication pathway, four key enzymes were selected: invertase (INV), sucrose synthesis (SUS), sucrose transporter (SUT), and hexokinase (HXK). Cluster information was then integrated by indicating an organ/tissue-preferred expression pattern: clusters A and B for leaf, cluster E for root, cluster H for pollen, cluster I for grain, and cluster J for ubiquitous expression patterns. Finally, the regulatory network between *GORI* and three cluster H genes was incorporated after adding qRT-PCR data between the *gori* knockout mutant and wild type anthers.

Rice plant images were downloaded from the International Rice Research Institute (https://www.flickr.com/photos/ricephotos/albums/72157643341257395) (accessed on 7 May 2021)), and the images were arranged using Adobe Illustrator CS6 (version 16.0.0).

3. Results

3.1. Identification of Genome-Wide Candidate Genes Related to Carbohydrate Metabolism

The MapMan toolkit and the KEGG database are useful information sources for the functional annotation of large-scale genes [20,21]. These two data sources were used to retrieve reliable carbohydrate metabolism-related genes. First, 266 genes involved in carbon metabolism pathways were found in the KEGG database. These genes were also searched in the MapMan toolkit, and genes in bincodes related to carbohydrate metabolism were identified: photosynthesis, major CHO, minor CHO, glycolysis, gluconeogenesis/glyoxylate cycle, OPP, TCA/organic acid transformation, and C1-metabolism. In addition to these bincodes, genes related to sugar or sucrose transport were added. Finally, 872 genes were collected using two public annotation sources. Because most expression data were available with locus IDs from the RGAP website (http://rice.plantbiology.msu.edu/ (accessed on 15 July 2021)), further analysis was performed on 866 candidate genes with RGAP locus IDs (Figure 1; Table S3).

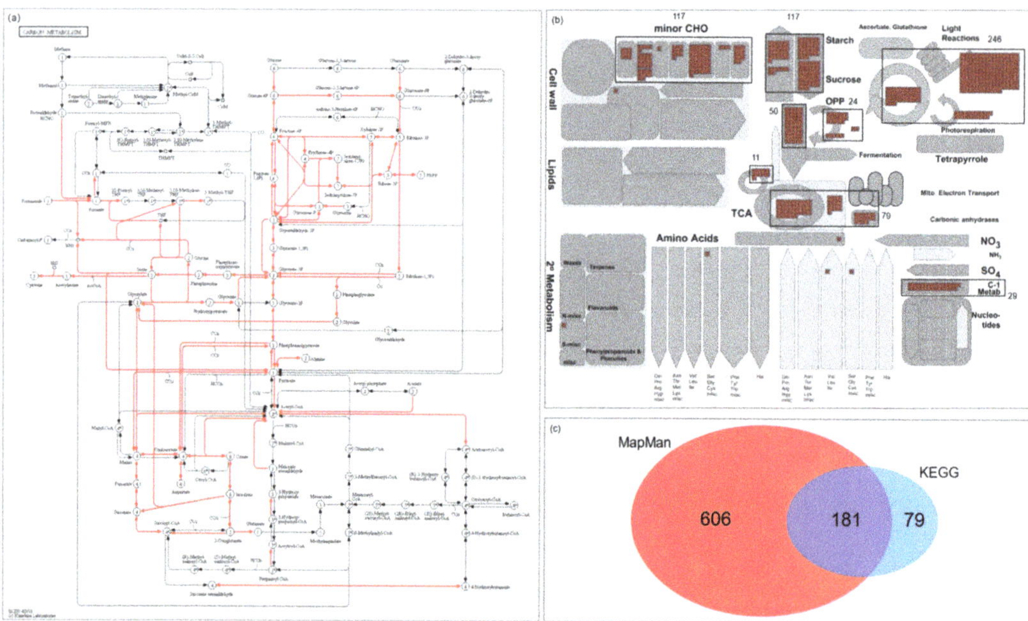

Figure 1. Schematic diagram summarizing 866 carbohydrate metabolism-related genes retrieved from the KEGG database and the MapMan toolkit. In the KEGG database, carbon metabolism pathway genes were selected and applied to the MapMan toolkit to find associated MapMan annotations. (**a**) There were eight bincodes related to 266 KEGG genes and with multiple members. (**b**) The bins with a black box and the total number of whole elements in each bin were indicated. Therefore, 681 MapMan genes associated with carbohydrate metabolism were selected. In addition, 106 carbohydrate transporters annotated in MapMan were included. A total of 787 genes from the MapMan toolkit and 266 from the KEGG database were collected. (**c**) Finally, 872 carbohydrate metabolism-related genes were collected. For further analysis, 866 genes with RGAP locus information were used. The detailed information of the 866 genes, including 181 overlapped genes, is listed in Table S3.

3.2. Functional Analysis of the Characterized Carbohydrate Metabolism-Related Genes

To analyze the functional significance of the 866 carbohydrate metabolism-related genes, functionally characterized genes were searched among them. To do this, the OGRO website was used [28]. Information of 1949 functionally characterized genes was then retrieved and classified according to major functional categories such as physiology, morphology, tolerance, or resistance. The functionally characterized roles for 76 of the 866 genes, including duplicate information about one locus, were identified (Figure 2). In the physiology category, eating quality was related to 22 genes, source activity was related to 19 genes, and flowering was related to three genes. In the morphology category, dwarf was related to five characterized genes, seed was related to four genes, and culm/leaf and root were related to two genes. Finally, regarding tolerance or resistance, salinity tolerance was related to five genes and cold and drought tolerance was related to two genes. As expected, the most frequently characterized functional category associated with carbon metabolism was eating quality in the physiology category, followed by source activity (Table 1). This result indicates that 866 carbohydrate genes might be useful candidates for enhancing the grain yield of rice associated with eating quality and source activity.

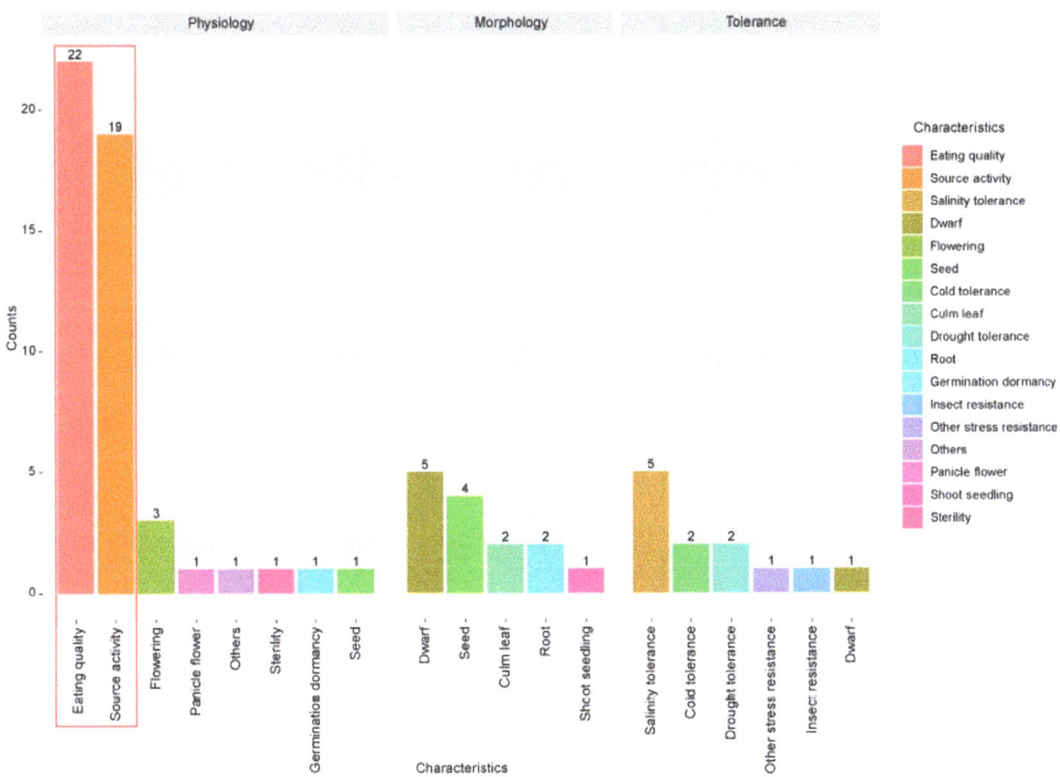

Figure 2. Distribution of functionally characterized genes from 866 carbohydrate metabolism-related genes according to phenotype. A literature search was performed to analyze the functional significance of the 866 carbohydrate metabolism-related genes. Functionally characterized gene information was retrieved from the OGRO database. To visualize the results of the 76 functionally characterized genes indicated in Table 1, three categories of major characteristic information from the OGRO database were used: physiology, morphology, and tolerance/resistance. The number of characterized genes was counted according to the minor characteristics within three major categories: physiology, morphology, and tolerance. The red box indicates the most enriched minor characteristics of the functions associated with the 76 functionally characterized genes.

Table 1. Summary of functionally characterized genes from the 866 carbohydrate metabolism-related genes.

LOC_id	Symbol	Character_major [1]	Character_minor	Method [2]	Detailed Functions	Reference
Os02g44230	OsTPP1	R or T	Cold tolerance	OX	Cold and salinity tolerance	[37]
Os05g44210	OsTPS1	R or T	Cold tolerance	OX	Cold, drought, and salinity tolerance	[38]
Os01g64660	moc2	MT	Culm leaf	KO	Tiller bud outgrowth; tillering	[39]
Os12g44380	ossut2	MT	Culm leaf	KO	Sugar export; growth retardation	[40]
Os05g44210	OsTPS1	R or T	Drought tolerance	OX	Cold, drought, and salinity tolerance	[38]
Os06g36560	OsMIOX	R or T	Drought tolerance	OX	Drought tolerance	[41]
Os03g07480	OsSUT1	MT	Dwarf	KD	Dwarfism; flowering time	[42]
Os04g56320	OsrcaA2	MT	Dwarf	KD/OX	Dwarfism	[43]
Os07g05820	OsGLO4	R or T	Dwarf	KD	Growth inhibition	[44]
Os11g05110	OsPK1	MT	Dwarf	KO	Elongation of uppermost internode; dwarfism	[45]
Os11g05110	ospk1	MT	Dwarf	KO	Dwarfism; seed color; internode color; regulation of gibberellin and ABA biosynthesis	[46]
Os12g44380	ossut2	MT	Dwarf	KO	Sugar export; growth retardation	[40]
Os01g44220	osagpl2	PT	Eating quality	KO	Starch biosynthesis	[47]
Os01g55540	OsAAT2	PT	Eating quality	OX	Seed amino acid and protein content	[48]
Os02g52710	Amy1A	PT	Eating quality	KD	Seed starch content; high temperature-triggered grain chalkiness	[49]
Os03g03720	OsGAPDHB	PT	Eating quality	KD	Fragrance rice	[50]
Os03g09250	RINO1	PT	Eating quality	KD	Seed phytic acid content	[51]
Os03g09250	RINO1	PT	Eating quality	KD	Seed phytic acid content	[52]
Os03g52760	lpaN15-186	PT	Eating quality	KO	Seed phytic acid content	[53]
Os03g55090	pho1	PT	Eating quality	KO	Seed starch content; grain maturation	[54]
Os05g32710	OsISA2	PT	Eating quality	OX	Seed starch content	[55]
Os05g33570	flo4	PT	Eating quality	KO	Seed protein and lipid content	[56]
Os06g04200	wx	PT	Eating quality	NV	Seed amylose content	[57]
Os06g06560	OsSSI	PT	Eating quality	KO	Seed amylopectin content	[58]
Os06g12450	SSIIa	PT	Eating quality	KD	Chalky kernel; amylose content	[59]
Os06g12450	ALK	PT	Eating quality	NV	Gelatinization temperature; gel consistency	[60]
Os08g09230	SSIIIa	PT	Eating quality	KD	Chalky kernel; amylose content	[59]
Os08g09230	flo5	PT	Eating quality	KO	Seed starch content	[61]
Os08g09230	ss3a	PT	Eating quality	KO	Seed starch content	[62]
Os08g25734	osagps2	PT	Eating quality	KO	Seed starch content	[47]
Os08g40930	OsISA1	PT	Eating quality	OX	Seed starch content	[55]
Os09g28400	Amy3A	PT	Eating quality	KD	Seed starch content; high temperature-triggered grain chalkiness	[49]
Os09g28420	Amy3B	PT	Eating quality	KD	Seed starch content; high temperature-triggered grain chalkiness	[49]
Os09g29404	isa3	PT	Eating quality	KO	Seed starch content	[63]
Os01g72090	se13	PT	Flowering	KO	Photoperiodic response	[64]
Os02g34560	Oscyt-inv1	PT	Flowering	KO	Root cell development; flowering time; fertility	[65]
Os03g07480	OsSUT1	PT	Flowering	KD	Dwarfism; flowering time	[42]
Os02g52710	AmyI-1	PT	Germination dormancy	KD/OX	Starch degradation; seed germination; seedling growth	[66]
Os03g07480	OsSUT1	R or T	Insect resistance	Others	Upregulation by aphid feeding; transfer sucrose	[67]
Os02g47020	SBPase	R or T	Other stress resistance	OX	Photosynthetic ability under high-temperature condition	[68]
Os02g01150	OsHPR1	PT	Others	KD/OX	Photorespiratory pathway	[69]
Os05g50380	LSU3	Others	Others	KO	Culm starch content	[70]
Os01g69030	OsSPS1	PT	Panicle flower	KO	Pollen germination through sucrose synthesis	[71]
Os02g34560	Oscyt-inv1	MT	Root	KO	Root cell development; flowering time; fertility	[65]
Os12g44380	ossut2	MT	Root	KO	Sugar export; growth retardation	[40]
Os02g17500	OsGMST1	R or T	Salinity tolerance	KD	Salinity tolerance	[72]
Os02g44230	OsTPP1	R or T	Salinity tolerance	OX	Cold and salinity tolerance	[37]

Table 1. *Cont.*

LOC_id	Symbol	Character_major [1]	Character_minor	Method [2]	Detailed Functions	Reference
Os03g07480	OsSUT1	R or T	Salinity tolerance	KD	Salinity tolerance	[73]
Os05g44210	OsTPS1	R or T	Salinity tolerance	OX	Cold, drought, and salinity tolerance	[38]
Os08g03290	OsGAPC3	R or T	Salinity tolerance	OX	Salinity tolerance	[74]
Os01g44220	OsAGPL2	MT	Seed	KO	Glassy/vitreous, shrunken grain	[75]
Os03g55090	pho1	MT	Seed	KO	Seed starch content; grain maturation	[54]
Os04g33740	GIF1	MT	Seed	NV	Grain filling; grain size	[12]
Os08g25734	OsAGPS2b	PT	Seed	KO	Seed weight; starch content; AGPase activities from developing endosperms of the seed	[75]
Os11g05110	ospk1	MT	Seed	KO	Dwarfism; seed color; internode color; Regulation of gibberellin and ABA biosynthesis	[46]
Os02g52710	AmyI-1	MT	Shoot seedling	KD/OX	Starch degradation; seed germination; seedling growth	[66]
Os01g11054	Osppc4	PT	Source activity	KD	Ammonium assimilation in leaves	[76]
Os01g64660	oscfbp1	PT	Source activity	KO	Photosynthetic sucrose biosynthesis; growth retardation	[77]
Os01g64960	qNPQ1-2	PT	Source activity	NV	Nonphotochemical quenching capacity; protection from photoinhibition	[78]
Os01g64960	PsbS	PT	Source activity	KD/OX	Nonphotochemical quenching capacity; photosynthetic rate in fluctuating light conditions	[79]
Os02g32660	BE2b	PT	Source activity	Others	Starch biosynthesis in endosperm; amylopectin biosynthesis; branch formation	[80]
Os03g57220	OsGLO1	PT	Source activity	KD	Photorespiration	[81]
Os04g56320	rca	PT	Source activity	KD	Rubisco activity	[82]
Os05g40180	OsSTN8	PT	Source activity	KO	Photosystem II repair during high light illumination	[83]
Os06g06560	SS1	PT	Source activity	Others	Starch biosynthesis in endosperm; chain elongation	[80]
Os06g51084	BE1	PT	Source activity	Others	Starch granule binding; amylopectin structure	[80]
Os07g05820	OsGLO4	PT	Source activity	KD	Rubisco activation; photosynthesis rate	[44]
Os08g45190	PGR5	PT	Source activity	KD	Photosynthetic capacity	[84]
Os10g37180	OsGDCH	PT	Source activity	KD	Leaf senescence induced by reactive oxygen species	[85]
Os12g17600	rbcS	PT	Source activity	KD	Rubisco content; photosynthetic capacity	[86]
Os12g17600	OsRBCS2	PT	Source activity	KD	Rubisco content	[87]
Os12g17600	rbcS	PT	Source activity	KD/OX	Rubisco content; photosynthetic capacity	[88]
Os12g19381	OsRBCS5	PT	Source activity	KD	Rubisco content	[87]
Os12g19470	OsRBCS4	PT	Source activity	KD	Rubisco content	[87]
Os12g44380	ossut2	PT	Source activity	KO	Sugar export; growth retardation	[40]
Os02g34560	Oscyt-inv1	PT	Sterility	KO	Root cell development; flowering time; fertility	[65]

[1] Major functional categories: R or T, resistance or tolerance; MT, morphological trait; PT, physiological trait. [2] Methods used to study: OX, overexpression; KO, knockout; KD, knockdown; NV, natural variation.

3.3. *Anatomical Dissection of Carbohydrate Metabolism-Related Genes via Meta-Expression Analysis*

To assess the functional roles of the 866 carbohydrate metabolism-related genes, meta-anatomical expression profiles consisting of 983 rice Affymetrix array anatomical sample data were first used [23]. Using the KMC algorithm, 729 genes with probes on the Affymetrix array were grouped into 11 anatomical clusters (Figure 3a; Table S4). This analysis could not be performed for 137 genes without probes on the Affymetrix array. Based on this analysis, carbohydrate metabolism-related genes may be involved in diverse morphological or physiological traits. For example, clusters A and B are closely associated with leaf and shoot development, cluster E is associated with root, clusters G and H are associated with pollen, and cluster I is associated with grain. In addition, cluster J, with ubiquitous expression patterns, might be related to the housekeeping function.

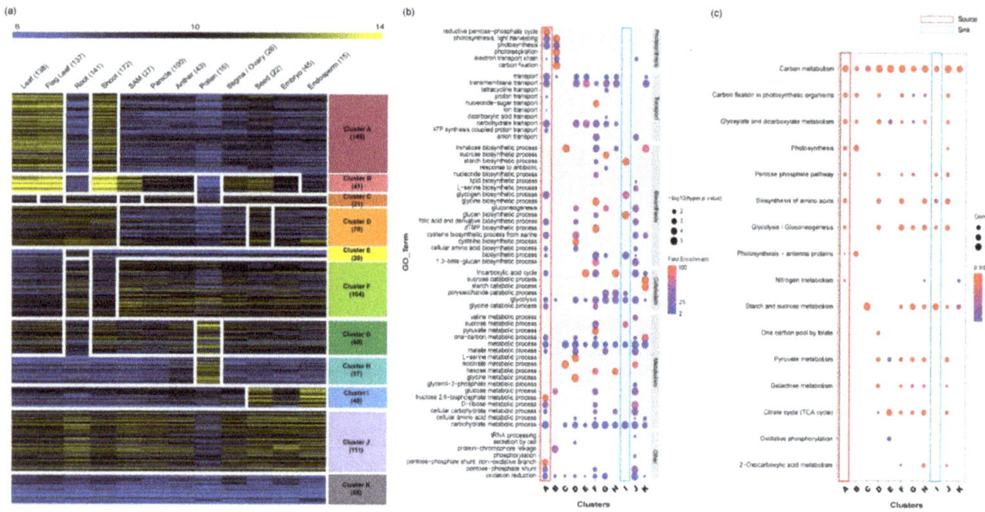

Figure 3. Dissection of the 866 carbohydrate metabolism-related genes using meta-anatomical expression profiles and functional enrichment analysis of 11 anatomical clusters. (**a**) Using large-scale microarray data, carbohydrate metabolism-related genes were visualized and dissected. The heatmap of the anatomical expression profiles of carbohydrate metabolism-related genes was grouped into 11 clusters via KMC clustering methods. The numbers after the name of each tissue/organ indicate the sample size, and the numbers below each cluster indicate the number of genes in the cluster. (**b**) GO enrichment analysis of 11 anatomical clusters. The GO enrichment assay revealed the characteristics of each cluster. GO terms were classified according to biological process GO terms. Dot color indicates the fold enrichment value (the blue color is 2, which is the minimum cutoff to select a significant fold enrichment value, and the red color indicates a higher fold enrichment value), and dot size indicates statistical significance (-log10(hyper *p*-value) is used, and a larger dot size means more significance). (**c**) KEGG enrichment analysis of 11 anatomical clusters. The enriched KEGG pathway indicated with the dot size represents the ratio of the selected genes to the total genes in the pathway, and the dot color illustrates the adjusted *p*-value. The numbers below the clusters indicate the number of mapped genes to each pathway. In addition, in GO and KEGG enrichment analyses, the source- and sink-related clusters are highlighted as red and blue boxes, respectively.

3.4. Functional Comparison and Enrichment Analysis of 11 Anatomical Clusters

The distribution of functionally characterized genes among clusters was searched to identify the relationships between anatomical expression patterns and the 11 clusters. Subsequently, 13 source activity genes were enriched in cluster A, which showed a leaf-preferred expression pattern. Similarly, 15 eating quality genes were in cluster I, which showed a grain-preferred expression pattern. Other clusters did not show a strong correlation between known functions and featured expression patterns (Figure S1).

Enrichment analysis of functional groups was also performed for each of the 11 anatomical clusters. To do this, GO and KEGG enrichment analyses were conducted. As a result, four GO terms associated with photosynthesis were enriched in cluster A: reductive pentose phosphate cycle (GO: 0019253), photosynthesis-light harvesting (GO: 0009765), photosynthesis (GO: 0015979), and electron transport chain (GO: 0022900). In cluster I, there were no photosynthesis-related GO terms. Instead, there were four GO terms related to sugar metabolism or biosynthesis: sucrose metabolic process (GO: 0005985), starch biosynthetic process (GO: 0019252), glucan biosynthetic process (GO: 0009250), and glycogen biosynthetic process (GO: 0005978; Figure 3b).

Consistent with the GO enrichment analysis results, KEGG enrichment also showed that photosynthesis pathways were enriched in cluster A, and starch and sucrose metabolism pathways were enriched in cluster I (Figure 3c). These results suggest that metabolic pathways might be further dissected by diverse developmental processes. Assigning expression patterns to each of the clusters will be useful for further functional analysis.

3.5. In Silico and In Vitro Expression Verification of Tissue-Preferred Genes

To validate the functional significance of gene clusters according to anatomical expression patterns, three genes (*Os10g26740*, *Os02g06540*, and *Os10g08022*) in cluster H associated with anther and pollen development were selected: sucrose transporter (*OsSUT3*), monosaccharide transporter, and fructose-bisphosphate aldolase isozyme, respectively. In silico analysis revealed that all these genes showed anther/pollen-preferred expression patterns (Figure 4a). This expression pattern was further confirmed by qRT-PCR, with samples in six tissues/organs (Figure 4b) matching the dissected model. Recently, a defect in the *GORI* gene changed the distribution of pectin in germinated pollen tubes and eventually led to the male sterile phenotype [35]. Thus, qRT-PCR was performed to analyze their regulatory roles by *GORI* in the *gori* mutant. Interestingly, all three genes were significantly downregulated in the *gori* mutant than in the wild-type (Figure 4c). It was speculated that these transporters and carbon metabolism-related enzymes regulated by *GORI* might be involved in the pollen tube growth process.

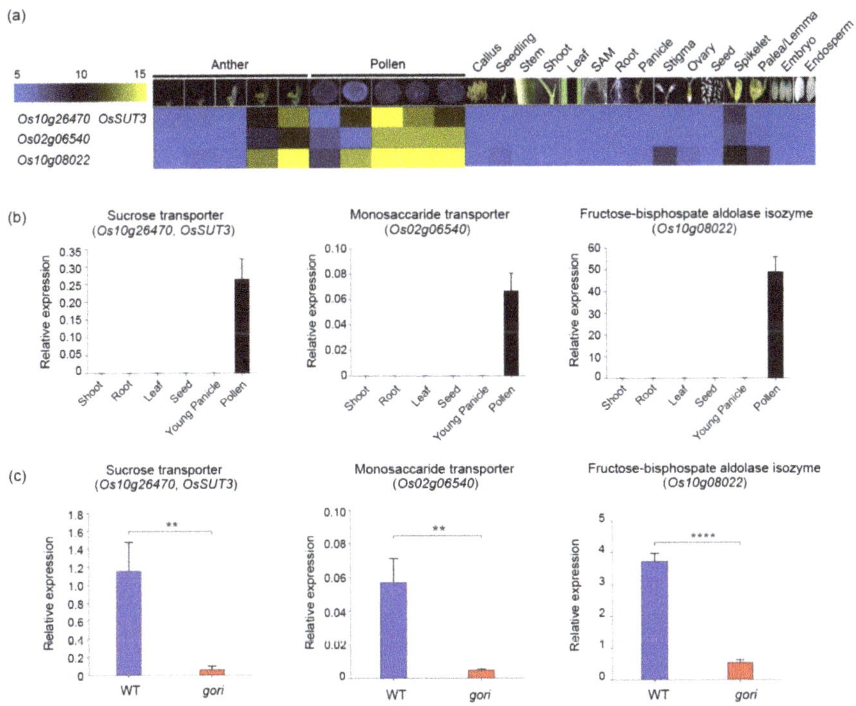

Figure 4. Validation of the dissected model and expression analysis of three carbohydrate metabolism-related genes in cluster H. (**a**) Heatmap analysis of three cluster H genes (*Os10g26740*, *Os02g06540*, and *Os10g08022*). Numeric values indicate an average of the normalized $\log_2$ intensity values. (**b**) Expression profiles of the three pollen-preferred genes based on qRT-PCR in various rice tissues: shoot, root, leaf, seed, young panicle, and pollen. (**c**) The expression of three pollen-preferred carbohydrate genes was significantly downregulated in the *gori* mutant compared to the wild-type plants. *OsUbi5* (*Os01g22490*) was used as an internal control. There were three biological replicates from the performed *t*-test on independent samples, with Bonferroni correction. ** $0.001 < p \leq 0.01$; **** $p \leq 0.0001$.

3.6. Construction of a Conceptual Carbohydrate Metabolism Model

In this section, a conceptual carbohydrate metabolism model that will help improve crop yield, as constructed previously for the rice endosperm, is proposed [89] (Figure 5). Notably, several studies fit well with this model. In pollen cluster H, *OsHXK10* is involved in anther dehiscent and pollen germination [90]. In addition, *CWINV3* mutation caused male sterility [91]. Regarding grain cluster I, *SUS3* overexpression increased cell wall polysaccharide deposition, resulting in enhanced biomass saccharification [92].

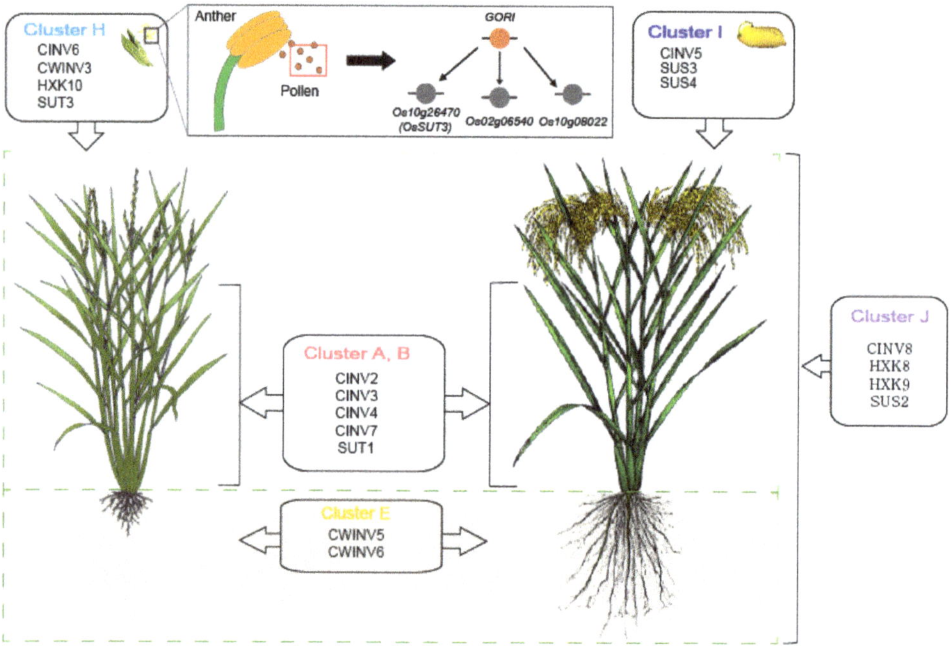

Figure 5. Conceptual carbohydrate model integrated with cluster information. A conceptual model was constructed by summarizing dissected carbohydrate metabolism-related genes according to the anatomical expression pattern. This model indicates the clusters associated with key enzymes (INVs, SUTs, SUSs, and HXKs) for source-sink communication. Clusters A and B are associated with leaf and flag leaf, cluster E is associated with root, cluster H is associated with anther and pollen, cluster I is associated with grain, and cluster J is associated with the whole rice plant based on ubiquitous expression patterns. The *GORI* regulatory model for pollen tissue was also combined.

This model, consistent with several functionally characterized genes, will be useful in providing guidelines for the spatial manipulation of carbohydrate metabolism-related genes in order to enhance crop yield.

4. Discussion

Improvement in crop yield is becoming more urgent due to the need to supply nearly 10 billion people. To maintain food security, many studies on carbohydrate metabolism in rice have been performed. However, these studies mostly focused on the source-sink mechanism and specific temporal samples such as the grain-filling stages. To complement this uneven interpretation of carbohydrate metabolism and provide new insights on rice productivity to enhance research associated with carbon metabolism, 866 carbohydrate metabolism-related genes were systemically dissected into 11 clusters according to meta anatomical expression data (Figure 3; Table S5).

As mentioned above, a functionally characterized gene search and functional group enrichment analyses indicated that most functionally characterized carbohydrate metabolism related genes were involved in the source-sink mechanism (Figure 2). Along with the results, this analysis showed some clusters showing root-, pollen-, and seed-preferred expression patterns (clusters E, H, and I). From among these, the expressions of three pollen-preferred carbon metabolism-related genes were confirmed, supporting the reliability of meta-anatomical expression data in this study (Figure 4).

Improving the seed setting rate through carbohydrate metabolism is a means of elevating rice productivity. One study reported that glycolysis could regulate pollen tube polarity in *Arabidopsis* [93]. Similarly, a recently characterized gene (*GORI*) involved in late pollen development in rice showed its connectivity to carbohydrate metabolism such as less pectin staining in the *gori* pollen tube [35]. Interestingly, three genes in cluster H showed significantly reduced expression when *GORI* was knocked out (Figure 4c). This result suggested that cluster H could be a suitable research candidate for further productivity improvement in the context of carbohydrate metabolism underlying late pollen development in rice.

Grain filling is also an important factor for rice yield increase [12]. In this analysis, the seed-preferred cluster I includes *SUS3* and *SUS4*, characterized as grain filling-related genes [94]. In addition, SWEET and glutamine synthetase, which were excluded in the analysis due to the limitations of the data source for functional categorization, play an important role in the grain-filling process related to sucrose transport and nitrogen metabolism, respectively [95–97]. Furthermore, when a hierarchical clustering of the *SWEET* genes with the 11 anatomical clusters was performed, *SWEET11* and *SWEET15* were close to cluster I (data not shown). Jointly, carbohydrate genes in cluster I could be useful genetic resources for further investigation of the rice grain-filling process and other metabolic processes.

Moreover, crop productivity can be affected by various factors, including environment, fertilizer, soil conditions, and even rhizobiome composition [2,98,99]. In particular, the interaction between crop root and the rhizobiome is related to root exudates, including amino acids, secondary metabolites, and carbohydrates [100]. Although the investigation of root exudates is understudied until now, the root-preferred cluster E could be an appropriate target for further studies on carbohydrate metabolism for generating exudates in the root.

5. Conclusions

This study aimed to improve the overall understanding of carbohydrate metabolism, which could provide some unknown clues for increasing rice productivity. For this, 866 carbohydrate metabolism-related genes were integrated into meta-anatomical expression data, and the significance of each cluster was shown using the functionally characterized roles in each cluster. Through an integrated analysis, carbohydrate metabolism-related genes were systemically dissected into 11 tissue-preferred clusters. Further functional enrichment analysis showed that two clusters (A and I) were strongly associated with source- and sink-preferred roles, respectively. In addition, the expression patterns of three pollen-specific cluster H genes were provided as examples of the reliability of the analysis. Furthermore, the reduced expression of the three cluster H genes in the *gori* mutant suggested that the tissue-preferred clusters could be suitable targets for further investigation. Collectively, a conceptual carbohydrate metabolism model summarizing the results was constructed, and it provided holistic insights on carbohydrate metabolism and suggested suitable candidates for improving crop productivity beyond source-sink mechanism-focused research.

Supplementary Materials: The following are available online at https://www.mdpi.com/article/10.3390/plants10081690/s1. Figure S1: Distribution of functionally characterized genes among anatomical clusters. Table S1: Characterized carbohydrate metabolism-related genes retrieved from the OGRO database. Table S2: Summary of the primers used in this study. Table S3: List and summary of the 866 carbohydrate metabolism-related genes in rice. Table S4: Classification of the carbon metabolism-related genes in rice using anatomical meta-expression data and KMC analysis. Table S5: Summarized cluster information of the 866 carbohydrate metabolism-related genes.

Author Contributions: Conceptualization, Y.-J.K., J.-S.J. and K.-H.J.; validation, W.-J.H., X.J. and S.-H.C.; formal analysis, W.-J.H., X.J. and Y.-J.K.; writing—original draft preparation, W.-J.H. and X.J.; writing—review and editing, W.-J.H., X.J. and K.-H.J.; visualization, W.-J.H. and X.J.; supervision, S.-T.K., J.-S.J. and K.-H.J. All authors have read and agreed to the published version of the manuscript.

Funding: This work was supported by grants from the New Breeding Technology Center Program (PJ01492703 to K.-H.J.) and the National Research Foundation, Ministry of Education, Science and Technology (2021R1A2C2010448 to K.-H.J.).

Data Availability Statement: The expression data presented in this study are available in the Supplementary Table S4.

Conflicts of Interest: The authors declare no conflict of interest.

References

1. Godfray, H.C.J.; Beddington, J.R.; Crute, I.R.; Haddad, L.; Lawrence, D.; Muir, J.F.; Pretty, J.; Robinson, S.; Thomas, S.M.; Toulmin, C. Food Security: The Challenge of Feeding 9 Billion People. *Science* **2010**, *327*, 812. [CrossRef]
2. Zhao, C.; Liu, B.; Piao, S.; Wang, X.; Lobell, D.B.; Huang, Y.; Huang, M.; Yao, Y.; Bassu, S.; Ciais, P.; et al. Temperature Increase Reduces Global Yields of Major Crops in Four Independent Estimates. *Proc. Natl. Acad. Sci. USA* **2017**, *114*, 9326. [CrossRef]
3. Yu, S.-M.; Lo, S.-F.; Ho, T.-H.D. Source–Sink Communication: Regulated by Hormone, Nutrient, and Stress Cross-Signaling. *Trends Plant Sci.* **2015**, *20*, 844–857. [CrossRef]
4. Sinclair, T.R.; Sheehy, J.E. Erect Leaves and Photosynthesis in Rice. *Science* **1999**, *283*, 1455. [CrossRef]
5. Sakamoto, T.; Morinaka, Y.; Ohnishi, T.; Sunohara, H.; Fujioka, S.; Ueguchi-Tanaka, M.; Mizutani, M.; Sakata, K.; Takatsuto, S.; Yoshida, S.; et al. Erect Leaves Caused by Brassinosteroid Deficiency Increase Biomass Production and Grain Yield in Rice. *Nat. Biotechnol.* **2006**, *24*, 105–109. [CrossRef]
6. Yoo, S.-C.; Cho, S.-H.; Zhang, H.; Paik, H.-C.; Lee, C.-H.; Li, J.; Yoo, J.-H.; Lee, B.-W.; Koh, H.-J.; Seo, H.S.; et al. Quantitative Trait Loci Associated with Functional Stay-Green SNU-SG1 in Rice. *Mol. Cells* **2007**, *24*, 83–94. [PubMed]
7. Lalonde, S.; Tegeder, M.; Throne-Holst, M.; Frommer, W.B.; Patrick, J.W. Phloem Loading and Unloading of Sugars and Amino Acids. *Plant Cell Environ.* **2003**, *26*, 37–56. [CrossRef]
8. Zhang, C.; Turgeon, R. Mechanisms of Phloem Loading. *Curr. Opin. Plant Biol.* **2018**, *43*, 71–75. [CrossRef]
9. Chen, L.-Q.; Qu, X.-Q.; Hou, B.-H.; Sosso, D.; Osorio, S.; Fernie, A.R.; Frommer, W.B. Sucrose Efflux Mediated by SWEET Proteins as a Key Step for Phloem Transport. *Science* **2012**, *335*, 207. [CrossRef] [PubMed]
10. Eom, J.-S.; Chen, L.-Q.; Sosso, D.; Julius, B.T.; Lin, I.; Qu, X.-Q.; Braun, D.M.; Frommer, W.B. SWEETs, Transporters for Intracellular and Intercellular Sugar Translocation. *Curr. Opin. Plant Biol.* **2015**, *25*, 53–62. [CrossRef] [PubMed]
11. Cho, J.-I.; Ryoo, N.; Ko, S.; Lee, S.-K.; Lee, J.; Jung, K.-H.; Lee, Y.-H.; Bhoo, S.H.; Winderickx, J.; An, G.; et al. Structure, Expression, and Functional Analysis of the Hexokinase Gene Family in Rice (*Oryza Sativa* L.). *Planta* **2006**, *224*, 598–611. [CrossRef] [PubMed]
12. Wang, E.; Wang, J.; Zhu, X.; Hao, W.; Wang, L.; Li, Q.; Zhang, L.; He, W.; Lu, B.; Lin, H.; et al. Control of Rice Grain-Filling and Yield by a Gene with a Potential Signature of Domestication. *Nat. Genet.* **2008**, *40*, 1370–1374. [CrossRef]
13. Li, B.; Liu, H.; Zhang, Y.; Kang, T.; Zhang, L.; Tong, J.; Xiao, L.; Zhang, H. Constitutive Expression of Cell Wall Invertase Genes Increases Grain Yield and Starch Content in Maize. *Plant Biotechnol. J.* **2013**, *11*, 1080–1091. [CrossRef]
14. Jeon, J.-S.; Jung, K.-H.; Kim, H.-B.; Suh, J.-P.; Khush, G.S. Genetic and Molecular Insights into the Enhancement of Rice Yield Potential. *J. Plant Biol.* **2011**, *54*, 1–9. [CrossRef]
15. Yang, J.; Zhang, J. Grain-Filling Problem in 'Super' Rice. *J. Exp. Bot.* **2010**, *61*, 1–5. [CrossRef]
16. Slewinski, T.L.; Zhang, C.; Turgeon, R. Structural and Functional Heterogeneity in Phloem Loading and Transport. *Front. Plant Sci.* **2013**, *4*, 244. [CrossRef]
17. Ruan, Y.-L.; Jin, Y.; Huang, J. Capping Invertase Activity by Its Inhibitor: Roles and Implications in Sugar Signaling, Carbon Allocation, Senescence and Evolution. *Plant Signal. Behav.* **2009**, *4*, 983–985. [CrossRef]
18. Zhu, T.; Budworth, P.; Chen, W.; Provart, N.; Chang, H.-S.; Guimil, S.; Su, W.; Estes, B.; Zou, G.; Wang, X. Transcriptional Control of Nutrient Partitioning during Rice Grain Filling. *Plant Biotechnol. J.* **2003**, *1*, 59–70. [CrossRef]
19. Yamakawa, H.; Hirose, T.; Kuroda, M.; Yamaguchi, T. Comprehensive Expression Profiling of Rice Grain Filling-Related Genes under High Temperature Using DNA Microarray. *Plant Physiol.* **2007**, *144*, 258–277. [CrossRef] [PubMed]
20. Kanehisa, M.; Goto, S. KEGG: Kyoto Encyclopedia of Genes and Genomes. *Nucleic Acids Res.* **2000**, *28*, 27–30. [CrossRef] [PubMed]
21. Thimm, O.; Bläsing, O.; Gibon, Y.; Nagel, A.; Meyer, S.; Krüger, P.; Selbig, J.; Müller, L.A.; Rhee, S.Y.; Stitt, M. MAPMAN: A User-Driven Tool to Display Genomics Data Sets onto Diagrams of Metabolic Pathways and Other Biological Processes. *Plant J.* **2004**, *37*, 914–939. [CrossRef]
22. Ouyang, S.; Zhu, W.; Hamilton, J.; Lin, H.; Campbell, M.; Childs, K.; Thibaud-Nissen, F.; Malek, R.L.; Lee, Y.; Zheng, L.; et al. The TIGR Rice Genome Annotation Resource: Improvements and New Features. *Nucleic Acids Res.* **2007**, *35*, D883–D887. [CrossRef]
23. Chandran, A.K.N.; Yoo, Y.-H.; Cao, P.; Sharma, R.; Sharma, M.; Dardick, C.; Ronald, P.C.; Jung, K.-H. Updated Rice Kinase Database RKD 2.0: Enabling Transcriptome and Functional Analysis of Rice Kinase Genes. *Rice* **2016**, *9*, 40. [CrossRef] [PubMed]

24. Cao, P.; Jung, K.-H.; Choi, D.; Hwang, D.; Zhu, J.; Ronald, P.C. The Rice Oligonucleotide Array Database: An Atlas of Rice Gene Expression. *Rice* **2012**, *5*, 17. [CrossRef] [PubMed]
25. Chandran, A.K.N.; Hong, W.-J.; Abhijith, B.; Lee, J.; Kim, Y.-J.; Park, S.K.; Jung, K.-H. Rice Male Gamete Expression Database (RMEDB): A Web Resource for Functional Genomic Studies of Rice Male Organ Development. *J. Plant Biol.* **2020**, *63*, 421–430. [CrossRef]
26. Saeed, A.I.; Sharov, V.; White, J.; Li, J.; Liang, W.; Bhagabati, N.; Braisted, J.; Klapa, M.; Currier, T.; Thiagarajan, M.; et al. TM4: A Free, Open-Source System for Microarray Data Management and Analysis. *Biotechniques* **2003**, *34*, 374–378. [CrossRef]
27. Moon, S.; Oo, M.M.; Kim, B.; Koh, H.-J.; Oh, S.A.; Yi, G.; An, G.; Park, S.K.; Jung, K.-H. Genome-Wide Analyses of Late Pollen-Preferred Genes Conserved in Various Rice Cultivars and Functional Identification of a Gene Involved in the Key Processes of Late Pollen Development. *Rice* **2018**, *11*, 28. [CrossRef]
28. Yamamoto, E.; Yonemaru, J.-I.; Yamamoto, T.; Yano, M. OGRO: The Overview of Functionally Characterized Genes in Rice Online Database. *Rice* **2012**, *5*, 26. [CrossRef]
29. Hong, W.-J.; Jung, K.-H. Comparative Analysis of Flanking Sequence Tags of T-DNA/Transposon Insertional Mutants and Genetic Variations of Fast-Neutron Treated Mutants in Rice. *J. Plant Biol.* **2018**, *61*, 80–84. [CrossRef]
30. Wickham, H. *Ggplot2: Elegant Graphics for Data Analysis*; Springer: New York, NY, USA, 2016; ISBN 978-3-319-24277-4.
31. Khatri, P.; Drăghici, S. Ontological Analysis of Gene Expression Data: Current Tools, Limitations, and Open Problems. *Bioinformatics* **2005**, *21*, 3587–3595. [CrossRef]
32. Jung, K.-H.; Dardick, C.; Bartley, L.E.; Cao, P.; Phetsom, J.; Canlas, P.; Seo, Y.-S.; Shultz, M.; Ouyang, S.; Yuan, Q.; et al. Refinement of Light-Responsive Transcript Lists Using Rice Oligonucleotide Arrays: Evaluation of Gene-Redundancy. *PLoS ONE* **2008**, *3*, e3337. [CrossRef] [PubMed]
33. Yu, G.; Wang, L.-G.; Han, Y.; He, Q.-Y. ClusterProfiler: An R Package for Comparing Biological Themes among Gene Clusters. *OMICS* **2012**, *16*, 284–287. [CrossRef] [PubMed]
34. Sakai, H.; Lee, S.S.; Tanaka, T.; Numa, H.; Kim, J.; Kawahara, Y.; Wakimoto, H.; Yang, C.; Iwamoto, M.; Abe, T.; et al. Rice Annotation Project Database (RAP-DB): An Integrative and interactive Database for Rice Genomics. *Plant Cell Physiol.* **2013**, *54*, e6. [CrossRef] [PubMed]
35. Kim, Y.-J.; Kim, M.-H.; Hong, W.-J.; Moon, S.; Kim, E.-J.; Silva, J.; Lee, J.; Lee, S.; Kim, S.T.; Park, S.K.; et al. GORI, Encoding the WD40 Domain Protein, Is Required for Pollen Tube Germination and Elongation in Rice. *Plant J.* **2021**, *105*, 1645–1664. [CrossRef] [PubMed]
36. Hong, W.-J.; Jiang, X.; Ahn, H.R.; Choi, J.; Kim, S.-R.; Jung, K.-H. Systematic Analysis of Cold Stress Response and Diurnal Rhythm Using Transcriptome Data in Rice Reveals the Molecular Networks Related to Various Biological Processes. *Int. J. Mol. Sci.* **2020**, *21*, 6872. [CrossRef]
37. Ge, L.-F.; Chao, D.-Y.; Shi, M.; Zhu, M.-Z.; Gao, J.-P.; Lin, H.-X. Overexpression of the Trehalose-6-Phosphate Phosphatase Gene OsTPP1 Confers Stress Tolerance in Rice and Results in the Activation of Stress Responsive Genes. *Planta* **2008**, *228*, 191–201. [CrossRef]
38. Li, H.-W.; Zang, B.-S.; Deng, X.-W.; Wang, X.-P. Overexpression of the Trehalose-6-Phosphate Synthase Gene OsTPS1 Enhances Abiotic Stress Tolerance in Rice. *Planta* **2011**, *234*, 1007–1018. [CrossRef]
39. Koumoto, T.; Shimada, H.; Kusano, H.; She, K.-C.; Iwamoto, M.; Takano, M. Rice Monoculm Mutation Moc2, Which Inhibits Outgrowth of the Second Tillers, Is Ascribed to Lack of a Fructose-1,6-Bisphosphatase. *Plant Biotechnol.* **2013**, *30*, 47–56. [CrossRef]
40. Eom, J.-S.; Cho, J.-I.; Reinders, A.; Lee, S.-W.; Yoo, Y.; Tuan, P.Q.; Choi, S.-B.; Bang, G.; Park, Y.-I.; Cho, M.-H.; et al. Impaired Function of the Tonoplast-Localized Sucrose Transporter in Rice, OsSUT2, Limits the Transport of Vacuolar Reserve Sucrose and Affects Plant Growth. *Plant Physiol.* **2011**, *157*, 109–119. [CrossRef]
41. Duan, J.; Zhang, M.; Zhang, H.; Xiong, H.; Liu, P.; Ali, J.; Li, J.; Li, Z. OsMIOX, a Myo-Inositol Oxygenase Gene, Improves Drought Tolerance through Scavenging of Reactive Oxygen Species in Rice (*Oryza Sativa* L.). *Plant Sci.* **2012**, *196*, 143–151. [CrossRef]
42. Ishimaru, K.; Hirose, T.; Aoki, N.; Takahashi, S.; Ono, K.; Yamamoto, S.; Wu, J.; Saji, S.; Baba, T.; Ugaki, M.; et al. Antisense Expression of a Rice Sucrose Transporter OsSUT1 in Rice (*Oryza Sativa* L.). *Plant Cell Physiol.* **2001**, *42*, 1181–1185. [CrossRef]
43. Sharma, A.; Komatsu, S. Involvement of a Ca(2+) Dependent Protein Kinase Component Downstream to the Gibberellin-Binding Phosphoprotein, RuBisCO Activase, in Rice. *Biochem. Biophys. Res. Commun.* **2002**, *290*, 690–695. [CrossRef]
44. Lu, Y.; Li, Y.; Yang, Q.; Zhang, Z.; Chen, Y.; Zhang, S.; Peng, X.-X. Suppression of Glycolate Oxidase Causes Glyoxylate Accumulation That Inhibits Photosynthesis through Deactivating Rubisco in Rice. *Physiol. Plant* **2014**, *150*, 463–476. [CrossRef]
45. Zhang, Y.; Xiao, W.; Luo, L.; Pang, J.; Rong, W.; He, C. Downregulation of OsPK1, a Cytosolic Pyruvate Kinase, by T-DNA Insertion Causes Dwarfism and Panicle Enclosure in Rice. *Planta* **2012**, *235*, 25–38. [CrossRef]
46. Zhang, Y.; Feng, F.; He, C. Downregulation of OsPK1 Contributes to Oxidative Stress and the Variations in ABA/GA Balance in Rice. *Plant Mol. Biol. Rep.* **2012**, *30*, 1006–1013. [CrossRef]
47. Lee, S.-K.; Hwang, S.-K.; Han, M.; Eom, J.-S.; Kang, H.-G.; Han, Y.; Choi, S.-B.; Cho, M.-H.; Bhoo, S.H.; An, G.; et al. Identification of the ADP-Glucose Pyrophosphorylase Isoforms Essential for Starch Synthesis in the Leaf and Seed Endosperm of Rice (*Oryza Sativa* L.). *Plant Mol. Biol.* **2007**, *65*, 531–546. [CrossRef] [PubMed]
48. Zhou, Y.; Cai, H.; Xiao, J.; Li, X.; Zhang, Q.; Lian, X. Over-Expression of Aspartate Aminotransferase Genes in Rice Resulted in Altered Nitrogen Metabolism and Increased Amino Acid Content in Seeds. *Theor. Appl. Genet.* **2009**, *118*, 1381–1390. [CrossRef] [PubMed]

49. Hakata, M.; Kuroda, M.; Miyashita, T.; Yamaguchi, T.; Kojima, M.; Sakakibara, H.; Mitsui, T.; Yamakawa, H. Suppression of α-Amylase Genes Improves Quality of Rice Grain Ripened under High Temperature. *Plant Biotechnol. J.* **2012**, *10*, 1110–1117. [CrossRef]
50. Lin, D.-G.; Chou, S.-Y.; Wang, A.Z.; Wang, Y.-W.; Kuo, S.-M.; Lai, C.-C.; Chen, L.-J.; Wang, C.-S. A Proteomic Study of Rice Cultivar TNG67 and Its High Aroma Mutant SA0420. *Plant Sci.* **2014**, *214*, 20–28. [CrossRef] [PubMed]
51. Kuwano, M.; Takaiwa, F.; Yoshida, K.T. Differential Effects of a Transgene to Confer Low Phytic Acid in Caryopses Located at Different Positions in Rice Panicles. *Plant Cell Physiol.* **2009**, *50*, 1387–1392. [CrossRef]
52. Kuwano, M.; Mimura, T.; Takaiwa, F.; Yoshida, K.T. Generation of Stable 'Low Phytic Acid' Transgenic Rice through Antisense Repression of the 1d-Myo-Inositol 3-Phosphate Synthase Gene (RINO1) Using the 18-KDa Oleosin Promoter. *Plant Biotechnol. J.* **2009**, *7*, 96–105. [CrossRef]
53. Kim, S.I.; Andaya, C.B.; Newman, J.W.; Goyal, S.S.; Tai, T.H. Isolation and Characterization of a Low Phytic Acid Rice Mutant Reveals a Mutation in the Rice Orthologue of Maize MIK. *Theor. Appl. Genet.* **2008**, *117*, 1291. [CrossRef]
54. Satoh, H.; Shibahara, K.; Tokunaga, T.; Nishi, A.; Tasaki, M.; Hwang, S.-K.; Okita, T.W.; Kaneko, N.; Fujita, N.; Yoshida, M.; et al. Mutation of the Plastidial α-Glucan Phosphorylase Gene in Rice Affects the Synthesis and Structure of Starch in the Endosperm. *Plant Cell* **2008**, *20*, 1833–1849. [CrossRef] [PubMed]
55. Utsumi, Y.; Utsumi, C.; Sawada, T.; Fujita, N.; Nakamura, Y. Functional Diversity of Isoamylase Oligomers: The ISA1 Homo-Oligomer Is Essential for Amylopectin Biosynthesis in Rice Endosperm. *Plant Physiol.* **2011**, *156*, 61–77. [CrossRef] [PubMed]
56. Kang, H.-G.; Park, S.; Matsuoka, M.; An, G. White-Core Endosperm Floury Endosperm-4 in Rice Is Generated by Knockout Mutations in the C4-Type Pyruvate Orthophosphate Dikinase Gene (OsPPDKB). *Plant J.* **2005**, *42*, 901–911. [CrossRef]
57. Wang, Z.-Y.; Zheng, F.-Q.; Shen, G.-Z.; Gao, J.-P.; Snustad, D.P.; Li, M.-G.; Zhang, J.-L.; Hong, M.-M. The Amylose Content in Rice Endosperm Is Related to the Post-Transcriptional Regulation of the Waxy Gene. *Plant J.* **1995**, *7*, 613–622. [CrossRef]
58. Fujita, N.; Yoshida, M.; Asakura, N.; Ohdan, T.; Miyao, A.; Hirochika, H.; Nakamura, Y. Function and Characterization of Starch Synthase I Using Mutants in Rice. *Plant Physiol.* **2006**, *140*, 1070–1084. [CrossRef]
59. Zhang, G.; Cheng, Z.; Zhang, X.; Guo, X.; Su, N.; Jiang, L.; Mao, L.; Wan, J. Double Repression of Soluble Starch Synthase Genes SSIIa and SSIIIa in Rice (*Oryza Sativa* L.) Uncovers Interactive Effects on the Physicochemical Properties of Starch. *Genome* **2011**, *54*, 448–459. [CrossRef]
60. Gao, Z.; Zeng, D.; Cheng, F.; Tian, Z.; Guo, L.; Su, Y.; Yan, M.; Jiang, H.; Dong, G.; Huang, Y.; et al. ALK, the Key Gene for Gelatinization Temperature, Is a Modifier Gene for Gel Consistency in RiceF. *J. Integr. Plant Biol.* **2011**, *53*, 756–765. [PubMed]
61. Ryoo, N.; Yu, C.; Park, C.-S.; Baik, M.-Y.; Park, I.M.; Cho, M.-H.; Bhoo, S.H.; An, G.; Hahn, T.-R.; Jeon, J.-S. Knockout of a Starch Synthase Gene OsSSIIIa/Flo5 Causes White-Core Floury Endosperm in Rice (*Oryza Sativa* L.). *Plant Cell Rep.* **2007**, *26*, 1083–1095. [CrossRef] [PubMed]
62. Fujita, N.; Satoh, R.; Hayashi, A.; Kodama, M.; Itoh, R.; Aihara, S.; Nakamura, Y. Starch Biosynthesis in Rice Endosperm Requires the Presence of Either Starch Synthase I or IIIa. *J. Exp. Bot.* **2011**, *62*, 4819–4831. [CrossRef]
63. Yun, M.-S.; Umemoto, T.; Kawagoe, Y. Rice Debranching Enzyme Isoamylase3 Facilitates Starch Metabolism and Affects Plastid Morphogenesis. *Plant Cell Physiol.* **2011**, *52*, 1068–1082. [CrossRef] [PubMed]
64. Saito, H.; Okumoto, Y.; Yoshitake, Y.; Inoue, H.; Yuan, Q.; Teraishi, M.; Tsukiyama, T.; Nishida, H.; Tanisaka, T. Complete Loss of Photoperiodic Response in the Rice Mutant Line X61 Is Caused by Deficiency of Phytochrome Chromophore Biosynthesis Gene. *Theor. Appl. Genet.* **2011**, *122*, 109–118. [CrossRef]
65. Jia, L.; Zhang, B.; Mao, C.; Li, J.; Wu, Y.; Wu, P.; Wu, Z. OsCYT-INV1 for Alkaline/Neutral Invertase Is Involved in Root Cell Development and Reproductivity in Rice (*Oryza Sativa* L.). *Planta* **2008**, *228*, 51–59. [CrossRef] [PubMed]
66. Asatsuma, S.; Sawada, C.; Itoh, K.; Okito, M.; Kitajima, A.; Mitsui, T. Involvement of α-Amylase I-1 in Starch Degradation in Rice Chloroplasts. *Plant Cell Physiol.* **2005**, *46*, 858–869. [CrossRef]
67. Ibraheem, O.; Botha, C.E.J.; Bradley, G.; Dealtry, G.; Roux, S. Rice Sucrose Transporter1 (OsSUT1) up-Regulation in Xylem Parenchyma Is Caused by Aphid Feeding on Rice Leaf Blade Vascular Bundles. *Plant Biol.* **2014**, *16*, 783–791. [CrossRef] [PubMed]
68. Feng, L.; Wang, K.; Li, Y.; Tan, Y.; Kong, J.; Li, H.; Li, Y.; Zhu, Y. Overexpression of SBPase Enhances Photosynthesis against High Temperature Stress in Transgenic Rice Plants. *Plant Cell Rep.* **2007**, *26*, 1635–1646. [CrossRef]
69. Ye, N.; Yang, G.; Chen, Y.; Zhang, C.; Zhang, J.; Peng, X. Two Hydroxypyruvate Reductases Encoded by OsHPR1 and OsHPR2 Are Involved in Photorespiratory Metabolism in Rice. *J. Integr. Plant Biol.* **2014**, *56*, 170–180. [CrossRef]
70. Cook, F.R.; Fahy, B.; Trafford, K. A Rice Mutant Lacking a Large Subunit of ADP-Glucose Pyrophosphorylase Has Drastically Reduced Starch Content in the Culm but Normal Plant Morphology and Yield. *Funct. Plant Biol.* **2012**, *39*, 1068–1078. [CrossRef]
71. Hirose, T.; Hashida, Y.; Aoki, N.; Okamura, M.; Yonekura, M.; Ohto, C.; Terao, T.; Ohsugi, R. Analysis of Gene-Disruption Mutants of a Sucrose Phosphate Synthase Gene in Rice, OsSPS1, Shows the Importance of Sucrose Synthesis in Pollen Germination. *Plant Sci.* **2014**, *225*, 102–106. [CrossRef]
72. Cao, H.; Guo, S.; Xu, Y.; Jiang, K.; Jones, A.M.; Chong, K. Reduced Expression of a Gene Encoding a Golgi Localized Monosaccharide Transporter (OsGMST1) Confers Hypersensitivity to Salt in Rice (*Oryza sativa*). *J. Exp. Bot.* **2011**, *62*, 4595–4604. [CrossRef] [PubMed]
73. Siahpoosh, M.R.; Sanchez, D.H.; Schlereth, A.; Scofield, G.N.; Furbank, R.T.; van Dongen, J.T.; Kopka, J. Modification of OsSUT1 Gene Expression Modulates the Salt Response of Rice *Oryza Sativa* Cv. Taipei 309. *Plant Sci.* **2012**, *182*, 101–111. [CrossRef] [PubMed]

74. Zhang, X.-H.; Rao, X.-L.; Shi, H.-T.; Li, R.-J.; Lu, Y.-T. Overexpression of a Cytosolic Glyceraldehyde-3-Phosphate Dehydrogenase Gene OsGAPC3 Confers Salt Tolerance in Rice. *Plant Cell Tissue Organ Cult.* **2011**, *107*, 1. [CrossRef]
75. Tuncel, A.; Kawaguchi, J.; Ihara, Y.; Matsusaka, H.; Nishi, A.; Nakamura, T.; Kuhara, S.; Hirakawa, H.; Nakamura, Y.; Cakir, B.; et al. The Rice Endosperm ADP-Glucose Pyrophosphorylase Large Subunit Is Essential for Optimal Catalysis and Allosteric Regulation of the Heterotetrameric Enzyme. *Plant Cell Physiol.* **2014**, *55*, 1169–1183. [CrossRef]
76. Masumoto, C.; Miyazawa, S.-I.; Ohkawa, H.; Fukuda, T.; Taniguchi, Y.; Murayama, S.; Kusano, M.; Saito, K.; Fukayama, H.; Miyao, M. Phosphoenolpyruvate Carboxylase Intrinsically Located in the Chloroplast of Rice Plays a Crucial Role in Ammonium Assimilation. *Proc. Natl. Acad. Sci. USA* **2010**, *107*, 5226–5231. [CrossRef]
77. Lee, S.-K.; Jeon, J.-S.; Bornke, F.; Voll, L.; Cho, J.-I.; Goh, C.-H.; Jeong, S.-W.; Park, Y.-I.; Kim, S.J.; Choi, S.-B.; et al. Loss of Cytosolic Fructose-1,6-Bisphosphatase Limits Photosynthetic Sucrose Synthesis and Causes Severe Growth Retardations in Rice (*Oryza sativa*). *Plant Cell Environ.* **2008**, *31*, 1851–1863. [CrossRef]
78. Kasajima, I.; Ebana, K.; Yamamoto, T.; Takahara, K.; Yano, M.; Kawai-Yamada, M.; Uchimiya, H. Molecular Distinction in Genetic Regulation of Nonphotochemical Quenching in Rice. *Proc. Natl. Acad. Sci. USA* **2011**, *108*, 13835–13840. [CrossRef]
79. Hubbart, S.; Ajigboye, O.O.; Horton, P.; Murchie, E.H. The Photoprotective Protein PsbS Exerts Control over CO_2 Assimilation Rate in Fluctuating Light in Rice. *Plant J.* **2012**, *71*, 402–412. [CrossRef]
80. Abe, N.; Asai, H.; Yago, H.; Oitome, N.F.; Itoh, R.; Crofts, N.; Nakamura, Y.; Fujita, N. Relationships between Starch Synthase I and Branching Enzyme Isozymes Determined Using Double Mutant Rice Lines. *BMC Plant Biol.* **2014**, *14*, 80. [CrossRef] [PubMed]
81. Xu, H.; Zhang, J.; Zeng, J.; Jiang, L.; Liu, E.; Peng, C.; He, Z.; Peng, X. Inducible Antisense Suppression of Glycolate Oxidase Reveals Its Strong Regulation over Photosynthesis in Rice. *J. Exp. Bot.* **2009**, *60*, 1799–1809. [CrossRef]
82. Jin, S.-H.; Hong, J.; Li, X.-Q.; Jiang, D.-A. Antisense Inhibition of Rubisco Activase Increases Rubisco Content and Alters the Proportion of Rubisco Activase in Stroma and Thylakoids in Chloroplasts of Rice Leaves. *Ann. Bot.* **2006**, *97*, 739–744. [CrossRef]
83. Nath, K.; Poudyal, R.S.; Eom, J.-S.; Park, Y.S.; Zulfugarov, I.S.; Mishra, S.R.; Tovuu, A.; Ryoo, N.; Yoon, H.-S.; Nam, H.G.; et al. Loss-of-Function of OsSTN8 Suppresses the Photosystem II Core Protein Phosphorylation and Interferes with the Photosystem II Repair Mechanism in Rice (*Oryza sativa*). *Plant J.* **2013**, *76*, 675–686. [CrossRef] [PubMed]
84. Nishikawa, Y.; Yamamoto, H.; Okegawa, Y.; Wada, S.; Sato, N.; Taira, Y.; Sugimoto, K.; Makino, A.; Shikanai, T. PGR5-Dependent Cyclic Electron Transport Around PSI Contributes to the Redox Homeostasis in Chloroplasts Rather Than CO_2 Fixation and Biomass Production in Rice. *Plant Cell Physiol.* **2012**, *53*, 2117–2126. [CrossRef] [PubMed]
85. Zhou, Q.; Yu, Q.; Wang, Z.; Pan, Y.; Lv, W.; Zhu, L.; Chen, R.; He, G. Knockdown of GDCH Gene Reveals Reactive Oxygen Species-Induced Leaf Senescence in Rice. *Plant Cell Environ.* **2013**, *36*, 1476–1489. [CrossRef]
86. Makino, A.; Shimada, T.; Takumi, S.; Kaneko, K.; Matsuoka, M.; Shimamoto, K.; Nakano, H.; Miyao-Tokutomi, M.; Mae, T.; Yamamoto, N. Does Decrease in Ribulose-1,5-Bisphosphate Carboxylase by Antisense RbcS Lead to a Higher N-Use Efficiency of Photosynthesis under Conditions of Saturating CO_2 and Light in Rice Plants? *Plant Physiol.* **1997**, *114*, 483–491. [CrossRef] [PubMed]
87. Ogawa, S.; Suzuki, Y.; Yoshizawa, R.; Kanno, K.; Makino, A. Effect of Individual Suppression of RBCS Multigene Family on Rubisco Contents in Rice Leaves. *Plant Cell Environ.* **2012**, *35*, 546–553. [CrossRef]
88. Makino, A.; Sage, R.F. Temperature Response of Photosynthesis in Transgenic Rice Transformed with 'Sense' or 'Antisense' RbcS. *Plant Cell Physiol.* **2007**, *48*, 1472–1483. [CrossRef]
89. Hong, W.-J.; Yoo, Y.-H.; Park, S.-A.; Moon, S.; Kim, S.-R.; An, G.; Jung, K.-H. Genome-Wide Identification and Extensive Analysis of Rice-Endosperm Preferred Genes Using Reference Expression Database. *J. Plant Biol.* **2017**, *60*, 249–258. [CrossRef]
90. Xu, F.-Q.; Li, X.-R.; Ruan, Y.-L. RNAi-Mediated Suppression of Hexokinase Gene OsHXK10 in Rice Leads to Non-Dehiscent Anther and Reduction of Pollen Germination. *Plant Sci.* **2008**, *175*, 674–684. [CrossRef]
91. Oliver, S.N.; Van Dongen, J.T.; Alfred, S.C.; Mamun, E.A.; Zhao, X.; Saini, H.S.; Fernandes, S.F.; Blanchard, C.L.; Sutton, B.G.; Geigenberger, P.; et al. Cold-Induced Repression of the Rice Anther-Specific Cell Wall Invertase Gene OSINV4 Is Correlated with Sucrose Accumulation and Pollen Sterility. *Plant Cell Environ.* **2005**, *28*, 1534–1551. [CrossRef]
92. Fan, C.; Wang, G.; Wu, L.; Liu, P.; Huang, J.; Jin, X.; Zhang, G.; He, Y.; Peng, L.; Luo, K.; et al. Distinct Cellulose and Callose Accumulation for Enhanced Bioethanol Production and Biotic Stress Resistance in OsSUS3 Transgenic Rice. *Carbohydr. Polym.* **2020**, *232*, 115448. [CrossRef]
93. Chen, W.; Gong, P.; Guo, J.; Li, H.; Li, R.; Xing, W.; Yang, Z.; Guan, Y. Glycolysis Regulates Pollen Tube Polarity via Rho GTPase Signaling. *PLoS Genet.* **2018**, *14*, e1007373. [CrossRef]
94. You, C.; Zhu, H.; Xu, B.; Huang, W.; Wang, S.; Ding, Y.; Liu, Z.; Li, G.; Chen, L.; Ding, C.; et al. Effect of Removing Superior Spikelets on Grain Filling of Inferior Spikelets in Rice. *Front. Plant Sci.* **2016**, *7*, 1161. [CrossRef]
95. Ma, L.; Zhang, D.; Miao, Q.; Yang, J.; Xuan, Y.; Hu, Y. Essential Role of Sugar Transporter OsSWEET11 During the Early Stage of Rice Grain Filling. *Plant Cell Physiol.* **2017**, *58*, 863–873. [CrossRef] [PubMed]
96. Yang, J.; Luo, D.; Yang, B.; Frommer, W.B.; Eom, J.-S. SWEET11 and 15 as Key Players in Seed Filling in Rice. *New Phytol.* **2018**, *218*, 604–615. [CrossRef] [PubMed]
97. Tabuchi, M.; Sugiyama, K.; Ishiyama, K.; Inoue, E.; Sato, T.; Takahashi, H.; Yamaya, T. Severe Reduction in Growth Rate and Grain Filling of Rice Mutants Lacking OsGS1;1, a Cytosolic Glutamine Synthetase1;1. *Plant J.* **2005**, *42*, 641–651. [CrossRef]
98. Samuel, A.D.; Bungau, S.; Tit, D.M.; Melinte, C.E.; Purza, L.; Badea, G.E. Effects of Long Term Application of Organic and Mineral Fertilizers on Soil Enzymes. *Rev. Chim.* **2018**, *69*, 2608–2612. [CrossRef]

99. Ramakrishna, W.; Yadav, R.; Li, K. Plant Growth Promoting Bacteria in Agriculture: Two Sides of a Coin. *Appl. Soil Ecol.* **2019**, *138*, 10–18. [CrossRef]
100. Preece, C.; Peñuelas, J. A Return to the Wild: Root Exudates and Food Security. *Trends Plant Sci.* **2020**, *25*, 14–21. [CrossRef] [PubMed]

plants

MDPI

Review

Seed Geometry in the Vitaceae

Emilio Cervantes [1,*], José Javier Martín-Gómez [1], Diego Gutiérrez del Pozo [2] and Ángel Tocino [3]

[1] Instituto de Recursos Naturales y Agrobiología del Consejo Superior de Investigaciones Científicas (IRNASA-CSIC), Cordel de Merinas, 40, 37008 Salamanca, Spain; jjavier.martin@irnasa.csic.es
[2] Departamento de Conservación y Manejo de Vida Silvestre (CYMVIS), Universidad Estatal Amazónica (UEA), Carretera Tena a Puyo Km, 44, Puyo 150950, Ecuador; diego.gutierrez.pozo@gmail.com
[3] Departamento de Matemáticas, Facultad de Ciencias, Universidad de Salamanca, Plaza de la Merced 1-4, 37008 Salamanca, Spain; bacon@usal.es
* Correspondence: emilio.cervantes@irnasa.csic.es

Abstract: The Vitaceae Juss., in the basal lineages of Rosids, contains sixteen genera and 950 species, mainly of tropical lianas. The family has been divided in five tribes: Ampelopsideae, Cisseae, Cayratieae, Parthenocisseae and Viteae. Seed shape is variable in this family. Based on new models derived from equations representing heart and water drop curves, we describe seed shape in species of the Vitaceae. According to their similarity to geometric models, the seeds of the Vitaceae have been classified in ten groups. Three of them correspond to models before described and shared with the Arecaceae (lenses, superellipses and elongated water drops), while in the seven groups remaining, four correspond to general models (waterdrops, heart curves, elongated heart curves and other elongated models) and three adjust to the silhouettes of seeds in particular genera (heart curves of *Cayratia* and *Pseudocayratia*, heart curves of the Squared Heart Curve (SqHC) type of *Ampelocissus* and *Ampelopsis* and Elongated Superellipse-Heart Curves (ESHCs), frequent in *Tetrastigma* species and observed also in *Cissus* species and *Rhoicissus rhomboidea*). The utilities of the application of geometric models for seed description and shape quantification in this family are discussed.

Keywords: endosperm; geometry; morphology; seed shape; Vitaceae

Citation: Cervantes, E.; Martín-Gómez, J.J.; Gutiérrez del Pozo, D.; Tocino, Á. Seed Geometry in the Vitaceae. *Plants* **2021**, *10*, 1695. https://doi.org/10.3390/plants10081695

Academic Editor: Eleftherios P. Eleftheriou

Received: 9 July 2021
Accepted: 10 August 2021
Published: 18 August 2021

Publisher's Note: MDPI stays neutral with regard to jurisdictional claims in published maps and institutional affiliations.

Copyright: © 2021 by the authors. Licensee MDPI, Basel, Switzerland. This article is an open access article distributed under the terms and conditions of the Creative Commons Attribution (CC BY) license (https://creativecommons.org/licenses/by/4.0/).

1. Introduction

The Vitaceae Juss. contains sixteen genera with ca. 950 species of lianas primarily distributed in the tropics with some genera in the temperate regions. The Leeaceae Dumort., with a single genus of 34 species, mostly shrubs and small trees rather than lianas, included in the family in the APG IV [1], was later recognized as a separate family [2,3]. Both families constitute the order Vitales, one of the basal lineages of Rosids, whose closest relative remains controversial [4,5].

The Vitaceae has been divided in five tribes [3] (Table 1): (I) Ampelopsideae J. Wen and Z. L. Nie (*Ampelopsis* Michx., *Nekemias* Raf., *Rhoicissus* Planch., *Clematicissus* Planch.); (II) Cisseae Rchb. (*Cissus* L.); (III) Cayratieae J.Wen and L.M.Lu (*Cayratia* Juss. Ex Guill, *Causonis* Raf., *Acareosperma* Gagnep., *Afrocayratia*, *Cyphostemma* (Planch.) Alston, *Pseudocayratia* J.Wen, L.M.Lu and Z.D.Chen, *Tetrastigma* Planch.); (IV) Parthenocisseae J.Wen and Z.D.Chen (*Parthenocissus* Planch.) and (V) Viteae Dumort (*Ampelocissus* Planch., and *Vitis* L.). *Cayratia* and *Cyphostemma* were included in *Cissus* by Linné and Planchon considered the former as a section of *Cissus* [6,7].

Cissus is the largest genus in the family with 300 species of complex classification [8]. *Cyphostemma* is second, with 200 species of an interesting diversity in their range of distribution as well as in growth habits (vines and lianas, herbs, stem succulents and a tree) [9]. *Vitis* has seventy-five inter-fertile wild species distributed in three continents under subtropical, Mediterranean and continental-temperate climatic conditions. *Vitis vinifera* L. is the species with highest economic importance in the family with some taxonomic uncertainty about the differentiation between *V. vinifera* L. subsp. *vinifera* and *V. vinifera* L. subsp. *sylvestris*

(Willd.) Hegi [10,11]. Thousands of cultivars of *V. vinifera* are used worldwide in Viticulture. Species of other genera are widely cultivated, such as *Parthenocissus quinquefolia* (L.) Planch., the Virginia creeper, in temperate areas, and *Cissus incisa* Des Moul., the grape ivy, in tropical areas. Species of the genus *Tetrastigma* are the only host plants for the parasitic plant *Rafflesia arnoldii* R.Br., Rafflesiaceae, which is native only to a few areas within the Malay Archipelago [12].

Table 1. A summary of the taxonomy of the Vitaceae. The approximate number of species in each tribe and genus is given between parentheses. Data adapted from [3].

Tribe	Genera
I. Ampelopsideae (47)	*Ampelopsis* Michx. (18)
	Nekemias Raf. (9)
	Rhoicissus Planch. (14)
	Clematicissus Planch. (6)
II. Cisseae (300)	*Cissus* L. (300)
III. Cayratieae (368)	*Cayratia* Juss. (25)
	Causonis Raf. (30)
	Acareosperma Gagnep. (1)
	Afrocayratia (7)
	Cyphostemma (Planch.) Alston (200)
	Pseudocayratia J. Wen, L.M.Lu and Z.D. Chen (5)
	Tetrastigma (Miq.) Planch. (100)
IV. Parthenocisseae (16)	*Parthenocissus* Planch. (14)
	Yua C.L.Li (2)
V. Viteae (190)	*Ampelocissus* Planch. (115)
	Vitis L. (75)

The taxonomy and phylogenetic relationships of the Vitaceae are far from complete and will benefit from an accurate description of seeds in the unambiguous terms of geometry. From a practical point of view, the classification based on geometric models may contribute to the distinction between wild and crop grapes of *Vitis vinifera* [11].

The main objective of this review is to provide a framework for the description of seed morphology in the Vitaceae based on geometric models. A recent review of this subject in the Arecaceae described morphological types in the seeds of this family based on the similarity of seed images to geometric figures, like ellipses, ovals and others [13]. Members of both families, the Arecaceae and the Vitaceae, were present in the Neotropical flora in the Eocene [14], and the description of their seeds may serve as a model to develop this work in other plant families.

2. Seed Morphology in the Vitaceae

2.1. Quantification of Seed Shape by Geometric Models

The silhouettes of bi-dimensional images of seeds often resemble geometric figures that can be used as models for the description and quantification of seed shape in plant families. A recent review of the geometry of seeds in the Arecaceae described a series of models useful for the analysis of seed shape in this family [13]. Geometric models included the ellipses (the circle is a particular type of ellipse), ovals, lemniscates, superellipses, cardioid and derivatives, lenses and the water drop curve [13]. The reader is referred to this review for the algebraic description of the models and their application in the morphometry of seeds in the Arecaceae. The application of geometric models in morphometry is based on the comparison of bi-dimensional images of well oriented seeds with these figures by means of image programs working in two layers (Adobe Photoshop, Corel Photo Paint ...). The two images (seed and model) can be superimposed searching for a maximum similarity and the ratio between shared and total surface areas, that we have termed *J* index, is calculated with the data obtained in ImageJ [15]. *J* index measures the percent of similarity between two images (the seed and the model) and provides information on overall seed shape [16,17]. Bidimensional seed images of many plant species adjust well to one of three morphological types: the ellipse, the oval and the cardioid [18]. The seeds of the model plant *Arabidopsis thaliana* (L.) Heynh., those of the model legumes *Lotus japonicus* (Regel) K.Larsen and *Medicago truncatula* Gaertn., as well as the seeds of *Capparis spinosa* L., in the Capparaceae and *Rhus tripartita* DC. in the Anacardiaceae adjust well to cardioids or modified cardioids [19–23]. The seeds of *Ricinus communis* L. and *Jatropha curcas* L. in the Euphorbiaceae and those of cultivars of *Triticum sp.* in the Poaceae adjust well to ellipses of varied x/y ratio [24–26]. Oval shaped seeds occur frequently in the Cucurbitaceae, Berberidaceae, Eupteleaceae and Lardizabalaceae [27,28], while the cardioid is more common in Papaveraceae [28]. A given geometric type is sometimes associated with other morphological or ecological characteristics. For example, cardioid-type seeds were observed to be more frequent in small-sized seeds, while elliptic shape is more frequent in larger seeds [18]. In the Malvaceae, cardioid type seeds are associated with small herbs of annual cycle [29].

2.2. Seed Morphology in the Vitaceae

The seeds of the Vitaceae share structural characteristics. The endosperm presents in transversal section a typical "M" shape coincident with a pair of ventral in-folds and a dorsal chalaza knot allowing the identification of fossil seeds in this family [30].

For the application of morphology in taxonomy, characters of the seeds are selected and compared between different taxonomic groups. Frequently, the data concern measurements of defined structural components and the distances between well referenced seed positions. The results depend on the characters selected and the method used for comparison. Chen and Manchester applied successive PCA based on 57 characters to 252 seeds representative of 238 species belonging to 15 genera [31]. The conclusions were: (1) The seeds of *Leea*, *Cissus*, *Cyphostemma*, *Tetrastigma*, *Rhoicissus*, and *Cayratia* have a long or linear chalaza, visible from the ventral side and terminated very near the beak at the dorsal side, a condition that was termed "perichalaza" [32], whereas the seeds in the rest of the family usually have an oval chalaza, central in the dorsal position; (2) *Tetrastigma* and species of *Rhoicissus* have peculiar characteristics in their linear chalaza, which is located near the apical notch and extends their beak; their long, narrow, sometimes divergent ventral in-folds and their rugose surface. The authors conclude that, by comparison of selected sets of characters, the seeds can be distinguished to the generic level [31].

Seed shape in the Vitaceae is variable and seeds resembling geometric figures are frequent in this family. The seeds of *Ampelocissus* are pyriform, oval, or round in dorsal or ventral view [30]. The seeds of *Cissus* species are often described as globose with a pointed base, elliptic in outline or oblong (see, for example, [33–35]). These adjectives and other like sub-globose or terete are also applied to seed descriptions in other genera suggesting two important points: (1) The seeds of the Vitaceae are suitable for the comparison with geometric figures used as models. (2) The comparison may be quantitative, yielding measures that contribute to taxonomy. Table 2 contains a list of 131 species in the Vitaceae whose seeds have been observed for this work.

Table 2. A summary of the 131 species for which seed shape has been analysed in this work.

Tribe (Species Observed/Total)	Genera (Species Observed/Total)	Species (References for the Images)
I. Ampelopsideae (15/47)	*Ampelopsis* Michx. (13/18)	*Ampelopsis aconitifolia* [36], *A. arborea* [37], *A. bodinieri* [36], *A. cantoniensis* [31,36], *A. cordata* [38], *A. chaffanjoni* [36], *A. delavayana* [31], *A. denudata* [30], *A. glandulosa* [39], *A. grossedentata* [31], *A. humulifolia* [36], *A. japonica* [36], *A. megalophylla* [31,36]
	Rhoicissus Planch. (2/14)	*Rhoicissus revoilii* [31], *R. rhomboidea* [31]
II. Cisseae (33/300)	*Cissus* L. (33/300)	*Cissus antarctica* [31], *C. aralioides* [35,40], *C. barbeyana* [40], *C. bosseri* [40], *C. cactiformis* [40], *C. campestris* [31,41,42], *C. cornifolia* [40], *C. descoingsii,* [31,41], *C. diffusiflora* [40], *C. elongata* [40], *C. erosa* [43], *C. floribunda* [40], *C. fuliginea* [31], *C. granulosa* [31], *C. hastata* [40], *C. hypoglauca* [31], *C. integrifolia* [40,42], *C. leucophlea* [40], *C. penninervis* [31], *C. petiolata* [40], *C. pileata* [40], *C. populnea* [40], *C. quadrangularis* [44], *C. reniformis* [31,41], *C. repens* [40], *C. sciaphila* [40], *C. smithiana* [40], *C. sterculiifolia* [31], *C. subtetragona* [40], *C. trianae* [31], *C. tuberosa* [42], *C. verticillata* [31,41,42,45], *C. willardii* [42],
III. Cayratieae (40/365)	*Causonis* Raf. (1/9)	*Causonis* sp. [46]
	Cayratia Juss. (7/60)	*Cayratia cheniana* [46], *C. geniculata* [31], *C. imerinensis* [47], *C. japonica* [31,48], *C. oligocarpa* [31], *C. saponaria* [31], *C.* sp. [African, [46]]
	Cyphostemma (Planch.) Alston (3/200)	*Cyphostemma elephantopus* [49], *C. laza* [31], *C. junceum* [31]
	Pseudocayratia J. Wen, L.M.Lu and Z.D.Chen (3/5)	*Pseudocayratia dichromocarpa* [50], *P. pengiana* [50], *P. speciosa* [50,51]
	Tetrastigma (Miq.) Planch. (26/100)	*Tetrastigma campylocarpum* [52], *T. cauliflorum* [52], *T. caudatum* [52], *T. delavayi* [52], *T. dichotomum* [51], *T. formosanum* [52], *T. harmandi* [31], *T. hemsleyanum* [31,52], *T. henryi* [52], T. *hypoglaucum* [52], *T. jinghongense* [52], *T. kwangsiense* [30,31], *T. lanceolarium* [30], *T. laoticum* [52], *T. obovatum* [51,52], *T. obtectum* [51,52], *T. pachyllylum* [52], *T. pedunculare* [31,51,52], *T. petraeum* [52], *T. retinervum* [52], *T. rumicispermum* [31,51,52], *T. serrulatum* [52], *T. sichouense* [52], *T. thorsborneorum* [52], *T. triphyllum* [31,52], *T. xishuangbannaense* [31,52]
IV. Parthenocisseae (11/16)	*Parthenocissus* Planch. (9/14)	*Parthenocissus dalzielii* [36], *P. heptaphylla* [31], *P. henryana* [36], *P. heterophylla* [36], *P. himalayana* [53], *P. laetevirens* [36], *P. quinquefolia* [54], *P. tricuspidata* [36,37,45], *P. vitacea* [31]
	Yua C.L.Li (2/2)	*Yua austro-orientalis* [31], *Y. chinensis* [31]
V. Viteae (32/190)	*Ampelocissus* Planch. (13/115)	*Ampelocissus acapulcensis* [30], *A. bombycina* [30], *A. bravoi* [42], *A. cavicaulis* [30], *A. erdvendbergiana* [30], *A. grantii* [30], *A. javalensis* [30,42], *A. latifolia* [30], *A. macrocirrha* [30], *A. martinii* [42], *A. obtusata* [30], *A. ochracea* [30], *A. robinsonii* [30]
	Vitis L. (19/75)	*Vitis aestivalis* [55], *V. amurensis* [45,56], *V. brandoniana* [54], *V. cinerea* [57], *V. eolabrusca* [54], *V. flexuosa* [54], *V. grayensis* [58], *V. labrusca* [45,54,59], *V. lanatoides* [58], *V. latisulcata* [58], *V. palmata* [60], *V. pseudorotundifolia* [54], *V. rostrata* [54], *V. rotundifolia* [31,41,54], *V. rupestris* [45], *V. tiliifolia* [61], *V. tsoi* [31,41], *V. vulpina* [59,62], *V. wilsoniae* [31,41]

Seven geometric models for the description and quantification of seed shape in the Vitaceae were described before [45]. Models 1 to 5 were derived from modifications in the equations of the heart curve [63], Model 6 was derived from the pyriform curve [64], and

Model 7 was obtained by the modification of two independent equations representing an ellipse, searching for the similarity with the seeds of cultivars of *V. vinifera*. In subsequent work, Model 7 revealed particularly useful, because models derived from it adjusted well to the shape of many cultivars of *V. vinifera* in the Spanish collection at IMIDRA [65]. The following section contains a description of new models obtained for this work and their examples in the Vitaceae. The first part presents geometric models shared by the Arecaceae and the Vitaceae, while the second part contains the description of new models based on a series of equations derived from the equation of an ellipse that apply to species in the Vitaceae.

3. Geometric Models for Seed Description and Quantification in the Vitaceae

3.1. Models Shared with the Arecaceae

The comparison of well-oriented seeds with geometric figures reveals the diversity of shapes in a family, adds precision to the description and permits quantification of seed shape. The geometric analysis of seed shape in the Vitaceae shows a variety of forms shared with the Arecaceae, such as lenses, superellipses and water drops [13], while their combination is not frequent in other plant families.

Lenses and superellipses [13] can be derived from the same formula:

$$\left|\frac{x}{a}\right|^p + \left|\frac{y}{b}\right|^q = 1$$

with $p, q > 2$ for superellipses and $p > 2$, $1 < q < 2$ for lenses [66,67]; (see Data Availability Statement section).

Figure 1 presents examples of seeds whose images resemble lenses of different length/width ratios: *Cissus sterculiifolia* [31], *Tetrastigma petraeum* [51] and *Cissus quadrangularis* [44].

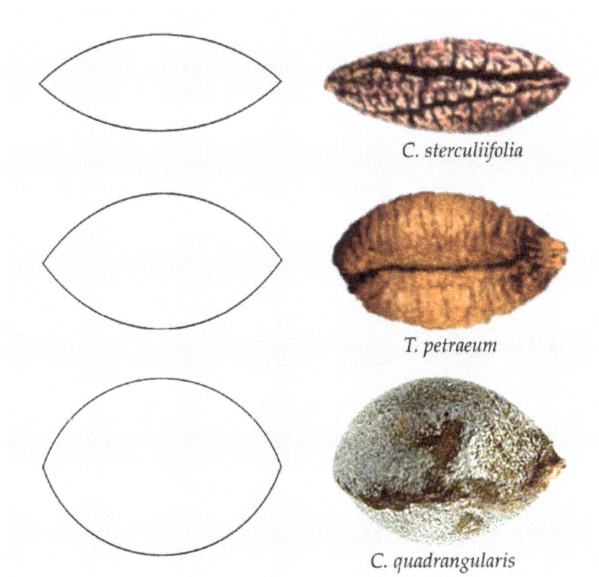

Figure 1. The images of seeds of *Cissus sterculiifolia* [31], *Tetrastigma petraeum* [51] and *Cissus quadrangularis* [44] resemble lenses of different proportions.

The seeds of *Cissus reniformis*, *Cyphostemma laza* and *Ampelocissus bravoi* adjust to superellipses of different proportions (not shown). Additionally, in some *Tetrastigma* species the seeds resemble superellipses, for example: *Tetrastigma henryi* [52], *Tetrastigma campylocarpum* [51] and *T. caudatum* [51] (Figure 2).

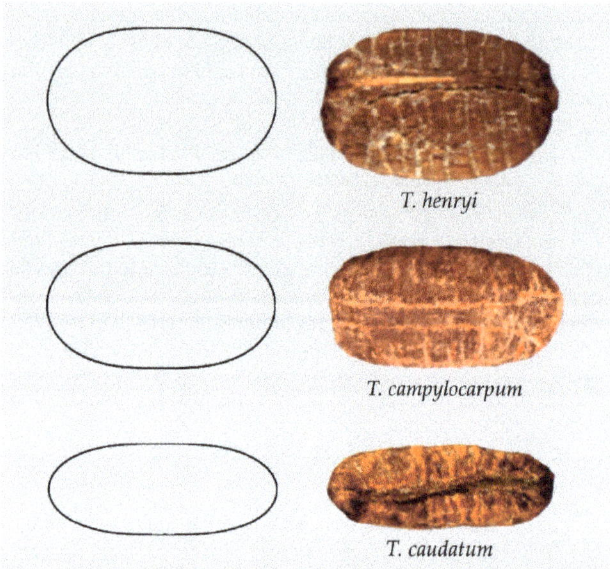

Figure 2. Examples of seed images resembling superellipses with different length/width ratios: First row: *Tetrastigma henryi* [51]. Second row: *Tetrastigma campylocarpum* [51]. Third row: *T. caudatum* [51].

Water drops and lemniscates described well the seeds of some species in the Arecaceae [13]. The former adapt well to the bi-dimensional shape of well-oriented images in the Vitaceae [45]. Figure 3 shows examples of seeds resembling elongated water drops see [45] for quantitative measurements in *Cissus verticillata*.

Figure 3. Examples of seed images resembling elongated water drops. From left to right: The model for an elongated water drop (Model 4 in [45]), seeds of *Cissus verticillata* and *Vitis vulpina*.

The equation describing the water drop is given in [13,45], while the equation describing the lemniscate is given in [13].

3.2. Geometric Models for the Vitaceae: New Models Obtained from a Series of Equations Derived from an Ellipse

A family of equations will be described that simplify the design of new models according to the variety of seed types found in the Vitaceae [45]. The task of finding a model adapted to a new shape will be easier knowing which term of an existing equation may give the changes needed to obtain a given figure. With this objective, the equations corresponding to all models described in this section derive from the ellipse of equation:

$$1 - x^2 - b^2y^2 = 0 \tag{1}$$

which can be expressed as:

$$\left(\sqrt{1-x^2} - b\,y\right)\left(\sqrt{1-x^2} + b\,y\right) = 0 \tag{2}$$

to remark the two explicit equations corresponding to the respective semi-ellipses that integrate it. Modifications in one of the factors, or in both, give equations of increasing complexity, whose graphic representations result in a variety of models.

A Water drop curve is obtained by the modification of Equation (2) to give:

$$\left(\sqrt{1-x^2} + b\,y\right)\left(\sqrt{1-x^2} + \frac{a}{50x^2+c} - b\,y\right) = 0 \tag{3}$$

with $a = b = 1,\ c = 2$. While the semi-ellipse corresponding to the left factor has not changed, the factor on the right determines the prominent part of the drop (beak). Increasing the value of a, increases the size of the beak (See Data Availability Statement section).

3.2.1. Water Drop Models

The three models represented in Figure 4 result from changing the values of a, b and c and modifying other terms in Equation (2): Model VAM1 $(a = 0.6; b = 1;\ c = 2)$ adjusts well to *Vitis amurensis*, *V. labrusca*, *V. rupestris* and *Cissus granulosa*; Model AGL1, a rounded Water drop, $(a = 0.3; b = 1.1;\ c = 1.6)$ adjusts well to seeds of *Ampelopsis glandulosa*, *Tetrastigma triphyllum* and *Cissus fuliginea*; Model AAR1, an elongated Water drop, $(a = 3; b = 1;\ c = 5)$ adjusts to seeds of *Ampelopsis arborea*, *Cissus campestris* and *C. willardii* (Data Availability Statement section). A slightly narrowed model adjusts well to seeds of *Tetrastigma hensleyanum* (not shown).

3.2.2. Heart Curves

Simultaneous modifications in both terms of Equation (2) result in a variety of heart curves. Heart curves are obtained with selected values of $a,\ b,\ c$ in:

$$\left(\sqrt{1-x^2} - \frac{a}{54x^2+9|x|+3} + y\right)\left(\sqrt{1-x^2} + \frac{b}{54x^2+9|x|+3} - y\right) = 0 \tag{4}$$

For example, Model ACO1 (see Figure 5) resulted from Equation (4) with $a = 1/2$; $b = 1/3$. Increasing a, reduces the size of the lower entry; increasing b, reduces the upper beak.

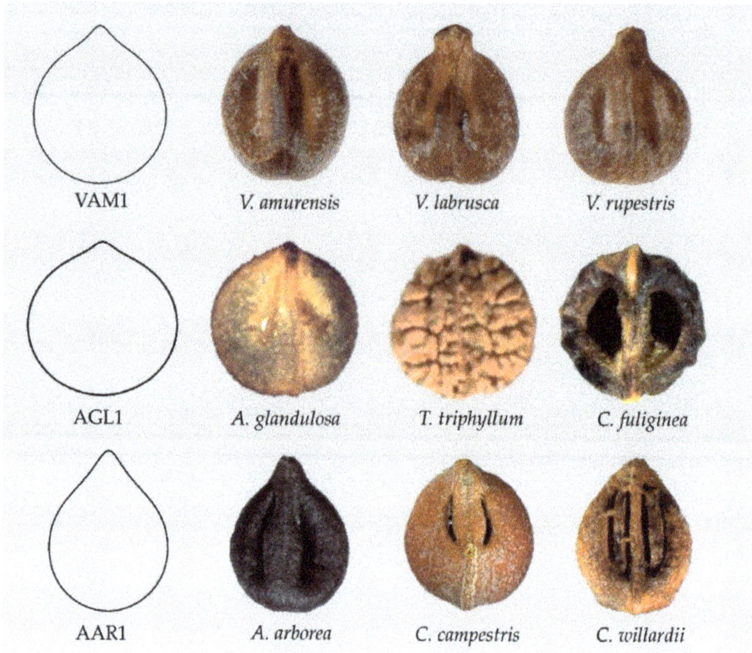

Figure 4. Seed images resembling water drops. From top to bottom: Model VAM1 (*Vitis amurensis*, *V. labrusca*, *V. rupestris*), Model AGL1 (*Ampelopsis glandulosa*, *Tetrastigma triphyllum*, *Cissus fuliginea*), Model AAR1 (*Ampelopsis arborea*, *Cissus campestris*, *C. willardii*).

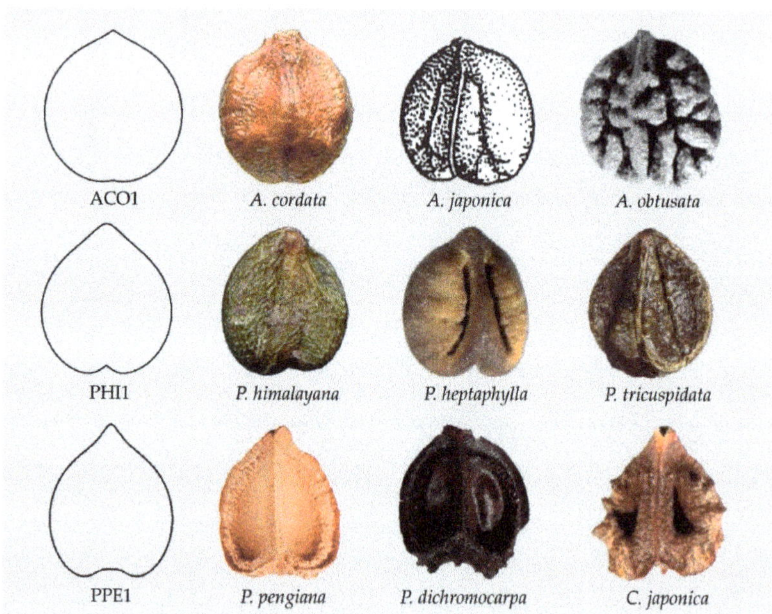

Figure 5. Seeds resembling heart curves. From top to bottom: Model ACO1 (*Ampelopsis cordata*, *A. japonica*, *A. obtusata*), Model PHI1 (*Parthenocissus himalayana*, *P. heptaphylla*, *P. tricuspidata*), Model PPE1 (*Pseudocayratia pengiana*, *P. dichromocarpa*, *Cayratia japonica*).

Model PHI1 (see Figure 5) was obtained by the following modification in Equation (4):

$$\left(\sqrt{1-x^2}+\frac{a}{10x^2+1}-y\right)\left(\sqrt{1-x^2}-\frac{b}{54x^2+9|x|+3}-y\right)=0 \quad (5)$$

with $a = 1/3$; $b = 1/2$.

Model PPE1 resulted from:

$$\left(\sqrt{1-x^2}+\frac{a}{5x^4+25x^2+1}-\frac{9}{10}y\right)\left(\sqrt{1-x^2}-\frac{b}{5x^4+25|x|^3+1}-\frac{9}{10}y\right)=0 \quad (6)$$

with $a = b = 1/3$. (Data Availability Statement section).

Changes in the above equations give modified water drop and heart curves. For example, changes in Equation (3) result in broadened heart curves, giving (i) Models ARO1, AJA1 and AER1 that describe the seeds of *Ampelocissus robinsonii, A. javalensis A. erdvendbergiana*, respectively; and (ii) Models AGR1 and ADE1 that resemble the seeds of *Ampelopsis grossdentata* and *A. denudata* (AGR1) and *A. delavayana* and *A. cantoniensis* (ADE1) respectively, see Figure 6 and Data Availability Statement section. The significance of differences between apparently similar models, such as PPE1 and AGR1, can be tested quantitatively on samples with a sufficient number of seeds. In principle, the difference at the basis of the figures (more flat and with a plane entry in PPE1) justifies the separation of the two models.

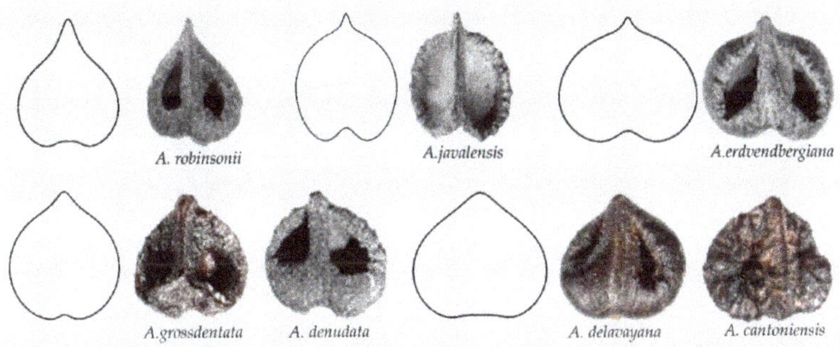

Figure 6. Models of broadened heart curves and seeds resembling each of them in *Ampelocissus* and *Ampelopsis*. (**Top**): Model ARO1, *Ampelocissus robinsonii*, Model AJA1, *Ampelocissus javalensis*, Model AER1, *A. erdvendbergiana*. (**Bottom**): Model AGR1, *Ampelopsis grossedentata, A. denudata*; Model ADE1, *A. delavayana, A. cantoniensis*.

Departing from the models described, it is possible to find new figures specific for seeds in other species; for example, seeds of *Ampelocissus cavicaulis, A. macrocirrha*, and *A. ochracea* share with *A. javalensis* the Squared Heart Curve (SqHC) type related to Model AJA1. Other models, such as AGR2, may fit better the shape of seeds of *A. grantii* and *A. latifolia* (Figure 7).

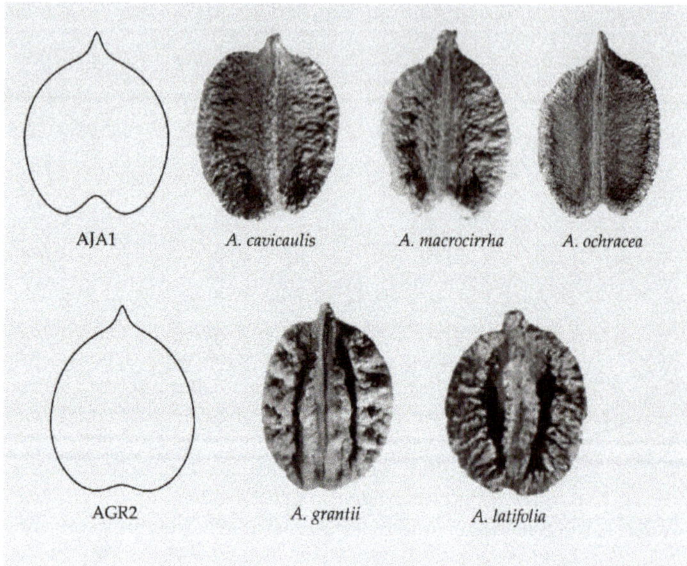

Figure 7. The seeds of many species of *Ampelocissus* are related with the Squared Heart Curve (SqHC). Model AJA1 adapts well to *Ampelocissus cavicaulis, A. macrocirrha* and *A. ochracea*. Model AGR2 represents better the shape of seeds of *A. grantii* and *A. latifolia*.

3.2.3. Pear Curves and Other Elongated Models

Some seed images resemble water drops in their overall shape; nevertheless, they show broader basis than waterdrops. Figure 8 shows the models YAU1, COL1, VAE1 and AME1 that correspond respectively to the shapes of *Yua austro-orientalis* and *Cissus trianae* (YAU1), *Cayratia oligocarpa* (COL1), *Yua chinensis* and *Vitis aestivalis* (VAE1) and *Ampelopsis megalophylla* (AME1). These four models share similar values of aspect ratio that justify the inclusion of model COL1 here with preference to the Squared Heart Curve (SqHC) group. Seeds of some species of *Ampelocissus* (e.g., *A. acapulcensis, A. bombycina, A. bravoi* can fit either model COL1 or models derived from it (not shown)).

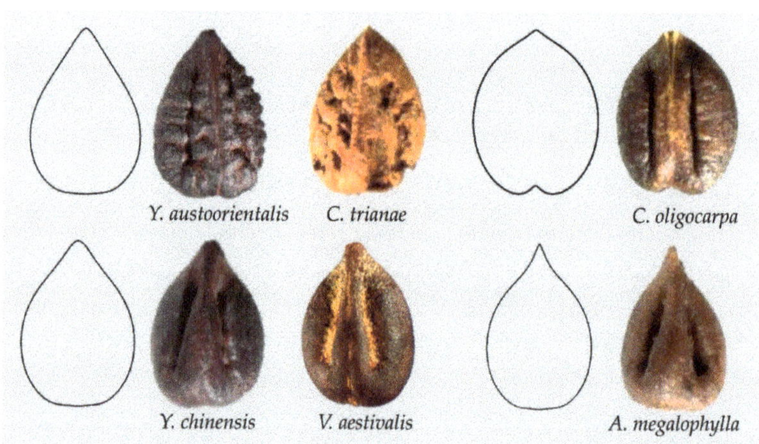

Figure 8. Models YAU1, COL1, VAE1 and AME1 with their respective seeds. *Yua austro-orientalis* and *Cissus trianae* (YAU1), *Cayratia oligocarpa* (COL1), *Yua chinensis* and *Vitis aestivalis* (VAE1) and *Ampelopsis megalophylla* (AME1).

The seeds of *Rhoicissus rhomboidea* and of many species of *Tetrastigma* resemble polarized ellipses with an end bi-lobulated and the other acute. We have termed this morphological group as the Elongated Superellipse-Heart Curve (ESHC) (Figure 9).

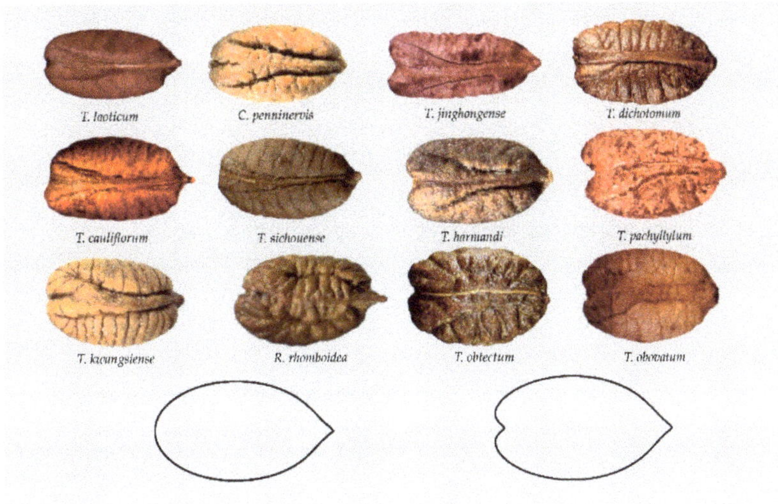

Figure 9. Seeds of *Rhoicissus rhomboidea* and many species of *Tetrastigma* resemble polarized ellipses with a side rounded or bi-lobulated and the other acute.

3.3. A Summary of Geometric Types in Seeds of the Vitaceae

Table 3 contains a summary of the morphological types for the Vitaceae according to the similarity of the seeds with geometrical figures. Ten morphological groups are described, three being present also in the Arecaceae [13] (lenses, superellipses and elongated water drops; termed respectively G I, G II and G III in Figure 10), three additional groups were described before for the Vitaceae [45] (water drops, heart curves and elongated heart curves; named G IV, G V and G VI in Figure 10), and four new groups are based on new models original from this work (G VII to G X). Three of the later models are particular for some genera and species. These are: (1) Heart curves of the *Cayratia* and *Pseudocayratia* types, with a marked entry at the basis and an acute protuberance (G VIII in Figure 10); (2) heart curves of the Squared Heart Curves (SqHCs) type in *Ampelocissus* and broadened models of *Ampelocissus* and *Ampelopsis* (G IX in Figure 10), and (3) Elongated Superellipse-Heart Curves (ESHCs), frequent in *Tetrastigma* species and observed also in *Cissus* species and *R. rhomboidea* (G X in Figure 10).

Table 3. A summary of groups based on morphological seed types for the analysed species in the Vitaceae. The number of cases found in each group is given between dashes.

Group (Geometric Model)	Examples
Group I (Lenses)-3-	*Cissus quadrangularis* [44], *C. sterculiifolia* [31], *Tetrastigma petraeum* [52]
Group II (Superellipses)-7-	*Ampelocissus bravoi* [42], *C. reniformis* [31,41], *Cyphostemma elephantopus* [49], *C. laza* [31], *Tetrastigma campylocarpum* [52], *T. caudatum* [52], *T. henryi* [52]
Group III (Elongated water drops)-15-	*Ampelopsis arborea* [37], *Cayratia imerinensis* [47], *Cissus aralioides* [35,40], *C. cornifolia* [40], *C. erosa* [43], *C. integrifolia* [40,42], *C. petiolata* [40], *C. pileata* [40], *C. populnea* [40], *C. verticillata* [31,41,42,45], *C. sciaphila* [40], *C. smithiana* [40], *C. willardii* [42], *Cyphostemma junceum* [31], *V. vulpina* [59,62]
Group IV (Water drops, normal or rounded)-14-	*Ampelopsis bodinieri* [36], *A. glandulosa* [36,39], *A. humulifolia* [36], *Cayratia cheniana* [46], *Cissus campestris* [31,41,42], *C. fuliginea* [31], *C. tuberosa* [42], *C. granulosa* [31], *Parthenocissus dalzielii* [36], *Tetrastigma triphyllum* [31,52], *Vitis amurensis* [45,56], *V. labrusca* [45,54,59], *V. palmata* [60], *V. rupestris* [45]
Group V (Heart curves normal or rounded)-19-	*Ampelopsis aconitifolia* [36], *A. chaffanjoni* [36], *A. cordata* [38], *A. japonica* [36], *Parthenocissus heptaphylla* [31], *P. heterophylla* [36], *P. henryana* [36], *P. himalayana* [52,53], *P. quinquefolia* [54], *P. vitacea* [31], *P. tricuspidata* [36,37,45], *Rhoicissus revoilii* [31], *T. lanceolarium* [30], *Vitis cinerea* [57], *V. flexuosa* [54], *V. lanatoides* [58], *V. latisulcata* [58], *V. tsoi* [31,41], *V. wilsoniae* [31,41]
Group VI (Elongated Heart curves)-6-	*Ampelocissus acapulcensis* [30], *Cissus oligocarpa* [31]. *V. eolabrusca* [54], *V. grayensis* [58], *V. pseudorotundifolia* [54], *V. tiliifolia* [61]
Group VII (Other elongated types)-11-	*Ampelopsis megalophylla* [31,36], *Causonis* sp. [46], *Cayratia saponaria* [31], *Cissus trianae* [31], *C. hypoglauca* [31], *Parthenocissus laetevirens* [36], *T. hypoglaucum* [52] *Vitis aestivalis* [55], *V. rotundifolia* [31,41,54], *Yua austro-orientalis* [31], *Y. chinensis* [31]
Group VIII (Heart curves of the *Cayratia* and *Pseudocayratia* types)-7-	*Cayratia japonica* [31,48], *Cayratia* sp. [African, [46]], *Pseudocayratia dichromocarpa* [50], *P. pengiana* [50], *P. speciosa* [50,52], *Tetrastigma formosanum* [51], *T. pedunculare* [31,51,52]
Group IX (Heart curves of the SqHC type of *Ampelocissus* and *Ampelopsis*)-15-	*Ampelocissus bombycina* [30], *A. cavicaulis* [30], *A. erdvendbergiana* [30], *A. grantii* [30], *A. javalensis* [30,42], *A. latifolia* [30], *A. macrocirrha* [30], *A. martinii* [42], *A. obtusata* [30], *A. ochracea* [30], *A. robinsonii* [30], *Ampelopsis cantoniensis* [31,36], *A. delavayana* [31], *A. denudata* [30], *A. grossedentata* [31]
Group X Elongated Superellipse-heart curves-16-	*Cissus elongata* [40], *C. penninervis* [31], *Rhoicissus rhomboidea* [31], *Tetrastigma hemsleyanum* [31,52], *T. jinghongense* [52], *T. laoticum* [52], *T. cauliflorum* [52], *T. dichotomum* [51], *T. harmandi* [31], *T. pachyllylum* [52], *T. kwangsiense* [30,31], *T. obovatum* [51,52], *T. obtectum* [51,52], *T. retinervum* [51], *T. serrulatum* [52], *T. sichouense* [52]
Undefined-18-	*Cayratia geniculata* [31], *Cissus antarctica* [31], *C. barbeyana* [40], *C. bosseri* [40], *C. cactiformis* [40], *C. descoingsii,* [31,41], *C. diffusiflora* [40], *C. floribunda* [40], *C. hastata* [40], *C. leucophlea* [40], *C. repens* [40], *C. subtetragona* [40], *T. delavayi* [52], *T. rumicispermum* [31,51,52], T. *thorsborneorum* [52], *T. xishuangbannaense* [31,52], *V. brandoniana* [54], *V. rostrata* [54]

In general, the distribution of morphological types is not in close agreement with the current taxonomic classification; nevertheless, some results may be summarized in this aspect. First, the seeds of the Elongated Superellipse-Heart Curves (ESHCs) type (Group X) are more frequent in *Tetrastigma* and have been observed in *Rhoicissus* and *Cissus*, but not in species of other genera. While many seeds in species of *Ampelopsis, Parthenocissus* and *Vitis* share the typical shapes of water drop and heart curves, the squared heart curve (SqHC) type (Group IX) has been predominantly observed in *Ampelocissus* and *Ampelopsis*. A number of species remain undefined due to one of these two reasons: First, their irregular seed shape making difficult the identification of an adequate model (*Cayratia geniculata, Cissus antarctica*) and, second, the seed images have geometric shapes but the identification of the model with the corresponding equation is pending (*Tetrastigma delavayi, T. rumicispermum*). In addition, further work will be done on the seeds of *Vitis* species.

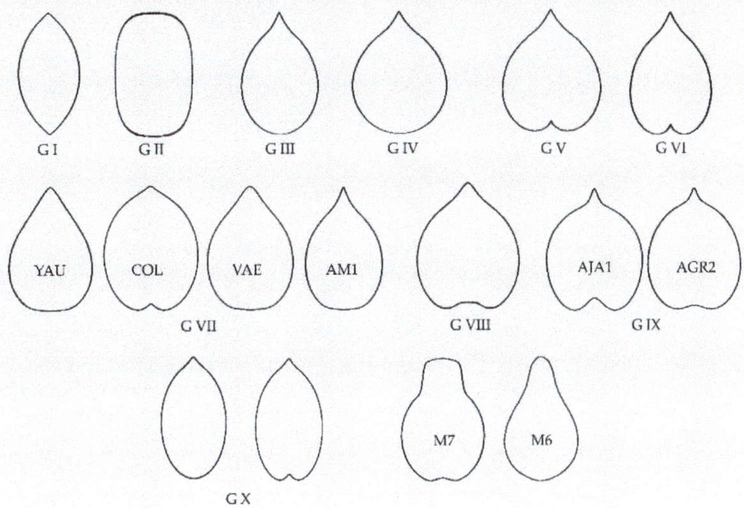

Figure 10. A summary of the models found for the description and quantification of seed shape in the Vitaceae. G I, G II and G III are lenses, superellipses and elongated water drops, respectively; G IV, G V and G VI correspond to water drops, heart curves and elongated heart curves, respectively; G VII contains four models corresponding to other elongated curves; G VIII presents an example of the heart curves of the *Cayratia* and *Pseudocayratia* types; G IX, heart curves of the Squared Heart Curves (SqHCs) type in *Ampelocissus* and broadened models of *Ampelocissus* and *Ampelopsis,* and G X, Elongated Superellipse-Heart Curves (ESHCs), frequent in *Tetrastigma* species and observed also in *Cissus* species and *R. rhomboidea.* Labelled as M7 and M6 are two models used in the description of seeds of grape varieties and as precursors for other models [45,65].

4. Discussion

Morphology has not received the attention due in recent decades due to the increased emphasis on molecular approaches, but the importance of descriptive aspects is rising [68]. Seed morphology in particular may provide the basis for developments in Ecology and Evolution. The morphological analysis shows a similarity in seed shape between two families from the Core Angiosperms that are not related traditionally by taxonomic criteria: the Arecaceae and the Vitaceae. Both families belong to very different clades, the Vitaceae to the Eudicot clade and the Arecaceae to the Monocot clade [1,6,69,70], and although the embryos of the former have two cotyledons and the latter have only one, their similarities in seed shape are in agreement with both families having endospermic seeds [3,4,70]. The seeds of the Arecaceae and the Vitaceae present a great diversity, including a combination of forms relatively infrequent in other plant families. Ellipses, ovals and cardioids are geometric forms frequent in plant families [18]; in contrast, superellipses are not so frequent. These adjust better to intermediate shapes between ellipses and rectangles. Other figures shared by the seeds of the Vitaceae and the Arecaceae are lenses and water drops of diverse proportions.

A distinctive aspect of seed morphology in the Vitaceae is the adjustment of their seeds to a diversity of water drops, heart curves and related figures. A series of variations derived from the equation of an ellipse have been described; their graphical representations give water drops and heart curves resembling with precision the seeds of diverse species of the Vitaceae. Both types of figures can be described as products from the modification of ellipses to obtain one pole acute and the other rounded or bi-lobulated.

The new models here described define the seed silhouettes of species in the diverse subfamilies of the Vitaceae. In addition to three groups based on models shared with the

Arecaceae, and three other groups described before (water drops, heart curves, elongated heart curves) [45], four new groups have been described. One of them corresponds to other types of elongated curves, and the remaining three are more specific. These correspond to: (1) Heart curves of the *Cayratia* and *Pseudocayratia* types, with a marked entry at the basis and an acute protuberance; (2) heart curves of the Squared Heart Curves (SqHCs) type in *Ampelocissus* and broadened models of *Ampelocissus* and *Ampelopsis*, and (3) Elongated Superellipse-Heart Curves (ESHCs), frequent in *Tetrastigma* species and observed also in *Cissus* species and *R. rhomboidea*.

The importance of seed morphology in taxonomy has been described for a long time, see, for example, [71]. Nevertheless, not all characters of seed morphology have the same relevance, and, in many instances, the lack of a morphological diagnostic key may be due to a high degree of homoplasy [72,73]. The similarity of the seed silhouette to a geometric model is the result of a complex process of development, and thus less submitted to homoplasy; in consequence, it may be a good character in taxonomy. In addition, the visualization of geometric figures that share the form of seeds may contribute to their classification complementing the results of artificial vision techniques [74–76].

In addition to taxonomy, classification based on seed shape acquires relevance in other research areas. Members of both families, the Arecaceae and the Vitaceae, were present in the Neotropical flora in the Eocene [14], and their fruits have been in the human diet for a long time [75,76]. Additionally, both families have been studied by means of phytoliths, microfossils useful in archaeobotany and archaeology [77,78].

In the first paragraph of the introduction to his book entitled *L'Évolution créatrice*, Henri Bergson recognized the importance of Geometry stating that *nôtre intelligence triomphe dans la géometrie, où se révelè la parenté de la pensée logique avec la matière inerte* [79]. Unfortunately, in the following pages of this text, the author abandoned the study of geometry, a model of precision, to enter the rhetorics of evolution. An approach to seed geometry in palms and grapes could support the words of P.B. Tomlinson (1990): "Palms are not then merely emblematic of the tropics, they are emblematic of how the structural biology of plants must be understood before evolutionary scenarios can be reconstructed" [80], quoted in [69].

5. Conclusions

Ten morphological types are described in the Vitaceae. Seven of them are general and three specific. Among the general types, three are shared with the Arecaceae and correspond to geometric figures well described (lenses, superellipses and elongated waterdrops). Four additional groups include waterdrops, normal or rounded, heart curves, normal or rounded, elongated heart curves and other elongated curves, respectively. Finally, the three specific types correspond to heart curves of the *Cayratia* and *Pseudocayratia* types, heart curves of the Squared Heart Curve (SqHC) type of *Ampelocissus* and *Ampelopsis*, and Elongated Superellipse-Heart Curves (ESHCs), frequent in *Tetrastigma* species and observed also in *Cissus* species and *R. rhomboidea*. All these groups are defined by geometric models obtained by the representation of algebraic equations. Modifications in the equations result in models adjusting to the shape of seeds for each species.

Author Contributions: Conceptualization, E.C., Á.T.; methodology, E.C., J.J.M.-G., D.G.d.P., Á.T.; software, J.J.M.-G., Á.T.; validation, E.C., J.J.M.-G., D.G.d.P., Á.T.; formal analysis, E.C., J.J.M.-G., D.G.d.P., Á.T.; investigation, E.C., J.J.M.-G., D.G.d.P., Á.T.; resources, E.C., J.J.M.-G., D.G.d.P., Á.T.; data curation, J.J.M.-G.; writing—original draft preparation, E.C.; writing—review and editing, E.C., J.J.M.-G., D.G.d.P., Á.T.; visualization, J.J.M.-G. All authors have read and agreed to the published version of the manuscript.

Funding: Project "CLU-2019-05-IRNASA/CSIC Unit of Excellence", funded by the Junta de Castilla y León and co-financed by the European Union (ERDF "Europe drives our growth").

Data Availability Statement: The Mathematica code for Geometric Models is given in: https://zenodo.org/record/4942111#.YMbkrfkzaM8.

Conflicts of Interest: The authors declare no conflict of interest.

References

1. The Angiosperm Phylogeny Group; Chase, M.W.; Christenhusz, M.J.; Fay, M.F.; Byng, J.W.; Judd, W.S.; Soltis, D.E.; Mabberley, D.J.; Sennikov, A.N.; Soltis, P.S.; et al. An update of the Angiosperm Phylogeny Group classification for the orders and families of flowering plants: APG IV. *Bot. J. Linn. Soc.* **2016**, *181*, 1–20. [CrossRef]
2. Ridsdale, C.E. A revision of the family Leeaceae. *Blumea* **1974**, *22*, 57–100.
3. Wen, J.; Lu, L.-M.; Nie, Z.-L.; Liu, X.-Q.; Zhang, N.; Ickert-Bond, S.; Gerrath, J.; Manchester, S.R.; Boggan, J.; Chen, Z.-D. A new phylogenetic tribal classification of the grape family (Vitaceae). *J. Syst. Evol.* **2018**, *56*, 262–272. [CrossRef]
4. Zhang, N.; Wen, J.; Zimmer, E.A. Congruent deep relationships in the grape family (Vitaceae) based on sequences of chloroplast genomes and mitochondrial genes via genome skimming. *PLoS ONE* **2015**, *10*, e0144701.
5. Zeng, L.P.; Zhang, N.; Zhang, Q.; Endress, P.K.; Huang, J.; Ma, H. Resolution of deep eudicot phylogeny and their temporal diversification using nuclear genes from transcriptomic and genomic datasets. *New Phytol.* **2017**, *214*, 1338–1354. [CrossRef] [PubMed]
6. Wen, J. Vitaceae. In *The Families and Genera of Vascular Plants*; Kubitzki, K., Ed.; Springer: Berlin, Germany, 2007; pp. 467–479.
7. Planchon, J.E. Monographie des Ampélidées vrais. In *Monographiae Phanaerogamarum 5*; de Candolle, A.P., de Candolle, A.C.P., Eds.; Masson: Paris, France, 1887; pp. 305–654.
8. Rossetto, M.; Jackes, B.R.; Scott, K.D.; Henry, R.J. Is the genus *Cissus* (Vitaceae) monophyletic? Evidence from Plastid and Nuclear Ribosomal DNA. *Syst. Bot.* **2002**, *27*, 522–533. [CrossRef]
9. Hearn, D.J.; Evans, M.; Wolf, B.; McGinty, M.; Wen, J. Dispersal is associated with morphological innovation, but not increased diversification, in *Cyphostemma* (Vitaceae). *J. Syst. Evol.* **2018**, *56*, 340–359. [CrossRef]
10. Terral, J.F.; Tabard, E.; Bouby, L.; Ivorra, S.; Pastor, T.; Figueiral, I.; Picq, S.; Chevance, J.-B.; Jung, C.; Fabre, L.; et al. Evolution and history of grapevine (*Vitis vinifera*) under domestication: New morphometric perspectives to understand seed domestication syndrome and reveal origins of ancient European cultivars. *Ann. Bot.* **2010**, *105*, 443–455. [CrossRef]
11. This, P.; Lacombe, T.; Thomas, M.R. Historical origins and genetic diversity of wine grapes. *Trends Genet.* **2006**, *22*, 511–519. [CrossRef]
12. Encyclopedia Britannica. Vitales. Available online: https://www.britannica.com/plant/Vitales (accessed on 25 July 2021).
13. Gutiérrez del Pozo, D.; Martín-Gómez, J.J.; Tocino, Á.; Cervantes, E. Seed geometry in the Arecaceae. *Horticulturae* **2020**, *6*, 64. [CrossRef]
14. Herrera, F.; Manchester, S.R.; Jaramillo, C. Permineralized fruits from the late Eocene of Panama give clues of the composition of forests established early in the uplift of Central America. *Rev. Palaeobot. Palynol.* **2012**, *175*, 10–24. [CrossRef]
15. Schneider, C.A.; Rasband, W.S.; Eliceiri, K.W. NIH Image to ImageJ: 25 years of image analysis. *Nat. Methods* **2012**, *9*, 671–675. Available online: http://imagej.nih.gov/ij/docs/guide (accessed on 20 January 2021). [CrossRef] [PubMed]
16. Cervantes, E.; Martín-Gómez, J.J.; Saadaoui, E. Updated methods for seed shape analysis. *Scientifica* **2016**, 5691825. [CrossRef] [PubMed]
17. Cervantes, E.; Martín-Gómez, J.J. Seed shape description and quantification by comparison with geometric models. *Horticulturae* **2019**, *5*, 60. [CrossRef]
18. Cervantes, E.; Martín-Gómez, J.J.; del Pozo, D.G.; Silva Días, L. An angiosperm species dataset reveals relationships between seed size and two-dimensional shape. *Horticulturae* **2019**, *5*, 71. [CrossRef]
19. Cervantes, E.; Martín-Gómez, J.J.; Ardanuy, R.; de Diego, J.G.; Tocino, Á. Modeling the *Arabidopsis* seed shape by a cardioid: Efficacy of the adjustment with a scale change with factor equal to the Golden Ratio and analysis of seed shape in ethylene mutants. *J. Plant. Physiol.* **2010**, *167*, 408–410. [CrossRef] [PubMed]
20. Martín Gómez, J.J.; Tocino, Á.; Ardanuy, R.; de Diego, J.G.; Cervantes, E. Dynamic analysis of *Arabidopsis* seed shape reveals differences in cellulose mutants. *Acta Physiol. Plant.* **2014**, *36*, 1585–1592. [CrossRef]
21. Cervantes, E.; Martín-Gómez, J.J.; Chan, P.K.; Gresshoff, P.M.; Tocino, Á. Seed shape in model legumes: Approximation by a cardioid reveals differences in ethylene insensitive mutants of *Lotus japonicus* and *Medicago truncatula*. *J. Plant Physiol.* **2012**, *169*, 1359–1365. [CrossRef]
22. Saadaoui, E.; Martín-Gómez, J.J.; Cervantes, E. Intraspecific variability of seed morphology in *Capparis spinosa* L. *Acta Biol. Cracov. Bot.* **2013**, *55*, 99–106. [CrossRef]
23. Saadaoui, E.; Martín-Gómez, J.J.; Tlili, N.; Khaldi, A.; Cervantes, E. Effect of climate in seed diversity of wild Tunisian *Rhus tripartita* (Ucria) Grande. *J. Adv. Biol. Biotechnol.* **2017**, *13*, 1–10. [CrossRef]
24. Martín-Gómez, J.J.; Saadaoui, E.; Cervantes, E. Seed shape of castor bean (*Ricinus communis* L.) grown in different regions of Tunisia. *J. Agric. Ecol. Res. Int.* **2016**, *8*, 1–11.
25. Saadaoui, E.; Martín, J.J.; Bouazizi, R.; Chokri, B.R.; Grira, M.; Abdelkabir, S.; Khouja, M.L.; Cervantes, E. Phenotypic variability and seed yield of *Jatropha curcas* L. introduced to Tunisia. *Acta Botánica Mex.* **2015**, *110*, 119–134. [CrossRef]
26. Martín-Gómez, J.J.; Rewicz, A.; Goriewa-Duba, K.; Wiwart, M.; Tocino, Á.; Cervantes, E. Morphological description and classification of wheat kernels Based on geometric models. *Agronomy* **2019**, *9*, 399. [CrossRef]
27. Cervantes, E.; Martín-Gómez, J.J. Seed shape quantification in the order Cucurbitales. *Mod. Phytomorphol.* **2018**, *12*, 1–13. [CrossRef]
28. Martín-Gómez, J.J.; Rewicz, A.; Cervantes, E. Seed shape diversity in families of the order Ranunculales. *Phytotaxa* **2019**, *425*, 193–207. [CrossRef]

29. Martín-Gómez, J.J.; del Pozo, D.G.; Cervantes, E. Seed shape quantification in the Malvaceae reveals cardioid-shaped seeds predominantly in herbs. *Botanica* **2019**, *25*, 21–31. [CrossRef]
30. Chen, I.; Manchester, S.R. Seed morphology of modern and fossil *Ampelocissus* (Vitaceae) and implications for phytogeography. *Am. J. Bot.* **2007**, *94*, 1534–1553. [CrossRef] [PubMed]
31. Chen, I.; Manchester, S.R. Seed morphology of Vitaceae. *Int. J. Plant. Sci.* **2011**, *172*, 1–35. [CrossRef]
32. Corner, E.J.H. *The Seeds of Dicotyledons*; Cambridge University Press: Cambridge, UK, 1976; Volume 1.
33. Trias-Blasi, A.; Parnell, J.A.N.; Chayamarit, K. *Cissus amplexicaulis* (Vitaceae), a new endemic species from Thailand. *Kew Bull.* **2010**, *65*, 487–490. [CrossRef]
34. Kew Science. Plants of the World Online. *Cissus aphylla* Chiov. Available online: http://powo.science.kew.org/taxon/urn:lsid:ipni.org:names:869770-1 (accessed on 9 July 2021).
35. Kew Science. Plants of the World Online. *Cissus aralioides* (Welw. ex Baker) Planch. Available online: http://powo.science.kew.org/taxon/urn:lsid:ipni.org:names:67514-1 (accessed on 9 July 2021).
36. Latiff, A. Studies in Malesian Vitaceae VII. The genus *Tetrastigma* in the Malay Peninsula. *Gard. Bull. Singap.* **1984**, *36*, 213–228.
37. Manchester, S.R.; Kapgate, D.K.; Wen, J. Oldest fruits of the grape family (Vitaceae) from the Late Cretaceous Deccan Cherts of India. *Am. J. Bot.* **2013**, *100*, 1849–1859. [CrossRef] [PubMed]
38. Native Plant Trust: Go Botany. *Ampelopsis cordata* Michx. Available online: https://gobotany.nativeplanttrust.org/species/ampelopsis/cordata/ (accessed on 9 July 2021).
39. Norfolk Cottage Garden Seeds & Plants. *Ampelopsis glandulosa* var. Brevipedunculata. Available online: https://www.norfolkcottagegarden.co.uk/images/seeds/macro/ampelopsis-glandulosa-var-brevipedunculata_seed-macro_img148_960px.jpg (accessed on 9 July 2021).
40. Adams, N.F.; Collinson, M.E.; Smith, S.Y.; Bamford, M.K.; Forest, F.; Malakasi, P.; Marone, F.; Sykes, D. X-rays and virtual taphonomy resolve the first *Cissus* (Vitaceae) macrofossils from Africa as early-diverging members of the genus. *Am. J. Bot.* **2016**, *103*, 1657–1677. [CrossRef] [PubMed]
41. Chen, I. History of Vitaceae Inferred from Morphology-Based Phylogeny and the Fossil Record of Seeds. Ph.D. Thesis, University of Florida, Gainesville, FL, USA, 2009.
42. Manchester, S.R.; Chen, I.; Lot, T.A. Seeds of *Ampelocissus*, *Cissus* and *Leea* (Vitales) from the Paleogene of Western Peru and their Biogeographic significance. *Int. J. Plant. Sci.* **2012**, *173*, 933–943. [CrossRef]
43. Smithsonian Tropical Research Institute. *Cissus Erosa* Subsp *Erosa*. Available online: https://biogeodb.stri.si.edu/bioinformatics/dfm/metas/view/7894 (accessed on 9 July 2021).
44. DELTA—Description Language for Taxonomy. *Cissus quadrangularis*. Available online: https://www.delta-intkey.com/famfs/images/vitace03.jpg (accessed on 9 July 2021).
45. Martín-Gómez, J.J.; Gutiérrez del Pozo, D.; Ucchesu, M.; Bacchetta, G.; Cabello Sáenz de Santamaría, F.; Tocino, Á.; Cervantes, E. Seed morphology in the Vitaceae based on geometric models. *Agronomy* **2020**, *10*, 739. [CrossRef]
46. Lu, L.; Wen, J.; Chen, Z. *Cayratia cheniana* (Vitaceae): An endangered new species endemic to the limestone mountains of Ninh Thuan Province, Vietnam. *Syst. Bot.* **2016**, *41*, 49–55. [CrossRef]
47. Descoings, B. *Flore de Madagascar et des Comores (Plantes Vasculaires) 124 Famille—Vitacées*; Museum National d'Histoire Naturelle: Paris, France, 1967. Available online: https://www.delta-intkey.com/angio/images/cayratia.gif (accessed on 9 July 2021).
48. Jiang, H.; Yang, J.; Liang, T.; Zhang, Z.; Wang, S.; Qi, X.; Sheng, P. Palaeoethnobotanical analysis of plant remains discovered in the graveyard of the Haihun Marquis, Nanchang, China. *Veg. Hist. Archaeobot.* **2021**, *30*, 119–135. [CrossRef]
49. The World of Plants: *Cyphostemma elephantopus*. Available online: https://public.fotki.com/PanosS/vitaceae-pages/cyphostemma/cyphostema_elephant.html (accessed on 9 July 2021).
50. Wen, J.; Lu, L.-M.; Hsu, T.-W.; Dang, V.-C.; Habib, S.; Boggan, J.K.; Okada, H.; Chen, I.-J.; Chen, Z.-D. *Pseudocayratia*, a new genus of Vitaceae from China and Japan with two new species and three new combinations. *J. Syst. Evol.* **2018**, *56*, 374–393. [CrossRef]
51. Habib, S.; Dang, V.-C.; Ickert-Bond, S.M.; Wen, J.; Chen, Z.-D.; Lu, L.-M. Evolutionary trends in *Tetrastigma* (Vitaceae): Morphological diversity and taxonomic implications. *J. Syst. Evol.* **2018**, *56*, 360–373. [CrossRef]
52. Habib, S.; Dang, V.-C.; Ickert-Bond, S.M.; Zhang, J.-L.; Lu, L.-M.; Wen, J.; Chen, Z.-D. Robust phylogeny of *Tetrastigma* (Vitaceae) based on ten plastid DNA regions: Implications for infrageneric classification and seed character evolution. *Front. Plant. Sci.* **2017**, *8*, 590. [CrossRef]
53. Norfolk Cottage Garden Seeds and Plants. *Parthenocissus himalayana*. Available online: https://www.norfolkcottagegarden.co.uk/images/seeds/macro/parthenocissus-himalayana-syn-semicordata_seed-macro_518_960px.jpg (accessed on 9 July 2021).
54. Tiffney, B.H.; Barghoorn, E.S. Fruits and seeds of the Brandon lignite. I. Vitaceae. *Rev. Palaeobot. Palynol.* **1976**, *22*, 169–191. [CrossRef]
55. Plants USDA. *Vitis aestivalis*. Available online: https://plants.sc.egov.usda.gov/home/plantProfile?symbol=VIAE (accessed on 9 July 2021).
56. Kirkbride, J.H., Jr.; Gunn, C.R.; Dallwitz, M.J. 2000 Onwards. Family Guide for Fruits and Seeds: Descriptions, Illustrations, Identification, and Information Retrieval; Version: 12th April 2021. Available online: https://www.delta-intkey.com/famfs/images/vitace04.jpg. (accessed on 9 July 2021).
57. Plants USDA. *Vitis cinerea*. Available online: https://plants.sc.egov.usda.gov/home/plantProfile?symbol=VICI2 (accessed on 9 July 2021).

58. Gong, F.; Karsai, I.; Liu, Y.-S. *Vitis* seeds (Vitaceae) from the late Neogene Gray Fossil Site, northeastern Tennessee, USA. *Rev. Palaeobot. Palynol.* **2010**, *162*, 71–83. [CrossRef]
59. Ardenghi, N.M.G.; Galasso, G.; Banfi, E.; Cauzzi, P. *Vitis* ×*novae-angliae* (Vitaceae): Systematics, distribution and history of an "illegal" alien grape in Europe. *Willdenowia* **2015**, *45*, 197–207. [CrossRef]
60. Plants USDA. *Vitis palmata*. Available online: https://plants.sc.egov.usda.gov/home/plantProfile?symbol=VIPA7 (accessed on 9 July 2021).
61. Smithsonian Tropical Research Institute. *Vitis tiliifolia*. Available online: https://biogeodb.stri.si.edu/bioinformatics/dfm/metas/view/10813?&lang=es (accessed on 9 July 2021).
62. Plants USDA. *Vitis vulgaris*. Available online: https://plants.sc.egov.usda.gov/home/plantProfile?symbol=VIVU (accessed on 9 July 2021).
63. Weisstein, E.W. "Heart Curve." From MathWorld—A Wolfram Web Resource. Available online: http://mathworld.wolfram.com/HeartCurve.html (accessed on 19 May 2021).
64. Weisstein, E.W. "Pear Curve." From MathWorld—A Wolfram Web Resource. Available online: http://mathworld.wolfram.com/PearCurve.html (accessed on 19 May 2021).
65. Cervantes, E.; Martín-Gómez, J.J.; Espinosa-Roldán, F.E.; Muñoz-Organero, G.; Tocino, Á.; Cabello-Sáenz de Santamaría, F. Seed morphology in key Spanish grapevine cultivars. *Agronomy* **2021**, *11*, 734. [CrossRef]
66. Gielis, J. A generic geometric transformation that unifies a wide range of natural and abstract shapes. *Am. J. Bot.* **2003**, *90*, 333–338. [CrossRef]
67. Gridgeman, N.T. Lamé Ovals. *Math. Gaz.* **1970**, *54*, 31–37. [CrossRef]
68. Christodoulou, M.D.; Clark, J.Y.; Culham, A. The Cinderella discipline: Morphometrics and their use in botanical classification. *Bot. J. Linn. Soc.* **2020**, *194*, 1–12. [CrossRef]
69. Judd, W.S.; Campbell, C.S.; Kellogg, E.A.; Stevens, P.F.; Donoghue, M.J. *Plant Systematics: A Phylogenetic Approach*, 2nd ed.; Sinauer Assoc.: Sunderland, MA, USA, 2002.
70. Dransfield, J.; Uhl, N.W.; Asmussen, C.B.; Baker, W.J.; Harley, M.M.; Lewis, C.E. *Genera Palmarum: The Evolution and Classification of Palms*; Kew Publishing; Royal Botanic Gardens of Kew: Kew, UK, 2008.
71. Rohrbach, P. *Monographie der Gattung Silene*; Verlag von Engelmann: Leipzig, Germany, 1869; pp. 1–249.
72. Jafari, F.; Zarre, S.; Gholipour, A.; Eggens, F.; Rabeler, R.K.; Oxelman, B. A new taxonomic backbone for the infrageneric classification of the species-rich genus *Silene* (Caryophyllaceae). *Taxon* **2020**, *69*, 337–368. [CrossRef]
73. Martín-Gómez, J.J.; Rewicz, A.; Rodríguez-Lorenzo, J.L.; Janoušek, B.; Cervantes, E. Seed morphology in *Silene* based on geometric models. *Plants* **2020**, *9*, 1787. [CrossRef]
74. Milanesi, C.; Costantini, L.; Firmati, M.; Antonucci, F.; Faleri, C.; Buracchi, A.; Cresti, M. Geometric morphometry and archaeobotany: Characterisation of grape seeds (*Vitis vinifera* L.) by analysis of form. *Open Access Libr. J.* **2014**, *1*, e634. [CrossRef]
75. Ucchesu, M.; Orrú, M.; Grillo, O.; Venora, G.; Paglietti, G.; Ardu, A.; Bacchetta, G. Predictive method for correct identification of archaeological charred grape seeds: Support for advances in knowledge of grape domestication process. *PLoS ONE* **2016**, *11*, e0149814. [CrossRef]
76. Ucchesu, M.; Orrú, M.; Grillo, O.; Venora, G.; Usai, A.; Serreli, P.F.; Bacchetta, G. Earliest evidence of a primitive cultivar of *Vitis vinifera* L. during the Bronze Age in Sardinia (Italy). *Veg. Hist. Archaeobotany* **2015**, *24*, 587–600. [CrossRef]
77. Morcote-Ríos, G.; Bernal, R.; Raz, L. Phytoliths as a tool for archaeobotanical, palaeobotanical and palaeoecological studies in Amazonian palms. *Bot. J. Linn. Soc.* **2016**, *182*, 348–360. [CrossRef]
78. Sharma, R.; Kumar, V.; Kumar, R. Distribution of phytoliths in plants: A review. *Geol. Ecol. Landsc.* **2019**, *3*, 123–148. [CrossRef]
79. Bergson, H. L'Évolution Créatrice. Édition Electronique (ePub, PDF) v.: 1.0: Les Échos du Maquis, avril 2013. Available online: https://philosophie.cegeptr.qc.ca/wp-content/documents/L%C3%A9volution-cr%C3%A9atrice.pdf (accessed on 16 August 2021).
80. Tomlinson, P.B. *The Structural Biology of Palms*; Clarendon Press: Oxford, UK, 1990.

plants

MDPI

Article

Arabidopsis thaliana Response to Extracellular DNA: Self Versus Nonself Exposure

Maria Luisa Chiusano [1,2,*], Guido Incerti [3], Chiara Colantuono [4], Pasquale Termolino [5], Emanuela Palomba [2], Francesco Monticolo [1], Giovanna Benvenuto [6], Alessandro Foscari [7], Alfonso Esposito [8], Lucia Marti [9], Giulia de Lorenzo [9], Isaac Vega-Muñoz [10], Martin Heil [10], Fabrizio Carteni [1], Giuliano Bonanomi [1] and Stefano Mazzoleni [1,*]

1 Department of Agricultural Sciences, University of Naples Federico II, Via Università 100, 80055 Portici, Italy; f.monticolo90@gmail.com (F.M.); fabrizio.carteni@unina.it (F.C.); giuliano.bonanomi@unina.it (G.B.)
2 Department of Research Infrastructures for Marine Biological Resources (RIMAR), Stazione Zoologica "Anton Dohrn", 80121 Napoli, Italy; emanuela.palomba@szn.it
3 Department of Agri-Food, Animal and Environmental Sciences, University of Udine, 33100 Udine, Italy; guido.incerti@uniud.it
4 Telethon Institute of Genetics and Medicine, via campi Flegrei, 34 Pozzuoli, 80078 Napoli, Italy; chiara.colantuono@gmail.com
5 Institute of Biosciences and Bioresources (IBBR), National Research Council of Italy (CNR), 80055 Portici, Italy; pasquale.termolino@ibbr.cnr.it
6 Biology and Evolution of Marine Organisms Department (BEOM), Stazione Zoologica "Anton Dohrn", 80121 Napoli, Italy; gio.benv@szn.it
7 Dipartimento di Scienze della Vita, University of Trieste, 34127 Trieste, Italy; alessandro.foscari@phd.units.it
8 Department of Cellular, Computational and Integrative Biology—CIBIO, University of Trento, 38123 Trento, Italy; alfonso.esposito@unitn.it
9 Department of Biology and Biotechnology "C. Darwin", Sapienza University of Rome, 00185 Rome, Italy; lucia.marti@uniroma1.it (L.M.); giulia.delorenzo@uniroma1.it (G.d.L.)
10 Departemento de Ingeniería Genética, CINVESTAV-Irapuato, Guanajuato 36821, Mexico; vega.aphelion@gmail.com (I.V.-M.); mheil@ira.cinvestav.mx (M.H.)
* Correspondence: chiusano@unina.it (M.L.C.); stefano.mazzoleni@unina.it (S.M.)

Citation: Chiusano, M.L.; Incerti, G.; Colantuono, C.; Termolino, P.; Palomba, E.; Monticolo, F.; Benvenuto, G.; Foscari, A.; Esposito, A.; Marti, L.; et al. *Arabidopsis thaliana* Response to Extracellular DNA: Self Versus Nonself Exposure. *Plants* **2021**, *10*, 1744. https://doi.org/10.3390/plants10081744

Academic Editors: Milan S. Stankovic, Paula Baptista and Petronia Carillo

Received: 27 July 2021
Accepted: 17 August 2021
Published: 23 August 2021

Publisher's Note: MDPI stays neutral with regard to jurisdictional claims in published maps and institutional affiliations.

Copyright: © 2021 by the authors. Licensee MDPI, Basel, Switzerland. This article is an open access article distributed under the terms and conditions of the Creative Commons Attribution (CC BY) license (https://creativecommons.org/licenses/by/4.0/).

Abstract: The inhibitory effect of extracellular DNA (exDNA) on the growth of conspecific individuals was demonstrated in different kingdoms. In plants, the inhibition has been observed on root growth and seed germination, demonstrating its role in plant–soil negative feedback. Several hypotheses have been proposed to explain the early response to exDNA and the inhibitory effect of conspecific exDNA. We here contribute with a whole-plant transcriptome profiling in the model species *Arabidopsis thaliana* exposed to extracellular self- (conspecific) and nonself- (heterologous) DNA. The results highlight that cells distinguish self- from nonself-DNA. Moreover, confocal microscopy analyses reveal that nonself-DNA enters root tissues and cells, while self-DNA remains outside. Specifically, exposure to self-DNA limits cell permeability, affecting chloroplast functioning and reactive oxygen species (ROS) production, eventually causing cell cycle arrest, consistently with macroscopic observations of root apex necrosis, increased root hair density and leaf chlorosis. In contrast, nonself-DNA enters the cells triggering the activation of a hypersensitive response and evolving into systemic acquired resistance. Complex and different cascades of events emerge from exposure to extracellular self- or nonself-DNA and are discussed in the context of Damage- and Pathogen-Associated Molecular Patterns (DAMP and PAMP, respectively) responses.

Keywords: exDNA; environmental DNA; DNA sensing; self-DNA inhibition; autotoxicity; plant response; DAMP; PAMP; EDAP

1. Introduction

Mazzoleni and co-workers [1] reported evidence that fragmented exDNA, accumulating in litter during the decomposition process, produces a concentration dependent, species-specific inhibitory effect, reducing root growth and seed germination of conspecifics. This discovery was also extended to different organisms other than plants, including microbes, fungi, protozoa and insects [2]. Such findings have relevant implications for plant-soil ecological theories, providing a chemical basis for autotoxicity [3] among the mechanisms of plant–soil negative feedback [4], and unexpected new functional roles of exDNA and its sensing at cellular level [5], in species interactions at community [6] and ecosystem [7] levels, with an impact on biomedical and biotechnological applications, and is thus deserving further investigation [8–10].

It is well known that exDNA is abundant in many habitats, including soil, sediments, oceans and freshwater [11–13]. In soil, it can persist over long periods of time [14] due to its binding to the mineral and humic fractions. The environmental DNA originates from the active release or decomposition and recycling of organic matter produced by the whole range of taxa inhabiting the belowground habitat (bacteria, archaea, fungi, protozoa, soil invertebrates and plants) [15]. The persistent nature of exDNA permits its exploitation for the assessment of microbial community composition, soil and plant biodiversity, as well as taxonomic and phylogenetic studies [11,16–18].

The exDNA evolutionary role has been discussed in relation to the well-reported process of horizontal gene transfer among microbial populations [15]. However, it also appears to have additional functional and ecological implications. In plants, it has been reported to be a relevant nutrient source, especially in conditions of low phosphate availability [15]. At plant root level, exDNA was found on mucilage surrounding the root tips, with a putative protective role [19–21] and it has been also shown to act as a signalling molecule, in association with altered expression of specific hormone genes [22].

The involvement of exDNA in signalling and self-recognition has been recently widely discussed for plants [1,7,9,23] and in the context of microbe- or damage-associated molecular patterns [24–26]. Moreover, the role and the cellular and molecular mechanisms underlying plant growth inhibition determined by extracellular self-DNA (i.e., DNA from the same or closely related species), as well as plant responses to extracellular nonself-DNA (i.e., DNA from phylogenetically unrelated species) are still poorly known and understood. Recent studies on the early plant response to fragmented exDNA [5] revealed a significant plasma membrane depolarization and an increased flux of intracellular calcium in Lima bean (*Phaseolus lunatus*) and maize (*Zea mays*) leaves after treatment with self-DNA, whereas nonself-DNA was unable to trigger such signalling events, thus confirming that plant responses to exDNA depends on the provenience of the DNA also at cellular level. More recently, [27], treating plants and suspension-cultured cells of common bean (*Phaseolus vulgaris*) with fragmented extracellular self-DNA, observed leaf generation of H_2O_2, activation of a mitogen-activated protein kinase (MAPK) and increase of extrafloral nectar secretion, that the authors commented as early immunity-related signalling responses. By contrast, nonself-DNA by lima bean and *Acacia farnesiana* exerted lower or no detectable effects. In analogy with mammals, that sense self- or nonself-exDNA as indicators of injury or infection, respectively [28], it was suggested that extracellular self-DNA acts as a damage-associated molecular pattern (DAMP) also in plants [27]. A growth inhibitory effect by self-DNA was demonstrated after a minimum of 48 h [1] and observed up to four days [27] after treatments, while immunity-related signals were observed up to 2 h [5,27]. Therefore, whether growth inhibition by self-DNA depends on the energetic cost of an immunity response [26], or it is a direct effect of the exposure to self-DNA [7], remains an open question. In addition, responses at the cellular level along the timeline preceding observable growth inhibition also deserve further clarification.

In this study, we analysed early effects after exposure to extracellular self- and nonself-DNA in the plant model *Arabidopsis thaliana* by whole-plant transcriptome profiling, exploiting the RNA-seq approach during the first 16 h post treatment (hpt), and studied the early exDNA spatial distribution at root and cell levels by microscopy confocal analysis. In order to avoid any confounding effect coming from genomic similarity, we used the

DNA of the animal *Clupea harengus* (common herring) as nonself treatment, i.e., a species phylogenetically very dissimilar to Arabidopsis.

In this context, specific questions and hypotheses addressed in this work are:

1—What are the early molecular responses to self-DNA before the inhibitory effect on root growth becomes evident?

2—Are such molecular responses different when compared to those from nonself-DNA?

3—Do extracellular self-DNA or nonself-DNA trigger a DAMP or another response?

4—Does early transcriptome evidence support the immunity-cost hypothesis and/or a direct mechanism triggered by self-DNA that exerts an inhibitory effect?

2. Results

2.1. Differential Gene Expression after Self and Nonself-DNA Exposure

The bioinformatics data processing (Table S1), revealed that, among the 32,678 genes reported in the reference Arabidopsis annotation file, 1473 and 5977 are differentially expressed genes (DEGs) in self and nonself treatments, respectively (Table S2a). Interestingly, the relative number of DEGs in common at each hpt is higher after the first hour than in the other two stages post treatment, with even more striking difference in the number of specific DEGs along the timeline following the exposure to DNA (Table S2b, Figure 1). Indeed, a very limited number of genes are differentially expressed compared to the control after exposure to self-DNA (always less than 2.5% of the total number of Arabidopsis genes). In nonself treatments, the number of DEGs is much higher, especially after 8 hpt, with expression shifts involving more than 15% of the total number of genes. The Venn diagrams of DEGs at each observation stage (Figure 1C) showed that most DEGs were specific per treatment.

After filtering by $|\log_2(FC)| \geq 1$, the total number of DEGs was reduced to 825 and 1949 in self and nonself treatments, respectively. Only 342 genes were in common between the two treatments (Table S2a), underlining that specific changes were indeed about threefold higher in nonself treatments than in self ones. Interestingly, the response to self-DNA involved 63% of the filtered DEGs which are specific, i.e., they do not appear in the list of DEGs responsive to nonself-DNA in all the stages.

The complete list of all DEGs and their fold changes, highlighting the filtered DEGs ($|\log_2(FC)| \geq 1$) and the expression levels (RPKM) at the corresponding hpt are reported for both self and nonself treatments (Table S3a and b, respectively).

2.2. GO Enrichment in Self and Nonself Treatments

Results of gene ontology (GO) enrichment analysis highlighted clear differences across treatments and exposure times for specific classes of GOs (Figure 2, Table S4). Major differences occur in GOs related to DNA transcription and RNA translation, the former showing significant enrichments in downregulated genes after 8 h of exposure exclusively in self-DNA, whereas the group of RNA translation GOs is enriched in upregulated genes at 8 and 16 h exclusively in nonself. At 1 h, and only in self treatments, one GO associated with signal transduction (GO:0007165) is enriched in downregulated genes. GOs related to hormones show relevant differences between self and nonself treatments at 16 h, with GOs related to brassinosteroids and cytokinins enriched by upregulated genes and abscisic acid (ABA) and gibberellin ones enriched by downregulated genes in self-DNA treatment. Nonself-DNA, instead, enriched GO:009751 (response to salicylic acid) by upregulated genes. On the other hand, the nonself-DNA at 8 h shows Brassinosteroids homeostasis, a response to ABA, and Auxin efflux GOs enriched by downregulated genes (Table S4).

GO enrichments by upregulated genes in the group Biotic stress is evident in nonself compared to self-DNA treatments, including the Induced systemic resistance GO term (GO:0009682), enriched at 1 and 8 h, and the systemic acquired resistance (GO:0009627), that is enriched at 16 h exclusively in the nonself-DNA treatment.

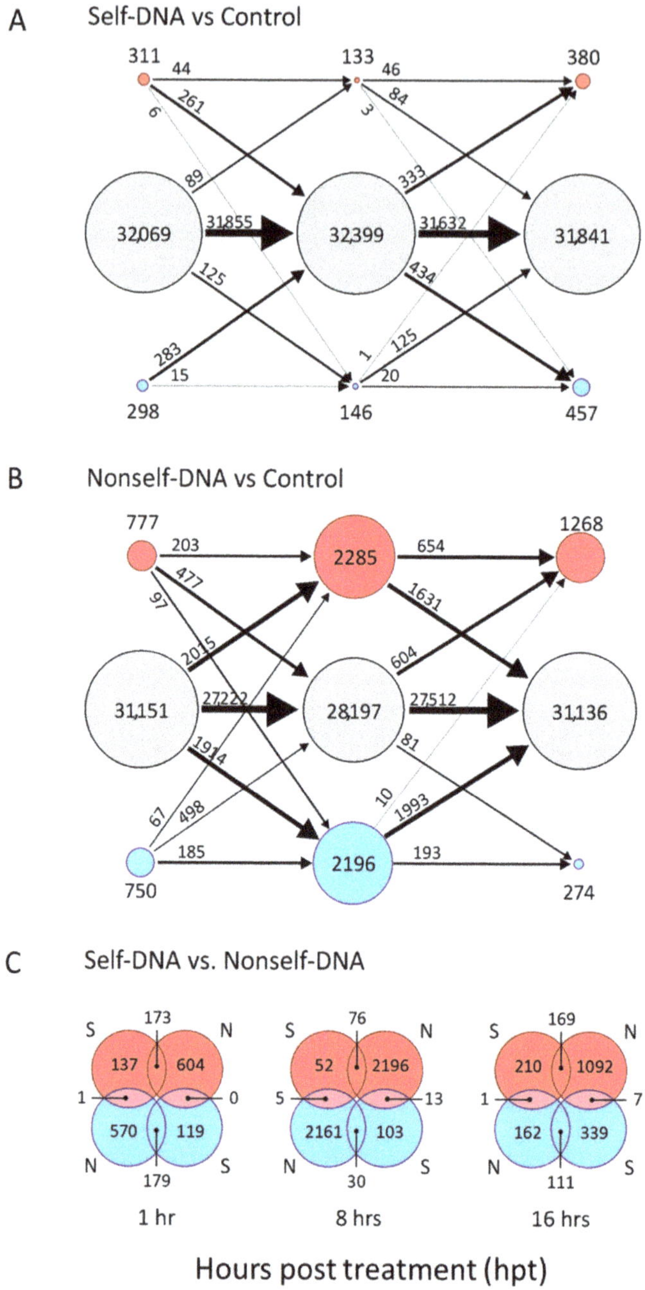

Figure 1. Number of differentially expressed genes (DEGs) (red circles: upregulated; blue circles downregulated, grey circles: not differentially expressed; circle size log-proportional to gene numbers) in treatment with self-DNA (**A**) and nonself-DNA (**B**) vs. control comparisons and transitions per stage (1, 8 and 16 hpt). The arrows indicate transitions over exposure time (line width log-proportional to the number of transitions displayed on each arrow). In (**C**), the Venn diagram for the self vs. nonself comparisons at each observation stage is reported.

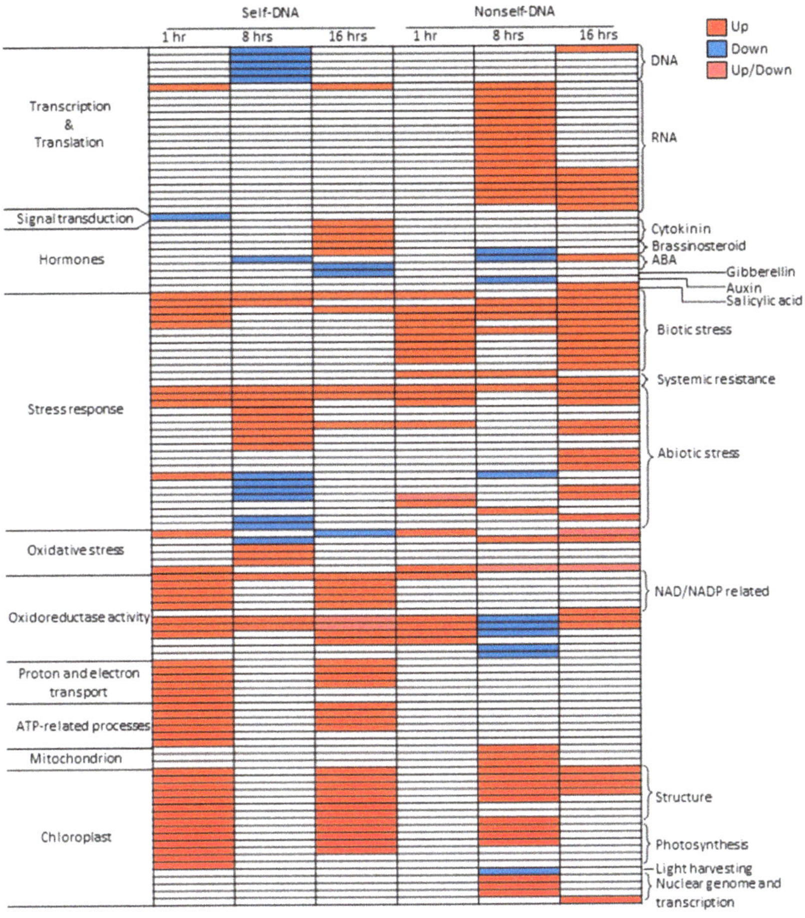

Figure 2. Summary of the GO enrichment analysis on filtered DEGs, with most enriched GOs (rows) grouped by functional process or cell compartment. The colour of each cell in the columns (indicating treatment type and stage) shows the pattern of expression of the enriching genes (full red: upregulated DEGs; blue: downregulated; light red: both up- and downregulated, with enrichment in upregulated DEGs showing lower *p*-value compared to the downregulated ones). In white, absence of enrichment is shown.

The Abiotic stress GO group also shows remarkable differences. In particular, exclusively in self-DNA treatment at 8 h, GOs related to responses to copper and cadmium ions (GO:0046688 and GO:0046686, respectively), as well as to ozone (GO:0010193) and light intensity (GO:0009642), are enriched by upregulated genes. This specificity is held in GOs related to the responses to water deprivation (GO:0009414), high light intensity (GO:0009644), heat (GO:0009408) and hyperosmotic salinity (GO:0042538), as well as to chitin and wounding (GO:0010200 and GO:0009611, respectively). These are all enriched by downregulated genes exclusively in self-DNA treatments (Figure 2, Table S5).

Interestingly, within the oxidative stress GOs group, an enrichment by upregulated genes, involving the response to superoxide radical activity (GO:0019430, GO:0006801 and GO:0004784), and by downregulated genes for the response to hydrogen peroxide (GO:0042542), is evident at 8 h exclusively in self-DNA treatment. A different pattern of enrichment by upregulated genes at 1 and 16 h in self-DNA treatments emerges for the GOs in the groups of proton and electron transport, ATP related processes, and Oxidoreductase activity (Figure 2, Table S4). Noticeably, the latter is instead enriched by downregulated genes in nonself treatments at 8 h.

It is noteworthy that the chloroplast related GOs groups (structure and photosynthesis) show a similar pattern due to enrichments by upregulated genes at 1 and 16 h in self-DNA treatment, while enrichment by upregulated genes at 8 h is evident for nonself-DNA treatments in the case of photosynthesis. Remarkably, light harvesting and the genome and transcription groups within these chloroplast related GOs do not show significant enrichment in self-DNA treatment, while being enriched in nonself-DNA treatments. Interestingly, also mitochondrion related GOs are enriched by upregulated genes exclusively at 8 h in nonself-DNA treatments.

2.3. Functional Response by Multivariate Analysis

K-means classification and Principal Component Analysis (PCA) ordination on all DEGs across experimental treatments showed 15 clusters that are clearly separated according to upregulation and downregulation patterns in the experimental treatments (Figure 3A). Details on DEGs in each cluster are reported in Table S5. In the PCA results (Figure 3B–E), the spreading of treatments and clusters in the bi-dimensional space defined by the first three principal components, highlighted several differences according to type and timing of exposure. Indeed, the early response to both treatments are located at the leftmost of the first component (accounting for 53.4% of the total variability), while the samples exposed to nonself-DNA at 8 and 16 h are separated from all other samples at the rightmost of the first component, and from each other along the second component (accounting for 17.4% of the total variability, Figure 3B). The third PCA axis (accounting for 14.0% of the total variability) is related to initial differences between gene expression patterns (at 1 h) in the response to self- versus nonself-DNA treatments (Figure 3D). Plotting the factorial scores of gene cluster centroids (Figure 3C,E) allows the identification of the genes most characterizing the treatments in the same component space (Figure 3B,D). Then, each centroid in Figure 3C is shown in Figure 3F labelled by the most frequently occurring GO keywords in the corresponding cluster (Table S5). Comparing the cluster ordination of Figure 3F with that of treatments in Figure 3B allows to highlight functional response differences among treatments and exposure times. In particular, two major groups of keywords clearly appear along the first principal component. On the left, oxidation–reduction processes and membrane/cell wall contribute to self-DNA treatment at all exposure times, and to nonself-DNA treatment at 1 h, consistent with gene upregulation in these treatments (clusters 8 and 12 in Figure 3A). On the opposite side, at the rightmost of the axis, ribosomal activity mostly contributes to nonself treatment at 8 h, associated with gene upregulation (see clusters 7 and 11 in Figure 3A). Along the second component, a clear distinction within nonself-DNA treatment, between 8 and 16 h, becomes evident, characterized by a defence response, in response to chitin, and a response to bacterium. Such a pattern, indicating a general biotic stress activity, corresponds to significant gene upregulation (clusters 4, 6 and 13 in Figure 3A) at 16 h in nonself-DNA treatment.

2.4. Differentially Expressed Genes Associated to Enriched GOs

The contribution at gene level of the trends highlighted by the GO enrichment analysis, was analysed in terms of number of DEGs in different GOs according to self- and nonself-DNA treatments (Figure 4). Additional details on up and downregulated DEGs are reported in Supplementary materials (Table S6a,b).

The analysis of DEGs within the GO groups highlighted the different responses between self- and nonself-DNA treatments (Figure 4). The number of specific DEGs after the treatment with self-DNA was higher at 1 h while decreasing at 8 h and rising again at 16 h. In particular, in this treatment, DEGs were observed enriching groups of DNA, RNA, proton and electron transport, ATP-related processes, NAD/NADP related processes and chloroplast related GOs (i.e., structure and photosynthesis) (Figure 4). Moreover, some other GOs (those related to biotic, abiotic stresses and mitochondrion) show a different pattern with a higher number of specific DEGs at the first hour, then progressively decreasing over time at 8 and 16 hpt.

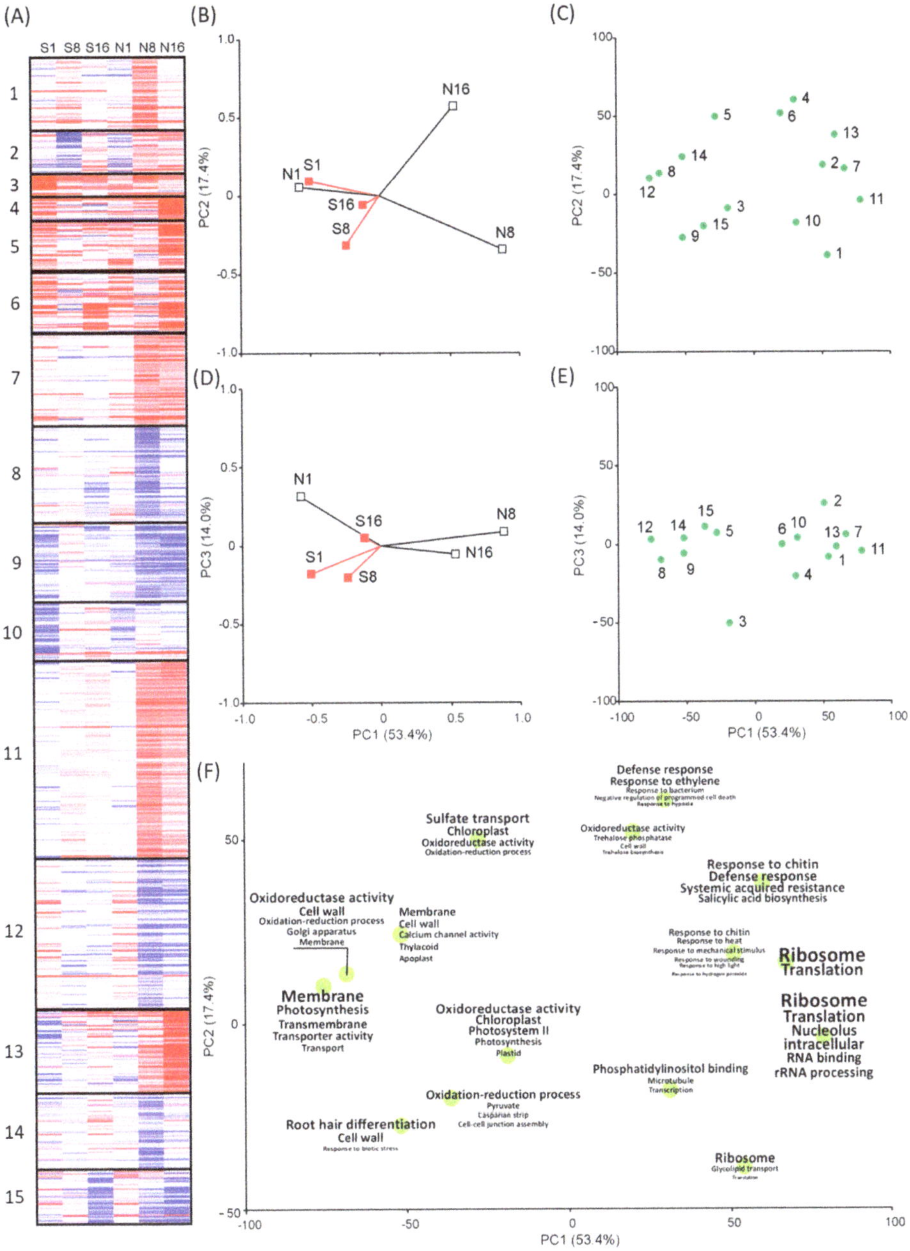

Figure 3. Classification and ordination of gene transcripts and samples based on RNA-seq data. Panels refer to heat map and K-means clustering based on fold changes in all stages of genes resulting differentially expressed (DEGs) in at least one of the six treatments (2 DNA sources × 3 observation times) (**A**). From red (high) to blue (low) the level of fold change values is shown. PCA ordination, with loading vectors of the samples (**B**,**D**) labelled by treatment (S: self-DNA, N: nonself-DNA) and time (1, 8, and 16 h), or factorial scores of cluster centroids (**C**,**E**) is shown. In (**F**), word cloud plot showing the position of each cluster in the first and second principal component space (as in (**C**)), overlapped by the keywords more frequently occurring in enriched gene ontologies (font size inversely log-proportional to enrichment *p*-value).

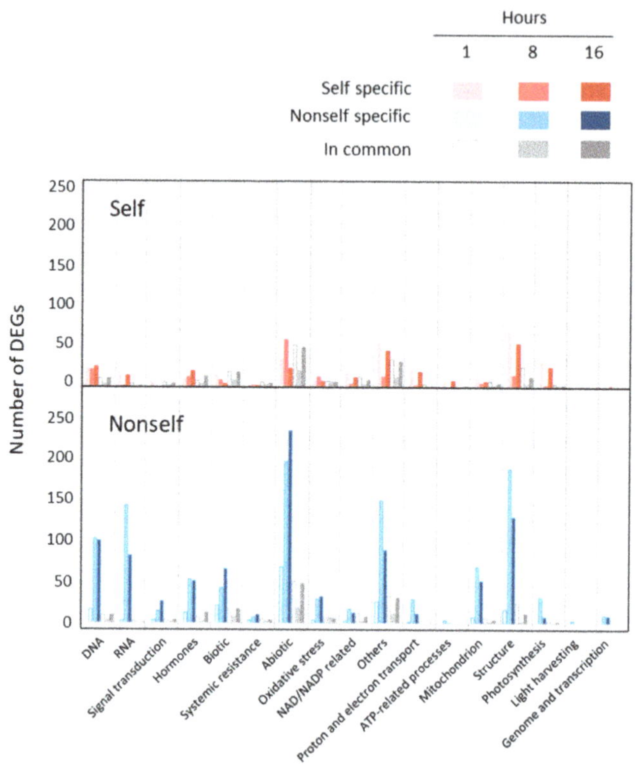

Figure 4. Number of either specific or common filtered DEGs in the main GO groups according to different self- and nonself-DNA treatments (at 1, 8 and 16 hpt). Specific DEGs after exposure to self-DNA are mostly shown at 1 and 16 hpt, whereas the nonself-DNA treatment is associated with an increased number of DEGs at 8 and 16 hpts.

On the contrary, after the nonself-DNA treatment, the number of specific DEGs increased at 8 hrs and persisted also at16 hpt for all considered GOs, with an evident increase in DEGs numbers for DNA, RNA, abiotic stress and chloroplast structure GOs.

2.4.1. One Hour Post Treatment

In the group of GOs related to DNA transcription and translation, the first hour does not report specific enrichments. However, DEGs counting show that downregulated genes prevailed, and the involvement of specific genes per each treatment highlights different reprogramming of the transcriptional asset in the two treatments. Interestingly, among the 12 upregulated genes in the self-DNA treatment, we identified the four RNA polymerases encoded by the chloroplast genome (ATCG00170; ATCG00180; ATCG00190; ATCG00740), all expressed over the $\log_2(FC) > 1$ cut-off (Table S3). In addition, self-specific upregulated DEGs include AT1G66600 (ABO3), a WRKY transcription factor involved in drought tolerance and in ABA mediated response [29]; AT5G47220 (ERF2), the ethylene responsive element and two transcription factors that are involved in ROS response (AT3G46080, AT1G52890). Noticeably, the higher expression of ERF2 is accompanied by the downregulation of specific genes involved in ethylene signal transduction (AT4G34410, AT5G52020) and ethylene responsive transcription factors (AT1G74930, AT1G28370). AT3G01220, exclusively downregulated in self-DNA treatment, is expressed during seed germination in the micropylar endosperm and in the root cap, and when mutated, increases seed dormancy and ABA sensitivity [30].

We identified 13 DEGs in response to self-DNA treatment associated with the enriched GOs among the RNA related GOs in the Transcription and translation group. All genes are encoded by the chloroplast genome (Table S3), with the only exception of one mitochondrial genome encoded gene (ATMG00090) that codes for a ribosomal protein S3 (RPS3), which is reported to contain a domain for resistance to pseudomonas syringae 3.

The GO:0007165 (signal transduction) was enriched by downregulated genes in self-DNA treatment, and the contribution of 5 specific genes encoding a CBL-interacting protein kinase (CIPK25) (AT5G25110), two Toll-Interleukin Resistance (TIR) proteins (AT5G44910, AT5G44920), an ankyrin repeat related gene (AT4G0346) and a phloem protein 2 A5 (PP2-A5) (AT1G65390), that is known to be induced upon and confers tolerance to spider mite attack [31] (Table S3).

Considering the 6 GOs NAD/NADP related processes in the group of oxidoreductase activity, 14 genes are exclusively upregulated in response to self-DNA treatment, of which 10 are encoded by the chloroplast genome (Table S3).

The group of proton and electron transport GOs included 6 GOs, with 21 genes exclusively upregulated in response to self-DNA treatment (Table S6a). These genes are mainly involved in the photosystem II reaction centre, in photosynthetic electron transfer and in the ATP synthase complex or membrane transporters. It is worth noting that when considering the GO group of ATP related processes, all the DEGs are self-specific and code for the ATP synthase complex, and all these genes are also contributing to the enrichment of the group of proton and electron transport related GOs.

Although the lack of enrichment in the GOs related to Mitochondrion in the first hour post both treatments, there are DEGs associated with these GOs. Considering the specific DEGs in each treatment, a higher contribution by self-DNA-related DEGs is reported (Table S6a). Interestingly, among these genes AOX1d (AT1G32350) encodes for one of the 4 AOX1 genes of *A. thaliana* genome [32].

When considering the chloroplast related GOs, where all the 4 groups of GOs are exclusively enriched in upregulated genes in the self-DNA treatment, the 7 GOs related to Structure (represented by a total of 2541 genes, Table S6a), reveal that the DEGs in response to self-DNA contributed more than the nonself treatment (94 versus 42 genes, and 69 versus 17 specific, respectively) (Figure 4). The major contribution was from upregulated genes (88 and 33 in self- and nonself-DNA treatments, respectively). Among the 94 DEGs in response to self-DNA treatment, 61 genes are encoded by the chloroplast genome and 57 of these chloroplast genes are specific of the self-DNA response at this stage.

In the case of the photosynthesis group (36 DEGs out of a total of 199 genes) all DEGs were from the self-DNA treatment (Figure 4), with 35 upregulated genes and 32 of them specific to this treatment, and 34 genes over 35 encoded by the chloroplast genome. In the nonself-DNA treatment, out of a total of 3 DEGs, 2 were upregulated and nonspecific DEGs (Table S6a).

The biotic stress GOs included 11 GOs associated with a total of 490 genes (Table S6a). These GOs were associated with 27 (10 specific) and 6 (5 specific) upregulated and downregulated genes, respectively, in self DNA response, while the corresponding figures for the nonself-DNA treatment were 40 (total DEGs), 34 (17 specific) upregulated and 6 (5 specific) downregulated DEGs, respectively, thus indicating remarkable difference in the response between the two treatments (Figure 4). Interestingly, among the DEGs, the self-DNA treatment shows a specific upregulation of BAG6 (AT2G46240) [33], one of the three genes belonging to the BAG family (AtBAG4, AtBAG6 and AtBAG7) that controls the induction of autophagy and have confirmed cytoprotective activities in response to cold, drought and heat stress.

The 2 enriched GOs within the Systemic resistance group corresponded to a total of 60 genes, 11 of which were DEGs at 1 h (Table S6a). PAD 4 (AT3G52430: phytoalexin deficient 4), that usually mediates TIR-NB-LRR signalling involved in the pathogen resistance response, as well as in root meristem growth arrest [34], is exclusively downregulated in self-DNA treatment and this is also confirmed by QRT-PCR (Table S7).

Considering the group of GOs associated with abiotic stress (20 enriched GOs for a total of 2697 genes), we identified 66 (24 specific) and 21 (11 specific) DEGs among upregulated and downregulated genes, respectively, in the self-DNA treatment. A total of 44 genes are upregulated and specific, while 35 are downregulated—25 of which are specific—in the nonself treatment, again sowing the differential responses between the two treatments (Table S6a, Figure 4).

In the group of oxidative stress (6 GOs for a total of 204 genes), 2 GOs were enriched at 1 h, corresponding to a total of 19 DEGs. Most of the genes (13 out of 19) in the self-DNA treatment are also included in DEGs involved in the abiotic stress GOs, indicating that the oxidative stress contribution is a key component of the early response elicited by self-DNA (Tables S3 and S6a).

2.4.2. Eight Hours Post Treatment

DEGs associated with the group of DNA enriched GOs, correspond to 25 and 108 total DEGs in self- and nonself-DNA treatments, respectively (Table S6a). Five genes are upregulated in response to self-DNA treatment at the 8 h (with 1 specific gene) and 65 (61 specific) are upregulated in response to nonself-DNA. All downregulated genes, 20 and 43 in response to self- and nonself-DNA, respectively, were specific of each treatment. Interestingly, although a higher number of DEGs is reported in nonself-DNA treatment, an exclusive GO enrichment (from downregulated genes) is evident in the self-DNA treatment, hence depicting a clear trend in the self-DNA response at 8 hpt (Figure 4). Among the genes that are downregulated in the self-DNA treatment, many follow the same trend already evident in the 1 hpt (Table S3). Among these genes, HSFA2A (AT2G26150), a transcription factor that is typically upregulated during stress response [35], and HSFC1 (AT3G24520), both confirmed by QRT-PCR, together with other AT-HSFA7B (AT3G63350) and 3 heat shock proteins (HSPs), all showing a significant negative $\log_2$(FC) in self-DNA treatments. The majority of the downregulated genes are ethylene related proteins or ethylene responsive transcription factors (AT1G12610: dwarf and delayed flowering 1 (DDF1); AT1G19210: Integrase-type DNA-binding superfamily protein; AT1G74930: ORA47; AT2G44840: ethylene-responsive element binding factor 13 (ERF13); AT4G11280: 1-aminocyclopropane-1-carboxylic acid synthase 6 (ACS6); AT4G25490:C-repeat/DRE binding factor 1 (CBF1); AT4G34410: redox responsive transcription factor 1 (RRTF1); AT5G05410: DRE-binding protein 2A (DREB2A); AT5G52020: Integrase-type DNA-binding superfamily protein; AT4G28110: myb domain protein 41 (MYB41)). In contrast, among the DEGs, those associated with salicylic acid or to the activation of the systemic acquired resistance highlight remarkable differences at transcription level between self- and nonself-DNA treatments at 8 h.

Considering the group of GOs associated with RNA, 18 GOs (related to cytoplasmic translation, biogenesis and/or assembly of ribosome and its subunits, rRNA cleavage, methylation and processing, rRNA binding) were enriched in upregulated genes exclusively in nonself-DNA treatment. Among the 146 DEGs, 145 genes were all specific to the nonself-DNA treatment, with 141 upregulated genes, among which 49 encode ribosomal proteins, indicating a consistent activation of the translation machinery in the nonself-DNA treatment at this stage (Table S6a, Figure 4).

Considering signal transduction, the reported GO:0007165 that is significantly enriched at 1 h, is not enriched at 8 h. However, at this stage we identified 17 DEGs associated with this GO (Table S6a, Figure 4), out of which only 2 genes are DEGs (upregulated) in response to self-DNA, both in common with the nonself-DNA treatment (AT4G11170 (Disease resistance protein (TIR-NBS-LRR class) family) and AT5G44990 (Glutathione S-transferase family protein)). To be noted, among the 13 upregulated DEGs in nonself-DNA treatment (Table S6a), 6 genes (AT1G57630, AT1G66090, AT1G72900, AT4G10170, AT5G41750, AT5G45000) are TIR domain containing proteins. Four downregulated genes were all specific in nonself-DNA treatment. A couple of them are involved in the signalling of phosphoinositides: AT3G55940 (Phosphoinositide-specific phospholipase C family protein) and AT5G58670 (phospholipase C1 (PLC1)). Interestingly, among the significant downregulated genes in nonself-DNA we also identified 4 further genes (AT3G03530: non-specific phospholipase C4 (NPC4); AT3G08510: phospholipase C 2 (PLC2); AT3G48610: non-specific phospholipase C6 (NPC6); AT3G51460: ROOT HAIR DEFECTIVE4 (RHD4)) all involved in the phosphoinositides signalling pathway, which is recognized as an early response involving membrane reorganization and lipid signalling in defence response [36].

In the group of biotic stress, we identified a total of 16 DEGs in the self-DNA treatment, 12 (4 specific) upregulated and 4 (all specific) downregulated (Table S6a). Among the upregulated genes, AT3G26830 (PHYTOALEXIN DEFICIENT 3 (PAD3)), AT5G40990 (GDSL lipase 1 (GLIP1)) and AT1G79680 (WALL ASSOCIATED KINASE (WAK)-LIKE 10 (WAKL10)) resulted upregulated at all observation stages in both DNA treatments, with the exception of the nonself-DNA treatment at 8 h. Differently, the self-specific downregulation of AT2G46240 (BAG6), AT1G80840 (WRKY40), AT2G27080 (Late embryogenesis abundant (LEA) hydroxyproline-rich glycoprotein family) and AT2G40000 (ortholog of sugar beet HS1 PRO-1 2 (HSPRO2)), putatively indicates a decrease of the biotic response to pathogen [37]. Accordingly, plants exposed to self-DNA showed a negligible representation of differential expression from biotic stress genes, when compared to abiotic stress related ones (Table S6a, Figure 4). Considering the GOs related to the systemic acquired resistance, we identified 13 DEGs, out of which 5 in the response to self-DNA (4 upregulated genes of which 2 specific), and 9 upregulated genes (7 specific) in the response to nonself DNA (Table S6a). Among these, AT4G12470 (AZI1), which is involved in the priming of salicylic acid induction and systemic immunity [38], was downregulated at 1 h and upregulated at 8 h in the nonself-DNA treatment.

Within the group abiotic stress, we found 78 DEGs in response to self-DNA treatment, out of which 35 were upregulated (18 specific), and 43 downregulated (40 specific) genes (Table S6a). In the case of nonself-DNA treatment, out of a total of 219 genes (Figure 4), 129 were upregulated (110 specific) and 90 downregulated (80 specific) genes (Table S6a). Among these genes, AT3G48360 is downregulated in self and is upregulated in nonself-DNA treatment. This gene has been shown to be downregulated in the presence of sugar while it is upregulated in the presence of nitrogen [39]. Moreover, the 35 upregulated genes in response to self-DNA included 5 GST genes (2 of which specific) and 5 peroxidases (1 specific), while among the 40 genes specifically down expressed in self-DNA treatment, 18 belong to the HSP protein superfamily (Table S3).

In the group of oxidative stress, we found 19 DEGs in self-DNA treatment, 6 of which upregulated (2 specific), and 13 were downregulated (10 specific) (Table S6a, Figure 4). Among the 2 genes exclusively upregulated in response to self-DNA, AT4G25100 codes for Fe superoxide dismutase 1 (FSD1), acting in plastidial, cytoplasmic and nuclear compartments with an anti-oxidative and osmoprotective role [40].

In the group of oxidoreductase activity, in the GOs concerning NAD/NADP related processes, we identified a total of 6 DEGs in the self-DNA treatment, all upregulated and 3 specific, while in the nonself-DNA treatment we found 15 upregulated genes (12 specific) and 6 downregulated genes, all specific (Table S6a).

In the group of proton and electron transport (Figure 4), we found a total of 32 DEGs, out of which 2 were differentially expressed in response to self-DNA (downregulated in this treatment), and 30 DEGs, with 17 and 13 specific genes upregulated and downregulated, respectively, in the nonself DNA treatment (Table S6a).

In the group of ATP-related processes, only 4 genes were differentially expressed, all exclusively in nonself-DNA treatment (Table S6a, Figure 4).

In the group of mitochondrion related GOs (3 GOs for a total of 1411 genes) we found 75 DEGs (Table S6a). It is noteworthy that among the three genes exclusively downregulated in response to self-DNA, AT4G25200 encodes for a mitochondrial localized small HSP that is regulated by HSFA2, that, as mentioned above, is also downregulated in the self-DNA treatment [41]. Differently (Figure 4), in nonself-DNA treatment, a total of 71 DEGs included 59 (57 specific) and 12 (all specific) upregulated and downregulated genes, respectively (Table S6a, Figure 4). Interestingly, one gene showed opposite behaviour in the two DNA treatments, being downregulated and upregulated in response to self- and nonself-DNA, respectively. This gene encodes for HSP26.5 (AT1G52560) that also is correlated in its expression with HSFA2 [41].

It is worthy to note that in the GO group of chloroplast (including structure, photosynthesis, light harvesting, and genome and transcription) no GO was enriched in self-DNA treatment at 8 h (Figure 2). Nevertheless, we still found specific responses in the self (17 DEGs including 10 (8 specific) upregulated and 7 (6 specific) downregulated genes) and in the nonself-DNA treatment (144 DEGs (142 specific) were upregulated and 49 (48 specific) were downregulated (Table S6a).

2.4.3. Sixteen Hours Post Treatment

After 16 hpt, in the group of DNA GOs, 135 genes are DEGs in self- and nonself-DNA treatments (Table S6a), 14 up and 10 down regulated genes are specific of the self-DNA treatment, while 94 up and 7 down regulated genes are specific of the nonself-DNA treatment (Figure 4). Among the 94 nonself- specific and upregulated genes, 13 belong to the MyB transcription factors family and 22 are WRKY transcription factors. Interestingly, among these latter genes, WRKY33 is involved in both the hypersensitive response and the systemic acquired resistance and WRKY70 is involved in the establishment of the systemic acquired resistance [42]. This indicates that nonself-DNA can be sensed as a PAMP by triggering the hypersensitive response and initiating a systemic acquired resistance.

In the group of RNA GOs, 97 were the total DEGs, including 14 DEGs in self-DNA treatment, 13 of which (12 specific) are upregulated, and 1 downregulated and specific, while 83 specific among 84 upregulated genes are DEGs in the nonself-DNA treatment (Table S6a, Figure 4). Interestingly, the 12 genes exclusively upregulated in response to self-DNA are encoded by the chloroplast genome and 8 of them are also upregulated at the 1 h post the same treatment. Differently, in the nonself-DNA treatment, the upregulated genes are more related to rRNA processing, maturation and stabilization, and less to ribosome biogenesis and structure (Table S3).

In the group of signal transduction GOs, with 1 GOs for a total of 432 genes (Table S6a), we report 31 total DEGs, all upregulated in nonself-DNA treatment (27 specific and 4 in common with the self-DNA treatment) (Table S6a, Figure 4). Among the 27 genes exclusively upregulated in response to nonself-DNA, 12 belongs to the group of disease resistance proteins (Table S3) and appear to be involved in the process of the hypersensitive response [43].

Concerning the stress response class of GOs, the group of biotic stress includes 88 DEGs in total all upregulated, with 21 DEGs (3 specific) in self and 85 DEGs (67 specific) in nonself treatments, respectively (Table S6a, Figure 4).

The group of systemic resistance showed a total of 16 DEGs, with 5 DEGs in self-DNA treatment (3 upregulated and in common with nonself, and 2 downregulated and one in common with nonself), and 15 genes, 13 (10 specific) upregulated and 2 (1 specific) downregulated in nonself (Table S6a). Among the 10 genes exclusively upregulated in response to nonself-DNA, AT3G52430 encodes for PAD4 which, as discussed above, when upregulated, mediates the TIR-NB-LRR signalling involved in the hypersensitive response [43].

In the group of abiotic stress, we identified 308 DEGs. Most of them (252), among which 213 specific DEGs, are upregulated in response to nonself-DNA, while 34 (24 specific) are downregulated in this treatment. In the self-DNA treatment, 51 genes (12 specific) are upregulated and 20 (10 specific) are downregulated (Table S6a, Figure 4).

In the group of oxidative stress, we found a total of 45 DEGs, with 12 total DEGs in response to self-DNA, including 4 (1 specific) upregulated and 8 (5 specific) downregulated genes, and 39 DEGs in nonself-DNA treatment, including 30 (27 specific) upregulated and 9 (6 specific) downregulated genes (Table S6a, Figure 4). In particular, 13 out of the 27 genes exclusively upregulated in response to nonself-DNA are peroxidases (Table S3).

Concerning the oxidoreductase activity GOs for the group of NAD/NADP related processes GOs, we identified a total of 32 DEGs. Considering the 10 DEGs exclusively upregulated in response to self-DNA treatment, 8 of them, coding for NADH dehydrogenase subunits, are encoded by the chloroplast genome and resulted upregulated also at 1 h (Table S3). In the nonself-DNA treatment, out of 21 DEGs, 17 (9 specific) were upregulated and 4 (all specific) were downregulated. Among the 9 genes exclusively upregulated in response to nonself-DNA treatment, none is encoded by the chloroplast genome, and 7 belong to the cytochrome p450 protein family.

In the group of proton and electron transport, we found 31 DEGs in total, with 19 DEGs, including 18 (17 specific) upregulated genes and 1 specific and downregulated gene, in self-DNA treatment (Table S6a). Among the 17 genes exclusively upregulated in response to self-DNA treatment, 16 are encoded by the chloroplast genome and all were also upregulated at 1 h.

For the group of ATP-related processes, exDNA treatments show nine DEGs in total. All of these are upregulated DEGs (Table S6a), including eight (seven specific) and two (one specific) DEGs in response to self- and nonself-DNA, respectively. Among the seven genes exclusively upregulated in the self-DNA treatment, six are encoded by the chloroplast genome and all are upregulated also at 1 h (Table S3).

For the group of mitochondrion, treatments showed a total of 42 DEGs, with 54 (50 specific) upregulated and 2 (all specific) downregulated genes in nonself-DNA treatment, and 8 upregulated (4 specific) and 2 downregulated and specific genes in response to self-DNA, the latter including AT5G24120 (Table S3). It is worth noting, such gene codes for SIGE, a transcriptional factors localized in both chloroplast and mitochondrion, which regulates the chloroplast transcriptional response to light intensity [44].

Considering the chloroplast group of GOs, the self-DNA treatment shows a trend similar to that observed at 1 h for the Structure and Photosynthesis subgroups (Table S6a, Figure 4). In the subgroup of structure, we found 196 DEGs, out of which 55 (46 specific) were upregulated and 10 (7 specific) were downregulated in self-DNA treatment. Among the 46 specific DEGs, 45 are encoded by the chloroplast genome and 40 of them are upregulated also at 1 h (Table S3). In nonself-DNA treatment, out of a total of 143 genes, 136 (127 specific) are upregulated and 7 (4 specific) are downregulated (Table S6a). Different from the self-DNA treatment, among the 127 genes upregulated and specific of the response to nonself-DNA, only 16 were upregulated also at 1 h and, interestingly, none is encoded by the chloroplast genome (Table S3). In the group of photosynthesis, we identified 34 DEGs (Figure 4). In self-DNA treatment, among a total of 26 DEGs, 23 were upregulated (21 specific) (Table S6a). Out of the 21 upregulated and specific genes, 20 are encoded by the chloroplast genome and were upregulated also at the 1 h (Table S3). In nonself-DNA treatment, out of a total of 10 DEGs (Figure 4), 9 were upregulated (7 specific) (Table S6a). For the group of light harvesting no DEGs are evident in both self- and nonself-DNA treatments.

In conclusion, in the self-DNA treatment, the genes involved in processes related to cell energy production and balance, oxidoreductases and chloroplast structure and photosynthesis show a recursive upregulation, since the pattern at 16 h resembles the one at 1 h.

2.5. Hormone Related DEGs after Self- and Nonself-DNA Treatments

The Hormones enriched GOs (10 GOs for a total of 610 genes) indicate processes related to cytokinins, brassinosteroids, ABA, gibberellins, auxins and salicylic acid.

To consider more details on possible DEGs associated with hormones, we considered the number of DEGs searching by each hormone as a keyword in GO categories of DEGs (Table S6b).

In the self-DNA treatment, among the upregulated DEGs at 1 h, we found (ranked by counts): nine genes (six specific) in the group of abscisic acid, seven genes, all self-specific, in the group of jasmonic acid (JA), five genes (three specific) for cytokinins, five (one specific) for salicylic acid, four for auxin, and one for the ethylene, while brassinosteroids and gibberellins were not involved in differential expression at this stage. Among these upregulated DEGs, it is worthy to mention AT5G15970, encoding for the protein KIN2, that is known to be induced by ABA and during water deficit stress. Additionally, AT1G15520 (PDR12) encodes for the pleiotropic drug resistance 12, an ABA related ABC transporter localized on the plasma membrane of guard cells and involved in ABA uptake and stomatal closure [45,46].

In the case of nonself-DNA treatments, among hormone with upregulated genes, we found: salicylic acid (five specific), abscisic acid (three specific), auxin (two specific), cytokines (one specific), ethylene and brassinosteroids (one specific). Among the downregulated DEGs, six were in the group of ABA (five specific), four in salicylic acid (four specific), three in cytokinins (two specific), two for the ethylene, both DEGs are nonself-specific, one specific in the group of brassinosteroids, and one in the group of auxin. We did not find DEGs for JA and gibberellins, neither among upregulated, nor among downregulated genes for nonself-DNA treatment at this stage. Overall, the data show a relevant role of ABA and JA in the response to self-DNA exposure at 1 h, while, in the case of nonself-DNA treatment, salicylic acid, ABA and auxin, play a major role at 1 h (Table S6b, Figure 4).

After 8 h of exposure to exDNA, we found a total of 85 DEGs. Among the 16 DEGs found in response to self-DNA, 7 were upregulated (6 specific) and 9 were downregulated (6 specific), whereas in nonself-DNA treatment, out of 73 DEGs in total, 25 were upregulated (23 specific) and 48 were downregulated (46 specific), indicating that the overall hormone reprogramming was active mainly in this specific treatment, at this stage. Hormone-related DEGs counts reveal a reduced involvement of ABA and JA associated DEGs in comparison with the first hour, in the response to self-DNA. In the case of nonself-DNA treatment, a remarkable specificity characterizes the response in comparison with the self-DNA treatment, with a clear involvement of salicylic acid, ABA and auxin activity related DEGs (Table S6b, Figure 4).

At 16 hpt, a total of 84 DEGs are evident (Table S6b, Figure 4). Among the upregulated genes in the self-DNA treatment, AT4G29740 and AT5G56970 encode for oxidases/dehydrogenases that catalyse the degradation of cytokinins, that are mainly involved in cell division processes and cell growth and differentiation [47], thus possibly revealing effects related to the growth inhibition caused by self-DNA exposure [1].

The high number of hormone-related DEGs in nonself-DNA treatment at 16 h, compared to the previous observation stages, may indicate a reprogramming of the hormonal activity in the nonself- compared to self-treatments (Table S6b, Figure 4).

A remarkable general difference on number of DEGs in self-DNA exposure when compared with nonself treatments, and also on their trends in different times post treatments is evident. Noticeable, the ABA, jasmonic acid, salicylic acid, ethylene and cytokinins related DEGs, increase in nonself-treatments in total and as specific upregulated DEGs during time, while the same trends are not evident in the self treatments.

2.6. DAMP and PAMP Associated Genes

To consider Arabidopsis genes involved in DAMPs or in PAMPs responses, we collected the list of known or putative receptors, mainly considering those responsive to extracellular nucleic acids, described in the literature [9,48–64]. Moreover, we also considered the expression patterns in both self and nonself-DNA treatments comparing the behaviour per stage (Table S8a). The summary of the total number of DEGs from self and nonself-DNA treatments at different time post exposure is reported in (Table S8b). Interestingly, it is evident that in both treatments, there are DEGs responsive genes in either the DAMP or the PAMP classes (Table S8b). In particular, in the DAMP class, the number of DEGs increases in both self and nonself treatments during time. In contrast, in the PAMP class, the number of DEGs remains almost stable in self treatments, while it is higher in the first and third stages, in comparison with the second stage post treatment, in nonself-DNA treatments. Interestingly, the number of DEGs showing a common behaviour in the two treatments increases during time for both classes, although the number of specific DEGs remains higher in nonself treatments, especially in the DAMP class. This may be due to the increase of DAMPs in the later stages of the nonself treatments, due to the cellular disruption revealed by the confocal analysis. It is worth noting, the very low numbers of specific DEGs from self-DNA treatments in both classes, which is even more striking in the PAMP class, and specifically at the 16 hpt. This may indicate that the differential sensing may be determined in the initial stages post treatment. Nevertheless, from this preliminary analysis, it is evident that the response to self-DNA is poorly characterized in terms of receptors of DAMPs or PAMPs (Table S8b). Considering the DEGs that are specific in the self-DNA treatment (Table S8a), it is worth mentioning AT1G57650, coding for an ATP-binding protein, and AT1G57630, coding for a TIR domain family protein, both upregulated and reported to respond to extracellular nucleic acids [9] and AT1G31540, coding for a TIR-NBS-LRR protein, which is down regulated, all belonging to the DAMP class. AT2G19190, coding for the Flagellin22-induced receptor-like kinase 1 and AT1G02900, coding for a Rapid alkalinization factor are both down regulated in the DAMP class.

At 8 hpt, specific DEGs from self-DNA treatments include defensins (three up regulated and one downregulated), and AT1G79680, coding for a cell wall associated kinase (WAK10), which is upregulated and reported to be a calcium receptor, and AT2G33580, in the PAMP class, coding for another membrane kinase, which is downregulated. Interestingly, among the three defensins that are classified as DAMPs and have DEGs in the nonself response, none is in common with the self-DNA response, and all are downregulated expect the one coded by AT3G24510, that is up regulated. It is of interest to note that the defensin pattern remains almost different in the two treatments also in the third stage. In particular, AT5G33355 remains upregulated also at the 16th hpt in self while AT3G24510 remains up in nonself treatments. AT1G34047 results a DEG at the 16 hpt only in the self-DNA response. Remarkably, defensins are the major class that is involved in the specific self-DNA response among the two classes (Table S8a).

2.7. QRT-PCR Results

Table S7 shows the results from QRT-PCR data of seven genes compared with the fold change of DEGs. The selected genes were chosen also to confirm some of the marker genes that could depict the behaviour in the two treatments. It is worth noticing the upregulation of the superoxide dismutase in self confirms the oxidative stress which is typical of this treatment. The expected general trend of AOX1d is confirmed in the two treatments per stage, together with the down regulation of HSFA2 (AT2G26150) and HSFC1 (AT3G24520) in the second stage of the self-DNA treatment, which is even more evident in the nonself one. Expression levels observed by RNAseq and QRT-PCR were well in accordance, as confirmed by the highly significant linear regression between the two series of data emerging from the comparison across seven gene transcripts, two DNA treatments and three observation time points (Figure S1, Pearson's $r = 0.814$, $P = 1.73 \times 10^{-10}$).

2.8. Differential Self- and Nonself-DNA Distribution by Confocal Analysis and Phenotypic Changes in Seedlings

Confocal microscopy of *A. thaliana* roots exposed to self-DNA revealed that labelled DNA (with both Cy3 and Alexa Fluor 555 dyes) was mainly visible outside the roots (Figure 5A–D) with an absence of fluorescence in the cytoplasm evident in all the images. At 1 h, the self-DNA fluorescence could be detected inside the root, but limited to the surface of the cells (Figure 5D).

In contrast, labelled nonself-DNA was clearly uptaken by the roots, in the cytoplasm and even at nuclear level (Figure 5F—I). The negative control showed no Cy3 fluorescence signal in any part of the root (data not shown). The exposure of the roots to FM4-64 dye after their treatments with either self- or nonself-DNA, showed a striking difference in the dye uptake and diffusive pattern inside the roots according to the type of extracellular DNA. Indeed, the dye remained outside the root that were previously exposed to self-DNA

(Figure 5E), whereas it massively entered the root cells when they had been previously treated with nonself-DNA (Figure 5L).

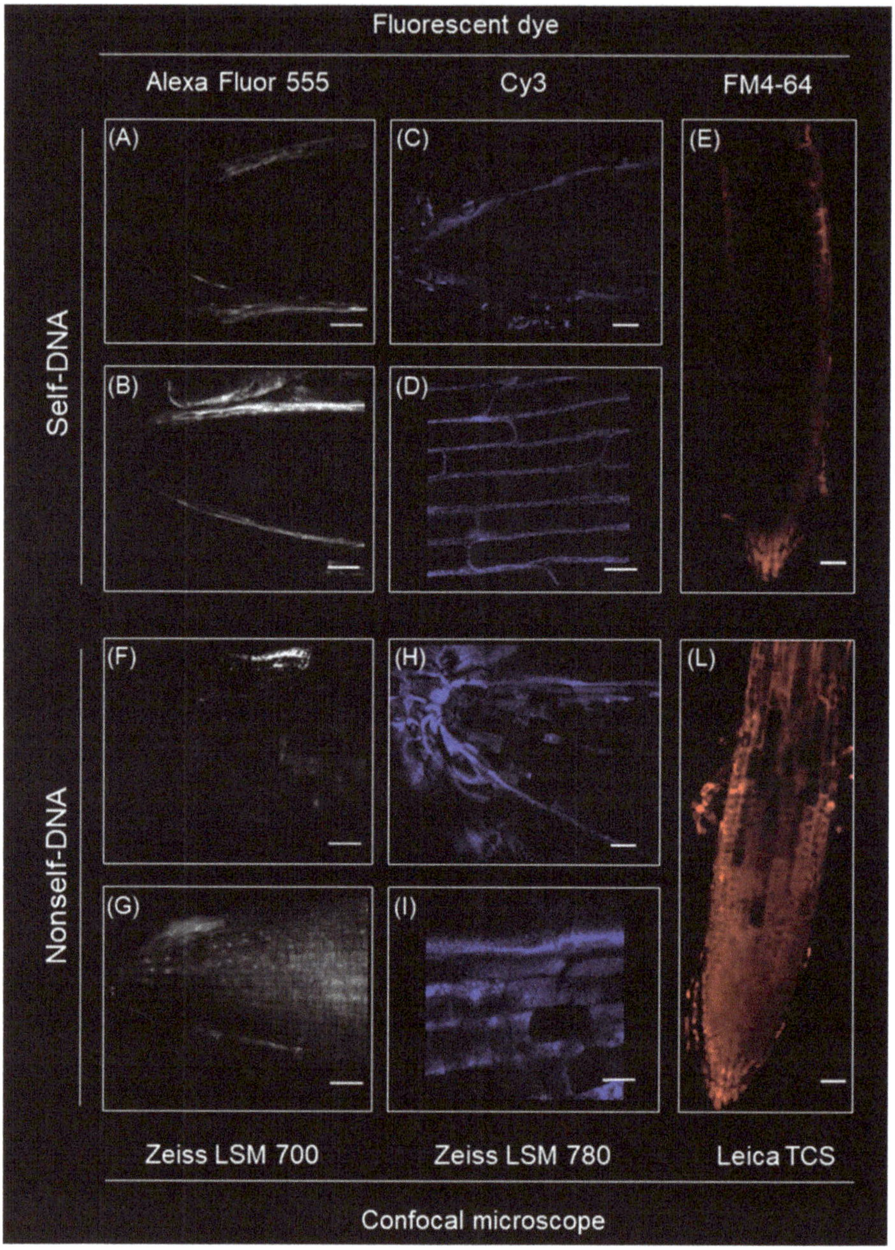

Figure 5. Fluorescence microphotography (confocal images) of early response (1 h) to self- vs. nonself-DNA treatments. From the left to the right, the following treatments are shown: Alexa Fluor 555 dye (in white) labelling self-DNA (**A**,**B**) and nonself-DNA (**F**,**G**); Cy3 dye (in blue) labelling self-DNA (**C**,**D**) and nonself-DNA (**H**,**I**); FM4-64 staining (in red) of A. thaliana roots after 1 h exposure to either self-DNA (**E**) or nonself-DNA (**L**). Scale bars 20 μm in (**A**–**C**,**E**–**G**,**I**,**L**); 10 μm in (**D**,**H**).

At the phenotypic levels, the main differential responses to self- and nonself-DNA are summarized in Figure 6. Of note, at macroscopic level, the exposure to self-DNA produced peculiar phenotypic effects: at 8 hpt with self-DNA, there is an increase in root hair density and a consistent root brownish; at a later stage 10 days post treatment, necrosis of root tips is accompanied by an inhibition of growth and leaf decolouring (Figure 6).

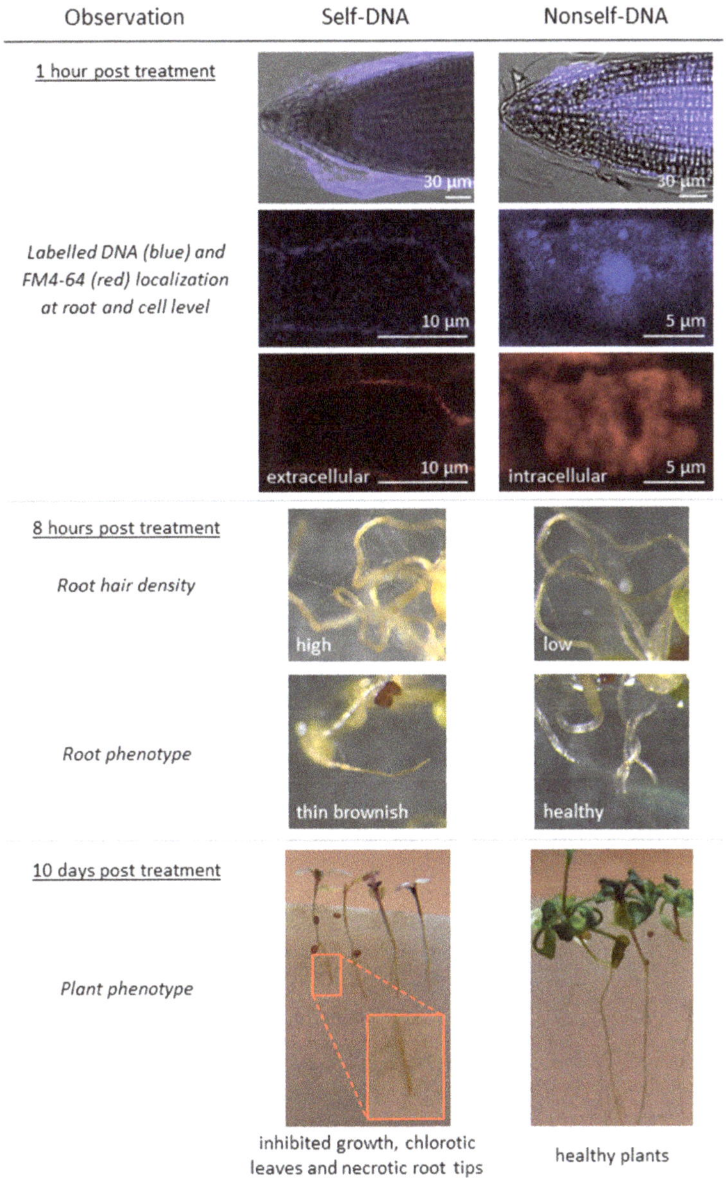

Figure 6. Summary of the main differences in response to extracellular self- and nonself-DNA in plants. A differential uptake according to treatments is clear after 1 h, morphological differences on thin roots and root apices appear after 8 h, full inhibition by the exposure to self-DNA is evident on the whole plants after 10 days.

3. Discussion

In Mazzoleni et al., 2015, it was reported that extracellular self-DNA, released in the soil during litter decomposition, or made available by experimental exposure, induced an inhibitory effect on root growth and seed germination in several plant species, without affecting heterospecifics [1]. Such findings have been ascribed to different putative mechanisms [8], including signalling and self-recognition [9,23], plant root defence [19] and microbe- or damage-associated molecular patterns [24–26].

The current study shows a clear-cut pattern in the plant transcriptomic response in the early stages after exposure and before evident phenotypic traits could be detected. Exposure to exDNA resulted in remarkable differences both between exposure to self- vs. nonself-DNA, and among different stages after exposure in each treatment. In parallel, remarkable differences were also highlighted in the early response by confocal microscopy showing that the root treated with self- and nonself-DNA have totally different physiological responses. A reduced root cell membrane permeability appears following the treatment with self-DNA, as indicated by the accumulation of both labelled self-DNA and FM4-64 in the outer layers of root cells. Conversely, after treatment with nonself-DNA, the labelled DNA diffused throughout the root reaching also the nuclei and, in this case, a clear diffusion of FM4-64 in the innermost part of the root was also evident.

The significantly different patterns of entrance of both labelled-DNA and of FM4-64 dye were consistent with the transcriptomic analysis results showing only in the case of nonself-DNA treatment the establishment of a hypersensitive response associated with cell wall and plasma membrane remodelling [65]. The uptake of nucleic acid macromolecules in roots was already reported [22]. However, to the best of our knowledge, this is the first evidence of exDNA entry in roots showing different distribution patterns between self and nonself-DNA.

In particular, the evident uptake of FM4-64 only after the exposure to nonself-DNA suggests an interesting activation of processes of endocytosis, vesicle dynamics and organelle organization as already reported in eukaryotic cells and plants [66].

At macroscopic effects peculiar phenotypic effects are revealed post self-DNA treatments, with an increased root hair density and a consistent root brownish evident already at 8 hpt. Only the exposure to self-DNA produced necrosis of root tips, inhibited growth and leaf decolouring at 10 days post treatment.

3.1. *Contrasting Transcriptome Dynamics in Response to Extracellular Self vs. Nonself-DNA*

The transcriptome analysis of plants treated with self-DNA showed a limited number of differentially expressed genes compared to the exposure to nonself-DNA, although remarkable GO enrichments could be revealed.

After one hour, a primary response to self-DNA sensing is the enrichment of upregulated genes involved in the chloroplast class (structure and photosynthesis groups) but with the lack of differential expression of nuclear genes related to chloroplast activity (genome and light harvesting groups). Significant enrichments are evident also in NAD/NADP related processes, proton and electron transport and ATP related processes, all due to upregulated DEGs, with the ATP related processes group evident only in self. Interestingly, there is no evidence of enrichment in the GOs related to mitochondrion. Nevertheless, among the 146 mitochondrial related genes, 7 are expressed in self and they are all upregulated, e.g., the mitochondrial gene RPS3 (ATMG00090) encoding a ribosomal protein related to pathogen resistance [64], and the nuclear gene AOX1d (AT1G32350) [32], as also confirmed by the QRT-PCR, which is typically upregulated in response to stress [65]. In particular, AOX1d contributes to the recovery from the inhibition of Complex III that is involved in the mitochondrial electron transport chain, thus indicating a block of the respiratory chain typical of the self-DNA treatment. The upregulation of AOX1D is coherent with the presence of nitric oxide (NO), as reflected by the upregulation of both NIA2 (Nitrate reductase: AT1G77760) and NIA1 (AT2G15620). The latter being related to the downregulation of AT2G28160 and AT3G25190 that are associated with the inhibition of ethylene production and generally associated with NO activation. NO is an alternative ROS product, determined by a drop of the oxygen concentration, also activating 2 oxoglutarate [66] and determining the inhibition of aconitase [67], which ends up with the upregulation of AOX1D. Noticeable, although the higher expression of ERF2 at the first hour, the down regulation of specific genes involved in ethylene signal transduction (AT4G34410, AT5G52020) and ethylene responsive transcription factors (AT1G74930, AT1G28370) is remarkable at 8 h in self treatments.

In addition, the hyper-activation of the chloroplast genome activity, in absence of a similar upregulation of chloroplast proteins encoded by the nuclear genome, reveals further specific peculiarities due to self-DNA exposure. The upregulation of chloroplast encoded genes, that does not meet a corresponding gene expression in nuclear encoded genes, could cause the overproduction of chloroplast related ROS, also caused by the lack of overexpression of genes like ascorbate peroxidases, namely the APX1 gene (AT1G07890). Ascorbate peroxidases acts as scavengers of H_2O_2 in the chloroplast, moreover suppressing the expression of H_2O_2-responsive genes under photo-oxidative stress [68]. This is also known to be accompanied by a downregulation of HSFA2 [41], which could explain the evidence here reported of an unexpected pattern of downregulated heat shock related proteins, as revealed for self-DNA exposure at 8 h.

However, further investigations should clarify the discrepancy here highlighted by the QRT-PCR that also shows a downregulation of the HSFA2 at the 8-h port treatment also in the nonself treatment that is not accompanied by the downregulation of heat shock proteins.

Interestingly, the upregulation of chloroplast related genes (involving proton and electron transport, ATP-related processes and structural components of chloroplast) observed at 1 h post exposure in self-DNA treatment, completely turned off after 8 h.

The enhanced ROS production starting at the first hpt in self, revealed by the specific signatures that witness these events and also reported as production of H_2O_2 in similar experiments of exposure to self-DNA [27], does not exclude the formation of singlet oxygen O^{2-}, because of the activation of superoxide dismutases in self. The O_2 drop down—maybe contributing to NO formation—is also accompanied by the downregulation of ethylene responsive transcription factors revealed at 8 hpt to self-DNA.

At 16 h, genes encoded by the chloroplast genome showed again upregulation in self-DNA treatments, indicating a recursive effect and a persisting stressing stimulus. It will be of interest, in future efforts, to monitor the hormonal as well as NO and H_2O_2 waves possibly accompanying the process. Moreover, photoinhibition should also occur, deteriorating the chloroplast machinery for longer exposure.

Considering the classes of genes specifically related to responses to stresses (biotic and abiotic stress, and systemic resistance) common DEGs are detected for both treatments. Nevertheless, the number of specific genes highlight initial milder differences between the two responses (Figure 4), while discrepancies become more evident in the two later stages. Indeed, self and nonself treatments, respectively, show along the three stages after treatment: 15-8-3 versus 22-44-67 DEGs in the biotic stress; 35-58-22 versus 68-199-237 for abiotic stress; 1-2-1 versus 4-8-11 in the systemic resistance.

In particular, genes related to heat, wounding and chitin response were downregulated, while responses to oxidative stress, toxic substances and ions were upregulated in self, involving genes encoding detoxification and anti-oxidation protective enzymes [69–72]. Such results clearly indicate that self-DNA triggered a response to oxidative stress and detoxification, while downregulating typical stress responsive genes, like HSPs, as it resulted evident at 8 h.

Downregulated genes in self at the first hpt include PAD4 (AT3G52430: phytoalexin deficient 4), as also confirmed by QRT-PCR, that usually, when upregulated, mediates TIR-NB-LRR signalling involved in the pathogen resistance response. The downregulation of PAD4, indeed, may be associated with the inhibition of TIR-NB-LRR signalling involved in the resistance responses mediated by TIR containing R proteins [43]. On the other hand, the upregulation of the systemic resistance and biotic stress responses is evident after exposure to nonself-DNA. Indeed, PAD4 (AT3G52430: phytoalexin deficient 4) is upregulated in nonself. Precisely, the self-specific downregulation at the first hour and the nonself-specific upregulation at 16 h, as also confirmed by QRT-PCR, also accompanied by the upregulation of genes involved in the hypersensitive response (AT3G52430, AT2G38470, AT1G91560, AT3G45640, AT5G07390, AT1G01480, AT4G11280 and of several TIR-NBS-LRR proteins), indicates the triggering of the related processes as a nonself-specific phenomenon. Moreover, an upregulation of genes related to systemic acquired resistance is also detected in this stage in nonself (AT3G45640, AT2G38470, AT1G19250, AT2G13810, AT3G56400, AT5G26920 and AT1G73810), consistent with the hypothesis that nonself-DNA acts as a PAMP triggering plant immune response [26], although we could not detect differential expression for enhanced disease susceptibility 1 (EDS1) and the senescence-associated gene 101 (SAG101) complex, that usually are also involved with PAD4 in triggering the two processes [43]. Overall, on one hand these findings suggest that the effects of nonself-DNA recall a PAMP-like response, as it is evolving towards a systemic acquired resistance, that is not revealed from our results from self-DNA treatments, the latter being possibly more related to a DAMP like response [23].

Additionally, the GO enrichment analysis indicates an evident upregulation of most genes involved in local or systemic response in nonself treatments, while self-DNA treatments highlight the absence of a hypersensitive response which is also accompanied by early upregulation of genes related to both ABA and jasmonic acid only in the first hpt.

Considering other hormone related responses, the results of GO enrichment analysis indicated a consistent upregulation of genes related to cytokinin and brassinosteroids and a downregulation of gibberellins in treatments with self-DNA. Differently, in the case of nonself-DNA, hormonal response trends revealed by the GO analysis were limited to an upregulation of genes related to ABA and salicylic acid. The upregulation of most of the genes belonging to the cytokinin oxidase/dehydrogenase (CKX) family in the self-DNA treatment indicates cytokinins breakdown and suggests that at 16 h cytokinin-mediated processes are negatively affected, possibly involving, among others, cell cycle regulation, cell proliferation and shoot and root development [73].

Self-DNA treated plants also showed downregulation of gibberellins transport. Interestingly, two loci (AT4G25010, AT5G50800), SWEET13 and SWEET14, are members of the SWEET family, known for including a major class of sugar membrane transporters in plants [74,75]. However, a recent study clarified an additional, interesting function of these two carriers, which can transport gibberellins at intra- and inter-cellular levels, thus possibly affecting plant development and growth [76]. In the same study, the highest levels of the proteins AtSWEET13 and AtSWEET14 were found in roots of 1-week-old seedlings Our finding of a downregulation exclusive of the self-DNA treatment after 16 hpt may be related to the growth inhibition observed at a later stage in our analyses (Figure 6) that confirmed what previously highlighted [1].

3.2. Hypotheses on the Mechanisms of Self-DNA Inhibition in Plants

The discovery of plant growth inhibition by self-DNA [1] could be the result of a mechanism resembling "processes of interference based on sequence-specific recognition of small-sized nucleotide molecules" [1], thus explaining the specificity and determining inhibition of the cell functionality [7]. A further hypothesis was suggested to explain the dosage-dependent growth-inhibition by self-DNA as the phenotypic consequence of a costly immune response [8,23]. However, it has been already underlined that "the molecular mechanism underlying growth inhibition by eDNA ... is uncharacterized" [9]

Self-DNA fragments that appear in the extracellular space have been suggested to act as DAMPs: i.e., endogenous signals of danger that indicate the disruption of cell integrity [27,77]. Mechanical damage, feeding by chewing herbivores (including hydrolases in their saliva) and even infection by necrotrophic pathogens cause the disruption of cells, and the subsequent release of ATP, small signalling peptides (AtPeps), or cell wall fragments, and all these DAMPs thus activating the plant response [49,51,78–80]. This signalling cascade comprises membrane depolarization, Ca^{2+} fluxes, ROS production and MAPK activation and the subsequent induction of a JA dependent broad spectrum immunity against chewing herbivores and necrotrophic pathogens [81]. The JA-dependent immune response causes dosage-dependent metabolic costs which, at the phenotypic level, may become apparent as stunted growth or a transient growth arrest [82–84]. Under this scenario, the general assumption is that the immunogenic properties of self-DNA and other DAMPs correlate with their dosage-dependent inhibitory effects on growth. Indeed, DAMPs trigger ROS, ethylene production and JA signalling in *A. thaliana* [85–90] with more than half of the DAMP-induced genes shared with JA-signalling [87,88,90]. Mechanical wounding of common bean (*Phaseolus vulgaris*) leaves resulted in enhanced levels of eATP, which triggered the production of ROS and the activation of catalase and polyphenol oxidase [91]. Correspondingly, fragmented self-DNA triggered membrane depolarization in maize (*Zea mays*) and lima bean (*Phaseolus lunatus*) [5], ROS production, MAPK activation and JA increase in common bean [27], and the expression of superoxide dismutase, catalase, and phenylalanine ammonia lyase in lettuce (*Lactuca sativa*) [25]. Intriguingly, herbivores might even secrete DNases to suppress the self-DNA-triggered plant immune response [92]. In accordance to the "cost" hypothesis, eATP and AtPeps strongly inhibited *A. thaliana* seedling growth in several studies, and this was suggested to be a direct and causal relation of the immunogenic and growth inhibiting effects of eATP or AtPeps [89,93,94]. Additionally, for self-DNA, growth inhibition correlates with immune responses in common bean [27] and *Lactuca sativa* [25].

However, consistent with our results on the upregulation of the response to biotic stress prevailing in the nonself treatments, together with the triggering of the hypersensitive response in the later stages of this treatment, nonself-DNA would be sensed as a PAMP and not a DAMP. PAMPs also trigger a growth arrest/inhibition that can be associated with their immunogenic effects [89,95–97]. However, growth reduction is not generally reported for nonself-DNA treatments, although few examples exist: nonself-DNA from bacteria triggered ROS and callose deposition in *A. thaliana* seedlings and strongly inhibited their growth, and nonself-DNA from herring triggered the same response, although to a lower degree [95]. In some cases, growth inhibition was also observed by heterospecific DNA from phylogenetically related plant species [1,25,27]. However, self-DNA always caused a stronger inhibitory effect than nonself, with the taxonomic distance between the exposed species and the species used as source of the DNA playing a significant role on the extent of the inhibition, thus giving origin to consider as self-DNA treatments, also those involving "homologous" DNA. In particular, exDNA from *A. thaliana* inhibited the growth of *Lepidium sativum* seedlings and vice versa, but DNA from *A. thaliana* did not inhibit *Acanthus mollis* growth [1]. Interestingly, *A. thaliana* and *Lepidium* belong to the same order (Brassicaceae), whereas the Acanthaceae belongs to a different order, the Lamiales. Similarly, DNA from *Capsicum chinense* inhibited *Lactuca sativa* (both Asterales), whereas DNA from *Acaciella angustissma* (Fabales) did not [25], and DNA from lima bean inhibited common bean growth whereas DNA from *Acacia farnesiana* did not [27].

ExDNA from phylogenetically unrelated species was even reported to promote growth, being used as a phosphorous source [22]. The easiest explanation for these observations would be that the growth-inhibitory effect of self-DNA could be linked to a mechanism recognizing self as well as "homologous" DNA, i.e., DNA from phylogenetically related species, although the latter is recognized to a lower extent [1,2,7,8,23,98]. Although generalizations may be limiting, it would have been more intuitive to expect a more costly response to foreign DNA, rather than to self or to closely related DNA. This could be also in agreement with the stronger molecular response, in terms of transcriptome changes, here revealed in nonself rather than in self treatments. However, this evidence appears in contrast with the hypothesis of the growth inhibition in self treatments as a consequence of the metabolic cost of the immunity response. As an alternative hypothesis, Mazzoleni et al., 2015 [1] and later Carteni et al., 2016 [7] suggested a different explanation based on a more direct effect, i.e., the possible "interference" of extracellular self- or "similar" DNA (e.g.: homologous, i.e., from phylogenetically related species or even similar, i.e., with convergent structure similarity, although not phylogenetically related) causing inhibition of the whole cell functionality, mediated by sequence-specific recognition of small-sized nucleotide molecules [99], which could hamper cell and gene expression functionality [100] or affect genome stability [101], inhibiting the growth. This could explain the self-DNA growth inhibition as a widely conserved property of living beings, and therefore justify its observation over a very wide range of organisms spanning from prokaryotes to metazoan [2]. On the other hand, it would be difficult to assume that specialized molecular machineries across several kingdoms, including immunity response at cell level, would have evolved and remained conserved to constantly produce similar but highly specific growth inhibition effects in all species (e.g., metabolic cost of immunity response). Consequently, the inhibitory effect of self-DNA has to be explained by a more general and basic recognition mechanisms, inhibiting cellular activities, leading to the production of ROS, causing cell or DNA damaging effects [102], determining cell cycle arrest and growth inhibition at the macroscopic level. In particular, it would be interesting to investigate how these mechanisms are associated with the more general frame of epigenetic responses to stress, including chromatin organization and its effect on genome stability, and related methylation profiles [103].

3.3. *EDAP: Extracellular DNA Associated Pathways*

We here demonstrate that the cell is able to sense exDNA distinguishing between self and nonself DNA, in plants. This is revealed by our observations from fluorescence microphotography, that indicates different patterns of exDNA localisation, with nonself-DNA (non-similar or phylogenetically distant) entering root tissues and cells, while self-DNA (conspecific and/or similar or "homologous") remaining outside. Furthermore, the transcriptome analysis reveals that specific and different molecular pathways are triggered by the early response to self- and nonself-DNA, respectively. We propose to define these pathways as extracellular DNA associated pathways, suggesting the new acronym EDAP. This acronym is useful to depict the differential response to self and nonself DNA exposure, since no specific pathway was described before that could explain the difference between the two categories of molecules (self and nonself DNA, respectively). Specifically, our findings show, on one hand, in the case of nonself-DNA a remarkable differential gene expression, involving both biotic and abiotic stress related genes, accompanied by the mounting of a hypersensitive response, putatively triggering a systemic acquired resistance. On the other hand, a minor differential expression is evident in the self-DNA response that is, however, remarkably associated with oxidative stress, and the activation of the chloroplast genes, with the down-regulation of stress responsive genes. The self-recognition is known to trigger an early intracellular Ca^{2+} spike signal and cell membrane depolarization, as observed 30 min after exposure to self-DNA [5]. This may be accompanied by the downregulation in signal transduction related genes, as here revealed by the transcriptome analysis. Among these, the CBL-interacting protein kinases, which are known to be active in the Ca^{2+} dependent signalling cascades [104] once bound to Calcineurin B-Like proteins (CBL), also regulate the response to oxygen deficiency and osmotic and salt stress [105,106]. This may suggest decreasing intracellular dynamics, possibly affecting the crosstalk between the chloroplast and the nuclear genome activities, after self-DNA exposure, that can be coupled with the activation of cation cotransporters, such as the vacuolar H^+/Ca^{2+} antiporter [107].

After the initial sensing, that can be in principle ascribed to the category of DAMP sensing, since self-DNA is a damage associated molecule, the signal can be rapidly propagated throughout the plant [108–111] and, once reaching photosynthetic organs, it can also induce the activation of the chloroplast genome (Figure 7), as clearly depicted by the transcriptome analysis and described here for the first time. The lack of the up-regulation of genes acting as chloroplast ROS scavenging, presumably due to the inhibition of the chloroplast-nuclear cross-talk here hypothesized, induces a positive feedback on ROS production which is associated with the MAPK activation and JA production that was reported in other experiments of exposure to self-DNA [27]. The abovementioned inhibition is also reported to induce ROS production. This is linked to O_2 drop down that may contribute to the activation of the NO pathway determining the downregulation of ethylene response after exposure to self-DNA. We hypothesize that the inhibition of the intracellular dynamics and crosstalk can be a typical initial cell response after the sensing of external self damage patterns, with the length of the exposure to the stressor agent strongly affecting cell fate and the extent of the damage to the whole organism.

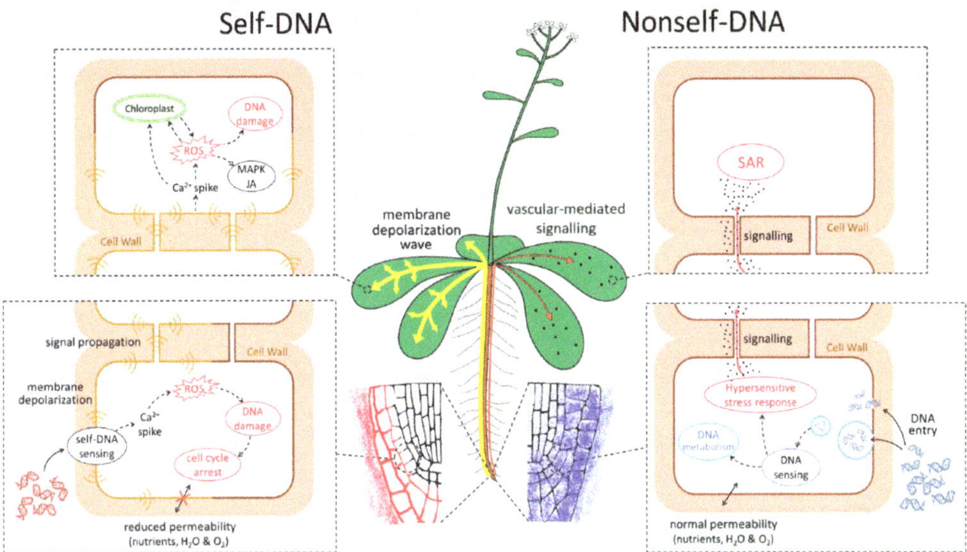

Figure 7. Model of Extracellular DNA Associated Pathways (EDAP) in plants. The exposure to either self- or nonself-DNA produces differential cellular responses. The self-DNA treatment triggers an electric response, starting with a sensing at membrane level, with calcium spikes followed by a reduced permeability of the roots, and a cascade of events involving the chloroplasts and inducing ROS production. On the other hand, nonself-DNA enters the cells where it is metabolized activating a cascade of events inducing a hypersensitive response.

To the best of our knowledge, self or nonself specific DNA receptors have not yet been described for plants so far [9,26,77]. However, among the putative eDNA/eRNA receptors described as DAMPs in *A. thaliana* [9,49–64], only three resulted upregulated at the first hpt in self, among which AT1G64160, that is a DEG also in the nonself treatment, in contrast with the 11 upregulated specific DEGS resulting from the nonself treatment.

DNA fragments may enter the cell, directly interfering with gene expression, as previously shown, e.g., for triplex-forming oligonucleotides (e.g., [112]) or even interacting at cytoplasmic level with redundant metabolic DNA [113]. However, our fluorescence microphotography experiments did not provide support to such hypothesis in the very early stage of the response to self. Indeed, the results here presented pave the way to consider that the sensing should occur in the extracellular environment, possibly on the extracellular surface of the plasma membrane or on its immediate surroundings, explaining the reasons of the accumulation of extracellular self-DNA on cell surfaces in contrast with the entrance and the active endocytosis stimulated by the nonself-DNA, that can also clarify the reasons of different molecular mechanisms associated with the different molecular responses, although further investigation should address possible explanations. On the basis of such logical reasoning, we speculate that nucleic acids themselves, known to be present in the extracellular environment could be involved in the self/nonself discrimination in plants and presumably in all living beings [21]. However, further validation on the location and the nature of the involved receptors or sensing mechanism still deserves further investigations to clarify if the sensing mechanism of exDNA, and specifically of self-DNA, could occur outside cells, possibly inhibiting successive processes like endocytosis or intracellular cross-talks to favour recovery from "damages" highlighted by the presence of "unexpected" self exDNA.

Our study clearly demonstrated that exposure to self-DNA produces a dramatic change of the biophysical state, directly determining the inhibitory effect, while no evidence of a costly activation of an immune response could be detected over our observation time frame.

Our analysis on possible DAMP and PAMP receptors or expression pathways, did not highlight clear trends in the context of known responses, but a more complex picture emerged for what concerns the response to extracellular DNA, clearly depicting the different response to self or nonself-DNA. This preliminary evidence surely needs more investigations and further assessments to better characterize what we here propose to describe as Extracellular DNA-Associated Pathways, introducing the acronym EDAP, that need to explain overlap and discrepancies in the sensing of self and nonself-DNA, highlighting novel perspectives on this subject.

4. Materials and Methods

4.1. DNA Extraction and Preparation

DNA was extracted from both *A. thaliana* young leaves and *Clupea harengus* (common herring) abdominal muscles with the same procedure. One gram of tissue was ground in liquid nitrogen and then mixed in CTAB buffer (0.1 M Tris-HCl pH 7.5, 0.7 M NaCl, 0.01 M EDTA, 1% CTAB). Samples were incubated 45 min at 65 °C and mixed periodically. Chloroform/Isoamylalcohol (24/1) was added to equal volume of sample mixed and centrifuged 20 min at 14,000× *g*. The upper phase was collected and transferred to a new tube for precipitation with 1 volume of Iso-propanol at −20 °C for 1 or more hours. Samples were centrifuged at max speed, at 4 °C for 20 min. The upper phase was discarded and pelleted DNA was washed with cold 70% ethanol in a centrifuge at max speed for 20 min. Ethanol was discarded and DNA was air dried and resuspended in 200 µL of sterile water. RNAse A (Thermo fisher, Third Avenue Waltham, MA, USA) was added at a final concentration of 0.25 mg/mL and incubated at 37 °C for 1 h. RNAse A was inactivated at 70 °C for 20 min.

The DNA was extracted with the same protocol from different tissue types of each species to randomize and overcome biases due to the methodology employed. Two independent DNA extractions were performed per sample (technical replicate). The DNA samples used as treatments were firstly checked by nanodrop standard quality parameters (260/280 and 260/230 ratio above 1.8) to evaluate DNA purity. The DNA quantity was evaluated using a QUBIT (Thermo Fisher, Third Avenue Waltham, MA, USA) fluorimeter and its integrity was evaluated by electrophoresis in 1% agarose gel. The DNA was sheared using a Bioruptor Plus (Diagenode, Seraing (Ougrée) Belgium, EU), for 12 min at high power, setting 60 s ON and 30 s OFF in order to reach an average size ranging from 200 to 700 bp. Such size was selected following previous evidence that this range was that found in decomposing litter and the most effective to produce inhibitory effects in vitro conditions [1].

4.2. Plant Materials and Treatments

A. thaliana (L.) Heynh. Col-0 (186AV) seeds were obtained from the "Centre de Ressources Biologiques" at the "Institut Jean Pierre Bourgin", Versailles, France (http://dbsgap.versailles.inra.fr/vnat/). Seeds were treated in 70% ethanol for 30 s, then transferred in a sodium hypochlorite (1:10 of commercial concentration) and 0.05% tween 20 solution for 10 min with occasional vortexing, washed with sterile milli-Q water 4 to

5 times, dried and resuspended in a sterile agar solution (0.2%). Sterilized seeds were vernalized for three days at 4 °C and prepared for sowing.

For transcriptome analyses, Petri dishes with a layer of thin Whatman paper were prepared in sterile conditions for each sample. The dishes were wetted with pure sterile water and an adequate number of Col-0 sterilized seeds were sown. The plates were put in the dark for 48 h in a growth chamber with 50% controlled humidity. After two days the plates were kept in 16h/8h light/dark cycles in controlled humidity conditions. If needed, sterile water was added to the plates from time to time, to all samples.

The experimental design for both transcriptome (Figure S2) and confocal analyses included two replicates of the following treatments: (1) control: with sterile distilled water; (2) self-DNA: with *A. thaliana* DNA; (3) nonself-DNA: with *Clupea harengus* DNA. Both DNA treatments were performed using a concentration of 200 ng/µL (as in [1]).

Before both transcriptome and confocal experiments, plants were grown for five days till the two true leaves stage.

For confocal analyses, *A. thaliana* plants (Col-0) were grown vertically on half strength MS basal medium. Five-days-old seedlings were placed on slides. They were treated with self and nonself-DNA and observed after 1 h.

For transcriptome analysis (summary in Figure S1), plants were grown until the appearance of the first true leaves. At this stage, the filter papers were imbibed with the control solution, 1.2 mL of sterile distilled water, or the same volume of 200 ng/µL of sonicated DNA (self or non self). Control and treated plates were harvested at 1, 8 and 16 h (two biological replicates per treatment), then immediately frozen with liquid nitrogen. Additional plants that were not harvested for RNA extraction were maintained for 15 days for final observations on longer terms phenotypic effects.

4.3. Transcriptome and Bioinformatics Analyses

Total RNA extraction from harvested plant material was performed using RNeasy micro kit from QIAGEN (Cat No./ID: 74004) following the standard extraction protocol and sent to a service provider for the RNA-seq analysis on Illumina Hiseq2500, by single read sequencing 1 × 15M.

Raw reads per sample were cleaned from adaptors and low-quality bases using the Trim Galore package (http://www.bioinformatics.babraham.ac.uk/projects/trim_galore/, access time: 3 September 2014), applying the default settings for single read sequencing. The cleaned reads were then mapped to the Arabidopsis nuclear and cytoplasmic genomes (version TAIR 10) using the STAR aligner software (version 2.4.2a) [114] allowing a maximum number of 10 mismatches.

Detailed results from the pre-processing are reported in Table S1. The mapped reads were counted per exon by featureCounts (version 1.4.6-p5) [115], using the strand specific count ("-s 2" option) and allowing the counting of read on overlapping features for each feature ("-O" option).

Assessment of replicates correlation in terms of RPKMs was performed by Pearson correlation and is reported as the correlation matrix and the associated dendrogram (Figure S3a,b). The principal component analysis of the same data per treatment per time is also shown (Figure S3c). DEGs call, comparing DNA treatments at each stage with the respective control, were made performing three different statistical approaches (FDR < 0.05): (i) DESeq2 [116]; (ii) edgeR and (iii) edgeR GLM The union of the three approaches was considered for subsequent analyses.

A K-means cluster analysis on log2 fold changes of all genes that resulted DEGs in at least one treatment/stage, was performed using MeV [115], by the Pearson Correlation as distance metric, setting at 15 the number of requested clusters according to the Figure of Merit (Figure S4). DEGs and samples were also submitted to PCA ordination, then plotting vector loadings for treatment combinations (timing and type of exposure to DNA) and factorial scores of cluster centroids in the multivariate space defined by the first three ordination axis.

GO enrichment analyses on DEGs filtered by $|\log_2(FC)| \geq 1$ were performed using the GOseq package [116] (FDR ≤ 0.05), and the reference GO annotation for Arabidopsis that was downloaded from Ensembl Plants (http://plants.ensembl.org/index.html, access time: 7 September 2018).

The GO enrichment analysis was also performed on the clusters obtained from the statistically significant DEGs, too. In order to better represent the prominent GO terms per cluster, selected keywords from the most enriched GOs were used to build word clouds (https://www.wordclouds.com/, access time: 11 November 2018), where the size of the represented words is directly proportional to the number of times they appear in the input data.

Lists of genes annotated with the most enriched GOs, showing different expression patterns between self-DNA and nonself-DNA treatments at each observation stage (i.e., 1, 8 and 16 h) were collected in order to quantitatively assess between-treatment differences and analyse their contribution in detail at single-gene level.

4.4. Confocal Microscopy Experiment

The confocal analyses were performed independently in two different institutes by using three different confocal microscopes (Zeiss LSM-700, Leica TCS and Zeiss LSM-780), in order to confirm and verify the reproducibility of the results.

The analyses were conducted treating roots by a DNA concentration of 100ng/μL DNA that included labelled DNA by Cy3 dye for an incubation period of 1 h. Seedlings were then transferred to a new slide and treated with 2 μM FM4-64 for 5 min. Before the confocal observations, seedlings were subjected to two sequential washing steps using tap water. Then, they were transferred to the slide for the confocal analyses. As a negative control, Arabidopsis seedlings were incubated for 1 h with the same amount of Cy3 (without DNA), then exposed to the same treatment with FM4-64. The images were analysed and edited by using the ImageJ free software.

4.5. QRT-PCR Validation

Seven target genes were selected considering their behaviour as DEGs in at least one stage/treatment and checked for duplication in the TAIR database (Arabidopsis.org). Using the RealTime qPCR online tool from IDT (https://eu.idtdna.com/scitools/Applications/RealTimePCR/, access time: 3 March 2021). One pair of primers was designed for amplification of the selected genes targeting the exon-exon junction, when possible, to exclude intronic regions arising from genomic DNA or not mature mRNAs. The 1st strand cDNA was synthesized with superscript III (ThermoScientific) starting from 600 ng of the same RNA samples used for the RNA-seq analysis and using manufacturer conditions. Each primer pair was verified in PCRs using first-strand cDNA as template. PCR conditions were as follows: denaturation at 94 °C for 4 min, cycling at 94 °C for 30 s, annealing at varying temperatures for 30 s, and extension at 72.0 °C for 30 s. Annealing temperatures ranged from 50 °C to 68 °C. Reaction products sizes and integrity were visualized by 1.2% agarose gel electrophoresis.

qPCR was performed in 20-μL reaction volumes containing 10 μL Power SYBR™ Green PCR Master Mix (Thermo-Fisher Scientific; catalog no. 4367659, Third Avenue Waltham, MA, USA), 500 nM forward and reverse primers and 1:10 dilution of 600 ng stock sample cDNA.

For each gene, we considered two biological replicates with two technical replicates per biological replicate. qPCR amplifications were performed on separate plates, where each plate contained primer pairs for the housekeeping gene GAPDH (AT1G13440.1, Jin et al., 2018). qPCR expression values were calculated for each gene as the difference between the quantification cycle of the gene and the reference gene, averaged over technical replicates (Table S7). Concordance between RNAseq and RT-qPCR results was assessed by testing the statistical significance of linear regression of the log2(FC) versus the log2(ΔΔCt) results, for those transcripts for which both metrices were available (n = 42, 7 transcripts × 2 DNA treatments × 3 time points). For each metric, the mean of replicated values was used for each combination of transcript, DNA treatment and observation time. Evident outliers were removed before fitting the regression model (Table S7).

5. Conclusions

In this work, we confirmed that the early molecular response to self-DNA is observable before the phenotypic effects of inhibition on root growth become evident, and that it can be clearly distinguished from the one to nonself-DNA. In the proposed model, self-DNA sensing at root level induces a membrane depolarization wave that rapidly propagates throughout the plant, leading to reduction of cell permeability and eventually to DNA damage and cell cycle arrest. Differently, nonself-DNA enters the cells eliciting a remarkable differential gene expression. This is associated with the activation of the hypersensitive response, possibly evolving into a systemic acquired resistance.

Mazzoleni and collaborators [1] reported that when plants are exposed to decomposing litter of phylogenetically similar species the inhibitory effect is still found, even if at a lower level. These observations, together with the need of deeper view on the mechanism through which the self- and nonself-DNA act, raise the interest for further investigations, both at transcriptomic and microscopic levels, of the A. thaliana response after the exposure to nonself-DNA at different degree of phylogenetic distance. Our results provide additional hints to the scenario concerning exDNA functions and its effects on plants, presumably justifying the same effects also in other species. They pave the way to additional investigations to demonstrate the mechanisms of exDNA sensing at extracellular level, and the specific recognition of self and nonself DNA in relation to the phylogenetic distance between the stimulating molecules and the structure features of the exposed organism genome. Finally, considering that self-DNA inhibition was demonstrated in many species across different kingdoms, further research should address the characterization of the EDAP in different model organisms, including prokaryotes and other eukaryotes.

Supplementary Materials: The following are available online at https://www.mdpi.com/article/10.3390/plants10081744/s1, Figure S1: Correlation analysis between log2FC RNA-seq and log2 ΔΔCt of QRT-PCR. The correlation was done using Person correlation, the R2 is also reported, Figure S2: Schematic representation of the experimental design, Figure S3: Correlation analysis (A) of gene expression (in RPKM) in all replicates (Control (C01, C08, C16), Self (S01, S08, S16), and Nonself (N01, N08, N16) indicate each stage per treatment. The associate cluster dendrogram (B) and the principal component analysis (PCA) (C) are also shown, Figure S4: Figure of Merit. The most suitable number of clusters expected per fold change of genes resulting in be DEGs in at least one stage per the two treatments, in a K-means clustering, is indicated by the red vertical bar, corresponding to 15 on the abscissa axis, Table S1: Summary of the RNA-seq pre-processing results showing, for each replicate, the total number of raw reads and the percentage of reads discarded after the cleaning steps and uniquely mapped on the Arabidopsis thaliana genome, Table S2: (a) Number of total genes and number of DEGs with assigned GO per treatment. (b) Number of common and specific DEGs in self-DNA and nonself-DNA treatments at 1, 8 and 16 h post treatment (hpt), Table S3: (a) List of statistically significant differentially expressed genes (DEGs) in self and nonself-DNA treatments and average RPKM values. Log2FC are shown for all DEGs. DEGs with a Log2FC ≥ | 1 | are indicated in colour cells: in red if upregulated and in blue if downregulated. (b) List of statistically significant differentially expressed genes (DEGs) in self and nonself-DNA treatments and average RPKM values. Log2FC are shown for all DEGs. DEGs with a Log2FC ≥ | 1 | are indicated in colour cells: in red if upregulated and in blue if downregulated, Table S4: Enriched Gene Ontologies (GOs) in self and nonself-DNA treatments organized in classes and groups. In red (UP) or in blue (DOWN) the GOs enriched by up or downregulate filtered DEGs (expression pattern) are shown, Table S5: Results of the cluster analysis based on fold changes in all stages of genes that are differentially expressed (DEGs) in at least one of the six treatments (2 DNA sources × 3 observation times). From red (high) to blue (low) the level of fold change values is shown, Table S6: (a) Number of filtered DEGs per class of GOs in self and nonself-DNA treatments at 1, 8 and 16 h post treatment (hpt). (b) Number of filtered DEGs per hormones in self and nonself-DNA treatments at 1, 8 and 16 h post treatment (hpt), Table S7: List of Genes that were significantly differentially expressed (DEGs) in at least one treatment/stage selected for validation by RT-qPCR, Table S8. (a) List of DAMPs and PAMPs in A. thaliana. (b) A summary of DEGs number per DAMP (525 genes in total) or PAMP (36 genes in total) classes of responsive genes.

Author Contributions: S.M. was the group coordinator; M.L.C. conceived the experiments and defined their design with P.T.; P.T. and E.P. performed the lab experiments; C.C., F.M. and E.P. performed the bioinformatic analysis; E.P., G.B. (Giovanna Benvenuto), L.M. and G.d.L. did the fluorescence microscopy experiments; C.C. did the RNA seq analysis and wrote the related methods; M.L.C. and S.M. supervised the analyses and interpreted the results; G.I., F.C., A.F., E.P. and C.C. prepared the figures; G.d.L., M.H., I.V.-M., G.B. (Giuliano Bonanomi) and A.E. contributed to the critical discussion and revision of the manuscript; S.M., M.L.C., G.I., and F.C. defined the model of the molecular mechanisms and wrote the manuscript. All authors have read and agreed to the published version of the manuscript.

Funding: This research received no external funding. The publication cost has been covered by a grant provided by the Museum Center "Musei delle Scienze Agrarie—MUSA" of the University of Napoli Federico II.

Data Availability Statement: Original datasets presented in this study can be found in the Sequence Read Archive (SRA), with the following accession identifier: BioProject ID: PRJNA707115.

Acknowledgments: EP is supported by a PhD fellowship funded by the Stazione Zoologica Anton Dohrn and by the NOSELF s.r.l (https://www.noself.it/). FM is supported by a PhD fellowship funded by the Department of Agricultural Sciences, Università degli Studi di Napoli Federico II The authors wish to thank technical suggestions from Rosa Paparo, Institute of Biosciences and Bioresources (IBBR), National Research Council of Italy (CNR) and Marco Miralto Department of Research Infrastructures for Marine Biological Resources (RIMAR), Stazione Zoologica "Anton Dohrn", Napoli.

Conflicts of Interest: The authors declare that the research was conducted in the absence of any commercial or financial relationships that could be construed as a potential conflict of interest.

References

1. Mazzoleni, S.; Bonanomi, G.; Incerti, G.; Chiusano, M.L.; Termolino, P.; Mingo, A.; Senatore, M.; Giannino, F.; Cartenì, F.; Rietkerk, M.; et al. Inhibitory and toxic effects of extracellular self-DNA in litter: A mechanism for negative plant-soil feedbacks? *New Phytol.* **2015**, *205*, 1195–1210. [CrossRef]
2. Mazzoleni, S.; Cartenì, F.; Bonanomi, G.; Senatore, M.; Termolino, P.; Giannino, F.; Incerti, G.; Rietkerk, M.; Lanzotti, V.; Chiusano, M.L. Inhibitory effects of extracellular self-DNA: A general biological process? *New Phytol.* **2015**, *206*, 127–132. [CrossRef]
3. Mazzoleni, S.; Bonanomi, G.; Giannino, F.; Incerti, G.; Dekker, S.C.; Rietkerk, M. Modelling the effects of litter decomposition on tree diversity patterns. *Ecol. Modell.* **2010**, *221*, 2784–2792. [CrossRef]
4. van der Putten, W.H.; Bradford, M.A.; Pernilla Brinkman, E.; van de Voorde, T.F.J.; Veen, G.F. Where, when and how plant–soil feedback matters in a changing world. *Funct. Ecol.* **2016**, *30*, 1109–1121. [CrossRef]
5. Barbero, F.; Guglielmotto, M.; Capuzzo, A.; Maffei, M.E. Extracellular Self-DNA (esDNA), but Not Heterologous Plant or Insect DNA (etDNA), Induces Plasma Membrane Depolarization and Calcium Signaling in Lima Bean (*Phaseolus lunatus*) and Maize (*Zea mays*). *Int. J. Mol. Sci.* **2016**, *17*, 1659. [CrossRef] [PubMed]
6. Zhang, N.; Van der Putten, W.H.; Veen, G.F.C. Effects of root decomposition on plant-soil feedback of early- and mid-successional plant species. *New Phytol.* **2016**, *212*, 220–231. [CrossRef] [PubMed]
7. Cartenì, F.; Bonanomi, G.; Giannino, F.; Incerti, G.; Vincenot, C.E.; Chiusano, M.L.; Mazzoleni, S. Self-DNA inhibitory effects Underlying mechanisms and ecological implications. *Plant Signal. Behav.* **2016**, *11*, e1158381. [CrossRef]
8. Veresoglou, S.D.; Aguilar-Trigueros, C.A.; Mansour, I.; Rillig, M.C. Self-DNA: A blessing in disguise? *New Phytol.* **2015**, *207*, 488–490. [CrossRef]
9. Bhat, A.; Ryu, C.-M. Plant Perceptions of Extracellular DNA and RNA. *Mol. Plant* **2016**, *9*, 956–958. [CrossRef]
10. Mazzoleni, S.; Cartenì, F.; Bonanomi, G.; Incerti, G.; Chiusano, M.L.; Termolino, P.; Migliozzi, A.; Senatore, M.; Giannino, F.; Rietkerk, M.; et al. New perspectives on the use of nucleic acids in pharmacological applications: Inhibitory action of extracellular self-DNA in biological systems. *Phytochem. Rev.* **2014**, *13*, 937–946. [CrossRef]
11. Nagler, M.; Insam, H.; Pietramellara, G.; Ascher-Jenull, J. Extracellular DNA in natural environments: Features, relevance and applications. *Appl. Microbiol. Biotechnol.* **2018**, *102*, 6343–6356. [CrossRef]
12. Pathan, S.I.; Arfaioli, P.; Ceccherini, M.T.; Ascher-Jenull, J.; Pietramellara, G. Preliminary evidences of the presence of extracellular DNA single stranded forms in soil. *PLoS ONE* **2020**, *15*, e0227296. [CrossRef]
13. Collins, R.A.; Wangensteen, O.S.; O'Gorman, E.J.; Mariani, S.; Sims, D.W.; Genner, M.J. Persistence of environmental DNA in marine systems. *Commun. Biol.* **2018**, *1*, 185. [CrossRef] [PubMed]

14. Agnelli, A.; Ascher, J.; Corti, G.; Ceccherini, M.T.; Pietramellara, G.; Nannipieri, P. Purification and isotopic signatures (δ 13C, δ 15N, Δ14C) of soil extracellular DNA. *Biol. Fertil. Soils* **2007**, *44*, 353–361. [CrossRef]
15. Levy-Booth, D.J.; Campbell, R.G.; Gulden, R.H.; Hart, M.M.; Powell, J.R.; Klironomos, J.N.; Pauls, K.P.; Swanton, C.J.; Trevors, J.T.; Dunfield, K.E. Cycling of extracellular DNA in the soil environment. *Soil Biol. Biochem.* **2007**, *39*, 2977–2991. [CrossRef]
16. Lejzerowicz, F.; Esling, P.; Majewski, W.; Szczuciński, W.; Decelle, J.; Obadia, C.; Arbizu, P.M.; Pawlowski, J. Ancient DNA complements microfossil record in deep-sea subsurface sediments. *Biol. Lett.* **2013**, *9*, 20130283. [CrossRef]
17. Ascher, J.; Ceccherini, M.T.; Pantani, O.L.; Agnelli, A.; Borgogni, F.; Guerri, G.; Nannipieri, P.; Pietramellara, G. Sequential extraction and genetic fingerprinting of a forest soil metagenome. *Appl. Soil Ecol.* **2009**, *42*, 176–181. [CrossRef]
18. Thomsen, P.F.; Willerslev, E. Environmental DNA—An emerging tool in conservation for monitoring past and present biodiversity. *Biol. Conserv.* **2015**, *183*, 4–18. [CrossRef]
19. Hawes, M.C.; Curlango-Rivera, G.; Wen, F.; White, G.J.; Vanetten, H.D.; Xiong, Z. Extracellular DNA: The tip of root defenses? *Plant Sci.* **2011**, *180*, 741–745. [CrossRef] [PubMed]
20. Wen, F.; Curlango-Rivera, G.; Huskey, D.A.; Xiong, Z.; Hawes, M.C. Visualization of extracellular DNA released during border cell separation from the root cap. *Am. J. Bot.* **2017**, *104*, 970–978. [CrossRef]
21. Monticolo, F.; Palomba, E.; Termolino, P.; Chiaiese, P.; de Alteriis, E.; Mazzoleni, S.; Chiusano, M.L. The Role of DNA in the Extracellular Environment: A Focus on NETs, RETs and Biofilms. *Front. Plant Sci.* **2020**, *11*, 2045. [CrossRef]
22. Paungfoo-Lonhienne, C.; Lonhienne, T.G.A.; Schmidt, S. DNA uptake by Arabidopsis induces changes in the expression of CLE peptides which control root morphology. *Plant Signal. Behav.* **2010**, *5*, 1112–1114. [CrossRef] [PubMed]
23. Duran-Flores, D.; Heil, M. Growth inhibition by self-DNA: A phenomenon and its multiple explanations. *New Phytol.* **2015**, *207*, 482–485. [CrossRef] [PubMed]
24. Panstruga, R. Bacterial RNA—A new MAMP on the block? *New Phytol.* **2016**, *209*, 458–460. [CrossRef]
25. Vega-Muñoz, I.; Feregrino-Pérez, A.A.; Torres-Pacheco, I.; Guevara-González, R.G. Exogenous fragmented DNA acts as a damage-associated molecular pattern (DAMP) inducing changes in CpG DNA methylation and defence-related responses in Lactuca sativa. *Funct. Plant Biol.* **2018**, *45*, 1065–1072. [CrossRef] [PubMed]
26. Heil, M.; Vega-Muñoz, I. Nucleic Acid Sensing in Mammals and Plants: Facts and Caveats. *Int. Rev. Cell Mol. Biol.* **2019**, *345*, 225–285. [CrossRef]
27. Duran-Flores, D.; Heil, M. Extracellular self-DNA as a damage-associated molecular pattern (DAMP) that triggers self-specific immunity induction in plants. *Brain Behav. Immun.* **2018**, *72*, 78–88. [CrossRef] [PubMed]
28. Toussaint, M.; Jackson, D.J.; Swieboda, D.; Guedán, A.; Tsourouktsoglou, T.-D.; Ching, Y.M.; Radermecker, C.; Makrinioti, H.; Aniscenko, J.; Bartlett, N.W.; et al. Host DNA released by NETosis promotes rhinovirus-induced type-2 allergic asthma exacerbation. *Nat. Med.* **2017**, *23*, 681–691. [CrossRef]
29. Ren, X.; Chen, Z.; Liu, Y.; Zhang, H.; Zhang, M.; Liu, Q.; Hong, X.; Zhu, J.-K.; Gong, Z. ABO3, a WRKY transcription factor, mediates plant responses to abscisic acid and drought tolerance in Arabidopsis. *Plant J.* **2010**, *63*, 417–429. [CrossRef]
30. Barrero, J.M.; Millar, A.A.; Griffiths, J.; Czechowski, T.; Scheible, W.R.; Udvardi, M.; Reid, J.B.; Ross, J.J.; Jacobsen, J.V.; Gubler, F. Gene expression profiling identifies two regulatory genes controlling dormancy and ABA sensitivity in Arabidopsis seeds. *Plant J.* **2010**, *61*, 611–622. [CrossRef]
31. Santamaría, M.E.; Martínez, M.; Arnaiz, A.; Rioja, C.; Burow, M.; Grbic, V.; Díaz, I. An Arabidopsis TIR-Lectin Two-Domain Protein Confers Defense Properties against *Tetranychus urticae*. *Plant Physiol.* **2019**, *179*, 1298–1314. [CrossRef] [PubMed]
32. Vanlerberghe, G.C.; McIntosh, L. Mitochondrial electron transport regulation of nuclear gene expression. Studies with the alternative oxidase gene of tobacco. *Plant Physiol.* **1994**, *105*, 867–874. [CrossRef] [PubMed]
33. Dickman, M.; Williams, B.; Li, Y.; de Figueiredo, P.; Wolpert, T. Reassessing apoptosis in plants. *Nat. Plants* **2017**, *3*, 773–779. [CrossRef] [PubMed]
34. Kim, T.-H.; Kunz, H.-H.; Bhattacharjee, S.; Hauser, F.; Park, J.; Engineer, C.; Liu, A.; Ha, T.; Parker, J.E.; Gassmann, W.; et al. Natural Variation in Small Molecule-Induced TIR-NB-LRR Signaling Induces Root Growth Arrest via EDS1- and PAD4-Complexed R Protein VICTR in Arabidopsis. *Plant Cell* **2012**, *24*, 5177–5192. [CrossRef] [PubMed]
35. Enomoto, T.; Tokizawa, M.; Ito, H.; Iuchi, S.; Kobayashi, M.; Yamamoto, Y.Y.; Kobayashi, Y.; Koyama, H. STOP1 regulates the expression of HsfA2 and GDHs that are critical for low-oxygen tolerance in Arabidopsis. *J. Exp. Bot.* **2019**, *70*, 3297–3311. [CrossRef] [PubMed]
36. Abd-El-Haliem, A.M.; Joosten, M.H.A.J. Plant phosphatidylinositol-specific phospholipase C at the center of plant innate immunity. *J. Integr. Plant Biol.* **2017**, *59*, 164–179. [CrossRef] [PubMed]
37. Murray, S.L.; Ingle, R.A.; Petersen, L.N.; Denby, K.J. Basal resistance against Pseudomonas syringae in Arabidopsis involves WRKY53 and a protein with homology to a nematode resistance protein. *Mol. Plant. Microbe Interact.* **2007**, *20*, 1431–1438. [CrossRef] [PubMed]
38. Dutton, C.; Hõrak, H.; Hepworth, C.; Mitchell, A.; Ton, J.; Hunt, L.; Gray, J.E. Bacterial infection systemically suppresses stomatal density. *Plant. Cell Environ.* **2019**, *42*, 2411–2421. [CrossRef] [PubMed]
39. Mandadi, K.K.; Misra, A.; Ren, S.; McKnight, T.D. BT2, a BTB Protein, Mediates Multiple Responses to Nutrients, Stresses, and Hormones in Arabidopsis. *Plant Physiol.* **2009**, *150*, 1930–1939. [CrossRef] [PubMed]
40. Dvořák, P.; Krasylenko, Y.; Ovečka, M.; Basheer, J.; Zapletalová, V.; Šamaj, J.; Takáč, T. FSD1: Developmentally-regulated plastidial, nuclear and cytoplasmic enzyme with anti-oxidative and osmoprotective role. *Plant. Cell Environ.* **2020**. [CrossRef] [PubMed]

41. Nishizawa, A.; Yabuta, Y.; Yoshida, E.; Maruta, T.; Yoshimura, K.; Shigeoka, S. Arabidopsis heat shock transcription factor A2 as a key regulator in response to several types of environmental stress. *Plant J.* **2006**, *48*, 535–547. [CrossRef]
42. Cao, F.Y.; DeFalco, T.A.; Moeder, W.; Li, B.; Gong, Y.; Liu, X.-M.; Taniguchi, M.; Lumba, S.; Toh, S.; Shan, L.; et al. Arabidopsis ETHYLENE RESPONSE FACTOR 8 (ERF8) has dual functions in ABA signaling and immunity. *BMC Plant Biol.* **2018**, *18*, 211. [CrossRef] [PubMed]
43. Coll, N.S.; Epple, P.; Dangl, J.L. Programmed cell death in the plant immune system. *Cell Death Differ.* **2011**, *18*, 1247–1256. [CrossRef] [PubMed]
44. Belbin, F.E.; Noordally, Z.B.; Wetherill, S.J.; Atkins, K.A.; Franklin, K.A.; Dodd, A.N. Integration of light and circadian signals that regulate chloroplast transcription by a nuclear-encoded sigma factor. *New Phytol.* **2017**, *213*, 727–738. [CrossRef] [PubMed]
45. Wang, X.; Wang, Y.; Wang, L.; Liu, H.; Zhang, B.; Cao, Q.; Liu, X.; Lv, Y.; Bi, S.; Zhang, S.; et al. Arabidopsis PCaP2 Functions as a Linker Between ABA and SA Signals in Plant Water Deficit Tolerance. *Front. Plant Sci.* **2018**, *9*, 578. [CrossRef]
46. Munemasa, S.; Hauser, F.; Park, J.; Waadt, R.; Brandt, B.; Schroeder, J.I. Mechanisms of abscisic acid-mediated control of stomatal aperture. *Curr. Opin. Plant Biol.* **2015**, *28*, 154–162. [CrossRef]
47. Avalbaev, A.M.; Somov, K.A.; Yuldashev, R.A.; Shakirova, F.M. Cytokinin oxidase is key enzyme of cytokinin degradation. *Biochemistry* **2012**, *77*, 1354–1361. [CrossRef]
48. Hou, S.; Liu, Z.; Shen, H.; Wu, D. Damage-Associated Molecular Pattern-Triggered Immunity in Plants. *Front. Plant Sci.* **2019**, *10*, 646. [CrossRef]
49. Bacete, L.; Mélida, H.; Miedes, E.; Molina, A. Plant cell wall-mediated immunity: Cell wall changes trigger disease resistance responses. *Plant J.* **2018**, *93*, 614–636. [CrossRef]
50. Li, Q.; Wang, C.; Mou, Z. Perception of Damaged Self in Plants. *Plant Physiol.* **2020**, *182*, 1545–1565. [CrossRef]
51. Pham, A.Q.; Cho, S.-H.; Nguyen, C.T.; Stacey, G. Arabidopsis Lectin Receptor Kinase P2K2 Is a Second Plant Receptor for Extracellular ATP and Contributes to Innate Immunity. *Plant Physiol.* **2020**, *183*, 1364–1375. [CrossRef] [PubMed]
52. Hornung, V.; Ablasser, A.; Charrel-Dennis, M.; Bauernfeind, F.; Horvath, G.; Caffrey, D.R.; Latz, E.; Fitzgerald, K.A. AIM2 recognizes cytosolic dsDNA and forms a caspase-1-activating inflammasome with ASC. *Nature* **2009**, *458*, 514–518. [CrossRef] [PubMed]
53. Greeff, C.; Roux, M.; Mundy, J.; Petersen, M. Receptor-like kinase complexes in plant innate immunity. *Front. Plant Sci.* **2012**, *3*, 209. [CrossRef] [PubMed]
54. Ferrusquía-Jiménez, N.I.; Chandrakasan, G.; Torres-Pacheco, I.; Rico-Garcia, E.; Feregrino-Perez, A.A.; Guevara-González, R.G. Extracellular DNA: A Relevant Plant Damage-Associated Molecular Pattern (DAMP) for Crop Protection Against Pests—A Review. *J. Plant Growth Regul.* **2020**, *40*, 451–463. [CrossRef]
55. Albert, I.; Hua, C.; Nürnberger, T.; Pruitt, R.N.; Zhang, L. Surface Sensor Systems in Plant Immunity. *Plant Physiol.* **2020**, *182*, 1582–1596. [CrossRef]
56. Cheung, A.Y.; Qu, L.-J.; Russinova, E.; Zhao, Y.; Zipfel, C. Update on Receptors and Signaling. *Plant Physiol.* **2020**, *182*, 1527–1530. [CrossRef]
57. Gravino, M.; Locci, F.; Tundo, S.; Cervone, F.; Savatin, D.V.; De Lorenzo, G. Immune responses induced by oligogalacturonides are differentially affected by AvrPto and loss of BAK1/BKK1 and PEPR1/PEPR2. *Mol. Plant Pathol.* **2017**, *18*, 582–595. [CrossRef]
58. De Lorenzo, G.; Ferrari, S.; Cervone, F.; Okun, E. Extracellular DAMPs in Plants and Mammals: Immunity, Tissue Damage and Repair. *Trends Immunol.* **2018**, *39*, 937–950. [CrossRef]
59. Choi, H.W.; Klessig, D.F. DAMPs, MAMPs, and NAMPs in plant innate immunity. *BMC Plant Biol.* **2016**, *16*, 232. [CrossRef]
60. Quintana-Rodriguez, E.; Duran-Flores, D.; Heil, M.; Camacho-Coronel, X. Damage-associated molecular patterns (DAMPs) as future plant vaccines that protect crops from pests. *Sci. Hortic.* **2018**, *237*, 207–220. [CrossRef]
61. He, K.; Wu, Y. Receptor-Like Kinases and Regulation of Plant Innate Immunity. *Enzymes* **2016**, *40*, 105–142. [CrossRef]
62. Bentham, A.R.; De la Concepcion, J.C.; Mukhi, N.; Zdrzałek, R.; Draeger, M.; Gorenkin, D.; Hughes, R.K.; Banfield, M.J. A molecular roadmap to the plant immune system. *J. Biol. Chem.* **2020**, *295*, 14916–14935. [CrossRef]
63. Erb, M.; Reymond, P. Molecular Interactions Between Plants and Insect Herbivores. *Annu. Rev. Plant Biol.* **2019**, *70*, 527–557. [CrossRef]
64. Bisgrove, S.R.; Simonich, M.T.; Smith, N.M.; Sattler, A.; Innes, R.W. A disease resistance gene in Arabidopsis with specificity for two different pathogen avirulence genes. *Plant Cell* **1994**, *6*, 927–933. [CrossRef]
65. Clifton, R.; Millar, A.H.; Whelan, J. Alternative oxidases in Arabidopsis: A comparative analysis of differential expression in the gene family provides new insights into function of non-phosphorylating bypasses. *Biochim. Biophys. Acta* **2006**, *1757*, 730–741. [CrossRef]
66. Selinski, J.; Hartmann, A.; Deckers-Hebestreit, G.; Day, D.A.; Whelan, J.; Scheibe, R. Alternative Oxidase Isoforms Are Differentially Activated by Tricarboxylic Acid Cycle Intermediates. *Plant Physiol.* **2018**, *176*, 1423–1432. [CrossRef] [PubMed]
67. Maruta, T.; Tanouchi, A.; Tamoi, M.; Yabuta, Y.; Yoshimura, K.; Ishikawa, T.; Shigeoka, S. Arabidopsis Chloroplastic Ascorbate Peroxidase Isoenzymes Play a Dual Role in Photoprotection and Gene Regulation under Photooxidative Stress. *Plant Cell Physiol.* **2009**, *51*, 190–200. [CrossRef] [PubMed]
68. Kliebenstein, D.J.; Monde, R.A.; Last, R.L. Superoxide dismutase in Arabidopsis: An eclectic enzyme family with disparate regulation and protein localization. *Plant Physiol.* **1998**, *118*, 637–650. [CrossRef] [PubMed]

69. Bechtold, U.; Murphy, D.J.; Mullineaux, P.M. Arabidopsis Peptide Methionine Sulfoxide Reductase2 Prevents Cellular Oxidative Damage in Long Nights. *Plant Cell* **2004**, *16*, 908–919. [CrossRef]
70. Simpson, P.J.; Tantitadapitak, C.; Reed, A.M.; Mather, O.C.; Bunce, C.M.; White, S.A.; Ride, J.P. Characterization of two novel aldo-keto reductases from Arabidopsis: Expression patterns, broad substrate specificity, and an open active-site structure suggest a role in toxicant metabolism following stress. *J. Mol. Biol.* **2009**, *392*, 465–480. [CrossRef]
71. Seki, M.; Narusaka, M.; Ishida, J.; Nanjo, T.; Fujita, M.; Oono, Y.; Kamiya, A.; Nakajima, M.; Enju, A.; Sakurai, T.; et al. Monitoring the expression profiles of 7000 Arabidopsis genes under drought, cold and high-salinity stresses using a full-length cDNA microarray. *Plant J.* **2002**, *31*, 279–292. [CrossRef] [PubMed]
72. Werner, T.; Motyka, V.; Laucou, V.; Smets, R.; Van Onckelen, H.; Schmülling, T. Cytokinin-deficient transgenic Arabidopsis plants show multiple developmental alterations indicating opposite functions of cytokinins in the regulation of shoot and root meristem activity. *Plant Cell* **2003**, *15*, 2532–2550. [CrossRef] [PubMed]
73. Chen, L.-Q.; Qu, X.-Q.; Hou, B.-H.; Sosso, D.; Osorio, S.; Fernie, A.R.; Frommer, W.B. Sucrose efflux mediated by SWEET proteins as a key step for phloem transport. *Science* **2012**, *335*, 207–211. [CrossRef]
74. Eom, J.-S.; Chen, L.-Q.; Sosso, D.; Julius, B.T.; Lin, I.W.; Qu, X.-Q.; Braun, D.M.; Frommer, W.B. SWEETs, transporters for intracellular and intercellular sugar translocation. *Curr. Opin. Plant Biol.* **2015**, *25*, 53–62. [CrossRef] [PubMed]
75. Kanno, Y.; Oikawa, T.; Chiba, Y.; Ishimaru, Y.; Shimizu, T.; Sano, N.; Koshiba, T.; Kamiya, Y.; Ueda, M.; Seo, M. AtSWEET13 and AtSWEET14 regulate gibberellin-mediated physiological processes. *Nat. Commun.* **2016**, *7*, 13245. [CrossRef]
76. Gallucci, S.; Maffei, M.E. DNA Sensing across the Tree of Life. *Trends Immunol.* **2017**, *38*, 719–732. [CrossRef] [PubMed]
77. Heil, M.; Ibarra-Laclette, E.; Adame-Álvarez, R.M.; Martínez, O.; Ramirez-Chávez, E.; Molina-Torres, J.; Herrera-Estrella, L. How plants sense wounds: Damaged-self recognition is based on plant-derived elicitors and induces octadecanoid signaling. *PLoS ONE* **2012**, *7*, e30537. [CrossRef]
78. Heil, M.; Land, W.G. Danger signals—Damaged-self recognition across the tree of life. *Front. Plant Sci.* **2014**, *5*, 578. [CrossRef]
79. Duran-Flores, D.; Heil, M. Sources of specificity in plant damaged-self recognition. *Curr. Opin. Plant Biol.* **2016**, *32*, 77–87. [CrossRef]
80. Wang, K.; Guo, Q.; Froehlich, J.E.; Hersh, H.L.; Zienkiewicz, A.; Howe, G.A.; Benning, C. Two Abscisic Acid-Responsive Plastid Lipase Genes Involved in Jasmonic Acid Biosynthesis in Arabidopsis thaliana. *Plant Cell* **2018**, *30*, 1006–1022. [CrossRef]
81. Vos, I.; Verhage, A.; Schuurink, R.; Watt, L.; Pieterse, C.; Van Wees, S. Onset of herbivore-induced resistance in systemic tissue primed for jasmonate-dependent defenses is activated by abscisic acid. *Front. Plant Sci.* **2013**, *4*, 539. [CrossRef]
82. Guo, Q.; Major, I.T.; Howe, G.A. Resolution of growth–defense conflict: Mechanistic insights from jasmonate signaling. *Curr. Opin. Plant Biol.* **2018**, *44*, 72–81. [CrossRef]
83. Heil, M.; Baldwin, I.T. Fitness costs of induced resistance: Emerging experimental support for a slippery concept. *Trends Plant Sci.* **2002**, *7*, 61–67. [CrossRef]
84. Pearce, G.; Yamaguchi, Y.; Barona, G.; Ryan, C.A. A subtilisin-like protein from soybean contains an embedded, cryptic signal that activates defense-related genes. *Proc. Natl. Acad. Sci. USA* **2010**, *107*, 14921–14925. [CrossRef] [PubMed]
85. Huffaker, A.; Pearce, G.; Veyrat, N.; Erb, M.; Turlings, T.C.J.; Sartor, R.; Shen, Z.; Briggs, S.P.; Vaughan, M.M.; Alborn, H.T.; et al. Plant elicitor peptides are conserved signals regulating direct and indirect antiherbivore defense. *Proc. Natl. Acad. Sci. USA* **2013**, *110*, 5707–5712. [CrossRef] [PubMed]
86. Choi, J.; Tanaka, K.; Cao, Y.; Qi, Y.; Qiu, J.; Liang, Y.; Lee, S.Y.; Stacey, G. Identification of a Plant Receptor for Extracellular ATP. *Science* **2014**, *343*, 290–294. [CrossRef]
87. Klauser, D.; Desurmont, G.A.; Glauser, G.; Vallat, A.; Flury, P.; Boller, T.; Turlings, T.C.J.; Bartels, S. The Arabidopsis Pep-PEPR system is induced by herbivore feeding and contributes to JA-mediated plant defence against herbivory. *J. Exp. Bot.* **2015**, *66*, 5327–5336. [CrossRef]
88. Poncini, L.; Wyrsch, I.; Dénervaud Tendon, V.; Vorley, T.; Boller, T.; Geldner, N.; Métraux, J.-P.; Lehmann, S. In roots of *Arabidopsis thaliana*, the damage-associated molecular pattern AtPep1 is a stronger elicitor of immune signalling than flg22 or the chitin heptamer. *PLoS ONE* **2017**, *12*, e0185808. [CrossRef]
89. Jewell, J.B.; Sowders, J.M.; He, R.; Willis, M.A.; Gang, D.R.; Tanaka, K. Extracellular ATP Shapes a Defense-Related Transcriptome Both Independently and along with Other Defense Signaling Pathways. *Plant Physiol.* **2019**, *179*, 1144–1158. [CrossRef]
90. Wang, Q.-W.; Jia, L.-Y.; Shi, D.-L.; Wang, R.; Lu, L.-N.; Xie, J.-J.; Sun, K.; Feng, H.-Q.; Li, X. Effects of extracellular ATP on local and systemic responses of bean (*Phaseolus vulgaris* L) leaves to wounding. *Biosci. Biotechnol. Biochem.* **2019**, *83*, 417–428. [CrossRef]
91. Huang, H.-J.; Cui, J.-R.; Xia, X.; Chen, J.; Ye, Y.-X.; Zhang, C.-X.; Hong, X.-Y. Salivary DNase II from *Laodelphax striatellus* acts as an effector that suppresses plant defence. *New Phytol.* **2019**, *224*, 860–874. [CrossRef]
92. Krol, E.; Mentzel, T.; Chinchilla, D.; Boller, T.; Felix, G.; Kemmerling, B.; Postel, S.; Arents, M.; Jeworutzki, E.; Al-Rasheid, K.A.S.; et al. Perception of the Arabidopsis danger signal peptide 1 involves the pattern recognition receptor AtPEPR1 and its close homologue AtPEPR2. *J. Biol. Chem.* **2010**, *285*, 13471–13479. [CrossRef]
93. Lim, M.H.; Wu, J.; Clark, G.; Roux, S. Apyrases (NTPDases) and extracellular nucleotides regulate plant responses to biotic and abiotic stresses. *Purinergic Signal.* **2014**, *10*, 712.
94. Yakushiji, S.; Ishiga, Y.; Inagaki, Y.; Toyoda, K.; Shiraishi, T.; Ichinose, Y. Bacterial DNA activates immunity in Arabidopsis thaliana. *J. Gen. Plant Pathol.* **2009**, *75*, 227–234. [CrossRef]

95. Niehl, A.; Wyrsch, I.; Boller, T.; Heinlein, M. Double-stranded RNAs induce a pattern-triggered immune signaling pathway in plants. *New Phytol.* **2016**, *211*, 1008–1019. [CrossRef] [PubMed]
96. Heidel, A.J.; Dong, X. Fitness Benefits of Systemic Acquired Resistance During *Hyaloperonospora parasitica* Infection in *Arabidopsis thaliana*. *Genetics* **2006**, *173*, 1621–1628. [CrossRef] [PubMed]
97. Ecker, J.R.; Davis, R.W. Inhibition of gene expression in plant cells by expression of antisense RNA. *Proc. Natl. Acad. Sci. USA* **1986**, *83*, 5372–5376. [CrossRef] [PubMed]
98. Hannon, G.J. RNA interference. *Nature* **2002**, *418*, 244–251. [CrossRef]
99. Gruenert, D.; Bruscia, E.; Novelli, G.; Colosimo, A.; Dallapiccola, B.; Sangiuolo, F.; Goncz, K. Sequence-specific modification of genomic DNA by small DNA fragments. *J. Clin. Invest.* **2003**, *112*, 637–641. [CrossRef]
100. Nisa, M.-U.; Huang, Y.; Benhamed, M.; Raynaud, C. The Plant DNA Damage Response: Signaling Pathways Leading to Growth Inhibition and Putative Role in Response to Stress Conditions. *Front. Plant Sci.* **2019**, *10*, 653. [CrossRef]
101. Kim, J.-H. Multifaceted Chromatin Structure and Transcription Changes in Plant Stress Response. *Int. J. Mol. Sci.* **2021**, *22*, 2013. [CrossRef]
102. Luan, S. The CBL-CIPK network in plant calcium signaling. *Trends Plant Sci.* **2009**, *14*, 37–42. [CrossRef]
103. Lee, K.-W.; Chen, P.-W.; Lu, C.-A.; Chen, S.; Ho, T.-H.D.; Yu, S.-M. Coordinated responses to oxygen and sugar deficiency allow rice seedlings to tolerate flooding. *Sci. Signal.* **2009**, *2*, ra61. [CrossRef] [PubMed]
104. Pandey, G.K.; Kanwar, P.; Singh, A.; Steinhorst, L.; Pandey, A.; Yadav, A.K.; Tokas, I.; Sanyal, S.K.; Kim, B.-G.; Lee, S.-C.; et al. Calcineurin B-Like Protein-Interacting Protein Kinase CIPK21 Regulates Osmotic and Salt Stress Responses in Arabidopsis. *Plant Physiol.* **2015**, *169*, 780–792. [CrossRef]
105. Cheng, N.-H.; Pittman, J.K.; Zhu, J.-K.; Hirschi, K.D. The protein kinase SOS2 activates the Arabidopsis H(+)/Ca(2+) antiporter CAX1 to integrate calcium transport and salt tolerance. *J. Biol. Chem.* **2004**, *279*, 2922–2926. [CrossRef] [PubMed]
106. Choi, W.-G.; Miller, G.; Wallace, I.; Harper, J.; Mittler, R.; Gilroy, S. Orchestrating rapid long-distance signaling in plants with Ca(2+), ROS and electrical signals. *Plant J.* **2017**, *90*, 698–707. [CrossRef] [PubMed]
107. Toyota, M.; Spencer, D.; Sawai-Toyota, S.; Jiaqi, W.; Zhang, T.; Koo, A.; Howe, G.; Gilroy, S. Glutamate triggers long-distance, calcium-based plant defense signaling. *Science* **2018**, *361*, 1112–1115. [CrossRef]
108. Salvador-Recatalà, V.; Tjallingii, W.F.; Farmer, E.E. Real-time, in vivo intracellular recordings of caterpillar-induced depolarization waves in sieve elements using aphid electrodes. *New Phytol.* **2014**, *203*, 674–684. [CrossRef] [PubMed]
109. Canales, J.; Henriquez-Valencia, C.; Brauchi, S. The Integration of Electrical Signals Originating in the Root of Vascular Plants. *Front. Plant Sci.* **2018**, *8*, 2173. [CrossRef]
110. Chin, J.Y.; Glazer, P.M. Repair of DNA lesions associated with triplex-forming oligonucleotides. *Mol. Carcinog.* **2009**, *48*, 389–399. [CrossRef] [PubMed]
111. Gahan, P.B.; Anker, P.; Stroun, M. Metabolic DNA as the origin of spontaneously released DNA? *Ann. N. Y. Acad. Sci.* **2008**, *1137*, 7–17. [CrossRef]
112. Dobin, A.; Davis, C.A.; Schlesinger, F.; Drenkow, J.; Zaleski, C.; Jha, S.; Batut, P.; Chaisson, M.; Gingeras, T.R. STAR: Ultrafast universal RNA-seq aligner. *Bioinformatics* **2013**, *29*, 15–21. [CrossRef] [PubMed]
113. Liao, Y.; Smyth, G.K.; Shi, W. featureCounts: An efficient general purpose program for assigning sequence reads to genomic features. *Bioinformatics* **2014**, *30*, 923–930. [CrossRef]
114. Robinson, M.D.; McCarthy, D.J.; Smyth, G.K. edgeR: A Bioconductor package for differential expression analysis of digital gene expression data. *Bioinformatics* **2010**, *26*, 139–140. [CrossRef] [PubMed]
115. Howe, E.A.; Sinha, R.; Schlauch, D.; Quackenbush, J. RNA-Seq analysis in MeV. *Bioinformatics* **2011**, *27*, 3209–3210. [CrossRef] [PubMed]
116. Young, M.D.; Wakefield, M.J.; Smyth, G.K.; Oshlack, A. Gene ontology analysis for RNA-seq: Accounting for selection bias. *Genome Biol.* **2010**, *11*, R14. [CrossRef]

plants

MDPI

Article

Agro-Alimentary Potential of the Neglected and Underutilized Local Endemic Plants of Crete (Greece), Rif-Mediterranean Coast of Morocco and Tunisia: Perspectives and Challenges

Mohamed Libiad [1,2,*], Abdelmajid Khabbach [2,3], Mohamed El Haissoufi [2], Ioannis Anestis [4], Fatima Lamchouri [2], Soumaya Bourgou [5], Wided Megdiche-Ksouri [5], Zeineb Ghrabi-Gammar [6,7], Vasileios Greveniotis [8], Ioannis Tsiripidis [9], Eleftherios Dariotis [10], Maria A. Tsiafouli [4] and Nikos Krigas [10,*]

1 Laboratory of Ecology, Systematics and Biodiversity Conservation (LESCOBIO), Department of Biology, Faculty of Sciences, Abdelmalek Essaâdi University, B.P. 2121, M'Hannech II, Tetouan 93000, Morocco
2 Laboratory of Natural Substances, Pharmacology, Environment, Modelling, Health and Quality of Life (SNAMOPEQ), Polydisciplinary Faculty of Taza, Sidi Mohamed Ben Abdellah University, B.P. 1223, Taza Gare, Taza 35000, Morocco; khamajid@hotmail.com (A.K.); mohamed.elhaissoufi1@usmba.ac.ma (M.E.H.); fatima.lamchouri@usmba.ac.ma (F.L.)
3 Laboratory of Biotechnology, Conservation and Development of Natural Resources (BCVRN), Department of Biology, Faculty of Sciences Dhar El Mahraz, Sidi Mohamed Ben Abdellah University, B.P. 1796, Fès-Atlas 30003, Morocco
4 Department of Ecology, School of Biology, Aristotle University of Thessaloniki, 54124 Thessaloniki, Greece; ganestis3@gmail.com (I.A.); tsiafoul@bio.auth.gr (M.A.T.)
5 Laboratoire des Plantes Aromatiques et Médicinales, Centre de Biotechnologie de Borj-Cédria, B.P. 901, Tunis 2050, Tunisia; bourgousoumaya@yahoo.com (S.B.); ksouriwided@yahoo.fr (W.M.-K.)
6 Institut National Agronomique de Tunisie, Université de Carthage, 43 Avenue Charles Nicolle, Cité Mahrajène, Tunis 1082, Tunisia; zghrabi@yahoo.fr
7 Laboratoire de Recherche Biogéographie, Climatologie Appliquée et Dynamiques Environnementales (BiCADE 18ES13), Faculté des Lettres des Arts et des Humanités de Manouba, Université de la Manouba, Campus Universitaire de la Manouba, Manouba 2010, Tunisia
8 Institute of Industrial and Forage Crops, Hellenic Agricultural Organization Demeter, 41335 Larisa, Greece; vgreveni@mail.com
9 Department of Botany, School of Biology, Aristotle University of Thessaloniki, 54124 Thessaloniki, Greece; tsiripid@bio.auth.gr
10 Institute of Plant Breeding and Genetic Resources, Hellenic Agricultural Organization Demeter, Thermi, P.O. Box 60458, 57001 Thessaloniki, Greece; eleftheriosdariotis@yahoo.com
* Correspondence: libiad001@gmail.com (M.L.); nikoskrigas@gmail.com (N.K.)

Citation: Libiad, M.; Khabbach, A.; El Haissoufi, M.; Anestis, I.; Lamchouri, F.; Bourgou, S.; Megdiche-Ksouri, W.; Ghrabi-Gammar, Z.; Greveniotis, V.; Tsiripidis, I.; et al. Agro-Alimentary Potential of the Neglected and Underutilized Local Endemic Plants of Crete (Greece), Rif-Mediterranean Coast of Morocco and Tunisia: Perspectives and Challenges. *Plants* **2021**, *10*, 1770. https://doi.org/10.3390/plants10091770

Academic Editors: Milan S. Stankovic, Paula Baptista and Petronia Carillo

Received: 5 August 2021
Accepted: 23 August 2021
Published: 25 August 2021

Publisher's Note: MDPI stays neutral with regard to jurisdictional claims in published maps and institutional affiliations.

Copyright: © 2021 by the authors. Licensee MDPI, Basel, Switzerland. This article is an open access article distributed under the terms and conditions of the Creative Commons Attribution (CC BY) license (https://creativecommons.org/licenses/by/4.0/).

Abstract: The neglected and underutilized plants (NUPs) could become alternative food sources in the agro-alimentary sector, enriching human and animal diets, offering the opportunity for sustainable exploitation, resilience to climate change, and production with resistance to pests and diseases. In the Mediterranean countries, these valuable resources are threatened by climate change, overexploitation, and/or monoculture. In this framework, we evaluated 399 local endemic NUPs of Crete (Greece), the Mediterranean coast, Rif of Morocco, and Tunisia, regarding their agro-alimentary potential, and assessed their feasibility and readiness timescale for sustainable exploitation with own previously published methodology. The methodological scheme was developed by experts in co-creative workshops, using point-scoring of seven attributes to evaluate the potential of the targeted NUPs in the agro-alimentary. Our results showed a diversity of promising local endemic NUPs of different families in the studied regions (Lamiaceae members are prominent), and we outlined the cases of 13 taxa with the highest optimum scores of agro-alimentary potential (>70%). Despite the diversity or the promising potential and current ex-situ conservation efforts to bridge gaps, our study indicated that only a few cases of Cretan local endemic NUPs can be sustainably exploited in the short-term. However, it is argued that many more local endemic NUPs can easily follow sustainable exploitation schemes if specific research gaps are bridged. Since NUPs can help to increased diversification of food production systems by adding new nutritional/beneficial species to human and animal diets, basic and applied research, as well as market and stakeholder attraction, is suggested as prerequisite to unlock the full potential of the focal endemic NUPs in the agro-alimentary sector.

Keywords: climate change; food security; Mediterranean countries; sustainable exploitation; phytogenetic resources

1. Introduction

Plants play a great role in human life. It is estimated that at least 31,128 plant species are currently used worldwide, and 17.79% are utilized for human nutrition [1]. Despite this diversity, only 30 plant species (including major staple crops) provide 95% of dietary energy or protein to feed the world [2,3]. However, regional studies in the Mediterranean countries highlight that more than 52% of the wild harvested plants can be used for agro-alimentary purposes [4,5]. Despite the high scale of use of plants in local Mediterranean economies, food resources are also affected by climate change, monocultures, and/or overexploitation [6]. To mitigate the effect of climate change and the degradation of land and water resources, it becomes urgent to engage improved crops and new species that are adapted to difficult environments and can increase the overall productivity and stability of agro-ecosystems [7]. Hence, there is need to change current agricultural practices and promote novel crops that are more resilient to climate change challenges, such as the neglected and underutilized plant species (NUPs). NUPs have an important role to play in diversifying and advancing agricultural development beyond the Green Revolution model by improving the yields of staple crops and by introducing new valuable crops to enrich human and animal nutrition [8]. These NUPs present tremendous opportunities for fighting poverty, hunger, malnutrition, and, at the same time, can increase the development of subsistence economies in local/regional scales [8–11].

The implication of citizens (or consumers), scientists, entrepreneurs, producers, market specialists and stakeholders, and the establishment of targeted or applied research and development programs can act as key drivers to promote sustainable food systems, especially regarding NUPs [8,12–14]. The instating of NUPs as alternative food sources in the agro-alimentary sector would depend on the availability of information describing their agronomical aspects, water-use, and possible drought tolerance [15]. There are some NUPs that could contribute to the food security of rural and urban people in India and some countries of South America, e.g., some Andean grains such as quinoa (*Chenopodium quinoa* Willd.) and cañahua (Bolivian name, known as cañihua in Peru) *Chenopodium pallidicaule* Aellen [8], and minor millets such as foxtail millet *Setaria italica* (L.) P. Beauv., proso millet *Panicum miliaceum* L., finger millet *Eleusine coracana* (L.) Gaertn, kodo millet *Paspalum scrobiculatum* L., little millet *Panicum sumatrense* Roth, and barnyard millet *Echinochloa colonum* (L.) Link [11]. In the Mediterranean Basin, the use of resistant NUPs to pests, diseases and extreme environmental factors (e.g., *Cistus ladanifer* L.) may be a viable solution for cultivations in poor and degraded soils. In addition, this species reveals interesting aptitudes that can be applied to food, pharmaceutical, phytochemical, and biofuel industries [16].

In both developed and developing countries, the increased demand from consumers for diversity and novelty in modern foods is currently creating new market niches for NUPs [7] due to their high content in vitamins, micronutrients, proteins, and other beneficial compounds [7,9,10]. Consequently, the introduction of NUPs in agro-alimentary production systems could play a strategic role in improving many dimensions of livelihoods and well-being, and therefore may represent an important source of household income, also encouraging women empowerment [7,9–11,17]. Conversely, better marketing and consumer awareness of the benefits associated with NUPs could play a critical role in their sustainable exploitation [12].

In an increasingly human-dominated world, conservation of species in the wild is (or should be) a top-priority [18], and the precedence should be given to the local endemic plants of biodiversity-rich regions [19,20]. Typically, the criteria adopted for conservation prioritization involve aspects of the geographic distribution, endemism, and the threats affecting each evaluated taxon [19,20], including harvesting from the wild. However, to date, some thousands of medicinal-aromatic plants, as well as the vast majority of the wild edible greens, are still collected directly from the wild [21], depleting their natural potential. In Mediterranean areas, the possibility of first genotypic selection, targeted and/or improved pilot cultivation, and new processing techniques for specific applications gives the NUPs the potential to be used as a valuable novel resource [16]. Furthermore, high education status and primary occupation of the household head may also have a major role for the conservation of NUPs at local scales [14].

Although NUPs are not intended to replace staple crops, crop diversification is envisaged as one of the best means to ensure sustainable agricultural production systems [3,17], by growing NUPs as part of crop rotation systems or by inter-cropping them with other crops. This practice could protect and enhance agro-biodiversity at field level and may disrupt the cycle of some pests and diseases [3]. The sustainable promotion of NUPs requires better understanding as well as improved marketing and high consumer awareness regarding the benefits associated with selected NUPs [12], while scientific research including agronomical aspects, breeding potential, post-harvest handling, and value chain establishment may bridge existing gaps, linking valuable local resources with farmers and markets [15,22]. This is the context of the current study. Our investigation aims at exploring and documenting for the first time the potential of the single-country/single-region endemic plants of Crete (Greece), Mediterranean Coast and Rif of Morocco, and Tunisia in the agro-alimentary sector, with the scope to identify the actions needed to remove barriers and bridge gaps regarding their sustainable exploitation.

2. Results

2.1. Cluster Analysis of Agro-Alimentary Attributes and of Focal Taxa

The results of the hierarchical cluster analyses of agro-alimentary attributes (Figure 1) showed that, in the case of Crete, the attribute 'food additive potential' was grouped together with the attribute 'wild edible greens' next to the 'bee attraction' attribute. The 'beverage potential' and 'spicy element' as attributes formed another subgroup, while these were clustered together with the subgroup 'aromatic properties' and 'type of aroma'.

In the case of Tunisia, as well as the Mediterranean coast and Rif of Morocco, a similar pattern was observed; the attributes 'beverage potential' and 'bee attraction' formed a subcluster, and they were clustered together with the attribute 'food additive potential'. The attributes 'type of aroma' and 'aromatic properties' were closely linked together and formed a cluster with the attribute 'spicy element', as in the case of Crete. Finally, the attribute 'wild edible greens' was not linked to any other attribute.

2.2. Diversity of Local Endemic NUPs

The evaluation of the potential of the focal endemic NUPs in the agro-alimentary sector showed that Lamiaceae family members (26 taxa) are mostly represented in the top 15 taxa of each of the three studied regions, followed by members of Asteraceae (5 taxa), Liliaceae (4 taxa in the case of Crete), Caryophyllaceae, Gentianaceae, Plumbaginaceae, and Rubiaceae (2 taxa each), as well as Apiaceae and Pinaceae (1 taxon each). Regarding the top 15 evaluated NUPs of each region, Lamiaceae members prevail again in the case of the Mediterranean coast and Rif of Morocco (11 taxa), followed by Crete (9 taxa) and Tunisia (6 taxa).

2.3. Agro-Alimentary Potential of the Focal Local Endemic NUPs

Supplementary Material S1 provides examples of scoring of taxa per attribute and the evaluation of the agro-alimentary potential (Level I) of the focal local endemic taxa of the studied regions is presented in Supplementary Material S2 as percentages of the maximum possible scores achieved.

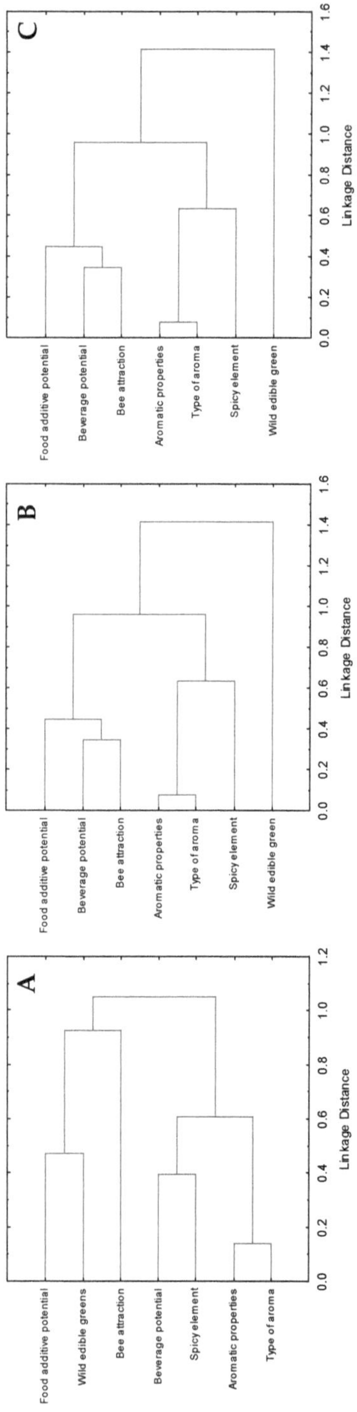

Figure 1. Graph of hierarchical clustering of the agro-alimentary attributes (complete linkage, 1-Pearson r distance) based on the score values of the local endemic plants of (**A**) Crete, (**B**) Mediterranean coast and Rif of Morocco, and (**C**) Tunisia.

2.3.1. Local Endemic Plants of Crete

Among the Cretan local endemics, the four highest evaluated taxa (Figure 2) were *Origanum dictamnus* L. (85.71%), *Origanum microphyllum* (Benth.) Vogel, *Sideritis syriaca* L. subsp. *syriaca* (80.95 % each), and *Thymbra calostachya* (Rech. f.) Rech. f. (71.42%); these plants showed a very interesting agro-alimentary potential. The scoring of *Origanum dictamnus* (85.71%) is illustrated in Figure 3. In total, three taxa [hierarchically: *Helichrysum doerfleri* Rech. f., *H. heldreichii* Boiss., and *Calamintha cretica* (L.) Lam.] ranked in above-average to high positions with scores 55–70%. Overall, nine taxa ranked above-average with scores 50–55%, i.e., the wild garlic or wild onion plants *Allium bourgeaui* Rech. f. subsp. *creticum* Bothmer, *A. circinnatum* Sieber subsp. *circinnatum*, *A. dilatatum* Zahar., and *A. platakisii* Tzanoud. and Kypr., *Micromeria hispida* Boiss. and Heldr. ex Benth. and *Micromeria sphaciotica* Boiss. and Heldr. ex Benth. (54.76% each). Moreover, another 20 taxa ranked in below-average to low positions with scores 35–50%. For 157 taxa, the scores ranked comparatively very low (<35%), and the lowest value was assigned to 30 taxa due to zero values, e.g., *Carex cretica* Gradst. and J. Kern (Supplementary Material S2).

Figure 2. Four top-evaluated Cretan endemic plants of Lamiaceae family, in terms of maximum possible score achieved regarding their agro-alimentary potential (Photos: N. Krigas). (**A**): *Origanum dictamnus*, (**B**): *Origanum microphyllum*, (**C**): *Sideritis syriaca* subsp. *syriaca*, (**D**): *Thymbra calostachya*. For these plants, propagation material has been collected from the wild to allow ex-situ conservation and propagation-cultivation trials at the Institute of Plant Breeding and Genetic Resources, Hellenic Agricultural Organization Demeter.

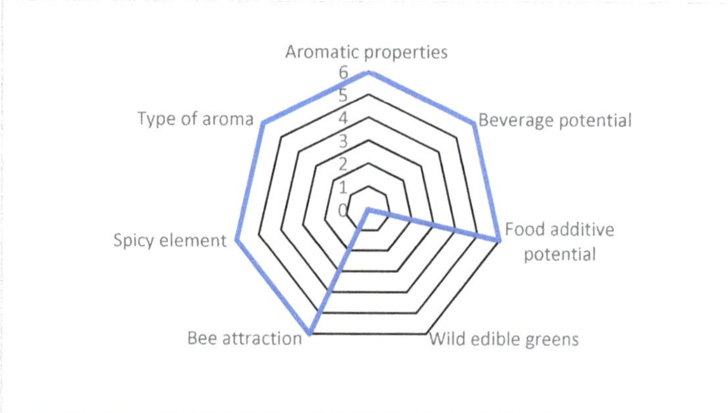

Figure 3. Evaluation example of *Origanum dictamnus* (Cretan endemic) scored for seven agro-alimentary attributes, reaching 85.7% of the optimum possible score. This example is hierarchically ranked in the highest (>70%) class. For attributes and scoring, see Section 4.

2.3.2. Local Endemic Plants of the Mediterranean Coast-Rif of Morocco

The six highest-evaluated North Moroccan taxa were *Centaurium erythraea* Rafn subsp. *bifrons* (Pau) Greuter, *Teucrium afrum* (Emb. and Maire) Pau subsp. *rubriflorum* (Pau and Font Quer) Castrov. and Bayon, *T. grosii* Pau, *T. gypsophilum* Emb. and Maire, *T. huotii* Emb. and Maire and *T. rotundifolium* Schreb. subsp. *sanguisorbifolium* (Pau and Font Quer) E.Cohen (each 71.43%), showing a very interesting agro-alimentary potential (Figure 4). The scoring of *Centaurium erythraea* subsp. *bifrons* (71.43%) is illustrated in Figure 5. In total, four taxa, namely *Teucrium chlorostachyum* Pau and Font Quer subsp. *chlorostachyum*, *T. chlorostachyum* subsp. *melillense* (Maire) El Oualidi, Mathez and T. Navarro, *T. rifanum* (Maire and Sennen) T. Navarro and El Oualidi, *Salvia interrupta* Schousb. subsp. *paui* (Maire) Maire, ranked in above-average to high positions, with scores > 55–70%. Overall, eight taxa ranked in lower to average positions with scores 35–50%, i.e., *Abies marocana* Trab. (47.62%), *Centaurium barrelieroides* Pau (45.24%), *Marrubium fontianum* Maire and *M. heterocladum* Emb. and Maire (42.86% each), *Anthemis mauritiana* Maire and Sennen subsp. *mauritiana*, *Origanum elongatum* (Bonnet) Emb. and Maire, *Stachys fontqueri* Pau and *Vicia cedretorum* Font Quer (38.10% each). For 67 taxa, the scores ranked comparatively very low (<35%), and the lowest positions were assigned to 9 taxa with the score zero values, e.g., *Hemicrambe fruticulosa* Webb (Supplementary Material S2).

Figure 4. Top-evaluated endemic plants of Mediterranean coast and Rif of Morocco in terms of maximum possible score achieved regarding their agro-alimentary potential. (**A**): *Teucrium huotii* (Photo: M. Rouviere, http://www.ville-ge.ch/musinfo/bd/cjb/africa/details.php?langue=fr&id=145076; accessed on 24 August 2021), (**B**): *Salvia interrupta* subsp. *paui* (Photo: A. Homrani Bakali, https://www.teline.fr/; accessed on 24 August 2021), (**C**): *Teucrium gypsophilum* (Left photo: F. Lamchouri; right photo: A. Khabbach). Propagation material has been collected from the wild for *Teucrium gypsophilum* to allow ex-situ conservation and propagation-cultivation trials at the Institute of Plant Breeding and Genetic Resources, Hellenic Agricultural Organization Demeter.

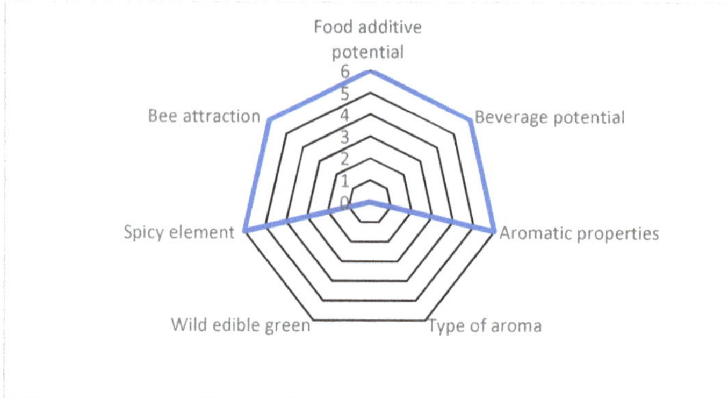

Figure 5. Evaluation example of *Centaurium eythraea* subsp. *bifrons* (endemic to the Mediterranean coast and Rif of Morocco) scored for seven agro-alimentary attributes, reaching 71.4% of the optimum possible score. This example is hierarchically ranked in the highest (>70%) class. For attributes and scoring, see Materials and methods.

2.3.3. Local Endemic Plants of Tunisia

Three Tunisian endemic taxa were ranked in the highest positions (>70%) regarding the general potential in the agro-alimentary sector, i.e., *Marrubium aschersonii* Magnus (76.19%), *Teucrium alopecurus* de Noë and *T. luteum* (Mill.) Degen subsp. *gabesianum* (S.Puech) Greuter and Burdet ex Greuter (71.43% each, Figure 6). The scoring of *Marrubium aschersonii* (76.19%) is illustrated in Figure 7. Only *Artemisia campestris* L. subsp. *cinerea* Le Houér was ranked in the above average to high positions (66.67%). Overall, eight taxa ranked in lower to average positions, with scores 40.48–47.62%, e.g., *Teucrium nablii* S Puech, *T. radicans* Bonnet and Barratte and *T. sauvagei* Le Houér. (47.62% each), *Calendula suffruticosa* Vahl subsp. *suffruticosa*, *Dianthus cintranus* Boiss. and Reut. subsp. *byzacenus* (Burollet) Greuter and Burdet, *D. rupicola* Biv. subsp. *hermaeensis* (Coss.) O. Bolòs and Vigo (42.86% each), *Galium afropusillum* Ehrend., and *G. olivetorum* Le Houér. (40.80% each). For another 70 taxa, the scores ranked comparatively very low (<35%). The lowest score was assigned to 11 taxa (2.38% each, see Supplementary Material S2).

(A) (B)

Figure 6. *Cont.*

(C)

Figure 6. Three top-evaluated Tunisian endemic plants of Lamiaceae family, in terms of maximum possible score, achieved regarding their agro-alimentary potential. (**A**): *Marrubium aschersonii* (Photo: G. Dakhlia), (**B**): *Teucrium alopecurus* (Photo: Z. Ghrabi-Gammar), (**C**): *Teucrium luteum* subsp. *gabesianum* (Left photos: Z. Ghrabi-Gammar; right photo: S. Bourgou). For these plants, propagation material has been collected from the wild to allow ex-situ conservation and propagation-cultivation trials at the Institute of Plant Breeding and Genetic Resources, Hellenic Agricultural Organization Demeter.

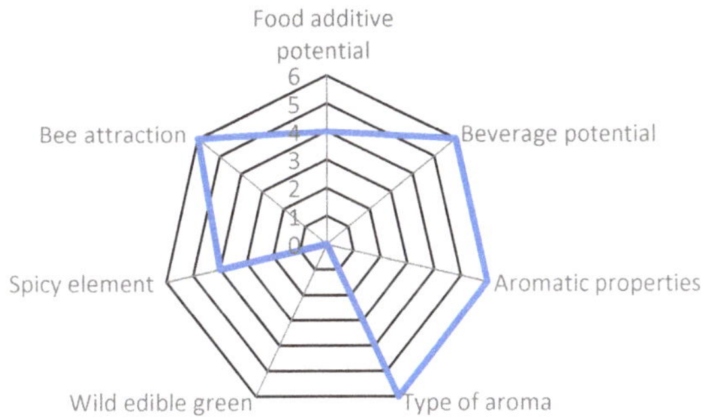

Figure 7. Evaluation example of *Marrubium aschersonii* (local Tunisian endemic) scored for seven agro-alimentary attributes, reaching 76.2% of the optimum possible scores. This example is hierarchically ranked in the highest (>70%) class. For attributes and scoring, see Materials and methods.

3. Discussion

3.1. Agro-Alimentary Potential of the Studied Local Endemic NUPs

Previous studies report that the unsustainable collection and trade directly from wild plant populations, when coupled with absence of knowledge on ex-situ conservation of propagation materials, may considerably decrease the availability of local phytogenetic resources, and this detrimental effect can lead in turn to increased prices of plant products [23,24]. Unfortunately, many plant materials, especially aromatic-medicinal plants and wild edible greens are still gathered in an unsustainable way; they are purchased directly from the wild as raw materials and are channelled for industrial use [21].

In the frame of the sustainable use of plant genetic resources as an essential component for food security and food diversity in the face of climate change, the Neglected and Underutilized Plant species (NUPs) are considered as promising alternative crops, if domesticated and sustainably used by marginalized farmers in local economies [8,11,22]. Previous studies estimate that the domestication and promotion of NUPs, such as the wild fennel in Morocco and amaranth in Ecuador, could increase the household annual income by 75% and 20%, respectively [8]. In addition, the NUPs could generate up to 62% of a farmer's annual income (1125 US$), e.g., *Gnetum* spp. in Nigeria and Cameroun [23]. Furthermore, the promotion of NUPs could contribute to their conservation and the maintenance of the associated indigenous knowledge, through wider use of their diversity, adoption of best cultivation practices, development of improved varieties, dissemination of high-quality seed, and capacity development [25].

To date, only some degree of attention has been given to NUPs, prioritizing members of Amaranthaceae or Poaceae [11,25]. However, the focus on local endemic NUPs of different regions is still very limited [2,3,22]. Local endemic NUPs are unfortunately plant species associated with limited and fragmented knowledge, usually attracting the attention of researchers or hobbyists but rarely that of citizens, politicians, and stakeholders [26,27]. The associated knowledge gaps and the research needs for most of these endemic NUPs are immense [9]. However, at least 43 local endemic NUPs of Crete (Greece), Rif and Mediterranean coast of Morocco, and Tunisia are currently traded in high prices worldwide, mainly due to their ornamental-horticultural value [22,28–30], and many of them have a very interesting potential in specific subsectors of the ornamental-horticultural industry [22]. The high market value and extant international trade of these NUPs suggest that at least some dozens of the focal NUPs studied herein are well-known, appreciated, and used mainly for ornamental-horticultural purposes. Hence, these commercial channels in place can be exploited to some extent to introduce adequately some of these NUPs (or other NUPs) in the agro-alimentary market. This could certainly be the case when such local endemic NUPs of ornamental-horticultural value also have a very interesting agro-alimentary potential. For example, the local Cretan endemic *Petromarula pinnata* (L.) A. DC. (Campanulaceae), is traditionally consumed in Crete as a wild edible green locally called 'petrofilia' (literally meaning rock-dwelling) or 'petromaroulida' (meaning rock-lettuce, thus alluring to its value as wild-growing fresh salad plant).

This study evaluated, for the first time, the agro-alimentary potential of the local endemic NUPs of Greece, Northern Morocco (Mediterranean Coast and Rif), and Tunisia and provided ranking of their potential (top-evaluated cases), thus allowing identification of the most interesting/promising cases of local endemic NUPs per country/region. Our study showed that the Lamiaceae local endemic NUPs (26 taxa) are most represented in the top 15 cases of each of the three studied regions, and thus should be considered quite promising in the agro-alimentary sector. Additionally, local endemic NUPs of Asteraceae (5 taxa), Liliaceae (4 *Allium* taxa in the case of Crete), or NUP members of another six families (Caryophyllaceae, Gentianaceae, Plumbaginaceae, Rubiaceae, Apiaceae, and Pinaceae) are also promising in the agro-alimentary sector. Due to their richness in volatile constituents and/or nutritional elements, these local endemic NUPs represent cases of taxa that could be used potentially as food additives and/or wild edible greens and/or for food flavouring as spicy elements such as members of the genera *Lactuca, Malva, Muscari, Rumex, Silene, Sanguisorba, Tragopogom, Thymbra* etc. (Supplementary material S1 and references therein), or are indeed used traditionally in local scales as food additives and/or wild edible greens and/or for food flavouring as spicy elements, e.g., the local endemic NUPs belonging to the genera *Allium, Campanula, Centaurea, Cotoneaster, Crepis, Hypochaeris, Origanum, Onopordum, Petromarula, Sonchus*, etc. (Supplementary material S1 and references therein) and/or for beverage preparations (e.g., *Sideritis* spp., *Origanum* spp., *Salvia* spp., *Teucrium* spp., *Thymbra* spp.). In the same fashion, many studies underline, nowadays, that the NUPs may have very high nutrient content or nutraceutical values, and consequently, they are often considered as 'superfoods' [7,12]. Recently, this has been the case of *Origanum dictamnus* (local Cretan endemic, with approved medicinal properties by the European Medicines Agency, www.ema.eu; accessed on 24 August 2021), for which a functional food potential has been thoroughly documented [31].

Apart from the above-mentioned, which refer to human nutritional/beneficial values, many of the evaluated local endemic NUPs can naturally attract pollinators and may feed bee populations (Supplementary Material S1), such as the NUP members of the families Lamiaceae, Dipsacaceae, and Asteraceae. Furthermore, many local endemic NUPs of the studied regions (Supplementary Material S2) belong to the top 10 families of the Mediterranean region with highly palatable plant species for grazing/foraging livestock [32,33], i.e., Asteraceae (30 Cretan, 15 north Moroccan, and 4 Tunisian local endemic NUPs), Poaceae (5 Cretan, 5 north Moroccan, and 4 Tunisian local endemic NUPs), Fabaceae (12 Cretan, 14 north Moroccan, and 5 Tunisian local endemic NUPs), Amaranthaceae (-), Brassicaceae (12 Cretan, 5 north Moroccan, and 2 Tunisian local endemic NUPs), Boraginaceae (1 north Moroccan and 6 Cretan local endemic NUPs), Caryophyllaceae (30 Cretan, 8 north Moroccan, and 3 local endemic Tunisian NUPS), Lamiaceae (14 Cretan, 16 north Moroccan, and 8 Tunisian local endemic NUPs), Apiaceae (7 Cretan, 4 north Moroccan, and 2 Tunisian local endemic NUPs), and Cistaceae (1 north Moroccan and 2 Tunisian local endemic NUPs). Since the local endemic NUPs of these families (n = 215 taxa; 53.88% of the focal NUPs) are naturally co-occurring in Cretan, Tunisian, or north Moroccan landscapes, together with other commonplace highly palatable members of the same families with more abundant populations, it is quite probable that these local endemic NUPs are also naturally preferred by foraging livestock at local scales due to their similar nutritional value for livestock feeding. This aspect brings into light another important agro-alimentary aspect related to the neglected foraging value of the local endemic NUPs. Certainly, the palatability of the local endemic NUPs of the studied regions could be further studied and documented appropriately, and selected local endemic NUPs could prove to be worth of propagation and cultivation at large-scales for forage/fodder of stabling livestock.

The diversity of unique Mediterranean NUPs with interesting agro-alimentary potential, as evaluated in this study, was further assessed in terms of estimated feasibility and readiness timescale for sustainable exploitation.

3.2. Sustainable Exploitation Feasibility of the Focal NUPs in the Agro-Alimentary Sector

Our previous study showed that there is only compromised feasibility in terms of sustainable exploitation [22] regarding the top 15 Moroccan (Mediterranean coast and Rif) endemics evaluated herein for their agro-alimentary potential, i.e., no taxon is evaluated in the highest class (>70%) or in above-average to high positions (>55–70%). *Abies marocana* (43.06%) and *Centaurium erythraea* subsp. *bifrons* (40.28%) are ranked below-average in terms of sustainable exploitation feasibility. Overall, the majority (13) of the top 15 taxa in agro-alimentary potential are ranked in lower positions in terms of exploitation feasibility (<35%). The same applies for the top 15 Tunisian endemics with interesting agro-alimentary potential; none of the Tunisian local endemic NUPs were evaluated as feasible in terms of sustainable exploitation (>70%) and no taxon is ranked in above-average to high positions (>55–70%), or even in average positions (>50–55%). Only *Artemisia campestris* subsp. *cinerea* is ranked marginally average (50%). The majority (13) of the top-evaluated taxa with agro-alimentary interest ranked in low positions (<35%) in terms of sustainable exploitation feasibility. The above findings mainly reflect the extant considerable research gaps such as absence of propagation and cultivation techniques in place, unavailability of propagation material, and compromised stakeholder interest, which actually hinder any kind of exploitation [22], see also Supplementary material S2. To justify this trend, a previous study [34] highlights, in a general way, the absence of horticultural experience regarding the local endemic NUPs of North Morocco and Tunisia, documenting that only a very small number of these local endemic plants are currently found under *ex-situ* conservation in botanic gardens and seed banks worldwide. This trend is in contrast with the comparatively higher number of local endemic NUPs of Crete that are currently in electronic trade worldwide [29].

Among the top 15 Cretan endemics taxa of the agro-alimentary sector, *Origanum dictamnus* is the most promising case, also achieving the highest score in terms of sustainable exploitation feasibility (91.67%). Thus, there are extant value chains and sustainable commercial exploitation at least in Crete where it is endemic and is locally cultivated [28,35,36], almost just as with any other crop [22]. In the same line, another five Cretan endemic taxa that ranked in above-average to high positions in terms of sustainable exploitation feasibility with scores >55.6–69.4% (hierarchically: *Calamintha cretica*, *Sideritis syriaca* subsp. *syriaca*, *Nepeta sphaciotica* P. H. Davis, *Helichrysum heldreichii*, *Thymbra calostachya*), as well as *Origanum microphyllum* that ranked in the above average class (50–55%) with score 52.78%, can also have the chance to become medicinal-aromatic crops in the future. Among them, *Sideritis syriaca* subsp. *syriaca* has already become a crop locally in Crete with established value chains [35,37,38]. All these cases of unique Cretan NUPs bring to light the fact that new agro-alimentary products can be potentially sourced from these local endemic NUPs, if research gaps are bridged, marketing is successfully engaged, and stakeholder attraction is carefully attained [22]. Additionally, these unique NUPs can possibly exploit extant value chains of related, but commonplace, plant products, namely those of Greek oregano [*Origanum vulgare* L. subsp. *hirtum* (Link) A. Terracc.], Turkish oregano (*Origanum onites* L.), Spanish oregano [*Thymbra capitata* (L.) Cav.], marjoram (*Origanum majorana*), Greek mountain tea or shepherds' tea (*Sideritis* spp.), everlasting or curry plant [*Helichrysum italicum* (Roth) G. Don], catmint (*Nepeta cataria* L.), etc.

The ability of modern technologies to transform harvested crops into a range of diverse products and uses with an extended shelf-life may create, nowadays, new opportunities to market novel products [7]. Local endemic NUPs such as those evaluated herein have a strong potential to shape a unique and solid product identity of local character (potential new products with protected designation of origin) which can also be exploited in terms of exclusive marketing, if well-protected legally [22]. In order to advertise new plant products based on, or sourced from, local endemic NUPs or popularize new uses for them, the development and promotion of user guides and recipe books, both in local and foreign languages, is required [8]. Top chefs, popular restaurants, TV shows, social media, and known food retailers can play a leading role in promoting and establishing new uses of NUPs, especially in nutrition, gastronomy, and food systems that occupy a great part of our daily livelihoods [8,28]. New culinary uses may also be documented and established even for plants never used before traditionally for strict culinary preparations, just as it was developed for the case *Origanum dictamnus* [28], as well as for *Origanum microphyllum* (a close relative of marjoram) and *Sideritis syriaca* subsp. *syriaca* [35]. For these Cretan local endemic NUPs, new culinary preparations have been introduced recently using their beneficial herbal teas, suggesting to incorporate them into standard Mediterranean meal preparations for an enhanced beneficial effect [28,35]. This trend actually represents a contemporary approach to the ancient and world famous Mediterranean nutrition, inspiring the enrichment of everyday food preparations with the beneficial health effects of EU approved traditional herbal medicines, such as *Origanum dictamnus* [36] and *Sideritis* spp. [37].

3.3. Readiness Timescale for Sustainable Exploitation of the Focal NUPs in the Agro-Alimentary Sector

Previous SWOT and gap analyses indicate that, in order to devise or to create new value chains in any economic sector for NUPs (such as the local endemic NUPs studied herein with respect to the agro-alimentary sector), five general conditions should first be accomplished as necessary prerequisites [22], i.e., extant high agro-alimentary potential unique product identity; availability of propagation material with Access and Benefit-Sharing (ABS) mechanisms already in place (Nagoya Protocol, EU Directive 511/2014), propagation and cultivation techniques in place and adequate research already conducted incorporated commercial interest (or triggered interest) able to attract stakeholders, and extant distribution channels. Among all the local endemics of all regions/countries (n = 399 taxa), the readiness timescale for sustainable exploitation was indeterminable in 67.67% of the cases (280 taxa) and determinable for only 119 taxa (29.82%) [22]. Among the top 15 of the local endemic NUPs with above average agro-alimentary potential in each of the three regions/countries, the readiness timescale for sustainable exploitation was determinable in only 33.33% of the cases, while for 66.67% of the taxa it was indeterminable.

Among the top 15 taxa evaluated as promising in the agro-alimentary sector, the readiness timescale was assessed as already achieved only in the case of *Origanum dictamnus* (Lamiaceae), a local Cretan endemic. The readiness timescale was designated as achievable in the short-term for 5 of the Cretan taxa (*Calamintha cretica, Helichrysum heldreichii, Nepeta sphaciotica, Sideritis syriaca* subsp. *syriaca, Thymbra calostachya*) and in the medium-term for *Origanum microphyllum*. The readiness timescale for *Sideritis syriaca* subsp. *syriaca* should also be considered as 'already achieved' based on the recently filled research gaps [38]. The best example-cases of local endemic Cretan plants with optimum evaluated agro-alimentary potential are illustrated in Table 1.

Table 1. Top cases of local endemic Cretan plants with strong agro-alimentary potential (Level I evaluation) associated with high feasibility and readiness timescale for sustainable exploitation (Level II and III evaluations, after [22]).

Taxon	Agro-Alimentary Potential (I)	Sustainable Exploitation Feasibility (II)	Readiness Timescale (III)
Origanum dictamnus	80.95%	91.67%	Already achieved
Sideritis syriaca subsp. *syriaca*	80.95%	66.67%	Short-term (or achieved)
Origanum microphyllum	80.95%	52.78%	Medium-term

The readiness timescale for the Moroccan (Mediterranean coast-Rif) *Centaurium erythraea* subsp. *bifrons* and *Abies marocana* as well as for the Tunisian *Artemisia campestris* subsp. *cinerea* was designated as achievable in the long-term (Supplementary Material S2). It seems that good chances are present for them if research gaps are filled promptly. *Argania spinosa* (L.) Skeels (Sapotaceae) represents a successful example of promotion of an endemic NUP of south-western Morocco and Algeria (however, not occurring in the Mediterranean coast and Rif studied herein). A. *spinosa* was traditionally used as food for centuries but was neglected and underutilized both locally and worldwide. After targeted research by scientists, and the documentation of its potential in the cosmetic sector, significant conservation and development efforts have been multiplied at local, regional and national scales, and these have opened the doors for the international markets [39]. At local scales, relevant studies [39] report the establishment of a local economic interest group for the development, preservation, and valorization of the argan forest of Morocco, promoting the optimization of women's work, the protection and maintenance of existing *A. spinosa* trees, the plantation of young trees, and the promotion of new and innovative products. Last, but not least, and acknowledging the current international value of the previously considered NUP *A. spinosa,* the United Nations recently decided to declare May 10 as the International Argan Day, which will be celebrated annually.

4. Materials and Methods

4.1. Study Area and Target-Plants

The study area of this work covers the island of Crete (Greece), the Mediterranean coast-Rif of Morocco and Tunisia (whole national territory). The catalogue of the local endemic plants studied herein (unique floristic elements of these areas thriving nowhere else) includes 399 taxa (species and subspecies), i.e., 223 single-island local endemic plant taxa of Crete (Greece) [29], 94 single-region endemic taxa of Rif and the Mediterranean coast of Morocco, as well as 82 single-country endemic taxa of Tunisia [34].

4.2. Methodological Scheme Applied

In the frame of the MULTI-VAL-END project (ARIMNet2), a group of 13 research scientists with complementary expertise see [22] from Greece, Morocco and Tunisia have conducted several workshops and meetings to develop a new methodology for the evaluation of NUPs in the agro-alimentary sector. This scheme was applied to the focal taxa of the study area. After detailed discussion and examination of the potential advantages and disadvantages related the attributes and their possible scoring, the members of the consortium adopted 19 attributes in total to be used for the evaluation of the targeted single-region and single-country endemic taxa (n = 399) in the agro-alimentary sector (Table 2). Among the 19 selected attributes, seven were assessed as sector-specific (Table 2), reflecting explicit interest concerning the specific potential of the target taxa in the agro-alimentary sector (Level I evaluation), while 12 of the attributes were employed as prerequisites of common interest across various economic sectors (e.g., agro-alimentary, ornamental-horticultural, medicinal-cosmetics sectors), thus facilitating the sustainable exploitation of the target-taxa (Level II evaluation) [22].

Up to four types of data sources per attribute were prioritized for the evaluation, i.e., literature survey, best expert judgment, survey over internet sources and interviews with elderly people (Supplementary Material S1). In five cases of attributes (food additive potential, beverage potential, aromatic properties, wild edible greens, and bee attraction), all four types of sources were used for the evaluation of taxa; in two cases of attributes (type of aroma and spicy element), three of them were consulted to score each taxon. The most common data source used was internet survey and best expert judgment. During the scoring of each attribute in the agro-alimentary sector, the experts of each country have reviewed and prepared in advance a list of selected data sources per attribute, thus facilitating the later stages of evaluation.

After consultation with the members of the consortium, the scaling for each attribute was defined (three-fold to five-fold), and this was based on the quality and quantity of extant information for every taxon and the concomitant possible score value. The scoring of each attribute was based on the relevance of the information obtained from the analysis of existing data. Therefore, one attribute allowed a three-grade scale (3 possible scores); two attributes were on four-grade scale and four attributes allowed five (5) possible scores (Table 2).

Through co-creation procedures [22], the directionality of attribute scaling and scoring values were designated, indicating the interesting and/or desired characteristics and/or the strong values per attribute for each studied plant taxon. According to best expert judgment, lower attribute score was always assigned to cases of taxa with an absence of data, undesired characteristics, and/or absence of values, while higher scores were assigned to cases of taxa with desired characteristics, and/or very interesting features. To apply the above-mentioned methodological scheme, three end-users with academic education (Bachelor and Master of Science) and/or PhD were recruited from the local academic environments in each country. During tutorials, they consulted the relevant information per attribute prepared by the task force following the guidelines given, and they scored independently the target-taxa of the three regions. The scoring procedure was completed in repetitive detached sessions, considering only one or few related attributes or one taxon at a time. In this way, all attributes and/or taxa of all three focal regions were progressively scored. By scoring completion per region, the datasets created were checked for consistency, and they were revised by the project's experts.

4.3. Evaluation Levels

Level I evaluation (L-I): At the first level, the agro-alimentary potential of each local endemic taxon was evaluated using a point scoring system with seven sector-specific attributes (Table 2). Examples of scoring of taxa along with guidelines and sources consulted are given in Supplementary Material S1. The sum of scorings for all attributes was calculated and it was expressed as relative percentage (%) of the maximum possible score that could be generated in the agro-alimentary sector, i.e., sum of maximum scores for all attributes. To illustrate the most interesting/promising taxa per country for the agro-alimentary sector, three lists of hierarchically ranked taxa per country were produced (see Supplementary Material S2).

Table 2. Sector-specific attributes and score values selected for the evaluation of the agro-alimentary potential (Level I evaluation, L-I) of the local endemic plants of Crete (Greece), Rif and Mediterranean coast of Morocco, and Tunisia, outlining the escalation of interest and the directionality of scoring. For examples on scoring of the studied taxa, guidelines and data sources used see Supplementary Material S1. For scoring of attributes of Levels II and III, see [22]).

Attribute	Short Description	Score 0	Score 1	Score 2	Score 3	Score 4	Score 5	Score 6	Possible Scores
Food additive potential	Use for alimentary or flavouring purposes	No	-	-	Under investigation	Possible	-	Yes	0, 3, 4, 6
Beverage potential	Use as tea infusion, tisane or decoction	No	-	-	Under investigation	Possible	-	Yes	0, 3, 4, 6
Aromatic properties	Perceived volatile constituents	No	Uncertain/Ambiguous	-	-	-	-	Yes	0, 1, 6
Type of aroma	Perceived basic aroma type	Bad or not smelling	Other	-	Indifferent/Light	Pungent	-	Pleasant	0, 1, 3, 4, 6
Wild edible green	Alimentary value as wild edible plant	No	Uncertain/Ambiguous	-	Under investigation	Possible	-	Yes	0, 1, 3, 4, 6
Spicy element	Savoury, flavouring and/or beneficial value	No	Uncertain/Ambiguous	-	Under investigation	Possible	-	Yes	0, 1, 3, 4, 6
Bee attraction	Apicultural nutritional value	No	Uncertain/Ambiguous	-	Under investigation	Possible	-	Yes	0, 1, 3, 4, 6

Level II evaluation (L-II): The feasibility for the sustainable exploitation of the endemic plants of Crete (Greece), Mediterranean coast-Rif of Morocco and Tunisia (Level II evaluation, L-II) is based on 12 attributes described in [22] as prerequisites of common interest across various economic sectors. Eight of these attributes represent the pre-conditions that should be met prior to any sustainable exploitation of the target taxa in any economic sector (also in the agro-alimentary sector), i.e., available initial plant material for propagation as well as species-specific propagation and cultivation techniques [22]. The remaining four attributes outline the special plant features and identity elements that could be exploited in product branding and marketing, thus facilitating trade exclusiveness, i.e., taxon's endemism or uniqueness, rarity, extinction risk, and protection statuses [22]. The sum of scorings for all these attributes outlines the most feasible cases for sustainable exploitation of taxa in an economic sector [22], thus also applying in the agro-alimentary sector.

Level III evaluation, L-III: The readiness timescale for value chain creation regarding the focal taxa (Level III evaluation) is based on the SWOT (Strengths, Weaknesses, Opportunities, Threats) and gap analyses performed in own previous research [22]. In brief, eight parameters are involved in the evaluations, i.e., feasibility ranking class (L-II), potential for up-scaling to address commercial demand, availability of propagation material, possibility to overcome legal restrictions on ABS (in relation to Interest), overview of extant research (research gaps), estimated attraction of new producers, and retailers, estimated difficulty for value chain creation, and estimated exploitation of distribution channels [22]. These criteria are applied in each case (focal taxon), and a single characterization is designated [22]. This evaluation allows determining if the sustainable exploitation in the agro-alimentary (or other) sector has already been achieved in some cases of taxa, whether this is indeterminable, or if it is achievable in short-term, medium-term, and long-term [22].

4.4. Statistical Analysis

To explore how the different agro-alimentary attributes and focal taxa are grouped in each study region, we performed complete linkage hierarchical cluster analyses with 1-Pearson r distance measure based on the individual scores of each of the endemic taxa for the seven selected agro-alimentary attributes. This type of analysis is aimed to examine possible patterns in the studied regions regarding the different dimensions of the agro-alimentary interest of the focal NUP taxa and how these are discerned or grouped together.

5. Conclusions

This study introduces a new methodological scheme for the multifaceted agro-alimentary evaluation of NUPs, focusing on 399 unique floristic elements (single-region or single-country endemic plants) of three Mediterranean regions (Crete, Greece; Mediterranean coast-Rif of Morocco; Tunisia). Although more research work and stakeholder attention is certainly needed to unlock the full potential of the evaluated herein local endemic NUPs, this study produced hierarchical ranking of their agro-alimentary potential and discussed feasibility and readiness timescale assessments for their sustainable exploitation.

In general, more effective uses of NUPs could support more nutrition-sensitive, resilient, and sustainable agro-alimentary systems. However, coordinated action, as well as basic and applied research, is needed to address many challenges such as domestication and ex-situ conservation concerns, breeding issues, poor consumer appeal, non-extant market niches or low market prices, unknown or difficult agro-processing, and compromised in-situ conservation of these NUPs [40,41], as these are often threatened by habitat degradation and human activities [12,32,33]. Yet, NUPs can help to increase the diversification of food production, adding new species to our diets with beneficial properties. To introduce local endemic NUPS in the agro-alimentary sector, the development of species-specific propagation and cultivation techniques is indispensable, improved cultivars should be aimed for in the future, and the development of new products that are able to attract stakeholders and extant distribution channels are required.

Supplementary Materials: The following are available online at https://www.mdpi.com/article/10.3390/plants10091770/s1, S1: Level I evaluation examples in the agro-alimentary sector related to local endemic plants of Crete, Greece (223 taxa), Mediterranean coast-Rif of Morocco (94 taxa) and Tunisia (92 taxa) with explanations, guidelines, data sources used, and scoring per attribute. S2: Hierarchically arranged percentages (%) of the maximum possible scores achieved regarding the agro-alimentary potential of the local endemic plants of Crete, Rif-Mediterranean coast of Morocco and Tunisia indicating blue cell cases (>70%), green cell cases (>55–70%), yellow cell cases (>50–55%), grey cell cases (>35–50%) and no colour cases (<35%) in combination with Level II (Sustainable exploitation feasibility) and Level III (Readiness timescale for sustainable exploitation) multifaceted evaluations [22].

Author Contributions: Conceptualization, N.K.; data curation, N.K., I.A., A.K., M.L., W.M.-K., I.T., M.A.T., M.E.H., and S.B.; formal analysis, N.K., I.A., I.T., and M.A.T.; funding acquisition, N.K., M.E.H., and S.B.; investigation, N.K., I.A., A.K., M.L., W.M.-K., Z.G.-G., F.L., E.D., V.G., M.E.H., and S.B.; methodology, N.K., I.A., A.K., M.L., W.M.-K., Z.G.-G., F.L., I.T., M.A.T., M.E.H., and S.B.; project administration, N.K.; resources, N.K., I.T. and S.B.; software, N.K., I.A., and M.A.T.; supervision, N.K., I.T., M.E.H., and S.B.; validation, I.A., A.K., M.L., W.M.-K., Z.G.-G., E.D., V.G., F.L., I.T., M.A.T., M.E.H., and S.B.; visualization, N.K., Z.G.-G., V.G., F.L., M.L., A.K., and M.A.T.; writing—original draft, M.L., N.K., V.G., I.A., and M.A.T.; writing—review and editing, N.K., A.K., M.L., W.M.-K., Z.G.-G., F.L., E.D., V.G., I.T., M.A.T., M.E.H., and S.B. All authors have read and agreed to the published version of the manuscript.

Funding: This research was supported by the ARIMNet2 2017 Transnational Joint Call through the MULTI-VAL-END project "Multifaceted Valorisation of single-country Endemic plants of Crete, Greece, Tunisia and Rif, Morocco for sustainable exploitation in the agro-alimentary, horticultural-ornamental and medicinal-cosmetic sectors" and was co-funded by the Hellenic Agricultural Organization Demeter of Greece, the State Secretariat for Higher Education and Scientific Research (SEESRS) of Morocco and the Ministry of Higher Education and Scientific Research (Ministère de l'Enseignement Supérieur & de la Recherche Scientifique, MESRS), Republic of Tunisia. ARIMNet2 (ERA-NET) has received funding from the European Union's Seventh Framework Programme for research, technological development and demonstration under grant agreement no. 618127.

Data Availability Statement: The data presented in this study are available on request from the corresponding authors.

Acknowledgments: The authors would like to thank the anonymous reviewers for their valuable comments and suggestions.

Conflicts of Interest: The authors declare no conflict of interest.

References

1. RBG. *The State of the World's Plants Report 2016*; Royal Botanic Gardens: Kew, UK, 2016. Available online: https://stateoftheworldsplants.org/2016/report/sotwp_2016.pdf (accessed on 6 February 2021).
2. Collins, W.W.; Hawtin, G.C. Conserving and using crop plant biodiversity in agroecosystems. In *Biodiversity in Agroecosystems*; Collins, W.W., Qualset, C.O., Eds.; CRC Press: London, UK, 1998; Volume 2, pp. 215–236.
3. FAO. Promoting Neglected and Underutilized Crop Species. 2017. Available online: http://www.fao.org/news/story/en/item/1032516/icode/ (accessed on 5 February 2021).
4. Khabbach, A.; Libiad, M.; Ennabili, A. Plant resources use in the Province of Taza (North of Morocco). *ProEnvironment* **2011**, *4*, 347–356.
5. Libiad, M.; Khabbach, A.; Ennabili, A. Exploitation of plants from upstream of the Sebou-wadi watershed (Province of Taounate, North of Morocco). *Biol. Divers. Conserv.* **2011**, *4*, 81–91.
6. Godfray, H.C.J.; Beddington, J.R.; Crute, I.R.; Haddad, L.; Lawrence, D.; Muir, J.F.; Pretty, J.; Robinson, S.; Thomas, S.M.; Toulmin, C. Food security: The challenge of feeding 9 billion people. *Science* **2010**, *327*, 812–818. [CrossRef] [PubMed]
7. IPGRI. *Neglected and Underutilized Plant Species: Strategic Action Plan of the International Plant Genetic Resources Institute*; International Plant Genetic Resources Institute: Rome, Italy, 2002; ISBN 92-9043-529-1.
8. Padulosi, S.; Thompson, J.; Rudebjer, P. *Fighting Poverty, Hunger and Malnutrition with Neglected and Underutilized Species (NUS): Needs, Challenges and the Way Forward*; Bioversity International: Rome, Italy, 2013; p. 56, ISBN 978-92-9043-941-7.
9. Dansi, A.; Vodouhe, R.; Azokpota, P.; Yedomonhan, H.; Assogba, P.; Adjatin, A.; Loko, I.Y.; Dossou-Aminon, I.; Akpagana, K. Diversity of the neglected and underutilized crop species of importance in Benin. *Sci. World J.* **2012**, *2012*, 1–19. [CrossRef] [PubMed]

10. Ebert, A.W. Potential of underutilized traditional vegetables and legume crops to contribute to food and nutritional security, income and more sustainable production. *Sustainability* **2014**, *6*, 319–335. [CrossRef]
11. Padulosi, S.; Mal, B.; King, I.O.; Gotor, E. Minor millets as a central element for sustainably enhanced incomes, empowerment, and nutrition in rural India. *Sustainability* **2015**, *7*, 8904–8933. [CrossRef]
12. Padulosi, S.; Cawthorn, D.M.; Meldrum, G.; Flore, R.; Halloran, A.; Mattei, F. Leveraging neglected and underutilized plant, fungi, and animal species for more nutrition sensitive and sustainable food systems. *Encycl. Food Secur. Sustain.* **2019**, *3*, 361–370. [CrossRef]
13. Pieroni, A.; Hovsepyan, R.; Manduzai, A.K.; Sõukand, R. Wild food plants traditionally gathered in central Armenia: Archaic ingredients or future sustainable foods? *Environ. Dev. Sustain.* **2020**, *23*, 2358–2381. [CrossRef]
14. Neupane, B.; Poudel, S. Documentation and on farm conservation of neglected and underutilized plant species in Lamjung district, Nepal. *Heliyon* **2021**, *7*, e05887. [CrossRef]
15. Chivenge, P.; Mabhaudhi, T.; Modi, A.T.; Mafongoya, P. The potential role of neglected and underutilised crop species as future crops under water scarce conditions in sub-Saharan Africa. *Int. J. Environ. Res. Public Health* **2015**, *12*, 5685–5711. [CrossRef]
16. Raimundo, J.R.; Frazão, D.F.; Domingues, J.L.; Quintela-Sabaris, C.; Dentinho, T.P.; Anjos, O.; Alves, M.; Delgado, F. Neglected Mediterranean plant species are valuable resources: The example of *Cistus ladanifer*. *Planta* **2018**, *248*, 1351–1364. [CrossRef] [PubMed]
17. Conti, M.V.; Campanaro, A.; Coccetti, P.; De Giuseppe, R.; Galimberti, A.; Labra, M.; Cena, H. Potential role of neglected and underutilized plant species in improving women's empowerment and nutrition in areas of sub-Saharan Africa. *Nutr. Rev.* **2019**, *77*, 817–828. [CrossRef] [PubMed]
18. Stockes, D.L. Why conserving species in the wild still matters. *Biodivers. Conserv.* **2018**, *27*, 1539–1544. [CrossRef]
19. Bacchetta, G.; Farris, E.; Pontecorvo, C. A new method to set conservation priorities in biodiversity hotspots. *Plant Biosyst.* **2012**, *146*, 638–648. [CrossRef]
20. Orsenigo, S.; Fenu, G.; Gargano, D.; Montagnani, C.; Abeli, T.; Alessandrini, A.; Bacchetta, G.; Bartolucci, F.; Carta, A.; Castello, M.; et al. Red list of threatened vascular plants in Italy. *Plant Biosyst.* **2021**, *155*, 310–335. [CrossRef]
21. Grigoriadou, K.; Sarropoulou, V.; Krigas, N.; Maloupa, E.; Tsoktouridis, G. GIS-facilitated effective propagation protocols of the Endangered local endemic of Crete *Carlina diae* (Rech. f.) Meusel and A. Kástner (Asteraceae): Serving ex-situ conservation needs and its future sustainable exploitation as an ornamental. *Plants* **2020**, *9*, 1465. [CrossRef]
22. Krigas, N.; Tsoktouridis, G.; Anestis, I.; Khabbach, A.; Libiad, M.; Megdiche-Ksouri, W.; Ghrabi-Gammar, Z.; Lamchouri, F.; Tsiripidis, I.; Tsiafouli, M.; et al. Exploring the potential of neglected local endemic plants of three Mediterranean regions in the ornamental sector: Value chain feasibility and readiness timescale for their sustainable exploitation. *Sustainability* **2021**, *13*, 2539. [CrossRef]
23. Ingram, V.; Ndumbe, L.N.; Ewane, M.E. Small scale, high value: *Gnetum africanum* and *buchholzianum* value chains in Cameroon. *Small-Scale For.* **2012**, *11*, 539–556. [CrossRef]
24. Nyadanu, D.; Aboagye, L.M.; Akromah, R.; Dansi, A. Agro-biodiversity and challenges of on-farm conservation: The case of plant genetic resources of neglected and underutilized crop species in Ghana. *Genet. Resour. Crop. Evol.* **2016**, *63*, 1397–1409. [CrossRef]
25. Padulosi, S.; Amaya, K.; Jäger, M.; Gotor, E.; Rojas, W.; Valdivia, R. A holistic approach to enhance the use of neglected and underutilized species: The case of Andean grains in Bolivia and Peru. *Sustainability* **2014**, *6*, 1283–1312. [CrossRef]
26. Rivera, D.; Obón, C.; Heinrich, M.; Inocencio, C.; Verde, A.; Fajardo, J. Gathered Mediterranean food plants—Ethnobotanical investigations and historical development. In *Local Mediterranean Food Plants and Nutraceuticals*; Heinrich, M., Müller, W.E., Galli, C., Eds.; Karger Publishing: Basel, Switzerland, 2006; Volume 59, pp. 18–74. [CrossRef]
27. Baldermann, S.; Blagojević, L.; Frede, K.; Klopsch, R.; Neugart, S.; Neumann, A.; Ngwene, B.; Norkeweit, J.; Schröter, D.; Schweigert, F.; et al. Are neglected plants the food for the future? *Crit. Rev. Plant Sci.* **2016**, *35*, 106–119. [CrossRef]
28. Krigas, N.; Menteli, V.; Vokou, D. Analysis of the ex-situ conservation of the Greek endemic flora at national European and global scales and of its effectiveness in meeting GSPC Target 8. *Plant Biosyst.* **2016**, *150*, 573–582. [CrossRef]
29. Menteli, V.; Krigas, N.; Avramakis, M.; Turland, N.; Vokou, D. Endemic plants of Crete in electronic trade and wildlife tourism: Current patterns and implications for conservation. *J. Biol. Res.* **2019**, *26*, 10–14. [CrossRef] [PubMed]
30. Khabbach, A.; Libiad, M.; El Haissoufi, M.; Lamchouri, F.; Bourgou, S.; Megdiche-Ksouri, W.; Ghrabi-Gammar, Z.; Menteli, V.; Vokou, D.; Tsoktouridis, G. Electronic commerce of the endemic plants of Northern Morocco (Rif region and Mediterranean coast) and Tunisia over the Internet. *Bot. Sci.* **2022**, *100*, 1–29. (in press). [CrossRef]
31. Maietta, M.; Colombo, R.; Corana, F.; Papetti, A. Cretan tea (*Origanum dictamnus* L.) as a functional beverage: An investigation on antiglycative and carbonyl trapping activities. *Food Funct.* **2018**, *9*, 1545–1556. [CrossRef]
32. Gamoun, M.; Louhaichi, M. Botanical composition and species diversity of arid and desert rangelands in Tataouine, Tunisia. *Land* **2021**, *10*, 313. [CrossRef]
33. Chebli, Y.; Otmani, S.E.; Chentouf, M.; Hornick, J.-L.; Bindelle, J.; Cabaraux, J.-F. Foraging behavior of goats browsing in Southern Mediterranean forest rangeland. *Animals* **2020**, *10*, 196. [CrossRef]
34. Libiad, M.; Khabbach, A.; El Haissoufi, M.; Bourgou, S.; Megdiche-Ksouri, W.; Ghrabi-Gammar, Z.; Sharrock, S.; Krigas, N. Ex-situ conservation of single-country endemic plants of Tunisia and northern Morocco (Mediterranean coast and Rif region) in seed banks and botanic gardens worldwide. *Kew Bull.* **2020**, *75*, 46. [CrossRef]

35. Stikoudi, M.; Maloupa, E.; Lazari, D.; Krigas, N. *Aromagarden and Cooking for Wellness: Discovering Secrets of Mediterranean Plants*; i-Print: Thessaloniki, Greece, 2016; ISBN 978-960-16-6011-0.
36. EMA/HMPC. *Final Assessment Report on Origanum dictamnus L., Herba (200431/2012)*; European Medicines Agency/Committee on Herbal Medicinal Products: London, UK, 2013. Available online: http://www.ema.europa.eu/docs/en_GB/document_library/Herbal_-_HMPC_assessment_report/2013/08/WC500147179.pdf (accessed on 18 February 2021).
37. EMA/HMPC. *Final Assessment Report on Sideritis Scardica Griseb.; Sideritis Clandestina (Bory & Chaub.) Hayek; Sideritis Raeseri Boiss. & Heldr.; Sideritis syriaca L., Herba (39455/2015)*; European Medicines Agency/Committee on Herbal Medicinal Products: London, UK, 2016. Available online: https://www.ema.europa.eu/en/documents/herbal-report/final-assessment-report-sideritis-scardica-griseb-sideritis-clandestina-bory-chaub-hayek-sideritis_en.pdf (accessed on 18 February 2021).
38. Kloukina, C.; Tomou, E.M.; Krigas, N.; Sarropoulou, V.; Madesis, P.; Maloupa, E.; Skaltsa, H. Non-polar secondary metabolites and essential oil of ex situ propagated and cultivated *Sideritis syriaca* L. subsp. *syriaca* (Lamiaceae) with consolidated identity (DNA Barcoding): Towards a potential new industrial crop. *Ind. Crop. Prod.* **2020**, *158*, 112957. [CrossRef]
39. Stussi, I.; Henry, F.; Moser, P.; Danoux, L.; Jeanmaire, C.; Gillon, V.; Benoit, I.; Charrouf, Z.; Pauly, G. *Argania spinosa*—How ecological farming, fair trade and sustainability can drive the reasearch for new cosmetic active ingredients. *SÖFW-J.* **2005**, *131*, 35–46. Available online: http://www.arganoils.com/articles/en/How%20Ecological%20Farming,%20Fair%20Trade%20and%20Sustainability%20Can%20Drive%20research%20for%20new%20cosmetic%20ingredients.pdf (accessed on 4 August 2021).
40. Dop, M.; Kefi, F.; Karous, O.; Verger, E.; Bahrini, A.; Ghrabi, Z.; El Ati, J.; Kennedy, G.; Termote, C. Identification and frequency of consumption of wild edible plants over a year in central Tunisia: A mixed-methods approach. *Public Health Nutr.* **2019**, *23*, 782–794. [CrossRef] [PubMed]
41. Karous, O.; Ben Haj Jilani, I.; Ghrabi-Gammar, Z. Ethnobotanical study on plant used by semi-nomad descendants' community in Ouled Dabbeb—Southern Tunisia. *Plants* **2021**, *10*, 642. [CrossRef] [PubMed]

plants

MDPI

Article

Quantitative Trait Locus Mapping for Drought Tolerance in Soybean Recombinant Inbred Line Population

Sanjeev Kumar Dhungana [1], Ji-Hee Park [1,*], Jae-Hyeon Oh [2], Beom-Kyu Kang [1], Jeong-Hyun Seo [1], Jung-Sook Sung [1], Hong-Sik Kim [3], Sang-Ouk Shin [1], In-Youl Baek [1] and Chan-Sik Jung [1]

1 Department of Southern Area Crop Science, Upland Crop Breeding Research Division, National Institute of Crop Science, Rural Development Administration, Miryang 50424, Korea; sanjeev@korea.kr (S.K.D.); hellobk01@korea.kr (B.-K.K.); next0501@korea.kr (J.-H.S.); sjs31@korea.kr (J.-S.S.); shinso32@korea.kr (S.-O.S.); baekiy@korea.kr (I.-Y.B.); jung100@korea.kr (C.-S.J.)

2 Department of Agricultural Biotechnology, Gene Engineering Division, National Institute of Agricultural Sciences, Rural Development Administration, Jeonju 54874, Korea; jhoh8288@korea.kr

3 Department of Central Area Crop Science, Crop Post-Harvest Technology Division, National Institute of Crop Science, Rural Development Administration, Suwon 16429, Korea; kimhongs@korea.kr

* Correspondence: heeya91@korea.kr; Tel.: +82-55-350-1234

Abstract: Improving drought stress tolerance of soybean could be an effective way to minimize the yield reduction in the drought prevailing regions. Identification of drought tolerance-related quantitative trait loci (QTLs) is useful to facilitate the development of stress-tolerant varieties. This study aimed to identify the QTLs for drought tolerance in soybean using a recombinant inbred line (RIL) population developed from the cross between a drought-tolerant 'PI416937' and a susceptible 'Cheonsang' cultivar. Phenotyping was done with a weighted drought coefficient derived from the vegetative and reproductive traits. The genetic map was constructed using 2648 polymorphic SNP markers that distributed on 20 chromosomes with a mean genetic distance of 1.36 cM between markers. A total of 10 QTLs with 3.52–4.7 logarithm of odds value accounting for up to 12.9% phenotypic variance were identified on seven chromosomes. Five chromosomes—2, 7, 10, 14, and 20—contained one QTL each, and chromosomes 1 and 19 harbored two and three QTLs, respectively. The chromosomal locations of seven QTLs overlapped or located close to the related QTLs and/or potential candidate genes reported earlier. The QTLs and closely linked markers could be utilized in maker-assisted selection to accelerate the breeding for drought tolerance in soybean.

Keywords: candidate gene; quantitative trait locus; recombinant inbred line; soybean drought tolerance; weighted drought coefficient

Citation: Dhungana, S.K.; Park, J.-H.; Oh, J.-H.; Kang, B.-K.; Seo, J.-H.; Sung, J.-S.; Kim, H.-S.; Shin, S.-O.; Baek, I.-Y.; Jung, C.-S. Quantitative Trait Locus Mapping for Drought Tolerance in Soybean Recombinant Inbred Line Population. *Plants* **2021**, *10*, 1816. https://doi.org/10.3390/plants10091816

Academic Editors: Milan S. Stankovic, Paula Baptista and Petronia Carillo

Received: 5 July 2021
Accepted: 30 August 2021
Published: 31 August 2021

Publisher's Note: MDPI stays neutral with regard to jurisdictional claims in published maps and institutional affiliations.

Copyright: © 2021 by the authors. Licensee MDPI, Basel, Switzerland. This article is an open access article distributed under the terms and conditions of the Creative Commons Attribution (CC BY) license (https://creativecommons.org/licenses/by/4.0/).

1. Introduction

Soybean (*Glycine max* [L.] Merr.) is one of the major commodity crops worldwide for food and feed sources (http://faostat.fao.org/). Increment in the production of major crops is crucial for global food security. However, the yield of many crops, including soybean, is challenged by global climate change [?]. Climate changes exacerbate the incidence of extreme weather patterns, such as erratic rainfall, elevated temperature, and the consequent drought stress, causing significant reductions in crop production [?]. Drought stress is a major abiotic stress that may cause more than 50% yield reduction in soybean [?]. Sensitivity of soybean plants to drought stress affects the global soybean yield because nearly 41% of the world's land is dryland [?], and unpredictable climatic variability, including increased drought events, is experienced in many parts [? ?]. Although the negative influence of drought on soybean depends on the severity, duration, and timing of the stress about the growth stage, the most susceptible stage to drought stress is the reproductive stage [? ?]. Therefore, acquisition of genetic information on drought tolerance at the reproductive stages of soybean is of great importance.

Low soil water availability brings several physiological and biochemical changes in soybean plants that may induce a wide range of injury symptoms, such as reduced photosynthesis [? ?], increased oxidative stress [?], and alterations in metabolism [?]. These changes are reflected in various visible traits, including reduced plant height, the number of nodes, branches and pods, biomass, and leaf area in soybean [? ? ?]. As drought tolerance is a complex quantitative trait controlled by multiple genes [?], it can be expected that several traits and loci are associated with the ability to tolerate water-deficit stress in soybean. Therefore, the quantitative trait locus (QTL) studies for drought tolerance comprising traits like plant height, the number of nodes, branches and pods, biomass, and leaf area could be of high significance.

Identification of the genomic regions associated with drought tolerance can help accelerate soybean genetic research and varietal improvement. A few linkage mapping studies have been carried out to identify QTLs related to drought tolerance in soybean considering different traits. For instance, QTLs have been detected using seed yield and drought susceptibility [?], leaf wilting coefficient, excised leaf water loss, relative water content and seed yield [?], the conditioning of fibrous roots that is related to drought avoidance [?], water use efficiency and leaf ash [? ?], beta and carbon isotope discrimination [?], canopy wilting [?], and plant height and seed yield [?]. Recently, Wang et al. [?] used a genome-wide association study to identify QTL for drought tolerance considering the relative plant height and plant weight.

One of the major limiting factors in the genetic study of drought tolerance was the availability of low-density markers, thereby reducing the efficiency and accuracy of QTL mapping. However, the rapid development of sequencing techniques has provided powerful tools like single nucleotide polymorphism (SNP) genotyping, enabling the development of the highest map resolution compared to other marker systems [? ?]. SNP markers have been used to discover QTL in many crops, including rice, maize, wheat, soybean, canola, barley, sugar beet, and cowpea [?]. Similarly, selection and measurements of relevant traits are equally important to precisely identify QTLs for stress tolerance. In this study, we considered a few vegetative as well as reproductive traits, such as plant height (PH), the number of nodes on the main stem (NN), branches (BN) and pods (PN), biomass (BM), and leaf area (LA) for phenotyping and SNP markers for genotyping the RIL population to identify QTL for drought tolerance. As these six traits are regarded as highly affected traits due to drought stress [? ? ?], this study provides valuable information on genetic understanding and breeding for drought tolerance in soybean.

2. Results

2.1. Soil Moisture Content

The soil moisture content of the control and treatment plots differed across three years according to the irrigation applied to the plots. On average, the control plots had 10–13% and the drought treatment plots had 3–10% soil moisture content. In 2017, the control plot showed an average of 11% and the treated plot showed an average of 7% soil moisture content. In 2018, the soil moisture content was 12.7 and 9.7% in the control and drought-treated plots, respectively. Similarly, the control plot showed 10% and the treated plot showed 3% moisture content in 2019.

2.2. Phenotypic Analysis of The Parents and 140 RILs

The drought-tolerant parent 'PI416937' had consistently higher weighted drought coefficient (WDC) than the susceptible parent 'Cheonsang' for all three combinations of traits (Table **??**). The mean WDC, calculated using two, three, and six traits, of 'PI416937' was 0.76, 0.80, and 0.79 and that of 'Cheonsang' was 0.42, 0.52, and 0.57, respectively. The highest WDC for 'PI416937' and 'Cheonsang' was found in 2019 and 2018, respectively. On the other hand, the highest WDC for the RILs was found in 2017. RIL distribution for WDC over three years showed normal distribution with transgressive segregation (Figure **??**).

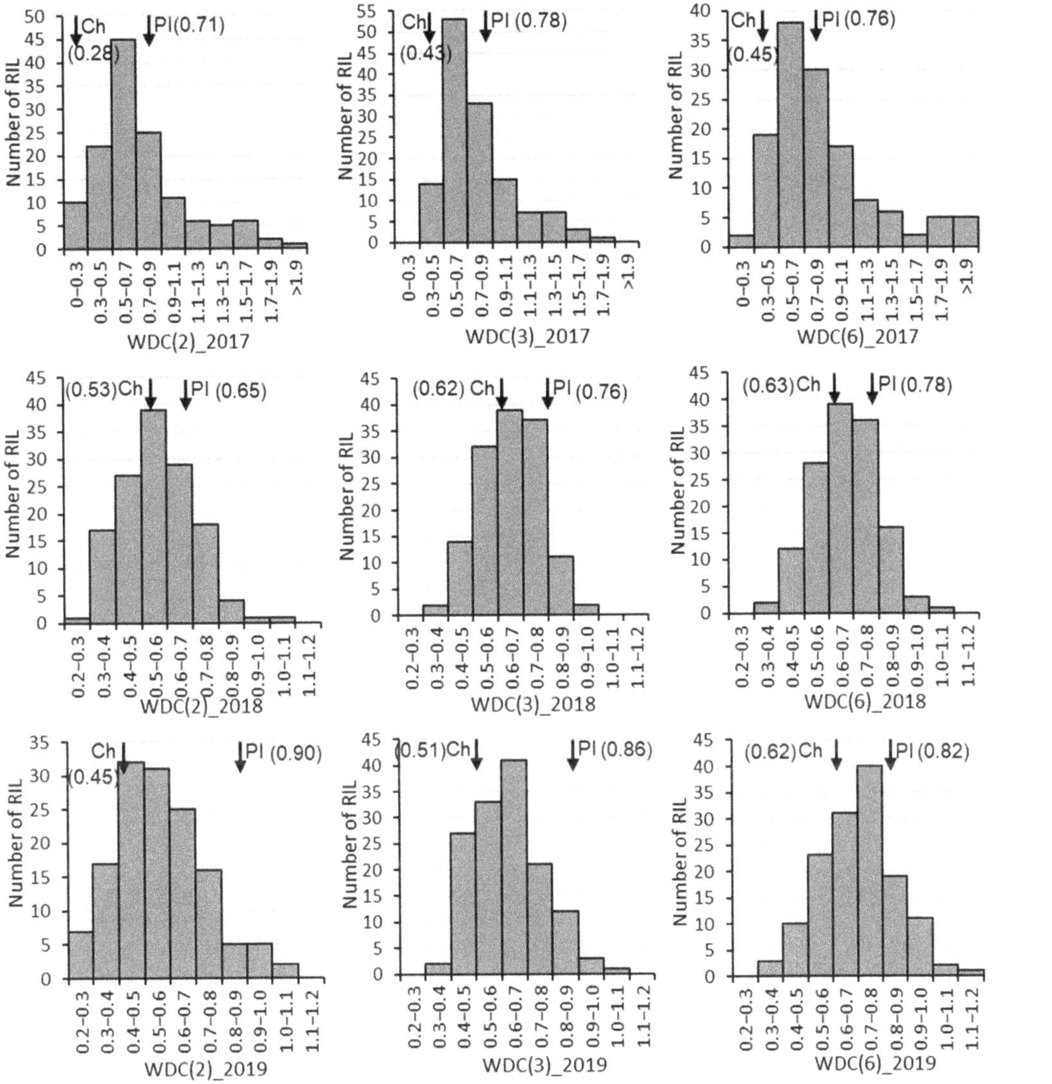

Figure 1. Frequencies in recombinant inbred line number for weighted drought coefficient (WDC) from 2017 to 2019. Ch and PI next to the inverted arrow (↓) with WDC value inside parentheses are abbreviated for the parents 'Cheonsang' and 'PI416937', respectively. The values in the parentheses after WDC indicate the number of traits considered to calculate WDC: 2 (biomass and leaf area), 3 (plant height, biomass, and leaf area), and 6 (plant height, node number, branch number, pod number, biomass, and leaf area).

Table 1. Weighted drought coefficient (WDC) of the parents and recombinant inbred lines (RILs) for three years (2017–2019) and their mean.

Trait	Year	Parents		RILs	
		PI416937	Cheonsang	Mean	Range
WDC (2)	2017	0.71	0.28	0.73	0.12–2.16
	2018	0.65	0.53	0.57	0.23–1.01
	2019	0.90	0.45	0.56	0.23–1.04
	Mean [1]	0.76	0.42	0.62	0.12–1.04
WDC (3)	2017	0.78	0.43	0.77	0.32–1.87
	2018	0.76	0.62	0.65	0.36–0.99
	2019	0.86	0.51	0.62	0.31–1.02
	Mean	0.80	0.52	0.68	0.31–1.87
WDC (6)	2017	0.76	0.45	0.86	0.22–3.06
	2018	0.78	0.63	0.67	0.38–1.01
	2019	0.82	0.62	0.70	0.33–1.15
	Mean	0.79	0.57	0.74	0.22–3.06

[1] Average value of three years. The values in the parentheses after WDC indicate the number of traits considered to calculate WDC: 2 (biomass and leaf area), 3 (plant height, biomass, and leaf area), and 6 (plant height, node number, branch number, pod number, biomass, and leaf area).

2.3. Linkage Mapping and QTL Analysis

The 19,259 polymorphic markers were binned (segregation distortion $p < 0.001$ and missing data with >15%) to eliminate the redundant markers. After binning, 2702 markers remained, out of which 54 markers with high map intervals and recombination frequencies were also eliminated. The 54 removed markers had as high as 63.34 cM map intervals and/or 0.6712 recombination frequencies. A total of 2648 SNPs were used to construct the linkage maps of 20 chromosomes (Supplementary Table S1) and QTL analysis. The total linkage maps spanned 3608.4 cM with a mean of 1.36 cM between markers. Chromosomes 13 (262.44 cM) and 15 (145.71 cM) had the largest and shortest linkage maps, respectively.

A total of 10 QTLs with a range of 3.52 to 4.71 LOD and 8.1 to 12.9% PVE were identified on seven chromosomes (1, 2, 7, 10, 14, 19, and 20). One QTL was found on five chromosomes 2, 7, 10, 14, and 20; two QTLs on chromosome 1; and three QTLs on chromosome 19 (Figure **??** and Table **??**). Five QTLs—*qWDC2-1*, *qWDC7-1*, *qWDC10-1* *qWDC19-1*, and *qWDC19-2* were detected on the different combinations of traits. These QTLs were considered to be stable QTLs for drought tolerance. Interestingly, *qWDC7-1* was detected on all three combinations of traits. *qWDC2-1* (LOD = 4.68, PVE = 10.6%), *qWDC7-1* (LOD = 4.44, PVE = 10.3%), and *qWDC19-2* (LOD = 4.57, PVE = 10.3%) which were identified on more than two trait combinations and had more than 10% PVE were considered to be stable and major QTL accounting for drought tolerance.

2.4. Candidate Gene Prediction

The potential candidate genes that resided within 200 kb of the QTLs were searched in Soybase (www.soybase.org, accessed on 20 April 2021), NCBI (https://www.ncbi.nlm.nih.gov/, accessed on 20 April 2021), and Phytozyme (https://phytozome.jgi.doe.gov, accessed on 20 April 2021).Twelve potential candidate genes were found within the 200 kb of the QTL regions (Table **??**). Four genes—*Glyma07g10321*, *Glyma07g10340*, *Glyma07g10440* and *Glyma07g11470*—reside in one of the major stable QTL *qWDC7-1*. They are related to myeloblastosis (MYB) transcription factor family, a leucine-rich repeat receptor-like protein kinase, calmodulin binding protein-like, and mitogen-activated protein kinase, respectively. Gene *Glyma01g04710* is related to glutathione S-transferase (GST). A few genes, such as *Glyma19g33750*, *Glyma19g34210*, and *Glyma20g22311* are found to be directly associated with a stress response.

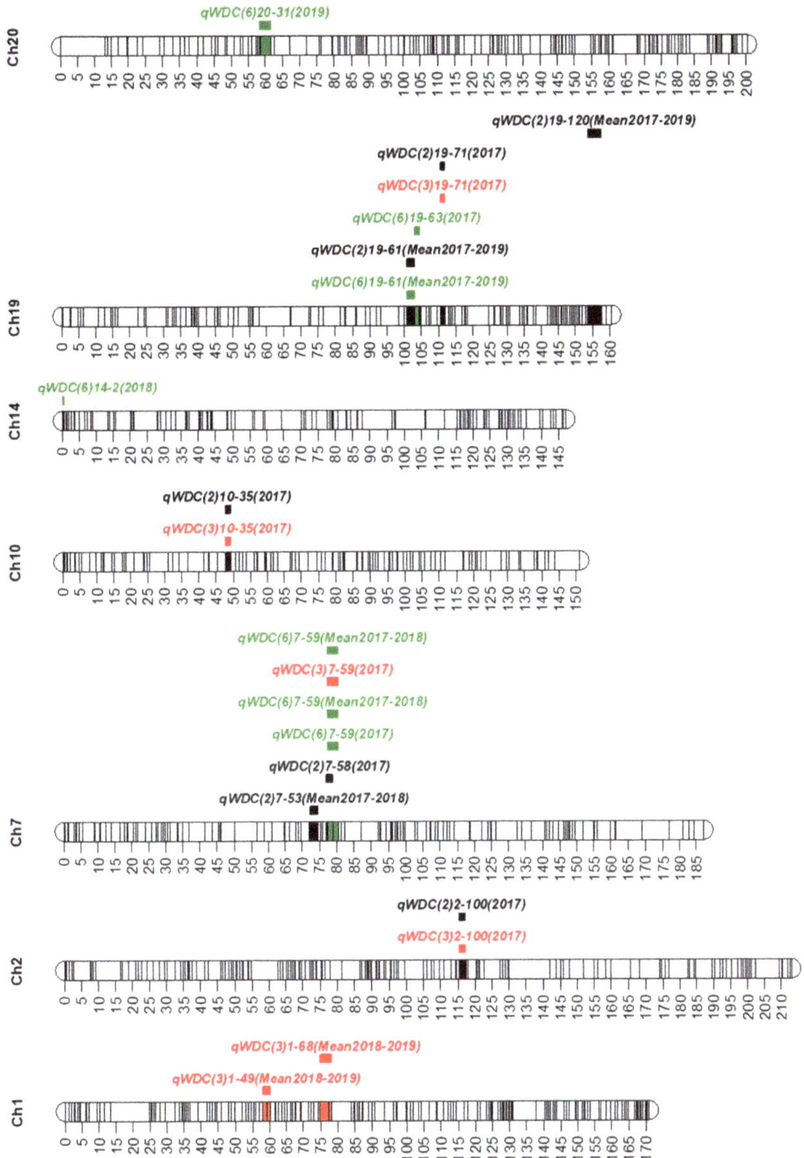

Figure 2. Positions of the QTLs for drought tolerance on seven chromosomes (Ch). q: QTL, WDC: weighted drought coefficient. In the QTL names, the first number in the parentheses after WDC represents the number of traits considered to calculate WDC (black, red, and green for 2, 3, and 6 traits, respectively; the second number is for chromosome name; the number after a dash (-) represents the sequential number of the marker on the linkage map; and mean denotes the average value of the traits in different years (2017–2019). The lines inside the chromosomes represent the position of markers used to construct the linkage map. The colored bars indicate the QTL regions. The scaled numbers next to chromosomes indicate the genetic length (cM) of the chromosome.

Table 2. QTLs for drought tolerance identified in a recombinant inbred line population derived from a drought-tolerant 'PI416937' and susceptible 'Cheonsang' parents.

QTL Name [1]	Traits [2]	Chr(LG) [3]	Genetic Position (cM)	Year	Marker Interval	Physical Position of Markers [4] (bp)	LOD [5]	PVE [6] (%)	Add [7]
qWDC1-1	3	1(D1a)	59.0	Mean (2018–2019)	AX-90430153–AX-90472468	4,523,676–5,311,697	3.59	9.0	0.0299
qWDC1-2	3	1(D1a)	77.5	Mean (2018–2019)	AX-90491463–AX-90348846	8,574,531–12,537,020	3.88	9.5	0.0303
qWDC2-1	3	2(D1b)	116.3	2017	AX-90446012–AX-90363541	14,349,986–14,561,578	4.53	10.3	0.0998
qWDC2-1	2	2(D1b)	116.3	2017	AX-90446012–AX-90363541	14,349,986–14,561,578	4.68	10.6	0.1276
qWDC7-1	2	7(M)	73.1	Mean (2017–2018)	AX-90361948–AX-90313028	8,428,091–8,490,557	3.51	8.2	−0.0539
qWDC7-1	2	7(M)	78.3	2017	AX-90524222–AX-90514687	8,678,861–9,630,217	3.90	8.7	−0.1102
qWDC7-1	6	7(M)	78.9	2017	AX-90395090–AX-90450726	9,458,480–10,232,736	3.58	8.5	−0.1403
qWDC7-1	6	7(M)	78.9	Mean (2017–2018)	AX-90395090–AX-90450726	9,458,480–10,232,736	3.57	8.7	−0.0711
qWDC7-1	3	7(M)	78.9	2017	AX-90395090–AX-90450726	9,458,480–10,232,736	4.44	10.0	−0.0935
qWDC7-1	3	7(M)	78.9	Mean (2017–2018)	AX-90395090–AX-90450726	9,458,480–10,232,736	4.42	10.3	−0.0489
qWDC10-1	3	10(O)	48.5	2017	AX-90408464–AX-90377420	38,465,737–38,164,877	4.14	9.2	−0.0910
qWDC10-1	2	10(O)	48.5	2017	AX-90408464–AX-90377420	38,465,737–38,164,877	4.47	9.9	−0.1189
qWDC14-1	6	14(B2)	0.4	2018	AX-90403945–AX-90517018	241,228–369,721	3.65	9.6	−0.0405
qWDC19-1	6	19(L)	101.5	Mean (2017–2019)	AX-90425812–AX-90334270	41,077,065–41,940,539	3.54	8.4	0.0467
qWDC19-1	2	19(L)	101.5	Mean (2017–2019)	AX-90425812–AX-90334270	41,077,065–41,940,539	3.52	8.1	0.0397
qWDC19-1	6	19(L)	104.2	2017	AX-90334270–AX-90311493	41,940,539–42,045,317	4.03	9.6	0.1491
qWDC19-2	3	19(L)	111.5	2017	AX-90480787–AX-90489545	43,473,467–43,030,013	4.45	10.0	0.0938
qWDC19-2	2	19(L)	111.5	2017	AX-90480787–AX-90489545	43,473,467–43,030,013	4.57	10.3	0.1203
qWDC19-3	2	19(L)	156.9	Mean (2017–2019)	AX-90403789–AX-90364479	49,436,986–49,727,405	3.91	9.1	−0.0425
qWDC20-1	6	20(I)	58.7	2019	AX-90318489–AX-90405719	30,153,192–32,595,196	4.71	12.9	0.0601

[1] QTL detected at the same, adjacent or overlapping marker intervals were considered the same QTL. [2] Number of traits (2: biomass and leaf area, 3: plant height, biomass, and leaf area, and 6: plant height, node number, branch number, pod number, biomass, and leaf area) considered to analyzed the QTL. [3] Chromosome (Chr) and linkage group (LG). [4] Physical position of the marker interval. The soybean reference genome (*Glycine max* Wm82.a1) was used to determine the physical position of the markers. [5] Logarithm of odds value at the peak likelihood of QTL. [6] Phenotypic variation explained by the QTL. [7] Additive effect, a positive value indicates that 'PI416937' contributed the allele, and negative value indicates that 'Cheonsang' contributed the allele for the PVE.

Table 3. Potential candidate genes related to stress tolerance that resided within 200 kb of the QTL regions.

SN	Gene Name	Physical Location (bp)	QTL	Gene Description
1	*Glyma01g04710*	4,323,774–4,325,439	*qWDC1-1*	Glutathione S-transferase, GST, Superfamily, GST domain containing
2	*Glyma01g04750*	4,367,349–4,370,072	*qWDC1-1*	Molecular chaperone (DnaJ superfamily)
3	*Glyma01g09515*	11,687,515–11,688,465	*qWDC1-2*	Growth factor activity
4	*Glyma07g10321*	8,632,573–8,633,965	*qWDC7-1*	MYB-like DNA-binding protein MYB
5	*Glyma07g10340*	8,640,435–8,642,843	*qWDC7-1*	Leucine-rich repeat receptor-like protein kinase
6	*Glyma07g10440*	8,726,024–8,730,618	*qWDC7-1*	Calmodulin-binding protein-like
7	*Glyma07g11470*	9,651,174–9,657,518	*qWDC7-1*	Mitogen-activated protein kinase
8	*Glyma19g33650*	41,237,434–41,241,804	*qWDC19-1*	Glutathione peroxidase
9	*Glyma19g33750*	41,345,217–41,346,003	*qWDC19-1*	Salt stress response/antifungal
10	*Glyma19g34210*	41,826,839–41,830,086	*qWDC19-1*	Heat shock transcription factor
11	*Glyma19g34550*	42,144,502–42,144,657	*qWDC19-1*	Golgi SNARE Bet1-related
12	*Glyma20g22311*	32,330,953–32,332,259	*qWDC20-1*	Stress responsive protein

The name and description of the drought stress-related potential candidate genes were searched in Soybase (www.soybase.org), NCBI (https://www.ncbi.nlm.nih.gov/), and Phytozyme (https://phytozome.jgi.doe.gov).

3. Discussion

The drought tolerance mechanism in plants is highly complex and is an outcome of complicated networks of multiple genes. Various physiological and biochemical alterations, due to drought stress, have been identified in soybean plants [? ? ? ?] that may visibly reflect in traits like PH, NN, BN, PN, BM, and LA [? ? ?]. Qi et al. [?] found a significant correlation between comprehensive drought resistance coefficient and WDC which was calculated by considering 35 morphological, physiological, and biochemical indicators including plant height and aboveground dry weight (biomass), which were also considered in the present study. These two traits (plant height and aboveground dry weight (biomass) incorporated in the previous report [?] were significantly correlated with other traits considered in the present study. As most of these six traits were significantly correlated (Supplementary Table S2), an integrated parameter WDC, derived from these traits, could appropriately represent them whilst analyzing the QTL for drought tolerance. Similarly, positive correlations of the number of nodes and pods with seed yield [?] as well as the associations of leaf area distribution with biomass and thereby with the number of pods, seed number, and seed yield [?] have been reported in soybean under low water availability, indicating the potential application of the QTL results of the present study in the soybean seed yield under drought condition.

The consistently higher WDC (Table **??**) value of 'PI416937' than that of 'Cheonsang' over three years showed the former parent is better drought-tolerant than the latter one. Wide range and continuous variations in WDC value of RILs across different environments (year) indicated a quantitative nature of WDC, suggesting the appropriateness of choosing these parents to develop the RIL population for QTL analysis. The transgressive segregation of the genotypes having WDC beyond either parent could be exploited in breeding for drought tolerance [?]. Although high broad-sense heritabilities for six traits were observed in individual years (up to 0.90), the mean year data showed relatively low heritability (up to 0.42) (Supplementary Table S3), suggesting a substantial influence of growing environment on the traits. The highly significant ($p < 0.0001$) genotype × year interaction also indicated the major influence of environment on the traits (Supplementary Table S4).

The chromosomal locations of seven QTLs identified in this study overlapped or positioned adjacent to related QTLs and/or potential candidate genes reported earlier, whereas two QTLs (*qWDC1-1* and *qWDC1-2*) on chromosome 1 and one QTL (*qWDC19-3*) on chromosome 19 were new. *qWDC2-1* was located nearby Satt266 (< 260 kb) that linked to a QTL for canopy wilting [?]. Another QTL *MPW2.2* (Gm02_14594196) for drought tolerance [?] was also located near to (< 33 kb) *qWDC2-1*. *qWDC7-1* was overlapped its position with the QTLs *qPH28-M-1* and *qPH-B2-1* for plant height [? ?] and the QTL *qPN-M-1* for pod number [?]. *qWDC10-1* was colocalized the physical position with the QTLs *MPW10.5* (Gm10_38212261) for drought tolerance [?], *qPH49-O-1* for plant height [?], and *qPN-O-1* for pod number [?]. A QTL *qPH-B2-1* for plant height [?] located within 300 kb from *qWDC14-1* identified on chromosome 14. *qWDC19-1* and *qWDC19-2* were colocalized with the QTLs *qPH07-L-1* and *qPH-L-2*, respectively, for plant height [? ?] and *qPN-L-1* for pod number [?], and located within the QTL *MPH19.2* for drought tolerance [?]. Similarly, the QTL *qWDC20-1* on chromosome 20 was overlapped with a QTL *qPN-I-1* for pod number [?].

Several biochemical mechanisms and genes might be involved in stress tolerance in soybean [?]. *Glyma01g04710* related to GST was found to be resided in the QTL *qWDC1-1*. GSTs play multiple roles in plants including drought stress response in *Arabidopsis* [?], rice [?], and soybean [?]. Over-expression of a GST gene, *GsGST*, from wild soybean (*Glycine soja*) enhances drought and salt tolerance in transgenic tobacco [?]. Overexpression of soybean BiP (binding protein), a molecular chaperon, similar to *Glyma01g04750* in QTL *qWDC1-1*, can enhance drought tolerance in soybean [?].

The products of four genes—*Glyma07g10321*, *Glyma07g10340*, *Glyma07g10440*, and *Glyma07g11470*—in the QTL region of chromosome 7 are related to the regulation of drought stress in soybean and other plants. For instance, *Arabidopsis* calmodulin-binding transcription factor CAMTA1 is involved in drought stress response [?]. GmMYB84, a novel MYB confers drought tolerance in soybean [?]. Overexpression of the leucine-rich receptor-like kinase gene *LRK2* increases drought tolerance and tiller number in rice [?]. Expression of a truncated ERECTA (a gene family encoding leucine-rich repeat receptor-like kinase) protein modified the growth and abiotic stress tolerance in soybean [?]. Morever, mitogen-activated protein kinase positively regulates drought stress in tomato [?].

In the QTL region of chromosome 19, four candidate genes were found. *Glyma19g33750* is associated with salt stress response and *Glyma19g34210* is related to a heat shock transcription factor. The other two genes—*Glyma19g33650* and *Glyma19g34550*—are linked with glutathione peroxidase and Golgi SNARE Bet1-related, respectively. Heat stress transcription factors play a crucial role in plants' response to several abiotic stresses by regulating the expression of stress-responsive genes, such as heat shock proteins [?] Overexpression of a glutathione peroxidase 5 (*RcGPX5*) gene increases drought tolerance in *Salvia miltiorrhiza* [?]. Furthermore, reactive oxygen species scavenging activities, including glutathione peroxidase, increased in soybean plants and were positively correlated with seed yield under drought stress [?]. Similarly, SNAREs are found to play a role in plant drought tolerance [?].

The QTLs for drought tolerance, which were identified considering up to six traits, were either colocalized or positioned adjacent to the previously reported QTLs and/or potential candidate genes associated with stresses and/or the traits of consideration. It increased the reliability of the QTL and the results could provide a valuable reference for the molecular marker-assisted selection and further fine-mapping of genes for drought tolerance.

4. Materials and Methods

4.1. Plant Material and Growing Conditions

A RIL population developed through the single seed descent method from a cross between a drought-tolerant 'PI416937' and susceptible 'Cheonsang' cultivar was used to analyze the QTL for drought tolerance. The parents and 140 RILs of $F_{6:7}$, $F_{6:8}$, and $F_{6:9}$ were grown in plastic houses at the Department of Southern Area Crop Science, Daegu (35°54′24″ N 128°26′51″ E) in 2017 and Miryang (35°29′32″ N 128°44′35″ E), Korea in 2018 and 2019. The plastic house was a kind of rain shelter with the ambient environmental condition. Soybean seedlings were grown in the seedling-growing plastic trays and then healthy uniform seedlings at the first trifoliate stage (V1) were transplanted in the plastic houses. Three to five plants of each genotype were transplanted in the plastic house at

30 cm row to row and plant to plant distance in two replications for control and drought stress each. Irrigation was applied through drip irrigation and drought stress was imposed from the V4 to R4 stages by withholding irrigation during the period. The plants in the control plots were regularly irrigated to avoid drought stress.

4.2. Measurement of Soil Moisture Content

The soil moisture content of the control and drought-stressed plots was measured using a soil moisture meter (TDR 300, Spectrum Technologies, Plainfield, IL, USA).

4.3. Measurement of Traits and Phenotyping

The plant height, number of nodes and branches on the main stem, number of pods, and leaf area were measured at the R6 stage, whereas the biomass (including seeds) was measured when plant was harvested at the R8 stage. The traits were measured in three to five plants of each replication. Leaf area was measured using the Easy Leaf Area software [?].

Each of drought coefficient (DC) value of six traits was calculated as the ratio of individual trait under the drought to control conditions as shown in the equation below.

$$DC = Trait_{Drought} / Trait_{Control}$$

The weighted drought coefficient (WDC) was calculated as follows [?]. This is one of the methods of comprehensive evaluation of drought tolerance in soybean that were identified from eight yield-related agronomic traits, and rigorous studies of different evaluation methods by establishing a relative correlation with the traits.

$$WDC = \sum_{i=1}^{n}[DC \times (|ri| \div \sum_{i=1}^{n}|ri|)]$$

where DC is mean drought coefficient of the traits considered, r is the correlation coefficient of the mean DC of the traits considered and the DC of individual traits.

The QTLs for drought tolerance were analyzed by considering the WDC values calculated from the combination of two (biomass and leaf area), three (plant height, biomass, and leaf area), and six (plant height, number of nodes, number of branches, number of pods, biomass, and leaf area) traits.

4.4. DNA Extraction and Genotyping

Genomic DNA was extracted from the young trifoliate leaves using a kit (Exgene™ Plant SV Miniprep Kit, GeneAll, Seoul, Korea) as described in a previous report [?]. The parents and RILs were genotyped using a 180K Axiom® SoyaSNP array [?].

4.5. Construction of Linkage Map and QTL Analysis

The polymorphic markers between the parents were separated from the 180K SNPs and subjected to screen for redundancy. In the genetic study, the redundant markers can make no additional information because they have identical segregation in the genetic population and show clustering at one genetic position in the linkage map construction [?]. Therefore, the redundant markers were separated out using the Bin function before the linkage map construction using the Map function in IciMapping V4.1 [?]. The algorithms set for the Bin function were as follows: significant distortion of $p < 0.001$ and missing data with >15%. The linkage map was constructed using the Kosambi mapping function following the manufacturer's instruction with the adjusted parameters: grouping by 3.0 logarithm of odds (LOD) threshold, ordering by nnTwoOpt, and rippling by the sum of adjacent recombination fractions. The SNPs with high map intervals and recombination frequencies were further removed.

QTLs were analyzed with the composite interval mapping (CIM) using QTL Cartographer V2.5 (available at https://brcwebportal.cos.ncsu.edu/qtlcart/WQTLCart.htm,

5 March 2021) following the manufacturer's instructions with adjusted parameters: Model 6, forward and backward regression, walk speed of 1.0 cM, and putative QTL with a window size of 10 cM. The number of control markers was 5, which was a default parameter. The LOD threshold for each trait was determined using a 1000 permutation test at $p < 0.05$. After the completion of the analysis, the QTL information was extracted by adjusting a minimum of 10 cM between QTL and 2-LOD support intervals. The graphical presentation of linkage maps with QTLs was done using MapChart 2.32 [?].

The QTLs were named by combing abbreviated letters *q* for QTL and *WDC* for weighted drought coefficient followed by the name of chromosome and nth QTL on the chromosome. For instance, *qWDC1-2* denotes the second QTL identified on chromosome 1.

4.6. Potential Candidate Genes Prediction

Potential candidate genes were searched within 200 kb regions of QTLs. The genes, which were directly linked to drought stress response and/or associated with the stress, were considered candidate genes. The name and function of drought stress-related potential candidate genes that resided in the QTLs were searched in Soybase (www.soybase.org), NCBI (https://www.ncbi.nlm.nih.gov/), and Phytozyme (https://phytozome.jgi.doe.gov). The Glyma1.1 gene version was used to collect the gene information.

4.7. Data Analysis

Analysis of variance (ANOVA) and Pearson's correlation were calculated in SAS9.4 using PROC GLM and PROC CORR, respectively. Broad-sense heritability (h^2) was determined as the ratio of genotypic variance ($\sigma^2{}_G$) to phenotypic variance ($\sigma^2{}_P$) as described earlier [?]. The genotypic variance ($\sigma^2{}_G$) component was estimated as: $M_3 - M_2/rY$ where M_3 is the mean square of genotype, M_2 is the mean square of genotype $\times$ year, r is the number of replications, and Y is the number of years. The phenotypic variance ($\sigma^2{}_P$) component was estimated using the equation $\sigma^2{}_P = \sigma^2_G + \sigma^2_{GY}/Y + \sigma^2_{e/rY}$ where $\sigma^2{}_{GY}$ and $\sigma^2{}_e$ are the components of genotype $\times$ year and error variances, respectively. The component of genotype $\times$ year variance ($\sigma^2{}_{GY}$) was estimated as: $M_2 - M_1/r$ where M_1 is the mean square of error ($\sigma^2{}_e$).

Supplementary Materials: The following are available online at https://www.mdpi.com/article/10.3390/plants10091816/s1, Table S1: Marker distribution and length of linkage maps of 20 chromosomes, Table S2: Correlation between different traits under control and drought conditions, Table S3: Plant height (PH), number of nodes on main stem (NN), number of branches on main stem (BN), number of pods (PN), biomass (BM), and leaf area (LA) under the control (C) and drought (D) in three years, Table S4: Analysis of variance for plant height, number of nodes and branched in main stem, number of pods, biomass, and leaf area of the recombinant inbred line (RIL) population derived from a drought-tolerant 'PI416937' and susceptible 'Cheonsang' parents.

Author Contributions: Conceptualization, B.-K.K., J.-H.S., S.-O.S. and H.-S.K.; investigation, J.-H.P., J.-H.O., B.-K.K., S.K.D., J.-H.S. and S.-O.S.; data curation, S.K.D., B.-K.K., J.-H.P., J.-H.O., J.-S.S., I.-Y.B. and C.-S.J.; writing—original draft preparation, S.K.D. and J.-H.P.; writing—review and editing, J.-S.S., I.-Y.B., C.-S.J. and H.-S.K.; funding acquisition, J.-H.P. All authors have read and agreed to the published version of the manuscript.

Funding: This research was funded by the Rural Development Administration Agenda Project (grant number PJ01186802), Republic of Korea.

Conflicts of Interest: The authors declare no conflict of interest.

References

1. Najafi, E.; Devineni, N.; Khanbilvardi, R.M.; Kogan, F. Understanding the Changes in Global Crop Yields through Changes in Climate and Technology. *Earths Future* **2018**, *6*, 410–427. [CrossRef]
2. Lesk, C.; Rowhani, P.; Ramankutty, N. Influence of Extreme Weather Disasters on Global Crop Production. *Nature* **2016**, *529*, 84–87. [CrossRef]

3. Wang, W.; Vinocur, B.; Altman, A. Plant Responses to Drought, Salinity and Extreme Temperatures: Towards Genetic Engineering for Stress Tolerance. *Planta* **2003**, *218*, 1–14. [CrossRef] [PubMed]
4. Reynolds, J.F.; Smith, D.M.S.; Lambin, E.F.; Turner, B.L.; Mortimore, M.; Batterbury, S.P.J.; Downing, T.E.; Dowlatabadi, H.; Fernandez, R.J.; Herrick, J.E.; et al. Global Desertification: Building a Science for Dryland Development. *Science* **2007**, *316*, 847–851. [CrossRef] [PubMed]
5. Cook, E.R.; Seager, R.; Cane, M.A.; Stahle, D.W. North American Drought: Reconstructions, Causes, and Consequences. *Earth-Sci. Rev.* **2007**, *81*, 93–134. [CrossRef]
6. Juenger, T.E. Natural Variation and Genetic Constraints on Drought Tolerance. *Curr. Opin. Plant Biol.* **2013**, *16*, 274–281. [CrossRef] [PubMed]
7. Westgate, M.E.; Peterson, C.M. Flower and Pod Development in Water-Deficient Soybeans (*Glycine Max* L. Merr.). *J. Exp. Bot.* **1993**, *44*, 109–117. [CrossRef]
8. Wijewardana, C.; Reddy, K.R.; Alsajri, F.A.; Irby, J.T.; Krutz, J.; Golden, B. Quantifying Soil Moisture Deficit Effects on Soybean Yield and Yield Component Distribution Patterns. *Irrig. Sci.* **2018**, *36*, 241–255. [CrossRef]
9. Ohashi, Y.; Nakayama, N.; Saneoka, H.; Fujita, K. Effects of Drought Stress on Photosynthetic Gas Exchange, Chlorophyll Fluorescence and Stem Diameter of Soybean Plants. *Biol. Plant.* **2006**, *50*, 138–141. [CrossRef]
10. Tang, Y.; Sun, X.; Wen, T.; Liu, M.; Yang, M.; Chen, X. Implications of Terminal Oxidase Function in Regulation of Salicylic Acid on Soybean Seedling Photosynthetic Performance under Water Stress. *Plant Physiol. Biochem.* **2017**, *112*, 19–28. [CrossRef]
11. Porcel, R.; Ruiz-Lozano, J.M. Arbuscular Mycorrhizal Influence on Leaf Water Potential, Solute Accumulation, and Oxidative Stress in Soybean Plants Subjected to Drought Stress. *J. Exp. Bot.* **2004**, *55*, 1743–1750. [CrossRef] [PubMed]
12. Du, Y.; Zhao, Q.; Chen, L.; Yao, X.; Zhang, W.; Zhang, B.; Xie, F. Effect of Drought Stress on Sugar Metabolism in Leaves and Roots of Soybean Seedlings. *Plant Physiol. Biochem.* **2020**, *146*, 1–12. [CrossRef] [PubMed]
13. Desclaux, D.; Huynh, T.-T.; Roumet, P. Identification of Soybean Plant Characteristics That Indicate the Timing of Drought Stress. *Crop Sci.* **2000**, *40*, 716–722. [CrossRef]
14. Cui, Y.; Jiang, S.; Jin, J.; Ning, S.; Feng, P. Quantitative Assessment of Soybean Drought Loss Sensitivity at Different Growth Stages Based on S-Shaped Damage Curve. *Agric. Water Manag.* **2019**, *213*, 821–832. [CrossRef]
15. Yan, C.; Song, S.; Wang, W.; Wang, C.; Li, H.; Wang, F.; Li, S.; Sun, X. Screening Diverse Soybean Genotypes for Drought Tolerance by Membership Function Value Based on Multiple Traits and Drought-Tolerant Coefficient of Yield. *BMC Plant Biol.* **2020**, *20*, 321. [CrossRef]
16. Maazou, A.-R.S.; Tu, J.; Qiu, J.; Liu, Z. Breeding for Drought Tolerance in Maize (*Zea Mays* L.). *Am. J. Plant Sci.* **2016**, *7*, 1858–1870. [CrossRef]
17. Du, W.; Wang, M.; Fu, S.; Yu, D. Mapping QTLs for Seed Yield and Drought Susceptibility Index in Soybean (*Glycine Max* L.) across Different Environments. *J. Genet. Genom.* **2009**, *36*, 721–731. [CrossRef]
18. Du, W.; Yu, D.; Fu, S. Detection of Quantitative Trait Loci for Yield and Drought Tolerance Traits in Soybean Using a Recombinant Inbred Line Population. *J. Integr. Plant Biol.* **2009**, *51*, 868–878. [CrossRef]
19. Abdel-Haleem, H.; Lee, G.-J.; Boerma, R.H. Identification of QTL for Increased Fibrous Roots in Soybean. *Theor. Appl. Genet.* **2011**, *122*, 935–946. [CrossRef] [PubMed]
20. Mian, M.A.R.; Bailey, M.A.; Ashley, D.A.; Wells, R.; Carter, T.E.; Parrott, W.A.; Boerma, H.R. Molecular Markers Associated with Water Use Efficiency and Leaf Ash in Soybean. *Crop Sci.* **1996**, *36*, 1252–1257. [CrossRef]
21. Mian, M.A.R.; Ashley, D.A.; Boerma, H.R. An Additional QTL for Water Use Efficiency in Soybean. *Crop Sci.* **1998**, *38*, 390–393. [CrossRef]
22. Specht, J.E.; Chase, K.; Macrander, M.; Graef, G.L.; Chung, J.; Markwell, J.P.; Germann, M.; Orf, J.H.; Lark, K.G. Soybean Response to Water: A QTL Analysis of Drought Tolerance. *Crop Sci.* **2001**, *41*, 493–509. [CrossRef]
23. Charlson, D.V.; Bhatnagar, S.; King, C.A.; Ray, J.D.; Sneller, C.H.; Carter, T.E.; Purcell, L.C. Polygenic Inheritance of Canopy Wilting in Soybean [*Glycine Max* (L.) Merr.]. *Theor. Appl. Genet.* **2009**, *119*, 587–594. [CrossRef]
24. Ren, H.; Han, J.; Wang, X.; Zhang, B.; Yu, L.; Gao, H.; Hong, H.; Sun, R.; Tian, Y.; Qi, X.; et al. QTL Mapping of Drought Tolerance Traits in Soybean with SLAF Sequencing. *Crop J.* **2020**, *8*, 977–989. [CrossRef]
25. Wang, W.; Zhou, B.; He, J.; Zhao, J.; Liu, C.; Chen, X.; Xing, G.; Chen, S.; Xing, H.; Gai, J. Comprehensive Identification of Drought Tolerance QTL-Allele and Candidate Gene Systems in Chinese Cultivated Soybean Population. *Int. J. Mol. Sci.* **2020**, *21*, 4830. [CrossRef] [PubMed]
26. Bhattramakki, D.; Dolan, M.; Hanafey, M.; Wineland, R.; Vaske, D.; Register, J.C., III; Tingey, S.V.; Rafalski, A. Insertion-Deletion Polymorphisms in 3′ Regions of Maize Genes Occur Frequently and Can Be Used as Highly Informative Genetic Markers. *Plant Mol. Biol.* **2002**, *48*, 539–547. [CrossRef] [PubMed]
27. Jones, E.S.; Sullivan, H.; Bhattramakki, D.; Smith, J.S.C. A Comparison of Simple Sequence Repeat and Single Nucleotide Polymorphism Marker Technologies for the Genotypic Analysis of Maize (*Zea Mays* L.). *Theor. Appl. Genet.* **2007**, *115*, 361–371. [CrossRef]
28. Mammadov, J.; Aggarwal, R.; Buyyarapu, R.; Kumpatla, S. SNP Markers and Their Impact on Plant Breeding. *Int. J. Plant Genom.* **2012**, *2012*, 728398. [CrossRef]
29. Qi, X.; Liu, Z.; Guan, R.; Wang, X.; Gou, Z.; Chang, R.; Qiu, L. Comparison of Evaluation Methods for Drought-Resistance at Soybean Adult Stage. *Acta Agron. Sin.* **2012**, *38*, 665–674. [CrossRef]

30. Rahman, M.M.; Hossain, M.M. Plant Density Effects on Growth, Yield and Yield Components of Two Soybean Varieties under Equidistant Planting Arrangement. *Asian J. Plant Sci.* **2011**, *10*, 278–286. [CrossRef]
31. Feng, Y.-Y.; He, J.; Turner, N.C.; Siddique, K.H.M.; Li, F.-M. Phosphorus Supply Increases Internode Length and Leaf Characteristics, and Increases Dry Matter Accumulation and Seed Yield in Soybean under Water Deficit. *Agronomy* **2021**, *11*, 930. [CrossRef]
32. Rieseberg, L.H.; Archer, M.A.; Wayne, R.K. Transgressive Segregation, Adaptation and Speciation. *Heredity* **1999**, *83*, 363–372. [CrossRef] [PubMed]
33. Xue, H.; Tian, X.; Zhang, K.; Li, W.; Qi, Z.; Fang, Y.; Li, X.; Wang, Y.; Song, J.; Li, W.-X.; et al. Mapping Developmental QTL for Plant Height in Soybean [*Glycine Max* (L.) Merr.] Using a Four-Way Recombinant Inbred Line Population. *PLoS ONE* **2019**, *14*, e0224897. [CrossRef] [PubMed]
34. Dong, Q.; Zhang, K.; Sun, X.; Tian, X.; Qi, Z.; Fang, Y.; Li, X.; Wang, Y.; Song, J.; Wang, J.; et al. Mapping QTL Underlying Plant Height at Three Development Stages and Its Response to Density in Soybean [*Glycine Max* (L.) Merri.]. *Biotechnol. Biotechnol. Equip.* **2020**, *34*, 395–404. [CrossRef]
35. Liu, S.; Xue, H.; Zhang, K.; Wang, P.; Su, D.; Li, W.; Xu, S.; Zhang, J.; Qi, Z.; Fang, Y.; et al. Mapping QTL Affecting the Vertical Distribution and Seed Set of Soybean [*Glycine Max* (L.) Merr.] Pods. *Crop J.* **2019**, *7*, 694–706. [CrossRef]
36. Ning, H.; Yuan, J.; Dong, Q.; Li, W.; Xue, H.; Wang, Y.; Tian, Y.; Li, W.-X. Identification of QTLs Related to the Vertical Distribution and Seed-Set of Pod Number in Soybean [*Glycine Max* (L.) Merri]. *PLoS ONE* **2018**, *13*, e0195830. [CrossRef]
37. Deshmukh, R.; Sonah, H.; Patil, G.; Chen, W.; Prince, S.; Mutava, R.; Vuong, T.; Valliyodan, B.; Nguyen, H.T. Integrating Omic Approaches for Abiotic Stress Tolerance in Soybean. *Front. Plant Sci.* **2014**, *5*, 244. [CrossRef]
38. Xu, J.; Zheng, A.-Q.; Xing, X.-J.; Chen, L.; Fu, X.-Y.; Peng, R.-H.; Tian, Y.-S.; Yao, Q.-H. Transgenic *Arabidopsis* Plants Expressing Grape Glutathione S-Transferase Gene (*VvGSTF13*) Show Enhanced Tolerance to Abiotic Stress. *Biochemistry* **2018**, *83*, 755–765. [CrossRef]
39. Kumar, S.; Asif, M.H.; Chakrabarty, D.; Tripathi, R.D.; Dubey, R.S.; Trivedi, P.K. Differential Expression of Rice Lambda Class GST Gene Family Members during Plant Growth, Development, and in Response to Stress Conditions. *Plant Mol. Biol. Report.* **2013**, *31*, 569–580. [CrossRef]
40. Hasan, M.S.; Islam, S.; Hasan, M.N.; Sajib, S.D.; Ahmed, S.; Islam, T.; Ghosh, A. Genome-Wide Analysis and Transcript Profiling Identify Several Abiotic and Biotic Stress-Responsive Glutathione S-Transferase Genes in Soybean. *Plant Gene* **2020**, *23*, 100239. [CrossRef]
41. Ji, W.; Zhu, Y.; Li, Y.; Yang, L.; Zhao, X.; Cai, H.; Bai, X. Over-Expression of a Glutathione S-Transferase Gene, GsGST, from Wild Soybean (*Glycine Soja*) Enhances Drought and Salt Tolerance in Transgenic Tobacco. *Biotechnol. Lett.* **2010**, *32*, 1173–1179. [CrossRef]
42. Valente, M.A.S.; Faria, J.A.Q.A.; Soares-Ramos, J.R.L.; Reis, P.A.B.; Pinheiro, G.L.; Piovesan, N.D.; Morais, A.T.; Menezes, C.C.; Cano, M.A.O.; Fietto, L.G.; et al. The ER Luminal Binding Protein (BiP) Mediates an Increase in Drought Tolerance in Soybean and Delays Drought-Induced Leaf Senescence in Soybean and Tobacco. *J. Exp. Bot.* **2009**, *60*, 533–546. [CrossRef] [PubMed]
43. Pandey, N.; Ranjan, A.; Pant, P.; Tripathi, R.K.; Ateek, F.; Pandey, H.P.; Patre, U.V.; Sawant, S.V. CAMTA1 Regulates Drought Responses in *Arabidopsis thaliana*. *BMC Genom.* **2013**, *14*, 216. [CrossRef]
44. Wang, N.; Zhang, W.; Qin, M.; Li, S.; Qiao, M.; Liu, Z.; Xiang, F. Drought Tolerance Conferred in Soybean (*Glycine Max.* L.) by GmMYB84, a Novel R2R3-MYB Transcription Factor. *Plant Cell Physiol.* **2017**, *58*, 1764–1776. [CrossRef] [PubMed]
45. Kang, J.; Li, J.; Gao, S.; Tian, C.; Zha, X. Overexpression of the Leucine-Rich Receptor-like Kinase Gene *LRK2* Increases Drought Tolerance and Tiller Number in Rice. *Plant Biotechnol. J.* **2017**, *15*, 1175–1185. [CrossRef] [PubMed]
46. Shanmugam, S.; Zhao, S.; Nandy, S.; Srivastava, V.; Khodakovskaya, M. Modification of Soybean Growth and Abiotic Stress Tolerance by Expression of Truncated ERECTA Protein from *Arabidopsis thaliana*. *PLoS ONE* **2020**, *15*, e0233383. [CrossRef]
47. Muhammad, T.; Zhang, J.; Ma, Y.; Li, Y.; Zhang, F.; Zhang, Y.; Liang, Y. Overexpression of a Mitogen-Activated Protein Kinase SlMAPK3 Positively Regulates Tomato Tolerance to Cadmium and Drought Stress. *Molecules* **2019**, *24*, 556. [CrossRef]
48. Guo, M.; Liu, J.-H.; Ma, X.; Luo, D.-X.; Gong, Z.-H.; Lu, M.-H. The Plant Heat Stress Transcription Factors (HSFs): Structure, Regulation, and Function in Response to Abiotic Stresses. *Front. Plant Sci.* **2016**, *7*, 114. [CrossRef]
49. Zhang, L.; Wu, M.; Teng, Y.; Jia, S.; Yu, D.; Wei, T.; Chen, C.; Song, W. Overexpression of the Glutathione Peroxidase 5 (*RcGPX5*) Gene from *Rhodiola Crenulata* Increases Drought Tolerance in *Salvia miltiorrhiza*. *Front. Plant Sci.* **2019**, *9*, 1950. [CrossRef]
50. Masoumi, H.; Masoumi, M.; Darvish, F.; Daneshian, J.; Nourmohammadi, G.; Habibi, D. Change in Several Antioxidant Enzymes Activity and Seed Yield by Water Deficit Stress in Soybean (*Glycine Max* L.) Cultivars. *Not. Bot. Horti Agrobot. Cluj-Napoca* **2010**, *38*, 86–94.
51. Kwon, C.; Lee, J.-H.; Yun, H.S. SNAREs in Plant Biotic and Abiotic Stress Responses. *Mol. Cells* **2020**, *43*, 501–508.
52. Easlon, H.M.; Bloom, A.J. Easy Leaf Area: Automated Digital Image Analysis for Rapid and Accurate Measurement of Leaf Area. *Appl. Plant Sci.* **2014**, *2*, 1400033. [CrossRef] [PubMed]
53. Dhungana, S.K.; Kim, H.-S.; Kang, B.-K.; Seo, J.-H.; Kim, H.-T.; Shin, S.-O.; Park, C.-H.; Kwak, D.-Y. Quantitative Trait Loci Mapping for Flooding Tolerance at an Early Growth Stage of Soybean Recombinant Inbred Line Population. *Plant Breed.* **2020**, *139*, 626–638. [CrossRef]
54. Lee, Y.-G.; Jeong, N.; Kim, J.H.; Lee, K.; Kim, K.H.; Pirani, A.; Ha, B.-K.; Kang, S.-T.; Park, B.-S.; Moon, J.-K.; et al. Development Validation and Genetic Analysis of a Large Soybean SNP Genotyping Array. *Plant J.* **2015**, *81*, 625–636. [CrossRef] [PubMed]

55. Meng, L.; Li, H.; Zhang, L.; Wang, J. QTL IciMapping: Integrated Software for Genetic Linkage Map Construction and Quantitative Trait Locus Mapping in Biparental Populations. *Crop J.* **2015**, *3*, 269–283. [CrossRef]
56. Wang, J.; Li, H.; Zhang, L.; Meng, L. *Users' Manual of QTL IciMapping*; The Quantitative Genetics Group, Institute of Crop Science, Chinese Academy of Agricultural Sciences (CAAS): Beijing, China; Genetic Resources Program, International Maize and Wheat Improvement Center (CIMMYT): Mexico City, Mexico, 2016.
57. Voorrips, R.E. MapChart: Software for the Graphical Presentation of Linkage Maps and QTLs. *J. Hered.* **2002**, *93*, 77–78. [CrossRef]
58. Toker, C. Estimates of Broad-Sense Heritability for Seed Yield and Yield Criteria in Faba Bean (*Vicia Faba* L.). *Hereditas* **2004**, *140*, 222–225. [CrossRef]

plants

MDPI

Review

γ-Aminobutyrate (GABA) Regulated Plant Defense: Mechanisms and Opportunities

Barry J. Shelp [1,*], Morteza Soleimani Aghdam [2] and Edward J. Flaherty [1]

[1] Department of Plant Agriculture, University of Guelph, Guelph, ON N1G 2W1, Canada; eflahert@uoguelph.ca

[2] Department of Horticultural Science, Imam Khomeini International University, Qazvin 34148-96818, Iran; soleimaniaghdam@eng.ikiu.ac.ir

* Correspondence: bshelp@uoguelph.ca; Tel.: +1-519-831-9878

Abstract: Global climate change and associated adverse abiotic and biotic stress conditions affect plant growth and development, and agricultural sustainability in general. Abiotic and biotic stresses reduce respiration and associated energy generation in mitochondria, resulting in the elevated production of reactive oxygen species (ROS), which are employed to transmit cellular signaling information in response to the changing conditions. Excessive ROS accumulation can contribute to cell damage and death. Production of the non-protein amino acid γ-aminobutyrate (GABA) is also stimulated, resulting in partial restoration of respiratory processes and energy production. Accumulated GABA can bind directly to the aluminum-activated malate transporter and the guard cell outward rectifying K^+ channel, thereby improving drought and hypoxia tolerance, respectively. Genetic manipulation of GABA metabolism and receptors, respectively, reveal positive relationships between GABA levels and abiotic/biotic stress tolerance, and between malate efflux from the root and heavy metal tolerance. The application of exogenous GABA is associated with lower ROS levels, enhanced membrane stability, changes in the levels of non-enzymatic and enzymatic antioxidants, and crosstalk among phytohormones. Exogenous GABA may be an effective and sustainable tolerance strategy against multiple stresses under field conditions.

Keywords: abiotic stress; antioxidants; biostimulants; biotic stress; GABA; metabolism; phytohormones; reactive oxygen species; signaling; tricarboxylic acid cycle

Citation: Shelp, B.J.; Aghdam, M.S.; Flaherty, E.J. γ-Aminobutyrate (GABA) Regulated Plant Defense: Mechanisms and Opportunities. *Plants* **2021**, *10*, 1939. https://doi.org/10.3390/plants10091939

Academic Editors: Milan S. Stankovic, Paula Baptista and Petronia Carillo

Received: 27 July 2021
Accepted: 14 September 2021
Published: 17 September 2021

Publisher's Note: MDPI stays neutral with regard to jurisdictional claims in published maps and institutional affiliations.

Copyright: © 2021 by the authors. Licensee MDPI, Basel, Switzerland. This article is an open access article distributed under the terms and conditions of the Creative Commons Attribution (CC BY) license (https://creativecommons.org/licenses/by/4.0/).

1. Introduction

The world population is predicted to be 9–10 billion people by 2050, so that a $60^{-1}10\%$ increase in global food production will be required as more marginal lands are being used for agricultural purposes [1]. Furthermore, at the current rate of global warming, the temperature is projected to increase by 1.5–2.4 °C [2]. With a 1.5 °C increase, heavy precipitation and associated flooding will intensify and be more frequent in most regions of Africa, Asia, North America and Europe. Additionally, more frequent and/or severe droughts will occur in a few regions on all continents except Asia. With further global warming, every region is projected to increasingly experience concurrent and multiple climatic changes (e.g., salinity, O_2 deprivation, acidity, heavy metals), which will adversely affect plant growth and development. These changing climatic conditions could facilitate the geographic expansion and aggressiveness of phytopathogens and modify host susceptibility [3]. Therefore, it is imperative to develop crop production systems that are more sustainable under stress conditions [4].

Under extreme environments, the overaccumulation of oxygen radicals and their derivatives (e.g., superoxide anion, $O_2^{\bullet-}$, hydroxyl radical, $^{\bullet}OH$; singlet oxygen, $^1O_2$; hydrogen peroxide, H_2O_2), known as reactive oxygen species (ROS), can lead to cellular damage, programmed cell death and lower plant productivity [5]. ROS are formed in many plant cell compartments, including chloroplasts, mitochondria, peroxisomes and plasma membrane. An optimum level of ROS is generally maintained by antioxidant defenses, so that a signal is transmitted to the nucleus through redox reactions, using the mitogen-activated protein kinase pathway in a variety of cellular mechanisms, to increase tolerance against diverse abiotic stresses [6].

There is considerable interest in improving stress tolerance using breeding and genetic engineering approaches, and exogenous application of natural compounds, including primary and secondary metabolites [4]. γ-Aminobutyrate (GABA) is a ubiquitous four-carbon, non-proteinogenic amino acid, which functions as both metabolite and signal in response to abiotic and biotic stresses [7–23]. The GABA shunt involves the activity of evolutionary-conserved enzymes that bypass two steps of the mitochondrial tricarboxylic acid cycle (TCAC), partially restore the stress-induced changes in respiratory processes, and alleviate oxidative injury. In addition, GABA accumulated during stress can bind directly to the aluminum-activated malate transporter (ALMT) and the guard cell outward rectifying K^+ (GORK) channel, thereby improving stress tolerance. Under acidic conditions, heavy metals activate malate efflux via root ALMT, resulting in heavy metal-malate complexes that are not readily absorbed [13].

Here, we review the pathways for, and the compartmentation of, GABA metabolism in plants, identifying some key gaps in our knowledge of the mechanisms involved, and then discuss stress-induced changes in flux, energy generation, redox balance and ROS. Second, we review evidence for GABA signaling in roots and stomata, and then discuss how the accumulation of GABA can influence K^+ and malate efflux, resulting in hypoxia, drought or aluminum tolerance. Third, we discuss the improvement of biotic stress resistance by the genetic enhancement of endogenous GABA. Finally, we discuss the improvement of abiotic stress tolerance using exogenous GABA to promote GABA, antioxidant, phytohormone and secondary pathways. These findings suggest that GABA application could be an appropriate treatment for dealing with different or simultaneous stresses in a field setting.

2. GABA Metabolism and Its Response to Abiotic Stress

2.1. Pathways and Compartmentation

Figure 1 shows an updated model of the metabolic and signaling pathways for GABA in plants, emphasizing vegetative organs. The model is based primarily on research with *Arabidopsis thaliana* (L.) Heynh., but where necessary, supporting evidence from other model plant systems such as petunia (Petunia x hybrida Juss.), tobacco (*Nicotiana tabaccum* L.), soybean (*Glycine max* (L.) Merr.), tomato (*Solanum lycopersicum* L.), rice (*Oryza sativa* L.), wheat (*Triticum aestivum* L.) and corn (*Zea mays* L.) is mentioned in this article [8,13,14,16] While the interaction of the GABA shunt (i.e., cytosolic glutamate (Glu) decarboxylase (GAD), mitochondrial GABA transaminase (GABA-T), and mitochondrial succinic semialdehyde (SSA) dehydrogenase (SSADH)) with the TCAC may be considered the central feature of this model, there is increasing evidence for the involvement of several branch pathways in the homeostasis of cellular GABA [17–19].

2.1.1. Biosynthesis of GABA from Glutamate, Polyamines or Proline

GAD is the direct source of GABA in the cytosol [24–28] (Figure 1). It is irreversible, pyridoxal 5′-phosphate-dependent, specific for L-Glu (K_ms for various plant GADs range from 3 to 32 mM), and maximally active at approximately pH 5.8 [15]. Many plant GADs possess a C-terminal domain that binds the Ca^{2+}/calmodulin (CaM) complex, thereby activating GAD activity at neutral pH [15,29–35]. *Arabidopsis* has five GAD genes in total, but only AtGAD1,2,4 possess the C-terminal domain [15,16]. Thus, stress-induced increases in cytosolic Ca^{2+}/CaM complexation or H^+ can activate/stimulate GAD activity [9,36–38]

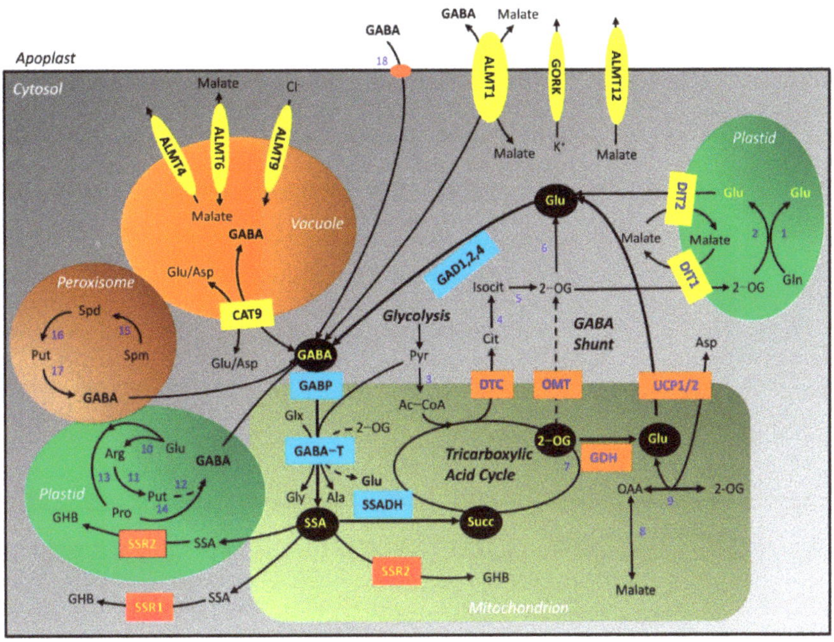

Figure 1. Model for stress-mediated GABA metabolism and signaling in *Arabidopsis*. The black ovals and blue squares represent important metabolites and enzymes/transporters, respectively, in the GABA shunt. Steps lacking convincing experimental support are shown as dashed lines. The orange squares represent enzymes/transporters that potentially link the tricarboxylic acid cycle back to the GABA shunt, the yellow squares represent transporters that link the GABA shunt to other pathways and organelles, and the red squares represent enzymes catalyzing the detoxification of SSA. The yellow ovals represent recently−identified transporters that are potentially regulated by GABA. Abbreviations: Ac−CoA, acetyl−CoA; Ala, alanine; ALMT, aluminum−activated malate transporter; Arg, arginine; Asp, aspartate; CAT, cationic amino acid transporter; Cit, citrate; DIT, dicarboxylate translocator; DTC, dicarboxylate/tricarboxylate carrier; GABA, γ−aminobutyrate; GABA−T, GABA transaminase; GABP, GABA permease; GAD, glutamate decarboxylase; GDH, glutamate dehydrogenase; GHB, γ−hydroxybutyrate; Gln, glutamine; Glu, glutamate; SSR, succinic semialdehyde reductase; Isocit, isocitrate; OAA, oxaloacetate; 2−OG, 2−oxoglutarate; OMT, 2−oxoglutarate/malate translocator; Pro, proline; Put, putrescine; Pyr, pyruvate; Spd, spermidine; Spm, spermine; Succ, succinate; SSA, succinic semialdehyde; SSADH; succinic semialdehyde dehydrogenase; UCP, uncoupling protein. Additional enzymes are indicated as numbers: 1, glutamine synthetase; 2, ferredoxin−dependent glutamate synthase; 3, pyruvate dehydrogenase complex; 4, aconitase; 5, isocitrate dehydrogenase; 6, glutamate:oxaloacetate (aspartate) transaminase or glutamate:pyruvate (alanine) transaminase; 7, 2−oxoglutarate dehydrogenase and succinyl−CoA ligase; 8, malate dehydrogenase; 9, aspartate transaminase; 10, urea cycle; 11, arginine decarboxylase, agmatine iminohydrolase and N−carbamoylputrescine amidohydrolase; 12, copper amine oxidase and aldehyde dehydrogenase 10A8; 13, Δ^{1-}pyrroline-5−carboxylate synthetase and Δ^{1-} pyrolline-5−carboxylate reductase; 14, spontaneous decarboxylation of proline to pyrrolidin^{-1}−yl, which is easily converted to Δ^{1-}pyrroline/4−aminobutanal and then GABA via aldehyde dehydrogenase (ALDH10A8); 15, polyamine oxidase (PAO2−4); 16, polyamine oxidase (PAO2,3); 17, uncertain copper amine oxidase and aldehyde dehydrogenase (ALDH10A9); 18, proline transporter (PROT1,2,3) or GABA transporter (GAT1) (modified from [17]). Important *Arabidopsis thaliana* gene names and identifiers are given in Supplementary Table S1.

The GABA level in the roots of the *Arabidopsis atgad1* mutant is 15% of the wild-type (WT), but the Glu level increases [39]. The GABA level in the shoot of the *atgad2* mutant is less than 25% of that in the WT, but the level in the roots is unaffected [40]. In contrast, the GABA level in the *atgad4* mutant is unaffected, compared to the WT [41]. Transgenic tobacco plants overexpressing a mutant petunia GAD lacking the autoinhibitory domain (*GADΔC*) exhibit severe morphological abnormalities, such as short stems, with high GABA and low Glu levels (18 and 8 mol% of total free amino acids, respectively, vs. 9 and 38% in the WT), as well as flowers that do not form pollen and abscise prematurely [29]. Notably, McLean et al. [42] have identified transgenic tobacco plants overexpressing *NtGADΔC* and exhibiting a normal phenotype, while the GABA levels are approximately 3-fold higher than in the WT.

Cytosolic GABA may also be derived indirectly from the metabolism of polyamines (PA) in organelles [43,44] (Figure 1). In *Arabidopsis*, the primary PA, putrescine (Put), is generated from the secondary PA, spermidine (Spd), and the tertiary PA, spermine (Spm), via FAD-dependent PA oxidases (PAO) in the peroxisome [45]. Put is produced in the plastid from arginine via arginine decarboxylase, agmatine imidohydrolase and carbamoylputrescine amidohydrolase [17,44]. 4-Aminobutanal (ABAL) can be derived in the peroxisome from both Spd and Put via two copper-dependent diamine oxidase activities (AtCuAOα3 and AtCuAOζ, based on the terminology in Tavladoraki et al. [46]) [47]. Terminal oxidation of Put in the plastid requires a plastidial CuAO, likely with a preference for diamines as substrates [17].

Early support for the existence of ABAL/pyrroline dehydrogenase in plants is available from the production of radiolabeled GABA from exogenously supplied radiolabeled Put, and the suppression of GABA production by the addition of aminoguanidine, a CuAO inhibitor [44,48]. More recent research demonstrated the conversion of ABAL into GABA via ABAL/pyrroline dehydrogenase activities (plastidial AtALDH10A8 and peroxisomal AtALDH10A9) [49,50]. Both AtALDH10A8 and AtALDH10A9 have strong alkaline pH optima, are NAD-dependent and use ABAL, as well as 3-aminopropanal and γ-trimethylaminobutyraldehyde (all with K_ms in the low micromolar range) as substrates to produce GABA, β-alanine and γ-butyrobetaine, respectively [50,51]. They are also prone to substrate inhibition. *ataldh10A8,9* seedlings are phenotypically normal, and the GABA and Glu levels are similar across the mutants and WT [50].

Proline is another potential indirect source of cytosolic GABA [52]. It is derived in the plastid/cytosol from Glu via NADP-dependent Δ^1-pyrroline-5-carboxylate synthetase and NADP-dependent Δ^1-pyrroline-5-carboxylate reductase [53]. Proline reacts with a hydroxyl radical, resulting in H–abstraction from the amine group, then spontaneous decarboxylation of proline and formation of pyrrolidin-1-yl [52]. Pyrrolidin-1-yl can easily be converted to Δ^1–pyrroline/ABAL and oxidized to GABA via ABAL/pyrroline dehydrogenase activity. To date, there is no direct evidence for the contribution of proline to GABA production in planta.

2.1.2. Conversion of GABA to Succinate or γ-Hydroxybutyrate

The mitochondrial-localized bidirectional amino acid transporter/GABA permease (AtBAT1/GABP) links the anabolic and catabolic portions of the GABA shunt [54] (Figure 1). However, GABA uptake by mitochondria isolated from the *atgabp* mutant is not eliminated, suggesting the possible existence of other mitochondrial GABA carriers with overlapping or redundant functions [16]. The substrate preference for AtGABP requires clarification (arginine, Glu and lysine, but not GABA and proline [55]; GABA, but not proline [54]; K_m Spd = 55 μM, K_m Put = 85 μM, K_m arginine = 1.4 mM [56]).

The pyruvate- and glyoxylate-dependent GABA-T (GABA-TP) catalyzes conversion of GABA to SSA in the mitochondrion [57–59] (Figure 1). It is reversible with a pH optimum of 9 and K_ms for GABA, pyruvate and glyoxylate of 0.18–0.34 mM, 0.14 mM and 0.11 mM, respectively [57,58]. This enzyme is often portrayed as possessing 2-oxoglutarate-dependent activity (GABA-TOG) [60–62]; however, an *AtGABA-TOG* gene has not yet been identified. In our opinion, a recent paper describing sugarcane GABA-TOG activity lacks rigor [63], and the existence of a plant GABA-TOG remains an open question [15].

Furthermore, the detection of GABA-TOG activity in crude extracts should be treated with skepticism [64–66]. The *atgaba-tp* mutant is phenotypically normal, except for lower seed production, and the leaf GABA level can increase up to 16-fold without an effect on the Glu level [58,67–70].

The conversion of SSA into succinate in the mitochondrion is catalyzed by SSADH activity [71,72] (Figure 1). The AtSSADH is reversible with a pH optimum of 9–9.5 and feedback-regulation by NADH and ATP (K_m SSA = 15 μM, K_m NAD = 130 μM, K_i NADH = 122 μM, K_i ATP = 8 mM). The *atssadh* mutant overaccumulates GABA (28 nmol g^{-1} FM) and H_2O_2 by 2- and 4-fold, respectively [73,74]. Succinate can contribute to the production of C skeletons and NADH via the TCAC and the generation of ATP via the mitochondrial electron transport chain (mETC), which in turn, prevents the accumulation of ROS [25,28,74]. Notably, SSADH and nearly every enzyme involved in the TCAC and mETC are succinylated, but the supply of succinate is limited during oxidative stress by lower OGDH and succinyl-CoA ligase activities [75,76]. These findings support the hypothesis that GABA shunt activity is necessary for the modification and regulation of respiratory activities to ensure an adequate ATP supply and minimize the generation of ROS (73).

SSA can be reduced to γ-hydroxybutyrate (GHB) via NADPH-dependent glyoxylate/SSA reductases in the mitochondrion/plastid (AtGLYR2/SSR2) and cytosol (AtGLYR1/SSR1) [77–83] (Figure 1). These proteins have affinity for SSA in the low millimolar range, glyoxylate in the low micromolar range, and NADPH in the low micromolar range, and are competitively inhibited by $NADP^+$ (K_i = 1–3 μM) [78,82–85]. The *atglyr1,2* mutants and *NAD kinase 1* overexpression line accumulate less and more GHB, respectively, with submergence than the WT [86]. The growth of plantlets or roots of various *Arabidopsis* lines with altered GLYR activity responds differentially to SSA or glyoxylate under chilling conditions [83]. Together, these findings are consistent with an elevated rate of SSA conversion to GHB with cold and low O_2, and suggest that AtGLYR1,2 are part of an adaptive response to stress-induced changes in redox balance [83]. Notably, the rice *osglyr1/2* double mutant displays stunted growth under photorespiratory conditions, compared to the WT [87], validating our earlier hypothesis that the GLYRs reduce both glyoxylate and SSA in planta [88]. The two GLYRs function in a redundant manner, which would be consistent with the diffusion of SSA and glyoxylate from their sites of origin. The fate of GHB in plants is uncertain; however, it could be linked to acetyl-CoA and fatty acid metabolism [12,16].

2.1.3. Biosynthesis of Glutamate from Succinate and 2-Oxoglutarate

Succinate is converted to citrate and 2-OG in the TCAC [25,28] (Figure 1). Two potential routes have been proposed for diverting these two important TCAC intermediates to the generation of cytosolic Glu [19]. The first involves the export of citrate from the mitochondrion via the dicarboxylate/tricarboxylate carrier (AtDTC) and its conversion to isocitrate and then 2-OG in the cytosol. The second involves the export of 2-OG via a 2-OG/malate translocator (AtOMT) [19]; contrary to a recent suggestion [89], we could find no evidence in the literature for a mitochondrial-specific OMT. In both cases, 2-OG would be converted to Glu via cytosolic transaminase activities. A third route is also possible, involving the direct synthesis of Glu from 2-OG via the mitochondrial glutamate dehydrogenase (GDH) and then export via the uncoupling proteins AtUCP1,2 [90,91]. The UCP is known to decrease the electrochemical gradient across the mitochondrial inner membrane, and prevent the over-reduction of the mETC [92].

Glu availability is an essential regulator of GAD activity, and the generation of cytosolic Glu from succinate bypasses two reactions of the TCAC (i.e., 2-OGDH and succinyl-CoA ligase). However, it does not exclude participation in the cytosolic GAD reaction of Glu originating in the plastid via the glutamine synthetase/glutamate synthase (GS/GOGAT) cycle [93] (Figure 1). The movement of 2-OG and Glu across the plastidial inner membrane is mediated by dicarboxylate translocators (AtDIT) [94].

2.1.4. GABA Transport

Arabidopsis grows efficiently on GABA as the sole N source, providing the first strong evidence for GABA uptake by plant cells [95] (Figure 1). Two types of plasma membrane-located amino acid transporters, amino acid permease 3 (AtAAP3) and proline transporter 1,2,3 (AtPROT1,2,3), transport GABA (K_m = 12.0 and 1.7–5 mM, respectively [96,97]). AtPROT2- and SlPROT1-mediated GABA transport is inhibited by proline and quaternary ammonium compounds [95,97]. AtGAT1 is a high-affinity plasma membrane-localized, proton-coupled transporter that is apparently specific for GABA (not transporting Glu or Asp) (K_m = 10 µM) [98,99]. Endogenous GABA in the *atgat1* mutant is unaffected by the addition of exogenous GABA, but it increases in the WT, confirming that AtGAT1 plays a role in GABA influx into the cell [99].

GABA is released from asparagus mesophyll cells via an unknown mechanism [100], but evidence is emerging for the bi-directional transport of GABA across the plasma membrane via the wheat root, aluminum-activated malate transporter (TaALMT1) [18,101,102] (Figure 1). AtALMT1 is highly homologous to TaALMT1, and there are plant ALMTs that encode channels with a preference for malate, chloride or nitrate [14]. However, their capacity to transport GABA has not yet been investigated.

To date, two organellar transporters for GABA have been described. The mitochondrial AtGABP was the first (see Section 2.1.2). Another one, SlCAT9, is localized to the tonoplast and links the cytosol with the vacuolar compartment, operating strictly in a stoichiometric exchange mode with Glu and aspartate so that the osmolarity of the vacuole does not change [103] (Figure 1).

2.2. Precursor–Product Relations and Flux in the GABA Shunt

Kaplan et al. [104] showed that exposure to 4 °C results in the sharp accumulation of both GABA and succinate from 12 to 24 h in the aerial portion of *Arabidopsis* plants. Subsequently, GABA declines, but succinate remains steady for 72 h. GHB accumulates sharply from 24 to 48 h and then declines to 96 h. Glu accumulates in a linear fashion from 12 to 96 h, whereas Put does so from 24 h. Furthermore, Espinoza et al. [105] demonstrated that exposure to 4 °C increases the GABA and alanine levels from 2 to 30 h; the GABA level remains steady for the remaining 28 h. Proline and Put begin to accumulate shortly thereafter (from 18–22 to 58 h), whereas the Spd level does not change. Interestingly, the succinate and malate levels decrease early in the time course and then remain steady. In contrast, the levels of 2-OG and glutamine (Gln) initially decline and then increase.

Some generalizations are possible from these two studies: (i) the cold-induced accumulation of GABA, alanine, succinate and GHB are not correlated with the availability of Put and proline; (ii) succinate and GHB may be concomitantly generated from SSA; (iii) SSA and succinate turnover may be restricted or an alternate source of succinate exists; and, (iv) Glu/Gln metabolism is altered under cold conditions. A definitive explanation for the increasing accumulation of Glu is not possible, despite the accumulation of three well-known products of Glu metabolism (i.e., GABA, Put and proline), though protein hydrolysis is known to be stimulated by low temperature [106].

Treatment with 50–150 mM NaCl for 6 d increases the GABA level in soybean (*Glycine max* (L.) Merr.) roots by 11- to 17-fold, compared to the control, as well as the diamine oxidase activity by 52–86%, but decreases the levels of Put, Spd and Spm [48]. Aminoguanidine inhibition of diamine oxidase activity increases the Put level from 28 to 51 nmol g^{-1} fresh mass (FM), but decreases the GABA level from 10.8 to 6.6 µmol g^{-1} FM. While these findings could be interpreted as support for the derivation of GABA from Put [48], the molar stoichiometry (ΔGABA/ΔPut) deviates markedly from the 1:1 ratio expected if Put is a major source of GABA, suggesting that aminoguanidine also interferes with the generation of GABA from Glu.

Other studies have modeled the stress-induced changes in flux through the GABA shunt using suspension cultures. For example, the mean GABA, alanine, proline, Glu and Gln pools in control tomato (*Solanum lycopersicum* L.) cells are 1.2, 2.0, 0.12, 0.84 and

6.77 μmol g^{-1} FM, respectively, over a 2-d period, whereas they are 13.0, 6.26, 31.2, 1.22 and 2.71 μmol g^{-1} FM, respectively, in cells adapted to water stress (25% polyethylene glycol 6000) [107]. Computer simulation of ^{15}N-labeling kinetics reveals that adaptation to water stress increases the N flux into GABA and alanine, suggesting high pyruvate availability and rapid turnover of both amino acids. The rate of GABA synthesis and catabolism, respectively, are 0.80 and 0.785 μmol h^{-1} g^{-1} FM in control cells, and 2.4 and 1.26 μmol h^{-1} g^{-1} FM in adapted cells. About 76% of the GABA is located in a metabolically inactive pool in unadapted cells, but only 38% in adapted cells. The proline pool increases by 300-fold due to greater synthetic rates from Glu and restricted oxidation, and the metabolically inactive Gln storage pool becomes depleted. Notably, the rate of nitrogen assimilation doubles, even though the total soluble protein (on a dry mass basis) decreases by 30%.

In related research, the mean GABA, alanine, proline, Glu and Gln pools in unadapted cowpea (*Vigna unguiculata* (L.) Walp) cells at 26 °C are 0.27, 5.37, 0.42, 1.79 and 4.11 μmol g^{-1} FM, respectively [108]. The GABA pool size in cells transferred to 42 °C increases to 1.85 and 3.24 μmol g^{-1} FM after 2 h and 1 d, respectively, and the other amino acids also accumulate, but less extensively than GABA. Total free amino acid levels increase approximately 1.5-fold after 1 d at 42 °C. The computer simulation suggests that heat shock induces a 63-fold increase in the rate of GABA synthesis over the first 2 h, without any change in the rate of GABA catabolism. The rate of GABA synthesis over the next 22 h increases 7-fold, and this is accompanied by a 3-fold increase in GABA catabolism. The rates of alanine and proline synthesis increase by 0.6 and 1.7-fold, respectively, over the 1-d period. The size of the free amino acid pool increases within 1 d, and the rate of protein synthesis decreases by 20–30%. An accelerated rate of protein degradation may also contribute to the effects of heat shock on the amino acid perturbations [108].

The stress-activated acceleration of flux through the GABA shunt can be explained by increases in GAD activity and GABA synthesis due to elevated levels of Glu (e.g., reduction in ATP availability, recycling of NH_3 and protein synthesis; increase in protein degradation) [93,109–111], and the Ca^{2+}/CaM complex or H^+ in the cytosol. The GOGAT- or GDH-mediated regeneration of Glu from 2-OG in the plastid and mitochondrion, respectively, is sufficient to sustain the increased formation of cytosolic GABA for a limited duration. However, the relative importance of PAs and proline to GABA accumulation remains controversial [35].

2.3. Respiration, Redox Balance and Reactive Oxygen Species

Recent studies have assessed the interaction between GABA catabolism and respiration. For example, increases in GAD activity and GABA level (1-fold) in leaves of tomato plants after 5 d of exposure to 200 mM NaCl, together with the 25% decrease in succinate [112], suggest that the SSADH-mediated production of succinate is insufficient to sustain the average rate of respiration. An increase in H_2O_2 level supports this interpretation. Che-Othman et al. [89] also found a 1-fold increase in the GABA level in wheat seedlings 3 to 11 d after exposure to 150 mM NaCl, compared to the control. Furthermore, increases in GAD activity (pH 5.8), and the levels of succinate, 2-OG, Glu and Gln, together with decreases in the activity and abundance of pyruvate dehydrogenase and OGDH, and the levels of citrate, aconitate, fumarate and malate suggest that: (i) elevated GABA shunt activity accounts for an approximate 20% increase in the respiration rate of salt-stressed leaves, despite the lower potential for pyruvate oxidation by TCAC; and (ii) GOGAT and GDH are likely indicators of N assimilation into Glu [89].

Respiration in soybean roots is reduced by 40%, after 6 h of hypoxia, but the ATP and pyruvate supply is enhanced via an activated glycolytic pathway, and the cytosolic NAD^+ is regenerated via fermentation reactions [113]. The direct flux of pyruvate into the TCAC is low, and the conversion of succinate to fumarate is markedly decreased due to restricted pyruvate dehydrogenase and succinate dehydrogenase activities. Pyruvate accumulation is minimized via the alanine transaminase- and GABA-TP-mediated formation of alanine, and the alanine transaminase reaction generates 2-OG for use by OGDH and succinyl CoA ligase to produce another ATP. The NAD^+ required for oxidation of 2-OG is apparently produced by the anti-clockwise operation of the TCAC malate dehydrogenase, which increases malate accumulation. The carbon flux from SSA to succinate is not eliminated, even though the $NADH/NAD^+$ ratio is presumably altered to some degree. GABA accumulates, at least in part, due to the stimulation of GAD activity by bound Ca^{2+}/CaM or lower cytosolic pH. Overall, both GABA and succinate appear to be temporary storage metabolites that readily supply the TCAC when hypoxia is mitigated [114].

The hypoxia-induced increase in $NADPH/NADP^+$ ratio might be attributed to increases in the activities of NAD kinases [86] and the oxidative pentose phosphate pathway [106]. In addition, there is evidence for NADPH-mediated reduction of alternative oxidases and then activation by either pyruvate or succinate [115], NADPH-mediated removal of SSA via SSR activity [78,83], and NADPH oxidase-mediated generation of H_2O_2 [116,117]. With the exception of hypoxia, abiotic stresses cause stomatal closure and decrease CO_2 fixation, leading to the underutilization of NADPH, over-reduction of the photosynthetic electron transport chain, and the generation of ROS. Stomatal closure also increases Rubisco activity, leading to the glycolate oxidase-mediated generation of H_2O_2 [88].

2.4. Genetic Manipulation of Endogenous GABA Modifies the Abiotic Stress Phenotype

Table 1 describes examples wherein the phenotype of GABA pathway mutants is modified under stress conditions. In general, the *atgad2* single and *atgad1/2* double mutants of *Arabidopsis* do not accumulate GABA in response to salinity, drought or hypoxia, but they become hypersensitive to the stress [40,69,70,118]. This suggests that AtGAD4-, PA- or proline-derived GABA is insufficient to counter the stress under consideration, though *ataldh10A8,9* single mutants have been shown to decrease the GABA level and confer a hypersensitive salinity phenotype [50]. In contrast, the GABA level in the *atgaba-t* mutant increases with salinity or hypoxia and the phenotype is more tolerant to the stress [70,118]. Furthermore, the loss in GABA accumulation in the *atgad1/2* mutant results in more RBOHF/NADPH oxidase-mediated H_2O_2 production and K^+ efflux via the GORK and Shaker-type outward rectifying K^+ channels than in the *atgaba-t* mutant [70,117,118]. Surprisingly, another study reported that the GABA level in the shoot of the *atgaba-t* mutant doubles with salinity (28 μmol g^{-1} DM), while the succinate and 2-OG levels decrease or are unchanged, but the mutant becomes hypersensitive to salinity [68]. The *atssadh* mutant overaccumulates GABA, H_2O_2 and GHB in the absence of stress, and displays hypersensitivity to heat stress [73,74].

Knockdown of tomato *SlGAD1-4* prevents the salinity-induced GABA accumulation and confers a hypersensitive salinity phenotype, whereas knockdown of *SlGABA-T1-3* slightly increases the GABA level, while conferring a hypersensitive salinity phenotype [112] (Table 1). In contrast, knockdown of *SlSSADH* increases the GABA level and confers a salinity tolerant phenotype. Furthermore, the GABA and H_2O_2 levels are higher and lower, respectively, than the corresponding levels in the *SlGAD1-4* and *SlGABA-T1-3* knockdown lines (12.5 and 0.5 μmol g^{-1} FM, respectively, vs. 6 and 0.75–0.8 μmol g^{-1} FM). Thus, the tolerance is inversely related with the H_2O_2 level.

Overall, these findings suggest that the capacity to tolerate stress is associated with H^+ consumption during GABA synthesis, pH regulation of H^+-ATPase, activation of the GABA shunt and TCAC, down-regulation of NADPH oxidase, and GABA inhibition of GORK-mediated K^+ efflux ([21,60,61,118]; also see Section 3.1). Notably, GABA is implicated in activating the transcription of 14-3-3 proteins, which are known activators of H^+-ATPases and GORK [119,120].

Table 1. Genetic manipulation of endogenous GABA modifies the abiotic stress phenotype.

Plant Species	Mutant	Abiotic Stress	Biochemical, Molecular and Physiological Responses	References
Arabidopsis (*Arabidopsis thaliana* (L.) Heyhn.)	*pop2-1 (gaba-t)*	150 mM NaCl for 4 d	• ↑ GABA (28 μmol g^{-1} DM), Pro (shoot) • ↓ DM, Succ, 2-OG, Gln; Glu unaffected (shoot) • ↑ GABA (up to 46 μmol g^{-1} DM), Glu, Pro (root) • ↓ Gln; Succ and 2-OG unaffected (root)	[68]
	gad1/2	100 mM NaCl for 3 wk	• ↓ FM and DM, Na^+ uptake (shoot) • ↑ GABA by 25% after 7 d (unlike the WT), loss of length • ↑ expression of *GORK*, *SKOR* and *RBOHF* (roots) • ↑ H_2O_2 (0.23 mmol g^{-1} FM), K^+ efflux (roots) • ↓ ATPase activity, MP (roots)	[70]
	pop2-5 (gaba-t).	100 mM NaCl for 3 wk	• ↓ FM and DM loss, compared to *gad1/2* (leaves) • ↑ GABA by 3.5-fold after 7 d (2.1 μmol g^{-1} FM), ATPase activity, MP, *SOS1* and *NHX1* expression more than in *gad1/2* (roots) • ↓ H_2O_2 level (0.17 mmol g^{-1} FM), *GORK* expression, K^+ efflux (roots)	[70]
	aldh10a8,9	150 mM NaCl for 2 d	• ↓ GABA from 65 nmol g^{-1} FM in WT to 28–40 nmol g^{-1} FM; Glu unaffected (shoot) • ↓ length (roots)	[50]
	gad1/2	Drought for 6 d	• ↓ RWC; delays dark-induced stomatal closure • ↑ stomatal conductance, stomatal number; GABA unaffected	[69]
	gad2	Drought for 3–7 d	• ↓ GABA • ↑ stomatal conductance, width of stomatal pore; ↓ RWC more quickly • *AtGAD2* Ox in guard cells ↑ GABA, ↓ stomatal conductance and ↑ RWC to WT level following 5 d drought	[40]
	gad1,2	Waterlogging for 1 wk	• ↓ shoot FM • ↓ GABA (0.8 μmol g^{-1} FM), H_2O_2 (0.3 μmol g^{-1} FM) (leaves) • ↑ H_2O_2 (0.30 μmol g^{-1} FM) within 1 d (roots) • ↑ K^+ efflux (roots), H_2O_2-mediated K^+ efflux, H_2O_2-mediated Ca^{2+} influx, OH-mediated K^+ efflux (roots) • ↓ MP, H+ efflux (roots)	[118]
	gaba-t	Waterlogging for 1 wk	• ↑ shoot FM, shoot DM, Chl, Chl fluorescence • ↑ GABA (6 μmol g^{-1} FM); ↓ H_2O_2 (0.15 μmol g^{-1} FM) (leaves), • ↓ H_2O_2 (0.15 μmol g^{-1} FM) within 1 d (roots) • ↓ K^+ efflux, H_2O_2-mediated K^+ efflux, H_2O_2-mediated Ca^{2+} influx, • OH-mediated K^+ efflux (roots) • ↑ MP, H+ efflux (roots)	[118]
	gaba-t	5% O_2 for 2 d	• ↑ GABA by 0–15% within 4 h, 15–80% within 48 h in both shoot and roots	[67]

Table 1. *Cont.*

Plant Species	Mutant	Abiotic Stress	Biochemical, Molecular and Physiological Responses	References
	ssadh	37 °C for 2 wk	• ↑ necrotic lesions, cell death under low fluence white light	[73,74]
Tomato (*Solanum lycopersicum* (L.))	VIGS *GAD1-4*	200 mM NaCl for 3 wk	• ↓ shoot FM, shoot height • ↓ Chl, P_N • ↑ H_2O_2 (0.8 µmol g^{-1} FM) • ↓ Glu; GABA (6 µmol g^{-1} FM), Succ, Gln unaffected	[112]
	VIGS *GABA-T1-3*	200 nM NaCl for 3 wk	• ↓ shoot FM; shoot height unaffected; ↓ Chl, P_N • ↑ H_2O_2 (0.75 µmol g^{-1} FM), GABA (6 µmol g^{-1} FM), Glu • ↓ Succ; Gln unaffected	[112]
	VIGS *SSADH*	200 mM NaCl for 4 wk	• ↑ shoot FM; shoot height unaffected; ↑ Chl, P_N • ↑ GABA (12.5 µmol g^{-1} FM), Gln • ↓ Glu; Succ, H_2O_2 unaffected (0.5 µmol g^{-1} FM)	[112]

Symbols: ↑, increases; ↓, decreases. Abbreviations: Chl, chlorophyll; DM/FM, dry/fresh mass; GABA, γ-aminobutyrate; GABA-T, pyruvate/glyoxylate-dependent GABA transaminase; GAD, glutamate decarboxylase; Gln, glutamine; Glu, glutamate; GORK channel, guard cell outward-rectifying K^+ channel; H_2O_2, hydrogen peroxide; MP, membrane potential; NHX and SOS, Na^+/H^+ exchanger; $O_2^{\bullet -}$, superoxide anion; •OH, hydroxyl radical; 2-OG, 2-oxoglutarate; Ox, overexpression; RBOHF, NADPH oxidase; RWC, relative water content; SKOR, stelar K^+ outward rectifying channel; SSADH, succinic semialdehyde dehydrogenase; Succ, succinate; VIGS, virus-induced gene silencing.

3. GABA Signaling and Its Response to Abiotic Stress

3.1. Root Malate and K^+ Efflux

Wheat root TaALMT1 is responsible for malate efflux, but this is inhibited by GABA binding to the cytosolic surface of the protein [18,101,102] (Figure 1). Indeed, both malate and GABA likely traverse the protein, though not simultaneously or through the same pore [18]. The GABA-binding domain is conserved in the GORK channels of the root epidermis [20]. The *atgork1* mutant displays an hypoxia (waterlogging)-tolerant phenotype, greater K^+ retention and hypoxia-induced Ca^{2+} signaling, and does not show any change in K^+ efflux in response to GABA [20,117] (Table 2). Therefore, one would expect the stress-induced accumulation of GABA to reduce the export of malate and K^+ from root cells [13,18,19,102]. Notably, the GABA-binding domain is conserved in most ALMTs [14] and further research is required to establish if they transport GABA, as well as anions.

3.2. Stomata Functioning

ALMTs and GORK facilitate the functioning of stomata [121] (Figure 1). During the opening, malate and Cl- fluxes into the guard cell vacuole are mediated by tonoplast AtALMT6 and malate-activated AtALMT9, respectively. During the closure, the efflux of malate and K^+ is mediated by plasma membrane-localized AtALMT12 and AtGORK, respectively. The drought-induced increase in abscisic acid (ABA) level upregulates GORK activity, resulting in stomatal closure and drought tolerance. The opening of AtALMT6 and AtALMT4 mediates malate efflux from the vacuole. Various Ca^{2+}-permeable channels contribute to the elevation of cytosolic Ca^{2+} during stress [121,122].

A role for GABA accumulation in drought resistance has been shown using *Arabidopsis* mutants with altered GABA levels. The *atgad1/2* double mutant has impaired stomatal closure and a drought-susceptible phenotype [69] (Table 1). Furthermore, the light-to-dark transition is slower, and GABA levels are lower than those in the WT without or with drought. Subsequent research with *atgad2* and *atalmt9* mutants demonstrated that the drought-induced accumulation of GABA suppresses light-induced stomatal opening, whereas it has no effect under constant light [40] (Tables 1 and 2). Stomatal opening in the *almt12* mutant showed WT sensitivity to GABA, whereas dark-induced stomatal closing is insensitive to GABA [40] (Table 2). These findings indicated that GABA accumulation in the cytosol of the guard cell reduces stomatal reopening and transpirational water loss, thereby improving drought tolerance. Further research is required to investigate iGABA inhibition of ALMT4,6 by cytosolic GABA contributes to the regulation of stomatal movement [123] (Table 2).

3.3. Overexpression of Malate Efflux Is Linked to Aluminum Tolerance

The accumulation of free aluminum (Al^{3+}) ions in the soil solution under low pH limits plant growth and productivity [124]. Ramesh et al. [101] have monitored malate efflux in roots of near-isogenic, Al^{3+} tolerant (ET8) and sensitive (ES8) lines of wheat. In the absence of Al^{3+}, GABA has no effect on malate efflux under acidic conditions. However, malate efflux increases in response to Al^{3+} treatment, and decreases in response to Al^{3+} and exogenous GABA in ET8, but not in ES8. While the GABA level is higher in ET8 than in ES8 under acidic conditions, Al^{3+} reduces the GABA level in both lines to similar levels. This suggests that high endogenous GABA inhibits the activity of TaALMT1 in the absence of Al^{3+}, and that both malate and GABA are exported from the cytosol when TaALMT1 is Al^{3+} activated, resulting in the sequestration of Al^{3+} and modulation of the plant sensitivity to Al^{3+} [102,124]. Table 2 summarizes various examples wherein ALMT1 overexpression enhances the efflux of malate in response to Al^{3+} treatment under acidic conditions [101,125–128]. Further research is required to establish the precise role of GABA efflux under acidic conditions in the presence of Al^{3+} [102]. It could be related to the alleviation of ammonium toxicity [129].

Table 2. Genetic manipulation of GABA receptors modifies the abiotic stress phenotype.

Plant Species	Mutant or Transgenic Line	Abiotic Stress	Biochemical, Molecular and Physiological Responses	References
Arabidopsis (*Arabidopsis thaliana* (L.) Heyhn.)	*almt4*	Drought for 2 wk	• ↑ whole-plant wilting • *AtALMT4* Ox complements *atalmt4* phenotype	[123]
	gork1	Waterlogging for up to 3 wk	• ↑ shoot FM, Chl; completes life cycle • ↑ K^+, Ca^{2+} retention in epidermal cells of mature and elongation zones of root • ↓ K^+ efflux rate	[117]
	almt9	Drought for up to 7 d	• GABA impairs dark-induced stomatal closure • *gad2/almt9* retains low GABA of *gad2* and restores stomatal conductance • *GADΔC* Ox does not alter stomatal conductance of *almt9* • stomatal conductance of *ALMT9*$^{F243C/Y245C}$ Ox similar to *gad2* and greater than *almt9*	[40]
	ZmALMT2 Ox	10 μM Al^{3+} (pH 4.5) for 1 d	• ↑ length, constitutive and Al^{3+}-activated Mal efflux (roots) • reverses the Al^{3+}-sensitive phenotype of *almt1/mate*	[125]
	BoALMT Ox	0.4 mM Al^{3+} (pH 4.5) for 2 d	• ↑ length, Al^{3+}-activated Mal and H+ efflux (roots)	[126]

Table 2. *Cont.*

Plant Species	Mutant or Transgenic Line	Abiotic Stress	Biochemical, Molecular and Physiological Responses	References
Arabidopsis and soybean (hairy roots) (*Glycine max* (L.) Merr.)	*GmALMT1* Ox	0.4 mM Al^{3+} (pH 4.5) for 2 d	• ↑ seedling growth, Mal efflux in root • ↓ Al^{3+} content	[127]
Tobacco (*Nicotiana tabaccum* L.)	*MsALMT1* Ox	30 μM Al^{3+} (pH 4.2) for 1 d	• ↑ root growth • ↓ Al^{3+} content	[128]
Barley (*Hordeum vulgare* L.)	*TaALMT1* Ox	0.1 mM Al^{3+} (pH 4.5) for 22 h	• ↑ root growth in Al^{3+}-sensitive line • restores Al^{3+}-activated, GABA-regulated Mal efflux in root	[101]

Symbols: ↑, increases; ↓, decreases. Abbreviations: Al^{3+}, aluminum; ALMT, Al^{3+}-activated malate transporter; Chl, chlorophyll; FM, fresh mass; GABA, γ-aminobutyrate; GAD, glutamate decarboxylase; Mal, malate; Ox, overexpression. Species abbreviations: Ms, *Medicago sativa* L. (alfalfa); Bo, *Brassica oleracea* var. *capitata* L. (cabbage); Zm, *Zea mays* L. (corn); Ta, *Triticum aestivum* L. (wheat).

4. GABA Metabolism and Its Response to Biotic Stress

4.1. Interaction between Plants and Other Organisms

Recent advances in our knowledge of the interactions between plants and other organisms have been reviewed in detail [16,61,130–132]. GABA inevitably accumulates in the host plant in response to bacterial and fungal infection, and infestation by invertebrate pests; however, the mechanism of action for GABA appears to differ. For example, the development of insect larvae and root-knot nematodes is delayed, presumably by disrupting the function of neuromuscular junctions [133–137], whereas in bacterial pathogens, quorum sensing is down-regulated, modulating the hypersensitive response in the host [138–144]. On the other hand, accumulated GABA may boost host endurance against fungal pathogens by sustaining TCAC activity and reducing oxidative damage [132,145–147].

4.2. Genetic Manipulation of Endogenous GABA Modifies the Biotic Stress Resistance

Table 3 briefly summarizes examples wherein genetic manipulation of endogenous GABA modifies biotic stress resistance. For example, transgenic tobacco and *Arabidopsis* plants with elevated GABA levels are more resistant to infection by *Agrobacterium* and *Pseudomonas* than WT plants [123,124], as well to predation by insect larvae and root-knot nematode [119–122]. In contrast, tomato plants with extremely low GABA levels are more susceptible to infection by *Ralstonia* [128]. Together, these findings provide strong support for the role of GABA in plant defense [114–116].

Recently, Deng et al. [148] reported that activation of the mitogen-activated protein kinase (MPK3/MPK6) signaling cascade greatly induces GABA biosynthesis in *Arabidopsis* (Table 3). The *gad1/2/4* triple and *gad1/2/4/5* quadruple mutants, in which the GABA levels are extremely low and the Glu and alanine levels are compromised, are more susceptible to both Pst and Pst-avrRpt2. Functional loss of AtMPK3/AtMPK6, their upstream AtMKK4/AtMKK5, or their downstream substrate, WRKY33, suppresses *AtGAD1* and *AtGAD4* expression after Pst-avrRpt2 treatment. These findings lend support for involvement of the MPK3/MPK6 signaling cascades in the induction of GAD and plant immunity against bacterial and fungal pathogens [149].

Table 3. Genetic manipulation of endogenous GABA modifies the biotic stress phenotype.

Plant Species	Mutant or Transgenic Line	Biotic Stress	Biochemical, Molecular and Physiological Responses	References
Tobacco (*Nicotiana tabaccum* L.)	*GADΔC* Ox	*Meloidogyne hapla* for 9 wk	• ↓ disease symptoms, nematode egg masses by 50–100%	[42]
		Heliothis virescens	• ↓ larval feeding by 80–90%	[135]
		Agrobacterium tumefaciens for 5 wk	• ↓ disease symptoms, stem tumor mass by 50% • ↑ lactonase *AttM* expression in *Agrobacterium*.	[138]
Arabidopsis (*Arabidopsis thaliana* (L.) Heynh)	*gaba-t*	*Spodoptera littoralis feeding* for 1 wk	• ↓ larval mass by ~40% • ↑ GABA 5-fold after brief feeding	[136,137]
	gad1/2/4	*Pseudomonas syringae* pv. *tomato* for 3 d	• ↓ infection-mediated GABA level • ↑ disease incidence • induction of *GAD1,4* regulated by *MKK4/MKK5-MPK3/MPK6* cascade and *WRKY33*	[148]
Tobacco or arabidopsis	*atpop (gaba-t)*, *NtGADΔC* Ox	*Pseudomonas syringae* pv. *tomato* for 2 d	• ↑ GABA, PR1 expression in plant • ↓ expression of bacterial Type III Secretion System with elevated GABA or exogenous GABA	[139,142]
Tomato (*Solanum lycopersicum* L.)	VIGS *GAD2*	*Ralstonia solanacearum* for 4 d	• ↑ disease incidence	[143]

Symbols: ↑, increases; ↓, decreases. Abbreviations: GABA, γ-aminobutyrate; GABA-T, GABA transaminase; GAD, glutamate decarboxylase; *GADΔC*, GAD lacking the C-terminal calmodulin binding domain; Ox, overexpression; PR, pathogen-related; VIGS, virus-induced gene silencing.

5. Exogenous GABA Improves Tolerance to Abiotic and Biotic Stresses

Table 4 briefly summarizes many examples from the literature wherein tolerance to hypoxia, drought, salinity, chilling, heat, osmotic stress and proton stress, as well as heavy metals (i.e., aluminum, arsenic and chromium) is improved in vegetative organs by the application of exogenous GABA [39,150–184]. The application of exogenous GABA typically increases the level of endogenous GABA, and elicits a diverse range of biochemical, molecular and physiological responses. The activity of the GABA shunt is increased to sustain the TCAC and energy production, though the precise response depends on the organ (shoot vs. roots) and the stress under consideration. Activities of N assimilation (including protein degradation) and PA pathways can also be modulated (also see [185]). Elevated endogenous GABA is also responsible for further increasing the stress-induced levels of non-enzymatic (ascorbic acid, GSH, phenols) and enzymatic (e.g., ascorbate oxidase, superoxide dismutase, ascorbate peroxidase, monodehydroascorbate reductase, glutathione reductase, glutathione peroxidase, glutathione S-transferase, catalase) antioxidants and osmolytes (sugars, organic and amino acids, including proline). These result in lower levels of ROS, malondialdehyde (a product of ROS-mediated peroxidation of membrane polyunsaturated fatty acids and marker for the depletion of antioxidant systems), and protein carbonylation (product of protein peroxidation and a marker of oxidative damage), lower activities of NADPH oxidase, lipoxygenase and polyphenol oxidase, and restoration of ion homeostasis (electrolyte leakage), which is indicative of membrane stability.

Table 4. Exogenous GABA modifies GABA, antioxidant, phytohormone and secondary pathways, and improves tolerance to abiotic stresses.

Plant Species	Abiotic Stress	Biochemical, Molecular and Physiological Responses	References
Arabidopsis (*Arabidopsis thaliana* (L.) Heyhn.)	Drought for up to 7 d	• ↓ light-stimulated increase in stomatal conductance • ↑ WUE	[40]
Tomato (*Solanum lycopersicum* L.)	2 °C for 2 d	• ↑ GABA; ↓ Pro, soluble sugars, MDA and EL	[150]
	75 mM NaCl for 6 d	• ↑ growth rate, DM, Chl • ↑ GABA, Glu, Pro, activities of GAD, SOD, POD and CAT, expression of *GAD1-3*, especially *GAD1* • ↓ ROS, MDA, Na^+, Na^+ flux from roots to leaves	[151]
Creeping bentgrass (*Agrostis stolonifera* L.)	35/30 °C or drought for 20–35 d	• ↑ Chl, RWC, cell membrane stability • ↑ Pro • ↓ ROS, MDA, EL • ↑ expression of *ABF3*, *POD*, *APX*, *HSP90*, *DHN3* and *MT1* during heat • ↑ expression of *CDPK26*, *MAPK1*, *ABF3*, *WRKY75*, *MYB13*, *HSP70*, *MT1*, *14-3-3*, *SOD*, *CAT*, *POD*, *APX* and DHAR during drought	[152,153]
	35/30 °C D:N for 35 d	• ↑ stomatal conductance, WUE, osmotic adjustment • ↑ amino acids (GABA), organic acids, sugars, ASC, GSH, GSSG, activities of SOD, APX, DHAR and GR • ↓ H_2O_2, MDA	[154]
	Drought for 25 d	• ↑ turf quality, RWC, osmotic adjustment, Chl, photochemical efficiency, amino acids (GABA), organic acids • ↓ EL	[155]
	38/33 °C D:N for 12 d	• ↑ Chl, P_N • ↑ GABA, expression of *CAD3*, *ACS*, *AER* and many *HSPs* • ↓ MDA, EL	[156]
	PEG (−0.52 MPa) for 12 d	• ↑ RWC, WUE, Chl, P_N • ↑ GABA, NO, activities of NR, GS, GOGAT, GDH, SOD, CAT, MDHAR and GR • ↓ Glu, GAD activity • ↓ photoinhibition, ROS, MDA, EL	[157]
Rice (*Oryza sativa* L.)	42/37 °C D:N for 10 d	• ↑ growth, survival, RWC, stomatal conductance • ↑ GABA, Pro, ASC, GSH, activities of CAT and APX • ↓ H_2O, MDA, EL	[158]
	100 µM As^{3+} (pH 4.4) for 22 d	• ↑ FM, shoot height, P_N, WUE • ↑ GABA, expression of *GAD2,3*, *GABA-T1,2* and *SSADH* (shoot) • ↓ activities of CAT, APX, POD, SOD and GR (shoot) • ↓ H_2O_2, GSH:GSSG (shoot) • ↓ expression of *GAD3*, *GABA-T2* and *SSADH* (roots)	[159]

Table 4. *Cont.*

Plant Species	Abiotic Stress	Biochemical, Molecular and Physiological Responses	References
	25 µM As^{3+} (pH 4.4) for 7 d	• ↑ Pro, Glu, linolenic acid, Put, expression of *ADC* and *SAMDC*, activities of MDHAR, GPX and POD (shoot) • ↓ Spd, Spm, ROS (shoot) • ↑ length, Put, expression of *ADC* and *SAMDC* (roots) • ↓ Spd, Spm, ROS	[160]
	30% PEG or/and 150 mM NaCl or 1 wk	• ↑ germination percent, vigor index, seedling growth, P_N, Chl, RWC, WUE • ↑ GABA, sugars, protein, starch • ↓ Pro, Na^+, H_2O_2, O_2•−, •OH, MDA • ↑ K^+, activities or/and expression of GR, SOD, CAT, APX, PAL, PPO, SKDH, CAD, GST, CHI, expression of many *CIPKs*,	[161]
White clover (*Trifolium repens* L.)	15% PEG for 17 d	• ↑ RWC, activities of GABA-T, OGDH, P5CS, PDH, ADC, ODC and SAMDC, *GAD* expression • ↑ GABA; transient ↑ Pro • ↓ EL, MDA, wilt • ↓ Glu, activities of GAD, PAO, CuAO and OAT • transient ↑ and ↓ in expression of *ADC*, *ODC*, *SAMDC*, *PAO* and *CuAO* • modifies free and conjugated PAs	[162]
	100 mM NaCl for 1 wk	• ↑ germination percentage, shoot DM and height. • ↑ GABA, activities of amylase, CAT, APX, POX, SOD, GR. DHAR and MDHR • ↑ expression of *CuZnSOD*, *MnSOD*, *FeSOD*, *CAT*, *APX*, *MDHR*, *GST*, *GPX*, *HKT1*, *HKT8*, *HAL2*, *H+-ATPase*, *SOS1* and genes encoding dehydrins • ↓ starch, soluble sugars, amino acids (including Pro) • ↓ H_2O_2, MDA, EL	[163]
	2 mg L^{-1} O_2 for 8 d	• ↑ GABA, Glu, expression and activities of *ADC*, *ODC* and *SAMDC* (roots) • ↓ expression and activities of *DAO* and *PAO* (roots) • modifies free and conjugated PAs (roots)	[164]
Melon (*Cucumis melo* L. var. *reticulatus*)	50–80 mM Ca(NO3)2 (pH 8.6) for 1 wk	• ↑ leaf area, P_N, chloroplast ATPase activities, ALA and improves chloroplast structure (seedling) • ↑ GABA, ASC, GSH (seedling) • ↑ activities of GAD, GABA-T, PAO, SOD, APX, GR, MDAR and DHAR, expression of *RBOHD* (seedling) • ↓ DAO activity, H_2O_2, MDA, Chl and its precursors (seedling) • ↑ Spd, Spm, activities of ADC, ODC and SAMDC, but ↓ Put (leaves and roots) • ↑ GABA-T activity (roots) • ↓ activities of GAD, PAO and DAO (roots) • Modifies levels of free and conjugated PAs	[165–167]
	75 mM NaCl/Na2SO4/NaHCO3/Na2CO3 (1:9:9:1 molar ratio, pH 8.6) for 24 h	• ↑ plant growth, NO and activities of NR, NOS, SOD, CAT, APX and GR • ↓ MDA, restores Na^+/K^+ balance	[168]

Table 4. *Cont.*

Plant Species	Abiotic Stress	Biochemical, Molecular and Physiological Responses	References
Mungbean (*Vigna radiata* (L.) R. Wilczek)	45/30 °C D:N for 30 d until maturity	• ↑ pod number, pod yield, seed yield, P_N • ↑ sugars, Pro, ASC, GSH • ↑ activities of Rubisco, P5CS, PDH, SOD, CAT, APX and GR • ↓ H_2O_2, MDA	[169]
Caragana intermedia Kuang et H.C. Fu	300 mM NaCl for 3 d	• ↑ GABA; ↓ H_2O_2 (roots) • transient ↑ in expression of *RBOHB, ACO1,2* and *ABA receptor*, ethylene production (roots) • ↓ *CuAO* expression (roots)	[170]
Wheat (*Triticum aestivum* L.)	100–200 mM NaCl for 21 d; 300 mM NaCl for 24 h	• ↑ germination, shoot DM, Chl, P_N, activities of SOD and CAT • ↓ MDA, EL	[171]
Corn (*Zea mays* L.)	300 mM NaCl for 48 h	• ↑ shoot height, leaf DM, area per plant, mitochondrial function. • ↑ GABA, Pro, sugars, activities of SOD and CAT activities. • ↓ GAD activity • ↓ MDA	[172]
	waterlogging for 14 d	• ↑ shoot and root DM, P_N, Chl, number of grana per chloroplast, GABA, activities of SOD, POD, CAT, GR and APX • ↓ MDA, H_2O_2, $O_2{\bullet}-$	[173]
Lettuce (*Latuca sativa* L.)	80 mM NaCl for 2 wk	• ↓ germination time • ↑ shoot and root DM, photosynthetic electron flow, activities of CAT, APX and SOD • ↓ Pro, H_2O_2, EL	[174]
Poplar (*Populus tomentosa* Carr.)	200 mM NaCl for 1 d	• ↓ H_2O_2, C_2H_4 (leaves) • ↑ ABA (leaves) • ↑ H_2O_2 (roots) • differential expression of various ABA/ethylene biosynthesis and receptor genes	[175]
Perennial ryegrass (*Lolium perenne* L.)	Drought for 14 d	• ↑ RWC, canopy temperature, turf quality, POD activity • ↓ wilt rating, MDA, EL	[176]
Black pepper (*Piper nigrum* L.)	10% PEG (*w/v*) for 15 d	• ↑ RWC, SOD activity • ↓ wilting %, Pro, MDA, impact on mitochondrial activity	[177]
Prunus rootstock (*Prunus avium* L.)	4 mg L^{-1} O_2 for 14 d	• transient ↓ and ↑ in H_2O_2 in roots and leaves, respectively. • ↑ P_N, stomatal conductance, Chl, Ala (leaves) • ↓ Glu (leaves) • transient ↑ in Glu, Ala, *GAD2,4* expression (roots)	[178]

Table 4. *Cont.*

Plant Species	Abiotic Stress	Biochemical, Molecular and Physiological Responses	References
Barley (*Hordeum vulgare* L.)	20 µM Al^{3+} (pH 4.5) for 1 d	• ↑ elongation, activities of SOD, CAT and POD (roots) • ↓ protein carbonylation, MDA, ROS	[179]
	20 mM NaCl for 6 d	• ↑ total phenolics and flavonoids (gallic acid, protocatechic acid, caffeic acid, quercetin), ABTS and DPPH radical scavenging activities • ↑ activities or expression of PAL, C4H, 4CL • ↓ MDA	[180]
Mustard (*Brassica juncea* L.)	0.15 or 0.3 mM Cr^{VI} for 5 d	• ↑ shoot height and DM, RWC, Chl, GSH, ASC, GSH/GSSG, ASC/DHA, activities of MDHAR, DHAR, GR, GPX, SOD and CAT • ↓ Pro, ROS, MDA, DHA, GSSG, MG, LOX activity	[181]
Tobacco (*Nicotiana tabaccum* L.)	200 µM Cd for 7 d	• ↑ Pro, GSH, stability of O_2-evolving complex, electron transfer rate in PSII, stomatal conductance, expression of *P5CS* and *P5CR* • ↓ H_2O_2, MDA	[182]
Sunflower (*Helianthus annuus* L.)	38 °C or drought 55 d	• ↑ shoot height and FM, root FM, Chl, transpiration, Pro and TSS • ↓ H_2O_2, MDA, EL • may ↑ activities of SOD and POD, expression of *Hsp70*, *MDHAR*, *Osmotin*, *GR* and *Aquaporin*	[183]
Carrot (*Daucus carota* L.)	Drought for 2 wk	• ↑ leaf and root growth, RWC, amino acids, CAT activity • ↓ MDA, EL	[184]

Symbols: ↑, increases/improves; ↓, decreases. Abbreviations: 14-3-3, regulatory protein; ABA, abscisic acid; ABF, transcription factor; ACS, acetyl-coenzyme A synthetase; AER, NADPH-dependent alkenal reductase P2; As^{3+}, arsenic; ASC, ascorbate; ADC, arginine decarboxylase; ALA, δ-aminolevulinic acid; Al^{3+}, aluminum; APX, ascorbate peroxidase; CAD3, cinnamyl alcohol dehydrogenase 3; CAT, catalase; Cd, cadmium; CDPK, calcium-dependent protein kinase; Cr^{VI}, chromium; CuAO, copper amine oxidase; D, day; DAO, diamine oxidase; DHA, dehydroascorbate; DHAR, dehydroascorbate reductase; DHN, dehydrin; DM/FM, dry/fresh mass; EL, electrolyte leakage; GABA, γ-aminobutyric acid; GABA-TP or GABA-TOG, pyruvate- or 2-oxoglutarate-dependent GABA transaminase; GAD, glutamate decarboxylase; Glu, glutamate; GOGAT, glutamate synthase; GPX, glutathione peroxidase; GR, glutathione reductase; GS, glutamine synthetase; GSH, reduced glutathione; GST, glutathione S-transferase; H_2O_2, hydrogen peroxide; HSP, heat shock protein; LOX, lipoxygenase; MAPK, mitogen-activated protein kinase; MDA, malondialdehyde; MDHAR, monodehydroascorbate reductase; miRNA, microRNA; MSI, membrane stability index; MT, metallothionein; MYB, transcription factor; N, night; NO, nitric oxide; NOS, NO synthase; NR, nitrate reductase; OAT, ornithine δ-aminotransferase; ODC, ornithine decarboxylase; P5CR, Δ^1-pyrroline-5-carboxylate reductase; P5CS, Δ^1-pyrroline-5-carboxylate synthetase; PA, polyamine; PAO, polyamine oxidase; PEG, polyethyleneglycol 6000; P_N, net photosynthesis; POD, peroxide dismutase; PPO, polyphenol oxidase; POX, peroxide oxidase; PP, phenylpropanoid; Pro, proline; Put, putrescine; PYL, pyrabactin resistance 1-like; RBOHD, respiratory burst oxidase homologue D/NADPH oxidase; ROS, reactive oxygen species; RuBisCo, ribulose bisphosphate carboxylase oxygenase; RWC, relative water content; SAMDC, S-adenosylmethionine decarboxylase; Spd, spermidine; Spm, spermine; SSADH, succinic semialdehyde dehydrogenase; TSS, total soluble sugars, WRKY, transcription factor; WUE, water use efficiency.

There is also evidence for the GABA-induced production of nitric oxide (NO) (Table 4), which could be associated with the enhancement of antioxidant defense, as well as regulation of epigenetic mechanisms and gene transcription [157,174,186]. The stress tolerance could also be related to GABA-induced changes in pathways associated with other phytohormones such as ABA (ABA receptors), ethylene (ACC oxidase, ACC synthase), PAs (arginine decarboxylase, free and conjugated forms, S-adenosylmethionine decarboxylase) and salicylate (Table 4) [60,156,172,175,187], which can regulate metabolic homeostasis and influence the expression of stress factors (miRNAs, transcription factors, heat shock proteins) with known and yet-to-be-determined functions [156,183,188]. It is known that the exogenous application of GABA, ABA and salicylate alleviate the drought-induced damage to membranes and leaf water status in creeping bentgrass by affecting similar metabolic pathways, yet cause differential changes in metabolite accumulation [155]. Additionally, NO and nitrate reductase are jointly needed for salicylate-induced water-stress tolerance in pepper plants [189]. The osmotin protein, which belongs to the PR-5 family of pathogenesis-related proteins, is known to inhibit the activity of defensive cell wall barriers in fungi [190]. Cinnamyl alcohol dehydrogenase is involved in lignin biosynthesis and alkenyl reductase and can detoxify cytotoxic substrates such as aldehydes [155]. Further research is required to investigate the effects of exogenous GABA on alternative respiratory pathways involved in the scavenging, regulation and homeostasis of ROS and NO [92,191], the action of other phytohormones [23,156,187,192–197], and post-translational modifications and epigenetic regulation of gene expression [147,198] during stress.

6. Concluding Remarks

Plants must endure a wide variety of abiotic and biotic stresses under field conditions. To prevent significant yield losses, many crop improvement programs strive to develop stress-tolerant cultivars. Plants may respond uniquely to different or simultaneous stresses, so breeding tolerance against one stress may be at the expense of tolerance to another. With climate change, plants will likely experience more extreme weather events or multiple stresses, including plant diseases. It would therefore be beneficial to develop a tolerance strategy against multiple stresses.

GABA metabolism has garnered considerable attention in recent years, in part because it often accumulates in response to a variety of abiotic (cold, heat, drought, salinity, salinity-alkalinity, osmotic, low O_2, heavy metal toxicity) and biotic (invertebrate pests, bacteria, fungi) stresses (Figure 2). These findings are typically attributed to stimulation of GABA anabolism or inhibition of GABA catabolism. However, there are clear cases in which GABA pathway activity is promoted to sustain respiration and the generation of energy, without the accumulation of GABA. On the other hand, accumulated GABA may bind to AMLT and GORK, interfering, respectively, with the transport of malate in stomatal guard cells and K^+ in root epidermal cells, thereby enhancing plant tolerance to drought and hypoxia. With very few exceptions, genetic manipulation of GABA metabolism and receptors, respectively, reveal positive relationships between GABA levels and abiotic/biotic stress tolerance, and between malate efflux from the root and heavy metal tolerance.

Common plant responses to avoid or tolerate abiotic and biotic stresses include stomatal closure and corresponding decreases in photosynthesis, and reduced leaf growth and root length, as well as greater ROS activity. These responses are coordinated by phytohormones such as ABA, NO, ethylene, salicylate and jasmonate. Thus, the enhancement of endogenous GABA by either genetic engineering or the application of exogenous GABA reduces the stress-induced ROS level and restores or partially restores the morpho-physiological features of the unstressed phenotype by promoting or modifying activities of the GABA shunt, TCAC, antioxidant, secondary metabolism and phytohormone pathways (Figure 2). Furthermore, elevated plant GABA adversely affects the activity of phytopathogens by various mechanisms. Therefore, exogenous GABA might function under field conditions as an effective and sustainable tolerance strategy against the multiple abiotic and biotic stresses that could be exacerbated by climate change.

Low temperature and controlled atmosphere conditions (low O_2, elevated CO_2) are extensively employed to extend the postharvest life of horticultural commodities, especially botanical fruits. However, these crops may suffer from chilling injury and other physiological disorders, as well as fungal decay. The use of exogenous GABA to improve the marketability of stored horticultural commodities will be described in a companion paper

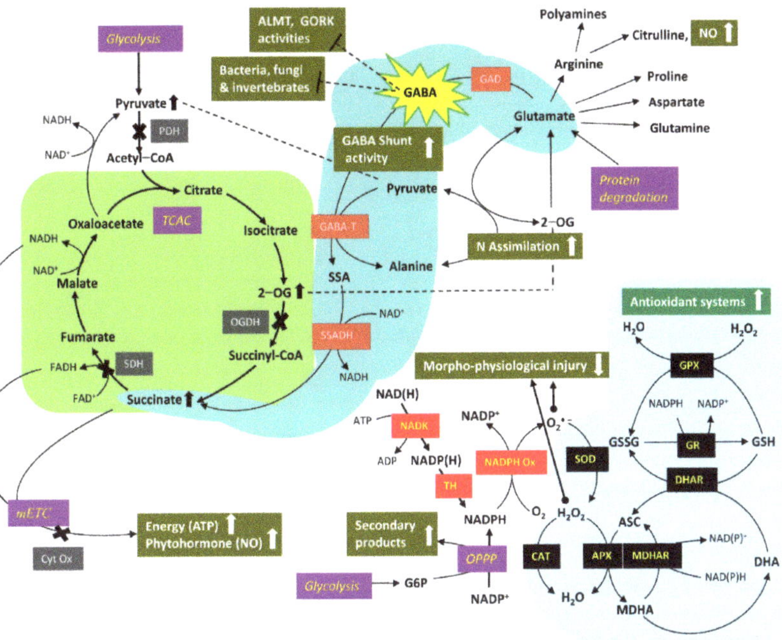

Figure 2. Abiotic and biotic stress-induced changes in morphological and physiological features are restored or partially restored by increasing endogenous GABA and GABA shunt activity. Depending upon the nature of the abiotic stress, the entry of pyruvate into the tricarboxylic acid cycle via pyruvate dehydrogenase, the catabolism of 2−oxoglutarate and succinate via 2−oxoglutarate dehydrogenase and succinate dehydrogenase, respectively, and activity of cytochrome oxidase in the mitochondrial electron transport chain may be restricted, thereby limiting the generation of molecules associated with energy transfer (NADH, FADH and ATP). These limitations are to some extent overcome by increasing the levels of endogenous GABA, by either GAD overexpression or application of exogenous GABA. The elevated endogenous GABA increases flux through the shunt, which in turn, increases the level and/or entry of succinate into the non−cyclic tricarboxylic acid cycle, mitochondrial electron transport chain, and accordingly, ATP generation. The elevated GABA also modifies the activity of stress-induced pathways involving enzymatic and non−enzymatic antioxidants, N assimilation, secondary products, and phytohormones (NO; ethylene and ABA are not shown) by uncharacterized mechanisms. The activity of phytopathogens is directly inhibited by accumulated GABA. Symbols: X, biochemical reaction potentially inhibited by stress; white filled arrows, increasing or decreasing activity of pathway(s) affected by GABA; black filled arrow, increasing level of metabolite affected by stress. Abbreviations: ALMT, aluminum−activated malate transporter; Cyt Ox, cytochrome oxidase; NADK, NAD kinase; OPPP, oxidative pentose phosphate pathway; PDH, pyruvate dehydrogenase; SAM, *S*−adenosylmethionine; SDH, succinate dehydrogenase; TCAC, tricarboxylic acid cycle; TH, transhydrogenase; see Table 4 legend for the remaining abbreviations.

Supplementary Materials: The following are available online at https://doi.org/10.5281/zenodo.5295721, Table S1: Important *Arabidopsis thaliana* genes associated with GABA metabolism and signaling.

Author Contributions: Conceptualization, B.J.S. and M.S.A.; writing, review and editing, B.J.S., M.S.A. and E.J.F.; preparation of the images; B.J.S. and E.J.F. All authors have read and agreed to the published version of the manuscript.

Funding: GABA research in the authors' laboratories is funded by the Natural Sciences and Engineering Research Council of Canada (B.J.S.) and the Imam Khomeini International University (M.S.A.).

Institutional Review Board Statement: Not applicable. This study did not involve humans or animals.

Informed Consent Statement: Not applicable. This study did not involve humans or animals.

Conflicts of Interest: The authors declare no conflict of interest.

References

1. Rockström, J.; Williams, J.; Daily, G.; Noble, A.; Matthews, N.; Gordon, L.; Wetterstrand, H.; DeClerck, F.; Shah, M.; Steduto, P.; et al. Sustainable intensification of agriculture for human prosperity and global sustainability. *Ambio* **2017**, *46*, 4–17. [CrossRef]
2. Intergovernmental Panel on Climate Change. Summary for Policymakers. In *Climate Change 2021: The Physical Science Basis. Contribution of Working Group I to the Sixth Assessment Report of the Intergovernmental Panel on Climate Change*; Masson-Delmotte, V., Zhai, P., Pirani, A., Connors, S.L., Péan, C., Berger, S., Caud, N., Chen, Y., Goldfarb, L., Gomis, M.I., et al., Eds.; Cambridge University Press: Cambridge, UK, 2021. Available online: https://www.ipcc.ch/report/sixth-assessment-report-working-group-i/ (accessed on 22 August 2021).
3. Saunders, D.G.O. Will yield gains be lost to disease? *Nat. Clim. Chang.* **2021**, *11*, 644–650. [CrossRef]
4. Godoy, F.; Olivos-Hernández, K.; Stange, C.; Handford, M. Abiotic stress in crop species: Improving tolerance by applying plant metabolites. *Plants* **2021**, *10*, 186. [CrossRef] [PubMed]
5. Hasanuzzaman, M.; Bhuyan, M.H.M.B.; Zulfiqar, F.; Raza, A.; Mohsin, S.M.; Mahmud, J.A.; Fujita, M.; Fotopoulos, V. Reactive oxygen species and antioxidant defense in plants under abiotic stress: Revisiting the crucial role of a universal defense regulator. *Antioxidants* **2020**, *9*, 681. [CrossRef] [PubMed]
6. Singh, A.; Kumar, A.; Yadav, S.; Singh, I.K. Reactive oxygen species-mediated signaling during abiotic stress. *Plant Gene* **2019**, *18*, 100173. [CrossRef]
7. Bown, A.W.; Shelp, B.J. The metabolism and functions of gamma-aminobutyric acid. *Plant Physiol.* **1997**, *115*, 1–5. [CrossRef] [PubMed]
8. Shelp, B.J.; Bown, A.W.; McLean, M.D. Metabolism and functions of gamma-aminobutyric acid. *Tr. Plant Sci.* **1999**, *4*, 446–452. [CrossRef]
9. Kinnersley, A.M.; Turano, F.J. Gamma aminobutyric acid (GABA) and plant response to stress. *Crit. Rev. Plant Sci.* **2000**, *19*, 479–509. [CrossRef]
10. Bouché, N.; Lacombe, B.; Fromm, H. GABA signalling: A conserved and ubiquitous mechanism. *Trends Cell Biol.* **2003**, *13*, 607–610. [CrossRef] [PubMed]
11. Bouché, N.; Fromm, H. GABA in plants: Just a metabolite? *Trends Plant Sci.* **2004**, *9*, 110–115. [CrossRef] [PubMed]
12. Fait, A.; Fromm, H.; Walter, D.; Galili, G.; Fernie, A.R. Highway or byway: The metabolic role of the GABA shunt in plants. *Trends Plant Sci.* **2008**, *13*, 14–19. [CrossRef]
13. Gilliham, M.; Tyerman, S.D. Linking metabolism to membrane signaling: The GABA–malate connection. *Tr. Plant Sci.* **2016**, *21*, 296–301. [CrossRef]
14. Ramesh, S.A.; Tyerman, S.D.; Gilliham, M.; Xu, B. γ-Aminobutyric acid (GABA) signalling in plants. *Cell. Mol. Life Sci.* **2017**, *74*, 1577–1603. [CrossRef] [PubMed]
15. Shelp, B.J.; Bozzo, G.G.; Trobacher, C.P.; Chiu, G.; Bajwa, V.S. Strategies and tools for studying the metabolism and function of γ-aminobutyrate in plants. I. Pathway structure. *Botany* **2012**, *90*, 651–668. [CrossRef]
16. Shelp, B.J.; Bown, A.W.; Zarei, A. 4-Aminobutyrate (GABA): A metabolite and signal with practical significance. *Botany* **2017**, *95*, 1015–1032. [CrossRef]
17. Shelp, B.J.; Zarei, A. Subcellular compartmentation of 4-aminobutyrate (GABA) metabolism in arabidopsis: An update. *Plant Signal. Behav.* **2017**, *12*, e1322244. [CrossRef] [PubMed]
18. Long, Y.; Tyerman, S.D.; Gilliham, M. Cytosolic GABA inhibits anion transport by wheat ALMT1. *New Phytol.* **2020**, *225*, 671–678. [CrossRef] [PubMed]
19. Bown, A.W.; Shelp, B.J. Does the GABA shunt regulate cytosolic GABA? *Tr. Plant Sci.* **2020**, *25*, 422–424. [CrossRef]
20. Adem, G.D.; Chen, G.; Shabala, L.; Chen, Z.-H.; Shabala, S. GORK channel: A master switch of plant metabolism? *Tr. Plant Sci.* **2020**, *25*, 434–445. [CrossRef]
21. Li, L.; Dou, N.; Zhang, H.; Wu, C. The versatile GABA in plants. *Plant Signal. Behav.* **2021**, *16*, 1862565. [CrossRef]
22. Fromm, H. GABA signaling in plants: Targeting the missing pieces of the puzzle. *J. Exp. Bot.* **2021**, *71*, 6238–6245. [CrossRef]
23. Xu, B.; Sai, N.; Gilliham, M. The emerging role of GABA as a transport regulator and physiological signal. *Plant Physiol.* **2021**. [CrossRef]
24. Wallace, W.; Secor, J.; Schraeder, L.E. Rapid accumulation of γ-aminobutyric acid and alanine in soyybean leaves in response to an abrupt transfer to lower temperature, darkness, or mechanical amnipulation. *Plant Physiol.* **1984**, *75*, 170–175. [CrossRef]
25. Tuin, L.G.; Shelp, B.J. In situ [^{14}C]glutamate metabolism by developing soybean cotyledons I. *Metabolic routes. J. Plant Physiol.* **1994**, *143*, 1–7. [CrossRef]
26. Tuin, L.G.; Shelp, B.J. In situ [^{14}C]glutamate metabolism by developing soybean cotyledons II. The importance of glutamate decarboxylation. *J. Plant Physiol.* **1996**, *147*, 714–720. [CrossRef]
27. Breitkreuz, K.E.; Shelp, B.J. Subcellular compartmentation of the 4-aminobutryate shunt in protoplasts from developing soybean cotyledons. *Plant Physiol.* **1995**, *108*, 99–103. [CrossRef]
28. Hijaz, F.; Killiny, N. Exogenous GABA is quickly metabolized to succinic acid and fed into the plant TCA cycle. *Plant Signal. Behav.* **2019**, *14*, e1573096. [CrossRef] [PubMed]
29. Baum, G.; Lev-Yadun, S.; Fridman, Y.; Arazi, T.; Katsnelson, H.; Zik, M.; Fromm, H. Calmodulin binding to glutamate decarboxylase is required for regulation of glutamate and GABA metabolism and normal development in plants. *EMBO J.* **1996**, *15*, 2988–2996. [CrossRef]

30. Ling, V.; Snedden, W.A.; Shelp, B.J.; Assmann, S. Analyses of a soluble calmodulin-binding protein from fava bean roots: Identification as glutamate decarboxylase. *Plant Cell* **1994**, *6*, 1135–1143. [PubMed]
31. Arazi, T.; Baum, G.; Snedden, W.A.; Shelp, B.J.; Fromm, H. Molecular and biochemical analysis of calmodulin interactions with the calmodulin-binding domain of plant glutamate decarboxylase. *Plant Physiol.* **1995**, *108*, 551–561. [CrossRef] [PubMed]
32. Snedden, W.A.; Arazi, T.; Fromm, H.; Shelp, B.J. Calcium/calmodulin regulation of soybean glutamate decarboxylase. *Plant Physiol.* **1995**, *108*, 543–549. [CrossRef]
33. Snedden, W.A.; Koutsia, N.; Baum, G.; Fromm, H. Activation of a recombinant petunia glutamate decarboxylase by calcium/calmodulin or by a monoclonal antibody which recognizes the calmodulin binding domain. *J. Biol. Chem.* **1996**, *271*, 4148–4153. [CrossRef] [PubMed]
34. Cholewa, E.; Cholewinski, A.J.; Shelp, B.J.; Snedden, W.A.; Bown, A.W. Cold-shock-stimulated γ-aminobutyric acid synthesis is mediated by an increase in cytosolic Ca^{2+}, not by an increase in cytosolic H^+. *Can. J. Bot.* **1997**, *75*, 375–382. [CrossRef]
35. Gut, H.; Dominici, P.; Pilati, S.; Astegno, A.; Petoukhov, M.V.; Svergun, D.I.; Grütter, M.G.; Capitani, G. A common structural basis for pH- and calmodulin-mediated regulation in plant glutamate decarboxylase. *J. Mol. Biol.* **2009**, *392*, 334–351. [CrossRef] [PubMed]
36. Knight, H.; Trewavas, A.J.; Knight, M.R. Calcium signaling in *Arabidopsis thaliana* responding to drought and salinity. *Plant J.* **1997**, *12*, 1067–1078. [CrossRef] [PubMed]
37. Bose, J.; Pottosin, I.I.; Shabala, S.S.; Palmgren, M.G.; Shabala, S. Calcium efflux systems in stress signaling and adaptation in plants. *Front. Plant Sci.* **2011**, *2*, 85. [CrossRef]
38. Behera, S.; Xu, Z.; Luoni, L.; Bonza, M.C.; Doccula, F.G.; De Michelis, M.I.; Morruis, R.J.; Schwarzländer, M.; Costa, A. Cellular Ca^{2+} signals generate defined pH signatures in plants. *Plant Cell* **2018**, *30*, 2704–2719. [CrossRef]
39. Bouché, N.; Fait, A.; Zik, A.; Fromm, H. The root-specific glutamate decarboxylase (GAD1) for sustaining GABA levels in *Arabidopsis*. *Plant Mol. Biol.* **2004**, *55*, 315–325. [CrossRef] [PubMed]
40. Xu, B.; Long, Y.; Feng, X.; Zhu, X.; Sai, N.; Chirkova, L.; Betts, A.; Herrmann, J.; Edwards, E.J.; Okamoto, M.; et al. GABA signalling modulates stomatal opening to enhance plant water use efficiency and drought resilence. *Nat. Commun.* **2021**, *12*, 1952. [CrossRef] [PubMed]
41. Zarei, A.; Chiu, G.Z.; Yu, G.; Trobacher, C.P.; Shelp, B.J. Salinity-regulated expression of genes involved in GABA metabolism and signaling. *Botany* **2017**, *95*, 621–627. [CrossRef]
42. McLean, M.D.; Yevtushenko, A.; Deschene, A.; Van Cauwenberghe, O.R.; Makhmoudova, A.; Potter, J.W.; Bown, A.W.; Shelp, B.J. Overexpression of glutamate decarboxylase in transgenic tobacco confers resistance to the northern root-knot nematode. *Mol. Breed.* **2003**, *11*, 277–285. [CrossRef]
43. Stitti, N.; Missihoun, T.; Kotchoni, S.S.; Kirch, H.H.; Bartels, D. Aldehyde dehydrogenase in *Arabidopsis thaliana*: Biochemical requirements, metabolic pathways, and functional analysis. *Front. Plant Sci.* **2011**, *2*, 65. [CrossRef]
44. Shelp, B.J.; Bozzo, G.G.; Trobacher, C.P.; Zarei, A.; Deyman, K.L.; Brikis, C.J. Hypothesis/review: Contribution of putrescine to 4-aminobutyrate (GABA) production in response to abiotic stress. *Plant Sci.* **2012**, *193–194*, 130–135. [CrossRef]
45. Moschou, P.N.; Wu, J.; Cona, A.; Tavladoraki, P.; Angelini, R.; Roubelakis-Angelakis, K.A. The polyamines and their catabolic products are significant players in the turnover of nitrogenous moleculses in plants. *J. Exp. Bot.* **2012**, *63*, 5003–5015. [CrossRef] [PubMed]
46. Tavladoraki, P.; Cona, A.; Angelini, R. Copper-containing amine oxidases and FAD-dependent polyamine oxidases are key players in plant tissue differentiation and organ development. *Front. Plant Sci.* **2016**, *7*, 824. [CrossRef] [PubMed]
47. Planas-Portell, J.; Gallart, M.; Tiburcio, A.F.; Altabella, T. Copper-containing amine oxidases contribute to terminal polyamine oxidation in peroxisomes and apoplast of *Arabidopsis thaliana*. *BMC Plant Biol.* **2013**, *13*, 109. [CrossRef] [PubMed]
48. Xing, S.G.; Jun, Y.B.; Hau, Z.W.; Liang, L.Y. Higher accumulation of gamma-aminobutyric acid induced by salt stress through stimulating the activity of diamine oxidases in *Glycine max* (L.) Merr. roots. *Plant Physiol. Biochem.* **2007**, *45*, 560–566. [CrossRef]
49. Missihoun, T.D.; Schmitz, J.; Klug, R.; Kirch, H.H.; Bartels, D. Betaine aldehyde dehydrogenase genes from *Arabidopsis* with different sub-cellular localization affect stress responses. *Planta* **2011**, *233*, 369–382. [CrossRef]
50. Zarei, A.; Trobacher, C.P.; Shelp, B.J. *Arabidopsis* aldehyde dehydrogenase 10 family members confer salt tolerance through putrescine-derived 4-aminobutyrate (GABA) production. *Sci. Rep.* **2016**, *6*, 35115. [CrossRef]
51. Jacques, F.; Zhao, Y.; Kopečná, M.; Končitíková, R.; Kopečný, D.; Rippa, S.; Perrin, Y. Roles for ALDH10 enzymes in γ-butyrobetaine synthesis, seed development, germination, and salt tolerance in *Arabidopsis*. *J. Exp. Bot.* **2020**, *71*, 7088–7102. [CrossRef]
52. Signorelli, S.; Dans, P.D.; Coitiño, E.L.; Borsani, O.; Monza, J. Connecting proline and γ-aminobutyric acid in stressed plants through non-enzymatic reactions. *PLoS ONE* **2015**, *10*, e0115349. [CrossRef]
53. Forlani, G.; Trovato, M.; Funck, D.; Signorelli, S. Regulation of proline accumulation and its molecular and physiological functions in stress defence. In *Osmoprotectant-Mediated Abiotic Stress Tolerance in Plants*; Hossain, M.A., Kumar, V., Burritt, D.J., Fujita, M., Mäkelä, P.S.A., Eds.; Springer Nature Switzerland: Cham, Switzerland, 2019; pp. 73–97. [CrossRef]
54. Michaeli, S.; Fait, A.; Lagor, K.; Nunes-Nesi, A.; Grillich, N.; Yellin, A.; Bar, D.; Khan, M.; Fernie, A.R.; Turano, F.J. A mitochondrial GABA permease connects the GABA shunt and the TCA cycle, and is essential for normal carbon metabolism. *Plant J.* **2011**, *67*, 485–498. [CrossRef]
55. Dündar, E.; Bush, D.R. BAT1, a bidirectional amino acid transporter in *Arabidopsis*. *Planta* **2009**, *229*, 1047–1056. [CrossRef]

56. Ariyaratne, M.; Ge, L.; Morris, P.F. Characterization of membrane transporters by heterologous expression in *E. coli* and production of membrane vesicles. *J. Vis. Exp.* **2019**, *154*, e60009. [CrossRef] [PubMed]
57. Van Cauwenberghe, O.R.; Shelp, B.J. Biochemical characterization of partially purified GABA: Pyruvate transaminase from *Nicotiana tabacum*. *Phytochemistry* **1999**, *52*, 575–581. [CrossRef]
58. Clark, S.M.; Di Leo, R.; Dhanoa, P.K.; Van Cauwenberghe, O.R.; Mullen, R.T.; Shelp, B.J. Biochemical characterization, mitochondrial localization, expression, and potential functions for an *Arabidopsis* γ-aminobutyrate transaminase that utilizes both pyruvate and glyoxylate. *J. Exp. Bot.* **2009**, *60*, 1743–1757. [CrossRef] [PubMed]
59. Shimajiri, Y.; Ozaki, K.; Kainou, K.; Akama, K. Differential subcellular localization, enzymatic properties and expression patterns of γ-aminobutyric transaminases (GABA-T) in rice (*Oryza sativa*). *J. Plant Physiol.* **2013**, *170*, 196–201. [CrossRef]
60. Podlešáková, K.; Ugena, L.; Spichal, L.; Dolezal, K.; De Diego, N. Phytohormones and polyamines regulate plant stress responses by altering GABA pathway. *Nat. Biotechnol.* **2019**, *48*, 53–65. [CrossRef]
61. Seifikalhor, M.; Aliniaeifard, S.; Hassani, B.; Niknam, V.; Lastochkina, O. Diverse role of γ-aminobutyric acid in dynamic plant cell responses. *Plant Cell Rep.* **2019**, *38*, 847–867. [CrossRef]
62. Bandehagh, A.; Taylor, N.L. Can alternative metabolic pathways and shunts overcome salinity induced inhibition of central carbon metabolism in crops? *Front. Plant Sci.* **2020**, *11*, 1072. [CrossRef] [PubMed]
63. Babu, G.G.; Naik, G.R. GABA: Pyruvate-dependent transaminase dominates GABA: 2-oxoglutarate dependent transaminase in sugarcane and their molecular characterization. *Int. J. Dev. Res.* **2013**, *3*, 9–16.
64. Shelp, B.J.; Walton, C.S.; Snedden, W.A.; Tuin, L.J.; Oresnik, I.J.; Layzell, D.B. GABA shunt in developing soybean seeds is associated with hypoxia. *Physiol. Plant.* **1995**, *94*, 219–228. [CrossRef]
65. Clark, S.M.; Di Leo, R.; Van Cauwenberghe, O.R.; Mullen, R.T.; Shelp, B.J. Subcellular localization and expression of multiple tomato γ-aminobutyrate transaminases that utilize both pyruvate and glyoxylate. *J. Exp. Bot.* **2009**, *60*, 3255–3267. [CrossRef] [PubMed]
66. Koike, S.; Matsukura, C.; Takayama, M.; Asamizu, E.; Ezura, H. Suppression of γ-aminobutyric acid (GABA) transaminases induces prominent GABA accumulation, dwarfism and infertility in the tomato (*Solanum lycopersicum* L.). *Plant Cell Physiol.* **2013**, *54*, 793–807. [CrossRef] [PubMed]
67. Miyashita, Y.; Good, A.G. Contribution of the GABA shunt to hypoxia-induced alanine accumulation in roots of *Arabidopsis thaliana*. *Plant Cell Physiol.* **2008**, *49*, 92–102. [CrossRef]
68. Renault, H.; Roussel, V.; El Amrani, A.; Arzel, M.; Renault, D.; Bouchereau, A.; Deleu, C. The *Arabidopsis pop2-1* mutant reveals the involvement of GABA transaminase in salt stress tolerance. *BMC Plant Biol.* **2010**, *10*, 20. [CrossRef] [PubMed]
69. Mekonnen, D.W.; Flügge, U.-I.; Ludewig, F. Gamma-aminobutyric acid depletion affects stomata closure and drought tolerance of *Arabidopsis thaliana*. *Plant Sci.* **2016**, *245*, 25–34. [CrossRef] [PubMed]
70. Su, N.; Wu, Q.; Chen, J.; Shabala, L.; Mithöfer, A.; Wang, H.; Qu, M.; Yu, M.; Cui, J.; Shabala, S. GABA operates upstream of H^+-ATPase and improves salinity tolerance in *Arabidopsis* by enabling cytosolic K^+ retention and Na^+ exclusion. *J. Exp. Bot.* **2019**, *70*, 6349–6361. [CrossRef]
71. Busch, K.B.; Fromm, H. Plant succinic semialdehyde dehydrogenase. Cloning, purification, localization in mitochondria, and regulation by adenine nucleotides. *Plant Physiol.* **1999**, *121*, 589–597. [CrossRef]
72. Busch, K.; Piehler, J.; Fromm, H. Plant succinic semialdehyde dehydrogenase: Dissection of nucleotide binding by surface plasmon resonance and fluorescence spectroscopy. *Biochemistry* **2000**, *39*, 10110–10117. [CrossRef] [PubMed]
73. Bouché, N.; Fait, A.; Bouchez, D.; Møller, S.G.; Fromm, H. Mitochondrial succinic-semialdehyde dehydrogenase of the γ-aminobutyrate shunt is required to restrict levels of reactive oxygen intermediates in plants. *Proc. Natl. Acad. Sci. USA* **2003**, *100*, 6843–6848. [CrossRef] [PubMed]
74. Fait, A.; Yellin, A.; Fromm, H. GABA shunt deficiencies and accumulation of reactive oxygen intermediates: Insight from *Arabidopsis* mutants. *FEBS Lett.* **2005**, *579*, 415–420. [CrossRef] [PubMed]
75. Jin, W.; Wu, F. Proteome-wide identification of lysine succinylation in the proteins of tomato (*Solanum lycopsicum*). *PLoS ONE* **2016**, *11*, e0147586. [CrossRef]
76. Zhang, K.; Xiong, Y.; Sun, W.; Wang, G.-L.; Liu, W. Global proteomic analysis reveals widespread lysine succinylation in rice seedlings. *Int. J. Mol. Sci.* **2019**, *20*, 5911. [CrossRef] [PubMed]
77. Breitkreuz, K.E.; Allan, W.L.; Van Cauwenberghe, O.R.; Jakobs, C.; Talibi, D.; André, B.; Shelp, B.J. A novel γ-hydroxybutyrate dehydrogenase. Identification and expression of an *Arabidopsis* cDNA and potential role under oxygen deficiency. *J. Biol. Chem.* **2003**, *278*, 41552–41556. [CrossRef]
78. Hoover, G.J.; Van Cauwenberghe, O.R.; Breitkreuz, K.E.; Clark, S.M.; Merrill, A.R.; Shelp, B.J. Characteristics of an *Arabidopsis* glyoxylate reductase: General biochemical properties and substrate specificity for the recombinant protein, and developmental expression and implications for glyoxylate and succinic semialdehyde metabolism *in planta*. *Can. J. Bot.* **2007**, *85*, 883–895. [CrossRef]
79. Simpson, J.P.; Di Leo, R.; Allan, W.L.; Clark, S.M.; Dhanoa, P.K.; Makhmoudova, A.; Hoover, G.J.; Mullen, R.T.; Shelp, B.J. Identification and characterization of a plastid-localized *Arabidopsis* glyoxylate reductase isoform: Comparison with a cytosolic isoform and implications for cellular redox homeostasis and aldehyde detoxification. *J. Exp. Bot.* **2008**, *59*, 2545–2554. [CrossRef]

80. Allan, W.L.; Simpson, J.P.; Clark, S.M.; Shelp, B.J. γ-Hydroxybutyrate accumulation in Arabidopsis and tobacco plants is a general response to abiotic stress: Putative regulation by redox balance and glyoxylate reductase isoforms. *J. Exp. Bot.* **2008**, *59*, 2555–2564. [CrossRef]
81. Ching, S.L.K.; Gidda, S.K.; Rochon, A.; van Cauwenberghe, O.R.; Shelp, B.J.; Mullen, R.T. Glyoxylate reductase isoform 1 is localized in the cytosol and not peroxisomes in plant cells. *J. Integr. Plant Biol.* **2012**, *54*, 152–168. [CrossRef]
82. Brikis, C.J.; Zarei, A.; Trobacher, C.P.; DeEll, J.R.; Akama, K.; Mullen, R.T.; Bozzo, G.G.; Shelp, B.J. Ancient plant glyoxylate/succinic semialdehyde reductases: GLYR1s are cytosolic, whereas GLYR2s are localized to both mitochondria and plastids. *Front. Plant Sci.* **2017**, *8*, 601. [CrossRef]
83. Zarei, A.; Brikis, C.J.; Bajwa, V.S.; Chiu, G.Z.; Simpson, J.P.; DeEll, J.R.; Bozzo, G.G.; Shelp, B.J. Plant glyoxylate/succinic semialdehyde reductases: Comparative biochemical properties, function during chilling stress, and subcellular localization. *Front Plant Sci.* **2017**, *8*, 1399. [CrossRef] [PubMed]
84. Hoover, G.J.; Prentice, G.A.; Merrill, A.R.; Shelp, B.J. Kinetic mechanism of a recombinant *Arabidopsis* glyoxylate reductase: Studies of initial velocity, dead-end inhibition and product inhibition. *Can. J. Bot.* **2007**, *85*, 896–902. [CrossRef]
85. Hoover, G.J.; Jørgensen, R.; Rochon, A.; Bajwa, V.S.; Merrill, A.R.; Shelp, B.J. Identification of catalytically important amino acid residues for enzymatic reduction of glyoxylate in plants, Biochim. *Biophys. Acta Proteins Proteom.* **2013**, *1834*, 2663–2671. [CrossRef]
86. Allan, W.L.; Breitkreuz, K.E.; Waller, J.C.; Simpson, J.P.; Hoover, G.J.; Rochon, A.; Wolyn, D.J.; Rentsch, D.; Snedden, W.A.; Shelp, B.J. Detoxification of succinate semialdehyde in *Arabidopsis* glyoxylate reductase and NAD kinase mutants subjected to submergence stress. *Botany* **2012**, *90*, 51–61. [CrossRef]
87. Zhang, Z.; Liang, X.; Lu, L.; Xu, Z.; Huang, J.; He, H.; Peng, X. Two glyoxylate reductase isoforms are functionally redundant but required under high photorespiration conditions in rice. *BMC Plant Biol.* **2020**, *20*, 357. [CrossRef] [PubMed]
88. Allan, W.L.; Clark, S.M.; Hoover, G.J.; Shelp, B.J. Role of glyoxylate reductases during stress: A hypothesis. *Biochem. J.* **2009**, *423*, 15–22. [CrossRef] [PubMed]
89. Che-Othman, M.H.; Jacoby, R.P.; Millar, A.H.; Taylor, N.L. Wheat mitochondrial respiration shifts from the tricarboxylic cycle to the GABA shunt under salt stress. *New Phytol.* **2020**, *225*, 1166–1180. [CrossRef] [PubMed]
90. Turano, F.; Dashner, R.; Upadhyaya, A.; Caldwell, C.R. Purification of mitochondrial glutamate dehydrogenase from dark-grown soybean seedlings. *Plant Physiol.* **1996**, *112*, 1357–1364. [CrossRef]
91. Monné, M.; Daddabbo, L.; Gagneu, D.; Obata, T.; Hielscher, B.; Palmieri, L.; Miniero, D.V.; Fernie, A.R.; Weber, A.P.M.; Palmieri, F. Uncoupling proteins 1 and 2 (UCP1 and UCP2) from *Arabidopsis thaliana* are mitochondrial transporters of aspartate, glutamate, and dicarboxylates. *J. Biol. Chem.* **2018**, *293*, 4213–4227. [CrossRef]
92. Popov, V.N.; Syromyatnikov, M.Y.; Fernie, A.R.; Chakraborty, S.; Gupta, K.J.; Igamberdiev, A.U. The uncoupling of respiration in plant mitochondria: Keeping reactive oxygen and nitrogen species under control. *J. Exp. Bot.* **2021**, *72*, 793–807. [CrossRef]
93. Scott-Taggart, C.P.; Van Cauwenberghe, O.R.; McLean, M.D.; Shelp, B.J. Regulation of γ-aminobutyric acid synthesis in situ by glutamate availability. *Physiol. Plant.* **1999**, *106*, 363–369. [CrossRef]
94. Renné, P.; Dreβ en, U.; Hebbeker, U.; Hille, D.; Flügge, U.I.; Westhoff, P.; Weber, A.P.M. The Arabidopsis mutant *dct* is deficient in the plastidic glutamate/malate translocator DiT2. *Plant J.* **2003**, *35*, 316–331. [CrossRef] [PubMed]
95. Breitkreuz, K.E.; Shelp, B.J.; Fischer, W.N.; Schwacke, R.; Rentsch, D. Identification and characterization of GABA, proline and quaternary ammonium compound transporters from *Arabidopsis thaliana*. *FEBS Lett.* **1999**, *450*, 280–284. [CrossRef]
96. Grallath, S.; Weimar, T.; Meyer, A.; Gumy, C.; Suter-Grotemeyer, M.; Neuhaus, J.M.; Rentsch, D. The AtProT family. Compatible solute transporters with similar substrate specificity but differential expression patterns. *Plant Physiol.* **2005**, *137*, 117–126. [CrossRef] [PubMed]
97. Schwacke, R.; Grallath, S.; Breitkreuz, K.E.; Stransky, E.; Stransky, H.; Frommer, W.B.; Rentsch, D. LeProT1, a transporter for proline, glycine betaine, and γ-aminobutyric acid in tomato pollen. *Plant Cell* **1999**, *11*, 377–391.
98. Meyer, A.; Eskandari, S.; Grallath, S.; Rentsch, D. AtGAT1, a high affinity transporter for γ-aminobutyric acid and Arabidopsis thaliana. *J. Biol. Chem.* **2006**, *281*, 7197–7204. [CrossRef]
99. Batushansky, A.; Kirma, M.; Grillich, N.; Pham, P.A.; Rentsch, D.; Galili, G.; Fernie, A.R.; Fait, A. The transporter GAT1 plays an important role in GABA-mediated carbon-nitrogen interactions in Arabidopsis. *Front. Plant Sci.* **2015**, *6*, 785. [CrossRef]
100. Chung, I.; Bown, A.W.; Shelp, B.J. The production and efflux of 4-aminobutyrate in isolated mesophyll cells. *Plant Physiol.* **1992**, *99*, 659–664. [CrossRef]
101. Ramesh, S.A.; Tyerman, S.D.; Xu, B.; Bose, J.; Kaur, S.; Conn, V.; Domingos, P.; Ullah, S.; Wege, S.; Shabala, S. GABA signalling modulates plant growth by directly regulating the activity of plant-specific anion transporters. *Nat. Commun.* **2015**, *6*, 7879. [CrossRef]
102. Ramesh, S.A.; Kamran, M.; Sullivan, W.; Chirkova, L.; Okamoto, M.; Degryse, F.; McLaughlin, M.; Gilliham, M.; Tyerman, S.D. Aluminum-activated malate transporters can facilitate GABA transport. *Plant Cell* **2018**, *30*, 1147–1164. [CrossRef]
103. Snowden, C.J.; Thomas, B.; Baxter, C.J.; Smith, J.A.C.; Sweetlove, L.J. A tonoplast Glu/Asp/GABA exchanger that affects tomato fruit amino acid composition. *Plant J.* **2015**, *81*, 651–660. [CrossRef]
104. Kaplan, F.; Kopka, J.; Sung, D.Y.; Zhao, W.; Popp, M.; Porat, M.; Guy, C.L. Transcript and metabolite profiling during cold acclimation of Arabidopsis reveals an intricate relationship of cold-regulated gene expression with modifications in metabolite content. *Plant J.* **2007**, *50*, 967–981. [CrossRef]

105. Espinoza, C.; Degenkolbe, T.; Caldana, C.; Zuther, E.; Leisse, A.; Willmitzer, L.; Hincha, D.K.; Hannah, M.A. Interaction with diurnal and circadian regulation results in dynamic metabolic and transcriptional changes during cold acclimation in *Arabidopsis*. *PLoS ONE* **2010**, *5*, e14101. [CrossRef] [PubMed]
106. Brikis, C.J.; Zarei, A.; Chiu, G.Z.; Deyman, K.L.; Liu, J.; Trobacher, C.P.; Hoover, G.J.; Subedi, S.; DeEll, J.R.; Bozzo, G.G.; et al. Targeted quantitative profiling of metabolites and gene transcripts associated with 4-aminobutyrate (GABA) in apple fruit stored under multiple abiotic stresses. *Hort. Res.* **2018**, *5*, 61. [CrossRef] [PubMed]
107. Rhodes, D.R.; Handa, S.; Bressan, R.A. Metabolic changes associated with adaptation of plant cells to water stress. *Plant Physiol.* **1986**, *82*, 890–903. [CrossRef] [PubMed]
108. Mayer, R.R.; Cherry, J.H.; Rhodes, D. Effects of heat shock on amino acid metbolism. *Plant Physiol.* **1990**, *94*, 796–810. [CrossRef] [PubMed]
109. Satya Narayan, V.; Nair, P.M. Metabolism, enzymology and possible roles of 4-aminobutyrate in higher plants. *Phytochemistry* **1990**, *29*, 367–375. [CrossRef]
110. Fontaine, J.-X.; Terce-Laforgue, T.; Armengaud, P.; Clement, G.; Renou, J.-P.; Pelletier, S.; Catterou, M.; Azzopardi, M.; Gibon, Y.; Lea, P.J.; et al. Characterization of a NADH-dependent glutamate dehydrogenase mutant of *Arabidopsis* demonstrates the key role of this enzyme in root carbon and nitrogen metabolism. *Plant Cell* **2012**, *24*, 4044–4065. [CrossRef]
111. Carillo, P. GABA shunt in durum wheat. *Front. Plant Sci.* **2018**, *9*, 100. [CrossRef]
112. Bao, H.; Chen, X.; Lv, S.; Jiang, P.; Feng, J.; Fan, P.; Nie, L.; Li, X. Virus-induced gene silencing reveals control of reactive oxygen species accumulation and salt tolerance in tomato by γ-aminobutyric acid metabolic pathway. *Plant Cell Environ.* **2015**, *38*, 600–613. [CrossRef]
113. António, C.; Päpke, C.; Rocha, M.; Diab, H.; Limami, A.M.; Obata, T.; Fernie, A.R.; van Dongen, J. Regulation of primary metabolism in response to low oxygen availability as revealed by carbon and nitrogen isotope redistribution. *Plant Physiol.* **2016**, *170*, 43–56. [CrossRef] [PubMed]
114. Diab, H.; Limami, A. Reconfiguration of N metabolism upon hypoxia stress and revcovery: Roles of alanine aminottransferase (AlaAT) and glutamate dehydrogense (GDH). *Plants* **2016**, *5*, 25. [CrossRef]
115. Vanlerberghe, G.C. Alternative oxidase: A mitochondrial respiratory pathway to maintain metabolic and signaling homeostasis duirng abiotic and biotic stress in plants. *Int. J. Mol. Sci.* **2013**, *14*, 6805–6847. [CrossRef]
116. Ludewig, F.; Hu, A.; Fromm, H.; Beauclair, L.; Bouché, N. Mutants of GABA transaminase (POP2) suppress the severe phenotype of *succinic semialdehyde dehydrogenase* (*ssadh*) mutants in Arabidopsis. *PLoS ONE* **2008**, *3*, e3383. [CrossRef]
117. Wang, F.; Chen, Z.-H.; Liu, X.; Colmer, T.D.; Shabala, L.; Salih, A.; Zhou, M.; Shabala, S. Revealing the roles of GORK channels and NADPH oxidase in acclimation to hypoxia in Arabidopsis. *J. Exp. Bot.* **2017**, *68*, 3191–3204. [CrossRef]
118. Wu, Q.; Su, N.; Huang, X.; Cui, J.; Shabala, L.; Zhou, M.; Yu, M.; Shabala, S. Hypoxia-induced increase in GABA content is essential for restoration of membrane potential and preventing ROS-induced disturbance to ion homeostasis. *Plant Commun.* **2021**, *2*, 100188. [CrossRef]
119. Lancien, M.; Roberts, M.R. Regulation of *Arabidopsis thaliana* 14-3-3 gene expression by γ-aminobutyric acid. *Plant Cell Environ.* **2006**, *29*, 1430–1436. [CrossRef] [PubMed]
120. van Kleeff, P.J.M.; Gao, J.; Mol, S.; Zwart, N.; Zhang, H.; Li, K.W.; de Boer, A.H. The *Arabidopsis* GORK K^+-channel is phosphorylated by calcium-dependent protein kinase 21 (CPK21), which in turn is activated by 14-3-3 proteins. *Plant Physiol. Biochem.* **2018**, *125*, 219–231. [CrossRef] [PubMed]
121. Saito, S.; Uozumi, N. Guard cell membrane anion transport systems and their regulatory components: An elaborate mechanism controlling stress-induced stomatal closure. *Plants* **2019**, *8*, 9. [CrossRef]
122. Demidchik, V.; Shabala, S.; Isayenkov, S.; Cuin, T.A.; Pottosin, I. Calcium transport across plant membranes: Mechanisms and functions. *New Phytol.* **2018**, *220*, 49–69. [CrossRef] [PubMed]
123. Eisenach, C.; Baetz, U.; Huck, N.V.; Zhang, J.; De Angeli, A.; Beckers, G.J.M.; Martinoia, E. ABA-induced stomatal closure involves ALMT4, a phosphorylation-dependent vacuolar anion channel of Arabidopsis. *Plant Cell* **2017**, *29*, 2552–2569. [CrossRef] [PubMed]
124. Kar, D.; Pradhan, A.A.; Datta, S. The role of solute transporters in aluminum toxicity and tolerance. *Physiol. Plant.* **2021**, *171*, 638–652. [CrossRef]
125. Ligaba, A.; Kochian, L.; Piñeros, M. Maize ZmALMT2 is a root anion transporter that mediates constitutive root malate efflux. *Plant Cell Environ.* **2012**, *35*, 1185–1200. [CrossRef] [PubMed]
126. Zhang, L.; Wu, X.-X.; Wang, J.; Qi, C.; Wang, X.; Wang, G.; Li, M.; Li, X.; Guo, Y.-D. BoALMT1, an Al-induced malate transporter in cabbage, enhances aluminum tolerance in *Arabidopsis thaliana*. *Front. Plant Sci.* **2018**, *8*, 2156. [CrossRef] [PubMed]
127. Liang, C.; Piñeros, M.A.; Tian, J.; Yao, Z.; Sun, L.; Liu, J.; Shaff, J.; Coluccio, A.; Kochian, L.V.; Liao, H. Low pH, aluminum, and phosphorus coordinately regulate malate exudation through *GmALMT1* to improve soybean adaptation to acid soils. *Plant Physiol.* **2013**, *61*, 1347–1361. [CrossRef] [PubMed]
128. Chen, Q.; Wu, K.-H.; Wang, P.; Yi, J.; Li, K.-Z.; Yu, Y.-X.; Chen, L.-M. Overexpression of MsALMT1, from the aluminum-sensitive *Medicago sativa*, enhances malate exudation and aluminum resistance in tobacco. *Plant Mol. Biol. Rep.* **2013**, *31*, 768–774. [CrossRef]
129. Ma, X.; Zhu, C.; Yang, N.; Gan, L.; Xia, K. γ-Aminobutyric acid addition alleviates ammonium toxicity by limiting ammonium accumulation in rice (*Oryza sativa*) seedlings. *Physiol. Plant.* **2016**, *158*, 389–401. [CrossRef]

130. Bown, A.W.; MacGregor, K.B.; Shelp, B.J. Gamma-aminobutyrate: Defence against invertebrate pests? *Trends Plant Sci.* **2006**, *11*, 424–427. [CrossRef]
131. Shelp, B.J.; Bown, A.W.; Faure, D. Extracellular γ-aminobutyrate mediates communication between plants and other organisms. *Plant Physiol.* **2006**, *142*, 1350–1352. [CrossRef]
132. Tarkowski, K.P.; Signorelli, S.; Höfte, M. γ-Aminobutyric acid and related amino acids in plant immune responses: Emerging mechanisms of action. *Plant Cell Environ.* **2020**, *43*, 1103–1116. [CrossRef]
133. Ramputh, A.; Bown, A.W. Rapid gamma-aminobutyric acid synthesis and the inhibition of the growth and development of oblique-banded leaf-roller larvae. *Plant Physiol.* **1996**, *111*, 1349–1352. [CrossRef] [PubMed]
134. Bown, A.W.; Hall, D.E.; MacGregor, K.B. Insect footsteps on leaves stimulate the accumulation of 4-aminobutyrate and can be visualized through increased chlorophyll fluorescence and superoxide production. *Plant Physiol.* **2002**, *129*, 1430–1434. [CrossRef]
135. MacGregor, K.E.; Shelp, B.J.; Peiris, S.E.; Bown, A.W. Overexpression of glutamate decarboxylase in transgenic tobacco deters feeding by phytophagous insect larvae. *J. Chem. Ecol.* **2003**, *29*, 2177–2182. [CrossRef]
136. Scholz, S.S.; Reichelt, M.; Mekonnen, D.W.; Ludewig, F.; Mithöfer, A. Insect herbivory-elicited GABA accumulation in plants is a wound-induced, direct, systemic, and jasmonate-independent defense response. *Front. Plant Sci.* **2015**, *6*, 1128. [CrossRef] [PubMed]
137. Scholz, S.S.; Malabarba, J.; Reichelt, M.; Heyer, M.; Mekonnen, D.W.; Ludewig, F.; Mithöfer, A. Evidence for GABA-induced systemic accumulation in Arabidopsis upon wounding. *Front. Plant Sci.* **2017**, *8*, 388. [CrossRef] [PubMed]
138. Chevrot, R.; Rosen, R.; Haudecoeur, E.; Cirou, A.; Shelp, B.J.; Ron, E.; Faure, D. GABA controls the level of quorum-sensing signal in *Agrobacterium tumefaciens*. *Proc. Natl. Acad. Sci. USA* **2006**, *103*, 7460–7464. [CrossRef]
139. Park, D.H.; Mirabella, R.; Bronstein, P.A.; Preston, G.M.; Haring, M.A.; Lim, C.K.; Collmer, A.; Schuurink, R.C. Mutations in γ-aminobutyric acid (GABA) transaminase genes in plants or *Pseudomonas syringae* reduce bacterial virulence. *Plant J.* **2010**, *64*, 318–330. [CrossRef]
140. Planamente, S.; Mondy, S.; Hommais, F.; Vigouroux, A.; Moréra, S.; Faure, D. Structural basis for selective GABA binding in bacterial pathogens. *Mol. Microbiol.* **2012**, *86*, 1085–1099. [CrossRef] [PubMed]
141. Kim, N.H.; Kim, B.S.; Hwang, B.K. Pepper arginine decarboxylase is required for polyamine and γ-aminobutyric acid signaling in cell death and defense response. *Plant Physiol.* **2013**, *162*, 2067–2083. [CrossRef] [PubMed]
142. McCraw, S.L.; Park, D.H.; Jones, R.; Bentley, M.A.; Rico, A.; Ratcliffe, R.G.; Kruger, N.J.; Collmer, A.; Preston, G.M. GABA (γ-aminobutyric acid) uptake via the GABA permease GabP represses virulence gene expression in *Pseudomonas syringae* pv. tomato DC3000. *Mol. Plant-Microbe Interact.* **2016**, *29*, 938–949. [CrossRef]
143. Wang, G.; Kong, J.; Cui, D.; Zhao, H.; Niu, Y.; Xu, M.; Jiang, G.; Zhao, Y.; Wang, W. Resistance against *Ralstonia solanacearum* in tomato depends on the methionine cycle and the γ-aminobutyric acid metabolic pathway. *Plant J.* **2019**, *97*, 1032–1047. [CrossRef]
144. Seifi, H.S.; Curvers, K.; De Vleesschauwer, D.; Delaere, I.; Aziz, A.; Hofte, M. Concurrent overactivation of the cytosolic glutamine synthetase and the GABA shunt in the ABA-deficient sitiens mutant of tomato leads to resistance against *Botrytis cinerea*. *New Phytol.* **2013**, *199*, 490–504. [CrossRef] [PubMed]
145. Gupta, K.; Sengupta, A.; Chakraborty, M.; Gupta, B. Hydrogen peroxide and polyamines act as double edged swords in plant abiotic stress responses. *Front. Plant Sci.* **2016**, *7*, 1343. [CrossRef] [PubMed]
146. Seifi, H.S.; Shelp, B.J. Spermine differentially refines plant defense responses against biotic and abiotic stresses. *Front. Plant Sci.* **2019**, *10*, 117. [CrossRef] [PubMed]
147. Seifi, H.S.; Zarei, A.; Hsiang, T.; Shelp, B.J. Spermine is a potent plant defense activator against gray mold disease on *Solanum lycopersicum*, *Phaseolus vulgaris*, and *Arabidopsis thaliana*. *Phytopathology* **2019**, *109*, 1367–1377. [CrossRef] [PubMed]
148. Deng, X.; Xu, X.; Liu, Y.; Zhang, Y.; Yang, L.; Zhang, S.; Xu, J. Induction of γ-aminobutyric acid plays a positive role to *Arabidopsis* resistance against *Pseudomonas syringae*. *J. Integr. Plant Biol.* **2020**, *62*, 1797–1812. [CrossRef]
149. Asano, T.; Nguyen, T.H.-N.; Yasuda, M.; Sidiq, Y.; Nishimura, K.; Nakashita, H.; Nishiuchi, T. Arabidopsis MAPKKK δ-1 is required for full immunity against bacterial and fungal infection. *J. Exp. Bot.* **2020**, *71*, 2085–2097. [CrossRef] [PubMed]
150. Malekzadeh, P.; Khara, J.; Heydari, R. Alleviating effects of exogenous gamma-aminobutiric acid on tomato seedling under chilling stress. *Physiol. Mol. Biol. Plants* **2014**, *20*, 133–137. [CrossRef] [PubMed]
151. Wu, X.; Jia, Q.; Ji, S.; Gong, B.; Li, J.; Lü, G.; Gao, H. Gamma-aminobutyric acid (GABA) alleviates salt damage in tomato by modulating Na^+ uptake, the *GAD* gene, amino acid synthesis and reactive oxygen species metabolism. *BMC Plant Biol.* **2020**, *20*, 465. [CrossRef]
152. Li, Z.; Peng, Y.; Huang, B. Physiological effects of γ-aminobutyric acid application on improving heat and drought tolerance in creeping bentgrass. *J. Am. Soc. Hort. Sci.* **2016**, *141*, 76–84. [CrossRef]
153. Li, Z.; Peng, Y.; Huang, B. Alteration of transcripts of stress-protective genes and transcriptional factors by γ-aminobutyric acid (GABA) associated with improved heat and drought tolerance in creeping bentgrass (*Agrostis stolonifera*). *Int. J. Mol. Sci.* **2018**, *19*, 1623. [CrossRef] [PubMed]
154. Li, Z.; Yu, J.; Peng, Y.; Huang, B. Metabolic pathways regulated by γ-aminobutyric acid (GABA) contributing to heat tolerance in creeping bentgrass (*Agrostis stolonifera*). *Sci. Rep.* **2016**, *6*, 30338. [CrossRef]
155. Li, Z.; Yu, J.; Peng, Y.; Huang, B. Metabolic pathways regulated by abscisic acid, salicylic acid and γ-aminobutyric acid in association with improved drought tolerance in creeping bentgrass (*Agrostis stolonifera*). *Physiol. Plant.* **2017**, *159*, 42–58. [CrossRef] [PubMed]

156. Li, Z.; Cheng, B.; Zeng, W.; Liu, Z.; Peng, Y. The transcriptional and post-transcriptional regulation in perennial creeping bentgrass in response to γ-aminobutyric acid (GABA) and heat stress. *Environ. Exp. Bot.* **2019**, *162*, 515–524. [CrossRef]
157. Tang, M.; Li, Z.; Luo, L.; Cheng, B.; Zhang, Y.; Zeng, W.; Peng, Y. Nitric oxide signal, nitrogen metabolism, and water balance affected by γ-aminobutyric acid (GABA) in relation to enhanced tolerance to water stress in creeping bentgrass. *Int. J. Mol. Sci.* **2020**, *21*, 7460. [CrossRef]
158. Nayyar, H.; Kaur, H.; Kaur, S.; Singh, R. γ-Aminobutyric acid (GABA) imparts partial protection from heat stress injury to rice seedlings by improving leaf turgor and upregulating osmoprotectants and antioxidants. *J. Plant Growth Regul.* **2014**, *33*, 408–419. [CrossRef]
159. Kumar, N.; Dubey, A.K.; Upadhyay, A.K.; Gautam, A.; Ranjan, R.; Srikishna, S.; Sahu, N.; Behera, S.K.; Mallick, S. GABA accretion reduces *Lsi-1* and *Lsi-2* gene expressions and modulates physiological responses in *Oryza sativa* to provide tolerance towards arsenic. *Sci. Rep.* **2017**, *7*, 8786. [CrossRef]
160. Kumar, N.; Gautam, A.; Dubey, A.K.; Ranjan, R.; Pandey, A.; Kumari, B.; Singh, G.; Mandotra, S.; Chauhan, P.S.; Srikrishna, S.; et al. GABA mediated reduction of arsenite toxicity in rice seedling through modulation of fatty acids, stress responsive amino acids and polyamines biosynthesis. *Ecotoxicol. Environ. Saf.* **2019**, *173*, 15–27. [CrossRef]
161. Sheteiwy, M.S.; Shao, H.; Qi, W.; Hamoud, Y.A.; Shaghaleh, H.; Khan, N.U.; Yang, R.; Tang, B. GABA-alleviated oxidative injury induced by salinity, osmotic stress and their combination by regulating cellular and molecular signals in rice. *Int. J. Mol. Sci.* **2019**, *20*, 5709. [CrossRef]
162. Yong, B.; Xie, H.; Li, Z.; Li, Y.-P.; Zhang, Y.; Nie, G.; Zhang, X.-Q.; Ma, X.; Huang, L.-K.; Yan, Y.-H.; et al. Exogenous application of GABA improves PEG-induced drought tolerance positively associated with GABA-shunt, polyamines, and proline metabolism in white clover. *Front. Physiol.* **2017**, *8*, 1107. [CrossRef] [PubMed]
163. Cheng, B.; Li, Z.; Liang, L.; Cao, Y.; Zeng, W.; Zhang, X.; Ma, X.; Huang, L.; Nie, G.; Liu, W.; et al. The γ-aminobutyric acid (GABA) alleviates salt stress damage during seeds germination of white clover associated with Na^+/K^+ transportation, dehydrins accumulation, and stress-related genes expression in white clover. *Int. J. Mol. Sci.* **2018**, *19*, 2520. [CrossRef] [PubMed]
164. Wang, I.; Fan, L.; Gao, H.; Wu, X.; Li, J.; Lv, G.; Gong, B. Polyamine biosynthesis and degradation are modulated by exogenous gamma-aminobutyric acid in root-zone hypoxia-stressed melon roots. *Plant Physiol. Biochem.* **2014**, *82*, 17–26. [CrossRef] [PubMed]
165. Hu, X.; Xu, Z.; Xu, W.; Li, J.; Zhao, N.; Zhou, Y. Application of γ-aminobutyric acid demonstrates a protective role of polyamine and GABA metabolism in muskmelon seedlings under $Ca(NO_3)_2$ stress. *Plant Physiol. Biochem.* **2015**, *92*, 1–10. [CrossRef] [PubMed]
166. Xiang, L.; Hu, L.; Xu, W.; Zhen, A.; Zhang, L.; Hu, X. Exogenous γ-aminobutyric acid improves the structure and function of photosystem II in muskmelon seedlings exposed to salinity-alkalinity stress. *PLoS ONE* **2016**, *11*, e0164847. [CrossRef] [PubMed]
167. Jin, X.; Liu, T.; Xu, J.; Gao, Z.; Hu, X. Exogenous GABA enhances muskmelon tolerance to salinity-alkalinity stress by regulating redox balance and chlorophyll biosynthesis. *BMC Plant Biol.* **2019**, *19*, 48. [CrossRef]
168. Xu, J.; Liu, T.; Qu, F.; Jin, X.; Huang, N.; Wang, J.; Hu, X. Nitric oxide mediates γ-aminobutyric acid-enhanced muskmelon tolerance to salinity–alkalinity stress conditions. *Sci. Hortic.* **2021**, *286*, 110229. [CrossRef]
169. Priya, M.; Sharma, L.; Kaur, R.; Bindumadhava, H.; Nair, R.M.; Siddique, K.H.M.; Nayyar, H. GABA (γ-aminobutyric acid), as a thermo-protectant, to improve the reproductive function of heat-stressed mungbean plants. *Sci. Rep.* **2019**, *9*, 7788. [CrossRef]
170. Shi, S.-Q.; Shi, Z.; Jiang, Z.-P.; Qi, L.-W.; Sun, X.-M.; Li, C.X.; Liu, J.F.; Xiao, W.F.; Zhang, S.-G. Effects of exogenous GABA on gene expression of *Caragana intermedia* roots under NaCl stress: Regulatory roles for H_2O_2 and ethylene production. *Plant Cell Environ.* **2010**, *33*, 149–162. [CrossRef] [PubMed]
171. Li, M.F.; Guo, S.J.; Yang, X.H.; Meng, Q.W.; Wei, X.J. Exogenous gamma-aminobutyric acid increases salt tolerance of wheat by improving photosynthesis and enhancing activities of antioxidant enzymes. *Biol. Plant.* **2016**, *60*, 123–131. [CrossRef]
172. Wang, Y.; Gu, W.; Meng, Y.; Xie, T.; Li, L.; Li, J.; Wei, S. γ-Aminobutyric acid imparts partial protection from salt stress injury to maize seedlings by improving photosynthesis and upregulating osmoprotectants and antioxidants. *Sci. Rep.* **2017**, *7*, 43609 [CrossRef] [PubMed]
173. Salah, A.; Zhan, M.; Cao, C.; Han, Y.; Ling, L.; Liu, Z.; Li, P.; Ye, M.; Jiang, Y. γ-Aminobutyric acid promotes chloroplast ultrastructure, antioxidant capacity, and growth of waterlogged maize seedlings. *Sci. Rep.* **2019**, *9*, 484. [CrossRef]
174. Kalhor, M.S.; Aliniaeifard, S.; Seif, M.; Asayesh, E.J.; Bernard, F.; Hassani, B.; Li, T. Enhanced salt tolerance and photosynthetic performance: Implication of ɤ-amino butyric acid application in salt-exposed lettuce (*Lactuca sativa* L.) plants. *Plant Physiol. Biochem.* **2018**, *130*, 157–172. [CrossRef]
175. Ji, J.; Yue, J.; Xie, T.; Chen, W.; Du, C.; Chang, E.; Chen, L.; Jiang, Z.; Shi, S. Roles of γ-aminobutyric acid on salinity-responsive genes at transcriptomic level in poplar: Involving in abscisic acid and ethylene-signalling pathways. *Planta* **2018**, *248*, 675–690. [CrossRef] [PubMed]
176. Krishnan, S.; Laskowski, K.; Shukla, V.; Merewitz, E.B. Mitigation of drought stress damage by exogenous application of a non-protein amino acid γ–aminobutyric acid on perennial ryegrass. *J. Am. Soc. Hort. Sci.* **2013**, *138*, 358–366. [CrossRef]
177. Vijayakumari, K.; Puthur, J.T. γ-Aminobutyric acid (GABA) priming enhances the osmotic stress tolerance in *Piper nigrum* Linn plants subjected to PEG-induced stress. *Plant Growth Regul.* **2016**, *78*, 57–67. [CrossRef]
178. Salvatierra, A.; Pimentel, P.; Almada, R.; Hinrichsen, P. Exogenous GABA application transiently improves the tolerance to root hypoxia on a sensitive genotype of *Prunus* rootstock. *Environ. Exp. Bot.* **2016**, *125*, 52–66. [CrossRef]

179. Song, H.; Xu, X.; Wang, H.; Wang, H.; Tao, Y. Exogenous γ-aminobutyric acid alleviates oxidative damage caused by aluminium and proton stresses on barley seedlings. *J. Sci. Food Agric.* **2010**, *90*, 1410–1416. [CrossRef]
180. Wang, M.; Zhu, Y.; Wang, P.; Gu, Z.; Yang, R. Effect of γ-aminobutyric acid on phenolics metabolism in barley seedlings under low NaCl treatment. *Antioxidants* **2021**, *10*, 1421. [CrossRef]
181. Mahmud, J.A.L.; Hasanuzzaman, M.; Nahar, K.; Rahman, A.; Hossain, M.S.; Fujita, M. γ-Aminobutyric acid (GABA) confers chromium stress tolerance in *Brassica juncea* L. by modulating the antioxidant defense and glyoxalase systems. *Ecotoxicology* **2017**, *26*, 675–690. [CrossRef]
182. He, G.; Zhang, H.; Liu, S.; Li, H.; Huo, Y.; Guoc, K.; Xu, Z.; Zhang, H. Exogenous γ-glutamic acid (GABA) induces proline and glutathione synthesis in alleviating Cd-induced photosynthetic inhibition and oxidative damage in tobacco leaves. *J. Plant Interact.* **2021**, *16*, 296–306. [CrossRef]
183. Razik, E.S.A.; Alharbi, B.M.; Pirzadah, T.B.; Alnusairi, G.S.H.; Soliman, M.H.; Hakeem, K.R. γ-Aminobutyric acid (GABA) mitigates drought and heat stress in sunflower (*Helianthus annuus* L.) by regulating its physiological, biochemical and molecular pathways. *Physiol. Plant.* **2021**, *172*, 505–527. [CrossRef]
184. Bashir, R.; Riaz, H.N.; Anwar, S.; Parveen, N.; Khalilzadeh, R.; Hussain, I.; Mahmood, S. Morpho-physiological changes in carrots by foliar γ-aminobutyric acid under drought stress. *Braz. J. Bot.* **2021**, *44*, 57–68. [CrossRef]
185. Chen, D.; Shao, Q.; Yin, L.; Younis, A.; Zheng, B. Polyamine function in plants: Metabolism, regulation on development, and roles in abiotic stress responses. *Front. Plant Sci.* **2019**, *9*, 1945. [CrossRef]
186. Ageeva-Kieferle, A.; Georgii, E.; Winkler, B.; Ghirardo, A.; Albert, A.; Hüther, P.; Mengel, A.; Becker, C.; Schnitzler, J.-P.; Durner, J.; et al. Nitric oxide coordinates growth, development, and stress response via histone modification and gene expression. *Plant Physiol.* **2021**, *187*, 336–360. [CrossRef]
187. Hijaz, H.; Nehela, Y.; Killiny, N. Application of gamma-aminobutyric acid increased the level of phytohormones in *Citrus sinensis*. *Planta* **2018**, *248*, 909–918. [CrossRef]
188. Renault, H.; El Amrani, A.; Palanivelu, R.; Updegraff, E.P.; Yu, A.; Renou, J.-P.; Preuss, D.; Bouchereau, A.; Deleu, C. GABA accumulation causes cell elongation defects and a decrease in expression of genes encoding secreted and cell wall-related proteins in *Arabidopsis thaliana*. *Plant Cell Physiol.* **2011**, *5*, 894–908. [CrossRef] [PubMed]
189. Kaya, C. Nitrate reductase is required for salicylic acid-induced water stress tolerance of pepper by upraising the AsA-GSH pathway and glyoxalase system. *Physiol. Plant.* **2021**, *172*, 351–370. [CrossRef]
190. Ullah, A.; Hussain, A.; Shaban, M.; Khan, A.M.; Alariqi, M.; Gul, M.; Jun, Z.; Lin, S.; Li, J.; Jin, S.; et al. Osmotin: A plant defense tool against biotic and abiotic stresses. *Plant Physiol. Biochem.* **2018**, *123*, 149–159. [CrossRef]
191. Patel, M.K.; Pandey, S.; Burritt, D.J.; Tran, L.-S.P. Plant responses to low-oxygen stress: Interplay between ROS and NO signaling pathways. *Environ. Exp. Bot.* **2019**, *161*, 134–142. [CrossRef]
192. Liu, T.; Xu, T.; Li, J.; Hu, X. NO is involved in JA- and H_2O_2-mediated ALA-induced oxidative stress tolerance at low temperatures in tomato. *Environ. Exp. Bot.* **2019**, *161*, 334–343. [CrossRef]
193. Prakash, V.; Singh, V.P.; Tripathi, D.K.; Sharma, S.; Corpas, F.J. Crosstalk between nitric oxide (NO) and abscisic acid (ABA) signalling molecules in higher plants. *Environ. Exp. Bot.* **2019**, *161*, 41–49. [CrossRef]
194. Bharath, P.; Gahir, S.; Raghavendra, A.S. Abscisic acid-induced stomatal closure: An important component of plant defense against abiotic and biotic stress. *Front. Plant Sci.* **2021**, *12*, 615114. [CrossRef] [PubMed]
195. Corpas, F.J.; González-Gordo, S.; Palma, J.M. Nitric oxide and hydrogen sulfide modulate the NADPH generating enzymatic system in higher plants. *J. Exp. Bot.* **2021**, *72*, 830–847. [CrossRef] [PubMed]
196. Cui, B.; Xu, S.; Li, Y.; Umbreen, S.; Frederickson, D.; Yuan, B.; Jiang, J.; Liu, F.; Pan, Q.; Loake, G.J. The Arabidopsis zinc finger proteins SRG2 and SRG3 are positive regulators of plant immunity and are differentially regulated by nitric oxide. *New Phytol.* **2021**, *230*, 259–274. [CrossRef] [PubMed]
197. Manrique-Gil, I.; Sánchez-Vicente, I.; Torres-Quezada, I.; Lorenzo, O. Nitric oxide function during oxygen deprivation in physiological and stress processes. *J. Exp. Bot.* **2021**, *72*, 904–916. [CrossRef]
198. Wurm, C.J.; Lindermayr, C. Nitric oxide signaling in the plant nucleus: The function of nitric oxide in chromatin modulation and transcription. *J. Exp. Bot.* **2021**, *72*, 808–812. [CrossRef] [PubMed]

plants

MDPI

Article

Tomato Domestication Affects Potential Functional Molecular Pathways of Root-Associated Soil Bacteria

Lisanne Smulders [1], Emilio Benítez [1,*], Beatriz Moreno [1], Álvaro López-García [2], María J. Pozo [2], Victoria Ferrero [3], Eduardo de la Peña [4,5] and Rafael Alcalá Herrera [1]

1 Department of Environmental Protection, Estación Experimental del Zaidín (EEZ), CSIC, C/Profesor Albareda 1, 18008 Granada, Spain; lisanne.smulders@eez.csic.es (L.S.); beatriz.moreno@eez.csic.es (B.M.); rafa.alcala@eez.csic.es (R.A.H.)
2 Department of Soil Microbiology and Symbiotic Systems, Estación Experimental del Zaidín (EEZ), CSIC, C/Profesor Albareda 1, 18008 Granada, Spain; alvaro.lopez@eez.csic.es (Á.L.-G.); mariajose.pozo@eez.csic.es (M.J.P.)
3 Department of Biodiversity and Environmental Management, Campus de Vegazana s/n, University of León, 24071 León, Spain; vferrv@unileon.es
4 Finca Experimental "La Mayora" CSIC, Instituto de Hortofruticultura Subtropical y Mediterránea (IHSM-UMA), CSIC, Algarrobo, 29750 Málaga, Spain; Eduardo.DeLaPena@ugent.be
5 Department of Plants and Crops, Faculty of Bioscience Engineering, Ghent University, Coupure Links 653, 9000 Ghent, Belgium
* Correspondence: emilio.benitez@eez.csic.es

Abstract: While it has been well evidenced that plant domestication affects the structure of the root-associated microbiome, there is a poor understanding of how domestication-mediated differences between rhizosphere microorganisms functionally affect microbial ecosystem services. In this study, we explore how domestication influenced functional assembly patterns of bacterial communities in the root-associated soil of 27 tomato accessions through a transect of evolution, from plant ancestors to landraces to modern cultivars. Based on molecular analysis, functional profiles were predicted and co-occurrence networks were constructed based on the identification of co-presences of functional units in the tomato root-associated microbiome. The results revealed differences in eight metabolic pathway categories and highlighted the influence of the host genotype on the potential functions of soil bacterial communities. In general, wild tomatoes differed from modern cultivars and tomato landraces which showed similar values, although all ancestral functional characteristics have been conserved across time. We also found that certain functional groups tended to be more evolutionarily conserved in bacterial communities associated with tomato landraces than those of modern varieties. We hypothesize that the capacity of soil bacteria to provide ecosystem services is affected by agronomic practices linked to the domestication process, particularly those related to the preservation of soil organic matter.

Keywords: bacterial functions; co-presence networks; metagenomics; microbial ecology; plant domestication

Citation: Smulders, L.; Benítez, E.; Moreno, B.; López-García, Á.; Pozo, M.J.; Ferrero, V.; de la Peña, E.; Alcalá Herrera, R. Tomato Domestication Affects Potential Functional Molecular Pathways of Root-Associated Soil Bacteria. *Plants* **2021**, *10*, 1942. https://doi.org/10.3390/plants10091942

Academic Editors: Petronia Carillo, Paula Baptista and Milan S. Stankovic

Received: 3 August 2021
Accepted: 14 September 2021
Published: 17 September 2021

Publisher's Note: MDPI stays neutral with regard to jurisdictional claims in published maps and institutional affiliations.

Copyright: © 2021 by the authors. Licensee MDPI, Basel, Switzerland. This article is an open access article distributed under the terms and conditions of the Creative Commons Attribution (CC BY) license (https://creativecommons.org/licenses/by/4.0/).

1. Introduction

The coevolutionary framework for analyzing interactions between plants and soil microorganisms has mainly been used for organisms involved in rhizosphere processes. Given that rhizosphere microbiomes are part of complex food webs affecting large numbers of nutrients released by the plant, it has been suggested that plants attract and select beneficial microbiomes by first releasing signals and then filtering species [1,2]. Rhizopshere microbiota are well known to play a critical role in both the adaptation of plants to the environment, but also contribute to a wide range of essential ecosystem services, such as carbon and nutrient cycling, plant growth promotion, soil structure stability, food web interactions and soil-atmosphere gas exchange, which ultimately affect soil productivity and sustainability [3].

In addition to plant genetics and developmental stage [4,5], other factors including soil management, agronomic practices, pathogen presence, soil pH, nutrient content, and moisture, have been suggested to affect root microbial community structure [6–8]. However, the question of how the host and its environment regulate microbiome assembly and co-occurrence in plant species has not been addressed yet. This is of particular interest for crops in the context of plant–soil feedback, where plants can change soil biology and chemistry in ways that could affect subsequent plant growth [9].

Crop genetic diversity is usually reduced during plant domestication, which is associated with the selection of certain morphological traits such as root architecture and exudate composition, leading to striking differences between crops and their wild relatives [10,11]. Therefore, domestication is expected to have a direct impact on the type and diversity of below-ground microorganisms [9]. Indeed, domestication and genetic selection have progressively differentiated the microbiota of modern crops from those of their wild progenitors. It has also been postulated that crops are more likely to display negative feedbacks as compared to wild relatives, as domestication potentially disrupted beneficial rhizosphere associations [12]. Previous studies of cultivated plants evidenced differences between bacteria associated with differing plant genotypes such as wheat (*Triticum aestivum* L.), rice (*Oryza sativa* L.), barley (*Hordeum vulgare* L.) and tomato (*Solanum lycopersicum* Mill.) [13–15], suggesting that traits selected during domestication could have a significant influence on rhizosphere microbiota composition.

Although the structure of root-associated microbial communities is widely accepted to depend, to a greater or lesser degree, on the plant genotype, little is known about whether domestication-mediated differences between rhizosphere microorganisms functionally affect microbial ecosystem services. In this scenario, an evaluation of functional soil microbial genes could help to determine the effect of domestication on functional redundancy or co-occurrence of basic metabolic capacity in the rhizospheres of crop varieties and their wild ancestors [16]. This is essential to identify agricultural practices that resulted in reduced trade-offs between agricultural productivity and the provision of ecosystem services.

This study aims to explore how plant domestication influences the assembly patterns of soil microbial communities by metagenomic analysis of bacterial communities and predicted functions in the rhizosphere of different tomato varieties along a domestication gradient.

2. Results

2.1. Bacterial Community Structure

Figure 1 shows the relative bacterial abundance of the tomato root-associated soils based on the 16S rRNA gene. Two main bacterial classes, Alphaproteobacteria and Actinobateria, dominated the total bacterial community with no differences observed between plant groups. Minority phyla such as Acidobacteria ($F = 7.152$, $p = 0.002$) and Gemmatimonadetes ($F = 4.720$, $p = 0.013$) were significantly less represented in the rhizosphere of wild tomato species than in tomato landraces and modern commercial cultivars. At the family level, the relative abundance of the *Gemmatimonadaceae* ($F = 4.133$, $p = 0.022$), *Microbacteriaceae* ($F = 5.419$, $p = 0.007$), and *Streptomycetaceae* ($F = 4.752$, $p = 0.022$) families decreased, while *Sphingomonadaceae* ($F = 7.887$, $p = 0.001$) increased in wild tomato relatives. Again, no differences between tomato landraces and modern commercial cultivars were detected.

The relative abundances of Acidobacteria_Gp16_unclassified ($F = 3.701$, $p = 0.031$), *Hyphomicrobiaceae* ($F = 6.736$, $p = 0.002$), and *Nocardioidaceae* ($F = 4.179$, $p = 0.021$) were different between wild and commercial cultivars, while landraces had intermediate values, generally not differing from the other two groups.

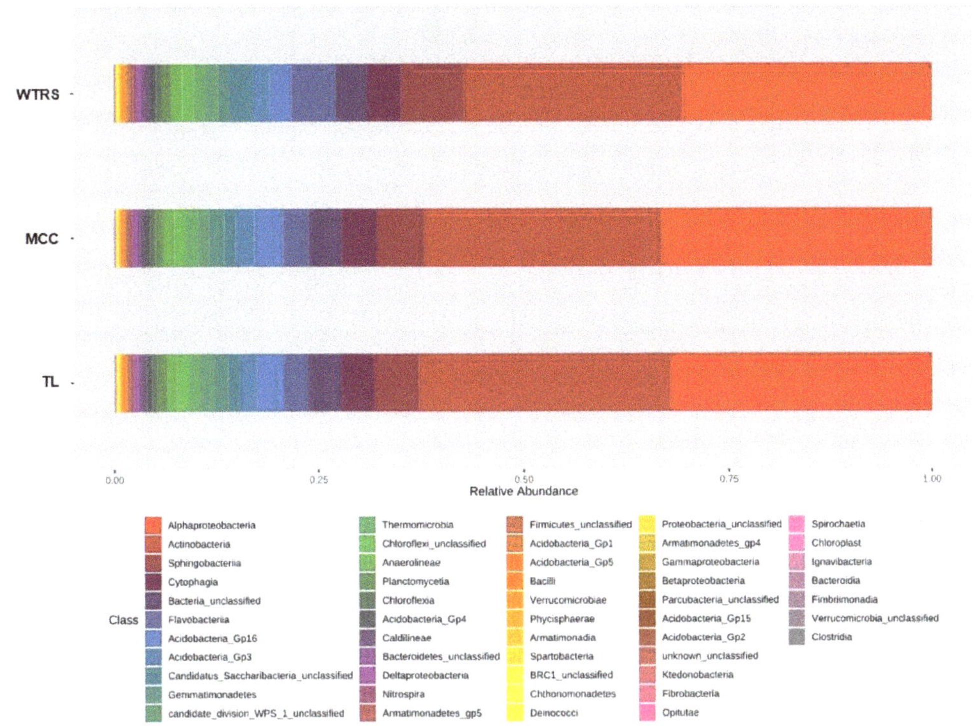

Figure 1. Relative abundance of bacteria of tomato rhizosphere soils. WTRS: wild tomato related species; TL: tomato landraces; MCC: modern commercial cultivars.

Linear discriminant analysis (LDA) at the genus level showed *Pedobacter* (*Sphingobacteriaceae*), *Rodococcus*, *Skermanella* and the proteobacterium *Microvirga* to be mainly responsible for the differences between the three tomato clusters (Figure 2). In addition, minor changes in bacterial diversity were observed at the OTU level (Table 1), as indicated by a significant decrease in the evenness of crop wild relatives ($F = 6.623$, $p = 0.003$).

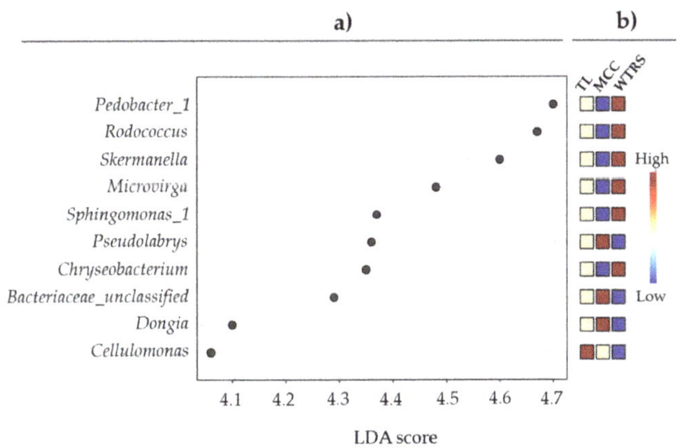

Figure 2. (**a**) Linear discriminant analysis (LDA) scores and, (**b**) heatmap from blue (low) via white to red (high) of genus relative abundances in root-associated soil of wild tomato related species (WTRS), tomato landraces (TL), and modern commercial cultivars (MCC).

Table 1. Richness estimates and diversity indices (means ± SE) for 16S rRNA libraries of tomato rhizosphere soils. Different letters indicate a significant difference among tomato varieties ($p < 0.05$, ANOVA, Dunn's post hoc-Bonferroni corrected p values) when exist. WTRS: wild tomato related species; TL: tomato landraces; MCC: modern commercial cultivars.

	WTRS	TL	MCC
Shannon–Wiener Diversity Index	5.79 ± 0.26	6.23 ± 0.06	6.18 ± 0.09
Shannon Entropy	8.35 ± 0.37	8.98 ± 0.08	8.92 ± 0.12
Species Richness	3260 ± 346	3727 ± 155	3471 ± 231
Total Abundance	52,838 ± 3338	59,144 ± 1702	57,100 ± 2964
Simpson Diversity Index	0.031 ± 0.013	0.010 ± 0.001	0.011 ± 0.002
Evenness	0.719 ± 0.021b	0.759 ± 0.004a	0.764 ± 0.005a
Chao-1	4453 ± 450	4983 ± 241	4619 ± 329

2.2. Bacterial Community Functional Analysis

We used metagenomics analysis to predict the functional potential of the bacterial community and to explore associated metabolic pathway networks using Kyoto Encyclopedia of Gene and Genome (KEGG) clusters.

At the level of functional units of gene sets, all tomato varieties shared all the 181 predicted functions related to soil bacteria (Table S1). However, 68 of them differed among tomato domestication types (Table 2). In general, wild tomatoes differed from modern cultivars and tomato landraces that usually showed similar values, but generally tomato landraces had intermediate values between modern cultivars and wild relatives. For example, the levels of the aromatic degradation metabolic pathway category, except for module M00541 (benzoyl-CoA degradation), tended to be significantly higher in bacteria growing in wild tomato accessions, indicating that tomato landraces drive bacterial communities with similar levels of predicted functions as modern commercial cultivars. Similarly, while the values for the metabolic categories of nitrogen, sulfur, cofactor/vitamin and purine were decreased in modern cultivars with respect to wild varieties, no differences between these wild and landrace cultivars were detected. By contrast, lipopolysaccharide and lipid metabolic pathway levels were clearly higher in both landrace and modern cultivars with respect to their wild relatives.

Table 2. Functional units of gene sets in metabolic pathways (KEGG modules) of tomato rhizosphere soils differentially represented among tomato varieties. Different letters indicate a significant difference among tomato varieties (ANOVA, Dunn's post hoc-Bonferroni corrected p values). WTRS: wild tomato related species; TL: tomato landraces; MCC: modern commercial cultivars.

Pathway Modules		WTRS	TL	MCC	p-Value
Amino acid metabolism; Arginine and proline metabolism	M00015_Proline biosynthesis, glutamate =>proline	1656a	1609b	1626b	0.02086
	M00023_Tryptophan biosynthesis, chorismate => tryptophan	3173b	3234a	3205a	0.008801
	M00037_Melatonin biosynthesis, tryptophan => serotonin => melatonin	69b	77a	74ab	0.04182
	M00040_Tyrosine biosynthesis, chorismate => arogenate => tyrosine	508a	464b	467b	5.579×10^{-6}
	M00042_Catecholamine biosynthesis, tyrosine => dopamine => noradrenaline => adrenaline	95b	106a	104ab	0.02289
	M00533_Homoprotocatechuate degradation, homoprotocatechuate => 2-oxohept-3-enedioate	491a	474b	473ab	0.04715
Amino acid metabolism; Aromatic amino acid metabolism	M00545_Trans-cinnamate degradation, trans-cinnamate => acetyl-CoA	1048a	1004b	1014b	0.001853

Table 2. *Cont.*

Pathway Modules		WTRS	TL	MCC	*p*-Value
Amino acid metabolism; Branched-chain amino acid metabolism	M00036_Leucine degradation, leucine => acetoacetate + acetyl-CoA	7057a	6938b	6874b	6.597×10^{-5}
Amino acid metabolism; Cysteine and methionine metabolism	M00017_Methionine biosynthesis, apartate => homoserine => methionine	5393b	5464a	5422a	0.006857
	M00035_Methionine degradation	2031b	2101a	2065b	6.836×10^{-5}
	M00338_Cysteine biosynthesis, homocysteine + serine => cysteine	233c	276a	257b	9.38×10^{-11}
Amino acid metabolism; Lysine metabolism	M00031_Lysine biosynthesis, mediated by LysW, 2-aminoadipate => lysine	81b	106a	98b	1.069×10^{-5}
Amino acid metabolism; Other amino acid metabolism	M00118_Glutathione biosynthesis, glutamate => glutathione	944a	876b	880b	7.781×10^{-8}
	M00027_GABA (gamma-Aminobutyrate) shunt	2018a	1942b	1921b	1.037×10^{-5}
Amino acid metabolism; Polyamine biosynthesis	M00133_Polyamine biosynthesis, arginine => agmatine => putrescine => spermidine	865b	895a	879ab	0.01599
	M00134_Polyamine biosynthesis, arginine => ornithine => putrescine	872a	838b	841b	0.008463
	M00136_GABA biosynthesis, prokaryotes, putrescine => GABA	677a	620b	636b	6.265×10^{-9}
Amino acid metabolism; Serine and threonine metabolism	M00555_Betaine biosynthesis, choline => betaine	1473a	1377b	1383b	4.206×10^{-11}
Carbohydrate metabolism; Central carbohydrate metabolism	M00006_Pentose phosphate pathway, oxidative phase, glucose 6P => ribulose 5P	1535ab	1523b	1546a	0.04953
	M00077_Chondroitin sulfate degradation	105b	118a	123a	6.408×10^{-5}
	M00008_Entner–Doudoroff pathway, glucose-6P => glyceraldehyde-3P + pyruvate	1993a	1905c	1928b	2.838×10^{-6}
	M00009_Citrate cycle (TCA cycle, Krebs cycle)	12,529a	12,667b	12,563a	0.0001383
	M00011_Citrate cycle, second carbon oxidation, 2-oxoglutarate => oxaloacetate	9185b	9286a	9207b	0.001811
	M00003_Gluconeogenesis, oxaloacetate => fructose-6P	5474b	5544a	5498b	0.008293
	M00633_Semi-phosphorylative Entner–Doudoroff pathway, gluconate/galactonate => glycerate-3P	85b	91ab	95a	0.03837
Carbohydrate metabolism; Other carbohydrate metabolism	M00061_D-Glucuronate degradation, D-glucuronate => pyruvate + D-glyceraldehyde 3P	1694a	1654b	1680ab	0.01455
	M00081_Pectin degradation	113b	129a	132a	6.318×10^{-5}
	M00114_Ascorbate biosynthesis, plants, glucose-6P => ascorbate	2958ab	2995a	2949b	0.0271
	M00131_Inositol phosphate metabolism, Ins(1,3,4,5)P4 => Ins(1,3,4)P3 => myo-inositol	1003a	969b	968ab	0.02686
	M00550_Ascorbate degradation, ascorbate => D-xylulose-5P	27a	19b	18b	1.195×10^{-5}
	M00554_Nucleotide sugar biosynthesis, galactose => UDP-galactose	199b	207ab	217a	0.005133
	M00565_Trehalose biosynthesis, D-glucose 1P => trehalose	3380b	3603a	3666a	2.2×10^{-16}

Table 2. *Cont.*

Pathway Modules		WTRS	TL	MCC	*p*-Value
Energy metabolism; Carbon fixation	M00170_C4-dicarboxylic acid cycle, phosphoenolpyruvate carboxykinase type	1302c	1357a	1321bc	1.883×10^{-5}
	M00172_C4-dicarboxylic acid cycle, NADP—malic enzyme type	3505a	3444b	3445b	0.03007
	M00173_Reductive citrate cycle (Arnon-Buchanan cycle)	10,778b	10,891a	10,850ab	0.004089
	M00374_Dicarboxylate-hydroxybutyrate cycle	7259b	7345a	7333a	0.01178
	M00620_Incomplete reductive citrate cycle, acetyl-CoA => oxoglutarate	2168b	2224a	2231a	0.0007637
Energy metabolism;; Methane metabolism	M00344_Formaldehyde assimilation, xylulose monophosphate pathway	913b	944a	942ab	0.03158
	M00345_Formaldehyde assimilation, ribulose monophosphate pathway	749b	808a	800a	6.137×10^{-7}
	M00346_Formaldehyde assimilation, serine pathway	3166b	3234a	3226ab	0.006593
	M00356_Methanogenesis, methanol => methane	22b	26ab	27a	0.03877
	M00358_Coenzyme M biosynthesis	177b	190a	198a	0.0004887
	M00378_F420 biosynthesis	82b	93a	89ab	0.05467
	M00563_Methanogenesis, methylamine/dimethylamine/trimethylamine => methane	465a	434b	464a	3.724×10^{-6}
Energy metabolism; Nitrogen metabolism	M00530_Dissimilatory nitrate reduction, nitrate => ammonia	1864a	1823b	1848a	0.01764
Energy metabolism; Sulfur metabolism	M00176_Assimilatory sulfate reduction, sulfate => H2S	2814a	2741b	2766ab	0.006395
Glycan metabolism; Glycosaminoglycan metabolis	M00076_Dermatan sulfate degradation	115b	129a	135a	2.073×10^{-5}
	M00077_Chondroitin sulfate degradation	105b	118a	123a	6.408×10^{-5}
	M00078_Heparan sulfate degradation	191b	215a	224a	2.272×10^{-7}
	M00079_Keratan sulfate degradation	475b	526a	547a	1.002×10^{-12}
Glycan metabolism; Lipopolysaccharide metabolism	M00060_KDO2-lipid A biosynthesis, Raetz pathway, LpxL-LpxM type	3058b	3132a	3124a	0.001684
	M00064_ADP-L-glycero-D-manno-heptose biosynthesis	692b	743a	771a	3.151×10^{-7}
Lipid metabolism; Fatty acid metabolism	M00082_Fatty acid biosynthesis, initiation	3785b	3861a	3842ab	0.01467
	M00083_Fatty acid biosynthesis, elongation	9121b	9218a	9214a	0.01719
	M00086_beta-Oxidation, acyl-CoA synthesis	1699b	1743a	1746ab	0.01575
Lipid metabolism; Lipid metabolism	M00113_Jasmonic acid biosynthesis	428b	454a	438b	0.002276
Metabolism of cofactors and vitamins; Cofactor and vitamin metabolism	M00116_Menaquinone biosynthesis, chorismate => menaquinol	943b	1026a	977b	1.104×10^{-10}
	M00117_Ubiquinone biosynthesis, prokaryotes, chorismate => ubiquinone	2772a	2703b	2707ab	0.01037
	M00122_Cobalamin biosynthesis, cobinamide => cobalamin	2143a	2105b	2153a	0.00244
	M00128_Ubiquinone biosynthesis, eukaryotes, 4-hydroxybenzoate => ubiquinone	74a	64b	67ab	0.01119
Nucleotide metabolism; Purine metabolism	M00546_Purine degradation, xanthine => urea	2126a	2089b	2125a	0.01079

Table 2. *Cont.*

Pathway Modules		WTRS	TL	MCC	*p*-Value
Xenobiotics biodegradation; Aromatics degradation	M00537_Xylene degradation, xylene => methylbenzoate	215a	199b	200ab	0.01542
	M00541_Benzoyl-CoA degradation, benzoyl-CoA => 3-hydroxypimeloyl-CoA	59b	67a	67ab	0.02392
	M00548_Benzene degradation, benzene => catechol	27a	20b	21b	0.0002254
	M00551_Benzoate degradation, benzoate => catechol/methylbenzoate => methylcatechol	124a	108b	110b	0.003958
	M00568_Catechol ortho-cleavage, catechol => 3-oxoadipate	445a	421b	433ab	0.01165
	M00569_Catechol meta-cleavage, catechol => acetyl-CoA/4-methylcatechol => propanoyl-CoA	466a	441b	430b	0.001843
	M00637_Anthranilate degradation, anthranilate => catechol	90a	73b	82b	1.726×10^{-5}

Amino acid metabolism pathways exhibited no clear tendency, although in modern commercial cultivars, cysteine and methionine pathway levels were higher and those of other amino acid pathways were lower. A similar variable pattern was observed with respect to both central carbohydrate and other carbohydrate metabolic pathways in the category of carbohydrate metabolism.

Finally, a marked increase in the carbon fixation and methane metabolic subfunctions and in the metabolic pathway categories glycan metabolism and lipid metabolism, respectively, was observed in the modern commercial cultivars.

2.3. Functional Networks of KEGG Orthologous Groups

Figure 3 and Table 3 show the co-presence networks and the topological properties of functional networks, respectively, for the modern:wild, landraces:wild and modern:landraces pairs. An increase in the average number of neighbors and a decrease in the characteristic path length were found in landraces:wild pairs (Table 3). Additionally, an increase in the network radius and diameter were detected in the pair modern:landraces. Finally, the pair modern:wild showed the largest number of KEGG-module nodes and the largest number of edges or inter-node connections in the network. The clustering coefficient, which reflects the tendency of organisms to form relatively high-density clusters, was zero. Co-occurrence networks are generated by connecting pairs of terms using a set of criteria defining co-occurrence. These networks connect across, rather than between, nodes. Every node, in which none of whose neighbors connect to each other, has a clustering coefficient of zero.

Highly connected clusters were retrieved for every network, four for the pair modern:wild and three for the other two pairs (Figures 4–6). On a closer analysis, we detected some links in highly connected clusters. Thus, bacterial functional units M00026 (histidine biosynthesis, PRPP => histidine), M00032 (lysine degradation, lysine => saccharopine => acetoacetyl-CoA), M00141 (C1-unit interconversion) and M00376 (3-hydroxypropionate bicycle) were highly connected in the soil of tomato landraces and wild relatives (Figure 4a), whereas modern varieties and wild relatives were connected by modules M00141, M00376, M00021 (cysteine biosynthesis, serine => cysteine) and M00089 (triacylglycerol biosynthesis) (Figure 5a–c). Finally, modules M00026 and M00141 were highly represented in modern varieties and tomato landraces (Figure 6a,b).

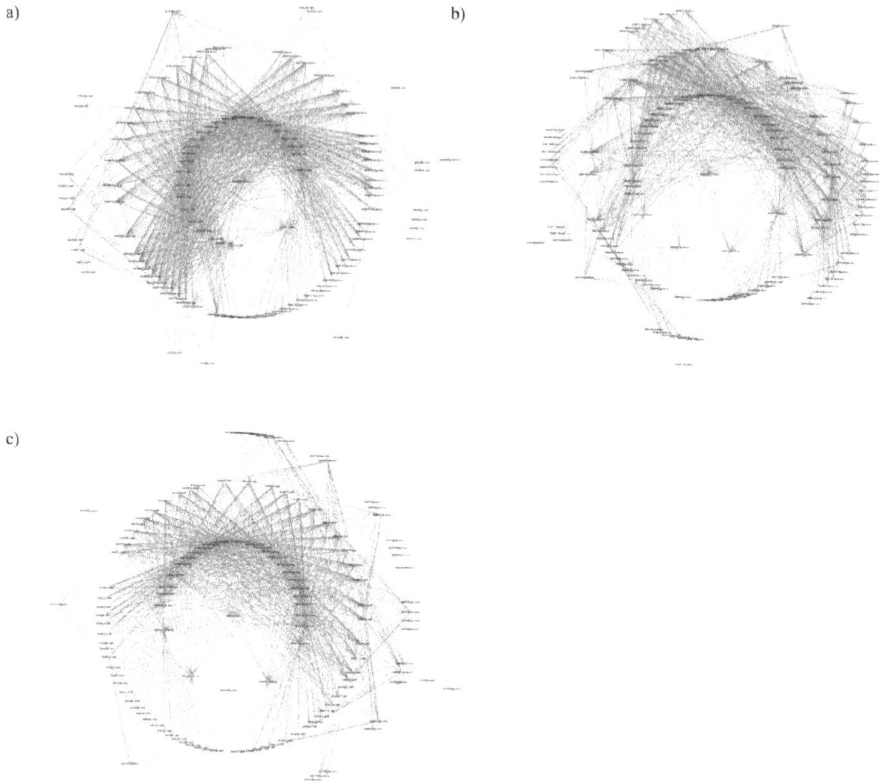

Figure 3. Co-presence network for the couples: (**a**) tomato landraces:wild tomato related species, (**b**) modern commercial cultivars:tomato landraces, (**c**) modern commercial cultivars:wild tomato related species. Node sizes reflect average relative abundance of each KEGG module. The line thickness is proportional to the edge weight. Node colors: green for wild tomato related species, blue for tomato landraces and red for modern commercial cultivars.

Table 3. Topological properties of pairwise functional networks. WTRS: wild tomato related species; TL: tomato landraces; MCC: modern commercial cultivars.

	MCC:WTRS	TL:WTRS	MCC:TL
Number of nodes	133	116	132
Number of edges	1005	1003	1001
Average number of neighbors	15,113	17,293	15,167
Network diameter	6	6	7
Network radius	3	3	4
Characteristic path length	2.542	2.371	2.577
Clustering coefficient	0.000	0.000	0.000
Network density	0.114	0.150	0.116
Network heterogeneity	0.850	0.780	0.869
Network centralization	0.230	0.325	0.309
Connected components	1	1	1

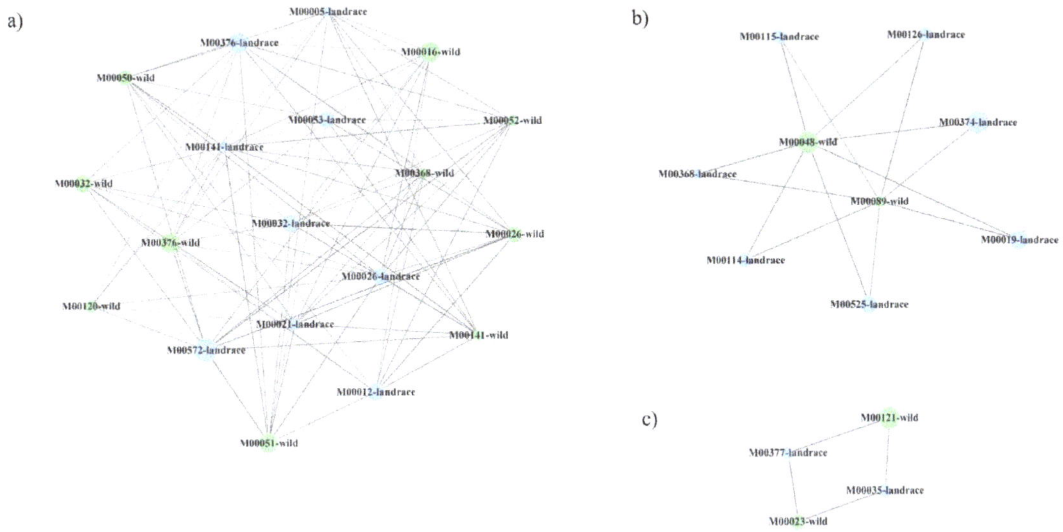

Figure 4. Three (**a**–**c**) most connected clusters in co-presence networks for the couple tomato landraces (blue):wild tomato related species (green). Node sizes reflect average relative abundance of each KEGG module. The line thickness is proportional to the edge weight.

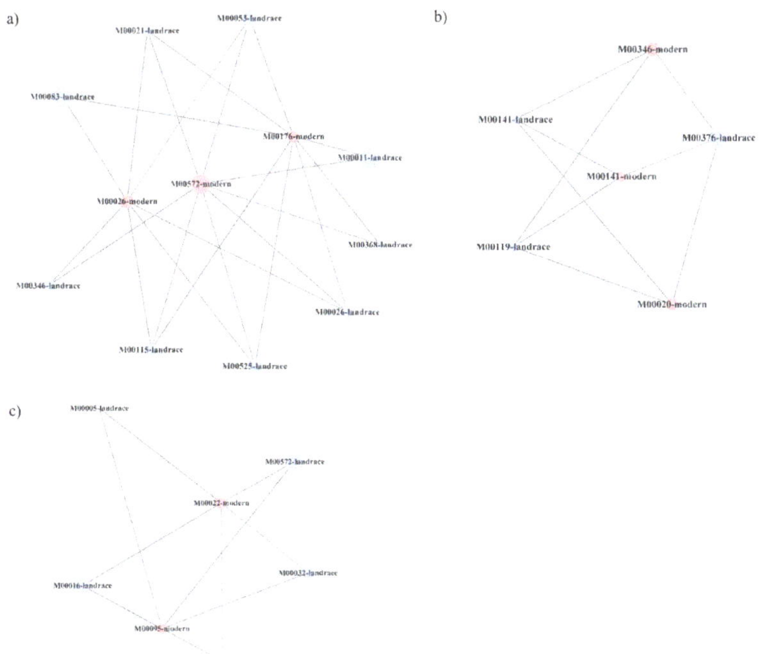

Figure 5. Three (**a**–**c**) most connected clusters in co-presence networks for the couple modern commercial cultivars (red):tomato landraces (blue). Node sizes reflect average relative abundance of each KEGG module. The line thickness is proportional to the edge weight.

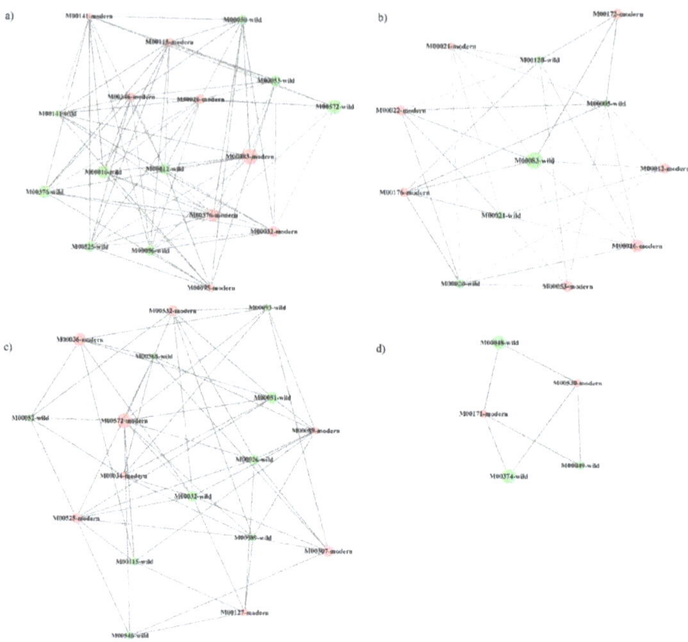

Figure 6. Four (**a–d**) most connected clusters in co-presence networks for the couple modern commercial cultivars (red):wild tomato relates species (green). Node sizes reflect average relative abundance of each KEGG module. The line thickness is proportional to the edge weight.

3. Discussion

Root traits selected during domestication were previously suggested to have a significant influence on the composition of the rhizosphere microbiome [13,17]. We found similar core bacterial microbiome members in tomato landraces and modern commercial cultivars, but detected small, though significant, differences in bacterial communities associated with both their rhizospheres and those of wild tomato relatives (Figure 1).

At family level, *Gemmatimonadaceae* (phylum Actinobacteria), *Microbacteriaceae* and *Streptomycetaceae* (Gemmatimonadetes) were represented less in the rhizosphere of wild tomato related species. At genera level, domestication gradually reduces the presence of the ubiquitous soil bacterium *Pedobacter*, the aromatic substrate metabolizer Rhodococcus and the alphaproteobacteria *Skermanella* and *Microvirga*, the latter considered a symbiotic nitrogen-fixing bacterium.

Previous studies highlighted the effect of plant species on the microbial composition and OTU abundance of the rhizosphere microbiome [5,18]. Domesticated crops often have shallow roots and shifts in traits such as leaf size and root architecture. Changes in these morphological traits results in increased litter quality, lower C:N ratio and root exudate composition, which could influence microbial community composition [2,9,19,20]. In this study, bacterial diversity at the OTU level was found to remain virtually unchanged along the domestication gradient, although evenness levels were significantly lower in the rhizosphere of tomato wild relatives. Evenness refers to the similarity of OTU frequencies in bacterial populations. Even though species evenness and richness are complementary, no differences were observed in the latter; the number of soil bacterial phyla recruited by wild type crops was similar to other tomatoes. Nevertheless, evenness does not necessarily translate into optimal diversity; ecosystem functions at the bacterial community level are more important than the bacterial species. As several species in an ecosystem may fulfill a similar function (redundancy), their even distribution is not essential as long as the function itself remains active. However, a more even species distribution within a bacterial community is assumed to make the ecosystem more resilient, as the risk of losing an essential component of the functional network would be much lower.

Using metagenomic analysis, the functional potential of the bacterial community was predicted and the associated metabolic pathway network explored (Table 2). The levels of the global metabolic pathway for aromatic degradation were significantly higher in bacteria associated to accessions of tomato wild relatives. The modules belonging to this pathway catalyze reactions involving various polyphenols such as catechol. Humification is known to involve biotic and abiotic transformations of soil litter layer materials into mature humic substances, where catechol and o-quinones derived from biotic activity in humic substance synthesis play a fundamental role [21]. In addition, the increase in catechol promotes the formation of humic substances through abiotic reactions in the catechol–Maillard system [22]. Thus, the observed decrease in the degradation of aromatic compounds to catechol indicates a loss of degradation capacity due to cultivation. Organic matter and humic substances play an important role in improving soil fertility and structure, water retention capacity and C sequestration in the environment [23], which diminishes along the domestication gradient. Another possible hypothesis is that plants affect microbial populations, and changes in environmental conditions, soils and cultivation techniques—with the gradual abandonment of organic materials in favor of agrochemicals—could reduce the degradation capacity of recalcitrant organic compounds associated with domestication and breeding. On the other hand, with respect to the carbon cycle, organic C taken up by microorganisms is partitioned into growth, metabolite excretion, and respiration [24]. We detected an increase in the Krebs cycle of wild tomato related species below-ground. After incorporation into the bacterial biomass, C is usually converted into stable organic matter or decomposed and respired as CO_2 depending on the chemical recalcitrance and degree of protection of the organic matter [25].

In this context, it has been suggested that crop wild relatives establish beneficial interactions with microbes more frequently than domesticated cultivars [26]. Given the abandonment of some agricultural practices related to exogenous organic matter inputs and the preservation of endogenous C, a concomitant loss of bacterial functions dealing with recalcitrant organic matter has been occurring for many years. It has also been evidenced that agronomic practices, such as tillage, irrigation and the use of other inputs such as pesticides and fertilizers influence the below-ground diversity and functions of soil microbes [27]. We therefore postulate that a loss in bacterial functions related to soil organic matter preservation occurs during tomato domestication.

A similar trend was detected in metabolic pathways related to biochemical cycles, such as the reduction in nitrates and sulphates and the formation of urea from purine metabolism. The decrease in these pathways that play a key role in plant growth could be attributed to the domestication process, or more precisely, to the emergence of modern commercial cultivars. Similar to the observations in the C-cycle, the increasing use of agrochemicals in modern agriculture may, in some way, be connected to the reduction on metabolic pathway levels caused by certain biochemical cycles.

Carbon fixation was more common in bacteria associated with modern commercial cultivars. This important process in soil carbon cycling is carried out by CO_2-fixing and CO-oxidizing bacteria and can reduce atmospheric CO_2 concentrations, thus indirectly mitigating global warming [24,28,29]. However, as no differences in the synthesis of ribulose 5 phosphate, an intermediary in the carbon fixation Calvin cycle, can be attributed to domestication, it is not possible to draw a clear picture of the effects of domestication on this ecosystem service.

On the other hand, pathways such as fatty acid and jasmonic acid biosynthesis were more commonly found in the rhizosphere of modern and landrace varieties. Fatty acids are involved in multiple functions, ranging from cell membrane constituents to cell signaling. Fatty acids have been used as indices of soil quality and even to describe food web connections [30], thus, positive feedback compared with their wild ancestors could be attributed to tomato crops. Jasmonic acid (JA) and its derivatives (collectively known as jasmonates) play an important role in regulating plant defenses against biotic stresses, and facilitating beneficial interactions between plants and microbes in the root zone [31,32]. JA signaling has been suggested to have evolved during land colonization by plants exposed to new biotic and abiotic stresses [33], and symbiotic relationships with microbes, including plant growth promoting bacteria and mycorrhizal fungi. Moreover, microbe induced systemic resistance to pathogens and pests involve JA signaling [34,35]. However, although JA production by bacteria and fungi in soil has been reported [36], its impact on plant–microbiome interactions remains unclear. Finally, regarding signalling, we detected a significant increase in the biosynthesis of gamma-aminobutyric acid (GABA) in wild tomato species compared to the groups that included cultivars. GABA is involved in inter-bacterial communication and interactions between bacteria and their host [37]. Furthermore, GABA production has been associated with bacterial overcoming of environmental stress [38].

Overall, these findings highlight the influence of tomato domestication on some molecular pathways of the associated soil bacteria, although all ancestral functional characteristics of bacteria have been conserved across time. However, we wonder whether there is a pattern of bacterial functional abundance associated with the tomato soil related to the domestication degree. To shed some light on this point, we calculated interactions between functional units of gene sets in metabolic pathways, which may help to address the question of how microbial genes work together to support specific microbiome functions [39]. In this study, we assessed pairwise relationships between bacterial functional units based on metagenomic sequencing of bacteria growing on tomato plants along a domestication gradient The highest connectance levels in bacterial communities were found in landraces:wild pairs due to an increase in network density as measured by the higher average number of connections established expressed by the average number of neighbors (Table 3). In addition, the increased connectance in the landraces:wild pairing with respect to the other two pairs was related to the decrease in the characteristic path length, defined as the average number of steps along the shortest paths for all possible pairs of network nodes. These changes suggest an intensification of microbial connectance relative to the pairs modern:landraces and modern:wild. Finally, an increase in the pair modern:landraces regarding the network radius and diameter measuring the longest of all the shortest calculated paths in the network, suggests a decrease in module-pathway connectance.

Highly connected clusters, or sets of nodes most of which are connected with one another, were then explored (Figures 4–6). Again, the highest connectance was detected for the pair landraces:wild varieties and nodes representing the same module in the two different types of tomato were recovered in a single cluster. For the rest of the pairs, even if they shared the same number of common modules, they were recovered in two or three different clusters. Overall, the above results suggest that certain functional groups such as the synthesis of certain amino acids or carbohydrate metabolism tend to be more evolutionarily conserved in bacterial communities associated with tomato landraces than those of modern varieties. However, we also found that most of the metabolic routes of bacteria associated to either landraces or modern cultivars with those associated to their ancestors were different. In this scenario, a possible process of divergent evolution in tomato lines, that is, the process by which groups of the same common ancestor evolve and accumulate differences in response to changes in both environmental conditions and biotic factors, could be debated. Nevertheless, further investigation is needed to clarify how tomato domestication has driven specific bacterial functions in root-associated soil.

4. Materials and Methods

4.1. Field Experiment

Seeds of 27 *Solanum lycopersicum* Mill., *S. habrochaites* and *S. pimpinellifolium* accessions were selected from La Mayora Institute of Subtropical and Mediterranean Horticulture (IHSM-UMA-CSIC) germplasm bank. Seeds were germinated and ten one-month-old seedlings per variety (n = 270) were randomly sown in an experimental field of La Mayora

IHSM (Málaga, Spain; 36.77° N, 4.04° W),) on 19th April 2018 in a Eutric Regosol soil [40]. They were grown until 16th July 2018. Just after transplanting, plants were watered with 15:15:15 solution (15% nitrogen, 15% phosphorus and 15% potassium) during 30 min adding-up a volume of 4 l per plant. During the course of the experiment, watering consisted of 30 min of water twice in a week (Mondays and Fridays) [41]. At harvest, the soil attached to the main and secondary roots was taken by shaking, placed in separate plastic bags, and kept at 4 °C. Then, samples from each variety were pooled, ground together using a mortar and pestle, sieved twice (2-mm mesh) and immediately stored at −20 °C until molecular analyses were performed.

In this study, cluster assays of the 27 tomato accessions, based on their degree of domestication, were carried out: (1) wild tomato species (accessions NR0407, NR1021, NR0136, NR0699, NR0937), (2) tomato landraces (accessions NR0025, NR0006, NR0044, NR0213, NR0275, NR0237, NR0469, NR0166, NR0063, NR0705, NR0612), and (3) modern commercial cultivars (accessions ABL104, ANL101, NR0071, NR0816, NR0080, NR1080, NR0504. COM1, COM2, COM3 and COM4 cultivars, which are protected under plant variety rights, have no accession number).

4.2. Chemical Characteristics of the Soil

Air-dried rhizosphere soil samples were used to determinate chemical properties. Total N and SOC were determined with the aid of the Leco-TruSpec CN elemental analyzer (LECO Corp., St Joseph, MI, USA). Total mineral content was determined by the digestion method with HNO3 65%:HCl 35% (1:3; v:v) followed by analysis using inductively coupled plasma optical emission spectrometry (ICP-OES) (ICP 720-ES, Agilent, Santa Clara, CA, USA). Detailed information on soil characteristics is given in Supplementary Material Table S2.

4.3. Molecular Analyses of Soil Bacteria

DNA was extracted from eight 1 g aliquots for each root-associated soil sample using the bead-beating method with the aid of a PowerSoil® DNA Isolation Kit (MoBio Laboratories, Solana Beach, CA, USA) according to the manufacturer's instructions. For each variety, two replicates were prepared by pooling four extractions and concentrating them at 35 °C to a final volume of 20 µL using a Savant Speedvac® concentrator (Fisher Scientific, Madrid, Spain). The V3-V4 hypervariable regions (ProV3V4 primers 5′ CCTACGGGNBGCASCAG 3′ and 5′ GACTACNVGGGTATCTAATCC 3′ [42,43]) of the 16S rRNA gene were used to characterize the bacterial communities of the two replicates per sample using the Illumina MiSeq platform (2 × 250 nucleotide paired-end protocol) at the genomic facilities of the López-Neyra Institute of Parasitology and Biomedicine (IPBLN-CSIC). Blockers were used to minimize amplification of mitochondria and chloroplasts. Raw sequences were preprocessed using the SEED2 platform [44] by first merging forward and reverse sequences. Quality filtering excluded sequences containing ambiguous bases (N) and those with a quality score of less than 30. Primers were removed and sequences trimmed to 400 bp length. The sequences were then clustered using the UPARSE method [45]: Operational Taxonomic Unit (OTU) radius set to 3% and sequence similarity to 97%. Singletons and chimeric sequences were removed. Taxonomic assignment of OTUs was performed using the classify.seqs algorithm in Mothur software (University of Michigan, Detroit, MI, USA) against the SILVA v132 database, after which no archaea were detected in the samples [46,47]. An abundance sample x OTU matrix was generated using OTU reads as a proxy of abundance using the Marker Data Profiling module in the MicrobiomeAnalyst tool (https://www.microbiomeanalyst.ca/ accessed on 2 August 2021). The most abundant sequence per OTU was selected as representative. Rarefaction curves were visualized using MicrobiomeAnalyst to confirm that all samples reached a plateau [48,49].

4.4. Predictive Metagenomics Profiling

To determine the potential functional metabolic capabilities of soil bacterial communities, we used Tax4fun, an open-source R package, which predicts the functional capabilities of these communities based on 16S datasets. Tax4Fun is applicable to output obtained from the SILVAngs web server [50]. Tax4fun was implemented in Shotgun Data Profiling (SDP) module of MicrobiomeAnalyst to predict functional pathways based on Kyoto Encyclopedia of Gene and Genome (KEGG, https://www.kegg.jp/ accessed on 2 August 2021) annotations [51,52]. KEGG functional annotations were based on modules, i.e., functional units of gene sets in the KEGG metabolic pathways database that can be linked to specific metabolic capacities and other phenotypic features [48].

4.5. Functional Networks

Similarities on functional profiles across tomato types were studied by looking for correlations in the abundance of modules. CoNet plug-in method [53] in Cytoscape software v.3.8.2 [54] was used to visualize these relationships by building co-occurrence networks. Thus, two nodes representing the same module in different tomato types should be connected in the case that both tomatoes have a similar pattern of abundance for that module. Thus, building networks by tomato type pairs gives an idea of the conservation of modules across domestication (i.e., the number of links between the same module in different tomato types). Co-occurrence networks were constructed based on the identification of significant positive associations, that is, co-presences of functional units in the tomato root-associated microbiome. Due to the different number of samples/tomato varieties in each domestication type, for arranging the construction of network, the number of samples in each domestication type was adjusted to the tomato type with the least number. The selection and order of samples was arranged randomly. This analysis was repeated 5 times by shuffling the input sample order to avoid spurious results. To run the analysis, KEGG modules having less than 20 reads were discarded from the analyses. KEGG module abundance was normalized by sample. A total of 2000 permutations were set up by keeping edge number constant. The significance of co-presences were evaluated by a combination of Spearman and Pearson correlations and Bray–Curtis dissimilarity (see e.g., [52,55], corrected for multiple testing using Bonferroni). Finally, the MCODE Cytoscape plugin [56] with default settings was then used to detect highly connected network modules. Only modules with an MCODE score greater than 2.0 were retained for analysis [39].

4.6. Statistical Analyses

OTU abundance information was normalized to the abundance value of the sample with the least number of sequences. Alpha diversity indices generated by SEED2 were used to compare bacterial richness and diversity in tomato accessions. Statistically significant differences in alpha diversity, the bacterial composition of the group of tomato varieties and predictive metagenomics profiling data were evaluated using generalized lineal model (GLM) with degree of domestication as fixed factor. We checked fixed factors for significance with Wald test from car package [57] and multiple comparisons between levels of the fixed factor were tested using Tukey's test with the package lsmeans and emmeans [58]. For each model, residuals were examined for model validation. Beta diversity, or species complexity differences between groups of tomato varieties, was determined by linear discriminant analysis (LDA) effect size (LEfSe) using the MicrobiomeAnalyst web server. Taxa with an LDA score > 4 were considered important biomarkers of each group given that a p value < 0.05 indicates significant differences between groups. Data were analyzed using R version 3.6.3 [59] and R Studio version 1.1.456 [60].

5. Conclusions

In our study we found that core bacterial microbiome is similar between tomato landraces and modern commercial cultivars with small differences with wild tomato. These findings highlight the influence of the host genotype on the potential functions of soil bacterial communities. Furthermore, we found that differences in eight biological metabolic pathways between wild tomatoes compared with tomato landraces and modern commercial. Thus, we conclude that all ancestral functional characteristics of bacteria have been conserved across time. In the light of these results, it becomes apparent that the capacity of soil bacteria to provide ecosystem services is affected by agronomic practices linked to the domestication process, particularly those related to the preservation of soil organic matter. We also assayed the relationships between functional units of bacteria growing on tomato plants along a domestication gradient, finding the highest levels of connection between bacterial communities driven by tomato landraces and their wild ancestors.

Supplementary Materials: The following are available online at https://www.mdpi.com/article/10.3390/plants10091942/s1, Table S1: Abundance of predicted functions related to root-associated soil bacteria in the 27 tomato accessions. Table S2: Chemical characteristics of the soil. Table S3: Tab-delimited taxonomy tables.

Author Contributions: Conceptualization, E.B., E.d.l.P. and L.S.; methodology, L.S., B.M., V.F. and R.A.H.; software, L.S., Á.L.-G., R.A.H. and E.B.; validation, E.B., R.A.H., L.S. and Á.L.-G.; formal analysis, E.B., L.S., R.A.H. and Á.L.-G.; investigation, E.B., L.S. and Á.L.-G.; resources, E.B., E.d.l.P., V.F. and M.J.P.; data curation, L.S., Á.L.-G., B.M., R.A.H. and E.B.; writing—original draft preparation, E.B., Á.L.-G. and L.S.; writing—review and editing, R.A.H., B.M., E.d.l.P. and M.J.P.; supervision, E.B. and M.J.P.; project administration; E.B.; funding acquisition, E.B. and M.J.P. All authors have read and agreed to the published version of the manuscript.

Funding: This research has received funding from the European Union's Horizon 2020 Research and Innovation programme under Grant agreement No 765290.

Institutional Review Board Statement: Not applicable.

Informed Consent Statement: Not applicable.

Data Availability Statement: All raw Illumina sequence data were deposited in the Sequence Read Archive (SRA) service of the European Bioinformatics Institute (EBI) database (BioProject ID: PRJNA693664).

Acknowledgments: Rafael Alcalá Herrera thanks the Junta de Andalucía (Spain) for funding his research at the EEZ-CSIC with a postdoctoral contract (Project P12-AGR-1419).

Conflicts of Interest: The authors declare no conflict of interest.

References

1. Reinhold-Hurek, B.; Bünger, W.; Burbano, C.S.; Sabale, M.; Hurek, T. Roots shaping their microbiome, global hotspots for microbial activity. *Annu. Rev. Phytopathol.* **2015**, *53*, 403–424. [CrossRef] [PubMed]
2. Cordovez, V.; Dini-Andreote, F.; Carrión, V.J.; Raaijmakers, J.M. Ecology and Evolution of Plant Microbiomes. *Annu. Rev. Microbiol.* **2019**, *73*, 69–88. [CrossRef] [PubMed]
3. Wall, D.H.; Bardgett, R.D.; Behan-Pelletier, V.; Herrick, J.E.; Jones, H.; Ritz, K.; Six, J.; Strong, D.R.; van der Putten, W.H. *Soil Ecology and Ecosystem Services*; Oxford University Press: Oxford, UK, 2012.
4. Houlden, A.; Timms-Wilson, T.M.; Day, M.J.; Bailey, M.J. Influence of plant developmental stage on microbial community structure and activity in the rhizosphere of three field crops. *FEMS Microbiol. Ecol.* **2008**, *65*, 193–201. [CrossRef]
5. Lei, S.; Xu, X.; Cheng, Z.; Xiong, J.; Ma, R.; Zhang, L.; Yang, X.; Zhu, Y.; Zhang, B.; Tian, B. Analysis of the community composition and bacterial diversity of the rhizosphere microbiome across different plant taxa. *Microbiol. Open* **2019**, *8*, e00762. [CrossRef]
6. Compant, S.; Samad, A.; Faist, H.; Sessitsch, A. A review on the plant microbiome, ecology; functions; and emerging trends in microbial application. *J. Adv. Res.* **2019**, *19*, 29–37. [CrossRef]
7. Liu, H.; Brettell, L.E.; Qiu, Z.; Singh, B.K. Microbiome-mediated stress resistance in plants. *Trends Plant. Sci.* **2020**, *25*, 733–743. [CrossRef]
8. Trivedi, P.; Leach, J.E.; Tringe, S.G.; Sa, T.; Singh, B.K. Plant–microbiome interactions, from community assembly to plant health. *Nat. Rev. Microbiol.* **2020**, *18*, 607–621. [CrossRef]
9. Pérez-Jaramillo, J.E.; Mendes, R.; Raaijmakers, J.M. Impact of plant domestication on rhizosphere microbiome assembly and functions. *Plant. Mol. Biol.* **2016**, *90*, 635–644. [CrossRef] [PubMed]
10. Schmidt, J.E.; Bowles, T.M.; Gaudin, A.C.M. Using ancient traits to convert soil health into crop yield, impact of selection on maize root and rhizosphere function. *Front. Plant. Sci.* **2016**, *7*, 1–11. [CrossRef]

11. Iannucci, A.; Fragasso, M.; Beleggia, R.; Nigro, F.; Papa, R. Evolution of the Crop Rhizosphere, Impact of Domestication on Root Exudates in Tetraploid Wheat (*Triticum turgidum* L.). *Front. Plant. Sci.* **2017**, *8*, 2124. [CrossRef]
12. Pérez-Jaramillo, J.E.; Carrión, V.J.; de Hollander, M.; Raaijmakers, J.M. The wild side of plant microbiomes. *Microbiome* **2018**, *6*, 143. [CrossRef]
13. Shenton, M.; Iwamoto, C.; Kurata, N.; Ikeo, K. Effect of Wild and Cultivated Rice Genotypes on Rhizosphere Bacterial Community Composition. *Rice* **2016**, *9*, 42. [CrossRef]
14. Carrillo, J.; Ingwell, L.L.; Li, X.; Kaplan, I. Domesticated tomatoes are more vulnerable to negative plant–soil feedbacks than their wild relatives. *J. Ecol.* **2019**, *107*, 1753–1766. [CrossRef]
15. Terrazas, R.A.; Balbirnie-Cumming, K.; Morris, J.; Hedley, P.E.; Russell, J.; Paterson, E.; Baggs, E.M.; Fridman, E.; Bulgarelli, D.A. A footprint of plant eco-geographic adaptation on the composition of the barley rhizosphere bacterial microbiota. *Sci. Rep.* **2020**, *10*, 12916. [CrossRef]
16. Jia, Y.; Whalen, J.K. A new perspective on functional redundancy and phylogenetic niche conservatism in soil microbial communities. *Pedosphere* **2020**, *30*, 18–24. [CrossRef]
17. Spor, A.; Roucou, A.; Mounier, A.; Bru, D.; Breuil, M.C.; Fort, F.; Vile, D.; Roumet, P.; Philippot, L.; Violle, C. Domestication-driven changes in plant traits associated with changes in the assembly of the rhizosphere microbiota in tetraploid wheat. *Sci. Rep.* **2020**, *10*, 12234. [CrossRef] [PubMed]
18. Mendes, R.; Garbeva, P.; Raaijmakers, J.M. The rhizosphere microbiome, significance of plant beneficial; plant pathogenic; and human pathogenic microorganisms. *FEMS Microbiol. Rev.* **2013**, *37*, 634–663. [CrossRef] [PubMed]
19. Martínez-Romero, E.; Aguirre-Noyola, J.L.; Taco-Taype, N.; Martínez-Romero, J.; Zuñiga-Dávila, D. Plant microbiota modified by plant domestication. *Syst. Appl. Microbiol.* **2020**, *43*, 126106. [CrossRef]
20. Milla, R.; García-Palacios, P.; Matesanz, S. Looking at past domestication to secure ecosystem services of future croplands. *J. Ecol.* **2017**, *105*, 885–889. [CrossRef]
21. Stevenson, F.J. *Humus Chemistry, Genesis, Composition, Reactions*, 2nd ed.; John Wiley and Sons: New York, NY, USA, 1994.
22. Zhang, Y.; Yue, D.; Ma, H. Darkening mechanism and kinetics of humification process in catechol-Maillard system. *Chemosphere* **2015**, *130*, 40–45. [CrossRef]
23. Tiessen, H.; Cuevas, E.; Chacon, P. The role of soil organic matter in sustaining soil fertility. *Nature* **1994**, *371*, 783–785. [CrossRef]
24. Gougoulias, C.; Clark, J.M.; Shaw, L.J. The role of soil microbes in the global carbon cycle, tracking the below-ground microbial processing of plant-derived carbon for manipulating carbon dynamics in agricultural systems. *J. Sci. Food Agric.* **2014**, *94*, 2362–2371. [CrossRef] [PubMed]
25. Six, J.; Frey, S.D.; Thiet, R.K.; Batten, K.M. Bacterial and Fungal Contributions to Carbon Sequestration in Agroecosystems. *Soil Sci. Soc. Am. J.* **2006**, *70*, 555–569. [CrossRef]
26. Pérez-Jaramillo, J.E.; de Hollander, M.; Ramírez, C.A.; Mendes, R.; Raaijmakers, J.M.; Carrión, V.J. Deciphering rhizosphere microbiome assembly of wild and modern common bean (*Phaseolus vulgaris*) in native and agricultural soils from Colombia. *Microbiome* **2019**, *7*, 114. [CrossRef]
27. Vries, F.; Wallenstein, M.D. Below-ground connections underlying above-ground food production: A framework for optimising ecological connections in the rhizosphere. *J. Ecol.* **2017**, *105*, 913–920. [CrossRef]
28. Cavicchioli, R.; Ripple, W.J.; Timmis, K.N.; Azam, F.; Bakken, L.R.; Baylis, M.; Behrenfeld, M.J.; Boetius, A.; Boyd, P.W.; Classen, A.T.; et al. Scientists' warning to humanity, microorganisms and climate change. *Nat. Rev. Microbiol.* **2019**, *17*, 569–586. [CrossRef] [PubMed]
29. Lynn, T.M.; Ge, T.; Yuan, H.; Wei, X.; Wu, X.; Xiao, K.; Kumaresan, D.; Yu, S.S.; Wu, J.; Whiteley, A.S.; et al. Soil Carbon-Fixation Rates and Associated Bacterial Diversity and Abundance in Three Natural Ecosystems. *Microb. Ecol.* **2017**, *73*, 645–657. [CrossRef] [PubMed]
30. de Carvalho, C.; Caramujo, M.J. The Various Roles of Fatty Acids. *Molecules* **2018**, *23*, 2583. [CrossRef] [PubMed]
31. Pieterse, C.M.; Leon-Reyes, A.; Van der Ent, S.; Van Wees, S.C. Networking by small-molecule hormones in plant immunity. *Nat. Chem. Biol.* **2009**, *5*, 308–316. [CrossRef]
32. Bari, R.; Jones, J.D. Role of plant hormones in plant defence responses. *Plant. Mol. Biol.* **2009**, *69*, 473–488. [CrossRef]
33. Kenrick, P.; Crane, P. The origin and early evolution of plants on land. *Nature* **1997**, *389*, 33–39. [CrossRef]
34. Pozo, M.J.; López-Ráez, J.A.; Azcón-Aguilar, C.; García-Garrido, J.M. Phytohormones as integrators of environmental signals in the regulation of mycorrhizal symbioses. *New Phytol.* **2015**, *205*, 1431–1436. [CrossRef] [PubMed]
35. Gruden, K.; Lidoy, J.; Petek, M.; Podpečan, V.; Flors, V.; Papadopoulou, K.K.; Pappas, M.L.; Martinez-Medina, A.; Bejarano, E.; Biere, A.; et al. Ménage à Trois, Unraveling the Mechanisms Regulating Plant-Microbe-Arthropod Interactions. *Trends Plant. Sci.* **2020**, *25*, 1215–1226. [CrossRef] [PubMed]
36. Eng, F.; Marin, J.E.; Zienkiewicz, K.; Gutiérrez-Rojas, M.; Favela-Torres, E.; Feussner, I. Jasmonic acid biosynthesis by fungi, derivatives; first evidence on biochemical pathways and culture conditions for production. *PeerJ* **2021**, *9*, e10873. [CrossRef] [PubMed]
37. Dagorn, A.; Chapalain, A.; Mijouin, L.; Hillion, M.; Duclairoir-Poc, C.; Chevalier, S.; Taupin, L.; Orange, N.; Feuilloley, M.G. Effect of GABA, a bacterial metabolite; on *Pseudomonas fluorescens* surface properties and cytotoxicity. *Int. J. Mol. Sci.* **2013**, *14*, 12186–12204. [CrossRef]

38. Somorin, Y.; Abram, F.; Brennan, F.; O'Byrne, C. The General Stress Response Is Conserved in Long-Term Soil-Persistent Strains of Escherichia coli. *Appl. Environ. Microbiol.* **2016**, *82*, 4628–4640. [CrossRef] [PubMed]
39. Li, L.; Wang, Z.; He, P.; Ma, S.; Du, J.; Jiang, R. Construction and Analysis of Functional Networks in the Gut Microbiome of Type 2 Diabetes Patients. *Genom. Proteom. Bioinform.* **2016**, *14*, 314–324. [CrossRef]
40. FAO. *World Reference Base for Soil Resources 2014: International Soil Classification System for Naming Soils and Creating Legends for Soil Maps*; Food and Agriculture Organization of the United Nations: Roma, Italy, 2015.
41. Ferrero, V.; Baeten, L.; Blanco-Sanchez, L.; Planello, R.; Diaz-Pendon, J.A.; Rodríguez-Echeverría, S.; Haegeman, A.; de la Peña, E. Complex patterns in tolerance and resistance to pests and diseases underpin the domestication of tomato. *New Phytol.* **2020**, *226*, 254–266. [CrossRef] [PubMed]
42. Takahashi, S.; Tomita, J.; Nishioka, K.; Hisada, T.; Nishijima, M. Development of a prokaryotic universal primer for simultaneous analysis of Bacteria and Archaea using next-generation sequencing. *PLoS ONE* **2014**, *9*, e105592. [CrossRef]
43. Lundberg, D.S.; Yourstone, S.; Mieczkowski, P.; Jones, C.D.; Dangl, J.L. Practical innovations for high-throughput amplicon sequencing. *Nat. Methods* **2013**, *10*, 999–1002. [CrossRef]
44. Vetrovský, T.; Baldrian, P.; Morais, D. SEED 2, a user-friendly platform for amplicon high-throughput sequencing data analyses. *Bioinformatics* **2018**, *34*, 2292–2294. [CrossRef]
45. Edgar, R.C. UPARSE, highly accurate OTU sequences from microbial amplicon reads. *Nat. Methods* **2013**, *10*, 996. [CrossRef] [PubMed]
46. Schloss, P.D.; Westcott, S.L.; Ryabin, T.; Hall, J.R.; Hartmann, M.; Hollister, E.B.; Lesniewski, R.A.; Oakley, B.B.; Parks, D.H.; Robinson, C.J.; et al. Introducing mothur, open-source; platform-independent; community-supported software for describing and comparing microbial communities. *Appl. Environ. Microbiol.* **2009**, *75*, 7537–7541. [CrossRef] [PubMed]
47. Quast, C.; Pruesse, E.; Yilmaz, P.; Gerken, J.; Schweer, T.; Yarza, P.; Peplies, J.; Glöckner, F.O. The SILVA ribosomal RNA gene database project, improved data processing and web-based tools. *Nucleic Acids Res.* **2013**, *41*, 590–596. [CrossRef]
48. Chong, J.; Liu, P.; Zhou, G.; Xia, J. Using MicrobiomeAnalyst for comprehensive statistical; functional; and meta-analysis of microbiome data. *Nat. Protoc.* **2020**, *15*, 799–821. [CrossRef]
49. Dhariwal, A.; Chong, J.; Habib, S.; King, I.L.; Agellon, L.B.; Xia, J. MicrobiomeAnalyst, a web-based tool for comprehensive statistical; visual and meta-analysis of microbiome data. *Nucleic Acids Res.* **2017**, *45*, 180–188. [CrossRef]
50. Aßhauer, K.P.; Wemheuer, B.; Daniel, R.; Meinicke, P. Tax4Fun, predicting functional profiles from metagenomic 16S rRNA data. *Bioinformatics* **2015**, *31*, 2882–2884. [CrossRef]
51. Breiman, L. Random forests. *Mach. Learn.* **2001**, *45*, 5–32. [CrossRef]
52. Segata, N.; Izard, J.; Waldron, L.; Gevers, D. Metagenomic biomarker discovery and explanation. *Genome Biol.* **2011**, *12*, R60. [CrossRef]
53. Faust, K.; Raes, J. CoNet app, inference of biological association networks using Cytoscape. *F1000Research* **2016**, *5*, 1519. [CrossRef] [PubMed]
54. Shannon, P.; Markiel, A.; Ozier, O.; Baliga, N.S.; Wang, J.T.; Ramage, D.; Amin, N.; Schwikowski, B.; Ideker, T. Cytoscape, a software environment for integrated models of biomolecular interaction networks. *Genome Res.* **2003**, *13*, 2498–2504. [CrossRef] [PubMed]
55. Weiss, S.; Van Treuren, W.; Lozupone, C.; Faust, K.; Friedman, J.; Deng, Y.; Xia, L.C.; Xu, Z.Z.; Ursell, L.; Alm, E.J.; et al. Correlation detection strategies in microbial data sets vary widely in sensitivity and precision. *ISME J.* **2016**, *10*, 1669–1681. [CrossRef] [PubMed]
56. Bader, G.D.; Hogue, C.W. An automated method for finding molecular complexes in large protein interaction networks. *BMC Bioinform.* **2003**, *4*, 2. [CrossRef] [PubMed]
57. Fox, J.; Weisberg, S. *An R Companion to Applied Regression*, 3rd ed.; Sage: Thousand Oaks, CA, USA, 2019.
58. Lenth, R.V. Least-squares means, The R package lsmeans. *J. Stat. Soft* **2016**, *69*, 1–33. [CrossRef]
59. R Development Core Team. *R, A Language and Environment for Statistical Computing*; R Development Core Team: Vienna, Austria, 2017.
60. RStudio Team. *RStudio, Integrated Development Environment for R*; RStudio Team: Boston, MA, USA, 2016.

plants

MDPI

Article

Enhancing Salt Tolerance in Soybean by Exogenous Boron: Intrinsic Study of the Ascorbate-Glutathione and Glyoxalase Pathways

Hesham F. Alharby [1], Kamrun Nahar [2], Hassan S. Al-Zahrani [1], Khalid Rehman Hakeem [1] and Mirza Hasanuzzaman [3,*]

[1] Department of Biological Sciences, Faculty of Science, King Abdulaziz University, Jeddah 21589, Saudi Arabia; halharby@kau.edu.sa (H.F.A.); hsalzahrani@kau.edu.sa (H.S.A.-Z.); kur.hakeem@gmail.com (K.R.H.)
[2] Department of Agricultural Botany, Sher-e-Bangla Agricultural University, Dhaka 1207, Bangladesh; knahar84@yahoo.com
[3] Department of Agronomy, Faculty of Agriculture, Sher-e-Bangla Agricultural University, Dhaka 1207, Bangladesh
* Correspondence: mhzsauag@yahoo.com

Abstract: Boron (B) performs physiological functions in higher plants as an essential micronutrient, but its protective role in salt stress is poorly understood. Soybean (*Glycine max* L.) is planted widely throughout the world, and salinity has adverse effects on its physiology. Here, the role of B (1 mM boric acid) in salt stress was studied by subjecting soybean plants to two levels of salt stress: mild (75 mM NaCl) and severe (150 mM NaCl). Exogenous B relieved oxidative stress by enhancing antioxidant defense system components, such as ascorbate (AsA) levels, AsA/dehydroascorbate ratios, glutathione (GSH) levels, the GSH and glutathione disulfide ratios, and ascorbate peroxidase, monodehydroascorbate reductase, and dehydroascorbate reductase activities. B also enhanced the methylglyoxal detoxification process by upregulation of the components of the glyoxalase system in salt-stressed plants. Overall, B supplementation enhanced antioxidant defense and glyoxalase system components to alleviate oxidative stress and MG toxicity induced by salt stress. B also improved the physiology of salt-affected soybean plants.

Keywords: trace element; plant nutrient; salinity; antioxidant defense system; glyoxalase system

Citation: Alharby, H.F.; Nahar, K.; Al-Zahrani, H.S.; Hakeem, K.R.; Hasanuzzaman, M. Enhancing Salt Tolerance in Soybean by Exogenous Boron: Intrinsic Study of the Ascorbate-Glutathione and Glyoxalase Pathways. *Plants* **2021**, *10*, 2085. https://doi.org/10.3390/plants10102085

Academic Editors: Milan S. Stankovic, Paula Baptista and Petronia Carillo

Received: 22 August 2021
Accepted: 28 September 2021
Published: 1 October 2021

Publisher's Note: MDPI stays neutral with regard to jurisdictional claims in published maps and institutional affiliations.

Copyright: © 2021 by the authors. Licensee MDPI, Basel, Switzerland. This article is an open access article distributed under the terms and conditions of the Creative Commons Attribution (CC BY) license (https://creativecommons.org/licenses/by/4.0/).

1. Introduction

Soybean (*Glycine max* L.) is a widely grown legume in America, Asia, Europe, and Africa and is consumed mostly as oil and soy protein [1]. However, soybean cultivation is now increasingly hampered due to various environmental stresses, including salinity, in many soybean growing areas. Climate change, anthropogenic activities, and poor agronomic management continue to increase the occurrence of salt stress in the crops growing in the field [2]. Salt-affected soils are distinguished by the presence of considerable amounts of soluble salts that are taken up by the plants growing there. This salt accumulation results in plant water stress due to disrupted osmotic potential, as well as interrupted uptake of essential nutrients and disturbance of the plant ionic balance [3]. Biochemical and physiological changes due to salt stress inhibit germination and post-germination development, with decreased root and shoot growth as common consequences of salt stress. The decreases in water absorption and mineral nutrients, the disruption of ionic and nutrient homeostasis, and the induction of ion toxicity lead directly and indirectly to growth inhibition [4]. Different physiological processes, including transpiration, photosynthesis, and translocation of assimilates, are hindered under salinity [5]. Therefore, altered developmental processes and yield reduction are key consequences of salt stress. Methylglyoxal generation can be increased 2- to 6-fold under abiotic stress [6].

Salt stress disrupts the equilibrium in osmotic potential as well as of different ions. Salt stress has injurious effects on chlorophyll (Chl) biosynthesis as well as on the efficiency of photosynthesis. The photon energy, therefore, cannot be appropriately used by photosystems, and plants suffer from oxidative stress due to excess production of reactive oxygen species (ROS) [3,7]. Ionic/nutrient imbalances alter the biosynthesis or metabolism of essential metabolites and disrupt the activity of enzymes involved in biochemical and physiological processes, further exacerbating ROS generation and oxidative damage in plants under salinity [4,8,9]. Many previous scientific papers provided evidence that oxidative stress tolerance is a prerequisite for enhancing overall stress tolerance, including salt stress [10].

Plants are equipped with antioxidant defense systems that protect against ROS generation, but these systems are only effective up to a certain limit. The antioxidant defense system includes both enzymatic and nonenzymatic components [10], and enhancement of this defense mechanism is a vital strategy for improving salt tolerance [10,11]. One metabolite that is involved in these defense systems is methylglyoxal (MG), a reactive substance produced as a by-product of different metabolic pathways in the chloroplast, mitochondria, and cytosol. Recently, the importance of the glyoxalase system has been prioritized for enhancing stress tolerance [8,10].

The effects of salt stress on mineral metabolism have received particular attention, but more study is required on micronutrients such as boron (B). Boron is a micronutrient that performs different physiological and developmental functions in plants. Boron deficiency decreases photosynthesis, pollen formation, nucleic acid biosynthesis, starch metabolism, and auxin function. Plant growth, including root elongation, is adversely affected by B deficiency. Plasmalemma-bound enzyme activities and the plasma membrane potential are altered due to B deficiency. Boron plays a role in protein configuration, the synthesis of phenolic compounds, and nitrogen metabolism [12–14]. A few recent research findings have demonstrated that supplementation with B can provide stress protection in plants by regulating physiological processes, improving the structural integrity of cytosolic organs, and enhancing antioxidant defense systems [3,9,15,16]. The aim of the present study was to investigate the antioxidant defense, the glyoxalase system, and the physiological attributes of salt-stressed soybean and the regulatory function of B in improving soybean performance under salt stress.

2. Materials and Methods

2.1. Growing Condition and Treatments

Soybean seeds were sown in plastic pots (63.5 cm depth and 26 cm diameter) in soil amended with organic manure and fertilizers at recommended doses [17]. After germination, five plants were allowed to grow in each pot. Two sets of pots were grown at the same time and under the same conditions. One set was used for morphological and growth studies, and the other was used for physiological and biochemical investigations. Twenty-day-old plants were treated with 75 mM NaCl (mild stress) and 150 mM NaCl (severe stress). A separate set of control plants was grown without added NaCl. The control and salt-treated seedlings were administered B (foliar application of 1 mM boric acid at one-week intervals up to 50 days after sowing [DAS]). A completely randomized design (CRD) was executed with three replications.

2.2. Growth Parameters

Plant height was measured using a measuring tape, and the average from five plants was considered for each replication.

Leaf photographs were captured with a digital camera, and Image-J software was used to calculate the leaf area [18].

Five sample plants were harvested from each pot (per replication) following completion of the treatments (45 DAS). Shoot fresh weight (FW) plant^{-1} was determined, and then the plants were oven-dried at 72 °C for 48 h for dry weight (DW) plant^{-1} determinations.

2.3. Measurement of Physiological Parameters

Five leaves were randomly selected from a plant from each pot of each replication to record the Soil Plant Analysis Development (SPAD) value with a SPAD meter (atLEAF, USA). Three SPAD readings were taken at different positions on a leaf. The Chl content was then calculated according to the conversion factor suggested by the user manual and following Zhu et al. [19].

The relative water content (RWC) [20] was recorded from whole leaf discs, which were weighed for FW and then floated in distilled water in petri plates and kept in the dark. The leaf weight was determined again after 8 h, after blotting excess water with tissue paper, and was treated as the turgid weight (*TW*). Dry weights were determined after drying at 80 °C (for 48 h), and the RWC was calculated using the following formula

$$\text{RWC } (\%) = \frac{FW - DW}{TW - DW} \times 100$$

Electrolyte leakage (EL) was recorded at 30 DAS by placing a 0.5 g leaf sample into a Falcon tube containing distilled water and placing the tubes in a water bath at 40 °C for 1 h. After cooling, the electrical conductivity (EC_1) was recorded with a conductivity meter. The samples were heated again at 121 °C in an autoclave for about 1 h, and the electrical conductivity (EC_2) was measured again after cooling the samples. The formula for computing the EL was [21]

$$\text{EL}(\%) = \frac{EC1}{EC2} \times 100$$

The proline (Pro) content was estimated by homogenizing leaf tissue with sulfosalicylic acid and centrifuging (11,500× *g*, 4 °C, 15 min). An aliquot of the supernatant was mixed with acid ninhydrin and glacial acetic acid and heated at 100 °C in a water bath for 60 min. After cooling, toluene was added, the contents were vortexed, and the toluene layer containing the Pro chromophore was optically assayed at 520 nm [22]. The Pro content was calculated using a standard curve made from known Pro concentrations (10, 20, 40, 60, and 80 μg mL^{-1}).

2.4. Estimation of Lipid Peroxidation and H_2O_2 Concentration

Lipid peroxidation was determined from the malondialdehyde (MDA) level. Leaves were homogenized in 5% (*w*/*v*) trichloroacetic acid (TCA) and centrifuged at 11,500× *g* for 15 min. The supernatant was mixed with thiobarbituric acid (TBA), heated at 95 °C for 60 min, cooled in an ice bath, and centrifuged again for 10 min. The absorbance of the supernatant was read at 532 nm to determine the malondialdehyde content. The measurements were corrected by deducting the absorbance at 600 nm, where the extinction coefficient was 155 mM^{-1} cm^{-1} [23].

The H_2O_2 was estimated by homogenizing leaves in potassium phosphate (K-P) buffer (pH 6.5) at 4 °C and centrifuging at 11,500× *g*. The supernatant was added to a mixture of $TiCl_4$ and H_2SO_4 and let stand at room temperature. After centrifuging at 11,500× g, the optical absorption of the supernatant was recorded spectrophotometrically at 410 nm to determine the H_2O_2 level [24].

2.5. Ascorbate and Glutathione Estimation

Leaves were homogenized in 5% meta-phosphoric acid containing 1 mM EDTA and centrifuged at 11,500× *g* (15 min; 4 °C) and the supernatant was analyzed for ascorbate [25,26] and glutathione [24,25,27]. Total AsA was estimated by reducing the oxidized portion with 0.1 M dithiothreitol (1 h at room temperature), and the supernatant was read at 265 nm using 1.0 unit ascorbate oxidase (AO). The ratio of oxidized AsA/dehydroascorbate (DHA) was calculated by deducting the reduced AsA from the total AsA. A standard curve was prepared from different concentrations of AsA (400, 600, and 800, and 1000 μM) to determine the AsA content of the plant samples.

GSH content was determined by a previously published method [24] with some modifications [27]. The supernatant was neutralized with K-P buffer (0.5 M) (pH 7). Then, after oxidizing GSH with 5,5-dithio-bis (2-nitrobenzoic acid) (DTNB) and reducing with nicotinamide adenine dinucleotide phosphate (NADPH) in the presence of glutathione reductase (GR), the supernatant was used to measure total GSH content spectrophotometrically at 412 nm. Oxidized GSH (GSSG) content was obtained after removing GSH with 2-vinylpyridine. The GSH level was obtained by subtracting GSSG from total GSH. Standard curves of known concentrations of GSH (20, 40, 60, and 80 μg mL^{-1}) and GSSG (12, 16, 20, and 24 μg mL^{-1}) were used to determine the contents in plant samples.

2.6. *Assays of Enzymes of the AsA-GSH Pathway and Glyoxalase System*

The protein contents of the plant samples were determined [28] by grinding leaf tissue in 50 mM K–P (potassium phosphate) buffer (pH 7.0) containing 100 mM KCl, 1 mM AsA, 5 mM β-mercaptoethanol, and 10% (*w*/*v*) glycerol and centrifuging at 11,500 ×*g*. The supernatants were utilized for enzyme activity determinations.

Ascorbate peroxidase (APX, EC: 1.11.1.11) activity was measured in K-P buffer (pH-7) containing H_2O_2, EDTA, and AsA. The absorbance was read at 290 nm, using 2.8 mM^{-1} cm^{-1} as the extinction coefficient [29].

The activity of monodehydroascorbate reductase (MDHAR, EC: 1.6.5.4) was measured in an assay buffer of Tris-HCl buffer, AsA, nicotinamide adenine dinucleotide (NADPH), and AO and recording the absorbance at 340 nm, using 6.2 mM^{-1} cm^{-1} as the extinction coefficient [25].

A previously published method [29] was used to determine dehydroascorbate reductase (DHAR, EC: 1.8.5.1) activity in an assay solution containing DHA, K-P buffer, and GSH and reading the absorbance at 265 nm using 14 mM^{-1} cm^{-1} as the extinction coefficient.

Glutathione reductase (GR, EC: 1.6.4.2) activity was measured in assay buffer containing NADPH, GSSG, K-P buffer pH 7, and EDTA and reading the absorbance at 340 nm using an extinction coefficient of 6.2 mM^{-1} cm^{-1} [25].

A previously published method [25] was used to determine the activity of glyoxalase I (Gly I, EC: 4.4.1.5) in an assay mixture containing GSH, K-P buffer, MG, and magnesium sulfate and reading the absorbance at 240 nm using 3.37 mM^{-1} cm^{-1} as the extinction coefficient. Glyoxalase II (Gly II, EC: 3.1.2.6) activity was determined using an assay buffer containing *S*-D-lactoyl glutathione, Tris-HCl buffer, and EDTA [25] and reading the absorbance at 412 nm using 13.6 mM^{-1} cm^{-1} as the extinction coefficient.

2.7. *Methylglyoxal Level*

Leaves were homogenized in 5% perchloric acid, centrifuged at 11,000× *g*, and the supernatant was decolorized and neutralized. The MG level was determined by adding sodium dihydrogen phosphate and *N*-acetyl-L-cysteine. The developed *N*-α-acetyl-*S*-(1-hydroxy-2-oxo-prop-1-yl)cysteine product was read after 10 min at 288 nm [4].

2.8. *Statistical Analysis*

Data were evaluated by analysis of variance (ANOVA) using CoStat v.6.400 software [30]. Means were separated according to Tukey's honestly significant difference (HSD) test at $p \leq 0.05$.

3. Results

3.1. *Growth Parameters*

Plant height was reduced by 13% and 40% by exposure to mild and severe salt stress, respectively, compared to unstressed control plants. Boron addition had no significant effect on the plant height of salt-treated plants (Figure 1A). Leaf area decreased at both levels of salt stress, and B addition had no significant effect on the leaf area of the salt-treated plants (Figure 1B).

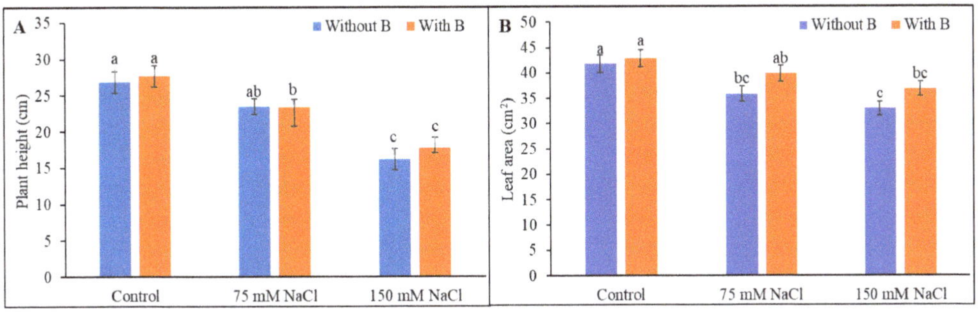

Figure 1. Plant height (**A**) and leaf area (**B**) of salt-stressed soybean following B supplementation. Twenty-day-old plants were subjected to two levels of salt (75 and 150 mM NaCl, 30 days) and supplemented with B (1 mM boric acid). Control treatments were grown without salt. Mean (±SD) was computed from three replicates for each treatment. Different letters over the bars indicate significant differences at $p \leq 0.05$ applying Tukey's HSD test.

Salt stress reduced shoot and root FW and DW compared to unstressed controls treatment. The decreased shoot FW under 150 mM salt stress was restored by B supplementation (Figure 2). Boron addition did not significantly mitigate the reductions in shoot dry weight, root fresh weight, or root dry weight induced by salt stress.

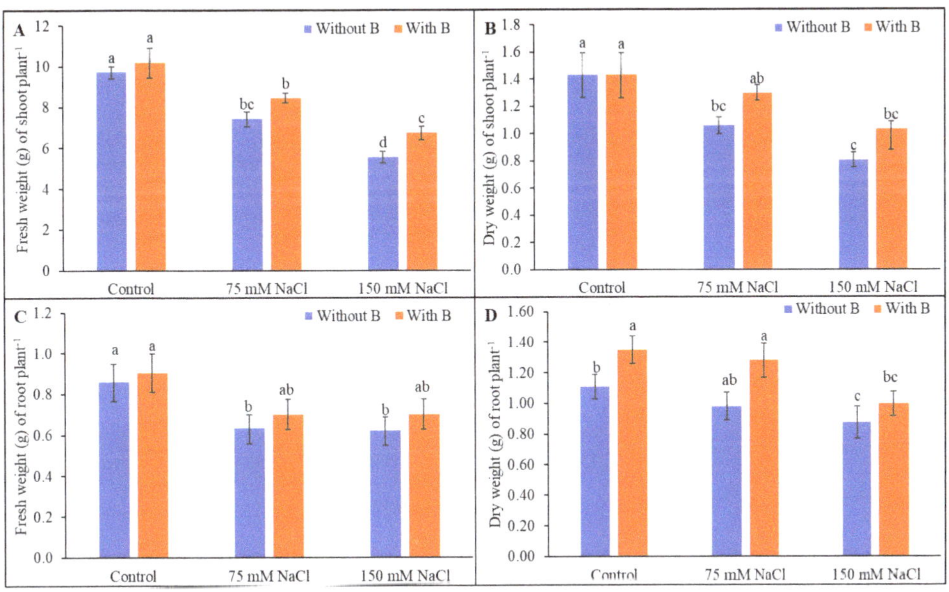

Figure 2. Shoot fresh weight (**A**), shoot dry weight (**B**), root fresh weight (**C**), and root dry weight (**D**) of soybean as affected by salt stress with or without B supplementation. Twenty-day-old plants were imposed with two levels of salt (75 and 150 mM NaCl, 30 days) and a set of plants were supplemented with B (1 mM boric acid). Control treatments were grown without salt. Mean (±SD) was computed from three replicates for each treatment. Different letters over the bars indicate significant differences at $p \leq 0.05$ applying Tukey's HSD test.

3.2. Physiological Parameters

Chlorophyll breakdown is common under salt stress, and the Chl content was decreased at both levels of salt stress compared with the unstressed controls. The addition of B to salt-treated seedlings increased the Chl level by 21% and 28% in the mild and severe salinity treatments, respectively, but these differences were not statistically significant compared to the salt-stressed seedlings without B supplementation (Figure 3A).

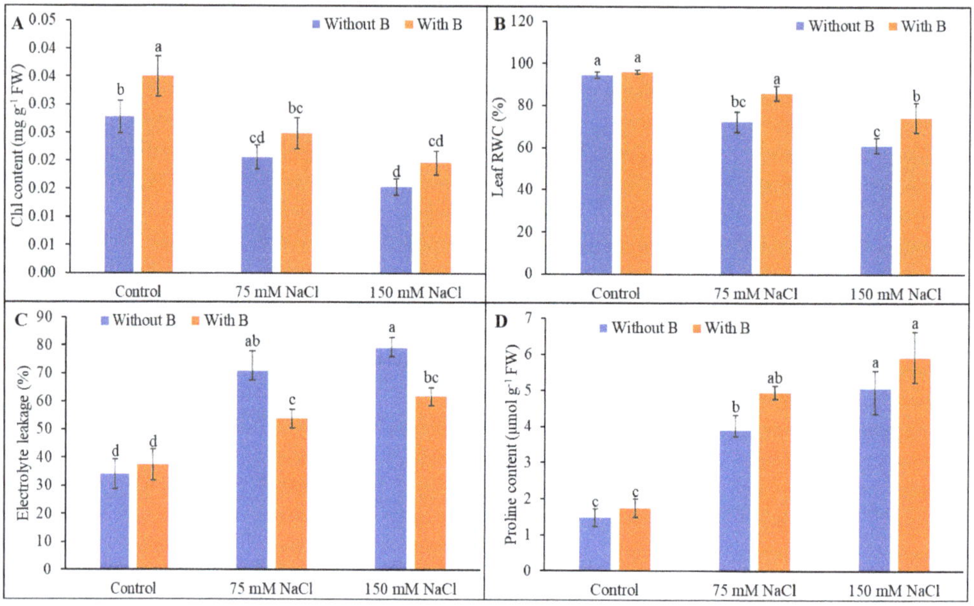

Figure 3. Chl content (**A**), leaf relative water content (**B**), electrolyte leakage (**C**) and proline content (**D**) of soybean as affected by salt stress with or without B supplementation. Twenty-day-old plants were imposed with two levels of salt (75 and 150 mM NaCl, 30 days) and a set of plants were supplemented with B (1 mM boric acid). Control treatments were grown without salt. Mean (±SD) was computed from three replicates for each treatment. Different letters over the bars indicate significant differences at $p \leq 0.05$ applying Tukey's HSD test.

Leaf RWC was decreased by salt treatment, but the relative water content was increased by 19% and 22% by B supplementation in plants treated with mild and severe salt stress (Figure 3B). Electrolyte leakage is a recognizable indicator of oxidative damage, and soybean plants showed a drastic rise in electrolyte leakage of 107% and 131% when subjected to mild and severe salinity treatments, respectively, compared to the unstressed controls. Supplementation with B noticeably diminished electrolyte leakage by 24% and 22% in the mild and severe salinity conditions, respectively, compared to salt stress alone (Figure 3C).

The Pro content increased upon exposure to salinity, and the level was further increased by B addition, but that increase was not statistically significant (Figure 3D).

3.3. *Malondialdehyde and H_2O_2 Levels*

Malondialdehyde content is a proxy for membrane lipid peroxidation and was increased by salt stress. The H_2O_2 levels showed a similar increase in salt-treated soybean seedlings. B addition decreased the MDA level by 38% (under severe salinity stress) and decreased the H_2O_2 level by 29% and 30% under mild and severe salt stress, respectively, compared to salt-treated plants without B supplementation (Figure 4).

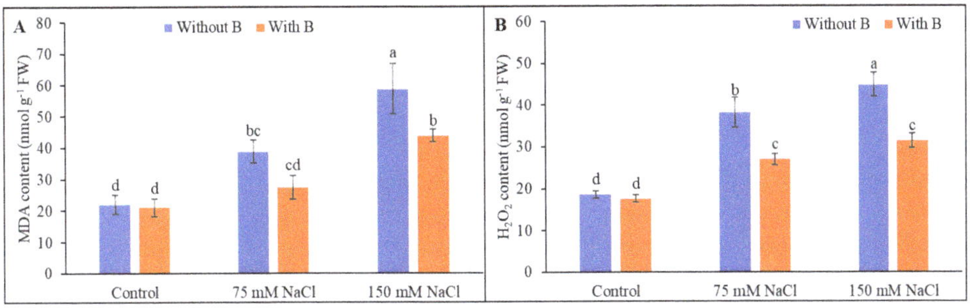

Figure 4. MDA (**A**) and H_2O_2 content (**B**) of soybean as affected by salt stress with or without B supplementation. Twenty-day-old plants were imposed with two levels of salt (75 and 150 mM NaCl, 30 days) and a set of plants were supplemented with B (1 mM boric acid). Control treatments were grown without salt. Mean (±SD) was computed from three replicates for each treatment. Different letters over the bars indicate significant differences at $p \leq 0.05$ applying Tukey's HSD test.

3.4. Ascorbate and Glutathione Pool

The AsA content was increased by exposure to mild salt stress but decreased under severe salt stress. The DHA level increased at both levels of salinity. Together, these changes resulted in a decrease in the AsA/DHA ratio in salt-treated soybean seedlings. However, the AsA/DHA ratio increased after B supplementation by 39% and 45% under mild and severe salt stress, respectively, largely due to the increase in AsA and decrease in DHA levels following B addition (Figure 5A–C).

The glutathione level was unaltered by salt treatment. The GSSG level was greatly increased by salt treatment and increased with the increase in the salinity level. The ratio of GSH/GSSG decreased in salt-treated soybean plants. B supplementation augmented the GSH level under mild and severe salt stress by 26% and 49%, respectively, and diminished GSSG by 22% and 28%, respectively, compared to salt-stressed plants not supplemented with B. These changes explained the maintenance of the GSH/GSSG ratio by B in the salt-treated plants (Figure 5D–F).

3.5. Activity of AsA-GSH Pathway Enzymes

The AsA metabolic and recycling enzymes differentially affected under salt stress. APX activity decreased under severe salt stress. The MDHAR and DHAR activity decreased under salt stress compared to unstressed controls. Boron supplementation increased the APX level under mild salinity but not under severe salinity. The MDHAR and DHAR activity was increased by B supplementation of salt-treated plants (Figure 6B,C). Glutathione reductase activity was unchanged by mild salt stress but was increased by severe salt stress compared to unstressed controls. Addition of B to salt-stressed plants increased GR activity compared to salt-stressed plants without B supplementation (Figure 6D).

3.6. Methylglyoxal Detoxification System

Methylglyoxal levels were increased strongly, by 36% and 118% following exposure to mild and severe salt stress, respectively, compared to unstressed controls. The methylglyoxal level was decreased by 21% in severely stressed plants following B supplementation. Glyoxalase I and Gly II activities were decreased by mild salt stress and further decreased by severe salt stress. Glyoxalase I and Gly II activities were increased in salt-stressed plants supplemented with B compared to salt-stressed plants without B supplementation (Figure 7).

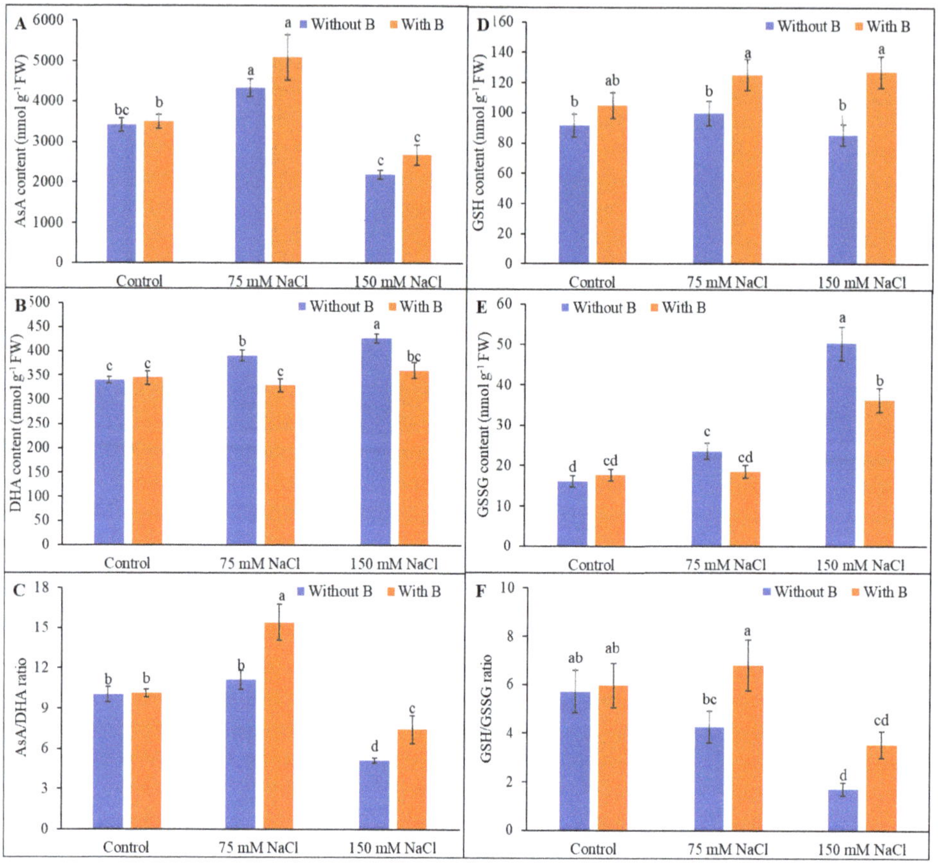

Figure 5. Ascorbate (**A–C**) and glutathione (**D–F**) pool of soybean as affected by salt stress with or without B supplementation. Twenty-day-old plants were imposed with two levels of salt (75 and 150 mM NaCl, 30 days) and a set of plants were supplemented with B (1 mM boric acid). Control treatments were grown without salt. Mean (±SD) was computed from three replicates for each treatment. Different letters over the bars indicate significant differences at $p \leq 0.05$ applying Tukey's HSD test.

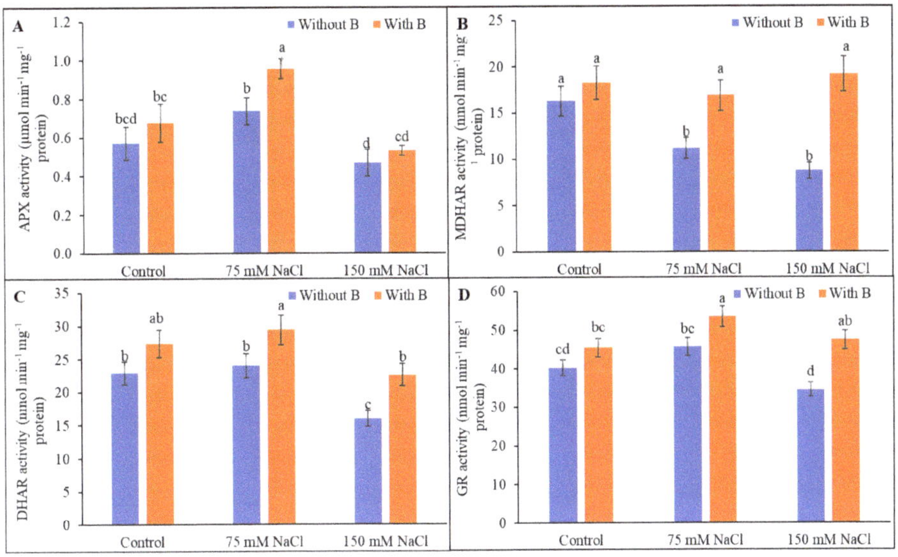

Figure 6. Activities of AsA-GSH pathway enzymes, APX (**A**), MDHR (**B**), DHAR (**C**), and GR (**D**) of soybean as affected by salt stress with or without B supplementation. Twenty-day-old plants were imposed with two levels of salt (75 and 150 mM NaCl, 30 days) and a set of plants were supplemented with B (1 mM boric acid). Control treatments were grown without salt. Mean (±SD) was computed from three replicates for each treatment. Different letters over the bars indicate significant differences at $p \leq 0.05$ applying Tukey's HSD test.

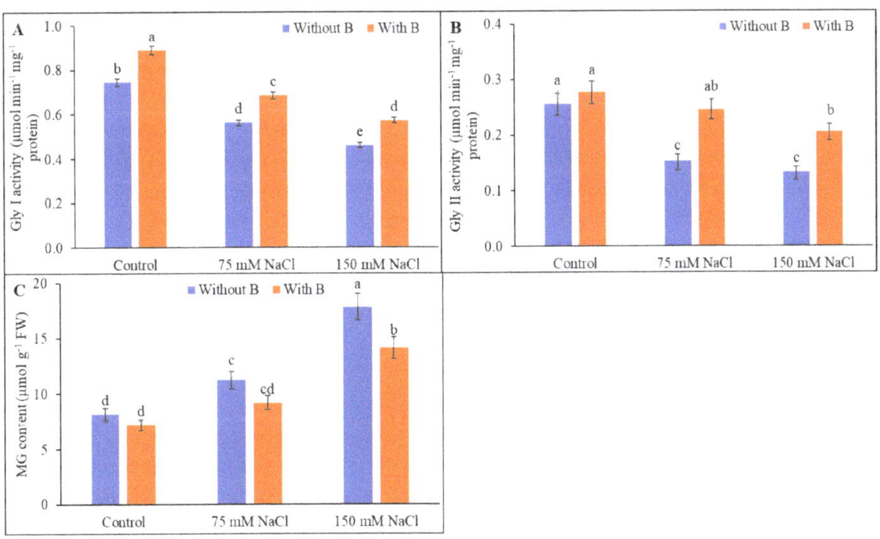

Figure 7. Activities of glyoxalase I (**A**), glyoxalase II (**B**), and level of MG (**C**) of soybean as affected by salt stress with or without B supplementation. Twenty-day-old plants were imposed with two levels of salt (75 and 150 mM NaCl, 30 days) and a set of plants were supplemented with B (1 mM boric acid). Control treatments were grown without salt. Mean (±SD) was computed from three replicates for each treatment. Different letters over the bars indicate significant differences at $p \leq 0.05$ applying Tukey's HSD test.

4. Discussion

Growth inhibition is the most common deleterious effect of salt exposure. The aim of the present study was to determine the various factors that contribute to this inhibition. Plants growing in saline conditions experience a reduction in the rate of water absorption due to altered osmotic potential. This is accompanied by a lower uptake of essential mineral nutrients, in part due to competition with Na [4]. Poor root growth under salt stress further hinders water and nutrient uptake, as well as transport. Photosynthesis is also hindered under salinity [5]. These alterations are the common reasons for salt-induced growth reduction [4,31].

Supplementation with B increased the shoot fresh weight of salt-treated plants. Several other studies have also demonstrated a B-induced improvement in growth parameters of plants growing under saline conditions [3,16]. However, the differences observed in the growth parameters of salt-stressed soybean plants in response to B application were not statistically significant in the present study. Further studies with different doses of B are needed to confirm the potential for a growth-enhancing role of B under salt stress in soybean.

Salt stress imposes oxidative stress that causes Chl pigment breakdown. Chl biosynthesis is also hampered during salt exposure due to nutrient imbalance and metabolic disruptions [3,4,11]. In the present study, salt stress reduced the Chl content in the soybean plants, but B supplementation did not mitigate the Chl losses significantly. Boron plays a role in cytoskeletal protein construction and nitrogen metabolism, which are both required for Chl synthesis [12–14]. Restoration of Chl levels has been reported in salt-stressed rose [16] and potato [3] plants following B application. Therefore, restoration of the Chl content in soybean may depend on the dosage, duration, or application method. Future studies should examine these possibilities to determine if better results can be achieved.

Pro accumulation and reduced RWC indicate the induction of osmotic stress in soybean plants under salt stress. A higher accumulation of Pro and other osmolytes is desirable to combat the osmotic stress induced by any environmental stress [4]. Increased Pro biosynthesis and accumulation have been reported in response to B in previous studies, which also showed improvements in plant water status [3,32]. Boron is involved in nitrogen metabolism and in the synthesis of some of the secondary metabolites connected to the biosynthesis of osmoprotectants such as Pro [14]. In the present study, B supplementation of salt-treated plants increased RWC but had no effect on Pro levels. This finding might reflect that plants have a number of endogenous osmoprotectant molecules that can regulate the water content of plants under stress conditions. Boron might, therefore, have a role in regulating the levels of osmoprotectants other than Pro and/or in regulating other physiological attributes that can improve the RWC of salt-stressed plants. Further investigation is needed to better understand the effects of B on osmoregulation.

Salinity causes osmotic stress that, in turn, creates oxidative stress and inhibits biomembrane functions. Ion toxicity also causes oxidative damage to biomembranes. Inhibition of antioxidant enzyme activity under salt stress increases ROS production and causes oxidative damage [10,33]. Soybean plants exposed to salinity showed increased generation of H_2O_2, which then damaged the membranes, as indicated by increased MDA levels. Electrolyte leakage was also higher during salt exposure, further confirming membrane damage. These common responses to oxidative stress have been documented in many different plant species during salt exposure [4,8,34]. In the present study, exogenous B supplementation decreased the salinity-induced oxidative stress, as confirmed by the decreased levels of H_2O_2, MDA, and electrolyte leakage from the salt-treated soybean plants.

Previous studies of B effects on salt stress responses have verified a role for B in oxidative stress mitigation. Decreases in lipoxygenase activity, electrolyte leakage, and MDA and H_2O_2 levels by B were reported in salt-stressed *Pistacia vera* leaves [7]. The MDA and H_2O_2 contents were diminished by B application in aluminum-stressed plants [35,36]. Downregulation of H_2O_2 production, as well as membrane lipid peroxidation, by B supplementation was confirmed in citrus plants under aluminum toxicity stress [9].

Salt stress generates a series of damaging secondary effects, including altered biosynthesis of metabolites and inhibition of enzyme activities [3,10,11,33]. In the present study, salt stress altered AsA and DHA levels, thereby decreasing the AsA/DHA ratio, whereas B supplementation reversed the damaging effects and restored the AsA/DHA ratio. Ascorbate peroxidase activity was decreased by severe salinity (but was unaltered under mild salt stress), and MDHAR and DHAR activities were decreased in salt-treated soybean plants, and these decreases were correlated with the decreases in AsA levels and the AsA/DHA ratio. Boron supplementation of the salt-treated soybean plants increased the MDHAR and DHAR activity and helped to restore the AsA level and decrease the DHA level, thereby increasing the AsA/DHA ratio above that observed in salt-stressed plants.

The GSH level was not altered in salt-treated soybean plants, but the GSSG level was greatly increased, resulting in a strongly decreased GSH/GSSG ratio under saline conditions compared to unstressed plants. B supplementation of the salt-stressed plants raised the GSH level and decreased the GSSG level, thereby increasing the GSH/GSSG ratio above that of salt-stressed plants. Glutathione reductase activity in salt-treated plants was increased by B supplementation, providing further improvement in the GSG/GSSG ratio.

Research that focuses on a potential regulatory role for B in antioxidant defense systems is sparse. The application of Bio-B fertilizer was shown to ameliorate freezing injury in grapevine through augmentation of activities of antioxidant enzymes, such as CAT, peroxidase (POD), and SOD [35]. The activity of APX and CAT and the levels of nonenzymatic antioxidant compounds, such as AsA and phenolic compounds, were enhanced considerably in *Pistacia vera* by B addition, and the trees showed better salinity tolerance following B supplementation [7]. Citrus plants under aluminum toxicity showed responses to B supplementation that included upregulated SOD activity and reduced activities of peroxidase, CAT, and polyphenol oxidase, as well as lower protein and proline levels [9]. Riaz et al. [15,36] confirmed that B supply decreased aluminum toxicity responses in trifoliate orange and increased the levels of antioxidant system components, including the activities of peroxidase, CAT, APX, phenylalanine ammonia lyase, and polyphenol oxidase, and the accumulation of Pro and secondary metabolites, while also improving components of the cell wall [36].

MG is produced through various abiotic stresses and has toxic effects on cell metabolism. Scavenging of MG through the glyoxalase system is a prerequisite for improving salt tolerance [4,8]. Gly I and Gly II activities were diminished and MG levels were increased in salt-treated soybean plants, and B application reversed these effects, leading to decreased MG accumulation. An increase in glyoxalase enzyme activity was also shown to decrease the MG content in salt-affected mung bean [4] and tomato (Parvin et al., 2020), in agreement with the present findings. However, a role for B in the maintenance of the glyoxalase system still requires investigation.

5. Conclusions

The present study revealed that B supplementation improved water relations of the salt-affected plants. Exogenous B also enhanced many components of the plant antioxidant defense system, including GSH, GSH/GSSG, AsA/DHA, and activities of APX, MDHAR, DHAR, and GR, to mitigate oxidative damage. Overall, B supplementation decreased the MDA level and electrolyte leakage in salt-treated soybean plants, while stimulating the methylglyoxal detoxification process through upregulation of the components of the glyoxalase system, including GSH level and Gly I and Gly II activities. B is a recognized fertilizer micronutrient with vital physiological and developmental functions [12–14], but few studies have probed the protective role of B under different abiotic stresses [3,9,16,36]. Therefore, many aspects remain to be revealed regarding the protective role of B in plants exposed to environmental stress. Combined application of B with other trace elements and their physiological roles also warrants further research.

Author Contributions: Conceptualization, H.F.A., K.R.H., M.H.; methodology, K.N., M.H.; formal analysis, M.H.; investigation, H.F.A., K.R.H., K.N.; resources, H.S.A.-Z.; writing—original draft preparation, K.N., M.H.; writing—review and editing, H.F.A., M.H.; supervision, H.F.A., H.S.A.-Z.; project administration, H.F.A., H.S.A.-Z.; funding acquisition, H.F.A., H.S.A.-Z. All authors have read and agreed to the published version of the manuscript.

Funding: This project was funded by the Deanship of Scientific Research (DSR) at King Abdulaziz University, Jeddah, under Grant No. RG-20-130-40.

Institutional Review Board Statement: Not applicable.

Informed Consent Statement: Not applicable.

Data Availability Statement: All data are available in this manuscript.

Acknowledgments: This project was funded by the Deanship of Scientific Research (DSR) at King Abdulaziz University, Jeddah, under Grant No. RG-20-130-40. The authors, therefore, acknowledge with thanks DSR for technical and financial support.

Conflicts of Interest: The authors declare no conflict of interest.

References

1. Živanović, L.; Popović, V. Soybean (*Glycine max*) production in world and in Serbia. *Zbornik Radova* **2016**, *21*, 129–135.
2. Gondek, M.; Weindorf, D.; Thiel, C.; Kleinheinz, G. Soluble salts in compost and their effects on soil and plants: A review. *Compost Sci. Util.* **2020**, *28*, 59–75. [CrossRef]
3. Mahmoud, A.W.M.; Abdeldaym, E.A.; Abdelaziz, S.M.; El-Sawy, M.B.I.; Mottaleb, S.A. Synergetic effects of zinc, boron, silicon, and zeolite nanoparticles on confer tolerance in potato plants subjected to salinity. *Agronomy* **2020**, *10*, 19. [CrossRef]
4. Nahar, K.; Hasanuzzaman, M.; Alam, M.M.; Rahman, A.; Suzuki, T.; Fujita, M. Polyamines confer salt tolerance in mung bean by reducing sodium uptake, improving nutrient homeostasis, antioxidant defense and methylglyoxal detoxification systems. *Front. Plant Sci.* **2016**, *7*, 1104. [CrossRef] [PubMed]
5. Pan, T.; Liu, M.; Kreslavski, V.D.; Zharmukhamedov, S.K.; Nie, C.; Yu, M.; Kuznetsov, V.V.; Allakhverdiev, S.I.; Shabala, S. Non-stomatal limitation of photosynthesis by soil salinity. *Crit. Rev. Environ. Sci. Technol.* **2021**, *51*, 791–825. [CrossRef]
6. Zhao, C.; Zhang, H.; Song, C.; Zhu, J.K.; Shabala, S. Mechanisms of plant responses and adaptation to soil salinity. *Innovation* **2020**, *1*, 100017. [CrossRef]
7. Tavallali, V.; Karimi, S.; Espargham, O. Boron enhances antioxidative defense in the leaves of salt-affected *Pistacia vera* seedlings. *Hortic. J.* **2018**, *87*, 55–62. [CrossRef]
8. Parvin, K.; Nahar, K.; Hasanuzzaman, M.; Bhuyan, M.B.; Mohsin, S.M.; Fujita, M. Exogenous vanillic acid enhances salt tolerance of tomato: Insight into plant antioxidant defense and glyoxalase systems. *Plant Physiol. Biochem.* **2020**, *150*, 109–120. [CrossRef]
9. Yan, L.; Riaz, M.; Wu, X.; Du, C.; Liu, Y.; Lv, B.; Jiang, C. Boron inhibits aluminum-induced toxicity to citrus by stimulating antioxidant enzyme activity. *J. Environ. Sci. Health Part C* **2018**, *36*, 145–163. [CrossRef]
10. Hasanuzzaman, M.; Bhuyan, M.H.M.; Parvin, K.; Bhuiyan, T.F.; Anee, T.I.; Nahar, K.; Hossen, M.; Zulfiqar, F.; Alam, M.; Fujita, M. Regulation of ROS metabolism in plants under environmental stress: A review of recent experimental evidence. *Int. J. Mol. Sci.* **2020**, *21*, 8695. [CrossRef]
11. Ahanger, M.A.; Qin, C.; Begum, N.; Maodong, Q.; Dong, X.X.; El-Esawi, M.; El-Sheikh, M.A.; Alatar, A.A.; Zhang, L. Nitrogen availability prevents oxidative effects of salinity on wheat growth and photosynthesis by up-regulating the antioxidants and osmolytes metabolism, and secondary metabolite accumulation. *BMC Plant Biol.* **2019**, *19*, 479. [CrossRef] [PubMed]
12. Blevins, D.G.; Lukaszewski, K.M. Boron in plant structure and function. *Annu. Rev. Plant Phys.* **1998**, *49*, 481–500. [CrossRef] [PubMed]
13. Saenz, J.L. Boron fertilization-A key for success. *Vine Vin. View* **2001**, *17*, 1–12.
14. Saleem, M.; Khanif, Y.M.; Ishak, F.; Samsuri, A.W.; Hafeez, B. Importance of boron for agriculture productivity: A review. *Int. Res. J. Agric. Sci. Soil Sci.* **2011**, *1*, 293–300.
15. Riaz, M.; Yan, L.; Wu, X.; Hussain, S.; Aziz, O.; Wang, Y.; Imran, M.; Jiang, C. Boron alleviates the aluminum toxicity in trifoliate orange by regulating antioxidant defense system and reducing root cell injury. *J. Environ. Manag.* **2018**, *15*, 149–158. [CrossRef] [PubMed]
16. Yousefi, H.; Dalir, N.; Rahnemaie, R.; Babaei, A. The alleviation of salinity-induced stress by using boron in soilless grown rose. *J. Plant Nutr.* **2019**, *43*, 1–12. [CrossRef]
17. BARC. *Fertilizer Recommendation Guide*; Bangladesh Agricultural Research Council: Dhaka, Bangladesh, 2018.
18. Ahmad, S.; Ali, H.; Rehman, A.; Khan, R.J.Z.; Ahmed, W.; Fatima, Z.; Abbas, G.; Irfan, M.; Ali, H.; Khan, M.A.; et al. Measuring leaf area of winter cereals by different techniques: A comparison. *Pak. J. Life Soc. Sci.* **2015**, *13*, 117–125.

19. Zhu, J.; Tremblay, T.; Liang, Y. Comparing SPAD and at LEAF values for chlorophyll assessment in crop species. *Can. J. Soil Sci.* **2012**, *92*, 645–648. [CrossRef]
20. Barrs, H.D.; Weatherley, P.E. A re-examination of the relative turgidity technique for estimating water deficits in leaves. *Aust. J. Biol. Sci.* **1962**, *15*, 413–428. [CrossRef]
21. Dionisio-Sese, M.L.; Tobita, S. Antioxidant responses of rice seedlings to salinity stress. *Plant Sci.* **1998**, *135*, 1–9. [CrossRef]
22. Bates, L.S.; Waldren, R.P.; Teari, D. Rapid determination of free proline for water stress studies. *Plant Soil* **1973**, *39*, 205–207. [CrossRef]
23. Heath, R.L.; Packer, L. Photoperoxidation in isolated chloroplast. I. Kinetics and stoichiometry of fatty acid peroxidation. *Arch. Biochem. Biophys.* **1968**, *125*, 189–198. [CrossRef]
24. Yu, C.W.; Murphy, T.M.; Lin, C.H. Hydrogen peroxide-induces chilling tolerance in mung beans mediated through ABA-independent glutathione accumulation. *Funct. Plant Biol.* **2003**, *30*, 955–963. [CrossRef] [PubMed]
25. Hasanuzzaman, M.; Hossain, M.A.; Fujita, M. Nitric oxide modulates antioxidant defense and the methylglyoxal detoxification system and reduces salinity-induced damage of wheat seedlings. *Plant Biotechnol. Rep.* **2011**, *5*, 353–365. [CrossRef]
26. Huang, C.; He, W.; Guo, J.; Chang, X.; Su, P.; Zhang, L. Increased sensitivity to salt stress in ascorbate-deficient *Arabidopsis* mutant. *J. Exp. Bot.* **2005**, *56*, 3041–3049. [CrossRef] [PubMed]
27. Paradiso, A.; Berardino, R.; de Pinto, M.; di Toppi, L.S.; Storelli, F.T.; de Gara, L. Increase in ascorbate–glutathione metabolism as local and precocious systemic responses induced by cadmium in durum wheat plants. *Plant Cell Physiol.* **2008**, *49*, 362–374. [CrossRef]
28. Bradford, M. A rapid and sensitive method for the quantitation of microgram quantities of protein utilizing the principle of protein–dye binding. *Anal. Biochem.* **1976**, *72*, 248–254. [CrossRef]
29. Nakano, Y.; Asada, K. Hydrogen peroxide is scavenged by ascorbate-specific peroxidase in spinach chloroplasts. *Plant Cell Physiol.* **1981**, *22*, 867–880.
30. *CoStat-Statistics Software Version 6.400*; CoHort Software: Monterey, CA, USA, 2008.
31. Naveed, M.; Sajid, H.; Mustafa, A.; Niamat, B.; Ahmad, Z.; Yaseen, M.; Kamran, M.; Rafique, M.; Ahmar, S.; Chen, J.T. Alleviation of salinity-induced oxidative stress, improvement in growth, physiology and mineral nutrition of canola (*Brassica napus* L.) through calcium-fortified composted animal manure. *Sustainability* **2020**, *12*, 846. [CrossRef]
32. Salim, Q.T.; Joody, A.T. Effect of silicon calcium, boron on Proline and relative water contents in apple leaves. *IOP Conf. Ser. Mat. Sci. Eng.* **2018**, *454*, 012179. [CrossRef]
33. Nahar, K.; Hasanuzzaman, M.; Fujita, M. Roles of osmolytes in plants adaptation to drought and salinity. In *Osmolytes and Plants Acclimation to Changing Environment: Emerging Omics Technologies*; Iqbal, N., Nazar, R., Khan, N.A., Eds.; Springer: New York, NY, USA, 2016; pp. 37–68. [CrossRef]
34. Hasanuzzaman, M.; Inafuku, M.; Nahar, K.; Fujita, M.; Oku, H. Nitric oxide regulates plant growth, physiology, antioxidant defense, and ion homeostasis to confer salt tolerance in the mangrove species, *Kandelia obovata*. *Antioxidants* **2021**, *10*, 611. [CrossRef] [PubMed]
35. Köse, C.; Güneş, A.; Kaya, Ö.; Kıtır, N.; Turan, M.; Sahin, F. Freeze injure and antioxidant enzyme activity of grapevine (*Vitis vinifera*) under bio-Boron fertilizer applications. *Erwerbs-Obstbau* **2018**, *60*, 3–10. [CrossRef]
36. Riaz, M.; Yan, L.; Wu, X.; Hussain, S.; Aziz, O.; Jiang, C. Boron supply maintains efficient antioxidant system, cell wall components and reduces aluminum concentration in roots of trifoliate orange. *Plant Physiol. Biochem.* **2019**, *137*, 93–101. [CrossRef] [PubMed]

plants

MDPI

Article

Biochar Amendments Improve Licorice (*Glycyrrhiza uralensis* Fisch.) Growth and Nutrient Uptake under Salt Stress

Dilfuza Egamberdieva [1,2,*], Hua Ma [3,*], Burak Alaylar [4], Zohreh Zoghi [5], Aida Kistaubayeva [2], Stephan Wirth [1] and Sonoko Dorothea Bellingrath-Kimura [1,6]

1 Leibniz Centre for Agricultural Landscape Research (ZALF), 15374 Müncheberg, Germany; swirth@zalf.de (S.W.); belks@zalf.de (S.D.B.-K.)
2 Department of Biotechnology, Al-Farabi Kazakh National University, Almaty 050040, Kazakhstan; kistaubayeva.kaznu@gmail.com
3 School of Life Sciences, Chongqing University, Chongqing 401331, China
4 Department of Molecular Biology and Genetics, Faculty of Arts and Sciences, Agri Ibrahim Cecen University, Agri 04100, Turkey; balaylar@agri.edu.tr
5 Faculty of Natural Resources and Marine Sciences, Tarbiat Modares University, Noor 46417-76489, Iran; zohre_zoghi@yahoo.com
6 Faculty of Life Science, Humboldt University of Berlin, 14195 Berlin, Germany
* Correspondence: Dilfuza.egamberdieva@zalf.de (D.E.); mh3660344@126.com (H.M.)

Citation: Egamberdieva, D.; Ma, H.; Alaylar, B.; Zoghi, Z.; Kistaubayeva, A.; Wirth, S.; Bellingrath-Kimura, S.D. Biochar Amendments Improve Licorice (*Glycyrrhiza uralensis* Fisch.) Growth and Nutrient Uptake under Salt Stress. *Plants* **2021**, *10*, 2135. https://doi.org/10.3390/plants10102135

Academic Editor: Milan S. Stankovic

Received: 10 September 2021
Accepted: 4 October 2021
Published: 8 October 2021

Publisher's Note: MDPI stays neutral with regard to jurisdictional claims in published maps and institutional affiliations.

Copyright: © 2021 by the authors. Licensee MDPI, Basel, Switzerland. This article is an open access article distributed under the terms and conditions of the Creative Commons Attribution (CC BY) license (https://creativecommons.org/licenses/by/4.0/).

Abstract: Licorice (*Glycyrrhiza uralensis* Fisch.) is a salt and drought tolerant legume suitable for rehabilitating abandoned saline lands, especially in dry arid regions. We hypothesized that soil amended with maize-derived biochar might alleviate salt stress in licorice by improving its growth, nutrient acquisition, and root system adaptation. Experiments were designed to determine the effect of different biochar concentrations on licorice growth parameters, acquisition of C (carbon), nitrogen (N), and phosphorus (P) and on soil enzyme activities under saline and non-saline soil conditions. Pyrolysis char from maize (600 °C) was used at concentrations of 2% (B2), 4% (B4), and 6% (B6) for pot experiments. After 40 days, biochar improved the shoot and root biomass of licorice by 80 and 41% under saline soil conditions. However, B4 and B6 did not have a significant effect on shoot growth. Furthermore, increased nodule numbers of licorice grown at B4 amendment were observed under both non-saline and saline conditions. The root architectural traits, such as root length, surface area, project area, root volume, and nodulation traits, also significantly increased by biochar application at both B2 and B4. The concentrations of N and K in plant tissue increased under B2 and B4 amendments compared to the plants grown without biochar application. Moreover, the soil under saline conditions amended with biochar showed a positive effect on the activities of soil fluorescein diacetate hydrolase, proteases, and acid phosphomonoesterases. Overall, this study demonstrated the beneficial effects of maize-derived biochar on growth and nutrient uptake of licorice under saline soil conditions by improving nodule formation and root architecture, as well as soil enzyme activity.

Keywords: biochar; licorice; soil enzymes; salinity; nutrients; root system

1. Introduction

Licorice is a perennial shrub widely distributed in South-Western Asia and the Mediterranean region, which consists of about 20 species. The most commonly grown species are *Glycyrrhiza glabra* L. and *Glycyrrhiza uralensis* Fisch., used as food and in traditional medicine to treat various health disorders [1–3]. The plant is well adapted to saline and arid soils because of its deep root system. Thus, licorice is used to restore soil fertility by increasing soil organic matter and soil biological activity [4,5]. However, exposure of licorice to excessive salinity may elicit nutrient deficiency and imbalances of N, K, P, and microelements, which are essential for plant growth [6,7]. Furthermore, salt stress disturbs the symbiotic performance of legumes with *Rhizobia*, thereby resulting in decreased nodulation and nitrogen fixation [8]. This process was explained by the failure of rhizobial colonization in the rhizosphere, and the formation of root nodules. There are several approaches to improving plant growth and development under drought and salinity, such as genetic engineering and the application of microbial inoculants [9]. Biochar amendment has also been repeatedly reported as an effective approach in restoring saline lands and increasing plant tolerance to salt stress; thus, it has gained attention in practical applications worldwide [6,10,11]. Biochar is a solid by-product of biomass-derived from forestry, agriculture, and the food industry, such as wood chips, crop residues, sewage sludge, or dairy manure. It is produced by pyrolysis or high temperatures under limited or complete absence of oxygen [12]. The positive effects of biochar on soil cation exchange capacity [13], water holding capacity [14] or soil organic matter content [15] were repeatedly demonstrated. Biochar also increased organic carbon in saline soil, supporting favourable conditions for soil microbes involved in nutrient cycles [16]. Tang et al. [17] also observed biochar-induced improvements in soil physical-chemical properties, which mitigated salt stress during seedling growth of *Brassica chinensis* L. Moreover, the soil under saline conditions amended with biochar experienced increased enzymatic activities such as of urease, invertase and phosphatase [18].

Biochar also plays a vital role in plant growth, providing nutrients and better nutrient availability [19]. The improved acquisition of nutrients such as N, P, K from soil amended with biochar was explained by enhanced bioavailability and increased microbial activity involved in nutrient cycling [20–22]. However, plant growth, nutrient acquisition, and soil biogeochemical processes after biochar application depend on the type and concentration of biochar. For example, Ibrahim et al. [23] observed that soil amendment with 2.5% biochar alleviated the harmful effects of salt stress in Sorghum (*Sorghum bicolor* L. Moench), whereas 5 and 10% biochar application rates harmed the plant growth under saline conditions. Our study hypothesized that the application of biochar mitigates the salt stress of licorice, improves soil nutrient acquisition through improving soil biological properties and root growth, and finally, that any beneficial effects depend on biochar concentration. Thus, our study could expand knowledge about the impact of increasing biochar concentrations on the improvement in the licorice root system (growth and architecture), symbiotic performance and nutrient availability, especially in salt-affected soils. Experiments were conducted in a greenhouse and included measurements of soil enzyme activities linked to carbon, nitrogen and phosphorus cycling.

2. Results

2.1. Plant Shoot and Root Growth

The licorice's root and the shoot biomass responded differently to the applied biochar concentrations of B2, B4 and B6 and soil salt stress. In non-saline soil amended with B2, the shoot and root growth of licorice significantly ($p < 0.05$) increased by 25 and 28% compared to plants grown in soil without biochar addition (Figures 1 and 2). However, there were no significant effects of B4 and B6 on licorice growth. The shoot and root biomass of plants even decreased after the amendment of B6 (Figure 1). Under saline soil conditions, biochar improved shoot and root biomass of licorice by 80 and 41%, respectively; however, B4 and B6 had no significant effects on shoot growth (Figure 1).

Notably, biochar improved the symbiotic performance of licorice, since the nodule number of plants grown in non-saline soil was 3.0 ± 1.0. In contrast, nodule numbers significantly increased threefold (8.6 ± 1.5) after the addition of B2 to the soil and twofold (6.0 ± 2.1) per plant at B4. Soil salinity inhibited nodule formation in the plant; no nodules were found on roots grown in saline soil without biochar addition. However, the soil which was amended with B2 and B4 increased nodule numbers to 4.1 ± 1.5 and 3.2 ± 1.0 per plant.

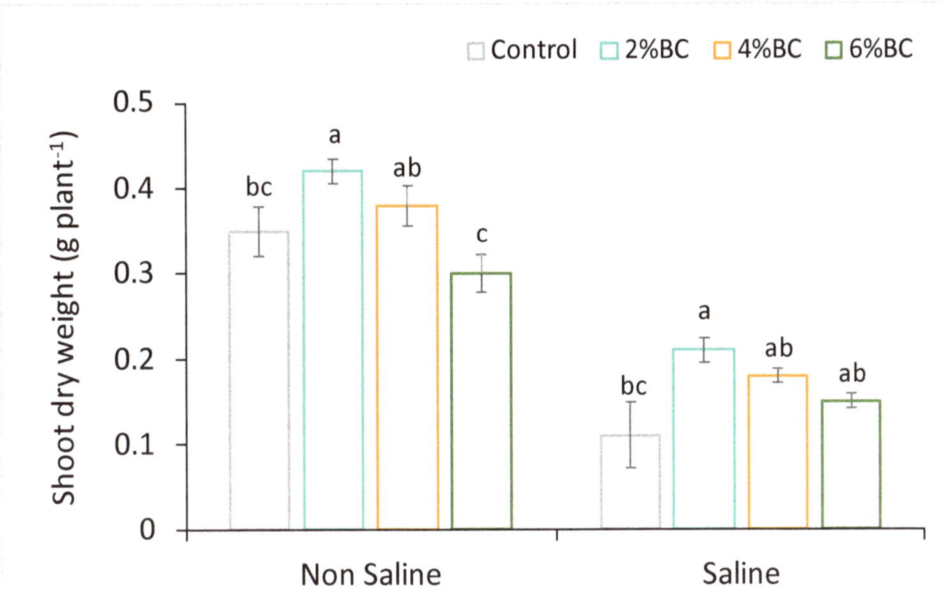

(A)

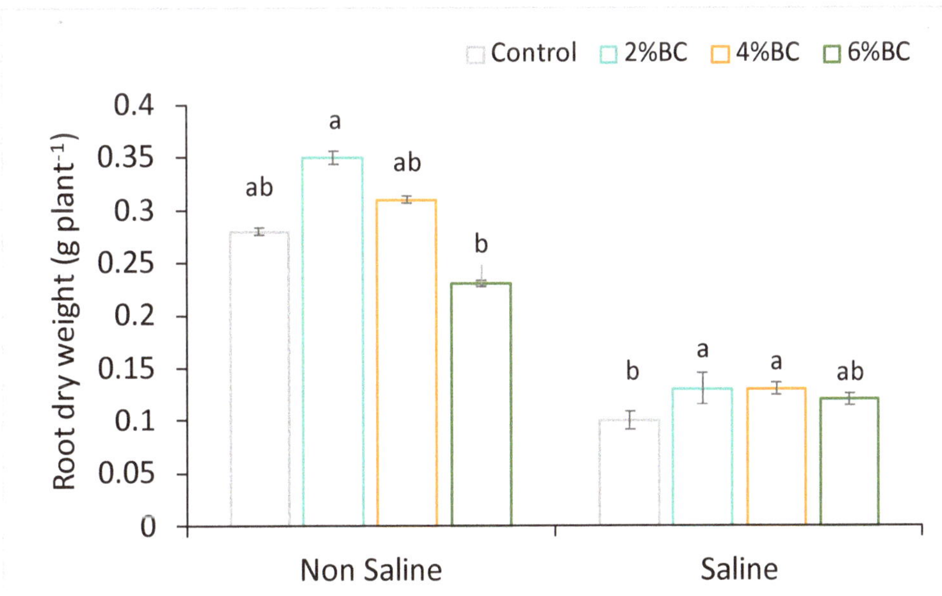

(B)

Figure 1. Shoot (**A**) and root (**B**) growth of licorice grown in soil amended with biochar at 2%, 4% and 6% concentrations under non-saline and saline soil conditions. Column means with different letters are significantly different based on Tukey's HSD test at $p < 0.05$.

Figure 2. Biochar addition (2%) enhanced whole-plant growth under saline soil conditions. (**A**): entire plants, (**B**): amplified root system of biochar-treated variant, (**C**): amplified root system of control plant.

2.2. Carbon and Nutrient Concentrations

The concentrations of C, N, and P in plant tissue were affected by different biochar concentrations applied under non-saline and saline soil conditions. In the biochar-amended soil, the nutrient concentration in the licorice plant tissues was higher than the plants grown in soil without the addition of biochar. The carbon content in the root tissue of licorice grown in non-saline soil amended with B2 slightly increased by 9% and by 7% under saline soil conditions, respectively (Figure 3A). Significant increases ($p < 0.05$) in N content of plant tissue over the controls were observed after B2 and B4 amendments under non-saline conditions, being 44 and 32% higher, respectively (Figure 3B). Under saline conditions, the soil amended with B2 and B4 showed an increased concentration of plant N content by 37 and 26%, respectively (Figure 3). Among other nutrients analyzed, only the P concentration of plants significantly ($p < 0.05$) increased after the amendment of B2 under non-saline conditions (Figure 3C).

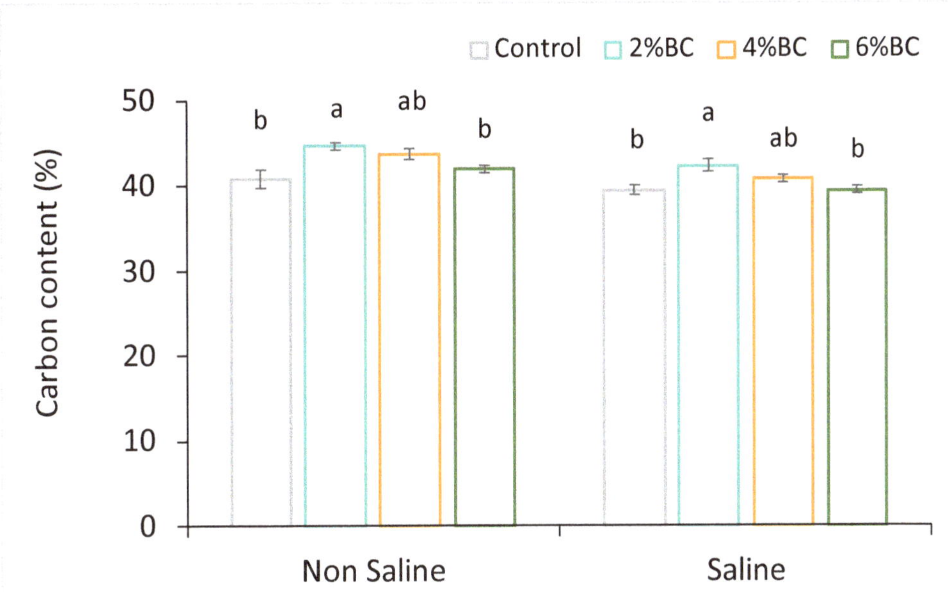

(**A**)

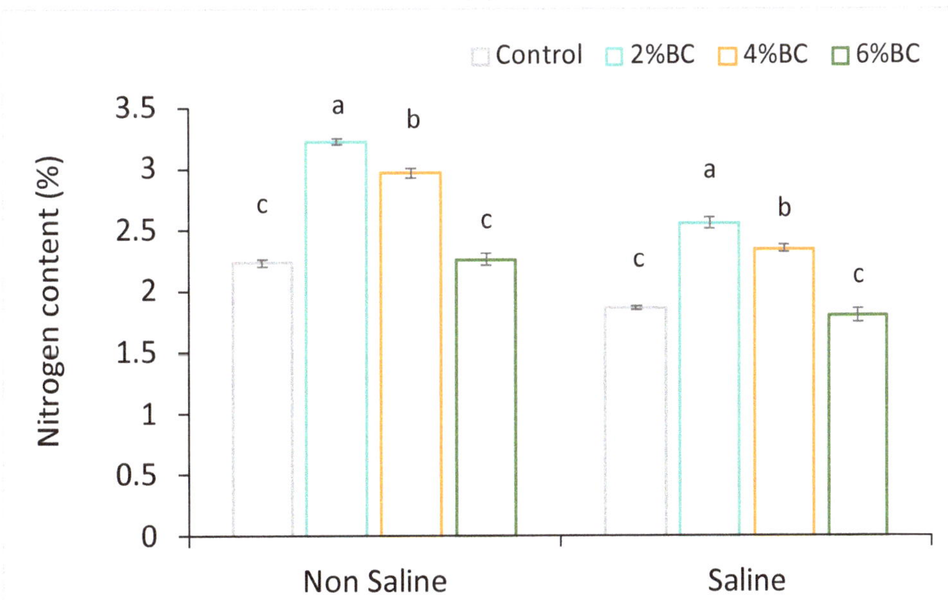

(**B**)

Figure 3. *Cont.*

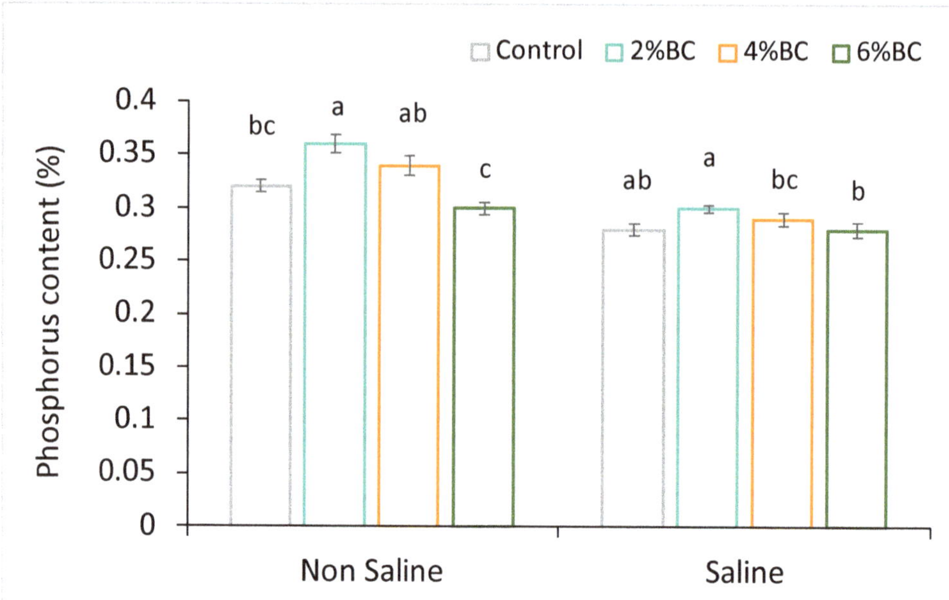

(C)

Figure 3. Concentrations (%) of carbon (**A**), nitrogen (**B**), and phosphorus (**C**) in licorice plant tissue grown after application of biochar at 2%, 4%, and 6% concentrations under non-saline and saline soil conditions. The different letters indicate significant differences based on Turkey's HSD test at $p < 0.05$.

2.3. Licorice Root Architecture

In general, licorice root architecture showed a differentiated response to different concentrations of biochar amendment. Interestingly, the total root length significantly increased by 26 and 57% under the B2 amendment under non-saline and saline conditions, respectively, as compared to the other treatments (Figure 4A). Root length similarly responded under the amendment of B4. The root length increased at both saline and non-saline soil conditions, although significant differences compared to the control were only observed under saline soil conditions (Figure 4A). Soil amendment changed the root project area, especially when grown under saline conditions (Figure 4B). The root project area significantly increased by 29–32% under non-saline and 37–48% under saline soil conditions compared to the control plant. By comparison, the surface area of the root significantly increased by 31% and 25% under non-saline and 51, and 36% under saline soil conditions compared to the control plants (Figure 4C). As shown in Figure 4D, the root diameter was not affected by biochar additions, except under B4, where an increase of 12% was observed under non-saline soil conditions (Figure 4D). The root volume responded differently to both biochar concentrations and soil conditions. Both B2 and B4 induced up to 52 and 43% higher root volumes under non-saline soil conditions than the control plants (Figure 4E). The root volume responded similarly to the addition of biochar under saline soil conditions, as it was increased by 73% under B2 amendment compared to the control plants (Figure 4E). There was no clear response from the root tips to different biochar additions, as they were significantly (45%) higher after B2 amendment compared to control plants under saline soil conditions only.

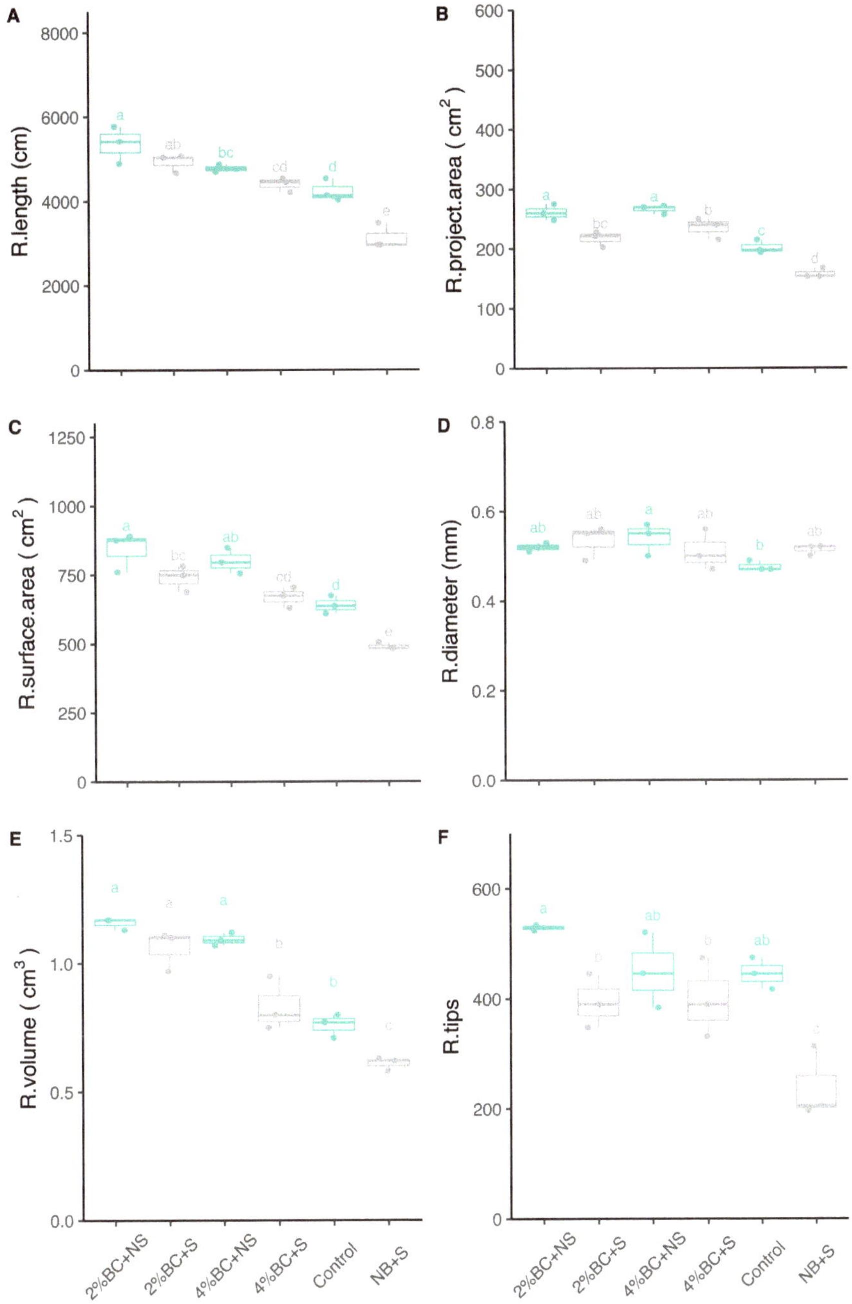

Figure 4. Root morphological traits, i.e., total root length (**A**), project area (**B**), surface area (**C**), root diameter (**D**), root volume (**E**), and root tips (**F**) of licorice grown after application of biochar at 2%, and 4% concentrations under non-saline and saline soil condition. The top and bottom of the box represent 75% and 25% quantile, respectively. The bars of the box represent maximum and minimum values of observations. The line in the box represents the median. The dot represents the value of each individual observation. Letters within each column mark significant differences at $p < 0.05$ based on Duncan's test. BC: biochar, NB: without biochar, S: saline, NS: non-saline, R.: root.

2.4. Soil Enzymes

The biochar amendment of B2 significantly increased fluorescein diacetate (FDA) hydrolase activity in non-saline and saline soil by 45 and 42%, respectively (Figure 5A); however, the activity decreased at the addition of B4 and B6 under both saline and non-saline conditions. The biochar amendments affected the acid and alkaline phosphomonoesterase activities under licorice (Figure 5B). An increase in acid phosphomonoesterase activity by 10 and 25% in non-saline soil and 13 and 32% in saline soil was observed after B2 and B4 amendments compared to the control, respectively. However, the effect was not significant (Figure 5C). In contrast, alkaline phosphomonoesterase activity decreased in soil amended with biochar at all of the applied concentrations (B2, B4, B6). Protease activity significantly increased after biochar additions (Figure 5D), i.e., an increase of 70 and 48% was observed in non-saline and of 96 and 96% in saline soil amended with B4 and B6 compared to the control, respectively (Figure 5D).

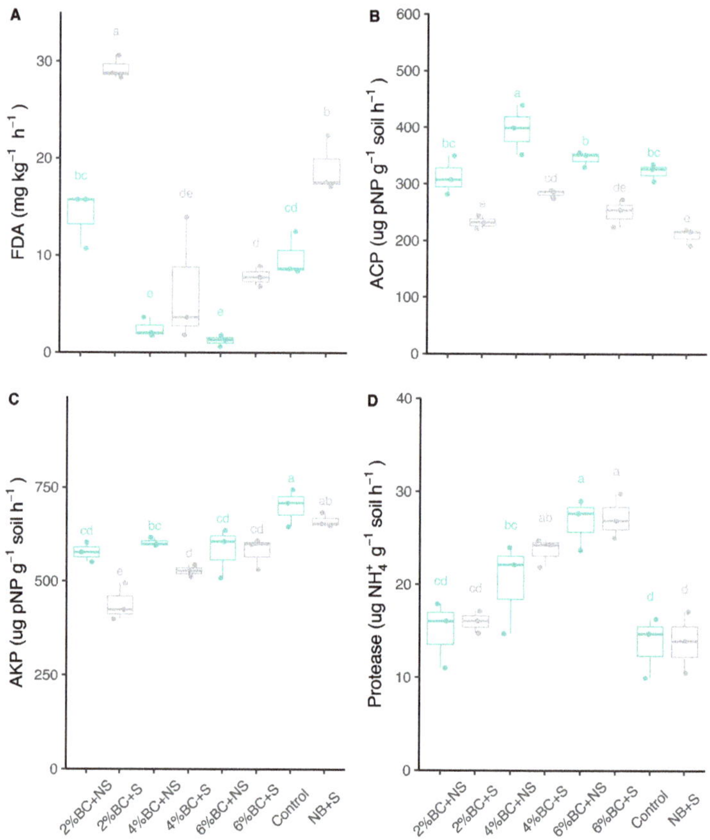

Figure 5. Effect of biochar amendment on FDA hydrolytic activity (**A**), soil acidic phosphomonoesterase activity (**B**), soil alkaline phosphomonoesterase activity (**C**), and soil protease activity (**D**) under non-saline and saline soil conditions. The top and bottom of the box represent 75% and 25% quantiles, respectively. The bars of the box represent maximum and minimum values of observations. The line in the box represents the median. The dot represents the value of each individual observation. Letters within each column mark significant differences at $p < 0.05$ based on Duncan's test. ACP: acidic phosphomonoesterase activity, AKP: alkaline phosphomonoesterase activity, FDA: fluorescein diacetate hydrolytic activity, Protease: protease activity. BC: biochar, NB: without biochar, S: saline, NS: non-saline.

2.5. Correlations and Redundancy Analysis

A cluster map of correlations was plotted to explore the relationship between measurements and retrace cause-effect dependencies, and a RDA was performed. The cluster map of correlations indicates four groups related to root morphological measurements, including root project area, root surface area, and root length. The root volume and tips were clustered with nodule numbers in one group (Figure 6) and significantly and strongly correlated.

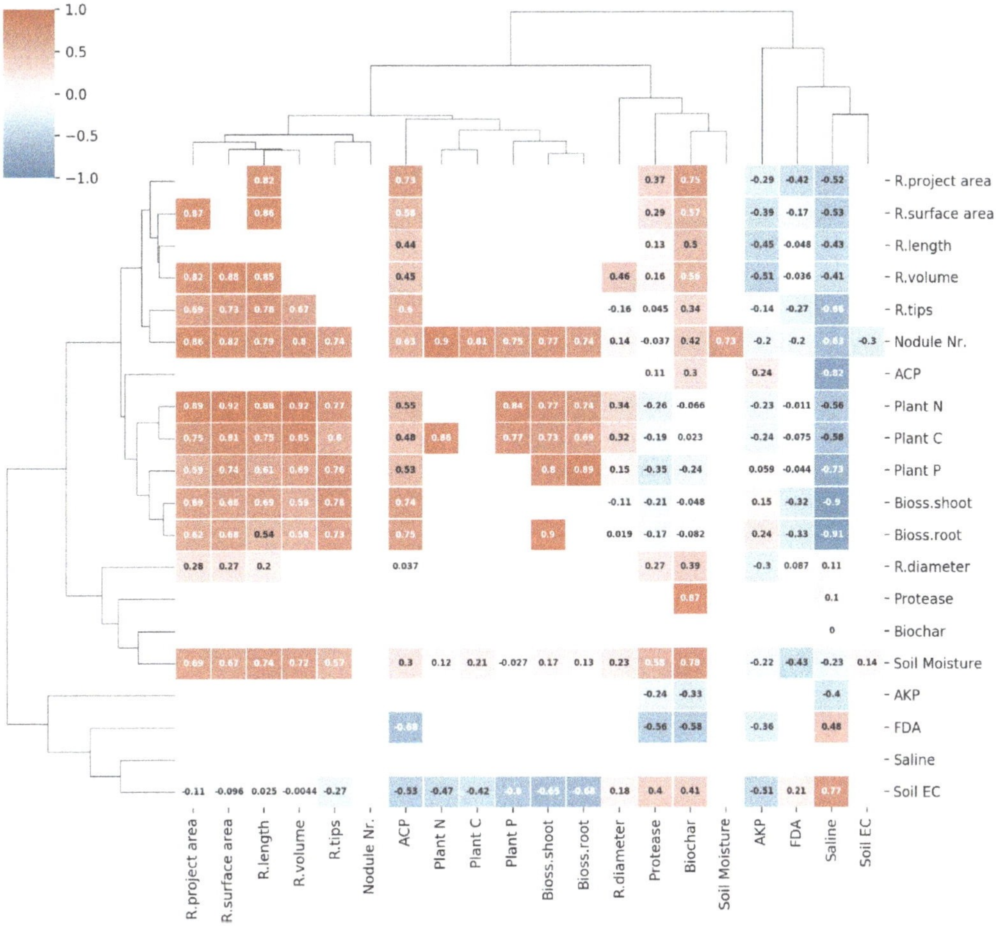

Figure 6. Cluster map of correlations of the root morphological measurements, plant measurements, nodule number, soil properties, saline conditions and biochar additions. Biochar: biochar additions, Bioss.root: root biomass, Bioss.shoot: shoot biomass, EC: soil electric conductivity, Nr.: number, Saline: saline conditions. The colour bar indicates Pearson correlation coefficient. The columns/rows of the data matrix are re-ordered according to the hierarchical clustering result; similar observations are close to each other. The blocks of 'high' and 'low' values are adjacent in the data matrix.

Plant tissue measurements, such as contents of N, C, and P, as well as biomass, were clustered with acidic phosphomonoesterase activity in one group. These measurements also significantly and positively correlated with each other. Soil moisture and biochar additions clustered together and showed strong correlations with most of the root morphological measurements apart from root diameter; however, only weak correlations with the plant measurements (Table 1 and Figure 6). Alkaline phosphomonoesterase activity, protease activity, and FDA hydrolytic activity showed weak correlations with root morphological and other plant measurements. Soil EC indicated a weak correlation with root morphological measurements and nodule number and showed a significantly negative correlation with shoot measurements. The saline soil condition indicated a significantly negative correlation with root morphological and aboveground plant parameters. Before the RDA was performed, data of the response variables (root project area, root surface area, root length, root volume, root tips, root diameter, shoot biomass, root biomass, plant N, P and K) as well as for the explanatory variables (biochar addition, saline condition, soil acidic and alkaline phosphomonoesterase activity, protease activity, FDA hydrolytic activity, soil moisture and EC) were normalized. As a result, the first three RDAs explained 70.57%, 9.63%, 2.37% of the total variance (Figure 7).

Table 1. Soil electric conductivity (EC) and moisture content during licorice growth in pot experiments.

Condition	Treatment	EC	Moisture
Saline	Control	0.165 ± 0.03	7.5 ± 0.86
	2% BC	0.385 ± 0.08	9.2 ± 0.76
	4% BC	0.421 ± 0.03	9.3 ± 0.95
	6% BC	0.750 ± 0.09	12.0 ± 1.10
Non-saline	Control	0.041 ± 0.01	7.2 ± 0.76
	2% BC	0.062 ± 0.01	11.5 ± 0.50
	4% BC	0.069 ± 0.02	9.5 ± 1.10
	6% BC	0.076 ± 0.01	13.1 ± 0.8

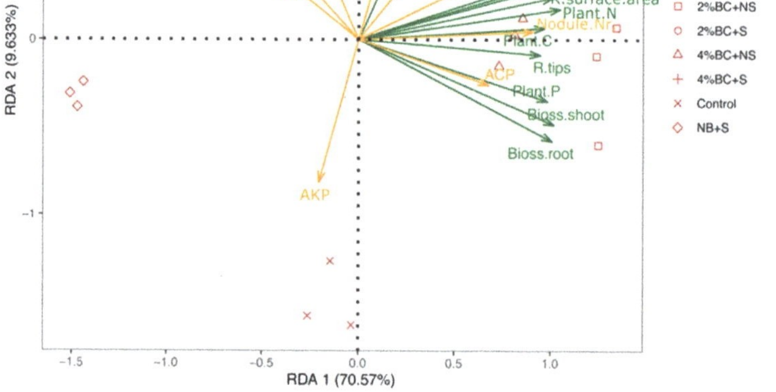

Figure 7. RDA-ordination triplot of variables. Response variables are shown by green arrows; explanatory variables are shown by yellow arrows. The angles between the arrows of the response and explanatory variables indicate correlations. The RDA1 and RDA2 axis explained 70.57% and 9.63% of the total variance, respectively.

A permutation test for RDA under a reduced model was performed to test the importance of the chosen model, the axis, and the explanatory variables (Table 2). The significance level of the model of RDA indicates that it can appropriately explain the data (Table 2). The RDA1 axis explained most of the variables and indicated the highest impact. In the explanatory variables, the acidic phosphomonoesterase activity contributes the most to explain the response variables.

Table 2. Permutation test for RDA under reduced model.

		Df	Variance	F	Pr(>F)	Signif.
Model		9	9.8537	5.6962	0.001	***
Axis	RDA1	1	8.0393	41.8259	0.001	***
	RDA2	1	1.0973	5.709	0.219	
	RDA3	1	0.2695	1.4023	0.981	
	RDA4	1	0.215	1.1185	0.987	
	RDA5	1	0.1404	0.7306	0.996	
	RDA6	1	0.038	0.1978	1	
	RDA7	1	0.0303	0.1574	1	
	RDA8	1	0.0204	0.1064	1	
	RDA9	1	0.0035	0.0181	1	
Explanatory variables	Protease	1	0.3749	1.9505	0.158	
	AKP	1	0.8829	4.5933	0.031	*
	ACP	1	4.9067	25.5281	0.001	***
	FDA	1	0.1533	0.7975	0.457	
	Soil.EC	1	1.6688	8.6824	0.002	**
	Soil.Moisture	1	0.5343	2.7798	0.079	.
	Nodule.Nr.	1	0.2909	1.5135	0.22	
	Saline	1	0.8186	4.2591	0.035	*
	Biochar	1	0.2233	1.1617	0.281	
Residual		8	1.5377			

Signif. codes: '***': 0.001, '**': 0.01, '*': 0.05, '.': 0.1.

The soil EC, alkaline phosphomonoesterase activity and the saline condition also affected the response variables. However, the biochar showed no significant contributions to explain the response variables. Thus, we traced back to the previous results, and a 2% biochar addition showed a positive effect on the root morphological and plant measurements; nevertheless, a 4% or a 6% biochar addition showed no significant impact on plant measurements. The overall contributions of three rates of biochar addition could not be significantly differentiated.

3. Discussion

Overall, this study showed that plant growth attributes, root architecture, nutrient acquisition and symbiotic performance of licorice under both non-saline and saline soil conditions improved after the amendment of biochar. However, the concentration of B2 had the most benefits for licorice root and shoot growth, as well as N and P uptake and C allocation under both soil conditions. Biochar dosage could be a crucial factor in determining the biochar potential in improving plant growth [24]. Liu et al. [25] observed a dose-dependent response of rice to biochar related to nutrient availability. They observed an enhancement of rice root grown under 0.05% biochar amendment, but not for 0.1% biochar. In another study, Batool et al. [26] reported improved physiological parameters of *Abelmoschus esculentus* L., such as leaf stomatal conductance and gas exchange rate at 1% biochar application compared 3%. Moreover, growth parameters of licorice were positively affected by 2% biochar treatments under salt stress, indicating alleviation of the adverse impact on plants. Several other reports demonstrated the positive impacts of soil amendments with biochar on plant growth and development under salt stress [27,28]. For example, Farooq et al. [29] observed an increased growth in seedlings and leaf area and reduced oxidative damage and Na+ accumulation in cowpea leaves after biochar application under salt stress conditions. A similar observation was reported for quinoa, where the amendment of biochar significantly increased plant height, shoot biomass, water use efficiency and yield under salt stress compared to control plants [30]. The effects of biochar on plant stress tolerance were also found to be dose-dependent. Ibrahim et al. [23] studied the impact of biochar rates (0%; 2.5%; 5%, 10%) on sorghum growth and physiological attributes under saline conditions. The plant height, leaf area, fresh and dry weight under stress were improved by 2.5 and 5% biochar amendment, whereas a 10% addition of biochar negatively influenced plant physiological properties. The effect has been explained by the regulation of the synthesis of anti-oxidant enzymes in plants induced by biochar [31], and also by a reduction in Na acquisition of plants [32].

The B2 concentration improved the root architecture of licorice under non-saline and saline conditions compared to the other treatments, indicating that a low biochar application rate strongly stimulates root system growth. In addition, the overall contributions of three rates of biochar addition on the root architecture traits were high according to the significant correlation among biochar and root architecture traits from the RDA result. Points for observations of the amendment of B2 and B4 under non-saline conditions can be projected perpendicular on the lines for response and explanatory variables, indicating the response values and explanatory variables at those sites on the triplot. The result also proves the root morphological growth was positively affected by B2 and B4 amendment under non-saline conditions. Similar findings were observed for the halophytes Sesbania (*Sesbania cannabina*) and Seashore mallow (*Kosteletzkya virginica*), where the root system, as well as shoot growth under salt stress, were improved by biochar application [33]. Several reports explain the possible mechanisms leading to the positive effects of biochar on plant growth. Laird et al. [34] observed reduced nutrient leaching and enhanced soil quality after applying biochar. The application rate of B2 increased C, N and P contents in plant tissues under both non-saline and saline conditions. The positive effect of biochar on plant growth was explained by the increased availability of essential nutrients for plant growth and development [35]. The nutrient uptake in plant tissue of corn such as N, P, K, and Mg was stimulated by the biochar amendment of alkaline soil [36]. Biochar is rich in organic carbon and minerals and supplies additional nutrients to the soil available for plant acquisition, improving plant nutritional status and plant development [37]. El-Naggar et al. [38] observed an increased cation exchange capacity (CEC) of soil after biochar amendment, associated with higher nutrient retention. Furthermore, reduced salt stress after biochar application was explained by improving stomatal conductance and water consumption of plants [32,39].

In our study, the nodule numbers decreased in licorice roots grown under saline soil conditions. In an earlier study, Zahran et al. [40] illustrated that abiotic stress might lead to an alteration in the *Rhizobium*-host plant recognition process. Remarkably, application rates of B2 and B4 improved the symbiotic performance of licorice under both non-saline and saline soil conditions. The result of RDA and the cluster map also indicated that biochar positively correlated with nodule numbers. However, only the observations of the B2 and B4 amendment under non-saline conditions can be projected on the line of nodule number on the triplot; apparently, saline conditions offset the effect of biochar to a certain degree. In earlier studies, several reports demonstrated increased nodule numbers in soybean roots after adding biochar [12,41]. Biochar is rich in carbon and several nutrients; some of these nutrients are directly available for microbes. Thus biochar can likely provide sources for bacterial survival and proliferation in soil and rhizosphere [42,43].

Soil enzyme activities responded differently to the biochar concentration applied under both non-saline and saline soil conditions. Fluorescein diacetate (FDA) hydrolase activity increased in soil amended with B2, whereas rates of B4 and B6 additions even inhibited enzyme activity under non-saline and saline conditions. Ma et al. [21] reported that applying 10 t ha^{-1} biochar produced from black cherrywood significantly increased soil FDA hydrolytic activity in a sandy field. Accordingly, an increase in FDA hydrolytic activity, which is an indicator of total microbial activity, was also reported by Chan et al. [15] after adding biochar into the soil. Furthermore, acid phosphomonoesterase activity was stimulated in soil amended with B2 and B4; however, alkaline phosphomonoesterase was inhibited under both soil conditions. In the explanatory variables of the RDA triplot, the acidic phosphomonoesterase activity contributed the most to explaining the response variables. The acidic phosphomonoesterase activity strongly correlated with plant P, shoot dry weight and root dry weight. It is likely that the P uptake may increase organic P mineralization [44] and thus, facilitated the association between plants and microorganisms [45,46]. Furthermore, P-solubilizing microorganisms are promoted to release organic acids to solubilize ortho-P [47,48]. A significant correlation between biochar and acidic phosphomonoesterase activity was also found, suggesting that biochar may have an indirect effect on P cycling for plant growth. Furthermore, soil protease activity strongly increased after applying B2, B4 and B6 under both saline and non-saline conditions. Correspondingly, increased enzyme activities involved in C and N cycles were also detected after adding maize biochar [41]. Thus, the increased enzyme activities in saline-alkaline soils amended with biochar in other studies demonstrated an improved microbial community related to central C, N, and P cycling activities [49,50]. Moreover, urease plays a crucial role in mineralizing soil organic nitrogen, as do phosphatases in transforming soil organic P forms [51,52].

4. Materials and Methods

4.1. Soil, Biochar and Plant

The soil used in the study was sandy loam, collected from the horizon (0–15 cm depth) of an experimental arable field under irrigation operated by the Experimental Field Station of the Leibniz Centre for Agricultural Landscape Research (ZALF), Müncheberg, Germany. The soil had the following contents: clay and fine silt (7%), coarse and medium silt (19%), sand (74%), C org (0.6%), total N (0.07%), P (0.03%), K (1.25%), and Mg (0.18%), the pH was 6.2. The biochar material was supplied from the Leibniz-Institute for Agrartechnik Potsdam-Bornim e.V. (ATB), Germany. The biochar was produced from maize by heating at 600 °C for 30 min and had the following properties (calculated per dry weight): dry matter (DM% fresh matter)—92.85; ash (%)—18.42; total organic carbon content (%)—75.47, N (%)—1.80; C/N ratio—41.93; Ca (g/kg $^{-1}$)—9.26; Fe (g/kg^{-1})—11.40; Mg (g/kg^{-1})—4.91; K (g/kg^{-1})—32.26; P (g/kg^{-1})—5.26; pH—9.89; EC—3.08 [53]. Licorice seeds were purchased from an online seed store of Chinese traditional medicine in China and were used for pot experiments.

4.2. Plant Growth Experiment

The experiment was conducted in a plant growth chamber at the Leibniz Centre for Agricultural Landscape Research (ZALF). Three concentrations of biochar 2, 4, and 6% (*w/w*) were used as a soil amendment. Pots (d = 0.16 m, 3 L) were filled with 1000 g air-dried soil and mixed with crushed chars (particle size < 3 mm). The seeds of licorice (*Glycyrrhiza uralensis* Fisch.) were surface-sterilized using 10% *v/v* NaOCl for 5 min and 70% ethanol for 5 min. After that, seeds were rinsed five times with sterile distilled water and transferred to paper tissue for germination in a dark room at 25 °C for 3–4 days. A total of three seeds were sown to each pot, and after one week, the seedlings were thinned to two plants per pot. The following treatments were set up: (i) plants grown in soil without biochar B0, (ii) plants grown in soil amended with 2% (B2), 4% (B4), and 6% (B6) biochar. Each treatment included four pots and was arranged in a randomized complete block design. The plants were grown under non-saline and saline (50 mM NaCl) conditions for 40 days at a temperature of 24 °C/16 °C (day/night) and in the humidity of 50–60%. Plants were irrigated with tap water containing 50 mM NaCl three times a week. The control treatment was only irrigated by tap water without NaCl. During plant growth, electrical conductivity of soil (EC) and moisture were measured every 3–4 days with UMP-2 BT+ sensor (UGT GmbH, Müncheberg, Germany). Average soil EC and soil moisture during plant growth are presented in Table 1. At harvest, the roots were separated from the shoots, and their biomass was oven-dried at 70 °C for 48 h. The dry weights of root and shoot and the number of nodules were determined from each plant.

4.3. *Plant and Soil Nutrient Analyses*

For the determination of carbon (C), nitrogen (N), and phosphorus (P) concentrations in plant tissues, oven-dried plants were homogenized by milling, and powders of shoots and roots were combined. The powder was analyzed with an inductively coupled plasma optical emission spectrometer (ICP-OES; iCAP 6300 Duo).

4.4. *Root Morphological and Architectural Traits*

Roots were separated from shoots and washed carefully with water. The entire root system was spread outward and analyzed using a scanner system (Expression 4990, Epson, Los Alamitos, CA, USA). Digital images of the root system were analyzed using Win RHIZO software (Régent Instruments, Quebec, Canada) for total root length, root volume, the number of root tips, root surface area, and average root diameter. The total number of nodules per plant root was counted under a stereomicroscope.

4.5. *Soil Enzyme Measurements*

Acid and alkaline phosphomonoesterase activities were assayed, according to Tabatabai and Bremner [54]. The concentration of p-nitrophenol (p-NP) produced in the assays by acid and alkaline phosphomonoesterase activities was calculated from a p-NP calibration curve after subtracting the absorbance of the control at 400 nm wavelength using a Lambda 2 UV-VIS spectrophotometer [55]. Protease activity was assayed using the method described by Ladd and Butler [56]. The ammonium released was calculated by relating the measured absorbance at 690 nm to a calibration graph containing 0, 1.0, 1.5, 2.0, and 2.5 µg of NH_4^+-N mL^{-1}. The assay of fluorescein diacetate (FDA) hydrolytic activity was performed according to Green et al. [57]. The fluorescein concentration was calculated by reference to a standard curve with 0, 0.001, 0.005, 0.05 and 0.15 mg of fluorescein.

4.6. *Statistical Analyses*

The data were subjected to one-way analysis of variance (ANOVA) using the software package SPSS-22 (SPSS Inc., Chicago, IL, USA). Multiple comparisons of the means were conducted by the least significant difference using Tukey's Honest Significant Difference (HSD) (LSD) ($p = 0.05$) test. Linear correlation analyses were performed with Pearson's correlation coefficients to clarify the relationship between various measurements by using python 3.8.1. A cluster map of the correlations was plotted to visualize the results. For further data exploration, a redundancy analysis (RDA) was performed to explain the dependent relationships between the explanatory variables (soil properties) and response variables (plant parameters) using the open-source statistical language R v1.3.1056 (R Studio, Boston, MA, USA). The results of the RDA were plotted on a triplot, on which the angles between the arrows of the response and explanatory variables indicate the correlations.

5. Conclusions

The results of our study revealed synergistic effects of concentration-dependent biochar amendments on licorice growth, nutrient uptake, and soil enzyme activities involved in the cycling of C, N and P in sandy loam soil under both non-saline and saline conditions. The improved acquisition of nutrients by licorice was explained by enhanced root growth, bioavailability of nutrients and increased soil microbial activity after biochar amendment. Remarkably, a medium-amount biochar amendment (B2) of the soil mostly improved the root system architecture and thus enabled improved nutrient uptake, and may support nitrogen fixation activity under salt stress. The use of excessive amounts of biochar, however, may result in unfavourable soil physico-chemical or soil ecological conditions such as over-critical aromatic C contents or unfavourable soil aggregation, which may negatively impact plant growth and microbial proliferation.

Overall, our findings underpin the notion of an elaborate interrelationship between biochar concentration and enhanced licorice growth, its root system architecture, symbiotic performance and nutrient acquisition under saline soil conditions.

Author Contributions: D.E. and S.D.B.-K. designed the experiments. Z.Z. and H.M. conducted the experiments. B.A. and H.M. analyzed the data. D.E., A.K., and S.W. wrote the manuscript. All authors have read and agreed to the published version of the manuscript.

Funding: This research was supported by a Georg Forster Research Fellowship (HERMES), Alexander von Humboldt Foundation, Bonn, Germany.

Data Availability Statement: Not applicable.

Conflicts of Interest: The authors declare no conflict of interest in the writing of the manuscript, or in the decision to publish the results.

References

1. Hayashi, H.; Hattori, S.; Inoue, K.; Sarsenbaev, K.; Ito, M.; Honda, G. Field survey of *Glycyrrhiza* plants in Central Asia (1). Characterization of *G. uralensis*, *G. glabra* and the putative intermediate collected in Kazakhstan. *Biol. Pharm. Bull.* **2003**, *26*, 867–871. [CrossRef]
2. Patil, S.M.; Patil, M.B.; Sapkale, G.N. Antimicrobial activity of *Glycyrrhiza glabra* Linn. Roots. *Int. J. Chem Sci.* **2009**, *7*, 585–591.
3. Sharma, V.; Agrawal, R.C.; Pandey, S. Phytochemical screening and determination of anti-bacterial and anti-oxidant potential of Glycyrrhiza glabra root extracts. *J. Environ. Res. Devel.* **2013**, *7*, 1552–1558.
4. Kushiev, H.; Noble, A.D.; Abdullaev, I.; Toshbekov, V. Remediation of abandoned saline soils using *Glycyrrhiza glabra*: A study for the Hungry steppes of central Asia. *Int. J. Agric. Sustain.* **2005**, 3, 102–113. [CrossRef]
5. Marui, A.; Nagafuchi, T.; Shinogi, Y.; Yasufuku, N.; Omine, K.; Kobayashi, T.; Shinkai, A.; Tuvshintogtokh, I.; Mandakh, B.; Munkhjargal, B. Soil physical properties to grow the wild licorice at semi-arid area in Mongolia. *J. Arid. Land. Stud.* **2012**, *22*, 33–36.
6. Egamberdieva, D.; Wirth, S.; Behrendt, U.; Allah, E.F.A.; Berg, G. Biochar treatment resulted in a combined effect on soybean growth promotion and a shift in plant growth promoting rhizobacteria. *Front. Microbiol.* **2016**, *7*, 209. [CrossRef] [PubMed]
7. Mwando, E.; Angessa, T.T.; Han, Y.; Li, C. Salinity tolerance in barley during germination-homologs and potential genes. *J. Zhejiang Univ. Sci. B.* **2020**, *21*, 93–121. [CrossRef]
8. Egamberdieva, D.; Berg, G.; Lindström, K.; Räsänen, L.A. Alleviation of salt stress of symbiotic *Galega officinalis* L. (goat's rue) by co-inoculation of *Rhizobium* with root colonizing *Pseudomonas*. *Plant Soil* **2013**, *369*, 453–465. [CrossRef]
9. Parray, J.A.; Jan, S.; Kamili, A.N.; Qadri, R.A.; Egamberdieva, D.; Ahmad, P. Current perspectives on plant growth-promoting rhizobacteria. *J. Plant Growth Regul.* **2016**, *35*, 877–902. [CrossRef]
10. Biederman, L.A.; Harpole, W.S. Biochar and its effects on plant productivity and nutrient cycling: A meta-analysis. *GCB Bioenergy* **2013**, *5*, 202–214. [CrossRef]
11. Saifullah-Saad, D.; Asif, N.; Zed, R.; Ravi, N. Biochar application for the remediation of salt-affected soils: Challenges and opportunities. *Sci. Total Environ.* **2018**, *625*, 320–335. [CrossRef]
12. Lehmann, J.; Rillig, M.C.; Thies, J.; Masiello, C.A.; Hockaday, W.C.; Crowley, D. Biochar effects on soil biota–a review. *Soil Biol. Biochem.* **2011**, *43*, 1812–1836. [CrossRef]
13. Novak, J.; Busscher, W.; Laird, D.; Ahmedna, M.; Watts, D.W.; Niandou, M. Impact of biochar amendment on fertility of a south eastern coastal plain soil. *Soil Sci.* **2009**, *174*, 105–112. [CrossRef]
14. Yu, O.Y.; Raichle, B.; Sink, S. Impact of biochar on the water holding capacity of loamy sand soil. *Int. J. Energy Environ. Eng.* **2013**, *4*, 1–9. [CrossRef]
15. Chan, K.Y.; Van Zwieten, L.; Meszaros, I.; Downie, A.; Joseph, S. Agronomic values of green waste biochar as a soil amendment. *Aust. J. Soil Res.* **2007**, *45*, 629–634. [CrossRef]
16. Manasa, M.R.K.; Katukuri, N.R.; Nair, S.D.; Yang, H.J.; Yang, Z.M.; Rong, B.G. Role of biochar and organic substrates in enhancing the functional characteristics and microbial community in a saline soil. *J. Environ. Manag.* **2020**, *269*, 110737. [CrossRef] [PubMed]
17. Tang, J.W.; Zhang, S.D.; Zhang, X.T.; Chen, J.H.; He, X.Y.; Zhang, Q.Z. Effects of pyrolysis temperature on soil -plant -microbe responses to *Solidago canadensis* L. derived biochar in coastal saline-alkali soil. *Sci. Total. Environ.* **2020**, *731*, 13. [CrossRef]
18. Lu, H.; Li, Z.; Fu, S.; Mendez, A.; Gasco, G.; Paz-Ferreiro, J. Effect of biochar in cadmium availability and soil biological activity in an anthrosol following acid rain deposition and aging. *Water Air Soil Pollut.* **2015**, *226*, 164. [CrossRef]
19. Cao, Y.; Ma, Y.; Guo, D.; Wang, Q.; Wang, G. Chemical properties and microbial responses to biochar and compost amendments in the soil under continuous watermelon cropping. *Plant Soil Environ.* **2017**, *63*, 1–7.
20. Ma, H.; Egamberdieva, D.; Wirth, S.; Li, Q.; Omari, R.A.; Hou, M.; Bellingrath-Kimura, S.D. Effect of biochar and irrigation on the interrelationships among soybean growth, root nodulation, plant P uptake, and soil nutrients in a sandy field. *Sustainability* **2019**, *11*, 6542. [CrossRef]

21. Ma, H.; Egamberdieva, D.; Wirth, S.; Bellingrath-Kimura, S.D. Effect of biochar and irrigation on soybean-rhizobium symbiotic performance and soil enzymatic activity in field rhizosphere. *Agronomy* **2019**, *9*, 626. [CrossRef]
22. Morales, M.M.; Comerford, N.; Guirinni, I.A.; Falkao, N.P.S.; Reeves, J.B. Sorption and desorption of phosphate on biochar and biochar—Soil mixtures. *Soil Use Manag.* **2013**, *29*, 306–314. [CrossRef]
23. Ibrahim, H.M.; Al-Wabel, M.I.; Usman, A.R.A.; Al-Omran, A. Effect of Conocarpus biochar application on the hydraulic properties of a sandy loam soil. *Soil Sci.* **2013**, *178*, 165–173. [CrossRef]
24. Gale, N.V.; Thomas, S.C. Dose-dependence of growth and ecophysiological responses of plants to biochar. *Sci. Total Environ.* **2019**, *658*, 1344–1354. [CrossRef] [PubMed]
25. Liu, M.; Lin, Z.; Ke, X.; Fan, X.; Joseph, S.; Taherymoosavi, S.; Liu, X.; Bian, R.; Solaiman, Z.M.; Li, L.; et al. Rice seedling growth promotion by biochar varies with genotypes and application dosages. *Front. Plant Sci.* **2021**, *12*, 580462. [CrossRef]
26. Batool, A.; Taj, S.; Rashid, A.; Khalid, A.; Qadeer, S.; Saleem, A.R.; Ghufran, M.A. Potential of soil amendments (Biochar and Gypsum) in increasing water use efficiency of *Abelmoschus esculentus* L. Moench. *Front. Plant Sci.* **2015**, *6*, 733. [CrossRef]
27. Hussain, M.; Farooq, M.; Nawaz, A.; Al-Sadi, A.M.; Solaiman, Z.M.; Alghamdi, S.S.; Ammara, U.; Ok, Y.S.; Siddique, K.H.M. Biochar for crop production: Potential benefits and risks. *J. Soils Sediments* **2017**, *17*, 685–716. [CrossRef]
28. Egamberdieva, D.; Ma, H.; Alimov, J.; Reckling, M.; Wirth, S.; Bellingrath-Kimura, S.D. Response of soybean to hydrochar-based. *Microorganisms* **2020**, *8*, 1674. [CrossRef]
29. Farooq, M.; Romdhane, L.; Rehman, A.; Al-Alawi, A.K.M.; Al-Busaidi, W.M.; Asad, S.A.; Lee, D.J. Integration of seed priming and biochar application improves drought tolerance in cowpea. *J. Plant Growth. Regul.* **2020**, *7*, 1–9. [CrossRef]
30. Yang, A.; Akhtar, S.S.; Li, L.; Fu, Q.; Li, Q.; Naeem, M.A.; He, X.; Zhang, Z.; Jacobsen, S.E. Biochar mitigates combined effects of drought and salinity stress in quinoa. *Agronomy* **2020**, *10*, 912. [CrossRef]
31. Thomas, S.C.; Frye, S.; Gale, N.; Garmon, M.; Launchbury, R.; Machado, N. Biochar mitigates negative effects of salt additions on two herbaceous plant species. *J. Environ. Manag.* **2013**, *129*, 62–68. [CrossRef]
32. Akhtar, S.S.; Andersen, M.N.; Liu, F. Residual effects of biochar on improving growth, physiology and yield of wheat under salt stress. *Agric. Water. Manag.* **2015**, *158*, 61–68. [CrossRef]
33. Zheng, H.; Wang, X.; Chen, L.; Wang, Z.; Xia, Y.; Zhang, Y.; Wang, H.; Luo, X.; Xing, B. Enhanced growth of halophyte plants in biochar-amended coastal soil: Roles of nutrient availability and rhizosphere microbial modulation. *Plant Cell. Environ.* **2017**, *41*, 517–532. [CrossRef] [PubMed]
34. Laird, D.A.; Fleming, P.D.; Wang, B.; Horton, R.; Karlen, D.L. Biochar impact on nutrient leaching from a midwestern agricultural soil. *Geoderma* **2010**, *158*, 443–449. [CrossRef]
35. Amini, S.; Ghadiri, H.; Chen, C.; Marschner, P. Salt-affected soils, reclamation, carbon dynamics, and biochar: A review. *J. Soils Sediments* **2016**, *16*, 939–953. [CrossRef]
36. Zhao, W.; Zhou, Q.; Tian, Z.; Cui, Y.; Liang, Y.; Wang, H. Apply biochar to ameliorate soda saline-alkali land, improve soil function and increase corn nutrient availability in the Songnen plain. *Sci. Total. Environ.* **2020**, *722*, 137428. [CrossRef] [PubMed]
37. Qayyum, M.F.; Steffens, D.; Reisenauer, H.P.; Schubert, S. Kinetics of carbon mineralisation of biochars compared with wheat straw in three soils. *J. Environ. Qual.* **2012**, *41*, 1210–1220. [CrossRef] [PubMed]
38. El-Naggar, A.H.; Usman, A.R.A.; Al-Omran, A.; Ok, Y.S.; Ahmad, M.; Al-Wabel, M.I. Carbon mineralisation and nutrient availability in calcareous sandy soils amended with woody waste biochar. *Chemosphere* **2015**, *138*, 67–73. [CrossRef] [PubMed]
39. Ghezzehei, T.A.; Sarkhot, D.V.; Berhe, A.A. Biochar can be used to capture essential nutrients from dairy wastewater and improve soil physico-chemical properties. *Solid Earth* **2014**, *5*, 953–962. [CrossRef]
40. Zahran, H.H.; Räsänen, L.A.; Karsisto, M.; Lindström, K. Alteration of lipopolysaccharide and protein profiles in SDS-PAGE of rhizobia by osmotic and heat stress. *World J. Microbiol. Biotech.* **1994**, *10*, 100–105. [CrossRef]
41. Wang, X.B.; Song, D.L.; Liang, G.Q.; Zhang, Q.; Ai, C.; Zhou, W. Maise biochar addition rate influences soil enzyme activity and microbial community composition in a fluvo-aquic soil. *Appl. Soil. Ecol.* **2015**, *96*, 265–272. [CrossRef]
42. Kolton, M.; Harel, Y.M.; Pasternak, Z.; Graber, E.R.; Elad, Y.; Cytryn, E. Impact of biochar application to soil on the root-associated bacterial community structure of fully developed greenhouse pepper plants. *Appl. Environ. Microb.* **2011**, *77*, 4924–4930. [CrossRef] [PubMed]
43. Prendergast-Miller, M.T.; Duvall, M.; Sohi, S.P. Localisation of nitrate in the rhizosphere of biochar-amended soils. *Soil. Biol. Biochem.* **2011**, *43*, 2243–2246. [CrossRef]
44. Masto, R.E.; Kumar, S.; Rout, T.K.; Sarkar, P.; George, J.; Ram, L.C. Biochar from water hyacinth (Eichornia Crassipes) and its impact on soil biological activity. *Catena* **2013**, *111*, 64–71. [CrossRef]
45. Warnock, D.D.; Lehmann, J.; Kuyper, T.W.; Rillig, M.C. Mycorrhizal responses to biochar in soil-concepts and mechanisms. *Plant Soil.* **2007**, *300*, 9–20. [CrossRef]
46. Blackwell, M.S.A.; Brookes, P.C.; de la Fuente-Martinez, N.; Gordon, H.; Murray, P.J.; Snars, K.E.; Williams, J.K.; Bol, R.; Haygarth, P.M. *Phosphorus Solubilisation and Potential Transfer to Surface Waters from the Soil Microbial Biomass Following Drying–Rewetting and Freezing–Thawing*, 1st ed.; Elsevier Inc.: Amsterdam, The Netherlands, 2010; Volume 106.
47. Vassilev, N.; Martos, E.; Mendes, G.; Martos, V.; Vassileva, M. Biochar of Animal Origin: A sustainable solution to the global problem of high-grade rock phosphate scarcity? *J. Sci. Food Agric.* **2013**, *93*, 1799–1804. [CrossRef] [PubMed]
48. Gul, S.; Whalen, J.K. Biochemical Cycling of Nitrogen and Phosphorus in Biochar-Amended Soils. *Soil Biol. Biochem.* **2016**, *103*, 1–15. [CrossRef]

49. Wu, H.; Zeng, G.; Liang, J.; Chen, J.; Xu, J.; Dai, J.; Li, X.; Chen, M.; Xu, P.; Zhou, X.; et al. Responses of bacterial community and functional marker genes of nitrogen cycling to biochar, compost and combined amendments in soil. *Appl. Microbiol. Biotechnol.* **2016**, *100*, 8583–8591. [CrossRef]
50. Shi, S.; Tian, I.; Nasir, F.; Bahadur, F.; Batool, A.; Luo, S.; Yang, F.; Wang, Z.; Tian, C. Response of microbial communities and enzyme activities to amendments in saline-alkaline soils. *Appl. Soil. Ecol.* **2019**, *135*, 16–24. [CrossRef]
51. Fu, Q.; Abadie, M.; Blaud, A. Effects of urease and nitrification inhibitors on soil N, nitrifier abundance and activity in a sandy loam soil. *Biol. Fertil. Soils* **2020**, *56*, 185–194. [CrossRef]
52. Margalef, O.; Sardans, J.; Fernández-Martínez, M.; Molowny-Horas, R. Global patterns of phosphatase activity in natural soils. *Sci. Rep.* **2017**, *7*, 1337. [CrossRef] [PubMed]
53. Reibe, K.; Götz, K.P.C.L.; Ross, T.F.; Doering, F.; Ellmer, L.; Ruess, R. Impact of quality and quantity of biochar and hydrochar on soil collembola and growth of spring wheat. *Soil. Biol. Biochem.* **2015**, *8*, 84–87.
54. Tabatabai, M.A.; Bremner, J.M. Use of p-nitrophenol phosphate for the assay of soil phosphatase activity. *Soil Biol. Bioch.* **1969**, *1*, 301–307. [CrossRef]
55. Acosta-Martinez, V.; Mikha, M.M.; Sistani, K.R.; Stahlman, P.W.; Benjamin, J.G.; Vigil, M.F.; Erickson, R. Multi-location study of soil enzyme activities as affected by types and rates of manure application and tillage practices. *Agriculture* **2011**, *1*, 4–21. [CrossRef]
56. Ladd, J.N.; Butler, J.H.A. Short-term assays of soil proteolytic enzyme activities using proteins and dipeptide derivatives as substrates. *Soil. Biol. Biochem.* **1972**, *4*, 19–30. [CrossRef]
57. Green, V.S.; Stott, D.E.; Diack, M. Assay for fluorescein diacetate hydrolytic activity: Optimalization for soil samples. *Soil Biol. Biochem.* **2006**, *38*, 693–701. [CrossRef]

plants

MDPI

Article

AIP1, Encoding the Small Subunit of Acetolactate Synthase, Is Partially Responsible for Resistance to Hypoxic Stress in *Arabidopsis thaliana*

Geunmuk Im and Dongsu Choi *

Department of Biology, Kunsan National University, Gunsan-si 54150, Korea; areas12@kunsan.ac.kr
* Correspondence: choid@kunsan.ac.kr

Abstract: Flooding is a significant stress to land plants, depriving them of essential oxygen. Plants have evolved diverse strategies with variable success to survive flooding. Similar strategies have been described in organisms from other kingdoms. Several fungal species can successfully survive a low-oxygen environment by increasing their branched-chain amino acid (BCAA) contents. BCAAs may act as alternative electron acceptors in the respiratory chain under an oxygen-limited environment. The key and first enzyme for BCAA biosynthesis is acetolactate synthase (ALS). We identified two homologous genes encoding the small subunit of ALS in Arabidopsis (*Arabidopsis thaliana*). We determined that *ALS INTERACTING PROTEIN1* (*AIP1*), which encodes the small subunit of ALS, is strongly expressed in all organs and highly expressed under submergence and low-oxygen stresses. We also showed that the overexpression of *AIP1* confers tolerance to low-oxygen stress. These results indicate that ALS may play an essential role under prolonged flooding or oxygen deficiency in Arabidopsis.

Keywords: ALS; BCAA; low oxygen; flooding; *AIP1*

Citation: Im, G.; Choi, D. *AIP1*, Encoding the Small Subunit of Acetolactate Synthase, Is Partially Responsible for Resistance to Hypoxic Stress in *Arabidopsis thaliana*. *Plants* **2021**, *10*, 2251. https://doi.org/10.3390/plants10112251

Academic Editors: Milan S. Stankovic, Paula Baptista and Petronia Carillo

Received: 29 September 2021
Accepted: 20 October 2021
Published: 22 October 2021

Publisher's Note: MDPI stays neutral with regard to jurisdictional claims in published maps and institutional affiliations.

Copyright: © 2021 by the authors. Licensee MDPI, Basel, Switzerland. This article is an open access article distributed under the terms and conditions of the Creative Commons Attribution (CC BY) license (https://creativecommons.org/licenses/by/4.0/).

1. Introduction

Due to the rapid progression of climate change, crop productivity has recently been declining in harsh environments [1]. As the world population continues to increase, a stable food supply is and will be a critical issue. By 2100, the average global temperature is predicted to rise by 2 °C, which will affect the climate in unprecedented ways [2]. Climate change is also responsible for sea levels rising and the flooding of farmlands, which will adversely influence agriculture and crop yields [3]. Frequent and prolonged flooding significantly reduces production of major staple crops such as rice (*Oryza sativa*), wheat (*Triticum aestivum*), and maize (*Zea mays*), which will threaten food security in the near future [1].

Flooding causes the depletion of oxygen inside plant cells, which blocks oxidative phosphorylation, dramatically reducing energy production and leading to severe hypoxic stress. Some plant species survive flooding by adopting special strategies, such as the formation of specialized tissue called aerenchyma that creates pockets of air [4,5]. Among such species, deepwater rice shows unique behaviors under prolonged flooding since it can escape complete submergence by promoting rapid stem growth to acquire enough oxygen supply from leaves that reach above the water line. By contrast, other flooding-tolerant species employ a breath-holding strategy, greatly reducing their metabolism until floodwaters recede. Plant adaptive strategies for flooding stress are diverse, complex, and remain mostly unexplained [4,6]. In recent years, various studies have focused on oxygen-sensing mechanisms that detect oxygen deficiency caused by flooding [7–9] and are the underlying transcriptional regulation. However, there are relatively few studies focusing on the factors that ultimately induce plant responses to hypoxia. Hypoxia and flooding induce changes in various metabolic processes [10–12], so certain enzymes and metabolites may play an important role in plant adaptive responses to hypoxia or flooding. Among diverse metabolic adaptations, the accumulation of amino acids can be essential to the survival of plants under the hypoxic conditions caused by floods [13–16].

In a microarray-based study using deepwater rice [11], we previously determined that the submergence-responsive gene (Os02g39570) encodes the putative regulatory subunit of acetolactate synthase (ALS) responsible for the biosynthesis of a certain group of amino acids. ALS has been the focus of studies pertaining to herbicide resistance. Since several amino acid biosynthetic pathways are targets of some herbicides, the metabolism of certain amino acids has been intensively studied [17,18].

Amino acids are the building blocks of protein biosynthesis, constituting a major organic form of transported nitrogen in plants by lying at the crossroads between carbon and nitrogen metabolism. They also participate in essential physiological processes, as they are the precursors of many plant secondary metabolites such as lignin, phytohormones, and flavonoids [19,20]. Of all essential amino acids, valine (Val), leucine (Leu), and isoleucine (Ile) have side chains and are collectively referred to as branched-chain amino acids (BCAAs) [21]. BCAAs are classified according to the small branching hydrocarbon residues responsible for their aliphatic character [21]. Plant BCAAs are essential compounds: They play an essential role in the biosynthesis of various secondary metabolites, in addition to their function as protein constituents [21,22]. Val and Ile are synthesized via two parallel pathways involving four common enzymes, which catalyze different products in the presence of different substrates [23]. ALS, the first common enzyme in the pathway, uses two pyruvate molecules to form acetolactate, leading to Val and Leu biosynthesis. Alternatively, the enzyme uses one pyruvate and one α-ketobutyrate substrate molecule to generate acetohydroxybutyrate as a precursor for Ile biosynthesis [23]. ALS has two subunits: a catalytic subunit and a regulatory subunit [24,25]. The regulatory subunit greatly enhances the reactivity of the catalytic subunit and is necessary for feedback inhibition by BCAAs [18,26–28].

Some fungi adapt to hypoxic environments through biosynthesis of BCAAs [29]. In this case, the fungus produces BCAAs to re-oxidize NAD(P)H and ALS strongly influences BCAA production in the hypoxic state [29]. Thus, as a result of the activation of ALS, the fungus gains an energy source in low-oxygen conditions that contributes to its survival. However, no hypoxic studies have been conducted on the genes encoding ALS regulatory subunits in seed plants.

In this study, we characterized two Arabidopsis (*Arabidopsis thaliana*) genes (*ALS INTERACTING PROTEIN1* [*AIP1*] and *AIP3*) orthologous to the rice ALS small subunit genes [27]. *AIP1* and *AIP3* encode ALS regulatory subunits and have diverse biological functions. However, their roles in hypoxia responses have not been studied in land plants. To understand their possible biological roles during hypoxia, we generated transgenic Arabidopsis plants overexpressing each gene. In this research, we provide the first report on the possible role of ALS in the low-oxygen response of Arabidopsis.

2. Results

2.1. Organ-Specific Expression Patterns of AIP1 and AIP3 Genes

AIP1 and *AIP3* genes were expressed in all organs tested. *AIP1* showed a relatively higher expression level in leaves, while *AIP3* was highly expressed in flowers, although the expression patterns of these two genes were very similar (Figure 1A,B). Overall, *AIP1* and *AIP3* were more highly expressed in leaves, siliques, and flowers compared to other tissues, but *AIP3* did not show statistically significant differences in its expression levels across different organs.

2.2. Expression Patterns of AIP1 and AIP3 under Submergence and Hypoxic Environments

AIP1 showed its strongest upregulation at 16 h after initiation of submergence, reaching levels ~2.5 times higher than air controls, before decreasing (Figure 2A).

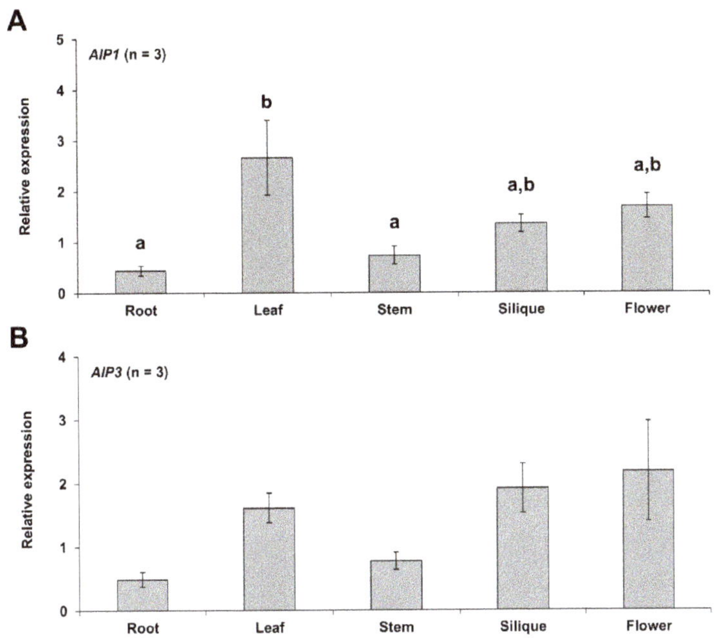

Figure 1. Expression patterns of *AIP1* and *AIP3* in various Arabidopsis tissues. RT-qPCR analysis of *AIP1* (**A**) and *AIP3* (**B**) relative transcript levels, performed using RNAs from various tissues. *ACT2* was used as an internal control. Error bars indicate standard errors; $n = 30$. Statistical significance was determined with one-way ANOVA followed by Tukey's honestly significant difference (HSD) test; different letters denote significant differences ($p \leq 0.05$).

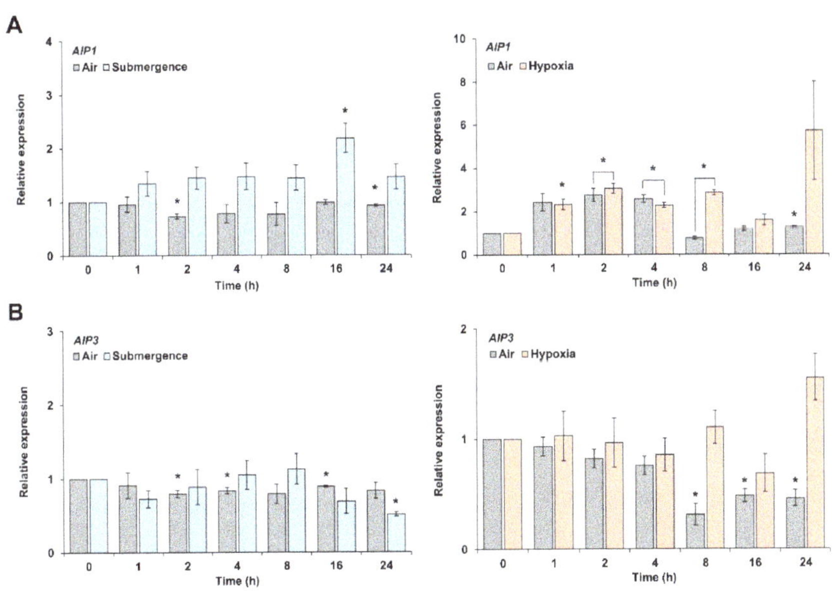

Figure 2. Expression patterns of *AIP1* and *AIP3* upon submergence or hypoxia. Seven-day-old seedlings were subjected to submergence or hypoxia in the dark. Control seedlings were maintained in normal air conditions in the dark and collected at the same time points. RT-qPCR analysis of *AIP1* (**A**) and *AIP3* (**B**) relative transcript levels from leaves. Error bars indicate standard errors. * $p \leq 0.05$ vs. 0 h, as determined by Student's *t*-test.

Relative *AIP3* transcript levels rose steadily from 2 to 8 h after initiation of submergence before gradually decreasing at later time points (Figure 2B). Under a hypoxic environment, *AIP1* was expressed the highest at 24 h after the beginning of treatment, while *AIP3* was expressed the highest at 8 h, then dropped at 16 h, before rising again at 24 h (Figure 2A,B). Since *AIP3* did not show significant differences under submergence and hypoxia, we focused on *AIP1*.

2.3. Assessment of the Tolerance to Hypoxia of AIP1-Overexpressing Lines by Recovery-Survival Rates

To assess the contribution of *AIP1* to tolerance against low-oxygen environments, we generated transgenic Arabidopsis lines overexpressing *AIP1* (fused to *GFP*). To evaluate the tolerance of these transgenic plants to 16 h of hypoxic stress, we measured their survival rate following a 3-day recovery period. Five independent GFP-tagged transgenic lines exhibited higher survival rates than those of the wild-type non-transgenic control group (Figure 3). We selected line *GFP-AIP1-1-2* overexpressing *GFP-AIP1* for further characterization (Figure S1).

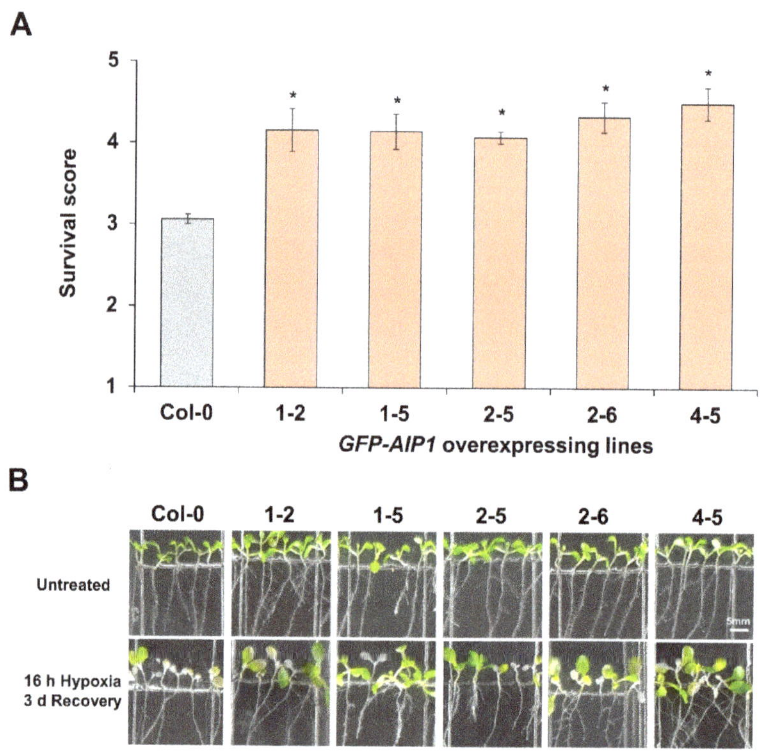

Figure 3. *GFP-AIP1* transgenic lines are more tolerant to hypoxic stress: (**A**) Survival scores of *GFP-AIP1*-overexpressing lines after a 16 h hypoxia treatment followed by a 3-day recovery period (Figure S2). Error bars indicate standard errors. * $p \leq 0.05$ vs. Col-0, as determined by Student's *t*-test. (**B**) Representative photographs of air- and hypoxia-treated (16 h) seedlings followed by a 3-day recovery period. Note the white seedlings for hypoxia-treated Col-0 plates.

2.4. Assessment of the Tolerance to Hypoxia of an AIP1-Overexpressing Line by Electrolyte Leakage

We next evaluated the electric conductivity of Col-0 and transgenic lines after hypoxic treatment. The electric conductivity of the *GFP-AIP1-1-2* line was significantly lower than that of the control Col-0 (Figure 4), in agreement with the higher survival score measured in the overexpression line (Figure 3).

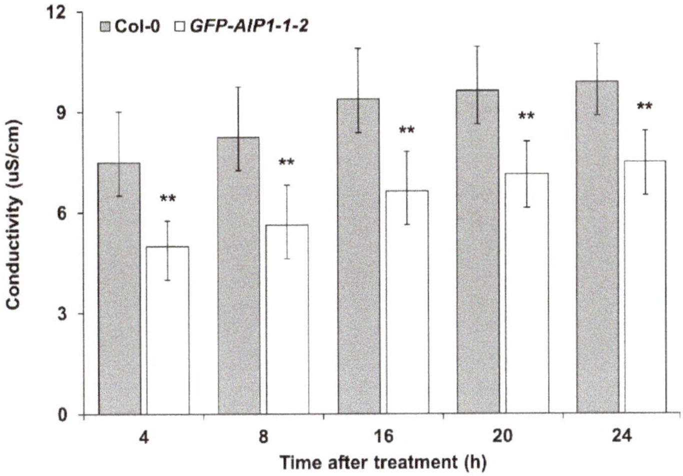

Figure 4. Electrolyte leakage from leaf disks between 4 and 24 h after hypoxic stress. Four-week-old plants were exposed to hypoxic conditions for 4 h. Electrolyte leakage of leaf disks was measured after 4–24 h of recovery under normoxia conditions. Error bars indicate standard error. ** $p \leq 0.01$ vs. Col-0, as determined by Student's *t*-test.

2.5. Measurement of BCAA Contents in a GFP-AIP1-Overexpressing Line

Since AIP1 is a subunit of ALS, we determined the free amino acid contents with a focus on BCAAs, in Col-0 and the *GFP-AIP1-1-2* line after hypoxic treatment. BCAA contents (especially those of Val and Leu) were significantly ($p \leq 0.05$) higher in the *GFP-AIP1-1-2* line relative to those in Col-0 (Figure 5).

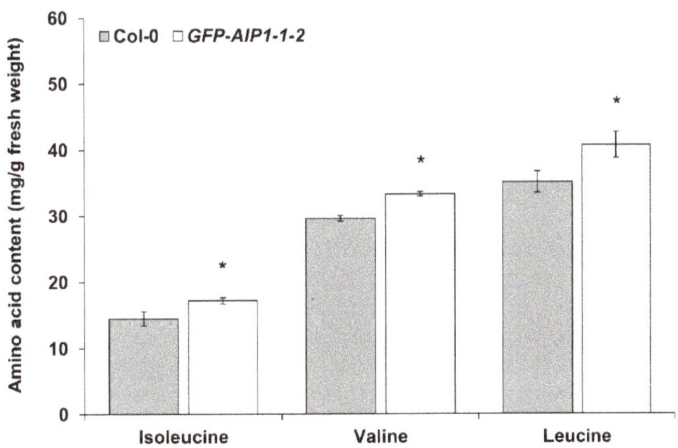

Figure 5. Measurements of BCAA contents in the *GFP-AIP1* overexpression line after hypoxic stress. Two-week-old plants were exposed to hypoxia for 4 h. The amounts of BCAAs were measured in leaves after hypoxic stress. Error bars indicate standard errors of two independent experiments, each with three technical replicates. * $p \leq 0.05$ vs. Col-0, as determined by Student's *t*-test.

3. Discussion

ALS is the enzyme catalyzing the first step in the biosynthesis of the BCAAs Val, Leu, and Ile [23,30]. As in bacteria and yeast (*Saccharomyces cerevisiae*), ALS in plants is composed of one catalytic and one regulatory subunit, forming a heterodimer whose activity is at least five times higher than homodimers of catalytic subunits [24,31]. A deeper understanding of the regulatory subunit is therefore very important to harness the catalytic power of ALS.

Roots and shoots may exhibit different tolerance mechanisms when faced with hypoxic stress [32]. In Arabidopsis, root and shoot responses to hypoxic stress are different [32,33]. In addition, several hypoxia-inducible genes, including a well-known hypoxic-response gene associated with fermentation, are abundantly expressed in roots [33,34]. For example, fermentation-related genes (*ALDEHYDE DEHYDROGENASE* [*ADH*], *PYRUVATE DECARBOXYLASE1* [*PDC1*], and *PDC2*) and several carbohydrate metabolism–related genes were more highly expressed in roots than in shoots after exposure to hypoxia for 1–24 h [10]. In the hypoxic state, Arabidopsis roots increase the expression of metabolism-related genes to overcome stress by creating a minimal level of ATP production [35]. However, *AIP1* and *AIP3* in this study were expressed to higher levels in the aboveground tissues and not in the roots. Considering that oxygen generated by active photosynthesis in leaves and stems mitigates the hypoxic state, we hypothesize that *AIP1* and *AIP3* are expressed mainly in oxygen-deficient tissues in leaves or stems, such as sieve tubes.

The inside of a sieve tube tissue is characterized by a relatively lower oxygen concentration than the surrounding tissues [36]. Measurements of the oxygen partial pressure using exudates in castor bean (*Ricinus communis*) revealed that that the oxygen concentration in the phloem was only 7% [36].

With a decrease in oxygen concentration, the content and current of sugar also decreased. There was a gradual increase in alanine, γ-aminobutyrate, methionine, and especially the BCAA Ile, as well as a gradual decrease in the carbon to nitrogen ratio, plus an increase in the ratio between succinate and malate in the phloem. These results suggest that metabolic functions and the functions of sieve tubes change adaptively with the oxygen concentration in the whole plant [36].

Flooding or submergence stress inevitably leads to hypoxic conditions by lowering intracellular oxygen partial pressure. *AIP1* expression increased after 16 h of submergence and after 24 h of hypoxia, and it was higher upon submergence or hypoxia stress compared to the control group (air treatment). However, *AIP3* displayed a lower expression level relative to the control group during the submergence treatment and followed a different pattern from that seen with *AIP1*. By contrast, *AIP3* exhibited a similar expression pattern as *AIP1* under hypoxic conditions, with a higher expression level after 24 h of hypoxia. *AIP1* and *AIP3* expression patterns were consistent under submergence and hypoxic conditions, prompting us to analyze the function of *AIP1*.

AIP1-overexpressing lines displayed better survival rates than the wild-type control during hypoxia-recovery tests (Figure 3), indicating that *AIP1* confers substantial resistance against hypoxic stress in Arabidopsis. We also confirmed the enhanced tolerance of *AIP1*-overexpressing lines to hypoxia by electrolyte leakage, as these transgenic lines showed less electrolyte leakage compared to the wild type (Figure 4). This improved tolerance may stem from higher BCAA contents, especially for Val and Leu (Figure 5).

BCAAs act as final receptor substitutes for electrons in fungi [29]. For instance, *Aspergillus nidulans* adapts to a hypoxic environment by employing a mechanism of NAD(P)H reoxidation to $NAD(P)^+$ through the biosynthesis of BCAAs. $NAD(P)^+$ can then be used for substrate-level phosphorylation to generate ATP. When applied to plants, this mechanism would produce $NAD(P)^+$ as a by-product of BCAA biosynthesis, leading to a minimal pool of ATP that plants can utilize through substrate-level phosphorylation to survive. In this context, we conclude that activation of the ALS small subunit enhances the tolerance of Arabidopsis plants to hypoxia by increasing BCAA levels. Although it is inefficient compared to the production of ATP through respiration, plants can nevertheless access a supply of ATP to support some growth, even when in a hypoxic environment, ultimately contributing to plant tolerance to hypoxic stress. Further study on the precise roles of ALS small subunits in hypoxia tolerance will be required, as will studies on the undefined biological role of the ALS holoenzyme.

4. Materials and Methods

4.1. Plant Materials and Growth Conditions

Arabidopsis (*Arabidopsis thaliana*) accession Columbia (Col-0) was used as the wild type [37]. The seeds were stratified at 4 °C for 3 days in the dark after surface sterilization, planted in pots filled with sterilized Horticulture nursery Sunshine Mix No. 5 (Sun Gro Horticulture, Vancouver, BC, Canada), and cultivated in a culture room under a long-day photoperiod (16 h light/8 h dark) and 60% humidity at 24 °C [38].

4.2. RNA Extraction and cDNA Synthesis

Total RNA was extracted as described previously [39]. RNA was quantified on a Nanodrop DS-11 (DeNovix, Wilmington, NC, USA) and only high-quality RNA with a ratio between OD_{260}/OD_{280} = 1.8–2.0 was used. The integrity and purity of the RNA were independently confirmed by electrophoresis on agarose gels. First-strand cDNAs were synthesized from 5 μg total RNA using a PrimeScript^{TM} 1st strand cDNA synthesis kit (TaKaRa, Kusatsu, Japan).

4.3. Tissue-Specific Expression of AIP1 and AIP3

Arabidopsis roots, stems, leaves, siliques, and flowers were collected for RNA extraction. Total RNAs were extracted from the roots and leaves of 7-day-old seedlings, 6-week-old stems, flowers from 8-week-old plants, and siliques from 9-week-old Arabidopsis plants [37].

RT-qPCR was conducted to check the organ-specific expression patterns of *AIP1* and *AIP3*. Primer Express v 3.0.1 (Applied Biosystems, Waltham, MA, USA) was used to design and build the primer pairs to amplify a PCR product between 100 and 150 bp. qPCR was performed with SYBR Premix Ex Taq^{TM} II(TaKaRa, Kusatsu, Japan) on a Thermal Cycler Dice Real Time instrument (TaKaRa, Kusatsu, Japan). The PCR conditions consisted of 40 consecutive cycles at 95 °C for 5 s and at 60 °C for 30 s.

4.4. Expression of AIP1 and AIP3 Genes upon Submergence and Hypoxia Conditions

For submergence treatment, Arabidopsis seedlings were first grown on plates with a half-strength Murashige and Skoog (MS) medium containing 0.5% (*w*/*v*) sucrose at 23 °C in a growth chamber for 1 week, before being moved to a glass box for submergence in the dark to minimize the production of oxygen from photosynthesis. After filling the glass box with distilled water to a height of 15 cm, the box was sealed and kept in the dark to maintain an oxygen-free submerged condition. At the same time, seedlings were incubated under normal air conditions without submergence as control. The seedlings were harvested for RNA extraction at seven time points (air: 0, 1, 2, 4, 8, 16, and 24 h; submergence: 1, 2, 4, 8, 16, and 24 h), as previously reported [38].

For each time point, the samples were divided into leaves and roots, and the leaves were used for RNA extraction. First-strand cDNAs were synthesized and subjected to qPCR as described above.

For hypoxic treatment, Arabidopsis were first grown on plates with half strength MS medium containing 0.5% sucrose at 23 °C in a growth chamber for 1 week, before being transferred to a glass box for hypoxic treatment in the dark to minimize oxygen production from photosynthesis. The glass box was sealed and supplied continuously with 99.99% argon gas at a rate of 500 cubic centimeters (cc)/min to maintain hypoxic conditions. In parallel, seedlings were incubated under a normal air supply as a control. The seedlings were harvested for RNA extraction at seven time points (air: 0, 1, 2, 4, 8, 16, and 24 h; hypoxia: 1, 2, 4, 8, 16, and 24 h).

First-strand cDNAs were synthesized and subjected to qPCR as above.

4.5. Generation of AIP1-Overexpressing Arabidopsis Transformants

The cDNA of *AIP1* was cloned into an ENTRY vector (pENTR 2B) and then recombined with a DESTINATION vector (pK7WGF2) in-frame and downstream of the green fluorescent protein (*GFP*) coding sequence to generate a GFP fusion as previously reported [39]. The resulting constructs were introduced into Agrobacterium (*Agrobacterium tumefaciens*) strain GV3101 and transformed into Arabidopsis Col-0 plants by the floral dip method. Homozygous T_3 lines were used for analyses after confirmation of constitutive overexpression of the genes.

4.6. Assessment of Hypoxia Tolerance of AIP1 Transformants by Measuring Recovery Rates

AIP1 transgenic seeds were sown onto square plates containing a half-strength MS medium supplemented with 0.5% sucrose and then incubated for 1 week in a growth chamber at 23 °C. As a control, Col-0 seeds were sown next to transgenic seeds on the same plates. The plates were then transferred to a glass box and placed in the dark to minimize oxygen production from photosynthesis. Argon gas (99.99%) was continuously flowed into the sealed glass box at a rate of 500 cc/min to maintain a hypoxic state. After 16 h of hypoxic treatment, plates were transferred to a 23 °C growth chamber and allowed to recover in air at a normal oxygen concentration for 3 days. After 3 days of recovery, the degree of leaf damage of Col-0 seedlings and *AIP1* transformants was evaluated with the standard five-step scale [40]. Fifteen transgenic lines were subjected to the hypoxia-survival test, and five lines (1-2, 1-5, 2-5, 2-6, and 4-5) were selected for further analysis according to consistent survival rate.

4.7. Assessment of Tolerance to Hypoxia by AIP1 Transformants Based on Electrolyte Leakage Assay

Seeds from Col-0 and *AIP1* transformants were sown in pots filled with sterilized Horticulture nursery Sunshine Mix No. 5 (Sun Gro Horticulture) and cultivated for 4 weeks in a culture room under a long day photoperiod (16 h light/8 h dark) with 60% humidity at 23 °C. After exposure to hypoxia or air for 4 h, two leaf disks (6.5-mm diameter) were excised from each plant (one disk per leaf) using a cork borer. Immediately after excision, the leaf disks were floated onto 2 mL sterilized distilled water in a well of a 12-well plate. The plates were then re-exposed to air at a normal oxygen concentration for 4, 8, 16, 20, or 24 h. From each well, 100 µL exudate was collected and applied to an electrical conductivity meter (HORIBA, LAQUAtwin-EC-22, Kyoto, Japan) to measure conductivity [41].

4.8. Measurement of BCAA Contents in AIP1 Transformants

AIP1 transgenic seeds were sown onto square plates containing a half-strength MS medium supplemented with 0.5% sucrose and incubated for 2 weeks in a growth chamber at 23 °C. As a control, Col-0 seeds were sown next to transgenic seeds on the same plates. The plates were then transferred to a glass box and placed in the dark to minimize oxygen production from photosynthesis. Argon gas (99.99%) was continuously flowed into the sealed glass box at a rate of 500 cc/min to maintain a hypoxic state. After 4 h of hypoxia or air treatment, the plant samples were divided into leaves and roots. Pulverized leaf tissues were hydrolyzed with 6 N HCL and then subsequently diluted with 0.02 N HCl. The Auto Amino Acid Analyzer (Hitachi, L-8900, Tokyo, Japan) was then used to quantify BCAA contents of the prepared samples.

Supplementary Materials: The following are available online at https://www.mdpi.com/article/10.3390/plants10112251/s1. Figure S1: Gene expression levels of *AIP1* in T_3 plants determined by RT-PCR plants. *ACTIN2* was used as an internal control. Figure S2: Survival scores represent non-damaged, 25% damaged, half-damaged, 75% damaged, and dead (5, 4, 3, 2, and 1, respectively).

Author Contributions: Investigation, G.I.; validation, G.I.; supervision, D.C.; writing—original draft preparation, G.I.; writing—review and editing, D.C. All authors have read and agreed to the published version of the manuscript.

Funding: This work was carried out with the support of "Cooperative Research Program for National Agricultural Genome Program (Project No. PJ013326012021)" Rural Development Administration, Republic of Korea.

Institutional Review Board Statement: Not applicable.

Informed Consent Statement: Not applicable.

Data Availability Statement: The data that support the findings of this study are available from the corresponding author upon reasonable request.

Conflicts of Interest: The authors declare that the research was conducted in the absence of any commercial or financial relationships that could be construed as a potential conflict of interest.

References

1. Mickelbart, M.V.; Hasegawa, P.M.; Bailey-Serres, J. Genetic mechanisms of abiotic stress tolerance that translate to crop yield stability. *Nat. Rev. Genet.* **2015**, *16*, 237–251. [CrossRef]
2. Malhi, G.S.; Kaur, M.; Kaushik, P. Impact of Climate Change on Agriculture and Its Mitigation Strategies: A Review. *Sustainability* **2021**, *13*, 1318. [CrossRef]
3. Fedoroff, N.V.; Battisti, D.S.; Beachy, R.N.; Cooper, P.J.M.; Fischhoff, D.A.; Hodges, C.N.; Knauf, V.C.; Lobell, D.; Mazur, B.J.; Molden, D.; et al. Radically rethinking agriculture for the 21st century. *Science* **2010**, *327*, 833–834. [CrossRef] [PubMed]
4. Drew, M.C. Plant injury and adaptation to oxygen deficiency in the root environment: A review. *Plant Soil* **1983**, *75*, 179–199. [CrossRef]
5. Jackson, M.B.; Colmer, T.D. Response and adaptation by plants to flooding stress. *Ann. Bot.* **2005**, *96*, 501–505. [CrossRef] [PubMed]
6. Fukao, T.; Barrera-Figueroa, B.E.; Juntawong, P.; Peña-Castro, J.M. Submergence and waterlogging stress in plants: A review highlighting research opportunities and understudied aspects. *Front. Plant Sci.* **2019**, *10*, 340. [CrossRef] [PubMed]
7. Choi, D. Molecular events underlying coordinated hormone action in submergence escape response of deepwater rice. *J. Plant Biol.* **2011**, *54*, 365. [CrossRef]
8. Gibbs, D.J.; Holdsworth, M.J. Every breath you take: New insights into plant and animal oxygen sensing. *Cell* **2020**, *180*, 22–24. [CrossRef]
9. Loreti, E.; Perata, P. The many facets of hypoxia in plants. *Plants* **2020**, *9*, 745. [CrossRef]
10. Hwang, J.H.; Lee, M.O.; Choy, Y.H.; Ha-Lee, Y.M.; Hong, C.B.; Lee, D.H. Expression profile analysis of hypoxia responses in Arabidopsis roots and shoots. *J. Plant Biol.* **2011**, *54*, 373. [CrossRef]
11. Hwang, S.T.; Kim, Y.L.; Sohn, S.H.; Choi, D. Gene expression profiling provides insight into the escape behavior of deepwater rice during submergence. *J. Plant Biol.* **2018**, *61*, 373–383. [CrossRef]
12. Minami, A.; Yano, K.; Gamuyao, R.; Nagai, K.; Kuroha, T.; Ayano, M.; Nakamori, M.; Koike, M.; Kondo, Y.; Niimi, Y.; et al. Time-course transcriptomics analysis reveals key responses of submerged deepwater rice to flooding. *Plant Physiol.* **2018**, *176*, 3081–3102. [CrossRef] [PubMed]
13. Streeter, J.G.; Thompson, J.F. Anaerobic accumulation of y-aminobutyric acid and alanine in radish leaves (*Raphanus Sativus*, L.). *Plant Physiol.* **1972**, *49*, 572–578. [CrossRef] [PubMed]
14. Reggiani, R.; Cantu, C.A.; Brambilla, I.; Bertani, A. Accumulation and interconversion of amino acids in rice roots under anoxia. *Plant Cell Physiol.* **1988**, *29*, 981–987.
15. De Sousa, C.A.F.; Sodek, L. Alanine metabolism and alanine aminotransferase activity in soybean (*Glycine Max*) during hypoxia of the root System and subsequent return to normoxia. *Environ. Exp. Bot.* **2003**, *50*, 1–8. [CrossRef]
16. Lee, D.J.; Chi, Y.T.; Kim, D.M.; Choi, S.H.; Lee, J.Y.; Choi, G.W. Ectopic Expression of CaRLK1 enhances hypoxia tolerance with increasing alanine production in *Nicotiana* Spp. *Plant Mol. Biol.* **2014**, *86*, 255–270. [CrossRef]
17. Singh, B.K.; Shaner, D.L. Biosynthesis of branched chain amino acids: From test tube to field. *Plant Cell* **1995**, *7*, 935–944. [CrossRef]
18. McCourt, J.A.; Pang, S.S.; King-Scott, J.; Guddat, L.W.; Duggleby, R.G. Herbicide-binding sites revealed in the structure of plant acetohydroxyacid synthase. *Proc. Natl. Acad. Sci. USA* **2006**, *103*, 569–573. [CrossRef]
19. Shi, Y. Characterization of the Arabidopsis *Glutamine Dumper1* Mutant Reveals Connections between Amino Acid Homeostasis and Plant Stress Responses. Ph.D. Thesis, Virginia Polytechnic Institute and State University, Blacksburg, VA, USA, 2015.
20. Jan, R.; Asaf, S.; Numan, M.; Kim, K.M. Plant secondary metabolite biosynthesis and transcriptional regulation in response to biotic and abiotic stress conditions. *Agronomy* **2021**, *11*, 968. [CrossRef]
21. Binder, S. Branched-chain amino acid metabolism in *Arabidopsis thaliana*. *Arab. Book* **2010**, *8*, e0137. [CrossRef]
22. Knill, T.; Schuster, J.; Reichelt, M.; Gershenzon, J.; Binder, S. Arabidopsis branched-chain aminotransferase 3 functions in both amino acid and glucosinolate biosynthesis. *Plant Physiol.* **2008**, *146*, 1028–1039. [CrossRef]
23. Chipman, D.; Barak, Z.; Schloss, J.V. Biosynthesis of 2-aceto-2-hydroxy acids: Acetolactate synthases and acetohydroxyacid synthases. *Biochim. Et Biophys. Acta—Protein Struct. Mol. Enzymol.* **1998**, *1385*, 401–419. [CrossRef]
24. Lee, Y.T.; Duggleby, R.G. Identification of the regulatory subunit of *Arabidopsis thaliana* acetohydroxyacid synthase and reconstitution with its catalytic subunit. *Biochemistry* **2001**, *40*, 6836–6844. [CrossRef]
25. Lee, Y.T.; Duggleby, R.G. Regulatory interactions in *Arabidopsis thaliana* acetohydroxyacid synthase. *FEBS Lett.* **2002**, *512*, 180–184. [CrossRef]

26. Gedi, V.; Yoon, M.Y. Bacterial acetohydroxyacid synthase and its inhibitors—A summary of their structure, biological activity and current status. *FEBS J.* **2012**, *279*, 946–963. [CrossRef] [PubMed]
27. Dezfulian, M.H.; Foreman, C.; Jalili, E.; Pal, M.; Dhaliwal, R.K.; Roberto, D.K.; Imre, K.M.; Kohalmi, S.E.; Crosby, W.L. Acetolactate synthase regulatory subunits play divergent and overlapping roles in branched-chain amino acid synthesis and Arabidopsis development. *BMC Plant Biol.* **2017**, *17*, 1–13. [CrossRef] [PubMed]
28. Duggleby, R.G. Identification of an acetolactate synthase small subunit gene in two eukaryotes. *Gene* **1997**, *190*, 245–249. [CrossRef]
29. Shimizu, M.; Fujii, T.; Masuo, S.; Takaya, N. Mechanism of de novo branched-chain amino acid synthesis as an alternative electron Sink in hypoxic *Aspergillus nidulans* cells. *Appl. Environ. Microbiol.* **2010**, *76*, 1507–1515. [CrossRef]
30. Binder, S.; Knill, T.; Schuster, J. Branched-chain amino acid metabolism in higher plants. *Physiol. Plant* **2007**, *129*, 68–78. [CrossRef]
31. Hershey, H.P.; Schwartz, L.J.; Gale, J.P.; Abell, L.M. Cloning and functional expression of the small subunit of acetolactate synthase from Nicotiana Plumbaginifolia. *Plant Mol. Biol.* **1999**, *40*, 795–806. [CrossRef] [PubMed]
32. Ellis, M.H.; Dennis, E.S.; Peacock, W.J. Arabidopsis roots and shoots have different mechanisms for hypoxic stress tolerance. *Plant Physiol.* **1999**, *119*, 57–64. [CrossRef]
33. Dolferus, R.; Ellis, M.; De Bruxelles, G.; Trevaskis, B.; Hoeren, F.; Dennis, E.S.; Peacock, W.J. Strategies of gene action in *Arabidopsis* during hypoxia. *Ann. Bot.* **1997**, *79*, 21–31. [CrossRef]
34. Ismond, K.P.; Dolferus, R.; De Pauw, M.; Dennis, E.S.; Good, A.G. Enhanced low oxygen survival in Arabidopsis through increased metabolic flux in the fermentative pathway. *Plant Physiol.* **2003**, *132*, 1292–1302. [CrossRef]
35. Geigenberger, P. Response of plant metabolism to too little oxygen. *Curr. Opin. Plant Biol.* **2003**, *6*, 247–256. [CrossRef]
36. Van Dongen, J.T.; Schurr, U.; Pfister, M.; Geigenberger, P. Phloem metabolism and function have to cope with low internal oxygen. *Plant Physiol.* **2003**, *131*, 1529–1543. [CrossRef] [PubMed]
37. Li, H. Characterization of *AtHIGD1* Gene in *Arabidopsis*. Master's Thesis, Kunsan National University, Gunsan-si, Jeollabuk-do (Miryong-dong), Korea, 2016.
38. Hwang, S.T. Molecular Physiological Characterization of *HIGD* Genes in Rice and *Arabidopsis* under Submergence or Hypoxia. Ph.D. Thesis, Kunsan National University, Gunsan-si, Korea, 2016.
39. Hwang, S.T.; Li, H.; Alavilli, H.; Lee, B.H.; Choi, D. Molecular and physiological characterization of *AtHIGD1* in Arabidopsis. *Biochem. Biophys. Res. Commun.* **2017**, *487*, 881–886. [CrossRef] [PubMed]
40. Gibbs, D.J.; Lee, S.C.; Isa, N.M.; Gramuglia, S.; Fukao, T.; Bassel, G.W.; Correia, C.S.; Corbineau, F.; Theodoulou, F.L.; Bailey-Serres, J.; et al. Homeostatic response to hypoxia is regulated by the N-end rule pathway in plants. *Nature* **2011**, *479*, 415–418. [CrossRef] [PubMed]
41. Hatsugai, N.; Katagiri, F. Quantification of plant cell death by electrolyte leakage assay. *Bio Protoc.* **2018**, *8*, e2758. [CrossRef]

plants MDPI

Article

Trait Variations and Probability Grading Index System on Leaf-Related Traits of *Eucommia ulmoides* Oliver Germplasm

Peng Deng [1], Xiangchen Xie [1], Feiyu Long [1], Liang Zhang [1], Yonghang Li [1], Zhangxu Zhao [2], Shiyao Yang [3], Yiran Wang [1], Ruishen Fan [1] and Zhouqi Li [1,*]

[1] College of Forestry, Northwest A&F University, Xianyang 712100, China; dengpeng@nwafu.edu.cn (P.D.); xxc@nwafu.edu.cn (X.X.); fly179@nwafu.edu.cn (F.L.); zhangliang2017@nwafu.edu.cn (L.Z.); yonghang@nwsuaf.edu.cn (Y.L.); wyr08@nwafu.edu.cn (Y.W.); fanruishen@nwsuaf.edu.cn (R.F.)
[2] College of Economics and Management, Northwest A&F University, Xianyang 712100, China; zhaozhangxu@nwafu.edu.cn
[3] College of Life Sciences, Northwest A&F University, Xianyang 712100, China; monster@nwafu.edu.cn
* Correspondence: lizhouqi@nwsuaf.edu.cn

Abstract: *Eucommia ulmoides* Oliver (EUO), an economic tree grown specifically in China, is widely used in various fields. To satisfy the requirements of industrial development, superior varieties need to be selected for different uses. However, there is no unified standard for breeders to reference. In this study, leaf-related traits were classified by a probability grading method. The results indicated there were significant differences between different planting models for the studied traits, and the traits in the Arbor forest model showed more abundant variation. Compared with genotype, the planting model accounted for relatively bigger variance, indicating that the standard should be divided according to planting models. Furthermore, the optimum planting model for different traits would be obtained by analyzing the variation range. Association analyses were conducted among traits to select the crucial evaluation indexes. The indexes were divided into three grades in different planting models. The evaluation system on leaf-related traits of EUO germplasm was established preliminarily, which considered planting models and stability across years for the first time. It can be treated as a reference to identify and evaluate EUO germplasm resources. Additionally, the study served as an example for the classification of quantitative traits in other economically important perennial plants.

Keywords: *Eucommia ulmoides* Oliver; trait variations; probability grading; quantitative traits; planting models; leaves

Citation: Deng, P.; Xie, X.; Long, F.; Zhang, L.; Li, Y.; Zhao, Z.; Yang, S.; Wang, Y.; Fan, R.; Li, Z. Trait Variations and Probability Grading Index System on Leaf-Related Traits of *Eucommia ulmoides* Oliver Germplasm. *Plants* **2021**, *10*, 2280. https://doi.org/10.3390/plants10112280

Academic Editors: Milan S. Stankovic, Paula Baptista and Petronia Carillo

Received: 23 August 2021
Accepted: 19 October 2021
Published: 25 October 2021

Publisher's Note: MDPI stays neutral with regard to jurisdictional claims in published maps and institutional affiliations.

Copyright: © 2021 by the authors. Licensee MDPI, Basel, Switzerland. This article is an open access article distributed under the terms and conditions of the Creative Commons Attribution (CC BY) license (https://creativecommons.org/licenses/by/4.0/).

1. Introduction

Eucommia ulmoiudes Oliver (EUO) is an economically important tree belonging to the monotypic family Eucommiaceae [1]. Unique to China, it has a high value for development and utilization. EUO has been used in China for more than 2000 years as a traditional Chinese medicine [2,3]. To date, 132 chemical compounds have been identified from it [4], and some of them contribute to treating hypertension, hyperlipemia, Alzheimer's disease, and aging [5]. Due to their health care function, the leaf and the bark were listed in the "Pharmacopia of China" from 2005 [6]. Surprisingly, the oil extracted from its cotyledon contains 66.4% of α-linolenic acid, 8–60 times greater than that in Oliver oil [7]. Additionally, EUO is a widely distributed tree species producing *Eucommia* rubber (*Eu*-rubber), a trans-polyisoprene (TPI) [8]. TPI performs better in insulation and corrosion resistance compared with cis-polyisoprene (CPI) produced by *Hevea brassiliensis* [9], which led to its use in insulated cables and medical instruments [10]. EUO is widely cultivated in 27 provinces in China, with a cultivation area of 0.35 million hectares [11]. Its better physical characteristics and extensive distribution make it a promising alternative or supplementary resource to *Hevea brassiliensis* [10,12]. A study reported that EUO orchards can produce 4000 kg fruit per hectare, about 520 kg of TPI, and the input–output ratio can reach up to 646.49% [7]. Furthermore, it is also used for landscaping and soil and water conservation as its peculiar fruit and a stronger adaption to the environment, respectively [13]. Therefore, it is called the "Chinese sacred tree" or "plants gold" with economic, social, and ecological benefits.

Almost all parts of EUO, including the bark, branches, leaves, flowers, and fruit, are of high value in various fields [14–17]. Compared with other organs, leaves are easy to access and abundant in yield [14]. They can be harvested throughout the entire growing season and are unconstrained by the reproductive growth stage. Therefore, it is an ideal tissue type to exploit. To date, leaves of EUO have been used in numerous fields. To satisfy the requirements of different industries, superior varieties or clones need to be selected according to target traits. For leaves, many traits are quantitative characteristics, and evaluation criteria are needed as a reference for breeders. However, the existing classification standard of leaf-related traits is not clear or comprehensive, which restricts the identification and evaluation of EUO germplasm.

There are various grading methods for the classification of quantitative traits, which include average classification, clustering analysis, and probability grading. Traditionally, the grading method has equal differences based on experience. Simple and feasible as it is, it cannot reflect the distribution of traits. Furthermore, it is difficult to reach an agreement [18]. Clustering analysis is utilized to classify the study objects into several clusters according to the similarities among them. This fits research objectives with large differences. For example, this method has been used to classify daylily flower color [19]. However, the effect of grading strongly depends on the study objects. Specifically, the sample must be representative and the sample capacity should be large enough to ensure the reliability of the standard. For example, a total of 36,000 *Panax notoginseng* seedlings were collected from 30 major producing areas to establish the grading standard of seedlings [20]. For the classification of the morphological traits, geometric morphometric analysis is more precise in describing the shape of leaves [21]. Additionally, the leaf shape fractal dimension (FD) is considered as the best index to reflect the complexity of leaf shape. However, it is more complicated and cannot be used for the other traits. A study reported that there was strong correlation between the FD and length-to-width ratio. Therefore, sufficient information on leaf shape can be provided by the length-to-width ratio, the simpler index [22]. Probability grading is an approach based on the probability distribution feature of quantitative trait values. It illustrates the average level and dispersion degree of traits and reflects the systematic position of the individual in the overall level [23]. Due to the objectivity in description, it contributes to the unification of standards among different breeders [18]. There are 3-grade and 5-grade standards for it [18,23]. Considering that the distribution of quantitative traits in nature obeys normal distribution [23], which shows the distribution characteristics of larger in the middle and smaller at both ends, the middle grade is allocated 40% of the occurrence probability. In addition, the values of middle grade are distributed around the average. Therefore, the probability grading method has a greater guidance value in practice. To date, it has been successfully used in numerous economically important trees, and ideal effects were obtained. Liu [24] proposed the probability grading method for the first time and successfully used it in the classification of economic traits of peach trees. Soon afterwards, Liu [23] and Liu [18] developed the method and applied it to the evaluation of quantitative traits in Chinese Jujuba. In recent years, the method has been used in the classification of important quantitative characteristics in mango [25], *Armeniaca vulgaris* [26], table grape [27], wild *Actinidia eriantha* [28], and apple fruit [29], among others. Surprisingly, it was also used to define the shape of the ray floret in large-flowered chrysanthemum to make the standard more accurate and objective than direct observation [30]. For EUO, it has been used in the classification of quantitative traits in male flowers [31] and fruit [32], but similar studies are rare in leaves of EUO. Studies on the classification of leaf-related traits of EUO mainly focused on the secondary metabolites of leaves [33,34], which is clearly not enough for the utilization of EUO. A more comprehensive probability grading system based on target traits is necessary.

EUO is a woody perennial plant; it usually takes 7 years after planting before it flowers and fruits [35], and then it turns from the vegetative to the reproductive growth stage. Unlike other generative organs, leaves can be obtained in all growing stages. Previous studies have reported significant differences in leaf-related traits between stages not only in EUO [36–38], but also in other species [39–41]. There are two kinds of planting models for EUO in production, including the Leaf-oriented cultivation model (LCM) and Arbor forest model (AFM) [42]. The difference between them in essence is that they are, respectively, in the vegetative and reproductive growth stages. To better guide production, the classification system should be formulated according to planting models.

In this study, the probability grading method was used to set up a standard for leaf-related traits of EUO. To satisfy the requirements of different industries, traits were classified according to target traits. The variations and distribution types of traits in different planting models were analyzed. In addition, this study revealed the impacts of planting models, genotypes, and tree age on different traits. To remove duplicated traits in evaluating germplasm, association analyses were analyzed. Considering the limited germplasm resources and generality of the standard, evaluation indexes were classified into three grades in separate models. This could serve as a reference for the evaluation, selection, and discrimination of EUO germplasm, which is beneficial for accelerating the process of EUO breeding.

2. Materials and Methods

2.1. Plant Materials

In this study, 56 Arbor forest model (AFM) trees and 48 Leaf-oriented cultivation model (LCM) clones were employed. Among these, 56 AFM trees comprised 15 EUO cultivars and 41 superior individuals. The 15 EUO cultivars were selected from different areas of China [43], and the 41 superior individuals were selected from the F1 generation of cross breeding of the superior cultivars according to the tree height and ground diameter. Forty-four in 56 of AFM trees and "Qinzhong NO. 1–4" were grafted to 48 LCM clones in August 2018. A field trial was set up at the nursery of the College of Forestry at Northwest Agricultural and Forestry University (108°05′ E, 34°24′ W) in Yangling, China. The 56 AFM trees were germinated in the EUO Germplasm Resources Collection Area at a spacing of 2 m × 2 m in March 2010. In addition, 48 LCM clones were laid out in randomized complete blocks at a spacing of 0.5 m × 0.5 m, including three blocks with 48 clones per block. There were 90 seedlings for each clone, with 30 seedlings in each block.

2.2. Studied Traits

The leaf-related traits of EUO in this study are listed in Table 1. Among these, T1 to T7 were used to describe the leaf morphology, and they can be used to distinguish and identify germplasms [44–46]. There are two main harvest products in *Eucommia ulmoides* Oliver: gutta-percha and chlorogenic acid. The total economic output depends on their contents and leaf yield. T8 and T9 depicted leaf yield, and T10 to T11 showed the content of secondary metabolites in leaves. Additionally, adaption must be considered when they are planted in different areas. T12 to T15 represented the water status of leaves, which were used to measure drought resistance [47–49].

Table 1. Leaf-related traits of *Eucommia ulmoides* Oliver in the current study.

No.	Traits	Abbreviation
T1	Leaf length (mm)	LL
T2	Leaf width (mm)	LW
T3	Leaf area (mm^2)	LA
T4	Leaf length–width ratio	LWR
T5	Thickness of a single leaf (mm)	TL
T6	Petiole length (mm)	PL
T7	Number of veins	NV
T8	Single leaf weight (g)	WL
T9	Total number of leaves	TNL
T10	Content of gutta-percha (%)	GP
T11	Content of chlorogenic acid (%)	CA
T12	Osmotic potential at turgor loss	$\Psi_\pi{}^0$
T13	Osmotic potential at full turgor	$\Psi_\pi{}^{100}$
T14	Critical water saturation deficit	RWD_0
T15	Relative bound water content at full turgor	*Ma*

2.3. Planting Models

The Leaf-oriented cultivation model (Figure 1a) and Arbor forest model (Figure 1b) of *Eucommia ulmoides* Oliver trees were displayed as follows.

Figure 1. Leaves in different planting models, including (**a**) the Leaf-oriented cultivation model and (**b**) the Arbor forest model.

2.4. Measurement of Leaf Water Status

For each LCM clone, a fresh sunshine leaf (the 8th leaf from the top) was taken from a healthy individual selected randomly from each block in August 2019. For every AFM tree, a fresh sunshine leaf (the 8th leaf from the top) was collected from similar-size branches in the direction of north, southeast, and southwest in August 2019. Thus, there were 3 replicates for every LCM clone or AFM tree. Leaves were collected in the evening and placed in water to absorb it throughout the night. The fresh leaf was weighed (W_f) and fastened to the PMS Model 1000 Pressure Chamber Instrument (PMS Instrument Company, Albany, OR, USA). By gradually applying pressure and recording the pressure and the weight of removed water, a fitted curve and straight line were acquired. The leaf was placed in a drying oven, and the dry weight (W_d) was recorded [47].

2.5. Sample Collections and Measurement of Leaf Morphological Traits

For each LCM clone, fresh leaves (the 10th leaf from the top) were collected from 10 healthy seedlings selected randomly from each block in September 2019, and the total fresh weight of them was about 20 g. For every AFM tree, the fresh leaves (the 10–15th leaf from the top) were collected from similar-size branches in the direction of north, southeast, and southwest in September 2017, 2018, and 2019, and the total fresh weight of them was about 60 g.

Leaf area (LA), leaf length (LL), and leaf width (LW) were measured by a Yaxin-1241 leaf area meter (Beijing Yaxin Technology Co., Ltd., Beijing, China). Petiole length (PL) and the thickness of a single leaf (TL) were measured using a vernier caliper. The single leaf weight (WL) was measured using an electronic balance.

2.6. Sample Processing and Measurement of the Secondary Metabolite Content in Leaves

The samples were placed in a drying oven at 60 °C. Then, they were stored at room temperature for later use. The samples were ground into powder using a DHS TL2020 tissue grinder apparatus (DHS Life Science & Technology, Beijing, China) before measurement.

The extraction and determination of gutta-percha were performed following the description of Ma [50]. Briefly, 100 mL NaOH solution (10%, m/m) was added to 5 g of powder in a water bath at 90 °C for 3 h, twice. After filtration, 60 mL HCI was added to the residue in a water bath at 40 °C for 2 h. Then, about 60 mL ethanol (60%, v/v) was added to the residue after filtration The solution was incubated for 1h and put into an ultrasonic cleaner (40 kHz, 40 °C) for 0.5 h. The content of gutta-percha (GP) could be obtained after filtration and drying at room temperature.

The extraction of chlorogenic acid was conducted with reference to Dong [51]. Separation and determination were performed using an Agilent Technologies (Santa Clara, CA, USA) HPLC system model 1260 following Ye [52]. The detection of chlorogenic acid was recorded at 320 nm. The 1260 DAD-chemstation (offline) software was utilized for data analysis (using peak area values) and the content of chlorogenic acid (CA) was determined using external calibration. The chlorogenic acid standard substance (HPLC $\geq$ 98%) was purchased from Shanghai Yuanye Bio-technology Co., Ltd (Shanghai, China).

2.7. Data Analyses

2.7.1. Chlorogenic Acid Content Determination

The chlorogenic acid content was determined by the following standard curve:

$$y = 2 \times 10^{-5}x + 0.0041 \left(R^2 = 0.9992\right) \tag{1}$$

where x is the peak area and y is the content.

2.7.2. Calculation of Water Status Parameters

According to the fitted curve and straight line, two formulas were obtained as follows:

$$1/\Psi_w = aV_1^b \tag{2}$$

$$1/\Psi_\pi = c + dV_2 \tag{3}$$

In this way, the following parameters were obtained:

$$\Psi_\pi^0 = \frac{1}{a}\left(\frac{d}{ab}\right)^{\left(\frac{b}{1-b}\right)} \tag{4}$$

$$\Psi_\pi^{100} = \frac{1}{c} \tag{5}$$

$$V' = \left(\frac{d}{ab}\right)^{\left(\frac{1}{b-1}\right)} \tag{6}$$

$$V_0 = -c/d \tag{7}$$

$$V_t = W_s - W_d \tag{8}$$

$$RWD_0 = \left(V'/V_t\right) \times 100\% \tag{9}$$

$$Ma = \left(V_t - V_0\right)/V_t \tag{10}$$

V': Water content of osmotic at turgor loss;
V_0: Water content of osmotic at full turgor;
Vt: Tissue-saturated water content.

2.7.3. Variance Analyses and Multiple Comparisons

Variance analyses and Tukey's HSD multiple comparisons were conducted using SPSS 22.0.

2.7.4. Normal Distribution Test and Grading Standard

The normal distribution test was performed using the "Shapiro test" protocol in R software. Traits incompletely obeying a normal distribution were treated as normal distribution only when the main part was normally distributed. Traits were separated into 3 grades by (X − 0.5246S) and (X + 0.5246S) if they obeyed a normal distribution ("X" indicates the mean and "S" indicates the standard deviation of the traits) [23].

3. Results

3.1. Leaf-Related Trait Variations

As shown in Table 2, leaf-related traits in the Arbor forest model showed a more abundant variation. It was likely that differences accumulated over time. For the Arbor forest model, the descending order of coefficient of variations (CV) were: leaf yield, secondary metabolite content, morphology, and the water status of leaves. The order of the Leaf-oriented cultivation model was similar to the Arbor forest model except for the order of the top two traits. The CV of secondary metabolite content was prominently larger than leaf yield.

For leaf morphological traits, whether in the Leaf-oriented cultivation model or the Arbor forest model, LA owned the largest range of variation, while NV owned the smallest. The CV of LA was 23.1 and 15.1% in the Arbor forest model and Leaf-oriented cultivation model, respectively. The counterpart of NV was 8.1 and 5.7% (Table 2).

For leaf yield traits, the WL contained broader variation (17.2%) in the Leaf-oriented cultivation model, while the TNL had broader variation (36.55%) in the Arbor forest model (Table 2).

For secondary metabolite content, the CA was more extensive in variation (35.99% for the Arbor forest model, 33.14% for the Leaf-oriented cultivation model); however, the GP was much narrower (23.57 and 19.87%, respectively) (Table 2).

Table 2. Variations and normal test for leaf-related traits in *Eucommia ulmoides* Oliver.

Traits	Planting Models [a]	Tree Ages	Germplasm Number	Min.	Max.	Range	Mean [b]	*SD* [c]	CV [d] (%)	*p* Value [e]
LL	L	1	48	192.4	272.7	80.3	232.8 A	17.0	7.3	0.7236
	A	8	56	92.7	221.3	128.6	147.1 B	27.3	18.5	0.2640
	A	9	56	107.0	160.3	53.3	134.5 C	11.4	8.5	0.9305
	A	10	56	120.7	209.3	88.6	147.6 B	14.7	10.0	0.1863
LW	L	1	48	93.0	138.2	45.2	115.2 A	11.3	9.8	0.6942
	A	8	56	39.0	94.9	55.9	63.8 B	12.5	19.6	0.1102
	A	9	56	46.2	80.1	33.9	60.8 B	8.3	13.7	0.0507
	A	10	56	51.2	93.3	42.1	66.9 B	8.0	12.0	0.1963
LA	L	1	48	12,314.8	22,737.1	10,422.3	17,877.1 A	2691.1	15.1	0.3470
	A	8	56	1971.4	12,558.9	10,587.5	5348.2 B	1664.0	31.1	0.1853
	A	9	56	3023.0	8191.3	5168.3	4893.3 B	1002.8	20.5	0.1302
	A	10	56	3383.1	11,749.5	8366.4	5905.9 B	1054.2	17.8	0.8026
LWR	L	1	48	1.62	2.46	0.84	2.07 B	0.19	9.1	0.9412
	A	8	56	1.61	3.21	1.60	2.35 A	0.29	12.3	0.2370
	A	9	56	1.80	2.84	1.03	2.24 A	0.24	10.7	0.3822
	A	10	56	1.60	2.81	1.21	2.25 A	0.26	11.4	0.3881
TL	L	1	48	0.43	0.59	0.16	0.51 A	0.04	7.7	0.6184
	A	8	56	0.14	0.40	0.26	0.25 C	0.05	18.9	0.0749
	A	9	56	0.18	0.36	0.18	0.26 BC	0.04	15.5	0.3698
	A	10	56	0.21	0.38	0.16	0.28 B	0.04	13.2	0.1244
PL	L	1	48	16.67	33.11	16.44	25.15 A	1.87	7.5	0.8569
	A	8	56	11.51	32.73	21.22	18.27 B	3.92	21.5	0.0681
	A	9	56	12.32	28.48	16.16	17.82 B	2.57	14.4	0.2036
	A	10	56	13.26	29.51	16.25	18.77 B	2.55	13.6	0.7481
NV	L	1	48	16.2	20.1	4.0	18.2 A	1.0	5.7	0.3619
	A	8	56	12.3	19.0	6.7	15.0 B	1.4	9.1	0.1447
	A	9	56	12.3	18.0	5.7	15.2 B	1.3	8.7	0.4919
	A	10	56	13.0	18.7	5.7	15.4 B	1.0	6.6	0.0721
WL	L	1	48	1.2328	2.6245	1.3918	1.8318 A	0.3144	17.2	0.3674
	A	9	56	0.1450	0.9157	0.7707	0.3766 B	0.1081	28.7	0.5583
	A	10	56	0.2027	0.9515	0.7488	0.4275 B	0.1110	26.0	0.5641
TNL	L	1	48	24	50	26	35 B	5	15.1	0.6885
	A	9	56	562	4060	3498	1656 A	550	33.2	0.1808
	A	10	56	347	3815	3468	1878 A	750	39.9	0.0843
GP	L	1	48	0.71	1.62	0.90	1.10 B	0.22	19.87	0.1229
	A	8	55	0.87	3.20	2.33	1.75 A	0.45	25.49	0.4512
	A	9	56	0.90	3.29	2.40	1.79 A	0.43	24.09	0.1148
	A	10	56	0.67	2.95	2.28	1.92 A	0.41	21.13	0.2740
CA	L	1	48	0.56	2.79	2.22	1.62 A	0.54	33.14	0.7486
	A	9	56	0.32	2.28	1.96	1.20 B	0.41	34.49	0.5308
	A	10	56	0.30	2.35	2.04	1.29 B	0.48	37.49	0.5547
Ψ_π^{0}	L	1	48	13.06	16.81	3.75	15.09 A	0.92	6.07	0.1170
	A	10	50	12.08	16.91	4.83	14.55 B	0.79	5.46	0.0754
Ψ_π^{100}	L	1	48	7.98	9.99	2.01	9.09 A	0.51	5.64	0.0647
	A	10	50	7.20	9.77	2.57	8.44 B	0.56	6.62	0.9538
RWD_0	L	1	48	6.32	14.27	7.94	9.66 A	1.71	17.69	0.2198
	A	10	50	4.33	11.50	7.17	7.63 B	1.65	21.59	0.4522
Ma	L	1	48	0.61	0.85	0.24	0.75 B	0.05	6.25	0.3185
	A	10	50	0.71	0.89	0.18	0.82 A	0.04	5.52	0.2169

[a] "L" indicates the Leaf-oriented cultivation model, "A" indicates the Arbor forest model. [b] The same letter for the same trait indicates a non-significant difference, while different letters indicate an extremely significant difference at "$p < 0.01$" level, according to Tukey HSD. [c] *SD* indicates standard deviation. [d] CV indicates coefficient of variation. [e] *p* value indicates the significance level of the normal distribution test. The same is below.

For leaf water status traits, the descending order of CV showed a consistent trend, whether in the Leaf-oriented cultivation model or the Arbor forest model. The descending order was as follows: RWD_0, Ψ_π^{0}, Ψ_π^{100}, and *Ma*. The CVs ranged from 6.8 to 25.3% in the Arbor forest model, the counterpart of which was 5.9 to 16.6% in the Leaf-oriented cultivation model (Table 2).

The CV shows the genetic potential of traits [23], and it could be used to describe and distinguish germplasm resources [53]. Characteristics owning a large range of variation could be better used to classify germplasm resources. Additionally, the degree of variation determined the effectiveness of selection. The leaf yield and the secondary metabolite content owned larger variation whether in the Leaf-oriented cultivation model or Arbor forest model. Accordingly, more attention should be given to them in superior germplasm selection.

3.2. Variance Analyses and Multiple Comparisons of Leaf-Related Traits

Table 3 indicates that there is a significant difference between different planting models for the studied traits. Compared with genotype, the planting model accounted for relatively bigger variance. Therefore, the grading standard should be divided according to planting models. For TL and leaf water status traits, there was a non-significant difference among genotypes. Considering their non-ideal effect in germplasm identification, they should be excluded in the evaluation system. On this basis, variance analyses were conducted among tree ages for the same trait. It showed there was a significant difference among the tree ages in different planting models, while differences were non-significant within the same planting model. Therefore, it was necessary to unify the standards in the same planting model. Furthermore, the optimum utilization model for different traits would be obtained by analyzing the variation range.

Table 3. Variance analyses of leaf-related traits in different planting models of *Eucommia ulmoides* Oliver.

Traits	Planting Model				Genotype			
	Df	MS	*F* Value	*p* Value	Df	MS	*F* Value	*p* Value
LL	1	828,917.83	1499.442	0.000	55	1855.53	3.356	0.000
LW	1	264,342.54	2093.847	0.000	55	588.84	4.664	0.000
LA	1	15,420,000,000	3019.958	0.000	55	16,918,968.05	3.314	0.000
LWR	1	3.030	35.737	0.000	55	0.359	4.229	0.000
TL	1	4.667	1089.642	0.000	55	0.004	0.899	0.678
PL	1	5023.425	495.826	0.000	55	91.702	9.051	0.000
NV	1	922.112	244.045	0.000	55	8.953	2.370	0.000
WL	1	117.398	1819.325	0.000	55	0.167	2.586	0.000
TNL	1	309,741,032.2	742.662	0.000	55	1,541,740.43	3.697	0.000
GP	1	43.648	415.807	0.000	55	0.857	8.162	0.000
CA	1	8.455	36.671	0.000	55	0.968	4.198	0.000
Ψ_{π}^{0}	1	17.075	7.829	0.006	46	2.296	1.053	0.391
Ψ_{π}^{100}	1	26.263	33.707	0.000	46	1.015	1.303	0.107
RWD_0	1	815.416	44.351	0.000	46	16.858	0.917	0.627
Ma	1	0.635	50.352	0.000	46	0.012	0.913	0.634

Df: Degrees of freedom, MS: Mean square.

For the leaf morphological traits shown in Table 2, the leaves in the Leaf-oriented cultivation model were remarkably larger than those in the Arbor forest model. This was consistent with another study. Du [36] reported that the leaves became smaller in size with increasing tree age in EUO. Compared with the leaves in the Arbor forest model, the leaves in the Leaf-oriented cultivation model were easy to access, which could save labor costs. Therefore, the leaves in the Leaf-oriented cultivation model were more likely to satisfy the demand for leaf use.

Interestingly, the two secondary metabolites appeared to have different features between different planting models. For GP, the Arbor forest model had a higher content than the Leaf-oriented cultivation model. This was consistent with a previous study. Du [36] found that the content of GP tended to increase gradually with tree age. However, it was the opposite for CA. This was consistent with previous research. Yang [37] and Zhang [38] pointed out that the CA content was reduced along with the tree age. It seemed that the leaves in the Leaf-oriented cultivation model were more suitable for medicinal use, while those in the Arbor forest model were more fit for rubber use (Table 2).

Leaf water status can be used to evaluate drought resistance [49], among which the $\Psi_{\pi}{}^{0}$ was considered as the best index to measure drought resistance [48] (p. 59). Many traits were negatively correlated with drought resistance, except *Ma*. The Arbor forest model had an advantage over the Leaf-oriented cultivation model in drought resistance (Table 2). Perhaps the seedlings in the Leaf-oriented cultivation model were more susceptible to the environment. In addition, these findings agreed with those of a preceding study [39]. Li found that $\Psi_{\pi}{}^{0}$, $\Psi_{\pi}{}^{100}$, and RWD_0 decreased significantly with the increase in age in the platform for *Populus simonii,* which indicated an increase in drought resistance with tree age.

In general, there was a significant difference between different planting models for all traits measured in this study. For traits in different models, the optimum utilization planting model can be obtained by analyzing the variation range. Consequently, the relationship between the target traits and planting models was expounded, which contributed to the improvement of EUO breeding.

3.3. Association Analyses among Leaf-Related Traits

To reduce repetitive traits in evaluating EUO germplasm, association analyses were conducted among leaf-related traits in different planting models (Table 4). There were (extremely) significant correlations among LL, LW, LWR, WL, and LA in the two planting models, which indicated LA provided more information. In addition, LA had larger variation compared with NV and PL. Therefore, LA can be the best index to show leaf morphology. Unlike WL, TNL was a relatively independent index, which can be used to represent leaf yield. Gutta-percha and chlorogenic acid were the main harvest products, and GP and CA represented the effective constituent content. In conclusion, LA, TNL, GP, and CA can be the evaluation indexes.

Table 4. Association analyses among leaf-related traits in different planting models of *Eucommia ulmoides* Oliver.

	LL	LW	LA	LWR	PL	NV	WL	TNL	GP	CA
LL	1	0.498 **	0.724 **	0.196	0.110	0.178	0.457 **	−0.296 *	0.277	−0.449 **
LW	0.701 **	1	0.933 **	−0.740 **	0.215	0.187	0.801 **	−0.211	0.006	−0.399 **
LA	0.873 **	0.940 **	1	−0.496 **	0.185	0.213	0.797 **	−0.279	0.125	−0.467 **
LWR	0.133	−0.604 **	−0.333 *	1	−0.151	−0.111	−0.545 **	−0.019	0.191	0.090
PL	0.387 **	0.127	0.234	0.289 *	1	0.218	0.229	−0.134	0.057	−0.168
NV	0.356 **	0.413 **	0.426 **	−0.155	0.236	1	0.055	0.026	0.108	−0.017
WL	0.447 **	0.462 **	0.524 **	−0.132	0.071	0.200	1	−0.080	0.153	−0.363 *
TNL	−0.023	0.118	0.062	−0.233	−0.201	0.078	−0.035	1	0.267	0.289 *
GP	−0.226	−0.223	−0.204	0.050	−0.152	0.139	−0.011	0.137	1	−0.315 *
CA	−0.174	−0.203	−0.181	0.068	−0.219	−0.015	−0.079	0.157	0.240	1

Association analyses among leaf-related traits in the Leaf-oriented cultivation model and Arbor forest model are, respectively, above and below diagonal. "*" and "**" indicate that the degree of correlation is significant ($p < 0.05$) or extremely significant ($p < 0.01$) among traits.

3.4. Distribution Type of Leaf-Related Traits

As seen in Table 2, all traits were normally distributed. The test for a normal distribution is the precondition of probability grading; it can indicate whether the samples are representative. If the samples obey a normal distribution, the germplasm resources studied can be treated as the result of random sampling. In this way, they can represent the overall to some extent, and the standard formulated by the germplasm can be more effective. The distribution of quantitative traits in nature is in accordance with a normal distribution. The reason why some traits do not obey a normal distribution is directional selection in breeding [23]. In addition, limited germplasm can affect it to some extent.

These will lead to some extreme points. Usually, they belong to the first or last grade of the total and can affect the distribution. However, the remaining part still obeys a normal distribution by removing the extreme values. A situation like this can be treated as a normal distribution [32]. In this way, the extreme values were divided into the first or last grade.

3.5. Probability Classification of Leaf-Related Traits

As shown in Figure 2 and Table S1, considering the stability across the years, dynamic measurements were obtained from 2017 to 2019 for the quantitative traits in the Arbor forest model, which corresponded to 8a to 10a of the tree ages. Grading points in different years could be unified within the Arbor forest model, as a non-significant difference was observed. However, the utilization of the Leaf-oriented cultivation model in production mainly focuses on 1a seedlings by coppicing every year. To better guide production, the standard in the Leaf-oriented cultivation model was formulated by 1a seedings. To promote and apply the standard, it should be easy to remember. Therefore, it was necessary to adjust the grading points slightly. Special attention was required for the part that was adjusted no more than 0.1 SD for corresponding traits [23].

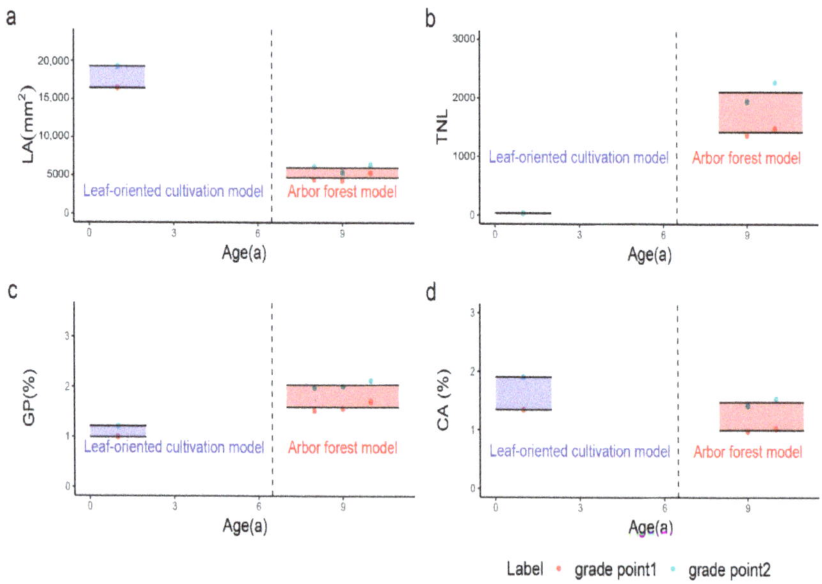

Figure 2. The classification criteria of the main evaluation indexes, including (**a**) LA, (**b**) TNL, (**c**) GP, and (**d**) CA. The black solid lines above and below the rectangle indicate the average level of grade points in each planting model. The lines at the top and bottom of the rectangle box represent the adjusted grade points. Traits were divided into three grades by the rectangle box: the 1st grade (the part below the rectangle), the 2nd grade (the part within the rectangle), and the 3rd grade (the part above the rectangle). The dashed line represents the boundary between the Leaf-oriented cultivation model and Arbor forest model.

Table 5 shows that the evaluation indexes were classified into three grades, respectively, in different planting models. Grade 3 was better in quality, Grade 2 was medium, and Grade 1 was worse.

Table 5. The criterion of leaf-related traits in *Eucommia ulmoides* Oliver.

Traits	Planting Models	Grade		
		1 (Worse)	2 (Medium)	3 (Better)
LA (mm^2)	L	≤16,465	16,465–19,230	≥19,230
	A	≤4730	4730–6035	≥6035
TNL	L	<32	32–37	≥37
	A	<1430	1430–2100	>2100
GP (%)	L	<1.00	1.00–1.20	>1.20
	A	<1.60	1.60–2.05	≥2.05
CA (%)	L	<1.35	1.35–1.90	≥1.90
	A	≤1.00	1.00–1.50	≥1.50

LA was used to distinguish leaf morphology. In the Leaf-oriented cultivation model, the variation range was 12,314.8–22,737.1 mm^2, with an average of 17,877.1 mm^2. The grading standard was as follows: grade 1 (smaller) ≤ 16,465 mm^2, 16,465 mm^2 < grade 2 (medium) < 19,230 mm^2, grade 3 (bigger) ≥ 19,230 mm^2; in the Arbor forest model, LA ranged from 1971.4 to 12,558.9 mm^2, with an average of 5382.5 mm^2. The classification criterion was as follows: grade 1 (smaller) ≤ 4730 mm^2, 4730 mm^2 < grade 2 (medium) < 6035 mm^2, grade 3 (bigger) ≥ 6035 mm^2.

TNL was used to represent leaf yield. In the Leaf-oriented cultivation model, the variation range was 24–50, with an average of 35. The grading standard was as follows: grade 1 (less) < 32, 32 ≤ grade 2 (medium) < 37, grade 3 (more) ≥ 37; in the Arbor forest model, the TNL ranged from 347 to 4060, with an average of 1767. The classification criterion was as follows: grade 1 (less) < 1430, 1430 ≤ grade 2 (medium) ≤ 2100, grade 3 (more) > 2100.

For GP, in the Leaf-oriented cultivation model, the variation range was 0.71–1.62%, with an average of 1.10%. The grading standard was as follows: grade 1 (lower) < 1.00%, 1.00% ≤ grade 2 (medium) ≤ 1.20%, grade 3 (higher) > 1.20%; in the Arbor forest model, the GP ranged from 0.67 to 3.29%, with an average of 1.82%. The classification criterion was as follows: grade 1 (lower) < 1.60%, 1.60% ≤ grade 2 (medium) < 2.05%, grade 3 (higher) ≥ 2.05%.

For CA, in the Leaf-oriented cultivation model, the variation range was 0.56–2.79%, with an average of 1.62%. The grading standard was as follows: grade 1 (lower) < 1.35%, 1.35% ≤ grade 2 (medium) < 1.90%, grade 3 (higher) ≥ 1.90%; in the Arbor forest model, the CA ranged from 0.30 to 2.35%, with an average of 1.25%. The classification criterion was as follows: grade 1 (lower) ≤ 1.00%, 1.00% < grade 2 (medium) < 1.50%, grade 3 (higher) ≥ 1.50%.

The quantitative traits were divided into five grades in most studies; the occurrence probability of Grades 1–5 was 10, 20, 40, 20, and 10%, respectively [30]. The selection of forest trees is a long-term process; it usually requires multiple selections, which is mainly due to the uncertainty of future performance. Therefore, the inclusion criterion should be reduced properly for forest tree selection. Therefore, it is appropriate to combine the first two and last two grades since the best grade is too narrow. In conclusion, it was preferable to divide the traits into three grades for this study; the occurrence frequency of Grades 1–3 was 30, 40, and 30%, respectively.

The probability grading index system on leaf-related traits of EUO germplasm was established preliminarily. It can be a standard for breeders to reference. Superior germplasm with a higher leaf yield and effective constituent content can be selected according to different target traits. They can greatly promote the economic performance in certain areas, which is quite critical to the large-scale production of industries.

4. Conclusions

In this study, to identify and evaluate EUO germplasm, leaf-related traits were classified by a probability grading method according to target traits. There was a significant difference between different planting models for the studied traits, and the traits in the

Arbor forest model owned abundant variations. The planting model accounted for a larger proportion of variance compared to the genotype, indicating the necessity to establish the grading standard according to planting models. Multiple comparisons analyses indicated there was a significant difference among the tree ages in different planting models, while a non-significant difference was observed within a planting model. Therefore, the grading points were unified in diverse years in the Arbor forest model. Considering the overlap among the traits in evaluating germplasm, association analyses were conducted to select the critical evaluation indexes. To make the standard more universal, the 3-grade standard was more suitable than the 5-grade standard. Germplasm in the Leaf-oriented cultivation model was more appropriate for leaf use and medical use, while germplasm in the Arbor forest model was more appropriate for rubber use and drought resistance. The evaluation system on leaf-related traits of EUO germplasm was established preliminarily, which considered target traits, planting models, and stability across the years. This can be a reference for the selection of superior germplasm for different target traits of EUO in different planting models. However, limited germplasm resources restrict the application of this standard to a wider scale. Therefore, additional studies should enlarge the germplasm resources.

Supplementary Materials: The following are available online at https://www.mdpi.com/article/10.3390/plants10112280/s1, Table S1: The grading standard of leaf-related traits of *Eucommia ulmoides* Oliver in different tree ages.

Author Contributions: Conceptualization, P.D. and Z.L.; Data curation, P.D., X.X., F.L., L.Z., Y.L., Z.Z., S.Y. and Y.W.; Formal analysis, P.D.; Funding acquisition, Z.L.; Investigation, P.D., X.X., F.L., L.Z., Y.L., Z.Z., S.Y., Y.W. and R.F.; Methodology, P.D. and Z.L.; Project administration, Z.L.; Software, P.D.; Supervision, Z.L.; Visualization, P.D.; Writing—original draft, P.D.; Writing—review and editing, R.F. and Z.L. All authors have read and agreed to the published version of the manuscript.

Funding: This research was funded by the Shaanxi Research and Development (R&D) Program (2019NY-012).

Data Availability Statement: The data presented in this study are available on request from the corresponding author.

Acknowledgments: The authors are thankful to Kang Gao for his assistance with the method of the classification standard of quantitative traits. We also thank Xiuping Yang and Hang Yu for their help in the usage and analyses in HPLC.

Conflicts of Interest: The authors declare no conflict of interest.

References

1. Tippo, O. The comparative anatomy of the secondary xylem and the phylogeny of the Eucommiaceae. *Am. J. Bot.* **1940**, *27*, 832–838. [CrossRef]
2. Zhu, M.; Wen, J.; Zhu, Y.; Su, Y.; Sun, R. Isolation and analysis of four constituents from barks and leaves of *Eucommia ulmoides* Oliver by a multi-step process. *Ind. Crops Prod.* **2016**, *83*, 124–132. [CrossRef]
3. Ouyang, D.; Wang, L.; Tang, T.; Feng, H. Genomic-Wide Identification and Characterization of the Uridine Diphosphate Glycosyltransferase Family in *Eucommia ulmoides* Oliver. *Plants* **2021**, *10*, 1934. [CrossRef] [PubMed]
4. Xing, Y.; He, D.; Wang, Y.; Zeng, W.; Zhang, C.; Lu, Y.; Su, N.; Kong, Y.; Xing, X. Chemical constituents, biological functions and pharmacological effects for comprehensive utilization of *Eucommia ulmoides* Oliver. *Food Sci. Hum. Wellness* **2019**, *8*, 177–188. [CrossRef]
5. He, X.; Wang, J.; Li, M.; Hao, D.; Yang, Y.; Zhang, C.; He, R.; Tao, R. *Eucommia ulmoides* Oliv.: Ethnopharmacology, phytochemistry and pharmacology of an important traditional Chinese medicine. *J. Ethnopharmacol.* **2014**, *151*, 78–92. [CrossRef] [PubMed]
6. China Pharmacopoeia Committee. *Chinese Pharmacopoeia of People's Republic of China*; Chemical Industry: Beijing, China, 2005.
7. Sun, Z.; Li, F.; Du, H.; Zhu, J.; Wang, Y.P. A novel silvicultural model for increasing biopolymer production from *Eucommia ulmoides* Oliver trees. *Ind. Crops Prod.* **2013**, *42*, 216–222. [CrossRef]
8. Bamba, T.; Fukusaki, E.; Nakazawa, Y.; Kobayashi, A. In-situ chemical analyses of trans-polyisoprene by histochemical staining and Fourier transform infrared microspectroscopy in a rubber-producing plant, *Eucommia ulmoides* Oliver. *Planta* **2002**, *215*, 934–939. [CrossRef]
9. Chen, R.; Namimatsu, S.; Nakadozono, Y.; Bamba, T.; Nakazawa, Y.; Gyokusen, K. Efficient regeneration of *Eucommia ulmoides* from hypocotyl explant. *Biol. Plant.* **2008**, *52*, 713–717. [CrossRef]

10. Wuyun, T.N.; Wang, L.; Liu, H.; Wang, X.; Zhang, L.; Bennetzen, J.L.; Li, T.; Yang, L.; Liu, P.; Du, L.; et al. The Hardy Rubber Tree Genome Provides Insights into the Evolution of Polyisoprene Biosynthesis. *Mol. Plant* **2018**, *11*, 429–442. [CrossRef]
11. Du, H. *China Eucommia Pictorial*; China Forestry Publishing House: Beijing, China, 2014.
12. Li, Y.; Wei, H.; Yang, J.; Du, K.; Li, J.; Zhang, Y.; Qiu, T.; Liu, Z.; Ren, Y.; Song, L.; et al. High-quality de novo assembly of the *Eucommia ulmoides* haploid genome provides new insights into evolution and rubber biosynthesis. *Hortic. Res.* **2020**, *7*, 183. [CrossRef]
13. Zhang, K. *Eucommia ulmoides Oliver*; China Forestry Publishing House: Beijing, China, 1990.
14. Zhang, K.; Dong, J.; Ma, B.; Gao, J.; Han, X. Studies on the distribution differences of the secondary metabolites in *Eucommia ulmoides*. *Sci. Silvae Sin.* **2002**, *38*, 12–16.
15. Zhu, M.; Sun, R. *Eucommia ulmoides* Oliver: A Potential Feedstock for Bioactive Products. *J. Agric. Food Chem.* **2018**, *66*, 5433–5438. [CrossRef]
16. Du, Q.; Wang, L.; Liu, P.; Qing, J.; Sun, C.; Sun, Z.; Du, H. Fatty Acids Variation in Seed of *Eucommia ulmoides* Populations Collected from Different Regions in China. *Forests* **2018**, *9*, 505. [CrossRef]
17. Wei, X.; Peng, P.; Peng, F.; Dong, J. Natural Polymer *Eucommia Ulmoides* Rubber: A Novel Material. *J. Agric. Food Chem.* **2021**, *69*, 3797–3821. [CrossRef]
18. Liu, P.; Liu, M.; Zhou, J.; Bi, P. Distribution and probability grading index system of quantitative character of Chinese Jujube. *Sci. Silvae Sin.* **2003**, *39*, 77–82.
19. Cui, H.; Zhang, Y.; Shi, X.; Gong, F.; Xiong, X.; Kang, X.; Xing, G.; Li, S. The numerical classification and grading standards of daylily (*Hemerocallis*) flower color. *PLoS ONE* **2019**, *14*, e0216460.
20. Chen, L.; Yang, Y.; Ge, J.; Cui, X.; Xiong, Y. Study on the grading standard of *Panax notoginseng* seedlings. *J. Ginseng Res.* **2018**, *42*, 208–217. [CrossRef]
21. Su, W.; Song, Y.; Qi, M.; Du, F. Leaf morphological characteristics of section Quercus based on geometric morphometric analysis. *Chin. J. Appl. Ecol.* **2021**, *32*, 2309–2315.
22. Shi, P.; Yu, K.; Niinemets, U.; Gielis, J. Can Leaf Shape be Represented by the Ratio of Leaf Width to Length? Evidence from Nine Species of Magnolia and Michelia (Magnoliaceae). *Forests* **2021**, *12*, 41. [CrossRef]
23. Liu, M. Study on probability grading of quantitative characters of Chinese Jujube. *Acta Hortic. Sin.* **1996**, *23*, 105–109.
24. Liu, M. Intraspecific variation of some economic traits in peach trees and its classification standard research. *J. Beijing Agric. Coll.* **1992**, *7*, 98–104.
25. Ni, Z.; Zhang, L.; Xie, D.; Luo, X.; Liu, C.; Wen, D.; Yu, Y. Variation and probability grading of main quantitative characteristics of mango germplasm resource in Nujiang Valley, Yunnan province. *J. Fruit Sci.* **2009**, *26*, 492–497.
26. Zhao, H.; Liu, W.; Liu, N.; Zhang, Y.; Zhang, Q.; Liu, S. Variation and probability grading of main quantitative traits of apricot (*Armeniaca vulgaris*) germplam. *J. Fruit Sci.* **2013**, *30*, 37–42.
27. Ma, X.; Zhao, Q.; Dong, Z.; Tang, X.; Wang, M.; Ren, R. Variation and Probability Grading of Main Quantitative Traits of Table Grape Resources. *J. Plant Genom. Resour.* **2013**, *14*, 1185–1189.
28. Lang, B.; Zhu, B.; Xie, M.; Zhang, W.; Seyrek, U.A.; Huang, C.; Xu, X. Variation and probability grading of the main quantitative characteristics of wild *Actinidia eriantha* germplasm resources. *J. Fruit Sci.* **2016**, *33*, 8–15.
29. Yan, Z.; Zheng, L.; Nie, J.; Li, Z.; Cheng, Y. Evaluation indices of sour flavor for apple fruit and grading standards. *J. Integr. Agric.* **2018**, *17*, 994–1002. [CrossRef]
30. Song, X.; Gao, K.; Fan, G.; Zhao, X.; Liu, Z.; Dai, S. Quantitative Classification of the Morphological Traits of Ray Florets in Large-flowered Chrysanthemum. *Hortscience* **2018**, *53*, 1258–1265. [CrossRef]
31. Du, Q.; Liu, P.; Qing, J.; Wei, Y.; Du, H. Variation and probability grading of main quantitative traits of male flowers for *Eucommia ulmoides* germplasm. *J. Beijing For. Univ.* **2016**, *38*, 42–49.
32. Du, Q.; Qing, J.; Wang, L.; Liu, P.; He, F.; Zhu, L.; Du, H. Variation and Probability Grading of Main Quantitative Traits of Fruits for *Eucommia ulmoides* Germplasm. *Bull. Bot. Res.* **2019**, *39*, 387–394.
33. Wei, Y.; Liu, P.; Du, Q.; Du, H. Comparison in Contents of Polyphenol and Flavonoid in Leaves of *Eucommia ulmoides* Germplasm. *For. Res.* **2016**, *29*, 529–535.
34. Qing, J.; Wei, Y.; Wang, D.; Wang, L.; Liu, P.; Du, H.; Du, Q. Study on Genetic Variation of Main Active Components in Leaves of *Eucommia ulmoides*. *Acta Bot. Boreali-Occident. Sin.* **2018**, *38*, 316–323.
35. Du, H. *Eucommia ulmoides Silviculture*; China Forestry Press: Beijing, China, 1996.
36. Du, H.; Du, L.; Xie, B.; Wuyun, T. Dynamic Accumulation of Gutta-percha Content in *Eucommia ulmoides* Oliv. Leaves. *J. Central South For. Univ.* **2006**, *26*, 1–6.
37. Yang, L.; Zhang, Z.; Mo, K.; Fu, Z. A Study of Variation Characteristics of Chlorogenic Acid Content in *Eucommia ulmoides* Leaves Under the Different Influence Factors. *J. Sichuan For. Sci. Technol.* **2012**, *33*, 50–52.
38. Zhang, W.; Zhu, W.; Zhang, L.; Fu, D.; Zhang, Z.; Zhang, L.; Zu, Y. Contents of main active ingredients and correlation with different ages and parts of *Eucommia ulmoides* Oliv. *Non-Wood For. Res.* **2020**, *38*, 46–57.
39. Li, J.; Li, Y.; Zhao, L.; Zuo, L. Effects of site conditions and tree age on *Robinia pseudoacacia* and *Populus simonii* leaf hydraulic traits and drought resistance. *Chin. J. Appl. Ecol.* **2012**, *23*, 2397–2403.
40. Li, Y.; Kong, D.; Huang, R.; Liang, H.; Xu, C.; Wu, H. Variations in essential oil yields and compositions of *Cinnamomum cassia* leaves at different developmental stages. *Ind. Crops Prod.* **2013**, *47*, 92–101. [CrossRef]

41. Damián, X.; Fornoni, J.; Domínguez, C.A.; Boege, K.; Baltzer, J. Ontogenetic changes in the phenotypic integration and modularity of leaf functional traits. *Funct. Ecol.* **2017**, *32*, 234–246. [CrossRef]
42. Zhu, M.; Xu, W.; Wen, J.; Zhu, Y.; Li, Y.; Su, Y.; Zhang, Q.; Sun, R. Dynamic changes of photosynthetic properties and chemical compositions of *Eucommia ulmoides* Oliver under two planting models. *Ind. Crops Prod.* **2017**, *96*, 46–56. [CrossRef]
43. Li, Y.; Wang, S.; Li, Z.; Jin, C.; Liu, M. Genetic diversity and relationships among Chinese *Eucommia ulmoides* cultivars revealed by sequence-related amplified polymorphism, amplified fragment length polymorphism, and inter-simple sequence repeat markers. *Genet. Mol. Res.* **2014**, *13*, 8704–8713. [CrossRef] [PubMed]
44. Wu, H.; Li, C.; Chen, D.; Li, J.; Huang, H.; Qiao, X.; Zuo, C. Clustering analysis of ancient tea plants in Yunnan Baiying Mountain based on morphological identification and leaf anatomical structure. *Chin. Agric. Sci. Bull.* **2014**, *30*, 114–121.
45. Kaplan, D.R. The science of plant morphology: Definition, history, and role in modern biology. *Am. J. Bot.* **2001**, *88*, 1711–1741. [CrossRef] [PubMed]
46. Samatadze, T.E.; Yurkevich, O.Y.; Hazieva, F.M.; Konyaeva, E.A.; Morozov, A.I.; Zoshchuk, S.A.; Amosova, A.V.; Muravenko, O.V. Agro-Morphological, Microanatomical and Molecular Cytogenetic Characterization of the Medicinal Plant *Chelidonium majus* L. *Plants* **2020**, *9*, 1396. [CrossRef]
47. He, X.; Cong, P.; Guo, Y.; Lu, J.; Wang, H.; Xue, P.; Zhang, X. Study on drought resistance of four herbs using pressure-volume curve. *Acta Sci. Nat. Univ. Nankaiensis* **2006**, *39*, 16–22.
48. Sun, Q.; Hu, J. *Research Technology of Plant Physiology*; Northwest Agricultural and Forestry University Press: Yangling, China, 2006.
49. Wang, A.; Di, B.; Repo, T.; Roitto, M.; Zhang, G. Responses of Parameters for Electrical Impedance Spectroscopy and Pressure-Volume Curves to Drought Stress in *Pinus bungeana* Seedlings. *Forests* **2020**, *11*, 359. [CrossRef]
50. Ma, B.; Wang, L.; Zhang, K.; Zhang, Y. Study on extraction method of *Eucommia ulmoides* rubber in laboratory. *J. Northwest For. Coll.* **1994**, *9*, 67–69.
51. Dong, J.; Ma, X.; Wei, Q.; Peng, S.; Zhang, S. Effects of growing location on the contents of secondary metabolites in the leaves of four selected superior clones of *Eucommia ulmoides*. *Ind. Crops Prod.* **2011**, *34*, 1607–1614. [CrossRef]
52. Ye, J.; Han, W.; Deng, P.; Jiang, Y.; Liu, M.; Li, L.; Li, Z. Comparative transcriptome analysis to identify candidate genes related to chlorogenic acid biosynthesis in *Eucommia ulmoides* Oliv. *Trees* **2019**, *33*, 1373–1384. [CrossRef]
53. Jing, S.; Li, B. Studies on description system of Hawthorn germplasm resources. *Acta Hortic. Sin.* **1990**, *17*, 81–88.

plants MDPI

Article

Plant Growth Regulators INCYDE and TD-K Underperform in Cereal Field Trials

Matthew J. van Voorthuizen [1], Jiancheng Song [1,2], Ondřej Novák [3] and Paula E. Jameson [1,2,*]

[1] School of Biological Sciences, University of Canterbury, Christchurch 8140, New Zealand; matthewmjvv@gmail.com (M.J.v.V.); jcsong88@163.com (J.S.)
[2] School of Life Sciences, Yantai University, Yantai 264005, China
[3] Laboratory of Growth Regulators, Institute of Experimental Botany of the Czech Academy of Sciences & Faculty of Science of Palacký University, CZ-783 71 Olomouc, Czech Republic; novako@ueb.cas.cz
* Correspondence: paula.jameson@canterbury.ac.nz

Abstract: Using plant growth regulators to alter cytokinin homeostasis with the aim of enhancing endogenous cytokinin levels has been proposed as a strategy to increase yields in wheat and barley. The plant growth regulators INCYDE and CPPU inhibit the cytokinin degrading enzyme cytokinin oxidase/dehydrogenase (CKX), while TD-K inhibits the process of senescence. We report that the application of these plant growth regulators in wheat and barley field trials failed to enhance yields, or change the components of yields. Analyses of the endogenous cytokinin content showed a high concentration of *trans*-zeatin (*t*Z) in both wheat and barley grains at four days after anthesis, and statistically significant, but probably biologically insignificant, increases in *cis*Z-*O*-glucoside, along with small decreases in *c*Z riboside (*c*ZR), dihydro Z (DHZ), and DHZR and DHZOG cytokinins, following INCYDE application to barley at anthesis. We discuss possible reasons for the lack of efficacy of the three plant growth regulators under field conditions and comment on future approaches to manipulating yield in the light of the strong homeostatic mechanisms controlling endogenous cytokinin levels.

Keywords: cytokinin; TD-K; thidiazuron; INCYDE; CPPU; isopentenyl transferase; IPT; cytokinin oxidase/dehydrogenase; CKX; wheat; barley; yield

Citation: van Voorthuizen, M.J.; Song, J.; Novák, O.; Jameson, P.E. Plant Growth Regulators INCYDE and TD-K Underperform in Cereal Field Trials. *Plants* **2021**, *10*, 2309. https://doi.org/10.3390/plants10112309

Academic Editors: Milan S. Stankovic, Paula Baptista and Petronia Carillo

Received: 7 October 2021
Accepted: 22 October 2021
Published: 27 October 2021

Publisher's Note: MDPI stays neutral with regard to jurisdictional claims in published maps and institutional affiliations.

Copyright: © 2021 by the authors. Licensee MDPI, Basel, Switzerland. This article is an open access article distributed under the terms and conditions of the Creative Commons Attribution (CC BY) license (https://creativecommons.org/licenses/by/4.0/).

1. Introduction

Food producers face a range of challenges in addressing global food security in the 21st century. These include continuing growth in food consumption in developing nations [1] and the effects of climate change, which will likely have significant and adverse effects on the environment and agriculture [2–4]. Increasing the yield of cereal crops, including wheat and barley, is fundamental to ensuring food security. In the 2019/2020 season, global production of wheat was more than 770 million tonnes, while for barley it was more than 150 million tonnes [5]. Several traits in cereals have been identified as important components of, and contributors to, overall yields, including having more productive tillers [6–8], a greater proportion of fertile grain-containing florets, larger grains, and leaf senescence occurring at an optimal time [9]. Notably, there can also be trade-offs between different components of yield, where increasing grain number can result in a decrease in grain weight [10–14]. Likewise, the production of more tillers is not necessarily beneficial, as small, unproductive tillers could direct resources away from productive tillers and negatively impact yield [15,16].

The cytokinins are a plant hormone group involved in many aspects of growth and development, including root and shoot growth [17–19], flower development [20,21], nitrogen signaling [22–24], senescence [25,26], stress response [27], seed yield components [28], and seed development [29–32], making them an important contributor to cereal yield.

Cytokinins are often grouped into three biologically active forms: naturally occurring, substituted adenines with either an N^6 isoprenoid side chain or an aromatic side chain; and the synthetic diphenyl ureas. Briefly, isopentenyl transferase (IPT) catalyses the first committed step towards the formation of the isoprenoid cytokinins. The first formed cytokinins are the nucleotides that are converted by LOG (LONELY GUY) to the active free base forms, *trans*-zeatin (*t*Z), N^6-isopentenyladenine (iP), *cis*-zeatin (*c*Z), and dihydrozeatin (DHZ), which are detected by a two-component signaling system. Cytokinin levels are controlled through destruction by cytokinin oxidase/dehydrogenase (CKX) or inactivation by cytokinin glucosyl transferase to *O*- or *N*-glucosides [32].

Previous attempts at manipulating yield and endogenous cytokinin have included the direct application of cytokinin itself to both wheat [33,34] and barley [35,36]. These approaches have involved direct injection into plant organs [34,37,38] or, more practically, through irrigation and spraying ([39], and references therein). However, success in field trials has been mixed, with findings in controlled experiments often hard to replicate in the field given the range of environmental factors and the complexity of analyzing their effects [12,39,40].

An alternative to the application of cytokinin has been the targeting of the enzymes that either deactivate cytokinin through glucosylation [41], or irreversibly degrade cytokinin via CKX [42,43]. Targeting *CKX* expression and/or activity has been suggested as a potential strategy to enhance yield [28,31,44–46], and *CKX* gene family members (GFMs) have been identified as being important for determining yield in both wheat and barley ([9], and references therein).

Given the challenge of increasing yield in the field using cytokinin [31], there has been a search for alternative compounds that might impact components of yield, including compounds that target CKX and compounds that might affect yield through other processes, including senescence. Such compounds include CPPU, TDZ, and the novel plant growth regulators (PGRs) INCYDE and TD-K [46–49]. These compounds became the focus for this research.

Thidiazuron is a substituted phenylurea (Figure 1a) that has been shown to inhibit CKX [50–52]. Thidiazuron has strong cytokinin activity [53–55]. It is able to activate cytokinin receptors [48,53,54,56] and has anti-senescence properties [46,57] that are stronger than *trans*-zeatin (*t*Z) and 6-benzylaminopurine (BAP) [48]. It is also able to promote shoot growth [58–61], increase fruit size [62], and produce ethylene when applied to leaves [48]. The latter property makes it desirable as a cotton defoliant [63].

CPPU (*N*-(2-chloro-4-pyridyl)-*N*′-phenylurea) is a diphenylurea derivative (Figure 1b) which is able to inhibit CKX [64,65] more strongly than TDZ [66]. Although it activates cytokinin receptors AHK3/AHK4, it does so more weakly than TDZ [52]. CPPU is also reported to be able to delay senescence [67], promote shoot formation [68], enhance fruit size [69–72], promote earlier flowering [73], and provide resistance to drought stress [74].

TD-K (*N*-furfuryl-*N*′-1,2,3-thiadiazol-5-yl-urea) is a diphenylurea thidiazuron derivative (Figure 1c) which has strong cytokinin activity comparable to BA in *Amaranthus* and tobacco callus bioassays [49]. TD-K has strong anti-senescence capacity, relative to TDZ and BA [48,49]. Compared to TDZ, it more weakly activates cytokinin receptors [49,53,54,56], is less able to promote ethylene production in mung bean hypocotyls [75], and, in contrast to TDZ, does not inhibit root growth [48].

INCYDE (2-chloro-6-(3-methoxyphenyl)aminopurine) is a substituted 6-anilinopurine derivative (Figure 1d). It is a stronger inhibitor of cytokinin oxidase/dehydrogenase than TDZ, while more weakly activating cytokinin receptors compared to TDZ and *t*Z [55]. It activated the cytokinin responsive reporter gene *ARR5:GUS* [76] in a dose-dependent manner. INCYDE was shown to enhance yield of Rapid Cycling *Brassica rapa* but only under specific, controlled conditions [49]. INCYDE increased shoot FW in *CKX1*-overexpressing *Arabidopsis thaliana* seedlings [76]. INCYDE application has been reported to increase flower production in tomatoes [77], shoot production when applied with BA [78], and has a dose-dependent inhibition of shoot and/or root growth in *Bulbine natalensis* and *Rumex* *crispus* [45] and micropropagated *Eucomis autumnalis* [78]. INCYDE is also reported to alleviate the effects of biotic [79] and abiotic stress [45,77]. Additionally, when applied in the field to barley, analogue INCYDE-F was responsible for altering the endogenous cytokinin content [80].

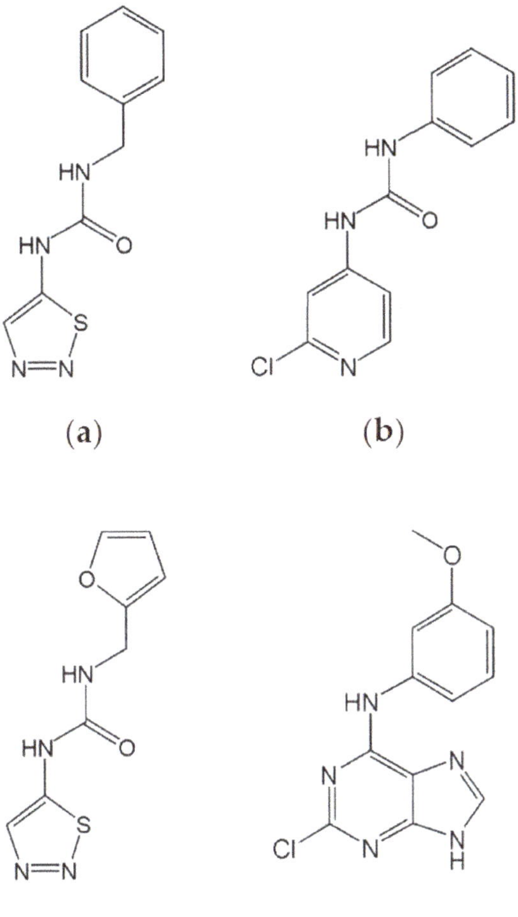

(a) (b) (c) (d)

Figure 1. Structures of plant growth regulators. (**a**) Thidiazuron. (**b**) CPPU. (**c**) TD-K. (**d**) INCYDE.

Three PGRs with different properties and modes of action were selected for this investigation: INCYDE, TD-K, and CPPU. These compounds were applied to wheat and barley in field trials and components of the yields were analyzed. The effects of these compounds on endogenous cytokinins was also examined.

2. Results

2.1. Field Trial Analyses

Analyses carried out on the harvested wheat and barley from the field trials did not reveal any statistically significant difference in the yield (T/ha), thousand grain weight (TGW) in grams (g), or protein composition between any of the treatments and the controls for either wheat or barley (Table 1). The Orator wheat (2013/14) field trial was broadly infected with *Septoria* during a critical time in development, which negatively impacted the yield. Given the lack of evidence for any change in yield, the field trials were discontinued.

Additional trials were carried out using outdoor pot trials where the same treatments and growth stages described for the field trials were used, but no statistically significant differences in yield or yield components were found for these trials either [81].

Table 1. Yield and protein composition in wheat (cv. Orator and cv. Torch) and barley (cv. Quench).

Treatment	Yield (T/ha)	TGW (g)	Protein (%)
Wheat Cultivar Orator (2013/14)			
Untreated Control	10.8 ± 0.2	46.0 ± 1.1	11.2 ± 0.2
DMSO 50 µM Control (GS 39, 51, 61, 65)	11.1 ± 0.1	46.6 ± 0.2	11.2 ± 0.1
DMSO 25 µM Control (GS 61, 65, 65 + 13 d)	11.2 ± 0.1	47.2 ± 0.7	11.0 ± 0.1
INCYDE 10 µM (GS 65)	11.4 ± 0.2	45.9 ± 1.1	11.4 ± 0
INCYDE 25 µM (GS 39, 51, 61, 65)	11.3 ± 0.2	46.2 ± 0.7	11.1 ± 0.1
INCYDE 25 µM (GS 39)	11.2 ± 0.2	45.0 ± 1.1	11.2 ± 0.1
INCYDE 25 µM (GS 51)	11.0 ± 0.1	45.4 ± 0.8	11.1 ± 0.2
INCYDE 25 µM (GS 61)	11.1 ± 0.2	45.6 ± 0.4	11.1 ± 0.1
INCYDE 25 µM (GS 65)	11.1 ± 0.2	45.1 ± 0.5	11.2 ± 0.1
INCYDE 50 µM (GS 61)	11.1 ± 0.1	45.5 ± 1.2	11.4 ± 0.1
INCYDE 50 µM (GS 65)	11.1 ± 0.2	46.9 ± 0.3	11.4 ± 0.1
TD-K 10 µM (GS 61, 65, 65 + 13 d)	11.1 ± 0.1	46.3 ± 1.1	11.1 ± 0.1
TD-K 25 µM (GS 61, 65, 65 + 13 d)	11.4 ± 0.1	46.0 ± 0.5	11.1 ± 0.1
Wheat Cultivar Torch (2014/15)			
Treatment	**Yield (T/ha)**	**TGW (g)**	**Protein (%)**
Untreated Control	14.5 ± 0.1	46.8 ± 1.1	9.8 ± 0.2
DMSO Control (GS 51, 61, 65, 65 + 15 d)	14.5 ± 0.3	48.1 ± 0.2	9.9 ± 0.2
TD-K 10 µM (GS 61, 65, 65 + 15 d)	14.5 ± 0.2	47.4 ± 0.5	10.0 ± 0.2
TD-K 50 µM (GS 61, 65, 65 + 15 d)	14.7 ± 0.3	49.1 ± 0.5	9.8 ± 0.05
CPPU 10 µM (GS 61, 65)	14.4 ± 0.3	47.1 ± 0.4	10.0 ± 0.2
CPPU 30 µM (GS 61, 65)	14.5 ± 0.3	48.8 ± 0.2	9.7 ± 0.1
CPPU 100 µM (GS 61, 65)	14.6 ± 0.2	48.4 ± 0.3	9.7 ± 0.3
CPPU 10 µM (GS 51, 65)	14.5 ± 0.2	47.8 ± 0.4	9.6 ± 0.1
CPPU 30 µM (GS 51, 65)	14.7 ± 0.1	47.2 ± 0.9	9.9 ± 0.1
CPPU 100 µM (GS 51, 65)	14.4 ± 0.2	48.1 ± 0.8	9.9 ± 0.2
Barley Cultivar Quench (2014/15)			
Treatment	**Yield (T/ha)**	**TGW (g)**	**Protein (%)**
Untreated Control	10.8 ± 0.2	52.3 ± 0.5	14.3 ± 0.1
DMSO Control (GS 51, 61, 65, 65 + 15 d)	11.0 ± 0.1	52.0 ± 0.4	13.9 ± 0.4
INCYDE 10 µM (GS 65)	11.0 ± 0.2	53.0 ± 0.2	13.8 ± 0.3
INCYDE 25 µM (GS 39, 51, 61, 65)	11.2 ± 0.2	51.8 ± 0.5	14.0 ± 0.2
INCYDE 25 µM (GS 39)	11.3 ± 0.2	52.5 ± 0.5	13.8 ± 0.2
INCYDE 25 µM (GS 51)	11.3 ± 0.1	52.4 ± 0.9	14.0 ± 0.4
INCYDE 25 µM (GS 61)	11.1 ± 0.1	52.2 ± 0.9	13.9 ± 0.5
INCYDE 25 µM (GS 65)	11.3 ± 0.1	52.5 ± 0.7	14.2 ± 0.4
INCYDE 50 µM (GS 61)	11.2 ± 0.1	52.5 ± 0.8	14.1 ± 0.4
INCYDE 50 µM (GS 65)	11.2 ± 0.3	52.3 ± 0.4	13.9 ± 0.5

Data were analyzed using an ANOVA, with protein percentage data logit-transformed prior to ANOVA. Data are presented as the means ± standard error (n = 4). Yield is provided in tonnes per hectare (T/ha), thousand grain weight (TGW) in grams (g) and protein as a percentage (%). Concentration of each treatment is given in µM, with growth stage (GS) indicating the growth stage (Zadoks scale [82]) targeted for treatment, and 'd' indicating the number of days after the respective growth stage. The dimethylsulfoxide (DMSO) controls list the GS targeted, with volumes equivalent to the DMSO used in the highest concentration within each field trial, with the exception of Orator (2013/14), where DMSO Control (GS 61, 65, 65 + 13 d) was provided at a volume equivalent to 25 µM applications.

2.2. LC–MS/MS Analyses in Grain

LC–MS/MS analyses of wheat and barley grains from control plants assessed four days after anthesis (4 DAA) show that the concentration of *t*Z was much greater than the concentration of the other free bases iP, *c*Z, or DHZ (Tables 2 and 3). Inactivation by glucosylation is clearly evident, as shown by the elevated levels of *c*Z- and *c*Z riboside-*O*-glucosides (*c*ZOG and *c*ZROG) in barley, and in wheat by elevated levels of *t*Z 9-glucoside (*t*Z9G), *c*ZOG, and *c*ZROG.

Table 2. LC–MS/MS analyses of the quantity of cytokinins in wheat (cultivar Torch, 2014/15) grains treated at anthesis with TD-K or CPPU. Measurements were made at four days after anthesis.

	Wheat Cytokinin Concentrations (pmol/g DW)		
Type	Control	TD-K 50 µM	CPPU 100 µM
*t*Z	794.5 ± 71.1	705.3 ± 75.7	889.2 ± 73.7
*t*ZR	60.0 ± 6.7	60.3 ± 3.2	63.5 ± 2.3
*t*ZOG	20.4 ± 2.8	24.7 ± 0.5	22.8 ± 1.3
*t*ZROG	5.8 ± 0.7	6.8 ± 0.2	6.5 ± 0.5
*t*ZRMP	115.9 ± 3.6	106.5 ± 2.8	103.1 ± 10.5
*t*Z7G	<LOD	<LOD	<LOD
*t*Z9G	247.3 ± 21.3	286.1 ± 3.5	268.7 ± 11.5
Total *t*Z types	1244.0 ± 104.8	1189.7 ± 85.2	1353.8 ± 57.0
iP	1.6 ± 0.2	1.3 ± 0.1	1.3 ± 0.1
iPR	2.2 ± 0.2	1.8 ± 0.1	2.5 ± 0.4
iPRMP	22.9 ± 1.5	22.1 ± 3.8	25.9 ± 4.0
iP7G	<LOD	<LOD	<LOD
iP9G	<LOD	<LOD	<LOD
Total iP types	26.8 ± 1.7	25.1 ± 3.9	29.7 ± 4.1
*c*Z	9.6 ± 0.5	7.8 ± 1.1	8.6 ± 1.2
*c*ZR	33.9 ± 3.9	24.7 ± 3.5	30.5 ± 4.0
*c*ZOG	114.3 ± 12.4	130.8 ± 11.7	116.1 ± 12.9
*c*ZROG	139.1 ± 13.2	154.5 ± 6.6	150.6 ± 8.9
*c*ZRMP	10.0 ± 0.8	7.7 ± 1.3	10.8 ± 1.2
*c*Z9G	<LOD	<LOD	<LOD
Total *c*Z types	306.5 ± 26.5	325.5 ± 12.2	316.5 ± 17.0
DHZ	0.23 ± 0.01	0.20 ± 0.03	0.20 ± 0.03
DHZR	2.9 ± 0.1	2.4 ± 0.3	2.6 ± 0.3
DHZOG	1.4 ± 0.2	1.7 ± 0.1	1.5 ± 0.1
DHZROG	9.5 ± 1.1	10.7 ± 0.6	10.0 ± 0.8
DHZRMP	<LOD	<LOD	<LOD
DHZ7G	15.0 ± 0.3	13.3 ± 0.9	14.6 ± 2.2
DHZ9G	0.07 ± 0.003	0.06 ± 0.01	0.07 ± 0.01
Total DHZ types	29.1 ± 1.5	28.2 ± 0.7	28.9 ± 2.6
Total CK bases	806.0 ± 71.4	714.6 ± 76.9	899.3 ± 74.9
Total CK ribosides	98.7 ± 6.8	89.1 ± 7.0	99.0 ± 7.0
Total CK nucleotides	148.9 ± 3.7	136.3 ± 2.0	139.8 ± 12.8
Total CK *O*-glucosides	290.5 ± 30.2	329.2 ± 18.8	307.5 ± 23.7
Total CK *N*-glucosides	262.3 ± 21.6	299.4 ± 3.9	283.4 ± 13.6
Total cytokinins	1606.4 ± 117.3	1568.5 ± 69.5	1728.9 ± 43.4

Treatments were compared to the control using a two-sided ANOVA. Data are presented as the means ± standard error (n = 3). LOD indicates below limit of detection. Treatments were made at anthesis (GS 60). Cytokinin abbreviations: CK (cytokinins), *t*Z (*trans*-zeatin), iP (N^6-isopentenyladenine), *c*Z (*cis*-zeatin), DHZ (dihydrozeatin), R (riboside), OG (*O*-glucoside), RMP (riboside-5′-monophosphate), 7G (7-*N*-glucoside), 9G (9-*N*-glucoside).

In wheat grains, neither TD-K nor CPPU treatment resulted in a significant change in any of the cytokinin metabolites compared to the control (Table 2). At four days following INCYDE treatment/anthesis in barley grains, there was a significant increase in the content of *c*Z *O*-glucoside (*c*ZOG), *c*Z-types overall and the total *O*-glucoside cytokinins (Table 3). Conversely, there were small but statistically significant decreases in the concentration of *c*ZR, DHZ, DHZR, DHZOG, and the total base and ribosides of *c*Z and DHZ cytokinins following INCYDE application.

Table 3. LC–MS/MS analyses of the quantity of cytokinins in barley (cultivar Quench, 2014/15) grains treated at anthesis with INCYDE. Measurements were made at four days after anthesis.

	Barley Cytokinin Concentrations (pmol/g DW)	
Type	Control	INCYDE 50 µM
*t*Z	759.5 ± 66.8	642.8 ± 30.0
*t*ZR	432.0 ± 44.3	458.9 ± 36.6
*t*ZOG	89.5 ± 2.7	97.6 ± 7.9
*t*ZROG	28.1 ± 1.8	30.2 ± 0.6
*t*ZRMP	455.3 ± 27.0	454.6 ± 13.0
*t*Z7G	<LOD	<LOD
*t*Z9G	46.2 ± 2.6	65.0 ± 9.1
Total *t*Z types	1810.6 ± 142.1	1749.1 ± 61.6
iP	2.0 ± 0.2	2.3 ± 0.3
iPR	4.5 ± 0.6	4.7 ± 0.3
iPRMP	86.8 ± 12.8	89.0 ± 11.9
iP7G	<LOD	<LOD
iP9G	<LOD	<LOD
Total iP types	93.3 ± 13.6	96.0 ± 12.2
*c*Z	3.6 ± 0.1	3.2 ± 0.1
*c*ZR	**23.4 ± 0.5**	**20.1 ± 0.5** *
*c*ZOG	**328.9 ± 11.8**	**417.1 ± 9.7** *
*c*ZROG	242.5 ± 8.4	256.8 ± 8.6
*c*ZRMP	20.0 ± 1.8	20.6 ± 1.9
*c*Z9G	<LOD	<LOD
Total *c*Z types	618.4 ± 18.8	**717.9 ± 12.8** *
DHZ	**0.9 ± 0.02**	**0.7 ± 0.04** *
DHZR	**6.5 ± 0.1**	**5.1 ± 0.3** *
DHZOG	**12.6 ± 0.6**	**10.6 ± 0.4** *
DHZROG	47.3 ± 3.1	43.0 ± 2.8
DHZRMP	<LOD	<LOD
DHZ7G	6.0 ± 0.3	5.0 ± 0.7
DHZ9G	0.06 ± 0.01	0.06 ± 0.01
Total DHZ types	73.0 ± 3.7	64.5 ± 3.8
Total CK bases	766.0 ± 67.0	649.0 ± 30.4
Total CK ribosides	466.3 ± 44.6	488.8 ± 37.1
Total CK nucleotides	562.2 ± 38.6	564.2 ± 11.1
Total CK *O*-glucosides	**748.9 ± 27.8**	**855.4 ± 25.2** *
Total CK *N*-glucosides	51.9 ± 2.6	70.1 ± 8.7
Total cytokinins	2595.2 ± 177.1	2627.5 ± 68.9

* Indicates a statistically significant ($p \leq 0.05$) difference for the treatment compared to the control using a two-sided ANOVA and *post hoc* two-sided Dunnett test (CI. 95%). Significant differences are provided in bold. Data are presented as the means ± standard error ($n = 3$). LOD indicates below limit of detection. Treatments were made at anthesis (GS 60). Cytokinin abbreviations: CK (cytokinins), *t*Z (*trans*-zeatin), iP (N^6-isopentenyladenine), *c*Z (*cis*-zeatin), DHZ (dihydrozeatin), R (riboside), OG (*O*-glucoside), RMP (riboside-5′-monophosphate), 7G (7-*N*-glucoside), 9G (9-*N*-glucoside).

3. Discussion

The region where our field trials were conducted, Canterbury, New Zealand, is known for world record cereal production (17.398 tonnes per hectare of wheat crop (Guinness World Records, 2020)). Our trials were conducted under optimal field conditions of water and fertilizer, which we recognize as a potentially challenging environment to assess PGR efficacy, a comment also made by Nisler et al. [66] with respect to their PGR field trials in the Czech Republic.

The lack of yield enhancement following INCYDE application (Table 1) suggests that this compound had little effect in our field trials on either wheat or barley. Positive trends in yield in field trials of wheat and barley treated with cytokinin derivatives similar to

INCYDE have been reported but these failed to reach statistical significance [80]. Consequently, our field data are not in conflict with this. While the Orator wheat (2013/14) field trial, where INCYDE was applied, was impacted by *Septoria*, there was no evidence of INCYDE ameliorating the effect of this disease, in contrast to the report by Reusche et al. [79]. This is not to imply that INCYDE is not efficacious under other conditions, as changes in gene expression occur following application [83], and responses are clearly evident under more controlled environments, including in bioassays [55], in in vitro culture settings [45,77–79], and in pot trials with Rapid Cycling *Brassica rapa* [49].

The statistically significant increase in *c*ZOG following INCYDE application to barley may show a mechanism in common with previous in vitro experiments, where INCYDE (with BA) enhanced *O*-glucoside accumulation in banana plantlets [84]. It is possible that active cytokinin forms may have been channelled into inactivated *O*-glucosides as a consequence of reduced inactivation by CKX, due to inhibition of CKX by INCYDE. Because of the activation of homeostatic mechanisms, and also because of the very high endogenous levels of active *t*Z immediately after anthesis, any transitory increases in active cytokinins, if they had occurred, are likely to be biologically insignificant.

Neither of the two diphenylurea-derivatives, TD-K or CPPU, enhanced yield (Table 1). This is in contrast with an increase of 120.9% for oilseed rape yield (6.038 vs. 4.99 T/ha), and 106% (7.02 vs. 7.49 T/ha) for spring barley reported in the TD-K patent for PGR application at BBCH50 (extension growth) [48]. Details of statistical significance are not, however, provided for the different crops. More recently, a diphenylurea derivative was applied to barley and wheat under field conditions in the Czech Republic [66]. Although these studies targeted earlier growth stages, including at BBCH 20–25, as well as seed treatments, they also targeted the emergence of the inflorescence (BBCH 51), and at a concentration range between 5 and 50 μM, which is comparable to that used in our study. However, the field data for wheat and barley treated with diphenylurea-derivative Compound 19 are only presented as percent of control without statistical analyses available [66]. The variability apparent between years (particularly in tiller number and 1000 grain weight) makes it essential that statistical analysis of the yield data (0.7 to 6.6% yield increase compared to control) is presented.

Likewise, CPPU, despite having success with enhancing fruit size, has not had much success when used to target cereals in the field ([32], and references therein). The difficulties of replicating findings from controlled environments onto the field have been reported [12,40], with field trials introducing a multitude of uncontrolled or difficult to control factors, many of which could affect cytokinin homeostasis.

An increased tiller number is not necessarily seen as desirable in wheat [9], so we specifically targeted the PGRs at later stages of development: for TD-K this was from anthesis onwards, due to its strong anti-senescence properties [48,49]; and for INCYDE and CPPU from GS39, when florets are being established, and/or GS51, when ears are particularly susceptible to stress [85,86], and or across anthesis, the latter chosen due to the rapidly changing cytokinin content and elevated *CKX* expression associated with this stage in development ([9], and references therein). Indeed, a high level of *t*Z cytokinin was identified in wheat four days after anthesis (DAA) (Table 2). This aligns with previous reports of high levels of zeatin in wheat early in grain development [87–91], and, moreover, confirms that this cytokinin is *t*Z. The transient nature of this narrow developmental window that is associated with cell division is also a possible reason for the lack of yield enhancement by cytokinins in cereal field trials ([31], and references therein), since in the field environment, anthesis is spread across several days, although we attempted to cover this by applications at GS61 and 65.

The high concentration of *t*Z in barley at 4 DAA (Table 3) has also been reported [92]. In contrast, the low concentration of *c*Z contrasts with the high peak of *c*Z reported previously in developing barley kernels [93]. This suggests that 4 DAA is possibly after the *c*Z peak. The high concentration of *c*ZOG suggests active deactivation of *c*Z within days post-anthesis.

Our research suggests that INCYDE, TD-K, and CPPU have little to no effect on components of harvestable yield in wheat and barley grown under optimal field conditions. Additionally, this research highlights some of the difficulties and issues of conducting field trials with PGRs, with any attempt to manipulate cytokinin made more difficult not only by strong homeostatic responses but also by the complex, pleiotropic nature of cytokinin [31,39]. Feedback responses following the disturbance of cytokinin homeostasis have been observed or suggested elsewhere in the form of an increase in *CKX* expression and/or activity [14,94–99]. An increase in cytokinin following CKX inhibition might also be responsible for an enhancement in the deactivation of cytokinins, which could explain the stronger production of *cis*-type *O*-glucosides seen in barley (Table 3). Feedback mechanisms might also involve *IPT* GFMs, with *HvIPT1* and *HvIPT2* both being downregulated in response to a local increase in cytokinin following the knockout of *HvCKX1* [100].

Despite these difficulties, targeting CKX is still an important strategy for manipulating cytokinin and yield [9,31,45,46], and arguably more suitable than alternative strategies, including the direct application of cytokinin, or targeting IPT, given that CKX is considered a more moderate or 'softer' regulator of cytokinin compared to IPT [101]. Future research could focus on determining if the endogenous changes in barley (Table 3) and, indeed, the lack of change in wheat, were the result of changes in expression of genes associated with cytokinin homeostasis, including biosynthesis (*IPTs*), degradation (*CKXs*), and glucosylation (*CGTs*), and whether these results could help explain the lack of yield in the field trials. Additionally, with the identification of the key *CKX* gene family members that affect yield in wheat (reviewed in [9]), and with interesting results in wheat [13,14,102], barley [12,87,103,104], and rice [44,105] trials, transgenic approaches hold significant potential for enhancing yield in cereals.

However, whether the resulting cereal is a result of genetic modification or gene editing, in some jurisdictions such plants are subject to legal and social restrictions which make their cultivation, processing, and marketing difficult or impossible [106–108]. In this context, non-transgenic approaches, such as the Targeting Induced Local Lesions In Genome (TILLING) strategy, offer numerous advantages, including overcoming the limits imposed by the lack of genetic variability in traditional breeding, the acceleration of breeding programs, and, above all, the possibility of developing new varieties that do not have the limitations that characterize transgenic organisms [107]. Both the CRISPR/Cas9-mediated gene editing technology and the TILLING approach have their own merits and demerits relating to the initial investment by researchers, the access to the requisite technology, the range of mutations that are either targeted (in gene editing) or identified (multiple point mutations in TILLING) and their use in breeding [106].

More recently, two in silico TILLING resources have been generated and made publicly available. These include the whole exome sequencing of over 1200 TILLING mutant lines of a well-known European bread wheat variety Cadenza [109,110]. Similarly, an in silico TILLING resource is being generated for the most widely grown Chinese bread wheat variety, Jimai 22. Within this population, multiple point mutants for not only all *CKX* GFMs but also the zeatin *O*-glucosyl transferase (*ZOGT*) GFMs have been identified [9,41]. Importantly, while *CKX* GFMs have been the target of much research [9], the high levels of cytokinin glucosides in wheat and barley, and the negative relationship of *ZOGT* gene expression with yield in wheat [41,91], indicate that the *ZOGT* GFMs warrant further investigation, which is beyond the tools offered by the CKX inhibiting PGRs.

4. Materials and Methods

4.1. Field Trials

Wheat and barley field trials were carried out over two seasons, near Lincoln, New Zealand (43°36′15.7″ S 172°25′56.0″ E and 43°37′04.7″ S 172°27′09.4″ E). Autumn-sown wheat (cultivar Orator) was grown in the 2013/14 season, while barley (cultivar Quench) and wheat (cultivar Torch) were grown in 2014/15. Sowing spacing was kept constant, to prevent any confounding effect on tiller number. Field trials were carried out in a farmer's paddock and subject to standard field management including regular irrigation, fertilizer application, and application of compounds, including herbicide, insecticide, and fungicides, where necessary. Field trials were planted in 10 m × 2.5 m plots, arranged in a randomized complete block design with four replicates for each treatment. Plant growth regulators INCYDE, TD-K, and CPPU were applied at concentrations between 10 and 100 μM at growth stages (GS), defined according to the Zadoks scale [82], including GS 39 (the appearance of flag leaf ligule), GS 51 (appearance of the spikelet), and at GS 61 to 69 (defined as anthesis). Plant growth regulators were applied at rates of 187 L/ha for the 2013/14 trial, and 170 L/ha for the 2014/15 trial.

INCYDE, TD-K and CPPU were prepared by dissolving compounds in dimethylsulfoxide (DMSO) (Scharlab), diluted with water and then, prior to application, mixed with surfactant (Yates Sprayfix, Yates) at 0.5% (*v/v*). Two controls were used in the field trials, 'untreated controls' where no application was made, and 'DMSO controls' where the amount of DMSO used was equivalent to the highest PGR concentration for each respective trial, unless stated otherwise in the results. Applications were made by New Zealand Arable using CO_2 pressurized hand-hand plot booms for applications rates between 170–190 L/ha.

4.2. Plant Material

4.2.1. Yield and Protein Composition Analyses

Once wheat and barley plants had senesced completely, plants were harvested with a Sampo combine harvester (Sampo Rosenlew Ltd., Pori, Finland) and protein content was analyzed by New Zealand Grainlab. Onboard weighing provided the analysis of yield (tonnes per hectare) and, using 20 g screened samples of grain, the TGW was calculated with a Numigral I seed counter (Sinar). Protein composition was analyzed using an Instalab® 700 NIR Analyzer (DICKEY-john). The thousand grain weight was calculated for each plot, using 20 g of screened grain samples.

4.2.2. LC–MS/MS Analyses

Grain material for LC–MS/MS analyses was sampled from the field trials, following anthesis-targeted application of either INCYDE (50 μM), TD-K (50 μM), CPPU (100 μM), or water + DMSO. Following treatment, whole heads were sampled at day 4 after anthesis, which was 4 days after treatment. Wheat and barley heads were frozen by immediately submerging the samples in liquid nitrogen and storing at −80 °C. Wheat and barley grains were dissected from the middle third section of the spike, with basal florets within the spikelet targeted in wheat [9]. Grains were then organized, based on the developmental stages as described in [111]. INCYDE-treated wheat grains were not sampled for LC–MS/MS analyses, given that this trial (wheat cv. Orator, 2013/14) was infected with *Septoria* at a critical time during grain development.

Grains were ground under liquid nitrogen and freeze dried with a Savant™ SPD131DDA SpeedVac™ Concentrator (Thermo Fisher Scientific) to produce samples weighing between 8 to 22 mg. For each treatment, three replicates were prepared. Samples were then analyzed according to [112]. Sample extraction was carried out with a modified Bieleski solution (60% MeOH, 10% HCOOH, and 30% H_2O), and [$^{13}C_5$]*c*Z, [$^{13}C_5$]*t*Z, [$^{2}H_5$]*t*ZR, [$^{2}H_5$]*t*Z7G, [$^{2}H_5$]*t*Z9G, [$^{2}H_5$]*t*ZOG, [$^{2}H_5$]*t*ZROG, [$^{2}H_5$]*t*ZMP, [$^{2}H_3$]DHZ, [$^{2}H_3$]DHZR, [$^{2}H_3$]DHZ9G, [$^{2}H_7$]DHZOG, [$^{2}H_3$]DHZMP, [$^{2}H_6$]iP, [$^{2}H_6$]iPR, [$^{2}H_6$]iP7G, [$^{2}H_6$]iP9G, [$^{2}H_6$]iPMP stable isotope-labelled standards (0.25 pmol of cytokinin bases, ribosides, *N*-glucosides, 0.5 pmol of cytokinin *O*-glucosides and nucleotides; Olchemim) were added to each sample to validate phytohormone determination. Sample purification was carried out with mixed-mode cation-exchange (MCX) cartridges (Oasis MCX, 30 mg/1 mL; Waters). Analytes were eluted by two-step elution using a 0.35 M NH_4OH aqueous solution and 0.35 M NH_4OH in 60% (*v/v*) methanol solution. The resulting eluate was subsequently evaporated to dryness and then dissolved in the mobile phase (15 mM ammonium formate pH 4.0 in 5% (*v/v*) methanol). LC–MS/MS analyses were carried out using a Acquity UPLC® System (Waters) and a triple quadrupole mass spectrometer Xevo™ TQ MS (Waters). The mass spectrometry data was then processed utilizing MassLynx™ Mass Spectrometry Software with TargetLynx™ (Waters).

4.3. Statistical Analyses

For yield and protein composition from the field trials, the mean was generated using four replicates for each treatment and the data presented with standard errors. Statistically significant differences, where $p \leq 0.05$, were determined between PGR treatments and the respective DMSO control using a two-way ANOVA. A logit transformation was made to protein composition data prior to ANOVA analysis. Similarly, statistically significant differences for LC–MS/MS data were determined between PGR treatments and the control using two-way ANOVA (significance level: 0.05), with a *post hoc* two-sided Dunnett test (Confidence Interval: 95%). To ensure the assumptions of the ANOVA were met, an examination of Q-Q plots of standardized residuals was made, and where necessary the equality of variances ensured through a Levene's test and plot of standardized residuals and predicted values.

Author Contributions: Conceptualization, P.E.J. and J.S.; methodology, J.S., P.E.J. and O.N.; validation, J.S., M.J.v.V. and O.N.; formal analysis, M.J.v.V. and P.E.J.; investigation, M.J.v.V., J.S. and P.E.J.; resources, P.E.J., J.S. and O.N.; data curation, M.J.v.V. and O.N.; writing—original draft preparation, M.J.v.V.; writing—review and editing, P.E.J., M.J.v.V. and J.S.; visualization, M.J.v.V.; supervision, P.E.J. and J.S.; project administration, P.E.J.; funding acquisition, P.E.J., J.S. and O.N. All authors have read and agreed to the published version of the manuscript.

Funding: This work was funded by the Foundation for Arable Research, New Zealand in the provision of a PhD scholarship (M.J.v.V.) and expendables (P.E.J., M.J.v.V. and J.S.); and by the Ministry of Education, Youth and Sports of the Czech Republic through the European Regional Development Fund-Project 'Plants as a tool for sustainable global development' (CZ.02.1.01/0.0/0.0/16_019/0000827, O.N.).

Institutional Review Board Statement: Not applicable.

Informed Consent Statement: Not applicable.

Data Availability Statement: All data are contained within the article.

Acknowledgments: The authors sincerely thank Hana Martínková and Petra Amakorová for their help with the analysis of phytohormones. We acknowledge Marek Zatloukal for INCYDE and Jaroslav Nisler for TD-K. We acknowledge New Zealand Arable for application of the PGRs and the New Zealand Grainlab for the yield and composition analyses.

Conflicts of Interest: The authors of this manuscript declare no conflict of interest.

References

1. Baldos, U.L.C.; Hertel, T.W. Global food security in 2050: The role of agricultural productivity and climate change. *Aust. J. Agric. Econ.* **2014**, *58*, 554–570. [CrossRef]
2. Schmidhuber, J.; Tubiello, F.N. Global food security under climate change. *Proc. Natl. Acad. Sci. USA* **2007**, *104*, 19703–19708. [CrossRef] [PubMed]
3. Nellemann, C.; MacDevette, M.; Manders, T.; Eickhout, B.; Svihus, B.; Prins, A.G.; Kaltenborn, B.P. *The Environmental Food Crisis: The Environment's Role in Averting Future Food Crises: A UNEP Rapid Response Assessment*; UNEP/Earthprint: Stevenage, UK, 2009.
4. Raza, A.; Razzaq, A.; Mehmood, S.S.; Zou, X.; Zhang, X.; Lv, Y.; Xu, J. Impact of climate change on crops adaptation and strategies to tackle its outcome: A review. *Plants* **2019**, *8*, 34. [CrossRef]
5. USDA Foreign Agricultural Service. *Grain: World Markets and Trade*; USDA Foreign Agricultural Service: Washington, DC, USA, 2021.
6. Langer, R.H.M.; Prasad, P.C.; Laude, H.M. Effects of kinetin on tiller bud elongation in wheat (*Triticum aestivum* L.). *Ann. Bot.* **1973**, *37*, 565–571. [CrossRef]
7. Darwinkel, A. Ear development and formation of grain yield in winter wheat. *Neth. J. Agric. Sci.* **1980**, *28*, 156–163. [CrossRef]
8. Harrison, M.A.; Kaufman, P.B. Hormonal regulation of lateral bud (tiller) release in oats (*Avena sativa* L.). *Plant Physiol.* **1980**, *66*, 1123–1127. [CrossRef] [PubMed]
9. Chen, L.; Zhao, J.; Song, J.; Jameson, P.E. Cytokinin dehydrogenase: A genetic target for yield improvement in wheat. *Plant Biotechnol. J.* **2020**, *18*, 614–630. [CrossRef] [PubMed]

10. Van Daele, I.; Gonzalez, N.; Vercauteren, I.; de Smet, L.; Inzé, D.; Roldán-Ruiz, I.; Vuylsteke, M. A comparative study of seed yield parameters in *Arabidopsis thaliana* mutants and transgenics. *Plant Biotechnol. J.* **2012**, *10*, 488–500. [CrossRef]
11. Guo, T.; Chen, K.; Dong, N.-Q.; Shi, C.-L.; Ye, W.-W.; Gao, J.-P.; Shan, J.-X.; Lin, H.-X. *GRAIN SIZE AND NUMBER1* negatively regulates the OsMKKK10-OsMKK4-OsMPK6 cascade to coordinate the trade-off between grain number per panicle and grain size in rice. *Plant Cell* **2018**, *30*, 871–888. [CrossRef] [PubMed]
12. Holubová, K.; Hensel, G.; Vojta, P.; Tarkowski, P.; Bergougnoux, V.; Galuszka, P. Modification of barley plant productivity through regulation of cytokinin content by reverse-genetics approaches. *Front. Plant Sci.* **2018**, *9*, 1676. [CrossRef]
13. Jabłoński, B.; Ogonowska, H.; Szala, K.; Bajguz, A.; Orczyk, W.; Nadolska-Orczyk, A. Silencing of *TaCKX1* mediates expression of other *TaCKX* genes to increase yield parameters in wheat. *Int. J. Mol. Sci.* **2020**, *21*, 4809. [CrossRef]
14. Jabłoński, B.; Szala, K.; Przyborowski, M.; Bajguz, A.; Chmur, M.; Gasparis, S.; Orczyk, W.; Nadolska-Orczyk, A. *TaCKX2.2* genes coordinate expression of other *TaCKX* family members, regulate phytohormone content and yield-related traits of wheat. *Int. J. Mol. Sci.* **2021**, *22*, 4142. [CrossRef] [PubMed]
15. Hendriks, P.W.; Kirkegaard, J.A.; Lilley, J.M.; Gregory, P.J.; Rebetzke, G.J. A tillering inhibition gene influences root–shoot carbon partitioning and pattern of water use to improve wheat productivity in rainfed environments. *J. Exp. Bot.* **2016**, *67*, 327–340. [CrossRef]
16. Kebrom, T.H.; Chandler, P.M.; Swain, S.M.; King, R.W.; Richards, R.A.; Spielmeyer, W. Inhibition of tiller bud outgrowth in the tin mutant of wheat is associated with precocious internode development. *Plant Physiol.* **2012**, *160*, 308–318. [CrossRef] [PubMed]
17. Werner, T.; Motyka, V.; Strnad, M.; Schmülling, T. Regulation of plant growth by cytokinin. *Proc. Natl. Acad. Sci. USA* **2001**, *98*, 10487–10492. [CrossRef]
18. Werner, T.; Motyka, V.; Laucou, V.; Smets, R.; Van Onckelen, H.; Schmülling, T. Cytokinin-deficient transgenic *Arabidopsis* plants show multiple developmental alterations indicating opposite functions of cytokinins in the regulation of shoot and root meristem activity. *Plant Cell* **2003**, *15*, 2532–2550. [CrossRef]
19. Brenner, W.G.; Schmülling, T. Transcript profiling of cytokinin action in *Arabidopsis* roots and shoots discovers largely similar but also organ-specific responses. *BMC Plant Biol.* **2012**, *12*, 112. [CrossRef] [PubMed]
20. Bartrina, I.; Otto, E.; Strnad, M.; Werner, T.; Schmülling, T. Cytokinin regulates the activity of reproductive meristems, flower organ size, ovule formation, and thus seed yield in *Arabidopsis thaliana*. *Plant Cell* **2011**, *23*, 69–80. [CrossRef]
21. D'Aloia, M.; Bonhomme, D.; Bouché, F.; Tamseddak, K.; Ormenese, S.; Torti, S.; Coupland, G.; Périlleux, C. Cytokinin promotes flowering of *Arabidopsis* via transcriptional activation of the FT paralogue TSF. *Plant J.* **2011**, *65*, 972–979. [CrossRef] [PubMed]
22. Takei, K.; Ueda, N.; Aoki, K.; Kuromori, T.; Hirayama, T.; Shinozaki, K.; Yamaya, T.; Sakakibara, H. *AtIPT3* is a key determinant of nitrate-dependent cytokinin biosynthesis in *Arabidopsis*. *Plant Cell Physiol.* **2004**, *45*, 1053–1062. [CrossRef]
23. Shtratnikova, V.Y.; Kudryakova, N.V.; Kudoyarova, G.R.; Korobova, A.V.; Akhiyarova, G.R.; Danilova, M.N.; Kusnetsov, V.V.; Kulaeva, O.N. Effects of nitrate and ammonium on growth of *Arabidopsis thaliana* plants transformed with the *ARR5::GUS* construct and a role for cytokinins in suppression of disturbances induced by the presence of ammonium. *Russ. J. Plant Physiol.* **2015**, *62*, 741–752. [CrossRef]
24. Gu, J.; Li, Z.; Mao, Y.; Struik, P.C.; Zhang, H.; Liu, L.; Wang, Z.; Yang, J. Roles of nitrogen and cytokinin signals in root and shoot communications in maximizing of plant productivity and their agronomic applications. *Plant Sci.* **2018**, *274*, 320–331. [CrossRef]
25. Gan, S.; Amasino, R.M. Cytokinins in plant senescence: From spray and pray to clone and play. *Bioessays* **1996**, *18*, 557–565. [CrossRef]
26. Zwack, P.J.; Rashotte, A.M. Cytokinin inhibition of leaf senescence. *Plant Signal. Behav.* **2013**, *8*, e24737. [CrossRef]
27. Cortleven, A.; Leuendorf, J.E.; Frank, M.; Pezzetta, D.; Bolt, S.; Schmülling, T. Cytokinin action in response to abiotic and biotic stresses in plants. *Plant Cell Environ.* **2019**, *42*, 998–1018. [CrossRef]
28. Schwarz, I.; Scheirlinck, M.T.; Otto, E.; Bartrina, I.; Schmidt, R.C.; Schmülling, T. Cytokinin regulates the activity of the inflorescence meristem and components of seed yield in oilseed rape. *J. Exp. Bot.* **2020**, *71*, 7146–7159. [CrossRef]
29. Riefler, M.; Novak, O.; Strnad, M.; Schmülling, T. *Arabidopsis* cytokinin receptor mutants reveal functions in shoot growth, leaf senescence, seed size, germination, root development, and cytokinin metabolism. *Plant Cell* **2006**, *18*, 40–54. [CrossRef] [PubMed]
30. Miransari, M.; Smith, D.L. Plant hormones and seed germination. *Environ. Exp. Bot.* **2014**, *99*, 110–121. [CrossRef]
31. Jameson, P.E.; Song, J. Cytokinin: A key driver of seed yield. *J. Exp. Bot.* **2016**, *67*, 593–606. [CrossRef] [PubMed]
32. Jameson, P.E.; Song, J. Will cytokinins underpin the second 'Green Revolution'? *J. Exp. Bot.* **2020**, *71*, 6872–6875. [CrossRef]
33. Wang, Z.; Cao, W.; Dai, T.; Zhou, Q. Effects of exogenous hormones on floret development and grain setting in wheat. *Plant Growth Regul.* **2001**, *35*, 225–231. [CrossRef]
34. Gupta, N.K.; Gupta, S.; Shukla, D.S.; Deshmukh, P.S. Differential responses of BA injection on yield and specific grain growth in contrasting genotypes of wheat (*Triticum aestivum* L.). *Plant Growth Regul.* **2003**, *40*, 201–205. [CrossRef]
35. Williams, R.H.; Cartwright, P.M. The effect of applications of a synthetic cytokinin on shoot dominance and grain yield in spring barley. *Ann. Bot.* **1980**, *46*, 445–452. [CrossRef]
36. Hosseini, S.M.; Poustini, K.; Ahmadi, A. Effects of foliar application of BAP on source and sink strength in four six-rowed barley (*Hordeum vulgare* L.) cultivars. *Plant Growth Regul.* **2008**, *54*, 231–239. [CrossRef]
37. Warrier, A.; Bhardwaj, S.N.; Pande, P.C. Effect of benzyladenine on grain growth in *Aestivum* wheat. *Plant Cell Physiol.* **1987**, *28*, 735–739. [CrossRef]

38. Sivakumar, T.; Srivastava, G.C. Effects of benzyl adenine and abscisic acid on grain yield and yield components in *Triticale* and wheat. *J. Agron. Crop. Sci.* **2001**, *186*, 43–46. [CrossRef]
39. Koprna, R.; De Diego, N.; Dundálková, L.; Spíchal, L. Use of cytokinins as agrochemicals. *Bioorg. Med. Chem.* **2016**, *24*, 484–492. [CrossRef] [PubMed]
40. Nagel, L.; Brewster, R.; Riedell, W.E.; Reese, R.N. Cytokinin regulation of flower and pod set in soybeans (*Glycine max* (L.) Merr.). *Ann. Bot.* **2001**, *88*, 27–31. [CrossRef]
41. Chen, L.; Zhao, J.; Song, J.; Jameson, P.E. Cytokinin glucosyl transferases, key regulators of cytokinin homeostasis, have potential value for wheat improvement. *Plant Biotechnol. J.* **2021**, *19*, 878–896. [CrossRef] [PubMed]
42. McGaw, B.A.; Horgan, R. Cytokinin oxidase from *Zea mays* kernels and *Vinca rosea* crown-gall tissue. *Planta* **1983**, *159*, 30–37. [CrossRef]
43. Galuszka, P.; Frébort, I.; Šebela, M.; Peč, P. Degradation of cytokinins by cytokinin oxidases in plants. *Plant Growth Regul.* **2000**, *32*, 315–327. [CrossRef]
44. Ashikari, M.; Sakakibara, H.; Lin, S.; Yamamoto, T.; Takashi, T.; Nishimura, A.; Angeles, E.R.; Qian, Q.; Kitano, H.; Matsuoka, M. Plant science: Cytokinin oxidase regulates rice grain production. *Science* **2005**, *309*, 741–745. [CrossRef]
45. Gemrotová, M.; Kulkarni, M.G.; Stirk, W.A.; Strnad, M.; Van Staden, J.; Spíchal, L. Seedlings of medicinal plants treated with either a cytokinin antagonist (PI-55) or an inhibitor of cytokinin degradation (INCYDE) are protected against the negative effects of cadmium. *Plant Growth Regul.* **2013**, *71*, 137–145. [CrossRef]
46. Nisler, J.; Kopečný, D.; Končitíková, R.; Zatloukal, M.; Bazgier, V.; Berka, K.; Zalabák, D.; Briozzo, P.; Strnad, M.; Spíchal, L. Novel thidiazuron-derived inhibitors of cytokinin oxidase/dehydrogenase. *Plant Mol. Biol.* **2016**, *92*, 235–248. [CrossRef]
47. Reynolds, A.G.; Wardle, D.A.; Zurowski, C.; Looney, N.E. Phenylureas CPPU and thidiazuron affect yield components, fruit composition, and storage potential of four seedless grape selections. *J. Am. Soc. Hortic. Sci.* **1992**, *117*, 85–89. [CrossRef]
48. Nisler, J.; Zatloukal, M.; Spíchal, L.; Koprna, R.; Doležal, K.; Strnad, M. 1,2,3-thiadiazol-5yl-urea Derivatives, Use Thereof for Regulating Plant Senescence and Preparations Containing These Derivatives. U.S. Patent 9,993,002, 12 June 2018.
49. Van Voorthuizen, M.J.; Nisler, J.; Song, J.; Spíchal, L.; Jameson, P.E. Targeting cytokinin homeostasis in rapid cycling *Brassica rapa* with plant growth regulators INCYDE and TD-K. *Plants* **2021**, *10*, 39. [CrossRef] [PubMed]
50. Chatfield, J.M.; Armstrong, D.J. Regulation of cytokinin oxidase activity in callus tissues of *Phaseolus vulgaris* L. cv Great Northern. *Plant Physiol.* **1986**, *80*, 493–499. [CrossRef]
51. Hare, P.D.; Van Staden, J. Inhibitory effect of thidiazuron on the activity of cytokinin oxidase isolated from soybean callus. *Plant Cell Physiol.* **1994**, *35*, 1121–1125. [CrossRef]
52. Kopečný, D.; Briozzo, P.; Popelková, H.; Šebela, M.; Končitíková, R.; Spíchal, L.; Nisler, J.; Madzak, C.; Frébort, I.; Laloue, M.; et al. Phenyl-and benzylurea cytokinins as competitive inhibitors of cytokinin oxidase/dehydrogenase: A structural study. *Biochimie* **2010**, *92*, 1052–1062. [CrossRef]
53. Yamada, H.; Suzuki, T.; Terada, K.; Takei, K.; Ishikawa, K.; Miwa, K.; Yamashino, T.; Mizuno, T. The *Arabidopsis* AHK4 histidine kinase is a cytokinin-binding receptor that transduces cytokinin signals across the membrane. *Plant Cell Physiol.* **2001**, *42*, 1017–1023. [CrossRef]
54. Spíchal, L.; Rakova, N.Y.; Riefler, M.; Mizuno, T.; Romanov, G.A.; Strnad, M.; Schmülling, T. Two cytokinin receptors of *Arabidopsis thaliana*, CRE1/AHK4 and AHK3, differ in their ligand specificity in a bacterial assay. *Plant Cell Physiol.* **2004**, *45*, 1299–1305. [CrossRef] [PubMed]
55. Zatloukal, M.; Gemrotová, M.; Doležal, K.; Havlíček, L.; Spíchal, L.; Strnad, M. Novel potent inhibitors of *A. thaliana* cytokinin oxidase/dehydrogenase. *Bioorg. Med. Chem.* **2008**, *16*, 9268–9275. [CrossRef]
56. Mok, M.C.; Mok, D.W.S.; Armstrong, D.J.; Shudo, K.; Isogai, Y.; Okamoto, T. Cytokinin activity of *N*-phenyl *N*′-1,2,3-thiadiazol-5-ylurea (Thidiazuron). *Phytochemistry* **1982**, *21*, 1509–1511. [CrossRef]
57. Ferrante, A.; Hunter, D.A.; Hackett, W.P.; Reid, M.S. Thidiazuron—A potent inhibitor of leaf senescence in *Alstroemeria*. *Postharvest Biol. Technol.* **2002**, *25*, 333–338. [CrossRef]
58. Malik, K.A.; Saxena, P.K. Regeneration in *Phaseolus vulgaris* L.: High-frequency induction of direct shoot formation in intact seedlings by N^6-benzylaminopurine and thidiazuron. *Planta* **1992**, *186*, 384–389. [CrossRef]
59. Singh, N.D.; Sahoo, L.; Sarin, N.B.; Jaiwal, P.K. The effect of TDZ on organogenesis and somatic embryogenesis in pigeonpea (*Cajanus cajan* L. Millsp). *Plant Sci.* **2003**, *164*, 341–347. [CrossRef]
60. Zhihui, S.; Tzitzikas, M.; Raemakers, K.; Zhengqiang, M.; Visser, R. Effect of TDZ on plant regeneration from mature seeds in pea (*Pisum sativum*). *Vitr. Cell. Dev. Biol. Plant* **2009**, *45*, 776. [CrossRef]
61. Guo, B.; Abbasi, B.H.; Zeb, A.; Xu, L.L.; Wei, Y.H. Thidiazuron: A multi-dimensional plant growth regulator. *Afr. J. Biotechnol.* **2011**, *10*, 8984–9000.
62. Piao, Y.L.; Wang, H.T.; Zhang, H.M. Effects of TDZ on grapes quality in cold area. *J. Agric. Sci. Yanbian Univ.* **2006**, *4*, 012.
63. Suttle, J.C. Involvement of ethylene in the action of the cotton defoliant thidiazuron. *Plant Physiol.* **1985**, *78*, 272–276. [CrossRef] [PubMed]
64. Hayata, Y.; Niimi, Y.; Iwasaki, N. Synthetic cytokinin-1-(2chloro4pyridyl)-3-phenylurea (CPPU)-promotes fruit set and induces parthenocarpy in watermelon. *J. Am. Soc. Hort. Sci.* **1995**, *120*, 997–1000. [CrossRef]
65. Bilyeu, K.D.; Cole, J.L.; Laskey, J.G.; Riekhof, W.R.; Esparza, T.J.; Kramer, M.D.; Morris, R.O. Molecular and biochemical characterization of a cytokinin oxidase from maize. *Plant Physiol.* **2001**, *125*, 378–386. [CrossRef]

66. Nisler, J.; Kopečný, D.; Pěkná, Z.; Končitíková, R.; Koprna, R.; Murvanidze, N.; Werbrouck, S.P.O.; Havlíček, L.; De Diego, N.; Kopečná, M.; et al. Diphenylurea-derived cytokinin oxidase/dehydrogenase inhibitors for biotechnology and agriculture. *J. Exp. Bot.* **2021**, *72*, 355–370. [CrossRef] [PubMed]
67. Wang, L.; Piao, Y.L.; Lv, H.; Bai, H.J.; Li, M.D. Effect of TDZ on fruiting and fruit growth of Xiaoxiangshui pear of Yanbian [J]. *J. Agric. Sci. Yanbian Univ.* **2009**, *3*, 007.
68. Guo, D.P.; Zhu, Z.J.; Hu, X.X.; Zheng, S.J. Effect of cytokinins on shoot regeneration from cotyledon and leaf segment of stem mustard (*Brassica juncea* var. tsatsai). *Plant Cell Tissue Organ. Cult.* **2005**, *83*, 123–127. [CrossRef]
69. Biasi, R.; Costa, G.; Giuliani, R.; Succi, F.; Sansavini, S. Effects of CPPU on kiwifruit performance. In *II International Symposium on Kiwifruit*; International Society for Horticultural Science: Korbeek-Lo, Belgium, 1991; Volume 297, pp. 367–374.
70. Flaishman, M.A.; Shargal, A.; Shlizerman, L.; Stern, R.A. The synthetic cytokinins CPPU and TDZ prolong the phase of cell division in developing pear (*Pyrus communis* L.) fruit. *Acta Hortic.* **2006**, *671*, 151–157. [CrossRef]
71. Stern, R.A.; Ben-Arie, R.; Neria, O.; Flaishman, M. CPPU and BA increase fruit size of 'Royal Gala' (*Malus domestica*) apple in a warm climate. *J. Hort. Sci. Biotechnol.* **2003**, *78*, 297–302. [CrossRef]
72. Ferrara, G.; Mazzeo, A.; Netti, G.; Pacucci, C.; Matarrese, A.M.S.; Cafagna, I.; Mastrorilli, P.; Vezzoso, M.; Gallo, V. Girdling, gibberellic acid, and forchlorfenuron: Effects on yield, quality, and metabolic profile of table grape cv Italia. *Am. J. Enol. Vitic.* **2014**, *65*, 381–387. [CrossRef]
73. Fukuda, N.; Oba, H.; Mizuta, D.; Yoshida, H.; Olsen, J.E. Timing of blue and red light exposure and CPPU application during the raising of seedlings can control flowering timing of *petunia*. *Acta. Hortic.* **2016**, *1134*, 171–178. [CrossRef]
74. Jianchang, H.; Yan, X.; Chunxiang, Z.; Hongbin, L. Protective effect of CPPU in papaya plant under drought stress. *J. Fruit Sci.* **2003**, *3*, 013.
75. Yip, W.K.; Yang, S.F. Effect of thidiazuron, a cytokinin-active urea derivative, in cytokinin-dependent ethylene production systems. *Plant Physiol.* **1986**, *80*, 515–519. [CrossRef]
76. Zatloukal, M.; Plihalova, L.; Klaskova, J.; Spíchal, L.; Koprna, R.; Dolezal, K.; Strnad, M. Substituted 6-Anilino-9-heterocyclylpurine Derivatives for Inhibition of Plant Stress. U.S. Patent 10,662,194, 26 May 2020.
77. Aremu, A.O.; Masondo, N.A.; Sunmonu, T.O.; Kulkarni, M.G.; Zatloukal, M.; Spichal, L.; Doležal, K.; Van Staden, J. A novel inhibitor of cytokinin degradation (INCYDE) influences the biochemical parameters and photosynthetic apparatus in NaCl-stressed tomato plants. *Planta* **2014**, *240*, 877–889. [CrossRef]
78. Aremu, A.O.; Stirk, W.A.; Masondo, N.A.; Plačková, L.; Novák, O.; Pěnčík, A.; Zatloukal, M.; Nisler, J.; Spíchal, L.; Doležal, K.; et al. Dissecting the role of two cytokinin analogues (INCYDE and PI-55) on in vitro organogenesis, phytohormone accumulation, phytochemical content and antioxidant activity. *Plant Sci.* **2015**, *238*, 81–94. [CrossRef]
79. Reusche, M.; Klásková, J.; Thole, K.; Truskina, J.; Novák, O.; Janz, D.; Strnad, M.; Spíchal, L.; Lipka, V.; Teichmann, T. Stabilization of cytokinin levels enhances *Arabidopsis* resistance against *Verticillium longisporum*. *Am. Phytopath. Soc.* **2013**, *26*, 850–860. [CrossRef]
80. Koprna, R.; Humplík, J.F.; Špíšek, Z.; Bryksová, M.; Zatloukal, M.; Mik, V.; Novák, O.; Nisler, J.; Doležal, K. Improvement of tillering and grain yield by application of cytokinin derivatives in wheat and barley. *Agronomy* **2021**, *11*, 67. [CrossRef]
81. Van Voorthuizen, M.J. Novel plant growth regulator effects on yield, senescence and cytokinin homeostasis in wheat, barley and rapid cycling *Brassica rapa*. Ph.D. Thesis, University of Canterbury, Christchurch, New Zealand, 2018.
82. Zadoks, J.C.; Chang, T.T.; Konzak, C.F. Konzak: "A decimal code for the growth stages of cereals". *Weed Res.* **1974**, *14*, 415–421. [CrossRef]
83. Berková, V.; Kameniarová, M.; Ondrisková, V.; Berka, M.; Menšíková, S.; Kopecká, R.; Luklová, M.; Novák, J.; Spíchal, L.; Rashotte, A.M.; et al. Arabidopsis response to inhibitor of cytokinin degradation INCYDE: Modulations of cytokinin signaling and plant proteome. *Plants* **2020**, *9*, 1563. [CrossRef] [PubMed]
84. Aremu, A.O.; Bairu, M.W.; Novák, O.; Plačková, L.; Zatloukal, M.; Doležal, K.; Finnie, J.F.; Strnad, M.; Van Staden, J. Physiological responses and endogenous cytokinin profiles of tissue-cultured "Williams" bananas in relation to roscovitine and an inhibitor of cytokinin oxidase/dehydrogenase (INCYDE) treatments. *Planta* **2012**, *236*, 1775–1790. [CrossRef] [PubMed]
85. Zhang, W.J.; Wang, J.Q.; Huang, Z.L.; Mi, L.; Xu, K.F.; Wu, J.J.; Fan, Y.H.; Ma, S.Y.; Jiang, D.G. Effects of low temperature at booting stage on sucrose metabolism and endogenous hormone contents in winter wheat spikelet. *Front. Plant Sci.* **2019**, *10*. [CrossRef]
86. Erena, M.F.; Lohraseb, I.; Munoz-Santa, I.; Taylor, J.D.; Emebiri, L.C.; Collins, N.C. The *WtmsDW* locus on wheat chromosome 2B controls major natural variation for floret sterility responses to heat stress at booting stage. *Front. Plant Sci.* **2021**, *12*. [CrossRef]
87. Jameson, P.E.; McWha, J.A.; Wright, G.J. Cytokinins and changes in their activity during the development of grains of wheat (*Triticum aestivum* L.). *Zeit. Pflanzenphysiol.* **1982**, *106*, 27–36. [CrossRef]
88. Lenton, J.R.; Appleford, N.E.J. *Cytokinins and Early Growth in Wheat*; Monograph, British Plant Growth Regulator Group: Wantage, UK, 1986; Volume 14, pp. 99–113.
89. Banowetz, G.M.; Ammar, K.; Chen, D.D. Postanthesis temperatures influence cytokinin accumulation and wheat kernel weight. *Plant Cell Environ.* **1999**, *22*, 309–316. [CrossRef]
90. Hess, J.R.; Carman, J.G.; Banowetz, G.M. Hormones in wheat kernels during embryony. *J. Plant Physiol.* **2002**, *159*, 379–386. [CrossRef]

91. Nguyen, H.N.; Perry, L.; Kisiala, A.; Olechowski, H.; Emery, R.N. Cytokinin activity during early kernel development corresponds positively with yield potential and later stage ABA accumulation in field-grown wheat (*Triticum aestivum* L.). *Planta* **2020**, *252*, 1–16. [CrossRef] [PubMed]
92. Faix, B.; Radchuk, V.; Nerlich, A.; Hümmer, C.; Radchuk, R.; Emery, R.J.N.; Keller, H.; Götz, K.-P.; Weschke, W.; Geigenberger, P.; et al. Barley grains, deficient in cytosolic small subunit of ADP-glucose pyrophosphorylase, reveal coordinate adjustment of C: N metabolism mediated by an overlapping metabolic-hormonal control. *Plant J.* **2012**, *69*, 1077–1093. [CrossRef] [PubMed]
93. Powell, A.F.; Paleczny, A.R.; Olechowski, H.; Emery, R.N. Changes in cytokinin form and concentration in developing kernels correspond with variation in yield among field-grown barley cultivars. *Plant Physiol. Biochem.* **2013**, *64*, 33–40. [CrossRef] [PubMed]
94. Motyka, V.; Faiss, M.; Strnad, M.; Kamínek, M.; Schmulling, T. Changes in cytokinin content and cytokinin oxidase activity in response to derepression of *ipt* gene transcription in transgenic tobacco calli and plants. *Plant Physiol.* **1996**, *112*, 1035–1043. [CrossRef] [PubMed]
95. Motyka, V.; Vaňková, R.; Čapková, V.; Petrášek, J.; Kamínek, M.; Schmülling, T. Cytokinin-induced upregulation of cytokinin oxidase activity in tobacco includes changes in enzyme glycosylation and secretion. *Physiol. Plant* **2003**, *117*, 11–21. [CrossRef]
96. Brugière, N.; Jiao, S.; Hantke, S.; Zinselmeier, C.; Roessler, J.A.; Niu, X.; Jones, R.J.; Habben, J.E. Cytokinin oxidase gene expression in maize is localized to the vasculature, and is induced by cytokinins, abscisic acid, and abiotic stress. *Plant Physiol.* **2003**, *132*, 1228–1240. [CrossRef]
97. Blagoeva, E.; Dobrev, P.I.; Malbeck, J.; Motyka, V.; Gaudinová, A.; Vaňková, R. Effect of exogenous cytokinins, auxins and adenine on cytokinin *N*-glucosylation and cytokinin oxidase/dehydrogenase activity in de-rooted radish seedlings. *Plant Growth Regul.* **2004**, *44*, 15–23. [CrossRef]
98. Hirose, N.; Takei, K.; Kuroha, T.; Kamada-Nobusada, T.; Hayashi, H.; Sakakibara, H. Regulation of cytokinin biosynthesis, compartmentalization and translocation. *J. Exp. Bot.* **2008**, *59*, 75–83. [CrossRef]
99. Vyroubalová, Š.; Václavíková, K.; Turečková, V.; Novák, O.; Šmehilová, M.; Hluska, T.; Ohnoutková, L.; Frébort, I.; Galuszka, P. Characterization of new maize genes putatively involved in cytokinin metabolism and their expression during osmotic stress in relation to cytokinin levels. *Plant Physiol.* **2009**, *151*, 433–447. [CrossRef] [PubMed]
100. Gasparis, S.; Przyborowski, M.; Kała, M.; Nadolska-Orczyk, A. Knockout of the *HvCKX1* or *HvCKX3* gene in barley (*Hordeum vulgare* L.) by RNA-Guided Cas9 nuclease affects the regulation of cytokinin metabolism and root morphology. *Cells* **2019**, *8*, 782. [CrossRef] [PubMed]
101. Zhao, J.; Bai, W.; Zeng, Q.; Song, S.; Zhang, M.; Li, X.; Hou, L.; Xiao, Y.; Luo, M.; Li, D.; et al. Moderately enhancing cytokinin level by down-regulation of *GhCKX* expression in cotton concurrently increases fiber and seed yield. *Mol. Breed.* **2015**, *35*, 60. [CrossRef] [PubMed]
102. Li, Y.; Song, G.; Gao, J.; Zhang, S.; Zhang, R.; Li, W.; Chen, M.; Liu, M.; Xia, X.; Risacher, T.; et al. Enhancement of grain number per spike by RNA interference of cytokinin oxidase 2 gene in bread wheat. *Hereditas* **2018**, *155*, 1–8. [CrossRef] [PubMed]
103. Zalewski, W.; Galuszka, P.; Gasparis, S.; Orczyk, W.; Nadolska-Orczyk, A. Silencing of the *HvCKX1* gene decreases the cytokinin oxidase/dehydrogenase level in barley and leads to higher plant productivity. *J. Exp. Bot.* **2010**, *61*, 1839–1851. [CrossRef]
104. Zalewski, W.; Gasparis, S.; Boczkowska, M.; Rajchel, I.K.; Kała, M.; Orczyk, W.; Nadolska-Orczyk, A. Expression patterns of *HvCKX* genes indicate their role in growth and reproductive development of barley. *PLoS ONE* **2014**, *9*, e115729. [CrossRef]
105. Yeh, S.Y.; Chen, H.W.; Ng, C.Y.; Lin, C.Y.; Tseng, T.H.; Li, W.H.; Ku, M.S. Down-regulation of cytokinin oxidase 2 expression increases tiller number and improves rice yield. *Rice* **2015**, *8*, 36. [CrossRef]
106. Uauy, C.; Wulff, B.B.; Dubcovsky, J. Combining traditional mutagenesis with new high-throughput sequencing and genome editing to reveal hidden variation in polyploid wheat. *Annu. Rev. Genet.* **2017**, *51*, 435–454. [CrossRef]
107. Gil, J.; Andrade-Martínez, J.S.; Duitama, J. Accurate, efficient and user-friendly mutation calling and sample identification for TILLING experiments. *Front. Genet.* **2021**, *12*, 624513. [CrossRef]
108. Lassoued, R.; Phillips, P.W.; Macall, D.M.; Hesseln, H.; Smyth, S.J. Expert opinions on the regulation of plant genome editing. *Plant Biotechnol. J.* **2021**, *19*, 1104–1109. [CrossRef] [PubMed]
109. Wheat Tilling. Available online: https://wheat-tilling.com/ (accessed on 25 October 2021).
110. Krasileva, K.V.; Vasquez-Gross, H.A.; Howell, T.; Bailey, P.; Paraiso, F.; Clissold, L.; Dubcovsky, J. Uncovering hidden variation in polyploid wheat. *Proc. Natl. Acad. Sci. USA* **2017**, *114*, E913–E921. [CrossRef] [PubMed]
111. Song, J.; Jiang, L.; Jameson, P.E. Co-ordinate regulation of cytokinin gene family members during flag leaf and reproductive development in wheat. *BMC Plant Biol.* **2012**, *12*, 78. [CrossRef]
112. Svačinová, J.; Novák, O.; Plačková, L.; Lenobel, R.; Holík, J.; Strnad, M.; Doležal, K. A new approach for cytokinin isolation from *Arabidopsis* tissues using miniaturized purification: Pipette tip solid-phase extraction. *Plant Methods* **2012**, *8*, 17. [CrossRef] [PubMed]

plants

MDPI

Article

Development of SNP Markers for White Immature Fruit Skin Color in Cucumber (*Cucumis sativus* L.) Using QTL-seq and Marker Analyses

D. S. Kishor [1], Hemasundar Alavilli [1], Sang-Choon Lee [2], Jeong-Gu Kim [3] and Kihwan Song [1,*]

[1] Department of Bioresources Engineering, Sejong University, Seoul 05006, Korea; kishoreflmes@gmail.com (D.S.K.); alavilli.sundar@gmail.com (H.A.)
[2] PHYZEN, Seongnam-si 13558, Gyeonggi-do, Korea; sclee0923@phyzen.com
[3] National Academy of Agricultural Science, Rural Development Administration, Jeonju 54874, Korea; jkim5aug@korea.kr
* Correspondence: khsong@sejong.ac.kr; Tel.: +82-2-3408-2905

Abstract: Despite various efforts in identifying the genes governing the white immature fruit skin color in cucumber, the genetic basis of the white immature fruit skin color is not well known. In the present study, genetic analysis showed that a recessive gene confers the white immature fruit skin-color phenotype over the light-green color of a Korean slicer cucumber. High-throughput QTL-seq combined with bulked segregation analysis of two pools with the extreme phenotypes (white and light-green fruit skin color) in an F_2 population identified two significant genomic regions harboring QTLs for white fruit skin color within the genomic region between 34.1 and 41.67 Mb on chromosome 3, and the genomic region between 12.2 and 12.7 Mb on chromosome 5. Further, nonsynonymous SNPs were identified with a significance of $p < 0.05$ within the QTL regions, resulting in eight homozygous variants within the QTL region on chromosome 3. SNP marker analysis uncovered the novel missense mutations in *Chr3CG52930* and *Chr3CG53640* genes and showed consistent results with the phenotype of light-green and white fruit skin-colored F_2 plants. These two genes were located 0.5 Mb apart on chromosome 3, which are considered strong candidate genes. Altogether, this study laid a solid foundation for understanding the genetic basis and marker-assisted breeding of immature fruit skin color in cucumber.

Keywords: cucumber; QTL-seq; SNP markers; white immature fruit skin color

Citation: Kishor, D.S.; Alavilli, H.; Lee, S.-C.; Kim, J.-G.; Song, K. Development of SNP Markers for White Immature Fruit Skin Color in Cucumber (*Cucumis sativus* L.) Using QTL-seq and Marker Analyses. *Plants* **2021**, *10*, 2341. https://doi.org/10.3390/plants10112341

Academic Editors: Milan S. Stankovic, Paula Baptista and Petronia Carillo

Received: 7 October 2021
Accepted: 28 October 2021
Published: 29 October 2021

Publisher's Note: MDPI stays neutral with regard to jurisdictional claims in published maps and institutional affiliations.

Copyright: © 2021 by the authors. Licensee MDPI, Basel, Switzerland. This article is an open access article distributed under the terms and conditions of the Creative Commons Attribution (CC BY) license (https://creativecommons.org/licenses/by/4.0/).

1. Introduction

Cucumber (*Cucumis sativus* L.; $2n = 2x = 14$) is a major economically important vegetable crop in the *Cucurbitaceae* family. Its total annual global production is 87,805,086 tons in an area of 2231,402 ha, and 80 % of global production comes from China with a yearly output of 70,338,971 tons (FAO, 2019). Cucumber fruits are usually consumed fresh or as processed pickles after 8–18 days of anthesis [1].

Cucumber fruits display wide phenotypic variation in immature fruit skin color from dark green to white appearance [2]. The green skin color of the cucumber fruit is directly associated with the accumulation of chlorophyll [1,3–5]. The external skin color of immature fruit is considered as an essential quality trait that decides consumer preference. In order to develop cucumber varieties with various skin colors, it is necessary to understand the inheritance and genes governing the fruit skin color.

Upon the availability of the genomic information of cucumber [6–8], several genes were identified, which led to the development of molecular markers for marker-assisted selection (MAS) of fruit skin color in cucumber. A single dominant gene *B* (R2R3-MYB) was identified for the orange mature fruit skin color on chromosome 4, and an Insertion-deletion (InDel) marker was developed based on the 1-bp deletion in the third exon of this gene, was co-segregated with fruit skin color [9]. High-resolution mapping for the dull fruit skin trait in cucumber revealed a single dominant gene *D* on chromosome 5 between markers SSR37 and SSR112, at a physical distance of 244.9 kb [10]. The *D* gene was reported to be closely linked to the genes governing fruit wart (*Tu*), uniform immature fruit color (*u*), and small spines (*ss*) [11,12].

Studies on green fruit skin trait have shown that SNPs in *ARC5* and *Ycf54* genes cause light-green immature fruit skin color in cucumber [3,13]. Yang et al. [14] studied the inheritance of the uniform immature fruit color in cucumber and identified a single recessive gene that confers uniform immature fruit color (*u*) phenotype. *u* gene was mapped to a 313.2 kb on chromosome 5 between co-dominant SSR markers SSR10 and SSR27 at a genetic distance of 0.8 and 0.5 cM, respectively [14], which further laid a strong foundation for marker-assisted breeding of uniform immature fruit color trait in cucumber.

Recent studies have shown that a single gene controls the external fruit color trait and white skin color is recessive over dark green skin color in cucumber [2,15,16]. A frameshift mutation, which leads to a premature stop codon in the *w* gene (*aprr2*) on chromosome 3, was reported to be a sole candidate gene responsible for white immature fruit color phenotype in cucumber and associated with chlorophyll biosynthesis [15]. Tang et al. [16] studied the genetics of white immature fruit color in cucumber and mapped a single recessive gene (w_0) for white immature fruit color on chromosome 3 to approximately 100.3 kb between two flanking markers, Q138 and Q193 [16]. In spite of several studies, Tang et al. [16] proposed that further study is necessary to understand the genetic basis of white immature fruit skin color in cucumber.

This study determined a novel genetic architecture for the white immature fruit skin color of cucumber by using QTL-seq and *SNP marker* analyses, which revealed that white fruit skin color trait is recessive over light-green fruit skin color of Korean slicer cucumber. Furthermore, this study identified a novel allelic variant of the *Chr3CG52930* gene and a new candidate gene, '*Chr3CG53640*', for white fruit skin color trait in cucumber. This finding would facilitate understanding of the genes involved in cucumber skin color and contribute to the development of cucumbers using marker-assisted breeding.

2. Materials and Methods

2.1. Genomic DNA Extraction and Pooling

Genomic DNAs were extracted from leaves of the two parental lines (Inbred line of Korean slicer cucumber 'MEJ' with light-green skin color and 'PI525075' with white skin color) and their F_2 plants using a cetyltrimethylammonium bromide (CTAB) method [17]. The integrity of the extracted genomic DNA was checked by an agarose gel electrophoresis. Genomic DNA (gDNA) was quantified using Quant-iT™ PicoGreen™ dsDNA Assay Kit (Invitrogen, Waltham, MA, USA) and Qubit fluorometer (Thermo Fisher Scientific, Waltham, MA, USA). Genomic DNAs of 16 F_2 plants with white immature skin color were mixed with equal amounts and used as white skin-pool, and genomic DNAs of 20 F_2 plants with light-green skin color were mixed with an equal quantity and used as light-green skin-pool for downstream analysis.

2.2. Whole-Genome Resequencing

Quality and quantity of gDNAs of two parental lines and pooled DNA samples were examined again using an agarose gel electrophoresis and Qubit fluorometer (Thermo Fisher Scientific, USA). The sequencing libraries were prepared according to the TruSeq DNA PCR-free Sample Preparation Kit (Illumina, San Diego, CA, USA). Briefly, fragmentation of 1µg of genomic DNA was performed using adaptive focused acoustic technology (AFA; Covaris, Woburn, MA, USA), and the fragmented DNA was end-repaired to create 5′-phosphorylated, blunt-ended dsDNA molecules. Following end-repair, DNA was size-selected with the bead-based method. These DNA fragments go through the addition of a single 'A' base and ligation of the Truseq indexing adapters. The purified libraries were quantified using quantitative PCR (qPCR) according to the qPCR Quantification Protocol

Guide (KAPA Library Quantification kits for Illumina Sequencing platforms) and qualified using the high-sensitivity DNA chip (Agilent Technologies, Santa Clara, CA, USA). The paired-end (2 × 150 bp) sequencing was performed using the HiSeq-X platform (Illumina, San Diego, CA, USA) by the Macrogen Co. (Seoul, Korea).

2.3. QTL-seq Analysis

QTL-seq analysis was performed using the QTL-seq program (version 2.1.3, https://github.com/YuSugihara/QTL-seq) (access on 7 October 2021) with default parameters as described previously [18]. Briefly, high-quality sequencing data were obtained by trimming the raw sequencing data using the Trimmomatic program [19] in the QTL-seq program. The high-quality sequencing data of the PI525075 parent with white skin color were aligned to the Korean cucumber genome (*Cucumis sativus* var. JEF, BioProject no: PRJNA732224, https://www.ncbi.nlm.nih.gov/bioproject/732224) (access on 7 October 2021) using BWA [20]. Then variants from PI525075 parent were used to generate the PI525075 reference genome by substituting the variant bases in the Korean cucumber genome. The high-quality sequencing data of two pooled DNA samples (white skin-pool and light-green skin-pool) were mapped to the PI525075 reference genome and variants (SNPs and InDels) were detected. The SNP index at each SNP position was calculated and then Δ (SNP index) was calculated using the formula: [SNP index (white skin-pool)—SNP index (light-green skin-pool)]. The average SNP index and Δ (SNP index) distribution were estimated in a given genomic interval using a sliding window approach with 2 Mb window size and 100 kb increment and plotted to generate SNP index plots for all chromosomes. The candidate genomic regions for the phenotype were determined based on the sliding window plots. The regions in which the average Δ (SNP index) was significantly greater than the surrounding region and exhibited an average $p < 0.05$ were considered as candidate QTLs.

Variants' information in the candidate QTLs was extracted from the variant calling file (VCF) generated by the QTL-seq program. Then, variants with $p < 0.05$ were further selected and used for further downstream analysis.

2.4. Annotation of Variants and Identification of Variants Causing Protein Sequence Change

The annotation of variants was performed using SnpEff software (version 5.0e, [21]). The variants present in gene regions (from 5′ UTR to 3′ UTR, including intron and exon) was annotated as genic, while other genomic regions were intergenic. Variants in coding sequences (CDSs) were further divided into synonymous and non-synonymous.

Variants that caused the change of amino acid in deduced protein sequence of the genes were identified among the selected variants with $p < 0.05$ in the QTL regions under the following criteria: (1) variants that were different between bulk1 (white skin) and bulk2 (light-green skin) were selected; (2) among the selected variants, only homozygous variants were selected; (3) homozygous variants that caused the amino acid change in protein sequences encoded by genes were finally selected based on variant annotation information as candidate variants associated with the phenotype.

2.5. Development of Molecular Markers

The flanking sequences of variants were extracted from the JEF Korean cucumber genome sequences and then used to design cleaved amplified polymorphic sequence (CAPS) and derived cleaved amplified polymorphic sequence (dCAPS) primers. CAPS and dCAPS primers were designed using in-house scripts modified from CAPS-finder.pl (https://github.com/mfcovington/CAPS-finder/blob/master/CAPS-finder.pl) (access on 7 October 2021) and dCAPS Finder (http://helix.wustl.edu/dcaps/dcaps.html) (access on 7 October 2021), respectively. Parameters of primer design were a primer size of 17– 25 mer, GC% of 50%, Tm of 50–60 °C, and amplicon size of 200–700 bp. BLASTN searches against reference genome sequences confirmed the specificity of designed primers. Designed molecular markers were validated by agarose gel electrophoresis after genomic

DNA PCR and restriction enzyme treatment with genomic DNAs of the parents and F_2 plants.

3. Results

3.1. Inheritance of Immature Fruit Skin Color

The phenotype of the two parental lines, MEJ and PI525075, along with their F_1 (MEJ/ PI525075), were shown in Figure 1a. The skin color of MEJ and PI525075 were evaluated for two skin color indices (one for white and two for light-green) at the immature fruit stage. The fruits of F_1 derived from a cross between MEJ and PI525075 were demonstrated to have a light-green immature fruit skin. In the F2 population, there were 106 and 30 plants with light-green skin and white skin, respectively, which was fit to a segregation ratio of 3:1. This implies the recessive nature of the gene for white immature fruit skin color (Table 1 and Figure 1b).

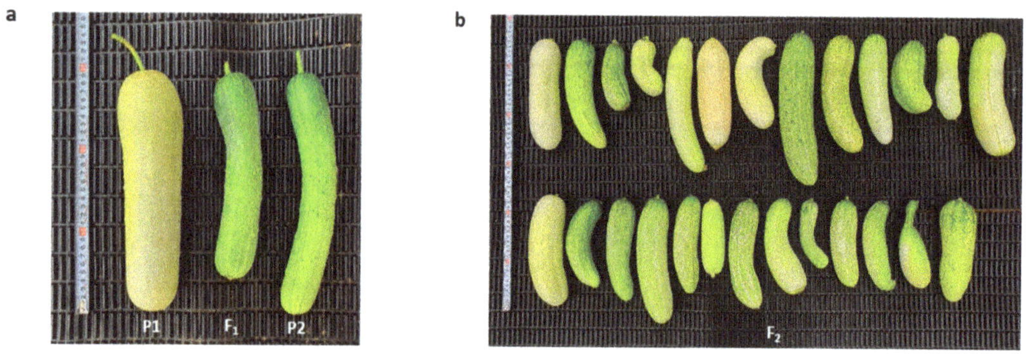

Figure 1. Phenotypic analysis of immature fruit skin color of immature cucumber fruits. (**a**) Fruit phenotype of PI525075 (P1), MEJ/PI525075 (F_1) and MEJ (P2); (**b**) fruit phenotype of F_2 derived from a cross between MEJ and PI525075.

Table 1. Segregation analysis of immature fruit skin color phenotype in cucumber.

Name	Population	Plant Number	Immature Fruit Skin Color Phenotype				
			White	Light-Green	Expected	χ^2	p Value †
PI525075	P1	10	10	0	-	-	-
MEJ	P2	10	0	10	-	-	-
MEJ/PI525075	F_1	10	0	10	-	-	-
MEJ/PI525075	F_2	136	30	106	3:1	0.62	0.42

† Not significant (p Value > 0.05).

3.2. Whole-Genome Re-Sequencing and QTL-seq Analysis

For whole-genome re-sequencing, 16 plants with white immature fruit skin and 20 plants with light-green immature fruit skin were selected among the 136 F_2 plants (Figure 1b) and their gDNAs were then pooled respectively to prepare the white fruit skin-pool and light-green fruit skin-pool along with their two parental lines, MEJ and PI525075. Whole-genome re-sequencing results showed that a total of 192,314,600 and 191,719,994 raw reads were generated for PI525075 and MEJ, respectively, while a total of 204,951,660 and 228,647,910 raw reads were obtained for white fruit skin-pool and light-green fruit skin-pool, respectively. Upon trimming, a total of 157,678,304 reads were generated from PI525075; 187,218,180 reads from MEJ; 175,957,140 reads from white fruit skin-pool, and 197,713,150 reads from light-green fruit skin-pool, each corresponding to more than 20 Gb read length, and more than 75% of the reads were clean reads (Table 2). Clean reads from the parents and bulks were compared to the estimated cucumber genome size of 350 Mb, which indicated that all genomes were sequenced at a depth ranging from 64.32X to 80.44X. In comparison with the PI525075 reference genome using QTL-seq program, we have identified a total of 160,392 SNPs for white fruit skin-pool and 120,899 SNPs for light-green fruit skin-pool across the chromosomes (Table 3).

Table 2. Summary of whole genome re-sequencing data used for QTL-seq analysis.

Samples	Raw Data		Trimmed Data			Coverage (X)
	Reads	Read Length (bp)	Reads	Read Length (bp)	%	
PI525075	192,314,600	29,039,504,600	157,678,304	22,512,072,535	77.52	64.32
MEJ	191,719,994	28,949,719,094	187,218,180	27,266,475,876	94.19	77.90
White-pool	204,951,660	30,947,700,660	175,957,140	25,010,424,993	80.82	71.46
Light-green-pool	228,647,910	34,525,834,410	197,713,150	28,154,853,131	81.55	80.44

Table 3. Polymorphisms identified in QTL-seq analysis.

Chromosome Number	No. of SNPs	
	White-Pool	Light-Green-Pool
1	11,896	8263
2	21,047	20,044
3	46,480	16,303
4	17,843	25,400
5	27,535	15,122
6	18,782	21,018
7	16,809	14,749
Total	160,392	120,899

To identify the genomic region governing immature fruit skin color trait between the white fruit skin-pool and light-green fruit skin-pool, the SNP index of individual SNPs were calculated using the parental PI525075 line as a reference genome and compared these with the bulks' sequences. An SNP index of zero represents entire short reads of the white fruit skin genome, whereas an SNP index of one indicates that the reads are from the light-green fruit skin genome. The average SNP index was estimated in a given genomic interval with 2 Mb window size and 100 kb increment and plotted to generate SNP index plots for white fruit skin and light-green fruit skin pools against all chromosomes (Figures S1 and S2). To identify the differences in the SNP-indices of the two pools, Δ (SNP index) was estimated by combining the SNP index information of white fruit skin and light-green fruit skin pools and a statistical confidence interval was plotted against the reference genome of cucumber (Figure S3). Furthermore, significant genomic regions were detected at a statistical significance of $p < 0.05$ according to the principle of SNP index estimation showed in QTL-seq analysis (Figure S3), resulting in two significant genomic regions harboring candidate QTLs for white fruit skin trait within the genomic region between 34.1 and 41.67 Mb on chromosome 3, and the genomic region between 12.2 and 12.7 Mb on chromosome 5 (Figure 2 and Table 4). The significant genomic region on chromosome 3 had an average SNP index value of 0.18 and 0.72 for white fruit skin-pool and light-green fruit skin-pool, respectively. Similarly, the second significant genomic region on chromosome 5 displayed an average SNP index value of 0.26 and 0.71 for white fruit skin-pool and light-green fruit skin-pool, respectively. The significant genomic region on chromosome 3 had an average Δ (SNP index) value of 0.53. In contrast, significant genomic region chromosome 5 had an average Δ (SNP index) value of 0.44 at the 95% confidence interval. These results indicated the presence of significant genomic regions conferring white immature skin color in cucumber.

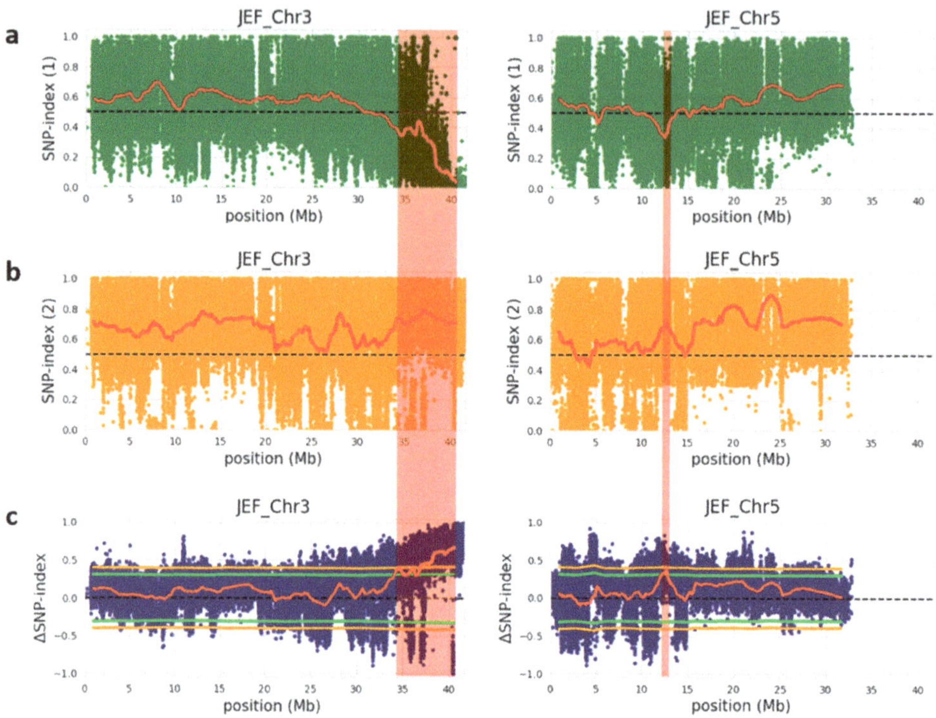

Figure 2. Identification of QTLs on chromosome 3 and 5 for immature fruit skin color based on QTL-seq analysis of F_2 population. (**a**) SNP index graph of white-bulk (green dots and red line represents SNP index and sliding window average of SNP index, respectively); (**b**) SNP-index graph of light-green-bulk (orange dots and red line indicates SNP index and sliding window average of SNP index, respectively); (**c**) Δ (SNP index) graph with statistical confidence intervals under the null hypothesis of no QTLs (green line, $p < 0.05$; orange line, $p < 0.01$) from QTL-seq analysis. The significant genomic regions with $p < 0.01$ are highlighted by red shaded bar. Blue dots and red line show Δ (SNP index) and sliding window average of Δ (SNP index), respectively. SNP and Δ (SNP indexes) were calculated based on 2 Mb interval with a 100 kb sliding window.

Table 4. Candidate QTLs associated with immature fruit skin color of cucumber based on QTL-seq analysis.

Chr.	Physical Position (Mb)	Size (bp)	SNPs	Genes
3	34.1–41.67	7,579,492	11,435	556
5	12.2–12.7	500,000	743	33

3.3. *Identification of SNPs and Candidate Genes via in Silico Analysis*

To identify the SNPs and potential candidate genes associated with immature fruit skin, homozygous variants that caused the change in amino acid in deduced protein sequence of genes in the QTL regions were mined with $p < 0.05$. As a result, a total of eight homozygous variants (six SNPs and two InDels) and seven potential candidate genes were identified within the QTL region between 34.1 and 41.67 Mb on chromosome 3 (Table 5). Among the eight homozygous variants, we detected a single base pair deletion in a gene encoding LSi6 (*Cucumis sativus*) aquaporin nodulin-26-like intrinsic protein (NIP), a single base insertion within a gene encoding hypothetical protein Csa_013022 (*Cucumis sativus*) PHT [solute carrier family 15 (peptide/histidine transporter) protein], and six missense mutations (non-synonymous SNPs) in five putative genes which are known to be involved in thioredoxin-related transmembrane protein 2 isoform X1 (*Cucumis sativus*)

(1 SNP); inactive poly [ADP-ribose] polymerase RCD1 (*Cucumis sativus*) (1 SNP); β-amyrin 11-oxidase (*Cucumis sativus*), AIT72036.1 cytochrome P450 (*Cucumis sativus*) (2 SNPs); Pyruvate kinase isozyme G, chloroplastic isoform X1 (*Benincasa hispida*), pyruvate kinase [EC:2.7.1.40] (1 SNP), and QWRF motif-containing protein 7 (*Cucumis sativus*) (1 SNP). By contrast, none of the homozygous variants caused the amino acid changes in the protein-coding genes between 12.2 and 12.7 Mb on chromosome 5.

3.4. Validation of Candidate Genes for Immature Fruit Skin Phenotype

To validate involvement of the candidate genes for immature fruit skin phenotype, a total of two CAPS and six dCAPS markers were developed based on the SNPs information available in the five putative genes (Table 6). These eight markers (M1 to M8) were firstly examined with two parental lines (MEJ and PI525075) and pooled gDNAs of F_2 plants with light-green and white fruit skin colors, identifying that all eight markers could discriminate two skin colors in tested samples (Figure 3). Since, M1, M2, and M5 markers were based on the same SNPs information's as M3, M4, and M6 markers, respectively. Therefore, M1, M2, M5, M7, and M8 markers were selected to be validated in 36 F_2 plants. Marker validation assay revealed that M7 and M8 markers designed for the SNPs located at *Chr3CG52930* and *Chr3CG53640* genes were co-segregated with the light-green and white fruit skin color phenotype in 36 F_2 plants (Figure 4). In contrast, none of the other markers (M1, M2 and M5) showed strong co-segregation among F_2 plants (Figure S4).

These results provided strong evidence showing that the *Chr3CG52930* gene-encoding pyruvate kinase isozyme G chloroplastic isoform X1 (*Benincasa hispida*) and *Chr3CG53640* gene-encoding QWRF motif-containing protein 7 (*Cucumis sativus*) are involved in the white immature fruit skin color phenotype in PI525075. Further, BLAST search showed that *Chr3CG52930* and *Chr3CG53640* genes from JEF Korean cucumber genome have 99 and 100 % sequence homology with the *Csa3G904080* and *Csa3G915140* genes of 'Chinese Long v2' cucumber genome, respectively.

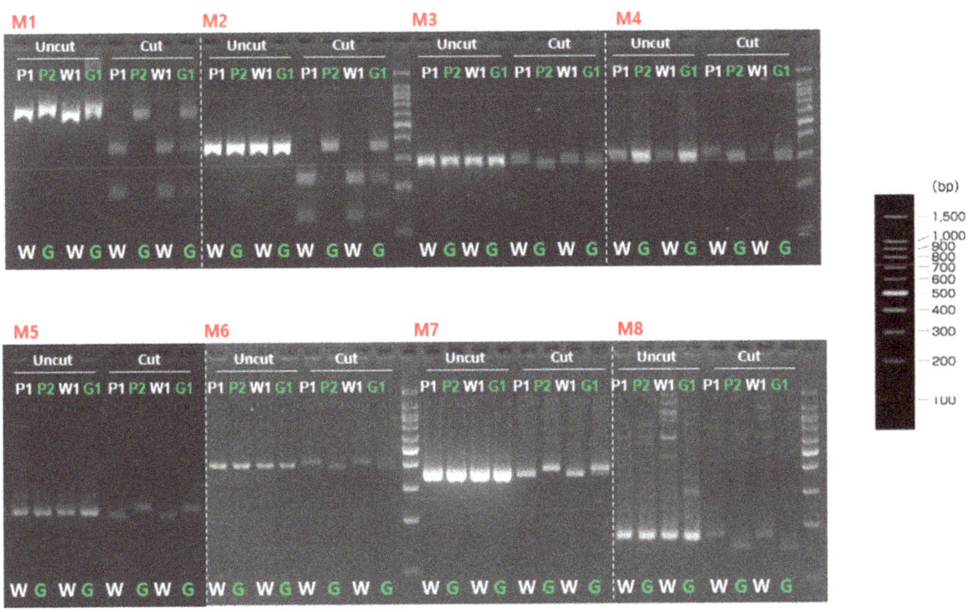

Figure 3. Marker validation with parental lines and bulked F_2 plants. P1, 'PI525075' (white); P2, 'MEJ' (light-green); W1, white pool, G1, light-green pool; W, white; G, light-green.

Table 5. Identification of sequence variations and candidate genes in the QTL region on chromosome 3.

Position	Ref	Alt	Effect	AA	Type	Gene ID	Description
34626707	TAAAAAAAAAAA	TAAAAAAAAAA	Deletion	ATG	InDel	*Chr3CG43770*	LSi6 [*Cucumis sativus* var. *sativus*], aquaporin NIP
40443941	T	C	Non_synonymous	I3V	SNP	*Chr3CG51850*	Thioredoxin-related transmembrane protein 2 isoform X1 [*Cucumis sativus*]
40660199	T	G	Non_synonymous	L457R	SNP	*Chr3CG52290*	Inactive poly [ADP-ribose] polymerase RCD1 [*Cucumis sativus*]
41075124	G	T	Non_synonymous	V250L	SNP	*Chr3CG52880*	β-amyrin 11-oxidase [*Cucumis sativus*], AIT72036.1 cytochrome P450 [*Cucumis sativus*]
41075140	T	G	Non_synonymous	I255R	SNP	*Chr3CG52880*	β-amyrin 11-oxidase [*Cucumis sativus*], AIT72036.1 cytochrome P450 [*Cucumis sativus*]
41102542	CTTTTTTTTTTTTT	CTTTTTTTTTTTTTT	Insertion	S123F	InDel	*Chr3CG52910*	Hypothetical protein Csa_013022 [*Cucumis sativus*], PHT; solute carrier family 15 (peptide/histidine transporter)
41106529	G	A	Non_synonymous	P338L	SNP	*Chr3CG52930*	Pyruvate kinase isozyme G, chloroplastic isoform X1 [*Benincasa hispida*], pyk; pyruvate kinase [EC:2.7.1.40]
41641571	G	A	Non_synonymous	A171T	SNP	*Chr3CG53640*	QWRF motif-containing protein 7 [*Cucumis sativus*]

Table 6. Primers used for genotyping the parents and F_2 plants.

Marker Name	Chr	Position	Gene ID	Marker Type	Forward/Reverse Sequence (5′–3′)	Restriction Site	Restriction Enzyme	Amplicon Size (Light-Green vs. White)
M1	3	40443941	*Chr3CG51850*	CAPS	TTCTCTAATGACGATGACGTTG/ATCCGGAATTTCTTTCTTCTTC	ACGT	*Hpy*CH4IV	691 vs. (264, 427)
M2	3	40660199	*Chr3CG52290*	CAPS	TCTGTGAACTTTGACAGTGGAG/TAGATCTCCACAAGTCTCCCAT	ACGT	*Hpy*CH4IV	360 vs. (226, 134)
M3	3	40443941	*Chr3CG51850*	dCAPS	TTTCCGGGATGAATTCCCGGATCGA/ATCCGGAATTTCTTTCTTCTTC	ATCGAT	*Cla*I	238 vs. 263
M4	3	40660199	*Chr3CG52290*	dCAPS	AAGATACCATCAAGTCCTCTTAAGC/GCAATTTATCTGCAACTGGTCT	AAGCTT	*Hind*III	272 vs. 297
M5	3	41075124	*Chr3CG52880*	dCAPS	GAAGATAGAGAATACAATTCAAGCT/TATGTAGAAGAGCCCAACAAGC	AAGCTT	*Hind*III	197 vs. 172
M6	3	41075140	*Chr3CG52880*	dCAPS	TGTTTTATCATTTTCAAATCTACGT/GCCTATTAAATTGGTGGATGAA	TACGTA	*Sna*BI	338 vs. 363
M7	3	41106529	*Chr3CG52930*	dCAPS	AACCTTGTGGACACTCGATGGACTT/ATGCGTGTTCCTCTAGTTTGTT	CTNAG	*Dde*I	327 vs. 302
M8	3	41641571	*Chr3CG53640*	dCAPS	ATGGAAGTCTCTGCTCTAACCTAAG/AAACTAGGCAGTCAACGAGGT	CCTNAGC	*Bpu*10I	113 vs. 138

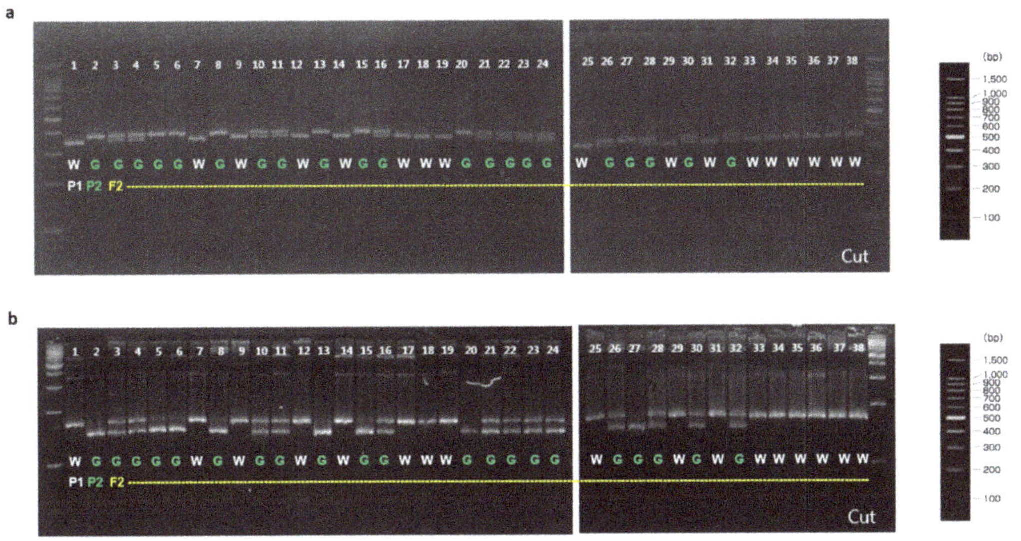

Figure 4. dCAPS analysis of novel alleles in the F_2 plants derived from a cross between MEJ and PI525075. P1, 'PI525075' (white); P2, 'MEJ' (light-green); W, white; G, light-green. (**a**) M7 marker for a novel allele of *Chr3CG52930*; (**b**) M8 marker for a novel of allele of *Chr3CG53640*.

4. Discussion

The immature fruit skin color is a major quality trait in cucumber, which is an important factor in determining consumer preference according to region. Likewise, chlorophyll is a primary natural pigment in the green peel of cucumber, resulting from chlorophyll biosynthesis [3,13]. A recent study suggests that white immature fruit skin color in cucumbers results from the lack of chlorophyll synthesis during fruit development [2], although several genes involved in chlorophyll biosynthesis are well known in flowering plants [22] but remain unclear in cucumber. Likewise, earlier studies have shown that white immature fruit skin color was a recessive trait and controlled mainly by a single recessive gene in cucumber [2,16]. In the present study, we have investigated the inheritance of white immature fruit skin color using an F_2 population derived from a cross between MEJ and PI525075, indicating that a single recessive gene controls the white immature fruit skin color of cucumber. Therefore, the present study's results are in accordance with the previous reports showing the single recessive gene inheritance for white immature fruit skin trait in cucumber [2,16].

With the growing popularity of next-generation sequencing (NGS), sequencing of the crop plants enables the detection of variants across the genome [23–26]. QTL-seq analysis allows rapid detection of the homozygous variants of a given phenotype by whole-genome resequencing of two bulked populations [18,27]. QTL-seq makes use of combined benefits of bulk-segregating analysis (BSA) and whole-genome resequencing, which can be used to identify the genomic regions responsible for the mutant phenotype in a single step [18,28–30]. The QTL-seq technique was recently applied to the cucumber population to identify the putative variants closely linked to causal genes responsible for subgynoecy, powdery mildew resistance, and light-green immature fruit skin traits in cucumber [13,31,32]. Although QTL-seq is a powerful tool, cucumber breeding has been largely dependent on conventional molecular mapping of putative genes, which is time consuming and requires large-scale DNA markers and many generations of advanced segregating populations [2,16,33]. Here we applied QTL-seq to detect the putative genes for the white immature fruit skin color trait in cucumber using an F_2 population derived from a cross between the MEJ (Korean type light-green skin color) and PI525075 (white skin color).

In the past, several genes controlling the immature fruit skin color have been reported in cucumber [2,3,13,14,16]. Liu et al. mapped and identified a single base pair insertion in the *APRR2* gene, resulting in a premature stop codon, which further disrupts the accumulation of chlorophyll and chloroplast development, leading to white immature fruit skin color in cucumber [2,15]. Furthermore, two markers based on InDel (LH392580) and SNP (ASPCR39250) were developed and validated within the 8.2 kb physical interval of the *APRR2* gene [15]. Similarly, the latest study in cucumber identified the w_0 gene for white immature fruit color on chromosome 3 to a 100.3 kb region containing 13 candidate genes between two flanking markers, Q138 and Q193 [16]. However, this study proposed that the *Csa3G904140* gene (w_0) is responsible for the white immature fruit skin color in cucumber. This study further highlighted that further study is required to validate whether the w_0 gene was the same as the *APRR2* gene reported in the previous study [15].

In contrast, our study identified the 1080 genes in the target genomic region based on QTL-seq analysis (Table S1); as a result, seven candidate genes were predicted for the immature fruit skin color within the QTL region on chromosome 3. Further, homozygous variants within these seven candidates were validated via SNP marker analysis, resulting in two SNP markers (M7 and M8) developed for the missense mutations showing co-segregation with the light-green and white fruit skin-colored F_2 plants. M7 marker designed at the nucleotide position of 6253 (P338L) of *Chr3CG52930* gene, whereas M8 marker designed at a nucleotide position of 511 (A171T) at *Chr3CG53640* gene on chromosome 3. Thus, it was likely that two genes located 0.5 Mb apart on chromosome 3 were solid putative genes for immature fruit skin color of cucumber, which laid a solid foundation for understanding the genetic basis of immature fruit skin color in cucumber.

Further, the *Chr3CG52930* gene-encoding pyruvate kinase isozyme G chloroplastic isoform X1 has been found to share 99% sequence homology with the *Csa3G904080* gene of the 'Chinese Long v2' cucumber genome. In the latest study, validation of three mutations in the *Csa3G904080* gene showed inconsistent results with the green and white cucumbers phenotype, which further concluded that the *Csa3G904080* gene was not a putative gene responsible for white pigmentation in cucumber [16]. Although Tang et al. [16] showed that expression of the *Csa3G904080* gene was higher in root and leaf than in fruit skin, the present study identified a novel missense mutation in the *Csa3G904080* gene and revealed a consistent result with the phenotype of green- and white-skinned cucumbers via SNP marker analysis. Therefore, these results indicated that the possible involvement of the *Csa3G904080* gene with white pigmentation requires further study due to their role in chloroplast biogenesis. Similarly, the *Chr3CG53640* gene encoding QWRF motif-containing protein 7 (*Cucumis sativus*) shows 100% sequence homology with the *Csa3G915140* gene of the 'Chinese Long v2' cucumber genome. A recent study has shown that a mutation in the gene-encoding QWRF motif-containing protein alters chlorophyll synthesis and reduces chlorophyll accumulation in Arabidopsis [34]. Therefore, we speculate that mutation in a *Chr3CG53640* gene could be responsible for the white pigmentation, resulting in reduced chlorophyll accumulation in immature fruit skin of the PI525075 cucumber line. Taken together, our study identified a novel allelic variant of the *Csa3G904080* gene and a new candidate gene, *Chr3CG53640*, for white immature skin color in cucumber via QTL-seq and SNP marker analyses. Hence, this study proposes a novel genetic resource controlling white immature skin color that has not been reported or validated in the previous studies. However, studying the gene function in vivo is necessary to confirm the association of *Csa3G904080* and *Chr3CG53640* genes with the white immature skin color in cucumber. Overall, this study provided a new genetic basis of immature fruit skin color trait in cucumber and is not limited to *APRR2* and w_0 genes, thus contributing to the development of cucumber cultivars by introgression of useful genes between the two variety groups with different skin colors.

Supplementary Materials: The following are available online at https://www.mdpi.com/article/10.3390/plants10112341/s1, Figure S1: SNP index graph of white-bulk. The green dots and the red line represent the SNP index and sliding window average of the SNP index, respectively. The SNP index was calculated based on a 2 Mb interval with a 100 kb sliding window. Figure S2: SNP index graph of light-green-bulk. The orange dots and the red line represent the SNP index and sliding window average of the SNP index, respectively. SNP index was calculated based on 2 Mb interval with a 100 kb sliding window. Figure S3: Plots of Δ (SNP index) of two bulks. Δ (SNP index) graph with statistical confidence intervals under the null hypothesis of no QTLs (Green line, $p < 0.05$; Orange line, $p < 0.01$) from QTL-seq analysis. Blue dots show Δ (SNP index). Red line indicates sliding window average of Δ (SNP index). Δ (SNP index) was calculated based on 2 Mb interval with a 100 kb sliding window. Figure S4: dCAPS analysis of M1, M2, and M5 markers in the F2 plants derived from a cross between MEJ and PI525075. P1, 'PI525075' (White); P2, 'MEJ' (Green); W, White; G, Green. (**a**) Genotyping result of M1 marker for a novel allele of *Chr3CG51850* gene. (**b**) Genotyping result of M2 marker for a novel of allele of *Chr3CG52290* gene. (**c**) Genotyping result of M5 marker for a novel allele of *Chr3CG52880* gene. Red arrow indicates no co-segregation with the phenotype. Table S1: Total number of genes found within the QTL region on chromosome 3.

Author Contributions: K.S. designed the study and conducted phenotyping. S.-C.L. performed genomic and marker analysis. J.-G.K. and K.S. supervised the study. D.S.K. performed markers validation and wrote the first draft of the manuscript, and edited; D.S.K., S.-C.L. and H.A. revised the manuscript. All authors have read and agreed to the published version of the manuscript.

Funding: This work was carried out with the support of "Cooperative Research Program for National Agricultural Genome Program (Project No. PJ01343201)" Rural Development Administration, Korea and (Project No. 20190455) Sejong University.

Data Availability Statement: All relevant raw sequence data generated in this study are available in the Sequence Read Archive (SRA) of the NCBI database with the following accession number: MEJ (SRX10261885) and PI525075 (SRR16093564).

Acknowledgments: We thank Hye Sik Kim, Hyo-Won Kim, and Junki Lee in Phyzen for providing bioinformatics analyses and Ho Bang Kim in Biomedic for providing experimental supports.

Conflicts of Interest: The authors hereby declare that they have no conflict of interest.

References

1. Che, G.; Zhang, X. Molecular basis of cucumber fruit domestication. *Curr. Opin. Plant Biol.* **2019**, *47*, 38–46. [CrossRef]
2. Liu, H.; Meng, H.; Pan, Y.; Liang, X.; Jiao, J.; Li, Y.; Chen, S.; Cheng, Z. Fine genetic mapping of the white immature fruit color gene w to a 33.0-kb region in cucumber (*Cucumis sativus* L.). *Theor. Appl. Genet.* **2015**, *128*, 2375–2385. [CrossRef] [PubMed]
3. Zhou, Q.; Wang, S.; Hu, B.; Chen, H.; Zhang, Z.; Huang, S. An ACCUMULATION AND REPLICATION OF CHLOROPLASTS 5 gene mutation confers light green peel in cucumber. *J. Integr. Plant Biol.* **2015**, *57*, 936–942. [CrossRef] [PubMed]
4. Sui, X.; Shan, N.; Hu, L.; Zhang, C.; Yu, C.; Ren, H.; Turgeon, R.; Zhang, Z. The complex character of photosynthesis in cucumber fruit. *J. Exp. Bot.* **2017**, *68*, 1625–1637. [CrossRef] [PubMed]
5. Wang, M.; Chen, L.; Liang, Z.; He, X.; Liu, W.; Jiang, B.; Yan, J.; Sun, P.; Cao, Z.; Peng, Q.; et al. Metabolome and transcriptome analyses reveal chlorophyll and anthocyanin metabolism pathway associated with cucumber fruit skin color. *BMC Plant Biol.* **2020**, *20*, 386. [CrossRef]
6. Yang, L.; Koo, D.-H.; Li, Y.; Zhang, X.; Luan, F.; Havey, M.J.; Jiang, J.; Weng, Y. Chromosome rearrangements during domestication of cucumber as revealed by high-density genetic mapping and draft genome assembly. *Plant J.* **2012**, *71*, 895–906. [CrossRef]
7. Huang, S.; Li, R.; Zhang, Z.; Li, L.; Gu, X.; Fan, W.; Lucas, W.J.; Wang, X.; Xie, B.; Ni, P.; et al. The genome of the cucumber, *Cucumis sativus* L. *Nat. Genet.* **2009**, *41*, 1275–1281. [CrossRef]
8. Li, Q.; Li, H.; Huang, W.; Xu, Y.; Zhou, Q.; Wang, S.; Ruan, J.; Huang, S.; Zhang, Z. A chromosome-scale genome assembly of cucumber (*Cucumis sativus* L.). *GigaScience* **2019**, *8*, guz072. [CrossRef]
9. Li, Y.; Wen, C.; Weng, Y. Fine mapping of the pleiotropic locus B for black spine and orange mature fruit color in cucumber identifies a 50 kb region containing a R2R3-MYB transcription factor. *Theor. Appl. Genet.* **2013**, *126*, 2187–2196. [CrossRef]
10. Yang, X.; Zhang, W.; Li, Y.; He, H.; Bie, B.; Ren, G.; Zhao, J.; Wang, Y.; Nie, J.; Pan, J.; et al. High-resolution mapping of the dull fruit skin gene D in cucumber (*Cucumis sativus* L.). *Mol. Breed.* **2014**, *33*, 15–22. [CrossRef]
11. Fanourakis, N.; Simon, P. Analysis of genetic linkage in the cucumber. *J. Hered.* **1987**, *78*, 238–242. [CrossRef]
12. Zhang, W.; He, H.; Guan, Y.; Du, H.; Yuan, L.; Li, Z.; Yao, D.; Pan, J.; Cai, R. Identification and mapping of molecular markers linked to the tuberculate fruit gene in the cucumber (*Cucumis sativus* L.). *TAG. Theor. Appl. Genetics. Theor. Und. Angew. Genet.* **2009**, *120*, 645–654. [CrossRef] [PubMed]

13. Lun, Y.; Wang, X.; Zhang, C.; Yang, L.; Gao, D.; Chen, H.; Huang, S. A CsYcf54 variant conferring light green coloration in cucumber. *Euphytica* **2016**, *208*, 509–517. [CrossRef]
14. Yang, X.; Li, Y.; Zhang, W.; He, H.; Pan, J.; Cai, R. Fine mapping of the uniform immature fruit color gene u in cucumber (*Cucumis sativus* L.). *Euphytica* **2014**, *196*, 341–348. [CrossRef]
15. Liu, H.; Jiao, J.; Liang, X.; Liu, J.; Meng, H.; Chen, S.; Li, Y.; Cheng, Z. Map-based cloning, identification and characterization of the w gene controlling white immature fruit color in cucumber (*Cucumis sativus* L.). *Theor. Appl. Genet.* **2016**, *129*, 1247–1256. [CrossRef]
16. Tang, H.-Y.; Dong, X.; Wang, J.-K.; Xia, J.-H.; Xie, F.; Zhang, Y.; Yao, X.; Xu, Y.-J.; Wang, Z.-J. Fine Mapping and Candidate Gene Prediction for White Immature Fruit Skin in Cucumber (*Cucumis sativus* L.). *Int. J. Mol. Sci.* **2018**, *19*, 1493. [CrossRef] [PubMed]
17. Doyle, J.J. Isolation of plant DNA from fresh tissue. *Focus* **1990**, *12*, 13–15.
18. Takagi, H.; Abe, A.; Yoshida, K.; Kosugi, S.; Natsume, S.; Mitsuoka, C.; Uemura, A.; Utsushi, H.; Tamiru, M.; Takuno, S.; et al. QTL-seq: Rapid mapping of quantitative trait loci in rice by whole genome resequencing of DNA from two bulked populations. *Plant J.* **2013**, *74*, 174–183. [CrossRef]
19. Bolger, A.M.; Lohse, M.; Usadel, B. Trimmomatic: A flexible trimmer for Illumina sequence data. *Bioinformatics* **2014**, *30*, 2114–2120. [CrossRef]
20. Li, H.; Durbin, R. Fast and accurate short read alignment with Burrows-Wheeler transform. *Bioinformatics* **2009**, *25*, 1754–1760. [CrossRef]
21. Cingolani, P.; Platts, A.; Wang, L.L.; Coon, M.; Nguyen, T.; Wang, L.; Land, S.J.; Lu, X.; Ruden, D.M. A program for annotating and predicting the effects of single nucleotide polymorphisms, SnpEff: SNPs in the genome of Drosophila melanogaster strain w1118; iso-2; iso-3. *Fly* **2012**, *6*, 80–92. [CrossRef] [PubMed]
22. Beale, S.I. Green genes gleaned. *Trends Plant Sci.* **2005**, *10*, 309–312. [CrossRef] [PubMed]
23. Kishor, D.S.; Seo, J.; Chin, J.H.; Koh, H.-J. Evaluation of Whole-Genome Sequence, Genetic Diversity, and Agronomic Traits of Basmati Rice (*Oryza sativa* L.). *Front. Genet.* **2020**, *11*, 86. [CrossRef]
24. Kishor, D.S.; Noh, Y.; Song, W.-H.; Lee, G.P.; Park, Y.; Jung, J.-K.; Shim, E.-J.; Sim, S.-C.; Chung, S.-M. SNP marker assay and candidate gene identification for sex expression via genotyping-by-sequencing-based genome-wide associations (GWAS) analyses in Oriental melon (*Cucumis melo* L.var.makuwa). *Sci. Hortic.* **2021**, *276*, 109711. [CrossRef]
25. Kishor, D.S.; Chung, S.-M. Development of SNP markers and validation assays in commercial Korean melon cultivars, using Genotyping-by-sequencing and Fluidigm analyses. *Sci. Hortic.* **2019**, *263*, 109113. [CrossRef]
26. Varshney, R.K.; Nayak, S.N.; May, G.D.; Jackson, S.A. Next-generation sequencing technologies and their implications for crop genetics and breeding. *Trends Biotechnol.* **2009**, *27*, 522–530. [CrossRef]
27. Abe, A.; Kosugi, S.; Yoshida, K.; Natsume, S.; Takagi, H.; Kanzaki, H.; Matsumura, H.; Yoshida, K.; Mitsuoka, C.; Tamiru, M.; et al. Genome sequencing reveals agronomically important loci in rice using MutMap. *Nat. Biotechnol.* **2012**, *30*, 174–178. [CrossRef]
28. Arikit, S.; Wanchana, S.; Khanthong, S.; Saensuk, C.; Thianthavon, T.; Vanavichit, A.; Toojinda, T. QTL-seq identifies cooked grain elongation QTLs near soluble starch synthase and starch branching enzymes in rice (*Oryza sativa* L.). *Sci. Rep.* **2019**, *9*, 8328. [CrossRef]
29. Yang, L.; Lei, L.; Li, P.; Wang, J.; Wang, C.; Yang, F.; Chen, J.; Liu, H.; Zheng, H.; Xin, W.; et al. Identification of Candidate Genes Conferring Cold Tolerance to Rice (*Oryza sativa* L.) at the Bud-Bursting Stage Using Bulk Segregant Analysis Sequencing and Linkage Mapping. *Front. Plant Sci.* **2021**, *12*, 402. [CrossRef]
30. Ramos, A.; Fu, Y.; Michael, V.; Meru, G. QTL-seq for identification of loci associated with resistance to Phytophthora crown rot in squash. *Sci. Rep.* **2020**, *10*, 5326. [CrossRef]
31. Zhang, C.; Badri Anarjan, M.; Win, K.T.; Begum, S.; Lee, S. QTL-seq analysis of powdery mildew resistance in a Korean cucumber inbred line. *Theor. Appl. Genet.* **2021**, *134*, 435–451. [CrossRef] [PubMed]
32. Win, K.T.; Zhang, C.; Silva, R.R.; Lee, J.H.; Kim, Y.-C.; Lee, S. Identification of quantitative trait loci governing subgynoecy in cucumber. *Theor. Appl. Genet.* **2019**, *132*, 1505–1521. [CrossRef] [PubMed]
33. Wang, Y.; Bo, K.; Gu, X.; Pan, J.; Li, Y.; Chen, J.; Wen, C.; Ren, Z.; Ren, H.; Chen, X.; et al. Molecularly tagged genes and quantitative trait loci in cucumber with recommendations for QTL nomenclature. *Hortic. Res.* **2020**, *7*, 3. [CrossRef] [PubMed]
34. Albrecht, V.; Šimková, K.; Carrie, C.; Delannoy, E.; Giraud, E.; Whelan, J.; Small, I.D.; Apel, K.; Badger, M.R.; Pogson, B.J. The Cytoskeleton and the Peroxisomal-Targeted SNOWY COTYLEDON3 Protein Are Required for Chloroplast Development in Arabidopsis. *Plant Cell* **2010**, *22*, 3423–3438. [CrossRef]

plants

MDPI

Article

Differences in Germination of ACCase-Resistant Biotypes Containing Isoleucine-1781-Leucine Mutation and Susceptible Biotypes of Wild Oat (*Avena sterilis* ssp. *ludoviciana*)

Fatemeh Benakashani [1], Jose L. Gonzalez-Andujar [2,*] and Elias Soltani [1]

[1] Department of Agronomy and Plant Breeding Sciences, College of Aburaihan, University of Tehran, Pakdasht 3391653755, Iran; benakashani@ut.ac.ir (F.B.); elias.soltani@ut.ac.ir (E.S.)

[2] Department of Crop Protection, Instituto de Agricultura Sostenible (CSIC), 14004 Córdoba, Spain

* Correspondence: andujar@ias.csic.es

Abstract: Herbicide resistance can affect seed germination and the optimal conditions required for seed germination, which in turn may impose a fitness cost in resistant populations. Winter wild oat [*Avena sterilis* L. ssp. *ludoviciana* (Durieu) Gillet and Magne] is a serious weed in cereal fields. In this study, the molecular basis of resistance to an ACCase herbicide, clodinafop-propargyl, in four *A. ludoviciana* biotypes was assessed. Germination differences between susceptible (S) and ACCase-resistant biotypes (WR_1, WR_2, WR_3, WR_4) and the effect of Isoleucine-1781-Leucine mutation on germination were also investigated through germination models. The results indicated that WR_1 and WR_4 were very highly resistant (RI > 214.22) to clodinafop-propargyl-contained Isoleucine to Leucine amino acid substitution. However, Isoleucine-1781-Leucine mutation was not detected in other very highly resistant biotypes. Germination studies indicated that resistant biotypes (in particular WR_1 and WR_4) had higher base water potentials than the susceptible one. This shows that resistant biotypes need more soil water to initiate their germination. However, the hydrotime constant for germination was higher in resistant biotypes than in the susceptible one in most cases, showing faster germination in susceptible biotypes. ACCase-resistant biotypes containing the Isoleucine-1781-Leucine mutation had lower seed weight but used more seed reserve to produce seedlings. Hence, integrated management practices such as stale seedbed and implementing it at the right time could be used to take advantage of the differential soil water requirement and relatively late germination characteristics of ACCase-resistant biotypes.

Keywords: ecological costs; germination models; herbicide resistance; hydrotime; target-site resistance

Citation: Benakashani, F.; Gonzalez-Andujar, J.L.; Soltani, E. Differences in Germination of ACCase-Resistant Biotypes Containing Isoleucine-1781-Leucine Mutation and Susceptible Biotypes of Wild Oat (*Avena sterilis* ssp. *ludoviciana*). *Plants* **2021**, *10*, 2350. https://doi.org/10.3390/plants10112350

Academic Editors: Milan S. Stankovic, Paula Baptista and Petronia Carillo

Received: 27 September 2021
Accepted: 27 October 2021
Published: 30 October 2021

Publisher's Note: MDPI stays neutral with regard to jurisdictional claims in published maps and institutional affiliations.

Copyright: © 2021 by the authors. Licensee MDPI, Basel, Switzerland. This article is an open access article distributed under the terms and conditions of the Creative Commons Attribution (CC BY) license (https://creativecommons.org/licenses/by/4.0/).

1. Introduction

Herbicide application is an effective and low-cost method for weed control throughout the world. Unfortunately, the extensive and widespread use of herbicides has resulted in the evolution of resistance in many weed species [1]. A great number of weed species (152 dicots and 111 monocots) present resistance to different families of herbicides [2]. Herbicide resistance in weeds is one of the most common problems, threatening human and animal food production [3].

Different mechanisms have been identified that are involved in the resistance of weed species to herbicides [4,5]. That resistance can evolve from variations in weed metabolism pathways and mutations [6], and many studies have shown that mutations in agroecosystems under herbicide selection may exhibit a competitive ability or adaptation cost relative to the susceptible wild-type, in herbicide untreated conditions [7,8]. For example, a single amino acid substitution (Isoleucine to Leucine) in an enzyme at herbicide site of action (Acetyl-CoA carboxylase; ACCase) could change the kinetics and function of the enzyme and cause herbicide resistance in winter wild oat (*Avena sterilis* ssp. *ludoviciana* (Durieu) Nyman) (hereafter referred to as *A. ludoviciana*) [9,10]. ACCase is the enzyme that catalyzes the first committed step in fatty acid synthesis, the carboxylation of acetyl-CoA to malonyl-CoA [11]. In the ACCase gene sequence, seven amino acid substitutions have been observed at different codon positions (Asn2078, Cys2088, Gly2096, Ile1781 and Ile2041) resulting in different herbicide resistance levels [12]. Among these amino acid substitutions, Ile-1781 (in ACCase) are the ones most abundantly found in plant species [12].

Weed biotypes with higher fitness produce more individuals; thus, the fitness difference between resistant biotypes (R) and susceptible biotypes (S) may be due to the difference in fertility, pollen and seed production, and the ability to compete [13].

Many studies have shown that herbicide resistance can affect seed germination and the range of optimal germination conditions [14–18]. Awareness of the dormancy and germination patterns of resistant weed seeds can also help in weed resistance management. Among environmental factors, temperature and water potential mainly impact seed dormancy and germination [19,20]. Hydrotime (HT) models are commonly used to describe seed germination response to water potential [21,22]. The hydrotime constant (θ_H) can be calculated by a multiplication between time to specific germination fraction and actual water potential (ψ) minus base water potential (ψ_b). Typically, different germination fractions have different values of $\psi_{b(g)}$ and the latter follow a normal bell curve in a seed population; the median of ψ_b ($\psi_{b(50)}$) is the base water potential and the standard deviation of the $\psi_{b(g)}$ ($\sigma_{\psi b}$) which shows germination uniformity. These three parameters (θ_H, $\sigma_{\psi b}$, $\Psi_{b(50)}$) may be used to explain germination fitness, but no study has proved such a claim before.

Low water potential has adverse effects on germination and seedling growth [23]. It has been reported that heterotrophic seedling growth is influenced by the weight of mobilized seed reserve (MSR) and the conversion efficiency of mobilized (CEM) seed reserve to seedling [24–26]. The weight of MSR can be divided into initial seed dry weight (ISDW) and the fraction of seed reserve (FSR), which is mobilized (i.e., the seed depletion ratio). Some authors use these components to investigate the impacts of water and salinity stress on seedling growth [24,26]. Soltani et al. [24] found that the most sensitive component of seedling growth (as affected by drought and salinity stress) is the weight of MSR. Cheng et al. [25] showed that seedling dry weight and MSR increased, while CEM declined during the seed germination process. Zheng and Ma [27] investigated heterotrophic seedling growth of *Bombax ceiba* as affected by seed aging and indicated that MSR and FSR significantly decreased with an increase in the duration of aging. However, their results showed no significant change in CEM with an increase in aging. There is no information on any changes in components of heterotrophic seedling growth between weeds that are either resistant or susceptible to herbicides.

A. ludoviciana is an annual member of the Poaceae family. This plant is a serious weed species in cereal fields around the world, whose geographic expansion is expected under climate change [10,28]. *A. ludoviciana* can severely reduce cereal yield [29]. Moderate winter wild oat densities, in the 20–80 panicles m^{-2} range, decreased barley yields by nearly 10% in experiments conducted in central Spain, with yield losses of up to 50% when densities reached 300 panicles m^{-2} [30].

Control of this weed is mainly based on acetyl-CoA carboxylase inhibitor herbicides. The increased application of ACCase herbicides and high initial frequency (6×10^{-10} plants) of resistant biotypes significantly affects resistance evolution [31]. To date, 263 resistant species have been reported worldwide [2]. The first cases of ACCase herbicide-resistant wild oat biotypes were found in Persia in 2006 [32,33]. Although resistance to ACCase inhibitors is numerically investigated, more research on the trait differences between resistant and susceptible biotypes would be necessary for resistant weed management. The objectives of this study were: (1) to evaluate the molecular basis for resistance of *A. ludoviciana* biotypes to clodinafop-propargyl as an ACCase herbicide; (2) to detect differences in germination and seedling growth between susceptible and resistant biotypes to ACCase inhibitor herbicide under different water potentials through germination models; and (3) to investigate the effect of Isoleucine-1781-Leucine mutation on germination.

2. Materials and Methods

2.1. Seed Source

A. ludoviciana seeds were collected from wheat fields of Khuzestan province in the Southwest region of Iran during July 2015 (Table 1). We used four populations that were suspected to be resistant to ACCase-inhibiting herbicides since they survived repeated post-emergence clodinafop-propargyl application at the recommended dosage (64 g ai ha^{-1}). Historical records showed that the fields had experienced previous applications of ACCase-inhibiting herbicides for more than 5 years. The populations were identified by the codes WR_1, WR_2, WR_3, and WR_4. Seeds of a susceptible population (S) were also collected from a wasteland in Khuzestan province where herbicide had never been used.

Table 1. Field Locations and History of Studied Wild Oat (*Avena sterilis* ssp. *ludoviciana*).

Table. *Cont.*	City	Location	The History of Grass Herbicide Application (Last 5 Years before Seed Sampling)
S	Andimeshk	32.45° N, 48.35° E	None
WR_1	Andimeshk	32.45° N, 48.35° E	Clodinafop-propargyl
WR_2	Shush	32.20° N, 48.25° E	Clodinafop-propargyl
WR_3	Ahvaz	31.32° N, 48.67° E	Clodinafop-propargyl
WR_4	Andimeshk	32.45° N, 48.35° E	Clodinafop-propargyl

Because the populations were collected from different areas, the collected seeds were first cultivated under the same conditions in Pakdasht (35.4669° N, 51.6861 E), Tehran, Iran, in 2015 and 2017 (one collection each year) in order to eliminate the environmental effects on seed production. To relieve dormancy, the seeds were stratified at 5 °C for 4 weeks. Then, they were planted in 2 × 3 cm^2 pots containing loam soil (30% sand, silt 35%, clay 35%) and decomposed manure; pH 7.5. The pots were irrigated to field capacity every 4 days. The plants were grown in a greenhouse with 25/18 °C of day/night temperature and natural photoperiod. To ensure that there was no chance of cross-pollination, spikes were covered with paper bags during the flowering stage. The produced seeds from each biotype were then used in whole plant dose response assay and first germination experiment conducted in 2015. To increase the accuracy of germination traits assessment, seeds were grown for another generation under the mentioned similar conditions. The germination characteristics of the seeds were re-evaluated in 2017.

2.2. Whole Plant Dose Response Assay

Eight seeds from each biotype were sown in 30 × 35 cm^2 plastic pots filled with loam soil (30% sand, silt 35%, clay 35%) and decomposed manure; pH 7.5). The pots were arranged in a completely randomized design with four replications. Thinning was applied to reduce seedlings number to four. Clodinafop-propargyl treatments were applied at 3–4 leaf stages with 0, 1, 2, 4, 8 and 16 times of the recommended field dose (64 g ai ha^{-1}). Four weeks after herbicide application, *A. ludoviciana* survival and aboveground biomass were recorded as a percentage of the untreated individuals. The four parameters log-logistic curve (Equation (1)) was fitted to the data using the R statistical software [34] with the add-on package *drc* [35].

$$Y = \frac{D - C}{1 + exp[b(loglog\ (X) - loglog\ (ED_{50}))]} \quad (1)$$

where Y is the biomass reduction, D is the upper limit, C is the lower limit, ED_{50} is the dosage (g ai ha^{-1}) that reduced fresh weight by 50%, b is the relative slope around ED_{50}, and X is the herbicide dose (g ai ha^{-1}). To describe the degree of resistance for biotypes, the ratio of absolute ED_{50} values of each resistant biotype to susceptible one was used to calculate the resistance index (RI) [36].

2.3. Investigating Molecular Basis of Resistance

The seeds (four seeds from each biotype) were sown in plastic pots (20 × 25 cm) filled with loam soil mixture, and the plants were maintained in a greenhouse at 18 and 22 °C in artificial light under a 16/8-h (day/light) photoperiod for leaf sampling. The greenhouse was located at Aburaihan Campus, University of Tehran (35°28′ N, 51°36′ E and 1020 masl), Iran. To assess the probability of mutation of the site of action being responsible for the resistance, genomic DNA was isolated from leaf tissues at 3-leaf to 4-leaf stage of three individual plants of each putative resistant and susceptible wild oat biotype using a modified cetyltrimethylammonium bromide (CTAB) extraction procedure [37]. The isolated DNA was quantified by measuring the absorbance at 260 nm by a spectrophotometer. Cleaved Amplified Polymorphic Sequences (CAPS) and derived Cleaved Amplified Polymorphic Sequences (dCAPS) molecular methods were used to identify the locations of four previously known mutations (Isoleucine-2041-Asparagine, Cysteine-2088-Arginine, Isoleucine-1781-Leucine, and Aspartic acid-2078-Glycine) responsible for target site-based herbicide resistance in the carboxyl transferase (CT) domain of the chloroplastic ACCase enzyme of the above-mentioned biotypes [4]. Primers and restriction enzymes used in CAPS and dCAPS methods are shown in Tables 2 and 3, respectively.

Table 2. Sequences of CAPS and D CAPS Primers.

Primer	Sequence 5′-3′	Usage	Reference
ACCF1	CACAGACCATGATGCAGCTC	CAPS for 2041and 2088	[4]
ACCR1	CTCCCTGGAGTTGTGCTTTC	-	
NsiI1781f	CTGTCTGAAGAAGACTATGGCCG	dCAPS for 1781	[1]
NsiI1781r	AGAATACGCACTGGCAATAGCAGCACTTCCATGCA	-	

Table 3. Restriction Sites of CAPS and D CAPS Restriction Enzymes.

Enzyme	Commercial Isoschizomers	Restriction Site	Technique	Reference
NsiI	*Ava*III, *Eco*T22I, *Mph*11031, *Zsp*2I	5′-ATGCA^T-3′ 3′-T^ACGTA-5′	dCAPS (1781)	[1]
EcoRI	*Fun*II	5′-G^AATTC-3′ 3′-CTTAA^G-5′	CAPS (2041)	[4]
EcoRV	*Eco*32I	5′-GAT^ATC-3′ 3′-CTA^TAG-5′	dCAPS (2078)	[4]
Eco47III	*Afe*I, *Aor*51HI, *Fun*I	5′-AGC^GCT-3′	CAPS (2088)	[4]

2.4. Germination and Seedling Growth

To investigate the germination differences of the biotypes, two-seed bioassay was conducted under different water potential conditions. As mentioned in the seed source section, the seeds produced under the same conditions in 2015 and the next generation in 2017 were studied in two separate experiments (Hereinafter referred to as the first and second experiments, respectively). Seed water content percentage was determined in four samples (100 seeds per sample) of each biotype by weighing fresh seeds (w1) and oven-dried seeds (w2) as follows (Equation (2)):

$$\text{Seed water content} = (\text{w1} - \text{w2})/\text{w1} \tag{2}$$

Then, ISDW was measured before starting the experiment by weighing fresh seeds minus seed water weight. Three replicates of 25 seeds for each biotype were germinated in Petri dishes of 150 mm in diameter on filter paper (Whatman No. 1) at 20 °C and dark, at five water potentials (0, −0.15, −0.30, −0.45, and −0.6 MPa). Polyethylene glycol 6000 (PEG) was used to maintain water potentials determined according to Michel and

Kaufmann [38]. Filter papers were soaked at the desired PEG solutions for 24 h, after which seeds were placed in Petri dishes and sealed. When the moisture of the Petri dishes decreased as affected by evaporation, seeds were moved into new Petri dishes and new solutions. Seed germination was assessed twice a day and the seeds with a radicle ≥2 mm long were considered germinated. Seeds were inspected for up to two weeks.

Heterotrophic seedling growth was evaluated after 14 days. The seedlings and seed remnants were separated first, and then weighed using an analytical balance with a milligram scale to determine the dry weight of seedlings (SLDW) and dry weight of seed remnants (FSDW). The weight of the mobilized seed reserve (MSR), the conversion efficiency of mobilized (CEM), and the fraction of mobilized seed reserve (FSR) were calculated as follows (Equations (3)–(5)) [4]:

$$\mathrm{MSR} = \mathrm{ISDW} - \mathrm{FSDW} \quad (3)$$

$$\mathrm{CEM} = \mathrm{SLDW}/\mathrm{MSR} \quad (4)$$

$$\mathrm{FSR} = \mathrm{MSR}/\mathrm{ISDW} \quad (5)$$

Data from the heterotrophic seedling growth test were analyzed as a combined analysis of multiple experiments (factors: experiment, biotype and water potential) and biotype × water potential interaction were compared by least significant difference (LSD) test if the interaction was significant by F test. Statistical analyses were performed with SAS software (Ver. 9.4.).

The hydrotime model was used to describe seed germination response to different water potentials (ψ; MPa) for each biotype [21,22]. The hydrotime constant (θ_H; MPa-hours) was obtained as the following equation (Equation (6)):

$$\theta_H = (\Psi - \Psi_{b(g)})tg \quad (6)$$

where $\psi_{b(g)}$ is the base water potential (MPa) for a specific germination percentage (g), and tg is the time (hours) to g percentage of germination for each biotype. Typically, variation in ψ_b follows a normal bell curve within a seed population [39]. Thus, the hydrotime model parameters were determined by repeated probit analysis using Equation (7), and the θ_H varied until the best fit was obtained for each biotype [20,23,39]:

$$\text{probit}\ (g) = [\Psi - (\theta_H/tg) - \Psi_{b(50)}]/\sigma_{\Psi b} \quad (7)$$

where $\psi_{b(50)}$ is the median, ψ_b, and $\sigma_{\psi b}$ is the standard deviations in ψ_b among the seeds within the biotypes. The calculations were performed for each replication separately to estimate standard errors of the parameters. The Excel software was used for all calculations.

3. Results and Discussion

3.1. Whole Plant Dose Response Assay

The resistant biotypes survived 4 weeks after all treatments while no susceptible biotype plant survived. The log-logistic model (Equation (1)) fitted adequately to the response of shoot fresh weight of biotypes to increasing rates of clodinafop-propargyl (Figure 1). The susceptible biotype was completely controlled at a rate lower than the recommended, suggesting that the S biotype is highly susceptible to clodinafop-propargyl rates (Table 4). The dose-results indicated that all identified resistant biotypes are classified as very highly resistant (RI > 100) to clodinafop-propargyl [40].

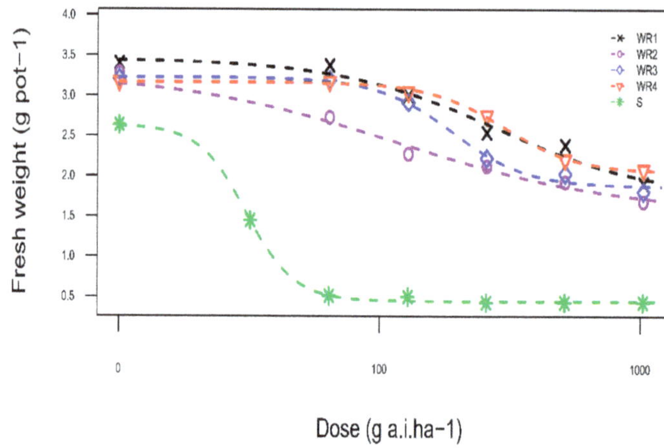

Figure 1. Effect of different concentrations of clodinafop- propargyl herbicide on aboveground fresh weight of susceptible and resistant biotypes. Symbols and lines represent actual and estimated response of resistant and susceptible biotypes, respectively. The symbols represent the mean of four replicates. The plants were grown in a greenhouse.

Table 4. Parameter estimates (SE) of four-parameter log-logistic model and resistance indices (RI) from whole plant bioassay of the suspected resistance (WR_1, WR_2, WR_3, WR_4) and susceptible (S) biotypes.

Biotype	B	C (g.pot^{-1})	D (g.pot^{-1})	Absolute ED_{50} (g a.i. ha^{-1})	RI
WR_1	1.45 (0.58)	1.74 (0.42)	3.45 (0.11)	>1024 ^{NA}	>214.22
WR_2	1.00 (0.58)	1.55 (0.40)	3.28 (0.13)	>1024 ^{NA}	>214.22
WR_3	3.01 (1.25)	1.87 (0.12)	3.23 (0.11)	>1024 ^{NA}	>214.22
WR_4	4.67 (3.13)	2.04 (0.16)	3.16 (0.09)	>1024 ^{NA}	>214.22
S	3.26 (2.93)	0.44 (0.06)	2.63 (0.13)	1.90	-

B: The relative slope around the parameter e, C is the lower limit, D is the upper limit, ED_{50} (absolute): Estimated by function *(ED (type="absolute"))* in the *drc* package of R software. RI = absolute ED_{50} Resistant population/absolute ED_{50} Susceptible population). ^{NA} Not possible to estimate the ED_{50} as plant fresh weight reduction was lower than 50% for all applied doses.

Resistance of *A. ludoviciana* to clodinafop-propargyl and other ACCase inhibitor herbicides had been reported in different countries in the world such as Australia, France [2] and Turkey [41]. The most common reason for weed resistance evolution is that herbicide application is the sole method of weed control combined with little or no variety in agronomic practices [42]. ACCase inhibitor herbicides have been extensively used by farmers for a decade as a practical selective herbicide to control weedy grasses in wheat production regions of Iran, especially in Khuzestan Province [43]. The results of screening studies confirmed the evolution of resistance in winter wild oat to ACCase inhibitors in Iran [32,44–46]. In a survey, Zand et al. [47] also characterized 52% of clodinafop-resistant *A. ludoviciana* populations in 50 farmer's fields in Khuzestan Province.

3.2. Molecular Basis for Resistance

The CAPS markers and dCAPS markers were amplified in all biotypes, along with the desired region of the ACCase enzyme (Figure 2). Results of CAPS and dCAPS detected the substitution of Isoleucine for Leucine at position 1781 in the CT domain of the acetyl-CoA carboxylase gene in WR_1 and WR_4 resistant biotypes. However, this amino acid substitution was not confirmed in the other resistant biotypes, WR_2, and WR_3 (Figure 2). Results of enzyme restriction with NsiI also showed WR_1 and WR_4 biotypes were heterozygous for the resistant 1781- Leucine (Figure 2).

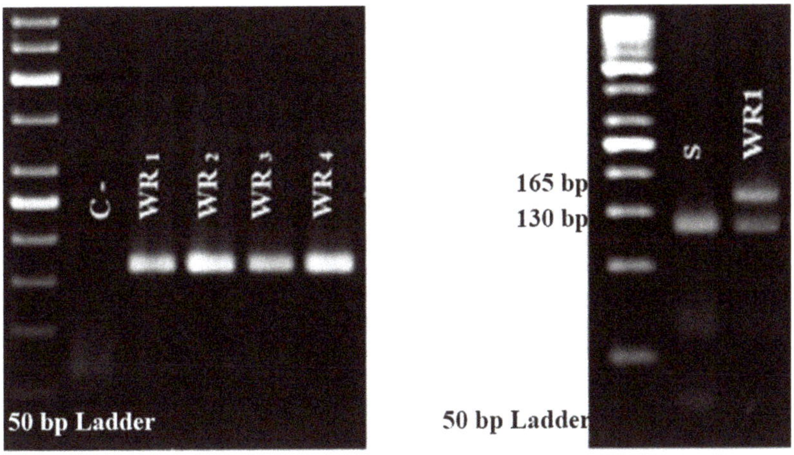

Figure 2. Polymerase chain reaction (PCR) (**Left**), and dCAPS analysis of individual *Avena sterilis* ssp. *ludoviciana* (**Right**). The size of the restriction enzyme (NsiI) digested fragment is 165 bp.

In most cases, resistance to ACCase inhibitors has been reported to be a result of target site mutations and insensitivity of ACCase [48,49]. It has been reported that I1781L substitution is the most frequent one conferring resistance to all three ACCase chemical herbicide families [12]. Yu et al. [4] found ACCase mutation in resistant *Lolium* populations. They detect 1781-Leu allele in many individuals (71%) of clethodim-resistant populations. These genotypes also exhibited cross resistance to aryloxyphenoxypropionate herbicides such as clodinafop, diclofop and fluazifop.

The resistance levels of biotypes containing I1781Le mutation were very high (RI > 214.22). Therefore, it was concluded that this substitution resulted in a high level of clodinafop-propargyl resistance in these populations. It was found that ACCase target site mutations conferred very high levels of resistance [50]. Resistance mechanism in other resistant biotypes (WR_2 and WR_3) that did not represent any point mutation in studied codons (Isoleucine-2041-Asparagine, Cysteine-2088-Arginine, Isoleucine-1781-Leucine, and Aspartic acid-2078-Glycine) was probably due to a mutation in other locations in the CT domain of the ACCase enzyme. Seven sites were reported to confer ACCase-inhibitor resistance in various weed species among the 13 conserved amino acid substitutions [6,51,52]. The results of a biochemical-based investigation of resistance to the acetyl-coenzyme A carboxylase (ACCase)-inhibiting herbicide diclofop-methyl in a resistant *Avena* population established that one or at least two independent resistance mechanisms (target-site ACCase resistance mutations and non–target-site enhanced rates of herbicide metabolism) can confer resistance in individual wild oat populations [53].

3.3. Germination and Seedling Growth

Results indicated that the total germination percentage differed among the biotypes (Figure 3). Water stress significantly decreased germination percentage (Figure 3) and seedling growth in all the biotypes (Table 5). The highest hydrotime constants (θ_H) were observed in WR_1 in both experiments (Table 6). The median base water potentials [$\psi_{b(50)}$] of the two experiments were significantly higher (less negative) in WR_1 and WR_4 biotypes as compared with other resistant biotypes (Table 6). The lowest median base water potential (−0.79 MPa in the first experiment and −0.91 MPa in the second experiment) was observed in the susceptible biotype. The values of $\sigma_{\psi b}$ for each biotype are indicated in Table 6.

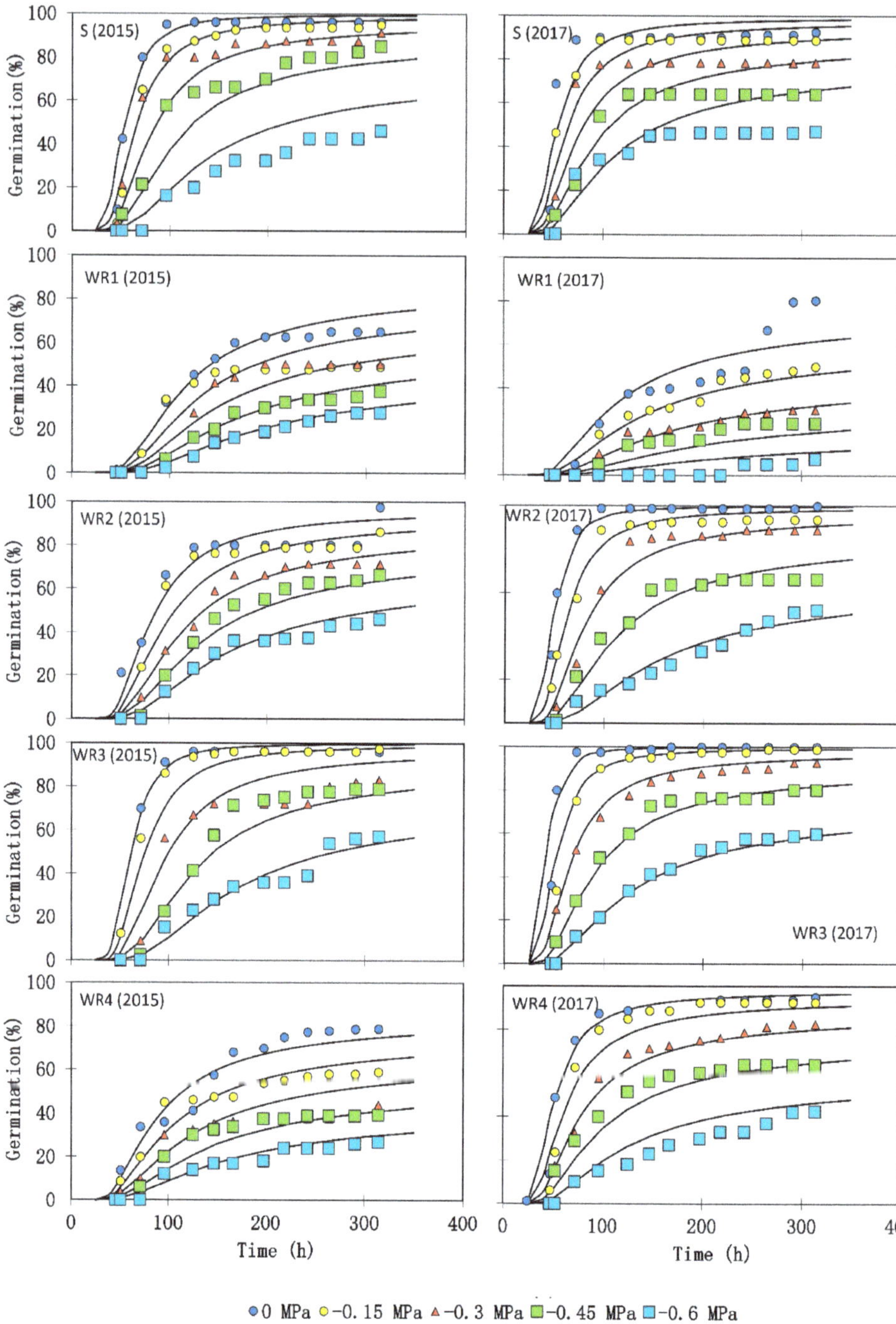

Figure 3. Germination time courses of five *Avena sterilis* ssp. *ludoviciana* biotypes (susceptible biotype (S) or resistant biotypes (WR_1–WR_4)). Symbols indicate interpolations of observed germination data and lines germination time courses predicted by the hydrotime model based on parameter estimates in Table 6. The symbols represent the mean of three replicates.

Table 5. Results of analysis of variance (mean squares) for initial seed dry weight (ISDW), seed remnants dry weight (FSDW), seedling dry weight (SLDW), the weight of mobilized seed reserve (MSR), the conversion efficiency of mobilized (CEM), and the fraction of seed reserve (FSR).

SOV	Df	ISDW	SLDW	FSDW	MSR	CEM	FSR
Experiment	1	0.0198	0.0035	0.220	0.373	0.0013	0.0019
Replication (Experiment)	2	0.0228	0.0909	0.335	0.387	0.0011	0.0024
Biotype	4	64.429 **	31.641 **	51.454 **	15.812 **	1.8749 **	0.0911 **
Water potential	4	0.0004	70.958 **	114.168 **	114.305 **	0.1843 **	0.8008 **
Bio × Exp	4	0.0004	0.0817	1.032	1.042	0.0155	0.0068
Bio × WP	16	0.0008	3.423 **	6.404 **	6.449 **	0.1286 **	0.0416 **
Exp × WP	4	0.0002	0.0551	0.898	0.893	0.0027	0.0052
Bio × Exp × WP	16	0.0005	0.0695	0.544	0.540	0.0071	0.0032
Residual	98	0.0074	0.2371	0.502	0.514	0.0114	0.0035

** Significant at 0.01 level of probability.

Table 6. Hydrotime constant (θ_H), median base water potential ($\Psi_{b(50)}$), standard deviation of the ψb(g) ($\sigma_{\Psi b}$). Coefficient of determination (R^2) and Root Mean Square Error (RMSE)for susceptible (S) and resistant biotypes (WR_1–WR_4).

Biotype	θ_H (MPa-Hours)	$\Psi_{b(50)}$ (MPa)	$\sigma_{\Psi b}$	RMSE	R^2
First experiment					
S	47.80 ± 6.34	−0.793 ± 0.012	0.268 ± 0.112	8.73	0.90
WR_1	87.42 ± 5.87	−0.609 ± 0.024	0.529 ± 0.164	9.28	0.84
WR_2	72.34 ± 4.45	−0.832 ± 0.008	0.431 ± 0.067	6.04	0.90
WR_3	48.01 ± 5.12	−0.778 ± 0.009	0.239 ± 0.082	7.43	0.80
WR_4	61.09 ± 6.78	−0.530 ± 0.010	0.502 ± 0.105	5.65	0.92
Second experiment					
S	48.99 ± 4.31	−0.908 ± 0.014	0.369 ± 0.089	10.56	0.81
WR_1	51.41 ± 3.44	−0.280 ± 0.018	0.393 ± 0.077	5.63	0.82
WR_2	34.41 ± 2.13	−0.701 ± 0.004	0.220 ± 0.084	4.71	0.95
WR_3	31.15 ± 2.87	−0.750 ± 0.008	0.222 ± 0.094	5.27	0.95
WR_4	36.31 ± 3.83	−0.687 ± 0.009	0.321 ± 0.075	6.05	0.93

As shown in Figure 4, ISDW significantly differed among biotypes, ranging from 13.56 mg (for WR_1) to 10.15 mg (for WR_3). Heterotrophic seedling growth test (except for ISDW) indicated significant interaction of the biotype and the water potential (Table 5). The FSDW ranged from 3.03 mg (for WR_3 in 0 water potential) to 11.58 mg (for WR_1 in −0.6 water potential) (Table 7). The SLDW changed significantly among biotypes in each water potential; and with the water potential decreasing, resistant biotypes, especially WR_1, lost seedling growth. Results indicated that biotypes used MSR variously, and had significantly different CEM seed reserve to seedling tissue (Table 7). The CEM values ranged from 0.00 (for WR_1) to 0.92 (for WR_3) mg mg^{-1}. In all water potential different resistant biotypes had the highest mobilized FSR and the S biotype the lowest value in 0 and −0.15 MPa (Table 7).

Resistant biotypes (in particular WR_1 and WR_4) had higher base water potential than the susceptible one, showing that resistant biotypes require more soil water for germination initiation. Hydrotime changes were different in the two experiments; in the first experiment, the hydrotime constant was higher in resistant biotypes than in the susceptible one, implying faster germination in the former, but in the second experiment, only WR_1 had higher hydrotime than the susceptible biotype. Thus, it seems that the changes of hydrotime are not affected by herbicide resistance. Opposite results, as observed in the first experiment, were reported before, in which faster germination was found in resistant biotypes of *Kochia scoparia* in comparison with the susceptible biotype [16]. In addition, the results of seed biology investigation of sulfonylurea-resistant prickly lettuce (*Lactuca serriola*) and susceptible biotypes showed that germination rate of the resistant biotype was 100% faster than the susceptible one [15], whereas slower germination had been detected in two resistant species of *Amaranthus* [14] and *Phalaris minor* [17]. The resistance mechanism and level of clodinafop-propargyl resistance are believed to account for the vast majority of the variability between resistant biotypes of *A. ludoviciana*. Since the biotypes were collected in a province with the same climatic conditions, there could not be any other important factor causing these large changes in biotypes.

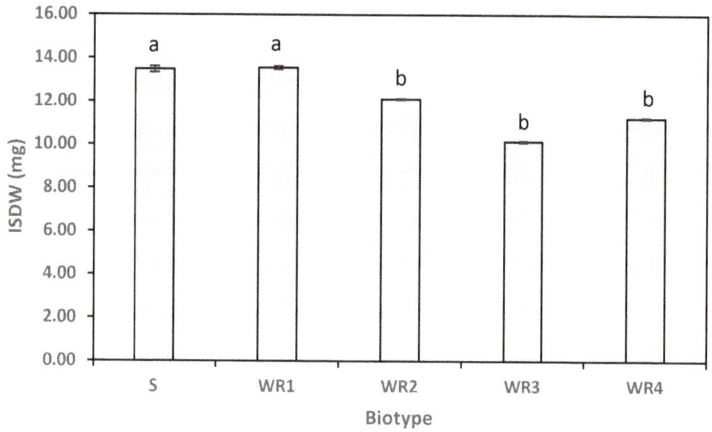

Figure 4. The average of initial seed dry weight (ISDW) among five *Avena sterilis* ssp. *ludoviciana* biotypes (susceptible biotype (S) or resistant biotypes (WR_1–WR_4)) produced in 2015 and 2017. The symbols represent the mean of three replicates. Error bars represent standard deviations. Means followed by the same letter are not significantly different according to the least significant difference (LSD) test.

Although *A. ludoviciana* grows in drylands and can survive and produce seeds under water stress [54], there are several advantages of a higher base water potential in the resistant biotype of *A. ludoviciana* than in the susceptible one as follows: Seedling emergence of winter annual weeds such as *A. ludoviciana* is not limited by soil moisture [54], and resistant biotypes require more water to germinate, so seedlings are more likely to grow under more moist conditions. In this condition, crop irrigation has a key impact on seedling emergence of weeds. If the base water potential is very low (e.g., −1.5 MPa), it is possible to have seedling emergence under a low soil moisture condition and a further reduction in soil moisture in the following days will cause the emerging seedlings to die. Due to their higher base water potential, seeds of resistant *A. ludoviciana* biotypes have to wait for the first irrigation; thus, they emerge simultaneously with the crop sowing date and do not experience water stress. Indeed, this is an avoidance mechanism to cope with water stress.

In our study, the susceptible biotype had a significantly higher grain weight than three resistant biotypes (WR_2, WR_3, and WR_4) but did not differ significantly from WR_1 (Figure 4). The grain weight of diclofop-methyl resistant individual plants of *Lolium rigidum* was significantly lower than that measured in susceptible plants. Early vigor of plants of resistant populations studied was also significantly lower than that measured in a susceptible population [55]. However, it was reported that there were no significant differences in one thousand seeds' weight of resistant *L. rigidum* populations containing Ile1781Leu and Ile 2041Asn mutations when compared to a sensitive population [56].

Table 7. Initial seedling dry weight (SLDW), seed remnants dry weight (FSDW), the weight of mobilized seed reserve (MSR), the conversion efficiency of mobilized (CEM), and the fraction of seed reserve (FSR) for susceptible biotype (S) or resistant biotypes (WR_1–WR_4).

Water Potential (MPa)	Biotype	SLDW (mg)	FSDW (mg)	MSR (mg)	CEM (mg mg^{-1})	FSR (mg mg^{-1})
−0.6	S	2.12 ^{a}	10.26 ^{b}	3.22 ^{a}	0.66 ^{b}	0.24 ^{a}
	WR_1	0.00 ^{c}	11.58 ^{a}	1.98 ^{bc}	0.00 ^{c}	0.15 ^{b}
	WR_2	2.00 ^{a}	8.82 ^{c}	3.27 ^{a}	0.53 ^{b}	0.27 ^{a}
	WR_3	1.22 ^{b}	8.75 ^{c}	1.40 ^{c}	0.89 ^{a}	0.14 ^{b}
	WR_4	1.53 ^{b}	8.76 ^{c}	2.47 ^{b}	0.63 ^{b}	0.22 ^{a}
−0.45	S	3.16 ^{a}	9.45 ^{b}	4.03 ^{a}	0.79 ^{a}	0.30 ^{b}
	WR_1	0.00 ^{c}	10.80 ^{a}	2.76 ^{b}	0.00 ^{c}	0.20 ^{c}
	WR_2	2.66 ^{a}	7.31 ^{d}	4.80 ^{a}	0.59 ^{b}	0.39 ^{a}
	WR_3	1.67 ^{b}	8.26 ^{c}	1.88 ^{c}	0.92 ^{a}	0.19 ^{c}
	WR_4	1.87 ^{b}	8.14 ^{c}	3.10 ^{b}	0.61 ^{b}	0.28 ^{b}
−0.3	S	3.41 ^{b}	8.83 ^{b}	4.65 ^{b}	0.74 ^{ab}	0.35 ^{bc}
	WR_1	0.00 ^{d}	11.10 ^{a}	2.44 ^{d}	0.00 ^{c}	0.18 ^{d}
	WR_2	4.66 ^{a}	6.63 ^{d}	5.49 ^{a}	0.86 ^{a}	0.45 ^{a}
	WR_3	2.90 ^{b}	6.17 ^{d}	3.96 ^{bc}	0.74 ^{ab}	0.39 ^{ab}
	WR_4	2.04 ^{c}	7.95 ^{c}	3.29 ^{c}	0.64 ^{b}	0.29 ^{c}
−0.15	S	3.97 ^{b}	8.51 ^{a}	4.98 ^{b}	0.80 ^{a}	0.37 ^{c}
	WR_1	3.27 ^{c}	7.31 ^{b}	6.26 ^{a}	0.52 ^{c}	0.46 ^{b}
	WR_2	5.37 ^{a}	5.68 ^{c}	6.42 ^{a}	0.84 ^{a}	0.53 ^{a}
	WR_3	3.93 ^{b}	5.50 ^{c}	4.67 ^{b}	0.84 ^{a}	0.46 ^{b}
	WR_4	2.87 ^{c}	6.85 ^{b}	4.37 ^{b}	0.67 ^{b}	0.39 ^{c}
0	S	4.80 ^{b}	8.11 ^{a}	5.36 ^{d}	0.89 ^{a}	0.40 ^{c}
	WR_1	5.03 ^{b}	3.84 ^{bc}	9.72 ^{a}	0.53 ^{c}	0.72 ^{a}
	WR_2	7.10 ^{a}	3.78 ^{bc}	8.33 ^{b}	0.85 ^{a}	0.69 ^{a}
	WR_3	5.04 ^{b}	3.03 ^{c}	7.12 ^{c}	0.71 ^{b}	0.70 ^{a}
	WR_4	3.79 ^{c}	4.50 ^{b}	6.74 ^{c}	0.56 ^{c}	0.59 ^{b}

Same letters within the same column for each water potential indicate no significant difference at $p = 0.05$.

Having a higher base water potential and lower seed weight in biotypes containing Il-1781-L mutation compared to non-mutant biotypes can be considered as one of the effects of this amino acid substitution. However, in the case of the hydrotime, since this trait in one of the resistant mutant-biotypes (WR_1) is not higher than two non-mutant resistant biotypes, this attribute cannot be related to the mentioned mutation.

The SLDW significantly varied among biotypes; WR_1 and WR_4 had the lowest SLDW among them. As indicated before, heterotrophic seedling growth is influenced by two components, MSR and CEM. Results showed that MSR and CEM changed significantly among the biotypes. In this regard CEM was lower in WR_1 and WR_4 than in the others and two components of MSR were significantly different. ISDW was significantly lower in resistant biotypes (except for WR_1) than in the S biotype, but mobilized FSR was significantly higher in resistant biotypes (except for WR_1 under water stress conditions) than in susceptible biotypes (Table 7). This shows that resistant biotypes (especially at the 0 MPa) used more seed reserve to produce seedling growth and needed more energy.

Our results revealed that all the suspected resistant biotypes of *A. ludoviciana* studied in this research were very highly resistant to clodinafop-propargyl. Two resistant biotypes contained Isoleucine to Leucine amino acid substitution and no mutations were found in two other biotypes. The herbicide-resistant biotypes had a higher base water potential and higher hydrotime (in WR_1 in both experiments and in all biotypes in the first experiment) for germination than the susceptible biotype. This shows that the latter can germinate, for a shorter time, at a lower soil water content than resistant biotypes. ACCase-resistant biotypes containing a mutation also used more seed reserve to start seedling growth. These results indicated that target site mutation at Ile1781 codon position could make ACCase-resistant biotypes less competitive than the S biotype; but they have become resistant to herbicides rather than growing faster.

Different methods can be used to control herbicide-resistant weeds, specifically ACCase-resistant *A. ludoviciana*. Crop management practices that lead to rapid stand establishment and canopy development minimize the effect of weeds. A number of management practices are necessary to control the growth of this weed, including crop rotation, planting certified seed, improving seedbed preparation, seeding at the correct rate, depth, and time of year. We believe that by designing and implementing appropriate management operations, such as stale seedbeds, at the right time, small differences between resistant and susceptible winter wild oats can be very useful in the control of resistant plants. A weed management program that includes monitoring weeds in the fields before and during the cultivation season is necessary to achieve success and does not use herbicides unless absolutely necessary. Using a herbicide over a long period of time increases herbicide resistance. To counteract this, it is recommended that herbicides be changed every few years.

Author Contributions: Conceptualization, E.S. and J.L.G.-A.; methodology, F.B.; software, F.B. and E.S.; validation, F.B., J.L.G.-A. and E.S.; formal analysis, F.B. and E.S.; investigation, F.B.; resources, F.B.; data curation, F.B. and E.S.; writing—original draft preparation, F.B. and E.S.; writing—review and editing, J.L.G.-A.; visualization, F.B. and E.S.; supervision, J.L.G.-A. and E.S.; project administration, E.S.; funding acquisition, F.B. All authors have read and agreed to the published version of the manuscript.

Funding: This research received no external funding.

Conflicts of Interest: The authors declare no conflict of interest.

References

1. Kaundun, S.S.; Windass, J.D. Derived cleaved amplified polymorphic, a simple method to detect a key point mutation conferring acethyl CoA carboxylase inhibitor herbicide resistance in grass weeds. *Weed Res.* **2006**, *46*, 34–39. [CrossRef]
2. Heap, I. The International Survey of Herbicide Resistant Weeds. Available online: http://www.weedscience.com (accessed on 22 August 2021).
3. Lonhienne, T.; Garcia, M.D.; Pierens, G.; Mobli, M.; Nouwens, A.; Guddat, L.W. Structural insights into the mechanism of inhibition of AHAS by herbicides. *Proc. Natl. Acad. Sci. USA* **2018**, *115*, 1945–1954. [CrossRef] [PubMed]
4. Yu, Q.; Collavo, A.; Zheng, M.Q.; Owen, M.; Sattin, M.; Powles, S.B. Diversity of acetyl-coenzyme A carboxylase mutations in resistant *Lolium* populations: Evaluation using clethodim. *Plant Physiol.* **2007**, *145*, 547–558. [CrossRef] [PubMed]
5. Delye, C.; Michel, S.; Berard, A.; Chauvel, B.; Brunel, D.; Guillemin, J.P.; Le Corre, V. Geographical variation in resistance to acetyl-coenzyme A carboxylase-inhibiting herbicides across the range of the arable weed *Alopecurus myosuroides* (blackgrass). *New Phytol.* **2010**, *186*, 1005–1017. [CrossRef] [PubMed]
6. Powles, S.B.; Yu, Q. Evolution in action: Plants resistant to herbicides. *Ann. Rev. Plant. Biol.* **2010**, *61*, 317–347. [CrossRef] [PubMed]
7. Coustau, C.; Chevillon, C.; French-Constant, R. Resistance to xenobiotics and parasites: Can we count the cost? *Trends Ecol. Evol.* **2000**, *15*, 378–383. [CrossRef]
8. Pavlicev, M.; Wagner, G.P. A model of developmental evolution: Selection, pleiotropy and compensation. *Trends Ecol. Evolution.* **2012**, *27*, 316–322. [CrossRef] [PubMed]
9. Jang, S.; Marjanovic, J.; Gornicki, P. Resistance to herbicides caused by single amino acid mutations in acetyl-C o A carboxylase in resistant populations of grassy weeds. *New Phytol.* **2013**, *197*, 1110–1116. [CrossRef] [PubMed]
10. Benakashani, F.; Zand, E.; Naghavi, M.R.; Sasanfar, H. Mutations in Acetyl-CoA Carboxylase enzyme, mechanism of cross-resistance in wild oat (*Avena ludoviciana* Deuri.) biotypes to ACCase inhibitor herbicides. *Iran J. Weed Sci.* **2015**, *10*, 179–190.
11. Sasaki, Y.; Nagano, Y. Plant acetyl-CoA carboxylase: Structure, biosynthesis, regulation, and gene manipulation for plant breeding. *Biosci. Biotechnol. Biochem.* **2004**, *68*, 1175–1184. [CrossRef] [PubMed]
12. Kaundun, S.S. Resistance to acetyl-CoA carboxylase-inhibiting herbicides. *Pest Manag. Sci.* **2014**, *70*, 1405–1417. [CrossRef]
13. Park, K.W.; Mallory-Smith, C.A. Multiple herbicide resistant in downy brome (*Bromus tectorum*) and its impact on fitness. *Weed Sci.* **2005**, *53*, 780–786. [CrossRef]
14. Weaver, E.S.; Thomas, G. Germination responses to temperature of atrazine-resistant and -S biotype of two pigweed (*Amaranthus*) species. *Weed Sci.* **1986**, *34*, 865–870. [CrossRef]
15. Alcocer-Ruthling, M.; Thill, D.C.; Shafii, B. Seed biology of sulfonylurea- resistant and -S biotype of prickly lettuce (*Lactuca serriola*). *Weed Technol.* **1992**, *6*, 858–864. [CrossRef]

16. Dyer, W.E.; Chee, P.W.; Fay, P.K. Rapid germination of sulfonylurea-resistant *Kochia scoparia* L. accessions is associated with elevated seed levels of branched chain amino acids. *Weed Sci.* **1993**, *41*, 18–22. [CrossRef]
17. Torres-Garcia, J.S.; Uscanga-Mortera, E.; Trejo, C.; Conde-Martinez, V.; Kohashi-Shibata, J.; Núñez-Farfán, J.; Martínez-Moreno, D. Effect of herbicide resistance on seed physiology of *Phalaris minor* (*Little seed canarygrass*). *Bot. Sci.* **2015**, *93*, 661–667. [CrossRef]
18. Osipitan, O.A.; Dille, J.A. Fitness outcomes related to glyphosate resistance in kochia (*Kochia scoparia*): What life history stage to examine? *Front. Plant Sci.* **2017**, *8*, 1090. [CrossRef]
19. Batlla, D.; Bench-Arnold, R.L. A predictive model for dormancy loss in *Polygonum aviculare* L. seeds based on changes in population hydrotime parameters. *Seed Sci. Res.* **2004**, *14*, 277–286. [CrossRef]
20. Soltani, E.; Gruber, S.; Oveisi, M.; Salehi, N.; Alahdadi, I.; Javid, M.G. Water stress, temperature regimes, and light control secondary dormancy induction and loss in *Brassica napus* L. seeds. *Seed Sci. Res.* **2017**, *27*, 217–230. [CrossRef]
21. Gummerson, R.J. The effect of constant temperatures and osmotic potentials on the germination of sugar beet. *J. Exp. Bot.* **1986**, *37*, 729–741. [CrossRef]
22. Bradford, K.J. Applications of hydrothermal time to quantifying and modeling seed germination and dormancy. *Weed Sci.* **2002**, *50*, 248–260. [CrossRef]
23. Soltani, E.; Farzaneh, S. Hydrotime analysis for determination of seed vigor in cotton. *Seed Sci. Technol.* **2014**, *42*, 260–273. [CrossRef]
24. Soltani, A.; Gholipoor, M.; Zeinali, E. Seed reserve utilization and seedling growth of wheat as affected by drought and salinity. *Environ. Exp. Bot.* **2006**, *55*, 195–200. [CrossRef]
25. Cheng, X.; Cheng, J.; Huang, X.; Lai, Y.; Wang, L.; Du, W.; Zhang, H. Dynamic quantitative trait loci analysis of seed reserve utilization during three germination stages in rice. *PLoS ONE* **2013**, *8*, e80002. [CrossRef] [PubMed]
26. Seyyedi, S.A.; Khajeh-Hosseini, M.; Moghaddam, P.R.; Shahandeh, H. Effects of phosphorus and seed priming on seed vigor, fatty acids composition and heterotrophic seedling growth of black seed (*Nigella sativa* L.) grown in a calcareous soil. *Ind. Crop. Prod.* **2015**, *74*, 939–949. [CrossRef]
27. Zheng, Y.L.; Ma, H.C. Effects of seed aging on seed germination and seed reserve utilization in mumian. *HortTechnology* **2014**, *24*, 471–474. [CrossRef]
28. Castellanos-Frías, E.; García de León, D.; Pujadas-Salva, A.; Dorado, J.; Gonzalez-Andujar, J.L. Potential distribution of *Avena sterilis* L. in Europe under climate change. *Ann. Appl. Biol.* **2014**, *165*, 53–61. [CrossRef]
29. Beckie, H.J.; Warwick, S.I.; Sauder, C.A. Basis for herbicide resistance in Canadian populations of wild oat (*Avena fatua*). *Weed Sci.* **2012**, *60*, 10–18. [CrossRef]
30. Torner, C.; Gonzalez-Andujar, J.L.; Fernandez-Quintanilla, C. Wild oat (*Avena sterilis* L.) competition with winter barley: Plant density effects. *Weed Res.* **1991**, *31*, 301–307. [CrossRef]
31. Vidal, R.A.; Fleck, N.G. The risk of finding herbicide resistant weed biotypes. *Planta Daninha* **1997**, *31*, 887–892.
32. Zand, E.; Benakashani, F.; Alizadeh, H.M.; Soufizadeh, S.; Ramezani, K.; Maknali, S.; Fereydounpoor, M. Resistance to aryloxyphenoxypropionate herbicides in wild oat (*Avena ludoviciana*). *Iran J. Weed Sci.* **2006**, *2*, 17–32.
33. Gherekhloo, J.; Oveisi, M.; Zand, E.; Deprado, R. A review of herbicide resistance in Iran. *Weed Sci.* **2016**, *64*, 551–561. [CrossRef]
34. R Core Team. *R: A Language and Environment for Statistical Computing*; R Foundation for Statistical Computing: Vienna, Austria, 2013; ISBN 3-900051-07-0. Available online: https://www.r-project.org/ (accessed on 10 September 2021).
35. Ritz, C.; Streibig, J.C. Bioassay analysis using R. *J. Stat. Softw.* **2005**, *12*, 1–22. [CrossRef]
36. Keshtkar, E.; Kudsk, P.; Mesgaran, M.B. Perspective: Common errors in dose–response analysis and how to avoid them. *Pest Manag. Sci.* **2021**, *77*, 2599–2608. [CrossRef] [PubMed]
37. Christoffers, M.J.; Berg, M.L.; Messersmith, C.G. An isoleucine to leucine mutation in acetyl-CoA carboxylase confers herbicide resistance in wild oat. *Genome* **2002**, *45*, 1049–1056. [CrossRef] [PubMed]
38. Michel, B.E.; Kaufmann, M.R. The osmotic potential of polyethylene glycol 6000. *Plant Physiol.* **1973**, *51*, 914–916. [CrossRef]
39. Dahal, P.; Bradford, K.J. Effects of priming and endosperm integrity on seed germination rates of tomato genotypes: II Germination at reduced water potential. *J. Exp. Bot.* **1990**, *41*, 1441–1453. [CrossRef]
40. Beckie, H.J.; Tardif, F.J. Herbicide cross resistance in weeds. *Crop. Prot.* **2012**, *35*, 15–28. [CrossRef]
41. Torun, H.; Uygur, F.N. Determination and mapping of resistant wild oat (*Avena sterilis* L.) populations to most commonly used herbicides in wheat fields for Osmaniye, Turkey. *Int. J. Agric. Sci.* **2018**, *11*, 8–14.
42. Beckie, H.J.; Hall, L.M.; Meers, S.; Laslo, J.J.; Stevenson, F.C. Management practices influencing herbicide resistance in wild oat. *Weed Technol.* **2004**, *18*, 853–859. [CrossRef]
43. Deihimfard, R.; Zand, E. Evaluating environmental impacts of herbicides on wheat agroecosystems in the provinces of Iran using EIQ model. *Environ. Sci.* **2005**, *6*, 1–9.
44. Benakashani, F.; Zand, E.; Alizadeh, H.M. Resistance of wild oat (*Avena ludoviciana*) biotypes to clodinafop-propargil herbicide. *Appl. Entomol. Phytopathol.* **2007**, *74*, 127–151.
45. Benakashani, F.; Rahimian Mashhadi, H.; Zand, E.; Alizadeh, H.; Naghavi, M.R. Investigation of the cross resistance to ACCase inhibitor herbicides in wild oat (*Avena ludoviciana* Durieu.) populations from Khuzestan province and chemical control of resistant populations. *Iran J. Weed Sci.* **2010**, *6*, 95–112.
46. Rastgou, M.; Rashed, M.H.; Zand, E.; Nasiri, M. Resistance of winter wild oat (*Avena ludoviciana* Durieu.) to aryloxyphenoxy propionate herbicides in wheat fields of Khuzestan province: First screening test. *Iran J. Weed Sci.* **2006**, *2*, 96–104.

47. Zand, E.; Benakashani, F.; Baghestani, M.A.; Maknali, A.; Minbashi, M.; Soufizadeh, S. Investigating the distribution of resistant wild oat (*Avena ludoviciana*) populations to clodinafop-propargil herbicide in South Western. *Iran. Environ. Sci.* **2007**, *3*, 85–92.
48. Delye, C. Weed resistance to acetyl coenzyme A carboxylase inhibitors: An update. *Weed Sci.* **2005**, *53*, 728–746. [CrossRef]
49. Kukorelli, G.; Reisinger, P.; Pinke, G. ACCase inhibitor herbicides—selectivity, weed resistance and fitness cost: A review. *Int. J. Pest Manag.* **2013**, *59*, 165–173. [CrossRef]
50. Gaines, T.A.; Duke, S.O.; Morran, S.; Rigon, C.A.G.; Tranel, P.J.; Küpper, A.; Dayan, F.E. Mechanisms of evolved herbicide resistance. *J. Biol. Chem.* **2020**, *295*, 10307–10330. [CrossRef] [PubMed]
51. Zhang, X.Q.; Powles, S.B. The molecular bases for resistance to acethyl co-enzyme A carboxylase (ACCase) inhibiting herbicides in two target based resistant biotypes of annual ryegrass (*Lolium rigidum*). *Planta* **2006**, *223*, 550–557. [CrossRef] [PubMed]
52. Delye, C.; Pernin, F.; Michel, S. 'Universal' PCR assays detecting mutations in acetyl-coenzyme A carboxylase or acetolactate synthase that endow herbicide resistance in grass weeds. *Weed Res.* **2011**, *51*, 353–362. [CrossRef]
53. Ahmad-Hamdani, M.S.; Yu, Q.; Han, H.; Cawthray, G.R.; Wang, S.F.; Powles, S.B. Herbicide resistance endowed by enhanced rates of herbicide metabolism in wild oat (*Avena* spp.). *Weed Sci.* **2013**, *61*, 55–62. [CrossRef]
54. Mahajan, G.; Loura, D.; Raymont, K.; Chauhan, B.S. Influence of soil moisture levels on the growth and reproductive behavior of *Avena fatua* and *Avena ludoviciana*. *PLoS ONE* **2020**, *15*, e0234648.
55. Matzrafi, M.; Gerson, O.; Rubin, B.; Peleg, Z. Different Mutations Endowing Resistance to Acetyl-CoA Carboxylase Inhibitors Results in Changes in Ecological Fitness of *Lolium rigidum* Populations. *Front. Plant Sci.* **2017**, *8*, 1078. [CrossRef] [PubMed]
56. Sabet Zangeneh, H.; Mohammaddust Chamanabad, H.R.; Zand, E.; Asghari, A.; Alamisaeid, K.; Travlos, I.S.; Alebrahim, M.T. Study of fitness cost in three rigid ryegrass populations susceptible and resistant to acetyl-CoA carboxylase inhibiting herbicides. *Front. Ecol. Evol.* **2016**, *4*, 142. [CrossRef]

plants

MDPI

Protocol

Machine Learning-Mediated Development and Optimization of Disinfection Protocol and Scarification Method for Improved In Vitro Germination of Cannabis Seeds

Marco Pepe, Mohsen Hesami and Andrew Maxwell Phineas Jones *

Department of Plant Agriculture, Gosling Research Institute for Plant Preservation, University of Guelph, Guelph, ON N1G 2W1, Canada; pepem@uoguelph.ca (M.P.); mhesami@uoguelph.ca (M.H.)
* Correspondence: amjones@uoguelph.ca

Abstract: In vitro seed germination is a useful tool for developing a variety of biotechnologies, but cannabis has presented some challenges in uniformity and germination time, presumably due to the disinfection procedure. Disinfection and subsequent growth are influenced by many factors, such as media pH, temperature, as well as the types and levels of contaminants and disinfectants, which contribute independently and dynamically to system complexity and nonlinearity. Hence, artificial intelligence models are well suited to model and optimize this dynamic system. The current study was aimed to evaluate the effect of different types and concentrations of disinfectants (sodium hypochlorite, hydrogen peroxide) and immersion times on contamination frequency using the generalized regression neural network (GRNN), a powerful artificial neural network (ANN). The GRNN model had high prediction performance ($R^2 > 0.91$) in both training and testing. Moreover, a genetic algorithm (GA) was subjected to the GRNN to find the optimal type and level of disinfectants and immersion time to determine the best methods for contamination reduction. According to the optimization process, 4.6% sodium hypochlorite along with 0.008% hydrogen peroxide for 16.81 min would result in the best outcomes. The results of a validation experiment demonstrated that this protocol resulted in 0% contamination as predicted, but germination rates were low and sporadic. However, using this sterilization protocol in combination with the scarification of in vitro cannabis seed (seed tip removal) resulted in 0% contamination and 100% seed germination within one week.

Keywords: hydrogen peroxide; sodium hypochlorite; generalized regression neural network; genetic algorithm; scarification; seed dormancy; plant tissue culture

Citation: Pepe, M.; Hesami, M.; Jones, A.M.P. Machine Learning-Mediated Development and Optimization of Disinfection Protocol and Scarification Method for Improved In Vitro Germination of Cannabis Seeds. *Plants* **2021**, *10*, 2397. https://doi.org/10.3390/plants10112397

Academic Editors: Milan S. Stankovic, Paula Baptista and Petronia Carillo

Received: 19 October 2021
Accepted: 5 November 2021
Published: 6 November 2021

Publisher's Note: MDPI stays neutral with regard to jurisdictional claims in published maps and institutional affiliations.

Copyright: © 2021 by the authors. Licensee MDPI, Basel, Switzerland. This article is an open access article distributed under the terms and conditions of the Creative Commons Attribution (CC BY) license (https://creativecommons.org/licenses/by/4.0/).

1. Introduction

For centuries, *Cannabis sativa* L. has been widely used around the world for various applications (e.g., textiles, food, cosmetics) [1]. These days, interest has been focused on medicinal and recreational facets, furthering commercial expansion. With Canada recently adopting the more globally appreciated view of cannabis, there exists an ever evolving, multi-billion-dollar industry focused on vegetative propagation [2]. Despite the reliance on clonal propagation, there is a continual need to germinate seeds to select new elite genotypes, perform pheno-hunting, as well as supporting breeding programs. To select new elite genotypes, plants are started from seed (technically achenes [3], but will be referred to as seed henceforth). During the vegetative phase of growth, a cutting is taken and maintained as a vegetative plant while the seedling is grown to maturity. Once the elite genotypes are selected, the cutting is then used as a source to propagate the clonal line. Maintaining the large population of cuttings during the phenotyping exercise represents a significant cost to producers and leaves the cutting derived mother plants exposed to insects and diseases.

To address the issues of insect, disease, and viral infections in mother plants, many producers use plant tissue culture to ensure that they are starting with clean material. For this process, nodal segments are disinfected and established in culture, a time consuming and relatively expensive endeavor. A potential alternative to the traditional approach is to first establish the seed in tissue culture. Once the seedlings are established and multiplied, micropropagated clones can be transferred into the growth facility and cultivated to maturity to identify elite genotypes. After selecting the elite genotypes, the in vitro parent material would be available for clonal propagation. This approach would greatly reduce the amount of space required for selecting new cultivars and provide a ready source of clean planting material once elite genotypes are identified. However, this approach requires an effective in vitro seed germination protocol with high germination speed and frequency. An efficient in vitro seed germination system would also support downstream biotechnologies (regeneration, transformation, etc.) in which seedling-derived tissues are preferred [4–6].

We previously reported the effect of different types and strengths of media in addition to carbohydrate types and levels as primarily important factors contributing to in vitro cannabis seed germination indices and morphological seedling traits [5]. Our results demonstrated that maximum germination percentage (82.67 ± 3.837%) was achieved with 0.43 strength mMS medium and 2.3% sucrose [5]. While the germination rate was over 80%, this was after 40 days of culture. Typically, the cannabis seed germinates within several days in the greenhouse/growth room, suggesting that something during the disinfection process was interfering with subsequent germination. To improve our previous protocol, we hypothesize that optimizing the disinfection protocol and seed scarification would increase the speed and frequency of seed germination.

As with most aspects of a tissue culture system, in vitro disinfection is a complex and non-linear process that is affected by numerous factors such as disinfectant and contaminant types and levels, media pH, immersion time, temperature, and their interactions (Figure 1) [7].

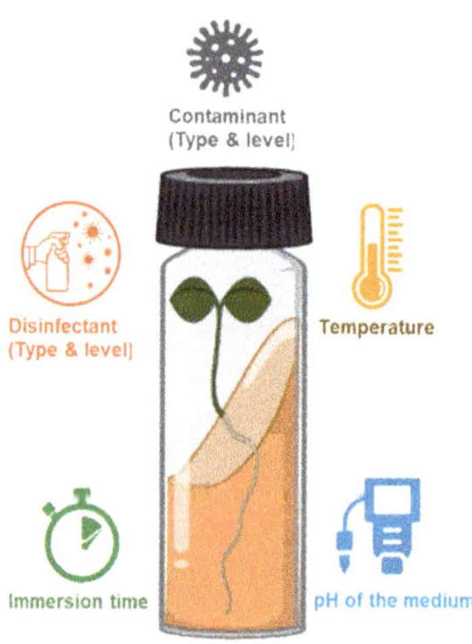

Figure 1. A schematic view of factors affecting the disinfection process.

In the disinfection process, the concentration of disinfectants plays a conflicting dual role relating to contamination frequency and seed viability [8]. Higher disinfectant concentrations generally lead to a greater control over contaminants; however, lower seedling viability is often the trade-off [9,10]. Therefore, it is necessary to optimize the disinfection process.

The disinfection process cannot be represented by a simple stepwise algorithm, especially when the datasets are highly imbalanced and noisy [10]. Therefore, artificial intelligence (AI) models combined with optimization algorithms (OAs) such as a genetic algorithm (GA) can be employed as an efficient and reliable computational method to interpret, forecast, and optimize this complex system [10–14]. This strategy (AI-OA) has been successfully used for modeling and optimizing different tissue culture systems, including in vitro decontamination, shoot proliferation, androgenesis, somatic embryogenesis, secondary metabolite production, and rhizogenesis [7,15,16]. Ivashchuk et al. [17] employed multilayer perceptron (MLP) and radial basis function (RBF) as two well-known artificial neural networks (ANNs) for modeling and predicting the effect of different disinfectants and immersion times for *Bellevalia sarmatica*, *Echinacea purpurea*, and *Nigella damascene* explant decontamination. They reported that both algorithms were able to accurately model and predict the disinfection process [17]. In another study, Hesami et al. [10] applied a hybrid MLP and non-dominated sorting genetic algorithm-II (NSGA-II) for the modeling and optimization of disinfectants and immersion times for chrysanthemum leaf segment decontamination. It was reported that MLP-NSGA-II had a high performance to predict and optimize the system [13]. A generalized regression neural network (GRNN) is another type of ANNs that has successfully been used for modeling and predicting different tissue culture processes [5,18–20]. Although there exist no reports using GRNN for the modeling and optimization of disinfection process, we previously showed that GRNN has a higher predictive performance than RBF, MLP, and the adaptive neuro-fuzzy inference system (ANFIS) for cannabis micropropagation [5,18]. Therefore, in the current study, we used GRNN-GA to model and optimize cannabis seed disinfection.

Mature seed germination can sometimes be more difficult than immature seed germination due to the increase in the seed coat's impermeability and the accumulation of inhibitors during seed maturation [21]. Hence, dormancy breaking plays a critical role relating to the speed and frequency of seed germination due to morpho-physiological dormancy [22,23]. Although there are no reports on the effects of scarification on cannabis seed germination, the positive impact of dormancy breaking by scarification has previously been suggested in several plants, such as *Limodorum* [24], *Salvia stenophylla* [25], and legumes [22]. Based on this evidence, studying the effect of scarification on cannabis seed germination can pave the way for devising an in vitro seed germination protocol with high speed and germination frequency. The current study uses GRNN-GA to model and optimize cannabis seed disinfection, and investigates the effect of scarification on seed germination. By combining these procedures, a superior in vitro cannabis seed germination protocol that limits contamination while allowing high germination rates in a short timeframe was established.

2. Results

2.1. *Effect of Different Disinfectants at Various Immersion Times on Contamination*

Based on our results (Table 1), different contamination rates were observed in various disinfection treatments.

As shown in Table 1, sodium hypochlorite was more successful than hydrogen peroxide in controlling contamination. Also, different levels of sodium hypochlorite at 15 min of immersion resulted in no contamination (Table 1).

2.2. *Data Modeling by Using GRNN*

According to our results, GRNN displayed an excellent performance for modeling and predicting contamination rates during in vitro sterilization (Table 2). Performance indices (RMSE and MBE) of the developed GRNN demonstrated that the obtained result was highly precise, and correlated in both training and testing sets (Table 2).

Additionally, R^2 was within the acceptable range for both training and testing sets, showing great prediction performance of the developed GRNN (Figure 2).

2.3. *Optimization via GA and Validation Experiment*

According to the optimization process (Table 3), 4.6% sodium hypochlorite along with 0.008% hydrogen peroxide for 16.81 min would result in no contamination. The results of the validation experiment verified that there was no contamination using this combination (Table 3).

Table 1. Effects of sodium hypochlorite and hydrogen peroxide (H_2O_2) concentrations and immersion times on contamination percentages of cannabis seeds.

Sodium Hypochlorite (%)	H_2O_2 (%)	Time (min)	Contamination (%) ± SE
0	0	0	100.0 ± 0.00
0	10	5	86.7 ± 6.67
0	10	10	86.7 ± 6.67
0	10	20	86.7 ± 6.67
0	20	5	86.7 ± 6.67
0	20	10	86.7 ± 6.67
0	20	20	73.3 ± 6.67
0	30	5	86.7 ± 6.67
0	30	10	80.0 ± 11.55
0	30	20	73.3 ± 6.67
5	0	5	53.3 ± 17.64
5	0	10	26.7 ± 6.67
5	0	15	0.0 ± 0.00
10	0	5	6.7 ± 6.67
10	0	10	0.0 ± 0.00
10	0	15	0.0 ± 0.00
15	0	5	6.7 ± 6.67
15	0	10	0.0 ± 0.00
15	0	15	0.0 ± 0.00

Table 2. Performance criteria of generalized regression neural network (GRNN) for contamination rate during cannabis seed disinfection.

Criteria	Training Set	Testing Set
Histogram	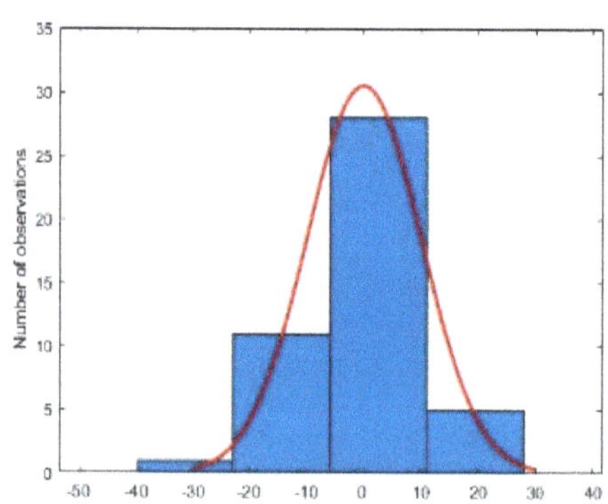	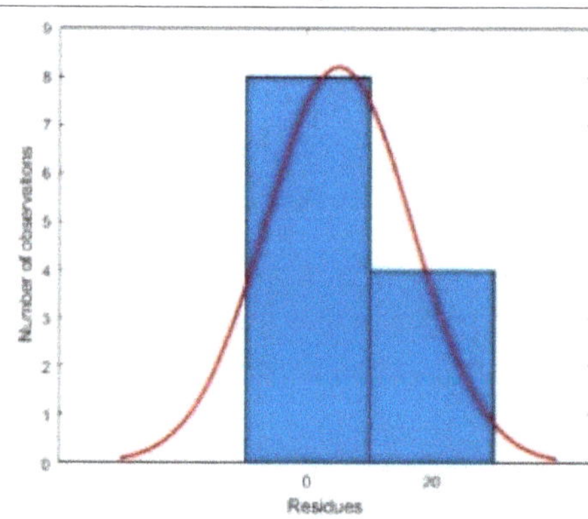
R^2	0.938	0.918
RMSE	9.888	12.247
MBE	−2.250	−5.000

R^2: Coefficient of determination; RMSE: Root mean square error; MBE: Mean bias error.

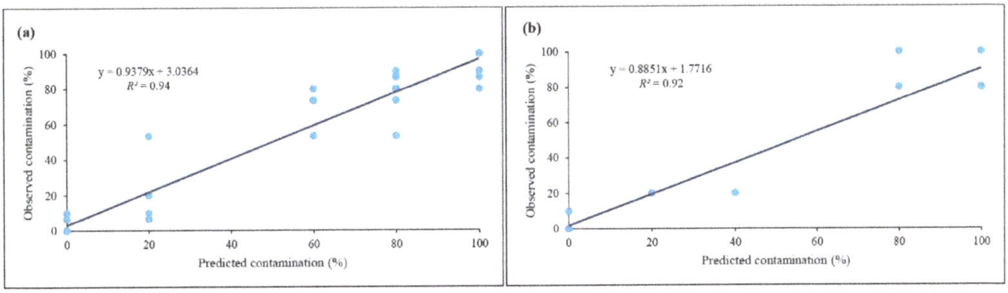

Figure 2. Scatter plot of observed data vs. predicted data of contamination percentage in (**a**) training and (**b**) testing sets.

Table 3. The results of optimization process via genetic algorithm (GA) and validation experiment.

Optimal Level of Input Variables	Predicted Contamination (%)	Contamination (%) in Validation Experiment
4.6% sodium hypochlorite + 0.008% hydrogen peroxide for 16.81 min	0	0 ± 0.0

2.4. Effect of Scarification on In Vitro Seed Germination

The seed scarification experiment (Figure 3) resulted in enhanced speed and frequency of in vitro germination of Finola seeds such that 100% germination was achieved within one week, while only 82.7 + 0.67% germination was observed in intact seeds (un- scarified seeds) after 40 days in our previous study. Moreover, we assessed the efficiency of the developed protocol on 10 drug-type cannabis genotypes (i.e., Bubba Island Kush, Glueberry OG, Critical Orange Punch, Frisian Dew, Banana Blaze, Blueberry, Durban Poison, Skunk #1, Passion #1, and Strawberry Cough; Dutch Passion, NL). The results showed more than 90% germination in these genotypes within one week when scarified.

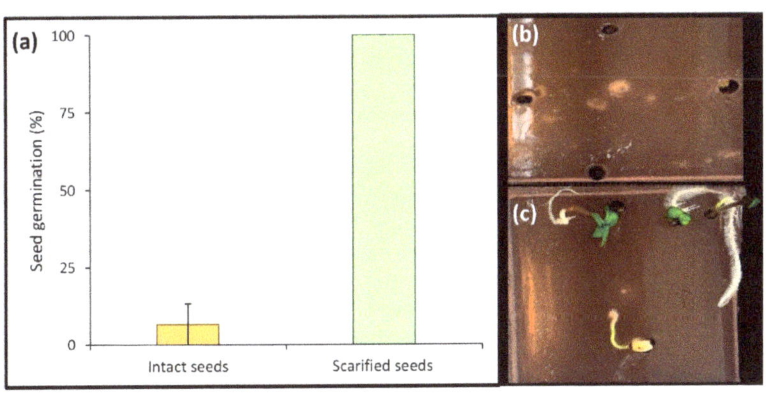

Figure 3. Effect of scarification on in vitro seed germination of cannabis after one week; (**a**) intact seeds vs. scarified seeds, (**b**) intact seeds, and (**c**) scarified seeds.

3. Discussion

In vitro seed germination of cannabis has great potential to improve the efficiency of elite cultivar selection, pheno-hunting, phenotyping, and to support various in vitro culture methods as initial explant materials [5]. In orthodox seeds, germination typically initiates with the passive uptake of water by the dry mature seed, and terminates with radicle protrusion through the seed envelope [23]. Different abiotic factors (e.g., temperature, light, medium composition, sterilization procedures, and scarification) affect seed germination, mainly through regulating the signaling and metabolism pathways of abscisic acid (ABA) and gibberellic acid (GA) [21]. Although cannabis seeds easily germinate within several days under greenhouse or field conditions, in vitro cannabis seed germination tends to be more difficult, with lower germination rates spread over a longer period of time. The cause of this difference is unknown, but is likely related to the disinfection protocol that may stress the developing embryo or potentially eliminate microbes that play a role in the germination process. As such, optimizing sterilization and scarification protocols can be considered the two most important procedures for successful in vitro seed germination [25].

The surface sterilization of initial source material, including seeds, is a prerequisite for the success of the culture [10]. Therefore, it is vital to optimize the sterilization protocol while allowing it to remain simple, cheap, environmentally friendly, and efficient [8]. Although various disinfectants and immersion times can be employed to sterilize the explants, each species and even explant type necessitates a particular sterilization protocol [10]. The hybrid of machine learning—optimization algorithm procedures offer promising computational methodology that is well suited to model and optimize in vitro culture systems such as sterilization [10].

Based on our results, GRNN-GA accurately predicted and optimized the in vitro surface sterilization of cannabis seed. According to the optimization process through GRNN-GA, 4.6% sodium hypochlorite along with 0.008% hydrogen peroxide for 16.81 min would result in no contamination. Similar to our results, previous studies showed that sodium hypochlorite was more successful than hydrogen peroxide in controlling contamination [10,25]. Additionally, the results of the validation experiment confirmed no differences between the optimized predicted and observed results, showing the robustness of GRNN-GA. In line with our results, previous studies showed that GRNN-GA can be considered a reliable computational method with high prediction performance for the modeling and optimizing of in vitro culture systems [5,18–20].

While the optimized seed disinfection protocol resulted in 0% contamination, germination was still slow and sporadic. The second experiment was performed to evaluate the effect of scarification on the speed and frequency of in vitro seed germination. The speed at which in vitro cannabis seeds germinate is remarkably slow in comparison to field germination. One possible explanation for this difference is the presence of different microbes (e.g., bacteria) that aid in the digestion of the seed coat or micropyle plug, thereby facilitating higher rates of imbibition, and thus, higher/quicker field germination rates. Since micropropagation is performed in sterile conditions, it seems that an additional step (i.e., seed scarification) should be considered to achieve a high germination rate [25]. When a viable seed is not able to germinate under appropriate conditions, the seed is considered to be dormant [26]. While cannabis seed is generally not thought to exhibit dormancy, it would appear that under these unusual circumstances, it demonstrates some physical dormancy. Following water uptake by the quiescent, dry, mature seed, germination occurs once the embryo can prevail over the constraints imposed by the testa and associated tissues [27]. As shown in Figure 4, the main constraints exerted by the covering seed structures include (i) the mechanical prevention of radicle protrusion, (ii) water uptake interference, (iii) interference with gas exchange, especially carbon dioxide and oxygen (iv) light filtration, and (v) inhibitor leakage restraint from the embryo [26–30]. The removal of the seed coat had an important role in breaking seed dormancy, which ultimately resulted in higher germination speed and frequency. Our results showed that seed scarification by removing the seed tips significantly increased the speed and frequency of seed germination. In line with our results, Pérez-Jiménez et al. [31] reported that seed scarification (i.e., removal of the seed coat) significantly increased the in vitro seed germination of *Prunus persica* L. Batsch. For the first time, we have demonstrated this method with respect to the micropropagation of cannabis.

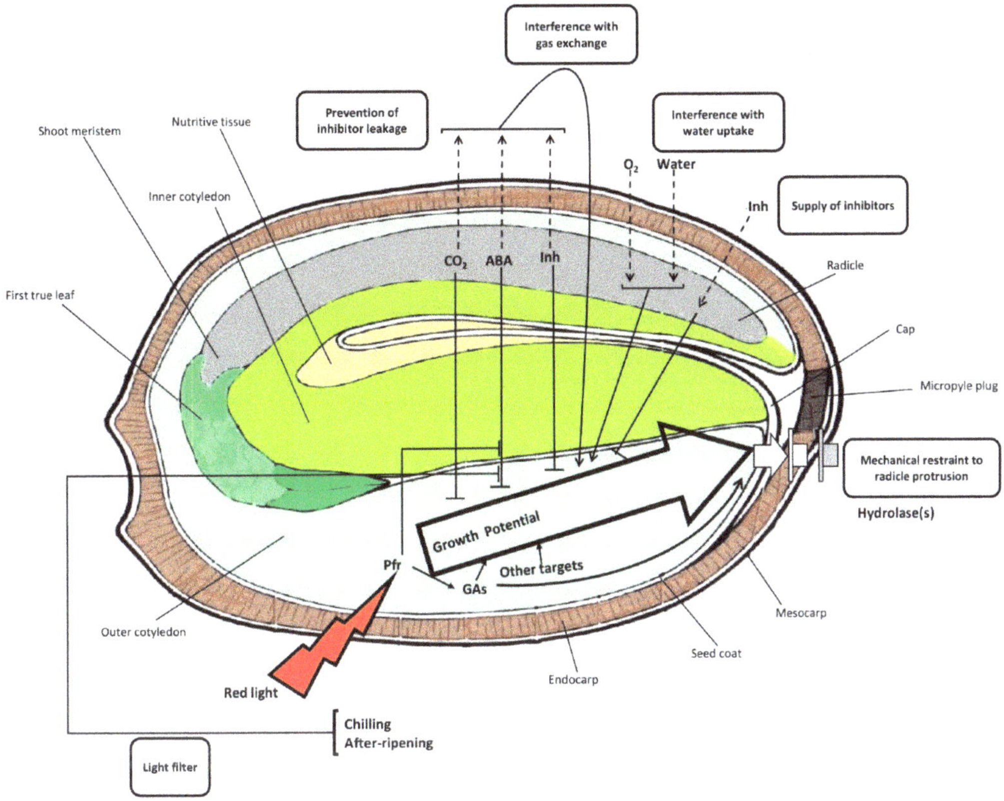

Figure 4. A schematic representation of potential interactions between the embryo and envelopes regulating cannabis seed germination and dormancy. GA: gibberellic acid; Pfr: far-red light photoreceptor phytochrome; ABA: abscisic acid; Inh: inhibitor.

4. Materials and Methods

4.1. Sterilization Procedure

Industrial hemp seed (*Cannabis sativa* cv. "Finola"; CSGA No.1 Certified seed, Lot #: 1908-18637-17-KKF-01) was employed in the first phase of this study. Different sterilants (sodium hypochlorite and hydrogen peroxide) at various immersion times (Table 1) were used for controlling contamination. The disinfection experiment was performed based on a completely randomized design with a factorial arrangement with 3 replications, each replication containing 5 seeds. For in vitro seed germination, the treated seeds were cultured in a previously optimized medium [5]. All media had 0.6% agar (Thermo-Fisher Scientific, Waltham, MA, USA) and the pH of the media was adjusted to 5.8 before autoclaving for 20 min at 120 °C. Thirty mL of media were poured into a Magenta GA7 box (Fisher Scientific, Hampton, NJ, USA). All culture boxes were placed in the growth chamber at 25 ± 2 °C under 16-h Photoperiod with 40 ± 5 µmol m^{-2} s^{-1} light intensity. Light was provided by multi-spectrum LEDs emitting only photosynthetically active radiation (400–700 nm). The obtained data was then used for modeling and optimizing the sterilization process using machine learning methods.

4.2. Modeling Procedure

In the current study, GRNN was used to develop a predictive model for contamination rate (Figure 5). To construct the model, sodium hypochlorite, hydrogen peroxide, and immersion time were considered as inputs, while the contamination rate was considered as the output (Figure 5).

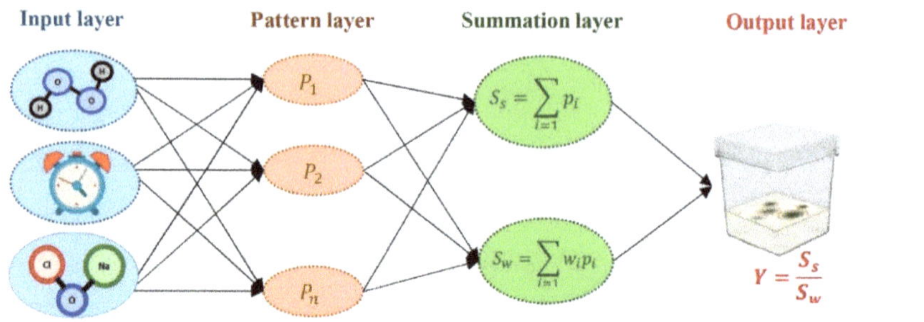

Figure 5. A schematic representation of generalized regression neural network (GRNN) used in this study.

Different performance criteria including Root mean square error (RMSE), mean bias error (MBE), and the coefficient of determination (R^2) were used to assess the efficiency of the developed predictive model.

4.3. *Optimization Procedure and Validation Experiment*

After data modeling, the developed GRNN model was linked to a GA (Figure 6) to find the optimal level of sterilants and immersion time for minimizing the contamination rate.

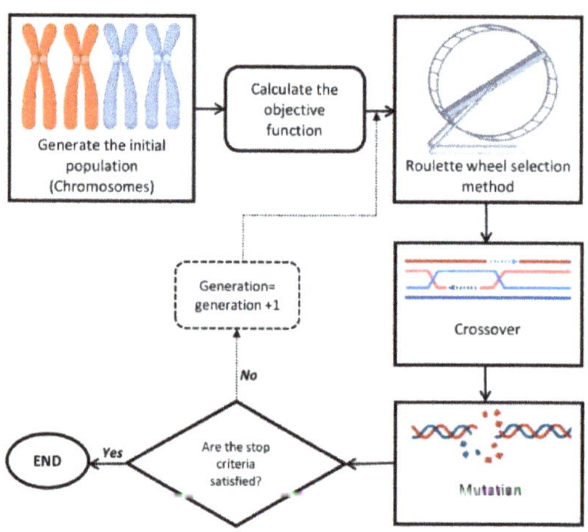

Figure 6. A schematic representation of genetic algorithm (GA).

In this study, initial population, generation number, mutation rate, mutation function, selection function, cross-over fraction, and cross-over function were, respectively, considered as 200, 1000, 0.05, uniform, Roulette Wheel, 0.6, and Two-point crossover.

To assess the performance of the developed model, the predicted-optimized result of GRNN-GA was experimentally evaluated with 3 replications, each replication containing 4 seeds.

4.4. *Scarification Procedure*

To assess the effect of seed scarification, the micropyle and its surrounding tissue were removed by scalpel, without injury to the embryo (Figure 7). This experiment was performed based on a completely randomized design with two treatments (scarified seeds and intact seeds) with three replications. Each replication contained four seeds.

Figure 7. A schematic representation of the scarification methodology for in vitro cannabis seed germination.

To assess the efficiency of the developed protocol, 10 drug-type genotypes of cannabis (i.e., Bubba Island Kush, Glueberry OG, Critical Orange Punch, Frisian Dew, Banana Blaze, Blueberry, Durban Poison, Skunk #1, Passion #1, and Strawberry Cough) were employed, and an in vitro seed germination rate on these genotypes was studied.

5. Conclusions

The present study was performed to establish an efficient in vitro seed germination protocol for *Cannabis sativa*. This was accomplished by optimizing an in vitro sterilization protocol for cannabis seeds using GRNN-GA, and to assess the effect of seed scarification on the speed and frequency of in vitro germination. The GRNN-GA model was able to precisely predict and optimize the disinfection process, but germination was still slow and sporadic. The result of the scarification experiment showed that seed scarification resulted in the reduction of in vitro germination time, while enhancing germination rate. Although we tested our protocol on different cannabis genotypes, future studies can evaluate the efficiency of the developed method on additional genotypes and further study the underlying mechanisms involved in seed dormancy and scarification in cannabis. Furthermore, we also suggest other methods, such as using sulfuric acid to evaluate scarification for in vitro cannabis seed germination; this development would improve the efficiency of the system. Due to the recalcitrant nature of *Cannabis* to in vitro culture, in addition to the much-needed refinement of in vitro seed disinfection techniques, the research presented offers opportunity to prepare micropropagated specimens with high efficiency. Our protocols can be implemented to reduce contamination and increase the germination rate of large-scale pheno-hunting and breeding programs for this economically important crop.

Author Contributions: Conceptualization, M.P., M.H. and A.M.P.J.; methodology, M.P., M.H. and A.M.P.J.; validation, M.P., M.H. and A.M.P.J.; formal analysis, M.P. and M.H.; investigation, M.P. and M.H.; resources, A.M.P.J.; data curation, M.P., M.H. and A.M.P.J.; writing—original draft preparation, M.P., M.H. and A.M.P.J.; writing—review and editing, M.P., M.H. and A.M.P.J.; visualization, M.P. and M.H.; supervision, A.M.P.J.; project administration, A.M.P.J.; funding acquisition, A.M.P.J. All authors have read and agreed to the published version of the manuscript.

Funding: This research was funded by the NSERC Discovery Grant, grant number RGPIN-2016-06252.

Institutional Review Board Statement: Not applicable.

Informed Consent Statement: Not applicable.

Data Availability Statement: All relevant data are within the paper.

Acknowledgments: Not applicable.

Conflicts of Interest: The authors declare no conflict of interest. The funders had no role in the design of the study; in the collection, analyses, or interpretation of data; in the writing of the manuscript, or in the decision to publish the results.

References

1. Hesami, M.; Pepe, M.; Alizadeh, M.; Rakei, A.; Baiton, A.; Phineas Jones, A.M. Recent advances in cannabis biotechnology. *Ind. Crop. Prod.* **2020**, *158*, 113026. [CrossRef]
2. Torkamaneh, D.; Jones, A.M.P. Cannabis, the multibillion dollar plant that no genebank wanted. *Genome* **2021**, *64*, 1–10. [CrossRef] [PubMed]
3. Small, E. Evolution and Classification of *Cannabis sativa* (Marijuana, Hemp) in Relation to Human Utilization. *Bot. Rev.* **2015**, *81*, 189–294. [CrossRef]
4. Adhikary, D.; Kulkarni, M.; El-Mezawy, A.; Mobini, S.; Elhiti, M.; Gjuric, R.; Ray, A.; Polowick, P.; Slaski, J.J.; Jones, M.P.; et al. Medical cannabis and industrial hemp tissue culture: Present status and future potential. *Front. Plant Sci.* **2021**, *12*, 627240. [CrossRef] [PubMed]
5. Hesami, M.; Pepe, M.; Monthony, A.S.; Baiton, A.; Phineas Jones, A.M. Modeling and optimizing in vitro seed germination of industrial hemp (*Cannabis sativa* L.). *Ind. Crop. Prod.* **2021**, *170*, 113753. [CrossRef]
6. Page, S.R.G.; Monthony, A.S.; Jones, A.M.P. DKW basal salts improve micropropagation and callogenesis compared with MS basal salts in multiple commercial cultivars of *Cannabis sativa*. *Botany* **2021**, *99*, 269–279. [CrossRef]
7. Hesami, M.; Jones, A.M.P. Application of artificial intelligence models and optimization algorithms in plant cell and tissue culture. *Appl. Microbiol. Biotechnol.* **2020**, *104*, 9449–9485. [CrossRef] [PubMed]
8. Cuba-Díaz, M.; Rivera-Mora, C.; Navarrete, E.; Klagges, M. Advances of native and non-native Antarctic species to in vitro conservation: Improvement of disinfection protocols. *Sci. Rep.* **2020**, *10*, 3845. [CrossRef]
9. Barampuram, S.; Allen, G.; Krasnyanski, S. Effect of various sterilization procedures on the in vitro germination of cotton seeds. *Plant Cell Tissue Organ Cult.* **2014**, *118*, 179–185. [CrossRef]
10. Hesami, M.; Naderi, R.; Tohidfar, M. Modeling and Optimizing in vitro Sterilization of Chrysanthemum via Multilayer Perceptron-Non-dominated Sorting Genetic Algorithm-II (MLP-NSGAII). *Front. Plant Sci.* **2019**, *10*, 282. [CrossRef]
11. Yoosefzadeh-Najafabadi, M.; Tulpan, D.; Eskandari, M. Application of machine learning and genetic optimization algorithms for modeling and optimizing soybean yield using its component traits. *PLoS ONE* **2021**, *16*, e0250665. [CrossRef] [PubMed]
12. Hameg, R.; Arteta, T.A.; Landin, M.; Gallego, P.P.; Barreal, M.E. Modeling and optimizing culture medium mineral composition for in vitro propagation of *Actinidia arguta*. *Front. Plant Sci.* **2020**, *11*, 554905. [CrossRef]
13. Niazian, M.; Niedbała, G. Machine learning for plant breeding and biotechnology. *Agriculture* **2020**, *10*, 436. [CrossRef]
14. Yoosefzadeh-Najafabadi, M.; Tulpan, D.; Eskandari, M. Using hybrid artificial intelligence and evolutionary optimization algorithms for estimating soybean yield and fresh biomass using hyperspectral vegetation indices. *Remote Sens.* **2021**, *13*, 2555. [CrossRef]
15. Zielińska, S.; Kępczyńska, E. Neural modeling of plant tissue cultures: A review. *BioTechnologia* **2013**, *94*, 253–268. [CrossRef]
16. Hesami, M.; Jones, A.M.P. Modeling and optimizing callus growth and development in *Cannabis sativa* using random forest and support vector machine in combination with a genetic algorithm. *Appl. Microbiol. Biotechnol.* **2021**, *105*, 5201–5212. [CrossRef]
17. Ivashchuk, O.A.; Fedorova, V.I.; Shcherbinina, N.V.; Maslova, E.V.; Shamraeva, E. Microclonal propagation of plant process modeling and optimization of its parameters based on neural network. *Drug Invent. Today* **2018**, *10*, 3170–3175.
18. Pepe, M.; Hesami, M.; Small, F.; Jones, A.M.P. Comparative analysis of machine learning and evolutionary optimization algorithms for precision micropropagation of *Cannabis sativa*: Prediction and validation of in vitro shoot growth and development based on the optimization of light and carbohydrate sources. *Front. Plant Sci.* **2021**, *12*, 757869. [CrossRef]
19. Salehi, M.; Farhadi, S.; Moieni, A.; Safaie, N.; Hesami, M. A hybrid model based on general regression neural network and fruit fly optimization algorithm for forecasting and optimizing paclitaxel biosynthesis in Corylus avellana cell culture. *Plant Methods* **2021**, *17*, 13. [CrossRef] [PubMed]
20. Hesami, M.; Condori-Apfata, J.A.; Valderrama Valencia, M.; Mohammadi, M. Application of artificial neural network for modeling and studying in vitro genotype-independent shoot regeneration in wheat. *Appl. Sci.* **2020**, *10*, 5370. [CrossRef]
21. Finkelstein, R.; Reeves, W.; Ariizumi, T.; Steber, C. Molecular aspects of seed dormancy. *Annu. Rev. Plant Biol.* **2008**, *59*, 387–415. [CrossRef]
22. Smýkal, P.; Vernoud, V.; Blair, M.W.; Soukup, A.; Thompson, R.D. The role of the testa during development and in establishment of dormancy of the legume seed. *Front. Plant Sci.* **2014**, *5*, 351. [CrossRef]
23. Tognacca, R.S.; Botto, J.F. Post-transcriptional regulation of seed dormancy and germination: Current understanding and future directions. *Plant Commun.* **2021**, *2*, 100169. [CrossRef] [PubMed]
24. Magrini, S.; De Vitis, M. In vitro reproduction of three *Limodorum* species (Orchidaceae): Impacts of scarification methods and nitrogen sources on mature seed germination and seedling development. *Plant Biosyst. Int. J. Deal. Asp. Plant Biol.* **2017**, *151*, 419–428. [CrossRef]
25. Musarurwa, H.T.; van Staden, J.; Makunga, N.P. In vitro seed germination and cultivation of the aromatic medicinal *Salvia stenophylla* (Burch. ex Benth.) provides an alternative source of α-bisabolol. *Plant Growth Regul.* **2010**, *61*, 287–295. [CrossRef]

26. Carrera-Castaño, G.; Calleja-Cabrera, J.; Pernas, M.; Gómez, L.; Oñate-Sánchez, L. An updated overview on the regulation of seed germination. *Plants* **2020**, *9*, 703. [CrossRef]
27. Han, C.; Yang, P. Studies on the molecular mechanisms of seed germination. *Proteomics* **2015**, *15*, 1671–1679. [CrossRef] [PubMed]
28. Alizadeh, M.; Hoy, R.; Lu, B.; Song, L. Team effort: Combinatorial control of seed maturation by transcription factors. *Curr. Opin. Plant Biol.* **2021**, *63*, 102091. [CrossRef]
29. Wu, Y.; Shen, Y.B. Seed coat structural and permeability properties of *Tilia miqueliana* seeds. *J. Plant Growth Regul.* **2021**, *40*, 1198–1209. [CrossRef]
30. Saffari, P.; Majd, A.; Jonoubi, P.; Najafi, F. Effect of treatments on seed dormancy breaking, seedling growth, and seedling antioxidant potential of *Agrimonia eupatoria* L. *J. Appl. Res. Med. Aromat. Plants* **2021**, *20*, 100282. [CrossRef]
31. Pérez-Jiménez, M.; Guevara-Gázquez, A.; Carrillo-Navarro, A.; Cos-Terrer, J. How carbon source and seedcoat influence the in vitro culture of peach (*Prunus persica* l. Batsch) immature seeds. *HortScience* **2021**, *56*, 136–137. [CrossRef]

plants

MDPI

Article

Leaf Area Calculation Models for Vines Based on Foliar Descriptors

Florin Sala [1], Alin Dobrei [2] and Mihai Valentin Herbei [3,*]

1 Department—Soil Sciences, Compartment of Soil Science and Plant Nutrition, Banat University of Agricultural Sciences and Veterinary Medicine "King Michael I of Romania", 300645 Timisoara, Romania; florin_sala@usab-tm.ro
2 Department—Horticulture, Compartment of Viticulture, Banat University of Agricultural Sciences and Veterinary Medicine "King Michael I of Romania", 300645 Timisoara, Romania; alin1969tmro@yahoo.com
3 Department—Sustainable Development and Environmental Engineering, Compartment Remote Sensing and GIS, Banat University of Agricultural Sciences and Veterinary Medicine "King Michael I of Romania", 300645 Timisoara, Romania
* Correspondence: mihai_herbei@yahoo.com

Abstract: In the case of foliar area studies on vines, with a large number of determinations, a simple, fast, sufficiently accurate and low-cost method is very useful. The typology of leaves on the vine is complex, characterized by several descriptive parameters: median rib; secondary venations of the first and second order; angles between the median rib and the secondary venations; sinuses; length and width of the leaf. The present study aimed to evaluate models for calculating the leaf area based on descriptive parameters and KA (KA as the surface constant used to calculate the leaf area) for six vine cultivars, 'Cabernet Sauvignon' (CS), 'Muscat Iantarnîi' (MI), 'Muscat Ottonel' (MO), 'Chasselas' (Ch), 'Victoria' (Vi) and 'Muscat Hamburg' (MH). The determined KA surface constants had subunit values (0.91 to 0.97), except for the cultivars 'Muscat Iantarnîi' and 'Muscat Ottonel' where the surface constant KA_2 (in relation to the second-order secondary venations) had supraunitary values (1.07 and 1.08, respectively). The determination of the leaf area was possible under different conditions of statistical accuracy ($R^2 = 0.477$, $p = 0.0119$, up to $R^2 = 0.988$, $p < 0.001$) in relation to the variety and parametric descriptors considered. The models obtained from the regression analysis facilitated a more reliable prediction of the leaf area based on the elements on the left side of the leaf, in relation to the median rib, compared to those on the right. The accuracy of the results was checked on the basis of minimum error (ME) and confirmed by parameters R^2, p and RMSE.

Keywords: foliar descriptors; leaf area; models; vine leaves

Citation: Sala, F.; Dobrei, A.; Herbei, M.V. Leaf Area Calculation Models for Vines Based on Foliar Descriptors. *Plants* **2021**, *10*, 2453. https://doi.org/10.3390/plants10112453

Academic Editors: Milan S. Stankovic, Paula Baptista and Petronia Carillo

Received: 21 October 2021
Accepted: 10 November 2021
Published: 13 November 2021

Publisher's Note: MDPI stays neutral with regard to jurisdictional claims in published maps and institutional affiliations.

Copyright: © 2021 by the authors. Licensee MDPI, Basel, Switzerland. This article is an open access article distributed under the terms and conditions of the Creative Commons Attribution (CC BY) license (https://creativecommons.org/licenses/by/4.0/).

1. Introduction

Foliar parameters are integral elements of the leaves, geometry, found in a certain proportionality with the leaf as a whole, and can be used to evaluate the leaf area, the indices dependent on the leaf area, as well as to study physiological, ecological and agricultural nature of plants [1–4]. Anatomical elements and descriptive parameters of the leaf lamina were used in the study and ampelographic characterization of genotypes in vines and in the evaluation of ecological plasticity in relation to certain environmental factors [5–10].

Some studies have evaluated changes at the molecular, cellular and topological levels of the leaves in relation to the plasticity of the respective genotypes [11,12]. Various other foliar studies have focused on the interception of solar energy [13], photosynthetic rate [14–16], nutritional status [17–19], water utilization ratio in relation to production [20,21], the peculiarities of growth and development of the vine [22,23], fruiting and production quality [24–29], quality of vines for human nutrition and phytopharmaceutical products [30,31], degree of attack of diseases and pests [32], the relationship of the vine with environmental factors and the reaction to stress condition [33–38].

Data on leaf parameters and in particular the leaf area, and indices targeting the leaf area (specific leaf weight—SLW, leaf area index—LAI, net assimilation rate—NAR, etc.) were used in assessing the relation of the vine with the ecological and technological factors [20,39]. The leaf area is highly correlated with leaf indices [40–42], with canopy cover [41,43–45], by the fraction of light intercepted by the canopy [46–48] and finally with crop coefficient Kc [49,50].

Descriptive parameters of the leaf lamina were important in the computerized reconstruction of leaves in different types and varieties of vines [51], with practical importance for the realization of study models and some specific foliar fingerprints. Some studies have focused on the seasonal cycle of growth and development in several varieties and varieties of vines, which is why it has been important to analyze the leaves in their dynamics [52,53]. The accurate determination of leaf parameters and leaf area (LA) is a key issue in crop growth analysis [54], as simple regression models relating LA and crop growth rate are commonly used to estimate crop yield [55].

Methods that are non-destructive, fast, low-cost and sufficiently accurate for determining leaf area are of interest due to their high efficiency in vine physiology and technology studies, especially in terms of leaf surface dynamics in relation to various factors.

The present study analyzed the leaf area of six vine cultivars in order to determine the leaf area by means of models developed based on the parameters of leaf lamina and KA surface constants specific to each cultivar.

2. Results

The vine leaf has a special typology and complexity that is characterized by a series of descriptive elements: the median rib, secondary venations of the 1st and 2nd order, lobes, sinuses, and angles between the median and secondary venations. Based on these descriptive elements and the general considerations presented, the present study aimed to evaluate some models of leaf surface calculation based on descriptive elements, KA surface constants and regression analysis. The surface constants are specific to each variety by leaf typologies, and the dimensions of the descriptive parameters of the leaves can be obtained with high accuracy by measurements (±0.5 mm). The study aimed to evaluate through a comparative analysis where foliar parameters and calculation method most easily facilitate the determination with high precision of the leaf area, to be promoted in the case of studies targeting a large number of leaf area determinations in the vine of life.

Measurements were made for each leaf on the median rib, the secondary venations of order 1 and 2, the distances between the terminations of the venations of order 1 and those of order 2, on the distances from the base of the sinuses to the base of the median rib, and, respectively, on angles α and β, the results being shown in Tables 1 and 2. At the same time, each leaf was scanned, resulting in the scanned leaf area (SLA) with high-precision (99.95–100.00), considered as a reference area for further comparisons with measured leaf area (MLA) in the study.

Table 1. Values of leaf areas and venations sizes at the level of vines in the studied cultivars.

Cultivar	MR	VL1	VR1	DV1	VL2	VR2	DV2
CS	9.21 ± 0.20	7.54 ± 0.19	7.92 ± 0.15	10.82 ± 0.24	5.80 ± 0.14	5.94 ± 0.10	11.52 ± 0.23
MI	9.16 ± 0.59	7.98 ± 0.39	7.86 ± 0.48	12.64 ± 1.10	6.00 ± 0.26	5.82 ± 0.24	11.10 ± 0.33
MO	9.30 ± 0.78	7.78 ± 0.60	8.04 ± 0.54	12.86 ± 0.71	5.60 ± 0.47	5.78 ± 0.31	10.86 ± 0.76
Ch	11.30 ± 0.49	9.64 ± 0.51	10.06 ± 0.39	14.98 ± 0.61	7.22 ± 0.32	7.62 ± 0.43	14.44 ± 0.63
Vi	13.56 ± 0.21	11.76 ± 0.51	12.18 ± 0.46	16.42 ± 0.84	8.88 ± 0.53	8.48 ± 0.48	17.04 ± 0.78
MH	15.46 ± 0.61	13.94 ± 0.70	12.88 ± 0.74	21.00 ± 1.73	10.56 ± 0.75	10.10 ± 0.48	20.46 ± 1.02

Note. CS—'Cabernet Sauvignon'; MI—'Muscat Iantarnîi'; MO—'Muscat Ottonel'; Ch—'Chasselas'; Vi—'Victoria'; MH—'Muscat Hamburg'; MR—median rib; VL1—venation left of order I; VR1—venation right of order 1; DV1—distance at the end of the venations VL1—VR1; VL2—second-order venation left; VR2—second-order venation right; DV2—distance at the end of the venations VL2–VR2.

Table 2. Values of the parameters related to the lobes and angles of the leaf venations in the studied cultivars.

Cultivar	DSL1	DSR1	DSL2	DSR2	α	β
CS	3.69 ± 0.17	3.49 ± 0.09	3.52 ± 0.10	3.57 ± 0.12	44.46 ± 0.77	54.91 ± 1.56
MI	5.54 ± 0.32	5.00 ± 0.44	4.98 ± 0.16	4.84 ± 0.18	51.94 ± 1.47	49.43 ± 1.44
MO	4.74 ± 0.39	4.64 ± 0.48	4.28 ± 0.31	4.48 ± 0.27	57.98 ± 1.83	54.52 ± 1.24
Ch	3.48 ± 0.20	3.82 ± 0.29	3.58 ± 0.27	3.88 ± 0.28	57.17 ± 1.63	52.97 ± 1.57
Vi	9.34 ± 0.55	9.46 ± 0.48	7.68 ± 0.38	7.68 ± 0.44	41.91 ± 1.22	44.76 ± 1.31
MH	6.14 ± 0.73	6.04 ± 0.46	6.20 ± 0.59	6.36 ± 0.45	50.43 ± 1.21	51.44 ± 1.54

Note. CS—'Cabernet Sauvignon'; MI—'Muscat Iantarnîi'; MO—'Muscat Ottonel'; Ch—'Chasselas'; Vi—'Victoria'; MH—'Muscat Hamburg'; DSL1—sinus base distance 1 left to lamina base; DSR1—sinus base distance 1 right to lamina base; DSL2—sinus base distance 2 left to lamina base; DSR2—sinus base distance 2 right to lamina base; α—the angle between the median rib (MR) and the right venation of the first order (VR1); β—the angle between the first-order straight venation (VR1) and the second-order straight venation (VR2).

Based on the leaf sizes obtained by measurement, the surface constants KA_1 and KA_2 were determined, the results being presented in Table 3. The optimal values for the surface constants were considered under the conditions of the minimum error between MLA and SLA (considered as reference) and of the statistical parameter RMSE.

Table 3. Values of area constant (KA) depending on leaf area and statistic safety parameters in the studied vine cultivars.

			KA_1					KA_2		
					'Cabernet Sauvignon'					
SLA (cm^2)			97.53					97.53		
KA	0.95	0.96	0.97	0.98	0.99	0.90	0.91	0.92	0.93	0.94
MLA (cm^2)	95.02	96.02	97.02	98.02	99.02	95.88	96.95	98.01	99.08	100.14
ME	−2.51	−1.51	−0.51	0.49	1.49	−1.65	−0.58	0.48	1.55	2.61
RMSE	4.9137	4.4959	4.2794	4.2948	4.5397	3.4282	3.0539	3.0338	3.3741	3.9835
					'Chasselas'					
SLA (cm^2)			154.07					154.07		
KA	0.89	0.90	0.91	0.92	0.93	0.91	0.92	0.93	0.94	0.95
MLA (cm^2)	151.29	152.98	154.69	156.39	158.09	150.19	151.84	153.49	155.14	156.79
ME	−2.79	−1.09	0.61	2.31	4.01	−3.88	-2.23	−0.58	1.07	2.72
RMSE	18.0852	17.7521	17.5824	17.5809	17.7476	8.9994	8.2622	7.8268	7.7444	8.0259
					'Muscat Hamburg'					
SLA (cm^2)			302.83					302.83		
KA	0.90	0.91	0.92	0.93	0.94	0.93	0.94	0.95	0.96	0.97
MLA (cm^2)	294.89	298.17	301.44	304.72	307.99	294.89	298.83	302.01	305.19	308.37
ME	−7.94	−4.67	−1.39	1.89	5.16	−7.18	−4.00	−0.82	2.36	5.54
RMSE	52.6654	52.3765	52.3025	53.1656	53.8714	41.2034	40.9519	40.9538	41.2089	41.7126
					'Muscat Iantarnîi'					
SLA (cm^2)			110.03					110.03		
KA	0.91	0.92	0.93	0.94	0.95	1.06	1.07	1.08	1.09	1.10
MLA (cm^2)	107.58	108.76	109.95	111.13	112.31	108.14	109.16	110.18	111.19	112.11
ME	−2.45	−1.27	−0.08	1.10	2.28	−1.89	0.87	0.15	1.17	2.19
RMSE	12.1725	12.0021	11.9556	12.4616	12.8615	16.8639	16.7988	16.7971	16.8588	16.9831
					'Muscat Ottonel'					
SLA (cm^2)			110.27					110.27		
KA	0.89	0.90	0.91	0.92	0.93	1.05	1.06	1.07	1.08	1.09
MLA (cm^2)	108.21	109.42	110.64	111.85	113.07	108.46	109.49	110.53	111.56	112.59
ME	−2.06	−0.85	0.34	1.58	2.80	−1.81	−0.78	0.25	1.29	2.32
RMSE	8.8879	8.5431	8.3732	8.3887	8.5888	6.2165	6.0236	6.0199	6.2057	6.5649

Table 3. *Cont.*

	KA_1					KA_2				
	'Victoria'									
SLA (cm^2)			216.17					216.17		
KA	0.95	0.96	0.97	0.98	0.99	0.91	0.92	0.93	0.94	0.95
MLA (cm^2)	211.59	213.82	216.05	218.28	220.50	210.63	212.95	215.26	217.58	219.89
ME	−4.57	−2.35	−0.12	2.11	4.33	−5.53	−3.22	−0.90	1.41	3.72
RMSE	13.8312	13.1502	12.8298	12.8971	13.3459	19.3721	18.8042	18.5139	18.5145	18.8057

Note. KA_1—valid for MLA = MR × DV1 × KA_1; KA_2—valid for MLA = MR × DV2 × KA_2; SLA—scanned leaf area (average leaf area for leaves, obtained by SLA method); MLA—measured leaf area; ME—minimal error, the lowest value is the best; RMSE—Root Mean Squared Error, the lowest value is the best. Area constant (KA) in each cultivar were considered depending on minimum values for ME and RMSE.

The determined KA surface constants had subunitary values, except for the cultivars 'Muscat Iantarnîi' and 'Muscat Ottonel', where the KA_2 surface constant had supraunitary values. Based on the KA values determined for each variety, the values of the leaf areas were calculated with high precision. From the comparative analysis of the obtained results, it was found that the leaf area was determined with greater precision based on the secondary venations of order 2 and KA_2, except for the cultivars 'Muscat Iantarnîi' and 'Victoria'.

The studied vine varieties presented a distinct foliar typology, genetically well-defined and characterized ampelographically [56], which was reflected in the clear differentiation of the results regarding the morphological and foliar surface elements. According to the values of the obtained leaf areas and the leaf descriptors [57], the leaves of the cultivars 'Cabernet Sauvignon', 'Muscat Iantarnîi', 'Muscat Ottonel' and 'Chasselas' were classified in class 8.5-Mesophyll (4500–18,225 mm^2) and the cultivars 'Victoria' and 'Muscat Hamburg' in class 8.6-Macrophyll (18,225–164,025 mm^2).

In the case of studies of vegetation dynamics in vines, with a high number of determinations, a simple, fast, sufficiently accurate and low-cost method is very useful. Based on the correlations identified between the values of the descriptive parameters of the leaf lamina, the prediction of leaf surfaces was evaluated only based on a known element at the level of the leaf lamina.

For this purpose, the regression analysis was used for each of the descriptive elements studied, the equations describing the prediction relations and statistical accuracy parameters being presented in Table 4. In the case of estimating the leaf area based on the angles α and β, no statistical certainty was registered, and the results were not taken into account.

Table 4. Equations for predicted leaf area (*PLA*) based on foliar parameters and statistical accuracy parameters.

Parameter x	Equation for Predicted Leaf Area (PLA)	Equation Number	Statistical Accuracy Parameters		
			R^2	p	RMSE
MR	$PLA = 1.2281x^2 - 0.1746x + 1.035$	(1)	0.965	$p << 0.001$	19.7330
VL1	$PLA = 0.3662x^2 + 22.575x - 92.945$	(2)	0.988	$p << 0.001$	17.1495
VR1	$PLA = -0.5495x^2 + 40.855x - 180.18$	(3)	0.739	$p << 0.001$	43.7254
DV1	$PLA = -0.1114x^2 + 22.493x - 141.23$	(4)	0.846	$p << 0.001$	34.8094
VL2	$PLA = 37.68x - 111.21$	(5)	0.997	$p << 0.001$	22.9111
VR2	$PLA = 1.489x^2 + 17.61x - 47.077$	(6)	0.917	$p << 0.001$	32.4634
DV2	$PLA = 19.713x - 115.45$	(7)	0.959	$p << 0.001$	27.4804
DSL1	$PLA = 18.38x + 64.325$	(8)	0.485	$p = 0.0103$	70.1936
DSR1	$PLA = 18.231x + 66.478$	(9)	0.477	$p = 0.0119$	70.3778
DSL2	$PLA = 31.157x + 7.935$	(10)	0.806	$p << 0.001$	61.7832
DSR2	$PLA = 31.23x + 4.4471$	(11)	0.752	$p << 0.001$	63.2706

Note. PLA—predicted leaf area; MR—median rib; VL1—venation left of order I; VR1—venation right of order 1; DV1—distance at the end of the venations VL1—VR1; VL2—second-order venation left; VR2—second-order venation right; DV2—distance at the end of the venations VL2—VR2; DSL1—sinus base distance 1 left to lamina base; DSR1—sinus base distance 1 right to lamina base; DSL2—sinus base distance 2 left to lamina base; DSR2—sinus base distance 2 right to lamina base.

The regression analysis based on descriptive parameters of the leaves led to the Equations (1)–(11), which predict the leaf area (PLA) in statistical accuracy conditions. Based on the values of the RMSE parameter and the correlation coefficient R^2, it was found that the elements on the left side of the median ribs (VL1, VL2, DSL1 and DSL2) facilitated the more accurate prediction of the leaf area compared to those on the right (VR1, VR2, DSR1 and DSR2). Such findings have not been found in the literature.

The analysis of statistical accuracy parameters (R^2, RMSE) found that the descriptive elements on the left side of the leaves facilitated a higher accuracy in determining the leaf area compared to the homologous elements on the right side (most likely due to leaf asymmetry, but the level of asymmetry has not been assessed), which recommends their use for calculating the leaf area in cultivars studied, when using only one known element of the leaf.

3. Discussion

Different methods can be used to determine the leaf area, classified into two broad categories, destructive and nondestructive, and direct and indirect, respectively [58]. Kvet and Marshall [59] concluded that the most appropriate method is the one in relation to the volume of plant material to be determined, the required accuracy, the time interval, the staff involved and the allocated costs, the planimetric determination or by scanning, providing the highest accuracy.

Direct methods for determining leaf area are based on measurements of leaf size and can be destructive, with greater accuracy [60–62], or non-destructive with portable devices or based on leaf size [60,63–66].

Destructive methods are generally more accurate but are more laborious, costly in terms of time, equipment and personnel. The simplest method is based on measuring the leaf area by planimetry or graph paper [58,67]. The gravimetric method, which is sufficiently accurate, is based on the exact determination of the weight of known surfaces (rectangular or circular) in a leaf to obtain a fit line, and the subsequent correlation with the weight of the leaves of interest to find the leaf surface [68]. However, this method is highly dependent on the cultivar, vegetation stages, plant age, leaf density, nutritional status and especially the hydration status of the leaves [69–72]. In some studies, the determination of leaf area was performed by combined non-destructive (scanning with portable devices) and destructive (gravimetric) methods [73,74]. Increasingly promoted are non-destructive methods that facilitate the repetitive study of leaves in the dynamics of growth and development processes in field conditions, for which portable scanners [17,53], imaging-based techniques [75–77], simple measurement methods based on leaf size [64,78,79] or mathematical and statistical models developed based on leaf size are used [80–82]. A number of other techniques have been proposed for estimating the leaf area in vines, based on indirect methods, such as imaging by measuring light extinction through the canopy [61,65,83–86], remote sensed imagery [87,88], ultrasonic-based method [89], remote sensing combined to Smart-App [90], or based on 3D point clouds resulted from UAV imagery [91]. In the case of such methods, a number of climatic, atmospheric parameters, or other external factors, can influence the accuracy according to which the leaf area is determined [92–94]. At the same time, these methods are very expensive because they require specialized equipment and certain calibration works, but they offer the possibility determining the leaf area and derived indices (leaf area index—LAI, leaf area duration—LAD, net assimilation rate—NAR, specific leaf area—SLA, specific leaf weight—SLW) over relatively large areas [84,95–97].

Indirect methods were used to determine the leaf area, canopy structure and leaf area index (LAI) in relation to different crops, climatic conditions, cropping systems and working techniques [84,98]. Williams and Ayards [20] found that the leaf area is in a linear relationship with LAI indices, water consumption and crop coefficient (Kc) in statistical accuracy conditions (R^2 = 0.89). Other research found the linearity relationship of the leaf surface with Kc and LAI [99]. The direct, non-destructive, in situ methods that use leaves dimensional parameters, relatively easy to measure, to leaf area estimation, are simple, fast, sufficiently accurate, with affordable costs and tools [58,100]. They are based on leaf length (L), maximum width (W), petiole length (Lp), leaf length x maximum width (LW), the square of the length (L2), the square of the width (W2) or some combination of these variables [101–104]. To determine the leaf area based on leaf size (L,W) in some studies, correction factors were used [104–106] or surface constants Kl or Kf [107] for the gravimetric method, which brought an extra precision to the calculation of the leaf area.

The estimation of the leaf area by using the leaf dimensions based on mathematical models was of interest due to its high speed and accuracy, certain parameters derived from statistical safety in calculations (R^2, p, RMSE) and the ability to estimate the accuracy level for subsequent comparisons with other results. However, when certain mathematical models were used to estimate leaf area in different crops, few models were used in vines to calculate leaf area [108]. The complexity of the vine leaf has led some models to develop based on the median vein [92,109], of lateral nerves of the first or second order [110–112], or based on the maximum length and width of the leaves [60,63,64,113]. To minimize errors, different leaf samples were proposed, such as number and position on the rope, then extrapolated to plant-level data, if necessary. Thus, Carbonneau [111] proposed measuring one leaf sample in each group of four contiguous leaves without losing accuracy, while Barbagallo et al. [114] proposed an empirical model to estimate primary leaf area per shoot based only on the measurement of three leaves: the largest leaf, the apical leaf and an intermediate leaf. These methods greatly reduce the workload if it is necessary to determine the leaf area for the whole plant and for many variants. Mabrouk and Carbonneau [115] proposed a model for determining the entire leaf area per shoot in the Merlot variety, based on the correlation between the total leaf area and the length of the primary and lateral shoots.

Good estimations of leaf area were found by using a model based on leaves in selected positions on the shoot [114]. Subsequent studies have shown that shoot length, however, is not always closely correlated with leaf area, especially for primary shoots [112,116]. Barbagallo et al. [117] found that cultivar and climatic and cultural factors affected linear and/or multiple regressions (using shoot length and leaf number as independent variables) to such an extent that it could not be used to accurately estimate leaf area per shoot. Another empirical model for estimating the leaf area per shoot has been proposed by Lopes and Pinto [112], which includes four variables: shoot length, number of primary leaves and the area of the largest and smallest leaves. Beslic et al. [118] considered that the method used depends on the cultivar and its leaf characteristics, such as shape, number of lobes, shape of sinuses, etc., and it always assumes the use of a large sample of leaves in order to produce the best prediction. Di Lorenzo et al. [119] found high correlations between shoot length and leaf area, and high correlations are also reported by Lopes and Pinto [80] for varieties 'Aragonez', 'Cabernet Sauvignon', 'Touriga Nacional', 'Jean' and 'Combined'. Complex, multi-variable models [112] provide greater accuracy but require more determination, while simpler methods have a higher margin of error. Based on the results obtained at cv. Blaufrankisch (*Vitis vinifera* L.), Beslic et al. [118] have considered that the original method proposed by Lopes and Pinto [112] is advantageous when it is difficult to determine the largest and the smallest leaf on a lateral shoot, as is the case with cultivars that have numerous and vigorous lateral shoots (which is not the case in cv. Blaufrankisch). Some studies require a large volume of determinations to find the leaf surface in dynamics or on the stem (LA per vine) and in the case of several variants [23,53,120].

Numerous studies have reported high accuracy in determining the leaf area in vines based on elements measured at the leaf level. Manivel and Weaver [121] found a high correlation between the length of vine leaves ('Grenache' cultivar) and their area (R^2 = 0.91). Carbonneau [111] and Carbonneau and Mabrouck [122] proposed a method using a number of linear parameters to estimate leaf area (LA). The best results were obtained by adding the lengths of the two main lateral veins. The coefficient of determination was 0.95 when 30% of the leaves on one stem were measured. Lopes and Pinto [112], when analyzing four grapevine cultivars (Fernão Pires, Vital, Touriga National, Periquita), they have obtained the predicted leaf area (PLA) under conditions of higher statistical accuracy when using first-order secondary venations compared to the median rib (the assessment being made on the basis of R^2). Montero et al. [108] determined the leaf area of the vine, 'Cencibel' cultivar, based on leaf size (leaf length and maximum width) obtained by simple regression analysis prediction relations with high accuracy (R^2 = 0.987 to 0.998). When they used maximum width (W), leaf length (L), petiole length (Lp) and dry weight of leaves (DML) as single variables in the regression equations were not as closely associated with total leaf area, although their R^2 values were also highly significant. Gutierrez and Lavín [79] determined the leaf area of the vine, Chardonnay variety, based on maximum length × maximum width for the shoot leaves and length between leaf apex and petiolar point × width between points of the superior lobules for the leaves of lateral shoots yielded the best linear mathematical indicators. Based on the determined foliar parameters, they obtained prediction relations of the leaf area with different accuracy levels (estimated based on the coefficient R^2), which suggests the differentiated contribution of the descriptive parameters of the leaves to the calculation of the leaf area and the need to know and choose those anatomical elements of the leaf that provide the greatest certainty in the calculation/prediction of the leaf area. High values for LA prediction based on median veins and maximum leaf width in two vine varieties (Niagara and DeChaunac) were also reported [113]. The accuracy and safety of the predictions were higher when based on the maximum width of the leaves than on their length. Tsialtas et al. [123] obtained high accuracy in predicting leaf area in the variety Cabernet Sauvignon (R^2 = 0.97). Similar results were also reported by Beslic et al. [81] to estimate leaf area in cv. Blaufrankisch.

Karim et al. [82] used linear regression models to estimate the leaf area of *Manihot esculenta* in parallel with gravimetric methods based on fresh and dry matter. They concluded that regression models obtained showed linear relationships when actual leaf area plotted against predicted leaf area of another one hundred leaves from different samples and that this confirmed accuracy of the developed models. Moreover, model selection indices had a high predictive ability (high R^2) with minimum error (low mean square error and percentage deviation). The selected models appeared accurate and rapid but unsophisticated, and they can be used for the estimation of LA in both destructive and non-destructive means in the Philippine Morphotype of Cassava.

Zufferey et al. [124], based on the length of each leaf lamina's two secondary lateral veins ('Chasselas', clone 14/33-4, rootstock 3309 C) and some allometric equations, obtained the leaf surface with statistically higher certainty in the case of secondary nerves based on R^2. Wang et al. [125] have performed geometric modeling based on B-spline for the study of leaves at Liriodendron. Tomaszewski and Górzkowska [126] have analyzed comparatively the variation of the shape of the leaves in fresh and dry states. Wen et al. [127] have used a multi-scale remashing method for leaf modeling.

In the case of the present study performed on six grape cultivars, the values of the R^2 coefficient for the prediction relations of the leaf area PLA had high values in the case of LA prediction based on MR, VL1, VL2, VR2 and DV2 (R^2 = 0.917 to 0.997) and reduced values in the case of prediction based on DSS1 and DSR1. Based on the leaf parameters MR and DV1 or DV2, four cultivars ('Cabernet Sauvignon', 'Chasslas', 'Muscat Hamburg', 'Muscat Ottonel') have recorded a higher accuracy and safety prediction of the leaf area based on the secondary venations of order 2 ($MR{\cdot}DV2{\cdot}KA_2$), and in two cultivars ('Muscat Iantarnîi' and 'Victoria'), a better prediction was obtained based on the first-order venations ($MR{\cdot}DV1{\cdot}KA_1$). Based on the models obtained from the regression analysis, the elements on the left side of the leaf, in relation to the median rib, facilitated a more reliable prediction of the leaf area compared to those on the right. The reliability of the results was checked on the basis of minimum error (ME) and confirmed by R^2, p and RMSE parameters.

4. Materials and Methods

4.1. Biological Material

The study on the determination of leaf area based on descriptive parameters of leaves and KA surface constants was performed on six grape cultivars with different leaf typologies: 'Cabernet Sauvignon', 'Muscat Iantarnîi', 'Muscat Ottonel', 'Chasselas', 'Victoria' and 'Muscat Hamburg'. The studied vine cultivars are cultivated in Arad and Timis counties, Romania, Figure 1.

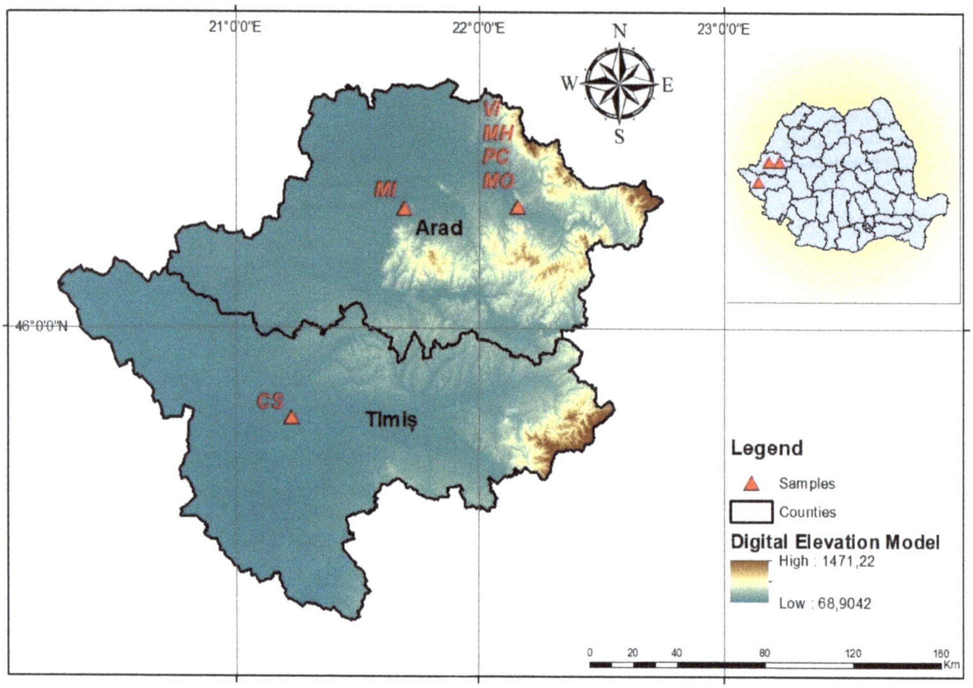

Figure 1. Cultivation area of the studied vine cultivars and leaf sampling locations, Arad and Timiș counties, Romania. CS—'Cabernet Sauvignon', MI—'Muscat Iantarnîi', MO—'Muscat Ottonel', Ch—'Chasselas', Vi—'Victoria' and MH—'Muscat Hamburg'. The map was made by the authors using ArcGIS software [128] and their own data.

4.2. Leaf Sampling

To determine the leaf area by scanning and based on the descriptive elements of the leaf lamina, 30 leaves from each variety were harvested and analyzed. The leaves were harvested in the grain-forming phenophase, BBCH 73–75 stage, and Principal growth stage 7: Development of fruits [129] from the main shoot, in the area of internodes 9–11, considered as typical leaves for characterization of grape cultivars. The leaves were immediately placed in plastic bags in the refrigerator and then transported to the laboratory for determination.

4.3. Measurement of Leaf Descriptive Parameters

At the level of the leaf lamina, specific descriptive parameters were determined for the vine, Figure 2: Median rib—Midrib (MR); left venation of order I (VL1); right venation of order I (VR1); distance at the end of the venations VL1-VR1 (DV1); second-order left venation (VL2); second-order right venation (VR2); distance at the end of the venations VL2-VR2 (DV2); sinus base distance 1 left to lamina base (DSL1); sinus base distance 1 right to lamina base (DSR1); sinus base distance 2 left to lamina base (DSL2); sinus base distance 2 right to lamina base (DSR2); the angle α between the median rib (MR) and the right venation of the first order (VR1); the angle β between the first-order straight venation (VR1) and the second-order straight venation (VR2).

Figure 2. Descriptive parameters determined at the level of the lamina of the vines.

The measurement of the length of the determined elements was done using a ruler, with an accuracy of ±0.5 mm. The determination of the angles α and β was done using the ImageJ software [130]. Based on the values of the obtained leaf areas and the leaf descriptors [57], the classification of the cultivars studied by leaf size classes was performed.

4.4. Determination of Leaf Area

The leaf area was determined for each leaf by scanning with ImageJ software (National Institutes of Health, USA [130]) (scanned leaf area—SLA). The scan was performed in a 1:1 ratio with the HP CM2320fxi MFP scanner(Hewlett-Packard, Boise, ID, USA), and the SLA was considered as a reference due to its high accuracy. At the same time, the leaf area was determined by measurement (measured leaf area—MLA) based on descriptive parameters at the leaf lamina (Figure 2). Regarding the software analysis of the leaf surface, numerous research articles have promoted such methods due to the facilities they present primarily related to the precision and accuracy of the analyses [131–133]. The measured leaf area was determined based on the parameters MR × DV1, MR × DV2 and KA (KA_1, KA_2) surface constants determined for each cultivar, Relation (12), as well as individually, based on each parameter by regression analysis.

$$MLA = MR \cdot DV \cdot KA \tag{12}$$

where: *MLA*—measured leaf area; *MR*—mid rib; *DV*—can be: DV1—distance to the end of the venation VL1–VR1; DV2—distance to the end of the venation VL2–VR2; *KA*—can be: KA_1—the corresponding surface constant DV1; KA_2—the corresponding surface constant DV2.

4.5. Statistical Analysis

All data were analyzed using variance analysis (ANOVA) and regression analysis. The assessment of the measurement accuracy and prediction of the leaf area was made by calculating the minimum measurement error (ME) related to the scanned leaf area (SLA) considered as a reference and based on the R^2 and RMSE parameters. Models were determined by regression analysis represented by polynomial functions of leaf surface prediction based on each determined leaf parameter. For statistical analysis of the results, the EXCEL application from the Microsoft Office 2007 package and the PAST software (University of Oslo, Norway) were used [134].

5. Conclusions

Surface constants (*KA*) were found for six vine cultivars and facilitated the determination of the measured leaf area (*MLA*) based on some foliar descriptor elements in conditions of high statistical safety (based on RMSE and ME). The elements on the left side of the median rib (VL1, VL2, DSL1 and DSL2) facilitated a more accurate prediction of the leaf area compared to those on the right (VR1, VR2, DSR1 and DSR2). Based on statistical safety parameters (R^2, RMSE), we found that the descriptive elements on the left side of the leaves facilitated a higher accuracy in determining the leaf area compared to the homologous elements on the right side, which recommends their use for calculating the leaf area in the cultivars studied when using only one known descriptor element of the leaf. In the case of estimating the leaf area based on the angles α and β, no statistical certainty was registered, and the results were not taken into account. The equations obtained for determining the foliar surface are based on the foliar parameters of the leaves in the six cultivars of vines studied. They can be tested/used in other varieties from the same group of leaf typology as those studied, but they can be adapted to other varieties, taking into account the specific values of the foliar parameters. The proposed method has the advantage of providing multiple ways to determine the leaf area of the vine based on the geometry elements of the leaves taken into account. It can be tested and adapted to other plant species, with leaves similar in geometric typology, to the vine.

Author Contributions: Conceptualization, F.S.; methodology, F.S. and A.D.; software, M.V.H.; validation, F.S., A.D. and M.V.H.; formal analysis, F.S.; investigation, A.D.; data curation, F.S. and M.V.H.; writing—original draft preparation, F.S.; writing—review and editing, M.V.H.; visualization, A.D.; supervision, F.S. All authors have read and agreed to the published version of the manuscript.

Funding: This research received no external funding.

Institutional Review Board Statement: Not applicable.

Informed Consent Statement: Not applicable.

Data Availability Statement: Not applicable.

Acknowledgments: The authors are grateful to the GEOMATICS Research Laboratory, BUASMV "King Michael I of Romania" from Timisoara, for the facility of the software used in this study. This paper is published from funds of the Banat's University of Agricultural Sciences and Veterinary Medicine from Timisoara.

Conflicts of Interest: The authors declare no conflict of interest.

References

1. Xiao, Y.; Tholen, D.; Zhu, X.-G. The influence of leaf anatomy on the internal light environment and photosynthetic electron transport rate: Exploration with a new leaf ray tracing model. *J. Exp. Bot.* **2016**, *67*, 6021–6035. [CrossRef] [PubMed]
2. Qu, M.; Zheng, G.; Hamdani, S.; Essemine, J.; Song, Q.; Wang, H.; Chu, C.; Sirault, X.; Zhu, X.-G. Leaf photosynthetic parameters related to biomass accumulation in a global rice diversity survey. *Plant Physiol.* **2017**, *175*, 248–258. [CrossRef]
3. Liu, H.; Ma, X.; Tao, M.; Deng, R.; Bangura, K.; Deng, X.; Liu, C.; Qi, L. A Plant leaf geometric parameter measurement system based on the Android platform. *Sensors* **2019**, *19*, 1872. [CrossRef] [PubMed]
4. Zhao, W.; Bai, Q.; Zhao, B.; Wu, Q.; Wang, C.; Liu, Y.; Yang, T.; Liu, Y.; He, H.; Du, S.; et al. The geometry of the compound leaf plays a significant role in the leaf movement of *Medicago truncatula* modulated by *mtdwarf4a*. *New Phytol.* **2021**, *230*, 475–484. [CrossRef]
5. Carneiro, L.C.; Lima, M.B. Ampelographic characterization of grapevine varieties using leaf shape. *Ciênc. Téc. Vitivin.* **1987**, *6*, 67–78.

6. Nicotra, A.B.; Atkin, O.K.; Bonser, S.P.; Davidson, A.M.; Finnegan, E.J.; Mathesius, U.; Poot, P.; Purugganan, M.D.; Richards, C.L.; Valladares, F.; et al. Plant phenotypic plasticity in a changing climate. *Trends Plant Sci.* **2010**, *15*, 684–692. [CrossRef]
7. Godoy, O.; Valladares, F.; Castro-Diez, P. The relative importance for plant invasiveness of trait means, and their plasticity and integration in a multivariate framework. *New Phytol.* **2012**, *195*, 912–922. [CrossRef] [PubMed]
8. Osunkoya, O.O.; Bayliss, D.; Panetta, F.D.; Vivian-Smith, G. Leaf trait co-ordination in relation to construction cost, carbon gain and resource-use efficiency in exotic invasive and native woody vine species. *Ann. Bot.* **2010**, *106*, 371–380. [CrossRef]
9. Fayolle, E.; Follain, S.; Marchal, P.; Chéry, P.; Colin, F. Identification of environmental factors controlling wine quality: A case study in Saint-Emilion Grand Cru appellation, France. *Sci. Total Environ.* **2019**, *694*, 133718. [CrossRef]
10. Cirillo, C.; Arena, C.; Rouphael, Y.; Caputo, R.; Amitrano, C.; Petracca, F.; De Francesco, S.; Vitale, E.; Erbaggio, A.; Bonfante, A.; et al. Counteracting the negative effects of copper limitations through the biostimulatory action of a tropical plant extract in grapevine under pedo-climatic constraints. *Front. Environ. Sci.* **2021**, *9*, 587550. [CrossRef]
11. Franks, P.J.; Beerling, D.J. Maximum leaf conductance driven by CO_2 effects on stomatal size and density over geologic time. *Proc. Natl. Acad. Sci. USA* **2009**, *106*, 10343–10347. [CrossRef]
12. Sack, L.; Scoffoni, C. Leaf venation: Structure, function, development, evolution, ecology and applications in the past, present and future. *New Phytol.* **2013**, *198*, 983–1000. [CrossRef]
13. Carbonneau, A. La surface foliaire exposee potentielle—Guide pour sa mesure. *Le Prog. Agric. Et Et Viticole* **1995**, *9*, 204–212.
14. Harrell, D.C.; Williams, L.E. Net CO_2 assimilation rate of grapevine leaves in response to trunk girdling and gibberellic acid application. *Plant Physiol.* **1987**, *83*, 457–459. [CrossRef] [PubMed]
15. Katerji, N.; Daudet, A.F.; Carbonneau, A.; Ollat, N. Étude à 1'échelle de la plante entière du fonctionnement hydrique et photosynthétique de la vigne: Comparaison des systèmes de conduite traditionnel et Lyre. *Vitis* **1994**, *33*, 197–203.
16. Petrie, P.R.; Trought, M.C.T.; Howell, G.S. Influence of leaf ageing, leaf area and crop load on photosynthesis, stomatal conductance and senescence of grapevine (*Vitis vinifera* L. cv. Pinot noir) leaves. *Vitis* **2000**, *39*, 31–36.
17. Bhat, Z.A.; Rashid, R.; Bhat, J.A. Effect of plant growth regulators on leaf number, leaf area and leaf dry matter in grape. *Not. Sci. Biol.* **2011**, *3*, 87–90. [CrossRef]
18. Shalan, A.M. Performance of *Vitis vinifera* cultivar Flame Seedless Grapevines under different node load per centimeter square of trunk cross-sectional area. *Asian J. Crop Sci.* **2013**, *5*, 139–152. [CrossRef]
19. Cuq, S.; Lemetter, V.; Kleiber, D.; Levasseur-Garcia, C. Assessing macro-element content in vine leaves and grape berries of *Vitis vinifera* by using near-infrared spectroscopy and chemometrics. *Int. J. Environ. Anal. Chem.* **2020**, *100*, 1179–1195. [CrossRef]
20. Williams, L.E.; Ayars, J.E. Grapevine water use and the crop coefficient are linear functions of the shaded area measured beneath the canopy. *Agric. For. Meteorol.* **2005**, *132*, 201–211. [CrossRef]
21. Van Leeuwen, C.; Tregoat, O.; Choné, X.; Bois, B.; Pernet, D.; Gaudillère, J.-P. Vine water status is a key factor in grape ripening and vintage quality for Red Bordeaux wine. How can it be assessed for vineyard management purposes? *J. Int. Sci. Vigne Vin* **2009**, *43*, 121–134. [CrossRef]
22. Hunter, J.J.; Visser, J.H. The effect of partial defoliation on growth characteristics of *Vitis vinifera* L. cv. Cabernet Sauvignon II. Reproductive growth. *S. Afr. J. Enol.Vitic.* **1990**, *11*, 26–32. [CrossRef]
23. Kadu, S.Y.; Tambe, T.B.; Patil, S.P. Studies on leaf morphology and vine vigour of vine grape wine varieties. *Asian J. Hort.* **2007**, *2*, 131–134.
24. Edson, C.E.; Howell, G.S.; Flore, J.A. Influence of crop load on photosynthesis and dry matter partitioning of Seyval grapevines. I. Single leaf and whole vine response pre- and post-harvest. *Am. J. Enol. Vitic.* **1993**, *44*, 139–147.
25. Poni, S.; Intrieri, C.; Silvestroni, O. Interactions of leaf age, fruiting, and exogenous cytokinins in Sangiovese grapevines under non-irrigated conditions. I. Gas exchange. *Am. J. Enol. Vitic.* **1994**, *45*, 71–78.
26. Kliewer, W.M.; Dokoozlian, N.K. Leaf area/crop weight ratios of grapevines: Influence on fruit composition and wine quality. *Am. J. Enol. Vitic.* **2005**, *56*, 170–181.
27. Bubola, M.; Janjanin, D.; Užila, Z.; Diklić, K.; Korlević, T. Impact of shoot trimming height on productive characteristics and fruit composition of Istrian Malvasia vines. In Proceedings of the 50th Croatian and 10th International Symposium on Agriculture, Zagreb, Opatija, Croatia, 16–20 February 2015; pp. 495–499.
28. King, P.D.; Smart, R.E.; McClellan, D.J. Timing of crop removal has limited effect on Merlot grape and wine composition. *Agric. Sci.* **2015**, *6*, 456–465. [CrossRef]
29. Parker, A.K.; Hofmann, R.W.; Van Leeuwen, C.; McLachlan, A.R.G.; Trought, M.C.T. Manipulating the leaf area to fruit mass ratio alters the synchrony of total soluble solids accumulation and titratable acidity of grape berries. *Aust. J. Grape Wine Res.* **2015**, *21*, 266–276. [CrossRef]
30. Akin, A. Effects of some growth regulating applications on leaf yield, raw cellulose and nutrient element content of the Müsküle table grape variety. *Afr. J. Biotechnol.* **2011**, *10*, 5601–5607. [CrossRef]
31. Oganesyants, L.; Panasyuk, A.; Kuzmina, H. Study of features of the biochemical composition of red vine leaves of autochthonous varieties in Russia. *BIO Web Conf.* **2015**, *5*, 02018-p1–02018-p4. [CrossRef]
32. Mohamed, N.; Lherminier, J.; Farmer, M.-J.; Fromentin, J.; Béno, N.; Houot, V.; Milat, M.-L.; Blein, J.-P. Defense responses in grapevine leaves against *Botrytis cinerea* induced by application of a *Pythium oligandrum* strain or its elicitin, oligandrin, to roots. *Phytopathology* **2007**, *97*, 611–620. [CrossRef]

33. Winkel, T.; Rambal, S. Influence of water stress on grapevines growing in the field: From leaf to whole-plant response. *Aust. J. Plant Physiol.* **1993**, *20*, 143–157. [CrossRef]
34. Köse, B. Effect of light intensity and temperature on growth and quality parameters of grafted vines. *Not. Bot. Horti Agrobot.* **2014**, *42*, 507–515. [CrossRef]
35. Hopper, D.W.; Ghan, R.; Cramer, G.R. A rapid dehydration leaf assay reveals stomatal response differences in grapevine genotypes. *Hortic. Res.* **2014**, *1*, 1–8. [CrossRef] [PubMed]
36. Bernardo, S.; Dinis, L.T.; Machado, N.; Moutinho-Pereira, J. Grapevine abiotic stress assessment and search for sustainable adaptation strategies in Mediterranean-like climates. A review. *Agron. Sustain. Dev.* **2018**, *38*, 66. [CrossRef]
37. Liu, G.T.; Jiang, J.F.; Liu, X.N.; Jiang, J.-Z.; Su, L.; Duan, W.; Li, R.-M.; Wang, Y.; Lecourieux, D.; Liu, C.-H.; et al. New insights into the heat responses of grape leaves via combined phosphoproteomic and acetylproteomic analyses. *Hortic Res.* **2019**, *6*, 100. [CrossRef]
38. Sánchez-Ortiz, A.; Mateo-Sanz, J.P.; Nadal, M.; Lampreave, M. Water stress assessment on grapevines by using classification and regression trees. *Plant Direct* **2021**, *5*, e00319. [CrossRef]
39. De la Riva, E.G.; Olmo, M.; Poorter, H.; Ubera, J.L.; Villar, R. Leaf mass per area (LMA) and its relationship with leaf structure and anatomy in 34 Mediterranean woody species along a water availability gradient. *PLoS ONE* **2016**, *11*, e0148788. [CrossRef]
40. Ritchie, J.T.; Johnson, B.S. Soil and plant factors affecting evaporation. In *Irrigation of Agricultural Crops—Agronomy Monograph; No. 30. ASA-CSSA SSSA*; Stewart, B.A., Nielson, D.R., Eds.; Publisher ASCE: Madison, WI, USA, 1990; pp. 363–390.
41. De Medeiros, G.A.; Arruda, F.B.; Sakai, E.; Fujiwars, M. The influence of crop canopy on evapotranspiration and crop coefficient of beans (*Phaseolus vulgaris* L.). *Agric. Water Manage.* **2001**, *49*, 211–224. [CrossRef]
42. Weraduwage, S.M.; Chen, J.; Anozie, F.C.; Morales, A.; Weise, S.E.; Sharkey, T.D. The relationship between leaf area growth and biomass accumulation in *Arabidopsis thaliana*. *Front. Plant Sci.* **2015**, *6*, 167. [CrossRef]
43. Heilman, J.L.; Heilman, W.E.; Moore, D.G. Evaluating the crop coefficient using spectral reflectance. *Agron. J.* **1982**, *74*, 967–971. [CrossRef]
44. Stevens, B.; Diels, J.; Vanuytrecht, E.; Brown, A.; Bayo, S.; Rujweka, A.; Richard, E.; Ndakidemi, P.A.; Swennen, R. Canopy cover evolution, diurnal patterns and leaf area index relationships in a Mchare and Cavendish banana cultivar under different soil moisture regimes. *Sci. Hortic.* **2020**, *272*, 109328. [CrossRef]
45. Tanioka, Y.; Cai, Y.; Ida, H.; Hirota, M. A spatial relationship between canopy and understory leaf area index in an old-growth cool-temperate deciduous forest. *Forests* **2020**, *11*, 1037. [CrossRef]
46. Ayars, J.E.; Johnson, R.S.; Phene, C.J.; Trout, T.J.; Clark, D.A.; Mead, R.M. Water use by drip irrigated late season peaches. *Irrig. Sci.* **2003**, *22*, 187–194. [CrossRef]
47. Duursma, R.A.; Falster, D.S.; Valladares, F.; Sterck, F.J.; Pearcy, R.W.; Lusk, C.H.; Sendall, K.M.; Nordenstahl, M.; Houter, N.C.; Atwell, B.J.; et al. Light interception efficiency explained by two simple variables: A test using a diversity of small- to medium-sized woody plants. *New Phytol.* **2012**, *193*, 397–408. [CrossRef]
48. Chen, H.; Zhao, X.; Han, Y.; Xing, F.; Feng, L.; Wang, Z.; Wang, G.; Yang, B.; Lei, Y.; Xiong, S.; et al. Competition for light interception in cotton populations of different densities. *Agronomy* **2021**, *11*, 176. [CrossRef]
49. Pereira, A.R.; de Camargo, M.B.P.; Nova, N.A.V. Coffee crop coefficient for precision irrigation based on leaf area index. *Bragantia Camp.* **2011**, *70*, 946–951. [CrossRef]
50. Vanino, S.; Pulighe, G.; Nino, P.; De Michele, C.; Bolognesi, S.F.; D'Urso, G. Estimation of evapotranspiration and crop coefficients of tendone vineyards using multi-sensor remote sensing data in a Mediterranean environment. *Remote Sens.* **2015**, *7*, 14708–14730. [CrossRef]
51. Martinez, M.-C.; Grenan, S. A graphic reconstruction method of an average vine leaf. *Agronomy* **1999**, *19*, 491–507. [CrossRef]
52. Snowball, A.M. Seasonal cycle of shoot development in selected *Actinidia* species. *N. Z. J. Crop Hortic. Sci.* **1997**, *25*, 221–231. [CrossRef]
53. Burg, P.; Vítěz, T.; Michálek, M. The evaluation of the vine leaves development dinamic. *Acta Univ. Agric. Silvic. Mendel. Brun.* **2013**, *61*, 17–23. [CrossRef]
54. Hunt, R. *Plant Growth Curves. The Functional Approach to Plant Growth Analysis*; Edward Arnold: London, UK, 1982; 260p.
55. Aase, J.K. Relationship between leaf area and dry matter in winter wheat. *Agron. J.* **1978**, *70*, 563–565. [CrossRef]
56. Constantinescu, G.; Negreanu, E.; Lăzărescu, V.; Poenaru, I.; Alexei, O.; Boureanu, E. *Ampelografia RPR*; Editura Academiei RPR: Bucuresti, Romania, 1960; Volume III, pp. 145–329.
57. Ellis, B.; Daly, D.C.; Huckey, L.J.; Johnson, K.R.; Mitchell, J.D.; Wilf, P.; Wing, S.L. *Manual of Leaf Architecture*; Cornell University Press: Ithaca, NY, USA, 2009; p. 190.
58. Marshall, J.K. Methods for leaf area measurement of large and small leaf samples. *Photosynthetica* **1968**, *2*, 41–47.
59. Kvet, J.; Marshall, J.K. Assessment of leaf area and other assimilating surfaces. In *Plant Photosynthetic Production: Manual of Methods*; Sestak, Z., Catsky, J., Jarvis, P.G., Eds.; Dr. W. Junk, N. V.: The Hague, The Netherlands, 1971; pp. 517–555.
60. Sepúlveda, G.; Kliewer, W.M. Estimation of leaf area of two grapevine cultivars (*Vilis vinifera* L.) using laminae linear measurements and fresh weight. *Am. J. Enol. Vitic.* **1983**, *34*, 221–226.
61. Sommer, K.J.; Lang, A.R.G. Comparative analysis of two indirect methods of measuring LAI as applied to minimal and spur pruned grape vines. *Aust. J. Plant Physiol.* **1994**, *21*, 197–206. [CrossRef]

62. Leblanc, S.G.; Fernandes, R.; Chen, J.M. Recent advancements in optical fi eld leaf area index, foliage heterogeneity, and foliage angular distribution measurements. *IEEE Int. Geosci. Remote. Sens. Symp.* **2002**, *5*, 2902–2904. [CrossRef]
63. Smith, R.J.; Kliewer, W.M. Estimation of Thompson Seedless grapevine leaf area. *Am. J. Enol. Vitic.* **1984**, *35*, 16–22.
64. Elsner, E.A.; Jubb, G.L. Leaf area estimation of Concord grape leaves from simple linear measurements. *Am. J. Enol. Vitic.* **1988**, *39*, 95–97.
65. Oliveira, M.; Santos, M. A semi-empirical method to estimate canopy leaf area of vineyards. *Am. J. Enol. Vitic.* **1995**, *46*, 389–391.
66. Sauceda-Acosta, C.P.; Gonzalez-Hernandez, V.A.; Sanchez-Soto, B.H.; Sauceda-Acosta, R.H.; Ramirez-Tobias, H.M.; Quintana-Quiroz, J.G. MACF-IJ, automated method for measuring color and leaf area through digital images. *Agrociencia* **2017**, *51*, 409–423.
67. Sestak, Z.; Catsky, J.; Jarvis, P.G. *Plant Photosynthesis Production, Manual of Methods*; Junk Publishers: The Hague, The Netherlands, 1971; p. 818.
68. Palaniswamy, K.M.; Gómez, K.A. Length-width method for estimating leaf area of rice. *Agron. J.* **1974**, *66*, 430–433. [CrossRef]
69. McKee, G.W. A coefficient for computing leaf area in hybrid corn. *Agron. J.* **1964**, *56*, 240–241. [CrossRef]
70. Epstein, E.; Robinson, R.R. A rapid method for determining leaf area of potato plants. *Agron. J.* **1965**, *57*, 515–516. [CrossRef]
71. Hoffman, G.J. Estimating leaf area from length measurements for hybrid granex onion. *Agron. J.* **1971**, *63*, 948–949. [CrossRef]
72. Palit, P.; Bhattacharayya, A.C. Measurement of leaf area per plant of white jute (*Corchorus capsularis* L.) and tossa jute (*C. olitorius* L.) using the average specific leaf weight value. *Trop. Agr.* **1984**, *61*, 59–62.
73. Castelan-Estrada, M.; Vivin, P.; Gaudillière, J.P. Allometric relationships to estimate seasonal above-ground vegetative and reproductive biomass of *Vitis vinifera* L. *Ann. Bot.* **2002**, *89*, 401–408. [CrossRef] [PubMed]
74. Bobeica, N.; Poni, S.; Hilbert, G.; Renaud, C.; Gomèz, E.; Delrot, S.; Dai, Z. Differential responses of sugar, organic acids and anthocyanins to source-sink modulation in Cabernet Sauvignon and Sangiovese grapevines. *Front. Plant Sci.* **2015**, *6*, 382. [CrossRef]
75. Costanza, P.; Tisseyre, B.; Huntter, J.J.; Deloire, A. Shoot development and non-destructive determination of grapevine (*Vitis vinifera* L.) leaf area. *S. Afr. J. Enol. Vitic.* **2004**, *25*, 43–47. [CrossRef]
76. Fuentes, S.; Poblete-Echeverría, C.; Ortega-Farias, S.; Tyerman, S.; De Bei, R. Automated estimation of leaf area index from grapevine canopies using cover photography, video and computational analysis methods. *Aust. J. Grape Wine Res.* **2014**, *20*, 465–473. [CrossRef]
77. Kalisperakis, I.; Stentoumis, C.; Grammatikopoulos, L.; Karantzalos, K. Leaf area index estimation in vineyards from uav hyperspectral data, 2d image mosaics and 3d canopy surface models. *ISPRS Int. Arch. Photogramm. Remote. Sens. Spat. Inf. Sci.* **2015**, *40*, 299–303. [CrossRef]
78. Ackley, W.B.; Crandall, P.C.; Rusell, T.S. The use of linear measurements in estimating leaf areas. *Amer. Soc. Hort. Sci.* **1958**, *72*, 327–330.
79. Gutierrez, A.T.; Lavín, A.A. Linear measurements for non destructive estimation of leaf area in Chardonnay vines. *Agric. Técnica* **2000**, *60*, 69–73.
80. Lopes, C.M.; Pinto, P.A. Easy and accurate estimation of grapevine leaf area with simple mathematical models. *Vitis* **2005**, *44*, 55–61.
81. Beslic, Z.; Todic, S.; Tesic, D. Validation of non-destructive methodology of grapevine leaf area estimation on cv. Blaufrankisch (*Vitis vinifera* L.). *S. Afr. J. Enol. Vitic.* **2010**, *31*, 22–25. [CrossRef]
82. Karim, M.R.; Fakir, M.S.A.; Mostafa, M.G.; Prodhan, A.K.M.A. Leaf area estimation by linear regression models in Cassava (Morphotype: Philippine). *J. Agrofor. Environ.* **2010**, *4*, 17–22.
83. Grantz, D.A.; Williams, L.E. An empirical protocol for indirect measurement of leaf area index in grape *Vitis vinifera* L. *HortScience* **1993**, *28*, 777–779. [CrossRef]
84. Ollat, N.; Fermaud, M.; Tandonnet, J.P.; Neveux, M. Evaluation of an indirect method for leaf area index determination in the vineyard: Combined effects of cultivar, year and training system. *Vitis* **1998**, *37*, 73–78.
85. Patakas, A.; Noitsakis, B. An indirect method of estimating leaf area index in cordon trained spur pruned grapevines. *Sci. Hortic.* **1999**, *80*, 299–305. [CrossRef]
86. Johnson, L.F.; Pierce, L.L. Indirect measurement of leaf area index in California North Coast vineyards. *HortScience* **2004**, *39*, 236–238. [CrossRef]
87. Dobrowski, S.Z.; Ustin, S.L.; Wolpert, J.A. Remote estimation of vine canopy density in vertically shoot-positioned vineyards: Determining optimal vegetation indexes. *Aust. J. Grape Wine Res.* **2002**, *8*, 117–125. [CrossRef]
88. Johnson, L.F.; Roczen, D.E.; Youkhana, S.K.; Nemani, R.R.; Bosch, D.F. Mapping vineyard leaf area with multispectral satellite imagery. *Comput. Electron. Agric.* **2003**, *38*, 33–44. [CrossRef]
89. Vitali, M.; Tamagnone, M.; La Iacona, T.; Lovisolo, C. Measurement of grapevine canopy leaf area by using an ultrasonic-based method. *OENO One* **2013**, *47*, 183–189. [CrossRef]
90. Orlando, F.; Movedi, E.; Coduto, D.; Parisi, S.; Brancadoro, L.; Pagani, V.; Guarneri, T.; Confalonieri, R. Estimating leaf area index (LAI) in vineyards using the PocketLAI Smart-App. *Sensors* **2016**, *16*, 2004. [CrossRef]
91. Comba, L.; Biglia, A.; Ricauda Aimonino, D.; Tirtia, C.; Mania, E.; Gay, P. Leaf Area Index evaluation in vineyards using 3D point clouds from UAV imagery. *Precision. Agric.* **2020**, *21*, 881–896. [CrossRef]

92. Schultz, H.R. An empirical model for the simulation of leaf appearance and leaf development of primary shoots of several grapevine (*Vitis vinifera* L.). *Sci. Hortic.* **1992**, *52*, 179–200. [CrossRef]
93. Bindi, M.; Miglietta, F.; Gozzini, B.; Orlandini, S.; Seghi, L. A simple model for simulation of growth and development in grapevine (*Vitis vinifera* L.). I. Model description. *Vitis* **1997**, *36*, 67–71. [CrossRef]
94. Fang, H.; Baret, F.; Plummer, S.; Schaepman-Strub, G. An overview of global leaf area index (LAI): Methods, products, validation, and applications. *Rev. Geophys.* **2019**, *57*, 739–799. [CrossRef]
95. Cohen, S.; Striem, M.J.; Bruner, M.; Klein, I. Grapevine leaf area index evaluation by gap fraction inversion. *Acta Hortic.* **2000**, *537*, 87–94. [CrossRef]
96. Radzali, M.H.; Kamal, N.A.M.; Diah, N.M. Measuring leaf area using Otsu segmentation method (LAMOS). *Indian J. Sci. Technol.* **2016**, *9*, 1–6. [CrossRef]
97. Zhang, W. Digital image processing method for estimating leaf length and width tested using kiwifruit leaves (*Actinidia chinensis* Planch). *PLoS ONE* **2020**, *15*, e0235499. [CrossRef] [PubMed]
98. Welles, J.M. Some indirect methods of estimating canopy structure. *Remote Sens. Rev.* **1990**, *5*, 31–43. [CrossRef]
99. Hatfield, J.L. Methods of estimating evapotranspiration. In *Irrigation of Agricultural Crops—Agronomy Monograph; No. 30. ASA-CSSA-SSSA*; Stewart, B.A., Nielson, D.R., Eds.; Publisher ASCE: Madison, WI, USA, 1990; pp. 435–474.
100. Kemp, C.P. Methods of estimating the leaf area of grasses from linear measurements. *Ann. Bot.* **1960**, *24*, 491–499. [CrossRef]
101. Robbins, N.S.; Pharr, D.M. Leaf area prediction models for cucumber from linear measurements. *HortScience* **1987**, *22*, 1264–1266.
102. Gamiely, S.; Randle, W.M.; Mills, H.A.; Smittle, D.A. A rapid and nondestructive method for estimating leaf area of onions. *HortScience* **1991**, *26*, 206. [CrossRef]
103. Payne, W.A.; Wendt, C.W.; Hossner, L.R.; Gates, C.E. Estimating pearl millet leaf area and specific leaf area. *Agron. J.* **1991**, *83*, 937–941. [CrossRef]
104. Rubin, S.S.; Danilevskaya, O.M. Opredelenie ploshchadi list'ev plodovykh dere'ev. (Leaf area determination of fruit trees.). *Bot. Zh.* **1957**, *42*, 728–730.
105. Gladyshev, N.P. K metodike opredelenya ploschady list'ev yablony. (On the methods of determination of the area of apple tree leaves). *Bot. Zh.* **1969**, *54*, 1571–1575.
106. Litschmann, T.; Vávra, R.; Falta, V. Non-destructive leaf area assessment of chosen apple cultivars. *VěDecké Práce OvocnářSké* **2013**, *23*, 205–212.
107. Nautiyal, M.C.; Singh, P.K.; Shukla, R.N.; Prakash, S.; Kumar, A. Correcting leaf area measurement by conventional methods: A new approach for apple (*Malus domestica* Borkh). *J. Hortic. Sci.* **1990**, *65*, 15–18. [CrossRef]
108. Montero, F.J.; de Juan, J.A.; Cuesta, A.; Brasa, A. Nondestructive methods to estimate leaf area in *Vitis vinifera* L. *HortScience* **2000**, *35*, 696–698. [CrossRef]
109. Junges, A.H.; Anzanello, R. Non-destructive simple model to estimate the leaf area through midvein in cultivars of *Vitis vinifera*. *Rev. Bras. Frutic.* **2021**, *43*, e-795. [CrossRef]
110. Carbonneau, A. Principes et methodes de mesure de la surface foliare. *Essai de caracterisation des types de feuilles dans le genre vitis. Ann. Amélio. Plantes* **1976**, *26*, 327–343.
111. Carbonneau, A. Analyse de la croissance des feuilles du sarment de vigne: Estimation de sa surface foliaire par echantillonnage. *Conn. Vigne Vin.* **1976**, *10*, 141–159. [CrossRef]
112. Lopes, M.A.; Pinto, P. Estimation of main and secondary leaf area of grapevine shoot. *Prog. Agric. Vitic.* **2000**, *117*, 160–166.
113. Williams, L.; Martinson, T. Nondestructive leaf area estimation of Niagara and DeChaunac grapevines. *Sci. Hortic.* **2003**, *98*, 493–498. [CrossRef]
114. Barbagallo, M.G.; Lorenzo, R.I.; Motisi, A.; Sottile, I. Estimation of leaf area changes along the primary shoot in grapevine. In Proceedings of the 9th GESCO Meeting, Budapest, Hungary; 1996; pp. 245–250.
115. Mabrouk, H.; Carbonneau, A. Une methode simple de determination de la surface foliaire de la vigne (*Vitis vinifera* L.). *Progr. Agric. Vitic.* **1996**, *113*, 392–398.
116. Tregoat, O.; Ollat, N.; Grenier, G.; Van Leeuwen, C. Survey of the accuracy and rapidity of several methods for vine leaf area assessment. *J. Int. Sci. Vigne Vin.* **2001**, *35*, 31–39. [CrossRef]
117. Barbagallo, M.G.; Collesano, G.; Di Lorenzo, R. *Valutazioni su un Metodo Diretto di Stima della Superficie Fogliare in Vitis vinifera L.*; V Giornate Scientifiche S.O.I.: Sirmione, Italy, 2000; pp. 263–264.
118. Beslic, Z.S.; Todic, S.R.; Tesevic, V.V.; Jadranin, M.B.; Novakovic, M.M.; Tesic, D. Pruning effect on content of quercetin and catechin in berry skins of cv. Blaufränkisch (*Vitis vinifera* L.). *Turk. J. Agric. For.* **2010**, *34*, 461–466. [CrossRef]
119. Di Lorenzo, R.; Costanca, P.; Pisciotta, A.; Vesco, G.; Barbagallo, M.G. Estimation of leaf area by means of linear regression equation in *Vilis vinifera* L: Limits and advantages. In Proceedings of the XIV International GESCO Viticulture Congress, Geisenheim, Germany, 23–27 August 2005; pp. 384–391.
120. De la Hera-Orts, M.L.; Martínez-Cutillas, A.; López-Roca, J.M.; Gómez-Plaza, E. Effects of moderate irrigation on vegetative growth and productive parameters of Monastrell vines grown in semiarid conditions. *Span. J. Agric. Res.* **2004**, *2*, 273–281. [CrossRef]
121. Manivel, L.; Weaver, R.J. Biometric correlations betwen leaf area and length measurement of 'Grenache' grape leaves. *HortScience* **1974**, *9*, 27–28.

122. Carbonneau, A.; Mabrouck, H. A simple method for determination of grapevine *Vitis vinifera* L. leaf area. *Progr. Agric. Vitic.* **1996**, *18*, 392–398.
123. Tsialtas, J.T.; Koundouras, S.; Zioziou, E. Leaf area estimation by simple measurements and evaluation of leaf area prediction models in Cabernet-Sauvignon grapevine leaves. *Photosynthetica* **2008**, *46*, 452–456. [CrossRef]
124. Zufferey, V.; Murisier, F.; Vivin, P.; Belcher, S.; Lorenzini, F.; Spring, J.L.; Viret, O. Carbohydrate reserves in grapevine (*Vitis vinifera* L. 'Chasselas'): The influence of the leaf to fruit ratio. *Vitis* **2012**, *51*, 103–110.
125. Wang, X.; Li, L.; Chai, W. Geometric modeling of broad-leaf plants leaf based on *B*-spline. *Math. Comput. Model.* **2013**, *58*, 564–572. [CrossRef]
126. Tomaszewski, D.; Górzkowska, A. Is shape of a fresh and dried leaf the same? *PLoS ONE* **2016**, *11*, e0153071. [CrossRef] [PubMed]
127. Wen, W.; Li, B.; Li, B.-j.; Guo, X. A leaf modeling and multi-scale remeshing method for visual computation via hierarchical parametric vein and margin representation. *Front. Plant Sci.* **2018**, *9*, 783. [CrossRef] [PubMed]
128. ESRI 2011. *ArcGIS Desktop: Release 10*; Environmental Systems Research Institute: Redlands, CA, USA, 2021.
129. Lorenz, D.H.; Eichhorn, K.W.; Bleiholder, H.; Klose, R.; Meier, U.; Weber, E. Phänologische Entwicklungsstadien der Weinrebe (*Vitis vinifera* L. *ssp.vinifera*). *Vitic. Enol. Sci.* **1994**, *49*, 66–70.
130. Rasband, W.S.; ImageJ. U. S. National Institutes of Health. Bethesda, MD, USA, 1997–2014. Available online: http://imagej.nih.gov/ij/ (accessed on 10 May 2013).
131. O'Neal, M.E.; Landis, D.A.; Isaacs, R. An inexpensive, accurate method for measuring leaf area and defoliation through digital image analysis. *J. Econ. Entomol.* **2002**, *95*, 1190–1194. [CrossRef]
132. Igathinathane, C.; Prakash, V.S.S.; Padma, U.; Ravi Babu, G.; Womac, A.R. Interactive computer software development for leaf area measurement. *Comput. Electron. Agric.* **2006**, *51*, 1–16. [CrossRef]
133. Easlon, H.M.; Bloom, A.J. Easy leaf area: Automated digital image analysis for rapid and accurate measurement of leaf area. *Appl. Plant Sci.* **2014**, *2*, 1400033. [CrossRef]
134. Hammer, Ø.; Harper, D.A.T.; Ryan, P.D. PAST: Paleontological statistics software package for education and data analysis. *Palaeontol. Electron.* **2001**, *4*, 1–9.

plants

MDPI

Article

Comparative Analysis of Genotyping by Sequencing and Whole-Genome Sequencing Methods in Diversity Studies of *Olea europaea* L.

James Friel [1], Aureliano Bombarely [1,2], Carmen Dorca Fornell [3], Francisco Luque [4] and Ana Maria Fernández-Ocaña [5,*]

[1] Dipartimento di Bioscienze, Università degli Studi di Milano, 20122 Milan, Italy; james.friel@unimi.it (J.F.); abombarely@ibmcp.upv.es (A.B.)
[2] Instituto de Biologıa Molecular y Celular de Plantas (IBMCP), CSIC, Universitat Politecnica de Valencia, 46011 Valencia, Spain
[3] Departamento de Didáctica de las Matemáticas y las Ciencias Experimentales, Facultad de Educación, Universidad Internacional de la Rioja (UNIR), 26006 Logroño, Spain; mariadelcarmen.dorcafornell@unir.net
[4] Instituto Universitario de Investigación en Olivar y Aceites de Oliva (INUO), Universidad de Jaén, 23071 Jaén, Spain; fjluque@ujaen.es
[5] Departamento de Biología Animal, Biologia Vegetal y Ecología, Facultad de Ciencias Experimentales, Campus de Las Lagunillas s/n, Universidad de Jaén UJA, 23071 Jaén, Spain
* Correspondence: amocana@ujaen.es

Citation: Friel, J.; Bombarely, A.; Fornell, C.D.; Luque, F.; Fernández-Ocaña, A.M. Comparative Analysis of Genotyping by Sequencing and Whole-Genome Sequencing Methods in Diversity Studies of *Olea europaea* L. *Plants* **2021**, *10*, 2514. https://doi.org/10.3390/plants10112514

Academic Editors: Luis Rallo, Fernando Pliego Alfaro, Pilar Rallo, Concepción Muñoz Díez and Carlos Trapero

Received: 6 August 2021
Accepted: 11 November 2021
Published: 19 November 2021

Publisher's Note: MDPI stays neutral with regard to jurisdictional claims in published maps and institutional affiliations.

Copyright: © 2021 by the authors. Licensee MDPI, Basel, Switzerland. This article is an open access article distributed under the terms and conditions of the Creative Commons Attribution (CC BY) license (https://creativecommons.org/licenses/by/4.0/).

Abstract: Olive, *Olea europaea* L., is a tree of great economic and cultural importance in the Mediterranean basin. Thousands of cultivars have been described, of which around 1200 are conserved in the different olive germplasm banks. The genetic characterisation of these cultivars can be performed in different ways. Whole-genome sequencing (WGS) provides more information than the reduced representation methods such as genotype by sequencing (GBS), but at a much higher cost. This may change as the cost of sequencing continues to drop, but, currently, genotyping hundreds of cultivars using WGS is not a realistic goal for most research groups. Our aim is to systematically compare both methodologies applied to olive genotyping and summarise any possible recommendations for the geneticists and molecular breeders of the olive scientific community. In this work, we used a selection of 24 cultivars from an olive core collection from the World Olive Germplasm Collection of the Andalusian Institute of Agricultural and Fisheries Research and Training (WOGBC), which represent the most of the cultivars present in cultivated fields over the world. Our results show that both methodologies deliver similar results in the context of phylogenetic analysis and popular population genetic analysis methods such as clustering. Furthermore, WGS and GBS datasets from different experiments can be merged in a single dataset to perform these analytical methodologies with proper filtering. We also tested the influence of the different olive reference genomes in this type of analysis, finding that they have almost no effect when estimating genetic relationships. This work represents the first comparative study between both sequencing techniques in olive. Our results demonstrate that the use of GBS is a perfectly viable option for replacing WGS and reducing research costs when the goal of the experiment is to characterise the genetic relationship between different accessions. Besides this, we show that it is possible to combine variants from GBS and WGS datasets, allowing the reuse of publicly available data.

Keywords: *Olea europaea* L.; olive; genotype by sequencing (GBS); single-nucleotide polymorphism (SNP); whole-genome sequencing (WGS); reference genome

1. Introduction

Olive tree (*Olea europaea*, ssp. *europaea*, var. *europaea*) is a member of the *Oleaceae* family which has an estimated 600 species of mostly small trees and shrubs [1,2]. Within the genus *Olea*, there are around 35 species and subspecies classified in three subgenera,

Olea, Paniculatae, and *Tetrapilus* [1,3], *Olea europaea* being the most widely cultivated and economically important species [1,4]. It is a long-lived, outcrossing species of fruit tree native to the Mediterranean basin. Nevertheless, its popularity as a commodity has extended its cultivation to other areas such as the coast of California (United States), the central coast of Chile, southern Africa, southwestern Australia, and Asia [1]. All subspecies are diploid (2n = 2x = 46), with the exception of two inter-subspecies polyploids [5]. As one of the world's oldest crops, the long history of cultivation and trade has made *O. europaea* culturally and economically significant to many countries in the Mediterranean basin [6]. The wild ancestor was domesticated around 6000 BC in the eastern region of the Mediterranean, but it has since spread across the world [6–9]. Though the exact origins of its domestication and distribution are unclear, it was likely spread east to west by humans through migration and the trade routes of the Levant area of the Mediterranean, contributing to its genetic diversity [10–12].

Over the past 50 years, agricultural intensification and global economics have contributed to a shift towards reliance on a small subset of high-performance cultivars, placing greater reliance on a limited supply of genetic resources [13–15]. Therefore, ex situ curation is becoming increasingly necessary for the maintenance and understanding of the genetic resources available for future breeding programs of insertion in the fields of new cultivars and their cultural conservation. To this end, in 1994 the International Olive Council (IOC) began to establish ex situ collections in national and international germplasm banks. Currently, there are three international germplasm banks located in Córdoba, Marrakech, and Izmir, as well as 19 national collections [14,16]. Since their establishment, limited molecular analysis has been carried out on olive germplasm banks to understand the genetic diversity, establish core collections, or suggest progenitors for future breeding programs.

There are around 1200 native cultivars, 3000 synonyms [8], and currently there are 5 published *Olea europaea* assemblies comprising 3 different cultivars, *Olea europaea* cv. 'Farga' (Oe6/Oe9) [17,18], *Olea europaea* cv. 'Picual' (Oleur0.6.1) [13] and *Olea europaea* cv. 'Arbequina' (Oe_Rao) [19], and a purported wild variety *Olea europaea ssp. sylvestris* (Oe451) [20]. Due to the prohibitive cost and effort involved in whole-genome sequencing, it was once an endeavour only available to consortia and solely focused on model organisms. However, the rapidly evolving sequencing technologies from companies such as Illumina, Pacific Biosystems, and Oxford Nanopore have consistently surpassed Moors law [21], allowing even small labs to afford to sequence their own species of interest. This affordable access to rapid high-throughput sequencing has been revolutionary for many areas of biology, with applications in de novo genome assembly, genotyping, gene–trait associations, metagenomics, transcriptomics, and epigenetics [22,23]. Given the economic and cultural importance of olive, it is not surprising that we have seen so many olive genome-sequencing projects in a short space of time. Each of the available olive genome assemblies uses different sequencing technologies, or a combination of methods. It is necessary when choosing a sequencing approach to consider the application of the genome, as the different types of errors, error rates, and biases that can come from a particular methodology will affect the overall quality and completeness of the assembly [24]. Furthermore, the different choices of assembly tool and pipeline will impact on the contiguity, accuracy, and handling of repeat regions in highly polyploid species [25].

Genetic profiling in many crops, as well as the analysis of genetic variation within and between their populations, has been achieved using cheap and effective biochemical and DNA markers such as random amplified fragment polymorphic DNA (RAPDs) [26,27], amplified fragment length polymorphisms (AFLPs) [28,29], SSRs [10,15,30,31], and single-nucleotide polymorphisms (SNPs) [32–34]. A sequence assembly of any quality is not required for many projects such as genetic profiling, establishing genetic relatedness, QTL mapping, or to perform a GWAS, although access to a complete and correctly annotated genome assembly provides the best account of individual genome variation and provides more information, increasing the potential resolution when using methods capable of recovering a higher density of variants, such as genotype by sequencing (GBS) [35], or whole-genome sequencing (WGS).

Both GBS and WGS could be used to call SNPs but differ massively in terms of missing data and their cost effectiveness. The main difference is that GBS is a reduced-representation approach to sequencing, that while quick and cost-effective, results in much more missing data due to its DNA fragmentation step. GBS uses restriction-site-specific digestion enzymes to fragment DNA samples. The DNA fragments then have unique barcode sequences ligated to the ends of the DNA fragments before fragment size selection is performed. The primary advantage of GBS is that by assigning sample-specific barcoded adapters, it is possible perform multiplexed sequencing in a single Illumina flow-cell lane for a large number of samples [35], making it much more cost effective than WGS. The number of SNPs that can be identified from within a WGS dataset can be significantly more than with GBS; however, this level of resolution is not always necessary in genetic-linkage-based research [36]. GBS has already been used in olives to generate genetic maps [20,37,38], study the diversity, and perform association analysis [39]. While the GBS library preparation itself is relatively simple, demultiplexing of the raw data is required to process the samples. This step can add extra difficulty for any researcher not already familiar with data processing.

It is likely that as the cost of sequencing and data processing continues to fall, we will see even more cultivar genomes assembled using differing technologies and assembly pipelines. It then becomes important to be able to assess the quality and functionality of a genome in order to choose the right assembly for a project's goal. Several tools and methods already exist to estimate genome completeness and contiguity [40]; however, the potential impact and bias that may arise from a reference genome's genetic background is not fully understood, nor to what extent this issue may affect different types of analysis.

In the context of a typical population genetics study, we wanted to understand if the genetic background would have a significant impact on the interpretation of group clustering and genetic relationships. Given the difference in data production and cost, we asked how comparable GBS datasets are to WGS datasets for this mode of research. Furthermore, we assessed the practicality of combining WGS and GBS datasets (Table 1), as this would allow for unrelated sequencing projects to access and combine publicly available data using two very different genotyping methods.

Table 1. List of *Olea europaea* cultivars sequenced using GBS and available WGS data.

Sequenced Cultivar	Country of Origin	Country Code	Available Data
Klon-14-1812	Albania	ALB	GBS/WGS
Kalamon	Greece	GRC	GBS/WGS
Koroneiki	Greece	GRC	GBS/WGS
Mastoidis	Greece	GRC	GBS/WGS
Mavreya	Greece	GRC	GBS/WGS
Myrtolia	Greece	GRC	GBS/WGS
Mari	Iran	IRA	GBS/WGS
Grappolo	Italy	ITA	GBS/WGS
Leccino	Italy	ITA	GBS/WGS
Arberquina	Spain	SPA	GBS/WGS
Manzanilla de Sevilla	Spain	SPA	GBS/WGS
Manzanillera de Huercal Overa	Spain	SPA	GBS/WGS
Menya	Spain	SPA	GBS/WGS
Morrut	Spain	SPA	GBS/WGS
Picual	Spain	SPA	GBS/WGS
Piñonera	Spain	SPA	GBS/WGS
Temprano	Spain	SPA	GBS/WGS

Table 1. *Cont.*

Sequenced Cultivar	Country of Origin	Country Code	Available Data
Verdial de Velez-Malaga-1	Spain	SPA	GBS/WGS
Abbadi Abou Gabra-842	Syria	SYR	GBS/WGS
Abou Kanani	Syria	SYR	GBS/WGS
Abou Satl Mohazam	Syria	SYR	GBS/WGS
Barri	Syria	SYR	GBS/WGS
Majhoj-1013	Syria	SYR	GBS/WGS
Uslu	Turkey	TUR	GBS/WGS
Removed Cultivars			
Chemlal de Kabylie	Algeria	DZA	GBS
Megaritiki	Greece	GRC	GBS
Shengeh	Iran	IRA	GBS
Barnea	Israel	ISR	GBS
Frantoio	Italy	ITA	GBS
Forastera de Tortosa	Spain	SPA	GBS
Llumeta	Spain	SPA	GBS/WGS
Picudo	Spain	SPA	GBS/WGS
Jabali	Syria	SYR	GBS
Maarri	Syria	SYR	GBS
Majhoj-152	Syria	SYR	GBS
Dokkar	Tunisia	TUN	GBS/WGS

2. Results and Discussion

2.1. Genome Assembly Comparison

In this study, we used five publicly available *Olea europaea* subsp. *europaea* assemblies (Table 2) comprising three different cultivars, 'Farga' (Oe6/Oe9) [17,18], 'Picual' (Oleur0.6.1) [13], and 'Arbequina' (Oe_Rao) [19], and a purported wild accession, *Olea europaea* subps. *sylvestris* (Oe451) [20]. Our full list of GBS samples and WGS samples is listed in Table 1. We removed 12 GBS samples from the initial 36-sample dataset that either failed to pass quality controls or for which no WGS data were available. The analysis was carried out on the remaining 24. The selection process is described in more detail in Section 4.5. These samples likely failed during the library preparation, although, there could be many reasons behind this problem. One of the most common is related to the quality of the DNA. Low DNA quality (e.g., due to impurities in the DNA extraction) reduces the efficiency of the restriction enzymes and leads to a partial digestion and reduction in the fragment population. Nevertheless, the low yield for the samples that failed (e.g., Barnea) indicates a bias in the amount of template used in the pool. Because the libraries were performed by an external service, it is difficult to assess where the problem was, but our guess is that it was related to the amplification step of some samples during the library preparation [35]. This would be an issue for a typical study of olive cultivar relatedness and would therefore need to be repeated. Fortunately, in this case the remaining samples were sufficient to compare GBS performance against WGS and the impact of assembly bias.

The assembly of GCA_002742605.1 (Oe451) used a whole-genome shotgun sequencing approach with the Illumina HiSeq 2000 platform to sequence *Olea europaea* var. *sylvestris* (wild olive). The assembly was performed using SOAPdenovo to generate a genome coverage of 220.0× [20]. For Oleur0.6.1, *Olea europaea* cultivar 'Picual' was selected as the genetic background [13]. The assembly process integrated Illumina HiSeq2500 and PacBio RSII sequencing to improve gap filling. *Olea europaea* cultivar 'Farga' based GCA_900603015.1/Oe6 [17] was assembled from Illumina HiSeq2500 with a genome coverage of 380×. Oe9 is an updated version of Oe6 [18], where a genetic map was used to anchor scaffolds to chromosomes. The most recently published is the *Olea europaea* cultivar 'Arbequina' (GWHAOPM00000000/Oe_Rao; NGDC) [19], which used Oxford Nanopore long-read sequencing and Hi-C data to construct chromosomes. The quality and completeness of each assembly was assessed before read mapping, SNP calling, and further analysis. Our comparison of genome assemblies was based on four metrics, (1) contiguity stats, (2) gene space completeness, (3) a k-mer completeness assessment using Merqury, and (4) the LTR Assembly Index (LTR) that evaluates the completeness of the genome using LTR retrotransposon elements (Table 2).

Table 2. Genome assembly and contiguity statistics.

Cultivar	**Wild Olive**	**Farga**	**Farga**	**Picual**	**Arbequina**
Assembly Stats	**Oe451 GCA_002742605.1 (Unver et al., 2017)**	**Oe6 GCA_900603015.1 (Cruz et al., 2016)**	**Oe9 GCA_902713445.1 (Julca et al., 2020)**	**Oleur0.6.1 (Jiménez-Ruiz et al., 2020)**	**Oe_Rao GWHAOPM00000962 (Rao et al., 2021)**
Chromosome assembly	Yes	No	yes	no	yes
Assembly size (Gb)	1.14	1.32	1.31	1.68	1.1
Scaffolds	41,256	11,038	9753	9174	962
Longest seq (Mb)	46.03	2.58	36.44	4.14	68.07
Shortest seq (bp)	452	500	500	1017	10,772
Average seq length (Kb)	27.69	119.47	135.00	183.12	1146.54
N90, number of seq	3410	3099	2019	4503	320
L90 (Kb)	23	111	116	86,918	279,924
N50, number of seq	23	901	162	1145	11
L50 (Mb)	12.57	0.44	0.73	0.41	42.60
% BUSCO complete	85.9	96.6	96.6	96.6	93.4
% BUSCO duplicated	13.4	23.6	23.6	51.5	18.3
% BUSCO fragmented	5.9	1.9	2.0	1.9	2.1
% BUSCO missing	8.2	1.5	1.4	1.5	4.5
Merqury completeness	72.62	78.39	78.73	89.41	NA
Merqury QV	43.00	34.78	35.68	33.04	NA
Merqury error	5.00867×10^{-5}	0.00033	0.00027	0.00050	NA
LTR_Retriever LAI	4.14	5.10	4.34	11.52	8.80

Assembly stats generated using custom script. Gene space completeness assessed using BUSCO v5 with eudicot_db10 dataset. Completeness, repeat regions, and sequencing data incorporation evaluated with Merqury v1.3 and Illumina short-read-derived k-mers. LTR Assembly Index (LAI) was estimated using the LTR_Retriever tool v2.9.0 with the default parameters.

With respect to (1), assembly contiguity statistics were generated using a custom script (see Section 4.4). First, we note that total assembly size varied between genetic backgrounds; Oleur0.6.1, at 1.68 Gb long, was by far the largest assembly, and over 360 Mbs longer than both Oe6 and Oe9 (Table 2). Indeed, it was over 500 Mbps longer than Oe451 or the 'Arbequina' genome (Oe_Rao). Jiménez-Ruiz et al. proposed that the bigger size of the 'Picual'-variety genome could be explained by the presence of a large number of duplicated DNA fragments coming from a very recent partial genome duplication event or artificially introduced repeat regions from assembly issues [18]. This comparison is interesting to observe because we are comparing the shotgun approach (Oe451, Oe6 and Oe9, Oleur0.6.1), where the genome is fragmented into reads of 250–800 bp [20], to sequencing with Pacbio first-generation RSII long reads (Oleur0.6.1), and the Oxford Nanopore third-generation long-read approach where reads are commonly 10–30 kb [41]. Shotgun approaches are very cost effective, but the process of sequence reassembly is more complicated than with long reads. The short reads make it difficult to correctly assemble repeat regions, particularly tandem repeats. This can also cause issues when using a genetic map to anchor to chromosomes, as the shorter read may not span a large enough section of the sequence to contain genetic markers. The trade-off is that on a per read basis, short reads are currently still more accurate and cheaper than long reads [41]. However, when only using short reads for an assembly, greater coverage is required, and this increases the number of errors in the dataset. Error correction and the filtering of low-quality reads is therefore an important step in the genome. We noted that Oe451, according to the Merqury assessment, had the highest QV score even with a 220× coverage, indicating careful management of the raw data during assembly. It is important to remember that each assembly will have its own unique set of errors introduced by sequencing or assembly issues, and thus any assembly should not be considered as a definitive sequence but rather, as stated in [42], only a working hypothesis.

Next, we checked (2) gene space completeness using Benchmarking Universal Single-Copy Orthologs (BUSCO) [43]. BUSCO genes are a set of ancestrally conserved genes used to estimate how complete a genome assembly is. Highly fragmented assemblies may contain greater percentages of missing or fragmented genes. The number of duplicated orthologs may be used as an indicator of possible errors that can occur when an assembly tool mistakenly assembles, or fails to assemble, reads into repeat regions. Nevertheless, polyploidy events and partial duplications may also lead to an increase in the number of duplicated genes detected by BUSCO, so often a deeper analysis needs to be carried out in order to distinguish a biological cause from a technical problem. In the publications for each of the olive tree genomes, BUSCO was used to evaluate assembly quality; however, the results were not comparable in terms of gene set database or BUSCO software version used. Thus, we ran BUSCO v.5 on all assemblies, using the eudicot_db10 gene set, which contains 2326 eudicot-specific genes. The results are different from those in the publications of the genomes used; however, this can be explained by the fact that the gene dataset's specificity can vary greatly depending on which is used, and older versions of the gene datasets may be less complete. Using the eudicot_db10 dataset, we obtained an identical completeness score of 96.6% for Oleur0.6.1, Oe6, and Oe9, which indicates that a high proportion of the core gene space was captured by these assemblies [44].

In addition to a recent genome duplication event in 'Picual', there are signs of at least two older whole-genome duplications in Olive [20], so it is unsurprising to see high levels of duplicated genes here. However, the 'Picual' genome has 51.5% duplicated genes, more than twice any other assembly. This could be an indication of a technical error such as a failure in the consensus calling due to high levels of heterozygosity, or in this case, there is also evidence of a biological origin for a recent genome duplication. BUSCO gene sets cannot have taken into account very recently uncovered duplications, and so this high duplication percentage in Oleur0.6.1 is difficult to evaluate. The number of missing BUSCO genes was low in the 'Picual' and 'Farga' assemblies (1.4–1.5%), as was the percentage of fragmented genes (1.9–2%), indicating an overall high level of completeness. Oe451 scored the lowest using this dataset, with a completeness score of only 85.9, an indication that a large portion of the genome is still unassembled despite its pseudo chromosomes; this may be an issue for use in particular studies, such as synteny, or the discovery of candidate genes by QTLs, but may not be an issue for SNP calling as we had intended if the GBS sites are not in the unassembled regions. Oe_Rao scored 93.4 for completeness and had a lower percentage of duplicated genes compared to the 'Farga' and 'Picual' assemblies.

We assessed each of the assemblies using (3) Merqury, a k-mer-based method able to evaluate the quality and completeness of a de novo assembly without the need for a reference genome. Merqury uses a similar method to KAT in which high-quality sequencing reads are decomposed into k-mers datasets, then the k-mer sets are compared to the genome assembly. Merqury summarises the quality assessment with two values: a completeness score that measures the completeness of the assembly based in the k-mer populations of the assembled and the unassembled reads, and a phred-scaled consensus quality (QV) score that measures the error produced during the haploid sequence consensus calling of the assembly. Additionally, Merqury's copy number spectra plots (Supplemental Figure S1) allow for a visual inspection of unassembled reads and artificial duplications [45]. The evaluation of Oe_Rao could not be carried out correctly as Merqury requires short reads and the Oxford nanopore reads used in the assembly are not compatible with the process, and no Illumina reads were publicly available at the time of this analysis. The two 'Farga' assemblies and the 'Picual' assembly had an identical BUSCO gene completeness score, yet, using Merqury, the overall completeness of the 'Farga' and 'Picual' assemblies was vastly different. This indicates a high amount of reads which were never used in the final 'Farga' assemblies (Oe6/Oe9). Indeed, in the Supplemental Figure S1 plots Oe6 and Oe9, we can see that this is the case. This result may be partially explained by the co-sequencing of the fungus genome *Aureobasisium pullulans* with the olive samples, which led to an assembly of 18 Mb [17], but the K-mer multiplicity indicates that there was also an important proportion of the repetitive content unassembled. However, these three scored an equally high QV, indicating highly accurate consensus calling during contig assembly. Discounting the Oe_Rao results, Oe451 scored the lowest in terms of completeness and had the most k-mers found only in the assembly, but had the highest QV score. Sequencing error types and error rates vary with the sequencing technology [36,41,46]; these may cause contig misassembles and scaffolding, and thus affect the reliability of a genome assembly for use in the development of genotyping markers, breeding programs, or population studies. A good example of the impact of the genome completeness on olive genetic studies can be found in Kaya et al.: 51% and 75% of the SNPs with strong association signals were mapped to the Oe451 and Oe6 reference genomes, respectively. Although these percentages are not correlated with the estimation of the completeness from Table 2, they are a good indication that the quality of the genome can influence the usability in other analyses.

Finally, the assemblies were evaluated using the LTR Assembly Index (LAI). Genome assemblies based on short-read sequencing technologies such as Oe451, Oe6, and Oe9 presented lower values than the assemblies based on long-read sequencing technologies, Oleur0.6.1 and Oe_Rao, as was expected and previously described in the use of LAI for the assessment of genome quality [47]. It is interesting to note that the updated version of Oe6, Oe9, has a lower LAI (4.34 compared with 5.10), meaning that some of the transposable element genome space was lost or fragmented during the improvement. From all the genome assemblies used, only the Oleur0.6.1 had a LAI value over 10, such that it is considered a standard value for high-quality genomes with a good contiguity.

Because every sequencing project used a different technological approach, genetic background, and assembly methodology, there is necessarily a great deal of difference between them and their overall quality. While the limitations of short read length can impact the handling of repeat regions during scaffolding with a single short-read mapping at the incorrect or multiple regions, long-read technologies such as Oxford Nanopore and Pacbio have much higher error rates. Further, the choice of assembly tool, corrections, and the use of an alignment tool with and without available references can impact the quality of a genome assembly. Considering the information collected in Table 2, Oleur0.6.1 appears to be the most complete assembly with the highest QV score of the four we could test. However, it remains unclear if the high percentage of duplicated genes is an error or true genome duplication. Furthermore, this genome is not yet anchored to chromosomes, limiting its use in some studies.

2.2. *Effect of Reference Genome Choice on Population Analysis*

We explored the potential effect genome selection could have on a typical genetic diversity study such as a population analysis by mapping the same GBS reads to each of the available reference genomes, and performed the same population analysis on all the resulting SNP datasets. This analysis began with a quality control assessement of our dataset, which consisted of unprocessed reads from all 36 of our selected Olea europaea cultivars, amounting to over 326 million raw reads. After pre-processing, mapping, and SNP calling (see Section 4.5 and Supplemental Figure S19) using three different genomes (Sylvestris/Oe451, Farga/Oe6, and Picual/Oleu r0.6.1), samples that had failed at one or more quality control checkpoints on all three genomes were removed from the rest of the analysis (see Sections 4 and 4.5) (Supplemental Table S1). Removed samples are listed in Table 1. As mentioned previously, the GBS samples may have failed during library preparation, or sequencing and would normally need to be repeated; however, 24 samples are sufficient for the focus of this work. The remaining 24 samples were then additionally mapped to the two remaining genomes (Farga/Oe9 and Arbequina/Oe_Rao), before variant calling, SNP filtering, and population analysis.

2.2.1. Analysis GBS Read Mapping and Variant Calling

Considering only the remaining 24 high-quality samples, the percentage of mapped reads did not vary much between assemblies; indeed, there was only a 2% difference separating the highest and lowest mapping genomes (Table 3). The average number of sites for each sample also showed very little variation, except for Oleur0.6.1, which had around 100,000 more sites, likely due to the increased number of duplicated genes in this assembly. It is important to highlight that a substantial difference was found in the number of variants called. Both Oe451 and Oe6 had double the number of variants called compared to Oe_Rao, and almost six times that of Oleur0.6.1. However, by far the most SNPs identified were from the improved 'Farga' genome, Oe9 with 23.7 million variants (Table 3). In terms of variants per loci, Oleur0.6.1 and Oe9 had, on average, almost triple the number of Oe451, Oe_Rao, and Oe6. The total number of SNPs called after filtering (see Section 4.5) was similar for all genomes. This massive drop off from Oe9 was most significantly explained by the 10,000 Kb thinning and 10,000 minimum quality score filtering steps. These two alone removed ~97% of total SNPs. Oleur0.6.1 was notable for having been left with half as many SNPs as the 'Sylvestris' and 'Farga' genomes. Perhaps the Oleur0.6.1 assembly contained several repeat regions which were incorrectly collapsed or, alternatively, reads may have been incorrectly assigned as a repeat(s), both of which could increase or decrease the number and frequency of SNPs, as multiple alleles from the same locus might be mistakenly identified as having come from different loci or vice versa. As can be seen in Table 2, there is indeed variation in the number of duplicated BUSCO genes and the k-mer duplications found across all of these genomes, indicating a difference in the assembly of repeat regions of each of the tested genomes. In such cases, it might be expected to also see variation in the levels of heterozygosity. However, heterozygosity per site was not significantly different, ranging from 0.31 to 0.36 (Table 3), making it difficult to interpret the source of this phenomenon. To test if collapsed regions were indeed the cause of this SNP count variation, we extracted the allele frequency of SNPs from three random samples ('Grappolo', 'Manzanilla de Sevilla', 'Piñonera') called from three of the genomes (Oe451, Oe6, Oleur0.6.1) and compared the frequency of alleles between 0.25 and 0.75 (Supplemental Figure S2). Allele frequencies for an individual different from 0, 0.5, and 1, such as 0.25 and 0.75, may be an indicative of the collapse of four copies into one during the genome assembly. Once adjusted by percentage, we observed similar profiles, with most alleles tending towards 0.25 and a small peak around 0.5. However, from our sub-sampling we did not see any significant difference and so it remains unclear why using Oleur0.6.1 resulted in half the number of SNPs and Oe451.

Table 3. GBS and WGS read mapping and variant calling.

Cultivar	*O europaea* var. *sylvestris*	Farga	Farga	Picual	Arbequina
	Oe451 GCA_002742605.1 (Unver et al., 2017)	Oe6 GCA_900603015.1 (Cruz et al., 2016)	Oe9 GCA_902713445.1 (Julca et al., 2020)	Oleur0.6.1 (Jiménez-Ruiz et al., 2020)	Oe_Rao GWHAOPM00000962 (Rao et al., 2021)
Total GBS reads mapped (M)	294.29	298.86	299.56	300.28	297.58
Total WGS reads mapped (M)	8767	NA	NA	8789	NA
% of total GBS reads	94.4	95.8	96.0	96.3	95.4
% of total WGS reads	100.4	NA	NA	107.14	NA
Avg number of sites GBS	325,680	350,581	348,652	432,644	340,715
Avg number of sites WGS	493,618	NA	NA	1,073,623	NA
GBS SNP before filtering	7,722,425	7,415,201	23,651,384	4,401,892	1,346,358
GBS SNP after filtering	13,343	12,507	13,410	7023	10,177

Table 3. *Cont.*

Cultivar	*O europaea* var. *sylvestris*	Farga	Farga	Picual	Arbequina
	Oe451 GCA_002742605.1 (Unver et al., 2017)	**Oe6 GCA_900603015.1 (Cruz et al., 2016)**	**Oe9 GCA_902713445.1 (Julca et al., 2020)**	**Oleur0.6.1 (Jiménez-Ruiz et al., 2020)**	**Oe_Rao GWHAOPM00000962 (Rao et al., 2021)**
GBS unfiltered SNPs per site	1.01	0.90	2.91	2.39	0.89
GBS filtered SNPs per site	0.02	0.02	0.02	0.01	0.02
Average Het/site GBS	0.36	0.34	0.35	0.31	0.35
Average Het/site WGS	0.42	NA	NA	NA	0.34
WGS SNP before filtering	144,579,296	NA	NA	128,172,089	NA
WGS SNP after filtering	119,190	NA	NA	114,169	NA
WGS/GBS SNP before filtering	150,952,260	NA	NA	11,679,534	NA
WGS/GBS SNP after filtering	9537	NA	NA	4004	NA

2.2.2. GBS Population Structure

The filtered SNPs obtained from each genome mapping were analysed using FASTSTRUCTURE (DARTR), ADMIXTURE (LEA), discriminate analysis of principle components (DAPC), and principal component analysis (PCA) to identify the distinct relationships among the cultivars (Supplemental Figures S3–S17).

In two previous studies, there were two distinct clusters identified among cultivars and a third cluster containing wild relatives [13,18]. In this work, the results from FASTSTRUCTURE identify two groups (K = 2, with cross entropy errors ranging 0.7 to 0.8) as being the most probable clustering for all GBS and WGS datasets (Supplemental Figure S8). As wild olive samples were not included in the analysis, two groups are in agreement with the previously published data [13]. When comparing the results of the model-based clustering from ADMIXTURE and genetic-distance-based PCA clustering, it was noted to be in agreement also with FASTSTRUCTURE. The first three principal components of the PCA only explained ~30% of the variation. However, it was sufficient to see a clear separation of the different clusters, a result observed constantly throughout all our SNP datasets (Supplemental Figures S3 and S4). In all cases, the first principal component (PC1) (~14%) separated the 24 cultivars into a group of 13 that could be described as primarily composed of eastern Mediterranean cultivars ('Barri', 'Abou Kanani', 'Abou Satl Mohazam', 'Majhol-1013', 'Temprano', 'Verdial de Velez-Malaga-1', 'Uslu', 'Kalamon', 'Morrut', 'Mari', 'Picual', 'Manzanilla de Sevilla', 'Abbadi Abou Gabra-842') and a group of 11 mostly northern Mediterranean cultivars ('Mastoidis', 'Klon-14-1812', 'Grappolo', 'Mavreya', 'Piñonera', 'Leccino', 'Myrtolia', 'Menya', 'Manzanillera de Huercal Overa', 'Koroneiki', and 'Arbequina'), reflecting the complex history of olive domestication [11,48], and aligning with previously published results [13,18]. ADMIXTURE at K = 2 produced a similar grouping as the PCA (Supplemental Figures S5–S7). The levels of estimated admixture in each individual cultivar did not appear to vary depending on the assembly used. It is also interesting to note the near identical results regardless of the number of SNPs. Oleur0.6.1 had almost half the number of filtered SNPs as Oe451 or Oe9, but it appears that this still provides sufficient resolution to evaluate genetic relatedness using these analyses.

In general, the DAPC posterior membership was also estimated to be the same regardless of assembly used during mapping. We could see some variation in group membership when samples were compared by country of origin. There was largely agreement while comparing all genomes; however, we saw some discrepancies, such as in the Oe6, Oe9, and Oe_Rao genomes, which gave a near identical result in which the cultivar 'Koroneiki' of Greece showed a much higher probability of belonging to the Syrian genetic group. Indeed, this was more in agreement with the comparison of principal components 1 and 2, seen in all datasets belonging to the Syrian genetic group.

We calculated, for each dataset, genetic distances and constructed neighbor-joining distance trees using the R package POPPR (Supplemental Table S4). We can see in all cases there are two clear operational taxonomic units (OTUs), or clusters, defined in each tree (Figure 1) with a smaller third cluster of cultivars, which is more variable depending on the reference genome used. Samples in the third group were those which had higher levels of admixture (Supplemental Figures S5 and S6) and were more difficult to assign. However, the two large groups were in agreement with the eastern Mediterranean and northern Mediterranean clusters identified in the PCA (Supplemental Figure S3). The topology of each grouping varied, but only slightly, and only within an out. There was no placement of individuals in another cluster; even those with greater admixture were either in a seperate group or with cultivars of the same region. The topography of the trees produced from Oe451 and Oe6 were near identical, and even closer matched than Oe6 to its updated Oe9 version. The bootstrapping values were high in all cases with only 2–3 nodes falling below 50 on any tree. Syrian and Italian cultivars were the most consistent in terms of clustering when using any other assemblies. The cultivar 'Mari' was seen in all trees to cluster with Syrian samples, but, interestingly, the DAPC results indicate this same grouping only when using a 'Farga' or 'Arbequina' genome. This might be due to some genetic relationship between the genomes and the 'Mari' sample, such as a possible introgression with some of the SNPs.

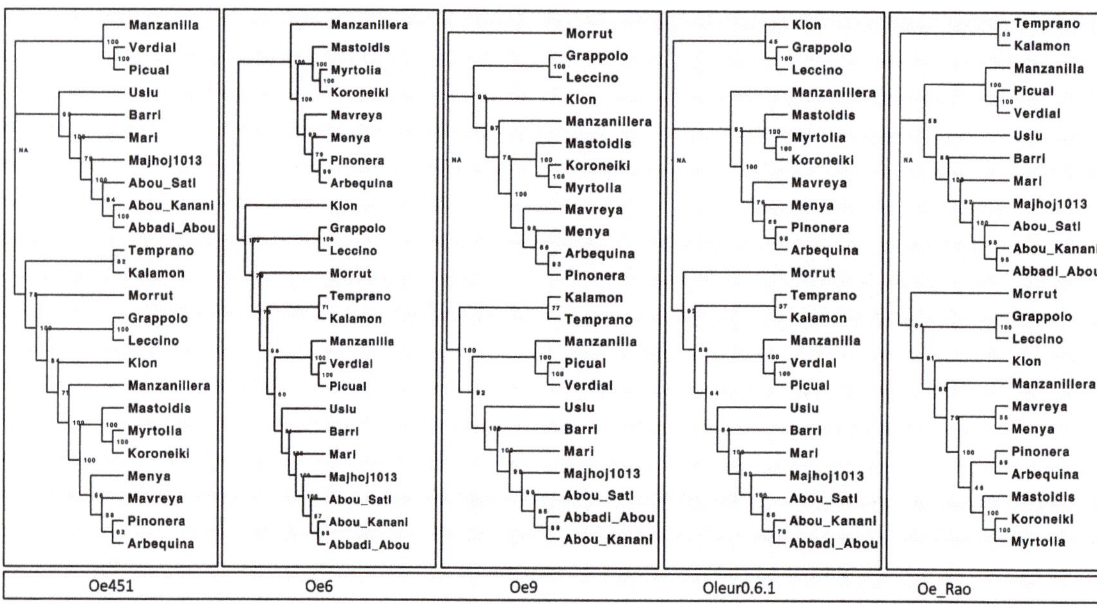

Figure 1. Neighbour-joining trees constructed using GBS data. Comparison of the neighbour-joining (NJ) trees constructed from SNPs called using GBS data mapped to each of the reference genomes. Using the R package Poppr, the genetic distance between each sample using Provesti's distance and NJ trees is plotted with a bootstrapping support of 100. The genome used to generate each is labelled above the corresponding tree.

Although there was great divergence in the number of SNPs associated with the use of each assembly, we saw little effect on the outcome of the population analysis. As observed in maize, reference genome selection can impact results [49,50]. Gage (2019) reported a reduction in the ability to robustly identify key loci of interest in a genome-wide association study (GWAS). Missing loci and inaccurate structural variations introduced during genome assembly may introduce reference bias, which may be more problematic in such cases. The impact of using any of the five selected assemblies in their current states appears to have been minimal in this population analysis, and indeed each reference genome provides confirmation of the results. However, it is still important to consider the vast differences in variants identified from each assembly and how they could impact other types of analysis. Using more than one reference assembly in this way can remove much of the bias introduced by many of the choices made during a genome sequencing project (see

Section 2.1) and is something that could be repeated in future olive breeding projects to improve the reliability of identified structural variations.

2.3. *Analysis of WGS Data*

As the population analysis using GBS data showed that the selection of different reference genomes can make a difference in the phylogenetic relationship of one sample to another, but not in the overall grouping of samples, and no significant difference was seen in the levels of estimated admixture, we wanted to see if this was also the case with WGS data. Since WGS would likely generate many more variants and sites, any small differences may be diluted or potentially exaggerated. To process the raw whole-genome sequencing (WGS) read set, we followed the same steps as used for the raw GBS read set (see Section 4.5).

The processed reads were mapped to two assemblies, the purported wild type (Oe451) and the genome that scored the highest overall in our analysis (Oleur0.6.1). First, looking at the number of reads mapped to each, we saw that the WGS read set had a much higher number of mapped reads than the GBS set, as expected (Table 3). In both cases, the percentage of processed reads that was mapped was over 100%. Some double mapping was detected but not enough to cause issue. Variant calling using WGS data produced significantly more variants than GBS; 144 million from WGS-Oe451 and 128 million called from WGS-Oleur0.6.1 before filtering. It was not surprising to have so many more variants called using WGS reads because of the lower amount of missing data [36], but it was interesting to note that when we compared the total number of variants called from Oe451 and Oleur0.6.1 with the GBS data, there was a difference of 3.3 million between the two datasets before filtering. The difference between GBS-Oe451 and GBS-Oleur0.6.1 was almost double after filtering, but there was only a ~10% difference between WGS-Oe451 and WGS-Oleur0.6.1.

We next ran the same population analysis script as previously used for the GBS datasets. The FASTSTRUCTURE, PCA, admixture, and DAPC results were again very similar regardless of the genetic background of the reference genome. As before, there was some difference in the topography of trees but overall, the same OTUs were reproduced. Bootstrapping values were more consistently higher using Oleur0.6.1. The results are near identical to the GBS results, suggesting that it was possible to merge GBS and WGS datasets.

Here, we also compared the genomes that produced the highest and lowest number of SNPs. It is worth noting that, again, despite the higher number of SNPs coming from Oe451, there appeared to be no difference in the outcome of the analysis when using any of the reference genomes or SNP datasets. The reasons for this might be that SNPs provide limited information and so a certain threshold, in terms of reliability and number, is needed for an SNP dataset to be useful. The quality of the SNPs called and filtered from each reference genome was sufficient, even with the lower number of SNPs from Oleur.0.6.1, to extract the same level of detail. We also must consider that while five reference sequences were used, the genetic background of one was a purported wild type and all the others were Spanish cultivars. Our results show no difference when using either sequence, but this may change if a cultivar from another region with a more distinct genetic history is used.

2.4. *Analysis of WGS/GBS Merged Data*

The ability of an SNP dataset to differentiate between groups can be quite low because of their biallelic nature [51] and so it could be expected that increasing the number of SNPs from WGS should significantly improve the power of differentiation. However, our results show that GBS and WGS data performed almost identically, despite a difference of roughly 10 times the number of SNPs called using WGS data (Table 3). Given the similarity of the population analysis results, we wanted to know if it was possible to combine variants from GBS and WGS datasets. This would allow for more collaborative opportunities and the supplementation of smaller datasets with already available public data.

The same two genomes (Oe451 and Oleur0.6.1) were chosen for this experiment. The vcf files of GBS-Oe451 and WGS-Oe45 were merged together using bcftools and a list of shared sites (see Section 4), as were GBS-Oleur0.6.1 and WGS-Oleur0.6.1. The merged WGS/GBS-Oe451 file consisted of over 150 million variants and after filtering, 9537 biallelic SNPs, less than 1% of the total variants. The WGS/GBS-Oleur0.6.1 merged file contained only around 11 million variants and less than 1% of total variants were high-quality SNPs. This is in stark contrast to the difference seen with GBS-Oe451 and GBS-Oleur0.6.1. We selected only the SNPs from shared sites and did not allow for any missing data, so this would explain the massive drop off compared with either WGS or GBS after only filtering.

Again, we ran the same population analysis script as with all previous SNP sets. FASTSTRUCTURE found the mostly likely number of clusters, based on lowest cross-entropy, to be K = 18–19 for WGS/GBS-Oe451 and K = 20 for WGS/GBS-Oleur0.6.1 (Supplemental Figure S8). This may be an overfit of samples due to there being two of each genotype (Supplemental Figure S11) and the WGS samples being more like their GBS counterpart than anything else, creating the highest probability clusters. However, we could see that the change in cross entropy decreased between possible clusters past K = 2 and K = 3. The sample clustering by PCA was near identical for the WGS/GBS-Oleur0.6.1 and WGS/GBS-Oe451 datasets (Figure 2). Importantly, they were similar to the GBS and WGS only analysis with all the reference genomes. We saw that GBS and WGS samples closely grouped together, forming the same clusters seen previously. The only notable exception was GBS Arbequina. It seems likely that this sample was mislabelled at some point in the process, as it is the only sample that does not pair up with its WGS counterpart. DAPC analysis was analogous to all previous results, with the same groups being identified. Analysis by NJ trees (Figure 3) revealed no major differences in terms of OTUs, but we observed that Oleur0.6.1 had some very low bootstrapping values and Oe451 had the better bootstrapping values, more closely resembling the WGS-only tree.

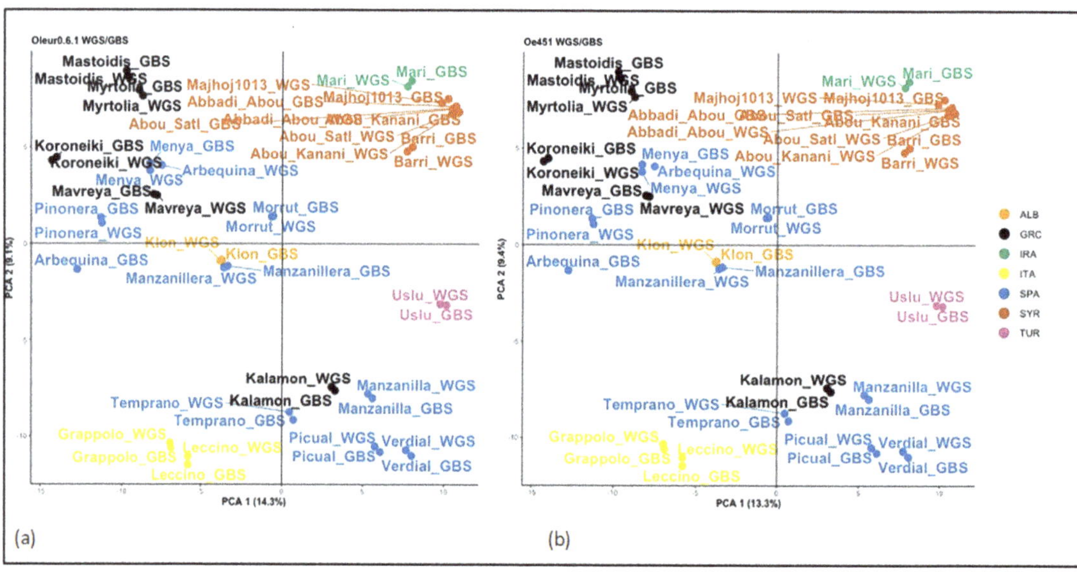

Figure 2. PCA clustering of WGS/GBS datasets. This shows the clustering of WGS and GBS merged samples. Principle components 1 and 2 are shown for each. (**a**) PCA of merged data with reference genome Oe451; (**b**) PCA of merged data with reference genome Oleur0.6.1.

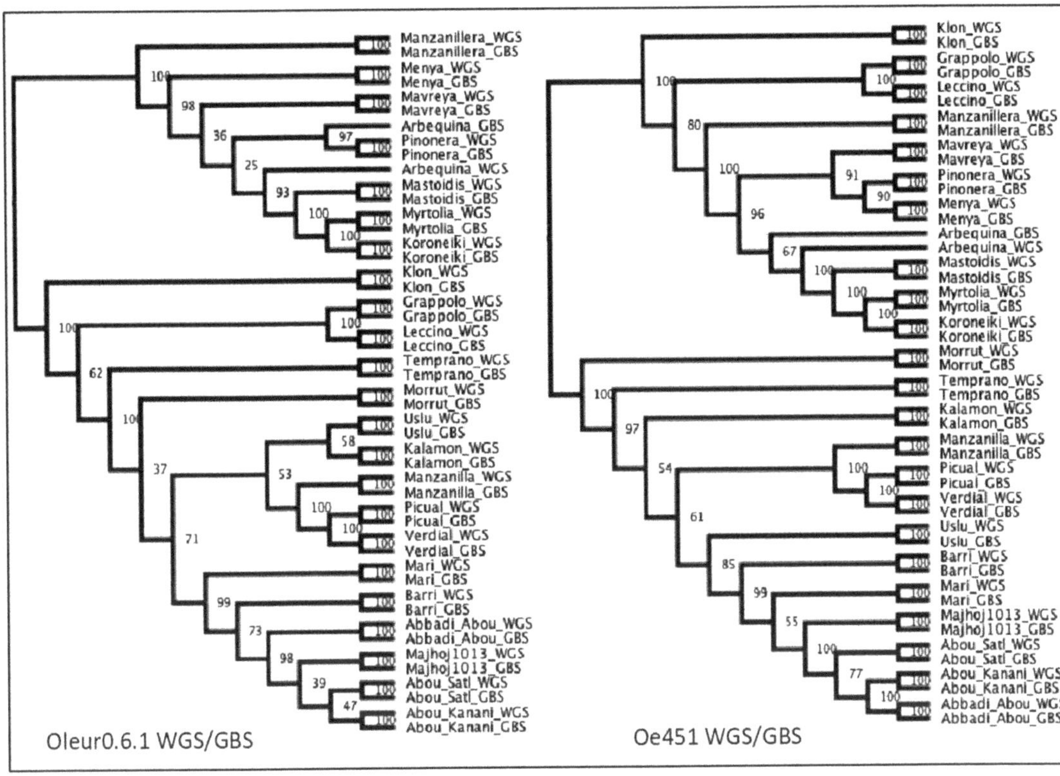

Figure 3. Neighbour-joining trees constructed from merged WGS and GBS data mapped to the Oleur0.6.1 and Oe451 reference genomes. Genetic distance between each sample using Provesti's distance and NJ trees plotted with a bootstrapping support of 100.

3. Conclusions

The results of this work indicate that GBS and WGS sequencing data are highly comparable for population structure analysis. Certainly, more information can be obtained from WGS, and great advances have been made to reduce its cost, but GBS remains much more affordable. For small sampling sizes WGS maybe be preferable, but a large-scale genotyping project can quickly become too expensive for most labs; we show that GBS is a cost-effective alternative capable of providing near identical results and identifying the same genetic relationships at a much lower cost. Further, the ability to successfully combine the two sequencing methods opens opportunities to mine data from a wider range of sources. This would also be a significant cost saving approach to consider for expanding existing olive tree WGS datasets and past genotyping projects, and allow for collaborations between research groups that have used either WGS or GBS genotyping methods. Although the sequencing cost is decreasing, it is not feasible to re-sequence hundreds of accessions to study the structure of the population or to identify some unknown accessions.

With respect to our comparative analysis between the different reference genomes used, the Oleur0.6.1 sequence may be the most accurately assembled, but lacking chromosomes can limit its application. However, we could not properly test the 'Arbequina' assembly using Mercury's best practices and therefore accurately compare it to the others in all aspects of our quality assessment. Any differences between the current existing genome assemblies are so small that we can say that for SNP-based genetic profiling at least, all are certainly suitable. Thus, the availability of genetic maps and chromosomes could potentially be a more important consideration when choosing which to use in future projects. Furthermore, we show that while the genetic background of the reference genome plays a role in the possible number of variants identified, there was little overall effect on population structure, genetic clustering, or analysis of genetic relationships. As more reference genomes of existing cultivars are rendered to the chromosome level, it might well be necessary to compare performance before choosing which to use. The current assemblies are of a purported wild type and three Spanish cultivars, so it is still possible that the genetic background may yet be an important factor if using cultivars from other regions as the reference.

Our results highlight the advantages of GBS while at the same time bringing to the table the possible limitations. GBS is a commonly used technique for other crops, but it has not been routinely implemented in olive breeding. Traditional breeding needs these types of tools to accelerate the development of new varieties able to face the important challenges olive production is currently facing.

4. Materials and Methods

4.1. Plant Material and Genomic DNA Extraction

In accordance with Belaj et al. (2012), and maximising the genetic diversity in a reduced number of genotypes, 36 olive tree cultivars were selected from the World Olive Germplasm Collection of the Andalusian Institute of Agricultural and Fisheries Research and Training (WOGBC) located in Córdoba Spain (Table 1). Total genomic DNA was extracted from fresh leaves using the Illustra DNA extraction kit Phytopure GE Healthcare (UK) in accordance with the protocol described in the manufacturer's instructions. To ensure high-quality DNA was used for sequencing, the purity was measured with a Qubit 2.0 Fluorometer Life Technologies NY (USA). DNA concentration was then normalised to 20 ng/µL. The DNA had a minimum 260/280 ratio of 1.8.

4.2. GBS Library Construction and Sequencing

The DNA were sent to DISMED, and the libraries were prepared by BGI. DNA samples were digested with ApeKI (New England Biolabs, Ipswitch MA) for 2 h at 75 °C, and T4 ligase (New England Biolabs, Ipswitch MA) was used to ligate to sticky ends, one of 36 unique "barcodes", and the "common adapter". Samples were incubated at 22 °C for 1 h and heated at 65 °C for 30 min. A set of 36 digested DNA samples, each with a different barcode adapter, were obtained. A total of 7 µL of each component of this set was combined in a unique sample and purified in a final volume of 50 µL with a commercial kit (QIAquick PCR Purification Kit; Qiagen group (Germany), according to the manufacturer's instructions. The result of this library was amplified in 50 µL, containing 5 µL of pool DNA fragments, 1× Taq Master Mix New England Biolabs (UK), and 12.5 pmol each of PCR primers, the sequences of which were: PCR primer: 5′AATGATACGGCGACCACCGAGATCTACACTCTTTCCCTACACGACGCTCTTCCGATCT and PCR primer: 5′ CAAGCAGAAGACGGCATACGAGATCGGTCTCGGCATTCCTGCTGAACCGCTCTTCCGATCT, containing complementary sequences for amplifying the fragments of DNA with ligated adapters. The PCR conditions were a primer step of 5 min at 72 °C; 98 °C for 30 s; 25 cycles of 98 °C for 30 s, 65 °C for 30 s, and 72 °C for 30 s; and a final extension step at 72 °C for 5 min. The library was purified as above (in a final elution of 30 µL) and 1 µL was used for the quality evaluation of fragment sizes. The library was considered suitable for sequencing if adapter dimers were minimal (~128 pb in length) and the majority of the other DNA fragments were between 170 and 350 bp. Paired-end sequencing of one 48-plex library per channel was performed on an Illumina HiSeq2000 Analyzer by BGI Genomes.

The sequences of the barcode adapter were: 5-ACACTCTTTCCCTACACGACGCTCTTCCGATCTxxxx and 5-CWGyyyyAGATCGGAAGAGCGTCGTGTAGGGAAAGAGTGT, where "xxxx" and "yyyy" are the barcode and barcode complement, respectively. The second, or "common" adapter sequence was shared among all samples and consisted of an

ApeKI-compatible sticky end: 5′-CWGAGATCGGAAGAGCGGTTCAGCAGGAATGCCGAG and 5′-CTCGGCATTCCTGCTGAACCGCTCTTCCGATCT.

4.3. WGS Library Construction and Sequencing

Raw sequencing data were obtained directly from the authors of Jiménez-Ruiz et al. (2020). Re-sequencing of all varieties was performed by 2 × 150 paired-end sequencing with Illumina HiSeq 4000, and all sequencing was carried out at the Duke Center for Genomics and Computational Biology (Durham, NC, USA). Raw data are available at NCBI BioProject ID: PRJNA556567.

4.4. Sequence Assembly Assessment

The five available assemblies used were *Olea europaea* subsp. *sylvestris* (version GCA_002742605.1/Oe451; NCBI), *Olea europaea* cultivar Picual (version Oleur0.6.1; https://genomaolivar.dipujaen.es/db/ (accessed on 2 February 2020)), the *Olea europaea* cultivar Farga (version GCA_900603015.1/Oe6; NCBI) [17] and its updated version (GCA_902713445.1/Oe9; NCBI) [18], and *Olea europaea* cultivar Arberquina (GWHAOPM00000000/Oe_Rao; NGDC) [19]. The same three forms of quality assessment were carried out on all the assemblies: contiguity, gene space completeness, and a k-mer-based evaluation.

Assembly and contiguity were calculated using a custom script; FastaSeqStats (https://github.com/aubombarely/GenoToolBox/blob/master/SeqTools/FastaSeqStats (accessed on 10 June 2013). N50 is a commonly used marker of sequence contiguity, counting the number of bases in the shortest fragments needed to span 50% of the genome. N90 is the same information at 90% of the total genome. Average sequence length provides similar information as the N50. L50 is the smallest number of contigs needed to cover 50% of the genome; L90 provides the same information at 90%.

Benchmarking Universal Single-Copy Orthologs (BUSCO) genes were used to evaluate the completeness of the assembly by searching for a list of known ancestrally conserved genes within the reference genome. All assemblies were assessed using BUSCO v.5 with the eudicot_db10 gene set containing 2326 eudicot-specific genes.

Merqury v1.3, (https://github.com/marbl/merqury (accessed on 10 January 2021) a k-mer-based method, was used to evaluate both quality and completeness. Using a similar method to KAT, high-quality sequencing reads were decomposed into k-mers datasets, then the k-mer sets were compared to the genome assembly. Merqury generates a k-mer dataset from the Illumina short-read sequencing data used in the genome sequence assembly. By decomposing the original sequencing reads into k-mers, Merqury can count how many times each k-mer appears in the assembly, as well as k-mers from the original Illumina reads not incorporated at all or that appear only in the assembly. These k-mer data were used to generate a completeness score and a phred-scaled consensus quality (QV) score, along with copy number spectra plots to visually inspect the assembly for unassembled reads and artificial duplications [45]. The Illumina short-read datasets were either downloaded from the same location as the assembly or kindly sent by the genome assembly curator. In the case of the 'Arbequina' assembly, only long-read sequencing was performed, which was unsuitable for use with Merqury.

LTR Assembly Index (LAI) was estimated using the LTR_Retriever tool v2.9.0 with the default parameters [47].

4.5. Read Processing, Mapping, Filtering, and Variant Calling

Raw reads were first processed and then mapped only to Oe451 to assess quality, using the pipeline described below and in Supplemental Material Figure S10. Samples with low SNP count can affect the overall number of SNPs available for analysis due to the percentage of missing data encountered during the VCF filtering (see Section 4 and Supplemental Figure S10). Before removing poorly performing samples, the full process was repeated with two other assemblies (Oleur0.6.1 and Oe6) to confirm the results (Supplemental Table S1). Total SNPs called from Oe451, Oe6, and Oleur0.6.1 after filtering were initially in the range 300–900. To increase this number further, low-quality samples were removed. The choice of samples to be included or excluded was based on performance at different stages of processing and the amount of missing data. To achieve this, the number of reads in each sample was counted and there were three clear groups, those with only 50,000 or fewer reads, a second group with 600,000 or fewer, and the majority of samples were within the range of 2–20 million reads. Samples with 50,000 or fewer reads were removed from further analysis, as this is too low to yield any useful information, along with any samples having less than 50% of the average number of sites per sample.

This reduced the sample set from 36 to the current 24, but significantly increased the number of total SNP loci shared across all samples. Sequence read mapping was then performed on all five genome assemblies to evaluate the influence that a particular assembly has on SNP discovery and on a population analysis using the following steps.

After sequencing, the GBS raw reads were demultiplexed by BGI Genomics. Next, the raw reads were processed with FASTQ_MCF v1.05 [52] to remove Illumina adapters and reads with a phred-scaled quality score of less than 30 and/or shorter than 50 bases. After trimming, the paired reads were aligned to each of the five genome assemblies; Oe451, Oe6, Oe9, Oleur0.6.1, and Oe_Rao, using BWA (Burrows–Wheeler Alignment Tool) v0.7.17-r1188t with default parameters. Prior to mapping, each reference genome was indexed using BWA. The output file of the mapping was in an unsorted SAM format, and these were converted to bam and sorted to save space and increase SNP calling efficiency using SAMTOOLS v1.7 [53]. Once sorted, the bam files were merged into a single bam file with BAMADDRG (https://github.com/ekg/bamaddrg (accessed on 14 April 2018).

Variants were called with FREEBAYES v1.3.1-16-g85d7bfc [54] using a custom script; MultiThreadFreeBayes, (https://github.com/aubombarely/GenoToolBox/tree/master/SNPTools/MultiThreadFreeBayes (accessed on 15 May 2018) this script allows FREEBAYES to run faster by using multiple threads on several scaffolds at the same time. Finally, the variant file (VCF) was filtered with VCFTOOLS v0.1.15 [55] using the following parameters: retain only biallelic SNPs, remove indels, a minimum read depth of 5 with a minimum mean depth of 20, a minimum SNP quality of 1000, no missing data, and an MAF of 0.05. Finally, SNPs were thinned to 1 per 10,000 Kb.

To merge the GBS and WGS datasets, we used the BCFTOOLS v1.7 [53] merge function. The WGS VCF file contained many more sites than the GBS, so a bed file containing all sites from the GBS dataset was supplied using the argument regions-file. This was carried out to reduce the final file size, as any site with missing data would eventually be removed during the filtering steps. After the files were merged, they were filtered using the same parameters as above.

4.6. Population Analysis

Each SNP dataset was analysed in the same way, with the same script to estimate population structure and genetic diversity from each of the datasets. The R script Olea_pop.R with commands and notations is available at https://github.com/frieljames/Olive_WGS_GBS (accessed on 1 April 2021). Supplemental Figure S19 shows an overview of the R programming pipeline. As a summary of the script, we cross analysed population structure in our datasets with the use of R and three different methods: (1) STRUCTURE [56], a Bayesian-based clustering method assuming Hardy–Weinberg equilibrium and linkage equilibrium between loci within populations, (2) discriminant analysis of principal components (DAPC) [57], which uses sequential K-means and model selection to infer genetic clusters, and (3) principal components analysis (PCA), a method that uses genetic distance to infer clusters. To begin, we used FASTSTRUCTURE [58], part of the DARTR package, a faster and more resource-efficient method of running STRUCTURE. The LEA package [59] performed a STRUCTURE analysis on the SNP data to infer the mostly likely genetic clusters based on allele frequency and clustering probability. ADMIXUTRE was estimated by running LEA's snmf functional analysis for the estimation of ancestral populations

(K) of 1–20. This estimates the admixture coefficient in the selected K range using sparse Non-Negative Matrix Factorisation algorithms. The output of this was used to estimate the admixture of groups at the largest K value. To perform a DAPC and a PCA, we used the R package ADGENETv.1.3-1 [60]. DAPC is different to STRUCTURE in that it does not rely on model assumptions or prior information; instead, it uses a multivariate method with sequential K-means and model selection to infer genetic clusters and assign individuals to clusters. To ensure that maximum variance was being used while attempting to avoid an overfit of the data, DAPC was performed using the suggested optimal number of principle components (PCs) for each dataset (Supplemental Figures S9–S11). These were identified as the number of PCs with the highest a-score, predicted by the optim.a.score() function, which selects an evenly distributed number of PCs in a pre-defined range, computes an a-score for each, and then interpolates the results using splines. For the PCA, the package ADGENET took a genlight object of SNP data to generate a genetic distance matrix for use in the PCA clustering.

Further estimation of possible ancestral populations was achieved with a neighbour-joining (NJ) approach, again based on allele frequencies. This was carried out with the R package POPPR [61]. The POPPR function aboot() allowed for the construction of a dendrogram with 100 bootstrapping support and obtained genetic distance using bitwise.dist, a method that calculates the fraction of different sites between samples equivalent to Provesti's distance. The tree was visualised in R with the package APE [62] using the function plot.phylo(). Afterwards, the tree was exported in the newick format to generate figures using FigTree v1.4.4 [63].

Finally, genetic diversity was calculated using the gl.basic.stats () function with the R package DARTR [64]. Selection and neutrality were estimated by calculating the number of segregating sites, nucleotide diversity, Watterson's theta, and Tajima's D. These values were generated using POPGENOME [65]. Additionally, population diversity was estimated using expected and observed heterozygosity, Fst and Fis values, for each of the assigned groups (POPGENOME) (Table S4).

Supplementary Materials: The following are available online at https://www.mdpi.com/article/10.3390/plants10112514/s1, Figure S1: Copy number spectrum plots (spectra-cn plot) of k-mer multiplicity from each set of x50 coverage Illumina reads. Histograms coloured by number of times each k-mer in the read set is found in a given assembly. The result is a set of histograms relating kmer counts in the read set to their associated counts in the assembly, Figure S2: Histograms of allele frequency sampled from 'Grappolo', 'Manzanilla de Sevilla', 'Piñonera' cultivars using vcftools, Figure S3: PCA clustering of GBS SNP sets. Plots show PCs 1 and 2 for SNPs called from a given reference genome, Figure S4: PCA clustering of WGS SNP sets. Plots show PCs 1 and 2 for SNPs called from a given reference genome, Figure S5: Estimated admixture for GBS SNP datasets Oe451, Oe6, Oe9, & Oleur.0.6.1. Ancestry co-efficient (0.0-1.0) estimated using STRUCTURE assigned populations, Figure S6: Estimated admixture for GBS Oe_Rao SNP dataset and WGS Oe451 & Oleur.0.6.1 SNP datasets. Ancestry co-efficient (0.0-1.0) estimated using STRUCTURE assigned populations, Figure S7: Estimated admixture for merged WGS/GBS Oe451 & Oleur.0.6.1 SNP datasets. Ancestry co-efficient (0.0-1.0) estimated using STRUCTURE assigned populations, Figure S8: Estimated number of populations using STRUCTURE for each dataset. Cross-entropy values along the y-axis, Figure S9: DAPC preliminary optimization of GBS data prior to running DAPC analysis. Reference genome SNP set indicated to the left of each row. Left most column contains the results the function optim.a.score() used to indicate best number of principal components to include. Center column the value of Bayesian Information Criterion (BIC) to assess the best supported model. Right-hand column contains the preassigned group (country of origin) representation using selected PCs and BIC model, Figure S10: DAPC preliminary optimization of WGS data prior to running DAPC analysis. Reference genome SNP set indicated to the left of each row. Left most column contains the results the function optim.a.score() used to indicate best number of principal components to include. Center column the value of Bayesian Information Criterion (BIC) to assess the best supported model. Right-hand column contains the preassigned group (country of origin) representation using selected PCs and BIC model, Figure S11: DAPC preliminary optimization of merged WGS and GBS data prior to running DAPC analysis. Reference genome SNP set indicated to the left of each row. Left most column contains the results the function optim.a.score() used to indicate best number of principal components to include. Center column the value of Bayesian Information Criterion (BIC) to assess the best supported model. Right-hand column contains the preassigned group (country of origin) representation using selected PCs and BIC model, Figure S12: DAPC group composition and clustering of GBS data set. Reference genome SNP set indicated to the left of each row. Left most column contains the membership probability of each individual sample using discriminant functions, proximal to admixture coefficients used by STRUCTURE. Right-hand column clustering using predefined groups (country of origin) with selected model, Figure S13: DAPC group composition and clustering of WGS data set. Reference genome SNP set indicated to the left of each row. Left most column contains the membership probability of each individual sample using discriminant functions, proximal to admixture coefficients used by STRUCTURE. Right-hand column clustering using predefined groups (country of origin) with selected model, Figure S14: DAPC group composition and clustering of merged WGS and GBS data set. Reference genome SNP set indicated to the left of each row. Left most column contains the membership probability of each individual sample using discriminant functions, proximal to admixture coefficients used by STRUCTURE. Right-hand column clustering using predefined groups (country of origin) with selected model, Figure S15: DAPC posterior membership probability of GBS data set. Reference genome SNP set indicated to the left of each row. Membership probability of each individual at K=2, Figure S16: DAPC posterior membership probability of WGS data set. Reference genome SNP set indicated to the left of each row. Membership probability of each individual at K=2, Figure S17: DAPC posterior membership probability of merged WGS and GBS data set. Reference genome SNP set indicated to the left of each row. Membership probability of each individual at K=2, Figure S18: Genotype-by-sequencing library preparation pipeline, Figure S19: Sequencing read processing and population analysis pipeline. Table S1: GBS read processing and variant calling, Table S2: GBS read processing and variant calling Stats for the raw GBS sequencing reads mapped to all 5 genome assemblies. Low quality individuals have been removed improving the overall number of SNPs available, Table S3: WGS read processing and variant calling. Stats for the raw WGS sequencing reads mapped to two genomes. Only samples matching the remaining GBS samples were selected, Table S4: Results of Popgenome analysis genetic diversity and neutrality.

Author Contributions: Conceptualisation, A.M.F.-O., A.B. and F.L.; Methodology, A.M.F.-O., A.B., and J.F.; formal analysis, J.F.; writing—original draft preparation, J.F.; writing—review and editing, J.F., C.D.F., A.B., F.L. and A.M.F.-O.; supervision, A.B. and A.M.F.-O.; project administration, A.M.F.-O.; funding acquisition, A.M.F.-O.All authors have read and agreed to the published version of the manuscript.

Funding: H2020 project NUMBER 101000427—GEN4OLIVE: Mobilization of Olive GenRes through pre-breeding activities to face the future challenges and development of an intelligent interface to ensure a friendly information availability for end users. Additionally, the project number UJA 2011/13/13 Detección y aplicabilidad de marcadores SNPs para estudios de diversidad en variedades de olivo y su mapeo genético. PLAN PROPIO DE LA UNIVERSIDAD DE JAEN.

Institutional Review Board Statement: Not applicable.

Informed Consent Statement: Not applicable.

Data Availability Statement: The novel GBS data produced for this study can be found under the NCBI Bioproject accession PRJNA750928. WGS data were already published, associated with the Bioproject accession PRJNA556567.

Acknowledgments: The authors want to acknowledge the technical and human support provided by CICT of Universidad de Jaén (UJA, MINECO, Junta de Andalucía, FEDER).

Conflicts of Interest: The authors declare no conflict of interest.

References

1. Green, P.S. A Revision of Olea L. (Oleaceae). *Kew Bull.* **2002**, *57*, 91–140. [CrossRef]
2. Wallander, E.; Albert, V.A. Phylogeny and Classification of Oleaceae Based on Rps16 and TrnL-F Sequence Data. *Am. J. Bot.* **2000**, *87*, 1827–1841. [CrossRef]
3. Besnard, G.; Rubio de Casas, R.; Christin, P.-A.; Vargas, P. Phylogenetics of Olea (Oleaceae) based on plastid and nuclear ribosomal DNA sequences: Tertiary climatic shifts and lineage differentiation times. *Ann. Bot.* **2009**, *104*, 143–160. [CrossRef]

4. Médail, F.; Quézel, P.; Besnard, G.; Khadari, B. Systematics, ecology and phylogeographic significance of Olea europaea L. ssp. maroccana (Greuter & Burdet) P. Vargas et al. a relictual olive tree in south-west Morocco. *Bot. J. Linn. Soc.* **2001**, *137*, 249–266. [CrossRef]
5. Besnard, G.; Garcia-Verdugo, C.; De Casas, R.R.; Treier, U.A.; Galland, N.; Vargas, P. Polyploidy in the Olive Complex (Olea europaea): Evidence from Flow Cytometry and Nuclear Microsatellite Analyses. *Ann. Bot.* **2008**, *101*, 25–30. [CrossRef]
6. Zohary, D.; Spiegel-Roy, P. Beginnings of Fruit Growing in the Old World. *Science* **1975**, *187*, 319–327. [CrossRef]
7. Newton, C.; Lorre, C.; Sauvage, C.; Ivorra, S.; Terral, J.-F. On the origins and spread of *Olea europaea* L. (olive) domestication: Evidence for shape variation of olive stones at Ugarit, Late Bronze Age, Syria—A window on the Mediterranean Basin and on the westward diffusion of olive varieties. *Veg. Hist. Archaeobotany* **2014**, *23*, 567–575. [CrossRef]
8. Terral, J.-F.; Alonso, N.; Capdevila, R.B.I.; Chatti, N.; Fabre, L.; Fiorentino, G.; Marinval, P.; Jordà, G.P.; Pradat, B.; Rovira, N.; et al. Historical biogeography of olive domestication (*Olea europaea* L.) as revealed by geometrical morphometry applied to biological and archaeological material. *J. Biogeogr.* **2004**, *31*, 63–77. [CrossRef]
9. Zohary, D.; Hopf, M.; Weiss, E. *Domestication of Plants in the Old World: The Origin and Spread of Domesticated Plants in Southwest Asia, Europe, and the Mediterranean Basin*, 3rd ed.; Oxford University Press: Oxford, UK, 2000; ISBN 978-0-19-181004-6.
10. El Bakkali, A.; Essalouh, L.; Tollon, C.; Rivallan, R.; Mournet, P.; Moukhli, A.; Zaher, H.; Mekkaoui, A.; Hadidou, A.; Sikaoui, L.; et al. Characterization of Worldwide Olive Germplasm Banks of Marrakech (Morocco) and Córdoba (Spain): Towards management and use of olive germplasm in breeding programs. *PLoS ONE* **2019**, *14*, e0223716. [CrossRef] [PubMed]
11. Besnard, G.; Khadari, B.; Navascués, M.; Fernández-Mazuecos, M.; El Bakkali, A.; Arrigo, N.; Baali-Cherif, D.; De Caraffa, V.B.-B.; Santoni, S.; Vargas, P.; et al. The complex history of the olive tree: From Late Quaternary diversification of Mediterranean lineages to primary domestication in the northern Levant. *Proc. R. Soc. B Boil. Sci.* **2013**, *280*, 20122833. [CrossRef] [PubMed]
12. Kaniewski, D.; Van Campo, E.; Boiy, T.; Terral, J.-F.; Khadari, B.; Besnard, G. Primary domestication and early uses of the emblematic olive tree: Palaeobotanical, historical and molecular evidence from the Middle East. *Biol. Rev.* **2012**, *87*, 885–899. [CrossRef] [PubMed]
13. Jiménez-Ruiz, J.; Ramírez-Tejero, J.A.; Fernández-Pozo, N.; Leyva-Pérez, M.d.l.O.; Yan, H.; Rosa, R.d.l.; Belaj, A.; Montes, E.; Rodríguez-Ariza, M.O.; Navarro, F.; et al. Transposon activation is a major driver in the genome evolution of cultivated olive trees (*Olea europaea* L.). *Plant Genome* **2020**, *13*, e20010. [CrossRef] [PubMed]
14. Rallo, L.; Barranco, D.; Castro-García, S.; Connor, D.J.; Campo, M.G.; del Rallo, P. High-Density Olive Plantations. In *Horticultural Reviews Volume 41*; John Wiley & Sons, Ltd.: Hoboken, NJ, USA, 2014; pp. 303–384. ISBN 978-1-118-70741-8.
15. Trujillo, I.; Ojeda, M.A.; Urdiroz, N.M.; Potter, D.; Barranco, D.; Rallo, L.; Diez, C.M. Identification of the Worldwide Olive Germplasm Bank of Córdoba (Spain) using SSR and morphological markers. *Tree Genet. Genomes* **2014**, *10*, 141–155. [CrossRef]
16. IOOC International Olive Oil Council. 2020. Available online: https://www.Internationaloliveoil.Org/2020 (accessed on 10 February 2021).
17. Cruz, F.; Julca, I.; Gómez-Garrido, J.; Loska, D.; Marcet-Houben, M.; Cano, E.; Galán, B.; Frias, L.; Ribeca, P.; Derdak, S.; et al. Genome sequence of the olive tree, Olea europaea. *GigaScience* **2016**, *5*, 29. [CrossRef] [PubMed]
18. Julca, I.; Marcet-Houben, M.; Cruz, F.; Gómez-Garrido, J.; Gaut, B.S.; Díez, C.M.; Gut, I.G.; Alioto, T.S.; Vargas, P.; Gabaldón, T. Genomic evidence for recurrent genetic admixture during the domestication of Mediterranean olive trees (*Olea europaea* L.). *BMC Biol.* **2020**, *18*, 148. [CrossRef]
19. Rao, G.; Zhang, J.; Liu, X.; Lin, C.; Xin, H.; Xue, L.; Wang, C. De novo assembly of a new Olea europaea genome accession using nanopore sequencing. *Hortic. Res.* **2021**, *8*, 64. [CrossRef]
20. Unver, T.; Wu, Z.; Sterck, L.; Turktas, M.; Lohaus, R.; Li, Z.; Yang, M.; He, L.; Deng, T.; Escalante, F.J.; et al. Genome of wild olive and the evolution of oil biosynthesis. *Proc. Natl. Acad. Sci. USA* **2017**, *114*, E9413–E9422. [CrossRef] [PubMed]
21. Muir, P.; Li, S.; Lou, S.; Wang, D.; Spakowicz, D.J.; Salichos, L.; Zhang, J.; Weinstock, G.M.; Isaacs, F.; Rozowsky, J.; et al. The real cost of sequencing: Scaling computation to keep pace with data generation. *Genome Biol.* **2016**, *17*, 53. [CrossRef]
22. Ansorge, W.J. Next-generation DNA sequencing techniques. *New Biotechnol.* **2009**, *25*, 195–203. [CrossRef] [PubMed]
23. Metzker, M.L. Sequencing technologies—The next generation. *Nat. Rev. Genet.* **2010**, *11*, 31–46. [CrossRef]
24. De Pristo, M.A.; Banks, E.; Poplin, R.; Garimella, K.V.; Maguire, J.R.; Hartl, C.; Philippakis, A.A.; Del Angel, G.; Rivas, M.A.; Hanna, M.; et al. A framework for variation discovery and genotyping using next-generation DNA sequencing data. *Nat. Genet.* **2011**, *43*, 491–498. [CrossRef]
25. Kyriakidou, M.; Tai, H.H.; Anglin, N.L.; Ellis, D.; Strömvik, M.V. Current Strategies of Polyploid Plant Genome Sequence Assembly. *Front. Plant Sci.* **2018**, *9*, 1660. [CrossRef] [PubMed]
26. Belaj, A.; Trujillo, I.; De La Rosa, R.; Rallo, L.; Giménez, M.J. Polymorphism and Discrimination Capacity of Randomly Amplified Polymorphic Markers in an Olive Germplasm Bank. *J. Am. Soc. Hortic. Sci.* **2001**, *126*, 64–71. [CrossRef]
27. Marita, J.M.; Rodriguez, J.M.; Nienhuis, J. Development of an algorithm identifying maximally diverse core collections. *Genet. Resour. Crop. Evol.* **2000**, *47*, 515–526. [CrossRef]
28. Angiolillo, A.; Mencuccini, M.; Baldoni, L. Olive genetic diversity assessed using amplified fragment length polymorphisms. *Theor. Appl. Genet.* **1999**, *98*, 411–421. [CrossRef]
29. Wachira, F.; Tanaka, J.; Takeda, Y. Genetic variation and differentiation in tea (Camellia sinensis) germplasm revealed by RAPD and AFLP variation. *J. Hortic. Sci. Biotechnol.* **2001**, *76*, 557–563. [CrossRef]

30. Belaj, A.; Šatović, Z.; Cipriani, G.; Baldoni, L.; Testolin, R.; Rallo, L.; Trujillo, I. Comparative study of the discriminating capacity of RAPD, AFLP and SSR markers and of their effectiveness in establishing genetic relationships in olive. *Theor. Appl. Genet.* **2003**, *107*, 736–744. [CrossRef]
31. Gómez-Rodríguez, M.V.; Beuzon, C.; González-Plaza, J.J.; Fernández-Ocaña, A.M. Identification of an olive (*Olea europaea* L.) core collection with a new set of SSR markers. *Genet. Resour. Crop. Evol.* **2021**, *68*, 117–133. [CrossRef]
32. Belaj, A.; del Carmen Dominguez-García, M.; Atienza, S.G.; Urdíroz, N.M.; De la Rosa, R.; Satovic, Z.; Martín, A.; Kilian, A.; Trujillo, I.; Valpuesta, V.; et al. Developing a core collection of olive (*Olea europaea* L.) based on molecular markers (DArTs, SSRs, SNPs) and agronomic traits. *Tree Genet. Genomes* **2012**, *8*, 365–378. [CrossRef]
33. He, J.; Zhao, X.; Laroche, A.; Lu, Z.-X.; Liu, H.K.; Li, Z. Genotyping-by-sequencing (GBS), an ultimate marker-assisted selection (MAS) tool to accelerate plant breeding. *Front. Plant Sci.* **2014**, *5*. [CrossRef]
34. Talavera, A.; Soorni, A.; Bombarely, A.; Matas, A.J.; Hormaza, J.I. Genome-Wide SNP discovery and genomic characterization in avocado (Persea americana Mill.). *Sci. Rep.* **2019**, *9*, 20137. [CrossRef]
35. Elshire, R.J.; Glaubitz, J.C.; Sun, Q.; Poland, J.A.; Kawamoto, K.; Buckler, E.S.; Mitchell, S.E. A Robust, Simple Genotyping-by-Sequencing (GBS) Approach for High Diversity Species. *PLoS ONE* **2011**, *6*, e19379. [CrossRef] [PubMed]
36. Wickland, D.P.; Battu, G.; Hudson, K.A.; Diers, B.W.; Hudson, M.E. A comparison of genotyping-by-sequencing analysis methods on low-coverage crop datasets shows advantages of a new workflow, GB-eaSy. *BMC Bioinform.* **2017**, *18*, 586. [CrossRef]
37. Marchese, A.; Marra, F.P.; Caruso, T.; Mhelembe, K.; Costa, F.; Fretto, S.; Sargent, D.J. The First High-Density Sequence Characterized SNP-Based Linkage Map of Olive ("*Olea Europaea*" L. Subsp. 'Europaea') Developed Using Genotyping by Sequencing. *Aust. J. Crop Sci.* **2016**, *10*, 857–863. [CrossRef]
38. Zhu, S.; Niu, E.; Shi, A.; Mou, B. Genetic Diversity Analysis of Olive Germplasm (*Olea europaea* L.) With Genotyping-by-Sequencing Technology. *Front. Genet.* **2019**, *10*, 755. [CrossRef] [PubMed]
39. Kaya, H.B.; Akdemir, D.; Lozano, R.; Cetin, O.; Kaya, H.S.; Sahin, M.; Smith, J.L.; Tanyolac, B.; Jannink, J.-L. Genome wide association study of 5 agronomic traits in olive (*Olea europaea* L.). *Sci. Rep.* **2019**, *9*, 18764. [CrossRef]
40. Whibley, A.; Kelley, J.L.; Narum, S.R. The changing face of genome assemblies: Guidance on achieving high-quality reference genomes. *Mol. Ecol. Resour.* **2021**, *21*, 641–652. [CrossRef]
41. Amarasinghe, S.L.; Su, S.; Dong, X.; Zappia, L.; Ritchie, M.E.; Gouil, Q. Opportunities and challenges in long-read sequencing data analysis. *Genome Biol.* **2020**, *21*, 30. [CrossRef]
42. Ekblom, R.; Wolf, J.B. A field guide to whole-genome sequencing, assembly and annotation. *Evol. Appl.* **2014**, *7*, 1026–1042. [CrossRef]
43. Simão, F.A.; Waterhouse, R.M.; Ioannidis, P.; Kriventseva, E.V.; Zdobnov, E.M. BUSCO: Assessing genome assembly and annotation completeness with single-copy orthologs. *Bioinformatics* **2015**, *31*, 3210–3212. [CrossRef]
44. Seppey, M.; Manni, M.; Zdobnov, E.M. BUSCO: Assessing Genome Assembly and Annotation Completeness. In *Gene Prediction: Methods and Protocols*; Kollmar, M., Ed.; Methods in Molecular Biology; Springer: New York, NY, USA, 2019; pp. 227–245. ISBN 978-1-4939-9173-0.
45. Rhie, A.; Walenz, B.P.; Koren, S.; Phillippy, A.M. Merqury: Reference-free quality, completeness, and phasing assessment for genome assemblies. *Genome Biol.* **2020**, *21*, 245. [CrossRef] [PubMed]
46. Schadt, E.E.; Turner, S.; Kasarskis, A. A window into third-generation sequencing. *Hum. Mol. Genet.* **2010**, *19*, R227–R240. [CrossRef] [PubMed]
47. Ou, S.; Chen, J.; Jiang, N. Assessing genome assembly quality using the LTR Assembly Index (LAI). *Nucleic Acids Res.* **2018**, *46*, e126. [CrossRef] [PubMed]
48. Besnard, G.; De Casas, R.R. Single vs multiple independent olive domestications: The jury is (still) out. *New Phytol.* **2016**, *209*, 466–470. [CrossRef]
49. Gage, J.L.; Vaillancourt, B.; Hamilton, J.P.; Manrique-Carpintero, N.C.; Gustafson, T.J.; Barry, K.; Lipzen, A.; Tracy, W.F.; Mikel, M.A.; Kaeppler, S.M.; et al. Multiple Maize Reference Genomes Impact the Identification of Variants by Genome-Wide Association Study in a Diverse Inbred Panel. *Plant Genome* **2019**, *12*, 180069. [CrossRef]
50. Della Coletta, R.; Qiu, Y.; Ou, S.; Hufford, M.B.; Hirsch, C.N. How the pan-genome is changing crop genomics and improvement. *Genome Biol.* **2021**, *22*, 3. [CrossRef] [PubMed]
51. Hakim, I.R.; Kammoun, N.G.; Makhloufi, E.; Rebaï, A. Discovery and Potential of SNP Markers in Characterization of Tunisian Olive Germplasm. *Diversity* **2010**, *2*, 17–27. [CrossRef]
52. Aronesty, E. Comparison of Sequencing Utility Programs. *Open Bioinform. J.* **2013**, *7*, 1–8. [CrossRef]
53. Li, H.; Handsaker, B.; Wysoker, A.; Fennell, T.; Ruan, J.; Homer, N.; Marth, G.; Abecasis, G.; Durbin, R. The Sequence Alignment/Map Format and SAMtools. *Bioinformatics* **2009**, *25*, 2078–2079. [CrossRef]
54. Garrison, E.; Marth, G. Haplotype-based variant detection from short-read sequencing. *arXiv* **2012**, arXiv:1207.3907.
55. Danecek, P.; Auton, A.; Abecasis, G.; Albers, C.A.; Banks, E.; DePristo, M.A.; Handsaker, R.E.; Lunter, G.; Marth, G.T.; Sherry, S.T.; et al. The variant call format and VCFtools. *Bioinformatics* **2011**, *27*, 2156–2158. [CrossRef] [PubMed]
56. Pritchard, J.K.; Stephens, M.; Donnelly, P. Inference of population structure using multilocus genotype data. *Genetics* **2000**, *155*, 945–959. [CrossRef] [PubMed]
57. Jombart, T.; Devillard, S.; Balloux, F. Discriminant analysis of principal components: A new method for the analysis of genetically structured populations. *BMC Genet.* **2010**, *11*, 94. [CrossRef]

58. Raj, A.; Stephens, M.; Pritchard, J.K. fastSTRUCTURE: Variational Inference of Population Structure in Large SNP Data Sets. *Genetics* **2014**, *197*, 573–589. [CrossRef] [PubMed]
59. Frichot, E.; François, O. LEA: An R package for landscape and ecological association studies. *Methods Ecol. Evol.* **2015**, *6*, 925–929. [CrossRef]
60. Jombart, T.; Ahmed, I. adegenet 1.3-1: New tools for the analysis of genome-wide SNP data. *Bioinformatics* **2011**, *27*, 3070–3071. [CrossRef]
61. Kamvar, Z.N.; Brooks, J.C.; GrãNwald, N.J. Novel R tools for analysis of genome-wide population genetic data with emphasis on clonality. *Front. Genet.* **2015**, *6*, 208. [CrossRef]
62. Paradis, E.; Schliep, K. Ape 5.0: An environment for modern phylogenetics and evolutionary analyses in R. *Bioinforma. Oxf. Engl.* **2019**, *35*, 526–528. [CrossRef]
63. Ramaut, A. FigTree. 2018. Available online: http://tree.bio.ed.ac.uk/software/figtree/ (accessed on 13 June 2020).
64. Gruber, B.; Unmack, P.J.; Berry, O.F.; Georges, A. dartr: An r package to facilitate analysis of SNP data generated from reduced representation genome sequencing. *Mol. Ecol. Resour.* **2018**, *18*, 691–699. [CrossRef]
65. Pfeifer, B.; Wittelsbürger, U.; Ramos-Onsins, S.E.; Lercher, M.J. PopGenome: An Efficient Swiss Army Knife for Population Genomic Analyses in R. *Mol. Biol. Evol.* **2014**, *31*, 1929–1936. [CrossRef]

plants

MDPI

Article

Morphological and Molecular Characterization of Some Egyptian Six-Rowed Barley (*Hordeum vulgare* L.)

Azza H. Mohamed [1,2,*,†], Ahmad A. Omar [2,3,*,†], Ahmed M. Attya [4], Mohamed M. A. Elashtokhy [5], Ehab M. Zayed [6] and Rehab M. Rizk [7]

1 Department of Agricultural Chemistry, College of Agriculture, Mansoura University, Mansoura 35516, Egypt
2 Citrus Research and Education Center, University of Florida (IFAS), Lake Alfred, FL 33850, USA
3 Biochemistry Department, Faculty of Agriculture, Zagazig University, Zagazig 44519, Egypt
4 Barley Research Department, Field Crops Research Institute, Agricultural Research Center, Giza 12619, Egypt; ahmed_1099@hotmail.com
5 Genetics Department, Faculty of Agriculture, Zagazig University, Zagazig 44511, Egypt; mmabdeltawab@zu.edu.eg
6 Cell Study Research Department, Field Crops Research Institute, Agricultural Research Center, Giza 12619, Egypt; ehabzayed977@gmail.com
7 Botany Department, Faculty of Science, Mansoura University, Mansoura 35516, Egypt; rehabrizk@mans.edu.eg
* Correspondence: azza@ufl.edu (A.H.M.); omar71@ufl.edu (A.A.O.); Tel.: +1-863-521-4886 (A.H.M.); +1-863-521-4569 (A.A.O.); Fax: +1-863-956-4631 (A.H.M. & A.A.O.)
† These authors contributed equally to this work.

Abstract: Barley production is essential in Egypt. In the present study, 15 different six-rowed Egyptian barley cultivars were studied. To differentiate between the different cultivars under study in terms of morphological characteristics and ISSR, molecular characterization reactions were carried out. Moreover, four cultivars (Giza 123, Giza 126, Giza 136, and Giza 138) were selected for further studies using scanning electron microscopy (SEM). Computational analysis of the DNA barcoding sequences of the two plastid markers *rbc*L and *mat*K was executed, and the results were deposited in the NCBI database. The morphological traits showed low statistical significance among the different cultivars under study via the data collected from two seasons, suggesting that the mean field performance of these Egyptian cultivars may be equal under these conditions. The results showed that the phylogenetic tree was divided into four groups, one of which contained the most closely related genotypes in the genetic distance, including Giza 124, Giza 130, Giza 138, Giza 136, and Giza 137, which converge in the indicative uses of farmers. The seed coat of the studied cultivars was "rugose". The elevation folding of the rugose pattern ranged from 11 ± 1.73 μm (Giza 126) to 14.67 ± 2.43 μm (Giza 123), suggesting variation in seed quality and its uses in feed and the food industry. According to the similarity matrix of ISSR analysis, the highest similarity value (93%) was recorded between Giza 133 and Giza 132, as well as between Giza 2000 and Giza 126. On the other hand, the lowest similarity value (80%) was recorded between Giza 130 and (Giza 133 and Giza 132), indicating that these cultivars were distantly related. Polymorphism information content (PIC) ranged from 0.26 for the primer ISSR UBC 835 to 0.37 for the primers ISSR UBC 814 and ISSR UBC 840. The current study showed that the *mat*K gene is more mutable than the *rbc*L gene among the tested cultivars.

Keywords: plastid markers; DNA barcoding; ISSR markers; Egyptian barley; agro-morphological traits; cluster analysis; genetic variation; biplot

Citation: Mohamed, A.H.; Omar, A.A.; Attya, A.M.; Elashtokhy, M.M.A.; Zayed, E.M.; Rizk, R.M. Morphological and Molecular Characterization of Some Egyptian Six-Rowed Barley (*Hordeum vulgare* L.). *Plants* **2021**, *10*, 2527. https://doi.org/10.3390/plants10112527

Academic Editors: Milan S. Stankovic, Paula Baptista, Petronia Carillo and Josefina Bota Salort

Received: 22 September 2021
Accepted: 18 November 2021
Published: 20 November 2021

Publisher's Note: MDPI stays neutral with regard to jurisdictional claims in published maps and institutional affiliations.

Copyright: © 2021 by the authors. Licensee MDPI, Basel, Switzerland. This article is an open access article distributed under the terms and conditions of the Creative Commons Attribution (CC BY) license (https://creativecommons.org/licenses/by/4.0/).

1. Introduction

Barley (*Hordeum vulgare* L.) is one of the main and oldest cereal crops on Earth. Worldwide, its grain production is ranked fourth after maize, rice, and wheat [1]. Barley is generally considered a poor man's crop because it is easy to cultivate, with few requirements, and has a high capacity for adaptation to harsh environments. Some literature estimates the age of barley at 11,000 years [2]. However, six-rowed barley did not arise until after 6000 BC [3]. Archaeological evidence has dated barley cultivation to 5000–6000 BC in Egypt [4–7]. Barley products, especially bread and beer, comprised a complete diet in ancient Egypt. Based on the health benefits of barley, as well as the need for agricultural development and reducing wheat imports, Egypt is currently examining the return of barley to the bread-making industry as a 30% ingredient. The global production volume of barley reached 142.37 million metric tons in the 2017/2018 crop year [1]. Furthermore, it is expected that barley production will decrease to 140.6 million metric tons in the next crop year [8]. Egypt's barley production has fluctuated substantially in recent years as it increased over the past two decades, ending at 108,000 tons in 2019 [1]. With the new policies for the sustainable development plan and reclaimed land expansion, the area and productivity are expected to increase.

There are approximately 38 Egyptian barley cultivars—two- and six-rowed—but the six-rowed barleys are the most famous and widely used in Egypt. Field evaluations have shown many differences between genotypes [9,10]. Furthermore, [11] found that all studied traits showed significant differences between genotypes, environments, and interactions. Moreover, Sharma, et al. [12] used Euclidean distances based on non-hierarchical cluster analysis to categorize total accessions into diverse clusters, and to determine and select accessions with decent yield and performance for other ancillary traits. The candidate breeding lines can be used in hybridization for barley improvement programs.

Molecular markers are an essential tool used to directly detect the differences between and within genetic materials at the DNA level; they provide a robust estimate of genetic similarity that is not often obtained using morphological data alone [13]. A comparison can also be made to determine the genetic distance between the Egyptian cultivars based on field characteristics and molecular parameters. The inter simple sequence repeats (ISSRs) technique has been successfully applied to many crop species [14–16]. ISSRs demonstrate the specificity of microsatellite markers, and require no specific sequence information for primer synthesis, using the advantage of random markers [17]; thus, they have been widely used for cultivar identification in different crops [18–20]. Moreover, Guasmi et al. [17] found that ISSR primers exhibited variations in the percentage of polymorphism, resolving power (Rp), and band informativeness (Ib); the rate of polymorphism was 66.67%, the Rp ranged from 0.74 to 1.16, and the average Ib ranged from 0.24 to 0.39, suggesting that ISSRs are robust molecular markers that can distinguish between Egyptian cultivars. On the other hand, according to Drine, et al. [21], ISSRs and random amplified polymorphic DNA (RAPD) markers identified 72.2% and 61% of polymorphic bands, respectively. Several parameters were used to compare the relative efficiency of these marker systems, including the effective multiplex ratio (EMR), marker index (MI), and polymorphic information content (PIC); the ISSR system showed higher values for all of the parameters examined. Moreover, Wang, et al. [22] used 10 ISSR primers to investigate the variation between Tibetan and Middle Eastern barley genotypes; the Tibetan genotypes contained 91 allelic variants, of which 79 were polymorphic (86.81%), while the Middle Eastern genotypes contained 82 allelic variants, of which 66 were polymorphic (80.49%). These results suggest that ISSRs are robust molecular markers that can be employed to distinguish Egyptian cultivars.

The morphology of barley grains observed via scanning electron microscopy (SEM) showed starch granules smaller than the standard ones; they appeared in abnormal shapes with a conspicuous peripheral groove and sunken cheeks [23,24]. When SEM was used to study the detailed structure, it proved that starch was degraded by both pitting and surface erosion [25]. The apparent shape of the seed under the electron microscope indicates the quality of its industrial and agricultural importance.

DNA barcoding is a genetic identification technology that uses a genetic region of short DNA sequences, called the DNA barcode [26]; this can be reliably characterized by similar morphological characteristics and chemical compositions [27]; it has two main objectives: identifying organisms, where an unknown sequence matches a known species sequence, and exploring species that are similar in terms of habitat delimitation and description of species [28]. A short DNA sequence obtained from established target regions of the chloroplast genome can be used to classify genera and/or species of plants with respect to orthologous databases, compared to conventional PCR-based markers [29]. DNA barcoding has been proposed as an essential tool for resolving the significant gaps in our current understanding of biodiversity. Furthermore, Barley and Thomson [30] demonstrated that the success of DNA barcoding varies broadly across DNA substitution models, and has a substantial influence on the number of operational taxonomic identified units. Moreover, using recent advances in combinatorial pooling and next-generation sequencing, Lonardi, et al. [31] proposed a new sequencing approach that addresses the challenge of de novo selective genome sequencing in a highly efficient manner. Barcodes can be employed to explain the relationships between Egyptian cultivars, and their relation to sequences within the database.

The main objective of this study was to measure and characterize the differences between the most economically important Egyptian barley cultivars, especially in making bread. The study investigated 15 Egyptian six-rowed cultivars at the field level, the molecular level, via scanning electron microscopic examination, and via DNA barcoding. The results obtained from the present study will potentially enhance breeding programs and lead to the development of new adaptive or high-yield barley cultivars with specific improved traits.

2. Results and Discussion

2.1. Field Experimental

2.1.1. Growing across Two Seasons

Figure 1 represents the average values of the field data during the two growing seasons, including grain filing period (day), maturity day (day), and hiding day (day) (Figure 1A); spike height (cm) and plant height (cm) (Figure 1B); the number of spikes per square meter and number of grains per spike (average of 10 spikes per square meter) (Figure 1C); biological yield (ton/ha) and grain yield (ton/ha) (Figure 1D); and weight of 1000 grains (g) (Figure 1E). The average mean values showed low statistical significance among the examined cultivars. There were no significant differences based on the least significant difference (LSD) for any of the studied traits except for biological weight, which showed a significant difference in values between the different genotypes (LSD = 1.78).

Figure 1 shows differences between genotypes in all studied traits, which were divided into three parts according to the convergence of the numerical values of the traits. There were indications of early cultivars being equal through the periods of maturity and seed fullness, and low statistical significance among them (Figure 1A). Plant height indicated vegetative solid growth (Figure 1B), which is sufficient for animal feed. These results are consistent with those of Amer, et al. [32], who found that the average yield of the new cultivar Giza 137 was 16.7 4.95 Ton/ha, while that of Giza 138 was 5.07 Ton/ha. These yields significantly exceeded the national checks Giza 123 and Giza 132 (3.88 Ton/ha). Giza 137 significantly out-yielded Giza 123 and Giza 132, by ~22.4 and ~20.7%, respectively. Furthermore, Giza 138 significantly exceeded the average of national checks Giza 123 (by ~25.6%) and Giza 132 (by ~23.9%).

On the other hand, Noaman, et al. [33] found that biological weight characterized new genotypes. Furthermore, Mariey, et al. [34] considered the Egyptian barley genotypes Giza 123, Giza 131, and Giza 136 to be salt tolerant. It is worthy of note that this study was performed under the conditions and climate of Giza Governorate, Egypt. Furthermore, the behavior of the varieties differs when studied under different environmental conditions—such as in the Sinai Peninsula or on the northern coast—even though they have the same genetic background; the same can be said of their other behavior under harsh conditions.

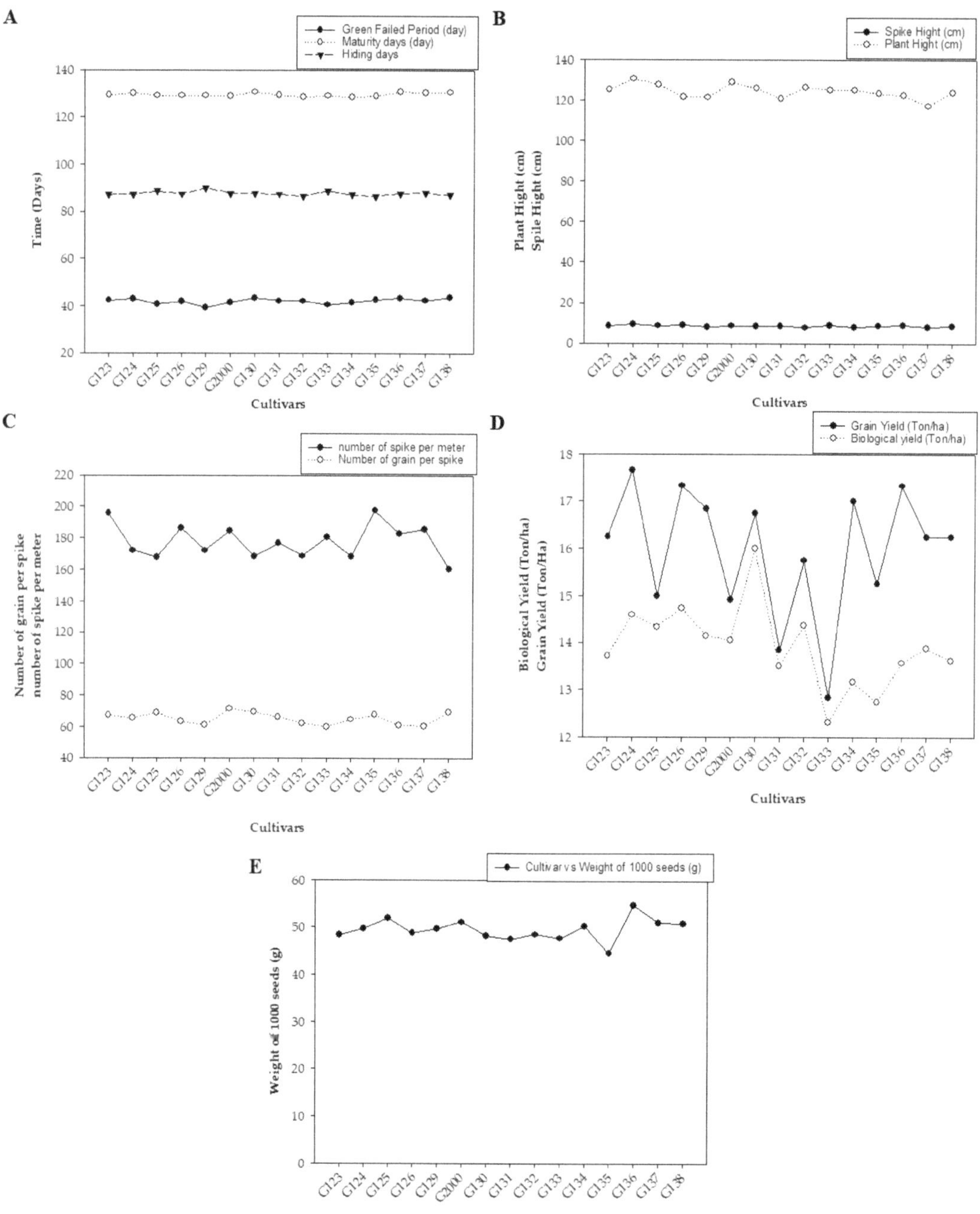

Figure 1. The averages of morphological traits in 15 barley genotypes grown in two seasons—2017/2018 and 2018/2019: (**A**) grain filing period (day), maturity day (day), and hiding day (day); (**B**) spike height (cm) and plant height (cm); (**C**) number of spikes per square meter and number of grains per spike; (**D**) biological yield (ton/ha) and grain yield (ton/ha); and (**E**) weight of 1000 grains (g).

2.1.2. Genetic Distance Dendrogram between Genotypes Based on Field Traits

The genetic tree of the genotypes was divided into four groups: Group I contained the most closely related genotypes in terms of genetic distance, including Giza 124, Giza 130, Giza 138, Giza 136, and Giza 1137 (Figure 2); members of this group were characterized by a high maturity day and high grain filing period (days), along with grain yield (Ton/ha) and biological yield (Ton/ha). Group II contained the two cultivars Giza 129 and Giza 133; this group could be described by a low number of grains per spike and high hiding days. Group III consisted of Giza 125 and Giza 2000 on one side of the group, and Giza 132 and Giza 134 on the other side (Figure 2); members of this group were characterized by high plant height and low-to-moderate maturity days. Group IV consisted of Giza 123, Giza 135, Giza 131, and Giza 126; this group could be characterized by height, a moderate number of spikes per square meter, spike height (cm), and the number of grains per spike, along with low-to-moderate weight of 1000 grain (g), hiding days, biological yield (Ton/ha), and maturity days (Figure 2). These results were consistent with the findings of Mariey and Khedr [35]. Moreover, based on their 10 agro-morphological traits, Mareiy, et al. [36] explored biplot and cluster analysis using Euclidean distance matrices and average linkage. According to PCA, all 15 genotypes fell into 4 groups. Cultivars in Group A tend to have higher yields, so they may be considered to be tolerant (Giza 16 and Giza 18). Nevertheless, Giza 124, Giza 132, and Giza 134 are among the cultivars in group D that produce lower grain yields. The characteristics of biological weight and biological yield are used to assess the production of grain in relation to the rest of the plant components, which are used as animal feed in the form of straw. Indeed, increasing the seed yield and decreasing the biological crop is beneficial to grain production, which is the goal of growing barley for nutrition and intensive production.

2.2. Scanning Electron Microscopy (SEM)

The term seed coat of barley caryopsis includes tissues from three separated organs: the pericarp, the testa, and the semipermeable membrane. Several unique compounds are synthesized in the seed coat, serving the plant's defense and control of its development in different ways. Additionally, many of these compounds are sources of industrial products and components for human consumption or animal feed [37]. The seed coat of the studied cultivars is "rugose" (Figure 3 and Table 1). The elevation folding of the rugose pattern ranges from 11 ± 1.73 μm (Giza 126) to 14.67 ± 2.43 μm (Giza 123). The extension of the rugose pattern (length) ranges from 16.00 ± 2.61 μm (Giza 126) to 18.67 ± 3.13 μm (Giza 136). The frequency pattern in 100 μm^2 ranges from 4.67 ± 0.51 (Giza 126) to 12.17 ± 1.69 (Giza 138). Thus, these cultivars could be promising for different purposes in service of contemporary Egyptian interests.

Table 1. The seed coat characteristics of four Egyptian six-rowed barley (*Hordeum vulgare* L.) cultivars (Giza 123, Giza 126, Giza 136, and Giza 138).

	Giza 138	Giza 136	Giza 126	Giza 123
Frequency pattern in 100 μm^2	12.17 ± 1.69	4.83 ± 0.52	4.67 ± 0.51	8.17 ± 0.99
The elevation folding of rugose (μm)	12.67 ± 2.04	12.67 ± 2.04	11.00 ± 1.73	14.67 ± 2.43
The extent of the rugose surface (length, μm)	18.00 ± 3.01	18.67 ± 3.13	16.00 ± 2.61	14.00 ± 2.27

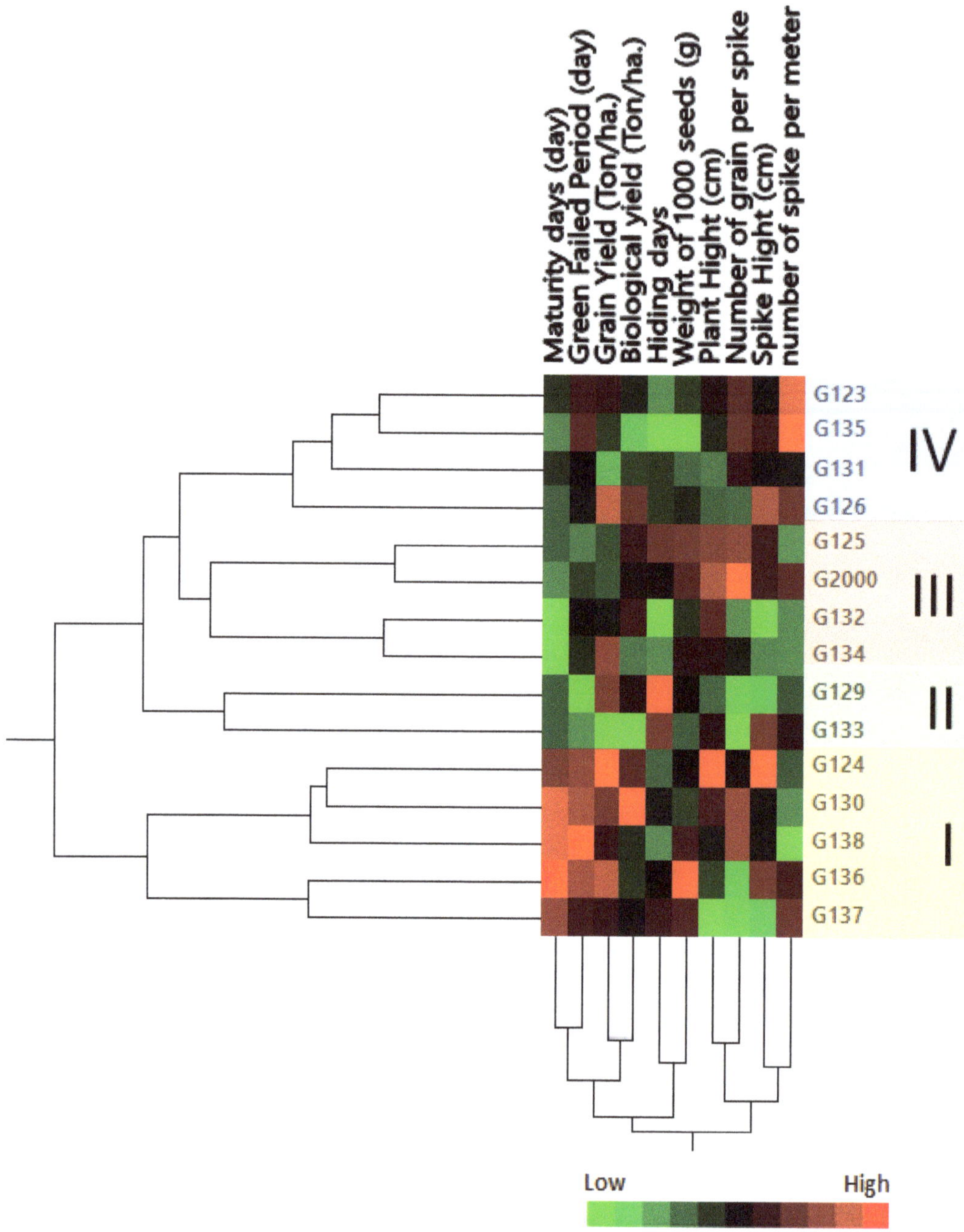

Figure 2. Cluster analysis and heatmap based on agro-morphological traits of 15 Egyptian six-rowed barley cultivars. The heatmap was constructed using JMP®, Version 15 (SAS Institute Inc., Cary, NC, USA, 1989–2019).

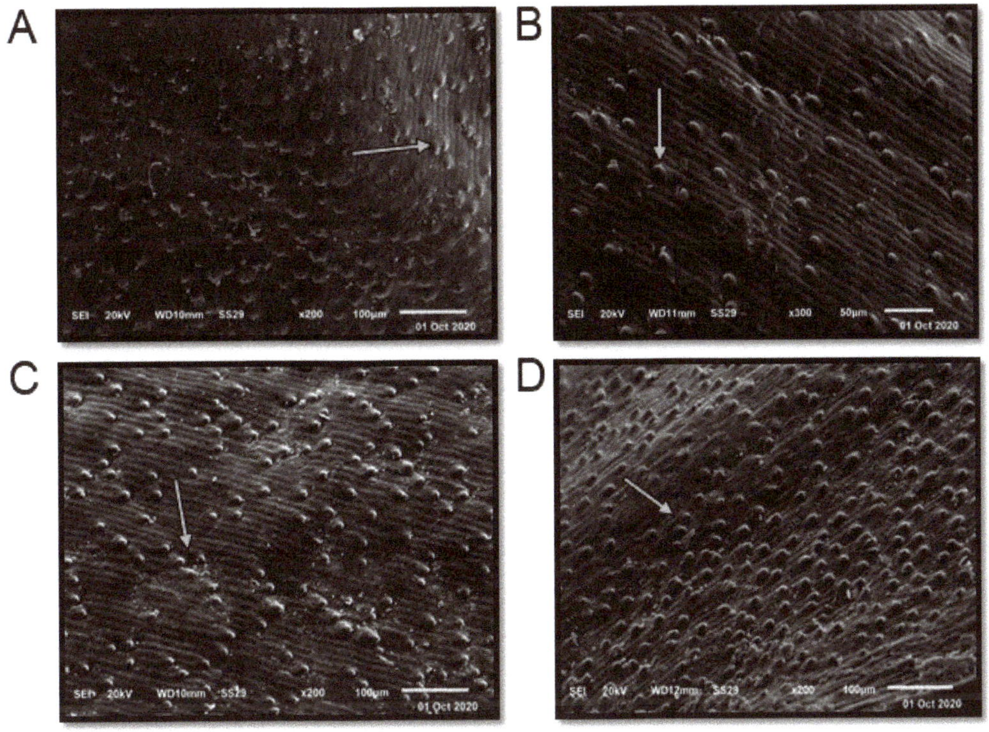

Figure 3. Scanning electron microscope (SEM) images of four Egyptian six-rowed barley (*Hordeum vulgare* L.) cultivars: (**A**) Giza 123, (**B**) Giza 126, (**C**) Giza 136, and (**D**) Giza 138. Scale bar = 100 µm. White arrows indicate the quality and shape of the rugose.

2.3. Molecular Characterization and Genetic Relationships as Revealed by ISSR Markers

The ability to effectively utilize genetic variability available to breeders is dependent upon an understanding of population diversity [38,39]. Thus, the primary benefit of cultivar differentiation at the molecular level is to explain with some accuracy the relationships between cultivars, in order to reduce selection costs within breeding programs and provide future breeders with molecular insights. The inter simple sequence repeats (ISSRs) fingerprinting profiles generated by 4 out of the 15 primers used in the present study, targeting 15 Egyptian six-rowed cultivars of barley, are displayed in Figure 4. The polymorphism generated by the 15 ISSR primers is summarized in Table 2. The 15 ISSR primers used in the present study produced a total number of bands (TNB) of 126, and 62 of those were polymorphic with uniqueness (PWU), with a polymorphism percentage (P%) of 50.07%. The TNB ranged from 5 for the ISSR UBC 844A and ISSR UBC 901 primers, to 14 for the ISSR UBC 835 primer. The number of PWU bands also varied, from two in ISSRs UBC 825 and UBC 901, to seven bands in the ISSR 857 and ISSR UBC 835 primers. The average number of PWU bands was 4.13 per primer (Table 2). The polymorphism information content (PIC) values varied between the ISSR primers. PIC ranged from 0.26 for the primer ISSR UBC 835 to 0.37 for the primers ISSR UBC 814 and ISSR UBC 840. Remarkably, some ISSR primers revealed distinct discrimination of 22–80% polymorphism, including ISSR UBC 825 and ISSR UBC 844A (Table 2). The ISSR primer UBC 844A recorded the lowest effective multiplex ratio (EMR) (6.20) and lowest marker index (MI) (0.02), whereas ISSR UBC 814 scored the highest in terms of PIC, resolving power (RP), and MI values (0.37, 12.67, and 0.05, respectively). Furthermore, ISSR UBC 835 scored the lowest values for PIC and MI (0.02 and 0.26, respectively) and the highest value in EMR (12.07).

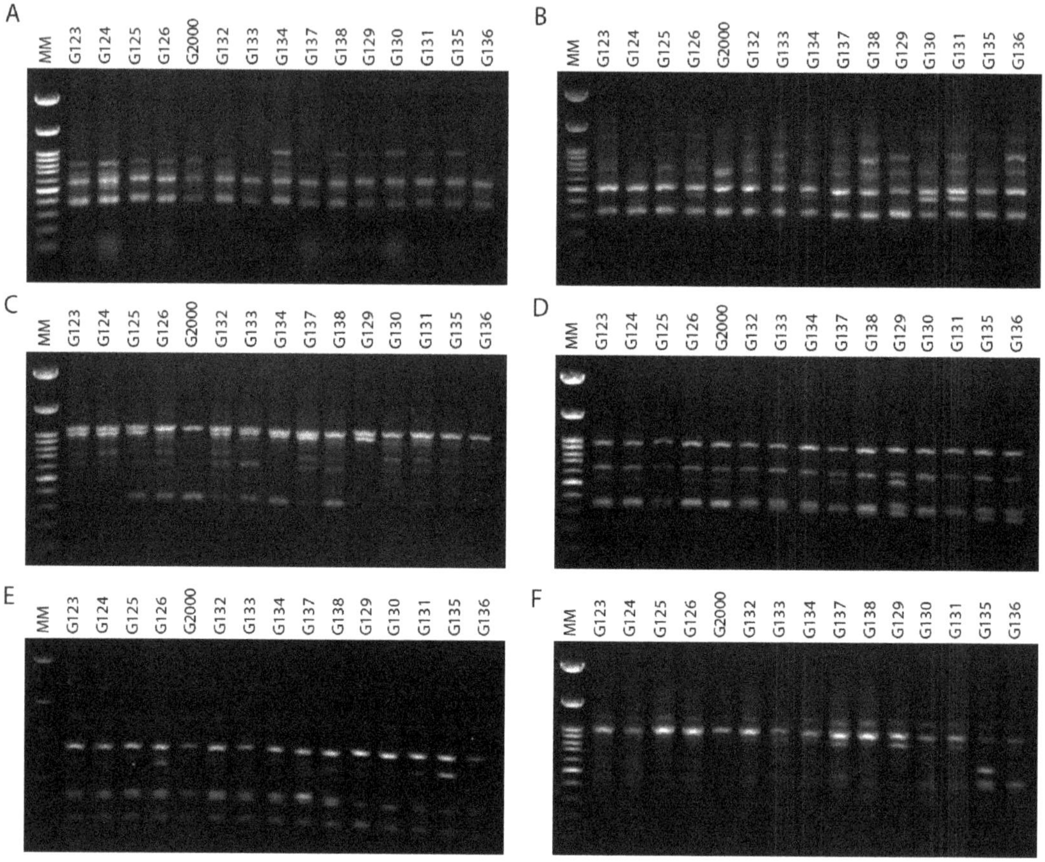

Figure 4. ISSR–PCR product profiles of 15 investigated samples of *Hordeum vulgare L*: (**A**) primer UBC 814, (**B**) primer UBC 826, (**C**) primer UBC 840, (**D**) primer UBC 808, (**E**) primer 807, and (**F**) primer 851. M: molecular size marker (100 bp).

Table 2. ISSR marker profiles for 15 Egyptian six-rowed barley cultivars.

Primer No.	Name	Sequence	MB	POU	UB	PWU	TNB	P%	MBF	PIC	RP	EMR	MI
1	UBC 825	(AC)$_7$ T	7	2	0	2	9	22	1.0	0.32	5.27	8.87	0.02
2	UBC 835	(Ag)$_8$ YC	7	7	0	7	14	50	0.8	0.26	5.86	12.07	0.02
3	UBC 814	(CT)$_7$ CAT	2	3	1	4	6	67	0.6	0.37	12.67	8.33	0.05
4	UBC 826	(AC)$_8$ C	6	4	0	4	10	40	0.8	0.30	7.40	11.30	0.03
5	UBC 827	(AC)$_8$ G	6	5	1	6	12	50	0.6	0.36	11.17	9.42	0.02
6	UBC 840	(gA)$_8$ TT	2	6	0	6	8	75	0.6	0.37	10.50	8.75	0.04
7	UBC 808	(Ag)$_8$ C	4	3	0	3	7	43	0.7	0.35	10.00	10.00	0.04
8	UBC 811	(gA)$_7$ gC	5	3	0	3	8	38	0.7	0.32	8.25	10.88	0.04
9	UBC 844A	(CT)$_8$ AC	1	4	0	4	5	80	0.4	0.37	10.40	6.20	0.04
10	UBC 901	(CA)$_8$ RY	3	2	0	2	5	40	0.8	0.27	6.00	12.00	0.05
11	807	(AG)$_8$ T	5	4	1	5	10	50	0.7	0.35	10.00	10.00	0.03
12	810	(GA)$_8$ T	4	6	0	6	10	60	0.7	0.33	8.60	10.70	0.03
13	841	(GA)$_8$ YC	3	4	0	4	7	57	0.7	0.33	9.14	10.43	0.04
14	857	(AC)$_8$ YG	5	2	0	2	7	29	0.8	0.28	6.29	11.86	0.04
15	851	(GT)$_8$ YG	4	3	1	4	8	50	0.6	0.36	11.25	9.38	0.04
	Total		64	58	4	62	126	-		-	-	-	-
	Mean		4.27	3.87	0.27	4.13	8.40	50.07		49	0.31	0.31	8.85

Each estimated parameter's minimum and maximum values are highlighted in yellow. MB: monomorphic bands; POU: polymorphic without uniqueness; UB: unique bands; PWU: polymorphic with uniqueness; TNB: total number of bands; P%: polymorphism (%); MBF: mean of band frequency; PIC: polymorphism information content; RP: resolving power; EMR: effective multiplex ratio; MI: marker index.

Genetic diversity in some six-rowed barley cultivars grown in Egypt was assessed using ISSR markers. The 15 ISSR primers produced 97 markers that were utilized to investigate the genetic diversity among the studied cultivars. A polymorphism percentage of 50.07%, with an average of 4.13 markers per primer, was found among the studied cultivars (Table 2). However, this number ranges from two for ISSR UBC 825 and ISSR 857, to seven for ISSR UBC 835. The ISSR primers produced single and unique bands, and four molecular primers had these bands (ISSR UBC 814, ISSR UBC 827, ISSR 807, and ISSR 851). The use of ISSR markers for fingerprinting previously resulted in high polymorphism between species, and reflected intraspecific variations within species [14,15,40]. In addition to the high level of polymorphism observed in the current study by ISSR, this may imply high insertional activity in the genome of the tested barley cultivars [21,41,42].

The genetic diversity parameter data revealed by ISSR markers were utilized to calculate the genetic diversity of the studied cultivars by using multivariate clustering, PCA, and heatmap analyses. In a PCA scatterplot, the ISSR markers reflect the robustness of the markers in categorizing the investigated cultivars. PCA analysis indicated that the four six-rowed Egyptian barley cultivars Giza 126, Giza 2000, Giza 125, and Giza 132 were distinct from the other cultivars (Figure 5). Neighboring affinity was also apparent between the Giza 135, Giza 136, and Giza 130 cultivars (Figure 5). Conversely, the rest of the cultivars—Giza 129, Giza 138, Giza 131, Giza 133, Giza 134, and Giza 137—were scattered at some distance from one another. The cultivars Giza 126 and Giza 2000 were the best foragers, as designated by cluster analysis (Figure 5), which also indicated a significant distance between Giza 123 and Giza 124 (Figure 5), and between Giza 132 and Giza 135, Giza 136, and Giza 130 (Figure 5). The differentiation of the studied cultivars in terms of years of release and pedigree may be due to previous alterations in production conditions. There is a possibility that these morphological characteristics can increase or decrease genetic variation between cultivars. Data from ISSR markers analyzed in this study might be explained by the instability of TNB insertion events, cultivar production, and behavior under environmental conditions [17,35]. There may be a correlation between the high degree of polymorphism observed in ISSR markers and genotype diversity [16,21,43]. Although there were differences between the dendrograms based on field characteristics, and in PCA results based on the molecular parameters, both sorted the cultivars into four groups closer to their uses in Egypt.

Multivariate compound similarity analysis is usually utilized to show more information about the genetic variance of plant breeds, which is detailed in heatmaps [40]. The multivariate compound similarities were presented as a heatmap constructed using R software. As indicated by the columns, 15 Egyptian barley cultivars were clustered into 5 clusters with at least 2 per cultivar (Figure 6). The first cluster included the Giza 134, Giza 133, and Giza 136 cultivars. The cultivars Giza 132, Giza 2000, and Giza 128 were discriminated as two neighboring pairs of cultivars. The third cluster consisted of Giza 126 and Giza 137, while Giza 135, Giza 131, Giza 129, and Giza 130 appeared as two neighboring clusters to make up the fourth cluster. The other cultivars—Giza 124, Giza 123, and Giza 125—were located in one group (Figure 6).

Based on the ISSR marker data for the studied cultivars, a genetic distance tree was constructed using Dice's genetic similarity matrix (Figure 7). In this tree, the two pairs (Giza 126 and Giza 2000) and (Giza 132 and Giza 133) were close to the other cultivars. In Egypt, these cultivars are used in human consumption and animal feed. In addition to the malt industry and the beer industry, the ancient Egyptian barley sector dates back to BC. Meanwhile, Giza 132 with Giza 130 and Giza 133 with Giza 130 were less similar to the rest of the barley cultivars, and have been nominated for a crossbreeding program for Egyptian barley breeders. On the other hand, Giza 129 was separated from the rest of the cultivars. All cultivars were distributed in the three clusters. According to the ISSR molecular marker polymorphism, a similarity matrix among the 15 cultivars was derived based on Dice's coefficient (Table 3). According to the similarity matrix of ISSR analysis, the highest similarity value (93%) was observed between (Giza 133 and Giza 132) and (Giza

2000 and Giza 126). Conversely, the lowest similarity value (80%) was recorded between Giza 130 and (Giza 133 and Giza132), indicating that these cultivars were distantly related, as shown in Table 3 and Figure 7. These distinctive cultivars could be expanded to improve soil properties, reduce fertilizer consumption, increase tolerance to drought and salinity, and facilitate growth in newly reclaimed lands. The results were nearly in agreement with those of previous studies [12,42–44].

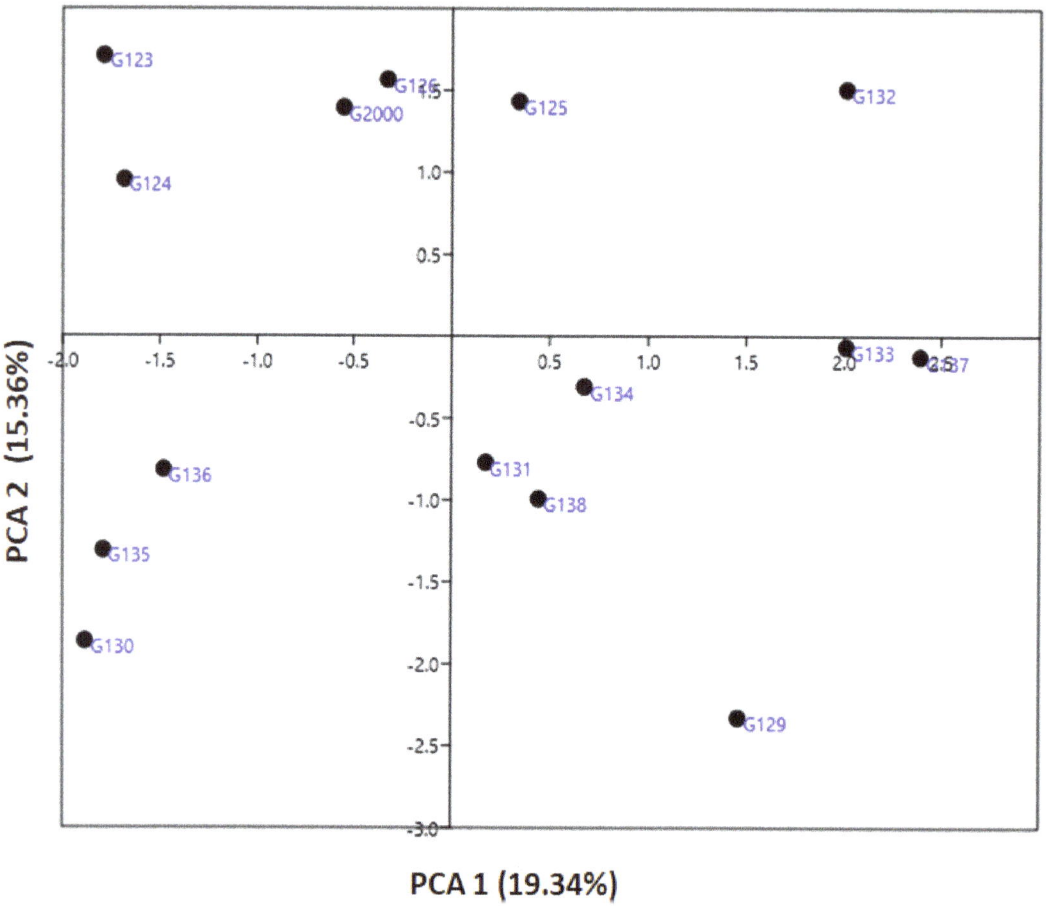

Figure 5. An illustration of the genetic diversity expressed in 15 Egyptian six-rowed barley cultivars, according to a principal component analysis (PCA) based on polymorphism of ISSR markers, using PAST software.

2.4. Biplots

Biplots were used to reflect the statistical values and their presentation in order to provide supportive information about all of the investigated parameters. Biplots have been used in previous studies to illustrate and present different types of data [45–47]. Through the different types of data, the information can be dispersed, but the biplot distributes the genotypes based on all of the traits under study, whether morphological data or molecular data. The biplot in Figure 8 shows the differences between the clusters in the morphological data and the clusters of the molecular data, as well as their interaction; it also clearly demonstrates the effects of each field trait on the genotypes, along with the effects of each initiator molecule on the Egyptian barley genotypes.

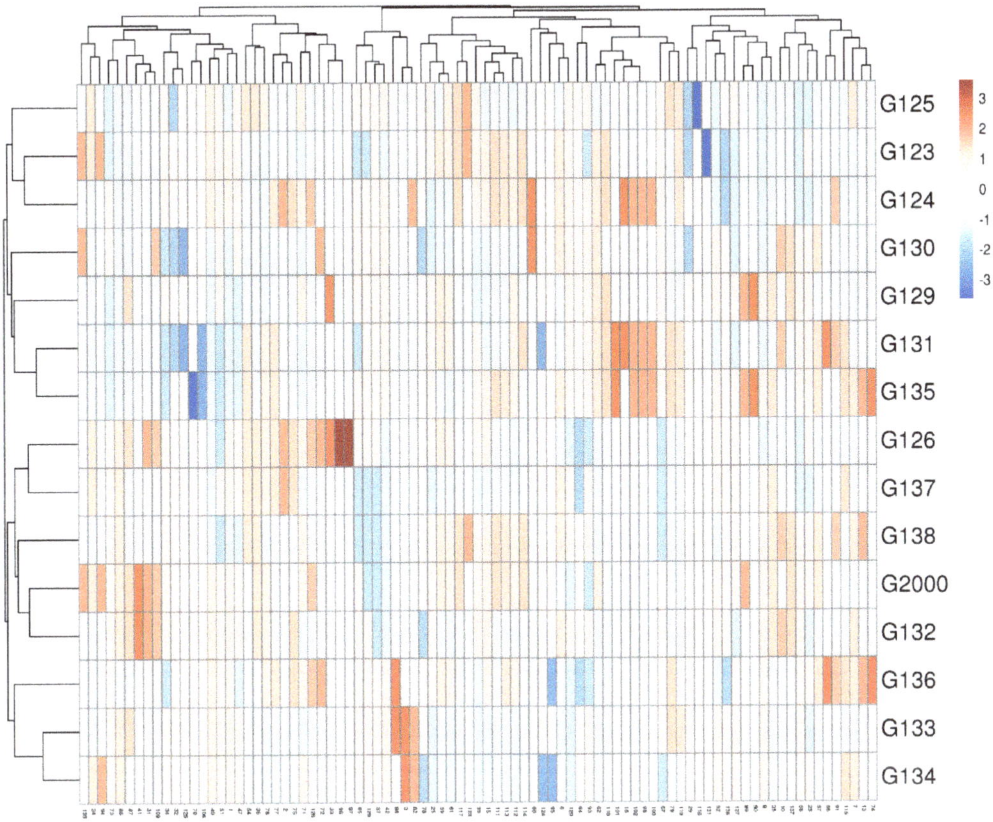

Figure 6. Multivariate heatmap illustrating the genetic diversity of 15 Egyptian six-rowed barley cultivars, based on the 15 ISSR primers for using the module of a heatmap of ClustVis—an online tool for clustering and visualizing of multivariate data [41].

To study the interaction between genotype and environment (GE), biplot analysis was utilized [48]. Using the constructed PCA biplot, it became clear which of the 10 morpho-agronomic traits and 15 ISSR primers contributed most to the discrimination of the examined cultivars (Figure 8). The 15 cultivars were divided into 3 groups based on 10 field traits and 15 molecular ISSR primers. The group that included Giza 130, Giza 136, Giza 138, and Giza 126 was the most influenced by the field and morphological characteristics, as shown in Figure 8. This group was established based on maturity day, biological yield, grain yield, weight of 1000 grains, grain field period, and the number of grains. At the same time, the genotypes Giza 129, Giza 137, and Giza 133 were more influenced by the molecular primers associated with age, including ISSRs 807, UBC 835, UBC 826, 851, and UBC 811, as well as hiding day. On the other hand, the third group was affected by plant height characteristics. The number of spikes per square meter, along with the remainder of the molecular parameters, characterized the cultivars Giza 134, Giza 131, Giza 132, Giza 126 Giza 135, Giza 123, Giza 125, and Giza 2000. Generally, when the cultivar falls on the adjective line, it is more impacted by it. Through the current data, we found that the genetic basis of the ISSR molecular markers is dominant over the morphological traits in the first and second groups, while the effect of field traits is predominant in the third and fourth groups, indicating the merging of the field cluster with the molecular cluster into one form in the biplot. Moreover, the contributions of the genes controlling the traits are shown through the molecular parameters, while the environment is shown by the field traits, and the differences in terms of environment and genetics in this study are united by environmental and genetic data [49,50].

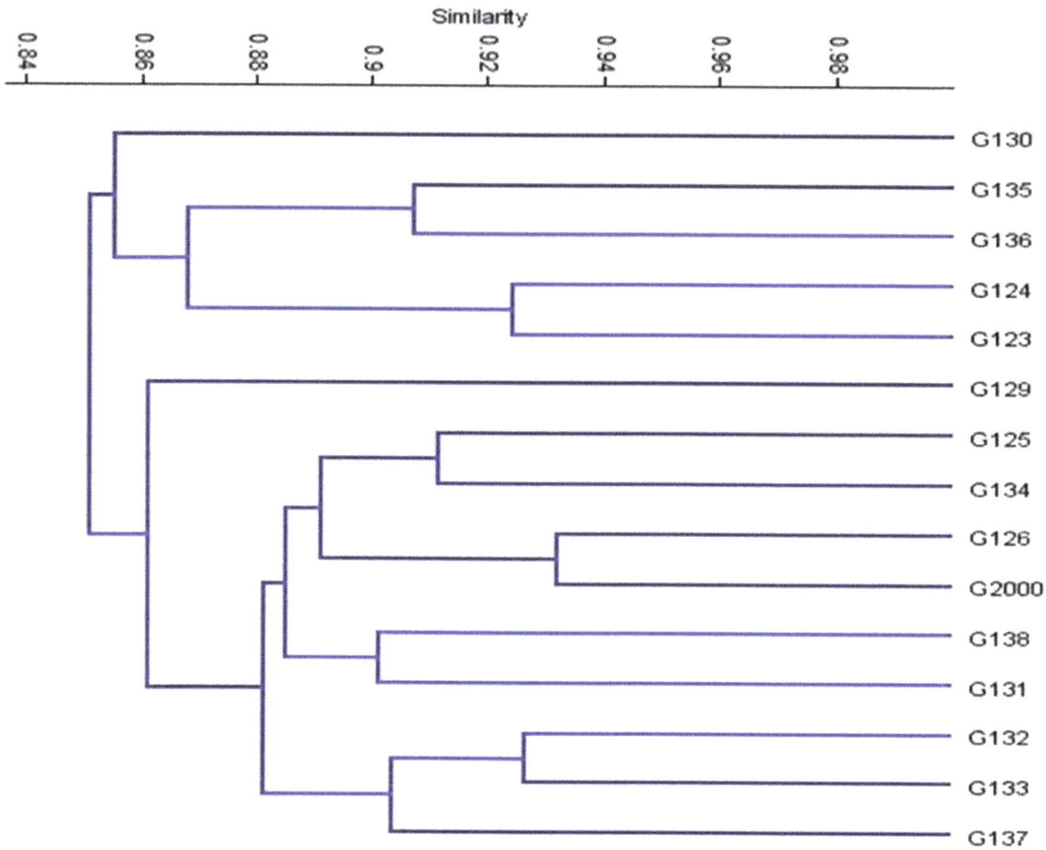

Figure 7. Cluster tree of genetic distance between 15 Egyptian six-rowed barley cultivars, based on the analysis of 15 ISSR primers according to Euclidean distance and the UPGMA algorithm in PAST software.

Table 3. Genetic similarity of the 15 Egyptian six-rowed barley cultivars, based on ISSR fingerprinting.

	G123	G124	G125	G126	G2000	G132	G133	G134	G137	G138	G129	G130	G131	G135	G136
G123	100														
G124	92	100													
G125	90	90	100												
G126	90	86	91	100											
G2000	87	88	91	93	100										
G132	86	86	91	91	89	100									
G133	84	83	89	87	86	93	100								
G134	86	87	91	86	88	87	91	100							
G137	83	83	87	86	84	91	90	86	100						
G138	84	84	90	90	89	87	88	89	89	100					
G129	80	82	85	82	82	86	88	89	86	87	100				
G130	84	86	84	84	84	80	80	83	82	89	83	100			
G131	86	88	89	89	86	90	87	86	89	90	88	88	100		
G135	87	85	83	86	86	82	85	86	82	85	83	86	88	100	
G136	87	88	87	88	87	84	87	87	81	87	86	86	86	91	100

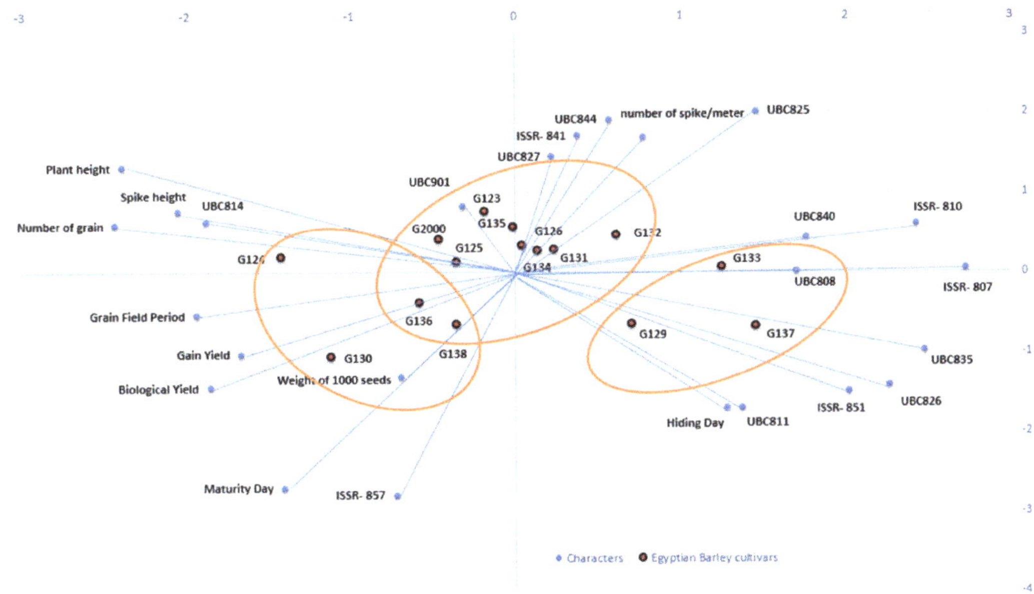

Figure 8. A biplot cluster tree illustrates the genetic distance between 15 Egyptian six-rowed barley cultivars, based on the analysis of 10 morpho-agronomic traits and 15 ISSR primers according to Euclidean distance and the UPGMA algorithm in PAST software.

2.5. DNA Barcoding Loci of *mat*K and *rbc*L Sequencing

DNA barcoding is an essential tool for species identification [51]. Genes from the chloroplast genome—such as *mat*K and *rbc*L—were used for DNA barcoding. The genetic diversity and phylogeny of the studied cultivars were determined by amplification and sequencing of both loci. Four barley cultivars were used for DNA barcoding. Two cultivars, marked by an asterisk (*), had a tough, inedible outer hull around the barley kernel (Giza 123 and Giza 138), while two cultivars marked by two asterisks (**) were characterized by sticks and sprouts that separate from the seed when ripe (Giza 126 and Giza 136) (Table 3). There was 100% amplification success with high specificity of PCR amplification of the *mat*K and *rbc*L regions for all four cultivars, as indicated by sharp DNA bands with no byproducts. The recorded size of the PCR product of the *mat*K region was 900 bp, while for the *rbc*L region it was 600 bp (data not shown). The GenBank accession numbers for *rbc*L in Giza 123, Giza 126, Giza 136, and Giza 138 are MW336986, MW391913, MW336987, and MW391914, respectively. The GenBank accession numbers for *mat*K in Giza 123, Giza 126, Giza 136, and Giza 138 are MW336988, MW336991, MW336990, and MW336989, respectively. To confirm the correct amplification of the *mat*K and *rbc*L sequences, a BLAST function was performed, identifying that all of the sequences were strongly coordinated with *mat*K and *rbc*L of the *Hordeum vulgare* sequences. Sun, et al. [52] assessed the possibility of using five intensively suggested regions (*rbc*L, *mat*K, *trn*H-*psb*A, internal transcribed spacer (*ITS*), and *ITS2*) as DNA barcode candidates to differentiate important species of Brassicaceae in China, in order to establish a new digital identification scheme for economic plants of Brassicaceae. They investigated 58 samples from 27 economic species of Brassicaceae for the success of PCR amplification, intra- and interspecific divergence, DNA barcoding gaps, and identification efficiency. Based on their results, the ITS showed superior discriminative ability, with a rate of 67.2% at the species level when compared with other markers.

Pairwise distances were calculated and evaluated based on the conserved *mat*K and *rbc*L gene sequences, using the WebLogo tool [53]. Additional information on the DNA barcoding regions of *mat*K and *rbc*L in four Egyptian six-rowed cultivars of *Hordeum vulgare* is provided in the Supplementary Materials (Supplementary Figures S1 and S2, respectively); this includes the alignment length, undetermined characters, missing percentages, and variable sites and their proportions, as well as parsimony-informative sites.

Figure 9A illustrates a phylogenetic tree of *mat*K sequence variation using the UPGMA algorithm to discriminate between the four investigated cultivars. To demonstrate the accuracy and efficacy of the created tree, 10 *mat*K sequences were obtained from NCBI and used as outgroups. The tree has three major clusters: the first includes the cultivars in two groups—Giza 123 and Giza 136—and the second includes Giza 126 and Giza 136 (Figure 9B). The third group comprises NCBI outgroup members (Figure 9B). There was a branch length of 1.5, and the bootstrap value displayed next to the branches designates the bootstrap value supporting the node. The Jukes–Cantor method was used to calculate the evolutionary distances based on the base substitutions per site. The bootstrap values were incredibly high (99%), confirming the validity of the tree branching. In the *rbc*L gene region, the four cultivars were distributed into two groups: the first included Giza 123, Giza 136, and Giza 138, while the other contained only Giza 126 (Figure 9C). Additionally, when 10 versions of the identical gene sequences from NCBI were added to GenBank, they resulted in 14 barley genotypes (Figure 9D). The 14 genotypes were distributed into 2 groups: the first group included the cultivar Giza 126 only, while the second group contained the other 13 cultivars (Figure 9D). There was a branch length of 3.5, and the bootstrap value displayed next to the branches designates the bootstrap value supporting the node. The Jukes–Cantor method was used to calculate the evolutionary distances based on the base substitutions per site. The ambiguous plant pairwise deletion option was used. In addition, the rest of the Egyptian cultivars were compared with GenBank's publications, where it was noted that the closest to Giza 123 were the HQ800432 and MN171390 versions. Using DNA barcoding, species could be classified quickly without relying on morphological characteristics. This technique uses DNA fragments of relatively small size as tags to describe or discover species [54].

On the other hand, the MN171392 version was close to Giza 136. Moreover, Giza 138 fell between two versions MN171388 and MN171387 (Figure 9D). After adding NCBI GenBank accession numbers, sequences had 24 genotypes for each region of the *rbc*L and *mat*K genes. The distribution of the GenBank NCBI accession numbers and the Egyptian cultivars did not differ from that of each gene separately from the regions of the *rbc*L and *mat*K genes; however, the similarity percentage was as follows in the *mat*K gene: The GenBank accession numbers of *rbc*L and *mat*K in Giza 123 and Giza 136 were distributed at 99% similarity, whereas the similarity rate of Giza 138 and Giza 126 reached 66% in the area of the genome. Meanwhile, in the *rbc*L gene region, the similarity rate was 56% for Giza 126, while the rest of the GenBank accession numbers and Egyptian cultivars were distributed at a similarity rate of 99%.

The results of the current study show that *rbc*L is less mutable than *mat*K in terms of sequence variability among the examined cultivars. Previous studies used the *mat*K region in many phylogenetic analyses of flowering plants, due to its conservative mode of evolution [55,56]. Four cultivars were differentiated in the present study according to *mat*K sequence variation, using 10 outgroup sequences from NCBI (Figure 9). The phylogenetic tree created using 10 NCBI-extracted *mat*K sequences of *Hordeum vulgare* confirmed the outstanding finding of separating the four cultivars Giza 123, Giza 126, Giza 136, and Giza 138, along with the *Hordeum vulgare* NCBI *mat*K sequence and its subspecies. Nevertheless, Giza 123 and Giza 136 were separated with the *Hordeum vulgare* NCBI *mat*K sequence of the NCBI accession numbers, suggesting sequence homology. In the second cluster, Giza 138 and Giza 126-super-supreme were in the same group, and shared high homology in *mat*K sequences. The *rbc*L region was used to distinguish between wild parents, as well as being used as precise sequences to distinguish between different degrees of biological diversity [57–59]. In addition to providing potentially helpful information for genome-assisted research, the present study also provides useful information for crop improvement.

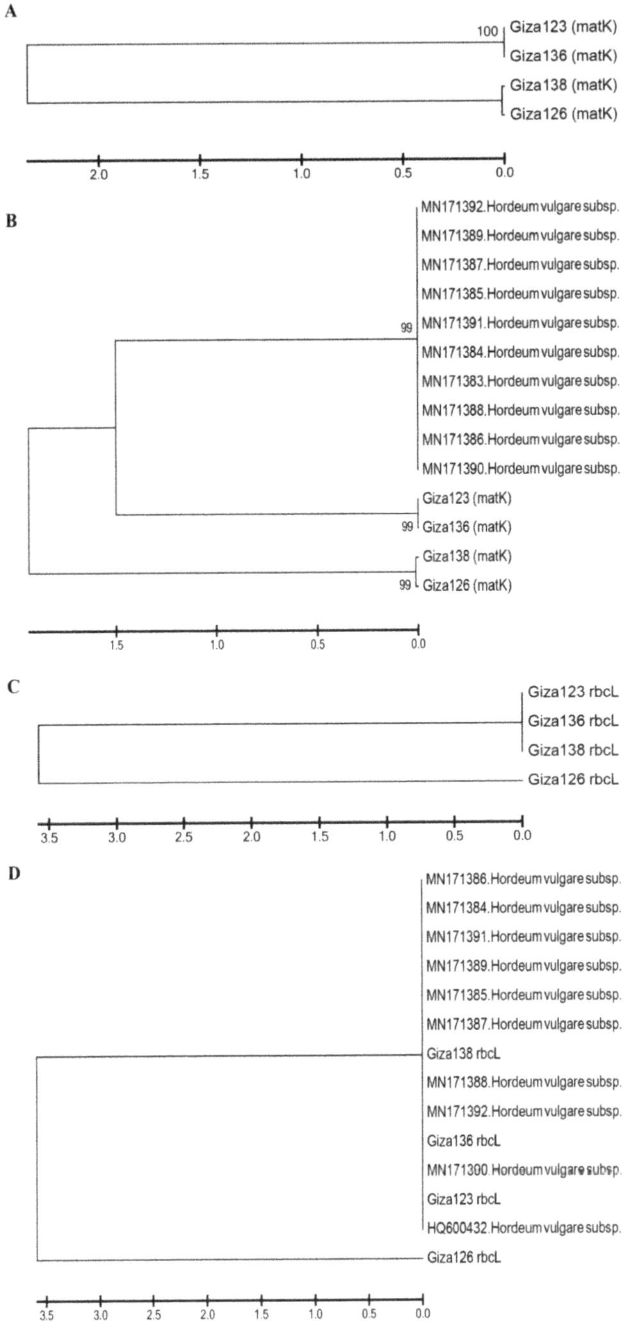

Figure 9. Phylogenetic trees based on (**A**) the *mat*K DNA barcoding region for four six-rowed barley cultivars (Giza 123, Giza 126, Giza 136, and Giza 138); (**B**) the *mat*K DNA barcoding region for four six-rowed barley cultivars, with 10 additional *mat*K sequences of *Hordium vulgaris* L. used as outgroups; (**C**) the *rbc*L DNA barcoding region for four six-rowed barley cultivars; and (**D**) the *rbc*L DNA barcoding region for four six-rowed barley cultivars, with 10 additional *rbc*L sequences of *Hordium vulgaris* L. used as outgroups, using MEGAX software.

3. Materials and Methods

3.1. Plant Materials

This study examined 15 Egyptian barley cultivars (all six-rowed). Those cultivars were selected because they are more critical to the Egyptian barley industry than the two-rowed lines. Viable grains of the studied cultivars were obtained from the Barley Research Department (BRD), Field Crop Research Institute (FCRI), Agricultural Research Center (ARC), Giza, Egypt, during two seasons: 2018/2019 and 2019/2020 (Table 4). These cultivars were chosen based on the recommendations of the barley breeders and the beer industry for their salinity and drought tolerance, high yield, and phytochemical characteristics—such as mineral elements and malt content.

3.2. Morphological Traits and Experimental Design

Two field experiments were carried out at El-Giza Agricultural Research Station (Giza, Egypt) during the successive winter seasons of 2018/2019 and 2019/2020 to study the morphological traits of the different cultivars. To differentiate between the studied cultivars based on morphological characteristics, the following parameters were recorded: days to 50% heading (HD), days to 50% maturity (MD), grain filling period (GFP) (days), plant height (PH) (cm), spike length (SL) (cm), number of grains per spike (average of 10 spikes per square meter), number of spikes per m^2 (No. Sp./m^2), weight of 1000 grains (g), biological yield (BY) (t/ha), and grain yield (GY) (Kg/ha). The grain filling period (GFP) was calculated using the following formula:

$$Grain\ filing\ period = maturity\ days - flowering\ day$$

A randomized complete block design (RCBD) with four replications was used. The plot size was 4 rows that were each 3 m long and 20 cm apart. Analysis of variance and least significant difference (LSD) at 5% were used for comparison between the cultivars.

Table 4. Name, origin, and year of release of the Egyptian six-rowed barley cultivars as recorded by the Barley Research Department, Field Crops Research Institute, Agricultural Research Center, Egypt, and the GenBank accession numbers for the *mat*K and *rbc*L genes for four six-rowed barley cultivars (Giza 123, Giza 126, Giza 136, and Giza 138).

No.	Cultivar	Origin (Year of Release)	*rbc*L GenBank	*mat*K GenBank	Kind	Pedigree	References
1	Giza 123	Egypt (1998)	MW336986	MW336988	Naked *	Giza117/FAO86	[60]
2	Giza 124	Egypt (1998)	NA	NA	Naked	Giza 117/Bahtim 52//Giza 118/ FAO86	[61]
3	Giza 125	Egypt (1995)	NA	NA	Naked	Giza 117/Bahtim52//Giza 118/ FAO86(2)	[62]
4	Giza 126	Egypt (1995)	MW391913	MW336991	Naked	Baladi Bahteem/S D729-Por12762-BC	[62]
5	Giza 129	Egypt (2003)	NA	NA	Hull-less **	Deir Alla106/Cel//As 46/Aths *2	[63]
6	Giza 2000	Egypt (2003)	NA	NA	Naked	Giza117/Bahteem52//Giza118/FAO86/3/Baladi16/Gem	[64]
7	Giza 130	Egypt (2003)	NA	NA	Hull-less	Comp Cross 229//Bco.Mr./DZ02391/3/Deir Alla 106	[63]
8	Giza 131	Egypt (2003)	NA	NA	Hull-less	CM67-B/CENTENO/CAM-B/ROW 906.73/4/GLORIA-BAR-COME-B/5/FALCON-BAR/6/LINO	[65]
9	Giza 132	Egypt (2006)	NA	NA	Naked	Rihane-05//As 46/Aths *2 Aths/Lignee 686	[66]
10	Giza 133	Egypt (2018)	NA	NA	Naked	Carbo/Gustoe	[67,68]
11	Giza 134	Egypt (2019)	NA	NA	Naked	Alanda-01/4/WI2291/3/Api/CM67//L2966-69	[67,68]
12	Giza 135	Egypt (2019)	NA	NA	Hull-less	ZARZA/BERMEJO/4/DS4931//GLORIA-BAR/COPAL/3/SEN/5/AYAROSA	[69]
13	Giza 136	Egypt (2019)	MW336987	MW336990	Hull-less **	PLAISANT/7/CLN-B/4/S.P-B/LIGNEE640/3/S.P-B/GLORIA-BAR/COME-B/5/FALCON-BAR/6/LINO	[70]
14	Giza 137	Egypt (2019)	NA	NA	Naked	Giza 118/4/Rhn-03/3/Mr25-//Att//Mari/Aths *3-02	[32]
15	Giza 138	Egypt (2019)	MW391914	MW336989	Naked *	Acsad1164/3/Mari/Aths *2//M-Att-73-337-1/5/Aths/lignee686/3/Deir Alla106//Sv.Asa/Attiki/4/Cen/Bglo."S"	[32]

NA: not available; *: has a tough, inedible outer hull around the barley kernel; **: hull-less—the sticks and sprouts separate from the bean when ripe, and the chromosome number is 2n = 2× = 14 (map view: "barley genome at ncbi.nlm.nih.gov"; retrieved 6 October 2014).

3.3. Scanning Electron Microscopy (SEM)

Four barley cultivars (Giza 123, Giza 126, Giza 136, and Giza 138) were studied using SEM. Those four cultivars were chosen based on the recommendations of the plant breeders. The chosen cultivars have high production demand and can withstand harsh conditions; they also have excellent synthetic qualities, which is the reason for their examination. For example, Giza 123 tolerates harsh conditions and high salinity levels; Giza 126 has excellent drought tolerance, and is grown under the rain on the northern coast of Egypt, while Giza 136 and Giza 138 are characterized by high yield under all conditions. Viable grains of the studied cultivars were obtained during the season of 2019. The clean and dry seed samples of the studied barley cultivars were placed on double-stick tape mounted on a copper electron microscope holder. The specimens were coated with gold, and then investigated and photographed with a JEOL JSM T200 at 25 kV, in the electron microscope unit of Mansoura University, Mansoura, Egypt. Seed coat technical terms were based on the works of Koul, et al. [71],Murley [72],and Stearn [73].

3.4. ISSR Molecular Markers

3.4.1. Extraction of Genomic DNA

Fresh leaf tissue (0.1 g of combined samples from three different plants) ground in liquid nitrogen with a mortar and pestle was used to extract genomic DNA using the cetyl trimethyl ammonium bromide (CTAB) protocol [74]. DNA concentration and purity for all samples were determined spectroscopically at 260 and 280 nm, respectively. DNA samples were stored at −20 °C for subsequent molecular analysis.

3.4.2. ISSR Amplification

ISSR amplification reactions were carried out in equal volumes (15 µL) containing 7.5 µL of 2× Master Mix (OnePCR™, GeneDireX, Inc., Taipei, Taiwan), 1 µL of DNA template (10 ng/µL), and 1 µL of primer. The names and sequences of the ISSR primers used in the current study are listed in Table 2. The amplification reaction was performed using a T100™ Thermal Cycler (Bio-Rad® Laboratories, Hercules, CA, USA). The polymerase chain reaction (PCR) program was as follows: initial denaturation at 94 °C for 4 min, followed by 30 cycles, with the first step at 94 °C for 30 s (denaturation), the second step varying between 46 and 52 °C—depending on the GC content of each primer—for 45 s (annealing), and the third step (extension) at 72 °C for 1 min, followed by a final extension step at 72 °C for 7 min. The reaction was stopped by maintaining the tubes at 4 °C for at least 30 min. Amplification products were separated via electrophoresis on 1.5% agarose gel in 1× TBE buffer (Tris-borate-EDTA). The gels were stained with 0.5 µg mL^{-1} ethidium bromide (EtBr) solution (Thermo Fisher Scientific, Carlsbad, CA, USA). Then, the gel was documented using a Bio-Rad ChemiDoc™ MP gel documentation system (Bio-Rad). The primers that gave reproducible results were used for data analysis. Polymorphism indices were calculated using iMEC (Online Marker Efficiency Calculator) (https://irscope.shinyapps.io/iMEC/) [75]. ClustVis, a web tool for visualizing clustering of multivariate data, was used to construct heatmaps (https://biit.cs.ut.ee/clustvis/) [41].

3.5. DNA Barcoding of Plastid Genes *rbc*L and *mat*K

DNA barcoding of sequences for the *rbc*L and *mat*K genes was performed using computational analysis. BioEdit software version 7.2.5 (https://bioedit.software.informer.com) was used to analyze and assemble the *rbc*L and *mat*K gene sequences for every cultivar. Using the BLAST function (https://www.ncbi.nlm.nih.gov), the sequences were compared with all accessible sequences in the database. The primers used for barcoding of the *rbc*L and *mat*K genes are listed in Table 5. The PCR program to amplify the two genes was as follows: initial denaturation at 94 °C for 4 min, followed by 40 cycles, with a denaturation step at 94 °C for 30 s, annealing step at 45 °C for 30 s, and elongation step at 72 °C for 30 s, followed by a final extension step at 72 °C for 7 min, after which it was maintained at 4 °C to stop the reaction. The PCR products were subsequently electrophoresed on 1.5%

w/v agarose, stained with 0.5 μg mL^{-1} EtBr solution (Thermo Fisher Scientific) in 1× TBE buffer, and visualized as described for the ISSR PCR amplification. The PCR products of the *mat*K and *rbc*L genes were recovered from agarose gel and purified using the Monarch DNA Gel Extraction Kit (New England Biolabs, Inc., Ipswich, MA, USA), according to the manufacturer's instructions. The purified *mat*K and *rbc*L amplicons were cloned into pGEM®-T Easy Vector Systems (Promega Corporation, Madison, WI, USA) before sequencing. After being transformed into the competent cells of the *E. coli* strain DH5α (Promega, Madison, WI, USA), the positive recombinants were identified via anti-ampicillin selection and verified by PCR screening. Three of the positive clones were sequenced using the ABI PRISM Big Dye Terminator Cycle Sequencing Ready Reaction Kit (Applied Biosystems, Waltham, MA, USA) in conjunction with ABI PRISM (3100 Genetic Analyzer, Macrogen DNA Sequencing Services, Seoul, Korea), as described by Badr, et al. [76]. Using Gblocks software version 0.91b, the revealed nucleotide sequence was assembled [77,78].

Table 5. Primer names, sequences, and product sizes for the *rbc*L and *mat*K genes' DNA barcoding.

Primer Names	Sequence	Product Size	References
*rbc*L-F	5′-ATGTCACCACAAACAGAGACTAAAGC-3′	600 bp	[79]
*rbc*L-R	5′-TCGCATGTACCTGCAGTAGC-3′		
*mat*K-F	5′-CGATCTATTCATTCAATATTTC-3′	900 bp	[80]
*mat*K-R	5′-TCTAGCACACGAAAGTCGAAGT-3′		

Online ClustalW2 software (https://www.ebi.ac.uk/Tools/msa/clustalw2/) was used to align multiple nucleotide sequences, which were double-checked using MEGAX (www.megasoftware.net). Gblocks version 0.91b [77,78] was used to review and assess the gaps in the positions. MEGAX software using the UPGMA algorithm was used to perform the phylogenetic analysis. Confidence of the clustering was attained using SEQBOOT (https://csbf.stanford.edu/phylip/seqboot.html). The sequence logos of the multiple sequence alignments were generated using the WebLogo tool [53]. Additionally, a principal component analysis (PCA) biplot based on the morpho-agronomic data matrix was constructed via multivariate analysis using PAST software versiong 4.02 (https://www.nhm.uio.no/english/research/infrastructure/past/).

3.6. Statistical Analysis

Standard analysis of variance (ANOVA) using least significant differences (LSD) was utilized to estimate the significant differences between the 15 cultivars of six-rowed barley [81]. Dendrogram cluster analysis was used to arrange a set of variables into clusters. A cluster analysis was performed using Euclidean distance and similarity levels [82,83]. ISSR markers that generated clear, distinct, and reproducible bands were recorded as (0) for absence or (1) for presence. The ability of ISSR primers to differentiate between investigated genotypes was analyzed by calculating the polymorphic information content (PIC) [84]. Resolving power (Rp) was measured following the formula of Gilbert, et al. [85]. Additionally, marker index (MI) and effective multiplex ratio (EMR) values were calculated. For the calculation of the coefficient of genetic similarity matrix, and for the construction of a distance tree illustrating the relationships between the tested genotypes, the ISSR marker matrices were used in combination with the unweighted pair group method with arithmetic mean (UPGMA) in PAST software version 4.02 [86]. Furthermore, by using PAST software version 4.02 [86], a PCA scatter diagram was constructed based on a Dice coefficient genetic similarity matrix. ClustVis, a web tool for visualizing clustering of multivariate data, was used to constructe heatmaps (https://biit.cs.ut.ee/clustvis/) [41].

4. Conclusions

Barley plays a vital role in Egypt in terms of agricultural development and added value, as it is used in new and marginal lands. Today, there is expansion in its cultivation due to its high adaptation to water scarcity and other harsh conditions. Thus, barley is added to wheat flour to increase the nutritional value of bread, and is also used in the manufacture of beer and malt. These cultivars are the most common forms of barley in Egypt, so they have been studied for their economic importance in terms of added value and sustainable development. Despite the differences at the molecular level, the examined Egyptian cultivars reflected similarities in terms of field performance under the optimal environment, exhibiting no differences in terms of field characteristics. These cultivars were closely distributed in a genetic tree, similar to the genetic tree based on the molecular description. These differences enable the breeders to choose the best of these cultivars from the most divergent, and to exclude the least different. Moreover, the electron microscope examination reflected differences in the seed surface characteristic, which helps in understanding the chemical content of the Egyptian barley grains and their economic importance. Interestingly, the sequencing results of four cultivars showed that the *rbc*L gene referred to the uniqueness of these four cultivars compared to the sequence database. Nevertheless, the second gene *mat*K revealed that these cultivars are very similar to the GenBank accession numbers. Additionally, the production of new sequences was added to the molecular information about the Egyptian barley cultivars, showing the differences between the Egyptian and European cultivars—especially since Egypt is one of barley's countries of origin. These results will potentially enhance breeding programs and aid in the development of new adaptive or high-yield barley cultivars with specific improved traits.

Supplementary Materials: The following are available online at https://www.mdpi.com/article/10.3390/plants10112527/s1: Figure S1: Sequence logos of the multiple sequence alignment of *mat*K; Figure S2: Sequence logos of the multiple sequence alignment of *rbc*L.

Author Contributions: Conceptualization, A.H.M., A.A.O. and E.M.Z.; methodology, all authors; software, A.H.M., A.A.O., A.M.A., R.M.R. and E.M.Z.; validation, A.H.M., A.A.O., M.M.A.E. and E.M.Z.; formal analysis, A.H.M., A.A.O., R.M.R. and E.M.Z.; investigation, A.H.M., A.A.O. and E.M.Z.; resources, A.M.A., M.M.A.E. and E.M.Z.; data curation, A.M.A., R.M.R. and M.M.A.E.; writing—original draft preparation, all authors; writing—review and editing, A.H.M., A.A.O. and E.M.Z.; visualization, A.A.O. and E.M.Z.; supervision, A.H.M., A.A.O. and E.M.Z. All authors have read and agreed to the published version of the manuscript.

Funding: This research received no external funding.

Institutional Review Board Statement: Not applicable.

Informed Consent Statement: Not applicable.

Data Availability Statement: All the obtained data presented in this the article or supplementary material.

Acknowledgments: The authors would like to thank Abdoallah Sharaf, Iva Mozgová's Lab, Institute of Plant Molecular Biology, Biology Center ASCR, Budweis, South Bohemia, Czechia for helping with the WebLogo analysis of the *mat*K and *rbc*L genes.

Conflicts of Interest: The authors declare no conflict of interest.

References

1. FAO. FAOSTAT. Food and Agriculture Organization of the United Nations. 2019. Available online: https://www.fao.org/faostat/en/#home (accessed on 30 October 2021).
2. Riehl, S. Barley in archaeology and early history. In *Oxford Research Encyclopedia of Environmental Science*; Oxford University Press: Oxford, UK, 2019.
3. Young, B. Barley: The Versatile Crop. *Ethnobot. Leafl.* **2001**, *2001*, 1–3. Available online: https://opensiuc.lib.siu.edu/cgi/viewcontent.cgi?article=1523&context=ebl (accessed on 25 October 2021).
4. Diamond, J.M.; Ordunio, D. *Guns, Germs, and Steel*; Vintage Publications: London, UK, 1999.
5. Kling, J. An introduction to barley-notes from css 330 world foods class. *Accessed April* **2004**, *18*, 2006.
6. Nesbitt, M.; Samuel, D. From staple crop to extinction? The archaeology and history of hulled wheat. In Proceedings of the Hulled Wheat: Promoting the Conservation and Use of Underutilized and Neglected Crops, Rome, Italy, 21–22 July 1995; pp. 41–100.

7. Zohary, D.; Hopf, M. *Domestication of Plants in the Old World: The Origin and Spread of Cultivated Plants in West Asia, Europe and the Nile Valley*; Oxford University Press: Oxford, UK, 2000.
8. Statista. Barley Production Worldwide 2008/2009–2018/2019 Published by M. Shahbandeh, April 27, 2020. 2020. Available online: https://www.statista.com/statistics/271973/world-barley-production-since-2008/ (accessed on 18 November 2021).
9. El-Banna, M.N.; Nassar, M.A.A.E.-G.; Mohamed, M.N.; Boseely, M.A.E.-A. Evaluation of 16 barley genotypes under calcareous soil conditions in Egypt. *J. Agric. Sci.* **2011**, *3*, 105.
10. Najafi, M.T. Evaluation of resistance to Sunn pest (*Eurygaster integriceps* Put.) in wheat and triticale genotypes. *Crop Breed. J.* **2012**, *2*, 43–48. Available online: https://www.sid.ir/EN/VEWSSID/J_pdf/1008420120106.pdf (accessed on 20 October 2021).
11. Al-Sayaydeh, R.; Al-Bawalize, A.; Al-Ajlouni, Z.; Akash, M.; Abu-Elenein, J.; Al-Abdallat, A. Agronomic Evaluation and Yield Performance of Selected Barley (*Hordeum vulgare* L.) Landraces from Jordan. *Int. J. Agron.* **2019**, *2019*, 1–12. [CrossRef]
12. Sharma, A.; Joshi, N.; Cheema, B.; Jindal, M.; Singh, S. Assessment of genetic diversity in barley (*Hordeum vulgare* L.). *J. Res. Punjab Agric. Univ.* **2014**, *51*, 105–108.
13. Bahieldin, A.; Ramadan, A.M.; Gadalla, N.O.; Alzohairy, A.M.; Edris, S.; Ahmed, I.A.; Shokry, A.M.; Hassan, S.M.; Saleh, O.M.; Baeshen, M.N.; et al. Molecular markers for salt tolerant wild barley *Hordeum spontaneum*. *Life Sci. J.* **2012**, *9*, 5838–5847.
14. Godwin, I.D.; Aitken, E.A.; Smith, L.W. Application of inter simple sequence repeat (ISSR) markers to plant genetics. *Electrophoresis* **1997**, *18*, 1524–1528. [CrossRef] [PubMed]
15. Reddy, M.P.; Sarla, N.; Siddiq, E. Inter simple sequence repeat (ISSR) polymorphism and its application in plant breeding. *Euphytica* **2002**, *128*, 9–17. [CrossRef]
16. Tanyolac, B. Inter-simple sequence repeat (ISSR) and RAPD variation among wild barley (*Hordeum. vulgare* subsp. spontaneum) populations from west Turkey. *Genet. Resour. Crop Evol.* **2003**, *50*, 611–614. [CrossRef]
17. Guasmi, F.; Elfalleh, W.; Hannachi, H.; Feres, K.; Touil, L.; Marzougui, N.; Triki, T.; Ferchichi, A. The use of ISSR and RAPD markers for genetic diversity among south tunisian barley. *ISRN Agron.* **2012**, *2012*, 1–10. [CrossRef]
18. Chahidi, B.; El-Otmani, M.; Jacquemond, C.; Tijane, M.h.; El-Mousadik, A.; Srairi, I.; Luro, F. Use of morphological and physiological characters, and molecular markers to evaluate the genetic diversity of three clementine cultivars. *C. R. Biol.* **2007**, *331*, 1–12. [CrossRef] [PubMed]
19. Hailu, F.; Merker, A.; Belay, G.; Johansson, E. Molecular diversity and phylogenic relationships of tetraploid wheat species as revealed by intersimple sequence repeats (ISSR) from Ethiopia. *J. Genet. Breed.* **2005**, *59*, 329–337.
20. Marotti, I.; Bonetti, A.; Minelli, M.; Catizone, P.; Dinelli, G. Characterization of some Italian common bean (*Phaseolus vulgaris* L.) landraces by RAPD, semi-random and ISSR molecular markers. *Genet. Resour. Crop Evol.* **2007**, *54*, 175–188. [CrossRef]
21. Drine, S.; Guasmi, F.; Ali, S.B.; Triki, T.; Boussorra, F.; Ferchichi, A. Genetic diversity analysis of different barley (*Hordeum vulgare* L.) genotypes from arid and humid regions using ISSR and RAPD markers. *J. New Sci.* **2016**, *34*, 1930–1939. Available online: https://www.jnsciences.org/index.php?option=com_attachments&task=download&id=238 (accessed on 19 November 2021).
22. Wang, A.; Yu, Z.; Ding, Y. Genetic diversity analysis of wild close relatives of barley from Tibet and the Middle East by ISSR and SSR markers. *C. R. Biol.* **2009**, *332*, 393–403. [CrossRef]
23. Johnson, P.E.; Patron, N.J.; Bottrill, A.R.; Dinges, J.R.; Fahy, B.F.; Parker, M.L.; Waite, D.N.; Denyer, K. A low-starch barley mutant, Risø 16, lacking the cytosolic small subunit of ADP-glucose pyrophosphorylase, reveals the importance of the cytosolic isoform and the identity of the plastidial small subunit. *Plant Physiol.* **2003**, *131*, 684–696. [CrossRef]
24. Patron, N.J.; Greber, B.; Fahy, B.F.; Laurie, D.A.; Parker, M.L.; Denyer, K. The lys5 mutations of barley reveal the nature and importance of plastidial ADP-Glc transporters for starch synthesis in cereal endosperm. *Plant Physiol.* **2004**, *135*, 2088–2097. [CrossRef]
25. Wijngaard, H.; Renzetti, S.; Arendt, E. Microstructure of buckwheat and barley during malting observed by confocal scanning laser microscopy and scanning electron microscopy. *J. Inst. Brew.* **2007**, *113*, 34–41. [CrossRef]
26. Chen, S.; Yao, H.; Han, J.; Liu, C.; Song, J.; Shi, L.; Zhu, Y.; Ma, X.; Gao, T.; Pang, X. Validation of the ITS2 region as a novel DNA barcode for identifying medicinal plant species. *PLoS ONE* **2010**, *5*, e8613. [CrossRef]
27. Wei, S.; Luo, Z.; Cui, S.; Qiao, J.; Zhang, Z.; Zhang, L.; Fu, J.; Ma, X. Molecular Identification and Targeted Quantitative Analysis of Medicinal Materials from Uncaria Species by DNA Barcoding and LC-MS/MS. *Molecules* **2019**, *24*, 175. [CrossRef]
28. DeSalle, R. Species discovery versus species identification in DNA barcoding efforts: Response to Rubinoff. *Conserv. Biol.* **2006**, *20*, 1545–1547. [CrossRef]
29. Viglietti, G.; Galla, G.; Porceddu, A.; Barcaccia, G.; Curk, F.; Luro, F.; Scarpa, G.M. Karyological Analysis and DNA Barcoding of Pompia Citron: A First Step toward the Identification of Its Relatives. *Plants* **2019**, *8*, 83. [CrossRef]
30. Barley, A.J.; Thomson, R.C. Assessing the performance of DNA barcoding using posterior predictive simulations. *Mol. Ecol.* **2016**, *25*, 1944–1957. [CrossRef]
31. Lonardi, S.; Duma, D.; Alpert, M.; Cordero, F.; Beccuti, M.; Bhat, P.R.; Wu, Y.; Ciardo, G.; Alsaihati, B.; Ma, Y. Barcoding-free BAC pooling enables combinatorial selective sequencing of the barley gene space. *arXiv* **2011**, arXiv:1112.4438.
32. Amer, K.A.; Abou El Enein, R.A.; El-Sayed, A.A.; Noaman, M.M.; Ahmed, I.A.; El-Moselhy, M.A.; Moustafa, K.A.; Abd El-Hamid, M.; Megahed, M.A.; El-Bawab, A.M.O.; et al. Giza 137 and Giza 138, new Egyptian six-rowed Barley cultivars for new land. *Egypt. J. Plant Breed.* **2017**, *21*, 380–395.
33. Noaman, M.; Asaad, F.; El-Sayed, A.; El-Bawab, A. Drought tolerant barley genotypes for rainfed areas in Egypt. *Egypt. J. Agric. Res.* **1997**, *75*, 1019–1036.

34. Mariey, S.; Farid, M.; khatab, I. Physiological and molecular characterization of some Egyptian barley (*Hordeum vulgare* L.) cultivars for salt tolerance. *Egypt. J. Genet. Cytol.* **2016**, *45*, 367–382. [CrossRef]
35. Mariey, S.; Khedr, R.A. Evaluation of some Egyptian barley cultivars under water stress conditions using drought tolerance indices and multivariate analysis. *J. Sustain. Agr. Sci.* **2017**, *43*, 105–114. [CrossRef]
36. Mareiy, S.A.; Farid, M.A.; Karima, A. Morphological and Molecular Characterization of Some Egyptian Barley Cultivars under Calcareous Soil conditions. *Middle East J. Agric. Res.* **2018**, *7*, 408–420. Available online: http://www.curresweb.com/mejar/mejar/2018/408-420.pdf (accessed on 25 October 2021).
37. Moïse, J.A.; Han, S.; Gudynaitę-Savitch, L.; Johnson, D.A.; Miki, B.L. Seed coats: Structure, development, composition, and biotechnology. *In Vitro Cell. Dev. Biol.-Plant* **2005**, *41*, 620–644. [CrossRef]
38. Jabbarzadeh, Z.; Khosh-Khui, M.; Salehi, H.; Saberivand, A. Inter simple sequence repeat (ISSR) markers as reproducible and specific tools for genetic diversity analysis of rose species. *Afr. J. Biotechnol.* **2010**, *9*, 6091–6095. Available online: https://www.ajol.info/index.php/ajb/article/view/92194/81648 (accessed on 25 October 2021).
39. Pérez de la Torre, M.; García, M.; Heinz, R.; Escandón, A. Analysis of genetic variability by ISSR markers in *Calibrachoa caesia*. *Electron. J. Biotechnol.* **2012**, *15*, 8. [CrossRef]
40. Alshehri, M.A. Genetic variation of Saudi Wheat Genotypes through ISSR and SCoT Assays. *Int. J. Pharm. Phytopharm. Res.* **2019**, *9*, 57–65.
41. Metsalu, T.; Vilo, J. ClustVis: A web tool for visualizing clustering of multivariate data using Principal Component Analysis and heatmap. *Nucleic Acids Res.* **2015**, *43*, W566–W570. [CrossRef] [PubMed]
42. Mariey, S.A.; Mohamed, A.M.; Ali, H.G. Effect of salinity stress on physiological and biochemical traits of barley cultivars. *Int. J. Environ.* **2018**, *7*, 65–77. Available online: http://www.curresweb.com/ije/ije/2018/65-77.pdf (accessed on 15 October 2021).
43. Samuel, A.D.; Bungau, S.; Tit, D.M.; Melinte, C.E.; Purza, L.; Badea, G.E. Effects of long term application of organic and mineral fertilizers on soil enzymes. *Rev. Chim.* **2018**, *69*, 2608–2612. [CrossRef]
44. Bungau, S.; Behl, T.; Aleya, L.; Bourgeade, P.; Aloui-Sossé, B.; Purza, A.L.; Abid, A.; Samuel, A.D. Expatiating the impact of anthropogenic aspects and climatic factors on long-term soil monitoring and management. *Environ. Sci. Pollut. Res.* **2021**, *28*, 1–23. [CrossRef]
45. Cao, Y.; Zhang, E.; Langdon, P.; Liu, E.; Shen, J. Spatially different nutrient histories recorded by multiple cores and implications for management in Taihu Lake, eastern China. *Chin. Geogr. Sci.* **2013**, *23*, 537–549. [CrossRef]
46. Mellor, L.F.; Nordberg, R.C.; Huebner, P.; Mohiti-Asli, M.; Taylor, M.A.; Efird, W.; Oxford, J.T.; Spang, J.T.; Shirwaiker, R.A.; Loboa, E.G. Investigation of multiphasic 3D-bioplotted scaffolds for site-specific chondrogenic and osteogenic differentiation of human adipose-derived stem cells for osteochondral tissue engineering applications. *J. Biomed. Mater. Res. Part B Appl. Biomater.* **2020**, *108*, 2017–2030. [CrossRef]
47. Mellor, L.F.; Huebner, P.; Cai, S.; Mohiti-Asli, M.; Taylor, M.A.; Spang, J.; Shirwaiker, R.A.; Loboa, E.G. Fabrication and evaluation of electrospun, 3D-bioplotted, and combination of electrospun/3D-bioplotted scaffolds for tissue engineering applications. *BioMed Res. Int.* **2017**, *2017*, 1–9. [CrossRef] [PubMed]
48. Yang, R.C.; Crossa, J.; Cornelius, P.L.; Burgueño, J. Biplot analysis of genotype× environment interaction: Proceed with caution. *Crop Sci.* **2009**, *49*, 1564–1576. [CrossRef]
49. Yang, C.; Tang, D.; Qu, J.; Zhang, L.; Zhang, L.; Chen, Z.; Liu, J. Genetic mapping of QTL for the sizes of eight consecutive leaves below the tassel in maize (*Zea mays* L.). *Theor. Appl. Genet.* **2016**, *129*, 2191–2209. [CrossRef]
50. Mohamed, N.E. Genotype by environment interactions for grain yield in bread wheat (*Triticum aestivum* L.). *J. Plant Breed. Crop. Sci.* **2013**, *7*, 150–157. Available online: https://academicjournals.org/journal/JPBCS/article-full-text-pdf/316DDD32668 (accessed on 15 November 2021). [CrossRef]
51. Khalil, S.; Ibrahim, S.; Ahmed, S.; Abdalla, K. DNA barcoding of jojoba (*simmondsia chinensis*) plants cultivated in Egypt using *rbc*L gene. *Egypt. J. Genet. Cytol.* **2020**, *49*, 245–254. Available online: http://www.journal.esg.net.eg/index.php/EJGC/article/view/333/324 (accessed on 15 November 2021).
52. Sun, X.; Qu, Y.; Yao, H.; Zhang, Y.; Yan, Q.; Hang, Y. Applying DNA barcodes for identification of economically important species in Brassicaceae. *Gen. Mol. Res.* **2015**, *14*, 15050–15061. [CrossRef] [PubMed]
53. Crooks, G.E.; Hon, G.; Chandonia, J.-M.; Brenner, S.E. WebLogo: A sequence logo generator. *Genome Res.* **2004**, *14*, 1188–1190. [CrossRef]
54. Selvaraj, D.; Park, J.-I.; Chung, M.-Y.; Cho, Y.-G.; Ramalingam, S.; Nou, I.-S. Utility of DNA barcoding for plant biodiversity conservation. *Plant Breed. Biotechnol.* **2013**, *1*, 320–332. [CrossRef]
55. Hilu, K.W.; Liang, g. The *mat*K gene: Sequence variation and application in plant systematics. *Am. J. Bot.* **1997**, *84*, 830–839. [CrossRef]
56. Zhang, X.; Fan, M.; Wang, D.; Zhou, P.; Tao, D. Top-k feature selection framework using robust 0–1 integer programming. *IEEE Trans. Neur. Net. Lear. Sys.* **2020**. [CrossRef]
57. Bafeel, S.; Arif, I.; Bakir, M.; Al Homaidan, A.; Al Farhan, A.; Khan, H. DNA barcoding of arid wild plants using *rbc*L gene sequences. *Gen. Mol. Res.* **2012**, *11*, 1934–1941. [CrossRef]
58. Kress, W.J.; Erickson, D.L. A two-locus global DNA barcode for land plants: The coding *rbc*L gene complements the non-coding *trn*H-*psb*A spacer region. *PLoS ONE* **2007**, *2*, e508. [CrossRef] [PubMed]

59. Sundari, K.; Jayali, A.; Sukamto, N. The application of barcode DNA *rbc*L gene for identification of medicinal plants: Red jabon and gofasa. *J. Phys. Conf. Ser.* **2019**, *1146*, 012030. [CrossRef]
60. Ahmed, I.A.; El-Gamal, A.S.; Abo El-Enein, R.A.; El-Sayed, A.A.; El-Hag, A.A.; El-Bawab, A.M.O.; El-Sherbiny, A.M.; El-Moselhy, M.A.; Asaad, F.A.; Megahed, M.A.; et al. Giza 123, a new barley variety for the newly reclaimrd areas in Egypt. *Egypt. J. Appl. Sci.* **1998**, *13*, 83–92.
61. Abo-Elenein, R.A.; Ahmed, I.A.; El-Sayed, A.A.; El-Gamal, A.S.; El-Sherbiny, A.M.; El-Bawab, A.M.O.; El-Hag, A.A.; Abd El-Hamid, M.M.; Asaad, F.A.; Megahed, M.A.; et al. Giza 124, a new cultivar for upper Egypt. *Egypt. J. Appl. Sci.* **1998**, *13*, 100–109.
62. Noaman, M.; El-Sayed, A.; Assad, F.; El-Sherbini, A.; El-Bawab, A.; El-Moselhi, M.; Rizk, R. Giza 125 and Giza 126, two new barley cultivars for rainfed areas of Egypt. *Egypt. J. Appl. Sci.* **1995**, *10*, 418–432.
63. El-Sayed, A.; Abo-Elenin, R.; El-Gamal, A.; Megahed, M.; El-Moselhy, M.; El-Sherbini, A.; El-Hag, A.; El-Bawab, A.; Abdel-Hameed, M.; Amer, K. Giza 129 and Giza 130, two newly released hulless barley varieties for irrigated lands in Egypt. *Egypt. J. Plant Breed.* **2003**, *7*, 387–398.
64. Ahmed, I.; El-Sayed, A.; Abo-El-Enin, R.; El-Gamal, A.; Noaman, M.; Rizk, A.; Mahfouz, H. Giza 2000, a new Egyptian barley variety for newly reclaimed lands and rainfed areas. *Zagazig J. Agri. Res.* **2003**, *30*, 2095–2112.
65. El-Sayed, A.; El-Enein, R.A.; El-Gamal, A.; El-Sherbiny, A.; El-Moselhy, M.; Megahed, M.; El-Hag, A.; El-Bawab, A.; Abdelhamid, M.; Amer, K. Two new food hull-less barley varieties for rainfed in Egypt. In Proceedings of the 4th International Crop Science Congress, Brisbane, Australia, 26 September–1 October 2004.
66. Noaman, M.M.; El-Sayed, A.A.; Abo El-Enein, R.A.; Ahmed, I.A.; El-Gamal, A.S.; El-Sherbiny, A.M.; Abd El-Hameed, M.m.; Megahed, M.A.; Moselhy, M.A.; El-Bawab, A.M.; et al. Giza 132, a new drought tolerant six-rowed barley cultivar. *Egypt. J. Appl. Sci.* **2006**, *21*, 46–58.
67. El-Bawab, A.; Anton, N.; Ashmawy, H.; Seham, M.M.; Abd El-Rahman, M. Evaluation of three barley cultivars under water deficit conditions of new reclaimed lands. *Egypt. J. Plant Breed.* **2014**, *18*, 687–699. [CrossRef]
68. El-Bawab, A.M.O.; El-Sayed, A.A.; El-Moselhy, M.A.; Mahmoud, M.A.; Amer, K.A.; Megahed, M.A.; Saad, M.F.; Ashmawy, H.A.; Eid, A.A.; Aboelenein, R.A.; et al. Giza 133 and Giza 134, two new varieties for new reclaimed areas in Egtpt. *Egypt. J. Plant Breed.* **2011**, *15*, 1–10.
69. El-Sayed, A.A.; El-Bawab, A.M.O.; El-Moselhy, M.A.; Amer, K.A.; Ashmawy, H.A.; Shendy, M.Z.; Abbas, S.I.; Eid, A.A.; Mahmoud, M.A.; Megahed, M.A.; et al. Giza 135, a new wide adapted hull-less barley variety for all barley production areas in Egypt. *Egypt. J. Plant Breed.* **2011**, *15*, 81–95.
70. El-Sayed, A.A.; El-Bawab, A.M.O.; El-Moselhy, M.A.; Amer, K.A.; Ashmawy, H.A.; Shendy, M.Z.; Abbas, S.I.; Eid, A.A.; Mahmoud, M.A.; Megahed, M.A.; et al. Giza 136, a new high yielding and stable hull-less barley variety for irrigated areas in Egypt. *Egypt. J. Plant Breed.* **2011**, *15*, 39–50.
71. Koul, K.; Nagpal, R.; Raina, S. Seed coat microsculpturing in Brassica and allied genera (subtribes Brassicinae, Raphaninae, Moricandiinae). *Ann. Bot.* **2000**, *86*, 385–397. [CrossRef]
72. Murley, M.R. Seeds of the Cruciferae of northeastern North America. *Am. Midl. Nat.* **1951**, *46*, 1–81. [CrossRef]
73. Stearn, W.T. *Botanical Latin*; Timber Press, Incorporated: Portland, OR, USA, 1973.
74. Doyle, J.J.; Doyle, J.L. A rapid DNA isolation procedure for small quantities of fresh leaf tissue. *Phytochem. Bull.* **1987**, *19*, 11–15.
75. Amiryousefi, A.; Hyvönen, J.; Poczai, P. iMEC: Online marker efficiency calculator. *Appl. Plant Sci.* **2018**, *6*, e01159. [CrossRef]
76. Badr, A.; El-Sherif, N.; Aly, S.; Ibrahim, S.D.; Ibrahim, M. Genetic diversity among selected medicago sativa cultivars using inter-retrotransposon-amplified polymorphism, chloroplast DNA barcodes and morpho-agronomic trait analyses. *Plants* **2020**, *9*, 995. [CrossRef]
77. Castresana, J. Selection of conserved blocks from multiple alignments for their use in phylogenetic analysis. *Mol. Biol. Evol.* **2000**, *17*, 540–552. [CrossRef] [PubMed]
78. Talavera, G.; Castresana, J. Improvement of phylogenies after removing divergent and ambiguously aligned blocks from protein sequence alignments. *Syst. Biol.* **2007**, *56*, 564–577. [CrossRef]
79. Bell, K.L.; Loeffler, V.M.; Brosi, B.J. An *rbc*L reference library to aid in the identification of plant species mixtures by DNA metabarcoding. *Appl. Plant Sci.* **2017**, *5*, 1600110. [CrossRef]
80. Barthet, M.M.; Hilu, K.W. Expression of *mat*K: Functional and evolutionary implications. *Am. J. Bot.* **2007**, *94*, 1402–1412. [CrossRef]
81. Steel, R.G.D.; Torrie, J.H. *Principles and Procedures of Statistics: A Biometrical Approach*; McGraw-Hill: London, UK, 1986.
82. Eisen, M.B.; Spellman, P.T.; Brown, P.O.; Botstein, D. Cluster analysis and display of genome-wide expression patterns. *Proc. Natl. Acad. Sci. USA* **1998**, *95*, 14863–14868. Available online: https://www.pnas.org/content/pnas/95/25/14863.full.pdf (accessed on 30 October 2021). [CrossRef] [PubMed]
83. Everitt, B.S. Cluster analysis. In *Multivariate Analysis for the Behavioral Sciences*; CRC Press: New York, NY, USA, 2018; pp. 341–363.
84. Ghislain, M.; Zhang, D.; Fajardo, D.; Huamán, Z.; Hijmans, R.J. Marker-assisted sampling of the cultivated Andean potato Solanum phureja collection using RAPD markers. *Genet. Resour. Crop Evol.* **1999**, *46*, 547–555. [CrossRef]

85. Gilbert, J.; Lewis, R.; Wilkinson, M.; Caligari, P. Developing an appropriate strategy to assess genetic variability in plant germplasm collections. *Theor. Appl. Genet.* **1999**, *98*, 1125–1131. [CrossRef]
86. Hammer, Ø.; Harper, D.A.; Ryan, P.D. PAST: Paleontological statistics software package for education and data analysis. *Palaeontol. Electron.* **2001**, *4*, 9. Available online: https://palaeo-electronica.org/2001_1/past/past.pdf (accessed on 1 November 2021).

plants

MDPI

Article

RWLMod—Potential Model to Study Plant Tolerance in Drought Stress Conditions

Florin Sala [1], Mihai Valentin Herbei [2,*] and Ciprian Rujescu [3]

1 Department—Soil Sciences, Compartment of Soil Science and Plant Nutrition, Banat University of Agricultural Sciences and Veterinary Medicine "King Michael I of Romania", 300645 Timisoara, Romania; florin_sala@usab-tm.ro

2 Department—Sustainable Development and Environmental Engineering, Compartment Remote Sensing and GIS, Banat University of Agricultural Sciences and Veterinary Medicine "King Michael I of Romania", 300645 Timisoara, Romania

3 Department—Management and Rural Development, Compartment Mathematics and Statistics, Banat University of Agricultural Sciences and Veterinary Medicine "King Michael I of Romania", 300645 Timisoara, Romania; rujescu@usab-tm.ro

* Correspondence: mihai_herbei@yahoo.com; Tel.: +40-722-751782

Abstract: Rationale: Water loss by evaporation is a normal physiological process, in order to regulate plant temperature. Under conditions of thermal and water stress, water loss is accelerated compared to normal conditions, and the response of plants is variable. In extreme cases, it can lead to wilting and death of plants. It was found that the phenomenon of water loss behaved as a pattern in different plant species, given by two functions, logistics (first part of water loss) and hyperbola (second part of water loss) in relation to a moment m, at which the rate of water loss (RWL) has reached its maximum value. Method: We studied the water loss process for a series of plant samples on different plant species (*Picea abies* L., H. Karst; *Juniperus communis* L.; *Pinus silvestris* L.; *Thuja occidentalis* L.; *Lamium purpureum* L.; *Veronica hederifolia* L.), measuring the rate of weight loss (RWL) in controlled conditions. The drying of the samples was done in identical conditions (thermo-balance, 100 °C, standard temperature for drying the plant samples) with the real-time recording of the drying time simultaneously with the water loss rate (RWL) from the plant samples. The exposure time varied, depending on each species sample, and was approximately 1000 s. Results: The experimental data was recorded at intervals of every 10 s, during the entire drying period. RWL values varied from 0.024 to 0.054 g/min at the beginning of the drying process and reached maximum values after 70–100 s, having values between 0.258 g/min and 0.498 g/min. During the drying period, this indicator presented different graphic evolutions, difficult to be described with a single function. The first segment was described by a logistic function, and the second was described by a hyperbola, resulting in a model (RWLMod) which described the real phenomenon. This model and theoretical calculation were used to quantify the water loss in a time interval and, compared with empirical dates, no significant differences were observed, which indicated an increased degree of accuracy regarding the use of this model. Recommendation and novelty of work: The novelty of the work is given by the obtained model (RWLMod), which makes possible the description of RWL over the entire time interval, and ensures a good fit with the real data. It recommends the method and model in studies of plant behaviour under stress in relation to different influencing factors.

Keywords: drought stress; drying processes; mathematical model; plant hydric stress tolerance; rate of weight loss; RWLMod; water evaporation

Citation: Sala, F.; Herbei, M.V.; Rujescu, C. RWLMod—Potential Model to Study Plant Tolerance in Drought Stress Conditions. *Plants* **2021**, *10*, 2576. https://doi.org/10.3390/plants10122576

Academic Editor: Milan S. Stankovic

Received: 24 September 2021
Accepted: 23 November 2021
Published: 25 November 2021

Publisher's Note: MDPI stays neutral with regard to jurisdictional claims in published maps and institutional affiliations.

Copyright: © 2021 by the authors. Licensee MDPI, Basel, Switzerland. This article is an open access article distributed under the terms and conditions of the Creative Commons Attribution (CC BY) license (https://creativecommons.org/licenses/by/4.0/).

1. Introduction

Water has a vital role in plant life, in relation to physiological and metabolic processes, plant nutrition, thermoregulation, plant growth and development, tolerance to stressors, etc. Knowing the dynamics of water loss in plants has multiple applications, beginning with a better understanding of the crop behaviour under conditions of water stress; we also identified technological issues on improving methods for the processing of plant products. The literature contains studies on the physiological mechanisms of water loss [1,2] on increasing tolerance regarding hydric stress [3–5], and on the drying process of some aromatic plants with economic possibilities [6]. In addition, the dynamics of plant humidity was studied under the influence of some external factors, such as air speed [7], or physiological and biochemical factors [8,9]. In the case of crop plants, there was a special interest regarding the increase of plant resistance to water stress and thermal factors; a number of studies have analysed physiological indices [10–12], water efficiency in plants [13,14], photosynthetic capacity and production quality [15,16], regarding the increased demands for food production, in the context of population growth and climate change [17,18]. There is a confirmed existence for the particular well-defined dynamic of some indicators describing humidity, and existing studies are using a mathematical characterization of these processes. Thus, the variation in humidity is behind the creation of some mathematical models, using the study of isothermal curves [19,20], or described using sigmoid curves [21]. In fact, sigmoidal curves that form a part of the mathematical model used in this paper are found in certain distinct classifications of nonlinear models used in agricultural sciences [22]. Phenomena with a downward asymptotic trend have been evaluated in studies regarding the behaviour of biological and biochemical processes [22]. Moreover, the necessity of using functions based on mathematical models which describe specific situations as accurately as possible require the adjustment and modification of classical models. Regarding old exponential models or those with limited growth, there are multiple concerns for setting functions to provide such an attribute. Power-Ricker and modified logistic function [23,24] are such examples, with applications in various fields of biology.

The problem of plant water loss was addressed in this study, in terms of rate of weight loss (RWL), under controlled conditions. By measuring the rate of water loss (RWL) in different plant species, the behaviour of the process was found according to a pattern, regardless of the species studied. A logistic function describes the first part of the process (RWL), and a hyperbolic function describes the second part of the process (RWL), in relation to the RWLmax value, recorded at all samples, but at different times. The typical approach in this study led to the finding of a mathematical model that described RWL throughout the drying process. This model (RWLMod) is the solution of this study in the description of RWL in plants. The model would facilitate the study and better understanding of the behaviour of plants (including crop plants) in conditions of water and heat stress, such approaches all the more necessary in the context of climate change.

2. Results

For each plant species studied, the series of data recorded in real time (Tables 1–6) on the rate of weight loss (RWL) expressed in grams/minute, respectively weight (g) and drying time, are presented. These values have been obtained for the corresponding time instants of the measurements, carried out at intervals of ten seconds. Similarly, for each sample, the maximum value of RWL has also been distinctly indicated, represented as RWL_{max}. The total duration of the process was specific to each sample, observing at the end of the tables the moment in time when the experiment stopped automatically, when the amount of water lost had become negligible. However, the upper limit of the time was about 1000 s. These data series were the basis for the statistical determination of the coefficients of the functional models.

Table 1. Statistical data series on water loss for the sample *Picea abies* L., H. Karst (Spruce).

Time (s)	Rate of Weight Loss (g/min)	Weight (g)	Time (s)	Rate of Weight Loss (g/min)	Weight (g)	Time (s)	Rate of Weight Loss (g/min)	Weight (g)
10	0.024	2.911	160	0.198	2.321	860	0.006	1.383
20	0.054	2.902	170	0.192	2.289	870	0.006	1.382
30	0.108	2.884	180	0.192	2.257	880	0	1.382
40	0.192	2.852	190	0.192	2.225	890	0.006	1.381
50	0.282	2.805	200	0.186	2.194	900	0.006	1.380
60	0.336	2.749	210	0.186	2.163	910	0.006	1.379
70	0.366 *	2.688	220	0.186	2.132	920	0	1.379
80	0.336	2.632	230	0.180	2.102	930	0.006	1.378
90	0.300	2.582	240	0.180	2.072	940	0.006	1.377
100	0.270	2.537	250	0.180	2.042	950	0	1.377
110	0.246	2.496	260	0.174	2.013	960	0.006	1.376
120	0.228	2.458	270	0.174	1.984	970	0.006	1.375
130	0.216	2.422	280	0.168	1.956	980	0	1.375
140	0.204	2.388	290	0.168	1.928	990	0	1.375
150	0.204	2.354	300	0.168	1.9	1000	0.006	1.374
			...	...	...			

* registered value of RWL_{max}.

Table 2. Statistical data series on water loss for the sample *Pinus silvestris* L. (Pine).

Time (s)	Rate of Weight Loss (g/min)	Weight (g)	Time (s)	Rate of Weight Loss (g/min)	Weight (g)	Time (s)	Rate of Weight Loss (g/min)	Weight (g)
10	0.024	3.017	160	0.174	2.501	1010	0.006	1.44
20	0.048	3.009	170	0.174	2.472	1020	0.006	1.439
30	0.078	2.996	180	0.168	2.444	1030	0.006	1.438
40	0.150	2.971	190	0.174	2.415	1040	0.006	1.437
50	0.252	2.929	200	0.168	2.387	1050	0	1.437
60	0.306	2.878	210	0.168	2.359	1060	0.006	1.436
70	0.324 *	2.824	220	0.162	2.332	1070	0.006	1.435
80	0.294	2.775	230	0.162	2.305	1080	0.006	1.434
90	0.258	2.732	240	0.168	2.277	1090	0.006	1.433
100	0.240	2.692	250	0.156	2.251	1100	0	1.433
110	0.216	2.656	260	0.162	2.224	1110	0.006	1.432
120	0.204	2.622	270	0.156	2.198	1120	0	1.432
130	0.192	2.59	280	0.156	2.172	1130	0.006	1.431
140	0.186	2.559	290	0.15	2.147	1140	0.006	1.43
150	0.174	2.53	300	0.15	2.122	1150	0	1.43
			...	...	...			

* registered value of RWL_{max}.

Table 3. Statistical data series on water loss for the sample *Juniperus communis* L. (Juniper).

Time (s)	Rate of Weight Loss (g/min)	Weight (g)	Time (s)	Rate of Weight Loss (g/min)	Weight (g)	Time (s)	Rate of Weight Loss (g/min)	Weight (g)
10	0.024	3.034	160	0.21	2.344	870	0.006	1.234
20	0.054	3.025	170	0.21	2.309	880	0.006	1.233
30	0.114	3.006	180	0.198	2.276	890	0.006	1.232
40	0.204	2.972	190	0.198	2.243	900	0.006	1.231
50	0.282	2.925	200	0.192	2.211	910	0.006	1.23
60	0.336	2.869	210	0.198	2.178	920	0	1.23
70	0.378	2.806	220	0.192	2.146	930	0.006	1.229
80	0.402	2.739	230	0.186	2.115	940	0.006	1.228
90	0.420 *	2.669	240	0.192	2.083	950	0.006	1.227
100	0.39	2.604	250	0.18	2.053	960	0	1.227
110	0.336	2.548	260	0.186	2.022	970	0.006	1.226
120	0.294	2.499	270	0.18	1.992	980	0	1.226
130	0.258	2.456	280	0.18	1.962	990	0.006	1.225
140	0.240	2.416	290	0.174	1.933	1000	0	1.225
150	0.222	2.379	300	0.174	1.904	1010	0.006	1.224
			...	...	...			

* registered value of RWL_{max}.

Table 4. Statistical data series on water loss for the sample *Thuja occidentalis* L. (Thuja).

Time (s)	Rate of Weight Loss (g/min)	Weight (g)	Time (s)	Rate of Weight Loss (g/min)	Weight (g)	Time (s)	Rate of Weight Loss (g/min)	Weight (g)
10	0.024	2.849	160	0.198	2.2	730	0.006	1.33
20	0.078	2.836	170	0.204	2.166	740	0.006	1.329
30	0.150	2.811	180	0.192	2.134	750	0.006	1.328
40	0.252	2.769	190	0.198	2.101	760	0.006	1.327
50	0.330	2.714	200	0.186	2.07	770	0.012	1.325
60	0.384	2.65	210	0.192	2.038	780	0	1.325
70	0.402 *	2.583	220	0.180	2.008	790	0.006	1.324
80	0.354	2.524	230	0.186	1.977	800	0.006	1.323
90	0.318	2.471	240	0.180	1.947	810	0.006	1.322
100	0.282	2.424	250	0.174	1.918	820	0	1.322
110	0.258	2.381	260	0.174	1.889	830	0.006	1.321
120	0.240	2.341	270	0.162	1.862	840	0.006	1.32
130	0.222	2.304	280	0.168	1.834	850	0	1.32
140	0.216	2.268	290	0.162	1.807	860	0	1.32
150	0.210	2.233	300	0.156	1.781	870	0.006	1.319
			...	...	...			

* registered value of RWL_{max}.

Table 5. Statistical data series on water loss for the sample *Lamium purpureum* L. (Nettle).

Time (s)	Rate of Weight Loss (g/min)	Weight (g)	Time (s)	Rate of Weight Loss (g/min)	Weight (g)	Time (s)	Rate of Weight Loss (g/min)	Weight (g)
10	0.054	2.702	160	0.252	1.862	1050	0.012	0.576
20	0.138	2.679	170	0.252	1.82	1060	0.006	0.575
30	0.240	2.639	180	0.240	1.78	1070	0.006	0.574
40	0.324	2.585	190	0.240	1.74	1080	0.006	0.573
50	0.390	2.52	200	0.234	1.701	1090	0.006	0.572
60	0.438	2.447	210	0.234	1.662	1100	0.006	0.571
70	0.468	2.369	220	0.228	1.624	1110	0.006	0.57
80	0.498 *	2.286	230	0.222	1.587	1120	0.006	0.569
90	0.432	2.214	240	0.216	1.551	1130	0.006	0.568
100	0.384	2.15	250	0.216	1.515	1140	0	0.568
110	0.336	2.094	260	0.210	1.48	1150	0.006	0.567
120	0.312	2.042	270	0.204	1.446	1160	0.006	0.566
130	0.288	1.994	280	0.198	1.413	1170	0	0.566
140	0.276	1.948	290	0.198	1.38	1180	0.006	0.565
150	0.264	1.904	300	0.186	1.349	1190	0	0.565
			...	...	...			

* registered value of RWL_{max}.

Table 6. Statistical data series on water loss for the sample *Veronica hederifolia* L. (Veronica).

Time (s)	Rate of Weight Loss (g/min)	Weight (g)	Time (s)	Rate of Weight Loss (g/min)	Weight (g)	Time (s)	Rate of Weight Loss (g/min)	Weight (g)
10	0.036	3.399	160	0.270	2.627	1280	0.006	0.49
20	0.066	3.388	170	0.264	2.583	1290	0.006	0.489
30	0.120	3.368	180	0.252	2.541	1300	0.006	0.488
40	0.198	3.335	190	0.252	2.499	1310	0.006	0.487
50	0.270	3.29	200	0.24	2.459	1320	0.006	0.486
60	0.324	3.236	210	0.246	2.418	1330	0	0.486
70	0.372	3.174	220	0.234	2.379	1340	0.006	0.485
80	0.414	3.105	230	0.240	2.339	1350	0.006	0.484
90	0.444	3.031	240	0.240	2.299	1360	0.006	0.483
100	0.462 *	2.954	250	0.228	2.261	1370	0	0.483
110	0.408	2.886	260	0.234	2.222	1380	0	0.483
120	0.366	2.825	270	0.228	2.184	1390	0.006	0.482
130	0.330	2.77	280	0.222	2.147	1400	0.006	0.481
140	0.306	2.719	290	0.228	2.109	1410	0	0.481
150	0.282	2.672	300	0.222	2.072	1420	0	0.481
			...	...	...			

* registered value of RWL_{max}.

Rate of weight loss has values between 0.024–0.054 g/min at the initial moment in time (t = 10 s), and the highest value was represented in the nettle. Afterwards, an increased growth rate was observed for this indicator, thus as approximately 70–100 s from the debut; RWL had values between 0.258 g/min for pine and 0.498 g/min for nettle. The limitation stage was observed at the end of this time interval and for a short period (10–20 s). The following stage regarded RWL variation; a rapid decrease on the unit of time, from the maximum values mentioned above until the values corresponding to the moment of time t of approximately 150 s, after which RWL had a slow trend of decrease until the end of the exposure period.

The following tables, respectively one for each sample (e.g., statistically calculated coefficients, corresponding to the functional model, sample "n" in Tables 7–12, show for each functional model the values of the coefficients, statistically calculated, then the values rsq. (R^2 coefficient of determination), and. sig. (significance probability) for testing the accuracy from the statistical point of view. Sample "n" indicates each plant species studied. For the first branch of the function, all values are superior to 0.990, indicating an almost perfect fitting, and all sig. values are smaller than 0.001, indicating a high degree of accuracy. In addition, for the second branch, the values rsq. are high, rsq. = 0.804 being, in fact, the lowest value. Further, for each individual sample, sig. < 0.001.

Table 7. Statistical coefficients calculated, corresponding to the functional model, sample "spruce".

a	b	u	**Spruce**	t_2	t_1
99.5066	0.9127	0.39		80	10
$I(t_2)$	$I(t_1)$	total branch 1 (integral calculation)	total branch 1 (measurement)	Sig.	Rsq.
31.30963	15.88585	0.25	0.27	0.000	1
α	β			t_2	t_1
−0.0031	31.1986			1000	80
$J(t_2)$	$J(t_1)$	total branch 2 (integral calculation)	total branch 2 (measurement)	Sig.	Rsq.
212.4123	136.4651	1.26	1.25	0.000	0.832

Note. Sig.—significance probability; Rsq.—R^2 coefficient of determination; a, b, α, β—coefficients of the function (8); u—upper bound; t_1 and t_2—time interval; I—values returned by equation (12) for the mentioned time moment; J—values returned by Equation (13) for the mentioned time moment.

Table 8. Statistical coefficients calculated, corresponding to the functional model, sample "pine".

a	b	u	**Pine**	t_2	t_1
112.688	0.9141	0.35		80	10
$I(t_2)$	$I(t_1)$	total branch 1 (integral calculation)	total branch 1 (measurement)	Sig.	Rsq.
28.11475	14.5556	0.22	0.24	0.000	0.99
α	β			t_2	t_1
0.0094	26.6			1150	80
$J(t_2)$	$J(t_1)$	total branch 2 (integral calculation)	total branch 2 (measurement)	Sig.	Rsq.
198.6827	117.5681	1.35	1.34	0.000	0.804

Table 9. Statistical coefficients calculated, corresponding to the functional model, sample "juniper".

a	b	u	**Juniper**	t_2	t_1
74.3438	0.9239	0.43		90	10
$I(t_2)$	$I(t_1)$	total branch 1 (integral calculation)	total branch 1 (measurement)	Sig.	Rsq.
38.83818	19.18521	0.32	0.36	0.000	0.996
α	β			t_2	t_1
−0.0129	41.5086			1010	90
$J(t_2)$	$J(t_1)$	total branch 2 (integral calculation)	total branch 2 (measurement)	Sig.	Rsq.
274.1153	185.6198	1.47	1.44	0.000	0.888

Table 10. Statistical coefficients calculated, corresponding to the functional model, sample "thuja".

a	b	u	Thuja	t_2	t_1
109.28	0.8973	0.41		80	10
$I(t_2)$	$I(t_1)$	total branch1 (integral calculation)	total branch 1 (measurement)	Sig.	Rsq.
32.82901	14.62778	0.3	0.32	0.000	0.997
α	β			t_2	t_1
−0.005	32.2814			870	80
$J(t_2)$	$J(t_1)$	total branch 2 (integral calculation)	total branch 2 (measurement)	Sig.	Rsq.
214.1464	141.058	1.21	1.2	0.000	0.874

Table 11. Statistical coefficients calculated, corresponding to the functional model, "nettle".

a	b	u	Nettle	t_2	t_1
27.6696	0.9262	0.51		80	10
$I(t_2)$	$I(t_1)$	total branch 1 (integral calculation)	total branch 1 (measurement)	Sig.	Rsq.
41.00061	18.55301	0.37	0.41	0.000	0.987
α	β			t_2	t_1
−0.0136	44.592			1190	80
$J(t_2)$	$J(t_1)$	total branch 2 (integral calculation)	total branch 2 (measurement)	Sig.	Rsq.
299.6035	194.3153	1.75	1.72	0.000	0.92

Table 12. Statistical coefficients calculated, corresponding to the functional model, sample "veronica".

a	b	u	Veronica	t_2	t_1
51.086	0.9342	0.47		110	10
$I(t_2)$	$I(t_1)$	total branch 1 (integral calculation)	total branch 1 (measurement)	Sig.	Rsq.
51.79227	22.49395	0.48	0.51	0.000	0.990
α	β			t_2	t_1
0.0147	49.2996			1420	110
$J(t_2)$	$J(t_1)$	total branch 2 (integral calculation)	total branch 2 (measurement)	Sig.	Rsq.
378.7108	233.3488	2.42	2.4	0.000	0.823

Moreover, we presented both values of water losses; on the one hand, those resulting from the theoretical integral calculations, and on the other hand, the real values obtained by real-time determinations. For each sample, separately, the results were presented distinctly, for each branch of the function. The data were statistically tested on the differences between the groups determined by the theoretical and empirical method, using the Mann–Whitney test in SPSS. The values obtained are U = 14.5, sig. = 0.575 for the time segment corresponding to the first branch (logistic model), respectively, and U = 15, sig. = 0.631 for the time segment corresponding to the second branch (hyperbole), indicating the acceptance of the null hypothesis. Therefore, it can be considered that the two data sets do not differ.

At the end of the presentation for each sample, the expression of the functional model is represented by a graph (rate of weight loss (g/min), sample, in the Figures 1–6. Here, the value of the theoretical maximum point (t = m) can also be observed, thus indicating the time (theoretical) when the maxim (theoretical) of RWL takes place. Thus, for the sample "spruce", the nonlinear Equation (1) was solved, resulting in the value: t = m = 81.2 s. Thus, for the sample "spruce", we have the Equation (2).

$$\frac{1}{\frac{1}{0.39} + 99.5066 \cdot 0.9127^{t}} = -0.0031 + \frac{31.1986}{t} \quad (1)$$

$$f(t) = \begin{cases} \frac{1}{\frac{1}{0.39} + 99.5066 \cdot 0.9127^{\,t}}, & t \leq 81.2 \\ -0.0031 + \frac{31.1986}{t}, & t > 81.2 \end{cases} \quad (2)$$

Similar calculations were made solving nonlinear equations corresponding to the other samples and resulting in the functions below. For pine, Equation (3) was found, with m = 80.6 s.

$$f(t) = \begin{cases} \frac{1}{\frac{1}{0.35} + 112.688 \cdot 0.9141^{t}}, & t \leq 80.6 \\ 0.0094 + \frac{26.658}{t}, & t > 80.6. \end{cases} \quad (3)$$

For the juniper sample, the resulted model is the Equation (4), and m = 92.5 s.

$$f(t) = \begin{cases} \frac{1}{\frac{1}{0.43} + 74.3438 \cdot 0.9239^{\,t}}, & t \leq 92.5 \\ -0.0129 + \frac{41.5086}{t}, & t > 92.5 \end{cases} \quad (4)$$

For thuja, Equation (5) was found, m = 74.8 s.

$$f(t) = \begin{cases} \frac{1}{\frac{1}{0.41} + 109.28 \cdot 0.8973^{t}}, & t \leq 74.8 \\ -0.005 + 32.2814, & t > 74.8 \end{cases} \quad (5)$$

Then, for nettle, the model is Equation (6), m = 86.6 s.

$$f(t) = \begin{cases} \frac{1}{\frac{1}{0.51} + 27.6696 \cdot 0.9262^{t}}, & t \leq 86.6 \\ -0.0136 + \frac{44.5920}{t}, & t > 86.6 \end{cases} \quad (6)$$

Respectively, Equation (7) was found for Veronica, and m = 109.8 s.

$$f(t) = \begin{cases} \frac{1}{\frac{1}{0.47} + 51.086 \cdot 0.9342^{t}}, & t \leq 109.8 \\ -0.0147 + \frac{49.2996}{t}, & t > 109.8 \end{cases} \quad (7)$$

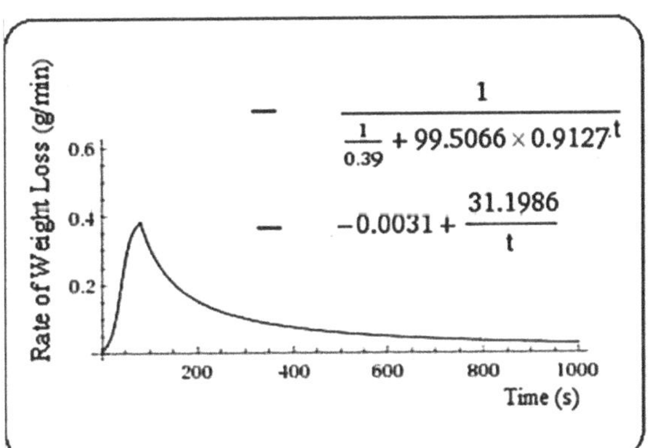

Figure 1. Rate of weight loss (g/min) in spruce samples.

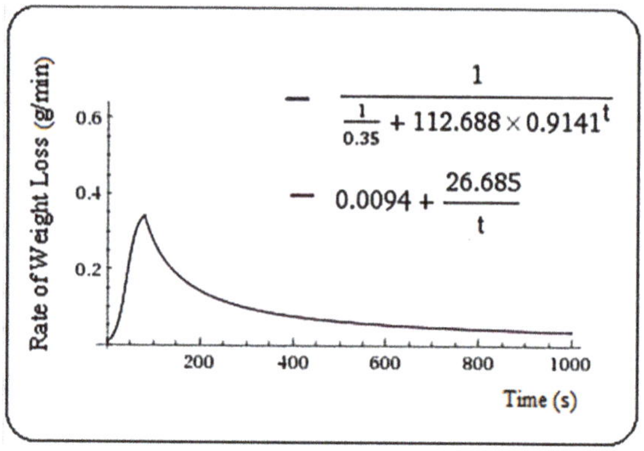

Figure 2. Rate of weight loss (g/min) in pine samples.

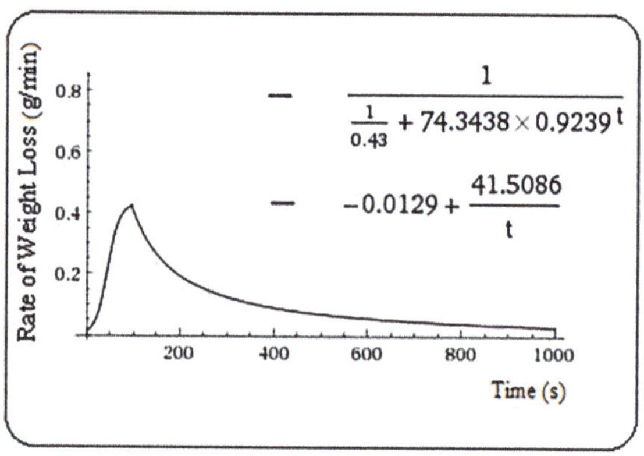

Figure 3. Rate of weight loss (g/min) in juniper samples.

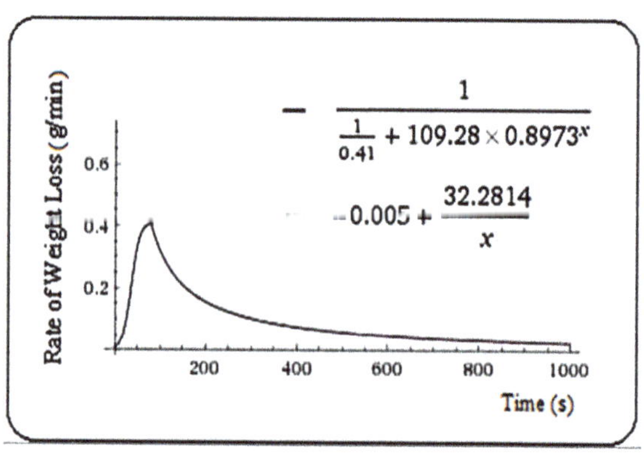

Figure 4. Rate of weight loss (g/min) in thuja samples.

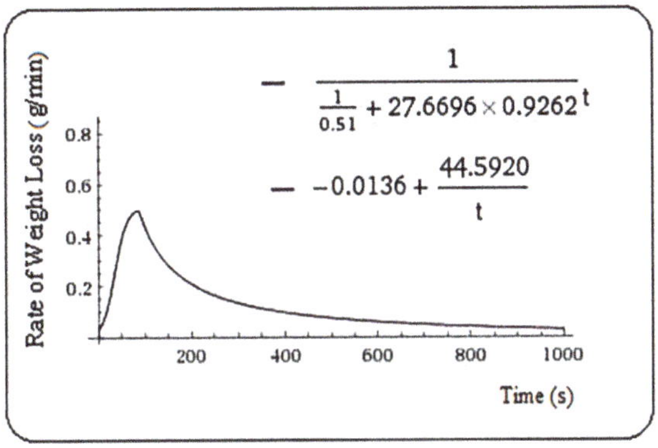

Figure 5. Rate of weight loss (g/min) in nettle samples.

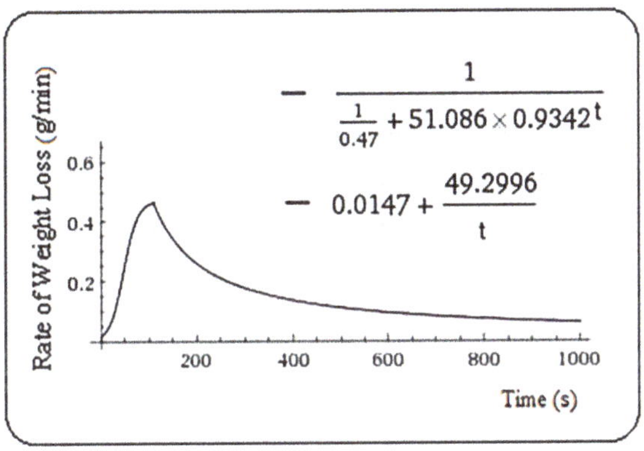

Figure 6. Rate of weight loss (g/min) in veronica samples.

3. Discussion

Multiple concerns have existed to study and describe aspects related to water in plants, or plant products, and the modeling approach has been the basis of many methods and techniques of investigation. The drying process is a method commonly used for conditioning plants. The generally accepted definition in the literature is reducing the moisture content of a certain product. The widespread use appears in preparation techniques for the medicinal herbs in beverages; in the beginning, fresh biomass (herba) will be subject to drying processes, a mandatory stage which reflects directly in the quality of the finished product, the period during which the product can be stored without quality depreciation, and also multiple economic aspects of generating profits [25].

Müller and Heindl [25] have studied the drying parameters for *Salvia officinalis*. One of the conclusions reported was related to the water activity (a_W). It correlates with the relative humidity (RH) of the air in the areas adjacent to the material to be analysed, therefore over the limit of RH > 70%, the development of some bacteria, fungus, residue (lees) has been noticed—issues that have a direct effect on the state of product quality.

The transformations occurring at a fixed temperature are also a topic often analyzed. Thus, sorption isotherms are described mathematically for *Artemisia dracunculus* with the help of the Halsey equation [25], but also with the help of other known models: BET, Caurie, GAB, Halsey, Henderson, Lewicki, Modified Mizrahi, Oswin, Peleg [26]. In addition, extensive isotherms for a series of plants were found, such as *Salvia officiinalis* [25], *Artemisia dracunculus*, *Mentha piperita*, *Thymus vulgaris* [26–28], *Mentha crispa* [29]), *Mentha viridis*, *Salvia officinalis*, *Lippia citriodora* [30], *Ficus deltoidea* [31]), *Melisa officinalis* [32]),

Chenopodium ambrosioides [33]), *Citrus sinensis* [34]), *Ziziphus spina-christi* [21]), *Phylanthus ambelica*, and *Zingiber officinale* [20]).

Furthermore, fruits are often subjected to drying processes. Simal et al. [35] and Kaya et al. [36] describe, using simulation methods, the dynamic water loss from kiwifruit; similar studies regarding various tropical fruits were conducted by Ceylan et al. [37] or Fernando and Amarasinghe [7].

This subject seems to be in full evolution at the moment, especially considering that modern food technology uses a high range of plants, for which the conducted studies are still not sufficiently detailed regarding water loss.

Rate of weight loss (RWL), defined as the amount of water lost in the time interval, can be discussed through physical analogy with the speed–vector size, thereby creating the possibility for it to be described with the help of mathematical physics equations. Jones and Sleeman [38] studied various biological models described by equations of this type.

Moreover, the rate of water loss is not constant; it is different from one species to another, from one organ of the plant to another, the way water content is present in the certain product (free water at intracellular level or present as links with various compounds).

The drying temperature, humidity and air velocity directly influence this parameter, and depending on a specific practical purpose, there are multiple studies that indicate the recommended parameters for drying in order to improve or protect some useful active principles.

Rate of weight loss of water in plants is less studied in a direct manner, but more indirectly by deduction from other calculations. However, here it has presented great interest mainly because of the different mathematical approaches that can be applied to this indicator. For example, determining the total quantity of water lost could be determined by using the properties of the definite integral.

At the same time, studying the RWL, there is the possibility of making direct observations in order to rapidly reduce the content of free water in plants. This fact can be used specifically in order to reduce the risk of developing microorganisms with undesirable effect over the quality of plant products.

In similar research to the one proposed in this study, Kaya and Aydin [39] investigated mint (*Mentha spicata* L.) and nettle (*Urtica dioica*) leaves in order to describe the loss of humidity in the presence of some variable external factors: air temperature, air velocity and relative humidity. Temperatures were lowered closer to the natural environment, with working values being 35, 45, 55 °C, and a longer drying period (30–50 min). Study regarding the leaves of lemon grass (*Cymbopogon citratus Stapf*) is also worth mentioning, although this was focused on the decrease of humidity, and various functional models that describe the trend shown by the experimental data were presented [40].

Theoretical mathematical models are most often made on simulated data, as a construction of the model, and will be subsequently verified in practice, on real data, in various case studies. The model proposed in this study (RWLMod) was built on real data, obtained from the six plant species studied. The model can be used in studies and in other plant species. It can also be used to describe RWL in relation to various influencing factors in plant life. The mathematical model was compared with the real distribution of the values obtained for each plant species. The fit between the mathematical model and the actual data series within each species, and the values for rsq. and sig. (as parameters of statistical safety of the fit), validates the model for the study conditions.

A direct overview on the experimental data from the present study indicated an initially ascending trend, followed by a descending part, characterized by right skewed distribution; an evolution which can be explained by physiological considerations [2]. Even if in practice mathematical modelling we find functions with approximately similar features, though in line with the trend shown by the experimental data which is the subject of this study, they present some particularities that make them different, and if these would be applied as models in this purpose, they could lead to significant errors. Even more RWL behaviour is different for the time period analysed, this being the reason why they chose to use a function consisting of two branches. The first branch, corresponding to the initial timeframe, when the plant samples show a rapid increase in the rate of water loss, is characteristic to an exponential model.

Basically, until the maximum level that occurs in a range of 80–100 s after the start, RWL increases by about 9–10 times compared to baseline. Because it is followed by a shorter range of time, a small limitation, we considered using a model based on sigmoid function, thus the phenomenon studied would be best approximated by a logistic function. For the second branch of the function, a hyperbole was used, in line with its graphic peculiarities. In the beginning, there is a rapid decrease of the water content on the time unit, in the vicinity of the maximum point, but towards the end, a very little amount of water begins to be lost even if the time interval is bigger. Specifically, in the case of each sample, towards the end of the time range studied, referring to the last 150 s of the drying process, RWL reveals low values, almost null. These aspects may be important for a better understanding of plant behaviour in extreme conditions of heat and water stress.

The chosen model (RWLMod) is a real function of a real continuous variable. In fact, the problem of continuity was placed only in one point, namely the separation point of the two time intervals. The continuity problem was solved after determining point (m) as being the intersection from the two branches of the function. In addition, this is a novelty brought by this study. The continuity immediately induces integrability of the function, and at the same time it implies the possibility to apply the formulas of integral calculation, more specifically the properties of the defined integrals regarding the determination of the area bounded by the graph of a function, to determine the amount of water lost between two given points in time. Moreover, this may be of interest for various other studies on the behaviour of plants in conditions of water and heat stress, especially in the context of climate change and the stress generated by it, for plants in general, and for crops in particular.

Performing direct comparisons on the total amount of water lost, on the one hand determined by the theoretical methods using formulas of integral calculation, and on the other hand using empirical data, there was little difference noticed, therefore insignificant. This was visible for all samples analysed, and statistical testing of the differences indicated that the values were close, and validated the proposed model (RWLMod).

4. Materials and Methods

Biological material was represented by different plants species, both woody trees (*Picea abies* L., H. Karst; *Juniperus communis* L.; *Pinus silvestris* L.; *Thuja occidentalis* L.), and herbaceous (*Lamium purpureum* L.; *Veronica hederifolia* L.). Fresh material, represented by leaves of species studied, was used for the determination. The tree species were about 20–25 years old. From herbaceous plants, leaf samples were taken at the flowering stage. The leaf samples were randomly harvested from the plant species studied, transported to the laboratory, and determinations were made. The humidity of the samples was variable (juniper, M = 59.71%; nettle, M = 79.16%; pine, M = 52.66%; spruce, M = 52.86%; thuja, M = 53.77%; veronica, M = 85.87%). Under the study conditions, a similar behaviour of RWL was recorded, regardless of plant species, and the samples' humidity.

The drying process was conducted under controlled conditions with a thermal balance AXIS (model ATS 60, Gdańsk, Poland), with an accuracy of determination of ± 0.001 g. Drying temperature was 100 °C (standard drying temperature), with automatic deactivation every five consecutive determinations with minimal differences, which automatically confirmed the end of the drying process. The drying temperature used (100 °C) does not represent the real living conditions of the plants; these conditions vary from one day to another, from one location to another. This temperature was chosen precisely from the perspective of capturing, in the mathematical model, the essence of the RWL phenomenon. There were between 87 and 142 data series recorded regarding drying parameters, distinctly for each sample, at every interval of 10 s. The data was automatically recorded on a computer using the software package PROMas version 2.2.0.0, and later processed mathematically and statistically.

For this study, there were recorded parameters regarding rate of weight loss due to water evaporation (RWL) in a process of controlled drying, representing the amount of water lost in the period of time and expressed in grams/minute. It was observed directly the maximum rate of weight loss due to water evaporation (RWL_{Max}), drying time (t) expressed in seconds, and weight (w) expressed in grams.

For the determination of the mathematical model (RWLMod) to describe the RWL from the plant samples considered in this study, the primary observation was that the general distribution of the RWL has several distinct graphic segments. The types of functions used to describe the phenomenon of water loss were determined starting from their particular shape which was resulted from the point clouds image. Values of the function coefficients were determined using procedure SPSS regression/curve estimation. Additionally, the values returned regarding the coefficient of determination, and also its level of statistical significance, were the basis for the confirmation of the chosen functional model. The graphic representations of the functions, determination of the intersection points between the two branches, were produced after solving some nonlinear equations, with the Wolfram Alpha application as a basis. The functions used to describe the two distinct branches of the graph of RWL are in Equation (8).

$$f(t) = \begin{cases} \frac{1}{\frac{1}{u}+a \cdot b^t}, & t \in [0, m] \\ \alpha + \frac{\beta}{t}, & t \in [m, \infty] \end{cases} \tag{8}$$

where a, b, α, β are the coefficients of the functions and u is the upper bound.

Thus, if for x = m the maximum value of the function is obtained, first branch (logistic function) specific for the initial phase, the time interval [0, m] describes the progress made before the maximum point ($m = RWL_{max}$). The time interval after the moment m is described by the hyperbola (Figure 7), presented in the second branch of the function (8). Continuity of the function is studied in the point t = m. Thus, to eliminate disadvantages of possible discontinuity points (points where the function is not integrable), the value of the point t = m was established as the intersection of the two functions, specifically by solving, for each sample individually, nonlinear Equation (9).

$$\frac{1}{\frac{1}{u} + a \cdot b^t} = \alpha + \frac{\beta}{t} \tag{9}$$

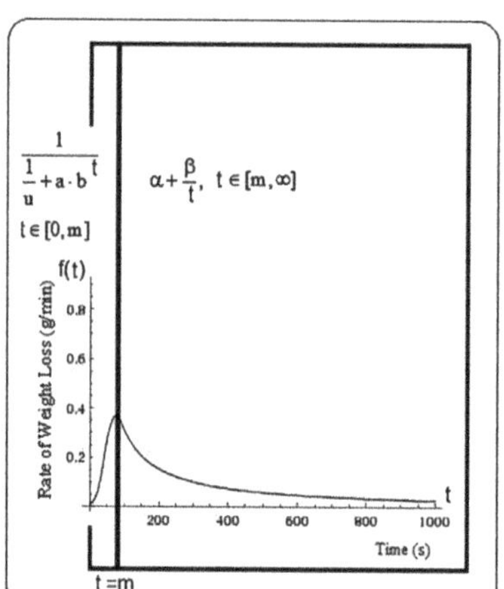

Figure 7. The general form of the function proposed to describe the trend of water loss.

The total quantity of water lost in a time interval $[t_1, t_2]$ represented by C (Figure 8), was determined using integral calculation, defined by the integral Equation (10).

$$C = \int_{t_1}^{t_2} f(t)dt \quad (10)$$

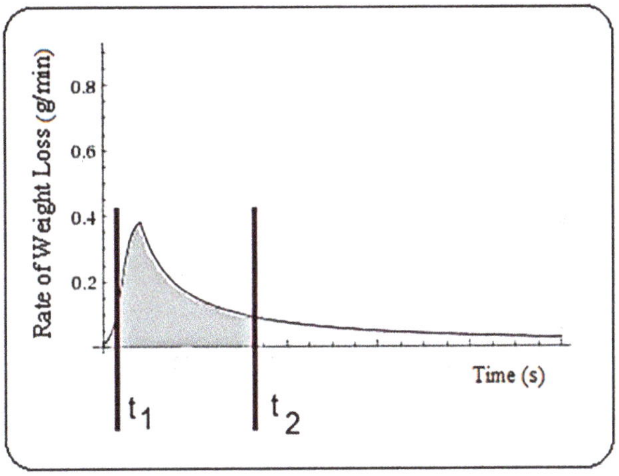

Figure 8. The integral calculation for the amount of water lost in the time interval t_1, t_2.

Primitive functions for each branch are represented by Equation (11).

$$I = \int f(t)dt = \int \frac{1}{\frac{1}{u} + a \cdot b^t} dt,\ J = \int \left(\alpha + \frac{\beta}{t}\right) dt \quad (11)$$

After performing the calculations, the results are described by Equations (12) and (13), respectively.

$$I = -\frac{u}{\ln b} \cdot \ln\left|b^{-t} + au\right| + C_1 \quad (12)$$

$$J = \alpha t + \beta \ln|t| + C_2,\ C_1, C_2 \in \Re \quad (13)$$

Regarding the positioning of the time interval compared to the maximum value, here are the cases I, II and III (Figure 9), and the quantity of water lost is described by Equations (14)–(16), respectively.

(I) $t_1 < t_2 < m$

$$C = \int_{t_1}^{t_2} f(t)dt = \int_{t_1}^{t_2} \frac{1}{\frac{1}{u} + a \cdot b^t} dt = \left(-\frac{u}{\ln b} \cdot \ln\left|b^{-t} + au\right|\right)\Big|_{t_1}^{t_2} = -\frac{u}{\ln b} \cdot \left(\ln \frac{\left|b^{-t_2} + au\right|}{\left|b^{-t_1} + au\right|}\right) \quad (14)$$

(II) $t_1 < m < t_2$

$$C = \int_{t_1}^{t_2} f(t)dt = \int_{t_1}^{m} \frac{1}{\frac{1}{u} + a \cdot b^t} dt + \int_{m}^{t_2} \alpha + \frac{\beta}{t} dt = -\frac{u}{\ln b} \cdot \left(\ln \frac{\left|b^{-m} + au\right|}{\left|b^{-t_1} + au\right|}\right) + \alpha(t_2 - m) + \beta \ln \frac{t_2}{m} \quad (15)$$

(III) $m < t_1 < t_2$

$$C = t = t_1 \int_{m}^{t_2} \alpha + \frac{\beta}{t} dt = \alpha(t_2 - t_1) + \beta \ln \frac{t_2}{t_1} \quad (16)$$

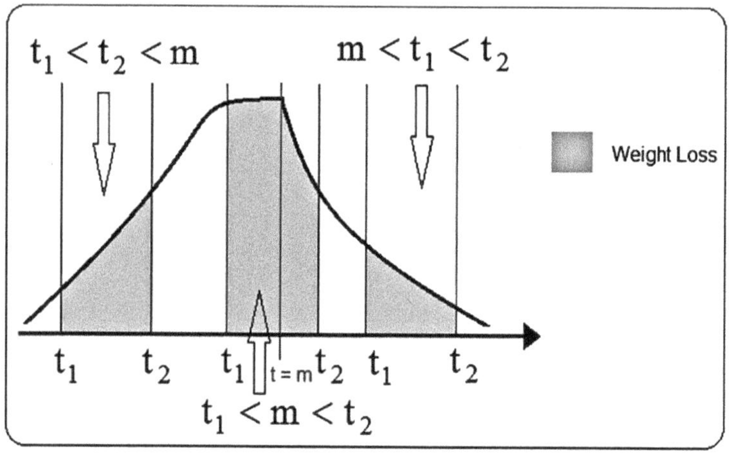

Figure 9. The amount of water lost in a period of time.

To evaluate the differences between the total amount of water lost, determined by the method of integral calculation described above and rate of weight loss determined by direct measurement, the time interval $[t_1, t_2]$ used is the entire time range from the first measurement, $t_1 = 0$, to the last measurement, corresponding to each sample individually. The maximum point t = m is found within the time interval $[t_1, t_2]$, so it was used in the formula (II). Determination by measuring the amount of water lost was actually determined by the type differences:$|\text{weight}(t_2) - \text{weight}(t_1)|$, using the primary test data.

5. Conclusions

The examples taken in the study, represented by the six different plant species, herbaceous and arboreal, showed that the obtained model (RWLMod) has a high level of safety and a high degree of robustness, being able to describe RWL with a high level of probability and fit, under controlled study conditions. Although only the sensitivity of the model in the description of RWL in some plant species was considered in this study, additional studies are possible to evaluate the sensitivity and accuracy of the model to other plant species, and in relation to possible factors influencing vegetation. There is also the possibility to evaluate other parameters, in relation to which the water regime in plants is influenced, and especially the loss of water in conditions of water and thermal stress. Although it exceeds the scope of this study, the proposed model would be useful in studies in crop plants, on water loss and tolerance to thermal and water stress in relation to soil conditions, minerals, plants nutritional status, biochemical composition of plants, and with crops health status. These aspects constitute, besides studies in other plant species, for the additional experimental validation of the model, and future directions of theoretical and applied research.

Author Contributions: Conceptualization, F.S. and C.R.; methodology, F.S.; software, C.R.; validation, F.S. and M.V.H.; formal analysis, C.R.; investigation, F.S.; data curation, C.R. and M.V.H.; writing—original draft preparation, F.S.; writing—review and editing, F.S. and C.R.; visualization, M.V.H. All authors have read and agreed to the published version of the manuscript.

Funding: This research received no external funding.

Institutional Review Board Statement: Not applicable.

Informed Consent Statement: Not applicable.

Data Availability Statement: The data presented in this study are available in article.

Acknowledgments: The authors thank the GEOMATICS Research Laboratory, BUASMV "King Michael I of Romania" from Timisoara, for the facility of the software use for this study. The authors also thank the Didactic and Experimental Resort, BUASMV "King Michael I of Romania" from Timisoara, for facilitating the experimental field of these studies. This paper is published from the own funds of the Banat's University of Agricultural Sciences and Veterinary Medicine from Timisoara.

Conflicts of Interest: The authors declare no conflict of interest.

References

1. Rand, R.H. Fluid mechanics of green plants. *Annu. Rev. Fluid Mech.* **1983**, *15*, 29–45. [CrossRef]
2. Sinha, R.K. *Modern Plant Physiology*; CRC Press: Boca Raton, FL, USA, 2004; p. 500.
3. Riederer, M.; Schreiber, L. Protecting against water loss: Analysis of the barrier properties of plant cuticles. *J. Exp. Bot.* **2001**, *52*, 2023–2032. [CrossRef]
4. Osakabe, Y.; Osakabe, K.; Shinozaki, K.; Tran, L.-S.P. Response of plants to water stress. *Front. Plant Sci.* **2014**, *5*, 86. [CrossRef]
5. Suguiyama, V.F.; Sanches, R.F.E.; Meirelles, S.T.; Centeno, D.C.; da Silva, E.A.; Braga, M.R. Physiological responses to water deficit and changes in leaf cell wall composition as modulated by seasonality in the Brazilian resurrection plant *Barbacenia purpurea*. *S. Afr. J. Bot.* **2016**, *105*, 270–278. [CrossRef]
6. Akpinar, E.K. Mathematical modelling of thin layer drying process under open sun of some aromatic plants. *J. Food Eng.* **2006**, *77*, 864–870. [CrossRef]
7. Fernando, J.A.K.M.; Amarasinghe, A.D.U.S. Drying kinetics and mathematical modeling of hot air drying of coconut coir pith. *SpringerPlus* **2016**, *5*, 807. [CrossRef] [PubMed]
8. Kissinger, M.; Tuvia-Alkalai, S.; Shalom, Y.; Fallik, E.; Elkind, Y.; Jenks, M.A.; Goodwin, M.S. Characterization of physiological and biochemical factors associated with postharvest water loss in ripe pepper fruit during storage. *J. Am. Soc. Hortic. Sci.* **2005**, *130*, 735–741. [CrossRef]
9. Dghim, F.; Abdellaoui, R.; Boukhris, M.; Neffati, M.; Chaieb, M. Physiological and biochemical changes in *Periploca angustifolia* plants under withholding irrigation and rewatering conditions. *S. Afr. J. Bot.* **2018**, *114*, 241–249. [CrossRef]
10. Bolat, I.; Dikilitas, M.; Ercisli, S.; Ikinci, A.; Tonkaz, T. The effect of water stress on some morphological, physiological, and biochemical characteristics and bud success on apple and quince rootstocks. *Sci. World J.* **2014**, *4*, 769732. [CrossRef]
11. Nemeskéri, E.; Helyes, L. Physiological responses of selected vegetable crop species to water stress. *Agronomy* **2019**, *9*, 447. [CrossRef]
12. Ru, C.; Hu, X.; Wang, W.; Ran, H.; Song, T.; Guo, Y. Evaluation of the crop water stress index as an indicator for the diagnosis of grapevine water deficiency in greenhouses. *Horticulturae* **2020**, *6*, 86. [CrossRef]
13. Boutraa, T.; Akhkha, A.; Al-Shoaibi, A.A.; Alhejeli, A.M. Effect of water stress on growth and water use efficiency (WUE) of some wheat cultivars (*Triticum durum*) grown in Saudi Arabia. *J. Taibah Univ. Sci.* **2010**, *3*, 39–48. [CrossRef]
14. Hembram, S.; Saren, B.K. Effect of water regime and plant geometry on growth, yield attributes and water use efficiency of rice (*Oryza sativa* L.). *J. Med. Plants Stud.* **2015**, *3*, 12–14.
15. Hatfield, J.L.; Prueger, J.H. Temperature extremes: Effect on plant growth and development. *Weather Clim. Extrem.* **2015**, *10*, 4–10. [CrossRef]
16. Fahad, S.; Bajwa, A.A.; Nazir, U.; Anjum, S.A.; Farooq, A.; Zohaib, A.; Sadia, S.; Nasim, W.; Adkins, S.; Saud, S.; et al. Crop production under drought and heat stress: Plant responses and management options. *Front. Plant. Sci.* **2017**, *8*, 1147. [CrossRef] [PubMed]
17. Calicioglu, O.; Flammini, A.; Bracco, S.; Bellù, L.; Sims, R. The future challenges of food and agriculture: An integrated analysis of trends and solutions. *Sustainability* **2019**, *11*, 222. [CrossRef]
18. Molotoks, A.; Smith, P.; Dawson, T.P. Impact of land use, population, and climate change on global food security. *Food Energy Secur.* **2021**, *10*, e261. [CrossRef]
19. Kouhila, M.; Kekhaou, N.; Otmani, M.; Fliyou, M.; Lahsasni, S. Experimental study of sorption isotherms and drying kinetics of Moroccan *Eucalyptus globulus*. *Dry. Technol.* **2002**, *20*, 2027–2039. [CrossRef]
20. Wijewardane, R.M.N.A.; Gunawardane, C.R.; Palipane, K.B.; Gunawardena, K.V.T.; Samarakoon, H.C.; Fernando, M.D. Evaluate the sorption behavior and identification of optimum drying conditions of *Phylanthus ambelica* (Nelli) and *Zingiber officinale* (Ginger). *J. Agric. Sci.* **2014**, *9*, 88–95. [CrossRef]
21. Saad, A.; Touati, B.; Draoui, B.; Tabti, B.; Abdenebi, A.; Benaceur, S. Mathematical modeling of moisture sorption isotherms and determination of isosteric heats of sorption of *Ziziphus* leaves. *Model Simulat. Eng.* **2014**, *2014*, 427842. [CrossRef]
22. Archontoulis, S.V.; Miguez, F.E. Nonlinear regression models and applications in agricultural research. *Agron. J.* **2015**, *107*, 786–798. [CrossRef]
23. Persson, L.; Leonardsson, K.; de Roos, A.M.; Gyllenberg, M.; Christensen, B. Ontogenetic scaling of foraging rates and the dynamics of a size-structured consumer-resource model. *Theor. Popul. Biol.* **1998**, *54*, 270–293. [CrossRef]
24. Bolker, B.M. *Ecological Models and Data in R*; Princeton University Press: Princeton, NJ, USA, 2008; p. 408.
25. Müller, J.; Heindl, A. Drying of medicinal plants. *Med. Aromat. Plant.* **2006**, 237–252. [CrossRef]
26. Dalgiç, A.C.; Pekmez, H.; Belibağlı, K.B. Effect of drying methods on the moisture sorption isotherms and thermodynamic properties of mint leaves. *J. Food Sci. Technol.* **2012**, *49*, 439–449. [CrossRef]

27. Soysal, Y.; Oztekin, S. Equilibrium moisture content equations for some medicinal and aromatic plants. *J. Agric. Eng. Res.* **1999**, *74*, 317–324. [CrossRef]
28. Arabhosseini, A.; Huisman, W.; van Boxtel, A.; Müller, J. Sorption isotherms of tarragon (*Artemisia dracunculus* L.). *Z. Arzn. Gew.* **2006**, *11*, 48–51.
29. Park, K.L.; Vohnikova, Z.; Brod, F.P.R. Evaluation of drying parameters and desorption isotherms of garden mint leaves (*Mentha crispa* L.). *J. Food Eng.* **2002**, *51*, 193–199. [CrossRef]
30. Kouhila, M.; Belghit, A.; Daguenet, M.; Boutaleb, B.C. Experimental determination of the sorption isotherms of mint (*Mentha viridis*), sage (*Salvia officinalis*) and verbena (*Lippia citriodora*). *J. Food Eng.* **2001**, *47*, 281–287. [CrossRef]
31. Johar, H.M.; Rukunudin, I.H.; Abdullah, S.; Kasim, F.H. Moisture sorption isotherms of *Ficus deltoidea* Jack leaves at 5 °C and 30 ° C. *J. Appl. Sci. Agric.* **2014**, *9*, 1–5.
32. Argyropoulos, D.; Alex, R.; Kohler, R.; Müller, J. Moisture sorption isotherms and isosteric heat of sorption of leaves and stems of lemon balm (*Melissa officinalis* L.) established by dynamic vapor sorption. *LWT-Food Sci. Technol.* **2012**, *47*, 324–331. [CrossRef]
33. Jamali, A.; Kouhila, M.; Mohamed, L.A.; Jaouhari, J.T.; Idlimam, A.; Abdenouri, N. Sorption isotherms of *Chenopodium ambrosioides* leaves at three temperatures. *J. Food Eng.* **2006**, *72*, 77–84. [CrossRef]
34. Bejar, A.K.; Mihoubi, N.B.; Kechaou, N. Moisture sorption isotherms–Experimental and mathematical investigations of orange (*Citrus sinensis*) peel and leaves. *Food Chem.* **2012**, *132*, 1728–1735. [CrossRef]
35. Simal, S.; Femenia, A.; Garau, M.C.; Rosselló, C. Use of exponential, Page's and diffusional models to simulate the drying kinetics of kiwi fruit. *J. Food Eng.* **2005**, *66*, 323–328. [CrossRef]
36. Kaya, A.; Aydin, O.; Dincer, I. Experimental and numerical investigation of heat and mass transfer during drying of Hayward kiwi fruits (*Actinidia deliciosa* Planch). *J. Food Eng.* **2008**, *88*, 323–330. [CrossRef]
37. Ceylan, I.; Aktaş, M.; Doğan, H. Mathematical modeling of drying characteristics of tropical fruits. *Appl. Therm. Eng.* **2007**, *27*, 1931–1936. [CrossRef]
38. Jones, D.S.; Sleeman, B.D. *Differential Equations and Mathematical Biology*; Chapman&Hall-CRC: New York, NY, USA, 2003; p. 408.
39. Kaya, A.; Aydın, O. An experimental study on drying kinetics of some herbal leaves. *Energy Convers. Manag.* **2009**, *50*, 118–124. [CrossRef]
40. Coradi, P.C.; de Castro Melo, E.; da Rocha, R.P. Mathematical modeling of the drying kinetics of the leaves of lemon grass (*Cymbopogon citratus* Stapf) and its effects on quality. *Idesia* **2014**, *32*, 43–56. [CrossRef]

plants

MDPI

Communication

Carboxylation Capacity Can Limit C_3 Photosynthesis at Elevated CO_2 throughout Diurnal Cycles

James Bunce †

Adaptive Cropping Systems Laboratory, USDA-ARS, Beltsville, MD 20705-2350, USA; buncejames49@gmail.com
† Retired.

Abstract: The response of carbon fixation in C_3 plants to elevated CO_2 is relatively larger when photosynthesis is limited by carboxylation capacity (V_C) than when limited by electron transport (J). Recent experiments under controlled, steady-state conditions have shown that photosynthesis at elevated CO_2 may be limited by V_C even at limiting PPFD. These experiments were designed to test whether this also occurs in dynamic field environments. Leaf gas exchange was recorded every 5 min using two identical instruments both attached to the same leaf. The CO_2 concentration in one instrument was controlled at 400 µmol mol^{-1} and one at 600 µmol mol^{-1}. Leaves were exposed to ambient sunlight outdoors, and cuvette air temperatures tracked ambient outside air temperature. The water content of air in the leaf cuvettes was kept close to that of the ambient air. These measurements were conducted on multiple, mostly clear days for each of three species, *Glycine max*, *Lablab purpureus*, and *Hemerocallis fulva*. The results indicated that in all species, photosynthesis was limited by V_C rather than J at both ambient and elevated CO_2 both at high midday PPFDs and also at limiting PPFDs in the early morning and late afternoon. During brief reductions in PPFD due to midday clouds, photosynthesis became limited by J. The net result of the apparent deactivation of Rubisco at low PPFD was that the relative stimulation of diurnal carbon fixation at elevated CO_2 was larger than would be predicted when assuming limitation of photosynthesis by J at low PPFD.

Keywords: photosynthesis; elevated CO_2; Rubisco; electron transport; light; diurnal cycle

Citation: Bunce, J. Carboxylation Capacity Can Limit C_3 Photosynthesis at Elevated CO_2 throughout Diurnal Cycles. *Plants* **2021**, *10*, 2603. https://doi.org/10.3390/plants10122603

Academic Editors: Milan S. Stankovic, Paula Baptista, Petronia Carillo and Hazem M. Kalaji

Received: 22 October 2021
Accepted: 25 November 2021
Published: 27 November 2021

Publisher's Note: MDPI stays neutral with regard to jurisdictional claims in published maps and institutional affiliations.

Copyright: © 2021 by the author. Licensee MDPI, Basel, Switzerland. This article is an open access article distributed under the terms and conditions of the Creative Commons Attribution (CC BY) license (https://creativecommons.org/licenses/by/4.0/).

1. Introduction

Photosynthesis by terrestrial vegetation is a large component of the annual global carbon balance, and predicting how the photosynthetic CO_2 assimilation (A) of terrestrial vegetation will respond to increased concentrations of CO_2 in the atmosphere is vital to understanding the future global carbon cycle [1–3]. Plants with C_3 photosynthetic metabolism are of predominant importance in terms of numbers of species, total CO_2 fixation, use as food for humans, and the responsiveness of photosynthesis to projected changes in atmospheric CO_2. Predictions of photosynthesis of C_3 plants at elevated CO_2 concentrations often use the Farquhar–von Caemmerer–Berry biochemical model of photosynthesis [4] and its recent modifications [5].

In the Farquhar–von Caemmerer–Berry model of C_3 photosynthesis [4], A at high photosynthetic photon flux density (PPFD) is limited by the maximum carboxylation capacity of Rubisco (V_{Cmax}) at low CO_2 concentrations, by maximum electron transport capacity (J_{max}) at higher CO_2, and sometimes by the rate of utilization of triose phosphates (TPU) at the highest CO_2 concentrations [5], with all of these rate-limiting parameters having different temperature dependencies. Because TPU limitation did not occur in these experiments, the focus here will be on V_{Cmax} and J_{max} at saturating PPFD, and V_C and J at limiting PPFD. The external CO_2 at the transition between limitation by V_{Cmax} and J_{max} at high PPFD varies among species, and with implementations of the FvCB model, from about 450 to 700 µmol mol^{-1} [3], so the issue is highly relevant to global change issues. Experiments exposing plants to elevated CO_2 have sometimes found photosynthesis at elevated CO_2 at high PPFD to be limited by V_{Cmax} and sometimes by J_{max}, depending upon species and level of elevated CO_2 (reviewed in [6]).

The relative increase in A caused by increased CO_2 concentration inside the leaf (C_i) is less at a given temperature when A is limited by J than when limited by V_C [3,4]. This is illustrated in Figure 1, which shows that for an increase in C_i from 250 to 375 μmol mol^{-1} (i.e., approximate C_i for ambient and 1.5 × ambient CO_2), A is stimulated about 20% more when A is limited by V_C than when limited by J, over a wide range of temperatures.

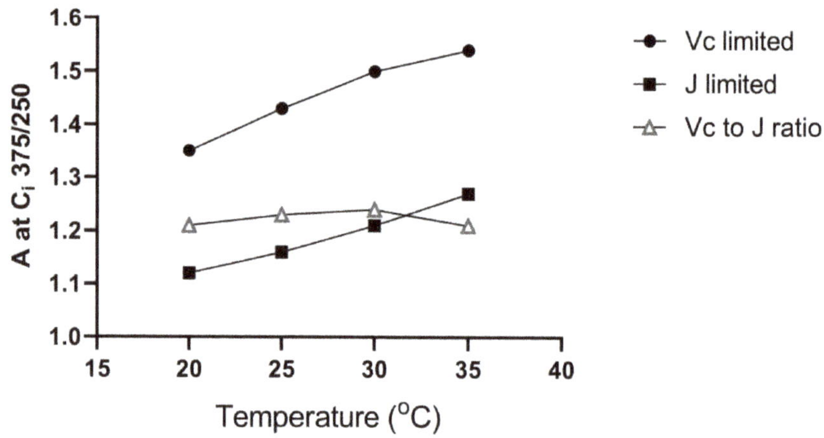

Figure 1. Hypothetical light-saturated rates of CO_2 assimilation (A) at internal CO_2 concentration (C_i) = 375 μmol mol^{-1} relative to those measured at C_i = 250 μmol mol^{-1} as a function of temperature, for leaves where A is limited by carboxylation capactity (V_c) or electron transport (J), and the ratio of these two values of A, using the Farquhar–von Caemmerer–Berry model [4].

At lower PPFD, it is often assumed that carboxylation capacity remains constant while electron transport rate decreases. This primarily lowers A at higher C_i (Figure 2) and would result in a smaller relative stimulation of A between a C_i of 250 and 375 μmol mol^{-1}, for example [7]. However, the assumption of constant carboxylation capacity at lower PPFD is incorrect, in the steady-state [8]. Bunce [8] found that the initial slope of A vs. C_i curves, an in situ measure of Rubisco carboxylation capacity, decreased as PPFD decreased in the steady-state, in a range of C_3 species measured, over a range of temperatures. The result was that the relative stimulation of A at elevated vs. ambient CO_2 did not decrease as PPFD decreased, at any temperature [8].

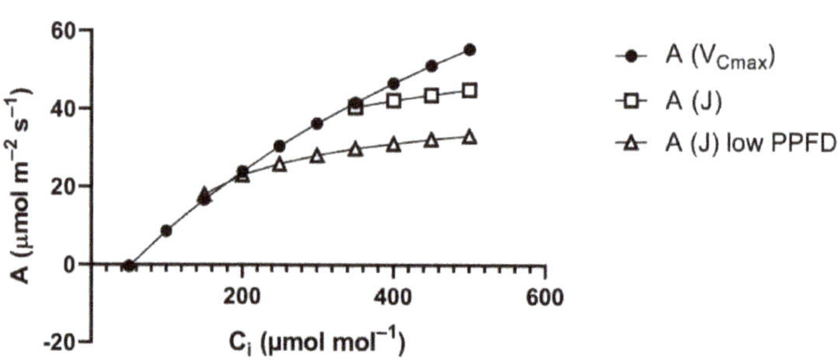

Figure 2. Hypothetical examples of assimilation rates (A) vs. internal CO_2 concentrations (C_i) at saturating and low PPFD, as predicted by the FvCB photosynthesis model [4]. Rates of A based on limitation by V_{Cmax} and by J at high and low PPFD are given. In all cases, actual A is the minimum of the A (V_{Cmax}) and A (J) curves.

The purpose of these experiments was to determine whether the same apparent decrease in V_C (deactivation of Rubisco) at low PPFD observed in the steady-state also occurred under naturally varying PPFD in dynamic field environments in C_3 species. This was tested by determining whether A at elevated CO_2 during different parts of diurnal cycles was better predicted from A at ambient CO_2 using the assumption of limitation of A at elevated CO_2 by Vc or by J.

2. Results

A representative diurnal time course of environment and leaf gas exchange for leaves of *G. max* at 400 and 600 μmol mol^{-1} CO_2 is presented in Figure 3. Representative diurnal time courses for the other two species are presented in Figures 4 and 5. Mean daytime stomatal conductances to water vapor were 500 and 450 mmol m^{-2} s^{-1} for *G. max* at 400 and 600 mmol mol^{-1} CO_2, respectively, and 120 and 100 mmol m^{-2} s^{-1} in *L. purpureus*, and 125 mmol m^{-2} s^{-1} at both CO_2 levels in *H. fulva*. Values of stomatal conductance at any time point can be obtained from the values of A and Ci at that point. Because photosynthesis was modeled based on measured values of Ci, impacts of stomatal conductance on photosynthesis are incorporated in the analysis.

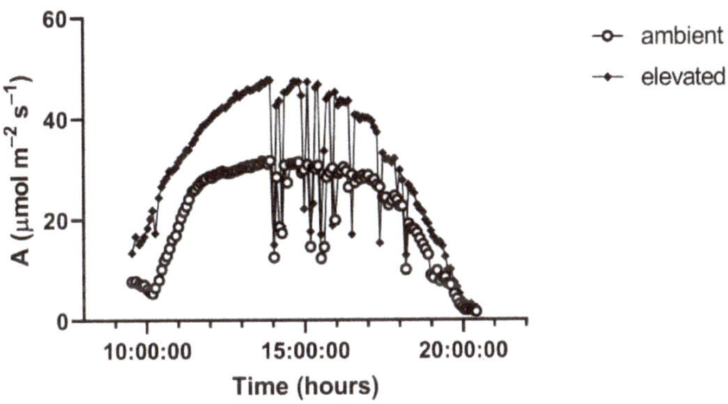

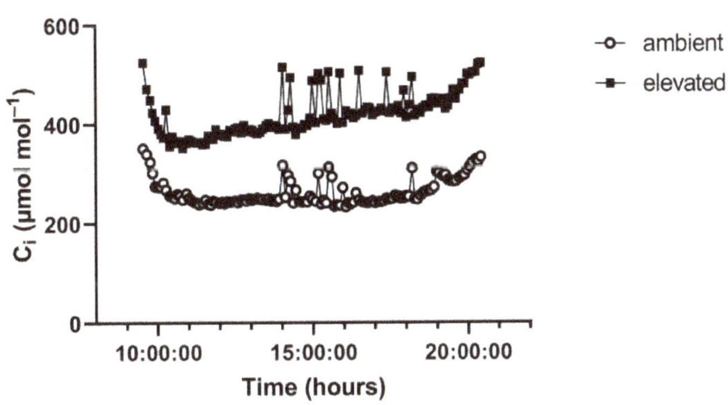

Figure 3. *Cont.*

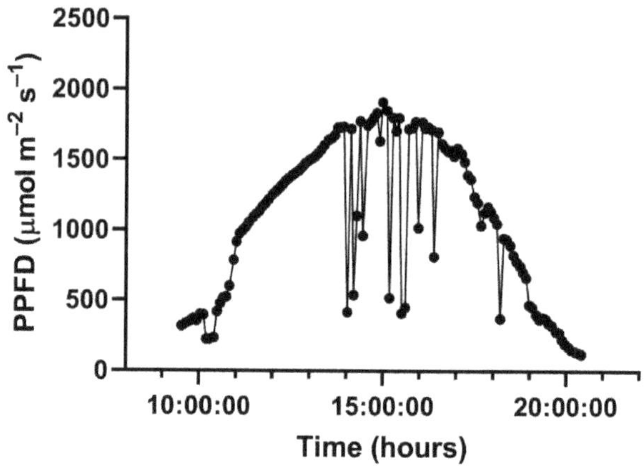

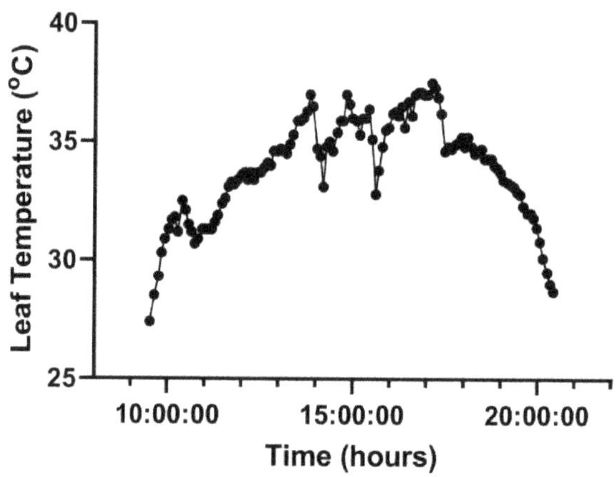

Figure 3. Diurnal patterns of PPFD, leaf temperature, A, and C_i of paired leaves of *G. max* kept at 400 (ambient) and 600 (elevated) μmol mol^{-1} CO_2. Environmental variables are given for only one of the two leaflets, for clarity. See text for details.

Because no differences occurred between morning and afternoon time periods in the outcome of the modeling of assimilation, either at the stable low or stable high PPFD periods, or after sudden decreases in PPFD due to clouds, values averaged over mornings and afternoons are presented in Table 1. The data reported in Table 1 for stable PPFD conditions represent means for two measurements (morning and afternoon) per day for three days in the cases of *G. max* and *H. fulva*, and four days in *L. purpurea*. The data for the sudden decrease in PPFD are means for two time points per day for each species, with three days in the case of *G. max* and *H. fulva*, and four days in *L. purpurea*. The results indicate that for all three species, when PPFD was stable, A at 600 μmol mol^{-1} CO_2 was accurately modeled by assuming limitation of A at both 400 and 600 μmol mol^{-1} CO_2 by Vc rather than J (Table 1). This was true for both high and low PPFD measurements, despite the rates of A being much lower at the low PPFD. In contrast, after sudden decreases in PPFD, A at 600 μmol mol^{-1} CO_2 was more accurately modeled by assuming limitation of A at both 400 and 600 μmol mol^{-1} CO_2 by J rather than by Vc (Table 1).

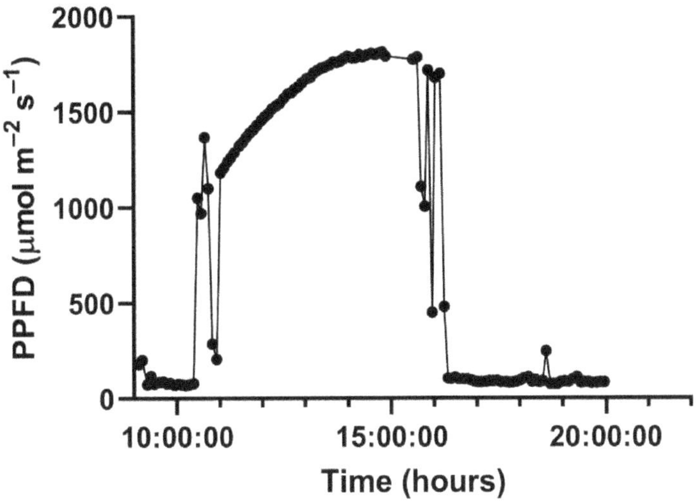

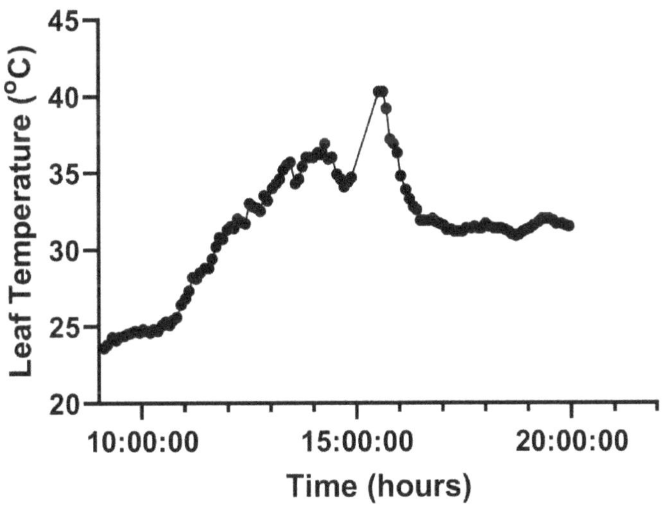

Figure 4. *Cont.*

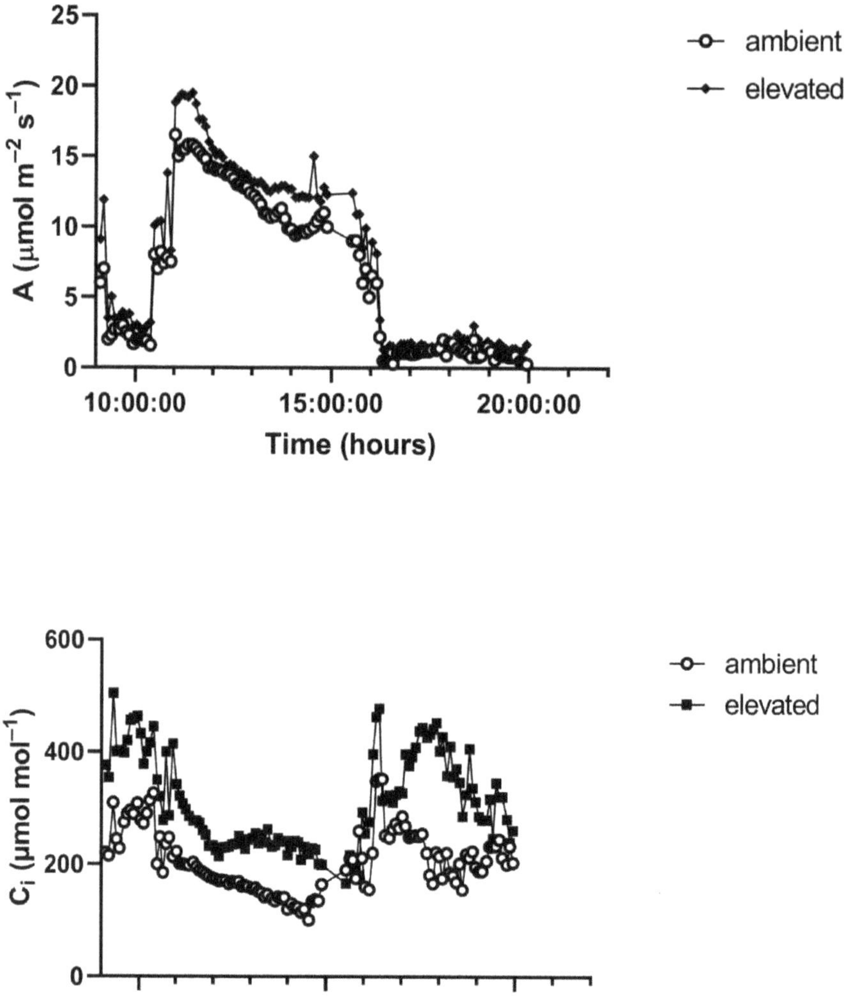

Figure 4. Diurnal patterns of PPFD, leaf temperature, A, and C_i of paired leaves of *L. purpureus* kept at 100 (ambient) and 600 (elevated) μmol mol^{-1} CO_2. Environmental variables are given for only one of the two leaflets, for clarity. See text for details.

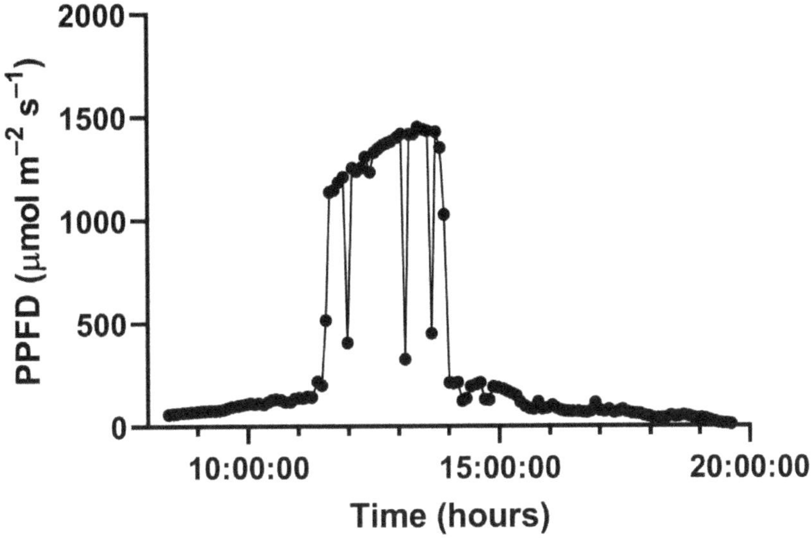

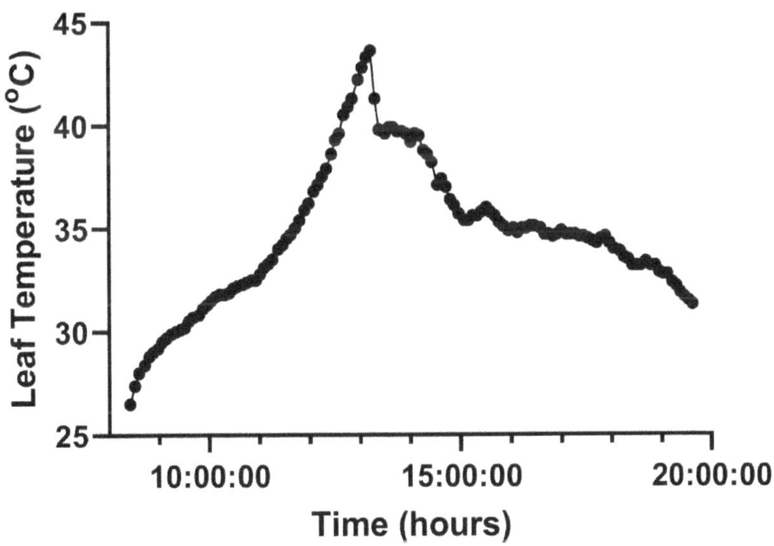

Figure 5. *Cont.*

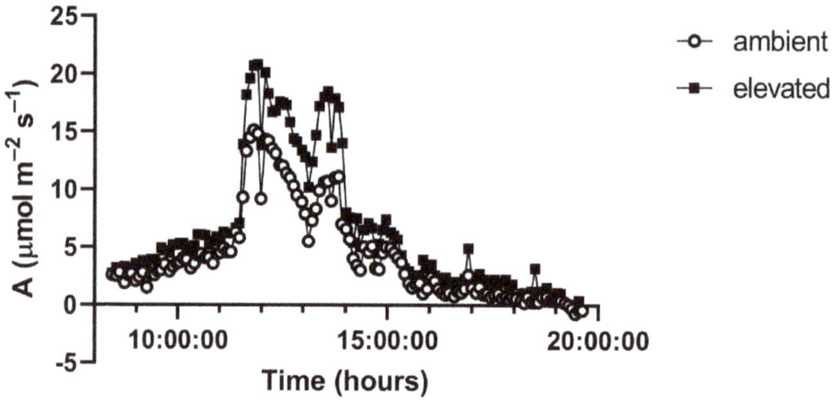

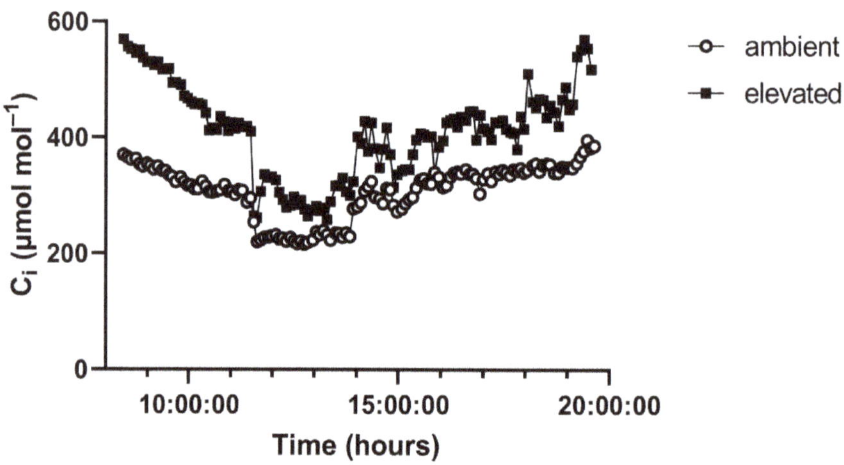

Figure 5. Diurnal patterns of PPFD, leaf temperature, A, and C_i of paired leaves of *H. fulva* kept at 400 (ambient) and 600 (elevated) μmol mol^{-1} CO_2. Environmental variables are given for only one of the two leaflets, for clarity. See text for details.

The Vc of all three species, measured at low PPFDs at 400 μmol mol^{-1} CO_2, had similar approximately linear increases in Vc with PPFD (Figure 6). Leaf temperatures ranged from 30 to 35 °C in this comparison. These similar values among species at low PPFD levels occurred despite large differences among species in V_{Cmax}, which ranged from about 160 μmol m^{-2} s^{-1} in *H. fulva* to 320 μmol m^{-2} s^{-1} in *G. max*, at 35 °C.

Table 1. CO_2 assimilation rates (A) of leaves measured at ambient (400 μmol mol^{-1}) and elevated (600 μmol mol^{-1}) CO_2 at stable, high PPFD (>1200 μmol m^{-2} s^{-1}), stable low PPFD (<400 μmol m^{-2} s^{-1}), or within 10 min after PPFD abruptly decreased by at least 700 μmol m^{-2} s^{-1} due to clouds. Also presented are assimilation rates at elevated CO_2 modeled using the model parameters fitted to the measured rates at ambient CO_2, assuming limitation of rates at elevated CO_2 either by V_C or by J at elevated CO_2. See text for details.

		Measured A (μmol m^{-2} s^{-1})		Modelled A (Elevated)		
Species	**PPFD**	**A Ambient**	**A Elevated**	**A V_C**	**A J**	**Mean T (°C)**
G. max	Stable high	29.3	43.5	43.3	36.1	37
	Stable low	8.7	14.1	14.2	12.0	33
	Sudden low	10.3	15.5	20.0	15.5	37
H. fulva	Stable high	11.2	20.4	20.6	16.1	37
	Stable low	5.1	6.6	6.7	6.0	33
	Sudden low	4.9	6.5	7.3	6.3	35
L. purpurea	Stable high	13.1	20.1	20.5	16.8	35
	Stable low	3.1	4.4	4.6	3.8	30
	Sudden low	2.5	3.0	4.0	3.2	37

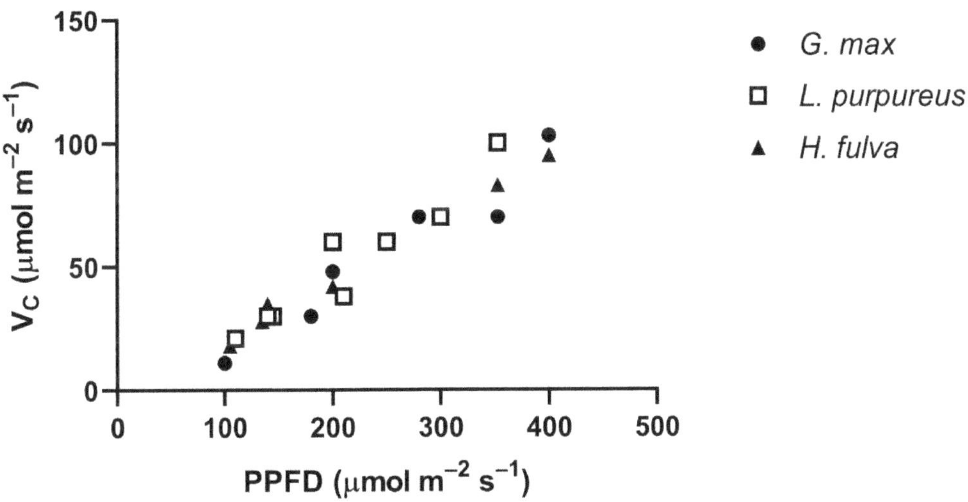

Figure 6. Vc measured at 30 to 35 °C at 400 μmol mol^{-1} CO_2 at a range of low PPFD values, for three species. There were two morning and two afternoon measurements on three days in *G. max* and *H. fulva*, and four days in *L. purpureus*.

3. Discussion

The results of this study indicated that for gradual changes in PPFD over the course of a day, caused by changes in the solar angle, leaf photosynthesis was always limited by Vc rather than by J in all three of these C_3 species, both at the approximate current ambient CO_2 concentration and at 1.5 times the current concentration. Models which assume limitation of C_3 photosynthesis by J at less than saturating PPFDs would underestimate the stimulation of daily CO_2 fixation at 1.5 times the current CO_2 concentration by approximately 50%.

The observed contrasting limitation of A by J after sudden decreases in PPFD caused by clouds obstructing direct solar radiation indicates that the assay of the type of limitation to photosynthesis used here was able to distinguish limitation by Vc from limitation by J. The change in type of limitation at low PPFD caused by the rate of decrease in PPFD is consistent with deactivation of Rubisco at low PPFD requiring a few to several minutes.

Conducting rapid A *vs.* C_i curves [9,10] under field conditions may be a new alternative method of determining the biochemical limitations to photosynthesis throughout a day, although to date these seem to only have been conducted at saturating PPFD. However, field-based rapid A vs. C_i measurements would be much more labor intensive than the method used here, and would also require information on the operational C_i throughout the day.

Several field experiments have indicated that V_{Cmax} corrected to a constant temperature may change during the course of a day [11–13], although the cause of the changes is not clear, and can occur even in shade [13]. In those cases, V_{Cmax} was assayed using steady-state A vs. C_i curves at high PPFD, although it was not clear whether the assays allowed sufficient time at high PPFD to fully re-activate Rubisco. Clearly, models of C_3 photosynthesis which assume constant V_{Cmax} throughout a day may be substantially in error. Furthermore, there is a notable lack of information on daily patterns of A vs. C_i curves at less than saturating PPFD, even at ambient CO_2. This is particularly important in predicting responses of photosynthesis to past and future changes in atmospheric CO_2, and may be one explanation for the observed insensitivity of biochemically based global photosynthesis models to past increases in atmospheric CO_2 [6,14].

4. Materials and Methods

Three C_3 species, *Glycine max* L. Merr., cultivar Clark, *Hemerocalus fulva* L., and *Lablab purpureus* L. Sweet were grown at the South Farm of the USDA Beltsville Agricultural Research Center, Beltsville, Maryland. The *G. max* and *L. purpureus* plants were grown from seed planted in the spring of 2019 in rows 70 cm apart, with plants thinned to about 7 cm between plants after emergence. The *H. fulva* plants were grown from tubers in rows 70 cm apart, with about 20 cm between tubers. The soil was a silt loam soil with a water table at about 1.5 m depth. Weeds were removed by hand. The prior crop was soybean, and no fertilizer was added to the soil after the well-fertilized soybean crop. Frequent precipitation prevented significant water deficits.

Two Ciras-3 portable photosynthesis systems (PP Systems, Amesbury, MA, USA) were used simultaneously each measurement day, with leaf chambers installed on opposite side leaflets of the same leaf in the case of *G. max* and *L. purpureus*, and both leaf chambers were on the same leaf in *H. fulva*, with about 5 cm between the chambers. Fully expanded, upper leaves were selected for measurement. Leaf chambers had circular windows 1.8 cm in diameter and were held horizontal. Leaf chambers were positioned so as to not be shaded by other leaves throughout the day. Leaf chambers were programmed to track ambient air temperature, and the external air temperature sensors were shaded at all times. Each chamber had internal PPFD sensors. The inlet air humidity setting was adjusted so that the chamber air had approximately the same water vapor content as that of the outside air. The CO_2 concentrations of air streams entering the two chambers were controlled at 400 and 600 μmol mol^{-1}. Leaf gas exchange data and environmental data were automatically recorded every 5 min from each leaf chamber throughout a 24 h period. Recording times of the two leaf chambers were not precisely synchronized, and the timing changed in each due to periodic automatic instrument self-tests, but recordings were synchronous within about 2 min. Measurements were begun soon after dew had evaporated from the leaves in the morning, and continued for 24 h. Measurements were continued for 24 h in order that rates in morning could be obtained without problems caused by dew on the leaves. Rates of respiration in darkness were not analyzed, for two reasons. First, flow rates were chosen based on obtaining accurate photosynthesis measurements, and created this very low, about 1 μmol mol^{-1}, differentials in darkness, so respiration measurements would be quite imprecise. Secondly, measuring small CO_2 differentials using clamp-on cuvettes has long been noted to produce artefactual apparent responses of respiration to CO_2 concentration because of leakage [15].

Measurements were made on three days each for *G. max* and *H. fulva*, and four days for *L. purpureus*, with days randomly assigned to the species, and with random assignment of the two gas exchange systems to the two CO_2 concentrations. All measurements were made in July and early August of 2019, on days selected as forecast to be mostly clear and without precipitation. All plants had flowered prior to the gas exchange measurements, but were plants were measured at least a month prior to reproductive maturity.

For each measurement day, detailed gas exchange analysis and modeling was applied to measurements during four periods: (1) periods of high (>1200 μmol m^{-2} s^{-1}), stable PPFD before mid-day, (2) periods of high, stable PPFD after mid-day, (3) periods of low (100–300 μmol m^{-2} s^{-1}), stable PPFD in early morning, and (4) periods of low, stable PPFD in late afternoon. Additionally, detailed analysis was made of periods of low (<500 μmol m^{-2} s^{-1}) PPFD which occurred after PPFD had decreased by at least 700 μmol m^{-2} s^{-1} within the last 10 min, because of clouds. This later type of data was available each measurement day, which is typical for this climate, where intermittent afternoon clouds are very common in mid-summer.

Assimilation rates for the 400 μmol mol^{-1} treatment under each of the above conditions were used to estimate values of Vc and J which would be consistent with the measured values of A, C_i, and temperature, using the FvCB model ([16], Equations (1) and (2)), with the temperature dependencies from Bernacchi et al. [17].

$$Anet = (1 - \Gamma^*/Ci) \times (Vc \times Ci)/(Ci + Kc\ (1 + O/Ko)) - Rd \quad (1)$$

$$Anet = (1 - \Gamma^*/Ci) \times (J \times Ci)/(4Ci + 8\Gamma^*) - Rd \quad (2)$$

where Ci is intercellular (CO_2), O is the oxygen concentration, γ^* is the photosynthetic CO_2 compensation point without dark respiration, Kc and Ko are the Michaelis–Menton constants for CO_2 and O_2, Vc is the carboxylation capacity of Rubisco, and J is the photosynthetic electron transport rate.

Mesophyll conductance was assumed to be infinite, so that the results would not be influenced by assumed finite values, given the lack of information about values of mesophyll conductance for *H. fulva* and *L. purpureus*, or its CO_2 dependence [10]. It was then tested whether values of A at 600 μmol mol^{-1} taken at the approximately the same time points were better fit by assuming limitation of A by the values of Vc or J estimated from the leaves at 400 μmol mol^{-1}. It could hypothetically occur that A at 400 μmol mol^{-1} would be limited by Vc while rates at 600 μmol mol^{-1} would be limited by J. That would lead to rates at elevated CO_2 in between those predicted by limitation by Vc or by J [6], but that did not occur in these experiments.

5. Conclusions

Field leaf gas exchange measurements on three herbaceous species indicated that, under clear sky conditions, photosynthesis was limited by the carboxylation capacity of Rubisco rather than by electron transport rate throughout diurnal cycles of solar radiation both at approximately the current atmospheric CO_2 concentration and at 1.5 times the current concentration. Abrupt decreases in radiation due to clouds temporarily resulted in limitation by electron transport capacity.

Funding: This research received no external funding.

Data Availability Statement: Data are available from the author upon request.

Conflicts of Interest: The author declares no conflict of interest.

References

1. Dusenge, M.E.; Duarte, A.G.; Way, D.A. Plant carbon metabolism and climate change: Elevated CO_2 and temperature impacts on photosynthesis, photorespiration and respiration. *New Phytol.* **2019**, *221*, 32–49. [CrossRef] [PubMed]
2. Lovenduski, N.S.; Bonan, G.B. Reducing uncertainty in projections of terrestrial carbon uptake. *Res. Lett.* **2017**, *12*, 044020. [CrossRef]

3. Rogers, A.; Medlyn, B.E.; Dukes, J.S.; Bonan, G.; Von Caemmerer, S.; Dietze, M.C.; Kattge, J.; Leakey, A.D.B.; Mercado, L.M.; Niinemets, Ü.; et al. A roadmap for improving the representation of photosynthesis in Earth system models. *New Phytol.* **2017**, *213*, 22–42. [CrossRef] [PubMed]
4. Farquhar, G.D.; von Caemmerer, S.; Berry, J.A. A biochemical model of photosynthetic CO_2 assimilation in leaves of C3 species. *Planta* **1980**, *149*, 78–90. [CrossRef] [PubMed]
5. Sharkey, T.D. O_2-insensitive photosynthesis in C_3 plants. Its occurrence and a possible explanation. *Plant Physiol.* **1985**, *78*, 71–78. [CrossRef] [PubMed]
6. Bunce, J.A. Limitations to soybean photosynthesis at elevated carbon dioxide in free-air enrichment and open top chamber systems. *Plant Sci.* **2014**, *226*, 131–135. [CrossRef] [PubMed]
7. Kirschbaum, M.U.F. The sensitivity of C_3 photosynthesis to increasing CO_2 concentration: A theoretical analysis of its dependence of temperature and background CO_2 concentration. *Plant Cell Environ.* **1994**, *17*, 747–754. [CrossRef]
8. Bunce, J.A. Light dependence of carboxylation capacity for C_3 photosynthesis models. *Photosynthetica* **2016**, *54*, 484–490. [CrossRef]
9. Stinziano, J.R.; Morgan, P.B.; Lynch, D.J.; Saathoff, A.J.; McDermitt, D.K.; Hanson, D.T. The rapid A-Ci responses: Photosynthesis in the phenomic era. *Plant Cell Environ.* **2017**, *40*, 1256–1262. [CrossRef] [PubMed]
10. Bunce, J. Three methods of estimating mesophyll conductance agree regarding its CO_2 sensitivity in the Rubisco-limited C_i range. *Plants* **2018**, *7*, 62. [CrossRef] [PubMed]
11. Kets, K.; Darbah, J.N.T.; Sober, A.; Riikonen, J.; Sober, J.; Karnosky, D.F. Diurnal changes in photosynthetic parameters of *Populus tremuloides*, modulated by elevated concentrations of CO_2 and/or O_3 and daily climatic variation. *Environ. Poll.* **2010**, *158*, 1000–1007. [CrossRef] [PubMed]
12. Singsass, E.L.; Ort, D.R.; DeLucia, E.H. Diurnal regulation of photosynthesis in understory saplings. *New Phytol.* **2000**, *145*, 39–49. [CrossRef]
13. Urban, O.; Klem, K.; Holisova, P.; Sigut, L.; Sprtova, M.; Teslova-Navratilova, P.; Zitova, M.; Spunda, V.; Marek, M.V.; Grace, J. Impact of elevated CO_2 concentration on dynamics of leaf photosynthesis in *Fagus sylvatica* is modulated by sky conditions. *Environ. Pollut.* **2013**, *185*, 271–280. [CrossRef] [PubMed]
14. Sun, Y.; Gu, L.; Dickinson, R.E.; Norby, R.J.; Pallardy, S.G.; Hoffman, F.M. Impact of mesophyll diffusion on estimated global land CO_2 fertilization. *Proc. Natl. Acad. Sci. USA* **2014**, *111*, 15774–15779. [CrossRef] [PubMed]
15. Bunce, J.A. Three new methods indicate that CO_2 concentration affects plant respiration in the range relevant to global change. *AoBP* **2021**, *13*, 1–5. [CrossRef] [PubMed]
16. Miao, Y.; Cai, Y.; Wu, H.; Wang, D. Diurnal and seasonal variations in the photosynthetic characteristics and the gas exchange simulations of two rice cultivars grown at ambient and elevated CO_2. *Front. Plant Sci.* **2021**, *12*, 651606. [CrossRef] [PubMed]
17. Bernacchi, C.J.; Pimentel, C.; Long, S.P. In vivo temperature response functions of parameters required to model RuBP-limited photosynthesis. *Plant Cell Environ.* **2003**, *26*, 1419–1430. [CrossRef]

plants

MDPI

Article

GIS-Facilitated Seed Germination and Multifaceted Evaluation of the Endangered *Abies marocana* Trab. (Pinaceae) Enabling Conservation and Sustainable Exploitation

Stefanos Hatzilazarou [1,†], Mohamed El Haissoufi [2,*,†], Elias Pipinis [3,†], Stefanos Kostas [1], Mohamed Libiad [2,4], Abdelmajid Khabbach [2,5], Fatima Lamchouri [2], Soumaya Bourgou [6], Wided Megdiche-Ksouri [6], Zeineb Ghrabi-Gammar [7,8], Vasiliki Aslanidou [1], Vasileios Greveniotis [9], Michalia A. Sakellariou [10], Ioannis Anestis [10], Georgios Tsoktouridis [10] and Nikos Krigas [10,*]

1 Laboratory of Floriculture, School of Agriculture, Aristotle University of Thessaloniki, 54124 Thessaloniki, Greece; hatzilaz@agro.auth.gr (S.H.); skostas@agro.auth.gr (S.K.); vaslanid@agro.auth.gr (V.A.)
2 Laboratory of Natural Substances, Pharmacology, Environment, Modelling, Health and Quality of Life (SNAMOPEQ), Polydisciplinary Faculty of Taza, Sidi Mohamed Ben Abdellah University, B.P. 1223, Taza Gare, Taza 35000, Morocco; libiad001@gmail.com (M.L.); khamajid@hotmail.com (A.K.); fatima.lamchouri@usmba.ac.ma (F.L.)
3 Laboratory of Silviculture, School of Forestry and Natural Environment, Aristotle University of Thessaloniki, 54124 Thessaloniki, Greece; epipinis@for.auth.gr
4 Laboratory of Ecology, Systematics and Biodiversity Conservation (LESCB), CNRST Labeled Research Unit N°18, Department of Biology, Faculty of Sciences, Abdelmalek Essaâdi University, B.P. 2121, M'Hannech II, Tetouan 93000, Morocco
5 Laboratory of Biotechnology, Conservation and Development of Natural Resources (BCVRN), Department of Biology, Faculty of Sciences Dhar El Mahraz, Sidi Mohamed Ben Abdellah University, B.P. 1796, Fès-Atlas 30003, Morocco
6 Centre de Biotechnologie de Borj-Cédria, Laboratoire des Plantes Aromatiques et Médicinales, B.P. 901, Tunis 2050, Tunisia; bourgousoumaya@yahoo.com (S.B.); ksouriwided@yahoo.fr (W.M.-K.)
7 Institut National Agronomique de Tunisie, Université de Carthage, 43 Avenue Charles Nicolle, Cité Mahrajène, Tunis 1082, Tunisia; zghrabi@yahoo.fr
8 Laboratoire de Recherche Biogéographie, Climatologie Appliquée et Dynamiques Environnementales (BiCADE 18ES13), Faculté des Lettres des Arts et des Humanités de Manouba, Campus Universitaire de la Manouba, Université de la Manouba, Manouba 2010, Tunisia
9 Institute of Industrial and Forage Crops, Hellenic Agricultural Organization Demeter, 41335 Larisa, Greece; vgreveni@mail.com
10 Institute of Plant Breeding and Genetic Resources, Hellenic Agricultural Organization Demeter, 57001 Thermi, Greece; michsakellariou@yahoo.com (M.A.S.); ganestis3@gmail.com (I.A.); gtsok1@yahoo.co.uk (G.T.)
* Correspondence: mohamed.elhaissoufi1@usmba.ac.ma (M.E.H.); nikoskrigas@gmail.com (N.K.); Tel.: +30-231-047-1110 (N.K.)
† Authors contributing equally.

Citation: Hatzilazarou, S.; El Haissoufi, M.; Pipinis, E.; Kostas, S.; Libiad, M.; Khabbach, A.; Lamchouri, F.; Bourgou, S.; Megdiche-Ksouri, W.; Ghrabi-Gammar, Z.; et al. GIS-Facilitated Seed Germination and Multifaceted Evaluation of the Endangered *Abies marocana* Trab. (Pinaceae) Enabling Conservation and Sustainable Exploitation. *Plants* **2021**, *10*, 2606. https://doi.org/10.3390/plants10122606

Academic Editors: Milan S. Stankovic, Paula Baptista and Petronia Carillo

Received: 29 October 2021
Accepted: 24 November 2021
Published: 27 November 2021

Publisher's Note: MDPI stays neutral with regard to jurisdictional claims in published maps and institutional affiliations.

Copyright: © 2021 by the authors. Licensee MDPI, Basel, Switzerland. This article is an open access article distributed under the terms and conditions of the Creative Commons Attribution (CC BY) license (https://creativecommons.org/licenses/by/4.0/).

Abstract: In the frame of the sustainable use of neglected and underutilized phytogenetic resources, and along with numerous studies in *Abies* spp. due to the innate conservation value of fir forests, this research focused on the Moroccan endemic fir, *Abies marocana*. The aim was triple-fold: to assess its potential and dynamics in economic sectors for sustainable exploitation; to determine the ecological conditions in which the species naturally thrives; and to find the appropriate requirements for its successful seed germination. We sourced multifaceted evaluations for three economic sectors performed in three levels, using 48 attributes and eight criteria from previous studies of our own, and the relevant species-specific assessments are overviewed herein in detail. The species' ecological profile was constructed using Geographical Information Systems (GIS) and open access data (Worldclim). Seed germination trials were performed to examine the effect of cold stratification (non-stratified, one- and two-months stratified seeds), the influence of four temperatures (10 °C, 15 °C, 20 °C, and 25 °C), and interactions thereof in relation to germination percentage (GP) and mean germination time (MGT). The experiments showed that the interaction of cold stratification and germination temperature has a strong effect on the GP and MGT of *A. marocana* seeds. A detailed GIS-derived ecological profile of the focal species was created in terms of precipitation and temperature natural regimes, enabling the interpretation of the seed germination results. The multifaceted evaluations reveal an interesting potential of the Moroccan fir in different economic sectors, which is mainly compromised due to extant research gaps, unfavorable conditions, and low stakeholder attraction. The findings of this study fill in extant research gaps, contribute to in situ and ex situ conservation strategies, and can facilitate the sustainable exploitation of this emblematic local endemic plant of northern Morocco.

Keywords: sexual propagation; cold stratification; in situ; ex situ; plant endemism; Morocco; phytogenetic resources

1. Introduction

The genus *Abies* Mill. (Pinaceae) includes about 40 recognized species distributed in the temperate regions of the northern hemisphere [1]. Historically, members of the genus *Abies* have been a valuable phytogenetic resource for humans. Fir trees have been long-exploited for their timber and the production of paper [2], and studies suggest that *Abies* spp. has also been used in traditional medicine to heal people's wounds [3]. In addition, *Abies* spp. has also been used widely in other sectors of the economy, such as the ornamental industry, with various fir species used traditionally as Christmas trees in different places [2].

The circum-Mediterranean firs form a distinct group of 10 native taxa and a natural hybrid, i.e., *Abies* × *borisii-regis* Mattf. [4]. Six fir species among them, i.e., *Abies cephalonica* Loundon, *A. cilicica* (Antoine and Kotschy) Carrière, *A. marocana* Trab., *A. nebrodensis* (Lojac.) Mattei, and *A. numidica* de Lannoy ex Carrière, are range-restricted in parts of the Mediterranean region. One of these, namely *A. marocana* (Moroccan fir), is a single-region local endemic species of Morocco which is confined to the Rif region, and it is classified as endangered, with a decreasing population [5]. This species could be considered as a relict tree species since it has endured a reduction of original distribution rather recently in Earth's history, due to climate changes and/or anthropogenic activities [6,7]. Currently, only two populations of *A. marocana* exist in the Rif Mountains of Morocco; the first one is confined to Mt. Tazaout, and the second spreads over the Chefchaouen Mountains (the mountains of Sfiha Tell, Tissouka, Lakraa, Talassemtane, Bouslimane, Taloussisse, Fahs, and Kharbouch) included in the Talassemtane National Park, an area belonging to the Intercontinental Biosphere Reserve of the Mediterranean [5,6,8,9]. In general, this species is adapted to the humid Mediterranean climate with cold winters, tolerates frosts over long periods, and grows on dolomitic limestone substrates, between 1400 m and 2100 m above sea level [9].

The extant Moroccan fir forests are affected by several abiotic factors, including droughts, the average temperature of the warmest quarter, the maximum temperature of the warmest month [9,10], annual average temperatures, annual average precipitation, soil moisture, and lithology [11], biotic factors, such as competition with less demanding species, as well as anthropogenic activities, including logging and habitat alterations [6,8,12]. Previous works have reported population declines and the poor natural regeneration of *A. marocana* [6,13]. Historically, the exploitation of the Moroccan fir forests began in the first half of the twentieth century [8]. Since then, the natural distribution of *A. marocana* was reduced by 70%, due to massive logging, land clearing, and fires [9]. This trend is still decreasing the area occupied by *A. marocana* stands, from 5000 ha [6] to currently less than 4000 ha [9]. To date, an absence of middle-aged plants is reported due to selective cutting, and the overgrazing which is practiced across its range limits the potential of natural regeneration. The situation is different in the untapped fir forest of Tazaot, where an abundance of young plants has been reported [13]. Currently, *A. marocana* is protected in situ within the Talassemtane National Park, where no logging is carried out. Forest fires and Indian hemp cultivation [8], however, may still threaten the extant species range.

In a wider context, the effects of climate change on *A. marocana* are, rather, perceived as moderate, due to the assumption that slow-growing conifers are able to regenerate under shelter if there are enough seedlings in mixed (*Abies marocana–Cedrus atlantica*) forests [10]. The structure, species composition, and competition in Moroccan fir forests could affect the response and resilience of *A. marocana* to climate change [10]. The competition with *Quercus ilex* L., *Q. faginea* Lam., and *Juniperus oxycedrus* L. could limit the regeneration of the Moroccan fir [12], while *A. marocana* and *Pinus nigra* J. F. Arnold-mixed forests seem to have a positive-to-neutral effect on *A. marocana* growth [10]. In the Tazaout forest, Moroccan firs regenerate more easily, and its stands are still dominant in the mixed formation with *Acer granatense* Boiss. and *Cedrus atlantica* (Endl.) Manetti ex Carrière, or when they co-occur with *Pinus pinaster* Aiton and *P. nigra*. However, 200 ha of Moroccan fir forest was burned in 2002, on the northern slope of Tazaout forest, and no regeneration has been observed there, probably due to the steep sloping and strong competition with other less-demanding species [8]. To resolve regeneration problems for this species in situ, deeper knowledge of *A. marocana* biology, as well as appropriate forest management actions, are required [10].

In general, to compile the necessary body of biological knowledge of threatened species (such as *A. marocana*), previous studies have highlighted the importance of seed germination studies both for the in situ and the ex-situ conservation of threatened plant resources [14–18]. Despite the fact that a large part of the Moroccan endemic flora includes hundreds of globally threatened plants [19], the number of local endemic taxa for which germination studies have been performed to date is still very low [20,21]. Currently, there are only a few germination studies related to some local endemic medicinal-aromatic plants of Morocco, such as *Thymus maroccanus* Ball., *Thymus broussonetii* Boiss [22,23], and *Origanum elongatum* (Bonnet) Emb. and Maire [15,24], and there is no relevant data for *A. marocana*.

Conservation-wise, germination and propagation trials need to be species-specific and focused on threatened local endemic species to support and facilitate long-term conservation strategies, as well as effective sustainable exploitation strategies [18]. In the absence of published data concerning the seed germination of *A. marocana*, and in the framework of diversified seed dormancy reports for different *Abies* species [25,26], the study herein investigates, for the first time, the seed germination requirements of *A. marocana*, an emblematic local endemic tree of Morocco which is threatened by extinction and thus should be prioritized. The absence of relevant information for the targeted fir species makes its propagation by domestic forest authorities difficult, hindering conservation plans and restoration programmes and hampering its possible exploitation in various economic sectors. Concerning the latter, the promising potential and the feasibility of creating value chains for *A. marocana* in different economic sectors make its sustainable exploitation achievable [18,27,28]. In this context, an understanding of the life cycle of *A. marocana* and its seed germination is crucial for achieving both its effective in situ conservation and its successful ex situ propagation for conservation purposes and sustainable exploitation strategies. To this end, an ecological profiling using Geographical Information Systems (GIS) can provide significant insight regarding the abiotic environmental conditions that *A. marocana* requires in its natural habitats, and may inform whether these conditions can be exploited or can be reproduced to some extent in the man-made environment of the ex-situ conservation efforts, thus facilitating propagation, germination trails, and acclimatization [29,30]. In turn, information stemming from GIS-facilitated germination studies on seed storage, the needed pre-treatments, and the optimal temperatures for seed germination is very useful for in situ conservation efforts as well as for sustainable exploitation strategies. Consequently, in the present study, we generated the GIS ecological profile of *A. marocana* with the aim to examine the effect of cold stratification, and to evaluate the impact of temperature on its seed germination.

2. Results

2.1. *Overview of the Potential of Abies marocana in Economic Sectors*

2.1.1. Ornamental-Horticultural Potential

Among the 94 Moroccan endemics comparatively examined [18], the highest-evaluated taxon was *Abies marocana* (72.5%), showing an interesting general potential in the ornamental-horticultural sector. The scoring of *Abies marocana* is illustrated in Figure 1, profiting from high-scoring in most of the attributes examined.

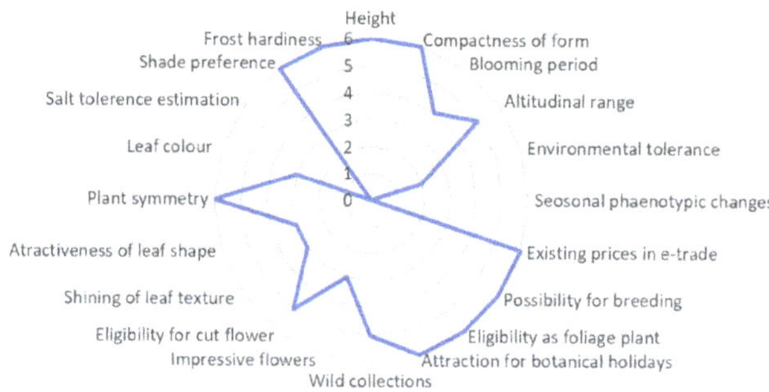

Figure 1. Evaluation of *Abies marocana* scored for 20 ornamental-horticultural attributes, reaching 72.5% of the optimum possible score, which is hierarchically ranked in the highest (>70%) general class; for details, see [18].

Concerning the potential in the subsectors of the ornamental-horticultural industry, *Abies marocana* scored 71.88% and ranked in the top three cases of local Moroccan endemics with the highest suitability for pot/patio plants (after *Salvia interrupta* subsp. *paui* and *Rhodanthemum hosmariense*). Although none of the local Moroccan taxa entered to the highest class of scores for home gardening suitability (>70%), *Abies marocana* received the highest score (68.83%) among them, and ranked first in above-average to high positions, followed by another seven Moroccan taxa. After *Salvia interrupta* subsp. *paui* and *Rhodanthemum hosmariense* ranked in top positions for landscaping eligibility (70.97% and 69.89% respectively), *Abies marocana* ranked in above-average to high positions (66.13%), followed by another eleven Moroccan taxa. Among Moroccan local endemics that are suitable for xeroscaping, *Abies marocana*, with 62.85%, ranked within the top 10 ones.

2.1.2. Medicinal-Cosmetic and Agro-Alimentary Potential

Abies marocana, scoring 44.44% in the medicinal-cosmetic sector (Figure 2A), was the third-highest evaluated taxon among the 94 local Moroccan endemics examined [27,28], showing a very interesting potential.

(A) (B)

Figure 2. Evaluation of *Abies marocana* scored for nine medicinal attributes (**A**) and for seven agro-alimentary attributes (**B**), reaching 44.44% and 47.62% of the optimum possible scores, respectively; these scores hierarchically ranked it in corresponding lower to average classes [20,28], respectively.

Although *Abies marocana* scored 47.62% in the agro-alimentary sector, it was included among the top 15 cases of local Moroccan taxa, with interesting agro-alimentary potential due to the strong aromatic properties of the pleasant aroma of the sourced resin (Figure 2B).

2.1.3. Feasibility and Readiness Assessments for Sustainable Exploitation

In terms of feasibility for sustainable exploitation (Level II evaluation based on 12 attributes) [18], *A. marocana* received the highest score (43.06%) among 94 local endemic plants of northern Morocco, which hierarchically ranked the taxon in the below-average to low class.

The readiness timescale for value chain creation regarding *A. marocana* (Level III evaluation upon completion of eight criteria) was assessed as achievable in the long-term [18].

2.2. Ecological Profiling of *Abies marocana*

In an attempt to facilitate both conservation efforts and sustainable exploitation [17], a GIS ecological profile was generated in terms of temperature and precipitation regimes across the original sites, where *A. marocana* is wild-growing in the natural environment (Figure 3).

2.2.1. Temperature-Related Attributes

The highest mean values of the average temperatures in the range of *A. marocana* (Figure 3) are peaked in mid- and late-summer, i.e., July (21.74 ± 1.66 °C) and August (21.29 ± 1.85 °C). During the autumn months, average temperatures start decreasing gradually, from 18.10 ± 2.12 °C in September to 9.35 ± 2.27 °C in November, and continue to reduce until mid-winter (5.00 ± 1.87 °C in January). By the end of winter (February), average temperatures start rising (5.62 ± 1.92 °C) and continue to increase gradually during the spring months, from 7.96 ± 1.93 °C in March to 13.91 ± 1.65 °C in May (Figure 3), reaching 18.02 ± 1.69 °C in early summer (June). Within this seasonal pattern, the minima of mean temperatures can be as low as −2.60 °C in January, and the maxima of mean temperatures can be as high as 32.70 °C in July (Figure 3). These temperature extremes represent the temperature limits in which the wild-growing *A. marocana* populations are adapted to thrive naturally, while mean diurnal range is 10.68 ± 0.44 °C and the annual temperature range is 12.61 ± 1.87 °C. These historical climatic data show no evidence for extreme low or high temperatures experienced by the *A. marocana* wild-growing populations. During the lowest average-temperature month (January), minimum mean temperature does not drop below 0 °C (T_{mean} of T_{min} = 0.64 ± 1.59 °C), while maximum mean temperature does not rise above 30 °C (T_{mean} of T_{max} = 28.7 ± 1.71 °C).

2.2.2. Precipitation-Related Attributes

Historical precipitation records suggest a strong seasonal raining pattern in the natural range of *A. marocana* (Figure 3). The highest mean values of precipitation occur naturally at the beginning of spring (March), with 171.98 ± 9.53 mm after a four-month period of high precipitation from the end of autumn and during winter (Figure 3), i.e., from November (141.30 ± 15.55 mm) to February (166.90 ± 13.59 mm). In mid-spring (April), precipitation starts declining from 87.28 ± 6.54 mm to 1.54 ± 0.78 mm during the driest month (July). After mid-summer, precipitation rises again (Figure 3), mainly in September (21.23 ± 3.40 mm) and October (74.16 ± 7.13 mm).

	(A) Minimum Temperature, Tmin (°C)				(B) Maximum Temperature, Tmax (°C)				(C) Average Temperature, Tavg (°C)				(D) Precipitation, P (mm)			
Months	Tmin	Tmax	Tmean	std	Tmin	Tmax	Tmean	std	Tmin	Tmax	Tmean	std	Pmin	Pmax	Pmean	std
Jan	-2.60	4.30	0.64	1.59	5.20	14.70	9.35	2.16	1.30	9.50	5.00	1.87	147.00	186.00	166.77	6.03
Feb	-2.30	4.70	0.99	1.59	5.70	15.90	10.24	2.26	1.70	10.30	5.62	1.92	123.00	205.00	166.90	13.59
Mar	-0.80	7.00	2.81	1.73	8.70	18.40	13.11	2.14	4.00	12.70	7.96	1.93	133.00	194.00	171.98	9.53
Apr	3.10	9.90	6.52	1.39	12.40	20.70	16.53	1.71	7.80	15.30	11.52	1.55	64.00	100.00	87.28	6.54
May	4.60	11.70	8.30	1.54	15.40	23.60	19.52	1.76	10.00	17.70	13.91	1.65	36.00	61.00	52.75	4.97
Jun	8.10	15.70	11.85	1.65	20.30	28.40	24.18	1.75	14.20	22.00	18.02	1.69	9.00	24.00	14.06	2.60
Jul	11.40	18.40	14.79	1.61	25.30	32.70	28.70	1.71	18.30	25.50	21.74	1.66	0.00	4.00	1.54	0.78
Aug	11.20	18.40	14.47	1.79	24.80	32.40	28.11	1.91	18.00	25.40	21.29	1.85	2.00	7.00	3.64	1.01
Sep	8.50	16.10	12.25	2.02	19.90	28.30	23.95	2.22	14.20	22.20	18.10	2.12	12.00	30.00	21.23	3.40
Oct	4.30	11.80	7.54	1.86	13.60	23.10	17.56	2.31	9.00	17.50	12.55	2.08	51.00	85.00	74.16	7.13
Nov	1.40	9.20	4.97	1.99	9.20	19.60	13.73	2.56	5.30	14.40	9.35	2.27	99.00	158.00	141.30	15.55
Dec	-1.10	5.80	2.07	1.68	6.20	15.90	10.41	2.28	2.50	10.90	6.24	1.98	91.00	182.00	123.22	20.44

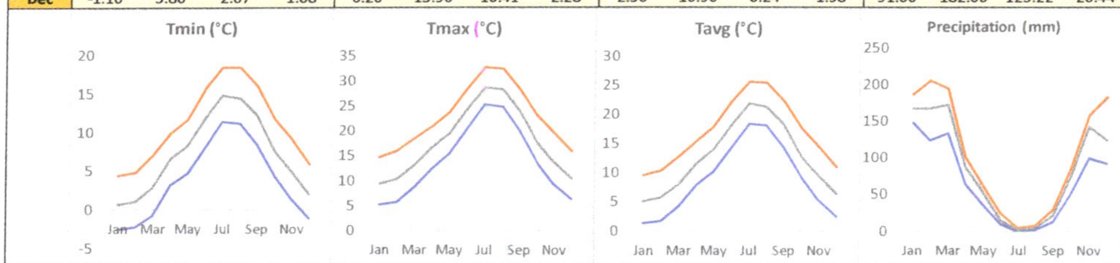

(E)

Nineteen Bioclimatic Variables	Min	Max	Mean	std
Annual Mean Temperature (Bio_1)	8.85	16.94	12.61	1.87
Mean Diurnal Range (Bio_2)	9.97	11.72	10.68	0.44
Isothermality (Bio_3)	35.98	41.26	38.07	1.25
Temperature Seasonality (Bio_4)	574.93	616.58	604.10	7.92
Max Temperature of Warmest Month (Bio_5)	25.30	32.70	28.70	1.72
Min Temperature of Coldest Month (Bio_6)	-2.60	4.30	0.64	1.59
Temperature Annual Range (Bio_7)	27.10	28.90	28.05	0.34
Mean Temperature of Wettest Quarter (Bio_8)	2.32	10.42	6.18	1.88
Mean Temperature of Driest Quarter (Bio_9)	16.85	24.33	20.35	1.73
Mean Temperature of Warmest Quarter (Bio_10)	16.85	24.38	20.44	1.84
Mean Temperature of Coldest Quarter (Bio_11)	1.85	10.22	5.62	1.92
Annual Precipitation (Bio_12)	816.00	1161.00	1024.81	54.88
Precipitation of Wettest Month (Bio_13)	147.00	205.00	173.67	9.03
Precipitation of Driest Month (Bio_14)	0.00	4.00	1.54	0.78
Precipitation Seasonality (Bio_15)	69.67	84.20	78.22	3.00
Precipitation of Wettest Quarter (Bio_16)	409.00	584.00	505.72	27.51
Precipitation of Driest Quarter (Bio_17)	11.00	35.00	19.23	4.24
Precipitation of Warmest Quarter (Bio_18)	14.00	40.00	21.19	4.58
Precipitation of Coldest Quarter (Bio_19)	404.00	573.00	456.92	33.08

Figure 3. GIS-derived ecological profile across the natural distribution range of *Abies marocana* wild-growing populations in Morocco, based on 313 occurrence records. In each case (**A–E**), minimum, maximum, average, and standard deviation is shown. Line graphs illustrate the minimum (blue lines), maximum (orange lines), and mean (grey lines) monthly values for temperature (°C) and precipitation (mm). (**A**) Minimum temperatures per month (°C), (**B**) maximum temperatures per month (°C), (**C**) average temperatures per month (°C), (**D**) precipitation per month (mm), (**E**) values for 19 bioclimatic variables. All data were extracted from WorldClim version 2.1.

2.3. Seed Germination Tests

The viability of the *A. marocana* seeds used in the experiments was low: 39 ± 8.24% (average of the four replications, S.D.), as indicated by the tetrazolium test.

Cold stratification (CS) and temperature, as well as the interaction thereof, significantly affected both the germination percentage (GP) and the mean germination time (MGT) of *A. marocana* seeds (Table 1). In the presence of a significant interaction, it was considered that the separate interpretation of the main effects was of inferior importance. In non-stratified seeds, the seeds incubated at 15 °C exhibited higher GP than seeds incubated at 10 °C, whereas no significant difference was observed in the GPs among seeds incubated at 10, 20, and 25 °C (Table 2). In cold-stratified seeds for one or two months, no significant difference among the incubation temperatures was observed in the GPs. Furthermore, in seeds incubated at 15, 20, or 25 °C, no significant difference was observed in the GPs between non-stratified seeds and seeds stratified for one or two months. However, in seeds incubated at 10 °C for the period of one month, CS resulted in a higher GP than non-stratified seeds, whereas no significant difference was observed in the GPs among seeds stratified for two months and non-stratified seeds (Table 2).

Table 1. Significance of factors and their interaction on the germination percentage and on the mean germination time of *Abies marocana* seeds, as estimated by ANOVA.

Attributes	Factors	Sum of Squares	df	Mean Square	F	Sig.
Germination Percentage (GP)	Cold Stratification (CS)	250.98	2	125.49	6.35	0.004
	Temperature (T)	367.82	3	122.60	6.20	0.002
	CS * T	380.42	6	63.40	3.21	0.013
Mean Germination Time (MGT)	Cold Stratification (CS)	14,770.65	2	7385.33	677.24	0.000
	Temperature (T)	2965.46	3	988.49	90.64	0.000
	CS * T	563.41	6	93.90	8.61	0.000

Table 2. Interaction of the factors "cold stratification" and "temperature incubation" on germination percentage and the mean germination time of *Abies marocana* seeds.

Cold Stratification (Months)	Incubation Temperature 10 °C	15 °C	20 °C	25 °C
Germination percentage (% ± S.D.)				
0	12.00 b [1] B [2] ± 5.66	35.00 a A ± 6.83	19.00 ab A ± 6.83	21.00 ab A ± 3.83
1	27.00 a A ± 6.83	32.00 a A ± 8.64	26.00 a A ± 6.93	23.00 a A ± 6.00
2	24.00 a AB ± 6.53	22.00 a A ± 5.16	16.00 a A ± 5.66	15.00 a A ± 6.83
Mean Germination Time (days ± S.D.)				
0	74.26 a [1] A [2] ± 4.50	50.15 b A ± 2.40	52.77 b A ± 3.70	48.18 b A ± 4.62
1	38.13 a B ± 5.99	17.72 b B ± 1.34	14.35 b B ± 1.01	14.75 b B ± 0.96
2	22.83 a C ± 3.55	18.35 ab B ± 2.48	15.46 b B ± 2.21	12.69 b B ± 2.54

[1] Within a row, percentages are statistically different at $p < 0.05$ when they do not share a common small letter. [2] Within a column, percentages are statistically different at $p < 0.05$ when they do not share a common capital letter. The comparisons were made using the R-E-G-WQ test.

According to Table 2, a significant effect of CS on MGT was observed. The CS for one or two months significantly hastened the germination of *A. marocana* seeds. Particularly, at 10 °C, seeds stratified for two months exhibited the lowest MGT. In seeds incubated at 15, 20, or 25 °C, a lower MGT was observed in stratified seeds (regardless of the period) compared to the non-stratified ones, whereas no significant difference was observed between the two periods of CS examined. Furthermore, the seeds incubated at 10 °C for germination exhibited the highest MGTs, regardless of CS period.

When incubated at temperatures ≥15 °C, following a one- or two-month period of CS, germinated seeds were recorded on the seventh day after the sowing and the germination was completed, at the end of third week for seeds incubated at 20 and 25 °C, and two weeks later for seeds incubated at 15 °C (Figure 4). In non-stratified seeds, the germination started at the end of the fourth week at temperatures ≥15 °C and at the end of the seventh week at 10 °C (Figure 4).

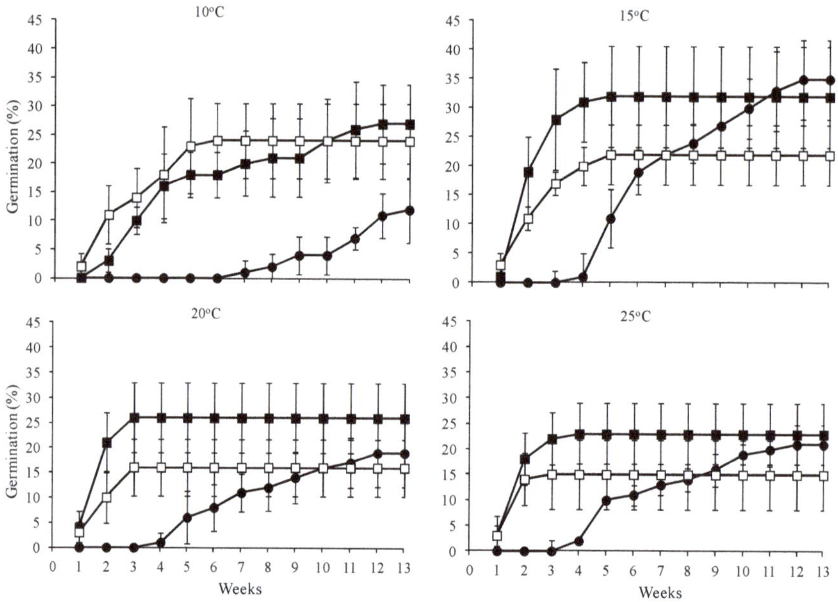

Figure 4. Effects of the cold-stratification period (CS) on the *Abies marocana* seed germination time course (• non-stratified, ■ one-month CS, and □ two-month CS) in seeds incubated at four constant temperatures (10, 15, 20, and 25 °C) under a 12-h light/12-h dark photoperiod.

3. Discussion

In the context of the sustainable use of phytogenetic resources and in the face of climate change, the Neglected and Underutilized Plant species (NUPs) are considered promising alternative crops if domesticated and sustainably used [18,31]. This study has explored, in a comprehensive way, and has assessed, in a multifaceted mode, whether the domestication procedure for *Abies marocana* can actually be achieved for its sustainable exploitation (see Section 4.1). Conservation-wise, this focal fir tree represents an emblematic local endemic of northern Morocco, which has been assessed as currently threatened by extinction, i.e., endangered with a decreasing population trend [5]. Therefore, intense research efforts are needed to obtain more information regarding its biological cycle if its extinction is to be prevented. In this way, the ecological profiling of *A. marocana* in terms of temperature and precipitation, as well as the first-ever germination tests presented herein, are considered contributions to fill such research gaps (see Section 4.2).

3.1. Potential, Feasibility, and Readiness for Value Chain Creation

Abies marocana represents an important NUP of Morocco, which is interesting for applications in different economic sectors. According to the multifaceted evaluations exercised [18,27,28] and overviewed herein, *A. marocana* represents a promising and highly suitable species for the ornamental-horticultural sector, especially as a pot plant as well as for landscaping and home gardening applications [18]. Currently, *A. marocana* has recognized ornamental value and is currently traded worldwide by three nurseries over the internet, and distinct value chains are already extant [21]. This kind of electronic commerce, however, for such a unique floristic element of Morocco, is performed with no official permission granted by domestic authorities; it is subjected to sovereign rights due to the Nagoya protocol and may be further associated with conservation implications [21].

Almost the same interest applies for the medicinal-cosmetic sector, where *A. marocana* scored third in comparison to other local endemic NUPs of Morocco [28]. To date, comparative studies in members of the genus *Abies* report at least 277 compounds that have been isolated and identified across 19 species, reporting mostly terpenoids, flavonoids, lignans, phenols, and steroids [32]. Members of such categories of natural ingredients often find commercial applications in cosmetic products. Modern investigations in close *Abies* relatives report the isolation and identification of more than 90 constituents from *A. alba* seeds and cone scales, concluding that its seeds can be potentially exploited by the perfume industry [33]. Preliminary studies in *A. marocana* report a variety of diterpenoids, cadinanes, and cholestanes [32]. Yet, with limited available data to date, intense efforts and targeted phytochemical investigations are required to unveil the potentially valuable diversity of molecules and bioactive compounds that are probably naturally produced by *A. marocana*.

A. marocana ranked almost average (43.06%) in terms of feasibility evaluation for sustainable exploitation [18]. However, when this score is partitioned in different attributes, two opposite trends became evident: (i) the relative tendency of score increases because of high attribute assessments (scores of five or six) related to endemism rarity, extinction risk status, effective ex situ conservation [20], and current extant horticultural experience with seed germination trials as presented herein, and (ii) the relative tendency of score decreases due to low attribute assessments (scores of zero or one) owing to the absence of protection status and few marketed commercial products [21], high water demands, the restricted availability of initial propagation materials, the absence of vegetative propagation techniques or cultivation protocols or extant cultivations [18]. If the above-mentioned gaps and weaknesses are filled by targeted applied research, there is no doubt that the feasibility assessment of *A. marocana* will be improved considerably, reaching those reported for some local endemics of Crete (Greece) that are currently under sustainable exploitation, i.e., *Sideritis syriaca* subsp. *syriaca* [34] and *Origanum dictamnus* [18].

The readiness timescale for the sustainable exploitation of *A. marocana* is assessed as achievable in the long-term [18] due to restrictions related with propagation materials, limited knowledge of propagation–cultivation techniques, compromised commercial interest, low stakeholder attraction, and discrepancies in implementation of the Nagoya Protocol in different countries. The absence of experience in the massive propagation and in the cultivation protocols for this species may strongly compromise any attempt at upscaling for its future exploitation and commercialization, and further implies that multidisciplinary and applied research is needed urgently (e.g., massive propagation, cultivation, cultivation practices, agro-processing, fertilization regimes, agronomical aspects, stakeholder and distribution channels attraction, etc.) to overcome existing barriers. Therefore, the value chain for *A. marocana* is considered as difficult to create in the short- or medium-term, and consequently, can only be achieved in long-term. This study, however, provides solid information regarding the seed propagation of *A. marocana*, thus facilitating the sustainable exploitation of this promising species in economic sectors. For many species, propagation from seeds is the most common and the cheapest method exercised in commercial nurseries [35]. In this fashion, it seems that there are good perspectives for sustainable exploitation if more research gaps are filled promptly and quickly. The speeding-along of sustainable exploitation strategies can also be achieved when similar successful paradigms at local scales are followed. For instance, *Argania spinosa* (L.) Skeels (Sapotaceae) has long been another NUP which—after targeted research by scientists, documentation, conservation efforts, and stakeholder attraction were all achieved within about a decade [27]—is currently receiving increased global appreciation in the international markets [36].

3.2. Seed Germination Requirements for Conservation and Sustainable Exploitation

For many species, propagation from seeds is the most common and the cheapest method used in commercial nurseries [35], creating heterogenous individuals due to species genetic diversity. Although this may be undesirable in some cases (e.g., the artificial selection of suitable genotypes for the production of specific metabolites), it is highly required during conservation efforts to maintain and promote the genetic diversity of the focal taxon. However, a major constraint to sexual propagation is the poor germination due to poor natural seed quality and seed dormancy. Indeed, high percentages of empty seeds are often observed in many *Abies* spp. [26]. A significant problem in such cases is the separation and the removal of seeds that do not contain an embryo (non-viable seeds), especially when these are filled in with resin. As frequently reported in the relevant literature, the seeds of most (if not all) members of the genus *Abies* exhibit some degree of dormancy, and therefore a period of CS is required to overcome it [25,37]. However, seed dormancy among different *Abies* spp. is quite variable, and the degree of dormancy determines the duration of CS needed to overcome this natural barrier [26]. Furthermore, a variation in seed dormancy among provenances and among different seedlots harvested in different years has been reported for some *Abies* spp. [25]. Due to this multiple variation in seed dormancy, a three-week period of stratification at 3–5 °C is suggested as a necessary step by ISTA [38] for seed germination in 10 *Abies* spp., whereas double germination tests are recommended for many species (including *A. pinsapo* Boiss., the closest species to *A. marocana*).

It is well-known that seeds of *Abies* spp. are usually of poor quality [39–41]. According to the results of the tetrazolium test performed in this study, the potential germination capacity of the seedlot of *A. marocana* sourced from wild habitats in Rif, Morocco and used in the germination experiments was as low as 39%. Possibly, the date of cone collection, the storage period (12 months) of seeds, or both of them resulted in a reduction in the viability and germinability of the seeds of *Abies marocana*. It has been reported that seed germinability of many *Abies* species is improved when seeds are collected as close as possible to seed dispersal [26]. Concerning the effect of seed storage on germination, the storage of *A. cephalonica* seeds in the dark, in sealed containers, at 7 ± 2 °C for one year resulted in a significant decrease in seed germinability [40]. Furthermore, the germination of *A.* × *borisii-regis* seeds is reduced after a one-year period of storage at 0 °C, compared to fresh seeds [41]. Similarly, low viability (<44%) has been reported in seeds from four Polish provenances of *A. alba* Mill. [42], and low germination percentages have been observed in seeds of four Indian provenances of *A. pindrow* (Royle ex D.Don) Royle [43].

In the present study, the germination and germination rate of *Abies marocana* non-stratified seeds varied along a temperature gradient, and its seed germination seemed to have a specific temperature requirement. *Abies marocana* seeds without any treatment germinated best at 15 °C, while a remarkable number of viable seeds failed to germinate at a lower (10 °C) temperature. In the literature, there are studies that recommend alternating temperatures of 30° with light for 8 h, and 20 °C with dark for 16 h, to achieve maximum seed germination in different *Abies* species [38], while some others note that species often have a specific temperature range for seed germination [37]. *Abies amabilis* (Douglas ex Loudon) J. Forbes seeds germinate better (although slower) under a 15:10 °C (day:night) temperature regime, regardless of the stratification treatment [44]. A constant temperature of 10 °C is probably the best temperature for the seed germination of *A. pindrow* from four provenances [43], as the highest GP and MGT is observed in this temperature (in contrast, the lowest GPs and MGTs were recorded at 25 °C). However, *A. cephalonica* seeds seem to germinate well over a wide temperature range (5–25 °C) without any pre-treatment, at both constant and alternating temperatures [45].

In our study, the CS of *A. marocana* seeds resulted in the loss of sensitivity to temperature. In cold-stratified seeds, regardless of duration, no significant difference in GPs among the incubation temperatures was observed. Apart from widening the range of temperatures for optimum germination, stratification resulted in more rapid germination of *A.*

marocana seeds, thus confirming similar results found in previous studies for other *Abies* spp. [26,46,47]. In practical horticulture for sustainable exploitation, apart from high seed germination, uniform and rapid seed germination is equally significant in order to avoid environmental hazards within the nursery [36]. In general, the fastest rate of germination is usually observed at temperatures between 15 and 25 °C for the seeds of many plant species. However, the germination percentages of the cold-stratified seeds of *A. marocana* studied herein were found to be lower than those of viable seeds. This could be attributed to mold development during the stratification period in the laboratory chambers since increased mold growth was observed on the two-month cold-stratified seeds of *A. marocana*. The same problem has been reported in germination studies involving seeds of other members of the genus *Abies* [40,44,46,48].

Most *Abies* species exhibit some degree of seed dormancy, and stratification is required to improve germination in terms of capacity and/or speed [25]. In the present study, the detected response to temperature and CS confirms the existence of some degree of dormancy in *A. marocana* seeds. CS is regarded as the most important treatment for breaking dormancy in the seeds of many species of the temperate zone [37,49–51]. It is known that when CS reduces the temperature requirement for germination or increases the speed of germination, the species should be listed [after 37] as having physiological dormancy rather than being with non-dormant seeds.

Temperature is an important environmental factor that regulates seed germination in many plant species [52]. The Talassemtane forest, where *A. marocana* occurs naturally, in Morocco receives an average precipitation of 1500 mm annually [5,6]. In our study, the GIS-derived ecological profile of *A. marocana* informed that the mean annual precipitation in the species range is 1024.81 ± 54.99 mm. Previous studies focusing on niche modelling report that droughts, the average temperature of the warmest quarter, the maximum temperature of the warmest month, the annual average temperatures, and the annual average precipitation are critical abiotic factors for this species [9,10]. Obviously, *A. marocana* has developed specific adaptations and a survival strategy to prevent seed germination during unfavorable environmental conditions for seedling growth, such as cold or dry periods. Cones of *A. marocana* ripen in autumn with subsequent seed dispersal in mid-autumn or early winter [9]. According to the results of this study, *A. marocana* seeds germinated without any CS; however, their germination started with a month-long delay at temperatures from 15–25 °C (see Figure 4). Moreover, at 10 °C, the germination started after the 50th day. Based on the results of this study, in vivo germination in the wild habitats does not seem to happen during October and November or early winter because the temperatures recorded in the distribution area of *A. marocana* are quite lower than their preferable temperature for the initiation of seed germination (15 °C for the germination of non-stratified seeds in laboratory conditions). According to the GIS-derived ecological profile of *A. marocana* (Figure 3), the mean monthly air temperatures (not ground or below-ground temperatures) for October, November, and December are 12.55, 9.35, and 6.24 °C, respectively. In parallel, in the wild habitats of *A. marocana*, the increased precipitation over the winter months (mean precipitation from 74.16 mm in October to 123.22 mm in December and 171.98 mm in March), in combination with the low temperatures during this period (see Figure 3), probably create the ideal conditions for natural seed stratification. As inferred by the germination results of cold-stratified *A. marocana* seeds, their exposure to humid and cold conditions during wintertime, which are the appropriate conditions for dormancy-breaking, results in a widening of the temperature range over which germination actually occurs in spring. Subsequently, the seeds can probably germinate in early spring when the average mean temperatures are still relatively low (7.96 °C and 11.52 °C in March and April, respectively) and the soil moisture is high or progressively close to the favorable germination temperature detected in germination tests (15 °C).

4. Materials and Methods

4.1. Multifaceted Evaluation in Economic Sectors

The evaluation of the potential of *A. marocana* in specific economic sectors (Level I evaluation) was overviewed based on twenty attributes assessing its ornamental-horticultural potential [18], (ii) seven attributes assessing its agro-alimentary potential [27], and (iii) nine attributes assessing its potential in the medicinal-cosmetic sector [28]. A multidisciplinary scientific consortium, including 13 experienced scientists, was engaged to develop a new methodological scheme for the multifaceted evaluation of target NUPs, deciding collectively after a case-by-case examination and consultation the following issues: individual attributes per economic sector to be used for the evaluation; typology of attributes used for evaluation (sector-specific or inter-sectorial); selection of data sources to be used for documentation (one to four types per selected attribute); scaling per attribute (two-fold to seven-fold), and directionality for the scoring of different attributes (possible scores and value definitions based on the quality and quantity of the extant information retrieved). This methodological scheme is fully described in detail (e.g., individual scores per attribute) along with guidelines and examples of scorings for 399 local endemic taxa of three Mediterranean regions (Crete, Mediterranean Coast–Rif, and Tunisia) in previous studies of our own for the ornamental sector [18], the agro-alimentary sector [27], and the medicinal-cosmetic sector [28]. The evaluation of the potential of *A. marocana* in specific economic sectors (Level I evaluations) was overviewed based on twenty attributes assessing its ornamental-horticultural potential [18], (ii) seven attributes assessing its agro-alimentary potential [27], and (iii) nine attributes assessing its potential in the medicinal-cosmetic sector [28]. After the scoring of individual attributes, the sum of scorings for all attributes per economic sector was calculated and expressed as the relative percentage (%) of the maximum possible score that could be generated in each sector [18,27,28]. In this way, the relevant data and information sourced from widely scattered sources were assessed to comprehensively illustrate the relative potential of *A. marocana* in different economic sectors.

Feasibility evaluation of *A. marocana* (Level II evaluation) involved point-scoring of 12 selected attributes, considered either as prerequisites of common interest across various economic sectors (eight attributes), or as unique identity elements or special features (four attributes) to be exploited in terms of product branding and marketing [18]. The designated readiness timescale evaluation for the sustainable exploitation for *A. marocana* (Level III evaluation) involved the completion of eight criteria, based on SWOT (Strengths, Weaknesses, Opportunities, Threats) and gap analyses. These evaluations were sourced from previous research or our own [18] and are presented herein in detail for *A. marocana*.

4.2. Distribution Mapping and GIS Ecological Profiling

In total, 313 natural distribution points (occurrence records) of current *A. marocana* wild-growing populations were taken from previously published studies [9,53,54]. Based on the natural distribution range of *A. marocana* in Morocco (Figure 5), respective historical climate data of pixel size 30 sec was downloaded from the WorldClim website (https://www.worldclim.org/data/worldclim21.html, accessed on 26 November 2021) regarding minimum, maximum, and average temperatures and precipitation, as well as the respective values regarding 19 bioclimatic variables, i.e., Annual Mean Temperature (Bio_1), Mean Diurnal Range (Bio_2), Isothermality (Bio_3), Temperature Seasonality (Bio_4), Max Temperature of Warmest Month (Bio_5), Min Temperature of Coldest Month (Bio_6), Temperature Annual Range (Bio_7), Mean Temperature of Wettest Quarter (Bio_8), Mean Temperature of Driest Quarter (Bio_9), Mean Temperature of Warmest Quarter (Bio_10), Mean Temperature of Coldest Quarter (Bio_11), Annual Precipitation (Bio_12), Precipitation of Wettest Month (Bio_13), Precipitation of Driest Month (Bio_14), Precipitation Seasonality (Bio_15), Precipitation of Wettest Quarter (Bio_16), Precipitation of Driest Quarter (Bio_17), Precipitation of Warmest Quarter (Bio_18), and Precipitation of Coldest Quarter (Bio_19). The following layers were imported in the GIS environment:

Figure 5. Natural global distribution range of the threatened local endemic Moroccan fir (*Abies marocana*) based on 313 occurrence records [9,53,54].

(a) WorldClim version 2.1 [55], containing minimum, maximum, and average temperatures (°C) as well as precipitation values (mm) and data for 19 bioclimatic variables for every month derived from 1970–2000, with a raster resolution of 1 km^2;

and (b) an *A. marocana* distribution raster file, including 313 occurrence records [9,53,54] (Figure 5).

4.3. Seed Collection and Storage

On 31 August 2019, mature cones of *A. marocana* were collected by hand from random wild-growing individuals in natural habitats of the Talassemtane National Park, Rif, Morocco (035°07′02″ N, 05°08′00″ W and altitude 1539 m a.s.l). The species habitat was characterized by the presence of limestone rock outcrops on a steep slope (Figure 6). During collection, every paper bag with cones and seeds was labeled separately (taxon, date and location). After harvesting, the cones were stored in paper bags to allow for gradual desiccation. To extract the seeds in the laboratory, the cones were fragmented by hand and then the seeds were manually separated from the scales and cone axes (Figure 7). The separated seeds were de-winged by hand and then stored in glass containers in a laboratory refrigerator (3–5 °C).

Then, to be gradually dehumidified, the seeds were deposited for 12 months in the Seed Bank of the Institute of Plant Breeding and Genetic Resources, Agricultural Organization Demeter, in Thessaloniki, Greece, at 5 °C in a walk-in fridge with low relative humidity (25%). Prior to the seed transfer that took place in the frame of the MULTI-VAL-END project ("Multifaceted Valorisation of single-country Endemic plants of Crete, Greece, Tunisia and Rif, Morocco for sustainable exploitation in the agro-alimentary, horticultural-ornamental and medicinal-cosmetic sectors"; ARIMNet2), a MTA (Mutual Transfer Agreement) was officially signed by the legal bodies of the donor and recipient parties to ensure the endorsement of provisions of the Nagoya Protocol, as imposed by the EU Regulation 511/2014. Then, the taxonomically identified stored seedlot of *A. marocana* (Figure 6) obtained a unique IPEN (International Plant Exchange Network) accession number MA-1-BBGK-20,428 (identifier and passport data illustrating origin, restrictions of use, storage institution, year of collection, and identity number), and it was photographed using a Ricoh WG-6 camera with an incorporated LED ring light.

Figure 6. (**A**) Forests and habitat of *Abies marocana* in Talassemtane National Park of the Rif region in Morocco; (**B**) Habitat of a mature *A. marocana* individual; (**C**) Trunk and ramification of a random *A. marocana* mature individual; (**D**) Male inflorescences; (**E**) Ripe female cones with seeds.

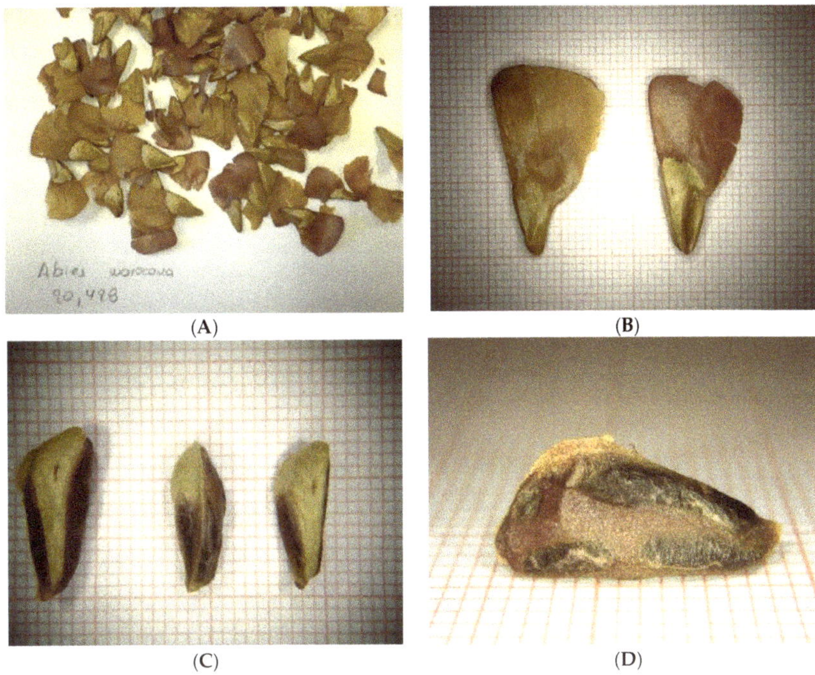

Figure 7. Seed morphology of *Abies marocana*: (**A**) Extracted seeds from cones; (**B**) Size variability of winged seeds; (**C**) Size variability of de-winged seeds; (**D**) Lateral view of de-winged seed.

4.4. Seed Treatment

Germination experiments were initiated in January 2021 and they were conducted in the Laboratory of Horticulture, School of Agriculture, Aristotle University of Thessaloniki, Greece. Before the experiments, a random sample of 100 seeds of *A. marocana* (four replications of 25 seeds) was subjected to a tetarazolium test in order to estimate seed viability. Pre-moistened seeds were cut longitudinally beside the embryo, and then were immersed in a 1% solution of 2,3,5-triphenyl tetrazolium chloride for 18 h in the dark [38]. Subsequently, the staining of cut seeds was examined (Figure 8).

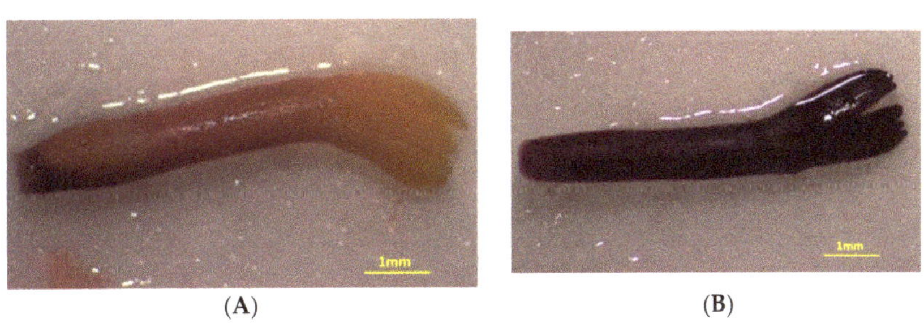

Figure 8. Non-viable (**A**) and viable (**B**) embryos during viability examination of *Abies marocana* seeds.

In order to determine the effect of cold stratification (CS) on germination, the seeds were mixed with moist sterilized river sand in plastic containers and were stratified for one or two months at 2–4 °C in separate plastic containers with 400 seeds each. In addition, non-stratified seeds (0 months) were subjected to germination.

4.5. Germination Tests

The non-stratified as well as the stratified seeds were placed for germination in growth chambers at the end of each CS period, and their germination response to four constant temperatures (10, 15, 20, and 25 °C) was evaluated. Experiments were maintained in a CRW-500SD growth chamber (Chrisagis, Athens, Greece) with relative humidity (RH) at 75 ± 1% and a light intensity of 82 μmol m^{-2} s^{-1} at the culture level from cool-white fluorescent tubes. The selection of temperature intervals was facilitated by the GIS ecological profile generated for *A. marocana* (see Section 2.2), based on 313 occurrence records located in nearby regions. For each treatment and temperature, there were four replications of 25 seeds. The seeds were placed on sterilized river sand in 12-cm glass Petri dishes, moistened with distilled water. The Petri dishes were randomly arranged on the shelves of the growth chambers under a 12-h light/12-h dark photoperiod regime, and were watered with distilled water for the whole experimental period. Attention was given to watering so as to avoid drying, as well as excess moisture, in the substratum (sand). The amount of water required to moisten the sand was not constant, and depended on the incubation temperature. Germinated seeds were counted each week for a period of 13 weeks. A seed was considered as germinated when an at least 2-mm-long radicle had emerged through the seed coat (Figure 9). Finally, for each temperature of each CS period, the germination percentage (GP) and the mean germination time (MGT) were calculated as the average of the four replications. The MGT was calculated for each replication per treatment, according to the following equation:

$$\mathrm{MGT} = \Sigma(\mathrm{Dn})/\Sigma\mathrm{n}$$

where n is the number of seeds which germinated on day D, counted from the beginning of the test [56].

Figure 9. Germinated seeds of *Abies marocana* with evident radicle protrusion.

4.6. Statistical Analysis

The experimental design was a completely randomized design with two factors. The factors were the duration of CS and the incubation temperature (3 × 4 factorial design). The GP data was transformed to arc-sine square root values before analysis [57,58]. The transformed data, as well as the MGT data, were checked for normality and homogeneity of variances, and they were analyzed using the ANOVA method [58] in the frame of the general linear model (GLM), while the comparisons of the means were performed using the R-E-G-WQ test [59,60]. All statistical analyses were carried out using SPSS 21.0 (SPSS, Inc., Armonk, New York, USA).

5. Conclusions

To date, there are only non-coordinated efforts for the ex-situ conservation of *Abies marocana* in five foreign countries, but no ex-situ conservation is offered in Moroccan facilities [20]. This fact is coupled with commercial demand, as illustrated by the global internet plant trade [21]. The current research explored comprehensively and documented substantially the existing potential of *Abies marocana* in different economic sectors (ornamental-horticultural, medicinal-cosmetic, and agro-alimentary), and overviewed the feasibility and readiness timescale regarding value chain creation for its sustainable exploitation. Furthermore, this study investigated the climatic conditions in which its wild-growing populations thrive naturally in Morocco, and revealed the germination requirements of *A. marocana* seeds (effects of temperature and cold stratification). Thus, the findings of this study fill in extant research gaps, may contribute to in situ and ex situ conservation strategies, and can facilitate the sustainable exploitation of this emblematic local endemic plant of northern Morocco.

Author Contributions: Conceptualization, S.H., N.K. and G.T.; data curation, N.K., I.A., A.K., M.L., W.M.-K., M.E.H., S.K., M.A.S., V.A., V.G. and S.B.; formal analysis, E.P., N.K., I.A., S.K. and S.H.; funding acquisition, N.K., M.E.H. and S.B.; investigation, N.K., S.H., I.A., V.G., E.P., S.K., A.K., M.L., W.M.-K., Z.G.-G., F.L., M.E.H., M.A.S., V.A., G.T. and S.B.; methodology, N.K., S.H., E.P., S.K., I.A., V.G., M.A.S., A.K., M.L., W.M.-K., Z.G.-G., F.L., M.E.H., G.T. and S.B.; project administration, N.K.; resources, N.K., M.E.H., V.G., S.H. and S.K.; software, N.K., I.A. and E.P.; supervision, N.K., S.H. and G.T.; validation, I.A., V.G., V.A., A.K., M.L., W.M.-K., M.A.S., Z.G.-G., F.L., G.T., M.E.H. and S.B.; visualization, N.K., A.K., E.P., M.L. and M.E.H.; writing—original draft, E.P., S.H., N.K., I.A., M.L., A.K. and M.E.H.; writing—review and editing, all authors. All authors have read and agreed to the published version of the manuscript.

Funding: This research was supported by the ARIMNet2 2017 Transnational Joint Call through the MULTI-VAL-END project "Multifaceted Valorisation of single-country Endemic plants of Crete, Greece, Tunisia and Rif, Morocco for sustainable exploitation in the agro-alimentary, horticultural-ornamental and medicinal-cosmetic sectors", and was co-funded by the Hellenic Agricultural Organization Demeter of Greece, the State Secretariat for Higher Education and Scientific Research (SEESRS) of Morocco, and the Ministry of Higher Education and Scientific Research (Ministère de l'Enseignement Supérieur et de la Recherche Scientifique, MESRS), Republic of Tunisia. ARIMNet2 (ERA-NET) has received funding from the European Union's Seventh Framework Programme for research, technological development, and demonstration, under grant agreement no. 618127.

Institutional Review Board Statement: Not applicable.

Informed Consent Statement: Not applicable.

Data Availability Statement: The data presented in this study are available on request from the corresponding authors.

Acknowledgments: The authors would like to thank Ioulietta Samartza for taking photographs of the seeds of *Abies marocana*, as well as the anonymous reviewers for their valuable comments and suggestions.

Conflicts of Interest: The authors declare no conflict of interest.

References

1. Vidakovic, V. *Abies* Mill.—Fir—Pinaceae. In *Conifers: Morphology and Variation*; Brekalo, B., Ed.; Graficki Zadov: Hrvatske, Yugoslavia, 1991; pp. 53–116.
2. Kohli, R.K.; Batish, D.R.; Singh, H.P. Important tree species: Forests and Forest Plants. In *Encyclopedia of Life Support System (EOLSS)*; Owens, J.N., Lund, H.G., Eds.; EOLSS Pub.: Oxford, UK, 2003; pp. 1–39.
3. Redžić, S.S. The ecological aspect of ethnobotany and ethnopharmacology of population in Bosnia and Herzegovina. *Coll. Antropol.* **2007**, *31*, 869–890. [PubMed]

4. Aussenac, G. Ecology and ecophysiology of circum-Mediterranean firs in the context of climate change. *Ann. For. Sci.* **2002**, *59*, 823–832. [CrossRef]
5. Alaoui, M.L.; Knees, S.; Gardner, M. *Abies pinsapo* var. *marocana*. The IUCN Red List of Threatened Species 2011; e.T34126A9841418. 2011. Available online: https://www.iucnredlist.org/species/34126/9841418 (accessed on 10 November 2021).
6. Baumer, M. Le sapin du Maroc. *Biol. Forêt* **1977**, *24*, 343–354. Available online: http://documents.irevues.inist.fr/bitstream/handle/2042/21153/RFF_1977_5_343.pdf (accessed on 6 July 2021). [CrossRef]
7. Hampe, A.; Jump, A. Climate relicts: Past, present, future. *Annu. Rev. Ecol. Evol. Syst.* **2011**, *42*, 313–333. [CrossRef]
8. Esteban, L.G.; De Palacios, P. *Pinsapo* forests: Past, present and future. *Bois Forêt Trop.* **2007**, *292*, 39–47. [CrossRef]
9. Alaoui, A.; Laaribya, S.; Ayan, S.; Ghallab, A.; López-Tirado, J. Modelling spatial distribution of endemic Moroccan fir (*Abies marocana* Trabut) in Talassemtane National Park, Morocco. *Austrian J. For. Sci.* **2021**, *138*, 73–94.
10. Navarro-Cerrillo, R.M.; Manzanedo, R.D.; Rodriguez-Vallejo, C.; Gazol, A.; Palacios-Rodríguez, G.; Camarero, J.J. Competition modulates the response of growth to climate in pure and mixed *Abies pinsapo* subsp. *maroccana* forests in northern Morocco. *For. Ecol. Manag.* **2020**, *459*, 117847. [CrossRef]
11. Ghallab, A. Modélisation de la Distribution Potentielle Actuelle des Espèces Forestières Majeures Dans le Rif Occidental du Maroc et Prédiction de L'impact du Changement Climatique. Doctoral Thesis, Université Abdelmalek Essaâdi, Tétouan, Marocco, 2018.
12. Navarro-Cerrillo, R.M.; Camarero, J.J.; Manzanedo, R.D.; Sánchez-Cuesta, R.; Lopez Quintanilla, J.; Sánchez Salguero, R. Regeneration of *Abies pinsapo* within gaps created by *Heterobasidion annosum*-induced tree mortality in southern Spain. *IForest* **2014**, *7*, 209. [CrossRef]
13. Boukil, A. Etude de l'évolution de la gestion de la sapinière de Talassemtane (Rif centro-occidental, Maroc). *Méditerranéenne* **1998**, *20*, 145–152.
14. Dadach, M. Recherche des Conditions Optimales de la Germination des Graines de Quelques Labiées du Mont de Tessala (Ouest Algérien) et Perspectives de Conservation. Ph.D. Thesis, Université Djillali Liabes de Sidi Bel Abbes, Sidi Bel Abbes, Algérie, 2013.
15. Belmehdi, O.; El Harsal, A.; Benmoussi, M.; Laghmouchi, Y.; Skali Senhaji, N.; Abrini, J. Effect of light, temperature, salt stress and pH on seed germination of medicinal plant *Origanum elongatum* (Bonnet) Emb. & Maire. *Biocatal. Agric. Biotechnol.* **2018**, *16*, 126–131. [CrossRef]
16. Flores-Enríquez, V.; Castillo, G.; Collazo-Ortega, M. Experimental seed germination for ex situ conservation of Mexican Podostemaceae. *Bot. Sci.* **2019**, *97*, 413–422. [CrossRef]
17. Grigoriadou, K.; Sarropoulou, V.; Krigas, N.; Maloupa, E.; Tsoktouridis, G. GIS-facilitated effective propagation protocols of the Endangered local endemic of Crete *Carlina diae* (Rech. f.) Meusel and A. Kástner (Asteraceae): Serving *ex-situ* conservation needs and its future sustainable exploitation as an ornamental. *Plants* **2020**, *9*, 1465. [CrossRef] [PubMed]
18. Krigas, N.; Tsoktouridis, G.; Anestis, I.; Khabbach, A.; Libiad, M.; Megdiche-Ksouri, W.; Ghrabi-Gammar, Z.; Lamchouri, F.; Tsiripidis, I.; Tsiafouli, M. Exploring the potential of neglected local endemic plants of three Mediterranean regions in the ornamental sector: Value chain feasibility and readiness timescale for their sustainable exploitation. *Sustainability* **2021**, *13*, 2539. [CrossRef]
19. Benzyane, M.; Aafi, A.; Sbay, H.; El Antry, S.; Yassine, M.; Ilmen, R.; Ghaïloule, D. *Les Ecosystèmes Naturels Marocains et les Changements Climatiques: La Résilience Ecologique à L'épreuve*; Collection Maroc Nature; Centre de Recherche Forestière: Rabat, Morocco, 2010; 88p.
20. Libiad, M.; Khabbach, A.; El Haissoufi, M.; Bourgou, S.; Megdiche-Ksouri, W.; Ghrabi-Gammar, Z.; Sharrock, S.; Krigas, N. *Ex-situ* conservation of single-country endemic plants of Tunisia and northern Morocco (Mediterranean coast and Rif region) in seed banks and botanic gardens worldwide. *Kew Bull.* **2020**, *75*, 46. [CrossRef]
21. Khabbach, A.; Libiad, M.; El Haissoufi, M.; Lamchouri, F.; Bourgou, S.; Megdiche-Ksouri, W.; Ghrabi-Gammar, Z.; Menteli, V.; Vokou, D.; Tsoktouridis, G. Electronic commerce of the endemic plants of Northern Morocco (Rif region and Mediterranean coast) and Tunisia over the Internet. *Bot. Sci.* **2021**, *100*, 1–29, in press. [CrossRef]
22. Belaqziz, R.; Romane, A.; Abbad, A. Salt stress effects on germination, growth and essential oil content of an endemic thyme species in Morocco (*Thymus maroccanus* Ball). *J. Appl. Sci. Res.* **2009**, *5*, 858–863. Available online: https://www.cabdirect.org/cabdirect/abstract/20093204278 (accessed on 15 September 2021).
23. Abbad, A.; Belaqziz, R.; Bekkouche, K.; Markouk, M. Influence of temperature and water potential on laboratory germination of two Moroccan endemic thymes: *Thymus maroccanus* Ball. and *Thymus broussonetii* Boiss. *Afr. J. Agric. Res.* **2011**, *6*, 4740–4745. [CrossRef]
24. Laghmouchi, Y.; Belmehdi, O.; Bouyahya, A.; Skali Senhaji, N.; Abrini, J. Effect of temperature, salt stress and pH on seed germination of medicinal plant *Origanum compactum*. *Biocatal. Agric. Biotechnol.* **2017**, *10*, 1–19. [CrossRef]
25. Edwards, D.G.W. Breaking dormancy in tree seeds with special reference to firs (*Abies* species)—The 1.4x solution. In Proceedings of the ISTA Forest Tree and Shrub Seed Committee Workshop Prague—Pruhonice, Czech Republic; Prochazkova, Z., Gosling, P.G., Sutherland, J.R., Eds.; Forestry and Game Management Research Institute Jiloviste—Strnady, CR and Forestry Commission Research Agency: Průhonice, Prague, Czech Republic, 2003; pp. 18–23.
26. Edwards, D.G.W. *Pinaceae—Pine* family: *Abies* P. Mill. fir. In *Woody Plant Seed Manual. Agriculture Handbook No. 727*; Bonner, F.T., Karrfalt, R.P., Eds.; USDA Forest Service: Washington, DC, USA, 2008; pp. 149–198.

27. Libiad, M.; Khabbach, A.; El Haissoufi, M.; Anestis, I.; Lamchouri, F.; Bourgou, S.; Megdiche-Ksouri, W.; Ghrabi-Gammar, Z.; Greveniotis, V.; Tsiripidis, I.; et al. Agro-Alimentary Potential of the Neglected and Underutilized Local Endemic Plants of Crete (Greece), Rif-Mediterranean Coast of Morocco and Tunisia: Perspectives and Challenges. *Plants* **2021**, *10*, 1770. [CrossRef]
28. Bourgou, S.; Jilani, I.B.H.; Karous, O.; Megdiche-Ksouri, W.; Ghrabi-Gammar, Z.; Khabbach, A.; Libiad, M.; El Haissoufi, M.; Lamchouri, F.; Greveniotis, V.; et al. Medicinal-cosmetic potential of the local endemic plants of Crete (Greece), Rif-Mediterranean Coast of Morocco and Tunisia: Priorities for sustainable exploitation. *Biology* **2021**. under review.
29. Krigas, N.; Mouflis, G.; Grigoriadou, K. Conservation of important plants from the Ionian Islands at the Balkan Botanic Garden of Kroussia, N Greece: Using GIS to link the in situ collection data with plant propagation and ex situ cultivation. *Biodivers Conserv.* **2010**, *19*, 3583–3603. [CrossRef]
30. Krigas, N.; Papadimitriou, K.; Mazaris, A.D. GIS and ex-situ plant conservation. In *Application of Geographic Information Systems*; Alam, B.M., Ed.; InTechopen.com: Rijeka, Croatia, 2012.
31. Padulosi, S.; Thompson, J.; Rudebjer, P. *Fighting Poverty, Hunger and Malnutrition with Neglected and Underutilized Species (NUS): Needs, Challenges and the Way Forward*; Bioversity International: Rome, Italy, 2013; p. 56. ISBN 978-92-9043-941-7.
32. Yang, X.W.; Li, S.M.; Shen, Y.H.; Zhang, W.D. Phytochemical and biological studies of *Abies* species. *Chem. Biodivers.* **2008**, *5*, 56–81. [CrossRef]
33. Wajs, A.; Urbańska, J.; Zaleśkiewicz, E.; Bonikowski, R. Composition of essential oil from seeds and cones of *Abies alba*. *Nat. Prod. Commun.* **2010**, *5*, 1291–1294. [CrossRef] [PubMed]
34. Kloukina, C.; Tomou, E.M.; Krigas, N.; Sarropoulou, V.; Madesis, P.; Maloupa, E.; Skaltsa, H. Non-polar secondary metabolites and essential oil of ex situ propagated and cultivated *Sideritis syriaca* L. subsp. *syriaca* (Lamiaceae) with consolidated identity (DNA Barcoding): Towards a potential new industrial crop. *Ind. Crop. Prod.* **2020**, *158*, 112957. [CrossRef]
35. Macdonald, B. *Practical Woody Plant Propagation for Nursery Growers*; Timber Press Inc.: Portland, Oregon, 2006; p. 669.
36. Stussi, I.; Henry, F.; Moser, P.; Danoux, L.; Jeanmaire, C.; Gillon, V.; Benoit, I.; Charrouf, Z.; Pauly, G. *Argania spinosa*—How ecological farming, fair trade and sustainability can drive the research for new cosmetic active ingredients. *SÖFW-J.* **2005**, *131*, 35–46. Available online: http://www.arganoils.com/articles/en/How%20Ecological%20Farming,%20Fair%20Trade%20and%20Sustainability%20Can%20Drive%20research%20for%20new%20cosmetic%20ingredients.pdf (accessed on 4 August 2021).
37. Baskin, C.C.; Baskin, J.M. *Seeds: Ecology, Biogeography, and Evolution of Dormancy and Germination*, 2nd ed.; Academic Press: San Diego, CA, USA, 2014.
38. ISTA (International Seed Testing Association). *International Rules for Seed Testing*; Seed Science and Technology: Wallisellen, Switzerland, 1999; p. 333.
39. Edwards, D.G.W. The stratification-redry technique with special reference to true fir seeds. In *National Proceedings, Forest and Conservation Nursery Associations*; General Technical Report, PNW-GTR-389; Landis, T.D., South, D.B., Eds.; U.S.D.A. Forest Service, Pacific Northwest Research Station: Portland, Oregon, 1996; pp. 172–182.
40. Politi, P.I.; Georghiou, K.; Arianoutsou, M. Reproductive biology of *Abies cephalonica* Loudon in Mount Aenos National Park, Cephalonia, Greece. *Trees* **2011**, *25*, 655–668. [CrossRef]
41. Varsamis, G.K.; Takos, I.A.; Merou, T.P.; Galatsidas, S.A.; Dimopoulos, P.D. Germination characteristics of *Abies* × *borisii-regis*. *Seed Technol.* **2014**, *36*, 51–59.
42. Skrzyszewska, K.; Chlanda, J. A study on the variation of morphological characteristics of silver fir (*Abies alba* Mill.) seeds and their internal structure determined by X-ray radiography in the Beskid Sadecki and Beskid Niski mountain ranges of the Carpathians (southern Poland). *J. For. Sci.* **2009**, *55*, 403–414. [CrossRef]
43. Rawat, B.S.; Khanduri, V.P.; Sharma, C.M. Beneficial effects of cold-moist stratification on seed germination behaviors of *Abies pindrow* and *Picea smithiana*. *J. For. Res.* **2008**, *19*, 125–130. [CrossRef]
44. Leadem, C.L. Stratification of *Abies amabilis* seeds. *Can. J. For. Res.* **1986**, *16*, 755–760. [CrossRef]
45. Daskalakou, E.N.; Koutsovoulou, K.; Mavroeidi, L.; Tsiamitas, C.; Kafali, E.; Radaiou, P.E.; Ganatsas, P.; Thanos, C.A. Interannual variability of germination and cone/seed morphometric characteristics in the endemic Grecian fir (*Abies cephalonica*) over an 8-year-long study. *Seed Sci. Res.* **2018**, *28*, 24–33. [CrossRef]
46. Jones, S.K.; Samuel, Y.K.; Gosling, P.G. The effect of soaking and prechilling on the germination of noble fir seeds. *Seed Sci. Technol.* **1991**, *19*, 287–294.
47. Blazich, F.A.; Hinesley, L.E. Low temperature germination of Fraser fir seed. *Can. J. For. Res.* **1984**, *14*, 948–949. [CrossRef]
48. Adkins, C.R.; Hinesley, L.E.; Blazich, F.A. Role of stratification, temperature, and light in Fraser fir germination. *Can. J. For. Res.* **1984**, *14*, 88–93. [CrossRef]
49. Pipinis, E.; Milios, E.; Kiamos, N.; Mavrokordopoulou, O.; Smiris, P. Effects of stratification and pre-treatment with gibberellic acid on seed germination of two *Carpinus* species. *Seed Sci. Technol.* **2012**, *40*, 21–31. [CrossRef]
50. Pipinis, E.; Milios, E.; Mavrokordopoulou, O.; Smiris, P. Effect of sowing date on seedling emergence of species with seeds enclosed in a stony endocarp. *J. Sustain. For.* **2018**, *37*, 375–388. [CrossRef]
51. Pipinis, E.; Stampoulidis, A.; Milios, E.; Kitikidou, K.; Akritidou, S.; Theodoridou, S.; Radoglou, K. Effects of seed moisture content, stratification and sowing date on the germination of *Corylus avellana* seeds. *J. For. Res.* **2020**, *31*, 743–749. [CrossRef]
52. Fenner, M.; Thompson, K. *The Ecology of Seeds*; Cambridge University Press: Cambridge, UK, 2005; p. 250.
53. Caudullo, G.; Welk, E.; San-Miguel-Ayanz, J. Chorological maps for the main European woody species. *Data Br.* **2019**, *12*, 662–666. [CrossRef]

54. Ben-Said, M.; Ghallab, A.; Lamrhari, H.; Carreira, J.A.; Linares, J.C.; Taïqui, L. Characterizing spatial structure of *Abies marocana* forest through point pattern analysis. *For. Syst.* **2020**, *29*, e016. [CrossRef]
55. Fick, S.E.; Hijmans, R.J. WorldClim 2: New 1-km spatial resolution climate surfaces for global land areas. *Inter. J. Climatol.* **2017**, *37*, 4302–4315. [CrossRef]
56. Hartmann, H.T.; Kester, D.E.; Davies, F.T.; Geneve, R.L. *Plant Propagation: Principles and Practices*, 6th ed.; Prentice-Hall: Upper Saddle River, NJ, USA, 1997; p. 770.
57. Snedecor, G.W.; Cochran, W.C. *Statistical Methods*, 7th ed.; The Iowa State University Press: Ames, IA, USA, 1980; p. 507.
58. Gomez, K.A.; Gomez, A.A. *Statistical Procedures for Agricultural Research*, 7th ed.; John Wiley and Sons Publications Inc.: New York, NY, USA, 1984; p. 680.
59. Klockars, A.; Sax, G. *Multiple Comparisons*; Sage Publications: Newbury Park, CA, USA, 1986; p. 87.
60. Toothaker, L. *Multiple Comparison Procedures*; Sage Publications: Newbury Park, CA, USA, 1993; p. 96.

plants

MDPI

Article

Molecular Authentication, Phytochemical Evaluation and Asexual Propagation of Wild-Growing *Rosa canina* L. (Rosaceae) Genotypes of Northern Greece for Sustainable Exploitation

Eleni Maloupa [1], Eleftherios Karapatzak [1], Ioannis Ganopoulos [1], Antonis Karydas [1], Katerina Papanastasi [1], Dimitris Kyrkas [2], Paraskevi Yfanti [2], Nikos Nikisianis [3], Anthimos Zahariadis [4], Ioanna S. Kosma [5], Anastasia V. Badeka [5], Giorgos Patakioutas [2], Dimitrios Fotakis [6] and Nikos Krigas [1,*]

1 Institute of Plant Breeding and Genetic Resources, Hellenic Agricultural Organization Dimitra, P.O. Box 60458, 57001 Thermi, Thessaloniki, Greece; maloupa@bbgk.gr (E.M.); ekarapatzak@gmail.com (E.K.); giannis.ganopoulos@gmail.com (I.G.); euripidis_1999@hotmail.com (A.K.); papanastasi@bbgk.gr (K.P.)

2 Department of Agriculture, University of Ioannina, 47100 Arta, Greece; dkyrkas@uoi.gr (D.K.); pyfanti@uoi.gr (P.Y.); gpatakiu@uoi.gr (G.P.)

3 Systada General Partnership, 55133 Kalamaria, Thessaloniki, Greece; nnikisia@gmail.com

4 Verus plus General Partnership, 55133 Kalamaria, Thessaloniki, Greece; zachariadis@verusplus.com

5 Laboratory of Food Chemistry, Department of Chemistry, University of Ioannina, 45110 Ioannina, Greece; i.kosma@uoi.gr (I.S.K.); abadeka@uoi.gr (A.V.B.)

6 Forest Research Institute, Hellenic Agricultural Organization Dimitra, 57006 Vassilika, Thessaloniki, Greece; fotakis@fri.gr

* Correspondence: nikoskrigas@gmail.com; Tel.: +30-23-1047-1110

Citation: Maloupa, E.; Karapatzak, E.; Ganopoulos, I.; Karydas, A.; Papanastasi, K.; Kyrkas, D.; Yfanti, P.; Nikisianis, N.; Zahariadis, A.; Kosma, I.S.; et al. Molecular Authentication, Phytochemical Evaluation and Asexual Propagation of Wild-Growing *Rosa canina* L. (Rosaceae) Genotypes of Northern Greece for Sustainable Exploitation. *Plants* **2021**, *10*, 2634. https://doi.org/10.3390/plants10122634

Academic Editors: Milan S. Stankovic, Paula Baptista and Petronia Carillo

Received: 1 November 2021
Accepted: 28 November 2021
Published: 30 November 2021

Publisher's Note: MDPI stays neutral with regard to jurisdictional claims in published maps and institutional affiliations.

Copyright: © 2021 by the authors. Licensee MDPI, Basel, Switzerland. This article is an open access article distributed under the terms and conditions of the Creative Commons Attribution (CC BY) license (https://creativecommons.org/licenses/by/4.0/).

Abstract: Dogroses belong to a taxonomically difficult genus and family and represent important phytogenetic resources associated with high ornamental, pharmaceutical-cosmetic and nutritional values, thus suggesting a potentially high exploitation merit. Triggered by these prospects, wild-growing *Rosa canina* populations of Greece were selected for investigation and evaluation of their potential for integrated domestication. We collected ripe rosehips from Greek native wild-growing populations (samples from seven genotypes) for phytochemical analysis (total phenolics, total flavonoids, antioxidant activity and vitamin C content), leaf samples for DNA analysis using the ITS2 sequence (nine genotypes) and fresh soft-wood stem cuttings for propagation trials (seven genotypes). After evaluation of these materials, this study reports for the first-time distinct DNA-fingerprinted genotypes from Greece with interesting phytochemical profiles mainly in terms of Vitamic C content (up to 500.22 ± 0.15 mg of ascorbic acid equivalents/100 g of sample) as well as effective asexual propagation protocols for prioritized *R. canina* genotypes via cuttings. The latter highlights the importance of the levels of external hormone application (2000 ppm of indole-3-butyric acid), the effect of season (highly-effective spring trials) and genotype-specific differences in rooting capacities of the studied genotypes. All inclusive, this study offers new artificially selected material of Greek native *R. canina* with a consolidated identity and interesting phytochemical profile. These materials are currently under ex-situ conservation for further evaluation and characterization in pilot field studies, thus facilitating its sustainable exploitation for applications in the agro-alimentary, medicinal-cosmetic, and ornamental sectors.

Keywords: biodiversity; ex-situ conservation; protocols; DNA barcoding; germplasm; phytogenetic resources; forest berries

1. Introduction

To date there are at least 373 recognized species worldwide in genus *Rosa* L. of Rosaceae family (www.theplantlist.org, accessed on 1 November 2021) and about 30,000 ornamental varieties [1,2]; the latter are probably derived from only seven species largely contributing to the creation of the modern commercial rose with further seven species providing some minor inputs [3]. Nevertheless, according to the Plant List database 19 rose species are still unplaced due to unresolved taxonomic status, and additionally >3100 species names of roses remain to be evaluated, thus outlining a notoriously complex taxonomy in genus *Rosa* [4]. Regarding the European dogroses of section *Caninae* (DC.) Ser., two basic trends currently prevail; they are either well-regarded by rose lovers for their attractiveness and pleasant scent or they are dreaded by scientists for their genetic complexity [5]. With about 60 rose species of Eurasian distribution, *Caninae* represents one of the largest sections of the genus *Rosa* [5]. The species included in this section cannot be well circumscribed by specific morphological traits due to multiple reproductive strategies (from apomixis to outcrossing including hybridization) and a unique meiotic system (the so-called canina meiosis), rendering dogroses as mostly pentaploids (rarely tetraploids or hexaploids; 2 n = 4 x, 5 x, 6 x = 28, 35, 42) with a base number of seven chromosomes [5]. The complex taxonomy of this family, genus, and section makes species identification difficult, and genetic studies including DNA barcoding may offer insight regarding relationships among closely related species [6].

The systematic genetic identification of model and non-model organisms through the use of modern molecular techniques based on the DNA sequence has become very popular in recent years. The DNA barcoding method is a widely used molecular tool which in combination with bioinformatic analysis has a wealth of applications in ecological, taxonomical, comparative biology, diversity, conservation, phylogenic and genetic studies amongst various plant species and taxonomic groups [7], but also in members of the Rosaceae family [6] and of the genus *Rosa* (e.g., [2,8–11]). Different DNA fingerprinting techniques were applied in members of the genus *Rosa* for effective characterization [12,13] using different molecular markers such as Inter Simple Sequence Repeats (ISSRs) [2,10], Random Amplified Polymorphic DNA (RAPD) markers [8], Internal Transcribed Spacer (ITS) markers [6] and Amplified Fragment Length Polymorphisms (AFLPs) markers [14]. From the plethora of markers that have been evaluated to date for barcoding (mainly of chloroplastic DNA, e.g., rbcL, matK, trnH-psbA, etc.), the nuclear internal transcribed spacer 2 (ITS2) is the predominant one due to its short length, high efficiency in species differentiation and clear PCR amplification results [15]. Studies including thousands of samples from 4800 medicinal plant species in 753 distinct genera [15] and of 893 Rosaceae members in 96 genera [6] suggest that ITS2 is the most suitable and effective DNA barcoding marker for identification purposes both in general and specifically in Rosaceae members including members of genus *Rosa*.

The first sequencing of the rose genome has been performed only recently [16] by sequencing of wild and heterozygous *Rosa multiflora* Thunb. genotypes. Two high quality reference genomes have been published which significantly enrich existing information, i.e., a high-density diploid SNP genetic map [17] and a high-density map for tetraploid rose crucial for anchoring pseudomolecules corresponding to the chromosomes [18]. These results were validated by HiC sequencing [19], reporting genes and transposable element annotation. During the last decade, next-generation sequencing has released transcriptome data from numerous tissues and developmental stages for various genotypes and rose species [20–22]. All these data accumulated to date are of paramount importance for gene annotation studies and such data and metadata may pave the way for the construction of a gene expression atlas anchoring candidate genes of interest through meta-analyses in several members of the genus *Rosa*.

Rose hips of several *Rosa* spp. of section *Caninae* are used in many regions worldwide (also in the Mediterranean region) for tea, soup, jam and jelly preparations [23,24]. Rose hips contain major biologically active components such as flavonoids, tannins, anthocyanins, phenolic compounds, fatty oils, organic acids as well as inorganic compounds [24]. Among dogroses, *R. canina* L. produces fruits (rose hips) of high pharmaceutical [24–26] and nutritional value [24]. *Rosa* spp. have been reported to be rich in several biologically active compounds such as high vitamin C content exceeding that of Citrus fruits. Other active compounds of *Rosa* spp. among others are phenolics, flavonoids, tannins, organic acids, etc. [27,28]. Due to their rich content in bioactive molecules associated with beneficial properties and few toxic or allergic reports and side effects [24], rose hips (including those of *R. canina*) are used in many countries for the treatment of many symptoms such as pain, gastroenteric ailments, cough, cold, inflammations, diarrhea etc., as well as for the prevention of many diseases including diabetes, hypertension, bronchitis, flu, arthritis etc. [24,28–31]. Previous targeted studies have shown that rose hips (including those of *R. canina*) demonstrate significant antioxidant, anticancer, anti-inflammatory, anti-obesity, anti-aging, antinociceptive, anti-Helicobacter pylori activities as well as gastro-, hepato-, nephro-, neuro- and cardioprotective activities [24,28–31]. Although *R. canina* is well-studied plant species in terms of phytochemisty, traditional uses and pharmacological profile [24], still there are very few studies examining wild-growing material from Greece, e.g., [28].

In the European context, Greece is quite rich in different *Rosa* spp. (http://portal.cybertaxonomy.org/flora-greece/intro accessed on 1 November 2021), including native wild-growing populations of 23 distinct species, thus almost half of the European *Rosa* species listed in Flora Europaea [32]. *R. canina* is widespread in Greece across a variety of habitat types (slopes, valleys, riverbanks, usually in thickets and open woodland along roads) from sea level up to 1600–1700 m (occasionally up to 2000 m). It flowers mainly in June and ripe fruits in wild-growing populations are usually observed from the end of August to the beginning of October [33]. For the successful utilization and exploitation of Greek native *R. canina* germplasm, the development of a distinct and solid identity based on DNA fingerprinting is aimed in the first place, elucidating relationships with other relevant materials. Furthermore, the development of effective asexual propagation protocol is crucial, safeguarding the steady transfer of desirable agronomic features and fruit traits to the offspring on one hand, and securing the production of uniform plant material on a commercial scale on the other hand. The use of cuttings is a very efficient and cost-efficient asexual propagation method that can serve these needs [34]. Although the propagation via cuttings of *R. canina* has been conducted on a variety of germplasm sources highlighting the enhancing effect on rooting of external application of indole-3-butyric acid (IBA) [35–37], there are no such studies however targeted to Greek native germplasm of *R. canina*.

In the frame of sustainable exploitation strategies [38–40] and coordinated research efforts to explore and evaluate the economic potential of neglected and underutilized phytogenetic resources which are native to Mediterranean regions [41–43], the aim of the current study was three-fold, focusing on: (i) the molecular authentication of Greek native *R. canina* germplasm (DNA fingerprinting of different genotypes); the evaluation of phytochemical content of selected Greek native genotypes; and (iii) the development of genotype-specific asexual propagation protocols thereof by cuttings. The study was conducted in two phases: Firstly, the implementation of preliminary propagation trials was performed from material collected directly from the wild using a variety of hormone levels and cutting types across different periods of the year, namely different growth stages of mother plants. This was done in order to be able to assess roughly the propagation potential of different *R. canina* genotypes. In parallel, wild-collected leaves were used for DNA barcoding, and wild harvested rose hips were used for phytochemical assessments of different genotypes. Consecutively, following-up the preliminary propagation results, and taking into account the overall assessment of molecular authentication and general phytochemical assessments of the studied Greek native genotypes, targeted propagation experiments were conducted on selected materials using ex-situ raised mother plants which originated in the wild. The consolidated identity of genotypes, the phytochemical profile and the development of a reliable, easy to implement and economically viable propagation protocols are aimed herein to further contribute to the sustainable utilization of the selected Greek native germplasm of *R. canina*.

2. Results

2.1. Authentication Efficiency of ITS2

To test the authentication efficiency of ITS2 for the selected Greek native *Rosa canina* genotypes, the BLAST1 and distance-based methods were selected. Using the BLAST1 method, the ITS2 barcode showed a high identification efficiency of 99% and 100% of the samples at the species and genus levels, respectively. Additionally, using the DISTANCE method, the ITS2 barcode showed identification efficiency of 98% at the species level. Figure 1A depicted that barcode ITS2 using the NJ tree method is able to distinguish Greek native *Rosa canina* genotypes from *R. canina* genotypes which are not native to Greece as well as from other *Rosa* spp. The neighbor-joining (NJ) phylogenetic tree resulting from DNA barcoding application using the ITS2 region classified all *Rosa* spp. samples into four different groups, and clearly discerned the Greek native *R. canina* samples collected in this study from all other samples of *R. canina* sourced from databases (Figure 1A). Bootstrap values further validate this classification. Although evolutionary relationships may be analyzed through the neighbor-joining tree, its key function herein is to repetitively evaluate bootstrap values to enlighten distinction of the Greek native germplasm (see distinct clades). Our results showed that the ITS2 gene is a valid choice with absolute (100%) efficiency for the distinction of the studied species (*R. canina*) among *Rosa* spp., classifying each *Rosa* species in separate monophyletic clusters.

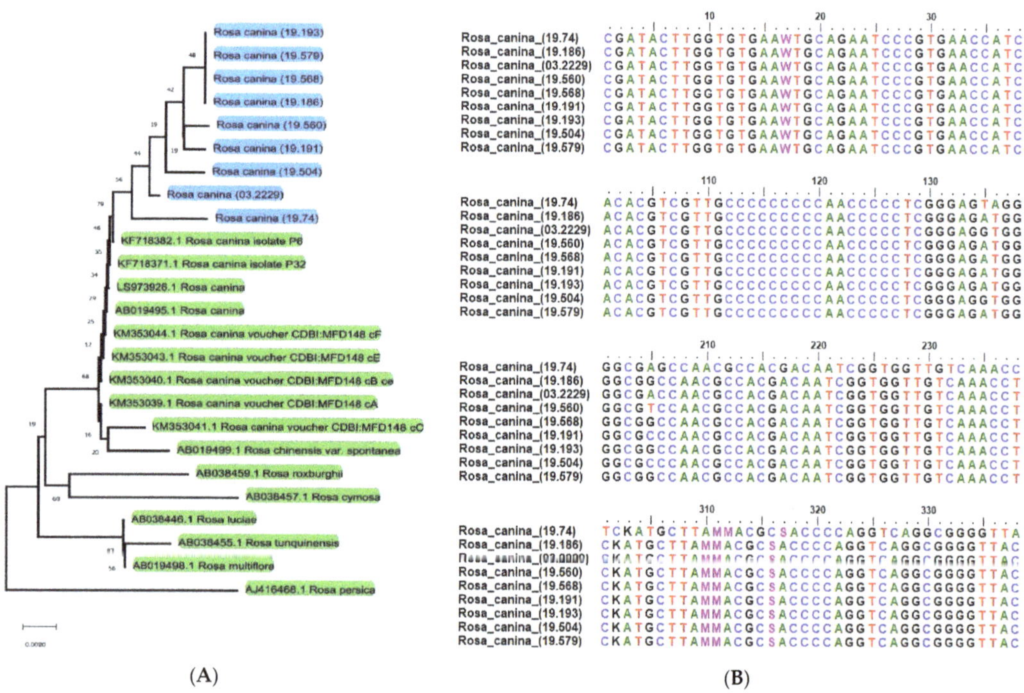

Figure 1. Phylogenetic tree (**A**) constructed on the basis of ITS2 regions of the Greek native *Rosa canina* genotypes contrasted with other *R. canina* and *Rosa* spp. genotypes retrieved from NCBI with multiple sequence alignment of the ITS2 bar-code region of the genotypes analyzed in this study (**B**). Results from neighbor-joining (NJ) bootstrap analyses with 500 replicates was used to assess the strength of the nodes. The node numbers indicated the bootstrap value of NJ. The distinct genotypes of this study are highlighted with blue.

The sequence of each amplicon and the alignment between *R. canina* Greek native genotypes is depicted in Figure 1B. The sequences of the *R. canina* genotypes studied herein possessed variation capable to produce differences in phylogenetic analysis. Specifically, we observed a number of single nucleotide polymorphisms (SNPs) which is responsible for the detected differences in the phylogenetic analysis. The evolutionary distances were computed using the Maximum Composite Likelihood method [44] and are presented in the units of the number of base substitutions per site. This analysis involved 25 nucleotide sequences. Codon positions included were 1st + 2nd + 3rd + Noncoding. All ambiguous positions were removed for each sequence pair (pairwise deletion option). There were 391 positions in total in the final dataset discriminating and classifying the Greek native *R. canina* genotypes from other genetically related genotypes or species. The constructed NJ tree of phylogenetic relationships included the Greek accessions of *R. canina* as a separate branch with 60% branch support. Therefore, with the use of the nuclear ITS2 barcoding sequence, the Greek native *R. canina* genotypes could be sufficiently distinguished from others which are not native to Greece, and from other closely related species genetically. Thus, using the ITS2 barcoding region, the nine *R. canina* genotypes studied herein were fingerprinted.

2.2. Phytochemical Analysis of Greek Native Rosa canina Rosehips

Table 1 shows the results as mean values and standard deviations (SD) of the total phenolic content (TPC), antioxidant activity (AA), total flavonoid (TF) and Vitamin C content of the Greek native *R. canina* samples analyzed (wild-growing plant material). Regarding TPC values, the lowest and highest values were recorded from GR-1-BBGK-O3,2229 (62.98 ± 0.01 mg GAE/100 g) and GR-1-BBGK-19,504 (215.46 ± 0.00 mg GAE/100 g), respectively, whereas statistically significant differences ($p < 0.05$) were found in most of the samples. On the other hand, AA showed limited variations as most samples presented AA values of the same order of magnitude, with the exception of GR-1-BBGK-19,568 which recorded the lowest AA (88.41% ± 0.46). Large variations with statistically significant differences ($p < 0.05$) were recorded in TF values; GR-1-BBGK-O3,2229 showed the lowest TF value (0.87 ± 0.01 mg CE/100 g) and GR-1-BBGK-19,674 the highest (2.46 ± 0.02 mg CE/100 g). Finally, Vitamin C content varied significantly ($p < 0.05$) among all studied samples. Most of the samples analyzed showed high content of Vitamin C as AAE/100 g as expected, with the exception of GR-1-BBGK-19,504 which showed the lowest value 71.85 ± 0.28 in relation to the other samples tested.

Table 1. Values of total phenolic content (TPC), antioxidant activity (AA), total flavonoids (TF) and Vitamin C content detected in samples of wild-growing genotypes of *Rosa canina* of northern Greece.

Population Sample	TPC (mg GAE/100 g)	AA (%RSA)	TF (mg CE/100 g)	Vitamin C (mg AAE/100 g)
GR-1-BBGK-19,191	83.56 ± 0.20 d	95.82 ± 0.50 a	1.44 ± 0.20 c	426.12 ± 0.52 b
GR-1-BBGK-03,2229	62.98 ± 0.01 f	95.36 ± 0.40 a	0.87 ± 0.01 d	350.38 ± 0.14 e
GR-1-BBGK-19,568	78.15 ± 0.02 e	88.41 ± 0.46 b	1.91 ± 0.02 b	500.22 ± 0.15 a
GR-1-BBGK-19,674	90.88 ± 0.02 c	95.31 ± 0.14 a	2.46 ± 0.02 a	398.32 ± 0.58 c
GR-1-BBGK-19,640	83.88 ± 0.03 d	95.37 ± 0.80 a	1.85 ± 0.03 b	390.30 ± 0.24 d
GR-1-BBGK-19,635	97.03 ± 0.30 b	95.71 ± 0.36 a	2.09 ± 0.30 ab	344.34 ± 0.55 f
GR-1-BBGK-19,504	215.46 ± 0.00 a	95.71 ± 0.00 a	2.00 ± 0.08 b	71.85 ± 0.28 g
Average	101.71 ± 48.65	94.53 ± 2.59	1.80 ± 0.50	354.50 ± 128.21

Values represent mean values ± Standard Deviation (S.D.) of samples analyzed in triplicate ($n = 3$); Values with different letters in the same column are statistically significant (Tukey post-hoc test, $p < 0.05$).

2.3. Preliminary Propagation Trials

A broad spectrum of rooting capacity was observed between the studied *R. canina* genotypes following external hormone application in preliminary trials (Table 2). Differences in rooting capacity were observed among genotypes at different seasons, in different vegetative stages and using different cutting types (Table 2). In particular, the genotype GR-1-BBGK-19,191 showed 44% rooting of hardwood cuttings in winter when treated with 10,000 ppm IBA, whereas softwood leafy cuttings at early growth in the following spring treated with 2500 ppm IBA powder presented 75.7% rooting.

Table 2. Results of preliminary propagation trials on different Greek native genotypes of *R. canina* utilizing the initial material collected directly from wild-growing populations. The table summarizes the most successful treatments used in terms of rooting frequencies. Original data are included in Table S1.

Population Sample (ACN)	Hormone Treatment (ppm IBA)	Mother Plant Development Stage **	Season of Year	Cutting Type	Rooting (%)
GR-1-BBGK-19,191	10,000	Dormancy	Winter	Hardwood cuttings	44.05
GR-1-BBGK-19,191	2500 *	Early growth (bud break)	Spring	Softwood cuttings	75.70
GR-1-BBGK-19,193	4000	Dormancy	Winter	Hardwood cuttings	10.71
GR-1-BBGK-19,193	2500 *	Early growth	Spring	Softwood cuttings	77.80
GR-1-BBGK-19,674	4000	Late growth	Autumn	Semi-hardwood cuttings	25.00
GR-1-BBGK-19,568	4000	Advanced growth	Summer	Softwood cuttings	1.30
GR-1-BBGK-19,579	2000	Advanced growth	Summer	Softwood cuttings	12.22
GR-1-BBGK-19,635	4000	Advanced growth	Summer	Softwood cuttings	28.00

* Hormone treatment applied through powdering, ** Early, advanced and late growth refer to the annual vegetative growth cycle.

2.4. Assessment of Greek Native Rosa canina Genotypes

Based hierarchically on: (i) the results of the molecular authentication achieved for nine Greek native *R. canina* genotypes (Figures 1 and 2), (ii) the success assessments regarding the preliminary trials on seven genotypes (Table 2) and (iii) the relative phytochemical interest in terms of comparative vitamin C content and total phenolic content detected in seven genotypes (Table 1), the genotypes GR-1-BBGK-19,191 and GR-1-BBGK-03,2229 were prioritized as most promising for sustainable exploitation strategies. The prioritization was based on the fulfillment of three criteria, i.e., effective DNA barcoding, high success of propagation trials as well as strong phytochemical interest (comparatively very high or high vitamin C content and high or low total phenolic content). Table 3 characterizes the results obtained from the molecular analysis as effective due to the fact that Greek genotypes were grouped independently from others when genetically compared; outlines the success of propagation trials in terms of rooting percentage as low (<40%) or high (>40%); and provides insight regarding the most important phytochemical interest in terms of comparative vitamin C content (very high: >400; high: >340–400; Low: <340 mg /100 g) and total phenolic content (very high: >90; high: 80–90; low: <80 mg of gallic acid/100 g).

Table 3. Multifaceted assessment of Greek native *Rosa canina* genotypes based on molecular authentication achieved, (Figure 1), success of preliminary propagation trials (see Table 2 for details), and relative phytochemical interest in terms of comparative vitamin C content (very high: >400 mg/100 g; high: >340–400; Low: <340) and total phenolic content (very high: >90 of gallic acid/100 g; high: 80–90; low: <80 mg).

IPEN Accession Number	DNA Barcoding	Success of Propagation Trials	Comparative Vitamin C Content	Comparative Total Phenolic Content
GR-1-BBGK-19,74	Effective	-	-	-
GR-1-BBGK-19,504	Effective	-	Low	Very high
GR-1-BBGK-19,560	Effective	-	-	-
GR-1-BBGK-19,568	Effective	Low	Very high	Low
GR-1-BBGK-19,635	-	Low	High	Very high
GR-1-BBGK-19,640	-	-	High	High
GR-1-BBGK-19,674	-	Low	High	Very high
GR-1-BBGK-03,2229	Effective	High	High	Low
GR-1-BBGK-19,186	Effective	-	-	-
GR-1-BBGK-19,191	Effective	High	Very high	High
GR-1-BBGK-19,193	Effective	High	-	-
GR-1-BBGK-19,579	Effective	Low	-	-

Figure 2. Representative photos of cutting propagation results (experiments in 2019 and 2020) on selected Greek native genotypes of *Rosa canina* originating form wild-growing material. (**A**) Softwood leafy cuttings (apical and sub-apical) of the prioritized genotype GR-1-BBGK-19,191; (**B**,**C**) Rooting results of apical and subapical cuttings of GR-1-BBGK-19,191, respectively, in 3:1 substrate across different hormone treatments tested; (**D**) Rooted cutting of the 4000 ppm IBA treatment in the 2020 experiment with the prioritized genotype GR-1-BBGK-03,2229; (**E**,**F**) Rooted cuttings of the 2020 experiment of the genotype GR-1-BBGK-19,674 (control and 2000 ppm IBA treatments) coming from mother plants with no fertilization (**E**) contrasted to those coming from mother plants with conventional fertilization in (**F**); (**G**) New plants of the genotype GR-1-BBGK-19,191 raised ex-situ under outdoor adaptation. Bars in photos A to F represent 1 cm.

2.5. Experimentation on Asexual Propagation of R. canina

Following the preliminary propagation results obtained (Table 1), when softwood cuttings of the more advanced growth of genotype GR-1-BBGK-19,191 were set again in a broader experiment during the summer, the rooting capacity reached 37.5% after 32 days under 2000–4000 ppm IBA (Table 4, Figure 2). However, this rooting capacity was reached through subapical cuttings treated with 2000 ppm IBA in 1:3 *v/v* peat/perlite and through apical cuttings treated with 4000 ppm IBA in 1:1 *v/v* peat/perlite (Table 4, Figure 2). Root number and root length on the other hand did not show to be significantly affected by hormone, cutting type or substrate (Table 2, $p < 0.05$). Similarly, in the genotype GR-1-BBGK-03,2229 which was studied across two consecutive years, rooting capacity took 42 days to reach 25% with soft wood cuttings of early growth during spring in 2019 both with 4000 ppm IBA in 1:1 *v/v* peat/perlite and 2500 ppm IBA powder in 1:3 *v/v* peat/perlite (Table 5, Figure 2). The same genotype during the following year, reached 66.7% rooting in 30 days with softwood cuttings taken when in advanced growth during the summer and treated with 4000 ppm IBA in 1:3 *v/v* peat/perlite, however without a statistically significant effect on root number or length (Table 5, $p < 0.05$).

Table 4. Rooting attributes of the prioritized *Rosa canina* genotype GR-1-BBGK-19,191 expressed as rooting percentage (%) and mean values (±SEM, $p < 0.05$) of root number and average root length (mm) of rooted cuttings for each hormone treatment (ppm IBA) and substrate type (rooting experiment of summer 2019). All cuttings were soft-wood, leafy sections of the first growth year. The two substrate type ratios shown refer to perlite/peat (*v/v*) under mist conditions. Values within each column that do not share the same letter are significantly different (Tukey HSD, $p < 0.05$). Original data are included in Table S2.

Substrate Type	Cutting Type	Hormone Treatment (ppm IBA)	Rooting (%)	Root Number	Root Length (mm)
3:1	Apical Cuttings	Control	0	0	0
		1000	12.5	4.00 (±0.00) *	28.50 (±0.00)
		2000	12.5	7.00 (±0.00)	43.57 (±0.00)
		4000	0	0	0
		6000	0	0	0
	Sub-apical Cuttings	Control	0	0	0
		1000	0	0	0
		2000	37.5 †	5.00 (±1.52) a	33.57 (±9.93) a
		4000	25.0	5.00 (±2.00) a	23.07(±9.07) a
		6000	0	0	0
1:1	Apical Cuttings	Control	0	0	0
		1000	0	0	0
		2000	0	0	0
		4000	37.5 †	6.33 (±2.60) a	51.79 (±9.31) a
		6000	0	0	0
	Sub-apical Cuttings	Control	0	0	0
		1000	12.5	1.00 (±0.00)	13.00 (±0.00)
		2000	25.0	3.50 (±2.50) a	40.91 (±13.08) a
		4000	0	0	0
		6000	12.5	22.00 (±0.00)	18.59 (±0.00)

The † symbol denotes the highest rooting frequency following pairwise comparisons of the observed rooting frequencies via Pearson X^2 tests. * In cases where only one replicate cutting managed to root, the standard error of the means for root number and length is 0.0 because they stem from a single value, as such those means are not included in the post-hoc test.

In another experiment during the summer of 2020 with genotype GR-1-BBGK-19,674 and mother plants grown in a cultivation trial under different fertilization regimes, rooting capacity of softwood cuttings coming from donor plants treated with conventional fertilization reached 50% under 2000 ppm IBA (Table 6).

Table 5. Rooting attributes of the prioritized *Rosa canina* genotype GR-1-BBGK-03,2229 in experiments of 2019 (A) and 2020 (B) expressed as rooting percentage (%) and mean values of root number (±SEM, $p < 0.05$) and average root length (mm) of rooted cuttings for each hormone treatment (ppm IBA) and substrate type. All cuttings were soft-wood, leafy sections of the first growth year. The two substrate type ratios shown refer to perlite/peat (*v/v*) under mist conditions. Values within each column that do not share the same letter are significantly different (Tukey HSD, $p < 0.05$, lowercase letters for 2019 and capital letters for 2020). Original data are included in Table S3.

Substrate Type	Treatment	Rooting (%)	Root Number	Root Length (mm)
A				
3:1	Control	6.25	1.00 (±0.00) **	30.00 (±00.00)
	1000	0	0	0
	2000	12.25	5.50 (±1.50) a	26.17 (±13.67) a
	4000	6.25	6.00 (±0.00)	21.33 (±00.00)
	2500 *	25.00	3.00 (±0.91) a	19.55 (±5.15) a
1:1	Control	6.25	3.00 (±0.00)	56.00 (±00.00)
	1000	12.25	6.00 (±1.00) a	43.14 (±12.85) a
	2000	12.25	7.00 (±4.00) a	80.08 (±27.25) a
	4000	25	8.25 (±1.18) a	52.54 (±8.88) a
	2500 *	12.25	7.00 (±3.00) a	42.72 (±9.47) a
B				
3:1	Control	0	0	0
	2000	16.7	1.00 (±0.16) A	6.00 (±4.83) A
	4000	66.7 †	3.00 (±1.71) A	14.75 (±6.65) A

* Hormone treatment applied through powdering. The † symbol denotes the highest rooting frequency following pairwise comparisons of the observed rooting frequencies via Pearson X^2 tests. ** In cases where only one replicate cutting managed to root, the standard error of the means for root number and length is 0.0 because they stem from a single value, as such those means are not included in the post-hoc test.

Table 6. Rooting attributes of the Greek native *Rosa canina* genotype GR-1-BBGK-19,674 of summer 2020 trials expressed as rooting percentage (%) and mean values (±SEM, $p < 0.05$) of root number and average root length (mm) of rooted cuttings for each fertilization status and hormone treatment (ppm IBA) of mother plants. All cuttings were soft-wood, leafy sections of first growth year. The substrate type used was 3:1 perlite/peat (*v/v*) under mist conditions. Values within each column that do not share the same letter are significantly different (Tukey HSD, $p < 0.05$). Original data are included in Table S4.

Mother Plant Fertilization Status	Hormone Treatment (ppm IBA)	Rooting (%)	Root Number	Root Length (mm)
No fertilization	Control	33.3	2.50 (±0.50) a	20.41 (±2.91) a
	2000	16.7	2.00 (±0.00) *	5.00 (±0.00)
Conventional	Control	33.3	2.00 (±1.00) a	67.50 (±42.52) a
	2000	50.0 †	3.33 (±0.33) a	87.77 (±11.39) a
Organic	Control	16.7	3.00 (±0.00)	86.67 (±0.00)
	2000	16.7	2.00 (±0.00)	92.50 (±0.00)

The † symbol denotes the highest rooting frequency following pairwise comparisons of the observed rooting frequencies via Pearson X^2 tests. * In cases where only one replicate cutting managed to root, the standard error of the means for root number and length is 0.0 because they stem from a single value, as such those means are not included in the post-hoc test.

3. Discussion

3.1. Molecular Authentication of Greek Native Genotypes of Rosa canina

DNA barcoding is a valid technique for the discrimination of *R. canina* genotypes since it is not affected by the stage of plant development and may further enhance the classical morphological identification offering insight regarding phylogenetic relationships of closely related species. In this study, the first-ever report regarding the molecular authentication of Greek native germplasm of *R. canina* is provided. The NJ (Neighbor-Joining) tree classification resulting from the use of barcoding technique in conjunction with ITS2 gene was in accordance with internationally accepted phylogenetic relations, and allowed the distinction of specific genotypes within the species *R. canina*. Using the ITS2 barcoding region, the nine Greek native *R. canina* genotypes studied herein were fingerprinted, and they were clearly separated from other genotypes of *R. canina* which are not native to Greece or other members in the genus Rosa. However, to further confirm the application of this barcoding technique using the ITS2 sequence, different species of genus *Rosa* from different habitats in different regions of Greece should be evaluated. Thus, ITS2 gene can be an effective and valid marker for the identification of the species and of different genotypes of *R. canina*, enriching the extant knowledge regarding the elucidation of evolutionary relationships and classification of Rosaceae and *Rosa* members.

3.2. Phytochemical Potential of Greek Native Genotypes of Rosa canina

The study herein represents the first comprehensive report of TPC, TF, AA and Vitamin C content of rosehips from Greek native genotypes of *Rosa canina*. The TPC values of the samples analyzed showed lower values compared to that reported previously in a study examining a single sample of Greek native plant material [28]; only GR-1-BBGK-19,504 showed TPC value in the same order of magnitude (215.46 ± 0.00 mg GAE/100 g) and this was the highest value detected among the studied Greek native genotypes. Previous studies [29] testing eight samples of *R. canina* from Transylvania, Romania have reported higher values than those detected in the present study, from 326.3 ± 5.65 to 575.1 ± 14.64 mg GAE/100 g of frozen rose hip pulp.

Regarding the TF content, [27] report average TF content of *R. canina* samples from Azerbaijan 2.02 ± 0.03 mg quercetin /100 g, which is slightly higher than that detected in samples of the present study (1.80 ± 0.50 mg CE/100 g). Among the studied Greek genotypes herein, GR-1-BBGK-19 showed above-average potential in terms of TF content and ranked comparatively high.

AA values reported previously in samples of *R. canina* from Azerbaijan [27] are lower than those detected herein for the Greek native samples (94.53% ± 2.59 vs. 83.41% ± 0.86). Among the studied Greek genotypes herein, GR-1-BBGK-19,191, GR-1-BBGK-03,2229 and GR-1-BBGK-19,504 were ranked comparatively above-average in terms of TF content.

Finally, Vitamin C content of the *R. canina* samples studied herein presented higher values (average 354.50 ± 128.21 mg AAE/100 g) compared to other investigations, e.g., from 112.20 ± 2.82 mg AA/100 g to 360.22 ± 2.87 mg AA/100 g [29]. Among the Greek genotypes of *R. canina* studied herein, the higher content was found in GR-1-BBGK-19,568 (500.22 ± 0.15 mg AA/100 g) followed by GR-1-BBGK-19,191.

3.3. Propagation Potential of Greek Native Genotypes

In the current study, rooting data of cutting propagation for Greek native *R. canina* germplasm are presented for the first time. From a commercial perspective, the rooting capacities observed herein are similar to those observed for *R. canina* genotypes in other studies [35,36,45] and can be considered above the threshold for commercial efficacy. With adequate mother plant growth, rooting capacity above 50% (namely for every two cuttings obtained from mother plants at least one successfully delivers a new plant) can be considered as commercially acceptable [34].

The Greek native *R. canina* (dogrose) genotypes studied herein have shown diversified propagation potential in terms of rooting capacity of cuttings. Different genotypes presented differences in propagation performance under different rooting factors, which suggests an interaction between genotype and the external factors involved in the rooting of cuttings. In previous propagation studies of *R. canina* genotypes with the use of IBA, a significant interaction between time of year and genotype was suggested [36], in the same fashion to the current results. Similarly, it has been suggested an effect of developmental stage of donor plants on hormone translocation and uptake by cuttings [36]. A significant effect of genotype on rooting of hardwood cuttings of *R. canina* and other rose hips has also been observed in earlier studies with indigenous eastern Mediterranean germplasm [46], including the promoting effect that IBA has on the rooting of cuttings [35]. In the current study, different types of cuttings of the same genotype showed varied rooting capacity at different annual growth stages of the donor material, consistently, however, requiring external hormone application. This observation is in agreement with similar results on the effects of donor plants' growth stage, cutting type and IBA on the rooting of cuttings in commercial germplasm of *R. hybrida* E. H. L. Krause grown in Mediterranean conditions [47]. Additionally, notable differences in cuttings' response to hormonal regimes between genotypes have also been observed in *R.* x *damascena* Herrm. [45]. Given that the cuttings conditions set for rooting were similar between different trials, this trend suggests that the effect of season is probably related to the growth developmental stage during which cuttings are taken. The different cutting types seem to indicate a genotype-varying response to the external hormone application. The effect of season on rooting of *Rosa* cuttings has been suggested as a significant factor by other investigators stemming from results of similar studies [48,49].

The effect of cutting type has been attributed by other studies to the differences in the nutrient status and carbohydrate concentration of the cuttings coming from different parts of the mother plant, as reported in *R. hybrida* commercial rootstock germplasm [50]. However, a clear conclusion cannot be drawn on the particular effect of cutting type in the current study since the nutritional status or carbohydrate balance of the Greek native *R. canina* germplasm studied herein have not yet been adequately studied and further research is needed. In addition, other studies dealing with rooting patterns across genotypes of native germplasm of *R. canina* suggest that the treatment of mycorrhizal fungi as well as growth promoting bacteria have possible synergistic effects on the rooting of cuttings [51,52]. Furthermore, grafting should be also examined as a method to overcome genetic differences at ease of propagation; this consists of grafting a desirable, difficult to root, genotype onto another genotype that is stronger in terms of rooting capacity. It is known that grafting is widely being performed in commercial roses [53]. However, there is evidence that grafting can affect plant physiology of interacting genotypes [54,55]. Consequently, caution should be taken when desirable fruit characteristics of the scion are involved such as natural content of vitamin C or total phenolics content. Undoubtedly, further research and experimentation is suggested regarding the potential implementation of grafting on Greek native *R. canina* germplasm.

4. Materials and Methods

4.1. Plant Populations of Rosa canina Sampled

Nine authorized botanical expeditions were organized in 2019 to explore different areas of Northern Greece (Epirus and North-central Greece) for wild-growing *R. canina* initial materials with vigorous growth and strong fruiting potential in the wild habitats. The collections were performed using the authorized special permit of the Institute of Plant Breeding and Phytogenetic Resources, Hellenic Agricultural Organization Demeter (Permit 82336/879 of 18/5/2019 & 26895/1527 of 21/4/2021). This permit is issued yearly by the Greek Ministry of Environment and Energy after detailed reporting of the applicant. The collections were performed in the frame of the research project "Highlighting local traditional varieties and wild native forest fruit trees and shrubs" (acronym: EcoVariety, T1EΔK-05434). In each expedition (Table 7, Figure 3, we collected from selected wild-growing *R. canina* populations (Greek native germplasm): (a) sets of fresh soft-wood stem cuttings as initial propagation material for propagation trials (in total, seven populations), (b) ripe rosehips sampled from three individuals for phytochemical analysis (in total, seven populations), and (c) leaf samples from 20 individuals destined for DNA analysis (in total, nine populations). The materials were taxonomically identified based on standard diagnostic keys for the European [32] and Greek *Rosa* material [33]. Consequently, each genotype was given a unique IPEN (International Plant Exchange Network) accession number by the Institute of Plant Breeding and Genetic Resources (IPB&GR) of the Hellenic Agricultural Organization Demeter.

Table 7. Selected *Rosa canina* genotypes sampled from various mountainous habitats of northern Greece assigned with different IPEN (International Plant Exchange Network) accession numbers.

IPEN Accession Number	Greek Prefecture	Area	Altitude (m)	Sampling
GR-1-BBGK-19,74	Central Macedonia	Mt Voras	862	LS
GR-1-BBGK-19,504	Central Macedonia	Kastaneri	780	RR, LS
GR-1-BBGK-19,560	Central Macedonia	Mt Vermio	1615	LS
GR-1-BBGK-19,568	Central Macedonia	Mt Tzena	1086	SWSC, RR, LS
GR-1-BBGK-19,635	Central Macedonia	Mt Kroussia	650	SWSC, RR
GR-1-BBGK-19,640	Central Macedonia	Mt Kroussia	700	RR
GR-1-BBGK-19,674	Western Macedonia	Ziaka	900	SWSC, RR
GR-1-BBGK-03,2229	Epirus	Ioannina	650	SWSC, RR, LS
GR-1-BBGK-19,186	Epirus	Mt Lakmos	1370	LS
GR-1-BBGK-19,191	Epirus	Anogeia	1081	SWSC, RR, LS
GR-1-BBGK-19,193	Epirus	Mt Xirovouni	1070	SWSC, LS
GR-1-BBGK-19,579	Epirus	Pades	1180	SWSC, LS

SWSC: Soft-wood stem cuttings for propagation; RR: Ripe rosehips for phytochemical analysis; LS: Leaf samples for DNA analysis.

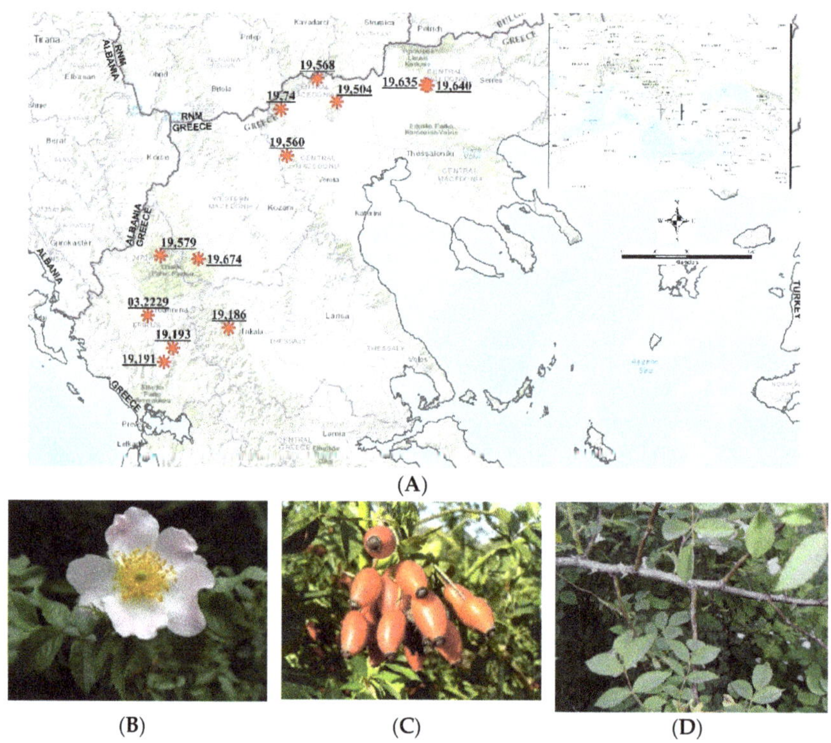

Figure 3. Overview of the collection sites of the *Rosa canina* Greek native germplasm analyzed (**A**), and morphology of flowers (**B**), fruits (**C**), and leaves (**D**) of *R. canina* GR-1-BBGK-03,2229 used for taxonomic identification, phytochemical analysis and DNA barcoding, respectively (for IPEN accession numbers see Table 7).

4.2. DNA Isolation

Approximately 30 mg of dried leaf sample was completely grounded in liquid nitrogen. Total DNA was isolated from leaf samples of *R. canina* using a Nucleospin Plant II (Macherey-Nagel) kit following the manufacturer's instructions.

4.3. Polymerase Chain Reaction (PCR) Amplification

One primer set of the nuclear ITS2 barcode region suggested by [56] was used for amplification and sequencing. The PCR amplification was performed according to [57].

4.4. Sequence Analysis

PCR products were directly sequenced in two directions of each fragment with a Big Dye terminator v3.1 Cycle sequencing kit (PE Applied Biosystems, Foster City, CA, USA) in an automated ABI 3730 sequencer (PE Applied Biosystems). The sequences were aligned using the CLUSTAL W program.

4.5. Molecular Data Analysis

Three methods were employed for molecular authentication of the selected *R. canina* genotypes: (1) Basic Local Alignment Search Tool (BLAST) search using the nucleotide database at NCBI [58]; (2) the genetic divergence method using maximum-likelihood models; and (3) tree topology analysis based on the neighbor-joining (NJ) method based on different loci in MEGAX [59] with the K2P distance model and 500 bootstrap replications. The sequences obtained after removing the primers used for PCR amplification were deposited to NCBI-Genbank BankIt (https://www.ncbi.nlm.nih.gov/BankIt/, accessed on 1 November 2021) under the accession numbers MK5334116 to MK5334124.

4.6. Phylogenetic Relationships

The phylogenetic relationships of different *Rosa* spp. were inferred using the Neighbor-Joining method [60]. The optimal tree with the sum of branch length = 0.07053901 is shown. The percentage of replicate trees in which the associated taxa clustered together in the bootstrap test (500 replicates) are shown next to the branches [61]. Phylogenetic analyses were conducted in MEGA X [59].

4.7. Phytochemical Analysis of Rosa canina Rosehips

The extracts were prepared according to the method described by [62] with some modifications. Part of the homogenized sample (2–5 g) was mixed with an appropriate amount of MeOH/H_2O (60:40), and the mixture was centrifuged (4 °C, 4000 rpm). The supernatant was collected and the volume was made up to 20 mL. This extract was used for the following analyses:

Total phenolic content (TPC): The determination of TPC was carried out using the method described by [62]. Phenolic extract 0.20 mL along with 2.3 mL of H_2O and 0.25 mL of Folin-Ciocalteu reagent were added in a volumetric flask. After 3 min, 0.50 mL of 20% Na_2CO_3 was added, and the volume was made up to 5 mL. The solution was stored in a dark place for 2 h. After 2 h the absorbance was measured at 725 nm against blank solution. Total phenolics were calculated using a standard curve of gallic acid at various concentrations. The results given are expressed as gallic acid equivalents (GAE)/100 g of sample. All analyses were carried out in triplicate.

Total Flavonoids (TF): The determination of TF was carried out according to the method of [29], with some modifications. Firstly, 5 μL of the above extract along with 3270 μL of H_2O and 75 μL of 5% $NaNO_2$ were added to a test tube, stirred, and stored in the dark for 5 min. After that, 150 μL of 10% $AlCl_3$-$6H_2O$ was added, mixed, and stored again in the dark for 6 min. Then, 500 μL of 1 M NaOH was then added, and the absorbance was measured at 510 nm against H_2O as blank. Total flavonoids were calculated using a standard curve of catechin at various concentrations. The results given are expressed as catechin equivalents (CE)/100 g of sample. All analyses were carried out in triplicate.

Antioxidant Activity (AA): The determination of AA was carried out according to the method described by [27], with some modifications. Phenolic extract 0.1 mL along with 2.9 mL 0.10 mM DPPH in MeOH were added in a 5 mL plastic cuvette, stored in a dark place for 15 min and then the absorbance was measured at 517 nm against MeOH as blank. The control sample was prepared daily using only 0.10 mM DPPH in MeOH. The percentage of radical scavenging activity (%RSA) was calculated using the following equation:

$$\%RSA = (Ao - As)/Ao \times 100,$$

where

Ao = Absorbance of control sample

As = Absorbance of the sample after 15 min of incubation

All analyses were carried out in triplicate.

Determination of Vitamin C: The determination of Vitamin C was carried out according to the method described by [63] after some modifications. A certain amount of homogenized sample (2–5 g) was added to a centrifuge tube with 5 mL of 4.5% *w/v* metaphosphoric acid (MPA) solution. The mixture was stirred and centrifuged at 8000 rpm at 4 °C for 20 min. Then, 1 mL of the supernatant was taken and diluted up to 10 mL with 4.5% MPA solution. This solution was filtered through 0.45 μm polyethersulfone filters. The vial was covered with aluminium foil to prevent oxidation of ascorbic acid and stored at 4 °C until HPLC-DAD analysis. HPLC-DAD conditions: Column (Agilent Eclipse XDB-C18) 4.6 mm × 150 mm, 5μm, elution solvent: aqueous 0.005 M H_2SO_4 solution at a flow rate of 0.5 mL/min (isocratic) and wavelength 245 nm. Vitamin C was calculated using a standard curve of ascorbic acid at various concentrations. The results given are expressed as ascorbic acid equivalents (AAE)/100 g of sample. All analyses were carried out in triplicate.

4.8. Preliminary Propagation Trials and Mother Plants' Growth Conditions

The soft-wood stem cuttings of the seven *R. canina* genotypes sampled (Table 7) were tested for rooting under various external hormone application treatments of indole-3-butyric acid (IBA) in propagation trays with peat: perlite at 1:3 *v/v* (Table 1); during experimentation, they were maintained at ambient temperature on mist bench in a plastic greenhouse with relative humidity (RH) maintained >85% where they were attended weekly to assess their rooting capacity. The produced mother plants were kept ex-situ at the grounds of IPB&GR under ambient conditions. The plants were watered regularly and were grown in 3 L pots using a mixture of peat and perlite (1:3 *v/v*). This allowed vigorous growth of mother plants, which enabled the raising of enough plant material for further experimentation during the next season.

4.9. Propagation Experimental Design, Cutting Types, Hormone Applications and Rooting Conditions

Following the preliminary observations (see Table 2 in results), two experiments were conducted in 2019 and another two experiments in 2020 regarding prioritized *R. canina* genotypes. In particular, an experiment was conducted on genotype GR-1-BBGK-19,191 in summer of 2019 which abided by a complete randomised design with five hormone application levels of Indole-3-butyric acid (IBA) (control; 1000 ppm; 2000 ppm; 4000 ppm; and 6000 ppm dissolved in 50% ethanol), two cutting types (1st year softwood apical and sub-apical cuttings) and two substrates (peat: perlite at 1:3 *v/v* and at 1:1 *v/v*), resulting in 20 treatments in total with eight replicate cuttings per treatment. Three fully developed leaves were kept in all cuttings.

Another experiment was conducted on genotype GR-1-BBGK-03,2229 in May 2019. This experiment was set in a complete randomized block design with two blocks each with five IBA application levels (control; 1000 ppm; 2000 ppm; 4000 ppm dissolved in 50% ethanol and 0.25% powder) and two substrates (peat: perlite at 1:3 *v/v* and at 1:1 *v/v*), resulting in 10 treatments in total with eight replicate cuttings per treatment in each block. Following the results of 2019 regarding GR-1-BBGK-03,2229, another experiment was conducted in July 2020 in a complete randomized block design with two blocks each consisted of three IBA levels (control; 2000 ppm; and 4000 ppm dissolved in 50% ethanol), resulting in three treatments with six replicate cuttings per treatment in each block. The substrate used was peat: perlite at 1:3 *v/v*. Cuttings in both years were of the same type (soft-wood leafy cuttings).

In addition, during 2020, further experimentation was performed on genotype GR-1-BBGK-19,674 using field established mother plants following the project's progress and taking into account the preliminary results. This experiment was conducted in a split-plot design having the fertilization status of mother plants in three main plots (no fertilization; conventional fertilization; organic fertilization) each having two sub-plots of IBA treatment (control and 2000 ppm dissolved in 50% ethanol). The six resulted treatments/sub-plots had six replicate cuttings each. The substrate used for rooting was peat: perlite at 1:3 *v/v*.

4.10. Cuttings' Performance and Growth Measurements

Observations on the progress of cuttings were taken weekly. When a treatment reached 100% rooting or after 40 days (whatever occurred first), the trays were taken out of mist and measurements were taken on rooting capacity, root number and root length per cutting. At the same time, rooted cuttings were transplanted in 0.5 L pots with peat: perlite 3:1 *v/v* substrate and were kept for the first two weeks within a greenhouse with shading and automated irrigation for plant establishment.

4.11. Statistical Analysis of Rooting Data and Phytochemical Data

The rooting data were subjected to analysis of variance (GLM-ANOVA) to establish overall treatment effects and the phytochemical data were analysed using one-way ANOVA. Consecutively, and following the results of the ANOVA, to dissect specific treatment effects (hormone level, cutting type, substrate type) on the variables measured, data were split and separate analyses of variance were conducted as appropriate. Means from rooting experiments and means of phytochemical data were compared separately using Tukey's HSD post hoc test Rooting frequencies were compared through pairwise Pearson Chi-Square tests. All analyses were conducted using the IBM-SPSS 23.0 software.

5. Conclusions

In the frame of sustainable exploitation strategies involving neglected and underutilized phytogenetic resources and domestication of wild-growing genotypes of native plants, the study herein focused on the exploration of the potential of Greek native *Rosa canina* germplasm. This study reported for the first time data regarding: (i) The effectiveness of fingerprinting distinct genotypes from Greece using the ITS2 sequence as molecular marker; (ii) The diverse phytochemical content in terms of total phenolics, total flavonoids, antioxidant activity and vitamin C content of different genotypes naturally occurring in northern Greece with interesting potential for applications; (iii) The effective propagation of selected and prioritized *R. canina* genotypes via cuttings, highlighting at the same time the importance of levels of external hormone application (IBA), the effect of season in terms of annual growth stage of donor plants, and genotype-specific differences in rooting capacities. The multifaceted documentation developed and assessed in this study offer new artificially selected plant material with consolidated identity and interesting phytochemical profile which is currently under ex-situ conservation for further evaluation and characterization in pilot field studies. In this way, the present work may pave the way for the sustainable exploitation of the selected Greek native genotypes of *R. canina*, facilitating future applications in the agro-alimentary, medicinal-cosmetic and ornamental sectors.

Supplementary Materials: The following are available online at https://www.mdpi.com/article/10.3390/plants10122634/s1, Table S1: Raw data used to create Table 4 regarding the number of roots and root length (mm) during the 2019 propagation experiment of the prioritized *Rosa canina* genotype GR-1-BBGK-19,191 for each substrate type, hormone treatment (ppm IBA, in a *w/v* solution dissolved in 50% ethanol) and cutting type, Table S2: Raw data used to create Table 5 regarding the number of roots and root length (mm) of the prioritized *Rosa canina* genotype GR-1-BBGK-03,2229 during propagation experiments of 2019 (A) and 2020 (B) for each hormone treatment (ppm IBA in a *w/v* solution dissolved in 50% ethanol), Table S3: Raw data used to create Table 6 regarding number of roots and root length (mm) of the Greek native *Rosa canina* genotype GR-1-BBGK-19,674 during the 2020 propagation experiment per mother plant fertilization status and hormone treatment (ppm IBA in a *w/v* solution dissolved in 50% ethanol), Table S4: Raw data used to create Table 2 regarding the observed rooting frequencies during the preliminary propagation trials on different Greek native genotypes of *Rosa canina* (initial material collected directly from wild growing populations). The Table S4 summarizes the most successful treatments in terms of total number of replicate cuttings set for rooting depending on the availability of the collected material along with details on mother plant developmental stage, season and cutting type, presenting the frequencies of rooted cuttings out of the total number of replicate cuttings set in each treatment.

Author Contributions: Conceptualization, N.K. and I.G.; methodology, E.M., E.K., I.G., A.K., K.P., D.K., P.Y., N.N., A.Z., I.S.K., A.V.B., G.P., I.G. and N.K.; software, I.G., D.F. and A.Z.; validation, E.K., I.G., A.K., K.P., D.K., P.Y., N.N., A.Z., I.S.K., A.V.B., G.P. and D.F.; formal analysis, E.K., I.G., I.S.K. and A.V.B.; investigation, E.K., I.G., A.K., K.P., D.K., P.Y., N.N., I.S.K., A.V.B. and G.P.; resources, E.M., I.G. and A.V.B.; data curation, N.K., E.K., K.P., I.G., N.N., I.S.K. and A.V.B.; writing—original draft preparation, E.K., I.G., D.F., A.V.B. and N.K.; writing—review and editing, E.M., E.K., I.G., A.K., K.P., D.K., P.Y., N.N., A.Z., I.S.K., A.V.B., G.P., D.F. and N.K.; visualization, N.K., E.K., I.G. and D.F.; supervision, E.M., A.V.B., G.P. and N.K.; project administration, E.M. and G.P.; funding acquisition, E.M., N.N., A.Z. and G.P. All authors have read and agreed to the published version of the manuscript.

Funding: This research has been co-financed by the European Union and Greek national funds through the Operational Program Competitiveness, Entrepreneurship and Innovation, under the call RESEARCH—CREATE—INNOVATE (project code: T1EDK-05434), entitled "Highlighting of local traditional and native wild fruit trees and shrubs".

Data Availability Statement: All data supporting the results of this study are included in the manuscript and datasets are available upon request.

Acknowledgments: The authors would like to thank the staff of the Institute of Plant Breeding and Genetic Resources, Hellenic Agricultural Organization-Demeter for administrative and technical support.

Conflicts of Interest: The authors declare no conflict of interest.

References

1. Young, M.A.; Schorr, P. *Modern Roses*, 12th ed.; Young, M.A., Ed.; American Rose Society: Shreveport, LA, USA, 2007; pp. 1–576, ISBN 1597250988.
2. Suprun, I.I.; Plugatar, S.A.; Stepanov, I.V.; Naumenko, T.S. Analysis of genetic relationships of genotypes of the genus *Rosa* L. from the collection of Nikita Botanical Gardens using ISSR and IRAP DNA markers. *Vavilovskii Zhurnal Genet. Sel.* **2020**, *24*, 474–480. [CrossRef]
3. Wylie, A.P. The history of garden roses, part 1. *J. R. Hortic. Soc.* **1954**, *79*, 555–571.
4. Tomljenovic, N.; Pejić, I. Taxonomic review of the genus *Rosa*. *Agric. Conspec. Sci.* **2018**, *83*, 139–147. Available online: https://hrcak.srce.hr/203011 (accessed on 1 November 2021).
5. Wissemann, V.; Ritz, C. Evolutionary patterns and processes in the genus *Rosa* (Rosaceae) and their implications for host-parasite co-evolution. *Plant Syst. Evol.* **2007**, *266*, 79–89. [CrossRef]
6. Pang, X.H.; Song, J.Y.; Zhu, Y.J.; Xu, H.X.; Huang, L.F.; Chen, S.L. Applying plant DNA barcodes for Rosaceae species identification. *Cladistics* **2010**, *27*, 165–170. [CrossRef]
7. Hebert, P.D.; Cywinska, A.; Ball, S.L.; De Waard, J.R. Biological identifications through DNA barcodes. *Proc. Royal Soc. B* **2003**, *270*, 313–321. [CrossRef]
8. Millan, T.; Osuna, F.; Cobos, S.; Torres, A.M.; Cubero, J.I. Using RAPDs to study phylogenetic relationships in *Rosa*. *Theor. Appl. Genet.* **1996**, *92*, 273–277. [CrossRef] [PubMed]
9. Bruneau, A.; Starr, J.R.; Joly, S. Phylogenetic relationships in the genus *Rosa*: New evidence from chloroplast DNA sequences and an appraisal of current knowledge. *Syst. Bot.* **2007**, *32*, 366–378. [CrossRef]
10. Oğraş, T.L.; Baştanlar, E.K.; Metin Karakaş, Ö.; Kandemir, I.; Özcelik, H. Assessment of genetic diversity of rose genotypes using ISSR markers. *Turk. J. Bot.* **2017**, *41*, 347–355. [CrossRef]

11. Agarwal, A.; Gupta, V.; Haq, S.U.; Jatav, P.K.; Kothari, S.L.; Kachhwaha, S. Assessment of genetic diversity in 29 rose germplasms using SCoT marker. *J. King Saud Univ. Sci.* **2019**, *31*, 780–788. [CrossRef]
12. Zhang, L.H.; Byrne, D.H.; Ballard, R.E.; Rajapakse, S. Microsatellite marker development in rose and its application in tetraploid mapping. *J. Am. Soc. Hortic. Sci.* **2006**, *131*, 380–387. [CrossRef]
13. Crespel, L.; Pernet, A.; Le Bris, M.; Gudin, S.; Saint-Oyant, L.H. Application of ISSRs for cultivar identification and assessment of genetic relationships in rose. *Plant Breed.* **2009**, *12*, 501–506. [CrossRef]
14. Koopman, W.J.M.; Wissemann, V.; De Cock, K.; Van Huylenbroeck, J.; De Riek, J.; Sabatino, G.J.H.; Visser, D.; Vosman, B.; Ritz, C.M.; Maes, B.; et al. AFLP markers as a tool to reconstruct complex relationships: A case study in *Rosa* (Rosaceae). *Am. J. Bot.* **2008**, *95*, 353–366. [CrossRef] [PubMed]
15. Chen, S.; Pang, X.; Song, J.; Shi, L.; Yao, H.; Han, J.; Leon, C. A renaissance in herbal medicine identification: From morphology to DNA. *Biotechol. Adv.* **2014**, *32*, 1237–1244. [CrossRef] [PubMed]
16. Nakamura, N.; Hirakawa, H.; Sato, S.; Otagaki, S.; Matsumoto, S.; Tabata, S.; Tanaka, Y. Genome structure of *Rosa multiflora*, a wild ancestor of cultivated roses. *DNA Res.* **2018**, *2*, 113–121. [CrossRef] [PubMed]
17. Saint-Oyant, L.H.; Ruttink, T.; Hamama, L.; Kirov, I.; Lakhwani, D.; Zhou, N.N.; Bourke, P.M.; Daccord, N.; Leus, L.; Schulz, D.; et al. A high-quality genome sequence of *Rosa chinensis* to elucidate ornamental traits. *Nat. Plants* **2018**, *4*, 473–484. [CrossRef] [PubMed]
18. Bourke, P.M.; Arens, P.; Voorrips, R.E.; Esselink, G.D.; Koning-Boucoiran, C.F.; van't Westende, W.P.; Santos Leonardo, T.; Wissink, P.; Zheng, C.; van Geest, G.; et al. Partial preferential chromosome pairing is genotype dependent in tetraploid rose. *Plant J.* **2017**, *90*, 330–343. [CrossRef]
19. Raymond, O.; Gouzy, J.; Just, J.; Badouin, H.; Verdenaud, M.; Lemainque, A.; Vergne, P.; Moja, S.; Choisne, N.; Pont, C.; et al. The *Rosa* genome provides new insights into the domestication of modern roses. *Nat. Genet.* **2018**, *50*, 772–777. [CrossRef]
20. Dubois, A.; Carrere, S.; Raymond, O.; Pouvreau, B.; Cottret, L.; Roccia, A.; Onesto, J.P.; Sakr, S.; Atanassova, R.; Baudino, S.; et al. Transcriptome database resource and gene expression atlas for the rose. *BMC Genom.* **2012**, *13*, 638. [CrossRef] [PubMed]
21. Liu, X.; Cao, X.; Shi, S.; Zhao, N.; Li, D.; Fang, P.; Chen, X.; Qi, W.; Zhang, Z. Comparative RNA-Seq analysis reveals a critical role for brassinosteroids in rose (*Rosa hybrida*) petal defense against *Botrytis cinerea* infection. *BMC Genet.* **2018**, *19*, 62. [CrossRef]
22. Neu, E.; Domes, H.S.; Menz, I.; Kaufmann, H.; Linde, M.; Debener, T. Interaction of roses with a biotrophic and a hemibiotrophic leaf pathogen leads to differences in defense transcriptome activation. *Plant Mol. Biol.* **2019**, *99*, 299–316. [CrossRef] [PubMed]
23. Pehlivan, M.; Mohammed, F.; Sevindik, M.; Akgul, H. Antioxidant and oxidant potential of *Rosa canina*. *Eurasian J. Forest Sci.* **2018**, *6*, 22–25. [CrossRef]
24. Ayati, Z.; Amiri, M.S.; Ramezani, M.; Delshad, E.; Sahebkar, A.; Emami, S.A. Phytochemistry, traditional uses and pharmacological profile of rose hip: A review. *Curr. Pharm. Des.* **2018**, *24*, 4101–4124. [CrossRef] [PubMed]
25. Paunović, D.; Kalušević, A.; Petrović, T.; Urošević, T.; Djinović, D.; Nedović, V.; Popović-Djordjević, J. Assessment of chemical and antioxidant properties of fresh and dried rosehip (*Rosa canina* L.). *Not. Bot. Hort Agrobot.* **2018**, *47*, 108–113. [CrossRef]
26. Gruenwald, J.; Uebelhack, R.; Moré, M.I. *Rosa canina*—Rose hip pharmacological ingredients and molecular mechanics counteracting osteoarthritis—A systematic review. *Phytomedicine* **2019**, *60*, 152958. [CrossRef]
27. Fattahi, S.; Jamei, R.; Hosseini Sarghein, S. Antioxidant and antiradical activities of *Rosa canina* and *Rosa pimpinellifolia* fruits from West Azerbaijan. *Iran. J. Plant Physiol.* **2012**, *2*, 523–529.
28. Kerasioti, E.; Apostolou, A.; Kafantaris, I.; Chronis, K.; Kokka, E.; Dimitriadou, C.; Tzanetou, E.N.; Priftis, A.; Koulocheri, S.D.; Haroutounian, S.A.; et al. Polyphenolic composition of *Rosa canina*, *Rosa sempervivens* and *Pyrocantha coccinea* extracts and assessment of their antioxidant activity in human endothelial cells. *Antioxidants* **2019**, *8*, 92. [CrossRef]
29. Roman, I.; Stanila, A.; Stanila, S. Bioactive compounds and antioxidant activity of *Rosa canina* L. biotypes from spontaneous flora of Transylvania. *Chem. Cent. J.* **2013**, *7*, 73. [CrossRef] [PubMed]
30. Guimarães, R.; Barros, L.; Calhelha, R.C.; Carvalho, A.M.; Queiroz, M.J.R.; Ferreira, I.C. Bioactivity of different enriched phenolic extracts of wild fruits from northeastern Portugal: A comparative study. *Plant Foods Hum. Nutr.* **2014**, *69*, 37–42. [CrossRef] [PubMed]
31. Tumbas, V.T.; Canadanovic-Brunet, J.M.; Cetojevic-Simin, D.D.; Cetkovic, G.S.; Ethilas, S.M.; Gille, L. Effect of rosehip (*Rosa canina* L.) phytochemicals on stable free radicals and human cancer cells. *J. Sci. Food Agric.* **2012**, *92*, 1273–1281. [CrossRef] [PubMed]
32. Klastersky, I. *Rosa* L. In *Flora Europaea*; Tutin, T.G., Heywood, V.H., Burges, N.A., Moore, D.M., Valentine, D.H., Walters, S.M., Webb, D.A., Eds.; Cambridge University Press: Cambridge, UK, 1968; Volume 2, pp. 25–31, ISBN 9780521066624.
33. Zieliński, J. *Mountain Flora of Greece*; Strid, A., Ed.; Cambridge University Press: Cambridge, UK, 1986; Volume 1, pp. 387–399, ISBN 9780748602070.
34. Blythe, E.K.; Sibley, J.L.; Tilt, K.M.; Ruter, J.M. Methods of auxin application in cutting propagation: A review of 70 years of scientific discovery and commercial practice. *J. Environ. Hortic.* **2007**, *25*, 166–185. [CrossRef]
35. Hoşafçi, H.; Arslan, N.; Sarihan, E.O. Propagation of dogrose (*Rosa canina* L.) plants by softwood cuttings. *Acta Hortic.* **2005**, *690*, 139–142. [CrossRef]
36. Kazankaya, A.; Yörük, E.; Doğan, A. Effect of IBA on rooting of *Rosa canina* hardwood cuttings from lake Van region, Turkey. *Acta Hortic.* **2005**, *690*, 153–158. [CrossRef]
37. Izadi, Z.; Zarei, H.; Alizadeh, M. Studies on vegetative propagation of *Rosa canina*. *Indian J. Hort.* **2012**, *69*, 598–601.

38. Maloupa, E.; Krigas, N.; Grigoriadou, K.; Lazari, D.; Tsoktouridis, G. Conservation strategies for native plant species and their sustainable exploitation: Case of the Balkan Botanic Garden of Kroussia, N. Greece. In *Floriculture Ornamental Plant Biotechnology*, 1st ed.; Teixeira da Silva, J.A., Ed.; Global Science Books: Middlesex, UK, 2008; Volume V, pp. 37–56, ISBN 9784903313122.
39. Grigoriadou, K.; Sarropoulou, V.; Krigas, N.; Maloupa, E.; Tsoktouridis, G. GIS-facilitated effective propagation protocols of the Endangered local endemic of Crete *Carlina diae* (Rech. f.) Meusel and A. Kástner (Asteraceae): Serving ex situ conservation needs and its future sustainable utilization as an ornamental. *Plants* **2020**, *9*, 1465. [CrossRef]
40. Krigas, N.; Mouflis, G.; Grigoriadou, K.; Maloupa, E. Conservation of important plants from the Ionian Islands at the Balkan Botanic Garden of Kroussia, N Greece: Using GIS to link the in situ collection data with plant propagation and ex situ cultivation. *Biodiver. Conserv.* **2010**, *19*, 3583–3603. [CrossRef]
41. Krigas, N.; Tsoktouridis, G.; Anestis, I.; Khabbach, A.; Libiad, M.; Megdiche-Ksouri, W.; Ghrabi-Gammar, Z.; Lamchouri, F.; Tsiripidis, I.; Tsiafouli, M.A.; et al. Exploring the potential of neglected local endemic plants of three Mediterranean regions in the ornamental sector: Value chain feasibility and readiness timescale for their sustainable exploitation. *Sustainability* **2021**, *13*, 2539. [CrossRef]
42. Libiad, M.; Khabbach, A.; El Haissoufi, M.; Anestis, I.; Lamchouri, F.; Bourgou, S.; Megdiche-Ksouri, W.; Ghrabi-Gammar, Z.; Greveniotis, V.; Tsiripidis, I.; et al. Agro-alimentary potential of the neglected and underutilized local endemic plants of Crete (Greece), Rif-Mediterranean coast of Morocco and Tunisia: Perspectives and challenges. *Plants* **2021**, *10*, 1770. [CrossRef] [PubMed]
43. Bourgou, S.; Jilani, I.B.H.; Karous, O.; Megdiche-Ksouri, W.; Ghrabi-Gammar, Z.; Khabbach, A.; Libiad, M.; El Haissoufi, M.; Lamchouri, F.; Greveniotis, V.; et al. Medicinal-cosmetic potential of the local endemic plants from of Crete (Greece), Rif-Mediterranean Coast of Morocco and Tunisia: Priorities for sustainable exploitation. *Biology* **2021**, under review.
44. Tamura, K.; Nei, M.; Kumar, S. Prospects for inferring very large phylogenies by using the neighbor-joining method. *Proc. Natl. Acad. Sci. USA* **2004**, *101*, 11030–11035. [CrossRef] [PubMed]
45. Nasri, F.; Fadakar, A.; Saba, M.K.; Yousefi, B. Study of indole butyric acid IBA effects on cutting rooting improving some of wild genotypes of damask roses *Rosa damascena* Mill. *J. Agric. Sci.* **2015**, *60*, 263–275. [CrossRef]
46. Ercişli, S.; Güleryüz, M. A study of the propagation of the hardwood cuttings of some rose hips. *Turk. J. Agric. For.* **1999**, *23*, 305–310. Available online: https://atif.sobiad.com/index.jsp?modul=makale-goruntule&id=R3mSOnsBYbO9RkQmEnlm (accessed on 1 November 2021).
47. Tawfik, A.A.; Ibrahim, O.H.; Abdul-Hafeez, E.; Ismail, S.A. Effect of cutting type, indol-3-butyric acid and the growing season on rooting of stem cuttings of *Rosa hybrida* cv Eiffel Tower. *J. Plant Prod.* **2018**, *9*, 537–542. [CrossRef]
48. Abbas, H.; Jaskani, M.J.; Hussain, Z.; Asif, M. Response of rose cuttings against root promoting hormones during spring and autumn. *Int. J. Biol. Biotech.* **2006**, *3*, 201–204. Available online: https://www.researchgate.net/publication/259465811_Response_of_rose_cuttings_against_root_promoting_hormones_during_spring_and_autumn (accessed on 28 October 2021).
49. Kashefi, M.; Zerei, H.; Bahadori, F. The regulating effect of the growth of indole butyric acid and the time of stem cutting preparation on propagation of damask rose ornamental shrub. *J. Ornam. Plants* **2014**, *4*, 49–55. Available online: https://www.sid.ir/en/journal/ViewPaper.aspx?id=438004 (accessed on 28 October 2021).
50. Otiende, M.A.; Nyabundi, J.O.; Ngamau, K.; Opala, P. Effects of cutting position of rose rootstock cultivars on rooting and its relationship with mineral nutrient content and endogenous carbohydrates. *Sci. Hortic.* **2017**, *225*, 204–212. [CrossRef]
51. Kinik, E.; Çelikel, F.G. Effects of plant growth promoting bacteria and auxin on cutting propagation of *Rosa canina* L. *Turkish J. Agric. Food Sci. Techn.* **2017**, *5*, 1714–1719. [CrossRef]
52. Kinik, E.; Çelikel, F.G. Cutting propagation of *Rosa canina* by mycorrhiza and auxin. *Int. J. Agric. Wildlife Sci.* **2020**, *6*, 1–7. [CrossRef]
53. Izadi, Z.; Zarei, H.; Alizadeh, M. Role of grafting technique on the success of stenting propagation of two rose (*Rosa* sp.) varieties. *Amer. J. Plant Sci.* **2013**, *4*, 41–44. [CrossRef]
54. Agbaria, H.; Bruria, H.; Zieslin, N. Effects of grafting on transpiration, CO_2 fixation and growth of rose plants (*Rosa* x hybrid cvs Liseta and Mercedes). *J. Hortic. Sci.* **1995**, *70*, 651–656. [CrossRef]
55. Nazari, F.; Khosh-Khui, M.; Salehi, H. Growth and flower quality of four *Rosa hybrida* L. cultivars in response to propagation by stenting or cutting in soilless culture. *Sci. Hortic.* **2009**, *119*, 302–305. [CrossRef]
56. Chen, S.; Yao, H.; Han, J.; Liu, C.; Song, J.; Shi, L.; Zhu, Y.; Ma, X.; Gao, T.; Pang, X.; et al. Validation of the ITS2 region as a novel DNA barcode for identifying medicinal plant species. *PLoS ONE* **2010**, *5*, e8613. [CrossRef] [PubMed]
57. Madesis, P.; Ganopoulos, I.; Ralli, P.; Tsaftaris, A. Barcoding the major Mediterranean leguminous crops by combining universal chloroplast and nuclear DNA sequence targets. *Genet. Mol. Res.* **2012**, *11*, 2548–2558. [CrossRef] [PubMed]
58. Ross, H.A.; Murugan, S.; Sibon, L.W.L. Testing the reliability of genetic methods of species identification via simulation. *Syst. Biol.* **2008**, *57*, 216–230. [CrossRef] [PubMed]
59. Kumar, S.; Stecher, G.; Li, M.; Knyaz, C.; Tamura, K. MEGA X: Molecular evolutionary genetics analysis across computing platforms. *Mol. Biol. Evol.* **2018**, *35*, 1547–1549. [CrossRef] [PubMed]
60. Saitou, N.; Nei, M. The neighbor-joining method: A new method for reconstructing phylogenetic trees. *Mol. Biol. Evol.* **1987**, *4*, 406–425. [CrossRef] [PubMed]
61. Felsenstein, J. Confidence limits on phylogenies: An approach using the bootstrap. *Evolution* **1985**, *39*, 783–791. [CrossRef] [PubMed]

62. Vavoura, M.V.; Badeka, A.V.; Kontakos, S.; Kontominas, M.G. Characterization of four popular sweet cherry cultivars grown in Greece by volatile compound and physicochemical data analysis and sensory evaluation. *Molecules* **2015**, *20*, 1922–1940. [CrossRef] [PubMed]
63. Lee, H.S.; Coates, G.A. Vitamin C in frozen, fresh squeezed, unpasteurized, polyethylene-bottled orange juice: A storage study. *Food Chem.* **1999**, *65*, 165–168. [CrossRef]

plants

MDPI

Article

Plant Response to Mechanically-Induced Stress: A Case Study on Specialized Metabolites of Leafy Vegetables

Jana Šic Žlabur [1], Sanja Radman [2,*], Sanja Fabek Uher [2], Nevena Opačić [2], Božidar Benko [2], Ante Galić [1], Paola Samirić [1] and Sandra Voća [1]

[1] Department of Agricultural Technology, Storage and Transport, University of Zagreb Faculty of Agriculture, Svetošimunska Cesta 25, 10000 Zagreb, Croatia; jszlabur@agr.hr (J.Š.Ž.); agalic@agr.hr (A.G.); paolasamiric@hotmail.com (P.S.); svoca@agr.hr (S.V.)

[2] Department of Vegetable Crops, University of Zagreb Faculty of Agriculture, Svetošimunska Cesta 25, 10000 Zagreb, Croatia; sfabek@agr.hr (S.F.U.); nopacic@agr.hr (N.O.); bbenko@agr.hr (B.B.)

* Correspondence: sradman@agr.hr

Abstract: Plants have evolved various adaptive mechanisms to environmental stresses, such as sensory mechanisms to detect mechanical stimuli. This plant adaptation has been successfully used in the production practice of leafy vegetables, called mechanical conditioning, for many years, but there is still a lack of research on the effects of mechanically-induced stress on the content of specialized metabolites, or phytochemicals with significant antioxidant activity. Therefore, the aim of this study was to determine the content of specialized metabolites and antioxidant capacity of lettuce and green chicory under the influence of mechanical stimulation by brushing. Mechanically-induced stress had a positive effect on the content of major antioxidants in plant cells, specifically vitamin C, total phenols, and flavonoids. In contrast, no effect of mechanical stimulation was found on the content of pigments, total chlorophylls, and carotenoids. Based on the obtained results, it can be concluded that induced mechanical stress is a good practice in the cultivation of leafy vegetables, the application of which provides high quality plant material with high nutritional potential and significantly higher content of antioxidants and phytochemicals important for human health.

Keywords: brushing; lettuce; chicory; phytochemicals; antioxidant capacity

Citation: Šic Žlabur, J.; Radman, S.; Fabek Uher, S.; Opačić, N.; Benko, B.; Galić, A.; Samirić, P.; Voća, S. Plant Response to Mechanically-Induced Stress: A Case Study on Specialized Metabolites of Leafy Vegetables. *Plants* **2021**, *10*, 2650. https://doi.org/10.3390/plants10122650

Academic Editors: Milan S. Stankovic, Paula Baptista and Petronia Carillo

Received: 8 November 2021
Accepted: 30 November 2021
Published: 2 December 2021

Publisher's Note: MDPI stays neutral with regard to jurisdictional claims in published maps and institutional affiliations.

Copyright: © 2021 by the authors. Licensee MDPI, Basel, Switzerland. This article is an open access article distributed under the terms and conditions of the Creative Commons Attribution (CC BY) license (https://creativecommons.org/licenses/by/4.0/).

1. Introduction

Consumer interest in cut leafy vegetables with distinct functional value is increasing, as is the need for a continuous supply to the market. Growing leafy vegetables in greenhouses allows them to be grown throughout the year, even in the cold months, thus ensuring the supply and availability of various leafy vegetables, especially lettuce, in the off-season [1,2]. Recently, the cultivation of leafy vegetables in greenhouses has been increasingly oriented towards hydroponic cultivation, mainly because of a number of advantages that this type of cultivation has over the conventional one: significant yield, high quality and healthy plant material, lower incidence of pathogens, less use of pesticides, less pollution, conservation of groundwater (closed hydroponic systems), high degree of automation, less physical labor, better control of water and nutrient supply to plants, fewer weeds, etc. [3,4]. Floating hydroponics (FH) is suitable for growing leafy vegetables such as lettuce, spinach, chicory, arugula, lettuce, medicinal, and also aromatic plants. The benefits are many and include faster growth, earlier harvest, more production cycles, and higher yield per unit area due to better control of plant nutrients [5,6]. Due to the higher plant density in FH and the lack of sunlight or the change in its spectrum during the winter season, plants compete for light, resulting in an undesirable elongation of the hypocotyl and internodes of the stem, leading to fragile plants, poor quality and uneven growth [2,7]. To produce the strongest and most resistant plant, different treatments can be combined to strengthen it. Mechanical conditioning (MC), i.e., stimulation of plants by tactile stimuli in the form of touching, brushing, or rubbing the plant material, is a successful, non-invasive, environmentally friendly, simple, and inexpensive measure to regulate plant growth that can reduce elongation and increase plant strength and resistance [8,9]. Indeed, during growth, plants are exposed to various abiotic and biotic stresses such as wind, rain, machinery, animals, pathogens, or plants themselves [10–12]. Plants have evolved numerous adaptations to these stresses and trigger a range of responses including anatomical, physiological, biochemical, and molecular, known as thigmomorphogenesis [13,14]. This plant adaptation is used to develop a successful production practice known as mechanical conditioning. This is based on the fact that mechanically-induced stress (MIS), applied naturally or under controlled conditions, impairs growth, thereby reducing the mass and size of major plant parts [15]. One of the most noticeable effects of MIS on plants is a reduction in the length of stems, leaves or petioles, resulting in plants that are smaller and more compact than those grown in unstressed controls, i.e., a reduction in primary shoot growth and an increase in secondary thickening [10,11,16–18].

However, it is important to emphasize that MIS is not necessarily associated with injury (wounding), but plants respond by activating its defense mechanism. Namely, the changes in the wax layer lead to the induction of so-called touch-sensitive genes and membrane receptors, whereupon the plant activates its defense mechanisms [19]. However, the response of plants to MIS in some variables other than morphological, such as the content of specific phytochemicals (chlorophyll, vitamins, phenols, etc.) is highly dependent on the species, and there are a number of unanswered questions [10,18].

Due to its positive effects on morphological properties [10–12,15,20], the MC technique has been successfully used for many years in winter production of various leafy vegetables and medicinal and aromatic plants, but there is still a lack of research on the effects of MIS on the chemical composition and content of specialized metabolites with significant antioxidant activity in plant material. Therefore, the aim of this study was to determine the content of specialized metabolites and antioxidant capacity of lettuce and green chicory under the influence of mechanical stimulation by brushing in a hydroponic growing system.

2. Results

2.1. Specialized Metabolites Content of Lettuce and Chicory

The results of the analyzed specialized metabolites of lettuce are presented in Tables 1 and 2. According to the results, the significant effect of MIS on ascorbic acid (AsA) content in lettuce (Table 1) is most pronounced in the first harvest period (L-MC10_1, L-MC20_1), with the highest AsA content (24.39 mg/100 g fw) determined in a sample treated with 20 brushings per day (L-MC20_1). In the first harvest period, an average of 57% higher AsA content was determined in treatment L-MC20_1 than in the treatment in which the procedure MC was not applied (L-MC0_1). In the same harvest period, the lettuce plants treated with MC with 20 passes per day also had the highest total phenol (TPC) content determined (102.46 mg GAE/100 g fw); even several times more than the control sample (L-MC0_1) and 14% more than the treatment with 10 passes per day (L-MC10_1). Total flavonoid (TFC) content in lettuce plants treated with 20 passes per day (L-MC20_2) was also highest; 87% higher than the control sample and 21% higher than the 10 passes per day treatment. In the second lettuce harvest period, the effect of mechanical stimulation on AsA content was not statistically justified, i.e., no significant differences were observed between treatments, although a slight increase in AsA content was observed in the treatment with 10 passes per day (L-MC10_2), about 7% compared to the sample without treatment (L-MC0_2). Significantly higher values of TPC, TFC and total non-flavonoid (TNFC) were observed in the second harvest period compared to the content of polyphenolic compounds in the first harvest period, with the highest TPC in the sample treated with 10 passes (L-MC10_2) and the highest TFC in the treatment with 20 passes (L-MC20_2). The highest levels of total chlorophyll (TCh) and carotenoids (TCA) in lettuce (Table 2) in the first harvest period were found in the samples treated with MC by 20 brushes per day, but in the second harvest the highest TCh and TCA were noted in plants MC treated by 10 brushes per day.

Table 1. Content of specialized metabolites in lettuce under the influence of brushing.

Treatment	AsA (mg/100 g fw)	TPC (mg GAE/100 g fw)	TFC (mg GAE/100 g fw)	TNFC (mg GAE/100 g fw)
		1st harvest		
L-MC0$_1$	15.53 ± 1.20 ^{b}	51.38 ± 5.67 ^{e}	35.59 ± 1.03 ^{f}	15.77 ± 6.50 ^{d}
L-MC10$_1$	23.71 ± 3.20 ^{a}	90.24 ± 0.82 ^{d}	55.19 ± 0.47 ^{d}	35.06 ± 1.27 ^{c}
L-MC20$_1$	24.39 ± 2.03 ^{a}	102.46 ± 1.25 ^{c}	66.72 ± 0.43 ^{b}	35.74 ± 1.14 ^{c}
		2nd harvest		
L-MC0$_2$	16.45 ± 2.35 ^{b}	107.47 ± 0.43 ^{b}	44.30 ± 0.29 ^{e}	90.34 ± 0.51 ^{a}
L-MC10$_2$	17.60 ± 2.52 ^{b}	150.94 ± 1.52 ^{a}	60.60± 0.1 ^{c}	63.17 ± 1.36 ^{b}
L-MC20$_2$	15.90 ± 1.97 ^{b}	86.89 ± 0.57 ^{d}	70.75 ± 0.67 ^{a}	16.13 ± 1.14 ^{d}
ANOVA	$p \leq 0.0008$	$p \leq 0.0001$	$p \leq 0.0001$	$p \leq 0.0001$

AsA—ascorbic acid content; TPC—total phenolics content; TNFC—total non-flavonoid content; TFC—total flavonoid content. Different letters show significant statistical differences between means.

Table 2. Pigment compounds of lettuce under the influence of brushing.

Treatment	Chl_a (mg/g)	Chl_b (mg/g)	TCh (mg/g)	TCA (mg/g)
		1st harvest		
L-MC0$_1$	0.2 ± 0.004 ^{d}	0.16 ± 0.002 ^{a}	0.36 ± 0.006 ^{e}	0.15 ± 0.002 ^{e}
L-MC10$_1$	0.17 ± 0.001 ^{e}	0.12 ± 0.002 ^{b}	0.29 ± 0.002 ^{f}	0.1 ± 0.001 ^{f}
L-MC20$_1$	0.28 ± 0.001 ^{c}	0.15 ± 0.001 ^{ab}	0.44 ± 0.001 ^{d}	0.20 ± 0.001 ^{d}
		2nd harvest		
L-MC0$_2$	0.28 ± 0.002 ^{c}	0.17 ± 0.001 ^{a}	0.45 ± 0.002 ^{c}	0.21 ± 0.001 ^{c}
L-MC10$_2$	0.37 ± 0.002 ^{a}	0.17 ± 0.045 ^{a}	0.57 ± 0.003 ^{a}	0.27 ± 0.001 ^{a}
L-MC20$_2$	0.35 ± 0.001 ^{b}	0.18 ± 0.001 ^{a}	0.53 ± 0.002 ^{b}	0.26 ± 0.001 ^{b}
ANOVA	$p \leq 0.0001$	$p \leq 0.0240$	$p \leq 0.0001$	$p \leq 0.0001$

Chl_a—chorophyll a content; Chl_b—chorophyll b content; TCh—total chlorophyll content; TCA—total carotenoids. Different letters show significant statistical differences between means.

Green chicory samples (Table 3) were again found to have higher AsA content in the first harvest, regardless of the treatment applied. The influence of MC showed a slightly different trend in AsA content than in the lettuce samples. Indeed, the highest AsA content (40.97 mg/100 g fw) was detected in the 10 brush strokes per day treatment (GC-MC10$_1$), with a 12% higher AsA content compared to the control sample (GC-MC0$_1$) and 19% higher compared to the 20 brush strokes per day treatment (GC-MC20$_1$). A significantly lower AsA content was observed in the second harvest period than in the first. However, in the second harvest period, the highest AsA content (33.23 mg/100 g weight) was detected in a sample treated with 20 brush strokes per day (GC-MC20$_2$), by an average of 32% compared to the treatment with 10 brush strokes (GC-MC10$_2$) and 41% compared to the control sample (GC-MC0$_1$). The highest levels of TPC (179.77 mg GAE/100 g fw), TFC (103.14 mg CTH/100 g fw) and TNFC (76.63 mg GAE/100 g fw) were found in the 20-pass treatment (GC-MC20$_2$) in the second harvest period. In general, as in lettuce samples, regardless of the treatment applied, the highest levels of polyphenolic compounds were determined in chicory in the second harvest period. Regarding the analyzed total chlorophylls content in green chicory (Table 4), no significant differences were found depending on the MC treatment (10 and 20 brushes per day) applied both in the first and second harvest period, while TCA content differed significantly considering the brushing treatment with the highest determined content in the first harvest period (0.30 mg/g) in plants treated with 10 passes per day (GC-MC10$_1$) and 0.48 mg/g in second harvest period. In general, the highest TCh and TCA contents were determined in the second harvest period with average values of 0.89 mg/g for TCh and 0.43 mg/g for TCA, which means about 62% higher TCh contents and about 65% higher TCA contents compared to the average values of TCh and TCA found in the first harvest period.

Table 3. Content of specialized metabolites in green chicory under the influence of brushing.

Treatment	AsA (mg/100 g fw)	TPC (mg GAE/100 g fw)	TFC (mg GAE/100 g fw)	TNFC (mg GAE/100 g fw)
		1st harvest		
GC-MC0$_1$	36.52 ± 2.75 ^{b}	104.85 ± 1.60 ^{d}	42.33 ± 1.95 ^{d}	62.51 ± 0.50 ^{c}
GC-MC10$_1$	40.97 ± 0.77 ^{a}	79.42 ± 1.13 ^{e}	26.80 ± 2.20 ^{e}	52.61 ± 1.27 ^{d}
GC-MC20$_1$	34.45 ± 4.47 ^{b}	50.03 ± 1.48 ^{f}	16.92 ± 1.84 ^{f}	33.09 ± 0.35 ^{e}
		2nd harvest		
GC-MC0$_2$	23.64 ± 1.97 ^{c}	172.64 ± 0.28 ^{b}	96.35 ± 0.72^b	76.28 ± 0.50 ^{a}
GC-MC10$_2$	25.31 ± 0.74 ^{c}	161.06 ± 0.69 ^{c}	92.45 ± 1.08^c	68.60 ± 0.39 ^{b}
GC-MC20$_2$	33.23 ± 0.75 ^{b}	179.77 ± 0.62 ^{a}	103.14 ± 1.33^a	76.63 ± 0.82 ^{a}
ANOVA	$p \leq 0.0001$	$p \leq 0.0001$	$p \leq 0.0001$	$p \leq 0.0001$

AsA—ascorbic acid content; TPC—total phenolics content; TNFC—total non-flavonoid content; TFC—total flavonoid content. Different letters show significant statistical differences between means.

Table 4. Pigment compounds of green chicory under the influence of brushing.

Treatment	Chl_a (mg/g)	Chl_b (mg/g)	TCh (mg/g)	TCA (mg/g)
		1st harvest		
GC-MC0$_1$	0.36 ± 0.001 ^{e}	0.15 ± 0.001 ^{f}	0.51 ± 0.001 ^{b}	0.25 ± 0.001 ^{e}
GC-MC10$_1$	0.43 ± 0.001 ^{d}	0.20 ± 0.001 ^{d}	0.63 ± 0.002 ^{b}	0.30 ± 0.001 ^{d}
GC-MC20$_1$	0.33 ± 0.001 ^{f}	0.18 ± 0.001 ^{e}	0.51 ± 0.001 ^{b}	0.23 ± 0.001 ^{f}
		2nd harvest		
GC-MC0$_2$	0.61 ± 0.002 ^{b}	0.29 ± 0.002 ^{c}	0.89 ± 0.004 ^{a}	0.42 ± 0.002 ^{b}
GC-MC10$_2$	0.71 ± 0.001 ^{a}	0.36 ^{a}	0.92 ± 0.26 ^{a}	0.48 ± 0.02 ^{a}
GC-MC20$_2$	0.56 ^{c}	0.30 ^{b}	0.86 ± 0.001 ^{a}	0.39 ± 0.001 ^{c}
ANOVA	$p \leq 0.0001$	$p \leq 0.0001$	$p \leq 0.0008$	$p \leq 0.0001$

Chl_a—chorophyll a content; Chl_b—chorophyll b content; TCh—total chlorophyll content; TCA—total carotenoids. Different letters show significant statistical differences between means.

2.2. Antioxidant Capacity of Lettuce and Chicory

The results of antioxidant capacity of lettuce and green chicory samples are shown in Figures 1 and 2. In general, relatively high values of antioxidant capacity were obtained in the first harvesting period, regardless of the treatment. The highest antioxidant capacity (2267.55 µmol TE/L) was determined in a lettuce sample treated with 20 brush passes per day (L-MC20$_1$), i.e., in plants with the most pronounced MIS. The same trend of antioxidant capacity of lettuce was observed in the second harvest period considering MC treatment; the highest antioxidant capacity was determined in the sample with 20 passes per day (L-MC20$_2$) (Figure 1). In the green chicory samples (Figure 2), the variations in antioxidant capacity are small compared to the lettuce samples, since no significant statistical differences were observed in either the first or second harvest period when treated with MC. However, regardless of MIS treatment and harvest period, green chicory samples are characterized by high antioxidant capacity, indicating a nutritionally valuable plant material.

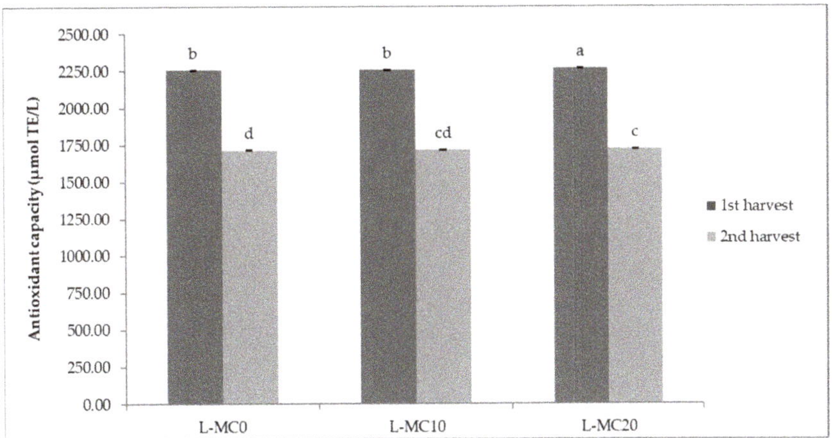

Figure 1. Antioxidant capacity of lettuce depending on different mechanical conditioning treatments. L—lettuce; MC—mechanical conditioning; 0, 10, 20—number of plants brushing per day. Different letters show significant statistical differences between means.

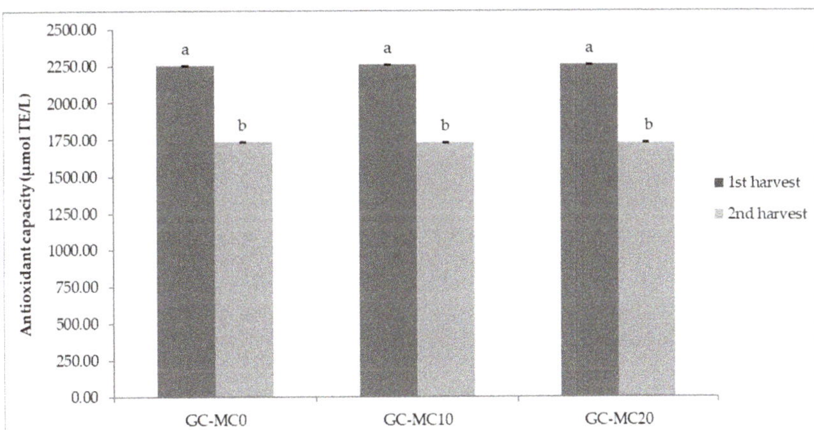

Figure 2. Antioxidant capacity of green chicory depending on different mechanical conditioning treatments. GC—green chicory; MC—mechanical conditioning; 0, 10, 20—number of plants brushing per day. Different letters show significant statistical differences between means.

3. Discussion

3.1. Specialized Metabolites Content

During growth, plants are exposed to many different stress factors (biotic and abiotic) and have developed numerous protective mechanisms, i.e., responses to them. The direct response of plants to environmental stress is increased production of reactive oxygen species (ROS). This is because ROS play an important role as a signaling molecule in regulating plant growth and enhancing plant responses to stress. In general, ROS are considered as by-products of aerobic metabolism in plants and are produced in various cellular organelles such as chloroplasts, mitochondria and peroxisomes. Plants require a threshold level of ROS for their vital functions, and change in concentration alters the physiology of plants, disrupting cell metabolism and causing irreversible damage to DNA material. Precisely to maintain the balance of ROS in cells, plants have also developed an antioxidant defense system (enzymatic and non-enzymatic mechanisms) that maintains a redox state of cells that helps to eliminate ROS [21–23].

Despite the lack of scientific research on the effect and correlation of MIS on AsA content, the effect of mechanical stimuli on AsA content can be explained by metabolic reactions that occur in plant tissues (cells) when the plant is exposed to stress. The non-enzymatic mechanisms of plant defense against ROS are mainly mediated by low molecular weight antioxidants such as glutathione, AsA and flavonoids, which are known to remove hydroxyl radicals and singlet oxygen. Ascorbic acid is indeed one of the best-known oxygen scavenging molecules, i.e., AsA protects metabolic processes from the harmful effects of H_2O_2 and other toxic oxygen-derived radicals. An important segment of enzymatic protection is the response of plants via enzymatic mechanisms (e.g., the enzyme ascorbate peroxidase, APX), such as another important ROS scavenging pathway, namely the Foyer–Halliwell–Asada pathway, also known as the ascorbate–glutathione cycle, which occurs in chloroplasts, mitochondria, apoplasts, cytosol, and peroxisomes. The glutathione–ascorbate cycle is a metabolic pathway for the de-toxification of hydrogen peroxide (H_2O_2), a reactive oxygen species produced as a byproduct in plant metabolism. The cycle involves the antioxidant metabolites ascorbate, glutathione, and NADPH, and the enzymes that link these metabolites [22,23]. Ascorbate peroxidase (APX), glutathione reductase (GR), monodehydroascorbate reductase (MDHAR) and dehydroascorbate reductase (DHAR) are key enzymes related to the non-enzymatic antioxidant metabolites AsA and glutathione. Dehydroxyascorbate (DHA), which is itself an oxidized form of ascorbate, is also formed from the eponymous cycle [24]. The ascorbate–glutathione cycle plays an important role in the reduction of dehydroxyascorbate to ascorbic acid [24]. Considering all this, it can be confirmed that AsA protects metabolic processes from the harmful effects of H_2O_2 and other toxic radicals from oxygen as well as membranes, either by directly removing toxic radicals such as $^1O_2$, O_2- and OH- or indirectly through the regeneration of the reduced forms of tocopherol or zeaxanthin. In addition, numerous studies show that plants exposed to various stress conditions, especially drought [25,26] and salinity [27,28], had increased AsA content. If we relate all this to the results of this study, it is worth noting that higher AsA content was found in lettuce in the first harvest period when plants were exposed to MIS, regardless of the brush strokes per day, while no significant changes were found in the second harvest period given the MIS treatment. In green chicory, in the first harvest period, the highest AsA content was found in plants treated with 10 brush strokes per day, while in the second harvest period, plants were treated with 20 brush strokes per day. (Tables 1 and 3). Comparing the AsA content in non-mechanically treated lettuce (control sample) with the results of the study by the authors Medina-Lozano et al. [29], we can generally conclude that slightly higher values were obtained in this study; however, compared to the study by van Treuren et al. [30], lower values were found in this study. Furthermore, the AsA content in the control green chicory samples ((GC-MC0), without mechanical treatment) analyzed in this study was higher than that reported in other literature data [31,32]. The reason for this wide dispersion of results is mainly due to the genetic characteristics of the varieties themselves, which are most commonly used for cultivation for commercial purposes [29–32].

The MIS, caused by the number of brushings, induces mechanical depolarization of the plant cell membrane, leading to the formation of various types of free radicals and initiating lipid peroxidation, thereby activating plant protective mechanisms against ROS, i.e., increased activity of antioxidant mechanisms and precursor enzymes of vitamin C synthesis (i.e., APX) [13,33]. Nowadays, among the best known and studied antioxidants present in cellular organelles and whose main function is detoxification of ROS are the polyphenolic compounds. Apart from being among the compounds with the most potent antioxidant activity, polyphenolic compounds form a chemically extremely large, diverse and highly significant group of plant secondary metabolites. Phenolic compounds are in fact the main products of the plant defense system united in secondary metabolism, i.e., the processes of biosynthesis of phenylpropanoids, anthocyanins, alkaloids, coumarins, terpenes, tannins, glucosinolates, flavonoids and isoflavonoids, lignans and lignans, among others. Secondary metabolites have several important ecological and physiological functions in the plant organism, ranging from structural roles (binding of cell wall polysaccharides), protection from attack by herbivores and microorganisms, attraction of pollinators, seed dispersal, communication of the plant with other plants and organisms, to the most important role, the responses of the plant organism to stress. Indeed, the phenylpropanoid biosynthetic pathway, leading to the accumulation of various phenolics compounds, is activated in plants under stressful environmental conditions. Thus, the accumulation of polyphenolic compounds can be regulated by various environmental factors such as water availability, wounding, herbicides, insect herbivory, nutrient deficiency and availability, light exposure, salinity, developmental stage of the plant, etc. [14,22,34,35]. Numerous scientific studies indicate increased accumulation of polyphenolic compounds when a plant organism is exposed to stress, as a direct response by the plant to defend itself. Mechanical conditioning also exposes the plant to mechanical stress, resulting in changes in metabolic pathways and increased activation of compounds whose main function is to remove excessive amounts of ROS, accumulated in plant cells [14]. Indeed, plants respond to MIS with stress responses such as the aforementioned increased ROS production and changes in calcium levels, followed by the activation of defense responses [12]. As mentioned above, polyphenols are one of the most effective compounds in the elimination of free radicals, and it is therefore expected that when the plant is exposed to stress, an increased accumulation of these compounds begins [36], which is confirmed by the results of this research. Lettuce and chicory plants that were more mechanically stimulated (Tables 1 and 2), i.e., exposed to MIS, also showed higher levels of polyphenolic compounds, especially total phenols and flavonoids. It should be emphasized that regardless of the mechanical treatment, i.e., MIS, the samples of lettuce and green chicory are rich in polyphenolic compounds and the results of total phenolics (TPC) are generally in agreement with other literature data, with the values of polyphenolic compounds obtained in this study for lettuce and green chicory not mechanically treated (control samples) similar or slightly lower than the data reported by other authors [36–40]. Based on the results obtained, it can be concluded that mechanical stimulation induces plant responses in direct response to stress exposure, as evidenced by the increased content of polyphenolic compounds in the lettuce and chicory samples in the MC treatments. The results of the higher content of polyphenolic compounds in plants exposed to stress are in agreement with the statements of other authors who also confirm that plants have the ability to biosynthesize higher amounts of polyphenols for defense purposes under stress than under normal growing conditions. Indeed, the biosynthesis of polyphenols in plants under stress conditions is regulated by different activities of specific enzymes in the metabolic pathways of polyphenol synthesis (shikimate/phenylpropanoid and mevalonate pathways) such as phenylalanine ammonia-lyase (PAL) and chalcone synthase (CHS). Under stress conditions, there are enhanced effects of enzymes associated with the regulation of gene transcription, which encode important biosynthetic enzymes of polyphenolic compounds [14,36].

Stress conditions to which plants are frequently exposed during their growth, especially those caused by abiotic factors such as heavy metals, drought, too much light (UV radiation), too low or too high temperatures, increased soil salinity, etc., usually have a negative effect on the performance of the photosynthetic process. These stress conditions reduce stomatal conduction, which leads to oxidative stress and reduces the activity of RUBISCO enzymes, which prevents the smooth process of photosynthesis by negatively affecting photosystem I and II, photosynthetic transport electron and finally chlorophyll biosynthesis [41]. The authors Benikhlef et al. [12] in their study on the induction of soft mechanical stress in plants of the species *Arabidopsis thalian* come to different conclusions. In fact, the aforementioned authors note that the application of MIS did not result in obvious damage to plant tissues, while chlorophyll content increased, which is directly related to the rapid change in calcium concentration and the release of ROS, accompanied by changes in cuticle permeability, induction of gene expression typically associated with mechanical stress, and the release of biologically active diffusers from the surface. The results of the pigment compounds analyzed in this study are mostly in agreement with the results of the mentioned study. In the lettuce samples (Table 3), both TCh and TCA content were highest in the first harvest period when MC was applied with 20 passes per day and 10 passes per day in the second harvest period. However, this effect of MIS on pigment compounds was not as pronounced in the green chicory samples, as there were no significant differences between the controls and those treated with MC. Considering that both species have a characteristic green color, high levels of total chlorophylls and carotenoids were also expected in the control samples, i.e., those not subjected to the MIS. The results of total chlorophylls and carotenoids content of the green chicory and lettuce samples from this study are significantly higher than those found in other research [32,42,43].

3.2. *Antioxidant Capacity*

Accumulation of antioxidants in plant cells is the expected response of the plant defense mechanism, wherein the bioactive compounds are those that are intensified to synthesize in order to detoxify ROS. From bioactive compounds, one that exhibits the most significant antioxidant activity in the plant cell are polyphenolic compounds, such as flavonoids [9,14,21–23,34–36]. It is also important to emphasize that antioxidants, in addition to the important effect and protection of plant cells, also have a beneficial effect on human health, given that their mechanisms of action, such as radical scavengers, chelators, quenchers, oxygen scavengers and antioxidant regenerators, effectively inhibit free radicals accumulated in human cells as a result of oxidative processes [44–46]. Based on the above mentioned and obtained results, it can be concluded that that cultivated leafy vegetables abound antioxidants and thus present a high value nutritional raw material. According to other studies, the lettuce and green chicory species analyzed in this study, regardless of MIS, are vegetables that have significant antioxidant properties, i.e., high levels of antioxidant capacity [32,37,47].

4. Materials and Methods

4.1. *Plant Material*

The research was conducted in 2019, at the Experimental Station of the Department of Vegetable Crops at the Faculty of Agriculture, University of Zagreb. Two types of green leafy vegetables were grown for research purposes: lettuce (*Lactuca sativa* L.) and green chicory (*Cichorium intibus* var. *foliosum*). Of the *Lactuca sativa* species, a mixed lettuce was sown in the following composition: 'Reggina di Maggio' 25%; 'Meraviglia delle Quatro Stagioni' 25%; 'Grunetta' 25%; 'Cavolo di Napoli' 25% (Hortus sementi, Longiano, Italy), while of the green chicory, the variety 'Zuccherina di Trieste' (Hortus sementi, Longiano, Italy) was sown.

4.2. *Floating Hydroponics*

Both leafy vegetables species were grown in floating hydroponics in an unheated greenhouse (45°49′33.208″ N, 16°1′42.832″ E). A total of 12 polystyrene boards filled with inert perlite substrate (Europerl d.o.o., Samobor, Croatia) were used for growing lettuce and chicory in the hydroponic system, of which six plates (two for each of the treatments, Figure 1) were sown with lettuce seeds and six with chicory seeds. Manual sowing of lettuce and chicory in an unheated greenhouse was carried out on 23 April. For both species, 30 seeds were sown per slot of the 0.96 m × 0.60 m board with a total of 102 slots (17 cm long and 0.5 cm wide). The amount of seed required depended on the seed size of the species, 22.77 g/m^2 for chicory and 10.35 g/m^2 for lettuce. After sowing, the seeds were covered with a finer granulation of perlite (0–3 mm) and the moisture of the substrate was maintained during germination. After emergence, the plates were placed in basins (4 m × 2 m × 0.25 m) filled with nutrient solution. The nutrient solution used in the experiment was adapted for growing lettuce and chicory [48]. The contents of the nutrient solution used for hydroponic cultivation of lettuce and chicory are shown in Table 5.

Table 5. Adapted nutrient solution for lettuce and green chicory cultivation.

Biogenic Element	Measuring Unit	Values
pH	mS/cm	5.5
EC		2.5
Na Cl HCO_3	mmol/L	<6 <6 <0.5
N-NH_4 K Ca Mg	mmol/L	<0.5 6 6 2
N-NO_3 S P	mmol/L	14 2 2
Fe Mn Zn B Cu Mo	μmol/L	40 8 8 50 1.5 1.5

4.3. Mechanical Conditioning

The mechanical conditioning process for both vegetable species began in early May, from the appearance of the cotyledon till harvests. The process was carried out by brushing with a burlap cloth in two mechanical conditioning treatments: 10 (MC10 treatment) and 20 (MC20 treatment) passes per day. Control plots were also established for lettuce and chicory, i.e., plots where the conditioning treatment was not applied (MC0 treatment). Brush treatments were performed at the same time each day, in the morning hours. The design of the mechanical conditioning experiment is shown in Table 6 and Figure 3.

Table 6. The design of the mechanical conditioning experiment.

Species	Brushing	Hydroponic System	Harvest Period	Treatment
Lettuce	0	FH	I.	L-$MC0_1$
			II.	L-$MC10_2$
	10	FH	I.	L-$MC20_1$
			II.	L-$MC0_2$
	20	FH	I.	L-$MC10_1$
			II.	L-$MC20_2$
Green chicory	0	FH	I.	GC-$MC0_1$
			II.	GC-$MC10_2$
	10	FH	I.	GC-$MC20_1$
			II.	GC-$MC0_2$
	20	FH	I.	GC-$MC10_1$
			II.	GC-$MC20_2$

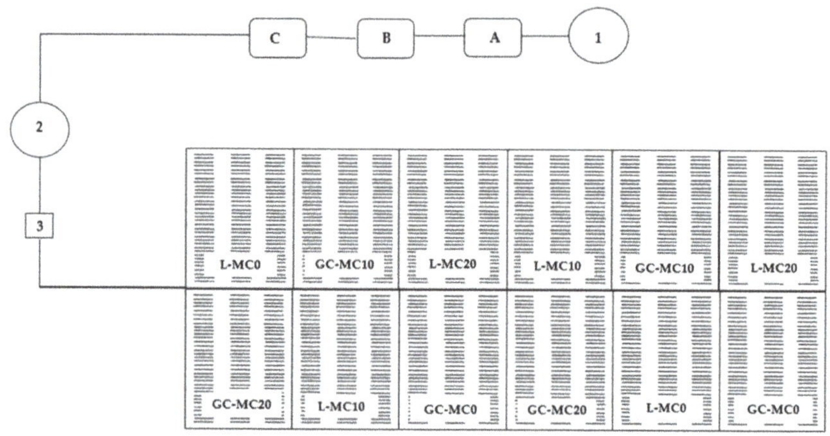

Figure 3. Graphical scheme of floating hydroponics cultivation of lettuce and green chicory. 1—water; A, B, C—tanks for concentrated nutrient solutions and injectors; 2—tanks for standard nutrient solutions; 3—pump; L—lettuce; GC—green chicory; MC—mechanical conditioning; 0, 10, 20—number of plants brushing per day.

4.4. Abiotic Parameters of Air and Nutrient Solution

The abiotic parameters of air (minimum and maximum temperature, relative humidity) and nutrient solution (pH, temperature, dissolved oxygen, pH and EC) were measured daily. The average minimum temperature measured in the greenhouses in May was 12.4 °C and the maximum temperature was 31.5 °C, while the relative humidity was 76%. The average pH of the solution was 5.4, while the average EC was 2.2 mS/cm. These conditions during growth resulted in favorable root development. To ensure the supply of sufficient oxygen to the roots, pumps were used which also mixed the nutrient solution. The amount of dissolved oxygen varied considerably during the growing cycle. The highest amount of dissolved oxygen in the nutrient solution (19.3 mg/L) was at the beginning of the growing cycle and the lowest (2.00 mg/L) at the end of the experiment (data not shown).

4.5. Harvest Period

Harvesting of lettuce and chicory was done twice during the growing season. The first harvest took place on 28 May, cutting the plants to avoid damaging the vegetative top of the plants and to ensure retro-vegetation. The second harvest took place after 16 days, on 10 June.

4.6. Determination of Specialized Metabolites Content

From the group of specialized metabolites following compounds were determined: ascorbic acid (AsA), i.e., vitamin C content, total phenolics (TPC), total flavonoids (TFC) and total non-flavonoids (TNFC). AsA was determined by titration with 2,6-dichlorindophenol according to the standard laboratory method available in AOAC [49]. AsA was isolated from the fresh leaves of lettuce and green chicory with 2% (*v*/*v*) oxalic acid; first 10 g ± 0.01 of fresh plant material was weighed and homogenized with 100 mL of 2% (*v*/*v*) oxalic acid. Prepared solution was filtered through Whatman filter paper and 10 mL of solution was used for titration with 2,6-dichlorindophenol till the appearance of a characteristic pink coloration. Namely, this method is based on the oxidimetric titrations using 2,6-dichlorindophenol as reducing agent. 2,6-dichlorindophenol is a solution of intense blue color that oxidizes L-ascorbic acid to dehydroascorbic acid until the color of the reagent changes, and also serves as an indicator for this redox reaction. The final AsA content was calculated according to Equation (1) and expressed as mg/100 g fresh weight.

$$AsA = (V\,(DKF) \times F)/D \times 100, \quad (1)$$

where V (DKF)—volume of DKF (mL); F—factor of DKF; D—sample mass used for titration.

The TPC, TFC and TNFC content was determined based on the colorimetric reaction, the development of blue color within phenols and reagent Folin–Ciocalteu measured spectrophotometrically (Shimadzu, 1900i, Kyoto, Japan) at 750 nm using dH_2O as a blank. The method was described by Ough and Amerine [50]. For the purpose of extraction of polyphenolic compounds from lettuce and green chicory leaves, 10 g ± 0.01 of fresh plant leaves were weighed into an Erlenmeyer flask, and 40 mL of 80% EtOH (v/v) was added and refluxed. The prepared sample was first heated to boiling point and additionally refluxed for 10 min. After 10 min, the sample was filtered through Whatman filter paper into a volumetric flask of 100 mL. After filtration, the remainder of the sample was transferred to the Erlenmeyer flask, another 50 mL of 80% EtOH (v/v) was added, and reflux was repeated for another 10 min. After the second reflux, the sample was filtered and the filtrates were combined while the flask was made up to the mark with 80% EtOH (v/v). The thus prepared sample was subjected to reaction with reagent Folin–Ciocalteu, according to the following procedure: to a volumetric flask of 50 mL, 0.5 mL of the ethanolic plant extract, 30 mL of distilled water (dH_2O), 2.5 mL of the freshly prepared reagent Folin–Ciocalteu (1:2 with dH_2O) and 7.5 mL of saturated sodium carbonate solution (Na_2CO_3) were added. The flask was made up to the mark with dH_2O and the reaction was allowed to stand at room temperature for 2 h with intermittent shaking. The same ethanolic extracts prepared for TPC were used for TFC determination. TFC separation was performed according to the following procedure: 10 mL of the ethanolic extract was added to the 25 mL volumetric flask, 5 mL HCl (1:4, v/v) and 5 mL formaldehyde were added. The prepared samples were treated with nitrogen (N_2) and left at room temperature for 24 h in a dark place. After 24 h, the samples were filtered and the same reaction was performed with Folin–Ciocalteu reagent as for TPC. Gallic acid was used as external standard and the final concentration of TPC, TFC and TNFC content was expressed as mg GAE/100 g fresh weight. TNFC content was mathematically expressed as the difference between total phenols and flavonoids.

From the plant pigments, the following have been identified: chlorophyll a (Chl_a), chlorophyll b (Chl_b), total chlorophylls (TCh) and total carotenoids (TCA), according to the method described by Holm [51] and Wettstein [52]. For the extraction of pigments from leaves of lettuce and green chicory, 0.2 g ± 0.01 of fresh plant leaves were weighed and a total volume of 15 mL of acetone (p.a.) was added, a total of three times. After each addition of acetone, the samples were homogenized using a laboratory homogenizer (IKA, UltraTurax T-18, Staufencity, Germany). The final solution was filtered and transferred to a 25 mL volumetric flask. The absorbance was measured spectrophotometrically (Shimadzu UV 1900i, Kyoto, Japan) at three wavelengths, 662, 644 and 440 nm, using acetone as a blank. The equations of Holm–Wettstein were used to quantify the individual pigments (2), and the final content was expressed in mg/g.

$$\begin{aligned} Chl\_a &= 9.784 \times A_{662} - 0.990 \times A_{644}\ [mg/L] \\ Chl\_b &= 21.426 \times A_{644} - 4.65 \times A_{622}\ [mg/L] \\ TCh &= 5.134 \times A_{662} + 20.436 \times A_{644}\ [mg/L] \\ TCA &= 4.695 \times A_{440} - 0.268 \times TCh\ [mg/L] \end{aligned} \quad (2)$$

4.7. Determination of Antioxidant Capacity

For the determination of antioxidant capacity, the ABTS assay was performed [53]; ABTS, 2,2′-azinobis (3-ethylbenzothiazoline-6-sulfonic acid), potassium persulfate, and Trolox were obtained from Sigma-Aldrich (St. Louis, MO, USA). Trolox (6-hydroxy-2,5,7,8-tetramethylchroman-2-carboxylic acid) was used as the antioxidant standard, and a stock standard Trolox (2.5 mM) was prepared in ethanol (80% v/v). To prepare the ABTS radical solution (ABTS+), 5 mL of ABTS solution (7 mM) and 88 mL of potassium persulfate solution (140 mM) were mixed and allowed to stand for 16 h in the dark at room temperature. On the day of analysis, a 1% ABTS+ solution (in 96% ethanol) was prepared. A total of 160 µL of ethanolic extract (prepared for phenol isolation) was directly injected into the cuvette and mixed with 2 mL of 1% ABTS+ while absorbance was measured at 734 nm (Shimadzu 1900i, Kyoto, Japan). The final results of antioxidant capacity were calculated from the calibration curve and expressed as µmol TE/L.

4.8. Statistical Analysis

Each sample (cultivar, mechanical conditioning treatment and control) of leafy vegetables cultivated in floating hydroponics was represented by two boards, while all chemical laboratory analyzes were performed in triplicate. The data obtained were averaged, expressed as mean ± standard deviation (SD), as shown in the figures and tables. The ANOVA and Duncan's multiple range tests (95% confidence limit) were performed to show the variations in the mean values among the samples. SAS statistical software ver. 9.4. was used for this purpose [54]. Different letters show significant differences between the means at $p \leq 0.0001$, while also the average deviation of the results from the mean for each parameter studied is expressed with the values of standard deviation.

5. Conclusions

The mechanically-induced stress in the form of brushing per day (10 and 20) did not cause damage to plant tissue and thus did not significantly affect the processes of primary metabolism, i.e., photosynthesis, as shown by the higher contents of total chlorophylls and carotenoids in lettuce in both harvest periods; with an average of 22% higher TCh in plants treated with 20 brushings per day in the first harvest period and 18% higher TCh in the second harvest period compared to the non-treated plants, and 33% higher TCA in the first and 24% higher in the second harvest period also in plants treated with 20 brushings per day compared to the non-treated plants. Moreover, the induced mechanical stimuli were sufficient to initiate plant signaling molecules for stress defense, as evidenced by higher levels of antioxidants such as ascorbic acid and polyphenolic compounds in the plants treated with mechanical conditioning. For lettuce in the first harvest period, AsA content was on average 55% higher in mechanically-stimulated plants compared to the non-treated plants, while for green chicory, a more pronounced effect of MIS was observed in the second harvest period, in which, on average, 24% higher AsA content was determined in treated plants. Polyphenolic compounds in the first harvest period in lettuce were on average 88% higher in MIS treated plants, regardless of the passes per day, while in the second harvest period, about 11% higher polyphenolic compounds were determined in comparison with non-treated plants. From all these, it can be concluded that implementation of induced mechanical stress is a good practice in the cultivation of leafy vegetables, the application of which produces high quality plant material with high nutritional potential, and significantly higher levels of antioxidants and phytochemicals important for human health. It should also be emphasized that most authors explain the effects of some abiotic stresses, mostly drought, high temperatures, salinity, etc., but there is still a lack of scientific data on the effects of mechanically-induced stress on the phytochemical status of plants. Individual compound determination is necessary to further explain the effects of mechanically-induced stress on the plant organism.

Author Contributions: Conceptualization, J.Š.Ž. and S.R.; methodology, J.Š.Ž., S.R. and S.F.U.; software, S.R. and N.O.; formal analysis, J.Š.Ž. and P.S.; resources, S.R.; data curation, P.S.; writing–original draft preparation, J.Š.Ž.; writing–review and editing, S.F.U., B.B. and S.V.; visualization, A.G. All authors have read and agreed to the published version of the manuscript.

Funding: This research received no external funding.

Conflicts of Interest: The authors declare no conflict of interest.

References

1. Radman, S.; Ćurko, J.; Toth, N.; Fabek, S.; Čoga, L.; Žutić, I.; Benko, B. Lamb's lettuce mineral content in floating system. *Acta Hortic.* **2014**, *1142*, 343–348. [CrossRef]
2. Toth, N.; Fabek, S.; Benko, B.; Žutić, I.; Stubljar, S.; Zeher, S. The effect of abiotic factors, sowing density and multiple harvest to arugula yield in floating hydroponic. *Glas. Zašt. Bilja* **2012**, *35*, 24–34.
3. Gonnella, M.; Serio, F.; Conversa, G.; Santamaria, P. Yield and quality of lettuce grown in floating system using different sowing density and plant spatial arrangements. *Acta Hortic.* **2002**, *614*, 687–692. [CrossRef]
4. Miceli, A.; Moncada, A.; Vetrano, F.; D'Anna, F. First results on yield and quality response of basil (*Ocimum basilicum* L.) grown in a floating system. *Acta Hortic.* **2003**, *609*, 377–381. [CrossRef]
5. Petropoulos, S.A.; Chatzieustratiou, E.; Constantopoulou, E.; Kapotis, G. Yield and Quality of Lettuce and Rocket Grown in Floating Culture System. *Not. Bot. Horti Agrobot.* **2016**, *44*, 603–612. [CrossRef]
6. Sharma, N.; Acharya, S.; Kumar, K.; Singh, N.; Chaurasia, O.P. Hydroponics as an advanced technique for vegetable production: An overview. *J. Soil Water Conserv.* **2018**, *17*, 364–371. [CrossRef]
7. Björkman, T. Dose and timing of brushing to control excessive hypocotyl elongation in cucumber transplants. *HortTechnology* **1999**, *9*, 40–42. [CrossRef]
8. Radman, S.; Bedek, M.; Židovec, V.; Toth, N.; Benko, B.; Žutić, I. The influence of mechanical stress on the morphological properties of thyme and coriander. In Proceedings of the 54th Croatian & 14th International Symposium on Agriculture, Vodice, Croatia, 17–22 February 2019; pp. 297–301.
9. Li, Z.G.; Gong, M. Mechanical Stimulation-Induced Cross-Adaptation in Plants: An Overview. *J. Plant Biol.* **2011**, *54*, 358–364. [CrossRef]
10. Biddington, N.L. The effects of mechanically-induced stress in plants—A review. *Plant Growth Regul.* **1986**, *4*, 103–123. [CrossRef]
11. Goodman, A.M.; Ennos, A.R. The effects of mechanical stimulation on the morphology and mechanics of maize roots grown in an aerated nutrient solution. *Int. J. Plant Sci.* **2001**, *162*, 691–696. [CrossRef]
12. Benikhlef, L.; L'Haridon, F.; Abou-Mansour, E.; Serrano, M.; Binda, M.; Costa, A.; Lehmann, S.; Métraux, J.P. Perception of soft mechanical stress in Arabidopsis leaves activates disease resistance. *BMC Plant Biol.* **2013**, *13*, 133. [CrossRef]
13. Saidi, I.; Ammar, S.; Demont-Caulet, N.; Thévenin, J.; Lapierre, C.; Bouzid, S.; Jouanin, L. Thigmomorphogenesis in *Solanum lycopersicum*, Morphological and biochemical responses in stem after mechanical stimulation. *Plant Signal. Behav.* **2010**, *5*, 122–125. [CrossRef] [PubMed]
14. Isah, T. Stress and defense responses in plant secondary metabolites production. *Biol. Res.* **2019**, *52*, 1–25. [CrossRef]
15. Potocka, I.; Szymanowska-Pułka, J. Morphological responses of plant roots to mechanical stress. *Ann. Bot.* **2018**, *122*, 711–723. [CrossRef]
16. Latimer, J.G. Mechanical conditioning for control of growth and quality of vegetable transplants. *HortScience* **1991**, *26*, 1456–1461. [CrossRef]
17. Latimer, J.G.; Beverly, R.B. Mechanical conditioning of greenhouse-grown transplants. *HortTechnology* **1993**, *3*, 412–413. [CrossRef]
18. Puijalon, S.; Bouma, T.J.; Douady, C.J.; van Groenendael, J.; Anten, N.P.R.; Martel, E.; Bornette, G. Plant resistance to mechanical stress: Evidence of an avoidance–tolerance trade-off. *New Phytol.* **2011**, *191*, 1141–1149. [CrossRef] [PubMed]
19. Braam, J. In touch: Plant responses to mechanical stimuli. *New Phytol.* **2005**, *165*, 373–389. [CrossRef] [PubMed]
20. Latimer, J.G. Mechanical conditioning to control height. *HortTechnology* **1998**, *8*, 529–534. [CrossRef]
21. Li, N.; Jia, Z.; Trush, M.A. Defining ROS in Biology and Medicine. *React. Oxyg. Species* **2016**, *1*, 9. [CrossRef]
22. Ahanger, M.A.; Singh Tomar, N.; Tittal, M.; Argal, S.; Agarwal, R.M. Plant growth under water/salt stress: ROS production; antioxidants and significance of added potassium under such conditions. *Physiol. Mol. Biol. Plants* **2017**, *23*, 731–744. [CrossRef] [PubMed]
23. Huang, H.; Ullah, F.; Zhou, D.X.; Yi, M.; Zhao, Y. Mechanisms of ROS Regulation of Plant Development and Stress Responses. *Front. Plant Sci.* **2019**, *10*, 800. [CrossRef]
24. Cuypers, A.; Vangronsveld, J.; Clisjters, H. The redox status of plant cells (AsA and GSH) is sensitive to zinc imposed oxidative stress in roots and primary leaves of *Phaseolus vulgaris*. *Plant Physiol. Biochem.* **2001**, *39*, 657–664. [CrossRef]
25. Alam, M.M.; Nahar, K.; Hasanuzzaman, M.; Fujita, M. Alleviation of osmotic stress in *Brassica napus*, *B. campestris* and *B. juncea* by ascorbic acid application. *Biol. Plant* **2014**, *58*, 697–708. [CrossRef]
26. Shafiq, S.; Akram, N.A.; Ashraf, M. Does exogenously-applied trehalose alter oxidative defense system in the edible part of radish (*Raphanus sativus* L.) under water-deficit conditions? *Sci. Hortic.* **2015**, *185*, 68–75. [CrossRef]
27. El-Sayed, O.M.; El-Gammal, O.H.M.; Salama, A.S.M. Effect of ascorbic acid, proline and jasmonic acid foliar spraying on fruit set and yield of Manzanillo olive trees under salt stress. *Sci. Hortic.* **2014**, *176*, 32–37. [CrossRef]
28. Shan, C.; Zhao, X. Effects of lanthanum on the ascorbate and glutathione metabolism of *Vigna radiata* seedlings under salt stress. *Biol. Plant* **2014**, *58*, 595–599. [CrossRef]
29. Medina-Lozano, I.; Bertolín, J.R.; Díaz, A. Nutritional value of commercial and traditional lettuce (*Lactuca sativa* L.) and wild relatives: Vitamin C and anthocyanin content. *Food Chem.* **2021**, *359*, 129864. [CrossRef]
30. van Treuren, R.; van Eekelen, H.D.L.M.; Wehrens, R.; de Vos, R.C.H. Metabolite variation in the lettuce gene pool: Towards healthier crop varieties and food. *Metabolomics* **2018**, *14*, 146. [CrossRef]

31. Mieszczakowska-Frąc, M.; Celejewska, K.; Płocharski, W. Impact of Innovative Technologies on the Content of Vitamin C and Its Bioavailability from Processed Fruit and Vegetable Products. *Antioxidants* **2021**, *10*, 54. [CrossRef]
32. Helaly, A.A.; Abdullah, H.M. Phytochemical Analysis and Yield Characterization of Eight *Cichorium intybus* L. Landraces. *J. Hortic. Sci. Ornam. Plants* **2017**, *9*, 39–51.
33. Locato, V.; Cimini, S.; De Gara, L. Strategies to increase vitamin C in plants: From plant defense perspective to food biofortification. *Front. Plant Sci.* **2013**, *4*, 152. [CrossRef]
34. Marchiosi, R.; dos Santos, W.D.; Polimeni Constantin, P.; de Lima, R.B.; Soares, A.R.; Finger-Teixeira, A.; Rodrigues Mota, T.; de Oliveira, D.M.; Foletto-Felipe, M.P.; Abrahão, J.; et al. Biosynthesis and metabolic actions of simple phenolic acids in plants. *Phytochem. Rev.* **2020**, *19*, 865–906. [CrossRef]
35. Šamec, D.; Karalija, E.; Šola, I.; Vujčić Bok, V.; Salopek-Sondi, B. The Role of polyphenols in abiotic stress response: The influence of molecular structure. *Plants* **2021**, *10*, 118. [CrossRef] [PubMed]
36. Sharma, A.; Shahzad, B.; Rehman, A.; Bhardwaj, R.; Landi, M.; Zheng, B. Response of phenylpropanoid pathway and the role of polyphenols in plants under abiotic stress. *Molecules* **2019**, *24*, 2452. [CrossRef] [PubMed]
37. Bayazid, A.B.; Park, S.H.; Kim, J.G.; Lim, B.O. Green chicory leaf extract exerts anti-inflammatory effects through suppressing LPS-induced MAPK/NF-κB activation and hepatoprotective activity in vitro. *Food Agric. Immunol.* **2020**, *31*, 513–532. [CrossRef]
38. Kim, H.-W.; Lee, S.-H.; Asamenew, G.; Lee, M.-K.; Lee, S.; Park, J.J.; Choi, Y.; Lee, S.H. Study on Phenolic Compounds in Lettuce Samples Cultivated from Korea Using UPLC-DAD-QToF/MS. *Korean J. Food Nutr.* **2019**, *32*, 717–729.
39. Zhou, W.; Liang, X.; Dai, P.; Chen, Y.; Zhang, Y.; Zhang, M.; Lu, L.; Jin, C.; Lin, X. Alteration of Phenolic Composition in Lettuce (*Lactuca sativa* L.) by Reducing Nitrogen Supply Enhances its Anti-Proliferative Effects on Colorectal Cancer Cells. *Int. J. Mol. Sci.* **2019**, *20*, 4205. [CrossRef]
40. Vidal, V.; Laurent, S.; Charles, F.; Sallanon, H. Fine monitoring of major phenolic compounds in lettuce and escarole leaves during storage. *J. Food Biochem.* **2018**, *43*, e12726. [CrossRef]
41. Sharma, A.; Kumar, V.; Shahzad, B.; Ramakrishnan, M.; Sidhu, G.P.S.; Bali, A.S.; Handa, N.; Kapoor, D.; Yadav, P.; Khanna, K.; et al. Photosynthetic Response of Plants Under Different Abiotic Stresses: A Review. *J. Plant Growth Regul.* **2020**, *39*, 509–531. [CrossRef]
42. Žnidarčić, D.; Ban, D.; Šircelj, H. Carotenoid and chlorophyll composition of commonly consumed leafy vegetables in Mediterranean countries. *Food Chem.* **2011**, *129*, 1164–1168. [CrossRef]
43. Derbie Assefa, A.; Choi, S.; Lee, J.-E.; Sung, J.-S.; Hur, O.-S.; Ro, N.-Y.; Lee, H.-S.; Jang, S.-W.; Rhee, J.-H. Identification and quantification of selected metabolites in differently pigmented leaves of lettuce (*Lactuca sativa* L.) cultivars harvested at mature and bolting stages. *BMC Chem.* **2019**, *13*, 56.
44. Wink, M. *Functions and Biotechnology of Plant Secondary Metabolites*; Blackwell Publishing Ltd.: West Sussex, UK, 2010.
45. Rasooli, I. *Phytochemicals—Bioactivities and Impact on Health*; InTech: Rijeka, Croatia, 2011.
46. Giusti, M.M.; Wallace, T.C. *Health-Promoting Components of Fruits and Vegetables in Human Health*; Nutrients: Basel, Switzerland, 2017.
47. Geneva, M.; Kostadinov, K.; Filipov, S.; Kirova, E.; Stancheva, I. Analysis of the antioxidant capacity of lettuce growth at different fertilizer regimes. *Agric. Sci.* **2021**, *74*, 145–154.
48. Consultant Plant Nutrition in Horticulture. Available online: https://cdnmedia.eurofins.com/corporate-eurofins/media/12142795/160825_manual_nutrient_solutions_digital_en.pdf (accessed on 3 April 2019).
49. Association of Officiating Analytical Chemists. *Official Methods of Analysis: Method 2002*, 17th ed.; AOAC International: Washington, DC, USA, 2002.
50. Ough, C.S.; Amerine, M.A. *Methods for Analysis of Musts and Wines*; John Wiley & Sons: Washington, DC, USA, 1988.
51. Holm, G. Chlorophyll mutations in barley. *Acta Agric. Scand.* **1954**, *4*, 457–471. [CrossRef]
52. Wettstein, D.V. Chlorophyll-letale und der submikroskopische Formwechsel der Plastiden. *Exp. Cell Res.* **1957**, *12*, 427–506. [CrossRef]
53. Miller, N.J.; Diplock, A.T.; Rice-Evans, C.; Davies, M.J.; Gopinathan, V.; Milner, A. A novel method for measuring antioxidant capacity and its application to monitoring the antioxidant status in premature neonates. *Clin. Sci.* **1993**, *84*, 407–412. [CrossRef] [PubMed]
54. *SAS Version 9.3*; SAS Institute Inc.: Cary, NC, USA, 2010.

plants

MDPI

Article

Chemical Composition and Biological Activities of Tunisian *Ziziphus lotus* Extracts: Evaluation of Drying Effect, Solvent Extraction, and Extracted Plant Parts

Touka Letaief [1,2,3], Stefania Garzoli [4,*], Valentina Laghezza Masci [3], Jamel Mejri [1], Manef Abderrabba [1], Antonio Tiezzi [3] and Elisa Ovidi [3]

[1] Laboratory of Materials Molecules and Applications (LMMA), IPEST, University of Carthage, BP 51, La Marsa 2070, Tunisia; touka.letaief@gmail.com (T.L.); jmejri2001@yahoo.fr (J.M.); manef.abderrabba@ipest.rnu.tn (M.A.)
[2] National Agronomic Institute of Tunisia (INAT), University of Carthage, Mahrajene 1082, Tunisia
[3] Department for the Innovation in Biological, Agrofood and Forestal Systems, Tuscia University, 01100 Viterbo, Italy; laghezzamasci@unitus.it (V.L.M.); antoniot@unitus.it (A.T.); eovidi@unitus.it (E.O.)
[4] Department of Drug Chemistry and Technology, Sapienza University, 00185 Rome, Italy
* Correspondence: stefania.garzoli@uniroma1.it

Citation: Letaief, T.; Garzoli, S.; Laghezza Masci, V.; Mejri, J.; Abderrabba, M.; Tiezzi, A.; Ovidi, E. Chemical Composition and Biological Activities of Tunisian *Ziziphus lotus* Extracts: Evaluation of Drying Effect, Solvent Extraction, and Extracted Plant Parts. *Plants* **2021**, *10*, 2651. https://doi.org/10.3390/plants10122651

Academic Editors: Milan S. Stankovic, Paula Baptista and Petronia Carillo

Received: 27 October 2021
Accepted: 28 November 2021
Published: 2 December 2021

Publisher's Note: MDPI stays neutral with regard to jurisdictional claims in published maps and institutional affiliations.

Copyright: © 2021 by the authors. Licensee MDPI, Basel, Switzerland. This article is an open access article distributed under the terms and conditions of the Creative Commons Attribution (CC BY) license (https://creativecommons.org/licenses/by/4.0/).

Abstract: The Tunisian *Ziziphus lotus* plant was investigated to determine its phytoconstituents and evaluate its biological activities. In particular, the GC/MS technique was used to describe the chemical composition of *Z. lotus* active extracts and fractions. Among the obtained extracts, the yields of the dried root methanolic extract (29.80%) and the fruit aqueous extract (48.00%) were the highest ones. The dried root methanolic extract exhibited the highest amount in the total phenolics (186.44 ± 0.26 mg GAE/g DW), total flavonoids (102.50 ± 3.53 mg QE/g DW), and tannins (60.714 ± 2.2 mg catechin/g DW). The root aqueous extracts revealed the highest antioxidant activity with an IC_{50} of 8.96 ± 0.38 mg/L and 16.46 ± 0.60 mg/L for the $ABTS^{\bullet+}$ and DPPH• assays, respectively. The total antioxidant capacity was accorded to the methanolic extract of the dried roots with a value of 304.07 ± 1.11 µg AAE/mg. The drying process was found to improve the qualitative and quantitative properties of the *Z. lotus* extracts. The evaluation of the cytotoxic activity against the SH-SY5Y cell line was carried out using MTT assay. The petroleum ether and dichloromethane extracts of the dried roots showed relevant cytotoxic activities. The thin layer chromatography and the GC-MS/GC-FID analysis led to the identification of the 13-epimanool as a potent cytotoxic compound.

Keywords: *Ziziphus lotus*; phenolics; antioxidant activity; SH-SY5Y cell line; chromatography

1. Introduction

Medicinal and aromatic plants represent an inexhaustible reservoir of secondary metabolites. These compounds do not play an essential role in plant growth, like the primary metabolites do, but rather serve to enable the plant to cope with extreme environmental conditions (drought, oxidative stress, pests, etc.), allowing better interaction between the plant and its surrounding environment [1]. Numerous classes of secondary metabolites exist according to their chemical structure, route of biosynthesis and solubility in water and organic solvents [2]. In order to explore such metabolites, an adequate extraction and isolation of the bioactive compounds, the screening of phytoconstituents, and the evaluation of their potential activities are needed [3]. Among the identified plants, only around 20% were studied in pharmaceutical and medicinal research [4], whereas many other plants are still not thoroughly explored [5]. In this context, the shrub *Z. lotus*, commonly named 'Sedra', is the subject of many current laboratory investigations. This plant belongs to the Rhamnaceae family which includes around 550 species spread over around 45 genera [6,7]. Being both a tropical and a subtropical plant, *Z. lotus* is commonly present in arid and semi-arid regions [8]. In Tunisia, this xerophytic plant can be found in the sand dunes of Saharan regions as well as in the arid and semi-arid zones where it occupies different types of soils [8,9]. *Z. lotus* is dormant from October to March and its fruits are harvested during the summer [10]. It forms clumps of a few meters in diameter and 2 to 5 m in height. Its thorny stems possess small deciduous leaves and tasty fruits called 'Nbeg' [6,8,11].

Z. lotus is a versatile shrub that is highly interesting especially for the inhabitants of dry areas [12]. It has been used in traditional medicine for the treatment of many diseases. In fact, it is used to treat diarrhea and regulate blood sugar levels [11], while its tonic febrifuge fruit is used for the dissolution of kidney stones [13]. Thanks to its emollient properties, the paste of this fruit has been used as an excellent pectoral remedy [6]; furthermore, it has been employed for the treatment of hazardous viral diseases like measles and smallpox [6].

Many findings in the research confirmed these traditional uses. Borgi et al. [14] outlined the anti-inflammatory, analgesic and antispasmodic activities of *Z. lotus* extracts. The aqueous extracts obtained from different parts of *Z. lotus* proved their effects as cytotoxic agents against T-cells, the major cause of autoimmune diseases [15]. Furthermore, *Z. lotus* extracts obtained by different solvents showed excellent antifungal activity against nine strains of pathogenic fungi [16]. Such endowments reflect the richness of *Z. lotus* in many active compounds, notably flavonoids and tannins [14], cyclopeptide alkaloids such as lotusine A and lotusine D [17], and vitamins (A, C, E) [15].

The aim of this work is to further investigate *Z. lotus* extract potencies. After the preparation of extracts from the dried and fresh parts (roots, leaves, fruits) of this plant by using different solvents, a screening of the phytoconstituents was carried out. The evaluation of the antioxidant activity was performed by $ABTS^{\bullet+}$, DPPH•, and TAC assays. In addition, the cytotoxic activity against the SH-SY5Y cell line was investigated. The extract showing a potent effect against this cancer cell line was subjected to GC-MS analysis.

2. Results

2.1. Extraction Yields

The yields of *Z. lotus* (root, leaf, and fruit) extracts obtained by different solvents were reported in Table 1. Extraction yields were between 0.50% and 48%.

Table 1. Yield of extracts of *Z. lotus* extracts.

Extracts		Extraction Yield (%)
Roots	PEE	0.50
	DE	1.30
	ME	29.80
	EE	5.58
	AE	7.07
Leaves	PEE	4.20
	DE	2.80
	ME	15.10
	EE	9.23
	AE	12.97
Fruits	ME	25.30
	AE	48.00

PEE: Petroleum Ether Extract; DE: Dichloromethane Extract; ME: Methanol Extract; EE: Ethanol Extract; AE: Aqueous Extract

2.2. Chemical Compositions of Z. lotus Extracts

Total phenolic content (TPC), total flavonoid content (TFC) and tannins of *Z. lotus* extracts were summarized in Table 2. Total phenolic content (TPC) of *Z. lotus* extracts, expressed as Gallic Acid Equivalent, ranged between 11.16 ± 0.13 and 186.44 ± 0.26 mg GAE/g DW. Among these phenolic compounds, flavonoids and tannins reached a maximum of

102.50 ± 3.53 mg QE/g DW and 60.71 ± 2.20 mg CE /g DW, respectively. These highest values were assigned to the methanolic extract of the roots.

Table 2. Chemical composition of *Z. lotus* extracts.

Samples		TPC (mg GAE/g DW)	TFC (mg QE/g DW)	Tannins (mg CE /g DW)
Roots	PEE	39.22 ± 0.62 ^{a}	19.63 ± 0.12 ^{a}	9.14 ± 0.90 ^{a}
	DE	30.15 ± 0.13 ^{a}	18.50 ± 0.88 ^{a}	6.90 ± 1.41 ^{a}
	ME	186.44 ± 0.26 ^{b}	102.50 ± 3.53 ^{b}	60.71 ± 2.20 ^{b}
	EE	100.13 ± 4.02 ^{c}	41.25 ± 0.63 ^{c}	27.70 ± 0.57 ^{b}
	AE	146.06 ± 2.50 ^{d}	36.70 ± 0.72 ^{a}	31.86 ± 1.49 ^{b}
Leaves	PEE	12.81 ± 0.10 ^{e}	3.06 ± 0.12 ^{d}	2.50 ± 0.37 ^{c}
	DE	11.16 ± 0.45 ^{e}	3.44 ± 0.06 ^{d}	3.44 ± 0.47 ^{c}
	ME	171.99 ± 1.14 ^{f}	21.35 ± 1.19 ^{e}	7.41 ± 0.68 ^{d}
	EE	41.70 ± 0.70 ^{g}	28.54 ± 1.89 ^{f}	1.66 ± 0.09 ^{d}
	AE	109.87 ± 2.07 ^{h}	17.10 ± 1.30 ^{d}	9.54 ± 0.26 ^{d}
Fruits	ME	26.12 ± 0.73 ^{i}	0.75 ± 0.13 ^{g}	1.00 ± 0.170 ^{e}
	AE	82.12 ± 1.70 ^{j}	13.40 ± 0.72 ^{h}	1.02 ± 0.10 ^{e}

TPC: Total Phenolics content; TFC: Total Flavonoids content; GAE: Gallic Acid Equivalent; QE: Quercetin Equivalent; CE: Catechin Equivalent; PEE: Petroleum Ether Extract; DE: Dichloromethane extract; ME: Methanol extract; EE: Ethanol Extract; AE: Aqueous Extract. Lowercase letters represent Tukey's Test comparison. Means within a column row with different letters were significantly different ($p < 0.05$).

2.3. *Antioxidant Activity*

The antioxidant potential of *Z. lotus* extracts was evaluated by using 2,2-azinobis (3-ethylbenzothiazoline-6-sulfonic acid) ($ABTS^{\bullet+}$), 2,2-diphenyl-1-picrylhydrazyl (DPPH•) and Total Antioxidant Capacity (TAC) methods. Table 3 reported the IC_{50} values. The $ABTS^{\bullet+}$ colorimetric assay determined IC_{50} values ranging from 8.96 ± 0.38 to 136.58 ± 0.41 mg/L for the root extracts and from 23.48 ± 0.63 to 249.37 ± 1.26 mg/L for the leaf extracts. The fruit extracts showed a low activity compared to the other part of the plants. The ME of the roots, with a concentration of 18.03 ± 0.61 mg/L, allowed a 50% inhibition of the free radical DPPH•. This methanolic extract possessed the potent total antioxidant capacity equivalent to 304.07 ± 1.11 mg ascorbic acid per mg of extract.

Table 3. Antioxidant activities (IC_{50} mg/L) of *Z. lotus* extracts.

Samples		Antioxidant Assay		
		$ABTS^{\bullet+}$ IC_{50} (mg/L)	DPPH• IC_{50} (mg/L)	TAC (mg AAE/mg Extract)
Roots	PEE	14.76 ± 0.02 ^{a}	101.06 ± 0.40 ^{a}	105.56 ± 0.37 ^{a}
	DE	136.58 ± 0.41 ^{b}	192.33 ± 0.60 ^{b}	91.11 ± 2.20 ^{a}
	ME	14.31 ± 0.13 ^{c}	18.03 ± 0.61 ^{c}	304.07 ± 1.11 ^{b}
	EE	27.42 ± 0.32 ^{d}	39.50 ± 0.49 ^{d}	167.41 ± 7.40 ^{c}
	AE	8.96 ± 0.38 ^{e}	16.46 ± 0.60 ^{e}	191.85 ± 0.00 ^{d}
Leaves	PEE	28.98 ± 0.06 ^{f}	NA	NA
	DE	29.51 ± 1.23 ^{g}	NA	154.44 ± 6.20 ^{e}
	ME	23.48 ± 0.63 ^{h}	33.66 ± 0.11 ^{f}	142.47 ± 0.85 ^{f}
	EE	249.37 ± 1.26 ^{i}	375.50 ± 1.50 ^{g}	173.09 ± 2.99 ^{g}
	AE	29.01 ± 0.44 ^{j}	64.80 ± 0.36 ^{h}	99.26 ± 4.62 ^{h}

Table 3. *Cont.*

Samples		Antioxidant Assay		
		ABTS•+ IC_{50} (mg/L)	DPPH• IC_{50} (mg/L)	TAC (mg AAE/mg Extract)
Fruits	ME	173.93 ± 0.88 ^{k}	343.00 ± 1.32 ^{i}	26.42 ± 2.26 ^{i}
	AE	342.25 ± 1.25 ^{l}	383.33 ± 0.29 ^{j}	40.74 ± 3.39 ^{j}

ABTS•+: 2, 2′-azinobis-3-ethylbenzothiazoline-6-sulfonate; DPPH•: 1, 1-diphenyl-2-picrylhydrazyl; TAC: Total Antioxidant Activity; AAE: Ascorbic Acid Equivalents; PEE: Petroleum Ether Extract; DE: Dichloromethane Extract; ME: Methanol Extract; EE: Ethanol Extract; AE: Aqueous Extract. Lowercase letters represent Tukey's Test comparison. Means within a column row with different letters were significantly different ($p < 0.05$); NA: Not Active. Ascorbic acid (IC_{50} = 2.74 ± 0.02 and 4.00 ± 0.00 mg/L for ABTS•+ and DPPH•, respectively) was used as antioxidant of reference to compare results.

2.4. Drying Effect on Phytochemical Composition and Antioxidant Activity

Quantitative evaluation of total TPC, TFC and tannins in *Z. lotus* extracts from the dried and fresh roots (DR and FR, respectively) and leaves (DL and FL, respectively) were summarized in Figure 1. Root extracts remained the highest in phytoconstituents. For all the extracting solvents, the shade-dried samples of *Z. lotus* roots presented higher values of phytochemical compounds in terms of phenolics, flavonoids, and tannins, than the fresh samples. As regards the leaves, the drying did not show a considerable effect on this chemical composition. The phosphomolybdenum assay for the total antioxidant capacity evaluation, confirmed the potent antioxidant capacity of the root extracts.

For both the ABTS•+ and DPPH• assays, the samples presenting the lowest IC_{50} values are the potent ones. Hence, the extracts of the dried roots are more efficient than the extracts of the fresh roots. The ethanolic extracts of the leaves (dried and fresh) show the lowest antioxidant activity (Figure 2).

2.5. Chemical Characterization of the Dried Root Petroleum Ether and Dichloromethane Extracts: GC-MS Analysis and Thin Layer Chromatography

The analysis of *Z. lotus* samples was carried out using a TurboMass Clarus 500 GC-MS/GC-FID. The main compound was n-hexadecanoic acid (90.6%) and Tetradecanoic acid, ethyl ester (72.8%), for the dried root petroleum ether extract (DR-PEE) and dried root dichloromethane extract (DR-DE), respectively. The compound 13-epimanool was a common compound of both petroleum ether and dichloromethane root extract fractions.

2.6. Cytotoxic Activity

The evaluation of the cytotoxic activity against the SH-SY5Y cell line was carried out by MTT assay. After 24 h of treatment, these human neuroblast cells revealed sensitive to the DR-PEE and DR-DE with an IC_{50} of 184.413 ± 4.77 and 16.148 ± 0.93. Increasing the treatment time to 48 h, improved considerably the cytotoxic potential of the extracts. The strongest activity (7.341 ± 1.98 µg/mL) was obtained by the DR-DE after 48 h of treatment (Table 4).

Table 4. MTT assay on SH-SY5Y cells. IC_{50} of *Z. lotus* DR-PEE and DR-DE extracts.

	DR-PEE		DR-DE	
	After 24 h of Treatment	After 48 h of Treatment	After 24 h of Treatment	After 48 h of Treatment
IC_{50} (µg/mL)	184.413 ± 4.77	20.941 ± 1.16	16.148 ± 0.93	7.341 ± 1.98

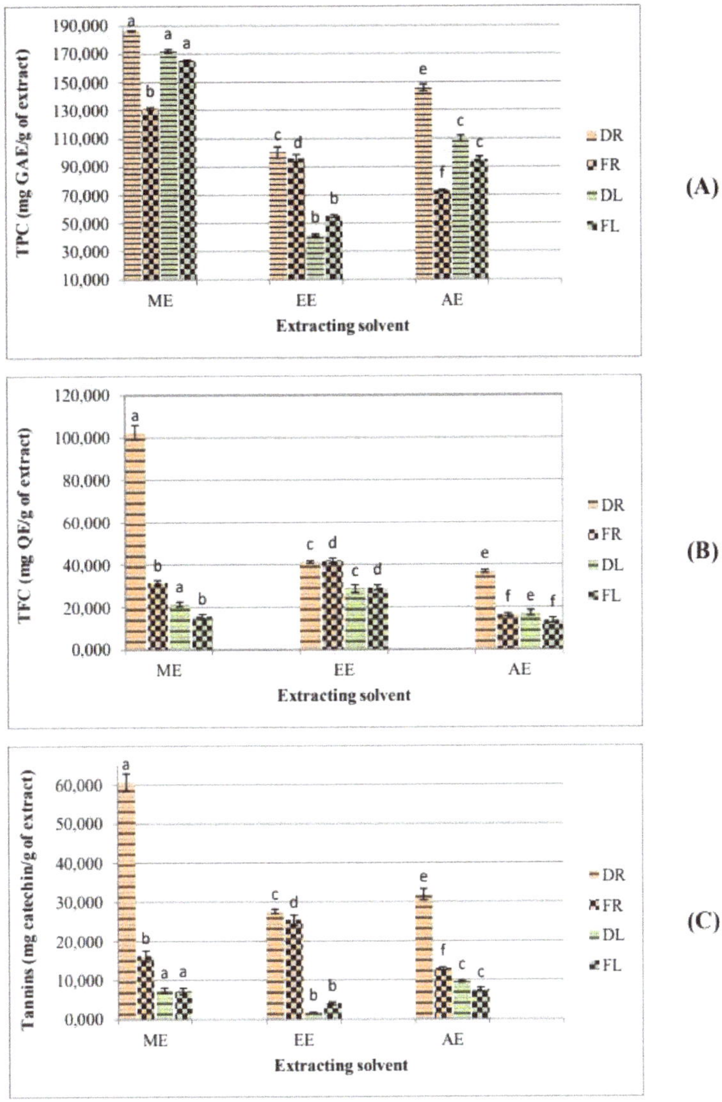

Figure 1. Drying effect on phytochemical composition of *Z. lotus* extracts. (**A**) Total phenolic content; (**B**) Total flavonoid content; (**C**) Tannin content. DR: Dried Roots; FR: Fresh Roots; DL: Dried Leaves; FL: Fresh Leaves. Histograms with the same colour marked with different letter were significantly different ($p < 0.05$).

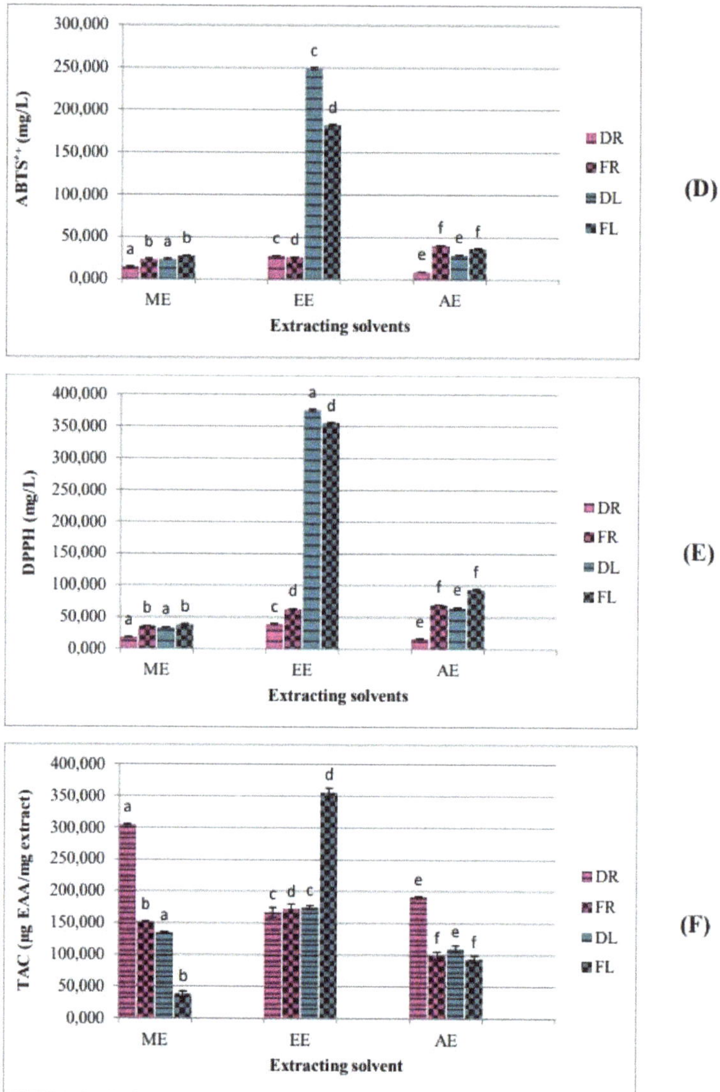

Figure 2. Drying effect on antioxidant activities of *Z. lotus* extracts. (**D**) ABTS$^{\bullet+}$ scavenging activity; (**E**) DPPH• scavenging activity; (**F**) Total antioxidant capacity. DR: Dried Roots; FR: Fresh Roots; DL: Dried Leaves; FL: Fresh Leaves. Histograms with the same color marked with different letter were significantly different ($p < 0.05$).

3. Discussion

The yield of the extraction depended on the solvent type and the extracted part of *Z. lotus*. The yield of methanolic extracts of roots and leaves resulted in 29.80% and 15.10%, respectively; and the water extract of the fruit allowed an extraction yield of 48.00%. The yields of the petroleum ether and dichloromethane extracts of the fruit were very low (data not shown). As suggested by Climati et al. [18], higher yields of the polar extracts (methanol, ethanol and aqueous) reflected the richness of *Z. lotus* samples in polar compounds.

The polar extracts of the roots and leaves showed the highest amount in phenolics ranging from 41.69 ± 0.70 mg GAE/g DW to 186.44 ± 0.26 mg GAE/g DW, which encompassed flavonoids, tannins, and other chemical compounds. This is probably due to the efficient interaction between the polar sites of the antioxidant compounds and the polar solvents (aqueous, methanolic, and ethanolic). The TPC of polar extracts was significantly ($p < 0.05$) affected by the extracting solvent. The methanolic extracts of roots and leaves presented the highest TPC amount; and the aqueous extract of the fruit was the richest one in phenolics. These TPC amounts might be affected by the interference of some other compounds that reduce the Folin-Ciocalteu reagent [19]. On the other hand, petroleum ether and dichloromethane extracts contained the lowest amounts of phytochemical compounds without a significant difference between the two extracting solvents.

The TFC values were significantly ($p < 0.05$) affected by the extracting solvent. For both roots and leaves, the highest values were assigned to the ME and EE regarding the PEE, DE, and AE. The tannin contents of roots and leaves ranged from 6.88 ± 1.41 to 60.71 ± 2.2 mg CE/g DW and from 1.66 ± 0.09 to 9.54 ± 0.26 mg CE /g DW, respectively. The ME and AE of the fruits presented the lowest tannin content with no significant difference between the two extracting solvents.

Concerning tissue phytochemical compositions as shown in Table 2, the TPC, TFC, and tannin contents varied significantly ($p< 0.05$) regarding the plant part. The root barks of *Z. lotus* presented the highest amount of phenolics (186.44 ± 0.26 mg GAE/g DW), flavonoids (102.50 ± 3.53 mg QE/g DW), and tannins (60.71 ± 2.20 mg CE/g DW); and these values were higher than those reported by Ghalem et al. [20]: TPC was 20.09 mg/g DW, TFC 0.02 mg/g, and tannins 1.56 mg/g DW, for the same species extracted with 70% acetone. In addition to the extracting solvent, other factors such as the harvesting period, the plant origin, and environmental conditions could explain this variability [21]. The leaf phenolics amount ranged from 11.16 ± 0.45 to 171.99 ± 1.14 mg GAE/g DW. The ME of the dried leaves was the richest one. These phenolic amounts were higher than those determined by Elaloui et al. [21] (21.98 mg GAE/g DW) and by Guirado et al. [9] (9.498 mg GAE/g DW). Taken all together, our findings confirmed the potent methanol capacity to extract the phytochemical compounds from *Ziziphus* leaves [21].

The flavonoid and tannin values of the leaves respectively varied between 3.06 ± 0.12 to 28.54 ± 1.89 mg QE/g DW and 1.66 ± 0.09 to 9.54 ± 0.26 mg CE/g DW. As demonstrated by Maraghni et al. [8], the plant endures extreme environmental conditions by the synthesis of flavonoids and tannins. This could clearly explain the richness in secondary metabolites of our samples, collected from a desert region in southern Tunisia.

The AE of the fruit was rich in phenolics (82.12 ± 1.70 mg GAE/g DW), flavonoids (13.40 ± 0.72 mg QE/g DW), and tannins (1.02 ± 0.10 mg CE/g DW). Samples investigated by Khouchlaa et al. [13] showed higher values of these bioactive compounds (phenolics 285.19 mg GAE/mg, flavonoids 2.66 mg QE/mg).

IC_{50} values obtained by $ABTS^{\bullet+}$ assay were significantly ($p < 0.05$) affected by the solvent used as well as the plant part. For almost all the extracting solvents, *Z. lotus* root extracts showed higher antioxidant activity than the leaves and fruits. The most active samples were AE (8.96 ± 0.38 mg/L) followed by the ME (14.31±0.13 mg/L) and the PEE (14.76 ± 0.02 mg/L) of the roots. For the leaves, the ME showed the best IC_{50} value (23.48 ± 0.63 mg/L). This IC_{50} value was higher than the value (3.82 mg/L) determined by Abderrahim et al. [22], studying the Algerian *Z. lotus* methanolic extract of the leaves. This variation could be due to the plant origin and the extracting method. In fact, in our study the leaves were successively extracted with solvents of increasing polarity, whereas in the mentioned study the ME was obtained by a direct extraction of the leaves by methanol.

Confirming the $ABTS^{\bullet+}$ assay, the highest DPPH• antioxidant levels were detected in root extracts. The extraction solvent and the plant part significantly ($p < 0.05$) influenced DPPH• IC_{50} values. Both the AE and the ME of this plant part were ranked as potent extracts with an IC_{50} of 16.46 ± 0.60 mg/L and 18.03 ± 0.61 mg/L, respectively. The ME of the leaves expressed the highest value among this plant organ extracts (33.66 ± 0.11 mg/L), almost as high as the one (28.19 mg/L) reported by Abderrahim et al. [22]

In both ABTS$^{•+}$ and DDPH• assays, the polar extracts (ME, EE, AE) showed higher quenching behaviour than the non-polar extracts (PEE, DE). This could be due to the richness of polar extracts in phenolics. The ABTS$^{•+}$ and DPPH reducing ability followed the same trend for root and leaf extracts (AE > ME > EE) and (ME > AE > EE), respectively. However, in the DPPH• assay IC_{50} values were higher; hence, the scavenging activity of the samples for this radical is lower. These results agree with many previous studies such as the investigations on the aqueous, acetone, and ethanol extracts of *Ziziphus mucronata* Willd. subsp.*mucronata* Willd [23]. Consequently, *Z. lotus* extracts react better with the ABTS$^{•+}$ assay which is based on rapid electron transfer reactions.

Z. lotus total antioxidant capacities (TAC) expressed in Equivalent of Ascorbic Acid (AAE/mg) showed a significant ($p < 0.05$) variability according to the solvent and tissue. Extracts from polar solvents as the ME and AE of the roots and the EE of leaves showed the highest total antioxidant capacity with values of 304.07 ± 1.11 mg AAE/mg, 191.85 ± 0.00 mg AAE/mg, and 173.09 ± 2.99 mg AAE/mg, respectively. Nevertheless, *Z. lotus* fruit extracts showed the lowest activity. Consequently, root and leaf extracts were significantly richer in phytochemicals and more active than the fruit extracts.

Removing water from the plant preserves the samples from deterioration and limits microbial multiplication [24]; however, it is crucial to determine possible effects of this drying process on the phytochemical compositions. The shade drying preserved the phytochemical composition of the extracts. In fact, for all the extracting solvents the dried roots showed significantly ($p < 0.05$) higher values of phenolic, flavonoids, and tannins than the fresh ones. As explained by Esparza-Martínez et al. [25], the drying process can ameliorate the cellular structure degradation leading to a better release of phenolic compounds. Nevertheless, for the leaves, no significant difference was observed in the phytochemical composition between the dried and fresh samples: the total phenolic composition might vary after a drying process depending on the plant tissue as well as the phenolic compound location in the cell [26].

Comparing the antioxidant activity of the dried samples to the fresh ones (Figure 2), the drying process significantly improved antioxidant capacity. Olufunmilayo et al. [27] attributed this difference to the richness of the dried samples in respect to fresh samples.

Exploring the chemical composition of the DR-DE extract, three compounds were identified (Table 5). The major compound was the tetradecanoic acid, ethyl ester (72.8%) belonging to the fatty acid ester class, known for its high hydrophobicity, and considered as a relatively neutral molecule used as a flavoring agent. This compound was previously identified in the fruit *n*-hexane fraction of *Z. lotus* [28]. The second major compound was the 13-epimanool (20.5%), a labdane-type diterpene. None of these compounds was previously identified in *Z. lotus* extracts or previously showed a cytotoxic potential.

Table 5. Chemical composition of *Z. lotus* active extracts and fractions.

Component	LRI [1]	LRI [2]	DR-PEE (%)	DR-DE (%)	[A] Fraction (%)	[B] Fraction (%)
Ethyl tridecanoate	1944	1943	0.2	6.7	-	14.9
2-pentadecanone	2026	2028	0.8	-	-	-
Tetradecanoic acid, ethyl ester	2055	2059	-	72.8	-	-
Pentadecanoic acid, ethyl ester	2178	2179	2.2	-	-	-
13-epimanool	2670	2676 *	0.8	20.5	100.0	85.1
Tetradecanoic acid	2680	2679	5.4	-	-	-
n-hexadecanoic acid	2943	2946	90.6	-	-	-
Total			100.0	100.0	100.0	100.0

[1] Linear Retention indices measured on polar column; [2] Linear Retention indices from literature; * Normal alkane RI; [A] DR-PEE active fraction, [B] DR-DE active fraction; DR-PEE: Dried Root-Petroleum Ether Extract; DR-DE: Dried Root-Dichloromethane extract.

The GC-MS/GC-FID analysis of DR-PEE (Table 5), revealed 6 compounds and palmitic acid (*n*-hexadecanoic acid), identified as the major compound (90.6%) of this extract, was reported as a potent cytotoxic agent against HCT-116 cells [29] and the human leukemic cells [30]. It has been also proved to be an apoptotic cell death inducer in human leukemic cell line MOLT-4. Hence, palmitic acid was suggested as an effective composite of anticancer remedies [30]. Additionally, 13-epimanool (0.8%) was detected as the common compound between the two active extracts, DR-DE and DR-PEE.

DR-PEE and DR-DE extracts were subjected to fractionation using the Thin Layer Chromatography (TLC) technique. Nine spots were isolated from the DR-PEE extract and six spots from the DR-DE, respectively named from the bottom E0 to E8 and D0 to D5.

Each isolated fraction was tested to evaluate its cytotoxic activity for 24 h. The spots named E6 (EC_{50} = 28.378 ± 0.47 μg/mL) and D4 (EC_{50} = 29.076 ± 1.39 μg/mL) were selected as the active fractions from the DR-PEE and DR-DE, respectively.

GC-MS/GC-FID analysis of the spots identified a unique compound 13-epimanool (100.0%) from the DR-PEE and two compounds from the DR-DE, 13-epimanool (85.1%) and ethyl tridecanoate (14.9%). Hence, 13-epimanool can be suggested to be responsible for the observed cytotoxic activity. Comparing the extracts to their active fractions, the activity of E6 (100% 13-epimanool) was approximately six times higher than the original extract DR-PEE. Consequently, 13-epimanool can be considered for the first time, to the best of our knowledge, as a potent cytotoxic compound expressing an affective antiproliferative activity. This gives the perspective to a deep study of this compound, extending from its isolation and purification to the determination of its mechanism of action.

Z. lotus extracts were tested to verify a possible cytotoxic activity on the SH-SY5Y human cell line and MTT assays were carried out. According to EC_{50} values on SH-SY5Y after 24 and 48 h of treatments, the dried root extracts of petroleum ether (DR-PEE) and dichloromethane (DR-DE) were selected and subsequent investigations were pursued. As reported by Rached et al. [31], the root barks are always expressing a potent cytotoxic effect. Consistent cytotoxic activities were also observed by testing the root bark extracts on hepatocellular HepG2 (48.3 μg/mL), breast MCF-7 (74 μg/mL) and cervical HeLa (69 μg/mL) [31].

The evaluation of the antiproliferative capacities of both extracts on neuroblastoma SH-SY5Y cells were reported in Table 5.

SH-SY5Y cells were more sensitive to DR-DE extract treatment in respect to the DR-PEE extract. In fact, after 24 h of treatment, the DR-PEE exhibited an EC_{50} of 16.148 ± 0.93 μg/mL regarding an EC_{50} of 184.413 ± 4.77 μg/mL for the DR-PEE. A comparative study investigating the *Neurolaena lobata* extracts, obtained by the increasing polarity extraction, reported the dichloromethane extract as the strongest antiproliferative agent against the human and murine anaplastic large cell lymphoma cell lines [32].

Furthermore, the evaluated activity of both samples showed time-dependent results since increasing the period of treatment to 48 h; EC_{50} values of 7.341 ± 1.98 μg/mL for DR-DE and 20.941 ± 1.16 μg/mL for DR-PEE were observed.

4. Materials and Methods

4.1. Preparation of Samples and Extracts

Z. lotus samples (leaves, fruits and roots) were collected in July 2017 from Oudhref-Gabes Region (South of Tunisia) (Figure 3). A part of the samples was cleaned and stored at 4 °C until use and the other part was shade-dried for two weeks at room temperature and stored in the absence of light and under dry conditions until use.

Figure 3. Different parts of *Ziziphus lotus*.

For both fresh and dried plant material organ (leaves, fruits, and roots), two extraction methods were processed: (1) Powdered plant material was successively extracted with increasing polarity solvents, namely petroleum ether, dichloromethane, and methanol. Starting with petroleum ether, 10 g of powder was macerated 3 times in 100 mL of the solvent for 30 min under constant agitation at room temperature. After each maceration, the obtained mixture was filtered using a filter paper. Then, the obtained powder was extracted by the subsequent solvent using the same protocol as for petroleum ether. (2) Plant material was extracted independently with two green solvents, namely ethanol and water. Next, 10 g of the powder was macerated in 300 mL of solvent during 90 min under constant agitation at room temperature. Then samples were filtrated and evaporated by a rotary vacuum evaporator at 35 °C to remove the solvent (Figure 4).

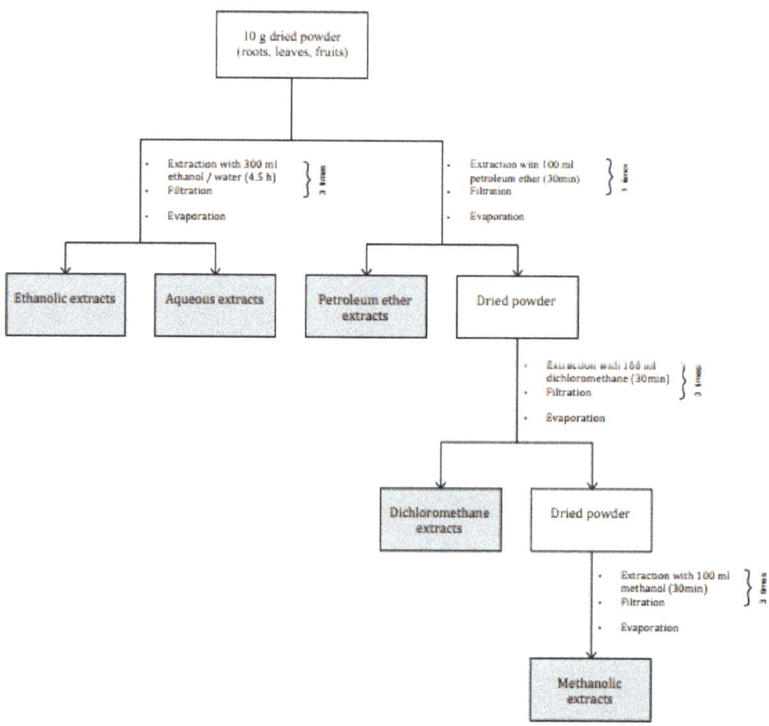

Figure 4. Preparation process of *Z. lotus* extracts.

The extraction yields were calculated:

$$\text{Yield percentage } (\%) = \frac{w}{W} \times 100 \quad (1)$$

w: the weight of residue in grams;
W: the weight of dried plant material in grams.

4.2. Total Phenolic Content

To determine phenolic content the Folin-Ciocalteu method was used [26]. A total of 100 µL of the diluted extract in DMSO was mixed with 500 µL of Folin Ciocalteu reagent (0.2 N). The mixture was kept in obscurity for 5 min at room temperature. Then, 400 µL of sodium carbonate solution (75 g/L, in water) was added; and after 30 min of incubation in the darkness, the absorbance was measured at 765 nm.

To plot a calibration curve, the gallic acid was used as standard and the amounts of phenolic content were expressed in milligrams of Gallic Acid Equivalents (GAE) per gram of Dry Weight (mg GAE/g DW). The measurements were carried out in triplicate.

4.3. Total Flavonoid Content

The evaluation of the flavonoid content was performed as outlined by Yahyaoui et al. [32] with some modifications. In brief, 250 µL of the solubilized extract was diluted in 2 mL of distilled water and mixed with 75 µL of a 15% sodium nitrite solution ($NaNO_2$). After 6 min, 75 µL of 10% aluminum chloride solution ($AlCl_3$) was added. After 6 min, 2 mL of sodium hydroxide solution (NaOH) (4%) was supplemented with a 100 µL of distilled water. After 15 min of incubation, the absorbance was measured at 510 nm.

The results were represented as milligrams of Quercetin Equivalent per gram of Dry Weight (mg QE/g DW) from the calibration relationship. The measurements were realized in triplicate.

4.4. Tannin Content

Tannins were estimated referring to the method cited by Ghazouani et al. [26], adapting some modifications. In an ice bath, 350 µL of each sample was mixed with 700 µL of vanillin (1% in 7 M H_2SO_4). After an incubation period of 15 min at 25 °C, the absorbance was measured at 500 nm. Results were expressed as mg of catechin equivalent per gram dry weight.

4.5. DPPH• Scavenging Activity

The antioxidant scavenging activity of *Z. lotus* extracts was evaluated using the 1,1-diphenyl-2-picrylhydrazyl (DPPH•) free radical method [26]. From each sample various dilutions were prepared, then a volume of 100 µL was mixed with 900 µL of freshly prepared methanolic DPPH• solution. After an incubation duration of 30 min in the darkness, the absorbance was measured at 520 nm using an UV-vis spectrophotometer.

Thus, the free radical-scavenging activity was expressed as the inhibition percentage defined as follow:

$$\% \text{ inhibition} = 100 \times (A(\text{blank}) - A(\text{sample}))/A(\text{blank}) \quad (2)$$

A (blank): the absorbance of the prepared DPPH• solution without the sample extract;
A (sample): the absorbance of the sample after the reaction with the DPPH• solution.

The DPPH• radical scavenging activity was expressed as the IC_{50} (mg/L), defined as the concentration of the test material able to reduce 50% of the initial DPPH• solution concentration. The IC_{50s} were graphically calculated using the linear regressions of the plotted lines; from the curves: %inhibition = f(treated extract concentrations).

Ascorbic acid was used as a standard. Each measurement was performed in triplicate.

4.6. ABTS•+ Scavenging Activity

The ABTS•+ (2,2′-azinobis-3-ethylbenzothiazoline- 6-sulphonate) scavenging activity of *Z. lotus* samples was assessed by referring to the experimental protocol described by Yahyaoui et al. [33]. Starting through the preparation of the 2, 2′-Azinobis-3-ethylbenzothiazoline-6-sulfonate by mixing the ABTS•+ aqueous solution (7 mM) with the potassium persulfate (2.5 mM) dissolved in water in the ratio of 1:1. The solution was used after 16 h of agitation in the darkness at room temperature. An amount of 100 μL of the sample was added to 900 μL of the diluted ABTS•+ solution. The absorbance was read at 734 nm after 6 min of incubation in the dark. The ABTS•+ radical scavenging activity was represented by the IC_{50} values (mg/L) estimated as the concentration expected to scavenge 50% of ABTS•+ radicals. The capacity of free radical scavenging IC_{50} was determined using the equation above using the DPPH• method. Measurements were performed in triplicate.

4.7. Total Antioxidant Capacity (TAC)

To determine the total antioxidant capacity of the samples, the phosphomolybdenum assay was processed [34]. The mixture of 0.1 mL of extract (0.5 mg/mL) and 1 mL of the reagent solution (0.6 M sulphuric acid, 28 mM sodium phosphate and 4 mM ammonium molybdate) was incubated for 90 min at 95 °C. The absorbance was measured at 765 nm. Results are expressed as milligram of ascorbic acid equivalent per gram of dry weight.

4.8. Thin Layer Chromatography (TLC)

TLC was conducted in a silica gel plate (20 cm × 20 cm, ICN Adsorbentien) and spots consisting of 80 μL of extract were processed [18]. The mobile phase was a mixture of petroleum ether-ethanol (7:3) and the run was proceeded for 3 h. At the end of the run, the plate was dried, the separated spots were viewed by using UV light and their outline marked with a pencil. Spots of the same line were scraped from the plate and the collected powders were extracted with ethanol for 30 min. After 5 min of centrifugation at 2500 rpm, the pellets were discarded and the supernatants concentrated in an evaporator.

4.9. GC-MS/GC-FID Analysis

Chemical analysis of *Z. lotus* samples was performed by a TurboMass Clarus 500 GC-MS/GC-FID from Perkin Elmer instruments (Waltham, MA, USA) equipped with a Stabilwax fused-silica capillary column (Restek, Bellefonte, PA, USA) (60 m × 0.25 mm, 0.25 μm film thickness). The operating conditions used were as follows: GC oven temperature was kept at 60 °C for 5 min and programmed to 220 °C at a rate of 5 °C/min, and kept constant at 220 °C for 25 min. Helium was used as a carrier gas at a flow rate of 1 mL/min. Solvent delay was 0–2 min and scan time was 0.2 s. Mass range was from 30 to 350 m/z using electron-impact at 70 eV mode. An amount of ~2 μg of each *Z. lotus* extract was diluted in 1 mL of methanol and 1 μL of the solution was injected into the GC injector at a temperature of 280 °C. The analysis was repeated twice. Relative percentages for quantification of the components were calculated by electronic integration of the GC-FID peak areas. The identification of the constituents was made by comparing the obtained mass spectra for each component with those reported in mass spectra Nist and Willey libraries. Linear retention indices (LRI) of each compound were calculated using a mixture of aliphatic hydrocarbons (C8–C30, Ultrasci) injected directly into GC injector at the same temperature program reported above.

4.10. Cytotoxic Activity

4.10.1. Cell Cultures

The evaluation of the cytotoxic activity was carried out using the stabilized human neuroblast cells (SH-SY5Y, ATCC® CRL-2266™) purchased from the American Type Culture Collection (ATCC-CRL2266). Cells were cultured in 75 cm^2 flasks in DMEM-F12 (Dulbecco's Modified Eagle's Medium: nutrient mixture F-12) culture medium supplemented with 10% fetal bovine serum (FBS), 1% glutamine, and 1% penicillin/streptomycin in a humidified incubator at 37 °C and under 5% CO_2. When the cells reached confluence, they were transferred into new flasks at a ratio of 1:20. Every four days, the medium was changed.

4.10.2. MTT

The MTT [3-(4,5-dimethyl-2-thiazolyl)-2,5-diphenyl-2H-tetrazolium bromide] assay was adopted to evaluate the cytotoxic activity of *Z. lotus* extracts [35]. Cells were seeded (2×10^5 cells/mL) in a 96 wells plate and incubated for 24 h before being treated with extracts diluted in DMSO. Treatment of the SH-SY5Y cells were carried out with different dilutions of samples ranging from 200 µg/mL to 0.39 µg/mL. Vinblastine sulfate (Merck KGaA, Darmstadt, Germany) and DMSO were used as positive and solvent controls, respectively. After the required incubation time, treatments were removed and a fresh medium with 0.5 mg/mL MTT was added. Plates were incubated for 4 h at 37 °C. The generated formazan crystals were solubilized by adding DMSO and absorbance was measured at 590 nm using a microplate reader (SunRise, TECAN, Inc, Boston, MA, USA). The percentage of viability was calculated as below:

$$\%\text{viable cells} = 100 \times [1 - (A(treated)/A(control))] \tag{3}$$

A(treated): absorbance mean of the treated cells
A(control): absorbance mean of the untreated cells

The EC50 was determined by using the linear regressions of the plotted curves describing the relation: %survived cells = f(extract concentrations).

Each measurement was performed in triplicate.

4.11. *Statistical Analysis*

Triplicate analyses were carried out to present results in means ± standard deviation. To compare the means, one way and two ways ANOVA was used. The statistical significance level was set up at $p < 0.05$. SPSS Statistics 28 software (IBM Corp., Armonk, NY, USA) was used to analyze the data.

5. Conclusions

Organic extracts prepared from the different parts of *Z. lotus* (leaves, fruit, roots) using solvents with different polarities were the subject of a phytochemical screening. The methanolic extract of the dried roots was the richest one in phytoconstituents, reflecting potent antioxidant activity. The petroleum ether and dichloromethane extracts of the dried roots showed notable cytotoxic activity against the SH-SY5Y cell line. The fractionation and the identification of these nonpolar extracts showed their richness in the 13-epimanool compound. Such findings justify the traditional use of this plant and encourage the investigation of *Z. lotus* in further studies.

Author Contributions: Conceptualization, T.L., J.M., M.A., S.G.; investigation, T.L., A.T., V.L.M., E.O., S.G.; data curation, T.L., S.G.; writing—original draft preparation, T.L.; writing—review and editing, T.L., A.T., V.L.M., E.O., J.M., M.A., S.G.; funding acquisition, T.L., M.A., S.G. All authors have read and agreed to the published version of the manuscript.

Funding: This research received no external funding.

Data Availability Statement: All generated data are included in this article.

Acknowledgments: The author T.L is thankful to Ali Letaief for the harvesting of *Z. lotus* samples.

Conflicts of Interest: The authors declare no conflict of interest.

References

1. Jain, C.; Khatana, S. Rekha Vijayvergia Bioactivity of secondary metabolites of various plants: A review. *Int. J. Pharm. Sci. Res.* **2019**, *10*, 494–504. [CrossRef]
2. Mera, I.F.G.; Falconí, D.E.G.; Córdova, V.M. Secondary metabolites in plants: Main classes, phytochemical analysis and pharmacological activities. *Rev. Bionatura* **2019**, *4*, 1000–1009. [CrossRef]
3. Abdullahi, R. Abubakar and Mainul Haque Preparation of Medicinal Plants: Basic Extraction and Fractionation Procedures for Experimental Purposes. *J. Pharm. Bioallied Sci.* **2020**, *12*, 1–10.
4. Altemimi, A.; Lakhssassi, N.; Baharlouei, A.; Watson, D.G.; Lightfoot, D.A. Phytochemicals: Extraction, isolation, and identification of bioactive compounds from plant extracts. *Plants* **2017**, *6*, 42. [CrossRef] [PubMed]
5. Yonbawi, A.R.; Abdallah, H.M.; Alkhilaiwi, F.A.; Koshak, A.E.; Heard, C.M. Anti-Proliferative, Cytotoxic and Antioxidant Properties of the Methanolic Extracts of Five Saudi Arabian Flora with Folkloric Medicinal Use: Aizoon canariense, Citrullus colocynthis, Maerua crassifolia, Rhazya stricta and Tribulus macropterus. *Plants* **2021**, *10*, 2073. [CrossRef] [PubMed]
6. Ghedira, K. *Zizyphus lotus* (L.) Desf. (Rhamnaceae): Jujubier sauvage. *Phytotherapie* **2013**, *11*, 149–153. [CrossRef]
7. San, B.; Yildirim, A.N.; Polat, M.; Yildirim, F. Mineral composition of leaves and fruits of some promising jujube (zizyphus jujuba miller) genotypes. *Asian J. Chem.* **2009**, *21*, 2898–2902.
8. Maraghni, M.; Gorai, M.; Neffati, M. The influence of water-deficit stress on growth, water relations and solute accumulation in wild Jujube. *J. Ornam. Hortic. Plants* **2011**, *1*, 63–72.
9. Guirado, E.; Tabik, S.; Alcaraz-Segura, D.; Cabello, J.; Herrera, F. Deep-learning Versus OBIA for scattered shrub detection with Google Earth Imagery: Ziziphus lotus as case study. *Remote Sens.* **2017**, *9*, 1220. [CrossRef]
10. Maraghni, M.; Gorai, M.; Neffati, M. Seed germination at different temperatures and water stress levels, and seedling emergence from different depths of Ziziphus lotus. *S. Afr. J. Bot.* **2010**, *76*, 453–459. [CrossRef]
11. Hammi, K.M.; Jdey, A.; Abdelly, C.; Majdoub, H.; Ksouri, R. Optimization of ultrasound-assisted extraction of antioxidant compounds from Tunisian *Zizyphus lotus* fruits using response surface methodology. *Food Chem.* **2015**, *184*, 80–89. [CrossRef] [PubMed]
12. Boussaid, M.; Taïbi, K.; Ait Abderrahim, L.; Ennajah, A. Genetic diversity of Ziziphus lotus natural populations from Algeria based on fruit morphological markers. *Arid L. Res. Manag.* **2018**, *32*, 184–197. [CrossRef]
13. Khouchlaa, A.; Talbaoui, A.; El Yahyaoui El Idrissi, A.; Bouyahya, A.; Ait Lahsen, S.; Kahouadji, A.; Tijane, M. Determination of phenol content and evaluation of in vitro litholytic effects on urolithiasis of moroccan *Zizyphus lotus* L. extract. *Phytotherapie* **2020**, *16*, s1. [CrossRef]
14. Borgi, W.; Bouraoui, A.; Chouchane, N. Antiulcerogenic activity of *Zizyphus lotus* (L.) extracts. *J. Ethnopharmacol.* **2007**, *112*, 228–231. [CrossRef]
15. Benammar, C.; Hichami, A.; Yessoufou, A.; Simonin, A.M.; Belarbi, M.; Allali, H.; Khan, N.A. *Zizyphus lotus* L. (Desf.) modulates antioxidant activity and human T-cell proliferation. *BMC Complement. Altern. Med.* **2010**, *10*, 54. [CrossRef] [PubMed]
16. Lahlou, M.; Mahi, J.H. Évaluation des activités antifongique et molluscicide de *Zizyphus lotus* (L.) Desf. du Maroc. *Ann. Pharm. françaises* **2002**, *60*, 410–414.
17. Ghedira, K.; Chemli, R.; Richard, B.; Nuzillard, J.M.; Zeches, M.; Men-Olivier, L. Le Two cyclopeptide alkaloids from *Zizyphus lotus*. *Phytochemistry* **1993**, *32*, 1591–1594. [CrossRef]
18. Climati, E.; Mastrogiovanni, F.; Valeri, M.; Salvini, L.; Bonechi, C.; Mamadalieva, N.Z.; Egamberdieva, D.; Taddei, A.R.; Tiezzi, A. Methyl carnosate, an antibacterial diterpene isolated from salvia officinalis leaves. *Nat. Prod. Commun.* **2013**, *8*, 429–430. [CrossRef] [PubMed]
19. Sánchez-Rangel, J.C.; Benavides, J.; Heredia, J.B.; Cisneros-Zevallos, L.; Jacobo-Velázquez, D.A. The Folin-Ciocalteu assay revisited: Improvement of its specificity for total phenolic content determination. *Anal. Methods* **2013**, *5*, 5990–5999. [CrossRef]
20. Ghalem, M.; Merghache, S.; Belarbi, M. Study on the antioxidant activities of root extracts of *Zizyphus lotus* from the western region of Algeria. *Pharmacogn. J.* **2014**, *6*, 32–42. [CrossRef]
21. Elaloui, M.; Ennajah, A.; Ghazghazi, H.; Ben Youssef, I.; Ben Othman, N.; Hajlaoui, M.R.; Khouja, A.; Laamouri, A. Quantification of total phenols, flavonoides and tannins from Ziziphus jujuba (mill.) and Ziziphus lotus (l.) (Desf). Leaf extracts and their effects on antioxidant and antibacterial activities. *Int. J. Second. Metab.* **2016**, *4*, 18. [CrossRef]
22. Abderrahim, B.; Harrar, A.; Gul, F.; Demirtas, I. Phenolic Compounds, Antioxidant and Antibacterial Activities of *Zizyphus lotus* L. Leaves Extracts. *Nat. Prod. J.* **2017**, *7*, 316–322.
23. Olajuyigbe, O.O.; Afolayan, A.J. Phenolic content and antioxidant property of the bark extracts of Ziziphus mucronata Willd. subsp. mucronata Willd. *BMC Complement. Altern. Med.* **2011**, *11*, 130. [CrossRef] [PubMed]
24. Buchaillot, A.; Caffin, N.; Bhandari, B. Drying of lemon myrtle (Backhousia citriodora) leaves: Retention of volatiles and color. *Dry. Technol.* **2009**, *27*, 445–450. [CrossRef]
25. Esparza-Martínez, F.J.; Miranda-López, R.; Guzman-Maldonado, S.H. Effect of air-drying temperature on extractable and non-extractable phenolics and antioxidant capacity of lime wastes. *Ind. Crops Prod.* **2016**, *84*, 1–6. [CrossRef]
26. Ghazouani, N.; Abderrabba, M.; Bouajila, J. Teucrium ramosissimum (Lamiaceae): Volatile Composition, Seasonal Variation, and Pharmaceutical Activity. *Anal. Lett.* **2016**, *49*, 1258–1271. [CrossRef]
27. Olufunmilayo, D.A.; Oyetola, O.; Shaid, R.O. Phytochemical Analysis and Antioxidant Activities of Dry and Fresh Leaves of Petivera alliacea and Ocimum. *Int. J. Sci. Basic Appl. Res.* **2015**, *24*, 1–13.

28. El Cadi, H.; Bouzidi, H.E.L.; Selama, G.; El Cadi, A.; Ramdan, B.; Oulad, Y.; Majdoub, E.; Alibrando, F.; Dugo, P.; Mondello, L.; et al. Physico-Chemical and Phytochemical Characterization of Moroccan Wild Jujube "*Zizyphus lotus* (L.)" Fruit Crude Extract and Fractions. *Molecules* **2020**, *25*, 5237. [CrossRef]
29. Ravi, L.; Krishnan, K. Cytotoxic Potential of N-hexadecanoic Acid Extracted from Kigelia pinnata Leaves. *Asian J. Cell Biol.* **2016**, *12*, 20–27. [CrossRef]
30. Harada, H.; Yamashita, U.; Kurihara, H.; Fukushi, E.; Kawabata, J.; Kamei, Y. Antitumor activity of palmitic acid found as a selective cytotoxic substance in a marine red alga. *Anticancer Res.* **2002**, *22*, 2587–2590.
31. Rached, W.; Barros, L.; Ziani, B.E.C.; Bennaceur, M.; Calhelha, R.C.; Heleno, S.A.; Alves, M.J.; Marouf, A.; Ferreira, I.C.F.R. HPLC-DAD-ESI-MS/MS screening of phytochemical compounds and the bioactive properties of different plant parts of: *Zizyphus lotus* (L.) Desf. *Food Funct.* **2019**, *10*, 5898–5909. [CrossRef]
32. Christine, U.; Ruxandra, P.; Benedikt, G.; Daniela, L.; Susanne, H.; Mareike, S.; Rene, D.; Bruno, W.; Gerda, E.; Melanie, H.; et al. The dichloromethane extract of the ethnomedicinal plant Neurolaena lobata inhibits NPM/ALK expression which is causal for anaplastic large cell lymphomagenesis. *Int. J. Oncol.* **2013**, *42*, 338–348.
33. Yahyaoui, M.; Ghazouani, N.; Saoudi, S.; Sifaoui, I.; Chammem, N.; Abderrabba, M. Experimental design methodologyapplication in the optimization of phytochemical compoundsextraction and antioxidant activity of Thymelaea hirsuta L. extracts. *J. Mater. Environ. Sci.* **2018**, *9*, 1551–1561. [CrossRef]
34. Tlili, N.; Mejri, H.; Yahia, Y.; Saadaoui, E.; Rejeb, S.; Khaldi, A.; Nasri, N. Phytochemicals and antioxidant activities of Rhus tripartitum (Ucria) fruits depending on locality and different stages of maturity. *Food Chem.* **2014**, *160*, 98–103. [CrossRef] [PubMed]
35. Ovidi, E.; Garzoli, S.; Laghezza Masci, V.; Turchetti, G.; Tiezzi, A. GC-MS investigation and antiproliferative activities of extracts from male and female flowers of Schinus molle L. *Nat. Prod. Res.* **2021**, *35*, 1923–1927. [CrossRef] [PubMed]

plants

MDPI

Article

Chemical Compounds, Antitumor and Antimicrobial Activities of Dry Ethanol Extracts from *Koelreuteria paniculata* Laxm

Tsvetelina Andonova [1,†], Yordan Muhovski [2,*,†], Hafize Fidan [3,†], Iliya Slavov [4], Albena Stoyanova [5] and Ivanka Dimitrova-Dyulgerova [1]

[1] Department of Botany and Methods of Biology Teaching, Faculty of Biology, University of Plovdiv, Paisii Hilendarski, 4000 Plovdiv, Bulgaria; ts_andonova@uni-plovdiv.bg (T.A.); ivadim@uni-plovdiv.bg (I.D.-D.)
[2] Life Sciences Department, Walloon Agricultural Research Centre, 5030 Gembloux, Belgium
[3] Department of Tourism and Culinary Management, Faculty of Economics, University of Food Technologies, 4000 Plovdiv, Bulgaria; hfidan@abv.bg
[4] Department of Biology, Faculty of Pharmacy, Medical University of Varna, 9000 Varna, Bulgaria; ijelev80@abv.bg
[5] Department of Tobacco, Sugar, Vegetable and Essential Oils, Faculty of Technology, University of Food Technologies, 4000 Plovdiv, Bulgaria; aastst@abv.bg
* Correspondence: y.muhovski@cra.wallonie.be
† Those authors participated equally to the manuscript.

Citation: Andonova, T.; Muhovski, Y.; Fidan, H.; Slavov, I.; Stoyanova, A.; Dimitrova-Dyulgerova, I. Chemical Compounds, Antitumor and Antimicrobial Activities of Dry Ethanol Extracts from *Koelreuteria paniculata* Laxm. *Plants* **2021**, *10*, 2715. https://doi.org/10.3390/plants10122715

Academic Editors: Milan S. Stankovic, Paula Baptista and Petronia Carillo

Received: 18 November 2021
Accepted: 8 December 2021
Published: 10 December 2021

Publisher's Note: MDPI stays neutral with regard to jurisdictional claims in published maps and institutional affiliations.

Copyright: © 2021 by the authors. Licensee MDPI, Basel, Switzerland. This article is an open access article distributed under the terms and conditions of the Creative Commons Attribution (CC BY) license (https://creativecommons.org/licenses/by/4.0/).

Abstract: *Koelreuteria paniculata* Laxm. is used in traditional medicine and has various established biological activities, however, the species is considered to be a potentially invasive alien tree species for Bulgarian flora. However, there is still much to be studied about the phytochemical and biological characteristics of the species. The present study aimed to determine the chemical composition of the ethanol extracts of aerial plant parts, by GC-MS analysis, and to thereby evaluate their in vitro antitumor and antibacterial properties. All three extracts were tested against the HT-29 and PC3 tumor cell lines using the MTT assay. Fifty-six components were identified from leaf, flower, and stem bark extracts, and over 10% were the following constituents: pyrogallol, α-terpinyl acetate, neryl acetate, and α-terpinyl isobutanoate. The oxygenated monoterpenes predominated in the extracts, followed by the oxygenated aliphatics and phenylpropanoids. Significant antiproliferative activity on the HT-29 cell line (IC_{50}–21.44 μg/mL and 23.63 μg/mL, respectively) was found for the flower and leaf extracts. Antibacterial activity was established for the following bacteria strains: *Bacillus subtilis* ATCC 6633, *Bacillus cereus* NCTC 10320, *Escherichia coli* ATCC 8739, *Pseudomonas aeruginosa* ATCC 6027, and *Proteus vulgaris* ATCC 6380. The stem bark and flower extracts showed better antimicrobial potential. *K. paniculata* could be considered as a potential source of biologically active substances with antitumor and antibacterial properties.

Keywords: *Koelreuteria paniculata*; dry ethanol extracts; GC-MS analysis; chemical compounds; antitumor and antimicrobial activities

1. Introduction

Nowadays, more and more authors are studying the chemical composition of various plant extracts in order to find new sources, of plant origin, to combat resistant human pathogens and cancers. Emerging allergies and the many side effects of synthetic drugs are grounds to look for their natural alternatives that are sufficiently effective and, at the same time, less harmful to human health [1–8]. Antibiotic-resistant bacteria are a significant modern problem. The key to solving it can be individual plant compounds, essential oils, or extracts containing some of the most active phytochemicals with antibacterial activity, such as polyphenols and terpenes [8]. Many natural plant compounds also exhibit potent anti-cancer activity. The therapeutic value of herbal sources in the fight against cancer has increased in recent years worldwide, as evidenced by the use of certain chemotherapeutic drugs isolated from medicinal plants [6].

The present study was focused on the evaluation of the biologically active components of possible natural sources, such as *Koelreuteria paniculata* Laxm. (belonging to the family of Sapindaceae), known as the golden rain tree. The species is native to China, but it is widely used for ornamental purposes in Europe, including Bulgaria. It is now considered to be an invasive alien tree, characterized by good overall adaptability, growth, development, and strong vitality characteristics [9,10].

K. paniculata has been the subject of some phytochemical studies concerning various plant extracts, such as ethanol [2,11–15], methanol [1,13–15], benzene/ethanol [13–15], and formaldehyde [13], as solvents establishing the different groups of natural components. Phenolic derivatives based on gallic acid, catechins, and phenolic acids are isolated and determined from ethanol (vacuum) extracts of fresh leaf parts [11]. Flavonoids, phenolic acids, and sterols were examined from the ethyl acetate fraction of the *K. paniculata* flowers [16]. Primary (fatty acids, carbohydrates) and secondary metabolites (methyl gallate, ethyl gallate, flavonoids, and their glycosides, sterols, and saponins) are established from 70% ethanol extract of air-dried powdered aerial parts [2,12].

In a few studies, GC-MS analysis was performed in order to identify the chemical composition of the extracts [1,13–15]. The authors have analyzed several extracts of bark, wood, branches, leaves, and roots using GC-MS analysis; however, it was not implemented for the identification of the chemical composition of the plant flowers. They found different groups of active components (fatty acids, phenols, mono and triterpenes, sterols, vitamin E, and others). In a previous study by our team, four essential oils from the aerial parts of *K. paniculata* have been isolated by hydrodistillation, for the first time, and they were identified by GC-MS analysis [17]. We found a rich content of different classes of compounds, as follows: aliphatic oxygenated compounds, oxygenated sesquiterpenes, sesquiterpene hydrocarbons, aliphatic hydrocarbons, and diterpenes as predominant groups.

The biological activities exhibited by plants are due to their composition, certain groups of compounds, and the possible interactions between them. The fractions of methanol leaf extract in *K. paniculata* were distinguished by antibacterial and antifungal activity against *Staphylococcus aureus, Bacillus subtilis,* and *Pyricularia grisea* [1]. The ethanol extracts obtained from the *K. paniculata* aerial parts showed antibacterial (against *E. coli*) and antimalarial (against chloroquine-sensitive and chloroquine-resistant *Plasmodium falciparum*) activities. Some compounds (such as methyl gallate and ethyl gallate) were among the biologically active secondary metabolites of the plant [2].

Galenic preparations of the *K. paniculata* leaf extracts exhibited high efficiency against six pathogenic microorganisms causing several diseases in humans, such as the following: *Enterococcus faecalis, Proteus mirabilis, Seracia marcescens, Salmonella typhimurium, Campylobacter jejuni,* and *Escherichia coli* [5].

The antitumor activity of *K. paniculata* extracts has been poorly studied. Kumar et al. [18] reported on the DNA protective effect of the methanol leaf extract and its hexane fraction. The antineoplastic activity of the carotenoid fraction of *K. paniculata* flowers was determined by Zhelev et al. [19]. The authors established low cytotoxicity against human hepatocarcinoma cell lines (HepG2) and human breast cancer cells (MDA-MB-231).

Considering the literature research, it is clear that the species may be a source of valuable biologically active compounds. Still, at the same time, it was evident that in this regard, there were gaps regarding their action and application. In addition, phytochemical research on this species in Bulgaria is scarce, which led us to expand our knowledge on the plant extracts from different parts of this tree species. The present study investigated the qualitative and quantitative composition of the ethanol extracts of the aerial parts of *K. paniculata* and evaluated their in vitro antitumor and antibacterial activities.

2. Results and Discussion

2.1. Chemical Compounds of Dry Ethanol Extracts from Aerial Parts of K. paniculata

The obtained dry (under vacuum) ethanol extracts' yields were 0.6521 (2.6084) g for stem bark, 0.624 (2.4960) g for leaves, and 0.516 (2.0640) g for flowers. The extracts were viscous liquids with a dark brown color and characteristic ointment. The three samples were analyzed via GC-MS analysis. The chemical compounds (with their peaks) are presented in Table 1 and Figure 1.

Table 1. Chemical compounds of the ethanol dry extracts of *Koelreuteria paniculata* aerial parts, (mean ± SD).

Peak	RT	RI_{calc}	RI_{lit}	Compound	*K. paniculata*, % of TIC			Compounds Identification
					Flower	Leaf	Stem Bark	
1	4.49	1174	1170	Furfuryl butanoate	nd	nd	nd	
2	4.68	1190	1186	(3Z)-Hexenyl butanoate	0.81 ± 0.0	0.49 ± 0.0	0.97 ± 0.01	GMD
3	4.79	1197	1199	γ-terpineol	3.78 ± 0.10	0.85 ± 0.01	0.55 ± 0.0	GMD
4	5.10	1206	1202	(2E,4E)-Hexadienol isobutanoate	0.13 ± 0.0	2.00 ± 0.07	2.66 ± 0.08	GMD
5	5.20	1218	1217	(2E,4E)-Nonadienol	0.20 ± 0.0	3.53 ± 0.09	4.68 ± 0.11	GMD
6	5.26	1225	1223	Methyl nonanoate	0.58 ± 0.0	0.22 ± 0.0	0.44 ± 0.0	GMD
7	5.30	1233	1228	Nerol	nd	0.49 ± 0.0	1.71 ± 0.05	GMD
8	5.41	1231	1229	(3Z)-Hexenyl 2-methyl butanoate	0.36 ± 0.0	3.78 ± 0.10	8.15 ± 0.20	GMD
9	5.60	1288	1288	Lavandulol acetate	0.47 ± 0.01	3.40 ± 0.09	2.02 ± 0.06	GMD
10	6.11	1316	1315	δ-terpinyl acetate	0.12 ± 0.0	0.71 ± 0.0	0.93 ± 0.01	GMD
11	6.23	1345	1346	α-terpinyl acetate	16.42 ± 0.51	20.24 ± 0.07	3.56 ± 0.09	GMD
12	6.35	1359	1360	Neryl acetate	0.15 ± 0.0	0.33 ± 0.0	12.37 ± 0.40	GMD
13	6.40	1362	1361	Pyrogallol	20.86 ± 0.70	2.02 ± 0.07	1.91 ± 0.05	NIST'08
14	6.55	1366	1363	α-methyl benzyl butyrate	nd	0.30 ± 0.0	0.82 ± 0.0	GMD
15	6.64	1375	1378	(3Z)-Hexenyl hexanoate	nd	0.79 ± 0.0	0.78 ± 0.0	GMD
16	6.82	1386	1386	Isobutyl phenylacetate	nd	1.13 ± 0.04	1.69 ± 0.04	GMD
17	6.94	1394	1395	Ethyl decanoate	5.85 ± 0.12	1.40 ± 0.05	1.47 ± 0.04	GMD
18	7.01	1401	1400	n-tetradecane	2.13 ± 0.07	1.08 ± 0.04	2.65 ± 0.07	GMD
19	7.08	1416	1417	β-caryophyllene	2.61 ± 0.07	3.86 ± 0.08	2.91 ± 0.07	GMD
20	7.12	1432	1433	α-*trans*-bergamotene	2.45 ± 0.07	0.95 ± 0.01	0.93 ± 0.01	GMD
21	7.16	1439	1438	Phenyl ethyl butanoate	3.89 ± 0.08	2.94 ± 0.08	2.89 ± 0.08	GMD
22	7.26	1453	1453	Geranyl acetone	nd	1.80±	2.53 ± 0.07	GMD
23	7.41	1465	1466	Linalool isovalerate	nd	0.92 ± 0.01	1.42 ± 0.03	GMD
24	7.47	1469	1468	n-dodecanol	2.21 ± 0.07	0.99 ± 0.01	1.90 ± 0.04	GMD
25	7.51	1472	1471	α-terpinyl isobutanoate	10.32 ± 0.40	4.77 ± 0.10	3.56 ± 0.09	GMD
26	7.55	1476	1479	α-curcumene	nd	0.45 ± 0.0	nd	GMD
27	7.64	1481	1482	γ-curcumene	nd	0.58 ± 0.0	nd	GMD
28	7.72	1486	1485	Phenyl ethyl 2-methylbutanoate	0.75 ± 0.0	3.21 ± 0.08	3.84 ± 0.09	GMD
29	7.76	1488	1489	β-selinene	3.01 ± 0.07	2.60 ± 0.07	2.63 ± 0.07	GMD
30	7.97	1497	1498	α-selinene	3.36 ± 0.08	0.66 ± 0.0	0.72 ± 0.0	GMD
31	8.15	1511	1514	β-curcumene	nd	0.33 ± 0.0	0.15 ± 0.0	GMD
32	8.26	1556	1552	Lauric acid	1.93 ± 0.03	0.88 ± 0.0	0.36 ± 0.0	NIST'08
33	8.41	1562	1560	Geranyl butanoate	0.81 ± 0.0	0.55 ± 0.0	0.78 ± 0.0	GMD
34	8.50	1568	1570	Octyl hexanoate	0.40 ± 0.0	0.62 ± 0.0	0.51 ± 0.0	NIST'08

Table 1. *Cont.*

Peak	RT	RI_{calc}	RI_{lit}	Compound	*K. paniculata*, % of TIC			Compounds Identification
					Flower	Leaf	Stem Bark	
35	8.56	1576	1573	Decyl butyrate	0.64 ± 0.0	0.37 ± 0.0	0.44 ± 0.0	NIST'08
36	8.71	1611	1611	Tetradecanal	0.20 ± 0.0	0.74 ± 0.0	0.82 ± 0.0	GMD
37	8.83	1623	1622	Isobutyl cinnamate	nd	1.22 ± 0.02	1.50 ± 0.02	GMD
38	9.07	1640	1639	Phenyl ethyl hexanoate	1.71 ± 0.03	9.05 ± 0.10	8.67 ± 0.07	GMD
39	9.17	1649	1650	β-eudesmol	0.42 ± 0.0	0.90 ± 0.01	0.69 ± 0.0	GMD
40	9.26	1654	1652	α-eudesmol	0.27 ± 0.0	0.72 ± 0.0	1.12 ± 0.02	GMD
41	9.37	1662	1661	Dihydro-eudesmol	0.35 ± 0.0	1.35 ± 0.02	1.93 ± 0.03	GMD
42	9.61	1671	1670	Epi-β-bisabolol	2.19 ± 0.07	3.03 ± 0.07	1.51 ± 0.02	GMD
43	9.67	1675	1674	β-bisabolol	2.88 ± 0.07	1.44 ± 0.02	1.66 ± 0.02	GMD
44	9.86	1690	1691	(Z)-α-*trans*-Bergamotol	2.00 ± 0.06	2.52 ± 0.07	1.00 ± 0.02	GMD
45	10.21	1699	1698	(2Z,6Z)-Farnesol	0.84 ± 0.0	0.51 ± 0.0	0.72 ± 0.0	GMD
46	10.50	1706	1704	δ-dodecalactone	0.97 ± 0.01	0.30 ± 0.0	0.57 ± 0.0	GMD
47	10.82	1718	1718	Methyl eudesmate	nd	0.45 ± 0.0	1.18 ± 0.02	GMD
48	11.93	1826	1825	(E)-Nerolidyl isobutyrate	nd	0.39 ± 0.0	nd	GMD
49	14.49	1953	1948	Phytol	nd	nd	0.14 ± 0.0	GMD
50	14.61	1971	1966	Palmitic acid	1.85 ± 0.05	0.86 ± 0.0	1.16 ± 0.02	GMD
51	14.73	1992	1990	Ethyl palmitate	0.29 ± 0.0	nd	nd	GMD
52	16.67	2018	2019	(6E,10Z)-Pseudo phytol	nd	nd	0.42 ± 0.0	GMD
53	16.85	2026	2024	Isopropyl hexadecanoate	nd	0.79 ± 0.0	0.25 ± 0.0	GMD
54	16.94	2029	2030	(6E,10E)-Pseudo phytol	nd	nd	0.81 ± 0.0	GMD
55	18.05	2133	2132	Linoleic acid	0.16 ± 0.0	4.32 ± 0.04	1.55 ± 0.02	GMD
56	18.28	2142	2141	Oleic acid	0.23 ± 0.0	nd	nd	GMD
57	21.58	2400	2400	Tetracosane	0.10 ± 0.0	nd	nd	GMD
				Total identified, %	**98.70**	**97.83**	**98.63**	

RT—Retention time; RI_{calc}—Kovats retention index, calculated by authors; RI_{lit}—Kovats retention index by literature data; TIC—Total ion current; nd—not detected; NIST'08—National Institute of Standards and Technology, Gaithersburg, MD, USA; GMD—Golm Metabolome Database.

Forty components were identified in the flower extract, representing 98.70% of the total content. Eighteen of them were in a concentration above 1%. The main components (over 3%) were as follows: pyrogallol (20.86%), α-terpinyl acetate (16.42%), α-terpinyl isobutanoate (10.32%), ethyl decanoate (5.85%), phenyl ethyl butanoate (3.89%), γ-terpineol (3.78%), α-selinene (3.36%), and β-selinene (3.01%). Qu et al. [16] first identified, in *K. paniculata* flowers, nine components in the ethyl acetate fraction by column chromatography and spectral analysis. The components are the following: sitosterol glucoside, gallic acid, kaempferol, luteolin, kaempferol-3-O-(6″-acetyl)-β-D-glucopyranoside, hyperoside-2″-O-acetyl, hyperoside-2″-O-galloyl, hyperoside, and kaempferol-3-O-D-glucopyranoside. In our previous study related to the essential oil composition of *K. paniculata* flowers, there were 38 phytochemicals identified, with twelve main compounds. The five common components for the two types of extracts have been proven, namely the following: β-caryophyllene, lauric acid, palmitic acid, oleic acid, and tetracosane [17].

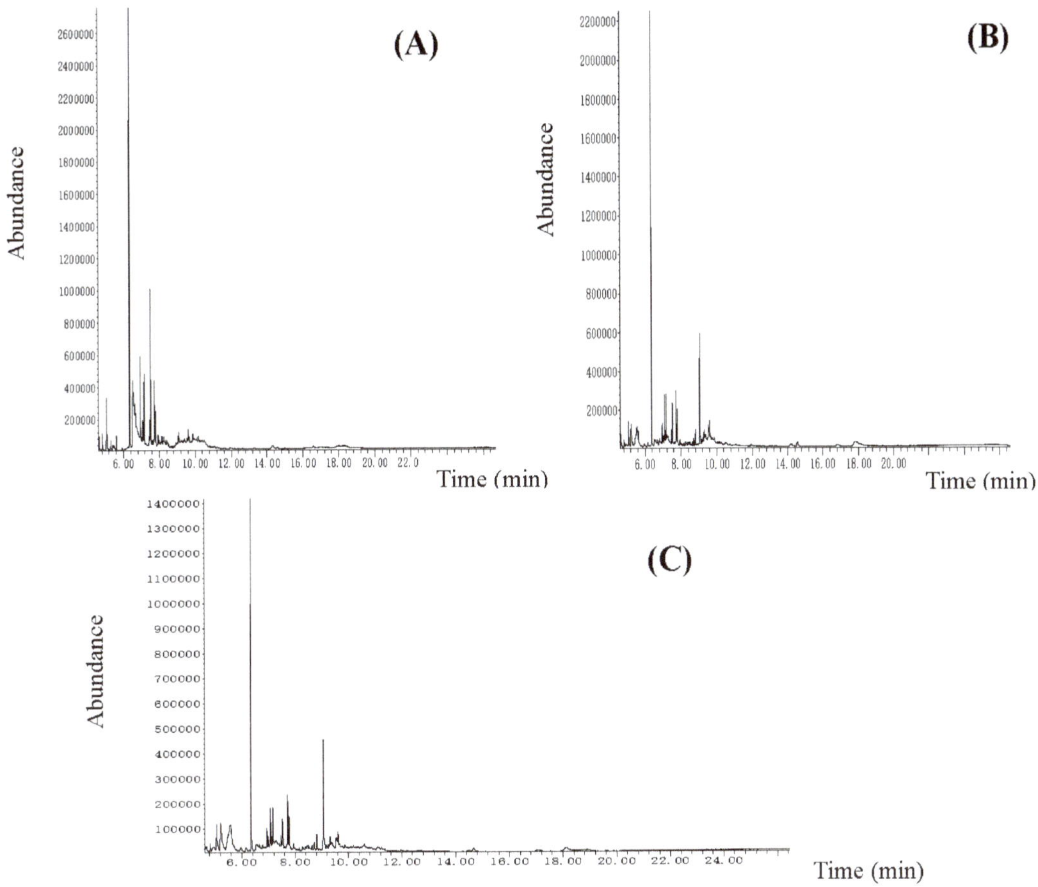

Figure 1. The GC-MS chromatograms of the analyzed dry ethanol extracts from *K. paniculata*: (**A**) flower extract, (**B**) leaf extract, (**C**) stem bark extract.

In the leaf extract, 50 components were identified, which represents 97.83% of the total content. Twenty-two of them were found in a concentration above 1% and the major ones (over 3%) were the following ten: α-terpinyl acetate (20.24%), phenyl ethyl hexanoate (9.05%), α-terpinyl isobutanoate (4.77%), linoleic acid (4.32%), β-caryophyllene (3.86%), (3Z)-hexenyl 2-methyl butanoate (3.78%), (2E,4E)-nonadienol (3,53%), lavandulol acetate (3.40%), phenyl ethyl 2-methylbutanoate (3.21%), and epi-β-bisabolol (3.03%). In the analyzed fractions of methanol extract from the dry leaves of *K. paniculata*, Ghahari et al. [1] found a smaller number (between two and seven) of the major components (over 3%), of which only linoleic acid (4.69%) was among those found by us. Wang et al. [14] isolated 13 active substances in both ethanol and methanol leaf extracts and 32 in benzene/ethanol leaf extract in which the best represented is ethyl gallate—a phenolic compound with antitumor activity. Andonova et al. [17] identified 49 components in the leaf essential oil from golden rain trees, of which six of them are the major ones (above 3%), different from those in the ethanol leaf extract. Only palmitic acid (2.89%), lauric acid (<1%), and β-caryophyllene (<1%) are common.

Fifty components were identified in the bark extract, representing 98.63% of the total content, and twenty-nine of them were in concentrations above 1%. The main components (over 3%) were as follows: neryl acetate (12.37%), (3Z)-hexenyl 2-methyl butanoate (8.15%), (2E, 4E)-nonadienol (4.68%), phenyl ethyl 2-methylbutanoate (3.84%), and α-terpinyl acetate (3.56%). Yang et al. [13] analyzed the ethanol bark extract of *K. paniculata* using GC-MS analysis and identified the components palmitic acid, linoleic acid, and ethyl oleate. The first two were also present in our extract in similar concentrations. The same authors reported data on the components of other types of bark extracts (formaldehyde, phenyl alcohol, and benzene alcohol extracts), where the concentrations of palmitic acid methyl ester (1.21%), vitamin E (0.31%), sorbitol (0.25%), and dihydrojasmone (0.18%) were identified. The authors found that the main components of the bark extracts were oleic acid (38.72%), lauric acid (5.90%), and acetic acid (4.41%), which were identified by TD-GC-MS analysis. GC-MS analysis of the bark essential oil of the golden rain tree conducted by Andonova et al. [17] identified thirty-six components with nine major ones. Six compounds of all identified were comparable with those in the ethanol extract-palmitic acid (3.20%), β-caryophyllene (1.81%), phytol (1.80%), oleic acid (1.03%), lauric acid (<1%), and tetracosane (<1%).

A comparative analysis of the chemical composition of the studied extracts showed that the one obtained from the flowers was dominated by pyrogallol. It is an odorless substituent and does not form the odor of the extract but instead determines its biological properties, mainly the antioxidant and antimicrobial potential [20]. The extracts of the flowers were also dominated by the monoterpene alcohol γ-terpineol, as well as its esters with acetic and isobutyric acid, which forms the smell of the extract as fresh bergamot-lavender-like (terpinyl acetate), floral (terpinyl isobutyrate), and pine with floral notes (γ-terpineol). According to our findings, the amount of these compounds was lower in the extracts obtained by the plant leaves and bark.

The differences in the identified components in our study, compared with those reported in the literature, are due to the plant's growing conditions, the technological parameters of the extraction, and the specificity of the used methodology.

The distribution of the components by chemical groups is presented in Figure 2. Oxygenated monoterpene (OM) derivatives predominated in all three of the extracts (flowers 32.49 $\pm$ 0.30%, leaves 35.21 $\pm$ 0.30% and stem bark 29.84 $\pm$ 0.25%), followed by aliphatic oxygen (AO) derivatives and phenylpropanoids (PP). The other groups were less represented, and their distribution can be seen in the figure, as the deviation of the values ranged from 0.07 to 0.1 for sesquiterpene hydrocarbons, from 0.15 to 0.20 for oxygenated aliphatics, 0.01 for aliphatic hydrocarbons, 0.08 to 0.09 for oxygenated sesquiterpenes, and from 0.20 to 0.26 for phenylpropanoids.

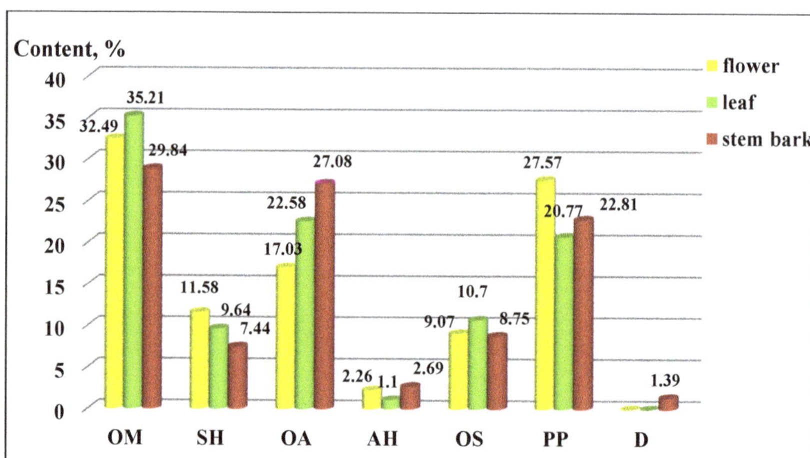

Figure 2. Composition by chemical groups from aerial parts in *Koelreuteria paniculata* ethanol extracts (%): AH—Aliphatic hydrocarbons; OA—Oxygenated aliphatics; OM—Oxygenated monoterpenes; SH—Sesquiterpene hydrocarbons; OS—oxygenated sesquiterpenes; PP—Phenylpropanoids; D—Diterpenes.

In our previous study, the distribution of the components in the different parts of *K. paniculata* showed some differences as aliphatic and oxygenated hydrocarbons, and sesquiterpenes represented the main part of the isolated essential oils [17].

The distribution of the functional groups concerning the total percentage content of *K. paniculata* ethanol extracts is presented in Figure 3. The group of esters had the highest percentage in all three of the studied extracts, followed by alcohols. An exception was the high content of phenols in the flowers (21.13 ± 0.20%), compared to the other two plant parts, which were present in a low percentage. The groups of acids, phenols, ketones, lactones, and aldehydes were very poorly represented in the examined extracts, as shown in Figure 3. The arrangement of the compounds by the functional groups was related to the manifested biological activities of the extracts. Gabrielli et al. [21] pointed out that phenols, followed by alcohols, aldehydes, ketones, ethers, and hydrocarbons, were of primary importance for the activity of the essential oils.

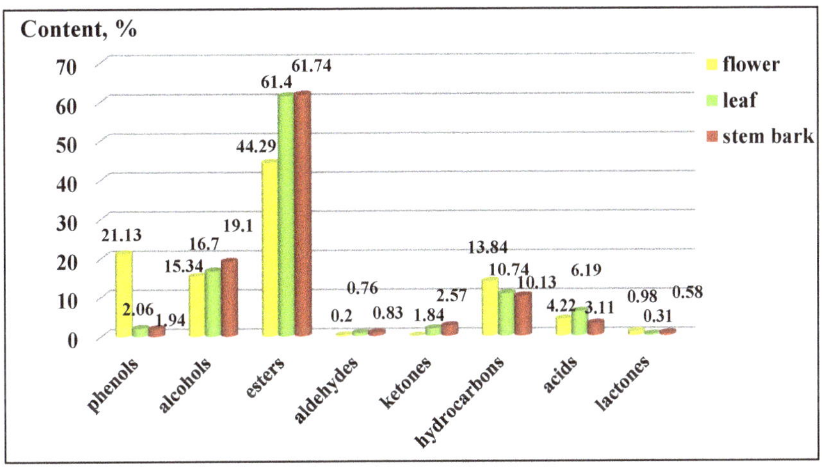

Figure 3. Composition by functional groups from *Koelreuteria paniculata* ethanol extracts (%): phenols; alcohols; esters; aldehydes; ketones; hydrocarbons, acids; lactones. The deviations in the values are in the range of 0.5 to 0.9% statistical error.

2.2. Antitumor Activity of the K. paniculata Ethanol Dry Extracts

The antiproliferative activity of the *K. paniculata* ethanol extracts obtained from the different plant parts was examined on two tumor cell lines—HT-29 and PC3. The two cell lines were not randomly selected. The human colon adenocarcinoma HT-29 cell line is widely used to study the biology of human colon cancers and showed many characteristics of mature intestinal cells [22]. Another cell line, PC3, is also valuable in carcinogenesis. Prostate cancer is the primary malignancy in men and the second leading cause of cancer-related deaths [23]. The obtained results are shown in Figure 4 and Table 2. Improved antiproliferative activity of the flower extract over the other two extracts (IC_{50}-21.44 µg/mL) on the cell line HT-29 was observed. Less pronounced activity (over two times) on the other cell line PC3 (IC_{50}-58.76 µg/mL) was also demonstrated. The leaf extract showed almost the same activity as the flower extract on the HT-29 cell line (IC_{50}-23.63 µg/mL), while prostate cancer cells were less sensitive to this extract (IC_{50}–80.56 µg/mL). The bark extract showed weak inhibition effects on the cell lines (IC50–339.4 µg/mL and 182.8 µg/mL for HT-29 and PC3 cell lines, respectively). As can be seen from the graphs (Figure 4E,F), the antiproliferative activity of the total bark extract was dose-dependent for both cell lines. It strongly resembled the antiproliferative effect of cisplatin, the antitumor standard in the present study. The total leaf and flower extracts affected cell growth only at low concentrations, and had almost the same values at higher concentrations at over 60 mg/mL for the HT-29 (Figure 4A,C) and over 125 mg/mL for PC3 (Figure 4B,D). As a possible reason for this, we can point out the differences in the chemical composition of the plant parts and the ethanol extracts obtained from them, especially the presence of the high content of pyrogallol in the composition of flowers. Pyrogallol is compared to antibiotics and also has antioxidant properties [20]. For example, Ahn et al. [24] reported the antitumor mechanisms of pyrogallol that showed significant cytotoxicity and reduced the number of colonies in Hep3B and Huh7 cells. Other authors revealed that phenols determined the antitumor effect of plant extracts on various tumor cell lines (including HT-29) [25,26].

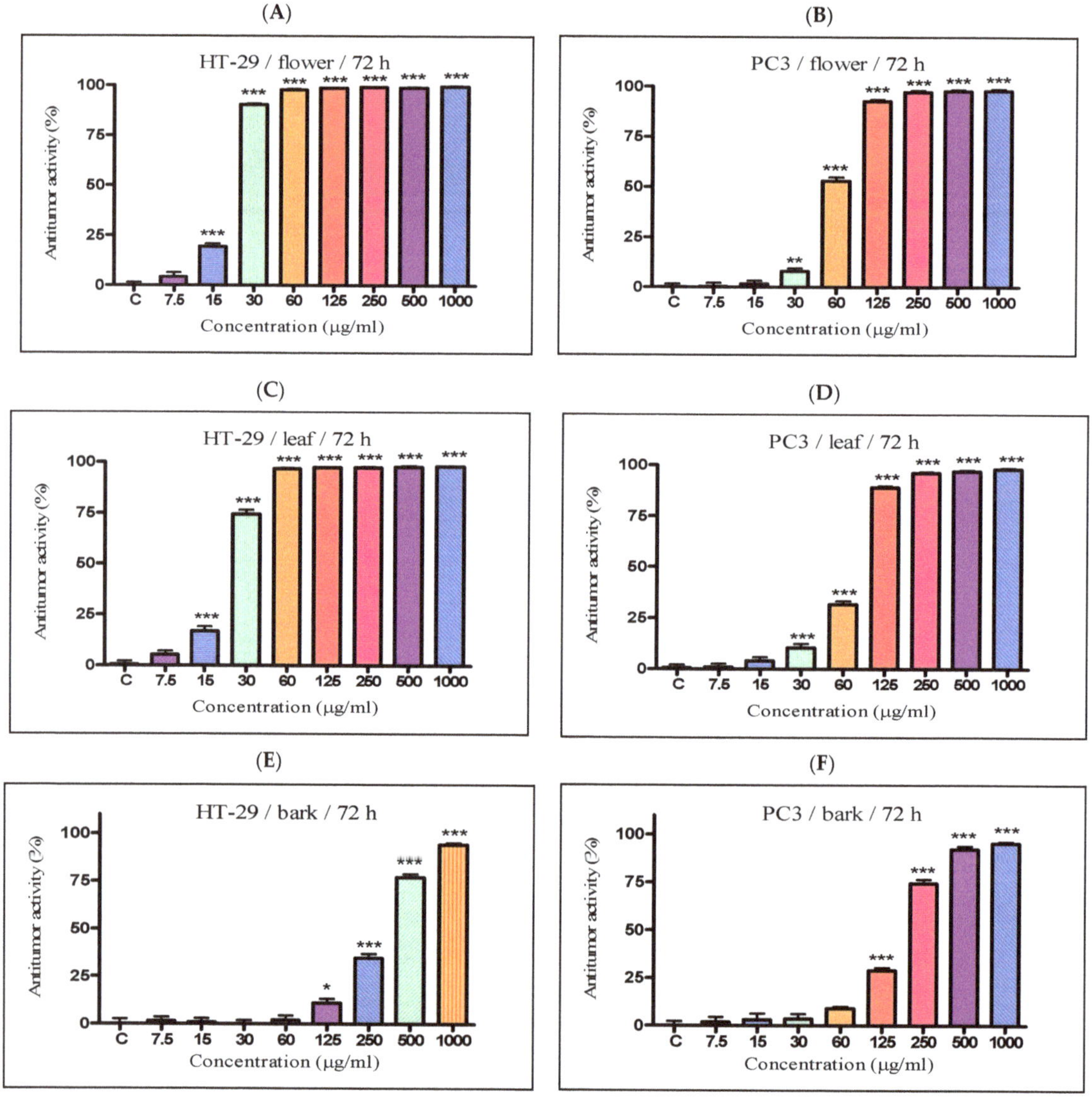

Figure 4. In vitro antiproliferative activity of ethanol extracts from three plant parts of *K. paniculata*. MTT assay was performed after 72 h. (**A**,**C**,**E**) show data for antiproliferative activity on cell line HT-29, of flower, leaf, and bark, respectively. (**B**,**D**,**F**) indicate results obtained on cell line PC3, of the same plant parts respectively. All samples were analyzed in triplicates. Values are represented as mean ± SD; One-way ANOVA followed by post hoc test using Tukey's multi-group comparison was performed: * $p < 0.05$, ** $p < 0.01$, *** $p < 0.001$.

Table 2. In vitro antiproliferative activity of the ethanol dry extracts of *Koelreuteria paniculata* aerial parts.

	IC_{50} of Mean ± SD (µg/mL)	
	HT-29	**PC3**
Flower extract	21.44 ± 0.20	58.76 ± 0.53
Leaf extract	23.63 ± 0.22	80.56 ± 0.75
Stem Bark extract	339.4 ± 1.31	182.8 ± 1.65
Cisplatin	2.5 ± 0.08	1.01 ± 0.03

IC_{50} determined following 72 h treatment with ethanol extracts. Antitumor activities were expressed as IC_{50} values (extract concentrations (µg/mL) required for 50% inhibition of cell growth), calculated using non-linear regression analysis (GraphPad Software, San Diego, CA, USA). Results were calculated from three measurements and expressed as mean ± SD. Cisplatin was used as standard to confirm the suitability of the used antitumor method.

Compared to the findings in our study, Zhelev et al. [19], using the MTT-test, found that the carotenoid fraction from *K. paniculata* flowers demonstrate relatively low cytotoxicity to HepG2 (human hepatocarcinoma) and MDA-MB-231 (human breast cancer cells), such as the HepG2 cell line, is more sensitive. The research, in this case, was related to the cytotoxicity of carotenoids and did not investigate their antiproliferative activity. Several articles have examined the ability of different *K. paniculata* extracts to protect various DNA structures from damaging factors. In the study by Kumar et al. [18,27], the methanol extracts and different fractions from the leaves showed a DNA protective effect in Calf thymus/pUC18, as authors associated its activity with the polyphenol constituents within it. In addition, Kumar and Kaur [28] established the potential of those extracts to inhibit lipid peroxidation and 4-nitroquinoline-1-oxide (4NQO)-induced genotoxicity. In vitro cytotoxicity assay on another *Koelreuteria* species (*K. elegans*) showed the promising anticancer activity of two phenols (from butanol fraction), methyl gallate and austrobailignan, against MCF-7 cell lines, which also reduced the cell proliferation of it [29].

2.3. Antimicrobial Activity of the K. paniculata Ethanol Dry Extracts

The results for the tested amounts of the extracts (100 µL, 150 µL) on nine pathogenic strains of microorganisms are presented in Table 3 and Figure 5. The bark extract was the most effective against the Gram-positive bacteria *Bacillus subtilis* ATCC 6633 (18 mm inhibition zone, IZ), *Bacillus cereus* NCTC 10,320 (14 mm IZ), and against the Gram-negative bacteria *Pseudomonas aeruginosa* ATCC 6027 (14 mm IZ) and *Proteus vulgaris* ATCC 6380 (8 mm IZ) at the higher tested concentration of the extract. The inhibitory zone of *K. paniculata* flower extract was quite similar against *P. vulgaris* (10 mm IZ), *B. subtilis* (14 mm IZ), and *B. cereus* (14 mm IZ). On the other hand, the *K. paniculata* leaf extract did not inhibit the test cultures against the Gram-negative bacterium *E. coli* ATCC 8739.

The differences in IZ values could be explained by the content of pyrogallol and terpineol esters. It is known that the activity on the main components of aromatic products (essential oils, extracts) was arranged in the following sequence: phenols > alcohols > aldehydes > ketones > esthers > hydrocarbons [30].

There was limited information concerning the antimicrobial properties of *K. paniculata*, as the reports were mainly about extract obtained from the plant's leaves. This is the first paper studying the antibacterial activity of extracts obtained from *K. paniculata* flowers and stem barks. Ghahari et al. [1] reported the antibacterial activity of *K. paniculata* methanol extract from the leaves against *B. subtilis* and *S. aureus*. Zazharskyi et al. [5] investigated the antimicrobial potential (with inhibition zone above 8 mm) of ethanol extracts from golden rain tree extracts against different pathogens, such as the following: *E. faecalis*, *P. mirabilis*, *S. marcescens*, *S. typhimurium*, *C. jejuni*, and *E. coli*; the last of which was the most sensitive microorganism. The authors did not find activity against the tested *P. aeruginosa* compared to the findings reported in our study. Ethyl and methyl gallate were the investigated phenols demonstrated in the study by Mostafa et al. [2]. They were reported as promising antimicrobial (against *E. coli*) and antimalarial (against chloroquine-sensitive plasmodia-*Plasmodium falciparum*) agents.

Table 3. Zones of growth inhibition (mm) of dry ethanol extracts of *Koelreuteria paniculata* aerial parts.

Test-Microorganism	QF	FE	LE	SBE	CH
		Gram-negative bacteria			
Escherichia coli ATCC 8739	150 µL	- *	9.02 ^{a} ± 0.01	-	28.05 ^{b} ± 0.01
	100 µL	-	6.02 ^{a} ± 0.01	-	
Salmonella enterica NCTC 6017	150 µL	-	-	-	24.04 ± 0.03
	100 µL	-	-	-	
Klebsiella (clinical isolate)	150 µL	-	-	-	21.03 ± 0.03
	100 µL	-	-	-	
Pseudomonas aeruginosa ATCC 6027	150 µL	-	-	14.02 ^{c} ± 0.01	21.03 ^{c} ± 0.03
	100 µL	-	-	5.03 ^{c} ± 0.01	
Proteus vulgaris ATCC 6380	150 µL	10.04 ^{d} ± 0.02	-	8.04 ^{d} ± 0.01	-
	100 µL	6.05 ^{d} ± 0.04	-	6.02 ^{d} ± 0.01	
		Gram-positive bacteria			
Staphylococcus aureus ATCC 25093	150 µL	-	-	-	25.03 ± 0.03
	100 µL	-	-	-	
Bacillus subtilis ATCC 6633	150 µL	14.03 ^{e} ± 0.03	-	18.04 ^{e} ± 0.02	39.04 ^{f} ± 0.03
	100 µL	11.03 ^{e} ± 0.04	-	11.03 ^{e} ± 0.02	
Bacillus cereus NCTC 10320	150 µL	14.0 6 ^{g} ± 0.03	-	14.03 ^{g} ± 0.02	-
	100 µL	-	-	6.02 ^{g} ± 0.01	
Listeria monocytogenes NCTC 11994	150 µL	-	-	-	21.06 ± 0.02
	100 µL	-	-	-	

Data are presented as means ± SD (standard deviation). $^{a–g}$ Means in a row not sharing the same superscript letter are significantly different at $p < 0.05$ (Tukey's HSD test). QF—Quantity of the filtrate; FE—Flower extract; LE—Leaf extract; SBE—Stem bark extract; CH—Chlorhexidine; *—No inhibitory activity was observed.

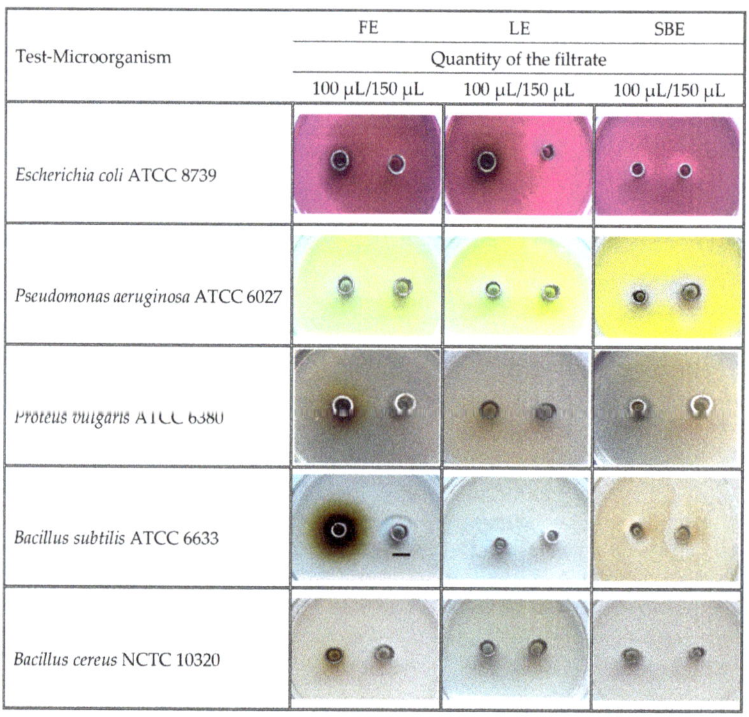

Figure 5. Images of the growth inhibition zones (over 6 mm) of dry ethanol extracts of *Koelreuteria paniculata* aerial parts; FE—Flower extract; LE—Leaf extract; SBE—Stem Bark extract. Scale bar indicated 10 mm.

The antimicrobial activity of different plants is influenced by the chemical composition of the plant and the concentration and conditions of obtaining the extracts. For example, Ham et al. [31] reported that neryl acetate had significantly strong and selective antibacterial activity against Gram-negative fish pathogens. Therefore, the presence of the component in the stem bark extracts could be the reason for its antimicrobial potential. Another study revealed the α-terpinyl acetate essential oil and extracts showed high antimicrobial effect against fungi, dermatophytes, bacteria and Candida yeasts [32]. The strain differences between the test cultures may also be relevant to the reported results [33].

3. Materials and Methods

3.1. Plant Material Collection and Identification

The samples from the aerial parts of *K. paniculata* (stem bark, leaves, and flowers—Figure 6) were collected between May and July 2020 in Plovdiv, Bulgaria (42°8′9.9492″ N, 24°44′31.8048″ E), and botanically identified by Prof. D-r. I. Dimitrova-Dyulgerova (Department of Botany, Faculty of Biology, University of Plovdiv "Paisii Hilendarski"). The voucher specimen (No 060436) has been deposited in the Herbarium of the Agricultural University, Plovdiv, Bulgaria (Herbarium SOA).

(**A**) (**B**) (**C**)

Figure 6. Aerial parts from *Koelreuteria paniculata*: (**A**) stem bark, (**B**) leaves, (**C**) flowers (photos taken by the authors).

3.2. Preparation of Dry Plant Extracts

Collected fresh and washed plant materials, after maceration (were ground into fine particles using a home grinder), were subjected to two serial extractions. For the elimination of non-polar compounds, chloroform was used (≥99% extra pure, Karl Roth, Germany) as the first solvent, and ethanol (96%, Ph. Eur., extra pure, Karl Roth, Germany) as a second solvent, to study the active and polar compounds. The extracts were obtained in a ratio of 1:10 (plant material:solvent) to complete exhaustion of the herb for 10 days with intermittent stirring. In this study, 400 g of fresh plant material was soaked in 4L of solvent. The supernatant from the chloroform extract was filtered using Whatman filter paper No. 1 (Sigma-Aldrich, Germany), and the residues were used for a second extraction. To concentrate the extracts, a rotary evaporator was used (Buchi, Rotavapor R-300) at 50 °C. Only the ethanol extracts were used for the present study. The dry extracts were collected in a vial and stored at 4 °C in the dark for further use for GC-MS analysis, antitumor, and antimicrobial tests.

3.3. *Cell Lines, Test-Microorganisms and Nutrient Media*

The PC3 (ATCC® CRL-1435™, human prostate adenocarcinoma) and HT-29 (ATCC® HTB-38™, human colon adenocarcinoma) cell lines were obtained from the American Type Culture Collection (ATCC, Manassas, VA, USA). Dulbecco's modified Eagle medium (DMEM), fetal bovine serum (FBS), antibiotics (penicillin and streptomycin), and the disposable consumables were supplied by Orange Scientific, Braine-l'Alleud, Belgium.

The following Gram-positive bacteria: *Listeria monocytogenes* NCTC 11994, *Staphylococcus aureus* ATCC 25093, *Bacillus subtilis* ATCC 6633, and *Bacillus cereus* NCTC 10320, and the following Gram-negative bacteria: *Escherichia coli* ATCC 8739, *Salmonella enterica* subsp. *enterica* serovar *abony* NCTC 6017, *Pseudomonas aeruginosa* ATCC 6027, *Proteus vulgaris* ATCC 6380, and *Klebsiella* (clinical isolate) were used in this study. The National Bank supplied the strains for industrial microorganisms and cell cultures. The following selective bacteriological media were used: Listeria Oxford agar base with an additive containing cycloheximide (Biolife); Endo agar (Sigma-Aldrich, Germany); Leifson agar (Merck); Baird-Parker agar base (Biolife) with yolk-tellurite additive and plate mount agar (Merck), Chlorhexidine (Sigma-Aldrich, Germany).

3.4. *Gas Chromatography-Mass Spectrometry (GC-MS) and GC-FID Analyses*

The GC-MS analysis was carried out with an Agilent 7890A gas chromatograph with an HP-5MS capillary column (30 m length, 0.32 mm in diameter, 0.25 μm film-coating thickness) coupled to a mass spectral detector Agilent MSD 5975C with helium as the carrier gas (1.0 mL/min). The temperature regime was in the range from 100 to 300 °C (100 °C, 2 min retention, increase to 180 °C with 15 °C/min, 1 min retention, increase to 300 °C with 5 °C/min, 10 min retention); injector and detector temperatures = 250 °C; mass-detector scan range-m/z = 50–550; injected sample volume-1 μL in flow split ratio 20:1. The compounds were identified by comparing retention times and relative Kovats (RI) indices with those of standard substances and mass spectral data from the Golm Metabolome Database (GMD) [34] and NIST'08 (National Institute of Standards and Technology, USA) (https://www.nist.gov/nist-research-library/reference-format-nist-publications, accessed on 10 February 2021). The experiment was carried out in triplicate.

3.5. *Antitumor Activity Assay*

The antitumor activity testing was performed on cell cultures from two human cell lines using the standard MTT-dye reduction assay, described by Mosmann [35]. The assay is based on the metabolism of the tetrazolium salt MTT to insoluble formazan by mitochondrial reductases. The formazan concentration can be determined spectrophotometrically. The measured absorption is an indicator of the cell viability and metabolic activity. The used cell lines were routinely grown as monolayer in 75 cm^2 tissue culture flasks in DMEM high-glucose (4.5 g/L), supplemented with 10% FBS and antibiotics. Cultures were maintained at 37.5 °C in a humidified atmosphere under 5% CO_2. Cells were plated at a density of 1×10^3 cells in 100 μL in each well of the 96-well flat-bottomed microplates and allowed to adhere for 24 h before treatment with the test compounds. A concentration range from 2 to 1000 μg/mL (double increasing manner) was applied for 72 h. The formazan absorption was registered using a microplate reader at λ = 540 nm. Cisplatin (Sigma-Aldrich, Germany) was used as a standard in the assay.

3.6. *Antimicrobial Activity Assay*

The antibacterial activity was determined by modifying the agar diffusion method by measuring the inhibition zones of pathogen growth around metal rings, into which a certain amount of test material was introduced. Selective media for the test cultures were inoculated with pathogen suspensions prepared from a 24-h culture on PCA. From a suitable ten-fold dilution of the suspension, the melted and cooled to 45–50 °C selective media was inoculated. After solidifying the media, sterilized metal rings with a diameter of Ø = 6 mm were placed on their surface, in which 0.10 and 0.15 μL of the extract were imported, respectively. Test cultures were incubated at 37 °C. The diameter (mm) of the growth inhibition zones of the test cultures was measured at 24 and 48 h, and a comparative assessment of their antibacterial activity was made. The final DMSO content was 5% (*v*/*v*), and this solution was used as a negative control. For positive control, chlorhexidine was used (100 µL). The experiments were performed in triplicate [36].

3.7. Statistical Analysis

The data of the antimicrobial activity test were analyzed and presented as mean values ± standard deviation (SD). Statistical analysis was carried out using Excel software. A one-way analysis of variance (ANOVA) was performed, and significant differences between samples were determined by applying the Tukey's honestly significant difference (Tukey "HSD") test, which is used to test differences among sample means for significance. Tukey "HSD" is considered to be a multiple comparison procedure that is used in order to test the significant differences between all possible pairs of mean values on a variable for groups of research samples. Antitumor activity was expressed as IC_{50} value (concentration required for 50% inhibition of cell growth), calculated using non-linear regression analysis (GraphPad Software, San Diego, CA, USA). The statistical analysis included the application of ANOVA, followed by Bonferroni's post hoc test. The lowest level of statistical significance was accepted as $p < 0.05$. The measurements in the GS/MS analysis were performed in triplicate and the results were presented as the mean value of the individual measurements with the corresponding standard deviation (SD), using Microsoft Excel.

4. Conclusions

In conclusion, the present study demonstrated the antitumor and antimicrobial potential of dry ethanol extracts of *K. paniculata* flowers, leaves, and stem bark. The antitumor activity against two cell lines (HT29-human colon adenocarcinoma and PC3-human prostate adenocarcinoma) and the antimicrobial potential against some pathogenic bacteria (*Pseudomonas aeruginosa* ATCC 6027, *Proteus vulgaris* ATCC 6380 and *Bacillus cereus* NCTC 10320) of *K. paniculata* ethanol extracts were investigated for the first time here. Significant antiproliferative activity was found for the flower and leaf ethanol extracts against the HT-29 cell line. The antibacterial activity (dose-dependent) was determined by the extracts of stem bark and flowers against Gram-positive strains of *Bacillus subtilis* ATCC 6633, and *Bacillus cereus* NCTC 10320, and Gram-negative strains of *P. vulgaris* ATCC 6380, and *P. aeruginosa* ATCC 6027. The leaf ethanol extracts inhibited only *E. coli* ATCC 8739 bacterial growth. Fifty-six components were identified in the studied aerial plant parts, among which the best represented, over 10% were pyrogallol (in the flowers), α-terpinyl acetate (in the leaves and flowers), neryl acetate (in the stem bark), and α-terpinyl isobutanoate (in the flowers). The oxygenated monoterpenes (by chemical groups) and the esters (functional groups) were the best-represented groups in all of the three extracts. Some of the compounds found in the extracts suggest a possible antioxidant potential. Future research should focus on the radical scavenging ability of extracts, as well as on the mechanism of action of proven antitumor activity. *K. paniculata* could be considered as a potential source of biologically active substances with application in pharmaceutical and food production. They would also be useful for the treatment of cancer, microbial infections, as well as for the production of natural preservatives to extend the shelf-life of food.

Author Contributions: Conceptualization, T.A.; I.D.-D.; Data curation, T.A.; Y.M.; H.F.; I.S.; A.S.; Formal analysis, T.A., Y.M., A.S., H.F., I.D.-D.; Methodology, T.A., I.S., H.F.; Supervision, I.D.-D.; Writing—original draft, T.A.; Writing—review & editing, T.A.; Y.M.; H.F.; I.S.; A.S.; I.D.-D. All authors have read and agreed to the published version of the manuscript.

Funding: This research received no external funding.

Institutional Review Board Statement: Not applicable.

Informed Consent Statement: Not applicable.

Data Availability Statement: Not applicable.

Conflicts of Interest: The authors declare no conflict of interest.

References

1. Ghahari, S.; Alinezhad, H.; Nematzadeh, G.A.; Ghahari, S. Phytochemical screening and antimicrobial activities of the constituents isolated from Koelreuteria paniculata leaves. *Nat. Prod. Res.* **2015**, *29*, 1865–1869. [CrossRef]
2. Mostafa, A.E.; El-Hela, A.A.; Mohammad, A.E.I.; Cutler, S.J.; Ross, S.A. Antimalarial compounds isolated from Koelreuteria paniculata growing in Egypt. *Planta Med.* **2015**, *81*, PC7. [CrossRef]
3. Chouhan, S.; Sharma, K.; Guleria, S. Antimicrobial activity of some essential oils—Present status and future perspectives. *Medicines* **2017**, *4*, 58. [CrossRef] [PubMed]
4. Manandhar, S.; Luitel, S.; Dahal, R.K. In vitro antimicrobial activity of some medicinal plants against human pathogenic bacteria. *J. Trop. Med.* **2019**, *2019*, 1–5. [CrossRef] [PubMed]
5. Zazharskyi, V.V.; Davydenko, P.O.; Kulishenko, O.M.; Borovik, I.V.; Zazharska, N.M.; Brygadyrenko, V.V. Antibacterial and fungicidal activities of ethanol extracts of 38 species of plants. *Biosyst. Divers.* **2020**, *28*, 281–289. [CrossRef]
6. Borah, P.; Banik, B.K. Medicinal plants and their compounds with anticancer properties. In *Green Approaches in Medicinal Chemistry for Sustainable Drug Design*; Elsevier: Amsterdam, The Netherlands, 2020; pp. 759–776. ISBN 978-0-12-817592-7.
7. Hadadi, Z.; Nematzadeh, G.A.; Ghahari, S. A study on the antioxidant and antimicrobial activities in the chloroformic and methanolic extracts of 6 important medicinal plants collected from North of Iran. *BMC Chem.* **2020**, *14*, 1–11. [CrossRef] [PubMed]
8. Álvarez-Martínez, F.J.; Barrajón-Catalán, E.; Herranz-López, M.; Micol, V. Antibacterial plant compounds, extracts and essential oils: An updated review on their effects and putative mechanisms of action. *Phytomedicine* **2021**, *90*, 153626. [CrossRef] [PubMed]
9. Petrova, A.; Vladimírov-Hurban, V.; Georgiev, V. *Invasive Alien Species of Vascular Plants in Bulgaria*; Institue of Biodiversity and Ecosystem Research, Bulgarian Academy of Sciences: Sofia, Bulgaria, 2013; ISBN 954-9746-30-5.
10. Ljubojević, M.; Tomić, M.; Simikić, M.; Savin, L.; Narandžić, T.; Pušić, M.; Grubač, M.; Vejnović, S.; Marinković, M. Koelreuteria paniculata invasiveness, yielding capacity and harvest date influence on biodiesel feedstock properties. *J. Environ. Manag.* **2021**, *295*, 113102. [CrossRef] [PubMed]
11. Lin, W.-H.; Deng, Z.-W.; Lei, H.-M.; Fu, H.-Z.; Li, J. Polyphenolic compounds from the leaves of Koelreuteria paniculata Laxm. *J. Asian Nat. Prod. Res.* **2002**, *4*, 287–295. [CrossRef] [PubMed]
12. Mostafa, A.E.; El-Hela, A.A.; Mohammad, A.-E.I.; Cutler, S.J.; Ross, S.A. New triterpenoidal saponins from Koelreuteria paniculata. *Phytochem. Lett.* **2016**, *17*, 213–218. [CrossRef] [PubMed]
13. Yang, Y.; Zheng, D.; Zhao, Y.; Wang, T.; Yang, Y.; Peng, W. Analysis of components in Koelreuteria paniculata. *Caribb. J. Sci.* **2017**, *50*, 13–18.
14. Wang, Y.; Zheng, D.; Zhao, Y.; Wang, T.; Yang, Y.; Ashraf, M.A.; Peng, W. Active components in branches and leaves of Koelreuteria paniculata. *Caribb. J. Sci.* **2018**, *51*, 11–21.
15. Wang, Y.; Liu, Q.; Zheng, D.; Zhao, Y.; Wang, T.; Yan, S.; Gu, H. Active constituents of Koelreuteria paniculata root. *Therm. Sci.* **2020**, *24*, 1737–1744. [CrossRef]
16. Qu, Q.-H.; Zhang, L.; Bao, H.; Zhang, J.-H.; You, X.-J.; Wang, J.-X. Chemical constituents of flavonoids from flowers of Koelreuteria paniculata. *Zhong Yao Cai Zhongyaocai J. Chin. Med. Mater.* **2011**, *34*, 1716–1719.
17. Andonova, T.; Dimitrova-Dyulgerova, I.; Slavov, I.; Muhovski, Y.; Stoyanova, A. A Comparative Study of Koelreuteria Paniculata Laxm. Aerial Parts Essential Oil Composition. *J. Essent. Oil Bear. Plants* **2020**, *23*, 1363–1370. [CrossRef]
18. Kumar, M.; Chandel, M.; Kumar, S.; Kaur, S. Studies on the antioxidant/genoprotective activity of extracts of Koelreuteria paniculata Laxm. *Am. J. Biomed. Sci* **2012**, *1*, 177–189.
19. Zhelev, I.; Georgiev, K.; Dimitrova-Dyulgerova, I. In-vitro antioxidant and antineoplastic activities of carotenoids from flowers of Koelreuteria paniculata. *World J. Pharm. Res.* **2016**, *5*, 53–60. [CrossRef]
20. Tinh, T.H.; Nuidate, T.; Vuddhakul, V.; Rodkhum, C. Antibacterial Activity of Pyrogallol, a Polyphenol Compound against Vibrio Parahaemolyticus Isolated from the Central Region of Thailand. *Procedia Chem.* **2016**, *18*, 162–168. [CrossRef]
21. Gabbrielli, G.; Loggini, F.; Cioni, P.L.; Giannaccini, B.; Mancuso, E. Activity of lavandino essential oil against non-tubercular opportunistic rapid grown mycobacteria. *Pharmacol. Res. Commun.* **1988**, *20*, 37–40. [CrossRef]
22. Baricault, L.; Denariaz, G.; Houri, J.-J.; Bouley, C.; Sapin, C.; Trugnan, G. Use of HT-29, a Cultured Human Colon Cancer Cell Line, to Study the Effect of Fermented Milks on Colon Cancer Cell Growth and Differentiation. *Carcinogenesis* **1995**, *16*, 245–252. [CrossRef] [PubMed]
23. Cooperberg, M.R.; Park, S.; Carroll, P.R. Prostate Cancer 2004: Insights from National Disease Registries. *Oncol. Williston Park NY* **2004**, *18*, 1239–1247.
24. Ahn, H.; Im, E.; Lee, D.Y.; Lee, H.-J.; Jung, J.H.; Kim, S.-H. Antitumor Effect of Pyrogallol via MiR-134 Mediated S Phase Arrest and Inhibition of PI3K/AKT/Skp2/CMyc Signaling in Hepatocellular Carcinoma. *Int. J. Mol. Sci.* **2019**, *20*, 3985. [CrossRef] [PubMed]
25. Kowalik, K.; Paduch, R.; Strawa, J.W.; Wiater, A.; Wlizło, K.; Waśko, A.; Wertel, I.; Pawłowska, A.; Tomczykowa, M.; Tomczyk, M. Potentilla Alba Extracts Affect the Viability and Proliferation of Non-Cancerous and Cancerous Colon Human Epithelial Cells. *Molecules* **2020**, *25*, 3080. [CrossRef]

26. Rajiv, C.; Roy, S.S.; Tamreihao, K.; Kshetri, P.; Singh, T.S.; Devi, H.S.; Sharma, S.K.; Ansari, M.A.; Devi, E.D.; Devi, A.K.; et al. Anticarcinogenic and Antioxidant Action of an Edible Aquatic Flora Jussiaea Repens L. Using In Vitro Bioassays and In Vivo Zebrafish Model. *Molecules* **2021**, *26*, 2291. [CrossRef] [PubMed]
27. Kumar, M.; Kumar, S.; Kaur, S. Investigations on DNA protective and antioxidant potential of chloroform and ethyl acetate fractions of Koelreuteria paniculata Laxm. *Afr. J. Pharm. Pharmacol.* **2011**, *5*, 421–427. [CrossRef]
28. Kumar, M.; Kaur, S. Golden rain tree leaf extracts as potential inhibitor of lipid peroxidation and 4-nitroquinoline-1-oxide (4-NQO)-induced DNA damage. *Afr. J. Biotechnol.* **2011**, *10*, 19758–19763. [CrossRef]
29. Abo-Elghiet, F.; Ibrahim, M.; Sleem, A.A. Anticancer activity of isolated compounds from twigs of Koelreuteria elegans cultivated in Egypt. *Azhar Int. J. Pharm. Med. Sci.* **2021**, *1*, 23–29. [CrossRef]
30. Baser, K.H.C.; Buchbauer, G. *Hand Book of Essential Oils: Science, Technology, and Applications*; Taylor and Francis Group, LLC CRC Pressisanimprintof Taylor &Francis Group, an Informa Business: Boca Raton, FL, USA, 2010. [CrossRef]
31. Ham, Y.; Yang, J.; Choi, W.-S.; Ahn, B.-J.; Park, M.-J. Antibacterial Activity of Essential Oils from Pinaceae Leaves Against Fish Pathogens. *J. Korean Wood Sci. Technol.* **2020**, *48*, 527–547. [CrossRef]
32. Vaičiulytė, V.; Ložienė, K.; Švedienė, J.; Raudonienė, V.; Paškevičius, A. α-Terpinyl Acetate: Occurrence in Essential Oils Bearing Thymus Pulegioides, Phytotoxicity, and Antimicrobial Effects. *Molecules* **2021**, *26*, 1065. [CrossRef]
33. Tajkarimi, M.M.; Ibrahim, S.A.; Cliver, D.O. Antimicrobial herb and spice compounds in food. *Food Control* **2010**, *21*, 1199–1218. [CrossRef]
34. Hummel, J.; Strehmel, N.; Selbig, J.; Walther, D.; Kopka, J. Decision tree supported substructure prediction of metabolites from GC-MS profiles. *Metabolomics* **2010**, *6*, 322–333. [CrossRef] [PubMed]
35. Mosmann, T. Rapid colorimetric assay for cellular growth and survival: Application to proliferation and cytotoxicity assays. *J. Immunol. Methods* **1983**, *65*, 55–63. [CrossRef]
36. Balouiri, M.; Sadiki, M.; Ibnsouda, S.K. Methods for in vitro evaluating antimicrobial activity: A review. *J. Pharm. Anal.* **2016**, *6*, 71–79. [CrossRef] [PubMed]

MDPI
St. Alban-Anlage 66
4052 Basel
Switzerland
www.mdpi.com

Plants Editorial Office
E-mail: plants@mdpi.com
www.mdpi.com/journal/plants

Disclaimer/Publisher's Note: The statements, opinions and data contained in all publications are solely those of the individual author(s) and contributor(s) and not of MDPI and/or the editor(s). MDPI and/or the editor(s) disclaim responsibility for any injury to people or property resulting from any ideas, methods, instructions or products referred to in the content.

www.ingramcontent.com/pod-product-compliance
Lightning Source LLC
Chambersburg PA
CBHW061451220926
51850CB00004B/65
9783036584188